P9-CQD-204

Aerosols Handbook

Measurement, Dosimetry, and Health Effects

Aerosols Handbook

Measurement, Dosimetry, and Health Effects

Edited by

Lev S. Ruzer and Naomi H. Harley

CRC PRESS

Boca Raton London New York Washington, D.C.

Library of Congress Cataloging-in-Publication Data

Aerosols handbook: measurement, dosimetry, and health effects / edited by Lev S. Ruzer and Naomi H. Harley.

p. cm.

Includes bibliographical references and index.

ISBN 1-56670-611-4 (alk. paper)

1. Aerosols—Toxicology—Handbooks, manuals, etc. I. Ruzer, Lev S. (Lev Solomonovich) II. Harley, Naomi H.

RA1270.A34A374A 2004
614.5′92—dc22

2004050336
CIP

Direct all inquiries to CRC Press, 2000 N.W. Corporate Blvd., Boca Raton, Florida 33431.

Visit the CRC Press Web site at www.crcpress.com

International Standard Book Number 1-56670-611-4
Library of Congress Card Number 2004050336
Printed in the United States of America 1 2 3 4 5 6 7 8 9 0
Printed on acid-free paper

Preface

Aerosols consist of particles in the very broad range of sizes from nanometers to hundreds of micrometers (4 to 5 orders of magnitude). Therefore, their behavior is complicated in the atmosphere, indoors, and especially in the lung.

Health effects associated with aerosols depend on the physical parameter that we call "dose." Dose depends on the quantity of aerosols in target cells. With the exception of some radioactive aerosols, it is practically impossible to measure dose directly. In practice, assessment of the dose is provided by measuring air concentration and calculating some known parameters.

According to the U.S. EPA, "in epidemiological studies, an index of exposure from personal or stationary monitors of selected pollutants is analyzed for associations with health outcomes, such as morbidity or mortality. However, it is a basic tenet of toxicology that the dose delivered to the target site, not the external exposure, is a proximal cause of a response. Therefore, there is increased emphasis on understanding the exposure–dose–response relationship. Exposure is what gets measured in the typical study and what gets regulated; dose is the causative factor."

In this book, we present a general, up-to-date overview of all aspects of aerosols, from their properties to the health outcomes. First, current issues related to aerosol measurement are detailed: standardization of measurements for different types of aerosols (indoor, medical and pharmaceutical, industrial, bioactive, and radioactive), with a special emphasis on breathing zone measurements. The handbook also discusses the problems of aerosol dosimetry, such as the definitions of aerosol exposure and aerosol dose, including the issue of nanometer particles, the mechanism of aerosol deposition in the lung, and modeling deposition with an emphasis on the corresponding uncertainty in risk assessment.

A separate part on radioactive aerosols includes aspects such as radon; natural and artificial aerosols; radioactive aerosols and the Chernobyl accident; dosimetry and epidemiology in miners, including direct dose measurement in the lungs; radon and thoron; and long-lived radionuclides in the environment.

It is especially important that the handbook includes an overview of nonradioactive and radioactive aerosols together, because behavior of radioactive aerosols in the lungs, including deposition and biokinetic processes, depends not on their activity, but on particle size distribution and breathing parameters. On the other hand, radioactivity of aerosols is the most useful tool for the study of their behavior in the lungs.

The handbook concludes with overviews of different aspects related to the health effects of diesel aerosols, health risks from ultrafine particles, and epidemiology to molecular biology.

Acknowledgments

Dr. Ruzer appreciates the support of William Fisk, head of the Indoor Environment Department (IED), Lawrence Berkeley National Laboratory, and staff scientists Richard Sextro, Lara Gundel, Michael Apte, Phillip Price, William Nazaroff, and Anthony Nero for their support. He would also like to acknowledge the work of the scientists and technicians of the Aerosol Laboratory at the All-Russian Scientific-Technical Institute for Physico-Technical and Radio-Technical Measurements (VNIIFTRI), Moscow, Russia.

Dr. Harley gratefully acknowledges the research support given by the U.S. Department of Energy EMSP Program DE FG02 03ER 63661 and the support of the staff at the New York University School of Medicine, Department of Environmental Medicine.

The editors thank the following individuals for their technical and administrative support: Joyce Cordell-Breckingridge, Olivia Salazar, Sondra Jarvis, Jeiwon Deputy, Eve Edelson, Julia Alter, Joanne Lambert, Marcy Beck, and Rita Labrie.

The editors also gratefully acknowledge the following individuals at CRC Press: Randi Cohen, who played an important role in preparing this book for publication from the beginning, and Matt Lamoreaux and Julie Spadaro, who worked hard during the final stages of production. Thanks and appreciation also go to the production team at Macmillan-India.

Editors

Lev S. Ruzer was born in Odessa, Ukraine, in 1922. His mother died when he was 5 years old. Five years later, his father was executed under Stalin's regime. After graduating from high school, he studied physics and mathematics at Odessa University until the beginning of World War II.

After being demobilized from the Soviet Army, he enrolled at a recently founded Department of Nuclear Physics at Moscow University. After graduation, he could not work as a scientist for political reasons: he had relatives in the United States and his father had been executed. For 8 years, he worked as a teacher in Moscow schools.

After Stalin's death, he found a job as a scientist in a medical institute. The research study included the assessment of dose to animals exposed to radon and its decay products. Based on this theoretical and experimental work, he defended his degree as a Candidate of Physico-Mathematical Sciences (equivalent to a Ph.D.) in 1961. He worked from 1961 to 1979 in the All-Union Institute of Physico-Technical and Radio-Technical Measurements as a founder and chairman of the Aerosol Laboratory.

In 1973, the Aerosol Laboratory became the State Standard on Aerosols in the Soviet Union. In addition to becoming the main center of aerosol measurement and metrology in the country, the laboratory focused on development and using methods of direct dosimetry on miners in both uranium and nonuranium mines.

In 1968, Dr. Ruzer published his book *Radioactive Aerosols* in Russian. In 1970, he defended his degree of Doctor of Technical Sciences, and in 1977 he became a professor. He also served as scientific supervisor to 8 Candidates of Science.

In 1977, he was discharged from his position for political reasons: his children were involved in dissident activity. He spent the following 8 years without work and unable to obtain permission to emigrate. He finally arrived in the United States in 1987. In 1989, his second book, *Aerosols R&D in the Soviet Union*, was published in English.

Since 1989, he has been working as a researcher in the Indoor Environment Department, Environmental Energy Technologies Division, at the Lawrence Berkeley National Laboratory.

Dr. Ruzer has published in more than 110 publications and holds three patents.

Naomi H. Harley received an undergraduate degree (B.S.) in electrical engineering from The Cooper Union, a masters degree (M.E.) in nuclear engineering, and a Ph.D. in radiological physics from the New York University Graduate School. She obtained an APC in management from the New York University Graduate Business School. Dr. Harley was elected a council member to the National Council on Radiation Protection and Measurements (NCRP) in 1982 and was made an honorary member in 2000. She is an advisor to the United Nations Scientific Committee on the Effects of Atomic Radiation (UNSCEAR).

Dr. Harley's major research interests are the measurement of inhaled or ingested radionuclides, the measurement of environmental radioactivity, the modeling of their fate within the human body, the calculation of the detailed radiation dose to cells specifically implicated in carcinogenesis, and risk assessment from exposure to internal radioactivity.

Dr. Harley has authored over 150 journal publications, six book chapters, and has four patents for radiation detection instrumentation. The most recent patent, issued in 2004, was for a miniature passive radon and thoron detector.

Contributors

Michael G. Apte
Lawrence Berkeley National Laboratory
Indoor Environment Department
Berkeley, California

Paul A. Baron
NIOSH
Cincinnati, Ohio

A.K. Budyka
Karpov Physico-Chemical Institute
Moscow, Russia

Beverly S. Cohen
Nelson Institute of Environmental Medicine
New York University
New York, New York

Daniel J. Cooney
Dispersed Systems Laboratory
University of North Carolina
Chapel Hill, North Carolina

D.E. Fertman
Scientific Engineering Centre
SNIIP
Moscow, Russia

Isabel M. Fisenne
USDHS Environmental Measurements Laboratory
New York, New York

Lucila Garcia-Contreras
Dispersed Systems Laboratory
University of North Carolina
Chapel Hill, North Carolina

Robert J. Garmise
Dispersed Systems Laboratory
University of North Carolina
Chapel Hill, North Carolina

James W. Gentry
Department of Chemical Engineering
University of Maryland
College Park, Maryland

Lara A. Gundel
Indoor Environment Department
Environmental Energy Technologies Division
Lawrence Berkeley National Laboratory
Berkeley, California

Max M. Häggblom
Department of Biochemistry and Microbiology
Rutgers University
New Brunswick, New Jersey

Naomi H. Harley
New York University School of Medicine
New York, New York

Maire S.A. Heikkinen
Department of Environmental Medicine
New York University School of Medicine
New York, New York

Anthony J. Hickey
Dispersed Systems Laboratory
University of North Carolina
Chapel Hill, North Carolina

William C. Hinds
Department of Environmental Health Sciences
Center for Occupational and Environmental Health
UCLA School of Public Health
Los Angeles, California

Mervi K. Hjelmroos-Koski
Environmental Health Sciences
School of Public Health
University of California
Berkeley, California

Philip K. Hopke
Department of Chemical Engineering
Clarkson University
Potsdam, New York

J.P. Johnson
Department of Mechanical Engineering
University of Minnesota
Minneapolis, Minnesota

Latarsha D. Jones
Transave, Inc.
Monmouth Junction, New Jersey

Kristin King Isaacs
National Health and Environmental Effects Research Laboratory
U.S. Environmental Protection Agency
Research Triangle Park, North Carolina

David B. Kittelson
Department of Mechanical Engineering
University of Minnesota
Minneapolis, Minnesota

V.L. Kustova
All-Russian Scientific Research Institute of Physico-Technical and Radio-Technical Measurements
VNIIFTRI
Moscow, Russia

Yu.V. Kuznetzov
All-Russian Scientific Research Institute of Physico-Technical and Radio-Technical Measurements
VNIIFTRI
Moscow, Russia

Janet M. Macher
California Department of Health Services
Environmental Health Laboratory Branch
Berkeley, California

Mark L. Maiello
Wyeth Research
R&D Environmental Health and Safety
Pearl River, New York

Ted B. Martonen
U.S. EPA
National Health and Environmental Effects Laboratory
Research Triangle Park, North Carolina
and
Department of Medicine
Division of Pulmonary Diseases
University of North Carolina
Chapel Hill, North Carolina

Andrew D. Maynard
National Institute for Occupational Safety and Health
Robert A. Taft Laboratory
Cincinnati, Ohio

B.I. Ogorodnikov
Karpov Physico-Chemical Institute
Moscow, Russia

I.V. Pavlov
VNIPI PT
Moscow, Russia

Phillip N. Price
Lawrence Berkeley National Laboratory
Indoor Environment Department
Berkeley, California

A.I. Rizin
Scientific Engineering Centre
SNIIP
Moscow, Russia

Charles E. Rodes
Center for Aerosol Technology
RTI International
Research Triangle Park, North Carolina

Jacky A. Rosati
U.S. EPA
National Risk Management Research Laboratory
Research Triangle Park, North Carolina

Lev S. Ruzer
Lawrence Berkeley National Laboratory
Indoor Environment Department
Berkeley, California

Jonathan M. Samet
Department of Epidemiology
Johns Hopkins Bloomberg School of Public Health
Baltimore, Maryland

Richard G. Sextro
Lawrence Berkeley National Laboratory
Indoor Environment Department
Berkeley, California

Hugh Smyth
Dispersed Systems Laboratory
University of North Carolina
Chapel Hill, North Carolina

Ira B. Tager
School of Public Health
University of California
Berkeley, California

Jonathan W. Thornburg
Center for Aerosol Technology
RTI International
Research Triangle Park, North Carolina

W.F. Watts
Department of Mechanical Engineering
University of Minnesota
Minneapolis, Minnesota

Contents

Chapter 1 Aspects of health-related aerosols 1
James W. Gentry

Chapter 2 Aerosol properties 19
William C. Hinds

Chapter 3 Advances in monitoring methods for airborne particles 35
Philip K. Hopke

Chapter 4 Ultrafine and nanoparticle emissions: A new challenge for internal combustion engine designers 47
D. B. Kittelson, W.F. Watts, and J.P. Johnson

Chapter 5 Breathing zone exposure assessment 61
Charles E. Rodes and Jonathan W. Thornburg

Chapter 6 Mechanisms of particle deposition 75
Kristin K. Isaacs, Jacky A. Rosati, and Ted B. Martonen

Chapter 7 Aerosol dose 101
Lev S. Ruzer, Michael G. Apte, and Richard G. Sextro

Chapter 8 Modeling deposition of inhaled particles 113
Ted B. Martonen, Jacky A. Rosati, and Kristin K. Isaacs

Chapter 9 Assessing uncertainties in the relationship between inhaled particle concentrations, internal deposition, and health effects 157
Phillip N. Price

Chapter 10 Aerosol chemistry and physics: Indoor perspective 189
Lara A. Gundel and Richard G. Sextro

Chapter 11 Aerosols in the industrial environment 225
Andrew D. Maynard and Paul A. Baron

Chapter 12 Medical and pharmaceutical aerosols 265
Hugh D.C. Smyth, Lucila Garcia-Contreras, Daniel J. Cooney, Robert J. Garmise, Latarsha D. Jones, and Anthony J. Hickey

Chapter 13 Bioaerosols 291
Maire S.A. Heikkinen, Mervi K. Hjelmroos-Koski, Max M. Häggblom, and Janet M. Macher

Chapter 14 Radioactive aerosols 343
Lev S. Ruzer

Chapter 15 Dosimetry and epidemiology of Russian uranium mines* 503
I.V. Pavlov

Chapter 16 Radioactive aerosols of the Chernobyl accident* 517
A.K. Budyka and B.I. Ogorodnikov

Chapter 17 Aerosol filtration (aerosol sampling by fibrous filters)* 541
A.K. Budyka and B.I. Ogorodnikov

Chapter 18 Radioactive aerosol standards 557
L.S. Ruzer, Yu.V. Kuznetzov, V.L. Kustova, D.E. Fertman, and A.J. Rizin

Chapter 19 Radon and thoron in the environment: Concentrations and lung cancer risk 569
Naomi H. Harley

Chapter 20 Risk from inhalation of the long-lived radionuclides uranium, thorium, and fallout plutonium in the atmosphere 585
Isabel M. Fisenne

Chapter 21 Health physics considerations of aerosols in radiosynthesis laboratories 595
Mark L. Maiello

Chapter 22 Diesel exhaust 601
Jonathan M. Samet

Chapter 23 Health effects of ambient ultrafine particles..607
Beverly S. Cohen

Chapter 24 Health effects of aerosols: Mechanisms and epidemiology 619
Ira B. Tager

Index 697

*Chapter translated from original Russian by Lev S. Ruzer.

chapter one

Aspects of health-related aerosols

James W. Gentry
University of Maryland

Contents

1.1 Overview 1
1.2 Total deposition 1
1.3 Regional deposition measurements and mucociliary clearance 4
1.4 Composition of aerosols 6
1.5 Principal laboratories 9
1.6 Journals and organizations 11
1.7 Awards 12
References 13

1.1 Overview

Over the last 40 years, major topics in health-related aerosols have been the preparation of suitable test aerosols, the measurement of total deposition, the measurement of regional deposition, the development of the bolus technique, and the study of clearance and retention. These are discussed in the sections below as well as several major laboratories, key organizations and journals, and awards. In this survey, we do not attempt to examine pulmonary biology or chemistry in detail nor do we discuss the important and growing area of nebulizers and drug delivery systems.

1.2 Total deposition

The first use of aerosols to examine pulmonary functions is attributed to B. Altshuler in his paper "Intrapulmonary Mixing of Gases Studied with Aerosols" at New York University (NYU) in 1959. In these measurements, a light scattering photometer was first used to measure the total particle deposition. A major limitation in this experiment is that the lower limit of the particles detectable is comparatively large. Two classes of optical instruments can be used to measure aerosol concentration. The first is a photometer that measures the total light scattered and can be used only with monodisperse particles and the second is a spectrometer that can be used with polydisperse aerosols. In theory, photometric methods for concentration measurements can be applied down to a particle size of 0.2 μm. For smaller particles, the light scattered by the air in the sensing volume will give a higher signal than the particles, and an alternative approach must be developed. In the late 1960s, the light source was replaced by a laser that allowed a decrease in the particle size that could be measured. (Jacobi et al., 1967; Gebhart, 1969). The 1970s was a period of increased use of aerosols to delineate

1-56670-611-4 / 05 / $0.00+$1.50

pulmonary function. Much of the work in the United States was carried out by E.D. Palmes and M. Lippmann at NYU, T. Mercer, P. Morrow, and W. Stöber at Rochester, and J.D. Brain at Harvard. In 1969, Lippmann and Albert (1969) published the paper that proposed a model based on rapid short time clearance from the upper airways. This and subsequent studies of the next decade used radioactively labeled aerosols to support this hypothesis.

The Harvard group (Brain et al. (1976)) pioneered studies of pulmonary deposition with aerosols. Chan and Lippmann (1980) utilized both modeling and experiments to provide an insight into regional deposition. However, the most fruitful approach to using aerosols was to use breath-holding measurements. This procedure was developed by E.D. Palmes in 1967 and 1975.

It was possible to measure both inhaled and exhaled air without a time delay. The aerosol could be delivered as a continuous flow or in boluses with small volumes. The air flow rate in the respiratory channel was measured with a pneumotachograph. On-line integration of the flow signals yielded the respired air volume. In Europe, Davies at the London School for Tropical Medicine is the person most responsible for the initial development of pulmonary research. It was in his laboratory that the methods for generation and measurement of the aerosols were first developed and where the experimental apparatus were first designed. Davies, who had worked at the Defense Laboratory in Porton Down, drew upon several developments of H. Walton and K. May there — notably the spinning disc generator and the cascade impactors. The generator used to produce monodisperse aerosols was the Sinclair-LaMer generator. Some of the co-workers of Davies in the period 1968–1970 were D. Swift, D. Muir, and J. Heyder. The procedures developed at London were adopted by the research group in Frankfurt.

The Frankfurt group contributed two special skills to the studies. They possesed expertise in the use and measurement of radioactive labeled isotopes and they had unsurpassed optical proficiency. This and the experimental thoroughness of the research group led to the most widely accepted data sets. The developments in measurement of total deposition discussed below are those carried out at Frankfurt (Heyder et al. 1972, 1975). The refinements of the measurements of the total deposition for oral breathing at Frankfurt can be used to highlight the development of the state of the science. The inhalation apparatus used at Frankfurt was a closed system, a modified version of the apparatus described by Muir and Davies (1967). Later on, a simpler open system was used. The aerosol number concentration and the volume of the respired air were measured simultaneously. The optical viewing chamber of the photometer was located directly between the mouthpiece and the mechanical controls of the flow. Originally, the particle size range was from 0.2 to 1 μm. In 1972, this range was increased to 3.0 μm and in 1974 to particles of 10 μm. Initially, the particle sizes were determined by Tyndall spectrometry with an "Owl," which was adapted from a design by Davies. The instrument was originally developed by D. Sinclair. Particles from 1 to 3 μm could be measured. Particle sizing of larger particles up to 10 μm was carried out with a sedimentation cell in air (Stahlhofen et al., 1975a). Later, Roth developed a single-particle spectrometer that could measure particles below 0.6 μm (Roth and Gebhart, 1978; Heyder et al., 1971). Using a focused argon laser having an intensity of the illuminating beam greater than 500 Wcm^{-2}, a scattering angle of 40°, and a sensitive photomultiplier, particles down to 0.08 μm were measured. This instrument, which was developed at Frankfurt, became the reference standard for commercial optical particle counters. For particles between 0.6 and 6 μm, Gebhart developed a low-angle scattering instrument with a white light source to obtain a monotonic calibration curve (Gebhart, 1969). In 1989, Brand developed a combined system comprising a differential mobility analyzer (DMA), a commercially available optical particle counter (LAS-X), and a white light optical counter. The instrument was used to obtain mass concentrations in Frankfurt at 15 min intervals (Brand et al., 1992). Fundamental studies on the efficiency of single-particle counters (Gebhart and Roth, 1986) and the application of optical counters for irregularly shaped particles (Gebhart, 1991) were also carried out at Frankfurt. A decade later, inhalation measurements were made for particles down to 0.005 μm (Heyder et al., 1986). The total deposition measurements with ultrafine silver particles for oral and nasal breathing were performed in a stationary open system for different breathing patterns (Schiller et al., 1988). The concentration of ultrafine particle was monitored with a commercially available condensation nucleus counter.

Although total deposition had been measured before, data of comparable quality had not been obtained. This improved quality is attributed to the superior optical instruments used. Deposition is at a minimum between 0.1 and 1.0 μm. The deposition of the small particles takes place by diffusion. The dimensionless group diffusion coefficient is inversely proportional to the particle diameter. As the particle size increases, the deposition efficiency increases as the appropriate dimensional groups associated with sedimentation and inertial impaction depend on the square of the diameter. In addition to particle characteristics — particle diameter and density (which affect both sedimentation and impaction but not diffusion) — particle deposition depends on the subject's breathing pattern, which determines the mean residence time of the inspired aerosol in the respiratory tract and the mean volumetric flow rate. Both mathematical and mechanical models have been developed to examine particle deposition. J. Gebhart and J. Heyder (1985) at Frankfurt used a granular bed to approximate lung deposition, while Cohen at NYU used a latex model of a dog's lung to examine deposition. These models were employed to examine deposition in the transition regime where neither sedimentation nor diffusion was dominant. The advantage of the models was that they could be extended over a wide range of different values. In interpreting the experimental results, the efficiency for small particles is correlated with a parameter inversely proportional to the diameter, and for large particles the measurements are correlated with a parameter proportional to the square of the diameter. This second group of experiments is said to be correlated with sedimentation, although both the sedimentation and the inertial parameters have the same functional relationship.

In mathematical modeling, two concepts have proved to be very useful. A paradigm of the lung is generated in which the lung is envisioned as an assembly of short capillary tubes of different diameters. This model creates a lung model where the effective diameters of the tubes have the same deposition efficiency as the measured values. These models gained more significance when regional deposition as well as total deposition values were obtained, although extensive investigations of mathematical and mechanical models to describe the deposition behavior of the lung in terms of effective airway diameters have been reported (Heyder and Gebhart, 1977). A second important area of research is the branched model of the lung. The initial Weibel model has been extensively reworked by the school of W. Hoffman at Salzburg. These models, which have a large number of branchings and allow the lung to be nonsymmetrical, are the state of the science. Models that assume that sedimentation and diffusion operate independently overestimate the deposition in the transition regime. Mathematical models of particle deposition were developed for the transition regime where neither sedimentation nor diffusion was dominant. Heyder et al. (1985) and Gentry et al. (1994) showed the most commonly used models agreed in the first approximation, and suggested a "vector model" as a more general approach.

An important aspect of the total deposition is the variability between subjects. First, it was noted that there was a remarkable intrasubject consistency in the measurements for subjects. The variation was found to be small even when the experiments were performed at different flow rates and breathing patterns (Heyder et al., 1982). Other observers have pointed out that although there was a small intrasubject variability there was a large intersubject variability. During the 1980s, a key aspect of total deposition studies was the experimental measurement of intersubject variability and their interpretation. Heyder reported the total deposition as a function of particle diameter for 20 subjects measured for different flow rates and breathing patterns. The variation was small and consistent with all the measurements. Measurements at Frankfurt on volunteers from Frankfurt and Bologna (Heyder et al., 1978) showed a similar consistency for nonsmokers. However, heavy smokers showed an increased particle deposition. The measurements demonstrated that certain types of behavior could dramatically increase deposition. These results suggested that the measurements could be used for early diagnosis of possible health problems, as increased deposition was observed before other symptoms appeared. They found that patients displayed much more variation in aerosol deposition than healthy subjects, even though the average deposition was very similar. They found that the overall deposition rate was lower for healthy patients at the fastest flow rates. These effects were linked with the effective size of the bronchial diameter in the lower respiratory tract.

1.3 Regional deposition measurements and mucociliary clearance

In the earliest measurements of regional deposition, a two-region model consisted of extrathoracic measurements and alveolar regions.

The Frankfurt group has been investigating regional deposition since 1974. They used a clearance, a collimated γ-ray detector system (whole-body counter) allowing independent detection of radioactively labeled aerosol particles that deposited in the extrathoracic and the intrathoracic region. The γ-ray detector gave a response independent of the distribution of radioactively labeled particles in the chest (Morsy et al., 1976). For the measurements used 4.7 μm iron oxide particles labeled with ^{198}Au, the deposition locations were measured using different detector and body positions. In subsequent measurements, regional depositions for different breathing patterns were measured (Stahlhofen et al., 1980). These studies required a monodisperse aerosol. Similar studies were carried out at the National Institute of Environmental Medicine in Stockholm. In typical measurements, they used 4.7 μm Teflon particles labeled with ^{111}In.

In 1981, a comparison was made between the two European groups using aerosols with diameters of 4.7 μm but differing in terms of the composition and radioactive isotopes used in labeling the aerosol. The participants from Sweden included P. Camner and K. Philipson, and the participants from Germany included W. Stahlhofen and J. Heyder, among others. The measurements were consistent with each other (Stahlhofen et al., 1981a). The clearance of the iron oxide particles was faster than the clearance of the insoluble Teflon particles. Possibly, there is an unknown mechanism involving particle solubility. Studies of biological variability of regional deposition in the human respiratory tract showed variations of less than 10% for the overall deposition, but showed variations of up to 40% in the extrathoracic regions. The variance in the alveolar region was smaller (Stahlhofen et al., 1981b).

Empirical expressions were used to describe deposition as a function of particle size in a three-compartment lung model (Rudolf et al., 1986; Heyder et al., 1986). The model was modified to account for differences in gender and age as well as other measures of biological variability.

Next, the tracheobronchial section was subdivided into two sections labeled tracheobronchial with mucociliary clearance and a bronchiolar section without mucociliary clearance. Large particles were deposited mainly in the extrathoracic region, very small particles mainly in the bronchiolar or lower bronchial region, and intermediate particles were retained mainly in the alveolar region. The deposition was strongly dependend on the breathing procedure. Currently, the model consists of five components, including a nasal as well as an extrathoracic region. This deposition model serves as a standard for the ICRP dosimetry model (Stahlhofen et al., 1989b; ICRP, 1994). The slow cleared fraction from the tracheobronchial region increased with decreasing particle size. This was confirmed by clearance measurements with ultrafine radioactively labeled ^{111}indium oxide particles (Roth et al., 1994). By 2001, the sensitivity of the lung counter had been extended to particles smaller than 30 nm.

An alternative method for probing effective air space dimension is the breath-holding method for lung research (Palmes et al., 1967). Since monodisperse aerosol particles settle with constant velocity, their deposition efficiency is dependent on the size of the airways, with the aerosol deposition being higher in the smaller airways. Along with the bolus technique, this method can be used to determine the effective airspace dimensions as a function of volumetric lung depth. Gebhart et al. (1981a,b) used single-breath inhalations of monodisperse aerosols to determine aerosol recovery from the human lung after periods of breath holding. He stated that the best conditions for sizing pulmonary air spaces of the human respiratory tract were obtained when boluses of the monodisperse aerosols having particles between 1 and 1.5 μm diameter were inspired. In addition, Heyder (1983) and Scheuch and Stahlhofen (1990) described the application of the bolus method to estimate the average airway diameter as a function of volumetric depth in the human respiratory tract. The number of particles decreased exponentially with the duration of breath holding, and an average airway radius was calculated for each lung depth from the slope of these exponential functions. Stahlhofen et al. (1989a) presented all three applications of the bolus method: the volumetric dispersion, the determination of averaged dimensions of the lungs, and the shallow bolus to clarify regional deposition from the clearance measurements.

The "bolus" technique has proved to be a very fruitful approach in estimating regional deposition. This is a procedure analogous to a pulse input in chemical engineering. A pulse of particles is introduced into the lung. The particle concentration is measured during an inhalation and expiration cycle. The origin of the bolus technique can be traced back to the work of Altshuler and Palmes at New York University. Altshuler (1969), Muir et al. (1970, 1971) described early applications of the bolus technique. The procedure requires that the subject breathe at regular intervals at a set time. The pulse is then introduced at a predetermined point within the breathing interval. Effectively, this determines how deeply the aerosol pulse penetrates into the lung. The subject then exhales with an exhalation rate equal to the inhalation rate. The initial pulse then exits the lung at the exact time of the input pulse. The fractional deposition is obtained by comparing the relative areas under the curves. The first description of the bolus technique for regional deposition came from Heyder (1983) and Heyder et al. (1988). Its advantage compared to the single-breath recovery method is that the depth to which the aerosol penetrates is known. In essence, one can pinpoint a small region in the lung. Also, the Frankfurt group developed the dispersion of the bolus to examine changes in lung ventilation. The deeper the bolus penetrates into the lungs, the higher the aerosol dispersion. An excellent description of the bolus technique was given by Blanchard (1996), in which he discussed the significance of distribution half-widths and their relationship with the dispersion (Blanchard, J., Aerosol bolus dispersion and aerosol-derived airway morphometry: assessment of lung pathology and response to therapy, *J. Aerosol Med.*, 9, 183–205).

The bolus techniques emphasize the necessity of having an excellent photometer. The Frankfurt group was quite fortunate in that they could utilize the photometer developed by J. Gebhart in 1988 for measurements with low concentrations. The instrument had two operating modes: a counting mode (particle concentrations below 10 cm^{-3}) and an analog mode (particle concentrations above 100 cm^{-3}) (Gebhart et al., 1988). The use of the instrument led to a technique now known as confocal microscopy. It was designed such that the laser beam passes through a slit forming a sheet of light with 80 μm thickness and a height of 15 mm configured at an angle of 60° relative to the aerosol stream. Only one particle was inside the sensing volume. This photometer design was flexible enough to be used with a number of different inhalation apparatus.

The application of the aerosol bolus method as a diagnostic tool in clinics was realized by using an inhalation apparatus and a photometer with 1 μm sebacate particles. Using this method, one can diagnose ventricular disturbances from aerosol dispersion measurements. Changes in lung function occur in the small airways that contribute less than 30% to the total airway resistance. Thus, changes in lung function are insensitive to measurements of airway resistance. However, convective gas mixing, which can be characterized by aerosol dispersion, is most pronounced in the peripheral regions. A typical application for aerosol morphometry or the determination of the airway diameters as a function of the lung depth is the early diagnosis of emphysema, which is insensitive to conventional lung function tests. R. Siekmeier and P. Brand at Frankfurt were able to identify and control the primary site of morphometric changes induced by pharmaceuticals. This has proved to be an exceptionally fruitful area of research. Key contributions to the effect of drugs and pulmonary afflictions have been made by P. Anderson, C. Schiller-Scotland, J. Blanchard, F. Rosenthal (modeling), and W. Bennett, among others.

When particles are inhaled, three possibilities may occur — they may be exhaled before deposition, they may be deposited and remain in the body, or they may be deposited and then subsequently cleared by mucociliary transport. For almost two decades from 1969 to 1988, the standard picture was rapid clearance from the ciliated airways followed by a slow clearance from the alveoli. This model was formulated by Lippmann and Albert (1969) and Albert et al. (1973) at NYU. As late as 1988, this was almost universally accepted (Oberdorster, 1988), with 80% of the particles removed by faster clearance and 20% cleared from the alveoli. However, since the 1990s there has been growing acceptance of a slow clearance component from the tracheobronchial region (Stahlhofen et al., 1986; Scheuch, 1994; Smaldone et al., 1988; Wolff et al., 1989; Bennett et al., 1993, 1998, 2003).

Scheuch and Stahlhofen (1972) found that the measured regional deposition data were less than that predicted from lung modeling. This indicated less tracheobronchial deposition and more alveolar deposition than predicted. The 24 hour convention for tracheobronchial clearance hypothesis was examined experimentally by Scheuch using a bolus technique and labeled iron oxide particles. They were able to direct where the particles were deposited. Even for very shallow boluses the fraction of particles that were retained was high (Stahlhofen et al., 1986). Scheuch (1994) and Scheuch et al. (1996) tested the assumptions that all the particles that deposit in the tracheobronchial tree are cleared by mucociliary transport within 24 hours (Lippmann and Albert, 1969). He found that there was a long-time clearance (i.e., more than 24 hour) as well as a short-time clearance (Scheuch and Stahlhofen, 1992). These results suggest that radioactive and toxic aerosols that deposit in the bronchi are retained longer and may potentially have a greater impact on health than previously believed. Moreover, they implied that the three-compartment lung model is inadequate and needed to be modified by subdividing the bronchi component into bronchi with and without mucociliary clearance. The slow cleared fraction from the tracheobronchial region increased with decreasing particle size. This was confirmed by clearance measurements with ultrafine radioactively labeled ^{111}indium oxide particles (Roth et al., 1994). By 2001, the sensitivity of lung counters had been extended to particles smaller than 30 nm.

1.4 Composition of aerosols

A major focus of medical aerosols is the generation of their size and chemical properties. In order to be respirable, the particles must be in a narrow size range from 0.01 to 12 μm. Particles in the smaller size range are sufficiently small that their diffusion for their diffusion coefficients are appreciable. This is the principal mechanism of collection. The larger particles are collected either by sedimentation or impaction on the points where the bronchi bifurcate.

For investigations of total deposition in the human respiratory tract a nontoxic, monodisperse aerosol is required. The deposition is determined by measuring the particle concentration before and after breathing, requiring a particle of a controllable size. The approach that has proven successful is based on the use of a condensation aerosol generator (La Mer-generator). The generator was developed by D. Sinclair and V. LaMer at Columbia University. The idea was to generate insoluble nuclei in the size range of 0.1 μm. Vapor condenses on the particle and it grows to several microns. The design of the generator requires some care if the vapor is to condense on the particles rather than on the walls of the chamber. Only a few layers of the condensable material are sufficient to yield an aerosol of uniform size even if the initial distribution of nuclei is polydisperse. A typical formulation uses di-2-ethylhexyl-sebacate (DEHS) condensing on NaC nuclei. This type of generator was used successfully in London by D. Muir and D. Swift in Davies's laboratory and in Frankfurt. Eventually the generator produced stable, reproducible aerosols in the size range from 0.08 to 2.0 μm. This range was extended to 12 μm using a special condensation chimney (Stahlhofen et al., 1975b, 1976).

A general procedure for the development of methods for producing radioactively labeled aerosols and magnetic particles using iron oxide particles was developed for regional deposition and clearance measurements. Monodisperse iron oxide particles with aerodynamic diameters ranging from 1 to 6 μm were obtained by atomizing an aqueous Fe_2O_3 colloid with a spinning top generator and drying the droplets in a furnace. For radioactive labeling, the Fe_2O_3 colloid was mixed with colloidal radioactive gold, ^{198}Au. These particles proved useful for definitive studies as the Fe_2O_3 particles retained their structural integrity. Leaching of ^{198}Au from particles suspended in body liquid was less than 1% (Stahlhofen et al., 1979). Later, magnetic particles were produced from a nonmagnetic colloidal Fe_2O_3 solution using a modified spinning top aerosol generator (Möller et al. 1988, 1990; Stahlhofen and Möller, 1992). The particles were then reduced at 800°C to Fe_3O_4 (magnetite). An alternative approach by P. Camner and K. Philipson for producing radioactively labeled particles used 4.7 μm Teflon particles labeled with ^{111}In. The spinning disc generator was developed by Walton and Prewett and perfected by May (1949) at Porton Down.

Monodisperse ultrafine particles can be generated by first producing particles in the size range between 5 and 100 nm by homogeneous condensation. The particles are then classified by selecting particles of a specific electrical mobility. The key step in production is to produce very small particles initially by using a source with a large surface area. The generation of silver aerosols produced from either silver wool or from silver membrane filters and NaCl particles produced from a saturated porous boiling chip have been particularly successful. In the second procedure, a vacuum is pulled off the boiling chips, which are then immersed into a saturated aqueous solution. The solution penetrates into the pores. After evaporation, the pores are lined with fine NaCl crystals. These methods are effective because of the large surface area. The mobility of an aerosol, depending on both the elementary charge and the diameter can be uniquely related to a specific size only for particles smaller than 100 nm. The production of ultrafine radioactively labeled aerosols is more complicated. The vapors must be produced in a high-temperature furnace below 1200°C to avoid contamination of the aerosols by the tube material. This represents a special problem for ^{111}In oxide, which is insoluble in body fluids, has a reasonable half-life, and is a γ-emitter with convenient energy. Roth proposed a method in which (^{111}In) chelate is vaporized and then degraded to indium oxide (Roth et al., 1989; Roth and Stahlhofen, 1990). This approach allowed the production of more detectable aerosols than ones using ^{198}Au. Studies using this approach in animal exposure studies were initiated at Neuherberg in 1995. Another method pioneered at Neuherberg was the use of a spark generator. Ultrafine aerosols of metal and metal oxide were generated by spark generation of a metal electrode in an inert atmosphere of partial oxygen. If the metal electrode contained a radioactively labeled component (i.e., indium), the resulting aerosol would be radioactively labeled. One problem with this method is that frequently the particles were produced as a chain aggregate rather than individual particles.

Although not suitable for human experiments, fibrous aerosols of asbestos have been produced for animal experiments. The methods are based on procedures developed by K. Spurny at the Fraunhofer in Grafschaft. Research on this method is being continued by H. Muhle at the Fraunhofer ITA in Hanover. The particles of fibers are shaken loose from a dry, powdered asbestos in a vibrating generator. The amplitude and frequency of the vibration influence the number of particles. Typically, the particles have a broad distribution of lengths and a narrow distribution of fiber diameters. The individual fibers are collected and allowed to settle in water for a period of several months. The resulting fibers are narrowly distributed. Based on these fibers, Pott (1987) developed a curve relating the degree of carcinogenity of the fibers to their length. Chan and Gonda (1993) developed a procedure for producing fibers from cromolynic acid and then labeling them with ^{99}Tc.

In magnetopneumography (MPG), magnetic particles usually consisting of iron oxide are used instead of particles labeled with a radioactive isotope. These techniques have been used to examine lung particle retention as well as lung macrophage function. Many of the key studies were developed by D. Cohen, P. Valberg, and J. Brain at Harvard. An extensive overview of the application of the method is given in a review article. (Valberg and Brain, 1988). The origin of the method dates back to the early 1970s (Cohen, 1973, 1975). MPG is a method used to measure the biomagnetic fields of a living organism, which can originate from ferromagnetic contamination or from magnetic fields arising from ion transport currents in the human body, such as in the heart (magnetocardiography, MCG) or in the brain (magnetoencephalography, MEG). These methods, which detect very weak magnetic fields, require very sensitive magnetic field sensors and magnetic shielding. SQUIDs are superconducting quantum interferometer devices and allow the detection of magnetic fields of some Tesla (10^{-15} T). MPG was first used to detect ferromagnetic contamination in welders and was later applied to measure the clearance of ferromagnetic tracer particles (Cohen et al., 1979). In 1985, Freedman et al. (1988) presented their study on alveolar clearance in health and disease. Previously, clearance studies for the airways were performed using radioactively labeled iron oxide particles and were limited to a period of 2–3 days. In contrast, clearance mechanisms in the lung periphery require half the time and are unsuitable for humans. Möller and Stahlhofen constructed a system for the detection of magnetic particles in the lungs (Stahlhofen and Möller, 1988). They adapted their system to study phagocytosis of inhaled iron particles. From the relaxation time of the particles in a magnetic field, they were able to induce the viscosity of the intracellular fluids *in situ* (Stahlhofen and Möller, 1992). Among the significant studies are measurements of viscosity and elasticity by magnetic relaxation by Valberg (Valberg and Albertini (1985), estimation of the rate of phagocytosis by I. Nemoto (Cohen et al., 1984a, b), studies of retention of welding fumes by Kalliomaki (Kalliomaki et al., 1980), and cytoplasmic motility by Gehr and Brain. (Brain et al., 1984).

In producing medicinal aerosols, an important objective is to be able to deliver particles in a significant volume. One problem is that the size of the particles is strictly limited to aerodynamic diameters of less than 12 μm. The aerodynamic diameter is proportional to the geometric diameter and the square root of the particle density. A brilliant idea was developed by Edwards and co-workers in which larger particles ~ 20–30 μm that are porous are utilized. Such particles can be used to carry drugs and have the potential for delayed drug delivery. A particle with a porosity of 3.0% and a size of 20 μm has an effective aerodynamic diameter of 3.5 μm, which has an appreciable pulmonary deposition. Along with the deposition curves determined from bolus experiments, the potential to target specific regions of the tracheobronchial tract exists (Batycky et al., 1997; Edwards et al., 1997).

It has been conjectured that aerosol therapy would have an improved effectiveness for smaller amounts of the drug necessary to attain the same therapeutic result (Roth et al., 1996). If one were able to measure droplet size and the inhaled mass of particles, control breathing maneuvers, and estimate the deposition and clearance of the particles in the respiratory tract, one would be able to determine an exact protocol for delivery of the inhaled dose for a drug. Since most of the drugs delivered via inhalation are soluble in water, it is essential to characterize the change in particle size due to solvent evaporation (Roth et al., 1996). Hygroscopic aerosols that undergo growth during inhalation have been a subject of considerable interest for almost 20 years. In Frankfurt, experiments were performed with monodisperse sodium chloride particles with an extended size range from 0.3 to 2.0 μm generated with a vibrating orifice in 1989 (Gebhart et al., 1990). The measurements were made with three subjects for breathing patterns corresponding to rest and light exercises. Theoretical computations were made between the equilibrium size of the aerosol in humid atmospheres typical of the human airways and a model developed by the aerosol group at Neuherberg (Ferron et al., 1988). The deposition efficiency in the lung for particles using the sizes of the wet aerosol showed close agreement with the lung deposition models.

1.5 *Principal laboratories*

The groups that carried out nonproprieteary medical aerosol research include those at New York University, Rochester, Harvard, Lovelace in Albequerque, and several related laboratories in the Research Triangle area of North Carolina. In Europe, these laboratories included the Fraunhofer Institute at Grafschaft and later Hanover, the GSF laboratory at Frankfurt, and the GSF Laboratory at Munich. Unfortunately, only a few of these laboratories and groups have been chronicled in the histories of aerosol research. Below, we mention several of the laboratories that are of particular historical significance with the caveat that there is a need for similar descriptions of those centered at New York University. Unfortunately, there is no documentation here of the work at Harvard and at Rochester, both of which played an essential role in American aerosols science. A good but short description of health research in the United States is provided by McClellan (2000) in the first volume of the history of aerosol science. The Defense laboratory at Porton Down in England, which contributed a number of key instruments, is also mentioned.

A significant number of innovative apparatuses as well as key personnel were developed at the British Laboratory at Porton Down. In 1940, a group euphemistically titled the Emergency Public Health Laboratory Service was established under the leadership of P. Fildes to investigate the practicality of biological warfare (Clark, 2003). More than 20 years earlier, Porton Down had specialized in research on defense against chemical weapons so that its choice for aerosol work was logical. Among the scientists who passed through the laboratory and made noteworthy contributions to aerosol science were D. Henderson, C.N. Davies, K. May, H. Walton, H. Green, W. Lane, and R. Dorman. Among the devices and instruments perfected at Porton Down were the inertial impactor, the spinning disc generator, the conifuge, and the Henderson protocol for generating aerosols from bacteria. In 1950, the conifuge, a device exploiting centrifugal force in order to obtain a size-dependent separation based on particle size, was developed. Another size classifier, an inertial impactor, proved to be more useful and had a much greater impact. This device was developed by K. May. The idea was to force the air stream carrying the aerosol to negotiate a sharp 90° bend. If the particles have sufficient inertia (proportional to particle density, velocity, and the square of the diameter), they will depart from the gas stream lines and impact on the collection surface. The device developed by May and also later impactors operate on the principle that after each stage the sizes of the openings of the device are reduced so that the velocity is increased. The inertial effect is magnified so that a smaller size of particles has the critical value of the inertial parameter (Stokes number) for collection. The particle concentrations are then obtained by weighing the mass collected at each stage. A major subarea of health-related aerosol research is the development of procedures for producing aerosols of a given size range and composition. Before the development of effective spectrometers over a wide size range, it was necessary to attempt to produce particles of a single size (monodisperse aerosols). One method of doing this was by using the spinning disc generator. This device, first developed by H. Walton and perfected by K. May, consists of a rotating cone. The fluid introduced at the center and apex of the cone then dribbles down the side of the cone in rivets. As the cone rotates at a controlled speed, droplets of a narrow distribution are shed. As these evaporate, particles with a narrow size range are generated. It has the advantage that a comparatively high volume of aerosols can be generated, and it is especially useful in generating radioactively labeled mineral aerosols.

In 1960, the Lovelace organization began to study inhalable radionuclecides. In 1966, R. McClellan assumed leadership of the organization. It has remained a center of aerosol-related health research in the United States to date. Originally, the aerosol group was strongly influenced by products of the Rochester group. Some notable aerosol scientists who worked at Lovelace include Mercer, Thomas, Phalen, Raabe, Newton, Yeh, Chen, and Chang. The group was noteworthy for its emphasis on aerosol physics.

Arguably, the center associated with the initiation of the most significant developments in the use of aerosols was the Department of Environmental Medicine at New York University. Unfortunately, the developments there have not yet been featured in one of the recent reports of the history of aerosol science. The innovation is especially associated with B. Altshuler, E. Palmes, R. Albert, and M. Lippmann. Since 1967, E. Palmes's group measured the total deposition of aerosols during breath holding in order to measure the pulmonary air spaces (Palmes et al., 1967; Palmes, 1973). They also studied the variability of the size of airspaces among humans (Lapp et al., 1975). Many of the ideas used to investigate regional deposition, clearance, and total deposition had their origin at NYU. The program has exerted a substantial influence on American aerosol research. In addition to the scientists mentioned above, B. Cohen, T. Chan, D. Yeats, and C.S. Wang have been associated with major contributions to pulmonary research. The use of aerosols in studies of pulmonary behavior at NYU dates from at least 1959 (Altshluler et al., 1959).

One of the most influential centers in pulmonary aerosol research was the Institute of Biophysical Radiation Research (an external Institute of the Research Center for Environment and Health) in Frankfurt. This institute was established after the retirement of Boris Rajewsky of the Max-Planck Institute for Biophysics in 1968. The core of the aerosol research comprised W. Stahlhofen, J. Heyder, J. Gebhart, and C. Roth. The latter two members coming from Battelle Frankfurt provided specialized expertise in optical measurements and instrumentation. The research at Frankfurt from 1968 to 1994 provided a comprehensive overview of pulmonary research with aerosols, including the development of procedures for estimating total deposition, regional deposition, clearance, and enabled detection of changes owing to lung diseases using radioactive isotopes, breath-holding techniques, the bolus procedure, and the use of magnetic particles. At least nine Ph.D. theses, eighteen diploma theses, and three habilitations were carried out at Frankfurt. The contributions of the group were recognized by the international aerosol community in that two members of the group received the career achievement award of ISAM. The members have also received the Thomas T. Mercer award of ISAM and AAAR, the Smoluchowski award of GAeF and the Whitby award of AAAR. Members of the institute played a key role in the founding of GAeF and in reconstituting ISAM as a more broadly based international group. An extended history of this group is to appear in the second volume of the History of Aerosol Science.

A major laboratory of the Institute of the Research Center for Environment and Health is located at Neuherberg, a suburb of Oberschleissheim close to Munich. In 1974, a group under the direction of W. Jacobi at the Hahn-Meitner Institute in Berlin moved to Munich. This group included W. Kreyling, B. Haidler, and G. Ferron. In 1986, J. Heyder moved from Frankfurt to direct Project Inhalation. In 1994, this group formed the Institute of Inhalation Biology. The group originally specialized in inhalation studies on animals — beagles, mice, and rats. The studies with beagles has since been suspended. In recent years, they have emphasized the inflammation properties of aerosols, especially ultrafine aerosols. Especially fruitful was the work with G. Ferron on hygroscopic growth of particles, and with H. Schulz and W. Kreyling on particle clearance from the lungs, their development of the spark generator for ultrafine aerosols, their work on the health effects of ultrafine aerosols (Peters, 1997), and the development of protocol for metering controlled breathing and inhalation volumes for animals as small as mice. They have focused on chemical changes in proteins and lipids occurring during lung infections. A particularly innovative approach was to examine studies contrasting the behavior of different mouse breeds and the use of a cell exposure facility. The institute provides the leadership for Projektfeld Aerosole, whose major focus is on the environmental aspects of aerosols.

A second major center of health-related aerosol research is associated with the Fraunhofer Institute. In 1972, the Institute for Aerobiologie at Grafschaft founded in 1957 began to emphasize aerosol research. In 1974, the name of the institute was changed to Fraunhofer Institute of Toxicology and Aerosol Research (Fh-ITA). A second major branch was located in Hanover. Eventually, aerosol research at Grafschaft was phased out while research continues to this day at Hanover. The leadership of the institute rested with H. Oldiges and W. Stöber, who became institute codirectors in 1973 at Grafschaft, and also W. Stöber, H. Marquardt, U. Mohr, and U. Heinrich (the current director). Initially, the major focus of the institution was on the development of a methodology for producing aerosols for animal inhalation studies especially asbestos fibers and cadmium oxide particles and instrumentation for particle measurements especially the development of the spiral centrifuge, which is associated with the laboratory. The work on measurements of asbestos in the ambient environment and several years after inhalation are especially noteworthy. In recent years, the emphasis of the research has shifted toward inhalation and toxicology studies with humans, whereas the initial work at Grafschaft was limited to animal studies.

1.6 Journals and organizations

The three principal journals dedicated to Aerosol Science are the *Journal of Aerosol Science (JAS)*, the *Journal of Aerosol Science and Technology*, and the *Journal of Aerosol Medicine*. At the time of writing this article in 2003, all of these journals were doing well, with comparatively strong impact factors between 1 and 2. These journals have close associations with the European Aerosol Assembly (originally GAeF), the American Association for Aerosol Research (AAAR), and the International Society for Aerosols in Medicine. The *Journal of Aerosol Medicine* is the only one of the three journals that publishes predominantly articles relating to the application of Aerosols in medicine. Below, the journals are discussed in connection with the organizations with which they are associated.

The International Society for Aerosols in Medicine (ISAM) has been holding one congress at approximately 2-year intervals since 1973. The journal is now publishing its 16th volume, suggesting initial publication around 1988. Also, in 1988 ISAM presented the career achievement award to Paul Morrow. A detailed description of the organization and the journal has not yet been archived. Its meetings are now held on a biannual basis. The *Journal of Aerosol Medicine* has been edited by G. Smaldone since its inception.

The JAS was the first journal whose objective was to exclusively publish papers in aerosol science and technology. The history of the founding of the journal is archived in Kasper (2000). During its first 14 years, the contents of the journal covered nearly all aspects of aerosol science and strongly reflected the editorial judgment of its editor and founder C.N. Davies. In 1982, the journal entered into a close and formal partnership with the Gesellschaft für Aerosolforschung (GAeF). Now, there is a partnership association with the European Aerosol Assembly (EAA), which is an umbrella organization including a number of European national and regional aerosol organizations. A description of the founding of these organizations is given in the History of Aerosol Science edited by O. Preining and E.J. Davis. Gebhart (2000) indicates that the organization was founded in 1972 following an annual meeting, Arbeitestagung Schwebstofftechnik, which had started in 1953. The new organization pursued an active policy of attempting to recruit medical papers in the meeting, showing an increase from 5 to more than 20%. From 1982 to 1995, the *JAS* was the official voice for the annual meeting of GAeF. In 1995, EAA took over this role, with the JAS being the official journal of all European Aerosol societies. Although the JAS continues to publish health-related papers, they represent a comparatively small fraction of the annual meeting and the journal. The editors of the journal since Davies were D. Hochrainer, J. Vincent, G. Kasper, and E.J. Davis.

In 1982, both the American Association of Aerosol Research and the *Journal of Aerosol Science and Technology* (*AS&T*) were founded. The two organizations have had a close association with each other since the founding of AAAR. A description of the founding and the first 18 years of the organization are given in Ensor (2000). The contents of the journal and the AAAR meeting are similar to that of the European organization, with medical papers ranging from 5 to 20% of the total papers. The editors of the journal were D. Shaw, B.Y.H. Liu, P. Hopke, and R. Flagan.

1.7 Awards

The number of awards received is an important indicator of the progress of aerosol research in medicine. The three organizations most directly associated with aerosol research are AAAR, International Society for Aerosols in Medicine (ISAM), and the European Aerosol Congress (and affiliated organizations). The AAAR and the ISAM sponsor the T.M. Mercer Award. ISAM sponsors several awards, one of which is the Career Achievement Award that recognizes scientific accomplishments over a lifetime. Following is a list of the 14 award recipients:

- T.M. Mercer Award
  1995 P. Morrow U.S.
  1996 J. Brain U.S.
  1997 R. McClellan U.S.
  1998 W. Stöber Germany
  2000 W. Stahlhofen Germany
  2001 A. Ben-Jebria U.S.
  2002 C. Pope U.S.
  2003 R. Wolff Canada
  2004 G. Oberdörster U.S.
- Career Achievement Award (ISAM)
  1988 P. Morrow U.S.
  1991 W. Stahlhofen Germany

1993	H. Matthys	Germany
1995	J. Ferin	U.S.
1997	Y. Sato	Japan
2001	J. Heyder	Germany
2003	P. Gehr	Switzerland

In addition, three recipients of the International Aerosol Fellow Award — W. Stöber (1994), K. Willeke (1996), and R. McClellan (1998) — received the award primarily for their work in health-related aerosols. Also, two recipients of the David Sinclair Award — M. Lippmann (1990) and J. Vincent (1994) — were recognized for their work on aerosols in medicine and in industrial hygiene, respectively. Three scientists W. Hofmann (Austria), H. Hauck (Austria) and G. Smaldone (U.S.) received the Juraj Ferrin Award of ISAM.

Of the two principal awards for aerosol scientists below or close to 40 years of age, three have been received by aerosol scientists primarily associated with health-related aerosols. G. Scheuch received the Whitby Award of AAAR in 1993, W. Möller and D. Edwards received the Smoluchowski Award of the Gesellschaft für Aerosolforschung in 1994 and 2001, respectively. ISAM also has an award for young investigators. The following is a list of award recipients:

- Young Investigator Award (ISAM)

1993	A. Clark.	U.S.
1995	K. Driscoll	U.S.
1997	M. Svartengren	Sweden
1999	P. Diot	France
2001	W. Finlay	Canada
2003	J. Brown	U.S.

References

Albert, R.E., Lippman, M., Peterson, H.D., Berger, J., Sanborn, K., and Bohning, D., Bronchial deposition and clearance of aerosols Arch. Intern. Med. 131, 115–127, 1973.

Altshuler, B., Behavior of airborne particles in the respiratory tract, in *Circulatory and Respiratory Mass Transport*, Wolstenholme, G.E.W. and Knight, J., Eds., Churchill, London, 1969.

Altshuler, B., Palmes, E.D., Yarmus, L., and Nelson, N., Intrapulmonary mixing of gases with aerosols, *J. Appl. Physiol.*, 14, 321–327, 1959.

Bennett, W.D., Scheuch, G., Zeman, K.L., Brown, J.S., Kim, C., Heyder, J., and Stahlhofen, W., Bronchial airway deposition and retention of particles in inhaled boluses: effect of anatomic dead space, *J. Appl. Physiol.*, 85, 685–694, 1998.

Bennett, W.D., Chapman, W.F., Lay, J.C., and Gerrity, T.R., Pulmonary clearance of inhaled particles 24 to 48 hours post deposition: effect of beta-adrenergic stimulation, *J. Aerosol Med.*, 6, 53–62, 2003.

Brain, J.D., Knudson, D.E., Sorokin, S.P., and Davis, M.A., Pulmonary distribution of particles given by intratracheal instillation or by aerosol inhalation, *Environ. Res.*, 11, 13–33, 1976.

Brain, J.D., Bloom, S.B., Valberg, P.A., and Gehr, P., Correlation between the behavior of magnetic iron oxide particles in the lungs of rabbits and phagocytosis, *Exp. Lung Res.*, 6, 115–131, 1984.

Brand, P., Ruoss, K., and Gebhart, J., Performance of a mobile aerosol spectrometer for an in-situ characterization of environmental aerosols in Frankfurt city. *Atmos. Environ.*, 26A, 2451–2457, 1992.

Batycky, R.P., Hanes, J., Langer, R., and Edwards, D.A., A theoretical model of erosion and macromolecular drug release from biodegrading microspheres, *J. Pharm. Sci.*, 86, 1464–1477, 1997.

Chan, T.L. and Lippmann, M., Experimental measurements and empirical modeling of the regional deposition of inhaled particles in humans, *J. Am. Ind. Hyg.*, 41, 399–408, 1980.

Chan, H.K. and Gonda, I.K., 1993, Preparation of radiolabeled materials for studies of deposition of fibers in the human respiratory tract, *J. Aerosol Med.*, 6, 241–247, 1993.

Clark, J.M., A history of the development of aerosol science for chemical and biological defense at Porton Down, personal communication, 2003.

Cohen, D., Ferromagnetic Contaminants in the lungs and other organs of the body, Science, 180, 745–748, 1973.
Cohen, D., Measurements of the magnetic fields produced by the human heart, brain, and lungs, *IEEE Trans. Mag.*, MAG-11, 694–700, 1975.
Cohen, D., Arai, S.F., and Brain, J.D., Smoking imparis long term clearance from the lungs, Science, 204, 514–517, 1979.
Cohen, D., and Nemoto, I., Ferrimagnetic particles in the lung. Part I.: the magnetizing process, *IEEE Trans. Biomed. Eng.*, BME-31, 261–273, 1984.
Cohen, D., Nemoto, I., Kaufman, L., and Arai, S., Ferrimagnetic particles in the lung. Part II.: the relaxation process, *IEEE Trans. Biomed. Eng.*, BME-31, 274–285, 1984.
Edwards, D.A., Hanes, J., Caponetti, G., BenJebria, A., Eskew, M.L., Mintzes, J., Deaver, D., Lotan, N., and Langer, R., Large porous particles for pulmonary drug delivery, Science, 276, 1868–1871, 1997.
Ensor, D.S., History of the American Association for Aerosol Research, in *History of Aerosol Science*, Preining, O., and Davis, E.J., Eds., Verlag der Österreichischen, Akademie der Wissenschaften, 2000, pp. 317–328.
Ferron, G.A., Haider, B., and Kreyling, W.G., Inhalation of salt aerosol particles — II, *J. Aerosol Sci.*, 19, 611–634, 1988.
Freedman, A.P., Robinson, S.E., and Street, M.R., Magnetopneumographic study of human alveolar clearance in health and disease, *Ann. Occup. Hyg.*, 32, 809–820, 1988.
Gebhart, J., A Particle Size Spectrometer for Aerosols by use of Small Angle Scattering in a Laser Beam, Arbeitstagung Schwebstofftechnik, Battelle Institute, eV., Frankfurt/M, September 1969.
Gebhart, J., Response of single particle counters to particles of irregular shape, *Part. Part. Syst. Charact.*, 8, 40–47, 1991.
Gebhart, J., History of the Gesellschaft für Aerosolforschung GAeF, in *History of Aerosol Science*, Preining, O. and Davis, E.J., Eds., Verlag der Österreichischen, Akademie der Wissenschaften, 2000, pp. 285–294.
Gebhart, J. and Heyder, J., Removal of aerosol particles from stationary air within porous media, *J. Aerosol Sci.*, 16, 175–187, 1985.
Gebhart, J. and Roth, C., Background noise and counting efficiency of single optical particle counters, in Aerosols: Formation and Reactivity, 2nd International Conference, Berlin, Pergamon Press, Oxford. 1986.
Gebhart, J., Heyder, J., and Stahlhofen, W., Use of aerosols to estimate pulmonary airspace dimensions, *J. Appl. Physiol.*, 51, 465–476, 1981a.
Gebhart, J., Heyder, J., and Stahlhofen, W., *J. Appl. Physiol. Respir. Environ. Exercise Physiol.*, 51, 465–476, 1981b.
Gebhart, J., Heigwer, G., Heyder, J., Roth, C., and Stahlhofen, W., The use of light scattering photometry in aerosol medicine, *J. Aerosol Med.*, 1, 89–111, 1988.
Gebhart, J., Anselm, A., Ferron, G., Heyder, J., and Stahlhofen, W., Experimental data on total deposition of hygroscopic particles in the human respiratory tract, in International Aerosol Conference, Kyoto, 1990.
Gentry, J.W., Gebhart, J., and Scheuch, G., A vector model for the total aerosol deposition efficiency under the simultaneous action of more than one mechanism, *J. Aerosol Sci.*, 25, 315–325, 1994.
Heyder, J., Charting human thoracic airways by aerosols, *Clin. Phys. Physiol. Meas.*, 4, 133–137, 1983.
Heyder, J. and Gebhart, J., Gravitational deposition of particles from laminar flow through inclined circular tubes, *J. Aerosol Science*, 8, 289–295, 1977.
Heyder, J., Roth, C., and Stahlhofen, W., A laser spectrometer for size analysis of small airborne particles, *J. Aerosol Sci.*, 2, 341–351, 1971.
Heyder, J., Gebhart, J., Heigwer, G., Roth, C., and Stahlhofen, W., Experimental studies of the total deposition of aerosol particles in the human respiratory tract, *J. Aerosol Sci.*, 4, 191–208, 1972.
Heyder, J., Armbruster, A., Gebhart, J., Greim, E., and Stahlhofen, W., Total deposition of aerosol particles in the human respiratory tract for nose and mouth breathing, *J. Aerosol Sci.*, 6, 311–328, 1975.
Heyder, J., Gebhart, J., Roth, C., Stahlhofen, W., Stuck, B., Tarroni, G., DeZaiacomo, T., Formignani, M., Melandri, C., and Prodi, V., Intercomparison of lung deposition data for aerosol particles, *J. Aerosol Sci.*, 9, 147–155, 1978.

Heyder, J., Gebhart, J., Stahlhofen, W., and Stuck, B., Biological variability of particle deposition in the human respiratory tract during controlled and spontaneous mouth breathing, *Ann. Occup. Hyg.*, 26, 137–147, 1982.

Heyder, J., Gebhart, J., and Scheuch, G., Interaction of diffusional and gravitational particle transport in aerosols, *Aerosol Sci. Technol.*, 4, 315–326, 1985.

Heyder, J., Gebhart, J., Rudolf, G., Schiller, C.F., and Stahlhofen, W., Deposition of particles in the human respiratory tract in the size range 0.005–15 μm, *J. Aerosol Sci.*, 17, 811–825, 1986.

Heyder, J., Blanchard, J.D., Feldman, H.A., and Brain, J.D., Convective mixing in the human respiratory tract: estimate with aerosol boli, *J. Appl. Physiol.*, 64, 1273–1278, 1988.

ICRP, Human Respiratory Tract Model for Radiological Protection, ICRP Publication 66, Pergamon Press, Oxford, 1994.

Jacobi, W., Eichler, J., and Stolterfoht, N., Particle Size Spectrometry of Aerosols by Lightscattering in a Laser Beam, Hahn-Meitner Institute B65, November 1967.

Kasper, G., A history of the journal of aerosol science, in *History of Aerosol Science*, Preining, O., and Davis, E.J., Eds., Verlag der Österreichischen, Akademie der Wissenschaften, pp. 393–402, 2000.

Kalliomaki, K., Kalliomaki, P.L., Kelha, V., and Vaaranen, V., Instrumentation for measuring the magnetic lung contamination of steel welders, *Ann. Occup. Hyg.*, 23, 174–184, 1980.

Lapp, N.L., Hanknson, J.L., Amandus, H., and Palmes, E.D., Variability in the sizes of airspaces in normal human lungs as estimated by aerosols, *Thorax*, 30, 293–299, 1975.

Lippmann, M., and Albert, R.E., The effect of particle size on the regional deposition of inhaled aerosols in the human respiratory tract, *Am. Ind. Hyg. Assoc. J.*, 30, 257–275, 1969.

Lippmann, M., Yeates, D.B., and Albert, R.E., Deposition, retention, and clearance of inhaled particles, *Br. J. Ind. Med.*, 37, 337–362, 1980.

McClellan, R.O., History of aerosol science and health reseach, in *History of Aerosol Science*,Preining, O. and Davis, E.J., Eds., Verlag der Österreichischen, Akademie der Wissenschaften, 2000, pp. 129–146.

May, K.R., An improved spinning top homogeneious spray apparatus, *J. Appl. Phys.*, 20, 932–938, 1949.

Möller, W., Stahlhofen, W., and Roth, C., Preparation of spherical magnetic aerosols for long time clearance and relaxation stuidies in the human lung, *J. Aerosol Sci.*, 21, S657–S660, 1988.

Möller, W., Roth, C., and Stahlhofen, W., Improved spinning top aerosol-generator for the production of highly concentrated ferrimagnetic aerosols, *J. Aerosol Sci.*, 21, S657–S660, 1990.

Morsy, S.M., Werner, E., Stahlhofen, W., and Pohlit, W., Parametrization of the operational conditions of whole body counting systems, *Atomkernenergie*, 28, 289–294, 1976.

Muir, D.C., The effect of airways obstruction on the single breath aerosol curve, in *Airway Dynamics*, Bouhuys, A. and Thomas, Charles C., Eds. 1970.

Muir, D.C.F. and Davies, C.N., The deposition of 0.5 μm diameter aerosols in the lung of man, *Ann. Occup. Hyg.*, 10, 161–174, 1967.

Muir, D.C., Sweetland, K., and Love, R.G., Inhaled aerosol boluses in man, in *Inhaled Particles III*, Walton, W.H., Ed., Unwin Brothers, 1971.

Oberdorster, G., Lung clearance of inhaled insoluble and soluble particles, *J. Aerosol Med.*, 4, 289–330, 1988.

Palmes, E.D., Measurements of pulmonary air spaces using aerosols, Arch. Intern. Med. 131, 76, 1973.

Palmes, E.D., Altshuler, B., and Nelson, N., Deposition of aerosols in the human respiratory tract during breath holding, *Inhaled Particles and Vapours II*, Davies, C.N. Ed., Pergamon Press, Oxford, pp. 339–347, 1967.

Palmes, E.D., Wang, C.S., Goldring, R.M., Altshuler, B., Effect of depth of inhalation on aerosol persistence during breath holding, *J. Appl. Physiol.*, 34, 356–360, 1973.

Peters, A., Wichmann, H.E., Tuch, T., Heinrich, J., Heyder, J., Respiratory effects are associated with the number of ultrafine particles, *Am. J. Respir. Crit. Care Med.*, 155, 1376–1383, 1997.

Pott, F., Problems in defining carcinogenic fibers, *Ann. Occup. Hyg.*, 31, 799–802, 1987.

Roth, C. and Gebhart, J., Rapid particle size analysis with an ultramicroscope, *Microsc. Acta*, 81, 119–129, 1978.

Roth, C. and Gebhart, J., Aqueous droplet sizing by inertial classification, *Part. Part. Syst. Charact.*, 13, 192–195, 1996.
Roth, C. and Stahlhofen, W., Radioactively labeled ultrafine particles for clearance measurements. *J. Aerosol Sci.*, 21, S443–S446, 1990.
Roth, C., Westenberger, S., and Kreyling, W.G., Production of ^{111}In-labelled monodisperse aerosol particles, *J. Aerosol Sci.*, 20, 1289–1292, 1989.
Roth, C., Scheuch, G., and Stahlhofen, W., Clearance measurements with radioactively labeled ultrafine particles, *Ann. Occup. Hyg.*, 38, 101–106, 1994.
Roth, C., Gebhart, J., Just-Nübling, G., von Eisenhart-Rothe, B., and Beinhauer-Reeb, I., Characterization of amphotericin B aerosols for inhalation treatment of pulmonary aspergillosis, *Infection* 24, 273–279, 1996.
Rudolf, G., Gebhart, J., Heyder, J., Schiller, Ch. F., and Stahlhofen, W., An empirical formula describing aerosol deposition in man for any particle size, *J. Aerosol Sci.*, 17, 350–355, 1986.
Rudolf, G., Köbrich, R., and Stahlhofen, W., Modelling and algebraic formulation of regional aerosol deposition in man, *J. Aerosol Sci.*, 21, S403–S406, 1990.
Scheuch, G., Particle recovery from human conducting airways after shallow aerosol bolus inhalation, *J. Aerosol Sci.*, 22, 957–973, 1994.
Scheuch, G. and Stahlhofen, W., Dispersion of aerosol boluses in the human tracheo-bronchial tract during period of breath-holding, *J. Aerosol Sci.*, 21, S431–S434, 1990.
Scheuch, G. and Stahlhofen, W., The recovery of 1 μm aerosol particles from large human airways, *J. Aerosol Sci.*, 23, S477–S481, 1992.
Scheuch, G. and Stahlhofen, W., Deposition and dispersion of aerosols in the airways of the human respiratory tract: the effect of particle size, *Exp. Lung Res.*, 18, 343–358, 1992.
Scheuch, G., Stahlhofen, W., and Heyder, J., An approach to deposition and clearance measurement in human airways, *J. Aerosol Med.*, 9, 35–41, 1996.
Schiller, Ch.F., Gebhart, J., Heyder, J., Rudolf, G., and Stahlhofen, W., Deposition of monodisperse insoluble aerosol particles in the 0.005–0.2 μm size range within the human respiratory Tract, *Ann. Occup. Hyg.*, 32, 41–49, 1988.
Smaldone, G.C., Perry, R.G., Bennett, W.D., Messina, M.S., Zwang, J., and Ilowite, J., Interpretation of "24 hour lung retention" in studies of mucociliary clearance, *J. Aerosol Med.*, 1, 11–20, 1988.
Stahlhofen, W., Human lung clearance following bolus inhalation of radioaerosols, in *Extrapolation of Dosimetric Relationships for Inhaled Particles and Gases*, Crapo, J.D. et al., Academic Press, San Diego, 1989, pp. 153–166.
Stahlhofen, W. and Möller, W., Description of a biomagnetic method for detection of the behavior of magnetic aerosols in the human lungs, *J. Aerosol Sci.*, 19, 1087–1091, 1988.
Stahlhofen, W., and Möller, W., *In vivo* magnetopneumography with spherical magnetite particles — analysis of shear-rate dependence of intracellular viscosity. *J. Aerosol Sci.*, 23, 515–518, 1992.
Stahlhofen, W., Armbruster, L., Gebhart, J., and Greim, E., Particle sizing of aerosols by single particle observation in a sedimentation cell, *Atmos. Environ.*, 9, 851-857, 1975a.
Stahlhofen, W., Gebhart, J., Heyder, J., and Roth, C., Generation and properties of a condensation aerosol of di-2-ethy-sebacate (DES) — I: description of the generator, *J. Aerosol Sci.*, 6, 161–167, 1975b.
Stahlhofen, W., Gebhart, J., and Roth, C., Generation and properties of a condensation aerosol of di-2-ethy-sebacate (DES) – III: Experimental investigations into the process of aerosol formation, *J. Aerosol Sci.*, 7, 223–231, 1976.
Stahlhofen, W., Gebhart, J., Heyder, J., and Stuck, B., Herstellung von monodispersen Fe_2O_3 — Testaerosolen mit Hilfe der Zentrifugalzerstaeubung, *Staub – Reinhaltung der Luft*, 39, 73–108, 1979.
Stahlhofen, W., Gebhart, J., and Heyder, J., Experimental determination of the regional deposition of aerosol particles in the human respiratory tract, *Am. Ind. Hyg. Assoc. J.*, 41, 385–398, 1980.
Stahlhofen, W., Gebhart, J., Heyder, J., Philipson, K., and Camner, P., Intercomparison of regional deposition of aerosol particles in the human respiratory tract and their long-term elimination, *Exp. Lung Res.*, 2, 131–139, 1981a.
Stahlhofen, W., Gebhart, J., and Heyder, J., Biological variability of regional deposition of aerosol particles in the human respiratory tract, *Am. Ind. Hyg. Assoc. J.*, 42, 348–351, 1981b.

Stahlhofen, W., Gebhart, J., Rudolf, G., and Scheuch, G., Measurement of lung clearance with pulses of radioactively labeled aerosols, *J. Aerosol Sci.* 17, 333–336, 1986.

Stahlhofen, W., Gebhart, J., and Scheuch, G., Aerosol boluses, in *ASTM STP 1024*, Utell, M.J. and Frank, R., Eds., American Society for Testing Materials, Philadelphia, 1989a, pp. 127–138.

Stahlhofen, W., Rudolf, G., and James, A.C., Intercomparison of experimental regional aerosol deposition data, *J. Aerosol Med.*, 2, 285–308, 1989b.

Valberg, P.A. and Albertini, D.F., Cytoplasmic motions, rheology, and structure probed by a novel magnetic particle method, *J. Cell. Biol.*, 101, 130–140, 1985.

Valberg, P.A. and Brain, J.D., Lung particle retention and lung macrophage function evaluated using magnetic aerosols, *J. Aerosol Med.*, 1, 331–349, 1988.

Wolff, R.K., Tillquist, H., Muggenburg, B.A., Harkema, J.R., and Mauderly, J.L., Deposition and clearance of radiolabeled particles from small ciliated airways in beagle dogs, *J. Aerosol Med.*, 2, 261–270, 1989.

chapter two

Aerosol properties

William C. Hinds
UCLA School of Public Health

Contents

2.1 Introduction 19
2.1.1 Concentration 21
2.2 Basic particle properties 21
2.2.1 Particle size 21
2.2.2 Particle density 23
2.2.3 Particle shape....................24
2.3 Kinetic properties of aerosols 24
2.3.1 Mechanical....................24
2.3.1.1 Settling velocity 24
2.3.1.2 Inertial 26
2.3.1.3 Diffusion 26
2.3.2 Electrical....................27
2.3.3 Optical 27
2.3.4 Growth processes 29
2.3.4.1 Coagulation....................29
2.3.4.2 Condensation 29
2.4 Chemical properties 30
2.5 Atmospheric particle size ranges 32
Glossary 32
References 33

2.1 Introduction

Aerosols can be defined as solid or liquid particles suspended in air. Particle sizes range from about 2 nm to more than 100 μm. An aerosol is a two-phase system, consisting of the particles and the gas they are suspended in. The term aerosol refers to the mixture, that is, both the particles and the suspending gas. At some level, they are unstable, that is, their properties change with time. From the standpoint of health effects, we consider only those aerosols that are sufficiently stable to be inhaled.

We distinguish different types of aerosols based on their method of generation, the size of the particles, and whether the particles are solid or liquid. Typical particle size ranges and other properties are illustrated in Figure 2.1.

1-56670-611-4 / 05 / $0.00+$1.50

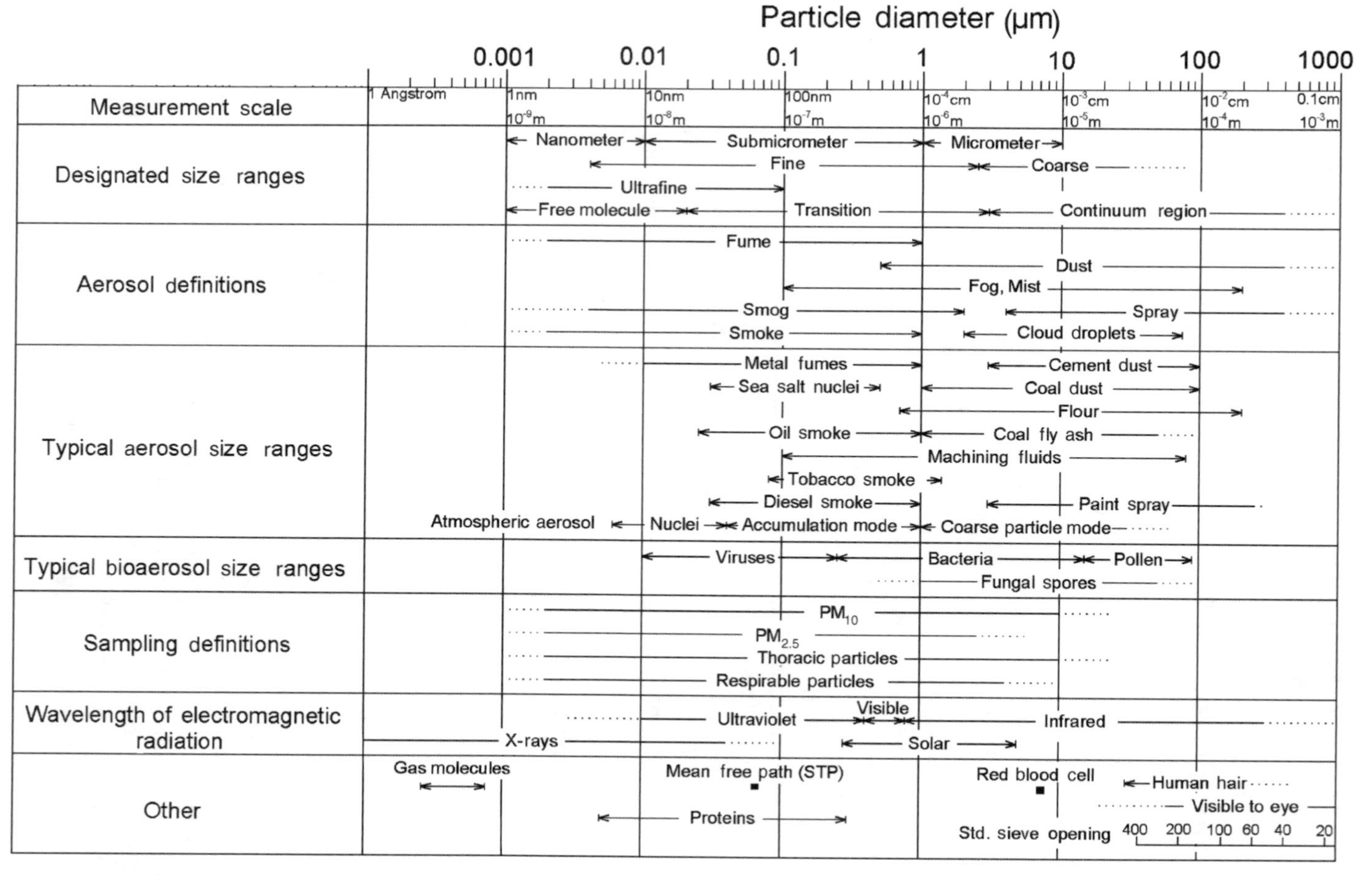

Figure 2.1 Typical aerosol particle size ranges (Hinds, 1999).

2.1.1 Concentration

The *mass concentration* of an aerosol is the mass of particles per unit volume of the mixture. Common units are mg/m^3, used for standards and measurement of occupational exposure aerosols, and $\mu g/m^3$, used for standards and measurements of air pollution. Aerosols may also be characterized by *number concentration*, the number of particles per unit volume. Number/cm^3 is often used. Number concentration is used for bioaerosols, asbestos, and ultrafine particles.

2.2 Basic particle properties

2.2.1 Particle size

Particle size is the most important attribute characterizing the properties and behavior of aerosols. Most aerosols have a wide range of sizes and their properties depend strongly on particle size. The symbols d, d_p, or D are commonly used for particle size. Particle size is most commonly expressed in micrometers (μm) but nanometer (nm) is also used for particles less than 0.1 μm. 1 μm is equal to 1000 nm, 10^{-3} mm, 10^{-4} cm, and 10^{-6} m.

As a result of the importance of particle size and the difficulty in measuring particle size distribution, several size-selective sampling schemes have been devised to sample a subset of all airborne particles based on particle size. Particles are separated aerodynamically while still airborne and before being collected. Such sampling collects all particles that are less than a specified size or range of sizes. Health effects-based, size-selective samplings have graded cutoffs in particle size over a well-defined size range. The cutoff size range is chosen to select those particles that can reach a particular region of the respiratory tract and potentially deposit there. Other sizes are removed from the sampled air in a precollector, and thus excluded from the sample.

The shapes of the cutoffs are defined by curves giving the transmission efficiency as a function of particle size of a precollector position immediately upstream of the sample filter. Figure 2.2 shows the size-selective criteria curves for the five most common definitions. Devices for carrying out this type of sampling are described in Chapter four.

As shown in Figure 2, the largest particle size definition is the *inhalable fraction*. It defines the particles that can enter the nose or mouth under average conditions. The ability of airborne particles of a given size to enter the nose or mouth depends on the direction and magnitude of the ambient wind. The upper limit for inhalable particles is not defined.

Next is the *thoracic fraction*, which selects those inhalable particles that can reach the thorax.

PM-10 is similar to thoracic particle mass but has a slightly sharper cutoff. It is usually referred to as particles less than 10 μm in size. Although not exactly correct, as shown in Figure 2, there is a decreasing fraction of particles larger than 10 μm that are included in PM-10, it is a useful way to consider PM-10.

Respirable fraction includes those particles that can reach the alveolar region of the respiratory system. The cutoff size for respirable particles is the particle size for 50% transmission through the precollector or 4.0 μm. Respirable particles are those that can reach the alveolar region, but do not necessarily deposit there. So the curves reflect the ability of particles of different sizes to traverse the preceding airways rather than deposit in the alveolar region.

PM-2.5 is only partly based on health effects and defines those particles that are in the accumulation mode. It consists of all the accumulation mode particles and a small fraction of the coarse particle mode particles.

The most common aerosol particle size distribution is the lognormal distribution. It usually fits the wide range and skew shape of aerosol particle sizes, especially those from

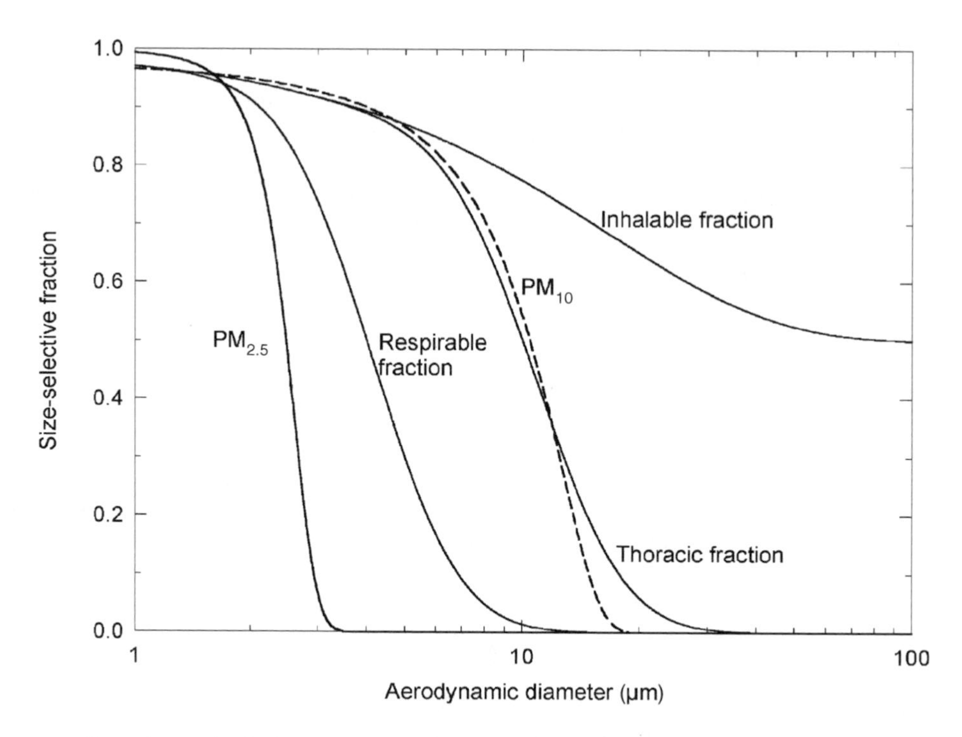

Figure 2.2 Sampling criteria for common size-selective sampling definitions (after Hinds, 1999).

a single source or formation process. Mixtures of aerosols with lognormal size distribution are usually not lognormally distributed and may be bimodal or trimodal. In addition to fitting the data, the mathematical form of the lognormal distribution is well suited for the calculation of many properties of aerosols. For a lognormal distribution, the particle diameters display a characteristic skew distribution (long tail for large sizes), but the distribution of the logarithms of particle diameters has a symmetrical normal distribution. In its simplest form a lognormal distribution is completely defined by two parameters, its geometric mean d_g and its geometric standard deviation, σ_g or GSD. A special property of lognormal distribution is that the median of the distribution is equal to the geometric mean and is usually used instead of the geometric mean. The frequency function for lognormal distribution is given by

$$df = \frac{1}{\sqrt{2\pi}d_p \ln\sigma_g} \exp\left(\frac{-(\ln d_p - \ln CMD)^2}{2\,(\ln \sigma g)^2}\right) dd_p \tag{1}$$

where *CMD* is the count median diameter,

$$\ln (CMD) = \ln d_g = \frac{\Sigma \ln d_i}{N} \tag{2}$$

where N is the total number of particles and

$$\ln \sigma_g = \left(\frac{\Sigma(\ln d_i - \ln CMD)^2}{N-l}\right)^{1/2} \tag{3}$$

For any aerosol there is a whole family of distributions, called moment distributions, that characterize different properties of the particles, such as the number, surface area, volume, or mass. The property must be proportional to a constant power of diameter. The two most common are the count distribution and the volume or mass distribution. The former describes how the particles are distributed among various sizes, while the latter describes how the mass of the particles is distributed among the various sized particles. The count or number distribution is characterized by a count median diameter (*CMD*) and a geometric standard deviation. The mass distribution is characterized by a mass median diameter (*MMD*) and a geometric standard deviation. Another special property of lognormal distribution is that for a given aerosol size distribution, the geometric standard deviation is the same for all moment distributions.

For a given aerosol the mass concentration C_m is linked to the number concentration C_N by the diameter of average mass, d_m, defined by

$$d_{\bar{m}} = \left(\frac{6}{\rho_p \pi N}\ \bar{m}\right)^{1/3} = \left(\frac{\Sigma d_p^3}{N}\right)^{1/3} \tag{4}$$

Thus,

$$C_m = C_N\,\bar{m} = C_N \frac{\rho\pi}{6} d_{\bar{m}}^3 \tag{5}$$

where $\bar{m}$ is the mass of the particle with average mass.

Another special property of lognormal distribution is that any of the variously defined average diameters can be calculated if a median size and a GSD are known or if any two diameters are known. For example, the *MMD* can be calculated from the *CMD* and σ_g by

$$MMD = CMD\ \exp(3\ \ln^2\sigma_g) \tag{6}$$

The diameter of average mass can be obtained by knowing only the *CMD* and σ_g by

$$d_{\bar{m}} = CMD\,\exp(1.5\ln^2\sigma_g) \tag{7}$$

These equations are called Hatch–Choate equations and are described in more detail in Hinds (1999).

Table 2.1 gives the mass and number of molecules for particles of different sizes, and the number and surface area per gram of particles for the range of spherical particle sizes most commonly encountered. It shows a large increase in specific area (surface area/g) as particle size becomes smaller.

2.2.2 *Particle density*

Particle density defines the mass per unit volume of a particle, not to be confused with the density of an aerosol, which, as discussed above, is referred to as mass concentration. By convention standard density is defined as 1000 kg/m^3 or the density of water. The density of droplets and particles from crushed material will be the same as the parent

Table 2.1 Some Properties of Standard Density Spheres

Particle diameter (μm)	Molecules/particle (H_2O)	Mass/particle (μg)	Number of particles/g (g^{-1})	Surface area/g (m^2/g)
0.001	1.8×10^{1}	5.2×10^{-16}	1.9×10^{21}	6000
0.01	1.8×10^{4}	5.2×10^{-13}	1.9×10^{18}	600
0.1	1.8×10^{7}	5.2×10^{-10}	1.9×10^{15}	60
1	1.8×10^{10}	5.2×10^{-7}	1.9×10^{12}	6
10	1.8×10^{13}	5.2×10^{-4}	1.9×10^{9}	0.6
100	1.8×10^{16}	5.2×10^{-1}	1.9×10^{6}	0.06

material. Agglomerated particles, including some smoke and fume particles, have densities less than their parent material because of the void space in their agglomerated structure.

2.2.3 Particle shape

The remaining physical property of particles is their shape. Liquid droplets and some condensation-formed particles are spherical; particles formed by crushing are irregular in shape; and some crystalline particles such as cubical sea salt particles exhibit regular geometric shapes. In the theoretical description of particle properties spherical shape is usually assumed. In the application of these theories to nonspherical particles, correction factors or equivalent diameters are used to estimate their properties. An equivalent diameter is the diameter of a spherical particle that has the same property or characteristic as the nonspherical particle. Aerodynamic diameter, described below, is an example of such an equivalent diameter. For reasonably compact particles, no more than three times longer than they are wide and without sharp edges, shape can often be neglected, that is, a spherical shape with a diameter equal to that of a sphere of equal volume (d_e) or some other approximation can be assumed. Under these conditions, shape seldom accounts for more than a 20% change in particle properties. For more extreme shapes, such as an asbestos fiber that is ten times longer than it is wide, the orientation of the particle must be taken into account and more complicated analysis is required.

2.3 Kinetic properties of aerosols

2.3.1 Mechanical

2.3.1.1 Settling velocity

The simplest and most important kind of aerosol particle motion is steady, straight-line motion resulting from a constant external force exerted on the particle. Under such conditions, aerosol particle velocity almost instantly reaches a constant terminal velocity that is proportional to the magnitude of the external force. When the external force is gravity, we call the resulting motion settling, and the constant velocity a given particle attains in this situation, the terminal settling velocity V_{TS}.

By equating the resisting force exerted by the air, given by Stokes law for spherical particles, to the force of gravity the settling velocity equation is obtained.

$$V_{TS} = \frac{\rho_p d_p^2 g C_C}{18\eta} \quad \text{for } Re_p < 1.0 \tag{8}$$

where ρ_p is the density of the particle (kg/m^3) d_p the diameter of the particle (m), g the acceleration of gravity (9.81 m/s^2) at sea level, C_C the slip correction or Cunningham's correction factor, a dimensionless correction factor for small particles discussed below, and η the viscosity of air (Pa s). Re_p is the Reynolds number that characterizes particles motion. It is given by

$$Re_p = \frac{\rho_a V d_p}{\eta} \tag{9}$$

where ρ_a is the density of the air (1.2 kg/m^3 at standard conditions) and V is the relative velocity between the particle and the air (V_{TS} in this case). Under standard conditions, where atmospheric pressure is 101 kPa (sea level) and the temperature is 293 K (20°C), standard density particles less than 80 μm in diameter will have Reynolds numbers less than 1.0 and their settling velocity is given by Equation (8).

For particles less than 1 μm in diameter, the slip correction factor C_C must be included in Equation (8). The slip correction factor corrects for the fact that Stokes law overestimates the drag force on a particle as its size approaches the gas mean free path (0.066 μm under standard conditions). The slip correction factor is 1.0 for large particles, but increases rapidly as particle size decreases below 1.0 μm. It depends on the ratio of particle size to the mean free path of the gas λ. The slip correction factor is given by Hinds (1999) as

$$C_C = 1 + \frac{\lambda}{d_p}\left[2.34 + 1.05 \exp\left(-0.39 \frac{d_p}{\lambda}\right)\right] \tag{10}$$

The mean free path is inversely related to the density of air, that is, λ equals $7.91\times10^{-8}/\rho_a$ for λ in meters and ρ_a in kg/m^3. Under standard conditions the mean free path is 0.066 μm and the slip correction factor is 1.15 for 1 μm particles, 2.93 for 0.1 μm particles, and 22.98 for 0.01μm particles. For particles greater than 3 μm in diameter, the slip correction factor in Equation (8) can be neglected with less than a 5% error.

The most important and widely used equivalent diameter for calculating particle motion is the aerodynamic diameter d_a. It is defined as the diameter of the aerodynamic equivalent sphere, that is, the spherical, standard density particle that has the same settling velocity as the particle of interest, as shown in Figure 2.3. The aerodynamic diameter standardizes shape (spherical) and density (1000 kg/m^3). Here, the settling velocity is a surrogate for all types of aerodynamic behavior. Thus, for a particular particle the aerodynamic diameter defines the size of the water droplet that has the same aerodynamic behavior as the particle.

Equation (8) for settling velocity can be written in terms of the aerodynamic diameter, d_a, as

$$V_{TS} = \frac{\rho_0 d_a^2 g C_C}{18\eta} \tag{11}$$

where ρ_0 is the standard density (1000 kg/m^3) and C_C is defined in terms of d_a.

In many situations, it is not necessary to know the true size, shape, or density of a particle if its aerodynamic diameter is known. Instruments that rely on aerodynamic separation, such as impactors, are used to measure aerodynamic diameter. For spheres, the particle

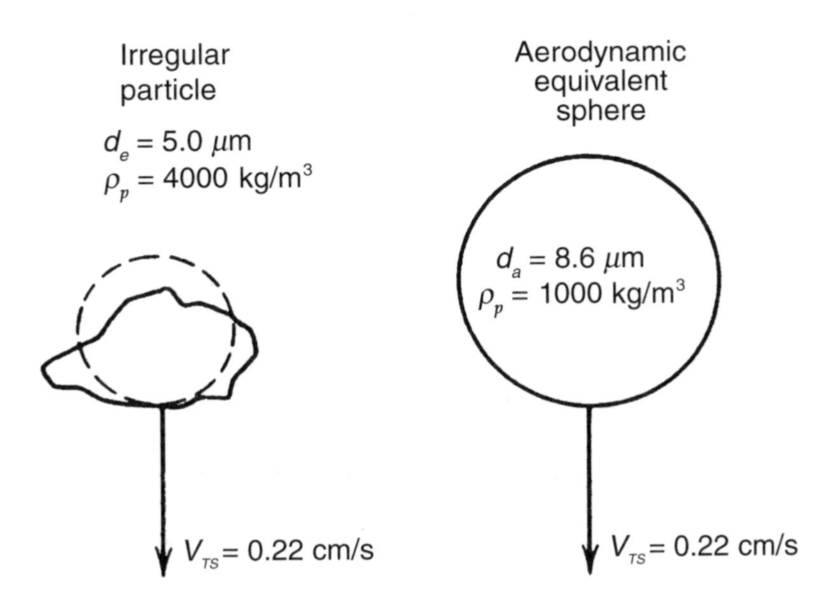

Figure 2.3 An irregular particle and its aerodynamic equivalent sphere (after Hinds, 1999).

diameter d_p and the aerodynamic diameter d_a are related by

$$d_a = d_p\left(\frac{\rho_p}{\rho_0}\right)^{1/2} \tag{12}$$

2.3.1.2 Inertial

The stopping distance of a particle is a measure of the ultimate distance it will travel in still air if it has a specified initial velocity V_0 and it is not acted on by any other force except air drag (resistance). The stopping distance S is given by

$$S = \frac{\rho_p d^2 C_C V_0}{18\eta} = \tau V_0 \tag{13}$$

where τ equals $\rho_p\, d_p^2\, C_C/18\eta$, which is a measure of the agility of a particle — the ease with which it can speed up, slow down, or turn corners.

One of the useful properties of the stopping distance is that it indicates how far a particle will travel in its original direction if the surrounding air is abruptly turned 90°. While this might occur in a tube or in the lung airways, the most important case is the flat plate impactor. Here a jet of air exits a nozzle and is directed normal to a flat plate where it is deflected 90°. Small and light particles follow the air stream to the side of the plate, while large and heavy particles continue toward the plate far enough to collide with it and stick. The parameter that determines whether a given size particle will collide or escape is the Stokes number, *Stk*. The Stokes number is defined for an impactor (Fuchs, 1964) as

$$Stk = \frac{\textit{stopping distance at jet velocity}}{\textit{radius of nozzle}} = \frac{\tau U}{D_j/2} = \frac{\rho_p d_p^2 U C_C}{9\eta D_j} \tag{14}$$

where D_j is the jet diameter, U is the velocity of the jet as it exits the nozzle, and $\rho_p d_p^2$ is usually expressed in terms of aerodynamic diameter $\rho_0 d_a^2$. Impactors are sharp cutoff devices, meaning there is a narrow range of particle sizes between 0 and 100% collection efficiency. Impactors are characterized by their cutoff size, the particle size for 50% collection d_{50}, and the best estimate of the ideal cutoff size. For impactors with round cross-section jets the cutoff size is given by Hinds (1999) and Marple *et al.* (2001),

$$d_{50}\sqrt{C_c} = \left[\frac{9\pi\eta D_j^3(Stk_{50})}{4\rho_p Q}\right]^{1/2} \tag{15}$$

or

$$d_{50}\sqrt{C_c} = \left[\frac{0.0068\eta D_j^3}{Q}\right]^{1/2} \quad \textit{for SI units} \tag{16}$$

where Q is the volumetric flow rate through the impactor nozzle in m^3/s. For an impactor with a rectangular slot nozzle, 0.0068 is replaced by 0.0167 in Equation (16) and D_j^3 is replaced by W^2L, where W and L are the width and length of the slot.

2.3.1.3 Diffusion

Aerosol particles exchange energy with the surrounding gas molecules and this interaction gives rise to particle Brownian motion and diffusion. Brownian motion is the irregular wiggling motion of aerosol particles and diffusion is the net transport of aerosol particles in a concentration gradient. Both processes increase as particle size decreases and are characterized by the particle diffusion coefficient. This diffusion coefficient is given by

$$D = \frac{kTC_C}{3\pi\eta d_p} \tag{17}$$

where k is Boltzmann constant, 1.38×10^{-23} J/K, and T is the absolute temperature in K. The diffusion coefficient determines the rate of particle transport in a concentration gradient J_D by Fick's first law of diffusion.

$$J_D = -D\frac{dN}{dx} \tag{18}$$

In the absence of a concentration gradient, a particle will still undergo Brownian motion. For this situation, the probability of finding a particle along a given axis at a given distance from its starting point after a time t has elapsed is described by a normal distribution with a mean of 0 (i.e., the starting point) and a standard deviation of

$$\sigma = \sqrt{2Dt} \tag{19}$$

2.3.2 *Electrical*

Most aerosol particles have some electrical charge and may be highly charged in some circumstances. When charged particles are subject to an electrostatic field, they experience a force that gives rise to particle motion at its terminal electrostatic velocity V_{TE}

$$V_{TE} = \frac{neEC_c}{3\pi\eta d_p} \quad \text{for } Re_p < 1.0 \tag{20}$$

where n is the number of excess elementary charges (positive or negative), e is the charge on an electron (1.60×10^{-19} C), and E is the strength of the electrostatic field at the location of the particle. The latter is the potential (voltage) gradient in a particular direction at the location of the particle. E can be determined for simple geometries. For example, the field strength between two large parallel plates, one at ground potential and one at 4000 V, separated by 10 mm is 4000/0.01 = 400,000 V/m.

Determining the charge on a particle is more difficult, so we usually estimate the charge from the conditions under which the particle acquired the charge. Most particles in the atmosphere will carry a low level of charge called Boltzmann equilibrium charge distribution. This results from collisions of the aerosol particles and bipolar ions naturally present in the atmosphere. There are about 10^9 ions/m^3 in the normal atmosphere with positive and negative ions approximately equally represented. Regardless of their initial charge, particles come to equilibrium with these ions over time and display bipolar Boltzmann equilibrium charge distribution. An average of 0.67 charges per particle are found on 0.1 μm particles and 2.34 on 1 μm particles. For particles larger than 0.05 μm, the fraction of particles having n charges on them when at Boltzmann equilibrium is given by

$$f_n = \left(\frac{5.31 \times 10^{-6}}{d_p T}\right)^{1/2} \exp\left(\frac{-1.67 \times 10^{-5} n^2}{d_p T}\right) \quad \text{for SI units} \tag{21}$$

Particles can acquire large bipolar charge by flame charging and large unipolar charge by static electrification processes. When particles are intentionally charged by diffusion charging or field charging with the unipolar ions, charge on a particle can reach 100 or 1000 times the Boltzmann equilibrium charge.

2.3.3 *Optical*

Aerosol particles interact strongly with light because their particle size is often of the same order of magnitude as the wavelength of light. This interaction affects visibility and is the basis for concentration and particle size-measuring instruments. Aerosol particles interact with light by scattering or absorbing the light that is incident on them. For homogeneous particles, absorption of light occurs when the material of the particle absorbs light. More important than absorption is the scattering of light by a particle, a result of the diffraction of light passing around the particle, reflection of light at the particle surface, and refraction of light as it passes through the particle. All particles scatter light. For particles < 0.05 μm, visible light scattering can be calculated by Rayleigh scattering. In this size range, scattering is a strong function of particle size, proportional to particle diameter raised to the 6th power. For large particles, greater than 5 or 10 μm, scattering is proportional to the square of particle size. In between these ranges light scattering is described by Mie theory, which is very complicated and requires calculation by computer.

The ability of an aerosol particle to scatter light can be thousands of times stronger than the same mass of molecules in the form of a vapor or in the form of millimeter size particles. This is one reason why aerosol particles have such a strong effect on atmospheric visibility. Visibility is expressed quantitatively as *visual range*, or how far you can see in a specified situation. The most commonly used situation is viewing a dark object silhouetted against a light background such as the daylight sky. For example, when viewing a distant mountain on the horizon against the daylight sky, the presence of aerosol particles between the observer and the mountain brightens the image of the mountain and thus reduces the apparent contrast between the mountain and the sky background. Taking into account the perception threshold of contrast required to distinguish an object from its background, the following equation describes the visual range L_V in terms of aerosol properties:

$$L_V = \frac{4.98}{N d_p^2 Q_e} \tag{22}$$

where L_V is the visual range in meters, N is the particle number concentration in m^{-3}, and Q_e is the extinction efficiency. The latter is a complicated function of particle size as shown in Figure 2.4 for visible light for materials with five refractive indices.

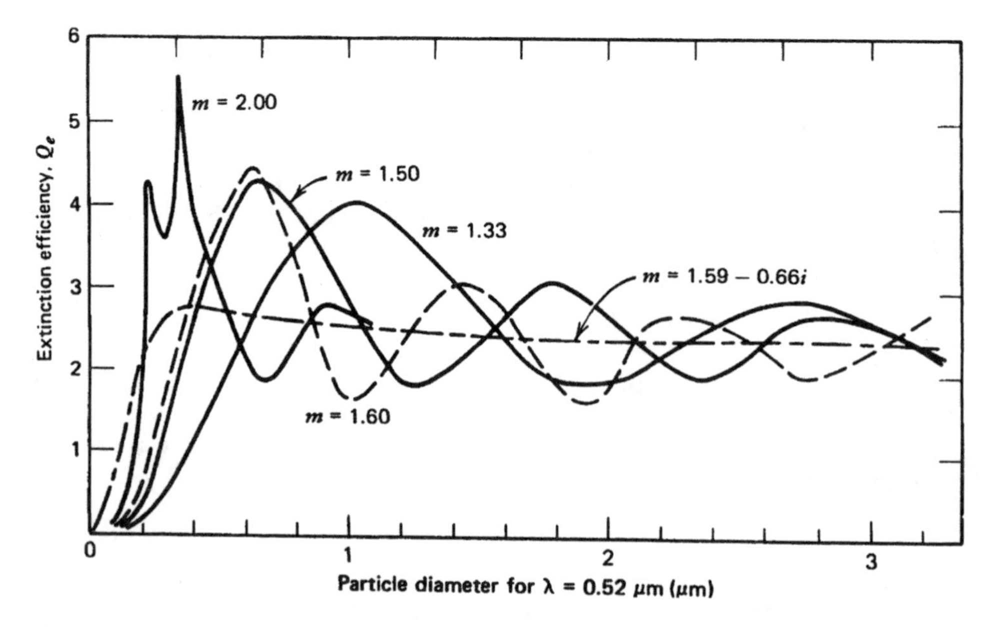

Figure 2.4 Extinction efficiency versus particle size for five refractive indices (after Hodkinson, 1966).

2.3.4 Growth processes

2.3.4.1 Coagulation

Micrometer-sized or smaller particles stick strongly to any surface they contact. As a consequence of this phenomenon along with Brownian motion, aerosol particles collide and stick together due to a process called thermal coagulation. The result is that at high-concentration conditions, the average particle size of aerosol increases and number concentration decreases with time. The rate of coagulation is proportional to number concentration squared and to the coagulation coefficient, K_0. For a monodisperse aerosol, the latter is given by

$$K_0 = 4\pi d_p D = \frac{4kTC_C}{3\eta} \quad \text{for } d_p > 0.1\ \mu\text{m} \tag{23}$$

For particles <0.1 μm, Equation (23) requires an additional correction factor; see Hinds (1999). The effect of coagulation on number concentration N over a period of time t is given by

$$N(t) = \frac{N_0}{1+N_0K_0t} \tag{24}$$

where N_0 is the starting number concentration at time t=0. In an ideal system, as the number concentration is reduced by coagulation the average particle size increases with an eightfold decrease in the number concentration causing an eightfold increase in the average particle volume, which corresponds to a twofold increase in the average particle size. Thus, the particle size changes with time due to coagulation according to

$$d(t) = d_0(1 + N_0K_0t)^{1/3} \tag{25}$$

where d_0 is the average particle diameter at time t=0.

Although Equations (24) and (25) apply to the ideal situation of monodisperse aerosols, they are accurate enough to be useful approximations for polydisperse aerosols and to provide an insight into the importance of coagulation in a wide variety of situations.

The situation for coagulation of polydisperse aerosols is more complicated and often requires computer analysis. An explicit expression for the average coagulation coefficient $\overline{K}$ for a lognormally distributed aerosol is given by Lee and Chen (1984)

$$\overline{K} = \frac{2kT}{3\eta}\left(1 + \exp(\ln^2\sigma_g) + \frac{2.49\lambda}{\text{CMD}}\left[\exp\left(0.5\ln^2\sigma_g\right) + \exp(2.5\ln^2\sigma_g)\right]\right) \tag{26}$$

where CMD and σ_g are the count median diameter and the geometric standard deviation of the lognormal size distribution. For typical polydisperse aerosols with lognormal size distributions coagulation can be up to six times faster than for a monodisperse aerosol of a size equal to the count median diameter.

2.3.4.2 Condensation

The formation and growth of aerosol particles by condensation of water vapor is an important mechanism for aerosol production in nature. It usually occurs when small aerosol particles called nuclei are present in a slightly supersaturated vapor environment. The saturation vapor pressure, p_s in kPa, for water vapor depends only on environmental temperature and is given by

$$p_s = \exp\left(16.7 - \frac{4060}{T-37}\right)\ \text{kPa} \quad \text{for } 273 < T < 373K \tag{27}$$

where T is the absolute temperature in K. The saturation vapor pressure and the actual vapor pressure p at a given temperature are used to define the saturation ratio S_R.

$$S_R = \frac{p}{p_s} \tag{28}$$

A saturation ratio >1.0 indicates that the environment is supersaturated and a saturation ratio <1.0 indicates that the environment is unsaturated.

Nuclei or small particles serve as sites for condensation. Insoluble nuclei participate passively by providing a surface of sufficient size on which condensation and growth can proceed. Soluble nuclei actively facilitate condensation and growth. For example, sodium chloride nuclei are common in the atmosphere and because of their affinity for water they will form concentrated salt droplets at high humidity. The affinity these droplets have for water is balanced by the Kelvin effect, the tendency of very small droplets to lose vapor molecules from their surface even under saturated conditions. When the supersaturation and initial droplets size are sufficient, growth by condensation will occur. For typical conditions, this requires an initial droplet size >0.1 μm and a saturation ratio >1.004.

Once conditions for growth are reached, growth proceeds at a rate given by Equation (29) as long as the supersaturation is maintained.

$$\textit{rate of growth} = \frac{d(d_p)}{dt} = \frac{4D_V M\,(p-p_s)}{\rho_p d_p RT} \quad \text{for } d_p > \lambda \quad \text{and} \quad 1 < S_R < 1.05 \tag{29}$$

where D_v is the diffusion coefficient for vapor molecules, M is the molecular weight of the vapor, and R is the universal gas constant. Thus, the rate of growth is proportional to the difference between the actual partial pressure of vapor and the saturated vapor pressure at the temperature of the system, $(p-p_s)$. Equation (29) provides satisfactory accuracy for the slow growth of droplets larger than 1 μm, but overestimates the rate by a factor of 2.4 for 0.1-μm droplets.

When p is less than p_s, the environment is unsaturated and droplet growth as given by Equation (29) will be negative; in other words, evaporation and droplet shrinkage occurs. Equation (29) can be used to predict the rate of evaporation for conditions of slow evaporation of water, $0.95 < S_R < 1.0$. When the surrounding environment is drier, $S_R<0.95$, evaporation takes place sufficiently fast such that self-cooling of the droplets occurs, slowing the rate of evaporation. Details on how to correct for rapid evaporation and condensation are given in Hinds (1999) and Sienfeld and Pandis (1998).

2.4 Chemical properties

The chemistry of aerosols reflects their sources and subsequent reactions. These reactions can include reactions between different compounds inside a particle or between the particle and compounds in the suspending gas. Most particles found in indoor or outdoor environments are complex chemical mixtures. The exceptions are aerosols formed by grinding pure compounds or aerosolizing liquids, such as may be found in some occupational exposure environments. In condensation from high-temperature processes, the high-molecular-weight compounds condense first to form a solid core onto which lower-molecular-weight compounds subsequently condense to form a surface layer.

Materials found in natural atmospheric aerosol include resuspended soil particles, sea salt particles, particles from dried botanical debris, forest fires, and volcanoes. Particles formed in the atmosphere by gas-to-particle conversion and photochemical processes are also found. The latter includes sulfates from SO_2 and H_2S, ammonium salts from NH_3, nitrates from NO_x, and photochemically formed particles from plant-derived terpenes or anthropometric volatile organic compounds.

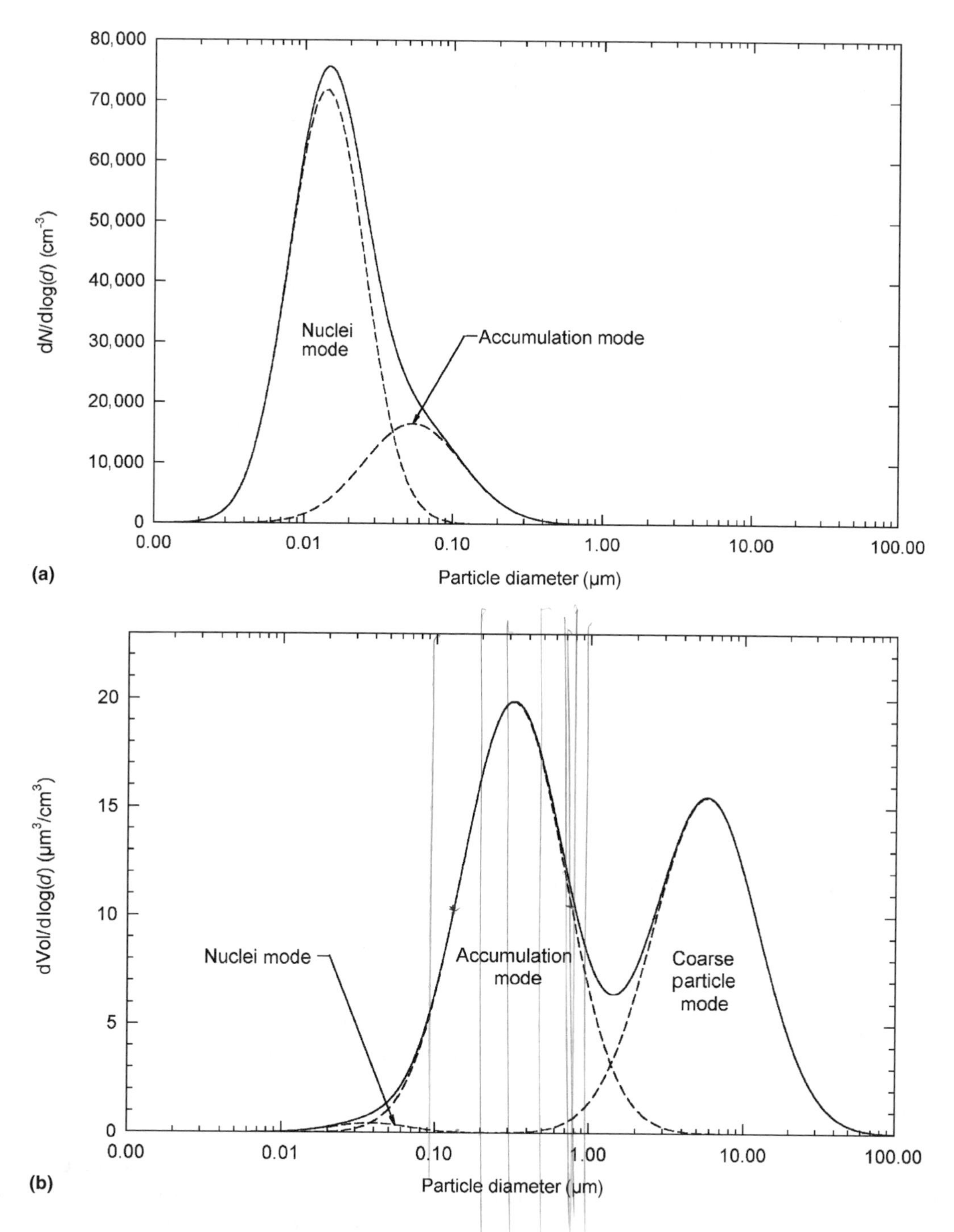

Figure 2.5 Typical trimodal urban aerosol particle size distribution: (a) number distribution and (b) volume distribution. Based on data from Whitby (1978) and Hinds (1999).

Semivolatile compounds are partitioned between the gas phase and the particle phase in an aerosol. This partitioning is sensitive to changes in temperature and reactions that change the vapor pressure of one or more of the ingredients and may be affected by particle size, with smaller particles favoring relatively more material in the vapor phase.

Aerosol particles have high specific surface, or surface area per unit mass of particles. Table 1 shows the surface area in m^2/g of particles of different sizes. The smallest particles have specific surfaces comparable to those of activated charcoal. When freshly formed, the surface of aerosol particles can readily adsorb vapor molecules and this facilitates reactions of various compounds.

2.5 Atmospheric particle size ranges

By convention, names are given to various particle size ranges. The term ultrafine particles refers to those smaller than 0.1 μm or 100 nm. Particles < 50 nm are called nanometer particles or nanoparticles. Particle size distributions in urban environments typically display three or two modes as shown in Figure 2.5. Particles in the smallest mode, called the nuclei mode, are < 100 nm with a modal peak in the 10–30 nm range. They consist of particles from combustion processes and those formed by gas-to-particle conversion. This mode has a relatively high number concentration, and particles coagulate with other nuclei and accumulation mode particles. The nuclei mode particles have a relatively short lifetime in the atmosphere and end up in the next larger mode, the accumulation mode.

The accumulation mode is typically from < 0.1 to about 2 or 3 μm, with a modal peak at about 0.3 μm. It includes combustion-formed particles, photochemically formed particles, and nuclei mode particles that have coagulated with accumulation mode particles. The nuclei and accumulation mode together form what is called fine particles or PM-2.5.

The coarse particle mode, those larger than 2.5 μm, are mechanically formed or resuspended particles such as windblown dust, large salt particles from sea spray, and automobile tire dust. Typically, the fine particle mode contains one third to two thirds of the mass of the urban aerosol, with the remainder in the coarse particle mode. As a result of their large size, coarse particles settle out readily with a lifetime ranging from hours to days. Fine and coarse particles have different chemical compositions reflecting their different sources. There is little mass exchange (coagulation) between the fine and coarse mode particles. The trimodal urban aerosol size distribution is an equilibrium size distribution that reflects the different sources, formation processes, growth and shrinkage processes, and removal processes occurring in the atmosphere.

Glossary

Bioaerosol An aerosol whose particles are of biological origin. Bioaerosols include viruses, viable and nonviable bacteria, and products of microorganisms, such as fungal spores and pollen.

Cloud A visible aerosol with a defined boundary. Atmospheric clouds are composed of water droplets.

Dust A solid particle aerosol formed by mechanical disintegration of its parent material.

Fume A solid particle aerosol formed by condensation of vapors or gaseous combustion products. Also, roughly used to refer to any noxious contaminant in the atmosphere.

Haze An atmospheric aerosol that affects visibility.

Mist and Fog Liquid particle aerosols formed by condensation or atomization.

Micrometer-sized aerosols Aerosols with particle sizes in the 1–10 μm size range.

Monodisperse Aerosols with all particles of the same size. These do not occur naturally but can be created in the laboratory for calibration and testing.

Nanometer size Aerosols with particles in the nanometer size range, usually 1–50 nm.

Polydisperse Aerosols with a range of particle sizes, the usual case. The range between the smallest and largest particles may be a factor of 100 or 1000.

Primary particles Particles emitted directly into the atmosphere.
Secondary particles Particles formed in the atmosphere by gas-to-particle conversion or photochemical processes.
Smog A usually visible aerosol, formed by photochemical processes in the atmosphere. Also, roughly used to describe any visible atmospheric pollution.
Smoke A visible aerosol formed by incomplete combustion.
Spray A droplet aerosol formed by mechanical breakup of a liquid.
Standard conditions For the phenomena discussed in this chapter, standard conditions refer to an atmospheric pressure of 101 kPa and a temperature of 293 K (20°C).
Submicrometer size An aerosol with particles less than 1 μm.

References

Fuchs, N.A., *The Mechanics of Aerosol*, Pergamon, Oxford, 1964. Republished, Dover Press, 1989.

Hinds, W.C., *Aerosol Technology: Properties, Behavior, and Measurement of Airborne Particles*, 2nd ed., Wiley, New York, 1999.

Hodkinson, J.R., The optical measurement of aerosols, in *Aerosol Science,* Davies, C.N., Ed., Academic Press, New York, 1966.

Lee, K.W. and Chen, H., Coagulation rate of polydisperse aerosols, *Aerosol Sci. Tech.*, 3, 327–334, 1984.

Marple, V.A., Rubow, K.L., and Olsen, B.A., Inertial, gravitational, centrifugal, and thermal collection techniques, in *Aerosol Measurement*, 2nd ed., Willeke, K. and Baron, P., Eds., Wiley, New York, 2001.

Sienfeld, J.H. and Pandis, S.N., *Atmospheric Chemistry and Physics*, New York, Wiley, 1998.

Whitby, K.J., The physical characteristics of sulfur aerosol, *Atmos. Environ.*, 12, 135–159, 1978.

chapter three

Advances in monitoring methods for airborne particles

Philip K. Hopke
Clarkson University

Contents

3.1 Introduction 35
3.2 Filter-based measurements 36
3.2.1 Mass 36
3.2.1.1 Fine particles 36
3.2.1.2 Coarse particles 37
3.2.2 Continuous mass measurements 38
3.2.2.1 Fine particles 38
3.2.2.2 Coarse particles 40
3.3 Particulate constituent measurements, fixed site 40
3.3.1 Filter-based integrated monitors 40
3.3.2 Continuous monitors 41
3.4 Supersites 44
3.5 Next steps 45
3.6 Conclusions 45
References 45

3.1 Introduction

The health effects of airborne particles have been extensively studied and strong correlations have been observed between human mortality and morbidity and particulate matter (PM) concentrations (e.g., Dockery et al., 1993). Excessive inhalation of pollutants such as PM can affect the functioning of the lungs, and may even cause asthma (Browner, 1996). Large quantities of compounds such as mercury, zinc, or lead can lead to a variety of chronic illnesses (Pimentel, 1997; Pimentel and Greiner, 1997). Organic constituents such as polycyclic aromatic hydrocarbons (PAHs) are also of concern due to their carcinogenic properties (Nielsen et al., 1996, 1999). In addition, fine PM also contributes to regional haze and a reduction in visibility. Thus, the sampling and analysis of airborne particles are critical tasks in determining whether there are potential environmental health problems, attaining air quality standards, and developing effective and efficient air quality management strategies to reduce exposure and adverse effects of airborne PM.

1-56670-611-4 / 05 / $0.00+$1.50

As a result of the reported correlations between PM and mortality with apparently higher risks arising from fine particles, a new United States National Ambient Air Quality Standard (NAAQS) for $PM_{2.5}$ was promulgated in 1997. Because of the limited data available on which to base this standard, it was agreed that implementation of the new standard would be deferred until after the next review of PM NAAQS. This review is currently underway, with the expectation that a decision will be made by the Environmental Protection Agency (EPA) Administrator on the standards in 2003 with any new NAAQS being promulgated in 2004.

The new $PM_{2.5}$ and PM_{10} NAAQS were challenged by the American Truckers Associations, who filed a lawsuit against the EPA Administrator claiming that EPA had exceeded its authority to regulate, that the basis of the standards was not sufficiently well defined, and that the regulations were arbitrary and capricious. With the exception of one part of the suit concerning PM_{10}, which EPA did not contest, the U.S. Supreme Court upheld EPA's position and thus, at the present time, $PM_{2.5}$ is in effect. The argument with respect to the PM_{10} standard was that it includes $PM_{2.5}$ and therefore is not the appropriate indicator for coarse particles. If EPA has a basis for the regulation of coarse particles, an appropriate standard, including a well-defined indicator, should be defined. As discussed below, a coarse particle standard is expected to be proposed in 2005. The existing 1987 PM_{10} standard remains in effect.

The $PM_{2.5}$ NAAQS includes both an annual average and a 24-h standard. The 24-h standard is set at $65\,\mu g\ m^{-3}$. The 98th percentile values are found for each of three consecutive years of sampling and averaged. If this average value exceeds $65\,\mu g\ m^{-3}$, the site is in nonattainment of the standard. The annual arithmetic average standard is set at $15\,\mu g\ m^{-3}$, with the average taken over three consecutive years of data.

In order to implement the new $PM_{2.5}$ standard, a network of samplers had to be deployed. Because of problems that EPA perceived with the performance standard set for PM_{10} samplers, a design standard was established for the reference sampler for $PM_{2.5}$. A major problem with the measurement of airborne particulate mass concentrations is that it is a complex mixture of nonvolatile and semivolatile materials. Although mass is thought to be a fundamental property of a material, in the case of particles, it is actually an operationally defined quantity. In other words, the mass concentration is what is obtained as a result of the measurement process. Since different measurement processes will give rise to different results, there is no absolute comparison standard against which to compare techniques. As part of the promulgation of the NAAQS for $PM_{2.5}$ in 1997, a Federal Reference Method (FRM) has been defined that provides a relatively high precision but a totally unknown accuracy. However, since it is defined by regulation to be the basis of attainment/nonattainment decisions with respect to the standard, it has become the basis of comparison with other methods.

In this chapter, measurement methods for airborne particulate mass and composition will be presented along with some of the background on how these devices are being deployed in the United States to provide improved information on the distribution of particle mass as well as compositional data that can be used to help develop implementation strategies to improve air quality.

3.2 *Filter-based measurements*

3.2.1 *Mass*

3.2.1.1 *Fine particles*

Mass measurements have traditionally involved the passage of air through a filter, resulting in a variety of positive and negative artifacts. The sampler is designed so that a particular fraction of the particle size distribution is separated aerodynamically from the ambient aerosol and transported to the filter. In the case of the $PM_{2.5}$ FRM, a design standard was adopted in which a set of machine drawings are provided in the regulations. If the instrument is constructed according to the tolerances given in the regulations by an appropriately certified workshop, it is by definition an FRM sampler.

There are quite stringent standards for developing a continuous equivalent sampler and, thus, there are currently no continuous $PM_{2.5}$ samplers that are approved for use in the compliance monitoring network.

PM samples are collected over a 24-h period and, because of the cost of equilibrating, weighing, deploying, retrieving, reequilibrating, and reweighing the filter, measurements in the United States are only required to be made every third day. Since we are interested in the mass primarily because there are efforts to protect public health, and water associated with the particles is not thought to pose a health risk, the "dry" mass concentration is the quantity of interest. However, removal of water without removal of any other semivolatile components is very difficult. For the filter-based methods, water is removed by equilibration of the same with air at 35% relative humidity and 23°C temperature for 24h. However, depending on particle composition, this may not permit the sample to be thoroughly dried. The FRM only requires 75% data capture. Hence, attainment decisions are made on <25% of the possible data that could be obtained if continuous measurement methods were available. There are then only limited data for health studies and implementation purposes.

Figure 3.1 illustrates the locations of the $PM_{2.5}$ FRM measurement sites, and Figure 3.2 shows the sites that, based on 2 years of data, are likely to be in violation of the NAAQS. This monitoring network was deployed to provide population-weighted measurements of $PM_{2.5}$ to ascertain compliance with the new NAAQS. However, it takes 3 years of data to declare nonattainment, and such designations will be made in November 2005.

3.2.1.2 *Coarse particles*

As a result of the decision by the Third Circuit of the U.S. Court of Appeals that was not disputed by the EPA, the proposed 1997 PM_{10} standard was invalidated. At present, the

Figure 3.1 Map of the United States showing the location at which the $PM_{2.5}$ FRM monitors have been deployed.

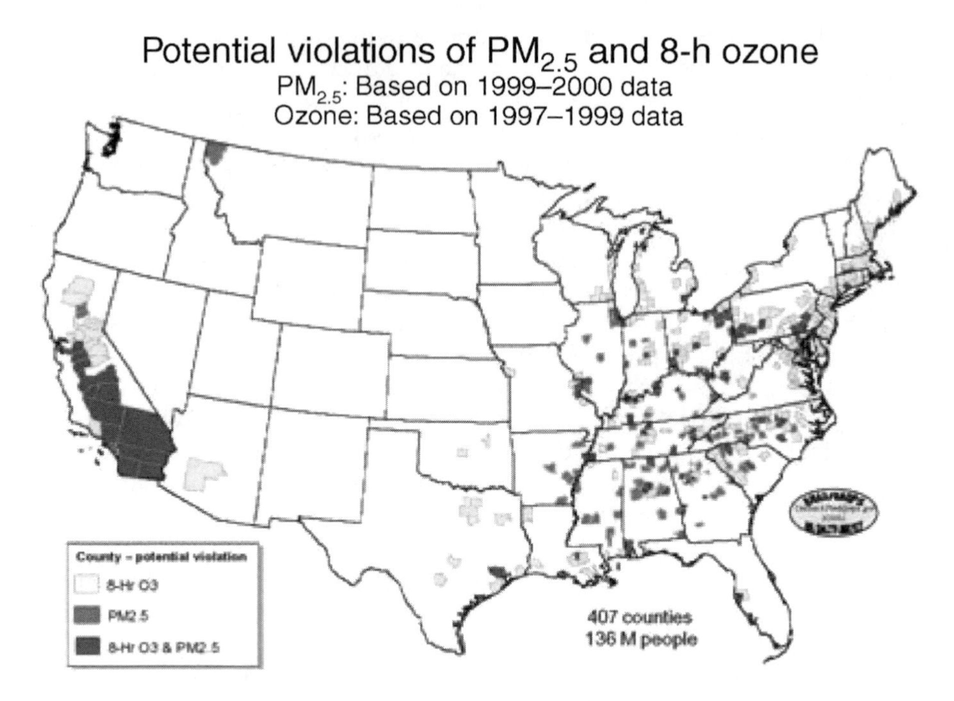

Figure 3.2 Map of the United States showing areas that may be in violation of the new NAAQS for ozone and $PM_{2.5}$.

1987 standard remains in place, but it is clear that the EPA intends to promulgate a coarse particle standard and has been discussing a $PM_{(10-2.5)}$ standard. The details of their proposals will not be available until the EPA releases its draft staff paper. This document was released in November 2003.

The initial proposal for an FRM for this size fraction was to utilize two side-by-side samplers. One sampler was the current FRM for $PM_{2.5}$ and the other was a $PM_{2.5}$ FRM with the WINS impactor removed to make it a PM_{10} sampler.

A new dichotomous sampler has been developed by Rupprecht and Patashnick Inc. (R&P) with volumetric flow control, which was tested in a recent study (Poor et al., 2002) and compared well with the $PM_{2.5}$ FRM in one city. This sampler has not been tested for its ability to measure $PM_{(10-2.5)}$. EPA has completed additional tests, but the results have not yet been published.

At this time, there is no proposal for the new PM coarse particle standard, and with very little available data, it is not possible to assess the locations likely to violate the standard. The initial proposal will be made in the Staff Paper that were made public in late 2003. The final standards are expected to be promulgated in 2005.

3.2.2 *Continuous mass measurements*

3.2.2.1 *Fine particles*

The development of various "continuous" mass monitor has taken place. These systems are actually semicontinuous since there are integration periods necessary to provide a measurement with an acceptable level of measurement uncertainty. The question for these monitoring systems then relates to the appropriate quantity to be measured. Should they (i) attempt to mimic the FRM, (ii) measure the mass under a well-defined set of thermodynamic conditions, or (iii) attempt to measure the actual particle mass concentration that was present in the atmosphere? Various manufacturers and academic researchers have adopted different approaches to address these various questions.

There are two direct approaches to the measurement of aerosol mass: the tapered element oscillating microbalance (TEOM®) and the beta gauge. Both these methods are based on basic physical principles. The TEOM operates on the theory of the vibrating reed. A vibrating reed that is anchored at one end vibrates with a single frequency and no harmonics. Thus, as mass is added to one end of the vibrating element, the resonant frequency changes and this change is directly proportional to the accumulated mass. In the beta gauge, the passage of electrons through a material is attenuated by the presence of the areal mass concentration between the radioactive source and the detector.

There are also indirect mass methods. In the continuous aerosol mass monitor (CAMM), the pressure drop across a pore-type filter increases with increasing mass loading within a limited range of particle sizes (Babich et al., 2000). Thus, after a period of time, the change in pressure drop across the filter can be related to the amount of accumulated particle mass. Another indirect method for fine particles is through the measurement of light scattering. Light scattering from a polydisperse aerosol is a complex phenomenon involving particle size and refractive index. However, many of these complexities average out sufficiently that reasonable estimates of the fine particle mass can be obtained from the amount of scattered light.

To use these mass measurement methods, the aerosol must be conditioned in some manner. For the continuous monitors, alternative methods have been used. The initial work with a TEOM made use of elevated temperatures to remove the water and any other material that would be volatile at 50°C. This temperature provides a well-defined set of thermodynamic conditions, but ensures the loss of any ammonium nitrate and most higher vapor pressure semivolatile organic constituents. To provide a sample to the sensing element without water at a lower temperature, a Nafion® drier is used. Nafion is a material that permits small, polar gas molecules like water to pass through a membrane. These driers have been used in a number of systems. However, heating the sensing element to 30°C will still remove semivolatile components, particularly when the ambient temperature is significantly below this value. The commercial beta gauge monitors typically heat the sample to a few degrees above ambient temperature to reduce associated water and minimize the loss of semivolatile compounds.

Research systems have been developed to try to address the shortcomings of the commercially available systems. Eatough and co-workers (Eatough et al., 2001; Pang et al., 2001, 2002a; Obeidi and Eatough, 2002; Obeidi et al., 2002) have developed the Real-Time Ambient Monitoring System (RAMS), which involved a very complicated system of denuders, driers, and a dual TEOM detection system. This system is currently undergoing evaluation in comparison to other conventional and research monitoring techniques (Lewtas et al., 2001). RAMS is very large and complex and does not represent a practical routine monitoring tool.

Although RAMS is very complex, it has led to the development of another dual TEOM system (Patashnick et al., 2001). This differential system uses dual electrostatic precipitators (ESPs) and an ambient-temperature TEOM connected to a common inlet and drier system. The air is split and passes through two channels, each with an ESP and a TEOM. One ESP is on and the other is off. The on/off states are switched frequently (every 5 min) so that in the on channel, the sampling artifacts are duplicated (adsorption of organic compounds on the filter, volatilization of materials from the deposit collected on the filter). Thus, the subtraction of the on channel from the off channel should provide a better estimate of the actual airborne particle masses. An alternative system using a filter is also being tested.

From this concept, an alternative system that utilizes a filter has been developed (Filter Dynamics Measurement Systems, R&P FDMS series 8500). The FDMS instrument computes its running PM mass concentration average based upon independent measurements of the volatile and nonvolatile fractions of the ambient particulate matter. To accomplish this, the FDMS unit constantly samples ambient air and uses a switching valve to change the path of the main flow every 6 min. The sampling process consists of alternate sample and purge (filtered) air streams passing through the exchangeable filter in the TEOM mass sensor. The purge filter in the FDMS main enclosure effectively removes both organic and inorganic aerosols at 4°C. The exchangeable purge filter can provide a time-integrated sample that can be used for subsequent chemical analysis. A standard R&P FRM-style molded filter cassette allows for the use of a variety of 47-mm-diameter filter media as the purge filter.

The sample and purge air flows alternately pass through the exchangeable filter in the TEOM microbalance, which generates a direct measurement of the collected mass. The system automatically adjusts the mass concentration from the particle-laden air stream by referencing it to the mass change that may occur during purging. For example, if the FDMS unit measures a decrease of filter mass during the 6-min purging period, this mass decrease is added back to the mass measurement obtained with particle-laden air.

At the moment, none of these continuous systems can be used to determine attainment or nonattainment of the PM NAAQS. Efforts are currently underway to develop new guidelines that will permit the development of local equivalence between the continuous monitors and the $PM_{2.5}$ FRM. There is considerable interest in obtaining the more complete data that a continuous monitor can provide as the well as the shorter time interval data that can be used to examine the relationship of short-term, peak concentrations with adverse health effects. Such guidance is expected to be provided by U.S. EPA to state and local air quality agencies in the near future.

3.2.2.2 *Coarse particles*

There is, at present, the initial development of a continuous coarse monitor based on the work of Misra et al. (2001). In this approach, a system similar to particle concentrators (used to provide concentrated airborne particles (CAPs) for toxicological studies in laboratory animals or people) and a TEOM is used. Because of the relatively low concentrations of coarse particle mass in most cases, a higher rate of aerosol sampling (50 lpm), with the particles concentrated by their inertia, is required to provide reasonable precision in the measurement. Extensive field testing of this system has not been carried out, but initial studies have been performed in Tampa, FL. EPA has conducted additional testing in 2002 and 2003 so that it could be considered as an FRM or FEM method for $PM_{(10-2.5)}$. Thus, the exact nature of the new FRM samplers that will be required for the determination of compliance with the new coarse particle NAAQS has not yet been determined.

3.3 *Particulate constituent measurements, fixed site*

3.3.1 *Filter-based integrated monitors*

As part of the monitoring network created in response to the needs generated by the promulgation of the $PM_{2.5}$ NAAQS, the U.S. EPA has initiated the deployment of a network of samplers to collect samples for chemical analyses. There will be a need to identify the particle sources, and the compositional data from this network are expected to provide data to the state and local air quality agencies to permit them to develop better air quality management plans (state implementation plans (SIPs)). There are 54 sites in this network that have been designated as "trends sites," which are expected to be operated over a long

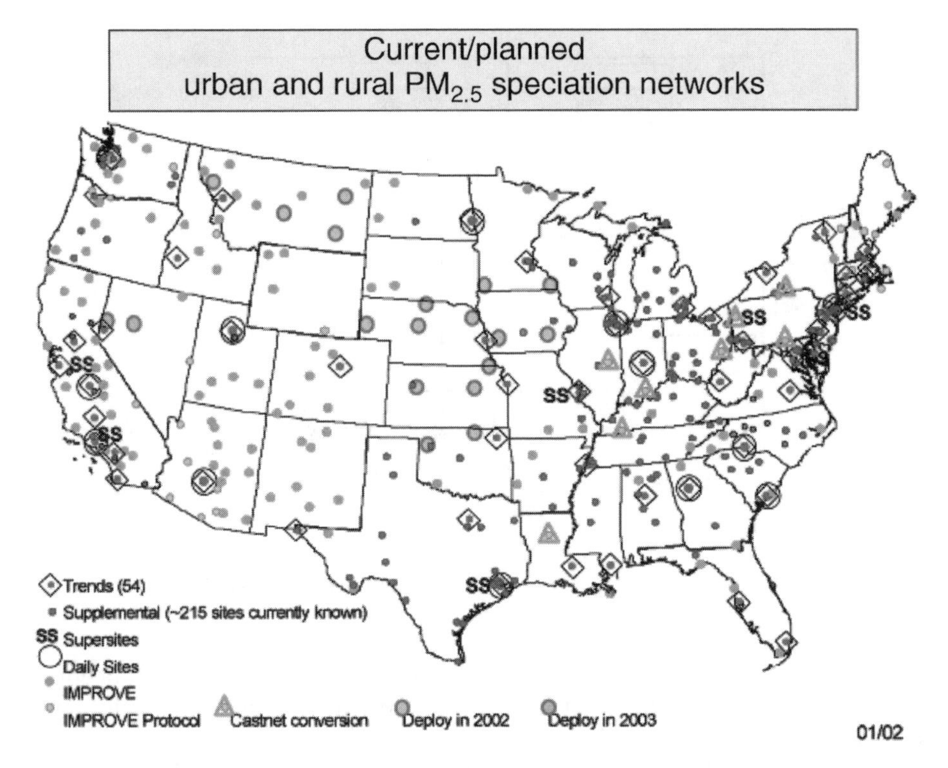

Figure 3.3 Map of the United States showing the planned deployment of the national speciation network including additional IMPROVE sites.

time period (at least 10 years) to provide trends in the $PM_{2.5}$ composition over this time interval. In addition to the speciation sites to be deployed in urban areas, there are also additional IMPROVE sites (Cahill et al., 1986) that will provide comparable data for rural areas. Figure 3.3 shows the complete planned speciation network deployment, which will have over 200 sampling sites across the United States.

The outline of a typical speciation network sampler is shown in Figure 3.4. It uses multiple channels to provide samples on appropriate filter media for a variety of analyses. One channel is used with a quartz filter to measure organic and elemental carbon using a modified NIOSH 5040 protocol (Birch and Cary, 1996). Annular denuders are used to separate gaseous HNO_3 from particulate nitrate. By capturing the particles on a base-impregnated fiber filter or a Nylasorb filter, the nitrate is retained on the filter until it is leached from the filter for analysis. Such a system is labor-intensive in terms of collection and analysis of the filters. A third channel collects the particles on a Teflon® filter for XRF analysis, and a comparable fourth channel provides a sample for the measurement of major ions through ion chromatography (IC). In some units, only a single Teflon filter is collected for sequential XRF and IC analyses.

3.3.2 Continuous monitors

One approach toward obtaining continuous information on airborne PM involves the development of measurement systems for major constituents (nitrate, sulfate, organic, and elemental carbon). An instrument for light-absorbing carbon measurements has been available for quite some time. The aethalometer was originally developed by Hansen et al. (1984). Initial models used a single-wavelength light source. Newer models have two to seven wavelengths to better discriminate among different types of light-absorbing carbon.

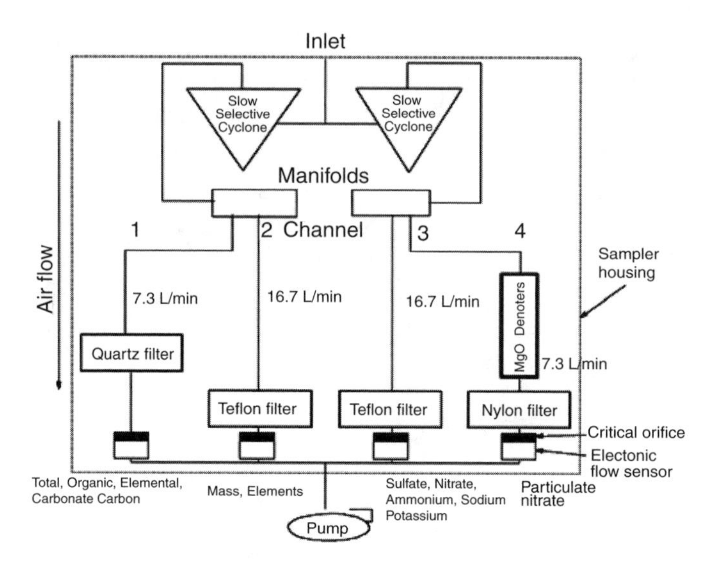

Figure 3.4 Schematic diagram of one of the speciation network samplers.

Absorbance at 370 nm appears to provide some indication of significant concentrations of PAHs, which is often associated with diesel particles. A new device developed in Japan has just come into the U.S. market introduced by TSI Inc. This instrument measures absorbance at 370 and 620 nm and assumes that in indoor air, the primary source of UV absorbance is PAH compounds associated with environmental tobacco smoke.

Another set of instruments has been developed to measure ambient PAH compounds (Wilson et al., 1994). Both field and hand-held monitors operate on the principle of photoionization. PAH compounds can be readily ionized by photons in the UV range and the current generated is measured. PAH compounds with three or more rings will be measured and an estimate of the total PAH concentration will be provided.

A set of instruments has been developed by R&P to measure nitrate, sulfate, and organic and elemental carbon. All these instruments depend on impaction onto a substrate that is subsequently heated to volatilize the material and, in some cases, catalytically convert the analyte species, which is then measured with a conventional gas monitor (Stolzenburg and Hering, 2000). Nitrate is converted to NO, sulfate is converted to SO_2, and carbon is oxidized to CO_2. Initial indications have been that the sulfate and nitrate instruments provide reasonable results when compared to side-by-side integrated samples. Much better characterizations of these instruments will be obtained from the supersite activities that are currently in progress.

Questions have been raised with the R&P OC/EC regarding the impactor cutoff of approximately 150 nm that may not capture all the diesel exhaust particles. Only two evolution temperatures are available: one for the inert gas cycle (OC) and the other for the oxygen-containing gas cycle (EC). In addition, there is no correction for pyrolysis during the OC evolution. Unpublished reports have suggested that this system provides a good estimate of the total OC+EC, but the fractions do not show good agreement with the laboratory measurements of integrated samples taken simultaneously with continuous measurements.

An OC/EC field instrument has been developed by Sunset Laboratory, which developed the laboratory instrument to implement the NIOSH 5040 method (Birch and Cary, 1996). This instrument collects particles on a filter and makes a pyrolysis correction using reflectance. It has been available for less than a year and is yet to be characterized. However, initial experience has demonstrated that it is quite promising as a useful continuous measure of OC and EC, indicating better agreement with the laboratory measurements.

Dr. George Allen, while he was at the Harvard School of Public Health, (HSPH), developed a continuous sulfate analyzer. It utilizes catalytic reduction of sulfate to SO_2 in a stainless-steel tube and detection with a conventional SO_2 monitor. Several variations of this system were made under licenses from HSPH, while the instrument is being commercialized by Thermo Andersen. Initial experience in the SEARCH monitoring network has been very encouraging.

Several groups have developed systems for individual particle characterization using laser ablation and time-of-flight mass spectrometry. Single-particle mass spectrometers have somewhat different characteristics. One of these systems has now been developed as a commercial instrument, but at a rather high cost per instrument.

Several groups have been working on the collection of fine PM by increasing the particle size through hygroscopic growth and analysis of the collected slurry. An initial size separation is obtained using a size-selective inlet. Steam is injected and the resulting droplets are collected in an impactor. The slurry from this impaction stage can then be introduced into an appropriate analytical instrument.

Weber et al. (2001) used ion chromatography to analyze major cations and anions. This system is a modification of the Dutch system developed by Slanina and co-workers (Khlystov et al., 1995), which is also being developed by Thermo Andersen. In this system, colorimetry is also used to determine the number of ammonium ions. In the Weber system, denuders are used to remove gaseous interferences such as HNO_3 and SO_2. It is being tested at several of the supersites.

Ondov has been developing a similar system for metal determination. The Semicontinuous Elements in Aerosol System (SEAS) is made entirely of glass to avoid contamination. Currently, there are no denuders on it, so the slurry contains both gaseous and particulate species. At this time, the only documentation on this system is the doctoral thesis of Kidwell (2000). The results from this system are available on the Baltimore Supersite webpage (http://www.chem.umd.edu/supersite/keyinstruments/SEASnew.htm).

Figure 3.5 shows the concentrations of selenium as measured in Baltimore in early September 2001. Figure 3.6 shows a map indicating the wind direction during the

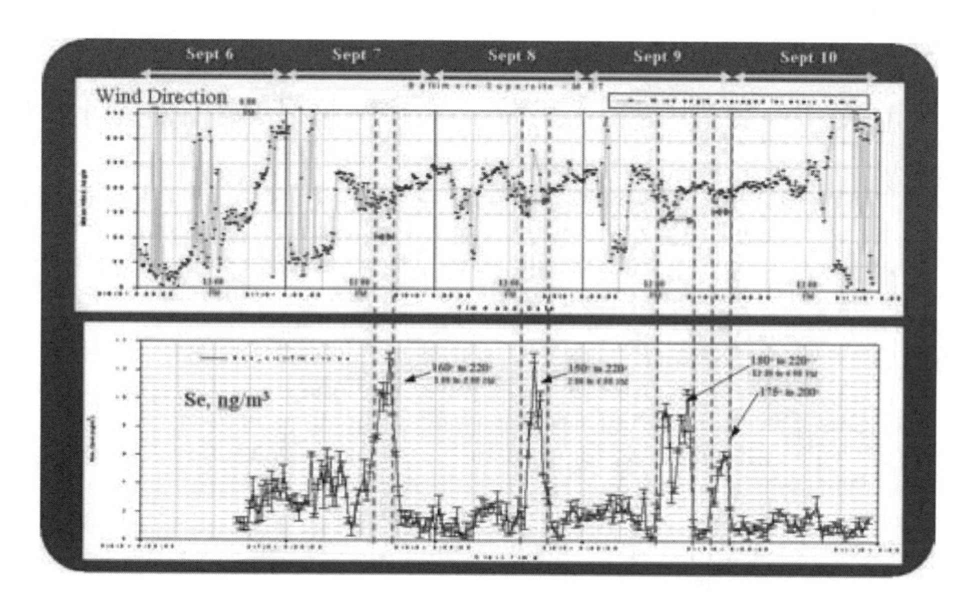

Figure 3.5 Top: Plot of wind direction as a function of time during early September 2001 as measured at the Clifton Park location of the Baltimore Supersite. Bottom: Plot of selenium concentrations as a function of time using the University of Maryland SEAS at the Baltimore Supersite.

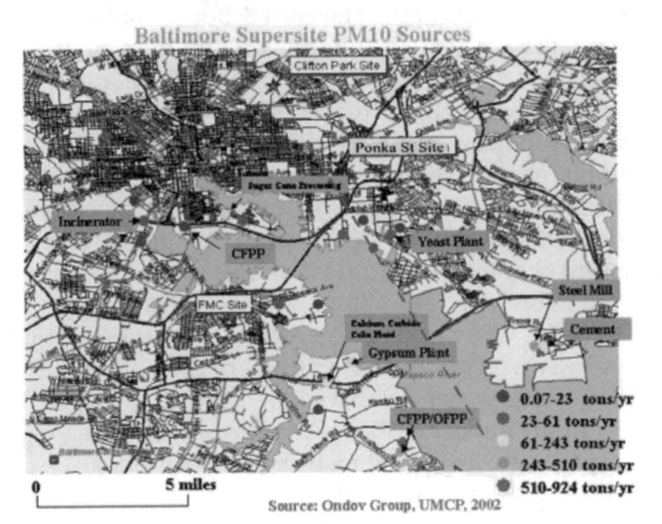

Figure 3.6 Map of Baltimore showing the location of Clifton Park and the likely source(s) of selenium as observed in Figure 3.5.

selenium peak and the location of likely sources. Along the direction of 160°, there is a large coal-fired power plant that is likely the source of the observed selenium. Similar results have been found for this system at the Pittsburgh Supersite, where even higher short-term selenium concentrations have been observed and attributed to a specific source. Results like this suggest that it is now possible to observe plumes that arise from specific point sources. It is likely that appropriate modeling will permit the identification of these sources and an estimation of the emission rates that give rise to the observed ambient concentrations. Such inverse modeling would provide useful information for air quality management.

3.4 Supersites

Seven sites have been established as the top end of the three tiered-monitoring program: FRM compliance monitors, speciation monitors, and supersites. The supersites are New York City, Pittsburgh, Baltimore, St. Louis, Houston, Fresno, and Los Angeles. These are limited duration programs to address specific hypotheses that are unique to each site, but with the general intent to test current-generation commercial and research-grade instruments. In general, only a single year of monitoring data are to be obtained and currently only very limited results are available from these studies, but it is anticipated that much of the data will be reported in the next 12–18 months.

The U.S. EPA has recently collaborated with the Baltimore supersites to create a publically available relational database that will house all of the supersites and related monitoring programs. During July 2001, there was a coordinated intensive monitoring effort involving the collaboration of 26 programs, including five of the supersites that cover the region from Houston (TX) to New Hampshire. A smaller effort was mounted during January 2002. An effort will be made to collect the data from all of the colloborating programs and incorporate them into the database. Initial operations of the website and database are planned for August 2004.

These efforts include conducting rigorous tests of the new commercial monitoring instruments as well as new state-of-the-art methods such as PILS, SEAS, and single-particle mass spectrometry systems.

3.5 Next steps

The U.S. EPA is currently reviewing its monitoring needs in a comprehensive planning effort. They have multiple networks that have been independently established at various times with different purposes. It is now appropriate to consider all of the monitoring needs, including NAAQS attainment determinations, epidemiology, atmospheric process research, regional haze, and acid rain management. The plan was released for public comment in 2003 (http://www.epa.gov/ttn/amtic/monitor.html) and its implementation will be reviewed by the Clean Air Scientific Advisory Committee in 2004. The results could lead to a significant redeployment of monitors to provide more utility of the data for the same level of expended resources.

As part of the monitoring network for the $PM_{2.5}$ standard, there is a movement to permit the use of continuous mass monitors. A major problem with the current network is that even when working perfectly, it only provides 24-h integrated sample data every third day. Thus, health effects hypotheses with respect to short-term, high exposures or multiple-day exposures can be examined. Also there will be a firmer basis for making attainment/nonattainment decisions. Lack of data can lead to major problems in the proper assessment of the attainment status of an area (Davidson and Hopke, 1984).

Data from the Speciation Network and the supersites are just beginning to become available; hence, it is not possible to report results at this time. These data should be able to provide additional information that can improve air quality management. There is an increased interest in using such data for source apportionment. The more refined information on particle compositions will make it possible to obtain more specificity in source identifications and a higher accuracy in source apportionment. These results will permit state and local air quality agencies to focus their control efforts on the specific sources that most likely lead to elevated concentrations and the nonattainment of air quality standards. New guidelines are being developed by the EPA's Office of Air Quality Standards and Planning (OAQPS), which are expected to be implemented in 2005 to permit state and local agencies to incorporate continuous monitors into the network. Such data will permit the consideration of other time intervals in future reviews of the standard.

It may then also be possible to associate the observed adverse health effects with specific source components in the particles. If such possibilities are realized, we can then refocus our energies from managing airborne particle mass to controlling only those sources that actually cause adverse effects.

3.6 Conclusions

Because of the interest in airborne PM arising from the promulgation of the $PM_{2.5}$ National Ambient Air Quality Standard in 1997 and the likely promulgation of a coarse particle standard in 2004, considerable efforts have been made to deploy monitoring systems and to develop and test new instruments, particularly continuous systems, to characterize the ambient aerosol. Much of the data that will be used to evaluate these instruments are just becoming available through the intensive monitoring campaigns (supersites) in the United States. These systems offer considerable hope for improved source identification and apportionment, and will offer valuable information to those responsible for the development of air quality management strategies.

References

Babich, P., Wang, P.Y., Allen, G., Sioutas, C., and Koutrakis, P., Development and evaluation of a continuous ambient $PM_{2.5}$ mass monitor, *Aerosol Sci. Technol.*, 32, 309–324, 2000.

Birch, M.E. and Cary, R.A., Elemental carbon-based method for monitoring occupational exposures to particulate diesel exhaust, *Aerosol Sci. Technol.*, 25, 221–241, 1996.

Browner, C., Environmental Health Threats to Children, EPA 175-F-96-001, 1996.

Cahill, T.A., Eldred, R.A., and Feeney, P.J., Particulate Monitoring and Data Analysis for the National Park Service 1982–1985, NPS contract number USDICX-3-0056, 1986.

Davidson, J.E. and Hopke, P.K., Implications of incomplete sampling on a statistical form of the ambient air quality standard for particulate matter, *Environ. Sci. Technol.*, 18, 571–580, 1984.

Dockery, D., Pope, A., and Xu, X., An association between air pollution and mortality in six US cities, *N. Engl. J. Med.*, 329, 1753–1759, 1993.

Eatough, D.J., Eatough, N.L., Obeidi, F., Pang, Y., Modey, W., and Long, R., Continuous determination of $PM_{2.5}$ mass, including semi-volatile species, *Aerosol Sci. Technol.*, 34, 1–8, 2001.

Hansen, A.D.A., Rosen, H., and Novakov, T., The aethalometer — an instrument for the real-time measurement of optical absorption by aerosol particles, *Sci. Total Environ.*, 36, 191, 1984.

Khlystov, A., Wyers, G.P., and Slanina, J., The steam-jet aerosol collector, *Atmos. Environ.*, 29, 2229–2234, 1995.

Kidwell, C.B., Sub-hourly airborne metal analysis by graphite furnace atomic absorption spectrometry after dynamic aerosol concentration, Doctoral thesis, UMCP, 2000.

Lewtas, J., Pang, Y., Booth, D., Reimer, S., Eatough, D.J., and Gundel, L.A., Comparison of sampling methods for semi-volatile organic carbon associated with $PM_{2.5}$, *Aerosol Sci. Technol.*, 34, 9–22, 2001.

Misra, C., Geller, M.D., Shah, P., Sioutas, C., and Solomon, P.A., Development and evaluation of a continuous coarse (PM_{10}–$PM_{2.5}$) particle monitor, *J. Air Waste Manage. Assoc.*, 51, 1309–1317, 2001.

Nielsen, T., Jorgensen, H.E., Larsen, J.C., and Poulsen, M., City air pollution of polycyclic hydrocarbons and other mutagens: occurrence, sources and health effects, *Sci. Total Environ.*, 189/190, 41–49, 1996.

Nielsen, T., Feilberg, A., and Binderup, M.-L., The variation of street air levels of PAH and other mutagenic PAC in relation to regulations of trace emissions and the impact of atmospheric processes, *Environ. Sci. Pollut. Res.*, 6, 133–137, 1999.

Obeidi, F. and Eatough, D.J., Continuous measurement of semivolatile fine particulate mass in Provo, Utah, *Aerosol Sci. Technol.*, 36, 191–203, 2002.

Obeidi, F., Eatough, N.L., and Eatough, D.J., Use of the RAMS to measure semivolatile fine particulate matter at riverside and Bakersfield, California, *Aerosol Sci. Technol.*, 36, 204–216, 2002.

Pang, Y., Ren, Y., Obeidi, F., Hastings, R., Eatough, D.J., and Wilson, W.E., Semi-volatile species in $PM_{2.5}$: comparison of integrated and continuous samplers for $PM_{2.5}$ research or monitoring, *Environ. Sci. Technol.*, 51, 25–36, 2001.

Pang, Y., Eatough, N.L., and Eatough, D.J., $PM_{2.5}$ semivolatile organic material at Riverside, California: implications for the $PM_{2.5}$ Federal Reference Method Sampler, *Aerosol Sci. Technol.*, 36, 277–288, 2002a.

Pang, Y., Eatough, N.L., Wilson, J., and Eatough, D.J., Effect of semivolatile material on $PM_{2.5}$ measurement by the 2.5 Federal Reference Method Sampler at Bakersfield, California, *Aerosol Sci. Technol.*, 36, 289–299, 2002b.

Patashnick, H., Rupprecht, G., Ambs, J., and Meyer, M.B., Development of a reference standard for particulate matter mass in ambient air, *Aerosol Sci. Technol.*, 34, 42–45, 2001.

Pimentel, D., Pest management in agriculture, in *Techniques for Reducing Pesticides, Environmental and Economic Benefits*, Pimentel, D., Ed., John Wiley & Sons, Chichester, England, 1997.

Pimentel, D. and Greiner, A., Environmental and socio-economic costs of pesticide use, in *Techniques for Reducing Pesticides, Environmental and Economic Benefits*, Pimentel, D., Ed., John Wiley & Sons, Chichester, England, 1997.

Poor, N., Clark, T., Nye, L., Tamanini, T., Tate, K., Stevens, R., and Atkeson,T., Field performance of dichotomous sequential PM air samplers, *Atmos. Environ.*, 36, 3289–3298, 2002.

Stolzenburg, M.R. and Hering, S.V., Method for the automated measurement of fine particle nitrate in the atmosphere, *Environ. Sci. Technol.*, 34, 907–914, 2002.

Weber, R.J., Orsini, D., Daun, Y., Lee, Y.N., Khotz, P.J., and Brechtel, F., A particle-into-liquid collector for rapid measurement of aerosol bulk composition, *Aerosol Sci. Technol.*, 35, 718–727, 2001.

Wilson, N.K., Barbour, R.K., Chuang, J., and Mukund, R., Evaluation of a real-time monitor for fine particle-bound PAH in air, *Polycyclic Aromatic Compd.*, 5, 167–174,1994.

chapter four

Ultrafine and nanoparticle emissions: A new challenge for internal combustion engine designers

D. B. Kittelson, W.F. Watts, and J.P. Johnson
Department of Mechanical Engineering
University of Minnesota

Contents

4.1 Introduction 47
4.2 Measurement of size and composition of diesel particles 49
4.2.1 Sampling issues....49
4.2.2 On-road and laboratory measurements of diesel particles 50
4.2.3 Particle structure and composition measurements 52
4.2.4 Particles from engines equipped with exhaust filters....55
4.3 Some future engine design issues....56
4.4 Discussion and conclusions 58
References 58

4.1 Introduction

Studies have linked environmental exposure to fine particles less than 2.5 μm in size to adverse health effects [7,28,31], although no causal mechanisms were identified. More recent studies have investigated the hypothesis that ultrafine particles <100 nm in size are causally involved in adverse health effects [8,9,25–27,34,]. The relationship between fine particles and health is logical, as the efficiency of particle deposition in the respiratory tract is a function of particle size.

The overall chemical composition of diesel particles from well-maintained modern engines is highly variable; however, the major constituents are always elemental carbon (EC, mainly from fuel), organic carbon (OC, mainly from lubricating oil), sulfates, and ash (mainly from lubricating oil). EC and OC are typically 80% or more of the particle mass. The OC/EC ratio generally decreases with load and is typically about one-third for highway cruise conditions. Figure 4.1 shows three weightings, number, surface, and mass of a typical diesel size distribution, and the relationships between the coarse, accumulation and nuclei modes and an alveolar plus tracheo-bronchial deposition curve [13]. The size distributions are modified versions of similar distributions published previously [35] and are representative of a diesel exhaust aerosol. The curves have a lognormal, trimodal form and the concentration in any size range is proportional to the area

1-56670-611-4 / 05 / $0.00+$1.50

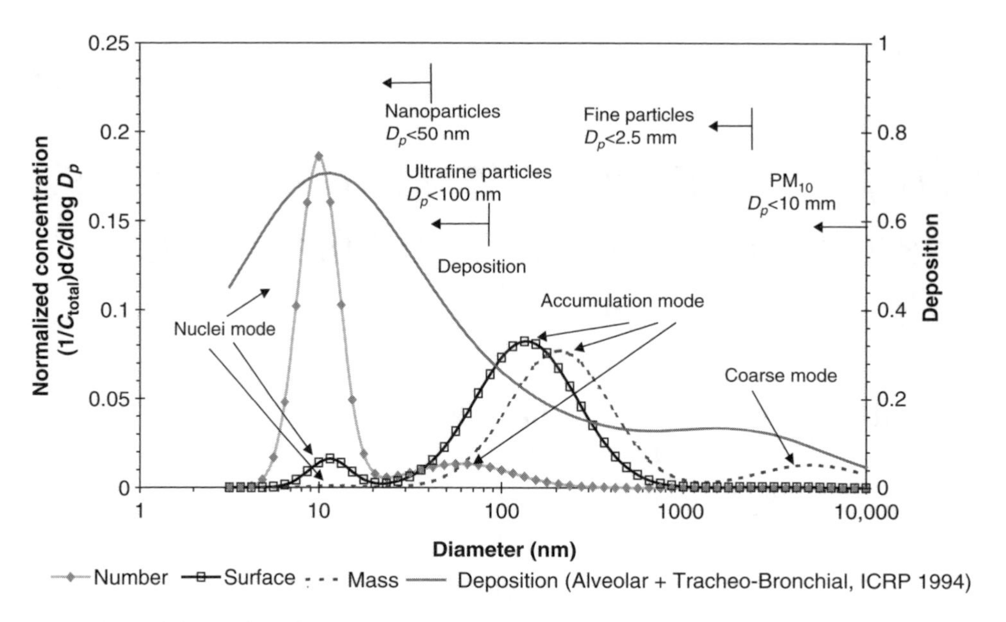

Figure 4.1 Trimodal size distribution.

under the corresponding curve in that range. Spark-ignition (SI) size distributions are similar, but have a smaller mass median diameter mainly due to fewer particles in the accumulation mode.

For diesel exhaust aerosols, the nuclei mode (roughly 3–30 nm) typically contains <10% of the particle mass but >90% of the particle number. Nuclei mode particles emitted by either diesel or SI engines are usually composed almost entirely of volatile organic material and sulfates, but may also contain metallic ash [1,10,21,24]. Most of the mass is found in the accumulation mode (roughly 30–500 nm), and it is composed of carbonaceous agglomerates and adsorbed materials. The coarse mode consists of particles larger than 1 μm and contains 5–20% of the mass. These relatively large particles are formed by reentrainment of particulate matter, mainly from the accumulation mode, that has been deposited on cylinder and exhaust system surfaces. Particles in crankcase fumes are also found in the coarse range. Since these modes consist of particles of different composition, which form at different times in the engine cycle, it is more meaningful to discuss size distributions in terms of these modes than in terms of arbitrary size boundaries such as nanoparticles or ultrafine particles.

Future diesel engines are likely to be equipped with highly efficient exhaust filters. These filters will remove nearly all of the solid carbon and ash particles, but will not necessarily remove volatile materials like OC and sulfates. As a result, it is likely that nearly all of the number and most of the mass emitted by these engines will be found in volatile nuclei mode particles.

Particulate matter emissions from internal combustion engines have traditionally been regulated solely on the basis of total particulate matter mass emissions; no reference is made either to the size or the number concentration of the emitted particles. In response to these regulations, modern engines emit much lower particulate matter mass concentrations. Particle number emissions remain unregulated. Recent measurements show that on-road number concentrations between 10^4 and 10^6 particles cm^{-3} are typical, even though ambient mass concentrations are well below regulated limits [4,17,18,32]. While on-road number concentrations are relatively high, they decay quickly downwind of the roadway. For typical urban conditions, Modeling [6] indicates that characteristic times and transit distances for 90% reduction of total number concentrations (nearly all in the ultrafine range) are on the order of a few minutes and 100–1000 m, respectively.

Submicron, on-road aerosols result mainly from the exhaust emissions of diesel and gasoline SI engines, but the relative contribution of each engine type is not well defined [12]. There is little correlation between the ambient particle number concentration and the ambient mass concentration, but there is a strong correlation between high-speed traffic and nanoparticle number concentration. Laboratory studies [3,10,11,22,23] have shown that nanoparticle emissions from SI engines are much more speed and load dependent than diesel engines. Our on-road studies [18,20,21] have shown that high speed and load conditions, such as freeway cruise and hard acceleration, produce number emissions approaching those of diesel engines; however, under less severe conditions, SI particle number emissions are much lower.

4.2 Measurement of size and composition of diesel particles

4.2.1 Sampling issues

The size and concentration of volatile and semivolatile particles emitted by diesel engines are strongly dependent upon dilution and sampling conditions. This is because the materials that comprise these particles (sulfur compounds and organics) are generally in the vapor phase in the tailpipe and undergo gas-to-particle conversion processes, nucleation, condensation, and adsorption, during dilution and cooling. These processes, especially nucleation, are highly nonlinear and extremely sensitive to conditions.

Figure 4.2 illustrates the principal processes that take place as particles are formed by combustion in a diesel engine and undergo dilution and cooling.

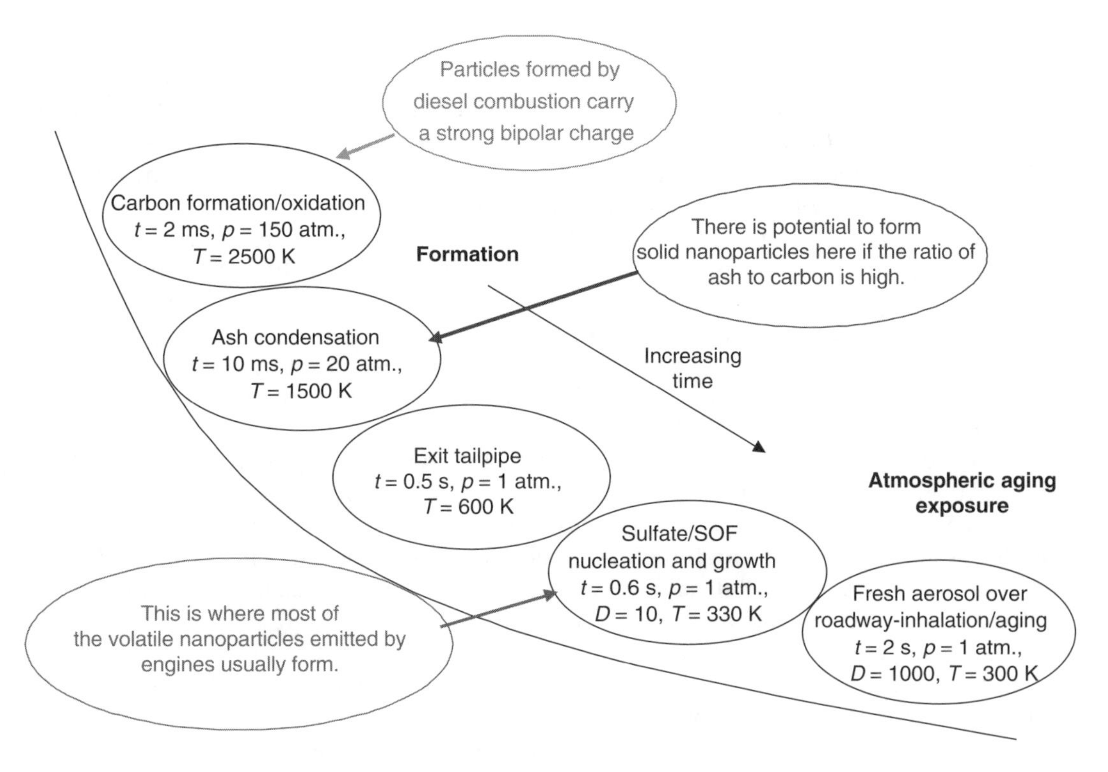

Figure 4.2 Particle formation history — most volatile nanoparticles form during dilution.

Carbonaceous soot particles are formed very early in the combustion process and most of them are oxidized at very high temperatures. Lubricating oil atomized and vaporized from cylinder walls is entrained and partially burned. Much of the organic carbon associated with diesel exhaust particles is associated with lubricating oil. Metallic additives in the lube oil may be converted to gas-phase compounds and then undergo gas-to-particle conversion as the exhaust dilutes and cools. Most of this material ends up as ash associated with accumulation mode particles, but ash nuclei may form if the ratio of metal to carbon particles is sufficiently high. Usually, however, most of the particles present in the tailpipe reside in the accumulation mode. As exhaust mixes with ambient air, volatile particle precursors, mainly sulfuric acid and hydrocarbons, become supersaturated and undergo gas-to-particle conversion forming volatile nuclei mode particles. Thus, most particles found in the nuclei mode form as the exhaust mixes with ambient air, not in the tailpipe. As a result, the concentration of particles in the nuclei mode is strongly influenced by the dilution conditions such as dilution rate, and temperature. The dependence of the nuclei mode on dilution conditions is particularly important in the laboratory where it is difficult to simulate on-road dilution conditions. On the other hand, solid particles, mainly carbon agglomerates and ash, are formed in the engine itself and are thus not influenced by dilution conditions.

Figure 4.3 shows the influence of dilution and sampling conditions on the size distributions measured for a medium-duty diesel engine running at medium speed and load [2]. The nuclei mode is strongly influenced by dilution conditions while the accumulation mode is not. Increasing residence time in the primary dilution chamber from 230 msec to 1 sec increases the size of the nuclei mode by two orders of magnitude. Decreasing the temperature in the primary dilution chamber from 66°C to 32°C increases the size of the nuclei mode by about one and a half orders of magnitude. Concerns about sampling and dilution led to the CRC E-43 program [21], in which laboratory and real-world, on-road samplings were compared.

4.2.2 On-road and laboratory measurements of diesel particles

The University of Minnesota mobile emissions laboratory (MEL) was used to collect gaseous and aerosol data while following heavy-duty trucks on rural roadways. The sample intake, located in front of the MEL, was set at a height of 4 m to sample the plumes heavy-duty trucks. The total sample-line length was less than 6 m, with a flow rate of 400 l min^{-1} through a 10.2 cm diameter tube. A manifold distributed the sample air to the instruments. The primary instrument used to determine the number, surface area, and volume size distributions was the TSI 3071 scanning mobility particle sizer (SMPS). It was operated in the high flow mode with 60 sec up-scan and 30 sec down-scan times. In this configuration, the SMPS measured particles from 8 to 300 nm. A stand-alone TSI 3025A ultrafine condensation particle counter (UCPC) was used to determine the total number concentration for particles ranging in size from about 3 to 1000 nm [15,36]. This UCPC has a maximum concentration of 100,000 particles cm^{-3}. Leaky-filter dilutors with dilution ratios ranging from 220:1 to 350:1 were used with the UCPC to prevent overranging with highly concentrated road way aerosols. Leaky-filter dilutors consist of a glass capillary tube placed inside an absolute capsule filter to create a leak through the filter. The diameter of the capillary tube determines the dilution ratio. Further details on the leaky-filter dilutor are available elsewhere [21].

Three gas analyzers were also used: a nondispersive infrared (NDIR) CO_2 analyzer, an NDIR CO analyzer, and a chemiluminescence NO_x analyzer. These instruments have a response time of about 1sec; they were used to determine on-road dilution ratios and background ambient gas concentrations.

Figure 4.4 shows size distributions measured in on-road chase experiments and on a chassis dynamometer with the same truck. The results shown are a composite of loaded

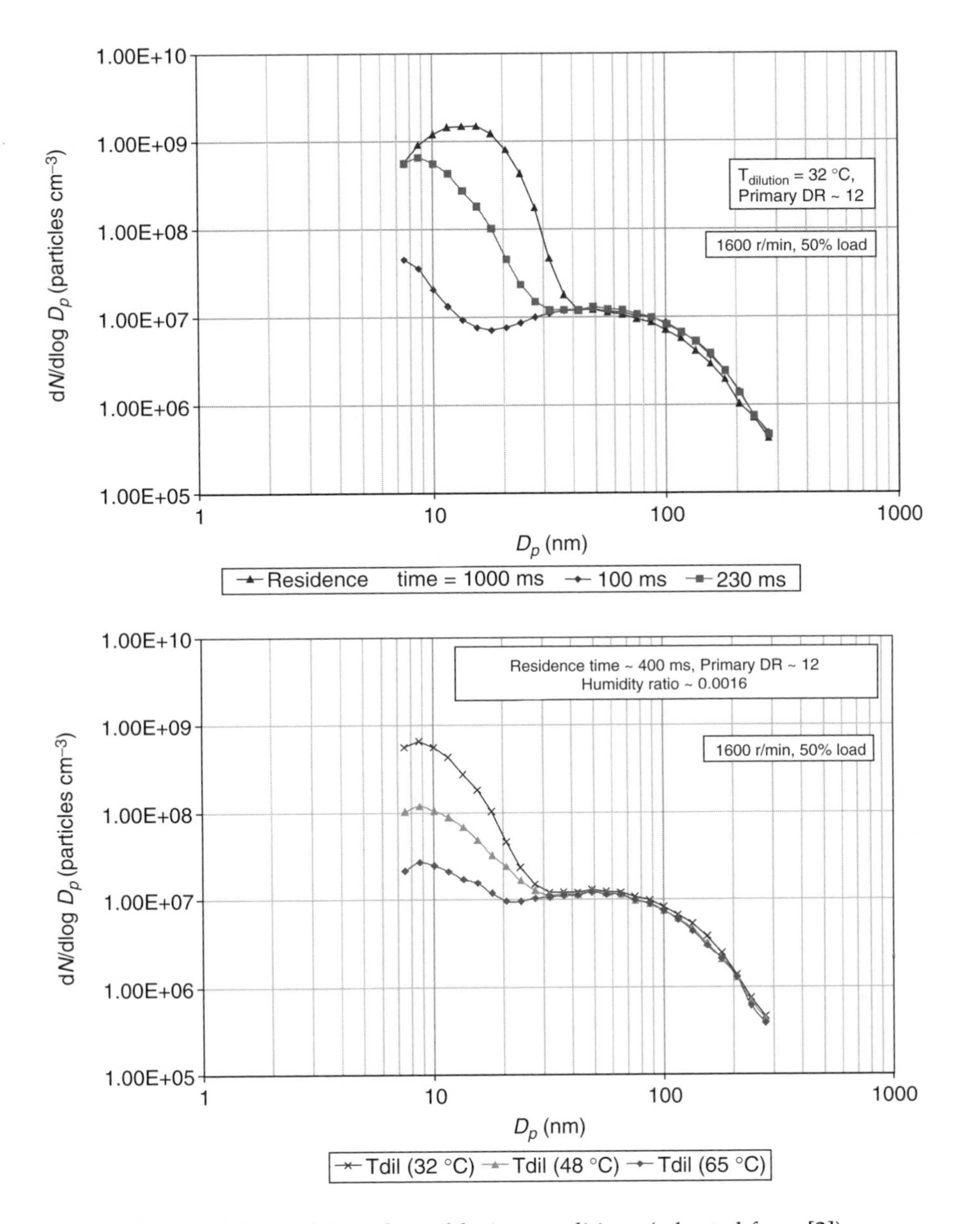

Figure 4.3 Sensitivity of the nuclei mode to dilution conditions (adapted from [2]).

and unloaded highway cruise under moderate summer conditions for a modern heavy-duty diesel engine with full electronic engine management.

A volatile nuclei mode was present both on-road and in the laboratory. The two-stage, porous tube/ejector dilutor system used in the laboratory tests on the chassis dynamometer could simulate on-road nuclei mode formation for composite, summer highway conditions. The relative sizes of the two modes are more significant than the absolute levels due to uncertainty in on-road dilution ratios. Results are shown for a CA fuel with about 50 ppm sulfur and an EPA fuel with about 350 ppm sulfur. In general, the CA fuel produced a smaller nuclei mode than EPA fuel. These results show that it is possible to simulate nuclei mode formation for carefully defined on-road conditions. However, at present, it is unclear as to which on-road conditions should be simulated. There are many variables, including temperature, previous operating history, road speed, exhaust system design, and others.

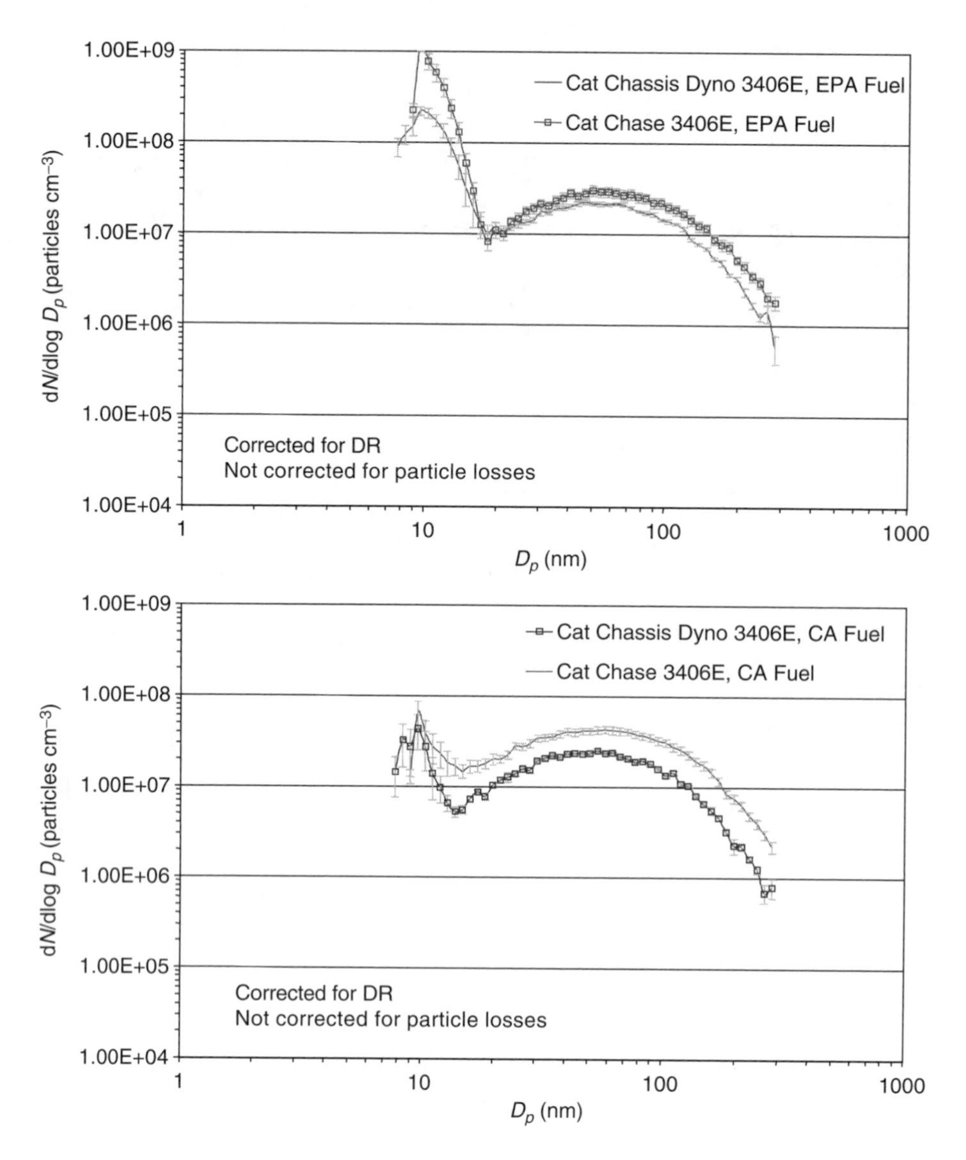

Figure 4.4 Size distributions measured on-road and on a chassis dynamometer (based on data from [19]).

4.2.3 Particle structure and composition measurements

Laboratory measurements of the structure and composition of diesel particles have been carried out under sampling conditions that gave nuclei mode formation similar to on-road conditions. The size of the nuclei mode is usually sensitive to fuel sulfur content, but a significant nuclei mode may form even with a very low content of fuel sulfur and lubricating oil. Figure 4.5 shows some typical size distribution measurements made as a Cummins Engine Company funded extension of the DOE/CRC E-43 program [21]. These measurements were made with a Cummins ISM engine (1999 technology) running on specially formulated low-sulfur lubricating oil and fuels with sulfur contents varying from 1 to 325 ppm sulfur. The same base fuel was used in all tests; fuel sulfur was varied with an additive. Four different operating conditions are shown: idle, light-medium load cruise (1400 r/min, 366 N m torque, 18% rated power), very light load cruise (1800 r/min, 154 N m,

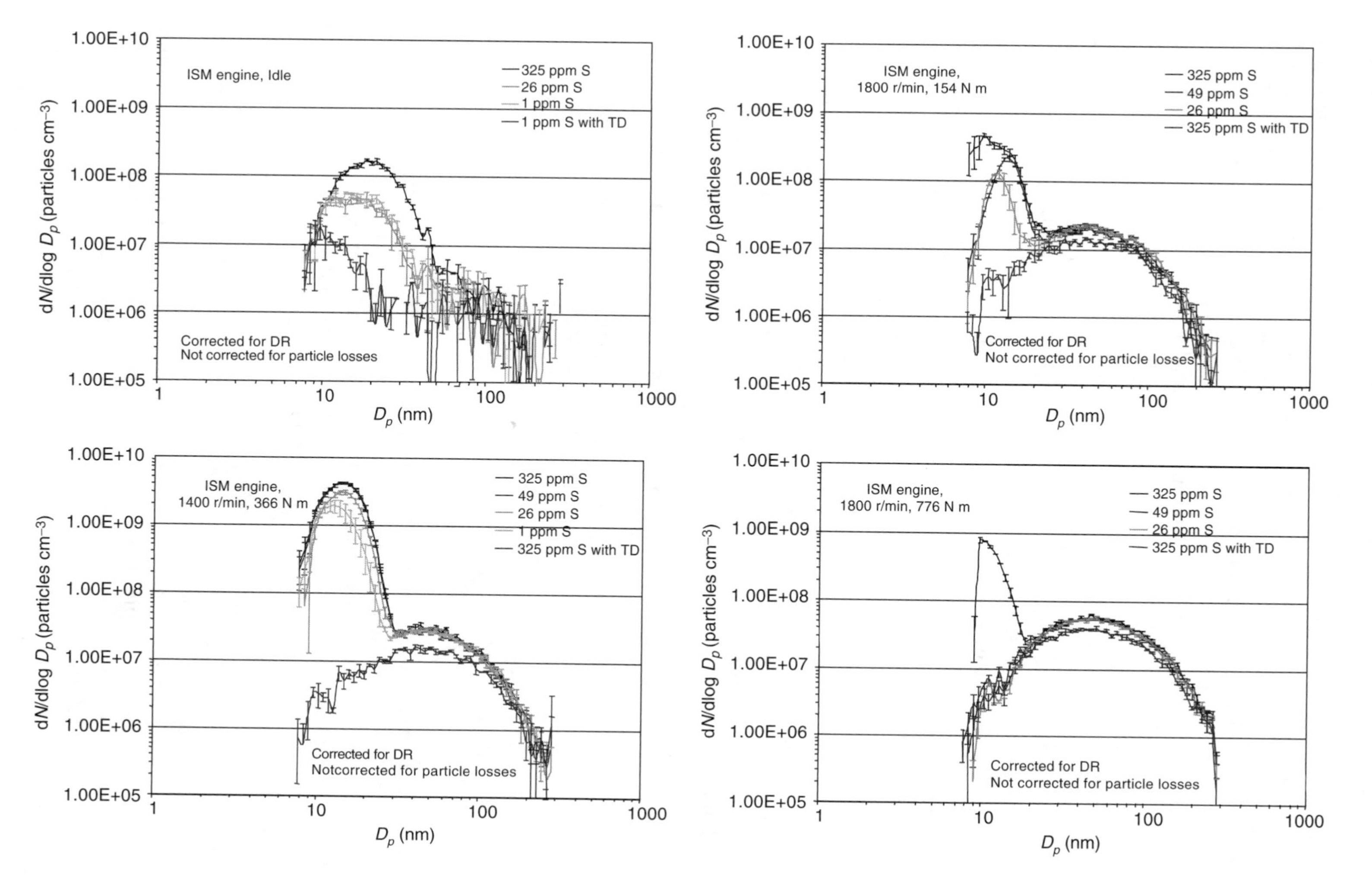

Figure 4.5 Size distributions measured with a modern heavy-duty diesel engine operating with very low sulfur lubricating oil and fuels with sulfur levels ranging from <1 to 325 ppm. TD denotes the thermal denuder that removes volatile particles (based on data from [19]).

10% rated power), and heavy load cruise (1800 r/min, 776 N m, 50% rated power). Measurements were made with and without a thermal denuder [5], which removed all particles that evaporate at 300°C or less. The size distributions clearly show the submicron bimodal size distribution typical of diesel exhaust particles. The size of the mainly volatile nuclei mode is influenced by the fuel sulfur concentration, but the mainly solid, accumulation mode is not. At idle, the largest nuclei mode was observed with the 325 ppm sulfur fuel. The 26 and 1 ppm sulfur fuels gave smaller nuclei modes of essentially the same diameter. The thermal denuder reduced but did not eliminate the nuclei mode, indicating the presence of nonvolatile particles. We believe that this solid nuclei mode is composed of ash associated with metals in the lubricating oil. At 1800 r/min, and 154 N m, the nuclei mode decreases but does not disappear as the sulfur content of the fuel is reduced. In this case, however, nuclei mode particles are nearly entirely removed by the thermal denuder, indicating that they are mainly volatile. At 1400 r/min, and 366 N m, a very large nuclei mode is present, which is not strongly influenced by the sulfur content of the fuel. It is completely removed by the thermal denuder, indicating that it consists of volatile material. At 1800 r/min, and 776 N m, only the highest fuel sulfur, 325 ppm, exhibits a nuclei mode. Again, this mode is volatile and is removed by the thermal denuder. This series of tests reveals a complex behavior with respect to the fuel sulfur content, ranging from the formation of a large nuclei mode regardless of the fuel sulfur to the formation of a nuclei mode only with the highest sulfur fuel.

A variety of physical and chemical methods were used to characterize the composition and size of diesel particles from modern engines as part of the CRC/DOE E-43 program [21,29,30,33,37]. A thermal desorption particle beam mass spectrometer (TDPBMS) was used to measure the volatility and mass spectra of the volatile fraction of all the particles in selected size ranges between 15 and 300 nm. Three different engines were tested, with fuels having sulfur contents ranging from 0 to 360 ppm. For these engines and fuels, the organic component of total diesel particles and nanoparticles appeared to be mainly unburned lubricating oil. The major organic compound classes found were alkanes, cycloalkanes, and aromatics. Low-volatility oxidation products and polycyclic aromatic hydrocarbons that have been found in previous GC-MS analyses were only a minor component of the organic mass. Nanoparticles formed with 360 ppm sulfur fuel contained small amounts of sulfuric acid, but those formed with less than 100 ppm sulfur fuel showed no evidence of sulfuric acid; the nanoparticles were nearly pure heavy organics.

The physical properties of the particles were also studied using tandem differential mobility analyzers (TDMAs). These experiments allowed size-resolved measurements of volatility and hygroscopicity to be made. In the volatility experiments, the particles were heated and the shrinkage was monitored. In the hygroscopicity measurements, the particles were humidified and their growth was monitored. The volatility measurements showed that diesel exhaust particles consist of an external mixture of more volatile and less volatile particles. At 30 nm, roughly the boundary between the nucleation mode and accumulation modes, both volatile and less volatile particles are found. For smaller sizes, volatile nuclei mode particles dominate; for larger sizes, less volatile carbonaceous agglomerates from the accumulation mode dominate. Volatile particles were found to evaporate in TDMAs like C24–C32 normal alkanes. These heavy alkanes are much more prevalent in lubricating oil than in fuel.

The hygroscopicity measurements were used to estimate the concentration of sulfuric acid in the particles. Hygroscopic particles were observed with the 360 ppm sulfur fuel. The growth increased with decreasing particle size, suggesting that the smallest particles were enriched with sulfuric acid. The growth observed with the smallest particles tested, 6.5 nm in diameter, was consistent with a sulfuric acid content of 20% by mass. The estimated mass fraction for 30 nm particles was only 5%. No detectable hygroscopic growth was observed for particles produced when the fuel sulfur content was less than 100 ppm.

Taken together, the size distribution, TDPBMS, and TDMA measurements lead to several conclusions about the formation and composition of the nuclei mode. The size distribution measurements show that, in general, increasing the fuel sulfur concentration increases the magnitude of the nuclei mode. On the other hand, the TDPBMS and TDMA measurements show that the nuclei mode consists mainly of heavy hydrocarbons, with a significant amount of sulfuric acid found only in the smallest particles when the fuel with the highest concentration of sulfur is used. Thus, it would appear that the presence of sulfur in the fuel facilitates the nucleation and growth of nuclei mode particles that consist mainly of heavy hydrocarbons. This same hypothesis was postulated based solely on size distribution measurements and physical arguments by Khalek et al. [16]. In most cases, the nuclei mode contains less than 1% of the particle mass. Since DPM often contains 10% or more sulfate and water and 20% or more OC, most of the sulfate and OC mass must reside in the accumulation mode, presumably adsorbed on the carbonaceous agglomerates formed by combustion. The nonvolatile material found in the nuclei mode at idle is likely to be ash formed metallic additives in the lubricating oil.

All of the tests described above were conducted on heavy- or medium-duty engines without after-treatment running under fully warmed-up, steady-state conditions. Under cold start and transient conditions, higher molecular weight hydrocarbons found in the fuel may play a more important role in nuclei mode formation. Many engines used in passenger cars and light trucks are equipped with flow-through exhaust catalysts. Under moderate operating conditions, these devices would be expected to remove much of the organic material associated with the nuclei mode and thus suppress its formation. However, under higher load conditions these devices, like the catalyzed filters discussed below, will tend to form sulfuric acid that may lead to nuclei mode formation.

4.2.4 Particles from engines equipped with exhaust filters

The emission standards for engines built for 2007 and beyond are so stringent that it is almost certain that it will be necessary to use exhaust filters. Figure 4.6 shows some typical size distributions measured upstream and downstream of a catalyzed filter system on a modern heavy-duty engine running on very low sulfur fuel, <1 ppm. There is little evidence of nuclei mode downstream of the filter and number concentrations are reduced by well over 99%. Changing dilution conditions influence size distributions upstream but had little influence downstream.

These filters are very effective at removing solid particles so that very little EC or ash is likely to leave the filters. Thus, particles emitted by future trap-equipped engines are likely to consist of mainly volatile materials like sulfuric acid and heavy hydrocarbons. Many exhaust filters are catalyzed to facilitate the combustion of collected particles. These catalysts are also effective at oxidizing SO_2 in the exhaust to SO_3, which reacts with water to form sulfuric acid. Consequently, the volatile particles emitted by such systems are likely to contain a higher sulfuric acid fraction than currently observed, although the absolute concentrations will be low. Figure 4.7 shows size distributions measured upstream and downstream of another catalyzed soot filter on a heavy-duty diesel engine. This time, the engine was operated at a very high load, rated torque condition, with a 26 ppm sulfur fuel. For this condition, a large nuclei mode is formed downstream of the filter. This is evident in the upper plot, which shows that the number concentration downstream is about ten times that upstream. The lower plot shows volume-weighted distributions for the same test. These results are shown on a linear scale so that the area under the curve is proportional to the total particle volume (roughly proportional to mass) in a given size range. The plot shows that the filter has essentially eliminated the accumulation mode where solid carbonaceous agglomerates are found, but has actually increased the

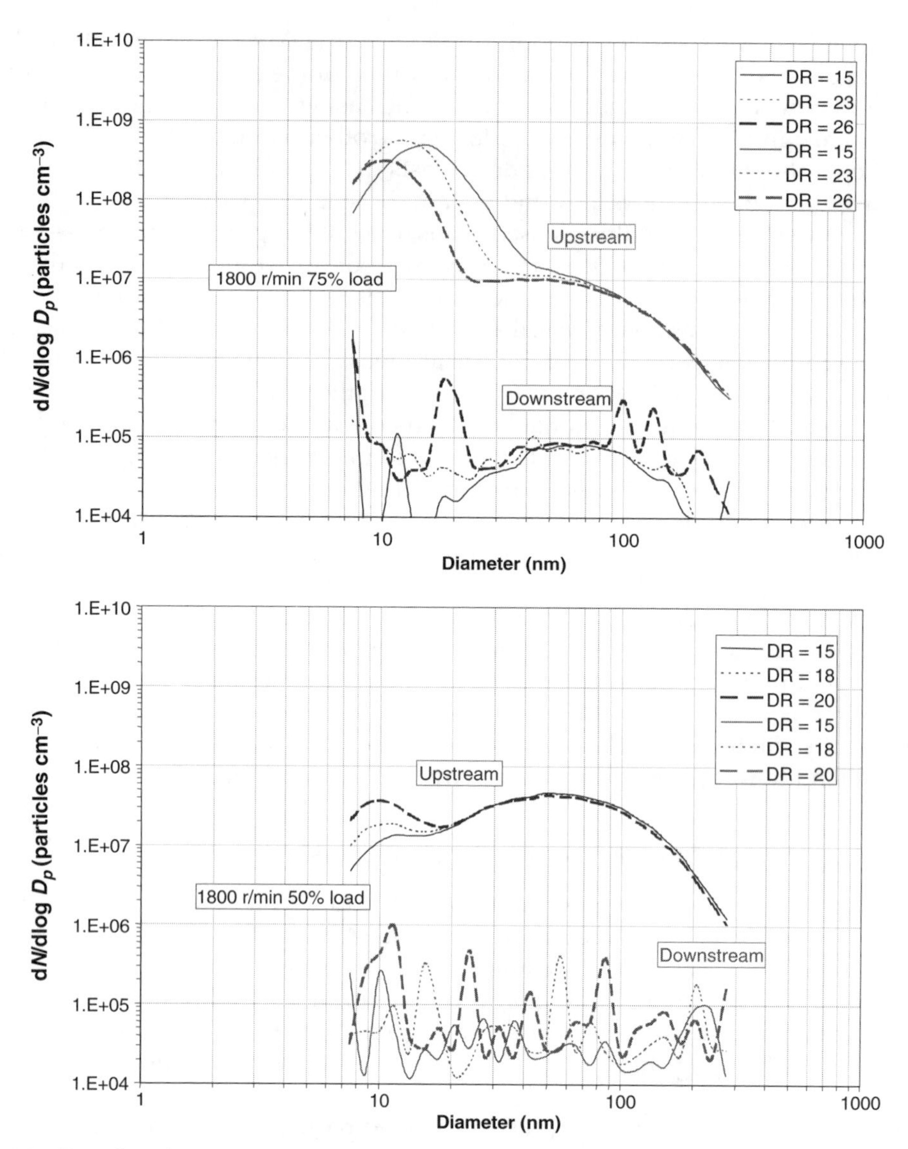

Figure 4.6 Size distribution measurements on a modern heavy-duty engine with a catalyzed exhaust filter.

concentration in the nuclei mode range due to particle formation downstream of the filter. However, the test with the thermal denuder shows that essentially all of this material is volatile. For this condition, the only significant particle emissions consist of volatile particles in the nuclei mode size range. Although chemical analysis was not performed, it is likely that most of this material is sulfuric acid.

4.3 Some future engine design issues

In the future, diesel engine designers will not only have to be concerned with direct emissions from engines, but also on how these emissions influence the performance of after-treatment systems. The engine and after-treatment system will have to be fully integrated. This has been an important aspect of the design of the SI engine for some time, but not for diesel engines.

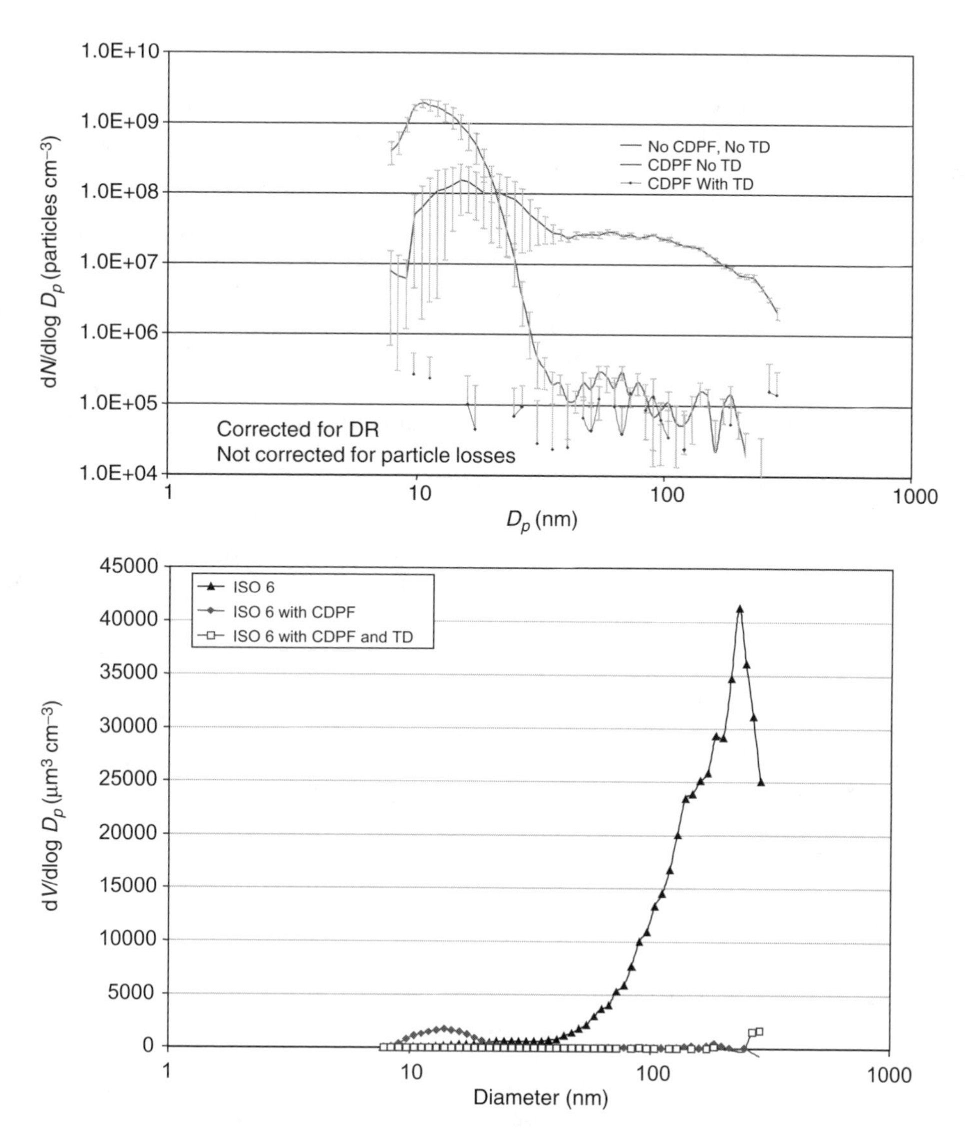

Figure 4.7 Size distribution measurements on a modern heavy-duty engine with a catalyzed exhaust filter. Left plot: number-weighted results; right plot: volume (mass)-weighted results. Results are shown with and without a thermal denuder.

A number of design issues will be associated with particle and nanoparticle control. For example, metals in the lube oil lead to solid ash particles. These particles cannot be oxidized in exhaust filters and over time the filters will become plugged. On the other hand, some of the constituents of the ash particles (e.g., Ca) may actually reduce soot ignition temperatures and aid in the regeneration of filters [14]. Some exhaust filtration systems trigger filter regeneration by injecting fuel into the cylinder late in the cycle. This raises the exhaust temperature and facilitates regeneration. However, some of this organic material may pass through the filter before regeneration is fully underway and produce puffs of nanoparticles.

Nanoparticle formation from engines running on gaseous fuels like natural gas and hydrogen may also pose a problem. While the combustion of the fuel in these engines leads to little or no particle formation, lube oil-related particles are still an issue. Such particles may be controlled by after-treatment, but this may make the engines uneconomical.

4.4 Discussion and conclusions

Current engines produce a bimodal size distribution in the submicron range, with a nuclei mode containing most of the particle number in the 3–30 nm diameter range and an accumulation mode containing most of the particle mass in the 30–500 nm range. Nuclei mode particles form mainly from volatile precursors, primarily heavy hydrocarbons and sulfuric acid. The formation of these volatile particles is strongly influenced by dilution and sampling conditions. Solid nuclei mode particles may form from metals in the lube oil or fuel. The accumulation mode consists primarily of solid carbonaceous agglomerates and adsorbed hydrocarbons. The solid carbon particles are formed within the cylinder, and the associated organic material results from partial combustion of fuel and lubricating oil. Solid particles may be nearly completely eliminated by filters expected to be used in future engines. Filters cannot directly remove the gas-phase precursors that lead to the formation of volatile, nuclei mode particles. Thus, under some conditions, engines equipped with high-efficiency filters are likely to emit primarily volatile particles, mainly in the nuclei mode region, unless the volatile precursors that lead to their formation are eliminated. Future diesel engine designers will not only have to consider exhaust concentrations, but also the influence of exhaust constituents on the performance of after-treatment systems.

References

1. Abdul-Khalek, I.S. and Kittelson, D.B., Real Time Measurement of Volatile and Solid Exhaust Particles Using a Catalytic Stripper, SAE Technical Paper Series, No. 950236, 1995.
2. Abdul-Khalek, I., Kittelson, D.B., and Brear, F., The Influence of Dilution Conditions on Diesel Exhaust Particle Size Distribution Measurements, SAE Paper No. 1999-01-1142, 1999.
3. Andersson, J. and Wedekind, B., DETR/SMMT/CONCAWE Particulate Research Program 1998–2001, Summary Report, Ricardo Consulting Engineers, 2001, 26 pp.
4. Booker, D.R., Urban Pollution Monitoring: Oxford City Centre, Research Report, AEA Technology, Aerosol Science Centre, Oxfordshire, U.K., 1997.
5. Burtscher, H., Baltensperger, U., Bukowiecki, N., Cohn, P., Hüglin, C., Mohr, M., Matter, U., Nyeki, S., Schmatloch, V., Streit, N., and Weingartner, E., Separation of volatile and non-volatile aerosol fractions by thermodesorption: instrumental development and applications, *J. Aerosol Sci.*, 32, 427–442, 2001.
6. Capaldo, K. and Pandis, S., Lifetimes of Ultrafine Diesel Aerosol, Subcontract Final Report Prepared for the University of Minnesota and the Coordinating Research Council under the E-43 Project Diesel Particulate Sampling Methodology, available from the Coordinating Research Council, Alpharetta, GA, 2001, 24 pp.
7. Dockery, D.W., Pope III, A., Xu, X., Spengler, J.D., Ware, J.H., Fay, M.E., Ferris Jr. B.G., and Speizer, F.E., An association between air pollution and mortality in six U.S. cities, *New Engl. J. Med.*, 329(24), 1753–1759, 1993.
8. Donaldson, K., Stone, V., Seaton, A., and MacNee, W., Ambient particle inhalation and the cardiovascular system: potential mechanisms, *Environ. Health Perspect.*, 109(Suppl. 4), 523–528, 2001.
9. Donaldson, K., Stone, V., Gilmour, P.S., Brown D.M., and MacNee, W., Ultrafine particles: mechanisms of lung injury, *Philos. Trans. R. Soc. Lond. A*, 358, 2741–2749, 2000.
10. Graskow, B.R., Kittelson, D.B., Abdul-Khalek, I.S., Ahmadi, M.R., and Morris, J.E., Characterization of Exhaust Particulate Emissions from a Spark Ignition Engine, SAE Technical Paper Series, No. 980528, 1998, 10pp.
11. Greenwood, S., Coxon, J.E., Biddulph, T., and Bennett, J., An Investigation to Determine the Exhaust Particulate Size Distributions for Diesel, Petrol, and Compressed Natural Gas Fuelled Vehicles, SAE Technical Paper Series, No. 961085, 1996.
12. Health Effects Institute., Improving Estimates of Diesel and Other Emissions for Epidemiologic Studies., Proceedings of an HEI Workshop, Baltimore, MD, December 4–6, 2002, HEI Communication 10, 2003, 162pp.

13. ICRP (International Commission on Radiological Protection)., Human Respiratory Tract Model for Radiological Protection, A Report of Committee 2 of the ICRP, Pergamon Press, Oxford, England, 1994.
14. Jung, Heejung, David B. Kittelson, and Michael R. Zachariah, The Influence of Engine Lubricating Oil on Diesel Nanoparticle Emissions and Kinetics of Oxidation, SAE Technical Paper Series, No. 2003-01-3179, 2003.
15. Kesten, J., Reineking, A., and Porstendoerfer, J., Calibration of a TSI model 3025 ultrafine condensation particle counter, *Aerosol Sci. Technol.*, 15, 107–111, 1991.
16. Khalek, I.A., Kittelson, D.B., and Brear, F., Nanoparticle Growth During Dilution and Cooling of Diesel Exhaust: Experimental Investigation and Theoretical Assessment, SAE Paper No. 2000-01-0515, 2000.
17. Kittelson, D.B., Engines and nanoparticles: a review, *J. Aerosol Sci.*, 29(5/6), 575–588, 1998.
18. Kittelson, D.B., Watts W.F., and Johnson, J.P., Nanoparticle emissions on Minnesota highways, *Atmos. Environ.* 38, 9–19, 2004.
19. Kittelson, D.B., Watts, W.F., and Johnson, J.P., Diesel Aerosol Sampling Methodology — CRC E-43 Final Report, Coordinating Research Council, Alpharetta, GA, 181pp. 2002, available at http://www.crcao.com/
20. Kittelson, D.B., Watts, W.F., Johnson, J.P., Remerowki, M.L., Ische, E.E., Oberdörster, G., Gelein, R.M., Elder, A.C., Hopke, P.K., Kim, E., Zhao, W., Zhou, L., and Jeong, C.-H., On road exposure to highway aerosols. 1. Aerosol and gas measurements, *Inhal. Toxicol.*, 16(Suppl. 1), 31–39, 2004.
21. Kittelson, D.B., Watts, W.F., Johnson, J.P., Zarling, D., Kasper, A., Baltensperger, U., Burtscher, H., Schauer, J.J., Christenson, C. and Schiller, S., Gasoline Vehicle Exhaust Particle Sampling Study, Contract final report U.S. Department of Energy Cooperative Agreement DE-FC04-01A166910, 2003, 152pp.
22. Maricq, M., Vehicle Particulate Emissions: A Comparison of ELPI, SMPS and Mass Measurement, International Seminar on Particle Size Distribution Measurement from Combustion Engines, Espoo, Finland, May 18–19, 1998.
23. Maricq, M.M., Podsiadlik, D.H., and Chase, R.E., Gasoline vehicle particle size distributions: comparison of steady state, FTP, and USO6 measurements, *Environ. Sci. Technol.*, 33, 2007–2015, 1999.
24. Mayer, A., Czerwinski, J., Matter, U., Wyser, M., Scheidegger, Kieser, D., and Weidhofer, VERT: Diesel Nano-Particles Emissions: Properties and Reduction Strategies, SAE Technical Paper Series. 980539, 1998, 12pp.
25. Oberdörster, G., Pulmonary effects of inhaled ultrafine particles, *Int. Arch. Occup. Environ. Health.*, 74, 1–8, 2001.
26. Oberdörster, G. and Utell, M.J., Ultrafine particles in the urban air: to the respiratory tract — and beyond? *Environ. Health Perspect.*, 110(8), A440–A441., 2002.
27. Peters, A., Wichmann, H.E., Th. Tuch, Heinrich, J., and Heyder, J., Respiratory effects are associated with the number of ultrafine particles. *Am. J. Respir. Crit. Care Med.*, 155, 1376–1383, 1997.
28. Pope III, C.A., Thun, M.J., Namboodiri, M.M., Dockery, D.W., Evans, J.S., Speizer, F.E., and Heath Jr. C.W., Particulate air pollution as a predictor of mortality in a prospective study of U.S. adults, *Am. J. Respir. Crit. Care Med.*, 151, 669–674, 1995.
29. Sakurai, H., Park, K., McMurry, P.H., Zarling, D.D., Kittelson, D.B., and Ziemann, P.J., Size-dependent mixing characteristics of volatile and non-volatile components in diesel exhaust aerosols, *Environ. Sci. Technol.*, 37, 5487–5495, 2003.
30. Sakurai, H., Tobais, H.J., Park, K., Zarling, D., Docherty, K.S., Kittelson, D.B., McMurry, P.H., Ziemann, P.J., On-line measurements of diesel nanoparticle composition, volatility, and hygroscopicity, *Atmos. Environ.*, 37, 1199–1210, 2003.
31. Seaton, A., MacNee, W., Donaldson, K., and Godden, D., Particulate air pollution and acute health effects, *Lancet*, 345, 1995, excerpt from section on — How particles may cause harm, p. 177.
32. Shi, J.P., Evans, D.E., Khan, A.A., and Harrison, R.M., Sources and concentration of nanoparticles (<10 nm diameter) in the urban atmosphere, *Atmos. Environ.*, 35, 1193–1202, 2001.
33. Tobias, H.J., Beving, D.E., Ziemann, P.J., Sakurai, H., Zuk, M., McMurry, P.H., Zarling, D., Waytulonis, R., and Kittelson, D.B., Chemical analysis of diesel engine nanoparticles using a

nano-DMA/thermal desorption particle beam mass spectrometer, *Environ. Sci. Technol.*, 35, 2233–2243, 2001.

34. Utell, M.J. and Frampton, M.W., Acute health effects of ambient air pollution: the ultrafine particle hypothesis, *J. Aerosol Med.*, 13(4), 355–359, 2000.
35. Whitby, K.T. and Cantrell, B.K., Atmospheric Aerosols: Characteristics and Measurement, in International Conference on Environmental Sensing and Assessment (ICESA), Institute of Electrical and Electronic Engineers (IEEE), IEEE 75-CH 1004-1, September 14–19, 1975, ICESA, Las Vegas, NV, paper 29-1, 1976, 6pp.
36. Wiedensohler, A., H.-C., Hansson, Keady, P.B., and Caldow, R., Experimental verification of the particle detection efficiency of TSI 3025 UCPC, *J. Aerosol Sci.*, 21(Suppl. 1), 617–620, 1990.
37. Ziemann, Paul, Sakurai, H., and McMurry, P.H., Chemical Analysis of Diesel Nanoparticles Using a NANO-DMA/Thermal Desorption Particle Beam Mass Spectrometer, Final Report, CRC Project No. E-43-4, April 2002, available at http://www.crcao.com

chapter five

Breathing zone exposure assessment

Charles E. Rodes and Jonathan W. Thornburg
RTI International

Contents

5.1 General sampling....61
5.1.1 Exposure assessment....61
5.1.2 Concentrations vs. personal exposures....62
5.1.3 Breathing zone exposures....63
5.1.3.1 Defining the breathing zone....64
5.1.3.2 Bluff body bias....67
5.1.4 BZE sampling applications....69
5.1.5 BZE sampling compliance....70
5.2 Sampling by contaminant type....70
References....71

5.1 General sampling

5.1.1 Exposure assessment

Concentrations measured at fixed locations can be dramatically different from those incorporated by personal exposure assessments made in the same space (Rodes et al., 1991). The concept that measurements of concentration do not necessarily quantify actual human exposures is illustrated by the classical risk paradigm (Rodes and Wiener, 2001):

sources→emissions→concentrations→exposures→doses→effects

where sources produce emissions that result in microenvironmental concentrations that result in exposures that are dependent on influencing factors, especially time-weighted proximity to the emissions from localized sources. The actual exposures produce body doses, dependent on uptake factors such as the inhalation rate, that can ultimately result in adverse health effects in susceptible populations.

While concentrations are produced by emissions from sources, exposures only occur if an individual is close enough (proximal) for a sufficiently extended period to result in a significant exposure level. Fixed-location monitors do not incorporate the element of proximity, nor do they account for the period of time that the person is actually nearby. The proximity to point sources produces concentration gradients that can be substantial. Since proximity changes as a person moves, personal exposure assessment provides the most complete and integrated picture of exposure. However, personal exposures are typically

1-56670-611-4 / 05 / $0.00+$1.50

more time consuming and expensive to obtain, and can be very burdensome to the individuals being studied. The issues of cost and burden must be weighed against the accuracy and representativeness that personal exposures provide.

The most applicable method of obtaining personal air exposures is through sampling directly in the breathing zone, and is specifically prescribed for most occupational scenarios to demonstrate compliance with permissible exposure limits (PELs) (e.g., OSHA, 2003; DOE, 2001; MSHA, 2003; NRC, 2003). There are no U.S. regulations requiring personal exposure monitoring for nonoccupational exposures. Personal exposure studies that integrate both occupational and nonoccupational periods should utilize breathing zone sampling approaches to allow the greatest flexibility in utilizing existing databases and models.

Personal exposure assessments can place an excessive burden upon study participants if the technology is not sufficiently miniaturized. Burden is discouraged by governmental and private Institutional Review Boards (IRBs) when dealing with human subjects in nonoccupational settings (e.g., NIH, 2003). Specifically obtaining breathing zone exposures (BZEs) can be the most burdensome approach to assessing personal exposures, but alternative approaches are available when this becomes an issue. Additionally, miniaturized personal exposure monitors typically have low collection rates, and often poor sensitivities (minimum detection limits) that may limit the comparability of the data with more robust, fixed-location technologies. This chapter will address the advantages and disadvantages of BZEs and alternative approaches from a number of perspectives, both occupational and residential.

5.1.2 *Concentrations vs. personal exposures*

The strategy of sampling of contaminants in or near the human breathing zone as most representative of inhalation exposure has been acknowledged as the most accurate approach (Cohen et al., 1984). A study by Sherwood (1966) on radioactive aerosol concentrations in a nuclear materials laboratory showed that collecting workers' personal exposure measures (PEMs) in the breathing zone provided levels of beta activity that exceeded single-location, room-average, microenvironmental[1] exposure measurement (MEM) values by a mean ratio of 7.7 (mean breathing-zone-to-room ratio). The primary explanation for the substantially elevated personal levels was the workers' closer proximity to the beta sources than the fixed-location, room monitor. In a similar study for both alpha and beta activity, Stevens (1969) similarly reported much higher breathing zone PEM levels compared with concurrent MEM levels, ranging from ratios of 2 to 3 when workers were near scattered, multiple point sources in the same room, and ratios from 5 to 15 when they were in close proximity to a single room source. Again the rationale was suggested to be driven primarily by the composite point-source-to-worker proximity during the 8-h exposure interval. Parker et al. (1990) studied the release of polydisperse 0.5 μm particles into a test room and reported that personal exposure measurements at the lapel of a manikin were 5 to 10 times higher 0.5 m from a point source than room average samplers a few meters away. They also reported that real-time measurements at the lapel and at the mouth, separated by only 0.3 m, showed very poor correlation for the concentration fluctuations. The potential for point sources to provide nonuniform concentrations in occupational and nonoccupational indoor settings was discussed by Nicas (1996) and Furtaw et al. (1996), respectively who provided models to estimate the influence of the room ventilation system on the room concentrations.

[1]Microenvironment is defined here to mean a localized, contained volume that generally defines the concentration — most often approximately bounded by the perimeter of the room when indoors.

Rodes et al. (1991) observed that activity pattern information during the integration period, in addition to source proximity, was critical to understanding nonoccupational personal exposures for aerosols. They suggested that the PEM-to-MEM ratios for residential exposure settings can be significantly different from occupational, due to typically weaker and more dispersed residential point sources and significant periods that include minimal or no sources. They reported median PEM-to-MEM ratios for nonoccupational aerosol exposure studies ranging from approximately 1.5 to 2.0 (much lower than typical occupational ratios), but still reported that the 95th percentile (most exposed) portion of the residential study populations exhibited ratios exceeding 4.0. Rea et al. (2001), using a personal real-time aerosol nephelometer, identified a wide variety of sources that contribute to daily nonoccupational personal exposures, in both outdoor locations and indoor microenvironments. They also reported that a significant portion of the aerosol personal aerosol exposures occurred in locations where the participants spent only 4 to 13% of their time. McBride et al. (1999) examined the dispersion of a gas tracer indoors in a private residence and noted a strong proximity effect (concentration decrease with distance) at distances up to 2 m from the source. Room background levels were typically measured within experimental error beyond 2 m from the source. They also reported that the room air exchange rate had little effect on the proximity effect within 0.5 m of the source. These data suggest that the room average MEM would need to be significantly closer to the PEM than 2 m in order to reflect the elevated breathing zone levels in close proximity to the sources. Rodes et al. (1995) also reported that the velocity profile near the body decreases sharply, defining a low-velocity boundary layer adjacent to the body only a few centimeters in thickness. Since the typical BZE inlet usually extends only a few centimeters from the body, particles will require a minimum size and inertia to penetrate across the flow streamlines to the inlet to be sampled. This bias is discussed subsequently in more detail in Section 5.1.3.2.

A number of strategies have been utilized to minimize the burden of conducting personal exposure assessments. In most scenarios, only the inlet system is placed in the breathing zone, with the pump located at the waist or in a backpack. The inlet can be clipped to the lapel, attached to special tabs on a vest, or attached to a shoulder strap adjacent to the lapel. Examples are shown in Figure 5.1. The swinging mass of a heavy inlet system can be annoying, and may affect the long-term comfort of the participant. This type of burden may result in two undesirable occurrences: (1) early dropout from the study (nonoccupational only) and/or (2) not wearing the sampler at all times (protocol compliance). Moving the inlet away from the breathing zone can reduce the perceived burden, but at the expense of reduced representativeness to BZE. Burden reduction approaches include utilizing a waistpack with the inlet mounted on the front, a backpack with the inlet mounted on the side, or for those participants who have physical disabilities, a luggage cart-mounted personal exposure system. Examples of these are shown in Figure 5.2. A comparison of sampling approaches by location is given in Table 5.1.

5.1.3 Breathing zone exposures

A relevant consideration here is the definition of "breathing zone." Occupational exposure studies almost always require that the inlet be placed in the breathing zone to capture BZEs. Baron (2003) recommends that personal exposure inlets sample "…air that would most nearly represent that inhaled by the employee" for occupational assessments. Fortunately, most participants in occupational studies (a) are healthy adults, (b) are required to participate as part of the job, and (c) only need to wear the monitor for 8-h periods. This simplifies the technologies applied from the perspective of the total burden (weight, size, obtrusiveness). Nonoccupational studies, on the other hand, often include unhealthy, elderly adults and children, utilize voluntary participants, and often focus on

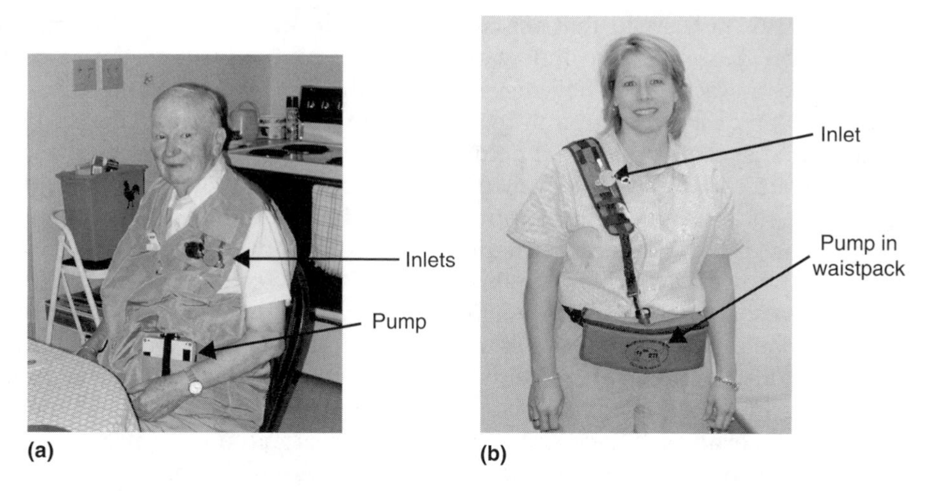

Figure 5.1 BZE monitoring using inlets fastened to (a) a specially made, low-lint vest (to distribute the weight) and (b) to a shoulder strap, (a) Low-lint vest used to support dual-channel pump and two aerosol sampling inlets near the breathing zone (b) PEM inlet mounted near breathing zone on a shoulder strap with pump inside a waistpack.

24-h assessment periods. This places a premium on the burden imposed on the study participants. Ideally, miniature passive badge samplers that can unobtrusively be attached to the lapel are most effective. These have been developed for radiation, a number of gases, and most recently efforts have been made to address aerosol collections passively (ORAU, 2004; Brown and Monteith, 2001). The size and weight of active (pumped) sampling systems for personal exposure pose substantial burden concerns for these classes of participants. The most burdensome version of personal exposure samplers places the inlet in or near the breathing zone by clipping it to a coat or vest lapel, or a shoulder strap.

5.1.3.1 *Defining the breathing zone*

Simplistically the air parcel immediately in front of the nose and mouth from which the inhalation volume is withdrawn could be construed as the breathing zone. Defining the boundaries of this parcel is not simple and varies with many factors, including the inhalation/exhalation rate, microenvironmental air velocity and turbulence, nose vs. mouth breathing, walking/running speed, etc. Often the location for obtaining breathing zone measures is defined simply by indicating that the inlet be attached to a "lapel" (e.g., NRC, 2003). Shapiro (1990) indicates that radiation personal monitors can be worn between the waist and the shoulders, also noting that breathing zone sampling should be conducted at the lapel location. In an effort to standardize measurement methodologies, while allowing a greater range of inlet types (many are simply too heavy to clip to a lapel), spatial definitions have been defined for the breathing zone, such as: " ...the hemisphere of 300 mm radius extending in front of the face and measured from the midpoint of an imaginary line joining the ears" (NOHSC, 2003). Rock (2001) similarly defines a 300 mm distance from the nose, and observes that such sampling is designated as "breathing zone," even when a respirator is worn. While such a rigidly defined space is useful for uniformity, limited data are available to define the concentration uniformity throughout such a volume, especially for aerosols.

Martinelli et al. (1983) documented BZE biases for selected elements, and suggested that resuspension of relatively large-particle clothing dust was a contributing factor. Bradley et al. (1994) reported that sampling efficiencies for respirable aerosol at various locations across the chest and back varied from 0.45 to 0.61. Part of the chest bias was surmised to result from the

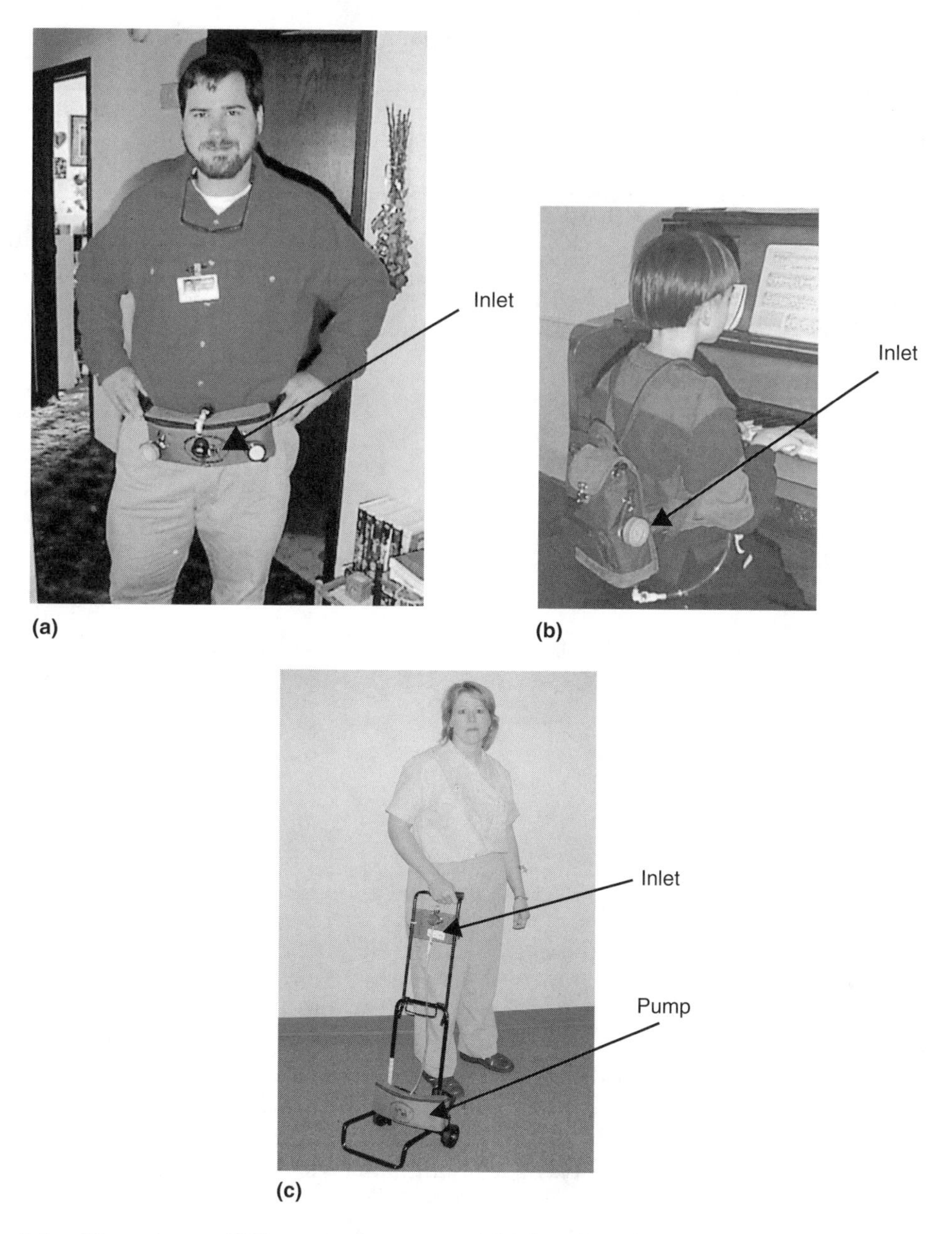

Figure 5.2 Alternate non-BZE personel exposure inlet locations for estimating BZEs. (a) Waistpack sampling arrangement showing inlet location with pump inside waistpack (b) Child's backpack showing inlet on side of pack and behind breathing zone (c) Exposure sampler mounted on luggage cart that can be pulled by participant.

inlet being in the downwash of cleaner exhaled breath from the nostrils. This observation lends credence to the placement of a breathing zone inlet to the left or right rather than directly under the head. Bull et al. (1987) generated 0.26 μm particles in 60-sec releases at point locations into a chamber and found that integrated samples collected in front of the nose and at the chest correlated to the same level (R = .84) as nose vs. the waist. The data suggested that, within experimental error, extending the breathing zone across the chest and as far down as the waist produced essentially equivalent data for these relatively small particles.

Table 5.1 Active (Pumped) Personal Exposure Sampling Approaches to Minimizing Participant Burden

Active Sampling Approach	PEM Inlet Location	Pump Location	Participant Burden Level	Target Cohort	Notes
Breathing zone exposure	Breathing zone — clipped to lapel or shoulder strap	Waistpack or backpack	High	Healthy adults	Assumes lapel concentrations same as BZ
Waistpack	Front of pack at waist level	Inside waistpack	Med. high	Healthy adults; older children	Okay for fine particles; may be susceptible to "ground cloud"
Backpack	Side of back pack at shoulder level	Inside backpack	Medium	Younger children	May interfere with book bags
Pull-cart	Pull-cart handle	Pull-cart platform	Low	Unhealthy adults; elderly	May be susceptible to "ground cloud"
Surrogate person	On trailing surrogate	On trailing surrogate	Minimal	Unhealthy adults; elderly; children	Requires continuous surveillance
Predictive models	None	None	None	Metro population	Requires fixed-location concentration data; may predict extremes inaccurately

A continuous tracer gas plume from a point source 1 m from a manikin-mounted PEM in a simulated indoor setting (equivalent air velocity and turbulence levels) was reported by Rodes et al. (1995) to be only 150 mm wide at the chest. This suggests that strong, nearby sources may produce significant differences between BZE and other personal sampling methods, if sufficiently long exposures occur. This scenario is most common in occupational settings. Similarly, Coker (1981) reported that workers' handedness (left vs. right) during daily activities (in this case, paint spraying) affected concentrations by as much as 50% across the chest, the right lapel concentrations being much higher than the left for right-handed individuals. Rodes et al. (2001) reported that body dander could be a significant contributor to the collected mass nonoccupational breathing zone aerosol. Importantly, this source and resuspended clothing dust are associated with the activities and character of the person being monitored, rather than external sources. The electrostatic charging of both a manikin (simulating the body) and the PEM were found to affect aerosol collection performance in a study by Smith and Bartley (2003). The efficiency was found to increase by ~10% for 7 μm particles when the charge could be effectively neutralized. Since participant clothing (e.g., sweaters) are surmised to periodically be highly charged, this may also influence breathing zone uniformity for aerosols.

The brief listing of spatial bias scenarios provided in the previous paragraph suggests that breathing zone assessments can be substantially different from personal exposure assessment measurements made at more distant body locations. Table 5.2 provides typical separation influences when the methodologies listed are used for personal exposure sam-

Table 5.2 Advantages and Disadvantages of the BZE Approach Compared to other Methods of Estimating Personal Exposures

Active Sampling Approach	Advantages	Disadvantages	Relevant Citations
Breathing zone exposure	Truest representation of inhaled exposure; 0 to ~30 cm from nose/mouth	Most burdensome	Cohen et al. (1984)
Waistpack	Removes inlet mass from lapel; eliminates need for vest or shoulder strap	~50 cm from nose/mouth; possible biases for particles <2.5 μm; probable biases for particles >2.5 μm from resuspension; "bluff body" effects at waist differ from head	Pellizzari et al. (2001)
Backpack	Easy for child to put on/take off; familiar (like a book bag); one of the few viable methods for studying children	~30 cm from nose/mouth; "trailing wake" sampling may bias particle collections >2.5 μm	Rodes and Wiener (2001), Eisner et al. (2002)
Pull-cart	Can utilize relative heavy hardware (e.g., 5 lb) with minimal burden; viable method for those who can not wear sampler	50–80 cm from nose/mouth;probable biases for particles >2.5 μm from resuspension	Evans et al. (2000)
Surrogate person	No physical burden except always having someone trailing	Can have significant separation distances of 100–300 cm; not easy to trail active children	Stevens et al. (2003)
Predictive models	Shown to predict median values exposure distributions of <10 μm aerosol	Unclear whether tails of distribution (e.g., most exposed) can be predicted reliably	Ott et al. (2000)

pling. This somewhat cursory review of the literature suggests that all of the methodologies (except predictive modeling) will probably provide measurements within about 20% of the true values for gases and total mass of particles <2.5 μm. As particles get larger above 2.5 μm, however, the spatial biases can increase to substantial values, especially if the potential exists for resuspended dust from clothing or the floor.

5.1.3.2 Bluff body bias

Studies of aerosol trajectories around objects have suggested that the presence or absence of the human body shape can affect particle capture methodologies (e.g., Wood and Birkett, 1979; Mark and Vincent, 1986; Botham et al., 1991). They all describe test methodologies that evaluate the performance of breathing zone samplers incorporating the presence of a manikin behind the inlet to simulate the human body. As mentioned previously, trajectories of particles approaching the body (and the PEM inlet) must navigate through the flow streamlines decelerating and diverging around the body shape. A mathematical study by Ingham and Yan (1994) attempted to estimate the level of sampling bias using a particle trajectory model, and suggested that the presence of the human "bluff" body behind an aspirated personal sampling inlet could bias the aerosol collection by as much as factors of two or more. The models showed the importance of influencing variables included factors such as (a) particle Stokes number (particle size), (b) level of particle bounce from inlet and clothing surfaces (and subsequent total capture by the inlet), (c) distance the sampling inlet projects away from the body, (d) sampling inlet velocity relative to the local air velocity, and (e) width of the inlet relative to that of the human body. The relative sampling efficiency data in Table 5.3 illustrate the predicted influences of particle size, air speed, and presence or absence of a bluff body behind the inlet (inlet alone in the freestream).

These data bolstered the notion that fixed-location MEM sampling without a bluff body simulating the human shape would potentially bias relationships with personal exposure sampling, even if the contaminant was uniformly distributed. The model also suggested that differences between personal sampling systems (size, shape, flowrate, etc.) could result in between-PEM-type biases. In order to minimize biases when intercomparing personal and fixed-location monitors, an elliptically shaped body such as that shown in Figure 5.3 can be used behind the inlet. This approach is not thought to require a physiologically correct manikin, but simply an approximate body shape to produce comparable flow streamlines around the body. The aluminum bluff body suggested by Rodes et al. (2003) is 40.6 cm tall, with a 2-to-1 elliptical cross-section of 40.6 cm wide by 20.3 cm deep.

Limited empirical data have are to validate these model projections for either between-PEM biases or biases between MEM with and without a bluff body shape, especially for nonoccupational settings. Heist et al. (2003) defined the flow streamline vectors around a heated, child-sized manikin in a simulated residential setting and provided data on scaling between body sizes. Preliminary data by Rodes et al. (2003) examining the influence of the presence or absence of a bluff body for fixed-location sampling in a controlled wind tunnel setting simulating private residences, and utilizing both fine Arizona Test Dust challenges. They found that the measured biases were less than 20% in all cases typically found in nonoccupational sampling. The data showed that for this aerosol type the samplers with a bluff body provided lower concentrations than those without a bluff body. The empirical data indicated that while the direction of the biases for selected parameters appears to be correctly predicted by the model, the degree of particle bounce and subsequent capture by the PEM inlet was much smaller than predicted by the total aspiration efficiency model of Ingham and Yan (1994). Further empirical studies are needed to

Table 5.3 Predicted Fractional *Overall Sampling Efficiency* from the Ingham and Yan (1994) Model Illustrating the Influence of Air Speed (cm/s) and Particle Size (μm) in Scenarios With and Without a Bluff Body Present.

Size (μm)	Indoor (15 cm/s)		Walking (100 cm/s)		Outdoor (400 cm/s)	
	Without body	With body	Without body	With body	Without body	With body
0.5	1.0	1.0	1.2	1.1	2.0	1.5
1.0	1.5	1.5	2.0	2.0	3.5	3.5
2.5	1.6	2.5	2.2	3.0	4.0	6.0
10	1.7	3.7	2.5	4.2	4.2	7.5

Note: Indoor and walking data interpolated by the authors from $\alpha = 5.0$ graph; outdoor data interpolated from $\alpha = 1.0$ graph.

Figure 5.3 Use of an elliptical aluminum body on which to mount a personal exposure inlet used as a fixed-location monitor indoors (at seated breathing zone elevation) or outdoors (at 2 m elevation). Bluff body with inlets located outdoors 2 m above the ground.

confirm these findings, given the potential for adding biases to BZE assessments, especially for particle sizes >2.5 μm.

5.1.4 BZE sampling applications

The most common application of breathing zone personal exposure assessments is to determine the distributions of exposure as related to a defined PEL. This requires that the methodologies be as accurate and representative of the true breathing zone concentrations. If the inlet cannot be physically placed continuously in the subject's breathing zone (e.g., small children), additional information is needed to subsequently allow any needed adjustment induced by the separation distance. As noted previously, gases and fine particles (≤2.5 μm) may be uniformly distributed in the sampled air. In this case, personal measurements made by all the approaches given in Table 5.1 may provide statistically equivalent results. Particles larger than ~2.5 μm, however, may exhibit significant microenvironmental concentration gradients that would result in detectable biases between inlet locations. The resuspended clothing dust and body dander inadvertently collected by the inlet may provide samples that are not representative of the air in the breathing zone. Utilizing a (low-lint) sampling vest can reduce clothing resuspension problems.

Not all personal exposure studies are conducted to estimate actual exposure levels. While occupational BZEs are most often conducted to make assessments relative to PELs, nonoccupational studies are often conducted to determine how well personal exposures associate with a dose or adverse health effect. In the case of associative studies, the absolute values are not nearly as important as the consistency and representativeness of the data. Rodes et al. (2002) observed that the lower filter face velocities of typical personal exposure samplers often result in significantly lower collected aerosol mass concentrations in warmer environments compared to high-flowrate, fixed-location samplers, due to reduced loss of volatile species. However, a Monte Carlo simulation analysis showed that either low-flow PEM or high-flow MEM mass concentrations would have produced statistically equivalent associations with concurrent adverse health effects.

5.1.5 BZE sampling compliance

As noted previously, the burden imposed on a study participant by BZE sampling can result in serious protocol compliance problems in nonoccupational studies where participation is voluntary. Occupational compliance is rarely of concern, since carrying a breathing zone monitor periodically is usually defined as a requirement of the job. A recent nonoccupational study reported by Lawless et al. (2003), using a specially designed activity sensor (Lawless, 2003) contained in the personal pump compartment, found that approximately 50% of the adult participants wore the monitor less than 60% of the protocol-defined sampling time. Two of the 35 participants wore the monitor less than 25% of the time. It was noted that burden level (weight, inconvenience) was a contributing factor, and that compliance was most important during periods of elevated exposures. While it is difficult to extrapolate broadly from these data, the findings strongly suggest that compliance can definitely be an important factor in personal exposure studies. When the monitor is not being worn as prescribed, it becomes a fixed-location monitor, resulting in data that are not representative of personal exposures. In settings with strong point sources, the distinction between personal and fixed-location monitors can be dramatically different. It is strongly suggested that efforts be made in any BZE study where burden might be construed to be a compliance factor, to assess the degree of compliance. While this may be as simple as asking follow-up questions, this approach may be inadequate for some participants.

5.2 Sampling by contaminant type

The application of specific contaminant categories (e.g., coarse aerosols) can accentuate specific performance areas such as accuracy, sampling artifacts, and representativeness. While this level of detail is beyond the scope of this chapter, it is appropriate to identify selected sources for additional information on BZE assessment. The sources of information describing specific sampling methodologies for gases, aerosols, and radiation are summarized in Table 5.4. The collection of personal exposure samples for gases and vapors, including those directly from the breathing zone, is described by Brown and Monteith (2001), including an array of active (pumped) and passive techniques. The active techniques require the air sample to be drawn through miniature absorber tubes clipped to the lapel containing specific substrates (e.g., charcoal, XAD-2) by a controlled flowrate pump. The pump is often attached to the waist. Low burden passive badge samplers are also described that attach to the lapel for a wide range of contaminants including organic vapor, amines, aldehydes, ozone, nitrogen dioxide, and mercury.

Particle-phase contaminants are almost universally sampled in an active mode, but typically require an aerodynamic sizing step that significantly increases the size and

Table 5.4 Selected Information Sources for BZE Technologies

Contaminant Category	Active (Pumped) Sampling	Passive Sampling	Notes
Gases	Brown and Monteith, (2001)	Brown and Monteith, (2001)	
Aerosols	Rodes and Wiener, (2001), Hering (2001)	Wagner and Leith, (2001)	Currently (5/03) only one passive aerosol device
Radiation	Ruzer (2004), Cohen and Heikkinen et al. (2001)	Ruzer (2004)	

complexity of the inlet. Often the resulting device is simply too cumbersome to attach directly to a lapel. The air stream is first sampled by a size-classifying device (e.g., cyclone, impactor) and the aerosol is then collected on either a filter or medium (e.g., impactor plate, agar (viable), or liquid bubbler). Hering (2001) describes many of the commonly used aerosol samplers for both fixed-location and personal exposures. Rodes and Wiener (2001) also summarize PEMs, including two miniature nephelometers.

Although radiation BZE assessment has some aspects similar to gases, the transport and fate properties of radiation require that it be considered separately. This topic is covered further in Chapter 14 (Ruzer, this volume).

References

Baron, P.A., NIOSH, Cincinnati, OH, personal communication to C.E. Rodes, RTI International, Research Triangle Park, NC, 6/4/03, 2003.

Botham, R.A., Hughson, G.L., Vincent, J.H., and Mark, D. Development of Test System for the Assessment of Personal Aerosol Samplers Under True Workplace Conditions, *AIHAJ*, 52: 423–427, 1991.

Bradley, D.R., Johnson, A.E., Kenny, L.C., Lyons, C.P., Mark, D., and Upton, S.L., The use of a manikin for testing personal aerosol samplers, *J. Aerosol Sci.*, 25 (Suppl. 1), S155–S156, 1994.

Brown, R.H. and Monteith, L.E., Gas and vapor sample collectors, in *Air Sampling Instrumentation of Atmospheric Contaminants*, 9th ed., Cohen, B.S. and McCammon, C., Eds., ACGIH, Cincinnati, OH, 2001, chap. 16, pp. 415–457.

Bull, R.K., Stevens, D.C. and Marshall, M. Studies of Aerosol Distribution in a Small Laboratory and Around a Humanoid Phantom, *JAS*, 18(3), 321–335, 1987.

Cohen, B.S., Harley, N. and Lippmann, M., Bias in air sampling techniques used to measure inhalation exposure, *AIHAJ*, 45(3), 187–192, 1984.

Cohen, B.S. and Heikkinen, M.S.A., in *Air Sampling Instrumentation of Atmospheric Contaminants*, 9th ed., Cohen, B.S. and McCammon, C., Eds., ACGIH, Cincinnati, OH, 2001, chap 21, pp. 623–660.

Coker, P.T., Recent Developments in Personal Monitoring for Exposure to Organic Vapours, International Environmental Safety, 13–15, 1981.

Department of Energy (DOE), DOE Occupational Radiation Exposure, 2001 Report, Report DOE/EH-0660, U.S. Department of Energy, Office of Safety and Health, Washington, DC, available from NTIS, Springfield, VA 22161, 2001.

Eisner, A.D., Heist, D.K., Drake, Z.E., Mitchell, W.J., and Wiener, R.W., On the impact of the human (child) microclimate on passive aerosol monitor performance, *Aerosol Sci. Tech.*, 36, 803–813, 2002.

Evans, G.F., Highsmith, R.V., Sheldon, L.S., Suggs, J.C., Williams, R.W., Zweidinger, R.B., Creason, J.P., Walsh, D., Rodes, C.E., and Lawless, P.A., The 1999 Fresno Particulate Matter Exposure Studies: comparison of community, outdoor, and residential PM mass measurements, *J. Air Waste Manage Assoc.*, 50, 1887–1896, 2000.

Furtaw, E.J., Pandian, M.D., Nelson, D.R., and Behar, J.V., Modeling indoor air concentrations near emission sources in imperfectly mixed rooms, *J. Air Waste Manage Assoc.*, 46, 861–868, 1996.

Heist, D.K., Eisner, A.D., Mitchell, W., and Wiener, R., Airflow around a child-size manikin in a low-speed wind environment, *Aerosol Sci. Tech.*, 37, 303–314, 2003.

Hering, S., Impactors, cyclones, and other particle collectors, in *Air Sampling Instrumentation of Atmospheric Contaminants*, 9th ed., Cohen, B.S. and McCammon, C., Eds., ACGIH, Cincinnati, OH, 2001, chap. 2., pp. 316–376.

Ingham, D.B. and Yan, B. The effect of a cylindrical backing body on the sampling efficiency of a cylindrical sampler, *J. Aerosol Sci.*, 25(3), 535–541, 1994.

Lawless, P.A., Portable air sampling apparatus including non-intrusive activity monitor and methods of using same, U.S. Patent 6,502,469B2, awarded January 7, 2003, U.S. Patent Office, 2003.

Lawless, P.A., Rodes, C.E., Thornburg, J.W., Williams, R.W., and Rea, A., Quantitative compliance monitoring in personal exposure sampling," *JAWMA*, in review, 2004.

Mark, D. and Vincent, J.H., A new personal sampler for airborne total dust in workplaces, *Ann. Occup. Hyg.*, 30(1), 89–102, 1986.

Martinelli, C.A., Harley, N.H., Lippmann, M., and Cohen, B.S., Monitoring real-time aerosol distributions in the breathing zone, *AIHAJ*, 44, 280–285, 1983.

McBride, S.J., Ferro, A.R., Ott, W.R., Switzer, W.R., and L.M. Hildemann, Investigations of the Proximity Effect for Pollutants in the Indoor Environment, *JEAEE*, 9(6), 602–621, 1999.

Mine Safety and Health Administration (MSHA), Verification of underground coal mine operators' dust control plans and compliance sampling for respirable dust; proposed rule, *Federal Register*, March 6, 2003, 30 CFR Parts 70, 75, and 90, 68, 10784–10872, 2003.

National Occupational Health & Safety Commission, Commonwealth of Australia, Glossary of Terms, website: www.nohsc.gov.au/OHSInformation/NOHSCPublications/fulltext/docs/h4/607.htm, 2003.

Nicas, M., Estimating exposure intensity in an imperfectly mixed room, *AIHAJ*, 57, 542–550, 1996.

NIH, Guidance on Reporting Adverse Events to Institutional Review Boards for NIH-Supported Multi-center Clinical Trials, website: http://grants.nih.gov/grants/guide/notice-files/not99-107.html, 2003.

Nuclear Regulatory Commission (NRC), Standards of protection against radiation, *Federal Register*, January 1, 2003, 10 CFR Part 20, 321–424, 2003.

Oakridge Associated Universities (ORAU) web site, http://www.orau.org/ptp/collection/dosimeters/ dosimeters.html, July, 2004.

Occupational Safety and Health Administration (OSHA), Permissible Exposure Limits, website: http://www.osha-slc.gov/SLTC/pel/, 2003.

Ott, W., Wallace, L., and Mage, D., Predicting particulate (PM_{10}) personal exposure distributions using a random component superposition statistical model, *J. Air Waste Manage Assoc.*, 50, 1390–1406, 2000.

Parker, R.C., Bull, R.K., Stevens, D.C., and Marshall, M., Studies of aerosol distributions in a small laboratory containing a heated phantom, *Ann. Occup. Hyg.*, 1, 35–44, 1990.

Pellizzari, E.D., Clayton, C.A., Rodes, C.E., Mason, R.E., Piper, L.L., Fort, B., Pfeifer, G., and Lynam, D. Particulate Matter and Manganese Exposures in Indianapolis, Indiana, *JEAEE*, 11: 423–440, 2001.

Rea, A.W., Zufall, M.J., Williams, R.W. and L. Sheldon, The Influence of Human Activity Patterns on Personal PM Exposure: A Comparative Analysis of Filter-Based and Continuous Particle Measurement, *JAWMA*, 51(9), 1271–1279, 2001.

Rock, J.C., Occupational air sampling strategies, in *Air Sampling Instrumentation of Atmospheric Contaminants*, 9th ed., Cohen, B.S. and McCammon, C., Eds., ACGIH, Cincinnati, OH, 2001, chap. 2, pp. 20–50.

Rodes, C.E. and Wiener, R.W., Indoor aerosols and exposure assessment, in *Aerosol Measurement*, Baron, P.A. and Willeke, K., Eds., Wiley Interscience, New York, 2001, p. 860.

Rodes, C.E., Kamens, R.M., and Wiener, R.W., The significance and characteristics of the personal activity cloud on exposure assessment measurements for indoor contaminants, *Indoor Air*, 2, 123–145, 1991.

Rodes, C.E., Kamens, R.M., and Wiener, R.W., Experimental considerations for the study of contaminant dispersion near the body, *AIHAJ*, 56, 535–545, 1995.

Rodes, C.E., Lawless, P.A., Evans, G.F., Sheldon, L.S., Williams, R.W., Vette, A.F., Creason, J.P., and Walsh, D., The relationships between personal PM exposures for elderly populations and indoor and outdoor concentration for three retirement center scenarios, *J. Expos. Anal. Environ. Epidemiol.*, 11, 1–13, 2001.

Rodes, C.E., Lawless, P.A., Thornburg, J.W., and Wallace, L., The characterization of biases in the assessment of personal PM exposures for non-occupational aerosols, *JAWMA*, in review, 2004.

Rodes, C.E., Lawless, P.A., Thornburg, J.W., Williams, R.W., Evans, G.F., Zweidinger, R., Norris, G., and McDow, S., The potential influence of face velocity on the loss of volatile species collected on Teflon filters, *JAWMA*, submitted, September, 2003.

Ruzer, L. (2004), Chapter 14, this volume.

Shapiro, J., *Radiation Protection — A Guide for Scientists and Physicians*, 3rd ed., Harvard University Press, Cambridge, MA, 1990.

Sherwood, R.J., On the interpretation of air sampling for radioactive particles, *Am. Ind. Hyg. Assoc. J.*, 27, 98–109, 1966.

Smith, J. and Bartley, D., Effect of sampler and manikin conductivity on the sampling efficiency of manikin-mounted personal samplers, *Aerosol Sci. Tech.*, 37, 79–81, 2003.

Stevens, D.C., The particle size and mean concentration of radioactive aerosols measured by personal and static air samples, *Ann. Occ. Hyg.*, 12, 33–40, 1969.

Stevens, C., Williams, R., Leovic, K., Chen, F., Vette, A., Seila, R., Amos, E., Rodes, C., and Thornburg, J., Preliminary Exposure Assessment Findings from the Tampa Asthmatic Children's Pilot Study (TACS), abstract prepared for the 2003 National Meeting of the AAAR, Anaheim, CA, May 2003.

Wagner, J. and Leith, D, Field Testing of a Passive Aerosol Sampler. *JAS*, 32: 33–48, 2001.

Wood, J.D. and Birkett, J.L., External airflow effects on personal sampling, *Ann. Occup. Hyg.*, 22, 299–310, 1979.

chapter six

Mechanisms of particle deposition

Kristin K. Isaacs
University of North Carolina at Chapel Hill

Jacky A. Rosati
U.S. EPA, National Risk Management Research Laboratory

Ted B. Martonen
U.S. EPA, National Health and Environmental Effects Laboratory and University of North Carolina at Chapel Hill

Contents

6.1 Introduction 76
6.2 Fundamentals of inhaled aerosols 76
6.2.1 Stokes's law 76
6.2.2 Terminal settling velocity and relaxation time....................77
6.2.3 Aerodynamic diameter 77
6.2.4 Modifications to the aerodynamic equations....................77
6.2.4.1 Correction for nonspherical particles 78
6.2.4.2 Correction for slip 78
6.2.4.3 Hygroscopicity 79
6.3 Domains of particle dynamics....................81
6.3.1 Free-molecule regime....................81
6.3.2 Continuum regime 81
6.3.3 Slip-flow regime 81
6.4 Fluid dynamics in airways....................82
6.4.1 Fundamentals of flow 82
6.4.1.1 Steady vs. unsteady flow 82
6.4.1.2 Laminar vs. turbulent flow....................82
6.4.2 Flow in idealized tubes 83
6.4.3 Flow in curved tubes 85
6.4.4 Flow in bifurcations and branching networks....................85
6.5 Particle motion....................86
6.5.1 Primary deposition mechanisms 86
6.5.1.1 Inertial impaction....................87
6.5.1.1.1 Laminar conditions 88
6.5.1.1.2 Turbulent conditions....................88
6.5.1.2 Sedimentation....................88
6.5.1.2.1 Laminar conditions 88
6.5.1.2.2 Turbulent conditions....................89

1-56670-611-4 / 05 / $0.00+$1.50

6.5.1.3 Diffusion 89
6.5.1.3.1 Laminar conditions 89
6.5.1.3.2 Turbulent conditions 89
6.5.2 Secondary deposition mechanisms 90
6.5.2.1 Interception 90
6.5.2.1.1 Laminar conditions 90
6.5.2.1.2 Turbulent conditions 91
6.5.2.2 Electrostatic charge 91
6.5.2.3 Cloud motion 92
6.6 Conclusions 93
Nomenclature 93
Disclaimer 94
Acknowledgement 94
References 95

6.1 Introduction

Particle deposition is an important topic of concern to diverse areas of aerosol science and technology, ranging from the design and manufacture of equipment for aerosol generation and characterization to the study of human health effects in aerosol therapy and inhalation toxicology. This chapter focuses on the application of aerosol science to particle deposition in the human respiratory system. Particle deposition in human airways is governed by multiple mechanisms. Inertial impaction, sedimentation, and diffusion are often considered the primary mechanisms of deposition, while interception, charging, and cloud motion may be important in some situations. These respective mechanisms are described and formulated in this chapter. The deposition efficiency of each mechanism is dependent upon interactions among aerosol characteristics, ventilatory parameters, and respiratory system morphologies.

In this chapter, factors governing both the motion and deposition of particles in the respiratory system are discussed. Then, the kinematic behavior of particles immersed in a fluid medium is reviewed. The motion of particles and fluids within tubular structures (e.g., airways) is next considered; finally, specific mechanisms of particle deposition in branching networks are discussed.

6.2 Fundamentals of inhaled aerosols

An aerosol is a suspension of particulate matter in a gaseous carrying medium. Several textbooks that address the general field of aerosol science and technology have been written. We refer the interested reader to the works of Fuchs,[1] Mercer[2], Reist,[3] Hinds,[4] and Friedlander.[5]

Particle size, density, and shape are important factors in the prediction of particle kinetics and aerosol deposition in human airways. In this section, particle characteristics, their effects on Stokes's law and terminal settling velocity, and their role in particle deposition are discussed.

6.2.1 Stokes's law

The interaction between an aerosol particle and the carrying gas is quantified by Stokes's law. Stokes's law provides a basis for the study of aerosol particle motion. The drag force on a particle moving through a fluid may be expressed as

$$F_D = 3\pi\mu d_p V_p \tag{6.1}$$

It is important to recognize that the application of Equation (6.1) requires the following assumptions:

1. the particle is a rigid sphere;
2. the carrying gas is incompressible;
3. the inertia of the particle is negligible compared to drag force;
4. there are no hydrodynamic interactions or boundary effects affecting the particle;
5. particle motion is constant;
6. the fluid velocity at the particle surface is zero.

In practice, these conditions must be checked for validity.

6.2.2 Terminal settling velocity and relaxation time

An aerosol particle will reach its terminal settling velocity when the drag force, F_D, on it is equal and opposite to the force of gravity, F_g:

$$F_D = F_g \tag{6.2}$$

By recognizing that $F_g = mg$ and substituting in Equation (6.1), this relationship may also be expressed as

$$3\pi\mu d_p V_p = mg \tag{6.3}$$

The equation for terminal settling velocity, V_{TS}, is derived by expressing the particle mass in terms of ρ_p and solving for V_p:

$$V_{TS} = \frac{\rho_p d_p^2 g}{18\mu} \tag{6.4}$$

The time it takes for a particle starting at rest to reach 63% of its terminal settling velocity is the relaxtion time, τ, expressed as

$$\tau = \frac{d_p^2 \rho_p}{18\mu} = \frac{d_{ae}^2 \rho_o}{18\mu} \tag{6.5}$$

Note that τ is dependent upon both particle properties (e.g., particle size and density) and fluid viscosity. When we substitute Equation (6.5) into Equation (6.4), we obtain V_{TS} expressed in terms of τ:

$$V_{TS} = \tau g \tag{6.6}$$

6.2.3 Aerodynamic diameter

The aerodynamic diameter of a particle (d_{ae}) is an important parameter used to relate particles of differing shapes and densities. The d_{ae} of a particle is defined as the diameter of a unit density sphere that has the same V_{TS} as the particle in question. The d_{ae} is commonly used to characterize the kinetic behavior of larger (>1 μm) aerosol particles, and can be calculated as

$$d_{ae} = d_g \sqrt{\rho_p} \tag{6.7}$$

where d_g is the geometric diameter of the particle.

6.2.4 Modifications to the aerodynamic equations

Modifications may be made to adapt Stokes's law for use in nonidealized situations. For example, Equation (6.1) assumes that a particle is spherical in shape, an assumption that may not always be appropriate (e.g., fibers). Without appropriate correction, calculations of aerodynamic properties and deposition probabilities for inhaled particles may be inaccurate.

6.2.4.1 *Correction for nonspherical particles*

When particles are nonspherical, a dynamic shape correction factor, χ, may be applied to Stokes's law. This quantity is defined as the ratio of the drag force of the nonspherical particle to the drag force of a sphere having the same volume and velocity. Stokes's law, corrected for shape, may then be expressed as

$$F_D = 3\pi\mu d_p V\chi \tag{6.8}$$

The correction factor χ can also be used to modify d_{ae} for shape irregularity:[6]

$$d_{ae} = d_g\sqrt{\frac{\rho_p}{\chi}} \tag{6.9}$$

The terminal settling velocity, corrected for shape, is given by

$$V_{TS} = \frac{\rho_p d_p^2 g}{18\mu\chi} \tag{6.10}$$

Moss,[7] Stober,[8,9] and Leith[10] have investigated the effects of shape on the aerodynamic properties of particles. Table 6.1 provides dynamic shape factors for common particle shapes and types.

6.2.4.2 *Correction for slip*

Stokes's law assumes that the fluid velocity at the particle surface is zero. This no-slip assumption is not valid for small particles; thus, the Cunningham slip correction factor, C_c, should be applied when particles smaller than 10 μm are being considered. C_c may be expressed as

$$C_c = 1 + Kn\left[A_1 + A_2 \exp - \left(\frac{A_3}{Kn}\right)\right] \tag{6.11}$$

where Kn is the dimensionless Knudsen number (defined below), and A_1, A_2, and A_3 are constants derived from experimental measurements. While the values of these constants vary in the literature, [1,12–14] the most commonly accepted values, as determined by Davies,[12] are A_1=1.257, A_2=0.4, and A_3=1.1.

Table 6.1 Dynamic Shape Factors for Common Particle Shapes and Types

Type of Particle	Dynamic Shape Factor, c	Reference
Particles of Regular Shape		
Sphere	1.00	11
Cube	1.08	11
Cylinder (vertical axis)	1.01	11
Cylinder (horizontal axis)	1.14	11
Cube octahedron	1.03	2
Octahedron	1.06	2
Tetrahedron	1.17	2
Other Particles		
Bituminous coal	1.05 – 1.11	11
Sand	1.57	11
Talc	1.88 – 2.04	11
Quartz (0.65–1.85 μm)	1.84	2
Quartz (>4 μm)	1.23	2

Kn is the ratio of the mean free path of the gas molecules (λ) to the particle diameter. It may be expressed as

$$Kn = \frac{2\lambda}{d_p} \tag{6.12}$$

The mean free path may be calculated using

$$\lambda = \frac{1}{\sqrt{2} n \pi d_m^2} \tag{6.13}$$

If Kn is very small ($Kn \ll 1$), a particle will decelerate due to gas molecule collisions upon the particle's surface. Conversely, if Kn is large ($Kn \gg 1$), a particle will not be affected by gas molecules.

When the Cunningham slip correction factor is included, the Stokes's drag force on a particle becomes

$$F_D = \frac{3\pi\mu V d_p \chi}{C_c} \tag{6.14}$$

The terminal settling velocity, corrected for shape and slip, is given by

$$V_{TS} = \frac{\rho_p d_p^2 g C_c}{18\mu\chi} \tag{6.15}$$

Table 6.2 demonstrates the effect of slip correction on V_{TS} for particles of various sizes. Note that there are significant differences between the corrected and uncorrected V_{TS} values for submicron particles.

6.2.4.3 *Hygroscopicity*

In human airways, particle deposition may depend on the physicochemical (hygroscopic) properties of inhaled substances. Hygroscopicity may be defined as a particle's propensity to absorb water from a warm, humid environment, thereby changing its diameter and density. The degree and rate of hygroscopic growth are influenced by many factors, including chemical composition, temperature, relative humidity, initial particle size, and duration of exposure.[16]

As the relative humidity within the lungs may approach saturation (99.5%),[17] hygroscopic effects can substantially affect inhaled aerosols. As particle size and density are important factors in determining deposition, hygroscopic growth is a critical concern in the risk assessment of inhaled aerosols (inhalation toxicology) and the development of inhaled

Table 6.2 Terminal Settling Velocity Values Calculated using the Cunningham Slip Correction Factor [15]

Particle Diameter (μm)	C_c	V_{TS}[a] (without C_c) (cm s^{-1})	V_{TS}[a] (with C_c) (cm s^{-1})
0.001	228.234	3.02E−09	6.61E−07
0.01	23.3443	3.02E−07	6.77E−06
0.1	2.97393	3.02E−05	8.71E−05
0.5	1.34743	7.54E−04	1.005E−03
1	1.17273	3.02E−03	3.516E−03
3	1.05757	2.71E−02	2.864E−02
5	1.03454	7.54E−02	7.790E−02
10	1.01727	3.02E−01	3.066E−01

[a] Values given are calculated for a unit density sphere of a given size at 20°C and 101 kPa using Davies's constants.[12]

pharmaceutics (aerosol therapy). Hygroscopic growth has been shown to occur in many aerosols, including a number of environmental pollutants and pharmacological agents.[18,19]

Theoretical models are useful in characterizing properties (i.e., growth rate, equilibrium size, aerodynamic diameter) of hygroscopic particles under different *in vivo* environmental conditions. Various theoretical models of particle hygroscopicity have been developed for cigarette smoke,[21] aqueous droplets,[22] and soluble particles.[23]

Martonen[20] derived equations for the density and aerodynamic diameter of a particle in different lung airway generations, provided the particle growth rate, r_g, is known. The density ρ_i of a particle entering airway generation i can be calculated as

$$\rho_i = \left[\frac{d_0}{d_i}\right]^3 [\rho_0 - \rho_{H_2O}] + \rho_{H_2O} \tag{6.16}$$

where ρ_0 is the initial particle density, d_0 is the initial particle diameter, and d_i is the particle diameter at the entrance to airway generation i, given by

$$d_i = d_0 + r_g\left(\sum_{j=0}^{j=i-1} \frac{L_j}{V_j}\right) \tag{6.17}$$

where L_j is the length of airway j and V_j is the mean velocity of the particle in airway j. The aerodynamic diameter of the particle in airway i is then given by

$$d_{ae,i} = \sqrt{\rho_i C_c}\, d_i \tag{6.18}$$

A particle's size and density may change while traveling through the respiratory system (Figure 6.1). It may not be prudent, therefore, to expect the deposition pattern of inhaled aerosols to be determined by the preinspired sizes and densities of its constituent particles. The dynamic process of hygroscopic growth must be accounted for during a breathing cycle. It is also interesting to conjecture that evaporation may occur when aerosols from the warm, moist, deep lung are cooled during expiration.

Both experimental and computational studies have been performed to determine deposition patterns for inhaled hygroscopic particles in human airway networks. Experimental studies have been performed for NaCl[24,25] and cigarette smoke.[26] Computational models of the deposition of hygroscopic particles have been developed for

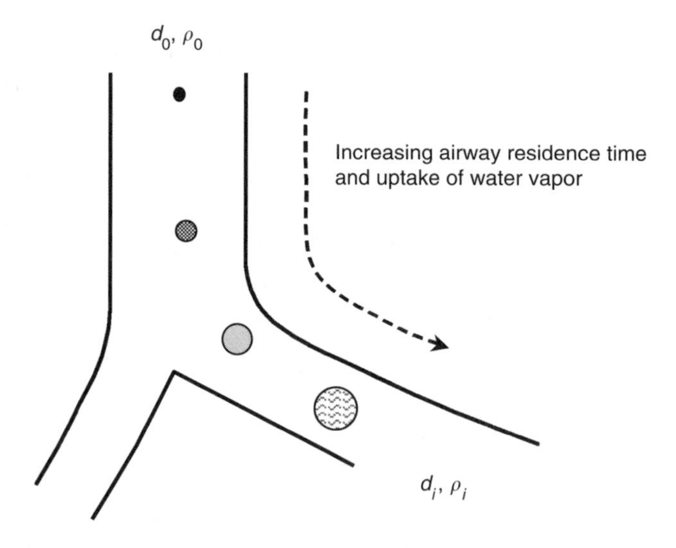

Figure 6.1 Variation in particle properties due to hygroscopic growth in a lung airway (not to scale).

sulfate aerosols,[27,28] cigarette smoke,[29,30] saline,[31–34] phosphoric acid aerosols,[35] atmospheric salts,[36] and several inhaled pharmaceutics.[37]

6.3 *Domains of particle dynamics*

The theory employed to quantify particle dynamics in a transporting medium will depend on the properties of both the particles and gases involved. Particle dynamics can be classified by size into three regimes of behavior. In the free-molecule regime, the behavior of very small particles is influenced by the motion of individual gas molecules. In the continuum regime, the surrounding gas acts on large particles as a continuum, or viscous fluid. In the slip-flow regime, a correction for fluid slip at the surface of the particle is employed. A fourth (transition) regime, situated between the slip-flow and free-molecule regimes, is sometimes considered,[38] although the behavior of particles within this regime is not independently defined. In some texts, the entire region between the free-molecule and continuum regimes is called the transition regime.[4]

The regimes are defined by *Kn* values, or (at constant temperature and pressure) by particle sizes. However, the actual values of the threshold *Kn* or d_p parameters for each regime vary in the literature. The regimes, their corresponding particle characteristics, and their associated dynamic theories are discussed below.

6.3.1 *Free-molecule regime*

The free-molecule regime has been defined to include particles having $Kn>10$,[38] or $Kn>20$.[6] In air at STP, these values correspond to particle diameters of 0.01 and 0.02 μm, respectively.

In the free-molecular regime, the motion of particles can be influenced by interactions with individual gas molecules. Particles and gas molecules in Brownian motion collide randomly, and after many collisions, the force exerted by the gas molecules will affect the direction of motion of the much more massive particles. In this case, particle motion must be quantified using gas dynamic theory.

Particle diffusion becomes a very important consideration in the study of particle dynamics in the free-molecule regime. In this regime, the diffusional velocity of the particle due to random Brownian motion is several orders of magnitude greater than the terminal settling velocity,[38] and thus, in this regime, the effects of gravitational forces may be neglected.

6.3.2 *Continuum regime*

In the continuum regime, the gas in which the particles are suspended can be assumed to act as a continuous, viscous, fluid. The continuum regime may also be called the Stokes's regime.[38] This regime is alternately defined as $Kn<0.1$,[38] $Kn<0.4$,[6] and $d_p<\lambda$ ($Kn<2$).[4] A value of $Kn<0.1$ corresponds to particle diameters of greater than 1.3 μm in air at standard temperatuer and pressure (STP).

In the continuum regime, particle motion within the gas is governed by the momentum (Navier–Stokes) equations for a viscous fluid, and the particle drag force and terminal settling velocity can be calculated from Stokes's law. In this region, no-slip conditions are assumed to exist at the surface of the particle.

6.3.3 *Slip-flow regime*

The slip-flow regime falls between the free-molecule and continuum regimes and has been variously defined as including particles having a value of *Kn* of 0.1–0.3 (with a transitional regime defined as having *Kn* of 0.3 – 10),[38] or 0.4 – 20.[4]

In the slip-flow regime, the assumption of no-slip conditions at the surface of the particle is not applicable. In this case, there exists a quantifiable velocity of the gas relative to the particle at its surface. In the slip-flow regime, the drag force exerted on a particle is overestimated (and the terminal settling velocity is underestimated) by Stokes's law. Therefore, within this regime Stokes's law must be corrected by the Cunningham slip correction factor (Equation 6.5).

6.4 Fluid dynamics in airways

Both motion and deposition of aerosols in the respiratory system are dependent upon airflow conditions within the airways. Particles will be affected by the nature of the velocity and pressure fields in which they are carried. Particles are entrained in the airflow, and are transported with both the bulk convective motion of the flow and any other secondary flow patterns initiated by airway geometries. Therefore, the fluid dynamic conditions (e.g., laminar vs. turbulent motion) of air in the bronchial tree are important considerations in both inhalation toxicology and aerosol therapy.

The dynamic behavior of air in the respiratory passages is governed by both morphological and respiratory parameters. Morphological considerations include airway dimensions (airway diameters and lengths), bifurcation angles, spatial arrangement of the branching network, and airway surface characteristics. Respiratory parameters describe the mechanics of ventilation, and include respiratory rates and tidal volumes. In this section, different lung airway morphologies and their corresponding airflow characteristics will be discussed. Both idealized and anatomically realistic morphologies will be introduced. A more advanced discussion of fluid flow modeling and its incorporation into aerosol deposition modeling will be provided in Chapter 8.

6.4.1 Fundamentals of flow

A discussion of the dynamics of airflow in the respiratory airways requires the introduction of some basic flow considerations.

6.4.1.1 Steady vs. unsteady flow

If the fluid properties (e.g., density, velocity) at each point in a prescribed flow field are time-invariant, then the flow is considered steady. In constrast, in unsteady flow the fluid properties at each point in the flow field vary with time. In some circumstances, when the time-dependent changes in fluid properties are very small, the flow may be assumed to be quasisteady. During normal respiration, the flow in the lung airways can be considered as quasisteady. However, under high-frequency breathing or forced expiration, flow in the airways is unsteady.[39] It is important for investigators to recognize the potential significance of transient, time-dependent, phenomena when analyzing airflow patterns in their particular studies.

6.4.1.2 Laminar vs. turbulent flow

Flow in airways may be either laminar or turbulent. Simply stated, in laminar flow, fluid motion occurs in smooth layers (laminae), and there is no fluid mixing between adjacent layers. In the turbulent regime, the flow structure is random and characterized by the chaotic motion of fluid particles within the larger structure of the mean fluid flow. In this case, mixing of fluid between layers may be initiated locally and propagated to downstream regions.

Flows in an idealized cylindrical passage can be characterized as either laminar or turbulent by calculating the value of dimensionless Reynolds number:

$$Re = \frac{\rho_a d_a V_a}{\mu} \tag{6.19}$$

Table 6.3 Reynolds Numbers by Airway Generation (Calculated at 1 atm and 37°C)

Airway Generation	Airway Cross-Section[42] (cm^2)	Airway Diameter[42] (cm)	Velocity[a] ($cm\ s^{-1}$)	Velocity[b] ($cm\ s^{-1}$)	Re[a]	Re[b]
0	2.54	1.800	98.43	393.70	1077.42	4309.68
1	2.33	1.220	107.30	429.18	796.07	3184.27
2	2.13	0.830	117.37	469.48	592.44	2369.76
3	2.00	0.560	125.00	500.00	425.70	1702.80
4	2.48	0.450	100.81	403.23	275.87	1103.49
5	3.11	0.350	80.39	321.54	171.10	684.41
6	3.96	0.280	63.13	252.53	107.50	430.00
7	5.10	0.230	49.02	196.08	68.57	274.26
8	6.95	0.186	35.97	143.88	40.69	162.76
9	9.56	0.154	26.15	104.60	24.49	97.97
10	13.40	0.130	18.66	74.63	14.75	58.99
11	19.60	0.109	12.76	51.02	8.46	33.82
12	28.80	0.095	8.68	34.72	5.02	20.06
13	44.50	0.082	5.62	22.47	2.80	11.21
14	69.40	0.074	3.60	14.41	1.62	6.49
15	117.00	0.050	2.14	8.55	0.65	2.60
16	225.00	0.049	1.11	4.44	0.33	1.32
17	300.00	0.040	0.83	3.33	0.20	0.81
18	543.00	0.038	0.46	1.84	0.11	0.43
19	978.00	0.036	0.26	1.02	0.06	0.22
20	1743.00	0.034	0.14	0.57	0.03	0.12
21	2733.00	0.031	0.09	0.37	0.02	0.07
22	5070.00	0.029	0.05	0.20	0.01	0.04
23	7530.00	0.025	0.03	0.13	0.01	0.02

[a] $15\ l\ min^{-1}$.
[b] $60\ l\ min^{-1}$.

In general, the flow in idealized tubes will be laminar for $Re<2300$.[40] Idealized tubes (i.e., those that are straight and have smooth, rigid walls, circular cross-sections, and constant diameters) are rarely observed *in vivo*.

Table 6.3 provides estimated Reynolds numbers in 24 different airway generations for two different respiratory rates. We note that the *Re* values predict laminar flow in every airway at a ventilation rate of $15\ l/min^{-1}$, but predict turbulent flow in the larger airways at a rate of $60\ l/min^{-1}$. However, it has been determined that due to disturbances in the inflow of air to the lungs caused by the larynx and the cartilaginous rings, turbulent flow is usually observed in the large airways,[39] and thus care must be exercised when attempting to relate *Re* values with *in vivo* flow conditions.[41]

6.4.2 Flow in idealized tubes

A straightforward way of simulating airflow within respiratory passageways is to consider airways as idealized tubes, as defined above. Flow in such an airway under both laminar and turbulent conditions is shown in Figure 6.2. As flow enters, the velocity of the fluid V_0 is uniform within the cross-section of the tube. This type of velocity pattern is called plug flow. As the flow proceeds, the velocity is retarded by the shear force the boundary surface of the tube imparts on the airflow. The no-slip condition exists at the tube surface, so the velocity of flow at the wall is zero. The axial distribution of flow velocity is called the velocity profile. The region of the flow where $V<V_0$ is called the boundary

layer. As the fluid moves along the tube, pressure and shear forces within the fluid will equilibrate, and a fully developed velocity profile will be attained. The distance from the entrance of the tube to the point where the laminar velocity profile is fully developed is called the entrance length, $L_{e,l}$. Under laminar conditions, the final velocity profile is parabolic, with the velocity increasing from zero at the tube walls to the free stream velocity V_0 at the center of the tube. The actual shape of the final parabolic profile is determined by both the absolute fluid viscosity and the pressure gradient driving the flow. In turbulent flow, the boundary layer grows more rapidly, the entrance length ($L_{e,t}$) is shorter, and the fully developed velocity profile is flatter.[40] In turbulent flow, a laminar sublayer will exist near the wall, and a transitional layer will lie between this laminar layer and the entirely turbulent flow that exists in the tube's center. These important details are shown in the detailed view of Figure 6.2.

Since inhaled particles are transported via airstreams, a knowledge of both flow development characteristics and velocity profiles is fundamental to predicting particle deposition

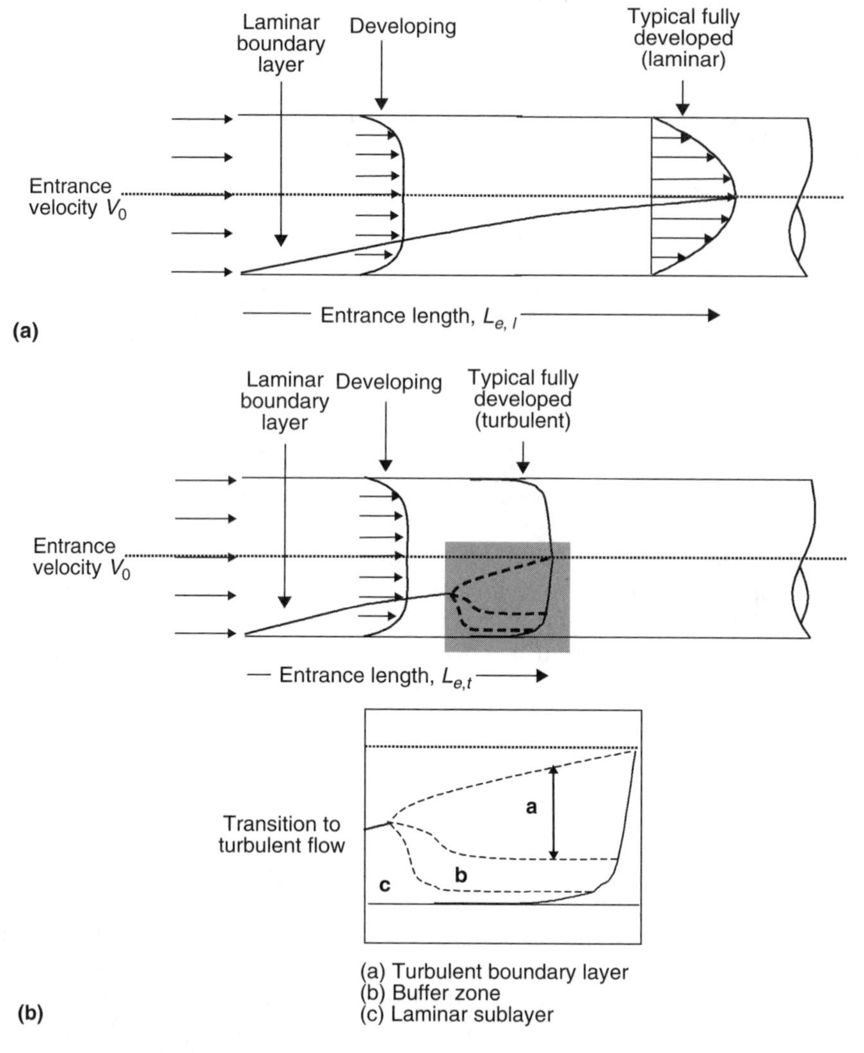

Figure 6.2 Developing flows and velocity profiles in an idealized airway for laminar and turbulent flow (not to scale).

patterns. However, the airflow patterns in human lungs are much more complex than the idealized situation described above. The human bronchial tree is a branching network of bifurcating, irregular tubes. This network also contains morphological surface features that affect flow patterns. A vast body of work has been generated in an attempt to characterize airflow in anatomically realistic human lung models. In this chapter, the discussion will be limited to a short overview of flow in curved tubes, bifurcations, and branching networks.

6.4.3 *Flow in curved tubes*

As a laminar, steady flow moves through a curved tube, secondary velocity patterns develop in the transverse plane of the tube.[39] These secondary patterns (Figure 6.3) are a result of the fast-moving fluid at the center of the tube being pushed toward the outside of the bend by inertia. Any distortion of the velocity profile can be predicted by the Dean number:

$$De = Re\left(\frac{d}{2r}\right)^{1/2} \tag{6.20}$$

where R is the radius of curvature of the bend and d is the tube diameter. The velocity profile will be affected for $De>25$, and large distortions will occur for $De>300$.[43] In curved tubes, velocity profiles are no longer radially symmetric. Instead, in laminar flow the profile becomes M-shaped in the transverse plane and skewed toward the outside wall in the plane of the bend.[44]

The flow in curved tubes has been studied experimentally,[45] theoretically,[43,46] and numerically.[47–49] Experimental[50, 51]and computational[52,53] studies of flow in curved airway configurations have also been performed.

6.4.4 *Flow in bifurcations and branching networks*

Because the lungs are a system of branching airways, much work has been done to characterize the airflow patterns within both single bifurcations and branching networks.[50, 51,54–57] Different types of bifurcations have been considered, incorporating two-dimensional, three-dimensional, symmetric, and asymmetric geometries. In general, experimental studies have indicated that a parabolic velocity profile will be skewed as it passes through a bifurcation, in a manner similar as in a curved tube.[58] Figure 6.4 provides a qualitative example of these skewed velocity profiles.

Flows in physical models of more complex branching networks (e.g., branching models derived from casts of human or animal lungs) have been studied experimentally.[50,51,56] However, the characterization of features such as secondary currents, particularly in downstream daughter branches, is difficult in such studies. Recently, computational fluid

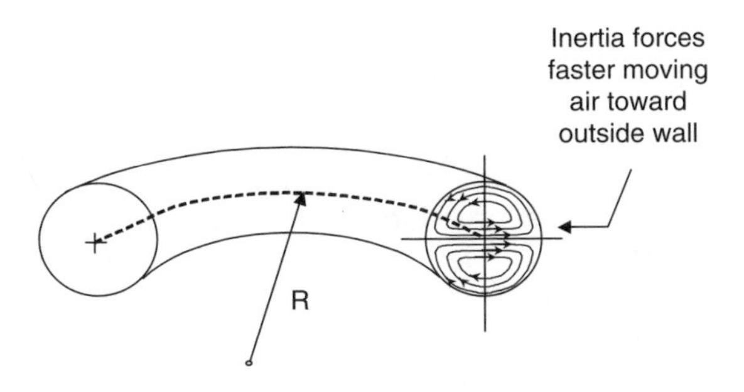

Figure 6.3 Secondary motion in a curved tube with radius of curvature *R*.

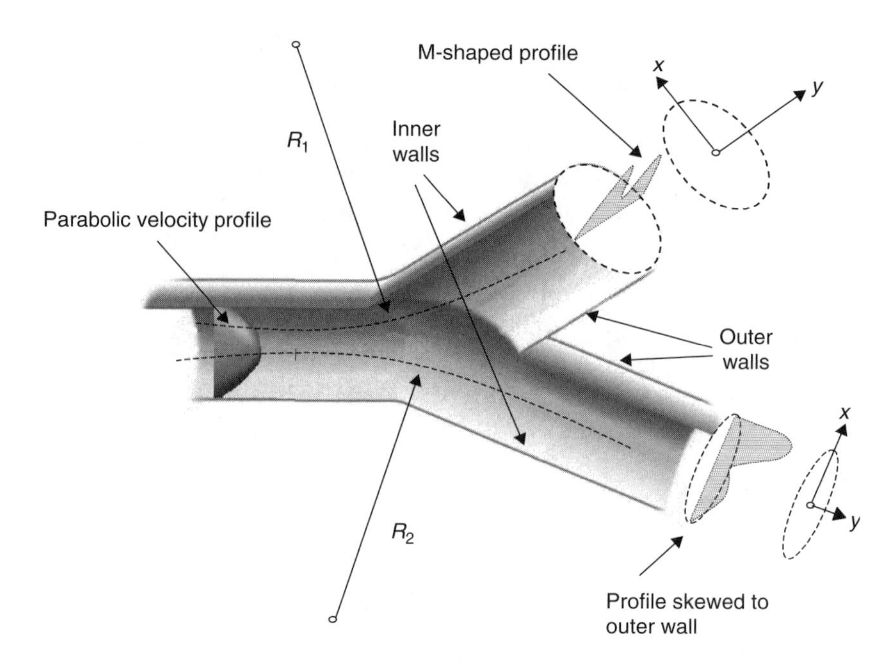

Figure 6.4 Qualitative description of velocity profiles in a bifurcation. Flow passing through a bifurcation behaves in a manner similar to flow passing through a curved tube. The radii of curvature R_1 and R_2 for the two daughter branches are shown. In the x plane (the plane of the bifurcation), the velocity profile becomes skewed toward the outer wall. In the y plane (which is perpendicular to the plane of the bifurcation), an M-shaped profile is observed.

dynamics (CFD) has been used to model airflow characteristics in a variety of branching models.[58–63] Such models, and their use in predicting particle deposition, will be reviewed in detail in Chapter 8.

6.5 Particle motion

The quantitative assessment of factors affecting inhaled particles was pioneered by Findeisen.[64] The subsequent work of Beeckmans[65] and Landahl[66] refined the analysis of particle deposition. The importance of these early efforts should be recognized.

Particle deposition within airways is governed by three primary mechanisms (inertial impaction, sedimentation, and diffusion), and several secondary mechanisms (interception, electric charge, and cloud motion). The respective deposition efficiencies of these mechanisms are dependent upon interactions among aerosol characteristics, ventilatory parameters, and lung morphologies.

6.5.1 Primary deposition mechanisms

Particle deposition efficiencies for inertial impaction, sedimentation, and diffusion are dependent upon fluid dynamics, airway geometries, and particle characteristics. It is therefore necessary to formulate expressions for deposition efficiencies that specifically consider these factors. Deposition efficiency is defined as the ratio of the number of particles deposited within a given respiratory system region or airway to the total number of entering (or inhaled) particles.

Many individual equations have been developed to determine particle deposition efficiencies within different regions of the respiratory system. The merits of various formulations have been reviewed by Martonen et al.[41] The equations most commonly used include those formulated by Martonen,[20,67] Beeckmans,[65] Landahl,[66] Landahl and Herrmann,[68,69] and Ingham.[70–72] The respective equations offered by the authors were derived using differing assumptions (e.g., turbulent vs. laminar conditions) and velocity fields (e.g uniform vs. parabolic velocity profiles). Unfortunately, other investigators have selected equations for use from the aforementioned authors without recognizing their incompatibilities. For instance, authors have used a sedimentation equation for a uniform velocity field while simultaneously using an impaction equation for a parabolic velocity field. This has led to confusion in the literature. To address this problem, Martonen[20,67] presented a consistent, compatible set of formulae, derived from well-defined conditions applicable to airways. These expressions are discussed below.

6.5.1.1 Inertial impaction

Inertial impaction occurs when particles have sufficient momentum to deviate from fluid streamlines and strike boundary airway surfaces (Figure 6.5a). Because momentum is the product of mass and velocity (mV), inertial impaction is an important deposition mechanism for large particles, usually greater than 1 μm in size. Inertial impaction is increasingly effective

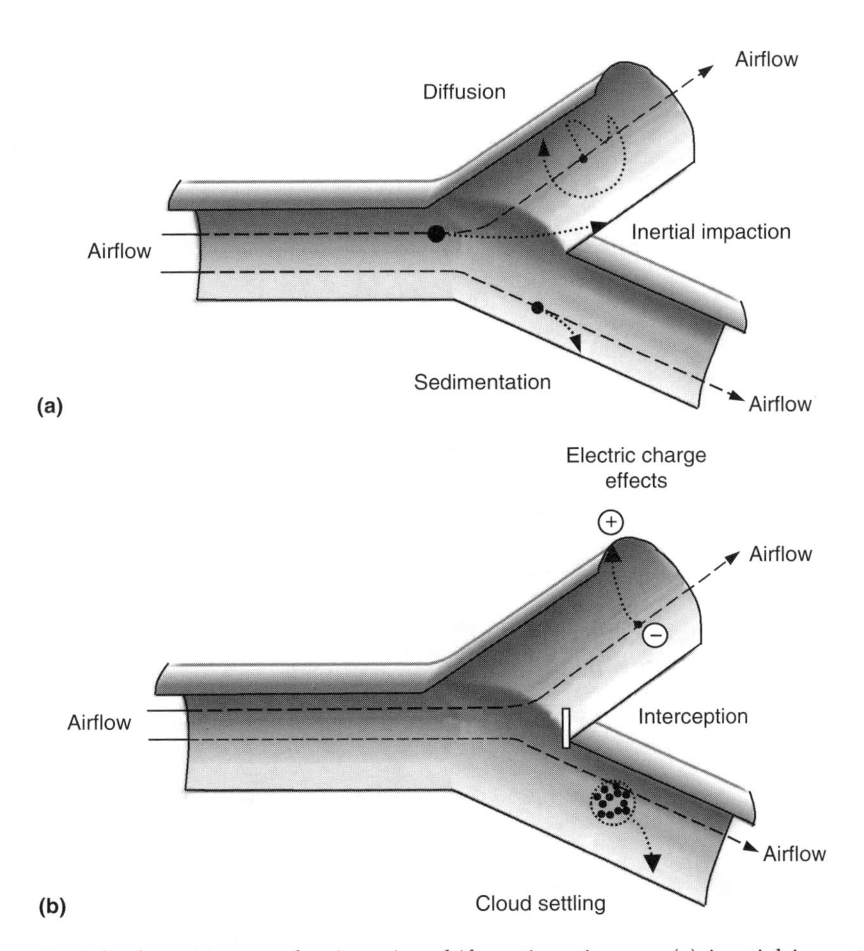

Figure 6.5 Particle deposition mechanisms in a bifurcating airway : (**a**) inertial impaction, sedimentation, and diffusion and (**b**) interception, electrostatic charge, and cloud motion. (Adapted from Martonen, T.B., Musante, C.J., Segal, R.A., Schroeter, J.D., Hwang, D., Dolovich, M.A., Burton, R., Spencer, R.M., and Fleming, J.S., *Respir. Care*, 45, 712, 2000. With permission.)

at higher velocities; thus, it occurs primarily in the upper airways of the tracheobronchial tree.[41] Inertial impaction as a mechanism of particle deposition has been widely studied.[73–77]

6.5.1.1.1 Laminar conditions. Martonen[20,67] expressed particle deposition efficiency from inertial impaction under laminar plug flow conditions as

$$P(I)=\frac{2}{\pi}\left[\beta(1-\beta^2)^{1/2}+\sin^{-1}(\beta)\right] \tag{6.21}$$

where

$$\beta=\frac{\tau V_a \theta_b}{d_a} \tag{6.22}$$

(Note that deposition efficiency is dependent upon τ, defined in Section 6.2.2.) These equations can be used to simulate particle deposition in the peripheral airways of the tracheobronchial tree. An equation applicable to fully developed (parabolic) velocity profile, appropriate for more distal airways, has been presented by Landahl.[66] However, airstream velocities are typically quite small in such areas, so the impaction deposition efficiency *per se* is negligible.

6.5.1.1.2 Turbulent conditions. Deposition efficiency for inertial impaction for turbulent flow conditions is[20,67]

$$P(I)=1-\exp\left[\frac{-4\tau V_a \theta_b}{\pi d}\right] \tag{6.23}$$

Turbulent flows, such as those requiring the use of Equation (6.23), may occur in the upper tracheobronchial airways of the human lung.

6.5.1.2 Sedimentation

Reduced flow velocities lead to increased particle residence times in airways. When this occurs (e.g., when the peripheral airways of the tracheobronchial tree or alveolar region are being considered), particles may be deposited by gravitational forces (Figure 6.5a). This is called sedimentation, and while particles as small as 0.1 μm may be affected, this mechanism increases in importance with increasing particle size.[78,79] Particle deposition from sedimentation, like inertial impaction, is directly dependent upon τ, the relaxation time of the aerosol particle. Sedimentation of particles has been studied by Yu and Thiagarajan[80] and Heyder and Gebhart.[81]

6.5.1.2.1 Laminar conditions. Particle deposition efficiency from sedimentation, *P(S)*, under laminar plug flow conditions can be expressed by[20,67]

$$P(S)=\frac{2}{\pi}\left[\beta(1-\beta)^{2(1/2)}+\sin^{-1}(\beta)\right] \tag{6.24}$$

The orientation of airways with respect to gravity must be considered; therefore, for upward flow

$$\beta=\frac{L_a V_{TS}\cos(\theta_b)}{d_a[V_a+V_{TS}\sin(\theta_b)]} \tag{6.25}$$

and for downward flow

$$\beta=\frac{L_a V_{TS}\cos(\theta_b)}{d_a[V_a-V_{TS}\sin(\theta_b)]} \tag{6.26}$$

These equations are suitable for calculating deposition in the peripheral airways of the tracheobronchial network. For downstream regions where flow is fully developed, and parabolic velocity profiles exist, the formulation of Beeckmans[65] may be employed.

6.5.1.2.2 Turbulent conditions Under turbulent flow conditions, deposition efficiency from sedimentation can be expressed as[20,65,67]

$$P(S)=1-\exp\left[\frac{-4\beta}{\pi}\right] \tag{6.27}$$

where β is given by either Equation (6.25) or Equation (6.26).

6.5.1.3 Diffusion

Deposition via diffusion (Figure 6.5a) occurs when particles exhibiting random Brownian motion collide with the airway surface. While deposition of particles between 0.1 and 1 μm in size occurs by both diffusion and sedimentation,[78,79] diffusion is the primary deposition mechanism for particles less than 0.1 μm in size,[82] since diffusion is governed by geometric diameter rather than aerodynamic diameter. Many studies of varying emphases have been performed to quantify the diffusion of aerosol particles.[70–72, 83–89]

6.5.1.3.1 Laminar conditions Under laminar plug flow conditions, deposition by diffusion may be described by[67]

$$P(D)=4\sqrt{\frac{K}{\pi}}-K \tag{6.28}$$

where

$$K=\frac{DL_a}{V_a d_a^2} \tag{6.29}$$

and the diffusion coefficient, D is given by

$$D=\frac{C_c kT}{3\pi\mu d_g} \tag{6.30}$$

Under parabolic laminar flow conditions, deposition by diffusion, $P(D)$, can be calculated from the expression derived by Ingham:[70]

$$P(D)=1-0.819e^{-3.66K}+0.0976e^{-22.3K}+0.0325e^{-57K}+0.0509e^{-49.9K^{2/3}} \tag{6.31}$$

The formulations given above have two major limitations: they are only valid for smooth-walled tubes and they ignore entrance affects. Regarding application to airways, the limitations may be quite serious because airways have natural surface features (e.g., cartilaginous rings). In addition, due to the serially branching quality of the respiratory network, velocity profiles may develop after each airway bifurcation. To address these issues, Martonen et al.[83,84,87] analyzed entrance effects in smooth-walled and rough-walled tubes. They determined that the deposition of ultrafine (~0.01 μm) particles would be underestimated by ~35% if airway surface structures were ignored.

6.5.1.3.2 Turbulent conditions
Particle deposition by diffusion under turbulent conditions can be calculated as[20,67]

$$P(D)=1-\exp\left[-\frac{8L_a v^{1/4}D^{3/4}Re^{7/8}}{57V_a d_a^2}\right] \tag{6.32}$$

which in air at STP reduces to

$$P(D)=1-\exp\left[-\frac{0.088L_a D^{3/4}Re^{7/8}}{V_a d_a^2}\right] \tag{6.33}$$

6.5.2 Secondary deposition mechanisms

6.5.2.1 Interception

Deposition by interception occurs when a particle contacts an airway surface while passing it at a distance less than or equal to its radius, without deviating from the flow streamline (Figure 6.5b). Interception is an important deposition mechanism for fibers, because as length increases, the likelihood of that fiber touching an airway surface increases.[90,91] Fiber orientation also plays a critical role in deposition by interception. While fibers oriented with the flow streamlines deposit similarly to particles of their equivalent diameter, fibers oriented away from the streamline, or fibers that are "tumbling," have an increased likelihood of deposition by interception.[91,92] Interception may also be important for nonfibrous particles having low density and large geometric diameters, because the larger geometric diameter increases the likelihood of a particle coming into contact with an airway surface.[2] It has been suggested that such "porous" particles may have important applications in aerosol therapy.[92,93] However, recent findings have indicated that certain issues remain unresolved concerning the use of porous particles as vehicles for drug delivery.[94]

Several studies have experimentally characterized fiber movement and deposition in human lung casts or bifurcated tubes.[95,96] Other studies have modeled fiber movement and deposition in the human lung,[91,97,98] while still others have modeled interception of both spherical particles and fibers.[99] The motion of particles, including fibers, in airway bifurcations has been systematically studied by Balásházy et al.[100–102] Harris and Fraser[97] generated simplified equations to model the deposition of fibers by interception under laminar and turbulent conditions. They determined that deposition by interception increases as fiber length increases and airway diameter decreases. Because of the potential importance of fibers in inhalation toxicology and aerosol therapy, their expressions for interception deposition efficiencies are presented below.

6.5.2.1.1 Laminar conditions Harris and Fraser[97] formulated expressions for the interception of fibers in laminar parabolic flows. For fibers that are tumbling in the flow, the deposition efficiency may be given by

$$P(Int)=\frac{f_t}{\pi r_a^4}\left[r_a^2\theta_s(L_f^2+\bar{L}_f^2)-\frac{\theta_s}{8}(L_f^4-\bar{L}_f^4)-\bar{L}_f^2(\tan\theta_s)\left(r_a^2+\frac{\bar{L}_f^2}{8}\right)+\frac{\bar{L}_f^4}{24}\tan^3\theta_s\right] \tag{6.34}$$

where the fraction of fibers tumbling is given by

$$f_i=\left(\frac{2}{\pi}\right)\tan^{-1}\left(\frac{\tan\varphi}{\beta_e}\right) \tag{6.35}$$

where φ has a value of 80°, and

$$\theta_s=\frac{\pi}{2}-\sin^{-1}\left(\frac{\bar{L}_f}{L_f}\right) \tag{6.36}$$

$$\bar{L}_f=L_f\cos\left[\tan^{-1}\left(0.34\beta_e+\frac{0.68L_a}{r_a}\right)\right] \tag{6.37}$$

and[91]

$$\beta_e=1.07\left(\frac{L_f}{d_f}\right)^{0.087} \tag{6.38}$$

Harris and Fraser[97] also present an equation for the interception of fibers that are oriented with the laminar flow streamlines.

6.5.2.1.2 Turbulent conditions Under turbulent flow conditions, the interception of fibers in the lung may be expressed as [97]

$$P(Int)=\frac{2}{\pi r_a^2}\left(\frac{1}{4}\sqrt{r_a^2-\frac{L_f^2}{16}}+r_a^2\sin^{-1}\frac{L_f}{4r_a}\right) \tag{6.39}$$

Asgharian and Yu[98] developed a more complex set of equations to characterize the deposition of large fibers by interception. They found that their equations agreed with the results of Harris and Fraser (at all aspect ratios and fiber sizes tested) in airways from the trachea to the 15th airway generation. However, they suggested that beyond the 15th generation, their equations are required to adequately quantify interception, while recognizing that the equations formulated by Harris and Fraser may still be valid for small fibers having $L_f/d_f \leq 10$ and $d_f < 1$ μm.

Interception can be an important deposition mechanism, particularly for fibers. However, its significance is often underemphasized. The link between the inhalation of asbestos fibers and lung disease has been well established.[103,104] Martonen and Schroeter[105] have described how the mechanism of interception, and its function in fiber deposition in the human lung, may play a significant role in respiratory health and disease.

Due to their high deposition and low clearance rates (which will be discussed in detail in Chapter 8), hollow fibers (microtubules) may be an effective and efficient vehicle for the delivery of inhaled pharmacological drugs. Techniques to generate liposomes as microtubules have been reported by Johnson et al.[106] The hollow fibers may be filled with pharmaceuticals of choice, and serve to transport the airborne drugs. The aerodynamic size of a fiber may be estimated as[9,107]

$$d_{ae}=2.19d_f\left(\frac{L_f}{d_f}\right)^{0.116} \tag{6.40}$$

We note that the d_{ae} of a fiber is relatively independent of length; therefore, the microtubule can be packed with a large quantity of drugs without adversely affecting its aerodynamic properties. Johnson and Martonen[107,108] have developed equations for the aerodynamic properties and behavior of tumbling fibers. The integration of these formulae into the aerosol delivery program of Martonen et al.[41] will produce an inhalation therapy model for the administration of aerosolized liposomes as microtubules.

6.5.2.2 *Electrostatic charge*

When, during aerosol generation, a friction or shear force is applied to the substance to be aerosolized, an electrostatic charge is often imparted to the particles.[3,109,110] In addition, aerosol particles may pick up charges as a result of collisions with atmospheric aerosols. These electrostatic charges will affect how particles, especially ultrafine particles, are deposited within human lungs.[111]

Two types of charging are important for particle deposition in the human lung: image charge and space charge. Image charge occurs when a charge-carrying particle induces the opposite charge in an airway surface, thus creating an attraction of the particle to the surface. Image charge occurs most often with highly charged particles. Space charge occurs when two particles of the same charge simultaneously repel, resulting in a particle colliding with the airway wall. Space charge occurs most often with high concentration, unipolar charged aerosols.[4] A number of studies have dealt with the various aspects of electric force effects.[112–121]

Experimental studies in human airway casts have been used to investigate how charge affects the deposition of particles in the tracheobronchial tree. Chan et al.[112] found that highly charged particles, between 2 and 7 μm in size, deposited significantly more than uncharged particles of the same size. Cohen et al.[113] found that charged ultrafine aerosols deposit at least three times more than charge-neutralized aerosols. Other experimental work has been carried out in human subjects. Scheuch et al.[115] found that negatively charged particles deposit more effectively than uncharged particles, with the greatest effects of charge occurring for submicron particles. Melandri et al.[118] found that unipolar charge on monodisperse aerosols resulted in an increase in deposition efficiency in the human lung. Several models of the effect of charge on particle deposition in the respiratory system have been developed.[119–121]

The published data from experimental and modeling studies, albeit limited, indicate that electric forces may have highly relevant effects on the deposition of inhaled particles. Therefore, the role of electric forces should not be ignored. However, their significance may be more of a concern for aerosol therapy (in which larger mass concentrations may be inhaled) than for inhalation toxicology.

6.5.2.3 Cloud motion

Under certain circumstances, an array of particles may behave as an entity rather than as individual constituent units. This type of behavior is called cloud motion (Figure 6.6). The motion of such an entity in a gas differs from that predicted by theory for individual particles. Cloud motion may play a role in the deposition of cigarette smoke in the human lung.[122]

Cloud motion occurs when the terminal settling velocity of a cloud of particles is much greater than the settling velocity for a single particle. The terminal settling velocity of a particulate cloud with a characteristic diameter, d_c, and drag coefficient, C_D, is given by[123]

$$V_c = \left(\frac{4\rho_c d_c g}{3C_D \rho_g} \right)^{1/2} \tag{6.41}$$

The cloud drag coefficient C_D is a function of cloud Reynolds number

$$Re_c = \frac{\rho_a V_c d_c}{\mu} \tag{6.42}$$

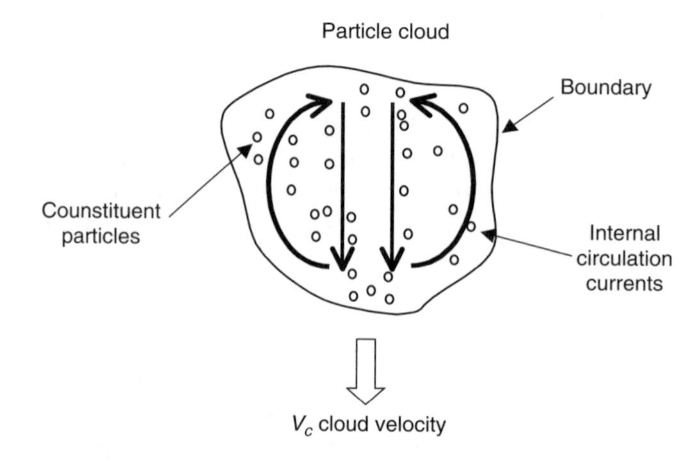

Figure 6.6 Cloud motion of concentrated aerosols. The settling velocity of the cloud is superimposed upon the motion of the constituent particles.

Using the relationship between C_D and Re recommended by Klyachko,[2] Martonen[123] derived the following expression for V_c (Equation (6.41)) for Re_c between 3 and 400:

$$V_c = \frac{\rho_c g d_c^2}{3\mu(6+Re_c^{2/3})} \tag{6.43}$$

The requirement of $V_c \gg V_p$ led directly to the following criteria for the existence of cloud motion:

$$\rho_c \gg \frac{1}{6}\rho_p C_c \left(\frac{d_p}{d_c}\right)^2 (6+Re_c^{2/3}) \tag{6.44}$$

Therefore, cloud motion may be a factor in airway deposition when the mass concentration of inhaled particles is very high.

6.6 Conclusions

The deposition of inhaled particles is a function of aerosol characteristics, ventilatory parameters, and respiratory system morphologies. Deposition in human airways is especially linked to air motion and fluid dynamics. We have provided the fundamental theory behind many of the factors that affect deposition, and have presented equations for the deposition efficiencies of several deposition mechanisms under different flow conditions. However, we stress that many of the basic formulae presented herein were derived for idealized or simplified conditions. The formulation of more advanced, simulation-based methods of predicting particle deposition in the respiratory system are currently being developed. We will present an overview of these methods in Chapter 8.

Nomenclature

Particle quantities

d_p particle diameter (μm)
d_g geometric diameter (μm)
d_{ae} aerodynamic diameter (μm)
V_p particle velocity (cm s^{-1})
V_{TS} particle terminal settling velocity (cm s^{-1})
ρ_p particle mass density (g cm^{-2})
ρ_u unit density (g cm^{-2})
m particle mass (g)
χ dynamic shape correction factor (dimensionless)
F_D drag force on a particle (g cm s^{-2})
τ relaxation time (s)
r_g particle hygroscopic growth rate (μm s^{-1})
C_c Cunningham slip correction factor (dimensionless)
D diffusion coefficient (cm^2 s^{-1})
n molecular density (molecules cm^{-3})
d_m molecular diameter (μm)
C_D particle drag coefficient (dimensionless)

Air characteristics

STP standard temperature and pressure: T=273 K (0° C), P=1 atm =1.01×10^6 g cm^{-1} s^{-2}
μ absolute air viscosity (g cm^{-1} s^{-1})
at STP, μ=1.82×10^{-4} g cm^{-1} s^{-1}
ρ_a air density (g cm^{-3})
at STP, ρ_g =1.2×10^{-3} g cm^{-3}

$\nu=\mu/\rho_g$ kinematic air viscosity ($cm^2\ s^{-1}$)
at STP, $\nu = 1.52\times10^{-1}\ cm^2\ s^{-1}$
λ mean free path of air (cm)
at STP, $\lambda=6.5\times10^{-6}$ cm
V_a mean velocity of air in an airway ($cm\ s^{-1}$)

Deposition efficiencies
$P(I)$ deposition efficiency for inertial impaction fraction
$P(S)$ deposition efficiency for sedimentation fraction
$P(D)$ deposition efficiency for diffusion fraction
$P(Int)$ deposition efficiency for interception fraction

Dimensionless parameters
$Re=\rho dV/\mu$ (Reynolds number)
$Kn=2\lambda/d_p$ (Knudsen number)
$De=Re(d/2R)^{1/2}$ (Dean number)

Fiber characteristics
L_f fiber length (μm)
$\overline{L}_f$ mean length of fiber projection into the plane normal to the airway axis (μm)
d_f fiber diameter (μm)

Particle cloud characteristics
d_c cloud terminal diameter (cm)
V_c cloud terminal settling velocity ($cm\ s^{-1}$)
$Re_c=\rho_a V_c d_c/\mu$(cloud Reynolds number [dimensionless])
ρ_c mass concentration of particles in a cloud ($g\ cm^{-3}$)

Airway properties
L_a airway length (cm)
d_a airway diameter (cm)
r_a airway radius (cm)
θ_b airway branching angle (°)

Other quantities
k Boltzmann's constant, $k = 1.38\times10^{-16}\ g\ cm^2\ s^{-2}\ molecule^{-1}\ K^{-1}$
T absolute temperature (K)
g gravitational acceleration rate, $g = 9.8\times10^2\ cm\ s^{-2}$
ρ_{H_2O} density of water at STP, $\rho_{H_2O} = 1\ g\ cm^{-3}$
d diameter of an idealized tube (cm)
R radius of curvature of an idealized tube (cm)

Disclaimer

The information in this document has been funded by the U.S. Environmental Protection Agency. It has been subjected to review by the National Health and Environmental Effects Research Laboratory and approved for publication. Approval does not signify that the contents necessarily reflect the views of the Agency, nor does mention of trade names or commercial products constitute endorsement or recommendation for use.

Acknowledgement

Kristin K. Isaacs was funded by the EPA/UNC DESE Cooperative Training Agreement CT827206, with the Department of Environmental Sciences and Engineering, University of North Carolina at Chapel Hill.

References

1. Fuchs, N.A., *The Mechanics of Aerosols*, Pergamon Press, New York, 1964.
2. Mercer, T.T., *Aerosol Technology in Hazard Evaluation*, Academic Press, New York, 1973.
3. Reist, P.C., *Aerosol Science and Technology*, McGraw-Hill, New York, 1993.
4. Hinds, W.C., *Aerosol Technology: Properties, Behavior and Measurement of Airborne Particles*, Wiley, New York, 1999.
5. Friedlander, S., *Smoke, Dust and Haze: Fundamentals of Aerosol Dynamic*, 2nd ed., Oxford University Press, New York, 2000.
6. Baron, P.A. and Willeke, K., Gas and particle motion, in *Aerosol Measurement: Principles, Techniques, and Applications*, 2nd ed., Baron, P. and Willeke, K., Eds., Van Nostrand Reinhold, New York, 2001, chap. 4.
7. Moss, O., Shape factors for airborne particles, *Am. Ind. Hyg. Assoc. J.*, 32, 221, 1971.
8. Stober, W.A., A note on the aerodynamic diameter and the mobility of non-spherical aerosol particles, *J. Aerosol Sci.*, 2, 453, 1971.
9. Stober, W.A., Dynamic shape factors of nonspherical aerosol particles, in *Assessment of Airborne Particles*, Mercer, T., Morrow, P., and Stober, W., Eds., Charles C. Thomas, Springfield, IL, 1972, chap. 14.
10. Leith, D., Drag on nonspherical object, *Aerosol Sci. Technol.*, 6, 153, 1987.
11. Davies, C.N., Particle fluid interaction, *J. Aerosol. Sci.*, 10, 477, 1979.
12. Davies, C.N., Definitive equations for the fluid resistance of spheres, *Proc. Phys. Soc.*, 57, 259, 1945.
13. Allen, M.D. and Raabe, O.G., Re-evaluation of Milikan's oil drop data for the motion of small particles in air, *J. Aerosol Sci.*, 6, 537, 1982.
14. Jennings, S.G., The mean free path in air, *J. Aerosol. Sci.*, 19, 159, 1988.
15. Crowder, T.M., Rosati, J.A., Schroeter, J.D., Hickey, A.J., and Martonen, T.B., Fundamental effects of particle morphology on lung delivery: predictions of Stokes's law and the particular relevance to dry powder inhaler formulation and development, *Pharm. Res.*, 19, 239, 2002.
16. Dennis, W.L., The growth of hygroscopic drops in a humid air stream, in *The Physical Chemistry of Aerosols*, Aberdeen University Press, Aberdeen, Scotland, 1961, chap. 2.
17. Ferron, G.A., Haider, B., and Kreyling, W.G., Inhalation of salt aerosol particles. I. Estimation of the temperature and relative humidity of the air in the human upper airways, *J. Aerosol Sci.*, 19, 343, 1988.
18. Morrow, P.E., Factors determining hygroscopic aerosol deposition in airways, *Physiol. Rev.*, 66, 330, 1986.
19. Hiller, F.C., Health implications of hygroscopic particle growth in the human respiratory tract, *J. Aerosol. Med.*, 4, 1, 1991.
20. Martonen, T.B., Analytical model of hygroscopic particle behavior in human airways, *Bull. Math. Biol.*, 44, 425, 1982.
21. Robinson, R.J. and Yu, C.P., Theoretical analysis of hygroscopic growth rate of mainstream and sidestream cigarette smoke particles in the human respiratory tract, *Aerosol Sci. Technol.*, 28, 21, 1998.
22. Finlay, W.H., Estimating the type of hygroscopic behavior exhibited by aqueous droplets, *J. Aerosol Med.*, 11, 221, 1998.
23. Ferron, G.A., The size of soluble aerosol particles as a function of the humidity of the air: application to the human respiratory tract, *J. Aerosol Sci.*, 8, 251, 1977.
24. Blanchard J.D. and Willeke, K., Total deposition of ultrafine sodium chloride particles in human lungs, *J. Appl. Physiol.*, 57, 1850, 1984.
25. Chan, H.K., Eberl, S., Daviskas, E., Constable, C., and Young, I., Changes in lung deposition of aerosols due to hygroscopic growth: a fast SPECT study, *J. Aerosol Med.*, 15, 307, 2002.
26. Hofmann, W., Morawska, L., and Bergmann, R., Environmental tobacco smoke deposition in the human respiratory tract: differences between experimental and theoretical approaches, *J. Aerosol Med.*, 14, 317, 2001.
27. Martonen, T.B. and Patel, M., Computation of ammonium bisulfate aerosol deposition in conducting airways, *J. Toxicol. Environ.Health*, 8, 1001, 1981.

28. Martonen, T.B., Barnett, A.E, and Miller, F.J., Ambient sulfate aerosol deposition in man: modeling the influence of hygroscopicity, *Environ. Health Perspect.*, 63, 11, 1985.
29. Muller, W.J., Hess, G.D., and Scherer, P.W., A model of cigarette smoke particle deposition, *Am. Ind. Hyg. Assoc. J.*, 51, 245, 1990.
30. Schroeter, J.D., Musante, C.J., Hwang, D., Burton, R., Guilmette, R., and Martonen, T.B., Hygroscopic growth and deposition of inhaled secondary cigarette smoke in human nasal pathways, *Aerosol Sci. Technol.*, 34, 1, 2001.
31. Ferron, G.A., Kreyling, W.G., and Haider, B., Inhalation of salt aerosol particles. II. Growth and deposition in the human respiratory tract, *J. Aerosol Sci.*, 19, 611, 1988.
32. Ferron, G.A. and Soderholm, S.C., Estimation of times for evaporation of pure water droplets and for stabilization of salt solution particles, *J. Aerosol. Sci.*, 21, 415, 1990.
33. Finlay, W.H. and Stapleton, K.W., The effect on regional lung deposition of coupled heat and mass transfer between hygroscopic droplets and their surrounding phase, *J. Aerosol Sci.*, 26, 655, 1995.
34. Eberl, S., Chan, H.K., Daviskas, E., Constable, C., and Young, I., Aerosol deposition and clearance measurement: a novel technique using dynamic SPET, *Eur. J. Nucl. Med.*, 28, 1365, 2001.
35. Martonen, T.B. and Clark, M.L., The deposition of hygroscopic phosphoric acid aerosols in ciliated airways of man, *Fundam. Appl. Toxicol.*, 3, 10, 1983.
36. Broday, D.M. and Georgopoulos, P.G., Growth and deposition of hygroscopic particulate matter in the human lungs, *Aerosol Sci. Technol.*, 34, 144, 2001.
37. Ferron, G.A., Oberdorster, G., and Hennenberg, R., Estimation of the deposition of aerosolised drugs in the human respiratory tract due to hygroscopic growth, *J. Aerosol. Med.*, 2, 271, 1989.
38. Hesketh, H.E., *Fine Particles in Gaseous Media*, Lewis Publishers, Inc., Chelsea, MA, 1986.
39. Pedley, T.J. and Kamm, R.D., Dynamics of gas flow and pressure-flow relationships, in *The Lung: Scientific Foundations,* Vol.1, Crystal, R.G. and West, J.B., Eds., Raven Press, New York, 1991.
40. Fox, R.W. and McDonald, A.T., *Introduction to Fluid Mechanics*, 3rd ed., Wiley and Sons, New York, 1985.
41. Martonen, T.B., Musante, C.J., Segal, R.A., Schroeter, J.D., Hwang, D., Dolovich, M.A., Burton, R., Spencer, R.M., and Fleming, J.S., Lung models: strengths and limitations, *Respir. Care*, 45, 712, 2000.
42. Weibel, E., Design of airways and blood vessels considered as branching trees, in *The Lung: Scientific Foundations,* Vol. 1 Crystal, R.G. and West, J.B., Eds., Raven Press, New York, 1991.
43. Dean, W.R., The streamline of motion of a curved pipe, *Philos. Mag.*, 5, 623, 1928.
44. Pedley, T.J. and Drazen, J.M., Aerodynamics theory, in *Handbook of Physiology, Section 3: The Respiratory System, Volume 3: Mechanics of Breathing,* Mackelm, P.T. and Mead, J., Eds., Williams and Wilkins, Baltimore, 1986.
45. Taylor, G.I., The criterion for turbulence in curved pipes, *Proc. R. Soc. London. Ser. A.*, 124, 243, 1929.
46. Yao, L.S. and Berger, S.A., Entry flow in a curved pipe, *J. Fluid Mech.*, 67, 177, 1975.
47. Greenspan, D., Secondary flow in a curved tube, *J. Fluid Mech.*, 57, 167, 1973.
48. McConalogue, D.J. and Srivastava, R.S., Motion of a fluid in a curved tube, *Proc. R. Soc. London. Ser. A.*, 307, 37, 1968.
49. Humphrey, J.A.C., Some numerical experiments on developing laminar flow in circular sectioned bends, *J. Fluid Mech.*, 154, 357, 1985.
50. Isabey, D. and Chang, H.K., A model study of flow dynamics in human central airways. Part II. Secondary flow velocities, *Respir. Physiol.*, 49, 97, 1982.
51. Chang, H.K. and El Masry, O.A., A model study of flow dynamics in human central airways. Part I. Axial velocity profiles, *Respir. Physiol.*, 49, 75, 1982.
52. Guan, X. and Martonen, T.B., Simulations of flow in curved tubes, *Aerosol Sci. Technol.* 26, 485, 1997.
53. Kleinstreuer, C. and Zhang, Z., Laminar-to-turbulent fluid-particle flows in a human airway model, *Int. J. Multiphase Flow*, 29, 271, 2003.
54. Schroter, R.C. and Sudlow, M.F., Flow patterns in models of the human bronchial airways, *Respir. Physiol.*, 7, 341, 1969.

55. Brech, R. and Bellhouse, B.J., Flow in branching vessels, *Cardiovasc. Res.*, 7, 593, 1973.
56. Pedley, T.J., Schroter, R.C., and Sudlow, M.F., Flow and pressure drop in systems of repeatedly branching tubes, *J. Fluid Mech.*, 46, 365, 1971.
57. Guan, X. and Martonen, T.B., Flow transition in bends and applications to airways. *J. Aerosol Sci., 31,* 833, 2000.
58. Martonen, T.B., Yang, Y., Xue, Z.Q., and Zhang, Z., Motion of air within the tracheobronchial tree, *Part. Sci. Tech.*, 12, 175, 1994.
59. Martonen, T.B., Guan, X., and Schrek, R.M., Fluid dynamics in airway bifurcations. I. Primary flows, *Inhal. Toxicol.*, 13, 261, 2001.
60. Martonen, T.B., Guan, X., and Schrek, R.M., Fluid dynamics in airway bifurcations. II. Secondary currents, *Inhal. Toxicol.*, 13, 281, 2001.
61. Martonen, T.B., Guan, X., and Schrek, R.M., Fluid dynamics in airway bifurcations. III. Localized flow conditions, *Inhal. Toxicol.*, 13, 291, 2001.
62. Yu, G., Zhang, G., and Lessman, R., Fluid flow and particle diffusion in the human upper respiratory system, *Aerosol Sci. Technol.*, 28, 146, 1998.
63. Calay, R.K., Kurujareon, J., and Hóldo, A.E., Numerical simulation of respiratory flow patterns within human lung, *Respir. Physiol. Neurol.*, 130, 201, 2002.
64. Findeisen, W., Über die von der molecularkinetischen theorie der wärmer geforderte bewegun von in ruhenden flüssigkeiten suspendierten teilchen, *Annalen der Physik*, 17, 549, 1935.
65. Beeckmans, J.M., The deposition of aerosols in the respiratory tract. I. Mathematical analysis and comparison with experimental data, *Can. J. Physiol. Pharmac.*, 43, 157, 1965.
66. Landahl, H.D., On the removal of air-borne droplets by the human respiratory tract. I. The lung, *Bull. Math. Biophys.*, 12, 43, 1950.
67. Martonen, T.B., Mathematical model for the selective deposition of inhaled pharmaceuticals, *J. Pharm. Sci.*, 82, 1191, 1993.
68. Landahl, H.D. and Herrmann, R.G., On the retention of airborne particulates in the human human lung, *J. Ind. Toxicol.*, 30, 181, 1948.
69. Landahl, H.D. and Herrmann, R.G., Sampling of liquid aerosols by wires, cylinders, and slides, and the efficiency of impaction of the droplets, *J. Colloid Sci.*, 4, 103, 1949.
70. Ingham, D.B., Diffusion of aerosols from a stream flowing through a cylindrical tube, *J. Aerosol Sci.*, 6, 125, 1975.
71. Ingham, D.B., Diffusion of aerosols from a stream flowing through a short cylindrical tube, *Aerosol Sci.*, 15, 637, 1984.
72. Ingham, D.B., Diffusion of aerosols in the entrance region of a smooth cylindrical pipe, *J. Aerosol Sci.*, 22, 253, 1991.
73. May, K.R. and Clifford, R., The impaction of aerosol particles in cylinders, spheres, ribbons, and discs, *Ann. Occup. Hyg.*, 10, 83, 1967.
74. Starr, J.R., Inertial impaction of particulates upon bodies of simple geometry, *Ann. Occup. Hyg.*, 10, 349, 1967.
75. Cheng, Y.-S. and Wang, C.-S., Inertial deposition of particles in a bend, *Aerosol Sci.*, 6, 139, 1975.
76. Reeks, M.W. and Skyrme, G., The dependence of particle deposition velocity on particle inertia in turbulent pipe flow, *J. Aerosol Sci.*, 7, 485, 1976.
77. Crane, R.I. and Evans, R.L., Inertial impaction of particles in a bent pipe, *J. Aerosol Sci.*, 8, 161, 1977.
78. Stahlhofen, W., Rudolf, G., and James, A.C., Intercomparison of experimental regional aerosol deposition data, *J. Aerosol Med.*, 2, 285, 1989.
79. Heyder, J., Gebhart, J., Rudolph, G., Schiller, C.F., and Stahlhofen, W., Deposition of particles in the size range 0.005–15 μm, *J. Aerosol Sci.*, 17, 811, 1986.
80. Yu, C.P. and Thiagarajan, V., Sedimentation of aerosols in closed finite tubes in random orientation, *J. Aerosol Sci.*, 9, 315, 1978.
81. Heyder, J. and Gebhart, J., Gravitational deposition of particles from laminar aerosol flow through inclined circular tubes, *J. Aerosol Sci.*, 8, 289, 1977.
82. Foster, W.M., Deposition and clearance of inhaled particles, in *Air Pollution and Health*, Holgate, S.T., Koren, H.S., Maynard, R.L., and Samet, J.M., Eds., Academic Press, New York, 1999, chap. 14.

83. Martonen, T.B., Zhang, Z., and Yang, Y., Particle diffusion from developing flows in rough-walled tubes, *Aerosol Sci. Technol.*, 26, 1, 1997.
84. Martonen, T.B., Zhang, Z., Yang, Y., and Bottei, G., Airway surface irregularities promote particle diffusion in the human lung, *Radiat. Prot. Dosim.*, 59, 5, 1995.
85. Shaw, D.T. and Rajendran, N., Diffusional deposition of airborne particles in curved bronchial airways, *J. Aerosol Sci.*, 8, 191, 1977.
86. Scheuch, G. and Heyder, J., Dynamic shape factor of nonspherical aerosol particles in the diffusion regime, *Aerosol Sci. Technol.*, 12, 270, 1990.
87. Martonen, T.B., Zhang, Z., and Yang, Y., Particle diffusion with entrance effects in a smooth-walled cylinder, *J. Aerosol Sci.*, 27, 139, 1996.
88. Hamill, P., Particle deposition due to turbulent diffusion in the upper respiratory system, *Health Phys.*, 36, 355, 1979.
89. Martin, D. and Jacobi, W., Diffusion deposition of small-sized particles in the brochial tree, *Health Phys.*, 23, 23, 1972.
90. Martonen, T. and Yang, Y., Deposition mechanisms of pharmaceutical particles in human airways, in *Inhalation Aerosols: Physical and Biological Basis for Therapy*, Hickey, A.J., Ed., Marcel Dekker, New York, 1996.
91. Sussman, R.G., Cohen, B.S., and Lippman, M., Asbestos fiber deposition in a human tracheobronchial cast. I. Empirical model, *Inhal. Toxicol*, 3, 161, 1991.
92. Edwards, D.A., Hanes, J., Caponetti, G., Hrkach, J., Ben-Jebria, A., Eskew, M.L., Mintzes, J., Deaver, D., Lotan, N., and Langer, R., Large porous particles for pulmonary drug delivery, *Science*, 276, 1868, 1997.
93. Edwards, D.A., Ben-Jebria, A., and Langer, R., Recent advances in pulmonary drug delivery using large, porous, inhaled particles *J. Appl. Physiol.*, 85, 379, 1998.
94. Musante, C.J., Schroeter, J.D., Rosati, J.A., Crowder, T.M., Hickey, A.J., and Martonen, T.B., Factors affecting the deposition of inhaled porous drug particles, *J. Pharm. Sci.*, 91, 1590, 2002.
95. Myojo, T., Deposition of fibrous aerosols in model bifurcations, *J. Aerosol Sci.*, 18, 337, 1987.
96. Myojo, T. and Takaya, M., Estimation of fibrous aerosol deposition in upper bronchi based on experimental data with model bifurcation, *Ind. Health*, 39, 141, 2001.
97. Harris, R.L. and Fraser, D.A., A model for deposition of fibers in the human respiratory system, *Am. Ind. Hyg. Assoc. J.*, 37, 73, 1976.
98. Asgharian, B. and Yu, C.P., A simplified model of interceptional deposition of fibers at airway bifurcations, *Aerosol Sci. Technol.*, 11, 80, 1989.
99. Cai, F.S. and Yu, C.P., Inertial and interceptional deposition of spherical particles and fibers in a bifurcating airway, *J. Aerosol Sci.*, 19, 679, 1988.
100. Balásházy, I., Martonen, T.B., and Hofmann, W., Fiber deposition in airway bifurcations, *J. Aerosol Med.*, 3, 243, 1990.
101. Balásházy, I., Martonen, T.B., and Hofmann, W., Inertial impaction and gravitational deposition of aerosols in curved tubes and airway bifurcations, *Aerosol Sci. Technol.*, 13, 308, 1990.
102. Balásházy, I., Martonen, T.B., and Hofmann, W., Simultaneous sedimentation and impaction of aerosols in two-dimensional channel bends, *Aerosol Sci. Technol.*, 13, 20, 1990.
103. Becklake, M.R., Asbestos-related diseases of the lung and other organs: Their epidemiology and implications for clinical practice, *Am. Rev. Respir. Dis.*, 114, 187, 1976.
104. Timbrell, V., Inhalation and biological effects of asbestos, in *Assessment of Airborne Particles*, Mercer, T.T., Morrow, P.E., and Stober, W., Eds., Charles C. Thomas, Springfield, IL, 1972, chap. 22.
105. Martonen, T.B. and Schroeter, J., Deposition of inhaled particles within human lungs, in *The Asbestos Legacy, Volume 23 of the Sourcebook on Asbestos Diseases: Medical, Legal, and Engineering Aspects*, Peters, G.A. and Peters, B.J., Eds., Matthew Bender and Company, Newark, NJ, 2001, chap. 3.
106. Johnson, D.L., Polikandritou-Lambros, M., and Martonen, T.B., Drug encapsulation and aerodynamic behavior of a lipid microtubule aerosol, *Drug Deliv.*, 3, 9, 1996.
107. Johnson, D.L. and Martonen, T.B., Behavior of inhaled fibers: potential applications to medicinal aerosols, *Part. Sci. Tech.*, 12, 161, 1994.
108. Johnson, D.L. and Martonen, T.B., Fiber deposition along airway walls: effects of fiber cross section on rotational interception, *J. Aerosol Sci.*, 24, 525, 1993.

109. Rosati, J.A., Leith, D., and Kim, C., Monodisperse and polydisperse aerosol deposition in a packed bed, *Aerosol Sci. Technol.*, 37, 1, 2003.
110. Kousaka, Y., Okuyama, K., Adachi, M., and Ebie, K., Measurement of electrical charge of aerosol particles generated by various methods, *J. Chem. Eng. Jpn.*, 14, 54, 1981.
111. Wilson, L.B., The deposition of charged particles in tubes with reference to the retention of therapeutic aerosols in the human lung, *J. Colloid Sci.*, 2, 271, 1947.
112. Chan, T.L., Lippman, M., Cohen, V.R., and Schlesinger, R.B., Effect of electrostatic charges on particle deposition in a hollow cast of the human larynx-tracheobronchial tree, *J. Aerosol Sci.*, 9, 463, 1978.
113. Cohen, B.S., Xiong, J.Q., Fang, C., and Li, W., Deposition of charged particles in lung airways, *Health Phys.*, 74, 554, 1998.
114. Grover, S.N. and Pruppacher, H.R., A numerical determination of the efficiency with which spherical aerosol particles collide with spherical water drops due to inertial impaction and phoretic and electrical forces, *J. Atmos. Sci.*, 34, 1655, 1977.
115. Scheuch, G., Gebhart, J., and Roth, C., Uptake of electrical charges in the human respiratory tract during exposure to air loaded with negative ions, *J. Aerosol Sci.*, 21, S439, 1990.
116. Chan, T.L. and Yu, C.P., Charge effects on particle deposition in the human tracheobronchial tree, *Ann. Occup. Hyg.*, 26, 65, 1982.
117. Hashish, A.H., Bailey, A.G., and Williams, T.J., Modelling the effect of charge on selective deposition of particles in a diseased lung using aerosol boli, *Phys. Med. Biol.*, 39, 2247, 1994.
118. Melandri, C., Tarroni, G., Prodi, V., De Zaiacomo, T., Formignani, M., and Lombardi, C.C., Deposition of charged particles in the human airways, *J. Aerosol Sci.*, 14, 657, 1983.
119. Yu, C.P. and Chandra, K., Precipitation of submicron charged particles in the human lung airways, *Bull. Math. Biol.*, 49, 471, 1977.
120. Chan, T.L. and Yu, C.P., Charge effects on particle deposition in the human tracheobronchial tree, *Ann. Occup. Hyg.*, 26, 65, 1982.
121. Hashish, A.H., Bailey, A.G., and William, T.J., Modelling the effect of charge on selective deposition of particles in a diseased lung using aerosol boli, *Phys. Med. Biol.*, 39, 2247, 1994.
122. Martonen, T.B. and Musante, C.J., Importance of cloud motion on cigarette smoke deposition in lung airways, *Inhal. Toxicol.*, 12, 261, 2000.
123. Martonen, T.B., The behavior of cigarette smoke in human airways, *Am. Ind. Hyg. Assoc. J.*, 53, 6, 1992.

chapter seven

Aerosol dose

Lev S. Ruzer, Michael G. Apte, and Richard G. Sextro
Lawrence Berkley National Laboratory

Contents

7.1 Introduction 101
7.2 Environmental dosimetry....102
7.3 Exposure and dose definitions 104
7.4 Uncertainty in dose assessment 105
7.5 Aerosol concentration safety standards....106
7.6 Nonuniformity of deposition and ultrafine/nanometer-sized particles 106
7.7 Characteristics of nanometer particles....108
7.8 Experimental data on nanometer particles 108
7.9 Radioactive markers 109
7.10 Conclusion 109
References 110

1493–1541, Paraselsus:
All substances are poisons,
there is none that is not a poison.
The right dose differentiates
a poison and a remedy.

7.1 Introduction

A well-defined pharmakinetic relationship exists between chemicals administered orally, dermally, or intravenously and the quantity of substance delivered to the specific target site. Certainly, some variability in dose exists due to interindividual differences in metabolism and transport kinetics, but it is relatively easy to define the final delivery of an agent to the target site. In the case of aerosols, the understanding of this relationship is complicated for several reasons:

1. Aerosols are a complex medium consisting of particles in a very wide size range, with diameters ranging from nanometers to tens of micrometers (4–5 orders of magnitude).
2. Diversity in particle size results in different mechanisms of interaction inside the three main branches of the lung: the extrathoracic, tracheobronchial, and alveolar regions.

1-56670-611-4 / 05 / $0.00+$1.50

3. The particle size distribution is altered inside the lung due to changes in humidity and temperature.
4. In each branch of the lung, the particle size distribution differs due to selective filtration.
5. Individual breathing characteristics and lung morphology vary widely.
6. The variability in clearance, translocation, and other biokinetic processes taking place after aerosol delivery to the site of the lung must be considered.

In the study of the pathway of aerosol exposure to dose to effect, three important conditions should be taken into account:

1. Dose, the physical parameter responsible for the health effect, should be defined correctly.
2. If, as in most practical cases, the dose is not measured directly, the correlation between the value of the measurable surrogate and the dose should be established.
3. The uncertainty related to the value of the dose surrogate should be assessed correctly.

The last condition is especially important in the case of aerosols where uncertainties are very high due to spatial and temporal variability, as are uncertainties associated with the interaction between aerosols and humans.

Figure 7.1 describes the particle dose pathway, indicating the complex set of physical parameters and mechanisms at play. The parameters and mechanisms shown in the figure are intended to be illustrative rather than comprehensive. Further details may be found in the chapters of this handbook as indicated in the figure.

7.2 Environmental dosimetry

Assessing the aerosol dose resulting from an environmental exposure has a particular set of issues. These problems vary depending upon environmental conditons, particle sources within environments, and the activities of the occupants and particle sources within them. The issues may vary greatly depending upon whether the exposed individuals are outdoors, indoors, and whether they occur in an occupational, residential, or another setting. Regardless of the environmental conditons, in order to accurately determine the burden of aerosol exposure on an individual, dose must be measured or estimated. Typically surrogates for actual dose such as aerosol concentration or *exposure* (time-weighted average concentration × duration of exposure) of particular aerosol species, or lung tissue dose estimates using inhalation models are used since true dose measurements are usually not possible. Other quantities such as the number or surface area of particles deposited in the whole lung, or to a specific lung site, can be defined as *intake* (included in Norms of Radiation Safety [3]). The use of any of these quantities in either epidemiological studies or control involves, explicitly or implicitly, assumptions about their relationships with one another. However, although it is the dose that is the cause of the biological effect, it is typically the airborne concentration that is measured, and it is the corresponding estimates of exposure that are used as correlates to health effects or predictors of risk.

It is the dose or intake to lung tissue that is studied to understand the actual insult leading to lung cancer or other diseases. If exposure is used as an index of dose, for example, in epidemiological studies, assumptions are made, typically implicitly, about a comparability or constancy of breathing rates, and of retention of aerosols in the lung.

It should be noted that, especially in industrial environments, the inhalation rates depend on the groups involved and on working conditions, specifically the physical load, which can vary substantially. Retention depends on the properties of the aerosols and also on the load of work, and is typically not known accurately.

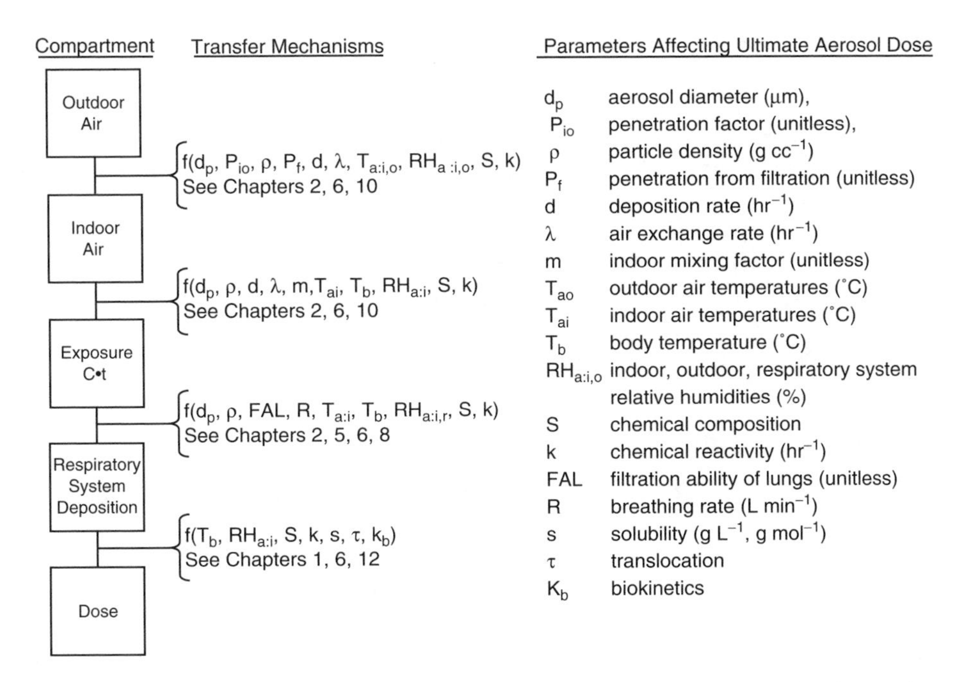

Figure 7.1 Parameters affecting ultimate inhaled particle dose, showing physical transfer mechanisms across macro to micro environmental compartments. The Chapters in this Handbook relating to the various transfer mechanisms are listed.

For example, in [4,5], a correlation was established between measured aerosol concentration in the breathing zone and gamma activity (dose) in the lungs of miners. The transfer coefficient called filtration ability of lungs (FAL) was defined and measured. FAL is a combination of a breathing rate and deposition coefficient. It was shown that FAL depends on physical activity and is different for different groups of miners. It was also shown [6,7] that intake — a combination of the concentration and FAL — is a more correct characteristic of exposure than the concentration time product, because it takes into account both concentration and physical activity. It is important to note that by applying FAL, it is possible to decrease the uncertainty in dose assessment.

The measured concentration is not complete enough to directly provide accurate exposure characterization, because the concentrations to which individuals are exposed vary substantially in time and space. For example, a variation of the building ventilation rate even for a short period of time can lead to a substantial change in concentrations. Moreover, the concentrations, and especially the particle size distribution of aerosols in the breathing zone (which is directly responsible for local deposition), may differ substantially from those measured by standard instruments [8–10].

Finally, in the occupation setting, the very concept of "work place"-associated concentration and work load are indefinite and job dependent, because workers are typically at a number of places during their working shift, with variable concentrations and nature of work. A similar situation exists with indoor exposure. Since the measurements of concentration in many environments may be performed only infrequently or even constructed, we cannot expect reliable correspondence between exposures estimated from spatially and temporally sparse measurements and actual personal exposures. Thus, the use of measurements of airborne concentration as a basis for estimating or comparing dose can lead to substantial errors. In aggregate, these errors may constitute as much as an order of magnitude or more, and therefore make risk assessment data unreliable. Unfortunately, even after the use of measured concentration for radioactive and nonradioactive aerosols was established as a measure of the dose to the lung more than 40 years ago, little discussion and experimental study were conducted to determine the degree of correlation between the measured concentration and actual dose for individuals, both workers and for the general population. Experimental data on the correlation between the measured concentration according to standard procedures (site measurement) and breathing zone measurements, not to mention dose, are very scarce. The only systematic study for radioactive aerosols provided in Polish mines [10] showed poor correlation between standard area measurement and breathing zone sampling.

Using direct measurement of naturally occurring radioactive radon decay products in the lungs of miners, it was shown that the ratio of the true dose and that based on measured concentration could be larger than 8 [4,5].

7.3 Exposure and dose definitions

It should be pointed out that the assessment of aerosol exposure presents substantial difficulties.

The term "exposure" came to the aerosol field from the study of the effect of gases. Aerosol science, including epidemiological studies of the health effect of aerosol exposure, was developed mainly in the second half of the 20th century.

In 1924, German chemist Fritz Haber proposed the following definition of exposure:

> A simple and practical measure for toxicity can be obtained that suffices for all practical purposes. For each war gas, the amount (c) present in one cubic meter of air expressed in milligrams and multiplied by the time (t) in minutes necessary for the experimental animal inhaling this air to obtain a lethal effect. The smaller this product ($c*t$) is, the greater is the toxicity of the war gas.

This was the original formulation of "Haber's law". The product of concentration and time was simply a measure of acute lethality in cats, nothing else. To expect a constant response to the same $c \times t$ product had been postulated in 1921 by experimental toxicologist and biochemist Flury. Haber's law did not take into account the difference between acute and chronic exposure.

The definition of exposure in the case of aerosols is much more complicated. The main difference in dosimetry between gases and aerosols is that aerosol deposited in the lungs is very nonuniform. The reason for this is that respirable fractions of aerosols consist of particles with diameters ranging from nanometers to 5–10 μm. Therefore, their particle size distribution and their deposition inside the lungs are very uneven.

According to the 1991 National Academy of Science report [11], the definition of exposure is

> An event that occurs when there is *contact at a boundary between humans and the environment* with a contaminant of a specific concentration for an interval of time; the units are concentration multiplied by time.

A statistical definition of exposure has been proposed [12]:

> An exposure at some instant of time is a joint occurrence of two events:

1. the pollutant of concentration C is present at a particular location in space at a particular time, and
2. the person is present at the same time and location in space

A later definition [13] addresses the notion that the target remains important, and also that different parts of the target can receive different exposures at the same time.

The last definition is more adequate for aerosols, because it takes into account the specific aerosol problem of nonuniformity.

In all these definitions, the key word is contact, which means that in the case of aerosols only breathing zone measurement should be used for the exposure and particle size measurement. If concentration (and particle size characterization) is provided at a distance from the breathing zone, the correlation should be established between breathing zone and sampling site measurement.

The next step is in defining the physical value that describes the delivery of aerosol particles to the lung. We will define this value — *intake* — using the analogy with radioactive aerosols. Intake is the product of concentration, breathing rate, and deposition coefficient. The deposition coefficient is a dimensionless value equal to the ratio of the quantity of aerosols deposited in the lung to the quantity that entered the lung. Intake is expressed in units of number of aerosol particles, particle surface area, or mass of the particles, depending on what units of concentration are used.

Based upon this discussion, we can formulate the definition of dose as a physical value responsible for a biological effect: Dose is the specific quantity of aerosols delivered to a target site that is directly responsible for a biological effect.

The term "quantity" is defined as follows:

1. In the case of radioactive aerosols, deposited energy per unit mass for alpha, beta, or gamma radiation is expressed in units of J/kg (Gray) or rads (100 erg/g) or the equivalent.
2. In the case of nonradioactive aerosols, quantity is the deposited number of particles, surface area, or mass of a discrete particle size.
3. The term "directly" means that dose is a quantity of the deposited amount of aerosol particles after the completion of all biokinetic processes.

According to EPA of the United States:

> In epidemiological studies, an index of exposure from personal or stationary monitors of selected pollutants is analyzed for associations with health outcomes, such as morbidity or mortality. However, it is a basic tenet of toxicology that the dose delivered to the target site, not the external exposure, is the proximal cause of a response. Therefore, there is increased emphasis on understanding the exposure–dose–response relationship. Exposure is what gets measured in the typical study and what gets regulated; dose is the causative factor [14].

Despite the fact that the importance of dosimetry is well understood, some uncertainties in interpretations exist even in terms of aerosol dose.

7.4 Uncertainty in dose assessment

It was shown [4,5] that in the case of natural radioactive aerosols, that is, the decay products of radon (in mines), it is possible to measure dose due to alpha radiation directly by measuring gamma activity from the lungs.

Such opportunities do not easily exist for nonradioactive aerosols; hence, the main method for nonradioactive aerosol dose assessment is through modeling. It should be noted, however, that calculated dose is very sensitive to input parameters [15]; in other words, the dose assessment based only on modeling can be unreliable. One of the main sources of uncertainty in lung dosimetry is in the assessment of particle size distribution in the breathing zone, the location where contact between aerosols and the organism occurs.

Reliable determination of the dose includes measurement of aerosol particle size distribution (concentration) in the breathing zone, a knowledge of the parameters of transformation of particle size distribution due to humidity, temperature, and other factors inside the respiratory tract, calculating or measuring directly (e.g., by using radioactive markers) aerosol deposition, and, finally, a knowledge or direct measurement of parameters of biochemical processes inside the lungs: translocation, clearance, absorption, etc.

Most current information on the clearance of aerosol from the respiratory tract comes from radioactive aerosol inhalation studies; much less is known about respiratory biokinetics of inhaled nonradioactive particles.

In terms of biokinetics, the main difference between radioactive and nonradioactive aerosols is that mass concentrations of radioactive aerosols are typically low in comparison with nonradioactive aerosols, which results in differences in clearance and translocation [1].

7.5 Aerosol concentration safety standards

In 1997, EPA established a new standard for particulate matter as an indicator of mass concentration: particle mass with aerodynamic diameters of 10 and 2.5 μm or less (PM_{10} and $PM_{2.5}$, respectively) [16].

In the light of the theoretical and experimental results of the last 5–7 years, it has become clear that quantifying aerosol mass deliveries to the lung is not always adequate for biological effects. One such very important example was mentioned in [17]. According to EPA, for a mass concentration of 50 $\mu g/m^3$ of ambient particulate aerosol, the daily deposition in the alveolar region was close to 50 μg [14].

It is easy to calculate that such a mass will produce, at the alveolar surface, only 1 particle per day per 1.5 mm^2 and 1 particle per 44 mm^2 for 1 and 3 μm particles, respectively. Such amounts will cover less than a millionth part of the lung surface. No known chemical constituents of ambient particulate matter have such a threshold of toxicity at this daily level.

However, the corresponding numbers of deposited particles per mm^2 and part of the covered surface of alveoli will be orders of magnitude higher for particles at the level of tens of nanometers. For the same mass, the number of particles is inversely proportional to the cube of the diameter. Therefore, in case of substantial nanometer particle concentration, even exposure to nontoxic chemicals can result, as experiments on animals have shown, in acute biological effect. Thus, in terms of dosimetry the concentration of nanometer particles should be measured correctly. It should be mentioned that the measurement of particles in this size present some difficulties In [18], it was shown that detection efficiency in this range of sizes decreases dramatically. For radioactive aerosols of such sizes, especially for the "unattached fraction" of radon decay products with a size around 1 nm (8–10 molecules), measurement techniques have been presented in many publications [19–28].

7.6 Nonuniformity of deposition and ultrafine/nanometer-sized particles

Many reasons can explain the discrepancy between dosimetric factors and health effects in the case of aerosols, the first being biological factors. However, from the point of view of dosimetry, two main factors can be used for the explanation:

1. nonuniformity in the distribution of aerosols both in breathing space and inside the lungs, making the use of average numbers inappropriate; and
2. the special role of ultrafine/nanometer aerosols in health effects [30–35].

Ultrafine particles contribute little mass to the fine fraction; however, they dominate particle surface area and particle number concentration.

The special role of nanometer-sized, ultrafine particles, and nonuniformity in aerosol dosimetry was demonstrated by Balashazy and Hoffman [35]. Present lung dosimetry models for radon decay products are based on deposition efficiencies for straight cylindrical airways, which are equivalent to the commonly accepted assumption that inhaled particles are uniformity deposited in these airways [36–38].

Because aerosol deposition depends only on particle size and not on radioactivity, we can assume that the same approach will be correct for nonactive aerosols. In the present models, depth–dose distributions on bronchial epithelium are obtained by integrating the surface activities over the surface of the cylindrical airways within the range of alpha particles, thereby assuming again that alpha particle sources are uniformly distributed on airway surfaces [39,40]. The assumption of uniformity is further supported by the theory that Brownian motion in straight cylindrical tubes *a priori* produces uniform deposition patterns.

In contrast, experimental studies about molecular-sized, ultrafine, and submicron particle deposition in single-pathway tracheobronchial models [41–43], in airway casts of the human tracheobronchial tree [44,45], and in single bronchial bifurcation models have demonstrated that particle deposition patterns are highly nonuniform. Experimental evidence exists to show that the main features of the deposition patterns within airways exist at bifurcation points in the lung [46].

Research showed that:

1. Deposition is enhanced at airway branching zones relative to cylindrical airway portions.
2. Deposition within a bifurcation is highest at the dividing spur.
3. Deposition is also enhanced at the inner sides of the daughter airways.

Computed enhancement factors indicate that the cells located at carinal ridges may receive localized doses that are 20–40 times (1 nm) and 50–115 times higher (10– 200 nm), respectively, than the corresponding average doses.

In current dosimetric models, for radon decay products the implicit assumption was made that epithelial cells on bronchial airway surfaces will receive the same average dose. In contrast, Balashazy and Hoffman [35] proposed that the target should be divided into two fractions:

1. the fraction of the surface without any particles being deposited there (i.e., the dose is zero); and
2. the remaining fraction, which may have many more particles deposited than indicated by the average deposition density.

Since it has been recently suggested that the number of multiple cellular hits may play a crucial role in the extrapolation of lung cancer from occupational to domestic environments [47], maximum enhancement factors may serve as a measure of the probability of multiple hits.

In spite of the fact that the nature of hits to lung cells from alpha particles and nanometer-sized and ultrafine particles is completely different, it is interesting to compare these numbers for a given concentration. The number of hits from alpha particles of radon decay products is given in Tables 2–14 of [48].

7.7 Characteristics of nanometer particles

It was shown by Preining [49] that particles in the size range of 5 nm contain only a few molecules with high proportions at the surface. For particles less than 5 nm in diameter, more than half of the molecules will be at the surface. Hence, the structure of such particles cannot be regarded as a part of continuum, and traditional concepts of particle volume and surface area are no longer prime parameters. The electronic state of the particle is determined by the electronic states of individual molecules. Their configuration plays an extremely important role in particle interactions with molecules, other particles, and biological cells. Nanoparticles are formed in very large numbers during nucleation, but their coagulation occurs very rapidly even at very low mass concentration. Thus, the lifetime of such particles is very short. One particularly important aspect of very small particles is the way in which they might interact with cells in biological systems.

7.8 Experimental data on nanometer particles

In relation to air pollution, for example, there is increasing concern about the role of small atmospheric particles in the observed increase in disease and mortality in human populations. It was demonstrated in [29] that rats exposed to ultrafine particles of, for example, titanium dioxide or carbon black sustain more lung injury and pathology than rats exposed to the same deposited mass of fine respirable particles composed of the same material.

The large number of deposited particles per unit mass of lung tissue may exceed the ability of macrophages to phagocytize them, and prolonged interaction with epithelial cells may be an important factor in stimulating inflammation and interstitial transfer of the particles. The large surface area of ultrafine particles in contact with the lung provides the opportunity for surface chemistry to have a profound effect. Ultrafine particles that penetrate the interstitium will make contact with interstitial macrophages and other sensitive cell populations, and this is likely to have a powerful inflammogenic effect that underlies the development of subsequent disease.

The potential importance of particles with a diameter of ~20 nm is illustrated by Oberdorster et al. [29] with a very low mass concentration of insoluble ultrafine particles and by Chen et al. [30] with a very low H^+ concentration on ultrafine particles surfaces.

The number of epidemiological studies conducted from the 1990s has demonstrated an association between ambient urban particulate air concentration down to below 100 $\mu g/m^3$ and increased mortality occurring in elderly people with preexisting cardiorespiratory diseases, and increased morbidity due to respiratory symptoms in children [31–34].

It has been proposed, based on results of the past and most recent inhalation studies with ultrafine particles in rats, that such particles (i.e., particles below ~50 nm in diameter) may contribute to the observed increased mortality and morbidity. It was demonstrated in the past that inhalation of highly insoluble particles of low intrinsic toxicity, such as TiO_2, results in significantly increased pulmonary inflammatory responses even when their size is in the ultrafine particle range, that is, ~20 nm in diameter. However, these effects were not of an acute nature and occurred only after prolonged inhalation exposure of the aggregated ultrafine particles at concentrations in the milligram per cubic meter range.

In contrast to the more recent studies with thermodegradation products of polytetrafluoroethylene (PTFE), it was found that freshly generated PTFE fumes containing single ultrafine particles (median diameter 26 nm) were highly toxic to rats at inhaled concentrations of $0.7–1.0 \times 10^6$ particles/m^3, resulting in acute hemorrhage pulmonary inflammation and death after 10–30 min of exposure. These results cannot be attributed to the gas-phase component. The calculated mass concentration of the inhaled ultrafine PTFE particles was less than 60 $\mu g/m^3$, a very low value to cause mortality in healthy rats. Aging of fumes with concomitant aggregation of the ultrafine particles significantly decreases their toxicity. Since ultrafine particles are always present in urban atmosphere, it was suggested that they play a role in causing acute lung injury in sensitive parts of the population.

7.9 Radioactive markers

In order to obtain more reliable results on deposition and dosimetry, radioactive markers were used. The main problem with such experiments on humans is ethical consideration and careful assessment of the risk associated with radiation. A Technegas aerosol suspension of ^{99m}Tc-labeled ultrafine carbon particles in the size range of 5 – 10 nm was used in experiments on humans. It was shown that particles of such size, which are similar to actual pollutant particles, diffuse rapidly into the blood stream and are deposited in substantial amounts in the abdomen, including the liver, stomach, and bladder [50].

Thus, the role of nanometer-sized particles should be considered not only from its relevance for lung dosimetry but also for the systemic effects. It should also be mentioned that radon decay products are good natural markers due to their short-lived nature and relatively high permissible concentration in comparison with most artificial aerosols. In mines, the primitive portable technique was used to measure the dose from radon decay products under especially difficult underground conditions [2].

For the purpose of studying deposition of aerosols in lungs with radon progeny as a marker in laboratory conditions and by using spectrometric detectors, a scanner, and three-dimensional tomographic techniques, it is possible to study experimentally, and not by modeling, the deposition of aerosol in different parts of the lung for different particle size distributions.

As a marker, radon decay products have some advantages:

1. Due to its nature, radon distributes uniformly in the air and produces with a known constant rate the marker itself – the "unattached fraction," consisting of an atom of nuclide surrounded by a group of 8–10 molecules.
2. Due to its extremely small size, such a marker does not change particle size distribution of the aerosols of interest.
3. Measurements in mines showed what kinds of concentrations are appropriate for the study in terms of ethical considerations.
4. Radon and its decay products have natural radioactive background. The risk of these nuclides is studied intensely. Based on the data provided in [51,52], the risk of its use in human study with an appropriate concentration can be hundreds of times smaller than the risk from radon and its decay products in open atmosphere.

One very interesting example in using radon decay products presented in [53] was the deposition of unattached radon progeny (nanometer-sized radioactive aerosols). This was studied on volunteers, after ethical approval by the special medical commission in Switzerland, by measuring deposited nanometer particles of radioactive Pb and Bi.

7.10 Conclusion

The main physical (dosimetric) characteristics related to the biological effect of aerosols are:
For radioactive aerosols (attached vs. unattached):

1. Concentration (Bq/m^3)
2. Exposure (Bq/m^3 s)
   (a) Dose (absorbed dose) (J/kg).

For nonradioactive aerosols:

1. Particle concentration (particle size distribution, PSD) (m^{-3})
2. Surface area concentration (m^{-1})
3. Mass concentration (g/m^3)
4. Exposure (concentration multiplied by time)
5. Intake (number of particles, surface area, mass)
6. Dose (number of particles, surface area, mass).

The concepts of dose for radioactive and nonradioactive aerosols are different. In the case of alpha, beta or gamma aerosols, we are talking about absorbed energy in the lung tissue at the target cells.

In the case of nonradioactive aerosols, dose represents the number of particles per mm^2 of the lung surface, surface of the aerosol particle, or mass of the particles, depending on the range of sizes. Special attention should be given to nanometer-sized particles and high deposition areas in the lung. The main sources of uncertainty in the dose assessment are nonradioactive characteristics: space and time distribution of aerosol concentration and PSD, especially differences in the breathing and sampling site, changes in humidity and temperature inside the lungs, and unknown characteristics of biokinetic processes.

References

1. Deposition, Retention and Dosimetry of Inhaled Radioactive Substances, NCRP Report No. 125, 1997.
2. Ruzer, L.S., *Radioaktivnie Aerozoli*, Energoatomizdat, Moscow, Russia, 2001, (in Russian).
3. NCRP Report No. 78, 1984.
4. Ruzer, L.S., Radioactive Aerosols, Determination of the Absorbed Dose, Doctoral dissertation, Moscow, 1970, (in Russian).
5. Ruzer, L.S., Nero, A.V., and Harley, N.H., Assessment of lung deposition and breathing rate of underground miners in Tadjikistan, *Radia. Prot. Dosim.* 58, 1995.
6. Ruzer, L.S., and Sextro, R.G., ANRI, No. 3, pp. 18–20 (in Russian).
7. Ruzer, L.S. and Sextro, R.G. SRA/ISEA Meeting, New Orleans, December 1996.
8. Schulte, H.F. Personal Air Sampling and Multiple Stage Sampling, Los Alamos Scientific Laboratory, Los Alamos, New Mexico, U.S.A., Symposium "Radiation Dose Measurements," Paris, 1967.
9. Sherwood, R.J. and Greenhalgh, D.M.S., Personal air sampler, *Ann. Occup. Hyg.*, 2, 127–132, 1960.
10. Domanski, T., Kluszynski, D., Olszewski, J., and Chrusciewski, W., Field monitoring vs individual miner dosimetry of radon daughter products in mines, *Pol. J. Occup. Med.*, 2, 147–160, 1989.
11. National Academy of Sciences (NAS), Human Exposure Assessment for Airborne Pollutants: Advances and Opportunities, Committee on Advances in Assessing Human Exposure to Airborne Pollutants, National Research Council, Washington, DC, 1991.
12. Ott, W.R., Concepts of human exposure to air pollution, *Environ. Int.*, 7, 179–196, 1966.
13. Duan, N. and Ott, W.R., Comprehensive Definitions of Exposure and Dose to Environment Pollution, in Proceedings of the EPA/A&WMA Specialty Conference on Total Exposure Assessment Methodology, November 1989, Las Vegas, NV, Air and Waste Management Association, Pittsburg, PA, 1990.
14. EPA, Review of the National Ambient Air Quality Standards for Particular Matter: Policy Assessment of Scientific and Technical Information, OAQPS Staff Paper, EPA/452/R-96-013, Office of Air Quality Planning and Standards, Research Triangle Park, NC, 1996.
15. Phalen, R.F., Schum, G.M., and Oldham, M.J.,*J. Aerosol Med.*, 3, 271–282, 1990.
16. Noble, C.A., *et al.*, *Aerosol Sci. Technol.*, 34, 457–464, 2001.

17. Valberg, P.A. and Watson, A.Y., *Inhal. Toxicol.*, 10, 641–662, 1998.
18. Banse, D.F. *et al.*, Particle counting efficiency of TSI CPC 3762, *J. Aerosol Sci.*, 32, 157–161, 2001.
19. George, A.C., Measurement of the uncombined fraction of radon daughters with wire screens, *Health Phys.*, 23, 390–392, 1972.
20. Holub, R.F. and Knutson, E.O., Measuring polonium-218 diffusion coefficient spectra using multiple wire screens, in *Radon and its Decay Products; Occurrence, Properties and Health Effects*, Hopke, P.K., Ed., ACS Symposium Series 331, American Chemical Society, Washington, DC, 1987, pp. 340–356.
21. Hopke, P.K., The initial behavior of ^{218}Po in indoor air, *Environ. Int.*, 15, 288–308, 1989.
22. Kartashev, N.P., *Atomnaya Energ.*, 24, 144, 1966.
23. Polev, N.M., Candidat dissertation, VNIIFTRI, Moscow, 1967.
24. Knutson, E.O., Tu, K.W., Solomon, S.B., and Strong, L., *Radiat. Prot. Dosim.*, 24, 261–264, 1988.
25. George, A.C. and Hinchliffe, L., *Health Phys.*, 23, 791–803, 1972.
26. George, A.C., Hinchliffe, L., and Sladowski, R., *Am. Ind. Hyg. Assoc. J.*, 36, 484–490, 1975.
27. Cooper, L.A., Jackson, P.O., Langford, L.C., Petersen, M.R., and Stuart, B.O., *Characteristics of Attached Radon-222 Daughters Under Both Laboratory and Field Conditions*, Battelle Pacific Northwest Laboratories, Richland, WA, 1973.
28. Raghavayya, M. and Jones, J.H., *Health Phys.*, 26, 417–430, 1974.
29. Oberdorster, G. *et al.*, Association of particulate air pollution and acute mortality: involvement of ultrafine particles, *Inhal. Toxicol.*, 7, 111–124, 1995.
30. Chen, L.C. *et al.*, Concentration and mass concentration as determinants of biological response to inhaled irritant particles. *Inhal. Toxicol.*, 7, 577–588, 1995.
31. Schwartz, J., Particulate air pollution and daily mortality in Detroit, *Environ. Res.*, 64, 26–35, 1991.
32. Schwartz, J. *et al.*, Increased mortality in Philadelphia associated with daily air pollution concentrations, *Ann. Rev. Respir. Dis.*, 145, 600–604, 1992.
33. Schwartz, J. *et al.*, Particulate air pollution and daily mortality in Stubenville, Ohio. Am. J. Epidemiol., 135, 12–19, 1992.
34. Pope, C.A. *et al.*, Daily mortality and PM-10 pollution in Utah Valley, *Arch. Environ. Health*, 47, 211–217, 1992.
35. Báláshazy, I. and Hoffman, W., Quantification of local deposition pattern of inhaled radon decay products in human bronchial airway bifurcations, *Health Phys.*, 78, 147–158, 2000.
36. National Research Council, *Comparative Dosimetry of Radon in Mines and Homes*, National Academy Press, Washington, DC, 1991.
37. Harley, N.H. *et al.*, The variability in radon decay product dose, *Environ. Int.*, 22 (Suppl. 1), S959–S964, 1996.
38. Hofmann, W. *et al.*, Comparison of different modeling approaches in current lung dosimetry models, *Environ. Int.*, 22(Suppl. 1), S965–S976, 1996.
39. Harley, N.H. *et al.*, Alpha absorption measurements applied to lung dose from radon daughters. *Health Phys.*, 23, 771–782, 1972.
40. Caswell, R.S. *et al.*, Systematic alpha-particle energy spectra and linear energy spectra for radon daughters,. *Radiat. Prot. Dosim.*, 52, 377–380, 1994.
41. Martin, D. *et al.*, Diffusion deposition of small-size particles in the bronchial tree, *Health Phys.*, 23, 23–29, 1972.
42. Martonen, T.B. *et al.*, Cigarette smoke and lung cancer, *Health Phys.*, 52, 213–217, 1987.
43. Shimo, M. *et al.*, Deposition of unattached RaA atoms in the tracheobronchial region, in Indoor Radon and Lung Cancer: Reality or Myth, Part 1, Cross, F.T., Ed., Battelle Press, Richland, WA, 1993, pp. 201–210.
44. Kim, C.S. *et al.*, Deposition of Ultrafine Particles in the Bifurcating Airways Models, Fourth International Aerosol Conference, Vol. 2, 1994, pp. 888–889.
45. Kinsara, A.A. *et al.*, Deposition of molecular phase radon progeny in lung bifurcations, *Health Phys.*, 68, 371–382, 1995.
46. Martonen, T.B. *et al.*, Cigarette smoke and lung cancer, *Health Phys.*, 52, 213–217.
47. Brenner, D.J., Radon: current challenges in cellular radiobiology, *Int. J. Radiat. Biol.*, 61, 3–13, 1992.

48. BEIR VI. Health effect of exposure to Radon, National Research Council, Washington, DC, National Academy Press, 1999.
49. Preining, O., The physical nature of very, very small particles and its impact on their behaviour, *J. Aerosol Sci.*, 29(5/6), 481–495, 1998.
50. Nemmar, A. *et al.*, Brief Rapid Communication, 2002.
51. Cole, L.A., Element of Risk, The Politics of Radon, Oxford, Oxford Univ. Press, 1993.
52. United State Environment Protection Agency, Home Buyer's and Seller's to Radon, EPA 402-K-00-008, July 2000.
53. Butterweck, G. *et al.*, *Radiat. Prot. Dosim.*, 94, 247–250, 2001.

chapter eight

Modeling deposition of inhaled particles

Ted B. Martonen
U.S. EPA, National Health and Environmental Effects Laboratory and University of North Carolina at Chapel Hill

Jacky A. Rosati
U.S. EPA, National Risk Management Research Laboratory

Kristin K. Isaacs
University of North Carolina at Chapel Hill

Contents

8.1 Introduction 114
8.2 Fluid dynamics in airways 115
8.2.1 Fundamental equations 115
8.2.2 Boundary conditions 116
8.2.3 Idealized velocity profiles 116
8.2.4 Computational fluid dynamics 117
8.3 Aerosol deposition models 117
8.3.1 Classes of models 117
8.3.1.1 Empirical models 118
8.3.1.2 Deterministic models 118
8.3.1.3 Stochastic models 119
8.3.1.4 Computational fluid-particle dynamics 119
8.3.2 Merits and limitations of deposition models 120
8.3.2.1 Scientific foundations 121
8.3.2.2 Biological realism 121
8.3.2.3 Hardware and software issues 121
8.3.2.4 Advantages of modeling and simulation 122
8.4 Factors influencing aerosol deposition patterns 122
8.4.1 Respiratory system morphology 122
8.4.1.1 Idealized models 122
8.4.1.2 Data-driven models 123
8.4.1.3 Surface features 123
8.4.2 Ventilatory conditions 125
8.4.2.1 Mode of respiration 125
8.4.2.1.1 Effect of oral or nasal breathing on particle delivery to lungs 125
8.4.2.1.2 Effect of oral or nasal breathing on total particle deposition within the respiratory tract 125

1-56670-611-4 / 05 / $0.00+$1.50

8.4.2.2 Breathing pattern 125
8.4.2.2.1 Spontaneous breathing 126
8.4.2.2.2 Regulated breathing 126
8.4.3 Respiratory system environment 126
8.4.4 Clearance 128
8.4.4.1 Mucociliary clearance 128
8.4.4.1.1 Effect of drugs and inhaled contaminants on mucociliary clearance 128
8.4.4.1.2 Effect of disease on mucociliary clearance 129
8.4.4.1.3 Effect of age and activity on mucociliary clearance 130
8.4.4.2 Macrophage clearance or phagocytosis 130
8.4.4.2.1 Effect of particle size/fiber length and shape on macrophage clearance 130
8.4.4.2.2 Effect of drugs and inhaled contaminants on macrophage clearance 130
8.4.4.3 Free particle uptake and translocation to the interstitium 131
8.4.4.4 Importance of clearance in particle deposition modeling 131
8.4.5 Disease 131
8.4.5.1 Chronic obstructive pulmonary disease 131
8.4.5.2 Asthma 132
8.4.5.3 Cystic fibrosis 132
8.4.5.4 Effect of obstructive disease on particle deposition and distribution 132
8.4.5.5 Modeling disease 132
8.4.6 Age 133
8.5 Theory and experiment 133
8.5.1 Predictions of particle deposition 133
8.5.2 Particle deposition measurements 134
8.5.2.1 Casts and models 134
8.5.2.2 Deposition patterns deduced from clearance studies 135
8.5.2.3 Light-scattering methods 135
8.5.2.4 Gamma scintigraphy 135
8.5.2.5 Microdosimetry 136
8.5.3 Comparison of modeling and data 139
8.5.3.1 Simulations of compartmental particle deposition 140
8.5.3.1.1 Extrathoracic 140
8.5.3.1.2 Tracheobronchial 140
8.5.3.1.3 Pulmonary 142
8.5.3.2 Simulations of particle distribution generation by generation 143
8.5.3.3 Simulations of local particle deposition 144
8.6 Summary 145
Disclaimer 145
Acknowledgment 145
References 145

8.1 Introduction

The mathematical modeling of the deposition and distribution of inhaled aerosols within human lungs is an invaluable tool in predicting both the health risks associated with inhaled environmental aerosols and the therapeutic dose delivered by inhaled pharmacological drugs. However, mathematical modeling of aerosol deposition requires a knowledge of the intricate geometry of the respiratory network and the resulting complex motion of air and particles within the airways. In this chapter, an overview of the basic engineering theory and respiratory morphology required for deposition modeling is covered. Furthermore, current deposition modeling approaches are reviewed, and many factors affecting deposition are discussed. Experimental methods for measuring lung deposition are presented, albeit briefly, and the comparison between experimental results and modeling predictions is examined for a selection of modeling efforts.

8.2 Fluid dynamics in airways

The deposition patterns associated with inhaled particulate matter are intrinsically linked to the airflow patterns within the respiratory system. Therefore, any effort to realistically model particle deposition requires an understanding of the fundamental fluid dynamics theory behind the motion of air in the extrathoracic and lung airways.

8.2.1 Fundamental equations

Fluid motion is governed by the conservation of mass (continuity) equation and the conservation of momentum (Navier–Stokes) equation. The flow of air in the respiratory airways is usually assumed to be incompressible.[1] For incompressible flow, the continuity equation is given by

$$\nabla \cdot V = 0 \tag{8.1}$$

and the Navier–Stokes equation is

$$\rho\left[\frac{\partial V}{\partial t} + (V \cdot \nabla)V\right] = \rho f - \nabla p + \mu \nabla^2 V \tag{8.2}$$

where ∇ and ∇^2 are the gradient and Laplacian operators, respectively (defined below), ρ is the fluid density, μ is the absolute fluid viscosity, and p is the hydrodynamic pressure. The parameter f is any externally applied volumetric force, such as gravity.

In studying fluid flow in airways, it is convenient to use the cylindrical coordinate representation of the motion equations. Noting that the gradient operator ∇ in cylindrical coordinates is

$$\frac{\partial}{\partial r} + \frac{1}{r}\frac{\partial}{\partial \theta_\theta} + \frac{\partial}{\partial z}$$

the continuity equation in cylindrical coordinates becomes

$$\frac{1}{r}\frac{\partial}{\partial r}(rV_r) + \frac{1}{r}\frac{\partial}{\partial \theta}V_\theta + \frac{\partial}{\partial z}V_z = 0 \tag{8.3}$$

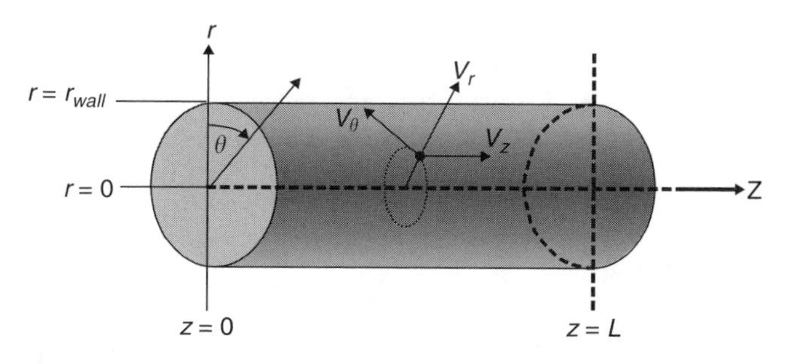

Figure 8.1 Cylindrical coordinate system for an arbitrary airway and corresponding velocity components at an arbitary point.

where V_r, V_θ, and V_z are the components of fluid velocity in the radial (r), circumferential (θ), and axial (z) directions, respectively (Figure 8.1). The corresponding equations for momentum in the r, θ, and z directions become

$$\frac{\partial V_r}{\partial t}+(V\cdot\nabla)V_r-\frac{1}{r}V_\theta^2=-\frac{1}{\rho}\frac{\partial p}{\partial r}+f_r+\frac{\mu}{\rho}\left(\nabla^2 V_r-\frac{V_r}{r^2}-\frac{2}{r^2}\frac{\partial V_\theta}{\partial \theta}\right) \tag{8.4}$$

$$\frac{\partial V_\theta}{\partial t}+(V\cdot\nabla)V_\theta+\frac{V_r V_\theta}{r}=-\frac{1}{\rho r}\frac{\partial p}{\partial \theta}+f_\theta+\frac{\mu}{\rho}\left(\nabla^2 V_\theta+\frac{V_\theta}{r^2}+\frac{2}{r^2}\frac{\partial V_r}{\partial \theta}\right) \tag{8.5}$$

$$\frac{\partial V_z}{\partial t}+(V\cdot\nabla)V_z=-\frac{1}{\rho}\frac{\partial p}{\partial z}+f_z+\frac{\mu}{\rho}\nabla^2 V_z \tag{8.6}$$

where

$$V\cdot\nabla=V_r\frac{\partial}{\partial r}+\frac{1}{r}V_\theta\frac{\partial}{\partial \theta}+V_z\frac{\partial}{\partial z} \tag{8.7}$$

and the Laplacian operator in cylindrical coordinates is defined as

$$\nabla^2=\frac{1}{r}\frac{\partial}{\partial r}\left(r\frac{\partial}{\partial r}\right)+\frac{1}{r^2}\frac{\partial^2}{\partial \theta^2}+\frac{\partial^2}{\partial z^2} \tag{8.8}$$

If the flow is steady, then the time derivatives in Equations (8.4)–(8.6) can be ignored.

8.2.2 Boundary conditions

The above system of four nonlinear partial differential equations can only be solved analytically when a number of assumptions are made for very simple flow geometries. Therefore, in most cases, numerical methods are required to determine a solution for the velocity and pressure fields. Numerical solution of the equations of motion requires a knowledge about the velocity or pressure at some or all boundaries of the flow geometry. The nature of the flow characteristics and geometry determine which boundary conditions need defining.

As flow has been studied in a vast variety of geometric lung models, it is difficult to recommend boundary conditions that are appropriate in every circumstance. In almost all cases, however, it is assumed that the no-slip condition (V at the wall $= 0$) exists along the walls of any airway. The definition of any inlet and outlet boundary conditions is not as straightforward. For instance, the flow velocity at the inlet may be designated as steady (time-invariant) or unsteady, and may be uniform over the diameter of the inlet or vary with radial direction. In addition, in different cases, the velocity, velocity gradient, pressure, or pressure gradient may be defined. In airway flow modeling, the velocity profile at the exit of an airway is often assumed to be parabolic, and the conducting airway is assumed to be sufficiently long enough for the flow to be fully developed.[2]

In some simplified cases, only the inlet conditions may need to be specified. For example, if one assumes that the pressure in the fluid varies only in the axial direction, then the mathematical character of the equations changes and only the inlet velocity profile is needed to obtain a solution.[3] In this case, the outlet velocity profile is determined as part of the model solution.

8.2.3 Idealized velocity profiles

The velocity profiles present in real lung airways are determined by a number of factors, including ventilatory conditions, lung morphology, and airway generation(s). Experimental studies[4–6] have shown that these profiles may be skewed and, in general, geometrically complex. However, by assuming idealized velocity profiles in individual airway generations, fundamental equations of fluid motion can be reduced and the development of simplified expressions for deposition via different mechanisms can be obtained. Idealized velocity profile types include laminar fully developed (parabolic), laminar undeveloped (plug), and turbulent. The type of velocity profile selected determines the appropriate expressions for particle deposition (e.g., via sedimentation, diffusion, or inertial impaction) in a given airway. Examples of such expressions are presented in detail in Chapter 6.

Martonen[7,8] presented recommendations for velocity profiles in different generations of lung airways, based on previously published experimental measurements. At the entrance to the trachea, it was assumed that the velocity profile was determined by the action of the laryngeal jet. In the upper tracheobronchial (TB) airways, the flow was considered turbulent, while in the lower TB airways the flow was assumed to be laminar undeveloped. In the pulmonary region, both laminar undeveloped and developed profiles were considered to serve as limiting cases, since the actual velocity profiles in this region are likely to lie in the transitional region between plug and parabolic flow. These flow conditions have been used in several particle deposition model studies.[2] A more recent modeling study[9] provided further evidence that the flow in the TB airways is not fully developed. Specifically, it was determined that an undeveloped flow model more accurately predicted total deposition of ultrafine particles than did a fully developed flow model.

8.2.4 Computational fluid dynamics

The field of study concerning the numerical solution of the equations of motion for fluids is known as *computational fluid dynamics* (CFD). CFD is an invaluable tool in the study of fluid and particle motion under circumstances that are difficult to simulate with a physical experiment or a simplified theoretical model. CFD involves segmenting a flow area (e.g., a system of airways and bifurcations) into many discrete elements or volumes, collectively called a *mesh*. In each element, the partial differential equations that describe the fluid motion are converted into algebraic equations by relating the fluid properties of the elements to the properties in adjacent elements. There exist several schemes for performing this *discretization*, including finite difference, finite element, and finite volume methodologies. In the finite difference method, the differential equations are approximated by Taylor series expansions. The theories behind finite element and finite volume discretizations are beyond the scope of this chapter.

Many commercial packages are available for performing CFD, including the finite-element-based FIDAP (Fluent, Inc.) and the finite-volume-based programs CFX-F3D (AFA Technology) and FLUENT (Fluent Inc.). FIDAP has been used to study respiratory system airflow in studies by Martonen et al.[10–15] and Kimbell et al.,[16,17] while CFX-F3D has been used by Yu et al.[18] and Martonen et al.[19] An example of a CFD simulation of the velocity patterns in an airway after eight bifurcations is shown in Figure 8.2.

8.3 Aerosol deposition models

8.3.1 Classes of models

The modeling of aerosol deposition in the lungs has been approached from several different conceptual directions. However, the vast majority of aerosol deposition models can be categorized as empirical, deterministic, or stochastic in nature. Empirical models are based on fitting numerical relationships to experimental data. In deterministic modeling, both the physical nature of the lung and the fluid and particle dynamics associated with respiration are quantified by simplified expressions, and the resulting particle motion is

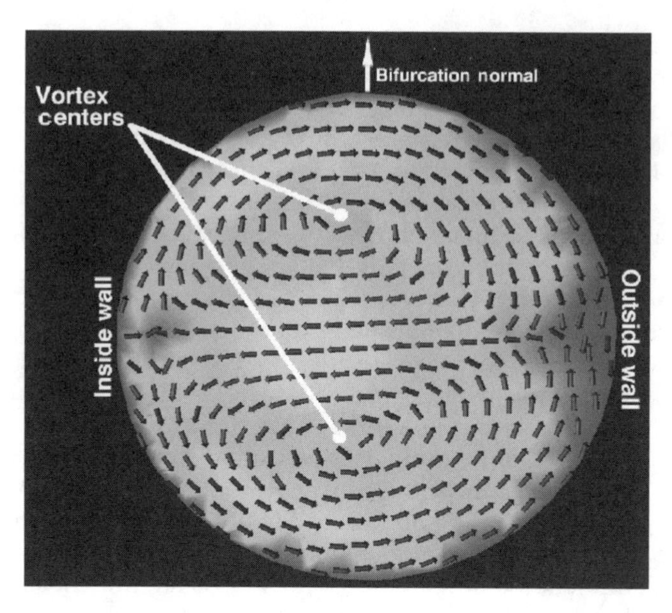

Figure 8.2 Image of a CFD simulation of secondary velocity currents in the eighth generation of a bifurcating airway, using FLUENT 6.0.

calculated. In stochastic modeling, the geometry of the airways is varied randomly to account for inter- and intra-subject variation.

Empirical and deterministic models were the first efforts designed to simulate and predict aerosol deposition in airways, and stochastic models were later developed to intrinsically incorporate biological variability into such simulations. However, as computer resources became more powerful, computational fluid particle dynamics (CFPD) models of particle motion and deposition (based on CFD analyses) began to be developed. While CFPD models do not comprise a distinct class by themselves (i.e., they may be either deterministic or stochastic in nature), they are also discussed below.

8.3.1.1 Empirical models

In empirical deposition models, regional or total aerosol deposition is described by equations derived by fitting algebraic relationships to experimental data. Stalhofen et al.[20] developed a semiempirical model for total and regional deposition, deriving equations for aerosol deposition in the extrathoracic (nasal, oral, and laryngeal), "fast-cleared" thoracic (ciliated airways), and "slow-cleared" thoracic (peripheral) regions as a function of respiratory parameters and aerodynamic diameter. The model, which was based on previous empirical models,[21,22] considered data from a wide variety of deposition experiments; theoretical relationships were used for particle sizes for which no data were available. Other empirical deposition models have been developed for the head,[23,24] tracheobronchial,[23–25] and alveolar[26] regions.

Empirical models have been combined with other types of models,[27] and modeling predictions derived from empirical relationships have been used to validate and confirm the results from more advanced deposition models.[9,28]

8.3.1.2 Deterministic models

Deterministic models are developed using an engineering approach to the simulation of air and particle motion. In deterministic models, simplifying assumptions about airway geometries and airflow conditions are made in order to derive expressions for particle trajectories from particle momentum equations. Such models vary in complexity; deterministic modeling efforts may range from simple analytical expressions that can be solved algebraically to systems of nonlinear ordinary or partial differential equations. In addition, deterministic models may describe particle deposition in a single airway, a bifurcation, or a complete branching network of respiratory airways.

Martonen[7,29] has developed deterministic models of particle deposition in human lungs. These models were formulated by modeling the airways of the lung as either straight or curved smooth-walled tubes, and assuming a fixed (laminar or turbulent) velocity profile in each airway generation. Other deterministic models of particle deposition in the respiratory system have been developed by Gradón and Orlicki,[30] Yu,[31] Yu and Diu,[32] Egan and Nixon,[33] Anjilvel and Asgharian,[34] and Phalen et al.[35,36]

In deterministic models, the simulated particle deposition patterns are determined solely by the input parameters to the model. Therefore, for any set of model input parameters, the same deposition pattern is found. This is not necessarily true of the next class of models—stochastic models.

8.3.1.3 Stochastic models

Models of particle deposition are categorized as stochastic if the morphological description of the lung is considered to vary in a random manner, within prescribed limits. The concept of stochastic deposition modeling, first introduced by Koblinger and Hofmann,[37] has been used by Hofmann and co-workers to simulate aerosol particle deposition in both human[37,38] and rat[39–41] lungs. Both radon progeny[42] and cigarette smoke[43] deposition have been specifically addressed.

In stochastic deposition modeling, the morphometric parameters describing the geometry of the lung are not given constant values, but are described instead by statistical (e.g., lognormal) distributions, which are in turn based on experimental measurements. These morphometric parameters may include airway diameters, lengths, and branching angles. Other, less obvious parameters (such as ratios of parent airway cross-sectional areas to the sum of daughter airway cross-sectional areas) may also be stochastically defined. Then, as each modeled particle enters the lungs, its pathway is determined by a random selection of values for each of the required morphometric parameters within their corresponding lognormal distribution. For example, the properties of the daughter airways are randomly assigned for each bifurcation.[44] The average resulting deposition in each airway is then calculated from the behavior of the entire ensemble of particles.

8.3.1.4 Computational fluid-particle dynamics

CFPD refers to the study of the motion of particles as determined by CFD simulations. In CFPD studies of particle deposition, CFD solutions of fluid velocities are coupled with the solution of particle trajectory equations developed from Newton's Second Law.[45] Particles are deposited when their trajectory intersects with an airway wall.

In recent years, CFPD methods have been used to examine the effects of complex flow patterns on particle motion and deposition in the respiratory system. Martonen and Guan[46,47] performed linked fluid dynamics and particle motion studies that examined the effects of tumors on the deposition of particles in an idealized two-dimensional (2D) airway bifurcation. Three-dimensional (3D) CFPD studies of the motion of particles in the extrathoracic and large tracheobronchial airways have also been performed.[19] Zhang et al.[48–50] and Comer et al.[45,51] have performed a series of CFPD studies aimed at predicting particle deposition in airway bifurcations networks. However, as noted by Martonen,[52] caution should be exercised when interpreting the aforementioned studies, as they did not take natural respiratory features such as the larynx, the cartilaginous rings, or carinal ridges into account. CFPD simulations of particle deposition have also been performed for the oral cavity.[53]

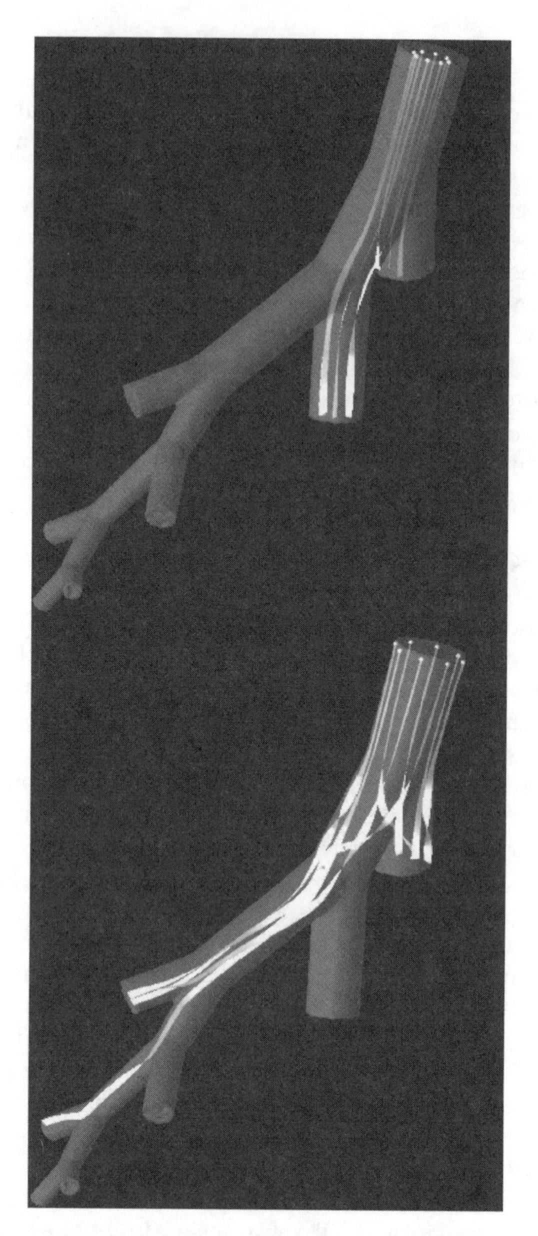

Figure 8.3 CFPD simulations of particle trajectories in a bifurcating airway using FLUENT 6.0. The flow rate is 120 l/min, and the two panels show trajectories for two different sets of initial particle locations.

An example of a CFPD simulation in a multiply bifurcating airway (performed in FLUENT) is shown in Figure 8.3. In this uncoupled case, the flow in the airway was first simulated, and then the trajectories of a group of particles of unit density were calculated.

8.3.2 Merits and limitations of deposition models

Each deposition model class has both scientific strengths and drawbacks. It is incumbent upon the researcher to choose an appropriate modeling technique for his or her needs. In selecting a suitable deposition model, one must consider many factors, including the scientific foundations of the models, the desired level of biological realism, and the available computer-related resources.

8.3.2.1 *Scientific foundations*

The scientific foundations of the different classes of models must be considered when selecting a model for the interpretation of experimental data or the prediction of particle deposition pattern. Empirical models may be valid for the physical system for which they were derived, but application to other systems may result in spurious conclusions. In addition, deterministic models are only as valid as the assumptions made in their derivation. One must be careful to be aware of these assumptions and use the models accordingly. Stochastic models are derived from observations of morphological variability; however, such formulations can be generated only if an appropriate amount of experimental data is available. One must be mindful of the limitations of stochastic models derived from limited experimental data. CFPD modeling is based on well-established methods that have been in use for years in a wide variety of scientific fields, especially in mechanical and aerospace engineering. However, the scientific validity of any CFPD simulation is based on the adequate application of these methods to the problem at hand. For example, the geometry of realistic respiratory system passages is more irregular than that in many industrial or engineering applications. Appropriate discretization of the complex geometry (for solution of the Navier–Stokes equations) may be very difficult, and an invalid mesh may result in specious computational results. However, more robust discretization (meshing) algorithms are continuously being developed and tested.

8.3.2.2 *Biological realism*

Particle deposition models vary greatly in their level of biological realism. For example, empirical models contain no information about particle motion or the physiology or anatomy of the respiratory system (i.e., the respiratory system is treated as a "black box"), yet they may be useful in interpreting data from experiments in which subjects breathe well-characterized aerosols in a known manner. Many deterministic models take into account both respiratory system morphology and the motion of individual particles. They are therefore able to model deposition in different respiratory system regions (extrathoracic, tracheobronchial, or pulmonary), in individual lung airway generations, or in individual airways.

Stochastic models may present a limited "anatomically realistic" model *per se* (i.e., surface features not considered); however, they have the important advantage of being able to model the realistic biological variability that is present among each lung pathway in a single subject and between individual subjects.[54] Stochastic models by their construction provide estimations of intra- and inter-subject variability in deposition.[54]

CFPD models are capable of simulating deposition in realistic airway configurations, considering flow conditions that arise from airway surface features. Therefore, CFPD models can predict *local* deposition (i.e., at cells, bifurcations, and carinal rings) caused by secondary flow currents, which cannot be predicted from simplified analytical models. CFPD can be used to predict deposition in complex anatomical geometries where the assumption of a smooth-walled cylinder should not be made, for example, in the larynx, mouth, or nasal passages. A particular advantage of CFPD models is that they provide the potential for coupling imaging studies with deposition modeling. Irregular respiratory system morphologies can be extracted from CT or MRI images, and these morphologies can provide a basis for CFPD studies.

8.3.2.3 *Hardware and software issues*

Deposition models of different classes require vastly different hardware and software resources. For example, empirical modeling requires no specialized computer programs, as the models are simple algebraic relationships. Deterministic models (those derived from particle motion and flow equations using various simplifying assumptions) vary greatly in computational efficiency, based on the nature of the assumptions made and the complexity of the airway geometry considered. However, CFPD modeling is almost always computationally intensive, requiring high-powered hardware with adequate processing and memory resources. In general, CFPD also requires either expensive third-party software or a large amount of complicated, challenging, in-house programming. As an example, Zhang and Kleinstreuer[55] used the CFD package CFX 4.3 (AEA Technology Inc.) to perform their studies of secondary flow patterns in an airway branching network. The software ran on a multiprocessor Silicon Graphics workstation, and the typical run time for a fluid flow and particle transport simulation for a single breathing cycle was approximately 72h. However, the use of CFPD models is becoming more frequent as the computing power of desktop personal computers and workstations increases.

8.3.2.4 Advantages of modeling and simulation

Up to this point we have focused on the merits and limitations of particular types of deposition models; we would also like to comment on the advantages and challenges of using modeling methods in the study of inhaled particle deposition. Modeling can be a powerful research tool, as it can be used to predict behaviors, phenomena, or physiological parameters that cannot be measured. In addition, modeling has the potential to help maximize both financial and animal resources, by aiding in the design of appropriate experiments for a given scientific hypothesis. As noted by Martonen,[56] modeling studies should be integrated in a complementary manner with human inhalation exposure studies.

One of the main challenges in simulation is that its valid and rigorous use may call for uniquely trained scientific personnel. For example, performing CFPD may call for appropriately trained interdisciplinary scientists who are capable of understanding the computational, mathematical, and physiological nuances of modeling complex biological systems. Therefore, rigorous modeling studies may require collaboration among physicians, toxicologists, and engineers.

8.4 Factors influencing aerosol deposition patterns

In the previous section, a brief overview of the different classes of lung deposition models was presented. Now, we will discuss some of the morphological, ventilatory, and situational factors that affect deposition patterns in the human respiratory system. These are factors that may be considered in both deterministic and stochastic models of particle deposition and distribution.

8.4.1 Respiratory system morphology

Any modeling of the deposition of aerosols in the human respiratory system requires a description of the morphology of the airway(s) being studied. Both the overall branching structure of the airway tree and dimensions (e.g., diameters and lengths) of individual airways must be considered. Both idealized morphology models and models based on specific experimental observations have been used in particle deposition modeling.

8.4.1.1 Idealized models

Many morphology models of the respiratory system have been derived from experimentally obtained morphometric data. Early morphometric models were simplified to provide idealized representations of the branching network of the human lung. The most widely used idealized model is the symmetric model of Weibel.[57,58] In his model, the lung is characterized as a symmetric and dichotomously branching tree of tubular airways. Idealized asymmetrical models of the airway tree have also been developed. These include Weibel's asymmetric "B" model[58] and the models of Horsfield[59] and Horsfield and Cumming.[60,61]

8.4.1.2 Data-driven models

Other more biologically realistic morphological models have also been derived from experimentally obtained morphometric data. Soong et al.[62] presented a stochastic model of the human tracheobronchial tree, using Weibel's symmetric model as the underlying average model, and incorporating probability distributions of lengths and airways based on several experimental studies (e.g., Jesseph and Merendino[63] and Parker et al.[64]). Horsfield et al.[65] developed a theoretical model of the bronchial tree based on measurements from human casts. Kitaoka et al.[66] also presented a model of the 3D branching structure of the lung based on morphometric studies. Their model was derived using an algorithm guided by several morphology- or physiology-based "rules," such as prescribed relationships between airway length and diameter and between airway diameter and flow rate. More recently, Spencer et al.[67] have derived a morphological model of the lung airways using data-driven surface modeling techniques. In this model, anatomical data were used to define airway lengths, diameters, and orientation angles. The surfaces of the resulting airway network were then realized using advanced graphics rendering techniques. Specifically, nonuniform rational B-splines (NURBS) were used to model smooth airway connections and realistic lung surface features.

Data-driven models are especially useful in modeling the extrathoracic airways, as these passages are not easily modeled as idealized tubes. Figure 8.4 depicts a geometric model of the morphology of the human nasal passages derived from MRI images of an adult male. The irregular, tortuous shape of the nasal passages (as seen in the cross-sections) will result in a distinct particle deposition pattern that cannot be predicted using simplified geometries.

8.4.1.3 Surface features

There is much evidence to suggest that surface features in the lungs should be considered when modeling particle deposition. Both the cartilaginous rings (which are a pronounced

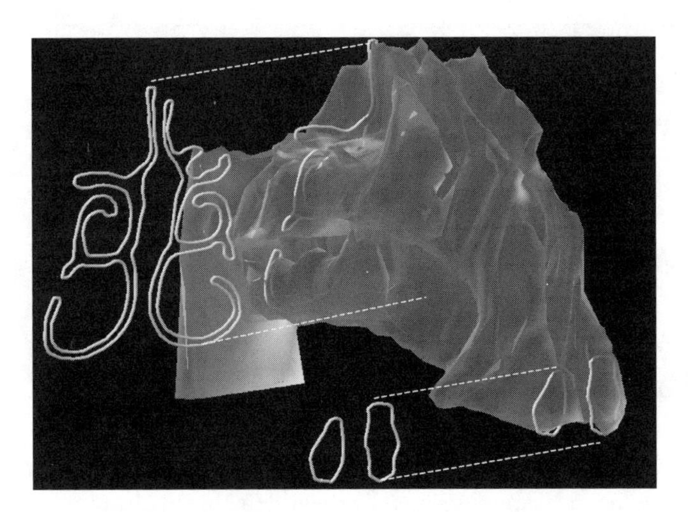

Figure 8.4 Morphological model of the human nasal passages derived from MRI imaging data. Note the complex cross-section of the nasal passage.

anatomical feature of the tracheobronchial airways) and the carinal ridges (which are situated at airway bifurcations) have been studied. Bronchoscopy images depicting these surface features are shown in Figure 8.5.

Cartilaginous rings affect airflow patterns in the large airways.[12,68] Using CFD modeling, Musante and Martonen[68] demonstrated that small eddies, produced between the rings, may increase localized particle deposition. They also predicted that flow instabilities produced by the rings could affect deposition in locations downstream from the rings themselves. In addition, errors in large airway deposition of up to 35% were possible if the rings were ignored.[69]

Martonen et al.[10] demonstrated via CFD modeling that localized deposition of particles at the carinal ridges could be explained by localized flow instabilities arising from the ridge geometry. They also predicted that the ridges could initiate flow effects that propagate to later generations, especially at high inspiratory flow rates. Using a numerical model, Balásházy and Hofmann[70] quantified deposition at carinal ridge sites for radon progeny of different sizes. They predicted that cells located at the ridge sites experienced deposition 20–115 times greater (depending on particle size) than the average airway dose.

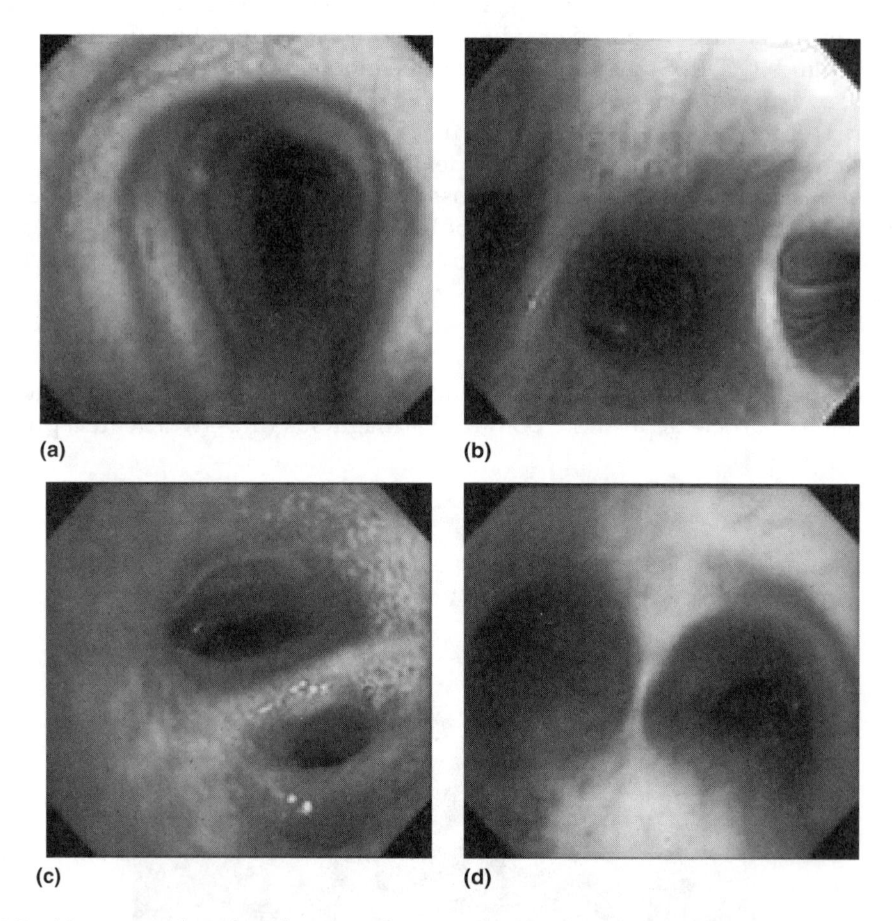

Figure 8.5 Airways and surface features photographed with videobronchoscopy. (**a**) Trachea and the main bronchi with the cartilaginous rings clearly visible. (**b**) Quadruple bifurcation, indicative of the complex branching pattern of the lung. (**c**) Blunt carinal ridges. (**d**) Sharp carinal ridges. University of Iowa, 1992–2003, reprinted with permission.

8.4.2 Ventilatory conditions

Ventilatory conditions have a distinct effect on aerosol deposition and distribution. Both the mode of respiration and breathing pattern must be considered when modeling particle deposition and clearance in the human lung.

8.4.2.1 Mode of respiration

Humans have the ability to breathe either nasally or orally. In contrast, rodents such as mice and rats are obligate nasal breathers. Thus, extrapolation of particle deposition data from these animals to human beings is difficult.

When sedentary, humans breathe through their nose, efficiently heating and humidifying the inhaled air. However, during exertion such as exercise, they switch over to oronasal breathing, or breathing through both mouth and nose. This switch is thought to occur when the breathing rate becomes so great that overcoming the relatively high pressure drop and resistance in the nasal passages is no longer an efficient means of respiration. In healthy adults, the switching point between nasal and oronasal breathing is thought to be approximately 35 l/min.[71,72] In children, however, this switching point is more variable.[73] Gender does not seem to play a role in switching point determination.[71]

8.4.2.1.1 Effect of oral or nasal breathing on particle delivery to lungs The route of breathing influences the quantity of inhaled contaminants or therapeutics delivered to human lungs. Particle penetration to the human lung is lower during nasal breathing (vs. oral breathing)[74] due to higher deposition efficiency in the nasal region, and there is thus more effective filtering of inhaled particles.[75,76] The higher deposition efficiency of nasal breathing is due to increased particulate matter removal by nasal hairs, impaction on pathway walls, and diffusion. Thus, it is less efficient to administer therapeutic aerosols to the human lung via the nose as opposed to the mouth.

8.4.2.1.2 Effect of oral or nasal breathing on total particle deposition within the respiratory tract Due to efficient particle removal by the nose,[77–79]total respiratory tract deposition is higher for nasal breathing than for oral breathing in nonsmokers.[43,80] As stated, the higher deposition efficiency in the nose vs. the mouth is due to increased particle removal by nasal hairs and inertial impaction. However, particulate matter removal in the nasal passages (and thus total deposition during nasal breathing) is highly dependent on particle size and inhalatory volume and flow, [76,78,79,81] as well as nasal passage morphology and development.[77,82,83]

8.4.2.2 Breathing pattern

Breathing pattern is an important factor in the respiratory deposition of therapeutic particles, as it may affect treatment efficiency.[84] Both spontaneous and regulated breathing patterns (Figure 8.6) have been used in human inhalation studies of particle deposition and clearance. The type of breathing pattern will affect the amount and pattern of particle deposition in the human lung. Studies comparing deposition data for spontaneous breathing patterns have been performed. [20,84,85–87]

Breathing pattern is typically described in terms of tidal volume (volume of air inhaled) and flow rate. In general, larger tidal volumes result in higher particle deposition in the human lung as particle-laden air penetrates deeper into the lung. Lower flow rates also result in higher particle deposition in the peripheral lung as velocities are slower and particles have more time to deposit by sedimentation or diffusion (see Chapter 6 for a discussion on deposition mechanisms).

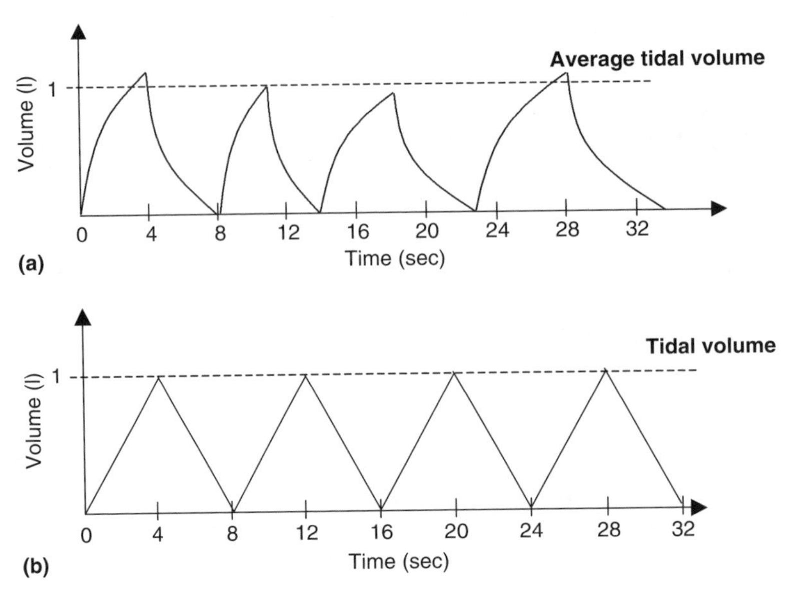

Figure 8.6 Spontaneous (**a**) and controlled (**b**) breathing patterns. Note that in spontaneous breathing, rate and tidal volume may vary, while these are constant in controlled breathing.

8.4.2.2.1 Spontaneous breathing Spontaneous breathing is unprescribed, with subjects breathing at their own pace, with only approximate tidal volume or flow requirements. Figure 8.6a illustrates the relationship between tidal volume and time for a typical spontaneous breathing pattern. Chan and Lippman[24] and Brown et al.[88] have used spontaneous breathing patterns in the study of the deposition of inhaled particles. Other studies have shown that spontaneous breathing of human subjects resulted in large inter-subject variability in the fractional deposition of inhaled particles.[85,89,90] Furthermore, Bennett and Smaldone[86] determined that it was differences in spontaneous breathing patterns (and not peripheral air space size or morphology) that influenced inter-subject variation in peripheral deposition.

8.4.2.2.2 Regulated breathing Regulated or academic breathing occurs when human subjects are required to follow a specified breathing pattern that varies from their own natural breathing rate. In this pattern, illustrated in Figure 8.6b, a constant flow rate is maintained. Regulated breathing has been used to attempt to mitigate the inter-subject variability introduced by spontaneous breathing.[86] Studies by Stahlhofen et al.,[91] Heyder et al.,[92] and Svartengren et al.[93] utilized regulated breathing in the determination of particle deposition in the human respiratory tract.

8.4.3 Respiratory system environment

Temperature and relative humidity (RH) in the human respiratory system vary with the mode of respiration and anatomical location. Table 8.1 provides temperature and humidity data[94–103] for different anatomical locations for both oral and nasal respiration. In general, a temperature of 37°C and an RH of 99.5% may be assumed for nasal respiration.[104] For oral respiration, 37°C and 90% RH may be assumed for air entering the trachea, with RH increasing by 1% per airway generation until it reaches 99.5% at the tenth generation.[104]

RH and temperature affect the growth of hygroscopic particles in the human lung. Hygroscopic growth occurs when the absorption of water from a humid environment

Table 8.1 Temperatures and Relative Humidities (RH) in the Human Respiratory Tract, for both Inspiration and Expiration

Anatomical Location	Nasal				Oral				References
	T_{insp} (°C)	T_{exp} (°C)	RH_{insp}	RH_{exp}	T_{insp} (°C)	T_{exp} (°C)	RH_{insp}	RH_{exp}	
Nasal Passages									
Distance from Nostril (cm)									
1.5	28.9 ± 2.3	—	69 ± 6.5	—	—	—	—	—	Keck et al.[94]
2.5	30.3 ± 1.6	—	78.7 ± 7.2	—	—	—	—	—	Keck et al.[94]
6.0	32.6 ± 1.5	—	90.3 ± 5.3	—	—	—	—	—	Keck et al.[94]
Laryngeal Cavity	32.3 ±0.8	36.4 ±0.2	98 –99	98-99	30.6 ±0.8	36.2 ±0.3	90	99	Ingelstedt[95]
Airway Generation, *i*									
i=0 (trachea)	34 – 35	36 – 37	—	—	33 - 34	36 – 37	—	—	Cole[96]
i=0 (trachea)	35.3	35.7	98	99.9	—	—	—	—	Perwitzschky[97]
i=0 (trachea)	35.4	36.2	—	—	34.5	35.8	—	—	Verzar et al.[98]
i=0 (trachea)	32.6	35.3	—	—	32.9	34.4	—	—	Herlitzka[99]
i=0 (trachea)	—	—	—	—	26.7	—	82.7	—	Dery et al.[100]
i=0 (trachea)	—	—	—	—	31.4 – 31.9	33.2 – 33.4	73.6 – 80.4	—	Dery[101]
i=0 (trachea)	—	—	—	—	31.2	32.6	—	—	McFadden et al.[102]
i=0 (trachea)	—	—	—	—	32	33	—	—	McFadden et al.[102]
i=0 (trachea)	—	—	—	—	32.2	33.4	—	—	Dery et al.[100]
i=0 (trachea)	36	36.2	98.3	99.2	—	—	—	—	McRae et al.[103]
i=1 (main)	—	—	—	—	30.6	—	85.8	—	Dery[101]
i=1 (main)	—	—	—	—	32.2	33.7	87	—	McFadden et al.[102]
i=2 (lobar)	—	—	—	—	33	34	—	—	McFadden et al.[102]
i=3 (segmental)	—	—	—	—	33.1	—	91.3	—	Dery et al.[100]
i=4 (subsegmental)	—	—	—	—	33.9	—	94.6	—	McFadden et al.[102]
i=4–5	—	—	—	—	33.9	35	—	—	Dery[101]
i=10–11	—	—	—	—	34.6	36	—	—	McFadden et al.[102]

Source: Adapted from Martonen, T.B., Hoffman, W., Eisner, A.D., and Ménache, M.G., The role of particle hygroscopicity in aerosol therapy and inhalation toxicology, *Extrapolation of Dosimetric Relationships for Inhaled Particles and Gases*, pp. 303–316, Copyright 1989, with permission from Elsevier.

causes changes in particle diameter and density. Since RH and temperature vary throughout the human lung, a particle's size and density may change while traveling through the respiratory system. Thus, a measured size distribution for hygroscopic particles is not likely to reflect actual particle sizes *in vivo*. A more extensive discussion of hygroscopicity and its effect on particle deposition in the human lungs may be found in Chapter 6.

Hygroscopic growth has been widely observed in many environmental and pharmaceutical aerosols[105–116] (Table 8.2). Therefore, it is often desirable to account for hygroscopic growth in particle deposition modeling. Deposition modeling studies[7,29] have accounted for hygroscopicity by incorporating the growth rate of aerosol particles as a function of residence time in the lung. Such experimental growth rate measurements for different aerosols have been presented.[117,118] The growth of aerosol particles has also been computationally predicted.[119–122]

8.4.4 Clearance

Inhaled foreign material is continually cleared from the respiratory tract. From a simplified perspective, inhaled insoluble particles are cleared from the human lung in two phases, mucociliary (or fast phase) clearance and phagocytosis (or slow phase) clearance. In addition, free particles may translocate out of the alveolar region of the lung into the lymphatic system or the lung interstitium. Depending on their lipophilicity, hydrophilicity, and/or size, soluble particles may be dissolved prior to physical clearance. Other particles, such as asbestos fibers and other biopersistent fibrous minerals, are often unable to be cleared from the lung; their retention may result in inflammation, tissue damage, and eventual disease. Oberdörster[123] provides a review of the clearance of both soluble and insoluble particles. Because the toxicity of a substance may be related to its time of residence in the respiratory system, clearance is an importance consideration in the risk assessment of inhaled particles. Thus, we will present a brief overview of its mediation by drugs, inhaled contaminants, age, and activity.

8.4.4.1 Mucociliary clearance

Mucociliary (or fast phase) clearance occurs in the tracheobronchial airways of the lung. Mucus is secreted by mucous glands in the bronchial walls and by goblet cells in the bronchial epithelium. This mucus is propelled by millions of cilia (collectively referred to as the mucociliary escalator) towards the pharynx, in the process transporting particles out of the conducting airways. At the pharynx, the mucus and particles are swallowed. The velocity of the mucus varies from a rate of 1 mm/min in the smaller airways to 2 cm/min in the trachea.[124] Mucociliary clearance of deposited particles generally occurs within 24 h after deposition in healthy individuals.[124–127] A comprehensive review of mucociliary clearance is presented by Yeates et al.[128]

8.4.4.1.1 Effect of drugs and inhaled contaminants on mucociliary clearance Numerous studies have shown that drugs can have a significant effect on mucociliary clearance. Beta-andrenergics, histamines, and amiloride have all been shown to increase mucociliary clearance in the human lung.[123,129–133] In contrast, cholinergic antagonists, aspirin, and anesthetics have all been shown to markedly decrease mucociliary clearance.[123,128,134,135]

Inhaled chemical contaminants such as sulfur dioxide, sulfuric acid, ozone, and tobacco smoke affect mucociliary clearance. At low concentrations ($<100\,\mu g/m^3$), sulfuric acid increases mucociliary clearance, while at high levels it seems to impair clearance by paralyzing the cilia.[136,137] Sulfur dioxide, ozone, and acute cigarette smoke exposures have all been shown to increase clearance in humans, whereas chronic cigarette smoking decreases mucociliary clearance.[123,138–142]

Table 8.2 Hygroscopic Growth of Various Environmental and Pharmacological Aerosols

Substance	Conditions		Diameter Increase (%)	Source
	T (°C)	RH (%)		
Mainstream Cigarette Smoke				
0.44 μm		100	65	Kousaka et al.[105]
~0.3 μm		99.5	~60	Li and Hopke[106]
~0.2 μm		99.5	~45	Li and Hopke[106]
—		100	~70	Hicks et al.[107]
NaCl				
0.1 μm	20	90	129	Gysel et al.[108]
0.3-0.5 μm	25	75	~90	Tang et al.[109]
0.3-0.5 μm	25	85	~110	Tang et al.[109]
0.3-0.5 μm	25	98	~280	Tang et al.[109]
$(NH_4)_2SO_4$				
0.05 μm	20	90	66	Gysel et al.[108]
0.1 μm	20	90	68	Gysel et al.[108]
$NaNO_3$				
0.05 μm	20	90	86	Gysel et al.[108]
0.1 μm	20	90	91	Gysel et al.[108]
NH_4HSO_4		98	~220	Tang and Munkelwitz [110]
Cromolyn sodium		98	31	Smith et al.[111]
Metaproterenol sulfate		98	29	Hiller et al.[112]
Isoproterenol sulfate		98	13	Hiller et al.[112]
Beclomethasone dipropionate		98	33	Hiller et al.[113]
Isoproterenol/phenylephedrine		90	24	Kim et al.[112]
Epinephrine		90	11	Kim et al.[112]
Metaproterenol		90	10	Kim et al.[112]
Albuterol		90	8	Kim et al.[112]
Isoetharine/phenylephedrine		90	10	Kim et al.[112]
Triamcinolone		90	17	Kim et al.[112]
Aerodur		98	37	Seemann et al.[115]
Bricanyl		98	144	Seemann et al.[115]
Cromolind		98	48	Seemann et al.[115]
Intal powder		97.4	30	Seemann et al.[115]
Intal composite		97.4	30	Seemann et al.[115]
Atropine sulfate		99.5	160	Peng et al.[116]
Isoproterenol hydrochloride		99.5	186	Peng et al.[116]
Isoproterenol hemisulfate		99.5	142	Peng et al.[116]
Disodium cromoglycate		99.5	26	Peng et al.[116]

8.4.4.1.2 Effect of disease on mucociliary clearance Mucociliary clearance is inhibited by numerous respiratory diseases. Acute respiratory infections such as pneumonia and influenza have been shown to impair clearance.[138,143,144] Chronic respiratory infections such as chronic bronchitis and bronchiectasis often result in an accumulation of mucus in the ciliary transport system, hindering mucociliary clearance.[145] While chronic obstructive pulmonary disease (COPD) patients generally have varied and erratic clearance rates,[127,128] asthma can result in reduced mucus transport rates and mucus plugging of the bronchi.[126,131,146,147] Patients with cystic fibrosis (CF) have been shown to have whole-lung clearance impairment as well as regional clearance impairment.[148] Small airway dysfunction or disease from chronic cigarette smoke also results in slowed mucociliary clearance.[140]

8.4.4.1.3 Effect of age and activity on mucociliary clearance Mucociliary clearance has been shown to decrease with age, starting at the age of 20, with large differences between adults >54 years old and adults 21–37 years.[149–152] Ho et al.[153] showed that adults over the age of 40 have decreased ciliary beat frequency, thus leading to slowed mucociliary clearance.

Increased physical activity, particularly aerobic exercise, has been shown to increase mucociliary clearance in humans.[123,147,154,155] Normal activities that do not require significant exertion have no effect on mucociliary clearance, while sleep significantly slows mucociliary clearance.[123,147]

8.4.4.2 Macrophage clearance or phagocytosis

Macrophage (or slow phase) clearance occurs in the alveolar region of the lung. As alveoli do not have cilia, deposited particles are engulfed by large bodies called macrophages (phagocytosis). These alveolar macrophages then migrate to the cilia surface and are cleared by mucociliary clearance, or move into the lymphatic system or blood stream for removal. This type of clearance may take months or years.[123]

8.4.4.2.1 Effect of particle size/fiber length and shape on macrophage clearance Particle size affects macrophage clearance rates, with clearance efficiency decreasing for particles smaller than 1 μm,[156] and optimal clearance occurring for particles between 1.5 and 3 μm.[123] Studies have shown that particle/fiber length affects alveolar clearance, with shorter fibers cleared more readily than longer fibers.[157–159] Particles such as asbestos and other nonsoluble fibers are often unable to be removed by phagocytosis because the macrophage is unable to engulf the entire fiber. This may result in incomplete phagocytosis, damage to the macrophage, and release of the macrophage's digestive enzymes. These digestive enzymes can cause extensive tissue damage. Thus, the persistence of these fibers in the lung can result in inflammation and disease.[160,161]

In addition, toxic nonfiber particles such as silica, which persist in the alveoli, provoke reactions that can lead to lung disease (i.e., silicosis). When a macrophage engulfs silica particles, the macrophage may release enzymes that cause fibroblast proliferation. This release of enzymes results from the crystalline structure of silica, the crystalline structure having been linked to the particle's fibrogenic potential.[162]

8.4.4.2.2 Effect of drugs and inhaled contaminants on macrophage clearance Inhaled contaminants such as ozone, nitrogen dioxide, and cigarette smoke affect alveolar macrophage activity. Ozone and nitrogen dioxide reduce phagocytosis as well as the bactericidal activity of macrophages, making it more difficult to fight bacterial lung infections.[163] Acute cigarette smoke exposures as well as high-level cigarette smoke exposures have been shown to inhibit macrophage action, while low levels of cigarette smoke exposure have been shown to increase macrophage action.[164,165] Exposure to nongaseous aerosolized contaminants such as nickel, cadmium, lead, manganese, chromium, and vanadium have been shown to damage alveolar macrophages.[162,163] Such exposures may result in disease (i.e., cadmium inhalation causing emphysema), potentially leaving the alveolar region unable to defend itself against or remove other inhaled contaminants, as well as causing the macrophages to release enzymes that damage the neighboring lung tissue. As discussed previously, toxic particulate contaminants such as silica and asbestos can cause macrophage impairment and damage, resulting in particle persistence and disease. The next section discusses inhaled particle overload in the alveolar region of the lung and resulting particle translocation.

8.4.4.3 *Free particle uptake and translocation to the interstitium*

Particles, particularly ultrafines, that are not rapidly cleared by macrophage action may persist in the alveoli or be taken up by epithelial cells and translocated from the alveolar region of the lung to the interstitial tissues and regional lymph nodes.[123,166–169] These particles may remain in the interstitium or regional lymph nodes for years, building up over time,[170] or may be removed by interstitial macrophages and/or penetrate into the postnodal lymph circulation.[169–171] It has been suggested that impaired clearance or a significant burden of particles in the lung (particle overload) increases translocation of particles to the interstitium.[166,167] Studies in animals have found that increasing particle number and dose rate, and decreasing particle size enhances the translocation of particles to the interstitium.[172–174] Enhanced interstitium translocation, particularly when toxic dusts (i.e., silica) are involved, has been linked with tissue damage, tumors, and fibrosis.[126,167,175]

8.4.4.4 *Importance of clearance in particle deposition modeling*

Several computational models of clearance mechanisms have been developed.[176–178] As clearance will affect both the residence time and local distribution of inhaled particles, the implementation of such models into particle deposition simulations is desirable.

Hofmann et al.[179] developed a model of particle clearance in which different clearance rates (derived from experimental studies) were associated with different generations of tracheobronchial airways. Furthermore, Martonen and Hofmann[180] described a model of clearance as a function of spatial location within airway branching sites. Specifically, they incorporated distinct clearance rates for tubular airway segments, bifurcation zones, and carinal ridges.

As discussed in the previous sections, age, activity level, drugs, inhaled contaminants, and disease can affect the efficiency of particle clearance, in turn influencing the number and local concentrations of inhaled particles. Particle deposition models that consider these overlapping factors could be of great use in both inhalation toxicology and aerosol therapy. Although we have discussed its influence as it relates to clearance, we will now provide a more complete discussion of the effects of disease on particle deposition and distribution.

8.4.5 *Disease*

Airway disease has a dual effect on particle deposition: influencing breathing pattern and physically changing airway morphology. In this section, we will discuss common respiratory diseases and how they affect airflow and airway morphology, thus affecting the deposition and distribution of inhaled aerosols.

8.4.5.1 *Chronic obstructive pulmonary disease*

COPD is a term that is generally applied to patients with emphysema and/or chronic bronchitis. Both of these diseases modify the structure of the human lung,[126] resulting in either obstructed airways and/or degeneration of the alveolar structure.

Emphysema may affect the respiratory or terminal bronchioles or the peripheral alveoli. Cigarette smoke and air pollution are the likely causes of emphysema, resulting in the destruction of elastin in the alveolar wall. This elastin destruction leads to enlarged airspaces and loss of alveolar structure, often resulting in intrapulmonary bronchi collapse during expiration.[75,126] In addition, the loss of the alveolar structure results in the loss of the capillary bed that transfers oxygen from the lungs to the blood.[126]

Chronic bronchitis is characterized by excessive mucus generation and alveolar wall thickening. The excessive mucus is a result of hypertrophied mucus glands, and causes the formation of mucus plugs that obstruct airways and may fully occlude small bronchi. Alveolar wall thickening results in reduced elasticity of alveolar walls, limiting regional ventilation.[75,126]

8.4.5.2 *Asthma*

Asthma is characterized by a reduction of the airway lumen due to constriction of the bronchial airways in response to a stimulus. This constriction may also in turn result in an increase in the mucus layer thickness. Chronic asthma can result in subepithelial fibrosis.[126] Asthma stimuli may include pollutants, allergens, or exercise. It has been noted that asthma has attained epidemic proportions on a global scale.[181]

8.4.5.3 *Cystic fibrosis*

CF is a genetic disease that causes the lung's epithelial cells to produce abnormally thick, excessive mucus. This slowly cleared mucus narrows airways and obstructs airflow, making tissue vulnerable to inflammation and recurrent infection. This inflammation and infection causes progressive respiratory disease, including bronchiectasis and chronic airway obstruction.[126,182] Impairment typically begins in the small airways and progresses proximally, with ventilation increasingly shifting from obstructed regions to healthy regions of the lung.[183]

8.4.5.4 *Effect of obstructive disease on particle deposition and distribution*

Several studies have investigated the effect of obstructive disease on particle deposition and distribution in the human lung. Exploring the effect on particle deposition, Kim and Kang[184] found a marked increase in the deposition of 1 μm particles in patients with COPD and asthma compared to normal subjects. Anderson et al.[185,186] found that the deposition of fine and ultrafine particles was increased in patients with CF and obstructive disease. Brown et al.[88] found that COPD patients had a greater dose rate for ultrafine particles than healthy subjects. Also, the deposition of particles increased with severity of obstruction or decrease in lung function. [183,187–189] Reasons for increased deposition in patients with obstructive disease include (1) reduction of airway diameter by constriction or mucus buildup, thus increasing inertial impaction on airway walls, (2) increased residence time of particles in the alveolar region resulting from non-uniform ventilation distribution, (3) collapse of airways due to flow limitation, and (4) flow perturbations or induced turbulence at sites of obstruction. [183,184,186]

Investigating the effect of obstructive disease on particle distribution, Brown et al.[190] found that a significant number of coarse particles deposit in the poorly ventilated tracheobronchial airways of CF patients, while these particles follow regional ventilation in healthy subjects. Other studies indicate that the deposition pattern of particles in patients with obstructive disease is heterogeneous, with an enhancement of deposition in various local regions.[191–193]

8.4.5.5 *Modeling disease*

Several models of particle deposition that specifically address disease have been developed. Segal et al.[194] modeled particle deposition in patients with COPD, using a modified deterministic model.[7,29] This work investigated the dependence of deposition pattern on the severity of disease. In addition, Martonen et al. simulated particle deposition in CF. This study found a proximal shift in particle deposition with severity of obstruction.[195–197] More recently, Martonen et al.[181] have simulated the effect of asthma on particle deposition patterns, comparing their results with data from imaging studies of asthma patients.

The United States Environmental Protection Agency (U.S. EPA) has identified people suffering from respiratory disease as a sensitive subpopulation needing particular consideration in the risk assessment of particulate matter and in the establishment of air pollution standards.[198] Therefore, more advanced models of particle deposition for a variety of diseases are needed.

8.4.6 Age

Age can be a significant factor influencing the deposition and distribution of particles in the human lung, probably due to the differences in airway geometry and ventilation between children and adults. Studies of airway geometry as a function of age are presented by Ménache and Graham,[199] Hofmann et al.,[201] and Martonen et al.[200] Age has also been shown to affect the percentage of nasal breathing,[202] thus affecting the amount of particulate matter that makes it to the human lung. Bennett et al.[203] showed that children have enhanced upper airway deposition of coarse particles when compared to adults, but that total deposition amounts are comparable.

Several studies have modeled aerosol deposition as a function of human subject age. Martonen et al.[200] found that total deposition within the human lung decreased with increasing age from 7 months to 30 years. Using the International Commission on Radiological Protection (ICRP) 66 model, Harvey and Hamby[204] found that for 1 μm particles, deposition and regional distribution varied by age, with extrathoracic deposition increasing significantly in younger age groups (young children and infants), who have smaller respiratory airway sizes. Hoffman et al.[201] showed that particle deposition does indeed depend on lung morphology and that dose per surface area decreases from 7 months of age to adulthood.

The U.S. EPA has also identified children as a sensitive subpopulation requiring additional consideration in the establishment of air quality standards.[198] Therefore, the development of particle deposition models for children is of particular importance. A model of particle deposition in a developing human lung has also been developed.[205] It was predicted that children may receive a localized particulate matter dose three times that of adults. Such models may be of great use in both inhalation toxicology and inhalation therapy.

8.5 Theory and experiment

Many different types of experimental protocols have been developed to measure particle deposition in the respiratory system. An overview of several of these methods, and a discussion of how the resulting data can be compared with particle deposition will be presented.

8.5.1 Predictions of particle deposition

Simulation studies can be used to predict the deposition of inhaled particles on differing spatial scales of resolution. Models can be developed that predict total respiratory system deposition or deposition in each of the regional (i.e., extrathoracic, tracheobronchial, or pulmonary) compartments. In addition, models can predict deposition efficiencies in each individual airway generation, or simulate the dose to a specific anatomical location (e.g., a carinal ridge or airway wall) within a respiratory passage. The level of detail desired and the type of experimental data available for model validation should govern the selection or development of an appropriate model for a given research purpose.

8.5.2 Particle deposition measurements

Much work has been done in attempting to quantify the deposition of particles in respiratory airways, encompassing a wide range of approaches and techniques. We provide an overview of aerosol deposition measurements that have been performed in casts, animal models, and human subjects. Comprehensive reviews of experimental measurements of respiratory particle deposition have been presented by Martonen,[8] Sweeney and Brain,[206] and Kim.[207]

8.5.2.1 Casts and models

Particle deposition measurements have been performed in models and replica casts of both human and animal lungs. Such studies provide a means of examining particle deposition airway by airway in realistic geometries. In addition, deposition studies in casts can be reproduced, and particles of different sizes and characteristics may be studied serially under the same conditions. Replica cast studies have been performed in the human tracheobronchial region,[24,208–210] in canine lungs,[209] and in nasal passages.[211,212] Other studies have been performed using laryngeal casts combined with a silicone rubber model of the tracheobronchial airways.[213,214] Deposition data obtained in the silicone model at a flow rate of 15 l/min are shown in Figure 8.7 for two particle sizes. Deposition in a TB replica cast[215] is also shown. In this figure, the ratio of the amount of aerosol deposited in a single generation to the amount entering the cast is plotted. Note that good agreement between the replica cast and the silicone model results was obtained. Table 8.3 summarizes localized (i.e., bifurcation) deposition from the two silicone model studies for several particle sizes at three different flow rates. Specifically, the table contains the bifurcation deposition ratio B_d, where

$$B_d = [(\text{aerosol mass deposited within a bifurcation})/(\text{bifurcation surface area})]/ [(\text{total aerosol mass deposited within the two airways of a generation, including the shared bifurcation})/(\text{total airway and bifurcation surface area})] \quad (8.9)$$

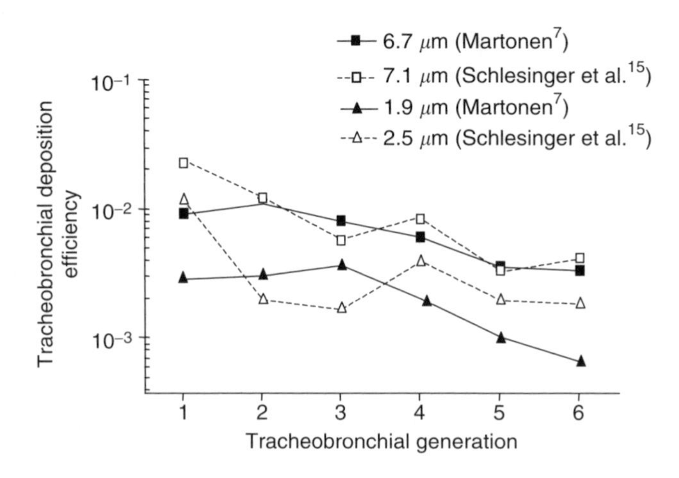

Figure 8.7 Deposition of particles in the tracheobronchial region as measured by Martonen[213] using a silicone rubber cast, compared with the replica cast data of Schlesinger et al.,[215] for two different particle sizes.

Table 8.3 Bifurcation Deposition Ratio B_d at Different Flow Rates

Tracheobronchial Model Generation	Q=15 l/min				Q=30 l/min			Q=60 l/min			
	1.9 μm[b]	2.1 μm[a]	3.0 μm[b]	6.7 μm[a]	3.3 μm[a]	3.6 μm[a]	6.8 μm[b]	1.9 μm[a]	6.1 μm[a]	8.7 μm[b]	10.6 μm[b]
1	0.80	0.42	0.51	0.83	0.17	0.25	0.65	0.50	0.84	1.10	1.20
2	1.15	1.08	1.15	0.92	1.08	1.19	1.35	1.30	1.14	1.40	1.52
3	1.50	1.38	1.62	1.36	2.29	1.83	2.31	1.93	1.84	3.05	3.16
4	0.91	0.90	0.87	1.12	1.34	1.47	1.57	1.74	1.46	2.10	1.94
5	1.10	0.85	0.91	1.18	1.48	1.63	1.50	1.63	1.78	1.75	1.60

[a] Martonen.[213]

[b] Martonen and Lowe.[214]

Note that in most cases the deposition in the bifurcation zone is enhanced in relation to the adjacent airways. For example, at a flow rate of 60 l/min the deposition of 8.7 μm particles at the third-generation bifurcation is three times greater than that in the adjacent airway segments. The values of B_d less than 1 may be due to large regions of the bifurcation site having 0 deposition, as observed by Schlesinger et al.[215]

8.5.2.2 Deposition patterns deduced from clearance studies

Experimental methods have been developed in which the regional deposition of particles is deduced from the measurement of the time course of clearance of particles from the thorax.[91] Specifically, radiolabeled particles are inhaled, and a whole body counter is used to measure the amount of radioactive activity in the stomach, chest, and extrathoracic regions. Since the removal of particles (by mucociliary clearance) in the tracheobronchial region occurs at a faster rate than removal (by macrophagic clearance) in the pulmonary region, deposition in the two regions can be deduced from the two-part slope of a normalized retention curve for the thorax.

8.5.2.3 Light-scattering methods

Traditionally, the total deposition of particles in the respiratory tract has been quantified using light-scattering photometry to compare the concentration of particles in inhaled and exhaled air.[91,207,216,217] When a monodisperse aerosol is used and ventilation is simultaneously measured, the deposition fraction in the respiratory system can be calculated. However, photometry cannot distinguish between differences in inspiratory and expiratory aerosol concentration and changes in aerosol size distribution; therefore, these methods are inappropriate for polydisperse or hygroscopic aerosols.[207] Rosati et al.[218] have developed a light-scattering, particle-sizing system that may be the best option for determining the total deposition of polydisperse aerosols in the respiratory tract. This system also has the potential to be applied to hygroscopic aerosols as it can determine particle sizes of inhaled and exhaled aerosols, and works well for different-sized polydisperse aerosols.[218,219]

8.5.2.4 Gamma scintigraphy

Radionuclide imaging has been widely used to measure both the concentration and spatial distribution of inhaled aerosols. In these studies, particles are tagged with a radioisotope (such as ^{99m}Tc) and then inhaled. 2D (planar) or 3D imaging modalities can then be used to measure the radioisotope emissions from specific locations within the body.

Planar gamma cameras can be used to obtain projections of the spatial distribution of inhaled radiolabeled aerosols.[221] These images may be useful in predicting total deposition

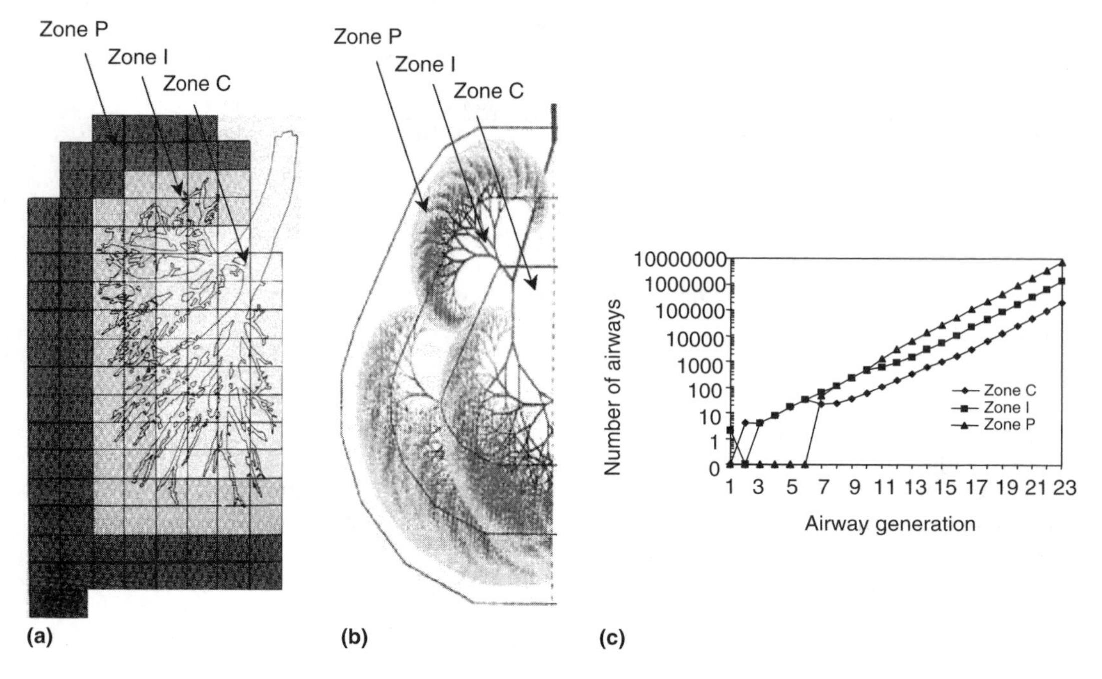

Figure 8.8 Association of the central, intermediate, and peripheral zones (**a**) of planar gamma camera images with a corresponding computerized lung branching network model (**b**). Panel (**c**) shows the resulting distribution of lung airways within each zone.

within the lung or extrathoracic passages, but the 2D nature of the images may obscure important deposition patterns.[222] Planar imaging studies have been performed for a variety of inhaled aerosols.[90,223,224] To assist in the interpretation of planar gamma camera data, Martonen et al.[223,224] have developed methods to associate regions of images (Figure 8.8a) with computer models of the human lung (Figure 8.8b). The computer model serves as a template to be superimposed on actual images, thus permitting the generational airway composition within the central (C), intermediate (I), and peripheral (P) zones of planar images to be predicted (Figure 8.8c).

Recently, 3D tomographic imaging modalities have been applied to the study of particle deposition patterns. Both 3D single-photon emission computed tomography (SPECT)[222,225,226] and positron emission tomography (PET)[227–229] have been employed in particle deposition studies. 3D methods provide a powerful means of associating particle deposition with distinct local regions within the respiratory system. In a series of papers,[222,225,230–233] Martonen and co-workers have recently presented computational methods for correlating the individual airways of a lung morphology model with the voxels of 3D SPECT images. These methods provide a means of validating 3D CFPD and deposition models using SPECT data while simultaneously providing a framework for predictive laboratory studies of targeted aerosol delivery. In Figure 8.9, a computer-generated 3D branching network is shown with an associated network of voxels. In practice, the computer model may be superimposed on the voxels of a SPECT image (Figure 8.9) to allow a quantification of particle deposition. Figure 8.10 shows an example of a SPECT image and a detailed view of the airway composition of an associated voxel.

8.5.2.5 Microdosimetry

In the experimental methods described above, particle deposition may be measured by region (e.g., by clearance studies) or airway by airway (e.g., in casts). However, in the

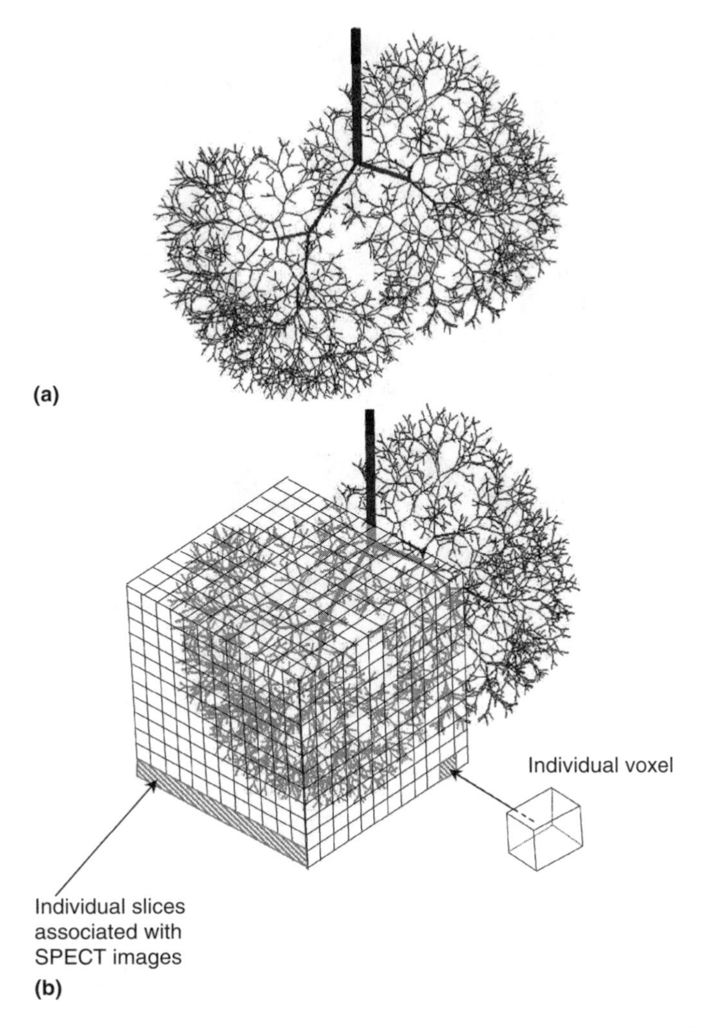

Figure 8.9 A branching lung network model (**a**) and an associated network of voxels (**b**, not to scale) for analysis of SPECT images of aerosol distribution.

study of the health risks associated with inhaled particles, it is desirable to obtain deposition measurements at ever finer levels of spatial resolution, measuring the dose to individual airway structures or cells. Unfortunately, there is little such data available. Schlesinger et al.[215] performed local deposition studies of 8 μm particles in a cast of an airway bifurcation. After the aerosols were deposited, the airway cast (Figure 8.11) was cut open and a microscope was used to count the number of particles in 1-mm^2 regions of the bifucation surface. Panels a and b of Figure 8.11 show their results for constant flow rates of 15 and 60 l/min, respectively. There is a definite "hot spot" of deposition at the carinal ridge at 15 l/min, and this area becomes wider with increasing flow rate. In addition, Martonen[234] presented a qualitative description of the local concentrations of 6.7 μm ammonium fluorescein particles in several generations of a bifurcating cast at a constant flow rate. Following the deposition of the particles, the cast was cut open and the distribution of the particles was imaged (Figure 8.12). Note the "hot spots" of particles at the carinal ridges. Additional studies of the dosimetry or microdosimetry of inhaled particles would be of great use in the validation of CFPD studies.

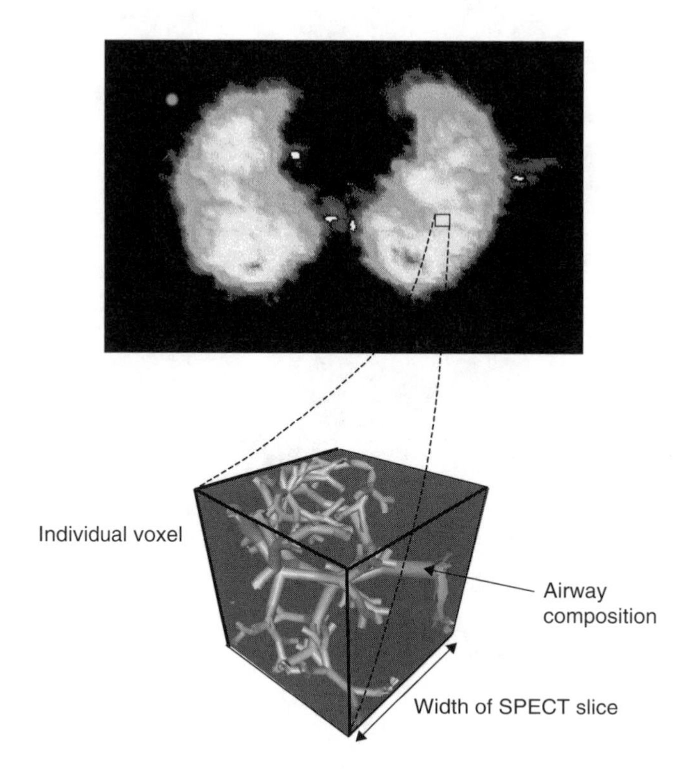

Figure 8.10 A SPECT image of radiolabeled aerosol distribution in the human lung and an associated voxel of the branching computer model; the composition of airways within the voxel is shown.

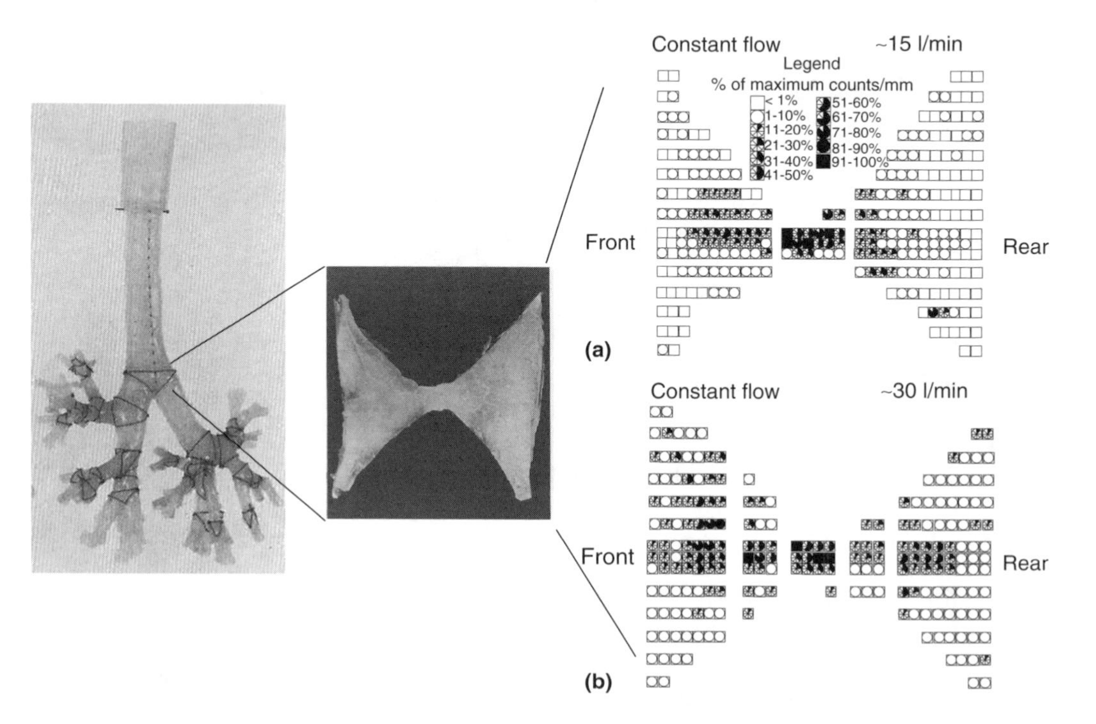

Figure 8.11 Determination of localized deposition by Schlesinger et al.[215] The replica cast, the expanded bifurcation zones, and the resulting measured local deposition are shown. Adapted from Schlesinger et al., *Ann. Occup. Hyg.*, Pergamon Press, 1982, with permission.

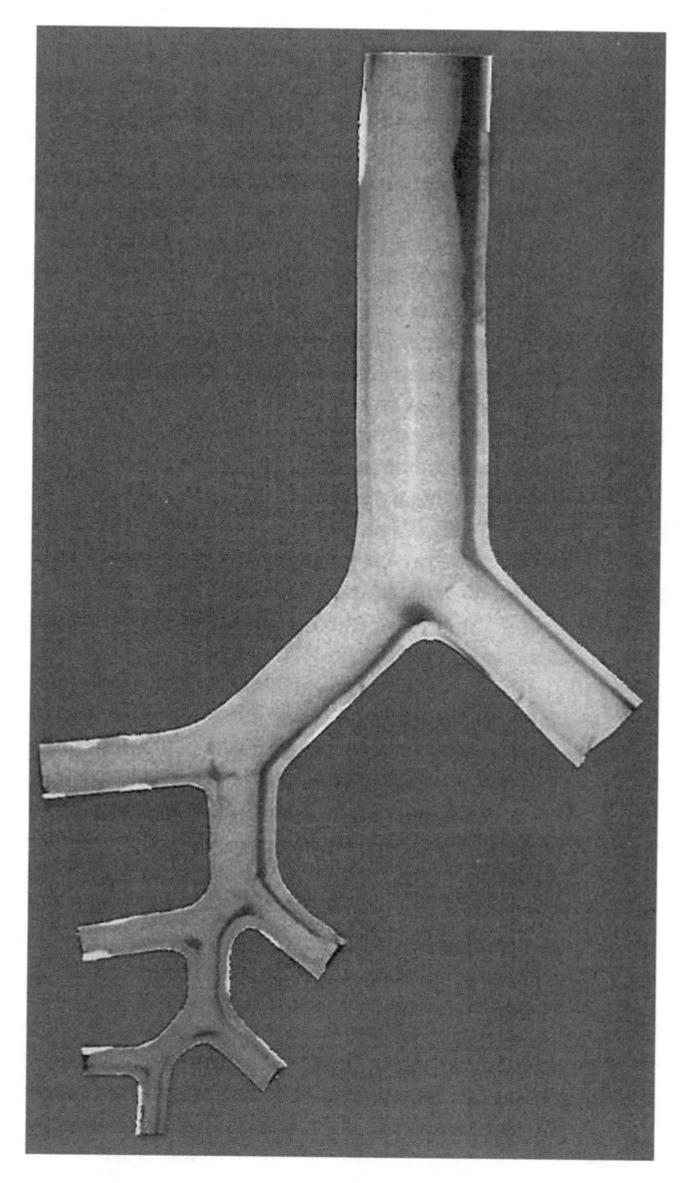

Figure 8.12 Enhanced deposition of 6.7 μm particles at different generations of bifurcations in a replica cast. Adapted from Martonen, T.B. et al. *Am. Ind. Hyg. Assoc. J.*, American Industrial Hygiene Association Press, 1992, with permission.

8.5.3 Comparison of modeling and data

Validation of deposition modeling results with experimental data is a crucial step in the modeling process. Studies comparing simulated particle deposition predictions with experimental results exist for a wide variety of models. In these studies, simulations of total and compartmental particle deposition have been examined,[237] with many particle sizes being considered. Stahlhofen et al.[20] presented a comprehensive overview and summary of a large amount of experimentally obtained particle deposition data, providing a resource for many subsequent modeling investigations.

Comparison of total deposition predictions with experimental data have been performed for both stochastic and deterministic models. Hoffmann and Koblinger[235] presented a comparison between their estimated total deposition, as predicted by their stochastic model, and that obtained from a variety of experimental sources as a function of particle size. They also considered mouth versus oral breathing and breathing pattern in their comparisons. In general, they found good agreement between the model and experiment at all particle sizes.

Theoretical predictions of total particle deposition obtained using the deterministic model of Martonen et al.[7] are compared with the experimental data of Heyder et al.[236] in Figure 13. The deposition formulae presented in Chapter VI form the foundation of this model. For the given ventilatory conditions, the predicted total deposition fractions are in relatively good agreement with the experimental data over a wide range of particle sizes. However, there are some systematic differences between the model predictions and the data, namely an overestimation of the total deposition fraction at larger particle sizes. Other comparisons of this deterministic model to human subject data have been performed by Segal et al.[237]

In the study of the fate of inhaled particles, one must be aware that inherent uncertainties exist in both experimental data and in model simulations. For example, uncertainty and error may be imparted into experimental data by uncertainties in the measurement of flow rates, particle size distributions, or experimental deposition measurements, or by intersubject variability in these measurements.[20] Uncertainty in model simulations may depend on one or more of the following:

(1) observational errors in any model input parameters, (2) natural (i.e., inter- or intra-subject) variation of model input parameters, (3) validity of the underlying theory of the model or any simplifying assumptions, (4) any approximation errors imparted by the computational numerical methods, or (5) round-off errors imparted by limitations of the computer hardware or software being used. Therefore, any comparisons of model and data should be undertaken with the sources of uncertainty in both the experimental system and modeling method in mind, and research aimed at explaining and/or controlling variability and uncertainty should be ongoing.

8.5.3.1 *Simulations of compartmental particle deposition*

8.5.3.1.1 Extrathoracic Due to the complex geometries present in the oral and nasal regions (e.g., the nasal geometry depicted in Figure 8.4), little theoretical modeling has been done in the extrathoracic compartment. Simple analytical models were developed early on for the nose,[238] but as the complexity of the nasal passages was recognized, and more experimental data were published,[239] it became apparent that more complex models were needed. Experimental deposition data have also been reported for the oral cavity,[24, 91,240] and analytical models have been developed.[241] However, most of the modeling that has been done for the nose[239,242,243] and mouth[242] region has been empirical in nature. In future, CFPD modeling, as introduced earlier in this chapter, may offer an alternative approach to the simulation of particle deposition in the extrathoracic passages.

8.5.3.1.2 Tracheobronchial Hofmann and Koblinger[235] compared their simulated tracheobronchial deposition values with the experimental data of Heyder et al.[236] For particle sizes of 0.05–1 μm, their model predicted deposition in the tracheobronchial region of 1–11%, while zero measured deposition was reported by Heyder et al. It was hypothesized that these differences were due to inherent limitations of the definition of different regions in the model.

Theoretical predictions of tracheobronchial deposition fraction for the model of Martonen[7,29] are plotted vs. corresponding experimental data in Figures 8.13 and 8.14.

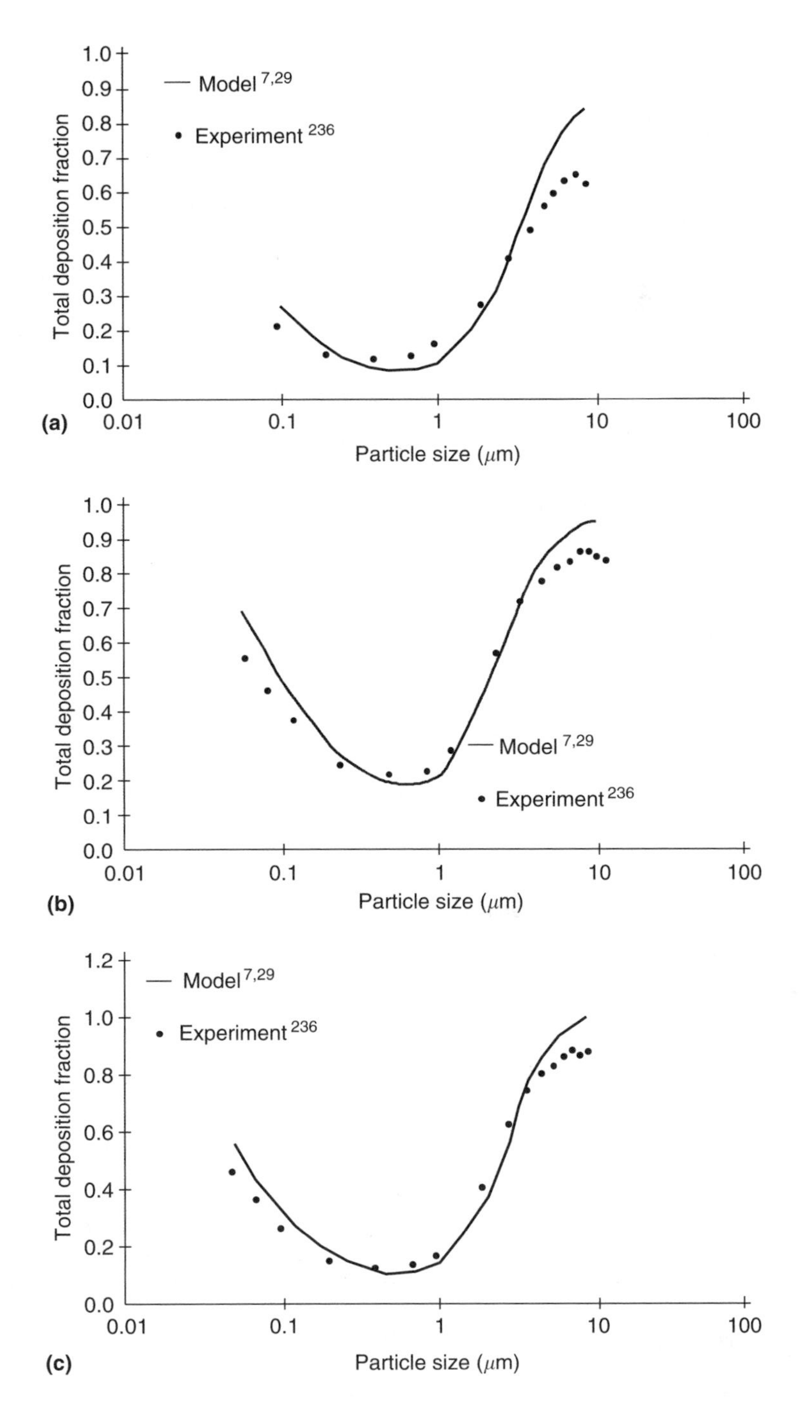

Figure 8.13 Model predictions of total deposition determined using the deterministic model of Martonen,[7,29] plotted against experimental data:[236] (**a**) breathing frequency=30 breaths/min, tidal volume=500 ml; (**b**) breathing frequency=15 breaths/min, tidal volume=1000 ml; (**c**) breathing frequency=30 breaths/min, tidal volume=1500 ml.

Again, relatively good agreement between theory and average experimental data was observed over the range of particle sizes investigated, suggesting that the physics of the inhaled particles are being adequately simulated.

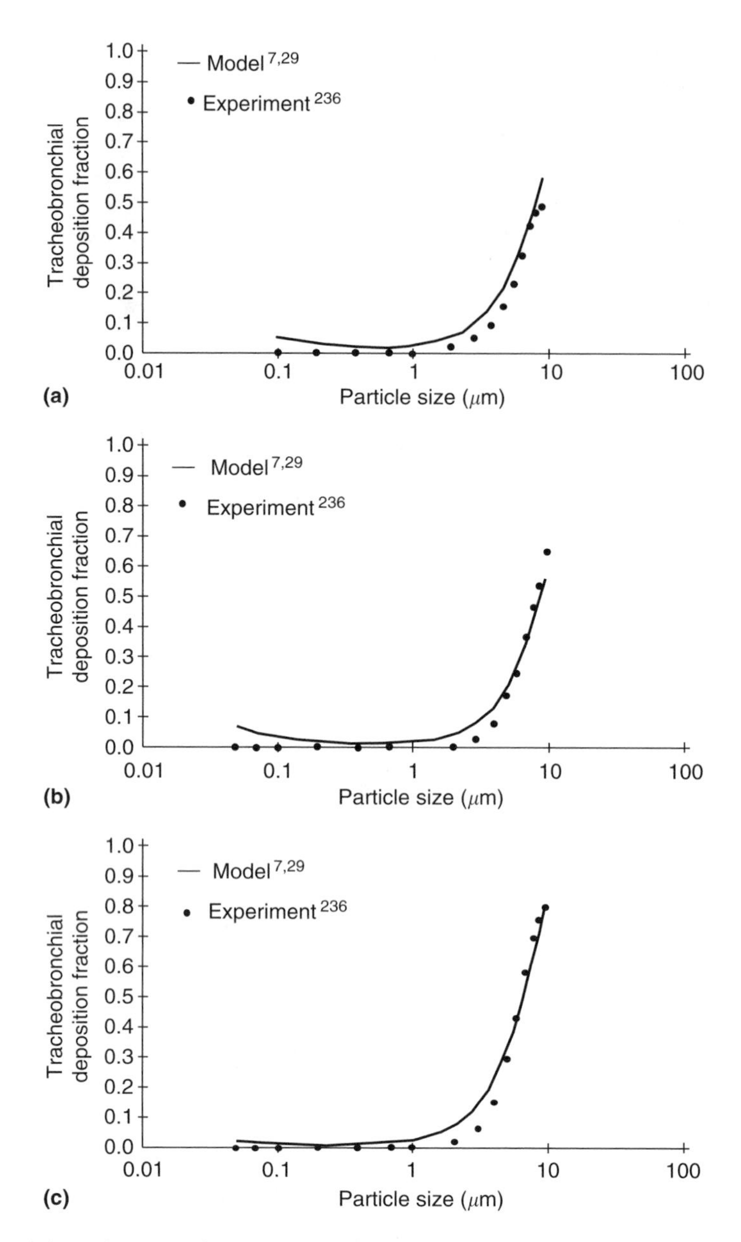

Figure 8.14 Model predictions of tracheobronchial deposition determined using the deterministic model of Martonen,[7,29] plotted against experimental data:[236] (**a**) breathing frequency=30 breaths/min, tidal volume=500 ml; (**b**) breathing frequency=15 breaths/min, tidal volume=1000 ml; (**c**) breathing frequency=30 breaths/min, tidal volume=1500 ml.

8.5.3.1.3 Pulmonary Hofmann and Koblinger[235] also compared their simulated pulmonary deposition values with the experimental data of Heyder et al.;[236] their model very closely predicted measured values in the pulmonary region. Results from the deterministic model of Martonen[7,29] for the pulmonary region are plotted against the experimental data of Heyder et al. in Figure 8.15. While generally good agreement was seen between theory and experiment over the particle sizes simulated, there was a noticeable shift of the predicted deposition with respect to the experimental data. Such a systematic trend in the model requires further investigation.

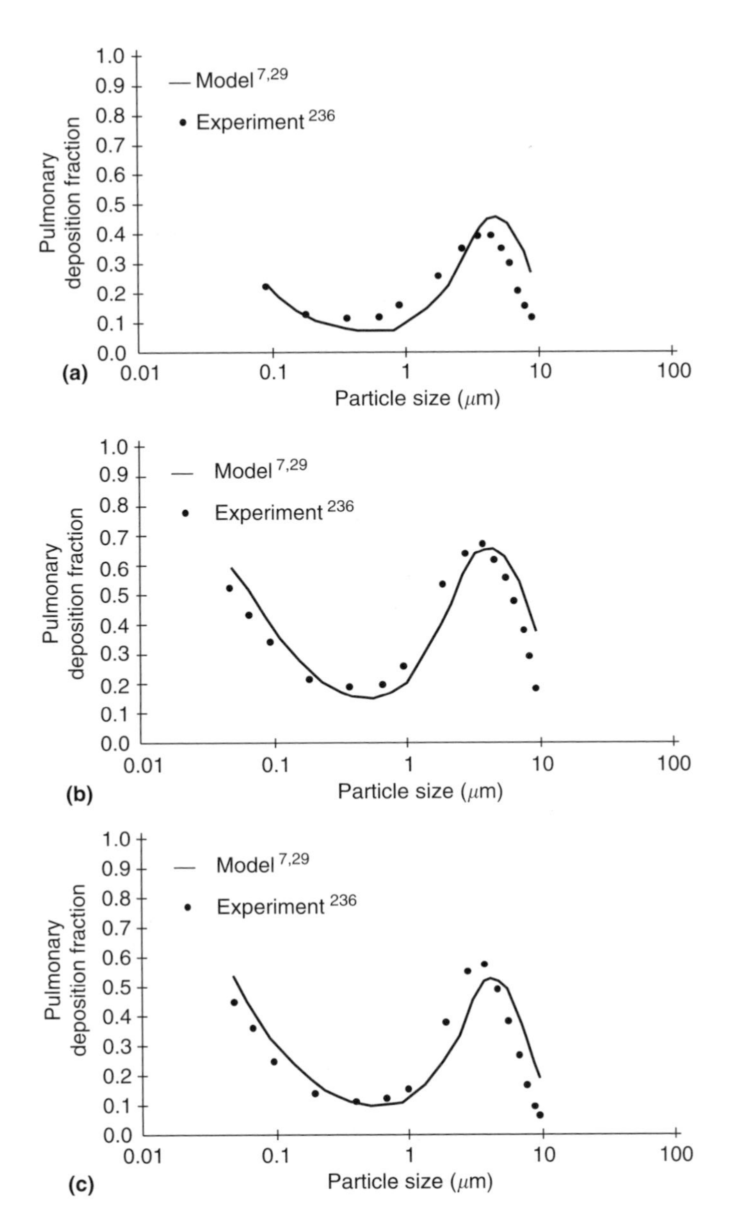

Figure 8.15 Model predictions of pulmonary deposition determined using the deterministic model of Martonen,[7,29] plotted against experimental data:[236] (**a**) breathing frequency=30 breaths/min, tidal volume=500 ml; (**b**) breathing frequency=15 breaths/min, tidal volume=1000 ml; (**c**) breathing frequency=30 breaths/min, tidal volume=1500 ml.

8.5.3.2 Simulations of particle distribution generation by generation

Figure 8.16 depicts particle deposition by generation as predicted by the model of Martonen[7,29] for a variety of ventilatory conditions. At higher flow rates (Figure 8.16c), the model predicts enhanced deposition of large particles in the tracheobronchial airways. Experimental measurements of deposition generation by generation are scarce, and additional accurate cast studies would be particularly useful in validating such simulations.

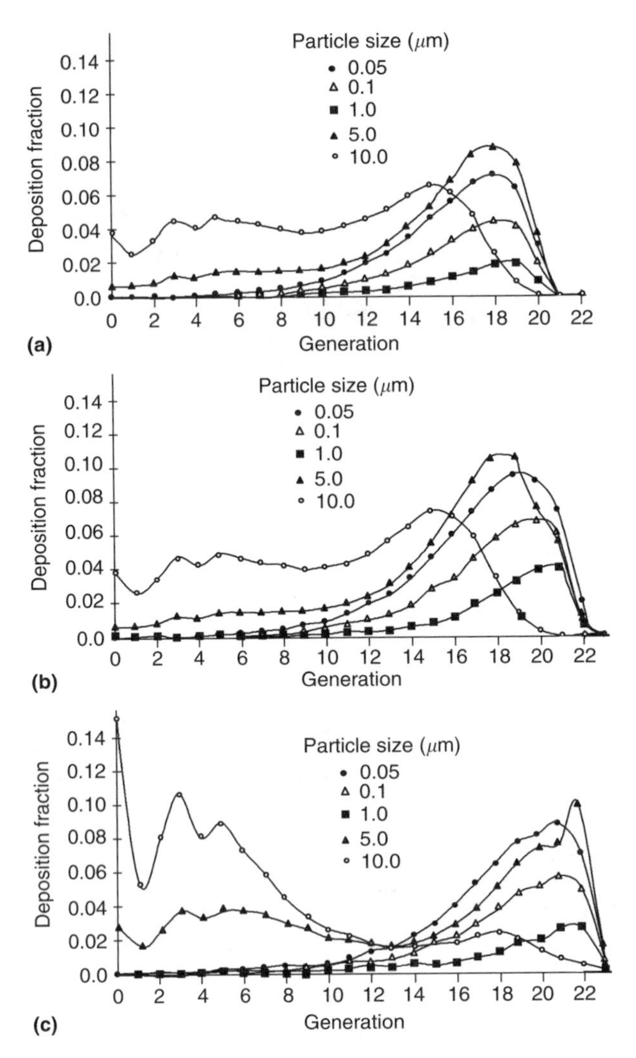

Figure 8.16 Model predictions of particle deposition generation by generation determined using the deterministic model of Martonen,[7,29] for a range of particle sizes: (**a**) breathing frequency=30 breaths/min, tidal volume=500 ml; (**b**) breathing frequency=15 breaths/min, tidal volume=1000 ml; (**c**) breathing frequency=30 breaths/min, tidal volume=1500 ml.

8.5.3.3 *Simulations of local particle deposition*

Experimental data describing the local airway concentrations of inhaled particles are relatively scarce, and are mainly derived from observations of cadaver airways and cast studies. Therefore, it is challenging to validate CFPD studies of local particle deposition in all except very simple geometries, although predicted accumulations of particles at carinal ridges have been shown to be consistent with experimental observations.[45] Recent efforts by Martonen and co-workers[232,233] present a methodology for associating individual airways of a branching lung morphology model with specific voxels of a corresponding SPECT image, thereby allowing for the comparison of simulated deposition with actual deposition measurements.

8.6 Summary

The modeling of particle deposition is of great use in both inhalation toxicology and inhalation therapy. In particular, modeling provides a means of predicting total, regional, and local respiratory system concentrations of inhaled particles, and offers a foundation for the development of targeted delivery protocols. In addition, modeling aids in interpreting experimental measurements and advances the understanding of events and variables that cannot be experimentally quantified.

Particle deposition in the human respiratory system is an extremely complex phenomenon, governed by a wide variety of overlapping and interacting factors. Development and validation of increasingly sophisticated computational models that address particle deposition on local and regional scales, and consider both biological variability and realism, will be instrumental in improving the prediction of both the health effects of inhaled particles and the therapeutic value of inhaled pharmaceutics.

Disclaimer

The information in this document has been funded wholly (or in part) by the U.S. Environmental Protection Agency. It has been subjected to review by the National Health and Environmental Effects Research Laboratory and approved for publication. Approval does not signify that the contents necessarily reflect the views of the Agency, nor does mention of trade names or commercial products constitute endorsement or recommendation for use.

Acknowledgment

Kristin K. Isaacs was funded by the EPA/UNC DESE Cooperative Training Agreement CT827206, with the Department of Environmental Sciences and Engineering, University of North Carolina at Chapel Hill.

References

1. Pedley, T.J. and Kamm, R.D., Dynamics of gas flow and pressure–flow relationships, in *The Lung: Scientific Foundations*, Vol. I, Crystal, R.G., West, J.B., and Barnes, P.J., Eds., Raven Press, New York, 1991.
2. Martonen, T.B. and Yang, Y., Deposition mechanics of pharmaceutical particles in human airways, in *Inhalation Aerosols: Physical and Biological Basis for Therapy*, Hickey, A.J., Ed., Marcel Dekker, Inc., New York, 1996.
3. White, F.M., *Viscous Fluid Flow*, McGraw-Hill, New York, 1991.
4. Chang, H.K. and El Masry, O.A., A model study of flow dynamics in human central airways. Part I. Axial velocity profiles, *Respir. Physiol.*, 49, 75, 1982.
5. Isabey, D. and Chang, H.K., A model study of flow dynamics in human central airways. Part II. Secondary flow velocities., *Respir. Physiol.*, 49, 97, 1982.
6. Schroter, R.C. and Sudlow, M.F., Flow patterns in the human bronchial airways, *Respir. Physiol.*, 7, 341, 1969.
7. Martonen, T.B., Mathematical model for the selective deposition of inhaled pharmaceuticals, *J. Pharm. Sci.*, 82, 1191, 1993.
8. Martonen, T.B., Surrogate experimental models for studying particle deposition in the human respiratory tract: an overview, in *Aerosols: Research, Risk Assessment, and Control Stategies*, Lee, S.D., Schneider, T., Grant, L.D., and Verkerk, P.J., Eds., Lewis Publishers, Chelsea, MI, 1986.

9. Zhang, Z. and Martonen, T.B., Deposition of ultrafine aerosols in human tracheobronchial airways, *Inhal. Toxicol.*, 9, 99, 1997.
10. Martonen, T.B., Yang, Y., and Xue, Z.Q., Effects of carinal ridge shapes on lung airstreams, *Aerosol Sci. Technol.*, 21, 119, 1994.
11. Martonen, T.B., Yang, Y., Xue, Z.Q., and Zhang, Z., Motion of air within the human tracheobronchial tree, *Part. Sci. Tech.*, 12, 175, 1994.
12. Martonen, T.B., Yang, Y., and Xue, Z.Q., Influences of cartilaginous rings on tracheobronchial fluid dynamics, *Inhal. Toxicol.*, 6, 185, 1994.
13. Martonen, T.B., Guan, X., and Schrek, R.M., Fluid dynamics in airway bifurcations. I. Primary flows, *Inhal. Toxicol.*, 13, 261, 2001.
14. Martonen, T.B., Guan, X., and Schrek, R.M., Fluid dynamics in airway bifurcations. II. Secondary currents, *Inhal. Toxicol.*, 13, 281, 2001.
15. Martonen, T.B., Guan, X., and Schrek, R.M., Fluid dynamics in airway bifurcations. III. Localized flow conditions, *Inhal. Toxicol.*, 13, 291, 2001.
16. Kimbell, J.S., Gross, E.A., Joyner, D.R., Godo, N.M., and Morgan, K.T., Application of computational fluid dynamics to regional dosimetry of inhaled chemicals in the upper respiatory tract of the rat, *Toxicol. Appl. Pharmacol.*, 121, 253, 1993.
17. Kimbell, J.S., Godo, N.M., Gross, E.A., Joyner, D.R., Richardson, R.B., and Morgan, K.T., Computer simulation of inspiratory airflow in all regions of the F344 rat nasal passages, *Toxicol. Appl. Pharmacol.*, 145, 388, 1997.
18. Yu, G., Zhang, G., and Lessman, R., Fluid flow and particle diffusion in the human upper respiratory system, *Aerosol Sci. Technol.*, 28, 146, 1998.
19. Martonen, T.B., Zhang, Z., Yue, G., and Musante, C.J., 3-D particle transport within the human upper respiratory tract, *J. Aerosol Sci.*, 33, 1095, 2002.
20. Stahlhofen, W., Rudolph, G., and James, A.C., Intercomparison of experimental regional aerosol deposition data, *J. Aerosol Med.*, 2, 285, 1989.
21. Rudolph, G., Gebhardt, J., Heyder, J., Scheuch, G., and Stahlhofen, W., An empirical formula describing aerosol deposition in man for any particle size, *J. Aerosol Sci.*, 17, 350, 1986.
22. Rudolph, G., Gebhardt, J., Heyder, J., Schiller, C.F., and Stahlhofen, W., Mass deposition from inspired polydisperse aerosols, *Ann. Occup. Hyg.*, 32, 919, 1988.
23. Gonda, I., A semi-empirical model of aerosol deposition in the human respiratory tract for mouth inhalation, *J. Pharm. Pharmacol.*, 33, 692, 1981.
24. Chan, T.L. and Lippmann, M., Experimental measurements and empirical modelling of the regional deposition of inhaled particles in humans, *Am. Ind. Hyg. Assoc. J.*, 41, 399, 1980.
25. Cohen, B.S. and Asgharian, B., Deposition of ultrafine particles in the upper airways: an empirical analysis, *J. Aerosol Sci.*, 21, 789, 1990.
26. Asgharian, B., Wood, R., and Schlesinger, R.B., Empirical modeling of particle deposition in the alveolar region of the lungs: a basis for interspecies extrapolation, *Fundam. Appl. Toxicol.*, 27, 232, 1995.
27. Carpenter, R.L. and Kimmel, E.C., Aerosol deposition modeling using ACSL, *Drug Chem. Toxicol.*, 22, 73, 1999.
28. Zhang, Z. and Martonen, T.B., Comparison of theoretical and experimental particle diffusion data within human airway casts, *Cell Biochem. Biophys.*, 27, 97, 1995.
29. Martonen, T.B., Analytical model of hygroscopic particle behavior in human airways, *Bull. Math. Biol.*, 44, 425, 1982.
30. Gradón, L. and Orlicki, D., Deposition of inhaled aerosol particles in a generation of the tracheobronchial tree, *J. Aerosol Sci.*, 21, 3, 1990.
31. Yu, C.P., A two-component theory of aerosol deposition in lung airways, *Bull. Math. Biol.*, 40, 693, 1978.
32. Yu, C.P. and Diu, C.K., A comparative study of aerosol deposition in different lung models, *Am. Ind. Hyg. Assoc. J.*, 43, 54, 1982.
33. Egan, M.J. and Nixon, W., A model of aerosol deposition in the lung for use in inhalation dose assessments, *Radiat. Prot. Dosim.*, 11, 5, 1985.
34. Anjilvel, S. and Asgharian, B., A multiple-path model of particle deposition in the rat lung, *Fund. Appl. Toxicol.*, 28, 41, 1995.

35. Phalen, R.F., Oldham, M.J., and Mautz, W.J., Aerosol deposition in the nose as a function of body size, *Health Phys.*, 57, 299, 1989.
36. Phalen, R.F. and Oldham, M.J., Methods for modeling particle deposition as a function of age, *Respir. Physiol.*, 128, 119, 2001.
37. Koblinger, L. and Hofmann, W., Analysis of human lung morphometric data for stochastic aerosol deposition calculations, *Phys. Med. Biol.*, 30, 541, 1985.
38. Koblinger, L. and Hofmann, W., Monte Carlo modeling of aerosol deposition in human lungs. Part I. Simulation of particle transport in a stochastic lung structure, *J. Aerosol Sci.*, 21, 661, 1990.
39. Koblinger, L. and Hofmann, W., Stochastic morphological model of the rat lung, *Anat. Rec.*, 221, 533, 1988.
40. Hofmann, W., Koblinger, L., and Martonen, T.B., Structural differences between human and rat lungs: implications for Monte Carlo modeling of aerosol deposition, *Health Phys.*, 57, 41, 1989.
41. Hofmann, W., Asgharian, B., Bergmann, R., Anjilvel, S., and Miller, F.J., The effect of heterogeneity of lung structure on particle deposition in the rat lung, *Toxicol. Sci.*, 53, 430, 2000.
42. Hofmann, W., Koblinger, L., and Mohamed, A., Incorporation of biological variability into lung dosimetry by stochastic modeling techniques, *Environ. Int.*, 22, S995, 1996.
43. Hofmann, W., Morawska, L., and Bergmann, R., Environmental tobacco smoke deposition in the human respiratory tract: differences between experimental and theoretical approaches, *J. Aerosol Med.*, 14, 317, 2001.
44. Hofmann, W. and Koblinger, L., Monte Carlo modeling of aerosol deposition in human lungs. Part II. Deposition fractions and their sensitivity to parameter variations, *J. Aerosol Sci.*, 21, 675, 1990.
45. Comer, J.K., Kleinstreuer, C., and Zhang, Z., Flow structures and particle deposition patterns in double-bifurcation airway models. Part 2. Aerosol transport and deposition, *J. Fluid Mech.*, 435, 55, 2001.
46. Martonen, T.B. and Guan, X., Effects of tumors on inhaled pharmacologic drugs. I. Flow patterns, *Cell Biochem. Biophys.*, 35, 233, 2001.
47. Martonen, T.B. and Guan, X., Effects of tumors on inhaled pharmacologic drugs. II. Particle motion, *Cell Biochem. Biophys.*, 35, 245, 2001.
48. Zhang, Z., Kleinstreuer, C., and Kim, C.S., Cyclic micron-sized particle inhalation and deposition in a triple bifurcation lung airway model, *J. Aerosol Sci.*, 33, 257, 2001.
49. Zhang, Z., Kleinstreuer, C., and Kim, C.S., Effects of curved inlet tubes on airflow and particle deposition in bifurcating lung models, *J. Biomech.*, 34, 659, 2001.
50. Zhang, Z., Kleinstreuer, C., and Kim, C.S., Flow structure and particle transport in a triple bifurcation airway model, *J. Fluids Eng.*, 123, 320, 2001.
51. Comer, J.K., Kleinstreuer, C., and Zhang, Z., Flow structures and particle deposition patterns in double-bifurcation airway models. Part 1. Air flow fields, *J. Fluid Mech.*, 435, 25, 2001.
52. Martonen, T., Commentary "Effects of asymmetric branch flow rates on aerosol deposition in bifurcating airways" by Z. Zhang, C. Kleinstreuer and C.S. Kim, *J. Med. Eng. Tech.*, 25, 124, 2001.
53. Zhang, Z., Kleinstreuer, C., and Kim, C.S., Micro-particle transport and deposition in a human oral airway model, *J. Aerosol Sci.*, 33, 1635, 2002.
54. Hofmann, W., Modeling techniques for inhaled particle deposition: the state of the art, *J. Aerosol Med.*, 9, 369, 1996.
55. Zhang, Z. and Kleinstreuer, C., Transient airflow structures and particle transport in a sequentially branching lung airway model, *Phys. Fluids*, 14, 862, 2002.
56. Martonen, T.B. and Schroeter, J.D., Risk assessment dosimetry model for inhaled particulate matter. I. Human subjects, *Toxicol. Lett.*, 138, 119, 2003.
57. Weibel, E.R., *Morphometry of the Human Lung*, Academic Press, New York, 1963.
58. Weibel, E.R., Design of airways and blood vessels as branching trees, in *The Lung: Scientific Foundations*, Vol. I, Crystal, R.G.,West, J.B., Barnes, P.J., Cherniak, N.S., and Weibel E.R., Eds., Raven Press, New York, 1991.
59. Horsfield, K., Pulmonary airways and blood vessels considered as confluent trees, in *The Lung: Scientific Foundations*, Vol. I, Crystal, R.G., West, J.B., Barnes, P.J., Cherniak, N.S., and Weibel E.R., Eds., Raven Press, New York, 1991.

60. Horsfield, K. and Cumming, G., Angles of branching and diameters of branches in the human bronchial tree, *Bull. Math. Biophys.*, 29, 245, 1967.
61. Horsfield, K. and Cumming, G., Morphology of the bronchial tree in man, *J. Appl. Physiol.*, 24, 373, 1968.
62. Soong, T.T., Nicolaides, P., Yu, C.P., and Soong, S.C., A statistical description of the human tracheobronchial tree geometry, *Respir. Physiol.*, 37, 161, 1979.
63. Jesseph, J.E. and Merendino, K.A., The dimensional interrelationships of the major components of the human tracheobronchial tree, *Surg. Gynecol. Obstet.*, 105, 210, 1957.
64. Parker, H., Horsfield, K., and Cumming, G., Morphology of the distal airways in the human lung, *J. Appl. Physiol.*, 31, 386, 1971.
65. Horsfield, K., Dart, G., Olson, D.E., Filley, G.F., and Cumming, G., Models of the human tracheobronchial tree, *J. Appl. Physiol.*, 31, 207, 1971.
66. Kitaoka, H., Takaki, R., and Suki, B., A three-dimensional model of the human airway tree, *J. Appl. Physiol.*, 87, 2207, 1999.
67. Spencer, R.M., Schroeter, J.D., and Martonen, T.B., Computer simulations of lung airway structures using data-driven surface modeling techniques, *Comput. Biol. Med.*, 31, 499, 2001.
68. Musante, C. and Martonen, T.B., Computational fluid dynamics in human lungs. I. Effects of natural airway features, in *Medical Applications of Computer Modeling: The Respiratory System*, Martonen, T.B., Ed., WIT Press, Boston, 2001.
69. Martonen, T.B., Zhang, Z., Yang, Y., and Bottei, G., Airway surface irregularities promote particle diffusion in the human lung, *Radiat. Prot. Dosim.*, 59, 5, 1995.
70. Balásházy, I. and Hofmann, W., Quantification of local deposition patterns of inhaled radon decay products in human bronchial airway bifurcations, *Health Phys.*, 78, 147, 2000.
71. Niinimaa, V., Cole, P., Mintz, S., and Shephard, R.J., The switching point from nasal to oronasal breathing, *Respir. Phys.*, 42, 61, 1980.
72. Niinimaa, V., Cole, P., Mintz, S., and Shephard, R.J., Oronasal distribution of respiratory airflow, *Respir. Phys.*, 43, 69, 1981.
73. James, D.S., Lambert, W.E., Mermier, C.M., Stidley, C.A., Chick, T.W., and Samet, J.M., Oronasal distribution of ventilation at different ages, *Arch. Environ. Health*, 52, 118, 1997.
74. Everard, M.L., Hardy, J.G., and Milner, A.D., Comparison of nebulised aerosol deposition in the lungs of healthy adults following oral and nasal inhalation, *Thorax*, 48, 1045, 1993.
75. Brain, J.D. and Sweeney, T.D., Effects of ventilatory patterns and pre-existing disease on deposition of inhaled particles in animals, in *Extrapolation of Dosimetric Relationships for Inhaled Particles and Gases*, Crapo, J.D., Miller, F.J., Smolko, E.D., Graham, J.A., and Hayes, A.W., Eds., Academic Press, Inc., New York, 1989, chap. 15.
76. Lennon, S., Shang, S., Lessmann, R., and Webster, S., Experiments on particle deposition in the human upper respiratory system, *Aerosol Sci. Tech.*, 28, 464, 1998.
77. Becquemin, M.H., Swift, D.L., Bouchikhi, A., Roy, M., and Teillac, A., Particle deposition and resistance in the noses of adults and children, *Eur. Respir. J.*, 4, 694, 1991.
78. Swift, D.L. and Strong, J.C., Nasal deposition of ultrafine ^{218}Po aerosols in human subjects, *J. Aerosol Sci.*, 27, 1125, 1996.
79. Schwab, J.A. and Zenkel, M., Filtration of particles in the human nose, *Laryngoscope*, 108, 120, 1998.
80. Morawska, L., Barron., W., and Hitchins, J., Experimental deposition of environmental tobacco smoke submicrometer particulate matter in the human respiratory tract, *Am. Ind. Hyg. Assoc. J.*, 60, 334, 1999.
81. Anderson, I., Lundquist, G.R., Proctor, D.F., and Swift, D.L., Human response to controlled levels of inert dust, *Am. Rev. Respir. Dis.*, 119, 619, 1979.
82. Cheng, K., Cheng, Y., Yeh, H., Guilmette, R.A., Simpson, S.Q., Yang, Y., and Swift, D.L., *In vivo* measurements of nasal airway dimensions and ultrafine aerosol deposition in the human nasal and oral airways, *J. Aerosol Sci.*, 27, 785, 1996.
83. Kesavan, J., Bascom, R., Laube, B., and Swift, D.L., The relationship between particle deposition in the anterior nasal passage and nasal passage characteristics, *J. Aerosol Med.*, 13, 17, 2000.

84. Brand, P., Friemel, I., Meyer, T., Schulz, H., Heyder, J., and Haubinger, K., Total deposition of therapeutic particles during spontaneous and controlled inhalations, *J. Pharm. Sci.*, 89, 724, 2000.
85. Heyder, J., Gebhart, J., Stahlhofen, W., and Stuck, B., Biological variability of particle deposition in the human respiratory tract during controlled and spontaneous mouth-breathing, *Ann. Occup. Hyg.*, 26, 137, 1982.
86. Bennett, W.D. and Smaldone, G.C., Human variation in the peripheral air-space deposition of inhaled particles, *J. Appl. Physiol.*, 62, 1603, 1987.
87. Schiller-Scotland, C.F., Hlawa, R., and Gebhart, J., Experimental data for total deposition in the respiratory tract of children, *Toxicol. Lett.*, 72, 137, 1994.
88. Brown, J.S., Zeman, K.L., and Bennett, W.D., Ultrafine particle deposition and clearance in the healthy and obstructed lung, *Am. J. Repir. Crit. Care Med.*, 166, 1240, 2002.
89. Bennett, W.D., Messina, M., and Smaldone, G.C., Effect of exercise on deposition and subsequention retention of inhaled particles, *J. Appl. Physiol.*, 59, 1046, 1985.
90. Messina, M.A. and Smaldone, G.C., Evaluation of quantitative aerosol techniques for use in bronchoprovocation studies, *J. Allergy Clin. Immunol.*, 75, 252, 1985.
91. Stahlhofen, W., Gebhart, J., and Heyder, J., Experimental determination of the regional deposition of aerosol particles in the human respiratory tract, *Am. Ind. Hyg. Assoc. J.*, 41, 385, 1980.
92. Heyder, J., Armbruster, L., Gebhart, J., Grein, E., and Stahlhofen, W., Total deposition of aerosol particles in the human respiratory tract for nose and mouth breathing, *J. Aerosol Sci.*, 6, 311, 1975.
93. Svartengren, M., Svartengren, K., Aghaie, F., Philipson, K., and Camner, P., Lung deposition and extremely slow inhalations of particles. Limited effect of induced airway obstruction, *Exp. Lung. Res.*, 25, 353, 1999.
94. Keck, T., Leiacker, R., Heinrich, A., Khunemann, S., and Rettinger, G., Humidity and temperature profile in the nasal cavity, *Rhinology*, 38, 167, 2000.
95. Ingelstedt, S., Studies on the conditioning of air in the respiratory tract, *Acta Otolaryngol.*, 131, 1, 1956.
96. Cole, P., Recordings of respiratory air temperature, *J. Laryngol. Otol.*, 68, 295, 1954.
97. Perwitzschky, R., Die temperatur and feuchtigkeitsverhaltnisse der atemluft in den luftwegen.1, *Mitt. Arch. Ohren Nasen Kehlkopfh*, 117, 1, 1928.
98. Verzar, F., Keith, T., and Parchet, V., Temperatur and feuchtigkeit der lugt in den atemwegen, *Pflugers Arch. Ges. Physiol.*, 257, 400, 1953.
99. Herlitzka, A., Sur la temperature tracheale de l'air inspire et expire, *Arch. Int. Physiol.*, 18, 587, 1921.
100. Dery, R., Pelletier, J., Jaques, H., Clavet, M., and Houde, J.J., Humidity in anaesthesiology. III. Heat and moisture patterns in the respiratory tract during anaesthesia with the semi-closed system, *Can. Anaesth. Soc. J.*, 14, 287, 1967.
101. Dery, R., The evolution of heat and moisture in the respiratory tract during anaesthesia with a non-rebreathing system, *Can. Anaesth. Soc. J.*, 20, 296, 1973.
102. McFadden, E.R., Pichurko, B.M., Bowman, F.H., Ingenito, E., Burns, S., Dowling, N., and Solway, J., Thermal mapping of the airways in humans, *J. Appl. Physiol.*, 58, 564, 1985.
103. McRae, R.D.R., Jones, A.S., Young, P., and Hamilton, J., Resistance, humidity and temperature of the tracheal airway, *Clin. Otolaryngol.*, 20, 355, 1995.
104. Martonen, T.B., Hoffman, W., Eisner, A.D., and Ménache, M.G., The role of particle hygroscopicity in aerosol therapy and inhalation toxicology, in *Extrapolation of Dosimetric Relationships for Inhaled Particles and Gases*, Crapo, J.D. and Smolko, E.D., Eds., Academic Press, Inc., New York, 1989.
105. Kousaka, Y., Okuyama, K., and Wang, C.S., Response of cigarette smoke particles to change in humidity, *J. Chem. Eng. Jpn.*, 15, 75,1982.
106. Li, W. and Hopke, P.K., Initial size distributions and hygroscopicity of indoor combustion aerosol particles, *Aerosol Sci. Tech.*, 19, 305, 1993.
107. Hicks, J.F., Pritchard, J.N., Black, A., and Megaw, W.J., Experimental evaluation of aerosol growth in the human respiratory tract, in: *Aerosols: Formation and Reactivity*, Schikarski W., Fissan H.J., and Friedlander, S.K., Eds., Pergamon Press, Oxford, 1986, 243.

108. Gysel, M., Weingartner E., and Baltensperger, U., Hygroscopicity of aerosol particles at low temperatures. 2. Theoretical and experimental hygroscopic properties of laboratory generated aerosols, *Environ. Sci. Technol.*, 36, 63, 2002
109. Tang, I.N., Munkelwitz, H.R., and Davis, J.G., Aerosol growth studies, II. Preparation and growth measurements of monodisperse salt aerosols, *J. Aerosol Sci.*, 8, 149, 1977.
110. Tang, I.N. and Munkelwitz, H.R., Aerosol growth studies, III. Ammonium bisulfate aerosols in a moist atmosphere, *J. Aerosol Sci.*, 8, 321, 1977.
111. Smith, G., Hiller, F.C., Mazumder, M.K., and Bone, R.C., Aerodynamic size distribution of cromolyn sodium at ambient and airway humidity, *Am. Rev. Respir. Dis.*, 121, 513, 1980.
112. Hiller, F.C., Mazumder, M.K., Wilson, J.D., and Bone, R.C., Effect of low and high relative humidity on metered-dose bronchodilator solution and powder aerosols, *J. Pharm. Sci.*, 69, 334, 1980.
113. Hiller, F.C., Mazumder, M.K., Wilson, J.D., and Bone, R.C., Aerodynamic size distribution, hygroscopicity and deposition estimation of beclomethasone dipropionate aerosol, *J. Pharm. Pharmacol.*, 32, 605, 1980.
114. Kim, C.S., Trujillo, D., and Sackner, M.A., Size aspects of metered-dose inhaler aerosols, *Am. Rev. Respir. Dis.*, 132, 137, 1985.
115. Seemann, S., Busch, B., Ferron, G.A., Silberg, A., and Heyder, J., Measurement of the hygroscopicity of pharmaceutical aerosols *in situ*, *J. Aerosol Sci.*, 26, 537, 1995.
116. Peng, C., Chow, A.H., and Chan, C.K., Study of the hygroscopic properties of selected pharmaceutical aerosols using single particle levitation, *Pharm. Res.*, 17, 1104, 2000.
117. Bell, K.A. and Ho, A.T., Growth rate measurements of hygroscopic aerosols under conditions simulating the respiratory tract, *J. Aerosol Sci.*, 12, 247, 1981.
118. Martonen, T.B., Bell, K.A., Phalen, R.F., Wilson, A.F., and Ho, A.T., Growth rate measurements and deposition modeling of hygroscopic aerosols in human tracheobronchial models, *Ann. Occup. Hyg.*, 26, 93, 1982.
119. Broday, D.M. and Georgopoulos, P.G., Growth and deposition of hygroscopic particulate matter in the human lungs, *Aerosol Sci. Tech.*, 34, 144, 2001.
120. Finlay, W.H., Estimating the type of hygroscopic behavior exhibited by aqueous droplets, *J. Aerosol Med.*, 11, 221, 1998.
121. Robinson, R.J. and Yu, C.P., Theoretical analysis of hygroscopic growth rate of mainstream and sidestream cigarette smoke particles in the human respiratory tract, *Aerosol Sci. Tech.*, 28, 21, 1998.
122. Ferron, G.A., The size of soluble aerosol particles as a function of the humidity of the air. Application to the human respiratory tract, *J. Aerosol Sci.*, 8, 251, 1977.
123. Oberdörster, G., Lung clearance of inhaled insoluble and soluble particles, *J. Aerosol Med.*, 1, 289, 1988.
124. West, J.B., *Pulmonary Pathophysiology, The Essentials*, 5th ed., Lippincott, Williams & Wilkins, Philadelphia, 1998.
125. Gehr, P., Schurch, S., Im Hof, V., and Geiser, M., Inhaled particles deposited in the airways are displaced towards the epithelium, in *Inhaled Particles VII, The Annals of Occupational Hygiene*, Walton, W.H., Critchlow, A., and Coppock, S.M., Eds., Pergamon Press, New York, 1994.
126. West, J.B., *Respiratory Physiology, The Essentials*, 6th ed., Lippincott, Williams & Wilkins, Philadelphia, 2000.
127. Yeates, D.B., Gerrity, T.B., and Garrard, C.S., Characteristics of tracheobronchial deposition and clearance in man, in *Inhaled Particles V, The Annals of Occupational Hygiene*, Walton, W.H., Critchlow, A., and Coppock, S.M., Eds., Pergamon Press, New York, 1982
128. Yeates, D.B., Gerrity, T.B., and Garrard, C.S., Particle deposition and clearance in the bronchial tree, *Ann. Biomed. Eng.*, 9, 577, 1981.
129. Fazio, F. and Lafortuna, C., Effect of inhaled salbutamol on mucociliary clearance in patients with chronic bronchitis, *Chest*, 80, 827, 1981.
130. Foster, W.M., Langenback, E.G., and Bergofsky, E.H., Respiratory drugs influence mucociliary clearance in central and peripheral ciliated airways, *Chest*, 80, 877, 1981.
131. Foster, W.M., Langenback, E.G., and Bergofsky, E.H., Lung mucociliary function in man, *Ann. Occup. Hyg.*, 26, 227, 1982.

132. Mortensen, J., Lange, P., Nyboe, J., and Groth, S., Lung mucociliary clearance, *Eur. J. Nucl. Med.*, 21, 953, 1994.
133. Weiss, T., Dorrow, P., and Felix, R., Effects of a beta adrenergic drug and a secretolytic agent on regional mucociliary clearance in patients with COLD, *Chest*, 80, 881, 1981.
134. Gerrity, T.R., Cotormanes, E., Garrard, C.S., Yeates, D.B., and Lourenco, R.V., The effect of aspirin on lung mucociliary clearance, *New Engl. J. Med.*, 308, 139, 1983.
135. Pavia, D., Sutton, P.P., Lopez-Vidriero, M.T., Agnew, J.E., and Clarke, S.W., Drug effects on mucociliary function, *Eur. J. Respir. Dis.*, 64, 304, 1983.
136. Leikauf, G., Yeates, D.B., Wales, K.A., Spektor, D., Albert, R.E., and Lippman, M., Effects of sulfuric acid aerosol on respiratory mechanics and mucociliary particle clearance in healthy non-smoking adults, *Am. Ind. Hyg. J.*, 42, 273, 1981.
137. Lippman, M., Schlesinger, R.B., Leikauf, G., Spektor, D., and Albert, R.E., Effects of sulfuric acid aerosols on respiratory tract airways, in *Inhaled Particles V, The Annals of Occupational Hygiene*, Walton, W.H., Critchlow, A., and Coppock, S.M., Eds., Pergamon Press, New York, 1982.
138. Camner, P., Clearance of particles from the human tracheobronchial tree, *Clin. Sci.*, 59, 79, 1980.
139. Kenoyer, J.L., Phalen, R.F., and Davis, J.R., Particle clearance from the respiratory tract as a test of toxicity: effect of ozone on short and long term clearance, *Exp. Lung. Res.*, 2, 111, 1981.
140. Weiss, T., Dorrow, P., and Felix, R., Regional mucociliary removal of inhaled particles in smokers with small airways disease, *Respiration*, 44, 338, 1983.
141. Foster, W.M., Costa, D.L., and Langenback, E.G., Ozone exposure alters tracheobronchial mucociliary function in humans, *J. Appl. Physiol.*, 63, 996, 1987.
142. Vastag, E., Matthys, H., Zsamboki, G., Kohler, D., and Daileler, G., Mucociliary clearance in smokers, *Eur. J. Respir. Dis.*, 68, 107, 1986.
143. Jarstrand, C., Camner, P., and Philipson, K., *Mycoplasma pneumoniae* and tracheobronchial clearance, *Am. Rev. Respir. Dis.*, 110, 415, 1974.
144. Camner, P., Jarstrand, C., and Philipson, K., Tracheobronchial clearance in patients with influenza, *Am. Rev. Respir. Dis.*, 108, 131, 1973.
145. Pavia, D., Sutton, P.P., Agnew, J.E., Lopez-Vidriero, M.T., Newman, S.P., and Clarke, S.W., Measurement of bronchial mucociliary clearance, *Eur. J. Respir. Dis.*, 64, 41, 1983.
146. Bateman, J.R.M., Pavia, D., Sheahan, N.F., Agnew, J.E., and Clarke, S.W., Impaired tracheobronchial clearance in patients with mild stable asthma, *Thorax*, 38, 463, 1983.
147. Pavia, D., Lung mucociliary clearance, in *Aerosols and the Lung: Clinical and Experimental Aspects*, Clarke, S.W. and Pavia, D., Eds., Butterworths, Boston, 1984, chap. 6.
148. Robinson, M., Everl, S., Tomlinson, C., Daviskas, E., Regnis, J.A., Bailey, D.L., Torzillo, P.J., Ménache, M., and Bye, P.T., Regional mucociliary clearance in patients with cystic fibrosis, *J. Aerosol Med.*, 13, 73, 2000.
149. Goodman, R.M., Yergin, B.M., Landa, J.F., Golinvaux, M.H., and Sackner, M.A., Relationship of smoking history and pulmonary function tests to tracheal mucous velocity in non-smokers, young smokers, ex-smokers, and patients with chronic bronchitis, *Am. Rev. Respir. Dis.*, 117, 205, 1978.
150. Puchelle, E., Sahm, J.M., Bertrand, A., Influence of age on bronchial mucociliary transport, *Scand. J. Respir. Dis.*, 60, 307, 1979.
151. Vastag, E., Matthys, H., Kohler, D., Gronbeck, L., and Daileler, G., Mucociliary clearance and airway obstruction in smokers, ex-smokers and normal subjects who never smoked, *Eur. J. Respir. Dis.*, 139, 93, 1985.
152. Incalzi, R.A., Maini, C.L., Fuso, L., Giordano, A., Carbonin, P.U., and Galli, G., Effects of aging on mucociliary clearance, *Compr. Gerontol.*, 3, 65, 1989.
153. Ho, J.C., Chan, K.N., Hu, W.H., Lam, W.K., Zheng, L., Tipoe, G.L., Sun, J., Leung, R., and Tsang, K.W., The effect of aging on nasal mucociliary clearance, beat, frequency, and ultrastructure of respiratory cilia, *Am. J. Respir. Crit. Care Med.*, 163, 983, 2001.
154. Wolff, R.K., Dolovich, M.B., Obminsky, G., Newhouse, M.T., Effects of exercise and eucapnic hyperventilation on bronchial clearance in man, *J. Appl. Physiol.*, 43, 46, 1977.
155. Salzano, F.A., Manola, M., Tricarico, D., Precone, D., and Motta, G., Mucociliary clearance after aerobic exertion in athletes, *Acta Otorhinolaryngol. Ital.*, 20, 171, 2000.

156. Adamson, I.Y. and Bowden, D.H., Dose response of the pulmonary macrophagic system to various particulates and its relationship to transepithelial passage of free particles, *Exp. Lung Res.*, 2, 165, 1981.
157. Timbrell, V., Deposition and retention of fibres in the human lung, *Ann. Occup. Hyg.*, 26, 347, 1982.
158. Coin, P.G., Stevens, J.B., and McJilton, C.M., Role of fiber length in the pulmonary clearance of amosite asbestos, *Am. Rev. Respir. Dis.*, 141, A521, 1990.
159. Coin, P.G., Roggli, V.L., and Brody, A.R., Persistence of long, thin chrysotile asbestos fibers in the lungs of rats, *Environ. Health Perspect.*, 102, 197, 1994.
160. Donaldson, K. and Tran, C.L., Inflammation caused by particles and fibers, *Inhal. Toxicol.*, 14, 5, 2002.
161. Oberdörster, G., Toxicokinetics and effects of fibrous and nonfibrous particles, *Inhal. Toxicol.*, 14, 29, 2002.
162. Witschi, H.R. and Last, J.A., Toxic responses of the respiratory system, in *Caserett and Doul's Toxicology, The Basic Science of Poisons*, 6th ed., Klaasen, C.D., Ed., McGraw-Hill, New York, 2001, chap. 15.
163. Hocking, W.G. and Golde, D.W., The pulmonary alveolar macrophage, *New Engl. J. Med.*, 301, 580, 1979.
164. Holt, P.G. and Keast, D., Environmentally induced changes in immunologic function. Acute and chronic effects of inhalation of tobacco smoke and other atmospheric contaminants in man and experimental animals, *Bacteriol. Rev.*, 41, 205, 1977.
165. Thomas, W.R., Holt, P.G., and Keast, D., Cigarette smoke and phagocyte function: effect of chronic exposure *in vivo* and acute exposure *in vitro*, *Infect. Immunol.*, 20, 468, 1978.
166. Morrow, P.E., Possible mechanisms to explain dust overloading of the lungs, *Fundam. Appl. Toxicol.*, 10, 369, 1988.
167. Oberdörster, G., Lung particle overload: implication for occupational exposures to particles, *Reg. Toxicol. Pharm.*, 27, 123, 1995.
168. Churg, A., The uptake of mineral particles by pulmonary epithelial cells, *Am. J. Respir. Crit. Care Med.*, 154, 1996.
169. Churg, A., Wright, J.L., and Stevens, B., Exogenous mineral particles in the human bronchial mucosa and lung parenchyma. I. Nonsmokers in the general population, *Exp. Lung Res.*, 16, 159, 1990.
170. Dumortier, P., De Vuyst, P., and Yernault, J.C., Comparative analysis of inhaled particle contained in human bronchoalveolar fluids, lung parenchyma and lymph nodes, *Environ. Health Perspect.*, 102, 257, 1994.
171. Geiser, M., Morphological aspects of particle uptake by lung phagocytes, *Microsc. Res.*, 57, 512, 2002.
172. Ferin, J., Oberdörster, G., and Penney, D.P., Pulmonary retention of ultrafine and fine particles in rats, *Am. J. Respir. Cell. Mol. Biol.*, 6, 535, 1992.
173. Ferin, J., Oberdörster, G., Soderhold, S.C., and Gelein, R., The rate of dose delivery affects pulmonary interstitialization of particles in rats, *Ann. Occup. Hyg.*, 38, 289, 1994.
174. Oberdörster, G., Finkelstein, J.N., Johnston, C., Gelein, R., Cox, C., Baggs, R., and Elder, A.C., Acute pulmonary effects of ultrafine particles in rats and mice, *Res. Respir. Health Eff. Inst.*, 96, 5, 2000.
175. Oberdörster, G., Ferin, J., and Lehnert, B.E., Correlation between particle size, *in vivo*, particle persistence and lung injury, *Environ. Health Perspect.*, 102, 173, 1994.
176. Sosnowski, T.R., Gradón, L., and Podgórski, A., Influence of insoluble aerosol depositions on the surface activity of the pulmonary surfactant: a possible mechanism of alveolar clearance retardation? *Aerosol Sci. Tech.*, 32, 52, 2000.
177. Gerrity, T.R., Garrard, C.S., and Yeates, D.B., A mathematical model of particle retention in the air-spaces of the human lung, *Br. J. Ind. Med.*, 40, 121, 1983.
178. Sanchis, J., Dolovich, M., Chalmers, R., and Newhouse, M., Quantitation of regional aerosol clearance in the normal human lung, *J. Appl. Physiol.*, 33, 757, 1972.
179. Hofmann, W., Ménache, M.G., and Martonen, T.B., Age-dependent lung dosimetry of radon progeny, in *Extrapolation of Dosimetric Relationships for Inhaled Particles and Gases*, Crapo, J.D., Smolko, E.D., Miller, F.J., Graham, J.A., and Hayes, A.W., Eds., Academic Press, San Diego, 1989.

180. Martonen, T.B. and Hofmann, W., Dosimetry of localised accumulations of cigarette smoke and radon progeny at bifurcations, *Radiat. Prot. Dosim.*, 38, 81, 1991.
181. Martonen, T.B., Fleming, J., Schroeter, J., Conway, J., and Hwang, D., In silico modeling of asthma, *Adv. Drug Deliv. Rev.*, 55, 829, 2003.
182. Davis, P.B., Drumm, M., and Knostan, M.W., Cystic fibrosis, *Am. J. Respir. Crit. Care Med.*, 154, 1229, 1996.
183. Brown, J.S., Regional Ventilation and Particle Deposition in the Healthy and Obstructed Lung, UMI Dissertation Services, Ann Arbor, MI, 2000.
184. Kim, C.S. and Kang, T.C., Comparative measurement of lung deposition of inhaled fine particles in normal subjects and patients with obstructive airway disease, *Am. J. Respir. Crit. Care Med.*, 155, 3, 1997.
185. Anderson, P.J., Blanchard, J.D., Brain, J.D., Feldman, H.D., McNamara, J.J., and Heyder, J. Effect of cystic fibrosis on inhaled aerosol boluses, *Am. Rev. Respir. Dis.*, 140, 1317, 1989.
186. Anderson, P.J., Wilson, J.D., and Hiller, F.C., Respiratory tract deposition of ultrafine particles in subjects with obstructive or restrictive lung disease, *Chest*, 97, 115, 1990.
187. Love, R.G. and Muir, D.C.F., Aerosol deposition and airway obstruction, *Am. Rev. Respir. Dis.*, 114, 891, 1976.
188. Siekmeier, R., Schiller-Scotland, C.H.F., Gebhart, J., and Kronenberger, H., Pharmacon-induced airway obstruction in healthy subjects: dose dependent changes of inspired aerosol boluses, *J. Aerosol Sci.*, 21, S423, 1990.
189. Anderson, P.J., Gann, L.P., Walls, R.C., Tennal, K.B., and Hiller, F.C., Utility of aerosol bolus behavior as a diagnostic index of asthma during bronchoprovocation, *Am. J. Respir. Crit. Care Med.*, 149, A1047, 1994.
190. Brown, J.S., Zeman, K.L., and Bennett, W.D., Regional deposition of coarse particles and ventilation distribution in healthy subjects and patients with cystic fibrosis, *J. Aerosol Med.*, 14, 443, 2001.
191. Ramana, L., Tashkin, D.P., Taplin, G.V., Elam, D., Detels, R., Coulson, A., and Rokaw, S.N., Lung imaging in chronic obstructive pulmonary disease, *Chest*, 68, 634, 1975.
192. Taplin, G.V., Tashkin, D.P., Chopri, S.K., Anselmi, O.E., Elam, D., Calvarese, B., Coulson, A., Detels, R., and Rokaw, S.N., Early detection of chronic obstructive pulmonary disease using radionuclide lung imaging procedures, *Chest*, 71, 567, 1977.
193. Ito, H., Ishii, Y., Maeda, H., Todo, G., Torizuka, K., and Smaldone, G.C., Clinical observations of aerosol deposition in patients with airway obstruction, *Chest*, 80, 837, 1981.
194. Segal, R.A., Martonen, T.B., Kim, C.S., and Shearer, M., Computer simulations of particle deposition in the lungs of chronic obstructive pulmonary disease patients, *Inhal. Toxicol.*, 14, 705, 2002.
195. Martonen, T.B., Katz, I., Hwang, D., and Yang, Y. Biomedical application of the supercomputer: targeted delivery of inhaled pharmaceuticals in diseased lungs, in *Computer Simulations in Biomedicine,* Power, H. and Hart, R.T., Eds., Computational Mechanics Publications, Boston, 1995.
196. Martonen, T.B., Katz, I., and Cress, W., Aerosol deposition as a function of airway disease: cystic fibrosis, *Pharm. Res.*, 12, 96, 1995.
197. Martonen, T.B., Hwang, D., Katz, I., and Yang, Y., Cystic fibrosis: treatment with a supercomputer drug delivery model, *Adv. Eng. Software*, 28, 359, 1997.
198. Environmental Protection Agency, 40 CFR Part 50, National Ambient Air Quality Standards for Particulate Matter (AD-FRL-5725-2, RIN 2060-AE66), *Federal Register*, 62, 38651, 1997.
199. Ménache, M.G. and Graham, R.C., Conducting airway geometry as a function of age, *Ann. Occup. Hyg.*, 41, 531, 1997.
200. Martonen, T.B., Graham, R.C., and Hofmann, W., Human subject age and activity level: factors addressed in biomathematical deposition program for extrapolation modeling, *Health Phys.*, 57, 49, 1989.
201. Hofmann, W., Martonen, T.B., and Graham, R.C., Predicted deposition of nonhygroscopic aerosols in the human lung as a function of subject age, *J. Aerosol Med.*, 2, 49, 1989.
202. Warren, D.W., Harifield, W.M., and Dalston, E.T., Effect of age on nasal cross-sectional area and respiratory mode in children, *Laryngoscope*, 100, 89, 1990.

203. Bennett, W.D., Zeman, K.L., Kang, C.W., and Schechter, M.S., Extrathoracic deposition of inhaled, coarse particles in children vs. adults, *Ann. Occup. Hyg.*, 41, 497, 1997.
204. Harvey, R.P. and Hamby, D.M., Age-specific uncertainty in particulate deposition for 1 micron AMAD particles using ICRP 66 lung model, *Health Phys.*, 82, 807, 2002.
205. Musante, C.J. and Martonen, T.B., Computer simulations of particle deposition in the developing human lung, *J. Air Waste Manage. Assoc.*, 50, 1426, 2000.
206. Sweeney, T.D. and Brain, J.D., Pulmonary deposition: determinants and measurement techniques, *Toxicol. Pathol.*, 19, 384, 1991.
207. Kim, C.S., Methods of calculating lung delivery and deposition of aerosol particles, *Respir. Care*, 45, 695, 2000.
208. Schlesinger, R.B. and Lippmann, M., Particle deposition in casts of the human upper tracheobronchial tree, *Am. Ind. Hyg. Assoc. J.*, 33, 237, 1972.
209. Cohen, B.S., Particle deposition in human and canine tracheobronchial casts: a determinant of radon dose to the critical cells of the respiratory tract, *Health Phys.*, 70, 695, 1996.
210. Phalen, R.F., Oldham, M.J., Beaucage, C.B., Crocker, T.T. and Mortensen, J.D., Postnatal enlargement of human tracheobronchial airways and implications for particle deposition, *Anat. Rec.*, 212, 368, 1986.
211. Gerde, P., Cheng, Y.S., and Medinsky, M.A., *In vivo* deposition of ultrafine aerosols in the nasal airway of the rat, *Fundam. Appl. Toxicol.*, 16, 330, 1991.
212. Kelly, J.T., Kimbell, J.S., and Asgharian, B., Deposition of fine and coarse aerosols in a rat nasal mold, *Inhal. Toxicol.*, 13, 577, 2001.
213. Martonen, T.B., Measurements of particle dose distribution in a model of a human larynx and tracheobronchial tree, *J. Aerosol Sci.*, 14, 11, 1983.
214. Martonen, T.B. and Lowe, J., Assessment of aerosol deposition patterns in human respiratory tract casts, in *Aerosols in the Mining and Industrial Work Environments*, Marple, V.A. and Liu, B.Y.H., Eds., Ann Arbor Science, Ann Arbor, MI, 1983.
215. Schlesinger, R.B., Bohning, D.B., Chan, T.L., and Lippmann, M., Particle deposition in a hollow cast of the human tracheobronchial tree, *J. Aerosol Sci.*, 8, 429, 1977.
216. Davies, C.N., Heyder, J., and Subba Ramu, M.C., Breathing of half-micron aerosols I. Experimental, *J. Appl. Physiol.*, 32, 591, 1972.
217. Heyder, J., Gebhardt, J., Heiger, G., Roth, C., and Stahlhofen, W., Experimental studies of the total deposition of aerosol particles in the human respiratory tract, *J. Aerosol Sci.*, 4, 191, 1973.
218. Rosati, J.A., Brown, J.S., Peters, T.M., Leith, D., and Kim, C.S., A polydisperse aerosol inhalation system designed for human studies, *J. Aerosol Sci.*, 33, 1433, 2002.
219. Rosati, J.A., Leith, D., and Kim, C.S., Monodisperse and polydisperse aerosol deposition in a packed bed, *Aerosol Sci. Tech.*, 37, 528, 2000.
220. Schlesinger, R.B., Gurman, J.L., and Lippmann, M., Particle deposition within bronchial airways: comparisons using cyclic and inspiratory flows, *Ann. Occup. Hyg.*, 26, 47, 1982.
221. Laube, B.L., *In vivo* measurements of aerosol dose and distribution: clinical relevance, *J. Aerosol Med.*, 9, S77, 1996.
222. Fleming, J.S., Halson, P., Conway, J., Moore, E., Nassim, M.A., Hashish, A.H., Bailey, A.G., and Holgate, S.T., Three-dimensional description of pulmonary deposition of inhaled aerosol using data from multimodality imaging, *J. Nucl. Med.*, 37, 873, 1996.
223. Martonen, T.B., Yang, Y., and Dolovich, M., Definition of airway composition within gamma camera images, *J. Thorac. Img.*, 9, 188, 1994.
224. Martonen, T.B., Yang, Y., Dolovich, M., and Guan, X., Computer simulations of lung morphologies within planar gamma camera images, *Nucl. Med. Comm.*, 18, 861, 1997.
225. Fleming, J.S., Sauret, V., Conway, J.H., Holgate, S.T., Bailey, A.G., and Martonen, T.B., Evaluation of the accuracy and precision of lung aerosol deposition measurements from single-photon emission computed tomography using simulation, *J. Aerosol Med.*, 13, 187, 2000.
226. Finlay, W.H., Stapleton, K.W., Chan, H.K., Zuberbuhler, P., and Gonda, I., Regional deposition of inhaled hygroscopic aerosols: in vivo SPECT compared with mathematical modeling, *J. Appl. Physiol.*, 81, 374, 1996.

227. Lee, Z., Berridge, M.S., Finlay, W.H., and Heald, D.L., Mapping PET-measured triamcinolone acetonide (TAA) aerosol distribution into deposition by airway generation, *Int. J. Pharm.*, 199, 7, 2000.
228. Dolovich, M.B., Influence of inspiratory flow rate, particle size, and airway caliber on aerosolized drug delivery to the lung, *Respir. Care*, 45, 597, 2000.
229. Dolovich, M.B., Measuring total and regional lung deposition using inhaled radiotracers, *J. Aerosol Med.*, 14, S53, 2001.
230. Fleming, J.S., Nassim, M., Hashish, A.H., Bailey, A.G., Conway, J., Holgate, S., Halson, P., and Moore, E., Description of pulmonary deposition of radiolabeled aerosol by airway generation using a conceptual three dimensional model of lung morphology, *J. Aerosol Med.*, 8, 341, 1995.
231. Fleming, J.S., Hashish, A.H., Conway, J., Hartley-Davies, R., Nassim, M.A., Guy, M.J., Coupe, J., and Holgate, S.T., A technique for simulating radionuclide images from the aerosol deposition pattern in the airway tree, *J. Aerosol Med.*, 10, 199, 1997.
232. Martonen, T.B., Hwang, D., Guan, X., and Fleming, J.S., Supercomputer description of human lung morphology for imaging analysis, *J. Nucl. Med.*, 39, 745, 1998.
233. Schroeter, J.D., Fleming, J.S., Hwang, D., and Martonen, T.B., A computer model of lung morphology to analyze SPECT images, *Comput. Med. Img. Graph.*, 26, 237, 2002.
234. Martonen, T.B., Deposition patterns of cigarette smoke in human airways, *Am. Ind. Hyg. Assoc. J.*, 53, 6, 1992.
235. Hofmann, W. and Koblinger, L., Monte Carlo modeling of aerosol deposition in human lungs. Part III. Comparison with experimental data, *J. Aerosol Sci.*, 23, 51, 1992.
236. Heyder, J., Gebhardt, J., Rudolf, G., Schiller, C.F., and Stahlhofen, W., Deposition of particles in the human respiratory tract in the size range 0.005–15 μm, *J. Aerosol Sci.*, 17, 811, 1986.
237. Segal, R.A., Martonen, T.B., and Kim, C.S., Comparison of computer simulations of total lung deposition to human subject data in healthy test subjects, *J. Air Waste Manage. Assoc.*, 50, 1262, 2000.
238. Swift, D.L. and Proctor, D.F., Access of air to the resiratory tract, in *Respiratory Defense Mechanisms*, Brain, J.D., Proctor, D.F., and Reid, L., Eds., Marcel Dekker, New York, 1977.
239. Heyder, J. and Rudolf, G., Deposition of aerosol particles in the human nose, *Inhal. Part.*, 4, 107, 1975.
240. Foord, N., Black, A., and Walsh, M., Regional deposition of 2.5–7.5 μm diameter inhaled particles in the healthy male non-smoker, *J. Aerosol Sci.*, 9, 343, 1978.
241. Cheng, K.H., Cheng, Y.S., Yeh, H.C., and Swift, D.L., Measurements of airway dimensions and calculation of mass transfer characteristics of the human oral passage, *J. Biomech. Eng.*, 119, 476, 1997.
242. Swift, D.L. and Proctor, D.F., A dosimetric model for particles in the respiratory tract above the trachea, *Ann. Occup. Hyg.*, 32, 1035, 1982.
243. Martonen, T.B. and Zhang, Z., Comments on recent data for particle deposition in human nasal passages, *J. Aerosol Sci.*, 23, 667, 1992.

chapter nine

Assessing uncertainties in the relationship between inhaled particle concentrations, internal deposition, and health effects

Phillip N. Price
Lawrence Berkeley National Laboratory

Contents

9.1 Introduction 158
9.1.1 Should we worry about dose or about exposure? 158
9.2 Epidemiological studies 160
9.2.1 Radon epidemiology 161
9.2.1.1 Background 161
9.2.1.2 Case–control studies of residential radon 162
9.2.1.3 Sources of error in exposure estimates in case–control studies of residential radon 164
9.2.1.4 Ecological studies of residential radon 165
9.2.1.5 Discussion of radon risk estimates 166
9.2.1.6 Implications for other aerosol exposure problems 168
9.2.2 Epidemiological studies of the effects of fine particle inhalation 169
9.2.2.1 Time-series studies of the health effects of fine particles 169
9.2.2.2 Cohort studies of the health effects of fine particle inhalation 171
9.2.2.3 Ecological epidemiological studies of the health effects of fine particle inhalation 171
9.3 Inhalation modeling and experiments 172
9.3.1 Motivation for experiments and modeling related to lung deposition of aerosols 172
9.3.2 Assessing uncertainties 172
9.3.3 General approach to assessing the effects of interpersonal variability 174
9.3.4 Respiratory tract morphology and other factors such as medical condition....175
9.3.5 Intersubject morphometric variability 175
9.3.6 Systematic morphometric variability 177
9.3.7 Breathing rate and inhalation details 178
9.3.8 Clearance of the lung 178
9.3.9 Estimating the dose-versus-exposure relationship in a population 180
9.4 Conclusion 182
Acknowledgments 183
References 183

1-56670-611-4 / 05 / $0.00+$1.50

9.1 Introduction

The question that ultimately motivates most aerosol inhalation research is: for a given inhaled atmosphere, what health effects will result in a specified population? To attempt to address this question, quantitative research on inhaled aerosols has been performed for at least 50 years (Landahl et al., 1951). The physical factors that determine particle deposition have been determined, lung morphology has been quantified (particularly for adults), models of total particle deposition have been created and validated, and a large variety of inhalation experiments have been performed. However, many basic questions remain, some of which are identified by the U.S. Committee on Research Priorities for Airborne Particulate Matter (NRC, 1998a) as high-priority research areas. Among these are: What are the quantitative relationships between outdoor concentrations measured at stationary monitoring stations and actual personal exposures? What are the exposures to biologically important constituents of particulate matter that cause responses in potentially susceptible subpopulations and the general population? What is the role of physicochemical characteristics of particulate matter in causing adverse health effects? As these questions show, in spite of significant progress in all areas of aerosol research, many of the most important practical questions remain unanswered or inadequately answered.

In this chapter, we discuss the sources and magnitudes of error that hinder the ability to answer basic questions concerning the health effects of inhaled aerosols. We first consider the phenomena that affect the epidemiological studies, starting with studies of residential radon and moving on to fine particle air pollution. Next, we discuss the major uncertainties in physical and physiological modeling of the causal chain that leads from inhaled aerosol concentration, to deposition in the airway, to time-dependent dose (i.e., the concentration of particles at a given point in the lungs as a function of time), to physiological effects, and finally to health effects.

Figure 9.1 illustrates, in greatly simplified form, the various factors that affect the relationship between exposure and health effects, as well as the measurements of these factors. For instance, the aerosol size and type distribution in the breathing zone is a factor that directly influences deposition in the respiratory tract, but parameters can be directly observed only at a sampling location, which may or may not sample the breathing zone. The network of rectangular boxes in the central column of the figure shows the causal chain that controls the relationship between aerosol exposure, dose, and ultimately health effects. To first order, all connections move down the chain: size distribution, inhalation details, and respiratory tract morphology control deposition, which, combined with clearance, affects dose, and so on. Connections can run the other direction too, as shown by the nonbold arrows in the figure: physiological effects can change the clearance rate and the inhalation rate. These effects can be quite important for large exposures to irritating compounds.

For a given inhaled atmosphere, what health effects will result in a specified population? The simplest attempts (at least conceptually) to address this question are epidemiological studies. Epidemiological studies directly investigate the statistical relationship between aerosol measurements and health outcomes is investigated directly, without attempting to model the causal chain that connects these parameters. Examples include studies on the relationship between radon (and its decay products) and lung cancer, both in miners and in the general population, and also studies on the relationship between outdoor aerosols and morbidity or mortality in the general population, as determined from hospital records or death certificates, respectively.

9.1.1 Should we worry about dose or about exposure?

We will not attempt a technical definition of "dose," but generally speaking we mean the biologically significant quantity of particles delivered to a given part of the respiratory

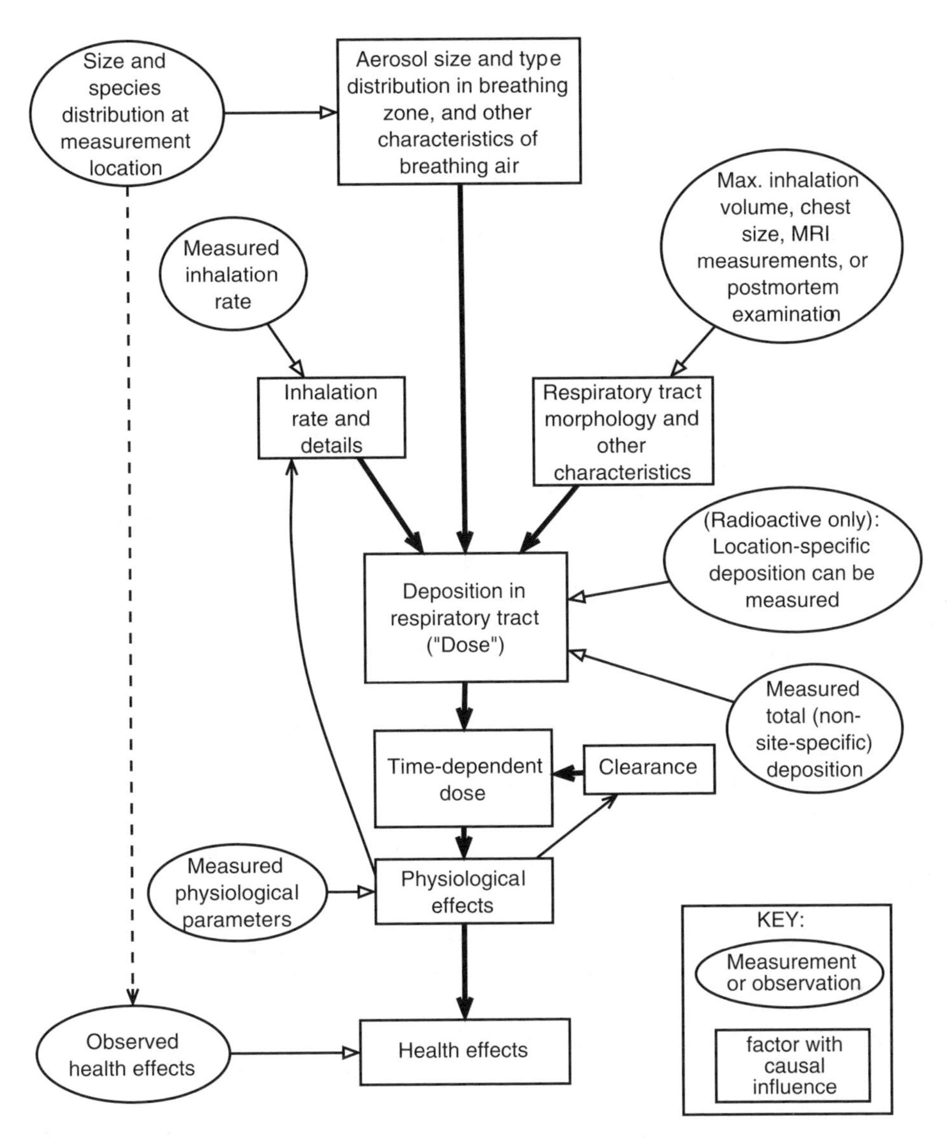

Figure 9.1 Schematic diagram showing the "causal chain" between aerosol inhalation and health effects (boxes), and the relationship of measurable parameters to physical or physiological quantities (ovals).

tract, so that dose may be a function of location. Determination of dose is very difficult, because dose is almost impossible to measure directly (except in the case of some radioactive particles). Most studies, including essentially all epidemiological studies, use a surrogate measure, "exposure," which is the time-integrated concentration of a pollutant in the subject's breathing air. Individuals with the same exposure might experience a very different dose: a large person performing exercise and breathing hard through the mouth will receive a different dose from that received by a small resting person breathing shallowly through the nose, even if they are breathing the same air.

Many questions in applied aerosol inhalation concern the relationship between *exposure* and health effects: what regulatory limits (if any) should be placed on pollutant emissions, what airborne pollutant concentration limits should be set by the Clean Air Act, at what indoor radon concentration should remediation be performed, and so on. From the point of view of a regulatory agency or an industrial hygienist, the knowledge that a particular quantity of a given pollutant, if deposited in the lower respiratory tract, will produce a particular health outcome is almost immaterial. What they really want to know is, what is an acceptable concentration of the pollutant in the air?

Since the relationship between exposure and health effects is often the question of interest, it is tempting to think that the difficult issue of true dosimetry (as opposed to exposure assessment) can be avoided altogether. Why not simply perform all experiments, case–control analyses, etc., in terms of exposure? The answer is that, because biological effects are caused by dose and not exposure, knowledge of the dose–response relationship can be used to answer questions that knowledge of the exposure–response relationship cannot resolve.

One situation in which an understanding of dose rather than merely exposure is required is the extrapolation of experimental results to different populations or environmental conditions. For example, data on miners exposed to high levels of radioactive aerosols produced by radon decay demonstrate quite convincingly that very high levels of exposure cause lung cancer, and that risk increases with exposure (see Lubin et al. [1995] for an overview, and Kusiak et al. [1993], Howe et al. [1986], Howe et al. [1987], Tomasek et al. [1994], and many others for original reports). What relevance, if any, do these results have for residential concentrations, which are typically lower by a factor of 100 or more? Working miners breathe much harder than most people, are simultaneously exposed to large concentrations of other (nonradioactive) aerosols, and experience a different relationship between radon concentrations (the quantity that is measured) and radon decay product concentrations (which actually provide the radioactive dose to the lung). Consequently, extrapolation to the general population is quite challenging. Similar issues arise in exposures to industrial compounds, for which extrapolation is usually required in order to apply observational results from exposed workers to the general population.

Another situation that requires an understanding of dose is airborne delivery of drugs such as asthma medicine: what particle size will be most efficient, how will the answer vary with health status (e.g., ability to take a deep breath), and so forth, are questions that require an understanding of the full relationship between exposure, dose, and health effects, not just exposure and health effect.

Generally speaking, observational data on the exposure–response relationship for a pollutant can be sufficient to roughly characterize the relationship in the regime in which health effects are very large, but such data cannot definitively address the relationship for much smaller exposures; for that, an understanding of some or all of the causal chain is required.

9.2 Epidemiological studies

Epidemiological approaches to determine the health effects from airborne particles attempt to avoid the "how and why" — the network of physical and physiological connections shown in Figure 9.1 — and instead directly investigate the matter of ultimate interest: what are the adverse outcomes (if any) from human exposure to aerosols?

Epidemiological studies of the health effects of aerosols fall into four categories:

1. Case–control studies that relate observed health outcomes to past particle exposures of cases (people with a particular health problem) and controls (people without the problem). Do sick people have higher exposure than healthy people? Or, equivalently as it turns out, are more highly exposed people more likely to become sick?
2. "Ecological" studies that compare long-term aerosol data, and long-term data on health outcomes, across cities or regions that have different airborne particulate

concentration. The essential question here is: Where airborne particle concentrations are higher, are people sicker?

3. Time-series studies that compare time-resolved aerosol data to time-resolved data on health outcomes, for a particular location or region. *When* airborne particle concentrations are higher, are people sicker?
4. Cohort studies that select a group of subjects and track them (ideally until death), and look for a difference in health effects as a function of exposure. Does the chance that someone will get sick depend on the particle concentrations to which they are exposed?

Unfortunately, all of these approaches have serious practical shortcomings. Case–control studies usually require retrospective assessment of exposures, often reaching far into the past, a task always subject to large errors. Ecological studies face an inherent problem known as the "ecological fallacy," which will be discussed below, and which renders such studies suitable for hypothesis testing and perhaps consistency testing rather than quantitative risk assessment. Time-series studies are probably the best choice for practical quantitative risk assessment of acute aerosol exposures, but as discussed below they can be subject to confounding variables, they often have low statistical power, and they cannot provide risk estimates for chronic exposure. Cohort studies are subject to confounding because highly exposed cohorts often differ from less-exposed cohorts in ways other than exposure, not all of which can be controlled. Also, cohort studies are limited in practice because they require a decades-long commitment to track the subjects after exposure. For this reason, cohort studies are normally possible only when exposure monitoring is performed as a matter of course (as was the case for some miner studies of radon), so that monitoring data can later be opportunistically used to perform a retrospective study long after the exposures occurred.

9.2.1 Radon epidemiology

9.2.1.1 Background

Inhalation of radioactive aerosols has been an area of intensive research. In the past 20 years or so, much of this research has been motivated by the study of indoor radon and its decay products. Radon is a naturally occurring radioactive gas whose decay products, which are themselves radioactive, provide most of the radiation to which people are exposed in their life, and there is considerable interest in quantifying the exposures and the dose–response relationship. The vast majority of research in this area uses radon measurements in an attempt to quantify risk, rather than measurements of radon decay products themselves, even though it is the decay products rather than radon itself that are responsible for the radiation dose. The ratio of total airborne activity concentration to radon activity concentration is known as the "equilibrium factor," and varies with time and from home to home (e.g., Huet et al., 2001; El-Hussein et al., 1998), increasing along with the particle concentration in indoor air. However, as discussed by James (1988), the dose delivered to the lung is controlled by the radon decay products that are "unattached" (not stuck to coarse aerosol particles), which varies inversely with equilibrium factor so that the dose for a given radon concentration is roughly independent of room conditions, for conditions typically found in homes.

Convincing evidence of radon-related lung cancer was found in miners, who were found to be at a substantial excess risk of lung cancer, and whose risk was found to increase with increasing cumulative exposure to radon and thus, presumably, its decay products. However, the relationship between airborne concentrations of radon to its inhalable decay products is much more complicated in mines than in homes for reasons involving the high airborne particle concentrations in mines. Also, estimates of radon concentrations to which miners were exposed are very uncertain. Therefore, although high levels of radon were convincingly shown to be associated with increased lung cancer risk, the relationship between cumulative exposure and increased risk was (and remains) uncertain by at least a factor of three. See Steinhausler (1988) for a discussion of many of these issues, with data.

In the past 25 years, concern has shifted from miners to the general population. Indoor radon concentrations are much lower than those in mines, but are not necessarily negligible. Extrapolation of risk from miner data is problematic, because miners were subject to very high radon exposure and were simultaneously exposed to other materials (such as dust, engine exhaust, etc.), in addition to having more, and longer, periods of deep breathing. Adjusting for these factors as well as possible, and (importantly) assuming a linear dose–response for inhaled radon decay products, led to predictions, albeit highly uncertain ones, for risk per unit dose for both smokers and nonsmokers in residences.

Indoor radon concentrations are highly variable. In the U.S., annual-average living-area radon concentrations are approximately lognormally distributed (Marcinowski et al., 1994) with a geometric mean of about 26 Bq/m^3 and a geometric standard deviation (GSD) of about 3.1. Even small areas are quite variable; for example, living-area average radon concentrations within most U.S. counties are approximately lognormally distributed with a GSD near 2.2.

When applied to the statistical distribution of radon and smoking in the U.S., model predictions based on extrapolated miner data suggested that between 10,000 and 20,000 people die per year due to inhaled radon decay products; most of the predicted deaths are among smokers, and these figures are derived from an unrealistic comparison to the number of people expected to die of lung cancer if no one were exposed to any radon (or radon decay products) whatsoever.

Epidemiological studies have attempted to determine the dose–response relationship (or, more correctly, the exposure–response relationship) for the range of radon exposures that occurs in residences. Radon research has included three of the four types of epidemiological studies mentioned above: (1) case–control studies, in which the radon exposures of individuals with lung cancer are compared with those who do not have lung cancer, (2) "ecological" studies that compare average lung cancer risk with average radon exposure, typically by county, and (3) cohort studies that attempt to follow until death cohorts of miners who were exposed to different radon concentrations, and look for a difference in lung cancer rates attributable to exposure. The challenges of quantifying miner exposures (and, even more difficult, doses) are very interesting, but we choose to focus here on residential exposures and will not discuss the miner studies.

9.2.1.2 Case–control studies of residential radon

Figure 9.2 shows the estimated odds ratios, with 95% confidence intervals, for different exposure bins, found by five case–control studies involving residential exposure to radon. Although these are arguably the best studies in terms of estimating exposures and avoiding other sources of error, this does not constitute a comprehensive list of studies of this type; see Field (2001) and Lubin and Boice (1997), which summarize the results of some more case–control studies. The statistical distribution of annual-average living-area mean radon concentrations in the U.S. is shown at the bottom of the figure (arbitrary vertical scale).

Even the seemingly straightforward task of plotting results from different radon case–control studies on a common scale can be tricky: each study used different exposure bins, some used different methods of determining how exposures would be binned (e.g., using average exposure over the past 25 years, or over the time period from 5 to 25 years ago), each study calculated the odds ratio relative to the lowest exposure bin (rather than to a group of people who were exposed to no radon, since such a group does not exist) and

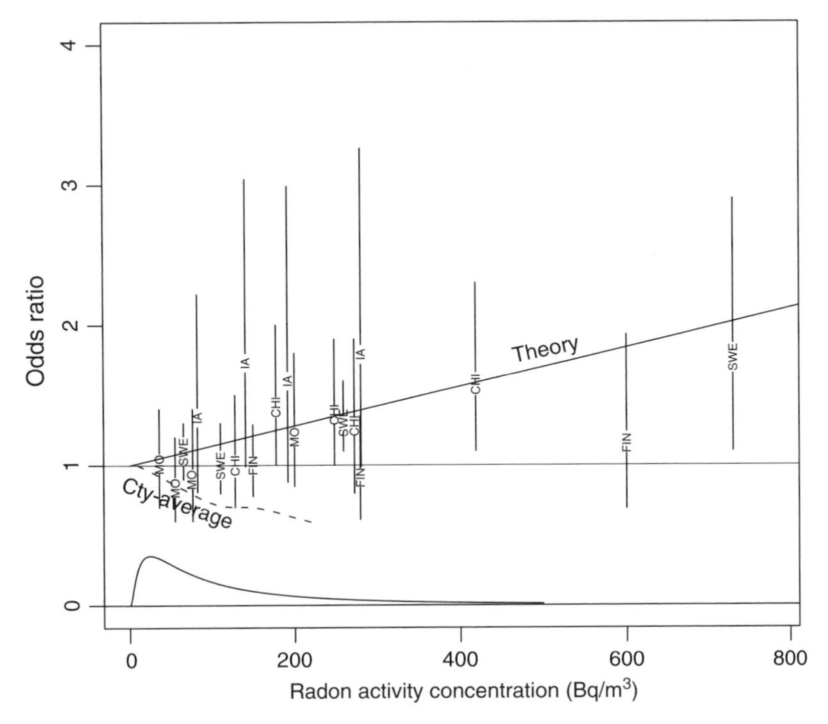

Figure 9.2 Estimated odds ratio (with 95% confidence bands) vs. mean radon activity concentration, showing data from several case–control studies. The odds ratio estimates are labeled according to the state or country in which the study was performed: IA=Iowa (Field et al., 2000), CHI=China (Wang et al., 2002), FIN=Finland (Auvinen et al., 1996), SWE=Sweden (Pershagen et al., 1994), and MO=Missouri (Alavanja et al., 1999). The "Theory" line is an approximation to the prediction based on the standard model for risk due to radon and smoking (NRC 1998b). The "Cty-average" curve is a smooth fit to Cohen's (2000) data on smoking-adjusted excess of county lung cancer deaths and county-mean radon concentrations. The probability density at the bottom shows the distribution of residential radon concentrations in the U.S. (arbitrary vertical units).

the exposure range for the lowest bin is different for each study, and so forth. Figure 9.2 does not attempt to adjust for these issues. For each study, the overall odds ratio (not separate odds for smokers and nonsmokers) is plotted for each bin at the mean radon concentration for subjects within that bin.

A line labeled "Theory" and representing approximately the extrapolation from miner studies (NRC, 1998b), which we will call the "standard model," is superimposed on the plot (actually the theory predicts relative risk, not odds ratio, a technical difference that should show a very slight downward curvature to the line, which we have not bothered to include). The plot shows the predicted odds ratio as a function of radon exposure for the U.S. proportion of smokers and nonsmokers; even if a linear response is assumed, the slope of this line is uncertain by at least a factor of three. All of the studies are consistent with the prediction from the standard model, but are also consistent, or nearly consistent, with radon having no effect on lung cancer (an odds ratio of unity) over the range of radon exposures tested. The studies that do find a "statistically significant" ($p < .05$) increase in lung cancer for increased radon concentration do so only for the highest exposure bin or two, and only by a bare margin. This does not, of course, indicate that there is no effect at lower concentrations — as the size of the error bars illustrates, each individual study has very low statistical power.

Lubin and Boice (1997) performed a meta-analysis of eight epidemiological studies, including several not shown in Figure 9.2, but failed to include recent studies from China

(Wang et al., 2002) and Iowa (Field et al., 2000), which were not available at the time they wrote their paper. They concluded that the best-fitting linear dose–response model is remarkably close to the best-guess extrapolation from miner data (the "Model" line in Figure 9.2), but that the 95% confidence intervals barely exclude "no effect." Looking at categorical rather than continuous effects, the relative risk at 150 Bq/m^3 is estimated to be 1.14, but with a 95% confidence interval from 1.01 to 1.30, thus barely excluding unity. The more recent results from the studies in Iowa and China narrow the error bars slightly without substantially changing the central estimates.

9.2.1.3 Sources of error in exposure estimates in case–control studies of residential radon

The following are some of the major sources of error for residential radon case–control studies.

1. *Population mobility*: People change residences, so monitoring in the current residence will not generally provide a good estimate of long-term exposure. Each study had some method for minimizing the influence of this fact, for example, by studying only subjects who had long lived in the same home for many years (Field et al., 2000), by monitoring in all past homes (Lagarde et al., 1997), or both (Wang et al., 2002). In all studies (other than Iowa), some homes could not be monitored and the missing data were imputed.
2. *Long-term temporal variability in radon concentrations*: Current monitoring in a home does not accurately estimate the past radon concentration in the home. Concentrations are known to vary from year to year; Steck (1994) found about a 25% year-to-year variation in a study of 100 homes in Minnesota. Two studies have attempted to avoid this problem. In addition to the conventional use of radon detectors to monitor individual homes, both the Iowa study (Field et al., 2000) and a Missouri study (Alavanja et al. [1999], not included in Figure 9.2) used conventional radon monitors and also used a novel exposure assessment technique based on the accumulation of a long-lived radon decay product, lead-210, in one or more glass objects belonging to a home's residents. These measurements are thought to reflect the long-term average radon concentration to which the object was exposed, which is assumed to be related to the owner's radon exposure. (In both studies, the result was a slight increase in estimated risk.)
3. *Spatial variability of radon concentrations within the home*: People spend time in different rooms on different floors of their house, so a house-average radon concentration will not accurately estimate personal exposure. The Iowa study (Field et al., 2000) attempted to avoid this problem by adjusting for the fraction of time subjects spent in each area of their home, at different periods of their lives. Most other studies neglect the effect of mobility within the home.
4. *Interviewer bias or respondent bias in questions regarding smoking or other risk factors*: Cases may respond differently from controls. Questions related to lifetime cigarette smoking, or other perceived risk factors such as time spent in the basement, may be more likely to draw biased answers from cases than controls, or *vice versa*.
5. *Uncontrolled or inadequately controlled confounding variables*: The populations of cases and controls can differ in many ways, due to both biases in selection procedures and to random chance. Information on known or suspected confounding variables is collected and used to attempt to control for such variables, for example, by adjusting for age, stratifying by smoking status, and so on, but may not completely remove the confounding effects.

Sources of exposure error such as items 1–4 can cause three problems. The first is the potential for bias: if any of the errors systematically over- or underestimate radon exposure for cases compared to controls, this will bias estimates of risk, a particular problem if the risk is small or the bias is large. The second is the problem of statistical power: the more noise, the less ability to distinguish a small signal, and the more cases and controls that are necessary. The third, and the most tractable if the magnitudes of the error can be estimated, is that random errors in exposure will yield an exposure–risk curve that is very flat: high-estimated-exposure categories will tend to contain subjects whose exposure is overestimated, and low-estimated-exposure categories will tend to contain subjects whose exposure is underestimated. If the magnitudes of the errors are known, however, the effect can be removed statistically (e.g., Field et al., 1997). Conceptually, the points to the right in Figure 9.2 need to be shifted slightly to the left, and the points at the left need to be shifted slightly to the right. This will yield a higher estimate of the slope, but will not improve the statistical power.

Sources of exposure error as discussed above can shift individual or group exposure estimates up or down, corresponding to moving to the left or right in Figure 9.2, usually tending to weaken the relationship between exposure and health effects. Confounding variables, on the other hand (item 5) can shift estimates up or down, and can thus totally change the exposure–response effect. The possibility of such confounding is ignored in the uncertainties quoted for the different studies, since after all the whole point of a case–control design is to try to eliminate this effect; however, in practice this is always a concern and the true uncertainty for each bin should be expanded somewhat.

Lubin et al. (1995) used a simulation method to determine the statistical power of case–control studies in the presence of population mobility and exposure estimate error. They concluded that for normal mobility and typical U.S. radon exposures, a case–control study would require somewhere in the range of 5000 to 13,000 cases (and about twice as many controls) in order to convincingly demonstrate increased radon risk at an exposure of 150 Bq/m^3, even if the standard model is correct. That number of cases would be vastly more than the number for any single case–control study that has been performed so far. Indeed, even the total number of cases in all of the case–control studies performed so far barely reaches the low end of this range, including 4236 cases in the eight studies examined by Lubin and Boice (1997), 413 in the Iowa study (Field et al., 2000), and 886 in the China study (Wang et al., 2002). Note, however, that the Iowa and China studies specifically selected low-mobility, highly exposed populations, thus decreasing exposure errors and increasing the expected risk, so that fewer cases would be needed to demonstrate an effect.

Overall, the case–control studies seem to indicate a slightly elevated risk of lung cancer for long-term exposure averaging over 150 Bq/m^3. The case–control results are consistent with the standard model, but they do not have sufficient statistical power to rule out a threshold below which there is no effect (or even a small protective effect), and are certainly consistent with risks much smaller than predicted by the standard model. In short, although case–control studies of radon can be informative for very high exposures and high risks, it seems unlikely that they will ever be able to provide reliable risk estimates below several hundred Bq/m^3.

9.2.1.4 *Ecological studies of residential radon*

A series of papers by Cohen (e.g., Cohen and Colditz, 1994; Cohen, 1995, 2000) examines excess lung cancer mortality (over what is expected from countywide smoking data) as a function of county mean indoor radon concentration, and finds a strong *negative* relationship. This relationship is quite robust, in the sense that it remains even if only a selected subset of counties is used: only urban counties, or only counties with above-average median family income, or only counties with near-median unemployment, or any of dozens of other categorizations. The nationwide relationship between county mean radon concentration and the county's excess lung cancer death rate is shown with a dashed line labeled "Cty-average" in Figure 9.2, which is based on a smooth curve through Cohen's county-mean data for females, scaled so as to equal unity at a radon concentration of 0.

Taken at face value, Cohen's data suggest a rather strong protective effect from radon at residential concentrations. However, it is well known (e.g., Robinson, 1950; Greenland and Morgenstern, 1989; Greenland and Robins, 1994; Gelman et al., 2001) that the "ecological fallacy" can lead to a spurious relationship in aggregated data. To give a classic example, Robinson (1950) found a positive correlation between statewide literacy (in English) and the fraction of residents who were foreign-born. This would suggest that the foreign-born were more likely to be literate, but in fact the reverse was true, and the observed effect in the aggregated data is due to the fact that more foreigners lived in states where the literacy of non-foreign-born people was high.

Cohen (1995, 2000) has argued that if lung cancer risk is linear in both smoking status and radon exposure, the deviation of the county–mean curve from the standard model *cannot*, mathematically speaking, be due to correlations between radon and smoking unless there is an almost perfect negative within-county correlation between these two variables. Lubin (2002) has shown that this is not true, and suggests that a within-county correlation between smoking and radon is the likely cause of the discrepancy between the aggregate data and the standard model. He has demonstrated that if the degree of correlation between radon and smoking is itself related to the county-mean radon concentration (in a very complicated and specific way, but not including extremely high negative correlation between radon and smoking status), then the county-aggregate results can be explained even if the individual-level risks are correctly predicted by the standard model.

Lubin's example demonstrates that confounding variables, even if not highly correlated with radon within counties, can cause effects of the correct magnitude to explain the county-aggregate data, even if the standard model is correct. However, the details of Lubin's example do not offer a plausible explanation, since his explanation requires very sharp shifts in the correlation between radon and lung cancer within counties, as a function of county-mean radon concentration. Lubin's example works mathematically, but not realistically. Cohen's work does seem to invalidate the standard model at some level, but this does not mean that it tells us anything about radon risk, since the problem may lie (for example) with the standard model's assumed multiplicative interaction between smoking and radon, or with one or more additional within-county confounding variables.

9.2.1.5 *Discussion of radon risk estimates*

There is no question that exposure to very high concentrations of radon (and thus its decay products) causes cancer: the miner studies are quite convincing. Linear extrapolation to low doses and to residential breathing rates suggests that even radon concentrations that are commonly experienced may involve significant increased risk. However, the expected risk is not high enough to easily test this hypothesis, given the large sources of uncertainty inherent in case–control and ecological studies.

Existing case–control studies, including over 5000 cases in all, are consistent with the standard model at 150 Bq/m^3, but even taken together are barely convincing (if that) in distinguishing the risk from zero, much less accurately quantifying it. Attempting to determine the exposure–risk curve at lower concentrations rapidly becomes even more problematic. Unless very significant resources are devoted to performing enormous case–control studies, epidemiological studies of radon risk will not yield a reliable exposure–risk curve for typical or even substantially elevated residential exposures, for example, in the range from 100 to 400 Bq/m^3. There is little chance that a massive case–control study will be performed to address this problem.

As for ecological studies such as Cohen's (2000): mindful of the recognized problems with ecological studies, we have previously stated (Price, 1995; Gelman et al., 2001) that ecological studies such as Cohen's examination of county-aggregate radon and lung cancer data "cannot be more than suggestive." We still feel this is true, but would now tack on an important addendum: extrapolations from much higher exposures, and case–control studies involving a few hundred to a few thousand cases, *also* can be no more than suggestive. The county-average data do indeed constitute a challenge to the standard model, since, as Lubin's (2002) example illustrates, there appears to be no reasonable way to reproduce Cohen's data using a dose–response model that is linear in both smoking and radon. However, Lubin's example also confirms that the county-mean exposure–response curve can be quite uninformative of the personal exposure–response curve. Given the fact that the case–control studies are incompatible with a strong protective effect near $100\,Bq/m^3$, the county-mean data do not seem to be useful for determining the risk from radon exposure.

As we have said before (Price, 1995), life is rarely simple and it seems implausible that the exposure–risk curve could be perfectly linear all the way to zero exposure for both smoking and radon. Figure 9.3 shows several alternative exposure–risk curves, all but one of which are more or less consistent with the available case–control data; certainly, more could be constructed. We are by no means proposing any of these as alternatives to the standard model, but rather wish to illustrate the nearly complete lack of convincing evidence about the shape of the curve below $200\,Bq/m^3$.

The distribution of personal exposures within each county is very wide. In consequence, if the exposure–response curve is highly nonlinear, the county mean curve will not track the individual exposure–response curve, even in the absence of confounding variables or within-county correlation between radon and other variables. For example, consider the lowermost fictional exposure–response curve in Figure 9.3, which has a strong protective effect at low exposures (and is therefore incompatible with the results of case–control studies). To examine what such an exposure–response curve would imply in terms of county-mean effects, in the absence of confounding, we assumed that (1) each person's exposure is a mixture of indoor exposure from their current residence, exposure from past residences, and outdoor exposure, (2) each individual source of exposure is lognormally distributed, and (3) there is mild correlation between current and past residential exposure, because people often move within the same county or region and there is some spatial correlation in radon levels. We used reasonable choices for the distributional parameters (which we will not discuss because we are merely trying to illustrate a point); results for predicted county-mean exposure and response are shown with open circles in Figure 9.3 for counties across a range of average concentrations. The county-mean results differ substantially from the individual exposure–response curve, showing a much smaller protective effect and very different behavior as a function of radon concentration.

As was recognized by Bogen and Layton (1998) in a cost–benefit analysis that considered a biologically plausible model for radon that includes a protective effect of low exposures, a protective effect at low concentrations could leave unchanged, or even increase, the expected benefit from radon reduction above $150\,Bq/m^3$. For instance, if the lowermost (protective) curve really did represent reality, then reducing the exposure from a premitigation value of $200\,Bq/m^3$ to a postmitigation value of $70\,Bq/m^3$ would provide a greater benefit than under the standard model. (If that curve really represented reality, it would also make sense for more than half the people in the country to try to increase their exposure to radioactive gas.)

Given the lack of useful data on the exposure–response relationship below $200\,Bq/m^3$, it seems reasonable that radon policy analyses should at least consider the possibility of a nonlinear relationship. Lin et al. (1999) noted this point in performing a decision analysis

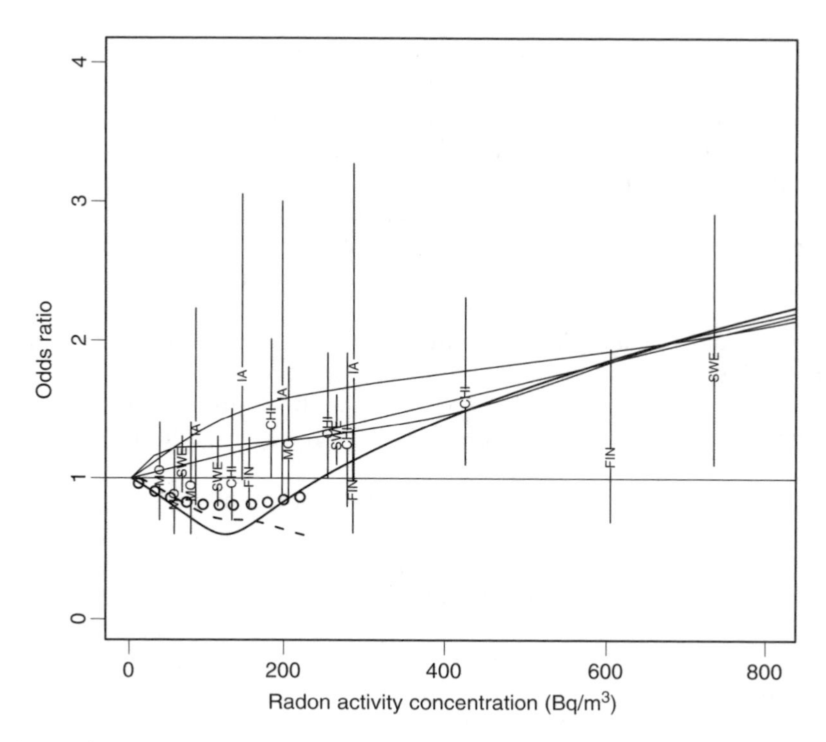

Figure 9.3 Lines show estimated odds ratio (with 95% confidence bands) vs. mean radon activity concentration, showing data from several case–control studies; see Figure 9.2 caption. The dashed curve is a smooth fit to Cohen's (2000) data on smoking-adjusted excess of county lung cancer deaths and county-mean radon concentrations. Curves show hypothetical nonlinear individual dose–response functions. Open circles show county-mean values if the individual dose–response follows the lowest curve.

concerning radon monitoring and mitigation. They first performed an analysis that assumed that the standard model holds, and then repeated the analysis under the assumption that there is a risk threshold so that exposure to concentrations below 150 Bq/m^3 holds no risk at all, quantifying the differences in benefits of a nationwide mitigation program in both scenarios. Such exercises can at least highlight what factors in the exposure–risk curve are important for making policy decisions.

9.2.1.6 Implications for other aerosol exposure problems

Case–control studies of any environmental hazard are subject to uncertainty in both exposure and outcome, with the latter uncertainty often being due to small-sample variation in the number of cases in each exposure category. In the presence of such uncertainty, there will always be some exposure below which a risk cannot be distinguished from zero even if it exists.

Many current problems in the broader field of aerosol exposure share common elements with the radon question as outlined above:

1. There is convincing evidence of adverse health effects at elevated exposures, but no certainty about how to extrapolate to lower exposures.
2. Case–control studies are difficult because estimating past exposures is error-prone.
3. Ecological studies are vulnerable to confounding by variables (smoking, in the case of radon) that have effects that may be much larger than the effect from the exposure at issue.

In spite of sharing these characteristics with the radon problem, some aerosol epidemiology studies differ from it in important ways. Specifically, the only (observed or expected) health effects from radon involve lung cancer, which has a long latency period. There is no possibility of detecting immediate health consequences from a short-term increase in radon exposure. In contrast, exposure to some aerosols can cause immediate observed consequences, a fact that allows time-series analysis to be brought to bear, as in studies of the relationship between fine particle concentrations and cardiopulmonary mortality (Schwartz, 1994; Tsai et al., 2000; Maynard and Maynard, 2002; and a great many others). We now briefly discuss epidemiological studies of the effects of fine aerosols.

9.2.2 *Epidemiological studies of the effects of fine particle inhalation*

9.2.2.1 *Time-series studies of the health effects of fine particles*

Time-series studies have the advantage of eliminating or greatly reducing many confounding effects found in other epidemiological studies. They can still be subject to confounding — for example, if more people die on days when airborne particle concentrations are high, this could be because high particle concentrations are associated with temperature, and high temperature is a risk factor for the frail. However, it is relatively easy to collect mortality and pollution data for many weeks, months, or even years, in which case confounding effects can be removed by stratifying on temperature, day of week, season, and so on.

Many studies have measured particle and gaseous pollutants, and looked for a relationship with health outcomes such as death or hospital admissions. These studies usually measure particulate concentration parameters such as PM_{10} and $PM_{2.5}$, along with some gaseous pollutants (such as ozone) and sometimes with some ability to resolve different types of particles, at one or more *outdoor* measurement locations. Summary statistics of these concentrations are then compared with health outcomes such as deaths due to heart or lung disease, hospital admissions, or other observables.

Particle measurements that are intended to characterize exposure are often subject to the major problem that the measurement location is spatially separated from the location(s) at which people are exposed. For example, pollutant measurements are often made outdoors at a few places in a city, whereas most people spend most of their time indoors (about 90%, in the United States). Thus, both the concentrations and size distributions of airborne particles to which people are exposed may be very different from the measurements. Given the reliance on remote ("ambient") outdoor concentration measurements, the relationship between such measurements and actual personal exposures is important (Wilson and Suh, 1997; Wislon et al., 2000). Table 9.1 summarizes results from several studies that investigated the relationship between ambient measurements (performed at one or more fixed monitoring locations in a city) and personal exposure measurements. In these studies, 24- or 48-h personal exposure measurements were compared to 24- or 48-h ambient monitoring data, and the correlation between the two types of measurement was determined. Some of the studies focused on populations believed to be at extreme risk, such as patients with chronic obstructive pulmonary disease or other cardiovascular disease; many of these are retirees and therefore not subject to workday exposures, so it seems plausible that they would experience a different relationship between ambient and personal exposures. Other studies examined the relationship for healthy elderly or healthy adults. All studies found very substantial intersubject variability in the correlation between ambient and personal exposure.

Several of the studies summarized in Table 9.1 have made the general claim that the correlation between ambient and personal exposures is high enough that ambient

Table 9.1 Correlation between One- or Two-Day Ambient Outdoor Measurements and Personal Exposure Measurements for Some Populations; "Patient" Refers to Treatment for Chronic Obstructive Pulmonary Disease

Study Population	Time Period	Median *r* for $PM_{2.5}$	Median *r* for SO_4
Elderly patients in Vancouver (a)	Late spring through early fall	0.48	0.96
Elderly patients in Amsterdam (b)	Winter and spring	0.79	
Elderly patients in Helsinki (b)	Winter and spring	0.76	
Healthy elderly in Baltimore (c)	Summer	0.76	0.88
	Winter	0.25	0.72
Healthy adults in Helsinki (d)	Workdays (throughout year)	0.43	
	Leisure time (throughout year)	0.48	
Healthy adults in Basel (g)	Throughout year	0.07	
Patients in Nashville (e)	Summer	0.0	
Patients in Boston (f)	Winter and summer	0.3	

(a) Ebelt et al. (2000), (b) Janssen et al. (2000), (c) Sarnat et al. (2000) (d) Kousa et al. (2002), (e) Rojas-Bracho et al. (1996), (f) Rojas-Bracho et al. (1998), (g) Oglesby et al. (2000).

exposures can be used as a surrogate for personal exposure in epidemiological studies. This is false. If it were possible to control for, or eliminate, *all* confounding variables, then *any* correlation whatsoever between ambient and personal exposure could be used for epidemiology. But in practice it is impossible to completely eliminate or control all confounding variables, and a low correlation greatly diminishes the ability to separate the signal due to particle inhalation from the effects of confounders. Although a few of the studies found substantial correlations between ambient and personal exposures to $PM_{2.5}$, several did not. The low correlations found in Basel and Nashville would prevent the success of time-series studies there, if these values are correct for the at-risk populations in those cities. The higher, but still low, correlations in Boston and wintertime Baltimore would greatly reduce the statistical power of studies in those cities, again assuming that those results hold for populations at risk from acute exposure to elevated air pollution.

The dismal ambient–personal correlations for some cities suggest that time-series analyses will not work everywhere. However, in cities in which the correlation is substantial, an effect of air-pollution-related mortality could be seen if it is large enough. An enormous number of such studies have been performed (a few recent ones are Moolgavkar et al., 1995; Samet et al., 2000 [with important corrections in Dominici et al., 2002b]; Lee et al., 2000; Dominici et al., 2002a; Le Tertre et al., 2002) and the evidence for increased risk of daily mortality associated with increased particulate air pollution is irrefutable. Elevated mortality is definitely associated with increased $PM_{2.5}$ levels, but the biologically important particle types cannot be conclusively determined because of the correlation between pollutants.

The time-series studies do not quantify personal exposure, and certainly not dose. At present this is not an important failing, because peak urban air particulate concentrations are high enough to demonstrate increased mortality, but regulations are likely to lead to reduced peak concentrations, and eventually the discernable health effects may be too small to separate from the noise. Almost certainly, researchers will then attempt to reduce uncertainties in exposure, and perhaps dose, in order to extrapolate to lower exposures, as was the case with radon.

The time-series studies, although conclusive, have some serious shortcomings. One problem is that they cannot be used to estimate chronic effects: the number of pollution-related deaths is caused by a combination of chronic (long-term) and acute (short-term) exposure, and time-series studies are sensitive only to the second of these. Since elevated peak concentrations of particulate air pollution are harmful, it seems likely that chronic exposure to somewhat lower concentrations also has an effect, but time-series studies cannot resolve this issue because they rely on relating the *change* in daily death rate with the *change* in daily exposure. If the change in daily exposure is greatly reduced, the statistical power of these studies will be too low to allow estimating its effect.

9.2.2.2 Cohort studies of the health effects of fine particle inhalation

A cohort study collects data on individuals, following them over a long time period, and looks for an association between health outcomes and exposure to putative risk factors. Like all epidemiological studies, cohort studies can be subject to confounding because exposure is not the only factor that differs between groups. Cohort studies have several advantages over case–control studies: they often include personal exposure measurements (not in the case of particle exposure, but in other areas of research), they can include information on timing of exposures and other risk factors, and, like case–control studies, they can include person-specific information on confounding factors. This latter characteristic makes cohort studies enormously better than ecological studies, because it is possible to directly adjust for the effects of confounding variables at the individual level, thus avoiding the kinds of issues with the "ecological fallacy" that are discussed above for the case of radon. Of course, there can still be confounding by a factor on which person-specific data were not collected.

Quite a few cohort studies of the effects of fine particle inhalation have been reported (Dockery et al., 1993 and Pope et al., 2002; for example). Almost all of these compare cohorts across cities — that is, they identify subjects who live in different cities, and follow them through time. These studies attempt to control for known risk factors, about which data are collected (examples include smoking, diet, age, obesity, fitness, and so on). However, there is still a large potential for error due to unknown confounding variables: lots of things vary between cities and it is hard to control for them all, or even to know what to control for. The reason why multiple cities must usually be included is that individual subject exposure estimates are not available: exposures are estimated for all subjects within a city, based on citywide monitoring data. These studies thus have one of the same major drawbacks as ecological regression — an inability to estimate individual exposures to the pollutant of interest — but with the key difference that known confounders can be controlled at the personal rather than group level.

Choosing subjects from within a single city or locale would be very advantageous for avoiding or minimizing the effects of confounding variables, but would require an ability to estimate exposure either for individuals or for highly- and less-exposed groups within the city.

9.2.2.3 Ecological epidemiological studies of the health effects of fine particle inhalation

Several studies, which we will not bother to cite, have attempted to estimate the effects of fine particle inhalation through ecological studies, comparing respiratory or cardiac mortality for high- and low-concentration cities. As with the example of ecological studies of radon, such results have very little quantitative value. No further studies of this type should be considered.

9.3 Inhalation modeling and experiments

9.3.1 Motivation for experiments and modeling related to lung deposition of aerosols

Any attempt to relate exposure to health effects through epidemiology is bound to fail at sufficiently low exposures: eventually the health effects will fall below the level at which they can be separated from the effects of confounding variables and statistical noise, even though the health effects may still have great practical significance at those levels. Moreover, even for health effects that can be convincingly identified or quantified through epidemiology, such quantification will often fail to answer important questions. For instance, epidemiological studies suggest that diesel engine exhaust may be carcinogenic at somewhat elevated exposures (McClellan, 1995). But diesel exhaust is a complex mixture of particles of varying chemical composition, and epidemiology cannot tell us which, or which combination, is responsible for causing cancer, an important question for regulators and for engineers attempting to design better engines or exhaust filtration systems.

Although epidemiological studies are unable to address many important questions concerning aerosol inhalation, there is an alternative: all or part of the causal chain in Figure 9.1 can be modeled or determined experimentally. The ideal experiment would expose a human subject with known nasal, tracheal, and lung morphology, and known breathing rate and breathing volume, to a known size distribution of aerosol particles, and would measure the deposition as a function of size and precise location in the respiratory tract. In practice, neither lung morphology nor deposition as a function of size and location can be precisely determined in a living subject; future improvements in radiological methods (Fleming et al., 2000) may help resolve both of these issues.

Gaining a complete understanding of the relevant parameters and phenomena is a rather ambitious task because the required experiments are difficult and many important physiological processes are not completely understood. Fortunately "ambitious" is not synonymous with "impossible," and thanks to decades of research, computational models of deposition can be used with confidence for certain tasks. For example, for controlled breathing rates and known subject parameters (such as tidal volume), total deposition for particles larger than 0.01 μm can be predicted rather accurately over a wide range of conditions, at least for typical healthy adult subjects. Figure 9.4 shows a typical example, from Hofmann and Koblinger (1992). This comparison of predictions from two models to experimental data (the mean deposition in three healthy adult subjects) shows excellent agreement for the total deposition, and rather good but imperfect agreement for bronchial deposition at small particle sizes; unfortunately, experimental values for the bronchial deposition of particles below 0.05 μm were not collected.

Although there are some differences between current deposition models that may have practical significance, most models currently in use agree with experiment and with each other rather well (e.g., see Bergmann et al., 1997; Segal et al., 2000) over the size range from 0.01 to 10 μm and for typical adult lung parameters. This makes sense because any model that does not agree with experiment would be abandoned. There are some differences between the ICRP-66 model (ICRP, 1994) and others for particles near and below 0.01 μm; Bergman et al. (1997) suggest that the ICRP model overestimates deposition for those sizes, compared to experimental data (Heyder et al., 1986).

9.3.2 Assessing uncertainties

There are two main causes of error in modeling aerosol deposition (or modeling anything, for that matter). The first is model misspecification: the model itself may simply be incorrect, failing to correctly include the effects of all significant parameters. The second is

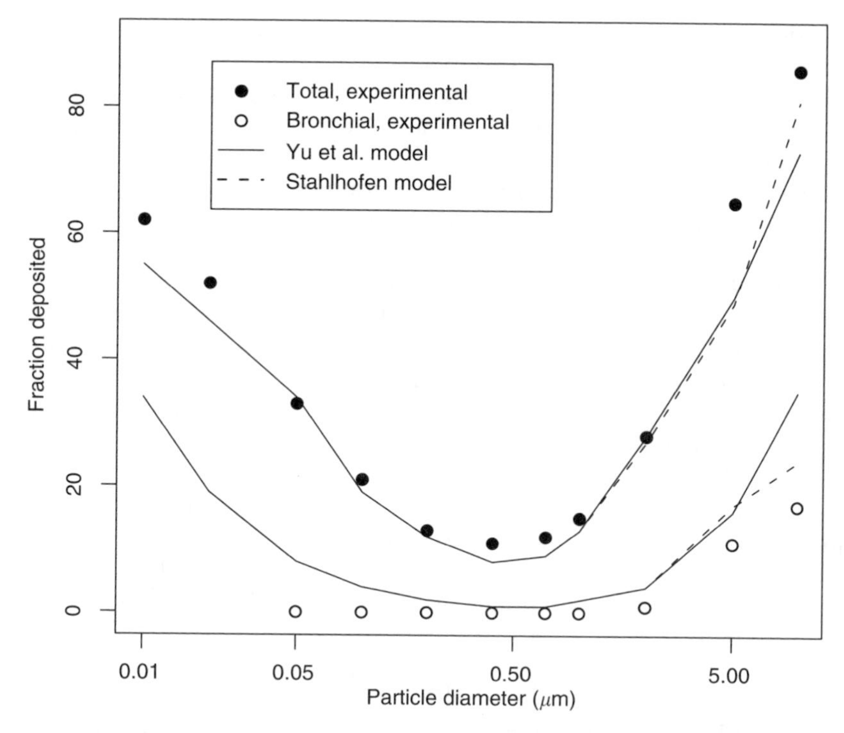

Figure 9.4 Points show total and bronchial deposition measured experimentally (values are from Hofmann and Koblinger, 1992); lines are fits from two models. Both models overpredict bronchial deposition in the range 0.05 to 0.50 μm, but otherwise the fit is very good.

parameter error: even if the model is correct, predictions will be wrong if the input parameters are wrong.

No computational model involving physiological phenomena is perfect. One can hope, however, that the magnitude of model misspecification error is small, at least for the range of parameter values over which the model is intended to apply. Barring that, one at least hopes that the approximate magnitude of the error is known.

Unfortunately, the magnitude of model misspecification error is notoriously difficult to evaluate. There are three basic approaches. The first is to compare the output of the model to experimental data. The error will be a combination of parameter error (discussed below) and model misspecification error. In performing this kind of test, it is important to compare the model to data that were *not* used to create it, or to estimate its parameters, in the first place; determining parameter values from one set of data and comparing the predictions to another set of data is called "cross-validation," and should be a standard procedure.

The second approach is to compare the model's output to that of another, superior model. For example, results from an analytical or semianalytical deposition model might be compared to those from a computational fluid dynamics (CFD) simulation. The CFD simulation will itself be imperfect, but the comparison between the two models will still be informative as to roughly the size of the error that can be expected. One might ask, why not simply discard the inferior model, if another is known to be superior. Reasons can include a regulatory requirement to use the inferior model, or computational impracticalities of always using the superior one.

Finally, if direct experimental comparison is not possible and a definitely superior model does not exist, the last resort is to compare the model's output to that of other plausible models that tackle the same problem in different ways. Different models are likely to share some of the same flaws so that errors estimated in this way may be understated. Still, if a researcher has several models at her disposal and they give different answers to the same problem, this is at least an indicator of the researcher's uncertainty, if not a measure of the model's intrinsic error.

As for the errors caused by incorrect input parameter values, the magnitudes of these errors are easy to evaluate if the uncertainty in the input parameters themselves is known. The most straightforward method, if the model is not too computationally burdensome, is to perform Monte Carlo simulation: run the model repeatedly, using input values drawn from distributions around their "best guess" values, and summarize the variation in the output. Latin Hypercube sampling of the parameters is another possibility that provides similar results at a lower computational burden. Sampling approaches such as these are routine in many areas of research, and are just beginning to see use in the field of aerosol inhalation (Molokanov and Badjin, 2000; Hofmann et al., 2002; Harvey and Hamby, 2001, 2002), where they should become routine.

Although total deposition and even regional deposition (i.e., deposition in the trachea, bronchi, and alveoli) can be predicted quite well for a healthy subject, other quantities of interest cannot be predicted nearly so accurately. We now discuss the elements of the causal chain and the extent to which they are or are not fully understood.

9.3.3 *General approach to assessing the effects of interpersonal variability*

For a laboratory subject, it is possible to directly measure some model input parameters (such as tidal volume, inspiratory capacity, and so on). Other parameters (such as details of lung structure) cannot be measured. Therefore, some parameters are known rather accurately and some will be estimated with error. Even in the absence of model misspecification, the error in input parameters will generate error in the model's prediction.

The range of the likely magnitude of the error, which is to say the uncertainty in the prediction, can be estimated from the uncertainty in the input parameters if the uncertainty is known. This uncertainty estimate can be made through analytical or semianalytical techniques for some analytical deposition models, or through a Monte Carlo or Latin Hypercube sampling procedure for more complicated models or when the uncertainty is itself a complicated function. In essence, we create "virtual" people who share the same values for measured parameters but have different values for the other parameters; our subject is one of these people, but we do not know which.

Now consider making predictions for a group of people who are not laboratory subjects. For each person, now *all* of the input parameters are uncertain, in contrast to the laboratory case in which some parameters are known. Thus, the same approaches can be taken as in the case of estimating the uncertainty for an individual; the only difference is in the number of parameters for which statistical sampling is required.

Even though the predicted deposition for an unknown individual will be subject to error due to uncertainty in the input parameters, it may be possible to predict the distribution of deposition values across a population. We might know that the tidal volumes in a given group of subjects are approximately normally (Gaussian) distributed with a particular mean and standard deviation, but do not know which individual has which tidal volume. Thus, the importance of intrapersonal variability depends on whether we are trying to make predictions for a specific individual, in which case our inability to determine the individual's input parameters is a source of error, or trying to make predictions for a population, in which case the inability to determine each individual's input parameters is irrelevant as long as we know their distribution across the population.

9.3.4 *Respiratory tract morphology and other factors such as medical condition*

There are three main issues concerning the effect of respiratory tract (especially the lungs) morphology on deposition:

1. Can current models predict important details of deposition for a given parametric description of the respiratory tract?
2. Do we have an acceptable parametric description of the respiratory tract?
3. Do we have a parametric description of the intersubject variability of respiratory tracts?

These questions are, of course, interrelated. If models failed to predict quantities such as total and regional deposition, we would not know whether the problem is with (1) or (2), and there would be no point in trying to address (3). In fact, though, as illustrated by Figure 9.4 and as documented in, for example, Hofmann (1996a) and Segal et al. (2000), the total and regional deposition, averaged over at least a few experimental subjects, can be predicted rather well over a wide range of particle sizes. This fact seems to jointly answer the first two questions in the affirmative for total and regional deposition, although local deposition (i.e., deposition for a given bronchial generation number or a particular location in the lung) is another story, as discussed below.

9.3.5 *Intersubject morphometric variability*

As for intersubject variability, models of the human respiratory tract are based on analysis of a relatively small number of human lung casts such as those summarized by Phalen et al. (1974), Yeh and Schum (1980), and Nikiforov and Schlesinger (1985). Available data show that there is significant intersubject variability in airway lengths, branching angles, and other relevant parameters (e.g., see Yu and Diu, 1982; Nikiforov and Schlesinger, 1985).

In order to evaluate the practical significance of morphometric variability, Hofmann et al. (2002) evaluated the variation in output from a model that predicts local, regional, and total deposition, for ten different realizations of a stochastic lung model (Koblinger and Hofmann, 1990), and for particles ranging from 0.01 to 10 μm. (In this context, we use "local" to mean the bronchial generation, not the actual spatial location in the lung.) The result is not a direct estimate of the effects of morphometric variability in real lungs. Instead, it is an estimate of the effects of morphometric variability in lungs created by the stochastic lung model, given the assumption that the deposition model is correct.

The model predicted rather small intersubject variability in *total* deposition at all sizes (of the order of 15%), and somewhat larger variability in regional deposition (around 5% for particles between 0.1 and 1 μm, increasing to about 30% outside that range). However, in spite of the generally low intersubject variability in results for total and regional deposition, the variability in predicted local deposition was quite high: Figures 9.5 and 9.6 show predicted deposition as a function of generation number, for 1 and 10 μm particles, respectively. As the figure shows, predictions varied by nearly a factor of ten for large (10 μm) particles and low generation numbers, and by more than a factor of two for small (1 μm) particles and high generation numbers.

As Hofmann et al. (2002) report, their estimates of intersubject variability in total deposition are well in line with the experimental data of Heyder et al. (1982) and Stahlhofen et al. (1981), but the estimated variability in regional deposition is somewhat less than that of Stahlhofen et al., a fact with several possible explanations (Hofmann et al., 2002). Two likely candidates are the following: (1) the stochastic lung model (Koblinger and Hofmann, 1985) may not include the full variation in human lung structures, and (2) the

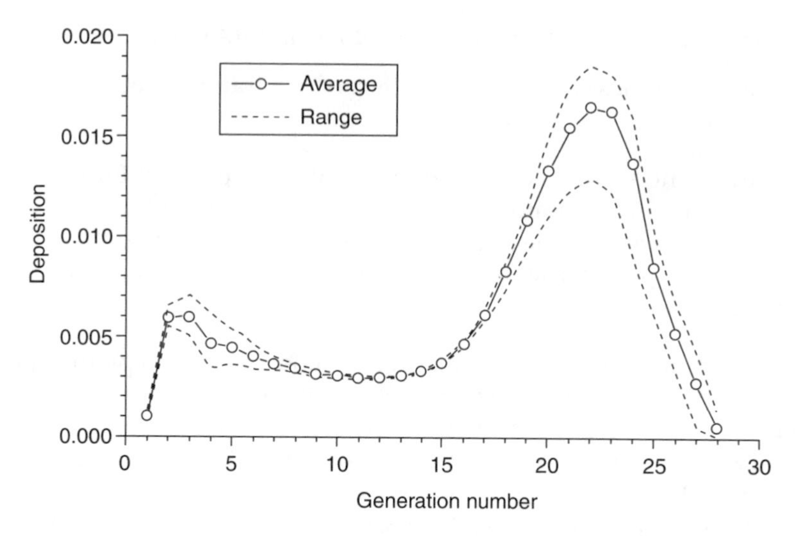

Figure 9.5 Predicted deposition vs. generation number for 1 μm particles, showing the average and range of values predicted for ten stochastic lung models that attempt to capture interpersonal morphometric variability. Moderate interpersonal variability is predicted for generation numbers 18 to 28. Figure from Hofmann et al. (2002).

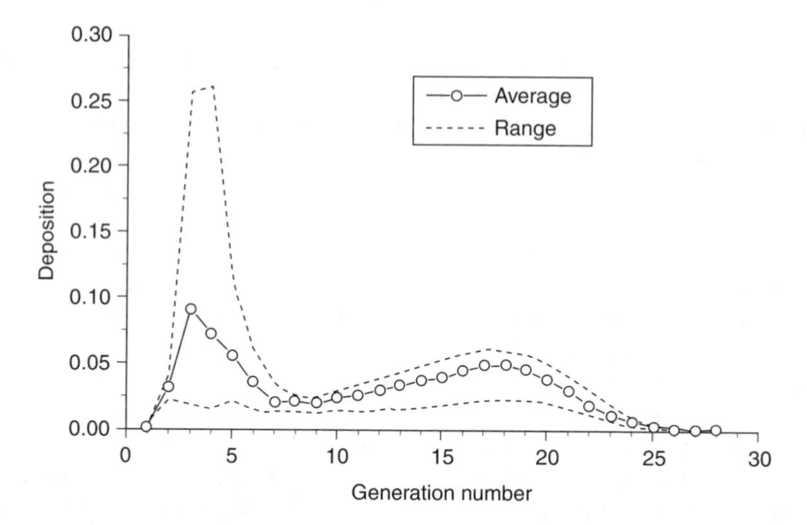

Figure 9.6 Deposition vs. generation number for 10 μm particles, showing the average and range of values predicted for ten stochastic lung models that attempt to capture interpersonal morphometric variability. The model suggests very high interpersonal variability in deposition as a function of generation number. Figure from Hofmann et al. (2002).

experiments of Stahlhofen et al. (1981) may not accurately measure local deposition, which was not directly observed but was estimated from particle clearance rates.

Modest intersubject variability in total and regional deposition had long been recognized (e.g., Heyder et al., 1982). It appears that at least some models of the variability of lung morphology (Koblinger and Hofmann, 1985) can predict this variability in deposition. It is much less clear that current morphological models allow correct prediction of the variability in *local* deposition (or, for that matter, predict its mean value). Satoh et al. (1996) have examined a normal human lung and suggest that there may be more *intra*subject variability in the number of bronchial generations than is present in current morphometric models. Further work on this subject is required if either the inter- or intrasubject variability in *local* deposition is to be accurately assessed.

Site-specific deposition (deposition to a particular group of cells) is yet another issue. Currently, no single model predicts total, regional, local, and site-specific deposition; site-specific models consider only a very small portion of the lung, and often use CFD techniques rather than the semiempirical or scaling-law-based techniques used by whole-lung models. Gradon and Pogorski (1996) and Hofmann (1996b) discuss the extent to which current computational models can predict site-specific deposition, which is estimated to be highly variable even within a short section of airway, with at least an order of magnitude variation in deposition probability.

Site-specific deposition could ultimately prove important for dosimetry: if cellular response is nonlinear with respect to the number of particles absorbed by the cell (or in contact with it), the nonuniformity of deposition even in a given lung generation will influence the physiological response.

9.3.6 Systematic morphometric variability

There is evidence of moderate sex differences (on the order of 10 to 30%) in the total deposition of aerosols in the range of 3 to 5 μm, even for the same flow rates and lung volumes (Kim and Hu, 1998). (In that study, "local" means a particular area of the lung, not a particular generation number).

Bennet et al. (1996) found that the mean *fractional* deposition of 2 μm particles was independent of sex for resting adults over the range from 18 to 80 years, but that because males had 45% higher minute ventilation, the deposition rate (deposition per unit time) was 30% greater in males than in females; this implies about the same deposition per unit surface area of the lung (on average) in men and women. Bennet and Zeman (1998) extended these experiments to children as young as age 4: the fractional deposition of 2 μm particles was the same for children, adolescents, and adults, but the rate of deposition normalized to lung surface area was substantially higher for children. The results suggest that for resting breathing rates, gender- and even age-related differences in the deposition of 2 μm particles are primarily due to breathing rates rather than morphological differences, although this may not be true at higher (nonresting) breathing rates. Chua et al. (1994) found no effect of age on local deposition in children with mild cases of cystic fibrosis, for children over 6 years old. Also, Nerbrink et al. (2002) report that using a standard lung deposition model predicted total deposition in asthmatic children with reasonable accuracy, for 1- to 2-μm particles.

Smith et al. (2001) report that the growth model for human lungs proposed by Phalen et al. (1985) accurately predicted airway parameters for subjects aged 3, 16, and 23 years (measured after autopsy). It seems that the characteristics of children's lungs that are important for total and regional deposition can be adequately predicted from the model of Phalen et al. or from scaling adult lung properties using the method of Habib et al (1994). This would probably not be true of infants, since the structure of the lung, and not just its scaling, changes substantially over the first two years of life.

Overall, it seems that current models should be sufficient for predicting deposition in healthy or nearly healthy children, but there is even less information about intersubject variability for children than for adults, so the distribution of deposition values over a group of children probably will not be predicted very well.

For the severely medically challenged, predictions will also be difficult. Brown et al. (2001) compared healthy adult subjects to adults with mild to moderate cystic fibrosis, and found substantial differences in regional deposition of 5-μm particles. Smaldone (2001) points out that patients with chronic obstructive pulmonary disease (COPD) have flow-limiting segments in the lung that will produce large local pressure drops and induce particle deposition in airways that do not experience deposition in healthy people. Kohlhaufl et al. (1999) found a modest (15%) increase in total deposition of 0.9 μm particles in women with airway hyperresponsiveness, compared to healthy women.

On the whole, there seems to be a tendency for lung disease to lead to increased deposition. As Smaldone (2001) noted, diseased lungs can be morphologically different from healthy lungs (in addition to other, nonmorphological differences), and it seems likely that accurate modeling of deposition in subjects with lung disease will require modifications to the morphometric models, for example, to allow for more within-subject variability in bronchial diameters in a given generation, due to partially blocked airways.

9.3.7 Breathing rate and inhalation details

The quantity of aerosol inhaled and deposited is of course strongly dependent on breathing rate (volume inhaled per unit time). Laboratory inhalation experiments often involve controlled breathing, and essentially always involve measurements of the breathing rate and tidal volume, so in such experiments there is seldom uncertainty in these important parameters.

Other inhalation details can also be important. For example, the short breath-hold time between inhalation and exhalation in normal breathing provides time for fine particles to settle gravitationally by a distance on the order of the size of an alveolar sac, so the length of breath-hold time (or its absence) can be an important parameter in some cases.

James et al. (1994) summarize and tabulate breathing data that were used in creating the ICRP-66 model and parameter input values (ICRP, 1994). Inhalation rates vary greatly with age and activity: volume per unit time varies by a factor of 20 from infants to adults, and by a factor of five to ten from resting to heavy exercise. In addition, for a given age and activity there is still substantial variation in both inhalation rate and in other inhalation-related parameters, such as fraction of air taken in through the mouth as opposed to the nose.

On the whole, it appears that current parameterizations of inhalation are adequate for modeling, and that the main uncertainties with respect to the details of air intake are associated with interpersonal variability and activity-dependent breathing.

9.3.8 Clearance of the lung

Once deposited, particles are removed from the respiratory tract by several mechanisms, including coughing and phagocytosis. The most important mechanism is mucociliary transport: a thin film of mucous is continuously created in the bronchioles and, propelled by ciliary action, carries deposited particles eventually to the throat, where they are swallowed or expectorated. Because the particles are carried by the mucous layer, clearance velocities are nearly independent of particle size and material, with some exceptions. However, clearance times vary substantially with particle size because the deposition location varies strongly with size, a fact that can be exploited for estimating clearance rates for different parts of the lung.

Most experiments do not directly measure clearance as a function of location (or generation number) in the lung, instead measuring parameters such as total clearance as a function of time, from which local or regional mucous velocities are derived via a model for velocity as a function of location. As a result, models can predict total clearance quite well even if the local mucous velocities are completely wrong. Various quantitative clearance models have been proposed, but unlike deposition models the clearance models are not in very good agreement with one another as to clearance velocities, particularly for high generation numbers.

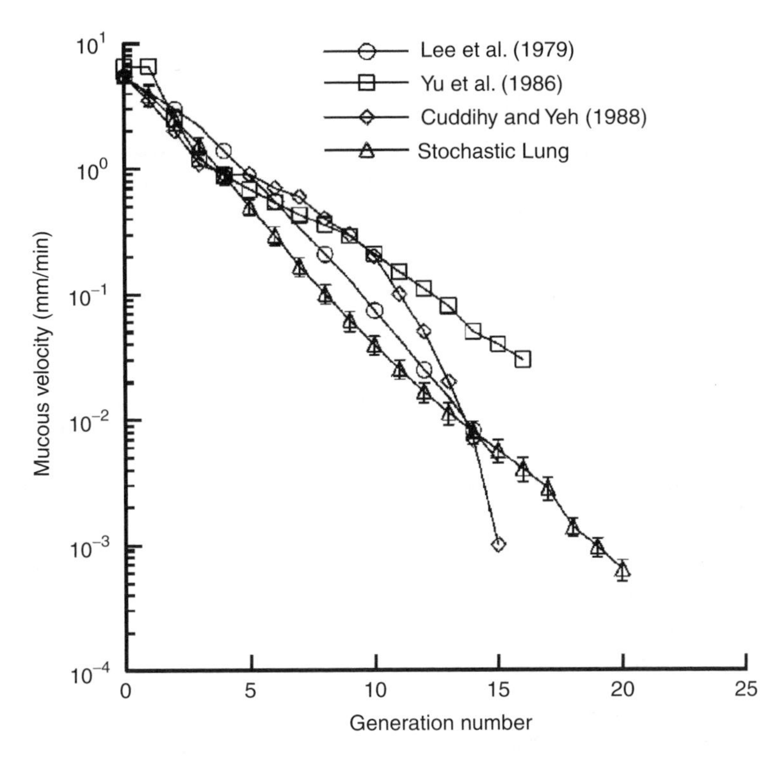

Figure 9.7 Predicted clearance velocity vs. generation number for four deposition models, showing a very large disagreement between models for generation numbers higher than five. The "stochastic lung" results, from Asgharian et al. (2001), show the mean and standard deviation of predictions for ten stochastic lung models. Figure from Asgharian et al. (2001).

Figure 9.7 compares four models for clearance velocity as a function of generation number (Lee et al., 1979; Yu and Xu, 1987; Cuddihy and, Yeh 1988; Asgharian et al., 2001). In each case, a clearance time was assumed and the model was solved to determine velocity in each airway generation. The stochastic lung model (Koblinger and Hofmann, 1985) used by Asgharian et al. (2001) differs from the other in having a lung morphology that is assumed to be more realistic, having pathways that differ in the number of generations.

The tour du force analysis by Asgharian et al., which produced their results shown in Figure 9.7, used ten stochastic lung models, which produced slightly different velocity predictions (which would correspond to an effect of interpersonal differences in lung morphology); the standard deviation of the velocity predictions for the ten stochastic lungs is plotted in the figure. The effect on velocities is rather small. However, because different lungs have different distributions of total airway lengths, and because lung morphology affects both deposition and clearance, the results on clearance time can be rather profound. Asgharian et al. simulated the deposition and clearance of distribution particles with a mass-mean aerodynamic diameter of 1 μm, in their ten stochastic lungs, as well as in a standard symmetric lung model and in a typical lung path. The results are shown in Figure 9.8. The differences in mass clearance profiles of the stochastic lungs suggest that even if there are no interpersonal differences in clearance physiology other than those caused by lung morphology, very large differences in mass removal rates are possible.

In addition to the obviously very large uncertainty in both typical clearance velocities and the amount of interpersonal variation in clearance velocities and times, other clearance-related parameters are very uncertain, and some important phenomena are not understood. A significant example is the presumed existence of a "slow clearance"

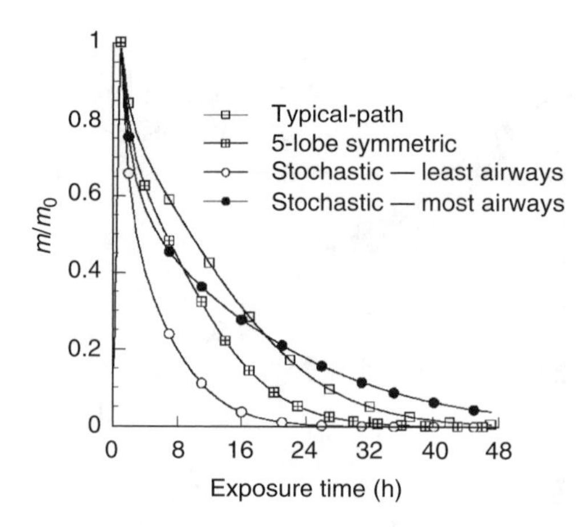

Figure 9.8 Predicted mass retained in the lung vs. time, for four different model lungs. Results from two of ten stochastic lungs generated from the same parametric lung description are shown; the difference in mass clearance shows the predicted effect of interpersonal morphometric variability even for individuals with the same gross lung description. Predictions from a symmetric lung model and a typical particle path can be very different from an individual prediction using the stochastic model. Figure from Asgharian et al. (2001).

mechanism to explain observed long-term (> 24 h) tracheobronchial retention of 1 to 3 μm particles, which seemed inconsistent with simple clearance models as discussed above. Explaining very slow clearance (>48 h) does require the introduction of a mechanism other than mucosal transport (e.g., macrophage uptake), but, as the results of Asgharian et al. show, some long-term retention may be explicable purely on morphological grounds, even without a separate long-term clearance mechanism.

Another important phenomenon that is poorly understood is "overloading," or dose-dependent decline of clearance. Overloading definitely occurs in rodents, but it is unclear whether it occurs in humans. Kuempel et al. (2001a, 2001b) examined this question by fitting a long-term clearance model to data on dust retention in miners, using dust loading data at autopsy and using estimated dust inhalation from work histories. The model fit was no better if an overloading mechanism was included.

Overall, quantitative modeling of clearance from the lung is very unreliable, and a great deal of progress is needed in order to improve predictions of clearance to the point where they can be used with confidence, as total and regional deposition models can, to make predictions for cases in which no experimental data are available. The degree of interpersonal variabiliy in the action of clearance mechanisms is even less certain. Bailey and Roy (1994) give a good discussion of these issues in an article in an annex to ICRP-66 that summarizes a prodigious volume of experimental literature.

9.3.9 Estimating the dose-versus-exposure relationship in a population

Molokanov and Badjin (2000) performed Monte Carlo simulation to evaluate parameter sensitivity of the ICRP-66 model for internal dose due to plutonium inhalation. Harvey and Hamby (2001, 2002) examined the same model in considerably more detail. They compiled data from a variety of sources to estimate the mean and standard deviation in the general population for all person-specific model parameters, as a function of age. Their recommended distributions are shown in Table 9.2. They used Latin Hypercube sampling to generate parameters for individuals, for whom the regional deposition of a specified (1 μm AMAD) particle distribution was calculated from the ICRP-66 model. By repeating

Table 9.2 Suggested Input Parameter Distributions Simulate the General Population, from Harvey and Hamby (2001, 2002)

	Parameter (units)	Adult Male	Adult Female	15 yr Male	15 yr Female	10 years	5 years	1 year	3 months
d_0	Diameter of trachea (cm)	1.65 (0.067)	1.53 (0.45)	1.59 (0.068)	1.52 (0.065)	1.31 (0.06)	1.06 (0.049)	0.75 (0.028)	0.62 (0.027)
d_9	Diameter of airway at gen. 9 (cm)	0.165 (0.067)	0.159 (0.006)	0.161 (0.007)	0.156 (0.007)	0.143 (0.007)	0.127 (0.006)	0.107 (0.004)	0.099 (0.004)
d_{16}	Diameter of airway at gen. 16 (cm)	0.051 (0.002)	0.048 (0.002)	0.047 (0.002)	0.045 (0.002)	0.039 (0.002)	0.031 (0.001)	0.022 (0.001)	0.020 (0.001)
BR	Breathing rate (m^3/h^1)	1.74 (0.67)	1.37 (0.45)	1.51 (0.51)	1.41 (0.45)	1.21 (0.41)	0.65 (0.19)	0.40 (0.11)	0.22 (0.06)
V_d	Anatomic dead space (ml)	146 (25.5)	124 (21.0)	130 (22)	114 (19)	78 (14.9)	46 (8.5)	20 (3.7)	14 (2.6)
FRC	Functional residual capacity (ml)	3301 (600)	2681 (500)	2677 (562)	2325 (488)	1484 (311)	767 (161)	244 (26)	148 (28)

Parameters show mean (standard deviation) of normal distributions. A proposed modification of the breathing rate parameterization is discussed in the text.

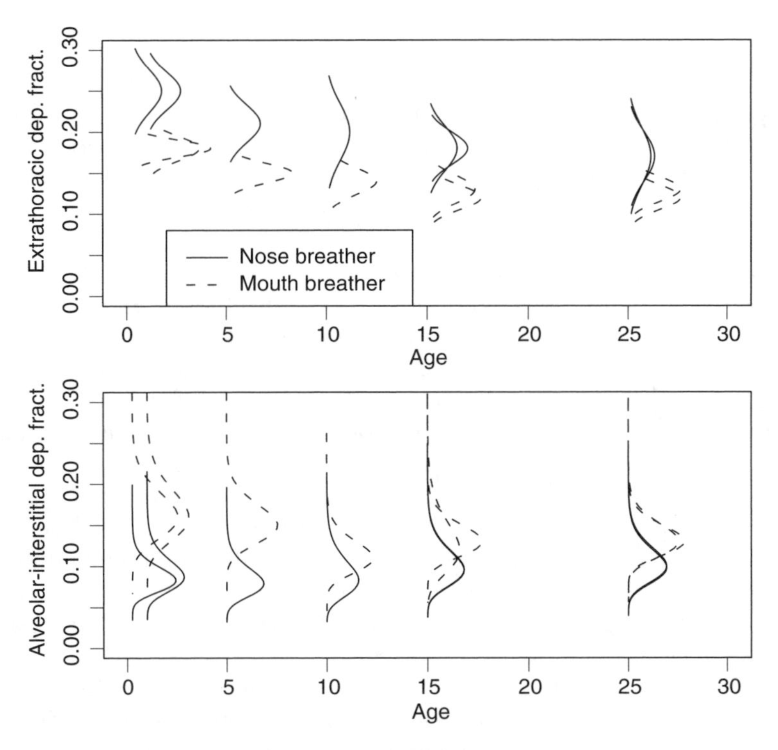

Figure 9.9 Predicted population distribution of extrathoracic and alveolar-interstitial deposition for 1 μm AMAD particles, as a function of age and breathing type, from Harvey and Hamby (2001, 2002). Much of the width of the distribution for each group is attributable to the wide distribution of assumed breathing rates. Separate plots are shown for males and females at ages 15 and 25 (adult), but are not labeled separately.

the sampling-and-calculation procedure many times, they created predicted distributions of regional deposition across the entire population of a given age and "breathing type" (mouth-breather or nose-breather). The results are shown in Figure 9.9 for two regions of the respiratory tract. Within each age group, distributions are very wide; most of the variability within each age group is due to the assumed interpersonal variation in breathing rate. Since the ICRP-66 model does not incorporate some known phenomena such as interpersonal morphological variation in the lung, the variability in results may be underestimated. Additionally, the sampling procedure assumes independence between the input parameters, whereas in fact they will be correlated (e.g., functional residual capacity will be positively correlated with diameter of trachea, since they are both correlated with body size); including such correlation could decrease or increase the widths of the predicted distributions, depending on details.

Figure 9.9 clearly illustrates the problem of using exposure as a proxy for dose. For instance, about 5% of male nose-breathers are predicted to have an alveolar-interstitial deposition fraction of less than 0.05, whereas about 5% of mouth-breathers are predicted to exceed 0.2, a factor of four difference which, as discussed above, is likely to be an underestimate. Of course, the distribution of physiological effects would be wider still, due to interpersonal variation in clearance, site-specific dose–response, and so on.

Of course, results such as these are only valuable to the extent that the model and assumptions on which they are based are trustworthy. The ICRP-66 model predicts regional deposition fairly well for a given set of parameter values, so the major issues in this particular case concern the accuracy of the statistical distributions of the input values.

9.4 Conclusion

A quantitative understanding of the relationship between exposure and dose is required in order to interpret epidemiological studies and to extrapolate to conditions or populations outside the studies, and for many other purposes such as optimizing aerosolized drug delivery, predicting the effects of drugs that affect airway constriction or other physiological parameters, and so forth.

Epidemiological studies definitively demonstrate increased mortality risk from exposure to very high concentrations of radon (several times higher than commonly occur in homes) and from surprisingly low concentrations of fine particles that are frequently experienced in cities. However, epidemiological studies based on exposure can only go so far in addressing questions related to the health effects of inhaled aerosols, because they are subject to substantial uncertainties due to errors in exposure, response recording, and to the presence of confounding variables.

Total and regional deposition in typical human subjects can be predicted very well for a laboratory subject, as indeed has been the case for many years. Local deposition (deposition at a particular location in the lung, or for a particular branching generation) is subject to larger predictive error; it is unclear whether this is due to shortcomings in the models or to shortcomings of parametric characterizations of the lung, particularly for high branching generation numbers.

Interpersonal variability in morphology and inhalation parameters seems to be reasonably well characterized for healthy noninfants, but more work is needed on characterizing these factors for infants and for people with lung disease. Since the latter group is a population of particular concern with regard to fine particle air pollution, this issue merits more attention.

The ability to predict clearance is still not very good, and predictions certainly cannot be trusted for high generation numbers. Even some basic clearance-related questions are unanswered, such as what level of particle loading (if any) induces an "overload" response. A much better understanding of the distribution of clearance-related parameters is needed for the general population and particularly for patients with lung disease.

Table 9.3 shows a qualitative evaluation of the adequacy of models, parameter estimates, and estimates of interpersonal variability in parameters, with regard to predicting

Table 9.3 Qualitative View of the Ability to Predict the Effects of Morphology, Inhalation Details, and Clearance on Total and Regional Dose and Time-Dependent Local Dose for Particles Larger than 0.1 μm Diameter, in an Experimental Subject (left columns) and Across the General Population (right columns)

	Healthy Adults	Children and Elderly	Diseased	Healthy Adults	Children and Elderly	Diseased
	Ability to Model the Effects in a Known Individual			**Ability to Model the Effects in the General Population**		
Morphology	++++	++++	+++	+++	+++	++
Inhalation details	++++	+++	+++	+++	+++	++
Clearance	++	++	+	+	+	+
	Ability to Predict Particle Deposition (dose)			**Ability to Predict Population Distribution of Deposition, if Exposure were Known**		
Total deposition	++++	+++	+++	+++	+++	++
Regional deposition	+++	++	++	++	++	+
	Ability to Predict Time-Dependent Dose, Including Clearance			**Ability to Predict Population Distribution of Time-Dependent Dose**		
Total	+++	+++	++	++	++	+
Regional	++	++	+	+	+	+
Local	+	+	+	+	+	+

++++ = very good, + = very poor.

time-dependent dose from the inhalation of particles in the range of 0.01 to 10 μm. Of course, the adequacy of a model ultimately depends on what level of accuracy is required; the table should therefore be interpreted as characterizing the relative rather than the absolute adequacy of models and parameter estimates. At least in the size range from 0.01 to 10 μm, the challenges of predicting total and regional deposition have been met, and the prediction of local dose is not bad. Clearance, interpersonal variability, and site-specific deposition constitute the new frontier, in the sense that uncertainties remain very high for predictions of these quantities.

Acknowledgments

This work was supported by the Assistant Secretary for Energy Efficiency and Renewable Energy, Building Technologies Program of the U.S. Department of Energy under Contract No. DE-AC03-76SF00098.

References

Alavanja, M.C., Lubin, J.H., Mahaffey, J.A., and Brownson, R.C., Residential radon exposure and risk of lung cancer in Missouri, *Am. J. Public Health*, 7, 1042–1048, 1999.

Asgharian, B., Hofmann, W., and Miller, F.J., Mucociliary clearance of insoluble particles from the tracheobronchial airways of the human lung, *J. Aerosol Sci.*, 32, 817–832, 2001.

Auvinen A., Makelainen I., Hakama M., Castren O., Pukkala E., Reisbacka H., and Rytomaa T., Indoor radon exposure and risk of lung cancer: a nested case-control study in Finland. *J. Nat. Cancer Inst.*, 88, 966–972, 1996. Errata in *J. Nat. Cancer Inst.*, 90, 401–402, 1998.

Bailey, M.R. and Roy, M., Clearance of particles from the respiratory tract. Human respiratory tract model for radiological protection, Annexe E, *Ann. ICRP*, 24, 301–380, 1994.

Bennet, W.D. and Zeman, K.L., Deposition of fine particles in children spontaneously breathing at rest, *Inhal. Toxicol.*, 10, 831–842, 1998.

Bennet, W.D., Zeman, K.L., and Kim, C., Variability of fine particle deposition in healthy adults: effect of age and gender, *Am. J. Respir. Crit. Care Med.*, 153, 1641–1647, 1996.

Bergmann, R., Hofmann, W., and Koblinger, L., Particle deposition modeling in the human lung: comparison between Monte Carlo and ICRP model predictions, *J. Aerosol Sci.*, 28, S433–S434, 1997.

Bogen, K.T. and Layton, D.W., Risk management for plausibly hormetic environmental carcinogens: the case of radon, *Belle Newslett.*, 7, 30–37, 1998.

Brown, J.S., Zeman, K.L., and Bennet, W.D., Regional deposition of coarse particles and ventilation distribution in healthy subjects and patients with cystic fibrosis, *J. Aerosol Med. — Deposition Clearance Effects Lung* 14, 443–454, 2001.

Chua, H.L., Collis, G.G., Newbury, A.M., Chan, K., Bower, G.D., Sly, P.D., and Lesouef, P.N., The influence of age on aerosol deposition in children with cystic fibrosis, *Eur. Respir. J.*, 7, 2185–2191, 1994.

Cohen, B.L., Test of the linear no-threshold theory of radiation carcinogenesis for inhaled radon decay products, *Health Phys.*, 68, 157–174, 1995.

Cohen, B.L., Updates and extensions to tests of the linear no-threshold theory, *Technology*, 7, 657–772, 2000.

Cohen, B.L. and Colditz, G.A., Tests of the linear no-threshold theory for lung cancer induced by exposure to radon, *Environ. Res.*, 64, 65–89, 1994.

Cuddihy, R.G. and Yeh, H.C., Respiratory tract clearance of particles and substances dissociated from particles, in *Inhalation Toxicology: The Design and Interpretation of Inhalation Studies and their Use in Risk Assessment*, Mohr, U., Ed., Springer-Verlag, Berlin, 1988, pp. 169–193.

Dockery, D., Pope, A., and Xu, X., An association between air pollution and mortality in six U.S. cities, *N. Engl. J. Med.*, 329, 1753–1759, 1993.

Dominici, F., Daniels, M., Zeger, S.L., and Samet, J.M., Air pollution and mortality: estimating regional and national dose–response relationships, *J. Am. Statist. Assoc.*, 97, 100–111, 2002a.

Dominici, F., McDermott, A., Zeger, S.L., and Samet, J.M., On generalized additive models in time series of air pollution and health, *Am. J. Epidemiol.*, 156: 3, 2002b.

Ebelt S.T., Fisher T.V., Petkan A.J., Vedal S., Braxer M., Exposure of chronic obstructive pulmonary disease (COPD) patients to particulate matter: relationship between personal and ambient air concentrations. *J. Air Waste Manage. Assoc.*, 50, 1081–1094.

El-Hussein, A., Ahmed, A.A., and Mohammed, A., Radiation dose to the human respiratory tract from inhalation of radon-222 and its progeny, *Appl. Radiat. Isot.*, 49, 783–790, 1998.

Field, R.W., A review of residential radon case–control epidemiologic studies performed in the United States, *Rev. Environ. Health*, 16, 151–167, 2001.

Field, R.W., Steck, D.J., and Nueberger, J.S., Accounting for random error in radon exposure assessment, *Health Phys.*, 73, 272–273, 1997.

Field, R.W., Steck, D.J., Smith, B.J., Brus, C.P., Fisher, E.L., Neuberger, J.S., Platz, C.E., Robinson, R.A., Woolson, R.F., and Lynch, C.F., Residential radon gas exposure and lung cancer: the Iowa lung cancer study, *Am. J. Epidemiol.*, 151, 1091–1102, 2000.

Fleming, J.S., Conway, J.H., Holgate, S.T., Bailey, A.G., and Martonen, T.B., Comparison of methods for deriving aerosol deposition by airway generation from three-dimensional radionuclide imaging, *J. Aerosol Sci.*, 31, 1251–1259, 2000.

Gelman, A., Park, D.K., Ansolabehere, S., Price, P.N., and Minnite, L.C., Models, assumptions and model checking in ecological regressions, *J. Rl Statist Soc. A*, 164, 101–118 (2001).

Gradon, L. and Podgorski, A., Deposition of inhaled particles: discussion of present modeling techniques, *J. Aerosol Med.*, 9, 343–355, 1996.

Greenland, S. and Morgenstern, H., Neither within-region nor cross-regional independence of exposure and covariate prevents ecological bias, *Int. J. Epidemiol.*, 18, 269–274, 1989.

Greenland, S. and Robins, J., Ecologic studies: biases, misconceptions, and counterexamples, *Am. J. Epidemiol.*, 139, 747–760, 1994.

Habib, R.H., Chalker, R.B., Suki, B., and Jackson, A.C., Airway geometry and wall mechanical properties estimated from subglottal input impedence in humans, *J. Appl. Physiol.*, 77, 441–451, 1994.

Harvey, R.P. and Hamby, D.M., Uncertainty in particulate deposition for 1 micrometer AMAD particles in an adult lung model, *Radiat. Prot. Dosim.*, 95, 239–247, 2001.

Harvey, R.P. and Hamby, D.M., Age-specific uncertainty in particulate deposition for 1 micrometer AMAD particles using the ICRP 66 lung model, *Health Phys.*, 82, 807–816, 2002.

Heyder, J., Gebhart, J., Stahlhofen, W., and Stuck, B., Biological variability of particle deposition in the human respiratory tract during controlled and spontaneous mouth-breathing, *Ann. Ocuup. Hygiene.*, 26, 137–147, 1982.

Heyder, J., Gebhart, J., Rodolf, G., Schiller, C.F., and Stahlhofen, W., Deposition of particles in the human respiratory tract in the size range 0.005–15 micrometers, *J. Aerosol Sci.*, 17, 811–825, 1986.

Hofmann, W., Lung Morphometry and Particle Transport and Deposition: Overview of Existing Models, in Aerosol Inhalation: Recent Research Frontiers, Proceedings of the International Workshop on Aerosol Inhalation, Lung Transport, Deposition and the Relation to the Environment, Marijnissen, J.C.M. and Gradon, L., Ed., Kluwer Academic Publishers, Boston, 1996a.

Hofmann, W., Modeling techniques for inhaled particle deposition: the state of the art, *J. Aerosol Med.*, 9, 369–388, 1996b.

Hofmann, W. and Koblinger, L., Monte Carlo modeling of aerosol deposition in human lungs. Part III: Comparison with experimental data, *J. Aerosol Sci.*, 23, 51–63, 1992.

Hofmann, W., Asgharian, B., and Winkler-Heil, R., Modeling intersubject variability of particle deposition in human lungs, *J. Aerosol Sci.*, 33, 219–235, 2002.

Howe, G.R., Nair, R.C., Newcombe, H.B. et al., Lung cancer mortality in relation to radon daughter exposure in a cohort of workers at the Eldorado Beaverlodge uranium mine, *J. Nat. Cancer Inst.*, 77, 357–362, 1986.

Howe, G.R., Nair, R.C., Newcombe, H.B. et al., Lung cancer mortality in relation to radon daughter exposure in a cohort of workers at the Eldorado Port Radium uranium mine: possible modification of risk by exposure rate, *J. Nat. Cancer Inst.*, 79, 1255–1260, 1987.

Huet, C., Tymen, G., and Bouland, D., Size distribution, equilibrium ratio and unattached fraction of radon decay products under typical indoor domestic conditions. *Sci. Total Environ.*, 272, 97–103, 2001.

ICRP (International Commission on Radiological Protection). *Human Respiratory Tract Model for Radiological Protection*, ICRP Publication 66, Pergamon Press, Oxford, 1994.

James, A., Lung Dosimetry, in *Radon and its Decay Products in Indoor Air*, Nazaroff, W.W. and Nero, A.V., Jr., Eds., John Wiley and Sons, New York, 1988.

James, A.C., Stahlhofen, W., Rudolf, G., Kobrich, R., Briant, J.K., Egan, M.J., Nixon, W., and Birchall, A., Annexe, D., Deposition of inhaled particles, in *International Commission on Radiological Protection: Human Respiratory Tract Model for Radiological Protection*, ICRP Publication 66, Pergamon Press, Oxford, 1994.

Janssen, N.A., de Hartog, J.J., Hoek, G., and Brunekreef, B., Personal exposure to fine particulate matter in elderly subjects: relation between personal, indoor, and outdoor concentrations, *J. Air Waste Manage. Assoc.*, 50, 1133–1143, 2000.

Kim, C.S. and Hu, S.C., Regional deposition of inhaled particles in human lungs: comparison between men and women, *J. Appl. Physiol.*, 84, 1834–1844, 1998.

Koblinger, L. and Hofmann, W., Analysis of human lung morphometric data for stochastic aerosol deposition calculations, *Phys. Med. and Biol.*, 30, 541–556, 1985.

Koblinger, L. and Hofmann, W., Monte Carlo modeling of aerosol deposition in human lungs. Part I: Simulation of particle transport in a stochastic lung structure. *J. Aerosol Sci.*, 21, 661–674, 1990.

Kohlhaufl, M., Brand, P., Scheuch, G., Meyers, T.S., Schulz, H., Haussinger, K., and Heyder, J., Increased fine particle deposition in women with asymptomatic nonspecific airway hyperresponsiveness, *Am. J. Respir. Crit. Care Med.*, 159, 902–906, 1999.

Kousa, A., Oglesby, L., Koistinen, K., Kunzli, N., and Jantunen, M., Exposure chain of urban air PM2.5 — associations between ambient fixed site, residential outdoor, indoor, workplace and personal exposure in four European cities in the EXPOLIS study, *Atmos. Environ.*, 36, 3031–3039, 2002.

Kuempel, E.D., Tran, C.-L., Smith, R.J., and Bailer, A.J., A biomathematical model of particle clearance and retention in the lungs of coal miners. II: Evaluation of variability and uncertainty. *Regul. Toxicol. Pharmacol.*, 34, 88–101, 2001a.

Kuempel, E.D., Tran, C.-L., Bailer, A.J., Smith, R.J., Dankovic, D.A., and Stayner, L.T., Methodological issues of using observational human data in lung dosimetry models for particulates, *Sci. Total Environ.*, 274, 67–77, 2001b.

Kusiak, R.A., Ritchie, A.C., Muller, J. et al., Mortality from lung cancer in Ontario uranium miners, *Br. J. Ind. Med.*, 50, 920–928, 1993.

Lagarde, F., Pershagen, G., Akerblom, G., Axelson, O., Baverstam, U., Damber, L., Enflo, A., Svartengren, M., and Swedjemark, G.A., Residential radon and lung cancer in Sweden: risk analysis accounting for random error in the exposure assessment, *Health Phys.*, 72, 269–276, 1997.

Landahl, H.D., Tracewell, T.N., and Lassen, W.H., On the retention of airborne particulates in the human lung, *AMA Arch. Ind. Hyg. Occup. Med.*, 3, 359–366, 1951.

Le Tertre, A., Quenel, P., Eilstein, D., Medina, S., Prouvost, H., Pascal, L., Boumghar, A., Saviuc, P., Zeghnoun, A., Filleul, L., Declercq, C., Cassadou, S., and Le Goaster, C., Short-term effects of air pollution on mortality in nine French cities: a quantitative summary, *Arch. Environ. Health*, 57, 311–319, 2002.

Lee, P.S., Gerrity, T.R., Hass, F.J., and Lourenco, R.V., A model for tracheobronchial clearance of inhaled particles in man and comparison with data, *IEEE Trans. Biomed. Eng.*, 26, 624–630, 1979.

Lee, J.T., Kim, H., Hong, Y.C., Kwon, H.J., Schwartz, J., and Christiani, D.C., Air pollution and daily mortality in seven major cities of Korea, 1991–1997, *Environ. Res.*, 84, 247–254, 2000.

Lin, C.Y., Gelman, A., Price, P.N., and Krantz, D.H., Analysis of local decisions using hierarchical modeling, applied to home radon measurement and remediation, *Statist. Sci.*, 14, 305–337, 1999.

Lubin, J.H., The potential for bias in Cohen's ecological analysis of lung cancer and residential radon, *J. Radiol. Prot.*, 22, 141–148, 2002.

Lubin, J.H. and Boice, J.D., Jr., Lung cancer risk from residential radon: meta-analysis of eight epidemiologic studies, *J. Nat. Cancer Inst.*, 89, 49–57, 1997.

Lubin, J.H., Boice, J.D., Jr., Edling, C., Hornung, R.W., Howe, G.R., Kunz, E., Kusiak, R.A., Morrison, H.I., Radford, E.P., Samet, J.M., Tirmarche, M., Woodward, A., Yao, S.X., and Pierce, D.A., Lung cancer in radon-exposed miners and estimation of risk from radon exposure, *J. Nat. Cancer Inst.*, 87, 817–827, 1995.

Marcinowski, F., Lucas, R.M., and Yeager, W.M., National and regional distribution of airborne radon concentrations in U.S. homes, *Health Phys.*, 66, 699–706, 1994.

Maynard, A.D. and Maynard, R.L., A derived association between ambient aerosol surface area and excess mortality using historic time series data, *Atmos. Environ.*, 36, 5561–5567, 2002.

McClellan, R.O., A mechanistic approach to assessing the lung cancer risk of diesel exhaust and carbon black, in *Aerosol Inhalation: Recent Research Frontiers*, Marijnissen, J.C.M. and Gradon, L., Eds., Kluwer Academic Publishers, The Netherlands, 1995.

Mills, P.K., Abbey, D., Beeson, W.L., and Petersen, F., Ambient air pollution and cancer in California Seventh-Day Adventists, *Arch. Environ. Health.*, 46, 271–280.

Molokanov, A.A. and Badjin, V.I., Parameter Uncertainty Analysis in the Task of Internal Dose Reconstruction Based on Am-241 Organ Activity Measurements, presented at 10th Congress of the International Radiation Protection Association, Hiroshima, Japan, 2000.

Moolgavkar, S., Leubeck, E.G., Hall, T.A., and Anderson, E.L., Particulate air pollution, sulfur dioxide and daily mortality: a reanalysis of the Steubenville data, *Inhal. Toxicol.*, 7, 35–44, 1995.

NRC (National Research Council), Research Priorities for Airborne Particulate Matter: I. Immediate Priorities and a Long-Range Research Portfolio, Committee on Research Priorities for Airborne Particulate Matter, National Academies Press, Washington, DC., 1998a.

NRC (National Research Council), Health Risks of Radon and Other Internally Deposited Alpha-emitters — BEIR VI, NRC Committee on the Biological Effects of Ionizing Radiations (BEIR), National Academy Press, Washington, DC., 1998b.

Nerbrink, O.L., Lindstrom, M., Meurling, L., and Svartengren, M., Inhalation and deposition of nebulized sodium cromoglycate in two different particle size distributions in children with asthma, *Pediatr. Pulmonol.*, 34, 351–360, 2002.

Nikiforov, A.I. and Schlesinger, R.B., Morphometric variability of the human upper bronchial tree, *Respir. Physiol.*, 59, 289–299, 1985.

Oglesby, L., Kunzli, N., Roosli, M., Braun-Fahrlander, C., Mathys, P., Stern, W., Jantunen, M., and Kousa, A., Validity of ambient levels of fine particles as surrogate for personal exposure to outdoor air pollution — results of the European EXPOLIS-EAS study (Swiss Center Basel), *J. Air Waste Manage. Assoc.*, 50, 1251–1261, 2000.

Phalen, R.F., Yeh, H.C., Raabe, O.G., Hulbert, A., and Crain, C.R., Comparative airway anatomy in the human beagle, dog, rat, and hamster and implications in inhaled particles, *Health Phys.*, 27, 634–635, 1974.

Pope, C.A., III, Burnett, R.T., Thun, M.J., Calle, E.E., Krewski, D., Ito, K., and Thurston, G.D., Lung cancer, cardiopulmonary mortality, and long-term exposure to fine particulate air pollution, *J. Am. Med. Assoc.*, 187, 1132–1141, 2002.

Price, P.N., Test of the linear no-threshold theory, *Health Phys.*, 69, 577–578, 1995 (letter to the editor).

Robinson, W.S., Ecological correlations and the behavior of individuals, *Am. Sociol. Rev.*, 15, 351–357, 1950.

Rojas-Bracho, L., Bahadori, T., Koutrakis, P., and Suh, H., Personal, Indoor, and Outdoor Exposures to Particulate Matter of COPD Patients Living in Private Residences, Abstract of meeting paper for Society for Risk Analysis Annual Meeting, 1996.

Rojas-Bracho, L., Koutrakis, P., and Suh, H., PM2.5 and PM10 Personal exposure and its Relationship with Outdoor Concentrations in a Group of COPD Patients Living in the Boston Area, Proceedings of the 10th Conference of the International Society for Environmental Epidemiology, Boston, MA, 1998.

Samet, J.M., Dominici, F., Curriero, F.C., Coursac, I., and Zeger, S.L., Fine particulate air pollution and mortality in 20 U.S. cities, 1987–1994. *N. Engl. J. Med.*, 343, 1742–1749, 2000.

Sarnat, J.A., Koutrakis, P., and Suh, H.H., Assessing the relationship between personal particulate and gaseous exposures of senior citizens living in Baltimore, MD, *J. Air Waste Manage. Assoc.*, 50, 1184–1198, 2000.

Satoh, K., Kobayashi, T., Mitani, M., Kawase, Y., Takahashi, K., Nishiyama, Y., Ohkawa, M., Tanabe, M., Koba, H., and Suzuki, A., Regular and irregular dichotomies of bronchial branching in the human lung, *Acad. Radiol.*, 3, 469–474, 1996.

Schwartz, J., Air pollution and daily mortality: a review and meta-analysis, *Environ. Res.*, 64, 36–52, 1994.

Segal, R.A., Martonen, T.B., and Kim, C.S., Comparison of computer simulations of total lung deposition to human subject data in healthy test subjects, *J. Air Waste Manage. Assoc.*, 50, 1262–1268, 2000.

Smaldone, G.C., Deposition and clearance: unique problems in the proximal airways and oral cavity in the young and elderly, *Respir. Physiol.*, 128, 33–38, 2001.

Smith, S., Cheng, Y.S., and Yeh, H.S., Deposition of ultrafine particles in human tracheobronchial airways of adults and children, *Aerosol Sci. Technol.*, 35, 697–709, 2001.

Stahlhofen, W., Gebhart, J., and Heyder, J., Biological variability of regional deposition of aerosol particles in the human respiratory tract, *Am. Ind. Hyg. Assoc. J.*, 42, 348–352, 1981.

Steck, D.J., Spatial and temporal indoor radon variations, *Health Phys.*, 62, 351–355, 1994.

Steinhausler, F., Epidemiological evidence of radon-induced health risks, in *Radon and its Decay Products in Indoor Air*, Nazaroff, W.W. and Nero, A.V., Eds., John Wiley and Sons, New York, 1988.

Tomasek, L., Darby, S.C., Fearn, T., Swerdlow, A.J., Placek, V., and Kunz, E., Patterns of lung cancer mortality among uranium miners in West Bohemia with varying rates of exposure to radon and its progeny, *Radiat. Res.*, 137, 251–261, 1994.

Tsai, F.C., Apte, M.G., and Daisey, J.M., An Exploratory Analysis of the Relationship between Mortality and the Chemical Composition of Airborne Particulate Matter, in Phalen, R.F. and Bell, Y.M., Eds., Proceedings of the Third Colloquium on Particulate Air Pollution and Human Health, Air Pollution Health Effects Laboratory, University of California, Irvine CA, 1999.

Wang, Z., Lubin, J.H., Wang, L., Zhang, S., Boice, J.D., Jr., Cui, H., Zhang, S., Conrath, S., Xia, Y., Shang, B., Brenner, A., Lei, S., Metayer, C., Cao, J., Chen, K., Lei, S., and Kleinerman, R., Residential radon and lung cancer risk in a high-exposure area of Gansu province, China, *Am. J. Epidemiol.*, 155, 554–564, 2002.

Wilson, W.E. and Suh, H.H., Fine particles and coarse particles: concentration relationships relevant to epidemiologic studies, *J. Air Waste Manage. Assoc.*, 47, 1238–1249, 1997.
Wilson, W.E., Mage, D.T., and Grant, L.D., Estimating separately personal exposure to ambient and nonambient particulate matter for epidemiology and risk assessment: why and how, *J. Air Waste Manage. Assoc.*, 50, 1167–1183.
Yeh, H.-C. and Schum, G.M., Models of human lung airways and their application to inhaled particle deposition, *Bull. Math. Biol.*, 42, 461–480, 1980.
Yu, C.P. and Diu, C.K., A comparative study of aerosol deposition in different lung models, *Am. Ind. Hyg. Assoc. J.*, 43, 54–65, 1982.
Yu, C.P. and Xu, G.B., Predicted deposition of diesel particles in young humans, *J. Aerosol Sci.*, 4, 419–429, 1987.

chapter ten

Aerosol chemistry and physics: Indoor perspective

Lara A. Gundel and Richard G. Sextro
Environmental Energy Technologies Division

Contents

Preface 190
10.1 Introduction 191
10.1.1 Importance of aerosol exposures 191
10.1.2 Significance of indoor environment to aerosol exposure 191
10.1.3 Outdoor particles 192
10.1.3.1 $PM_{2.5}$ composition overview 192
10.1.3.2 $PM_{2.5}$ source apportionment 192
10.1.4 Differences between indoor and outdoor environments 193
10.1.5 Evidence for indoor aerosol processing and generation 194
10.1.5.1 The building envelope 194
10.1.5.2 Indoor materials 195
10.1.5.3 Indoor aerosol generation 195
10.2 Composition of indoor particles and aerosol precursors 196
10.2.1 Combustion sources 196
10.2.1.1 Cooking with oils and meat 196
10.2.1.2 Fireplaces 199
10.2.1.3 Tobacco 200
10.2.1.4 Candles 201
10.2.1.5 Incense 201
10.2.1.6 Unvented kerosene heaters 202
10.2.2 Human activities and consumer products 202
10.2.2.1 Pesticides 202
10.2.2.2 Air fresheners 202
10.2.2.3 Walking 202
10.2.2.4 Cleaning 203
10.2.2.5 Cleaning products 203
10.2.2.6 Renovation 204
10.2.3 Indoor aerosol source apportionment 204
10.3 Physical properties of indoor aerosols 205
10.3.1 Size distributions 205
10.3.1.1 Number 206
10.3.1.2 Surface area 206
10.3.1.3 Volume and mass 206

1-56670-611-4 / 05 / $0.00+$1.50

10.3.2 Particle dynamics 208
10.3.2.1 Particle motion 208
10.3.2.2 Particle formation and phase partitioning 209
10.3.2.2.1 Nucleation....209
10.3.2.2.2 Condensation 209
10.3.2.2.3 Sorption/desorption 209
10.3.3 Surface area in the indoor environment 211
10.3.3.1 Particles 211
10.3.3.2 Indoor materials 211
10.3.3.3 Adsorption indoors 211
10.4 Fate and transport 213
10.4.1 Model of indoor aerosol behavior 213
10.4.2 Airflow and aerosol transport through penetrations in building envelopes ..215
10.4.3 Indoor aerosol deposition rates 216
10.4.4 Resuspension rate of particles on carpets/floors 217
References 218

When Sherlock Holmes mystified his friend Dr. Watson by his amazing deductions, he was utilizing to the fullest degree the data available to him — the muddy boot, the ash of a cigar, the torn ticket. This is a practice which is also basic in the highest forms of scientific research.

E.B. Wilson, Introduction to Scientific Research, 1952

Preface

Research in indoor aerosol physics and chemistry requires the type of careful observations and astute deductions that the fictional Sherlock Holmes used so well. Holmes's perspective assists in moving past the preconceptions based on outdoor chemistry and the presumption that indoor environments provide refuge from the assaults of air pollution.

The built environment provides shelter from the weather, but not from exposure to airborne particles, especially fine particles. Indoor aerosol chemistry and physics have become exciting and fruitful areas of investigation, inspired by the pioneering work of Weschler and Shields (1997, 1999), who showed that ozone reactions with volatile organic compounds indoors can produce secondary organic aerosols. Wolkoff et al. (2000) subsequently demonstrated that the products of these reactions are irritating to humans, even in low concentrations. In addition to pollutant specific indoor transport and fate models the future holds the promise of comprehensive models to predict the behavior of indoor pollutants in the indoor environment, as illustrated in the recent work of Bennett and Furtaw (2004).

This overview focuses mainly on indoor fine particle matter (PM) in urban homes and offices located in temperate zones. The objective is to summarize an understanding (as of late 2003) of the chemical composition of indoor PM from various sources, as well as the chemical and physical processes that contribute to its behavior and fate. Selected recent examples illustrate the promise and challenge of this area of research, and point to fruitful directions for future investigation.

In addition to the other chapters in this book, relevant background material on indoor particles is found in recent books edited by Morawska and Salthammer (2003) and Salthammer (1999). A recent issue of the journal *Atmospheric Environment*, vol 37(39 and 40) 2003 is devoted to indoor air chemistry and physics. For more insight about the fundamental chemical and physical processes that affect both outdoor and indoor air, readers can consult reference texts on atmospheric chemistry such as Finlayson-Pitts and Pitts (1999) and Seinfeld and Pandis (1998). Relevant environmental engineering concepts are clearly presented in the text by Nazaroff and Alvarez-Cohen (2000).

10.1 Introduction

10.1.1 Importance of aerosol exposures

Epidemiologists have established an association between concentrations of airborne particles and human morbidity and mortality. The mechanisms for the effects of particles on health have not yet been established, but currently the U.S. EPA supports a number of investigations in an effort to find out.

10.1.2 Significance of indoor environment to aerosol exposure

People spend 80–90% of their time indoors, so most human exposure to particles of outdoor origin takes place indoors. Nazaroff et al. (2003) compare mass flow rates in urban and indoor atmospheres to the human breathing rate (Table 10.1, from Nazaroff et al., 2003) and conclude that a nonreactive compound present indoors is about a thousand times more likely to be inhaled than if the same species is emitted or formed outdoors.

Indoor particulate matter has not yet been investigated as intensely or as systematically as outdoor PM. In the early 1990s, the EPA's PTEAM study compared personal exposure to PM_{10} mass in Riverside, CA, to indoor and outdoor concentrations (PM_{10} refers to particles smaller than 10 μm in diameter.). The resulting data were used to model PM infiltration and removal processes (Ozkaynak et al., 1996). Over half the indoor particle mass originated outdoors, even in homes with smoking and cooking. The PTEAM study found that personal exposures were higher than (<2.5 μm diameter) indoor or outdoor concentrations, and this excess exposure is now called the "personal cloud." Recently, Williams et al. (2003a, b) reported $PM_{2.5}$ exposures based on data from a longitudinal study in North Carolina that tracked indoor exposure to ambient PM. These reports also cite many related earlier studies. The personal cloud effect was confirmed, and mean personal $PM_{2.5}$ exposures were only moderately correlated with ambient $PM_{2.5}$ concentrations. These studies raise concern about the representativeness of central monitoring site data for exposure assessment.

With growing attention to the health effects of inhaled particles, more and more research has been directed to characterizing what people actually breathe indoors. A chemical tracer source apportionment approach (Schauer et al., 1996) is being used for outdoor particle source apportionment in the southern California children's health study (Manchester-Neesvig et al., 2003). While such projects are major contributions, they work best for well-characterized outdoor PM sources. They do not yet consider the indoor environment as a chemical reservoir that can change the composition of infiltrating particles

Table 10.1 Atmospheric Mass and Flow Rates in Global, Urban and Indoor Atmospheres[a]

Environment	Mass (kg)	Flow, F ($kg\,d^{-1}$)	Mass Breathed, Q[b] ($kg\,d^{-1}$)	Ratio $Q{:}F$
Global atmosphere	5×10^{18}	—	$\sim10^{11}$	—
Urban atmospheres[c]	$\sim10^{15}$	$\sim3\times10^{15}$	$\sim4\times10^{10}$	$\sim10^{-5}$
Indoor atmospheres[d]	$\sim10^{12}$	$\sim10^{13}$	$\sim8\times10^{10}$	$\sim10^{-2}$

[a]From Nazaroff et al. (2003).

[b]Includes air inside and outside of buildings.

[c]Sum of all urban environments (globally).

[d]Sum of all indoor environments (globally).

and generate PM. Improved indoor source apportionment methods such as the positive matrix factor approach (Yakovleva et al. (1999) for PTEAM data) are currently being applied to personal exposure data. An example, based on positive matrix factorization (Yakovleva et al., 1999) will be discussed later.

10.1.3 Outdoor particles

10.1.3.1 $PM_{2.5}$ composition overview

Since at least half of indoor PM has infiltrated from outdoors (e.g., Wallace et al., 2003), an overview of outdoor PM provides a useful starting point for the investigation of indoor aerosol chemistry. Figures 10.1 and 10.2 show the 2002 annual average $PM_{2.5}$ concentrations for urban and rural areas in the U.S., respectively, as reported by the Speciation Trends Network for urban areas and the Interagency Monitoring of Protected Visual Environments Network for rural areas (U.S. EPA 2003, p. 14).

Urban areas experienced higher PM concentrations than rural areas, but the difference was smaller in the eastern U.S. Throughout the country, carbonaceous material accounted for more of the PM mass than any other species. In the eastern and western U.S., sulfate and nitrate ions, respectively, accounted for the next greatest fractions of urban PM mass. Particulate ammonium was ubiquitous and originated from biogenic ammonia. These differences in ionic composition can be traced to regional differences in concentrations of gas-phase precursors SO_2 (from stationary sources, eastern U.S.) and NO_x (primarily from vehicles, western U.S.).

10.1.3.2 $PM_{2.5}$ source apportionment

Although the U.S. monitoring networks do not conduct detailed speciation of the carbonaceous components of $PM_{2.5}$, hundreds of individual compounds from a variety of combustion sources have been identified (U.S. EPA, Rogge et al., 1991–1997; Schauer et al., 1999–2002a,b). Vehicle exhaust is ubiquitous and usually the chief source of $PM_{2.5}$. Tire

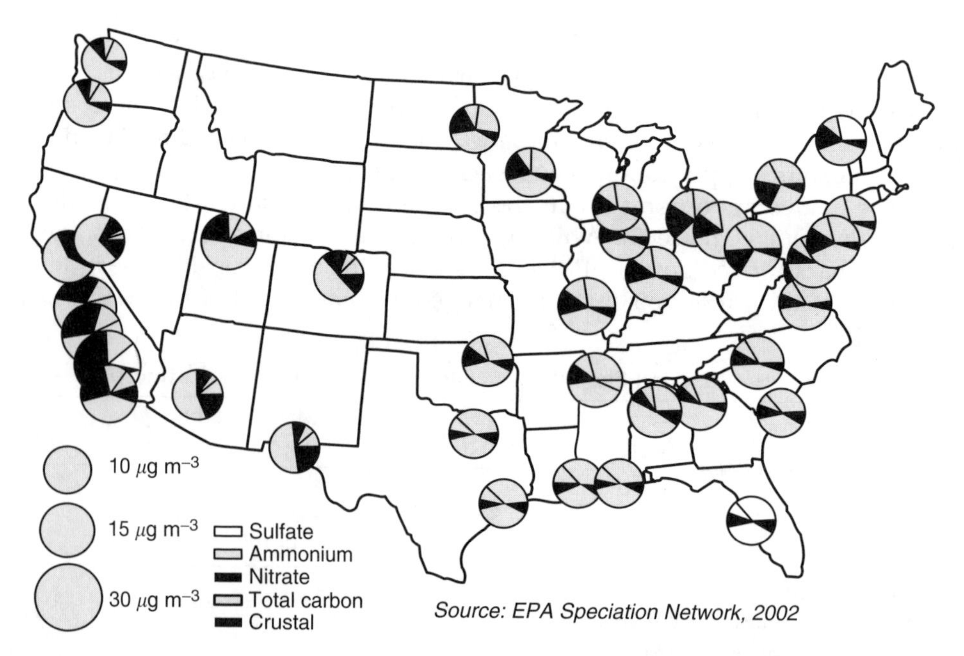

Figure 10.1 Average annual $PM_{2.5}$ concentrations for urban areas of the U.S. in 2002.

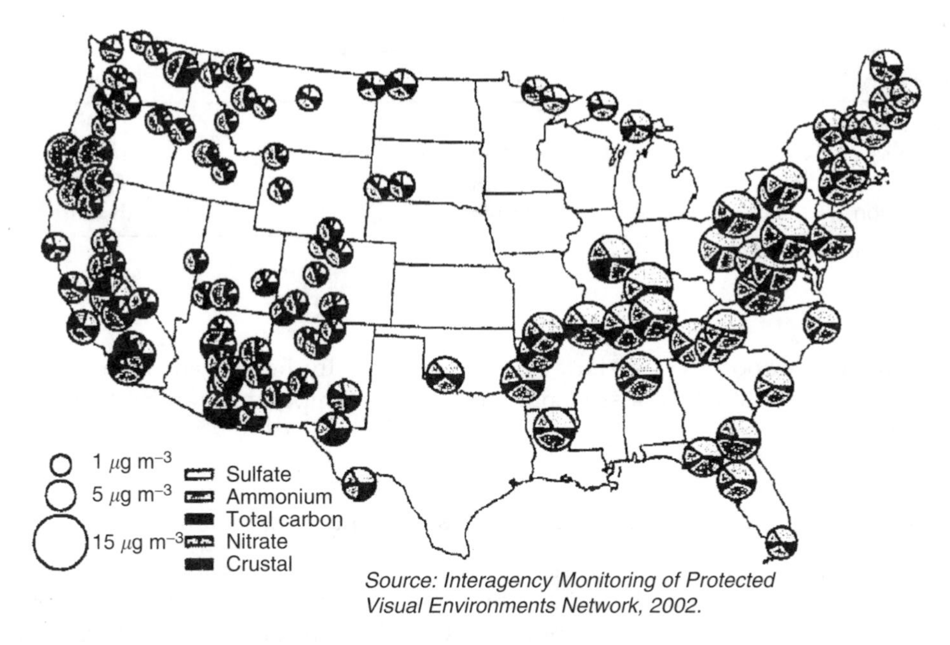

Figure 10.2 Average annual $PM_{2.5}$ concentrations for rural areas of the U.S. in 2002.

wear, plant detritus, spores, halogenated organic compounds from pesticides, PCBs, and dioxin-like species also contribute to airborne PM. Secondary organic aerosol (SOA) contributions typically follow seasonal patterns of ozone concentrations. SOA forms when ozone reacts with unsaturated hydrocarbon gases, leading to high-molecular-weight oxygenated compounds that condense on available particles.

More than half of the ambient fine carbonaceous mass has not yet been traced to individual compounds. Even so, emission profiles that resulted from the detailed source characterization studies of Rogge et al. (1991–1997) and Schauer et al. (1999–2002a, b) have been incorporated into molecular marker receptor source apportionment models. Schauer et al. (2002c) recently constructed and applied such a model to a recent smog episode in California to derive the contributions of 11 unique sources from measured concentrations of volatile, semivolatile, and particulate organic pollutants. Vehicle exhaust was the most important source of particles, among others.

10.1.4 Differences between indoor and outdoor environments

Table 10.2 compares some key physical parameters in outdoor and indoor environments. From a physical chemist's perspective, a building is a leaky reaction vessel with active surfaces. Whereas most outdoor atmospheric processes are not constrained by partitions or macroscopic surfaces, the indoor environment is defined by walls and covered with building materials and furnishings. Many of these substances can contribute to the air quality by adsorbing or emitting compounds that are chemically active. Indoor activities also generate gases and particles, leading to substantially higher indoor concentrations of many volatile organic compounds (VOCs) than outdoors. Communication between the indoor and outdoor environments takes place via air exchange through doors, windows, and cracks in the building envelope. The residence time of a particle in a room can be shorter or longer than the same particle in an air pocket of the same volume outdoors. Because of much lower levels of ultraviolet light, photochemistry does not play an

Table 10.2 Key Physical Parameters for Outdoor and Indoor Air[a]

Parameter	Urban Atmosphere	Indoor Atmosphere
Residence time	~10 h	~1 h
Light-energy flux	~1000 W m^{-2} (daytime)	~1 W m^{-2}
Surface–volume ratio	~0.01 $m^2 m^{-3}$	~3 $m^2 m^{-3}$
Precipitation	~10–150 cm yr^{-1}	Absent

[a]From Nazaroff et al. (2003).

important role indoors. Thus, dark reactions, especially heterogeneous processes, predominate indoors. Much less seasonal variation in temperature and relative humidity occurs indoors.

10.1.5 Evidence for indoor aerosol processing and generation

Figure 10.3 shows important processes involving indoor particles and gases. Airborne particles are shown entering and leaving a building (by infiltration and exfiltration, respectively). Soil particles are tracked indoors as hitchhikers on shoes and clothing. They deposit on floors and carpets, where they can be crushed to smaller size by foot traffic. Compounds can undergo phase change indoors, evaporating from or condensing on particles, depending on vapor pressure and temperature. Many indoor materials act as both sources and sinks for volatile and semivolatile pollutants, and in typical homes and offices, the building materials and furnishings have much higher exposed surface area than indoor aerosols. Figure 10.3 does not include sorption and desorption of volatile organic compounds, but these processes are discussed later on, in sections 10.3 and 10.4. Ultrafine particles form during indoor combustion (cooking, heating, smoking, etc.) and by chemical reactions of gas precursors such as cleaning products and ozone. Inevitably, indoor-generated ultrafine particles coagulate by collision with each other and with infiltrated particles of outdoor origin. Particles are removed by deposition to surfaces, and those larger than 0.8 μm can also be resuspended by indoor activities of the occupants (walking, vacuuming).

Although no U.S.-wide indoor PM speciation trends network operates, recent studies show intriguing differences between indoor and outdoor PM, as discussed below. The first example shows how an ordinary unoccupied home conditions infiltrating aerosol, the second points to the influence of indoor materials on attempts to characterize indoor PM, while the third provides evidence for the counterintuitive notion that fine particles are produced indoors by practices that residents often consider protective and health promoting.

10.1.5.1 The building envelope

Using real-time concentration and ventilation measurements in an unoccupied house in Fresno, CA, Lunden et al. (2003) found that indoor particulate sulfate and soot (black carbon) acted as conservative tracers for infiltration of outdoor $PM_{2.5}$, but indoor $PM_{2.5}$ had much less ammonium nitrate than predicted from the penetration factors for elemental carbon. They also found that indoor ammonia and nitric acid concentrations were usually lower than outdoors. The observations are consistent with disruption of the ammonium nitrate gas/particle equilibrium as the indoor surfaces take up nitric acid and ammonia.

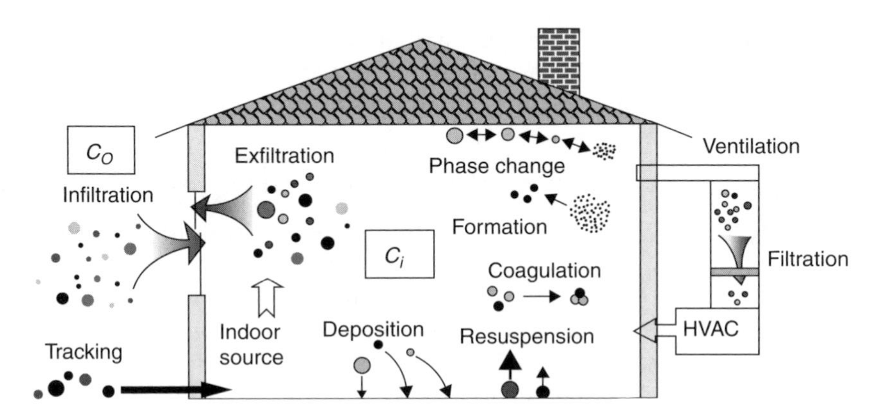

Figure 10.3 Processes affecting indoor aerosol concentrations; adapted from Thatcher et al., 2003. C_o and C_i represents the outdoor and indoor concentrations, respectively.

The affinity of indoor surfaces for these species led to evaporative dissociation of particulate ammonium nitrate.

10.1.5.2 Indoor materials

Several groups of investigators have recently found that conventional measurements of concentrations of indoor carbonaceous particles can exceed indoor $PM_{2.5}$ mass concentrations (Landis et al., 2002; Pang et al., 2002). Pang et al. showed that the conservation of mass was not actually violated and that the measured indoor particulate carbon concentrations were reduced substantially when denuders adsorbed semivolatile organic gases upstream of the particle collection medium (quartz fiber filter).

Meanwhile, indoor building materials and furnishings have been found to emit organic gases with a wide range of volatilities. Hodgson et al. (2000, 2003) and Rudel et al. (2003) reported high indoor concentrations of phthalate esters (from vinyl flooring). Other recent indoor air measurements include perfluoroalkyl sulfonamides from surface coatings (Shoeib et al., 2004), fire retardants from polyurethane foam (polybrominated diphenyl ethers: Sjodin et al., 2001; triethylphosphate: Salthammer et al., 2003), and consumer electronics (triethylphosphate: Carlsson et al., 2000). In many indoor environments, the gas-phase concentrations of semivolatile organic compounds can reach supersaturation, and condensation can occur onto walls, windows (Butt et al., 2004), textiles, airborne particles, and even quartz fiber filter sampling media (Weschler, 2003).

10.1.5.3 Indoor aerosol generation

Although the contribution of indoor combustion sources to respirable particles has been investigated for a long time, only recently has secondary aerosol formation been observed indoors. Using a pair of matched offices with the same ambient ozone concentrations and ventilation rates, Weschler and Shields (1999) tracked increased numbers of fine particles in the room into which *d*-limonene had been released. Limonene is a biogenic terpene, a cyclic alkene that is found in lemon and pine oils that are common components of cleaning products. Concerns about the irritancy and other potential health impacts of the gas and aerosol products have led to an explosion of work in this area.

The stage is now set for deeper exploration of indoor aerosol chemistry and physics. The three examples above illustrate important current research areas in indoor aerosol chemistry and building science. After reviewing common sources of indoor PM, aerosol precursors, and the chemical and physical processes that determine the fate of indoor particles, a discussion of fate and transport will attempt to show how the three examples are being incorporated into indoor aerosol modeling approaches. The chapter closes with an overview of indoor aerosol measurement and characterization techniques.

10.2 Composition of indoor particles and aerosol precursors

The PTEAM study identified outdoor air as the greatest contributor to indoor particle mass concentrations in Riverside, CA, in the fall of 1990 (Ozkaynak et al., 1996). Smoking and cooking were the next most significant contributors to $PM_{2.5}$ and PM_{10}. Other important indoor combustion sources included wood burning and the use of candles, incense, and unvented kerosene heaters. Table 10.3 lists these sources of indoor particles, along with compositional highlights and references. Not included are biomass fuels that are used for cooking in Asia and Africa, although some recent information is available in conference proceedings (Indoor Air, 2002). Biogenic aerosols are on the list, but much more information is available elsewhere (Macher et al., 1999). Some human activities that lead to indoor particle generation and resuspension are also included in Table 10.3. Indoor aerosol formation is discussed more thoroughly in Section 4 of this chapter. Table 10.4 lists characteristic compounds or chemical classes that are potential chemical tracers for each source.

10.2.1 Combustion sources

Indoor combustion sources typically operate less efficiently than vehicle engines or industrial boilers, leading to higher emission factors for unburned fuel and partially oxidized products. Semivolatile combustion-generated gases cool, or nucleate and condense onto existing particles and coagulate with existing particles. The aerosol community is indebted to the late Glenn Cass for leading comprehensive particle characterization efforts at the California Institute of Technology, and many studies from his laboratory are cited below. Starting in the early 1990s, Rogge et al. (1991–1998) began reporting detailed chemical characterization of PM from individual combustion sources that had been operated under carefully controlled conditions, as described by Hildemann et al. (1991). Schauer et al. (1999–2002) expanded these investigations by characterizing both the gas and particulate phases of many of the same combustion sources. In spite of this large body of work and the contributions of many other investigators, identified components typically account for less than half of the fine particulate mass emitted by each combustion source.

10.2.1.1 Cooking with oils and meat

Schauer et al. (2002a) found that frying vegetables in seed oils led primarily to the formation of fine particulate alkenoic and alkanoic acids, whereas the semivolatile gases contained mostly saturated and unsaturated aldehydes. Emission factors for $PM_{2.5}$ ranged from 13 to 30 mg kg^{-1} of stir-fried vegetables. Andrejs et al. (2002) found that particles larger than about 4 μm were composed primarily of triglyceride components of the original oil. Hildemann et al. (1991) found that charbroiling meat yielded more fine PM (40 g kg^{-1} meat) than Schauer et al. (2002a) found from cooking with oils. Rogge et al. (1991) and Schauer et al. (1999) reported more classes of organic compounds from meat cooking than from the use of seed oils. However, only about 10% of the particulate organic carbon from meat cooking could be traced to individual compounds. Of those, saturated and unsaturated fatty acids were the most abundant, as shown in Figure 10.4.

Aldehydes and ketones were the next most abundant. Carbonyl-containing compounds and unidentified organic compounds dominated the semivolatile gases from meat cooking, but olefins and alkanes were also appreciable contributors. Rogge (1991)

Table 10.3 Overview of Chemical Composition of Particles of Indoor Origin

	$PM_{2.5}$	PM_{10}	Processes	References
Outdoor particles	OC, NH_4^+, SO_4^{2-}, NO_3^-, EC, trace metals	Crustal components: Fe, Ca, Si	Infiltration, deposition	Ozkaynak et al. (1996), Williams et al. (2003a,b)
Combustion	OC, EC		Pyrolysis, evaporation, combustion, nucleation, condensation	Daisey and Gundel (1991), Manchester-Neesvig et al. (2003)
Cooking	Organic acids >> aldehydes, ketones			Rogge et al. (1991), Schauer et al. (1999, 2002), Andrejs et al. (2002)
Wood — fireplaces	Levoglucosan > methoxyphenols >resin acids			Rogge et al. (1998), McDonald et al. (2000), Schauer et al. (2001)
Tobacco smoking	Organic acids >alkanes > *N*-heterocycles			Rogge et al. (1994)
Candles	High MW alkanes >aldehydes >alkanoic acids >alkenes, >esters; EC; Pb		Evaporation, condensation; pyrolysis	Fine et al. (1999), van Alphen (1999), Nriagu and Kim (2000), Wasson et al. (2002)
Incense	PAH; aldehydes; $PM_{2.5}$			Cheng et al. (1995), Jetter et al. (2002)
Kerosene heaters	$PM_{2.5}$, SO_4^{2-}, HONO			Leaderer et al. (1990), Leaderer et al. (1999)
Biogenic aerosols				Macher et al. (1999)
Pets		Allergens in dander, endotoxins	Entrainment	Song and Liu (2003), Erwin et al. (2003)
Microbes		Endotoxins	Entrainment	Song and Liu (2003), Erwin et al. (2003)
Mites		Allergens	Entrainment	Loan et al. (2003)
Human activities				
Pesticide use		Flea powder	Entrainment	Nishioka et al. (1996), Becker et al. (2002), Rudel et al. (2003)
Air fresheners+O_3	Organic aldehydes and acids		Nucleation, condensation	Weschler et al. (1999)
Walking	Minor contributor	Crustal elements, mite wastes	Abrasion, resuspension	Abt et al. (2000)
Cleaning activities	Indoor and outdoor PM	PM	Resuspension	Ferro et al. (2002)
Cleaning prod+O_3	Organic aldehydes and acids		Nucleation, condensation	Weschler et al. (1997)
Renovation		Building materials	Abrasion, resuspension	Harrison and Brooke (1999)

Table 10.4 Potential Chemical Tracers for Indoor Sources of PM

PM Source	$PM_{2.5}$	PM_{10}	References
Outdoor particles	SO_4^{2-}, EC (if no indoor combustion sources)	Crustal components: Fe, Ca, Si	Ozkaynak et al. (1996), Williams et al. (2003a,b)
Combustion			
Cooking			
Meat	Cholesterol		Rogge et al. (1991)
Seed oils, meat	Triglycerides		Andrejs et al. (2002)
Plant biomass	Levoglucosan		Simoneit (2002)
Softwood (pine)	Retene, guaiacols, diterpenoids		Rogge et al. (1998), McDonald et al. (2000), Simoneit (2002), Schauer et al. (2001)
Hardwood (oak)	Syringols		Rogge et al. (1998), McDonald et al. (2000), Schauer et al. (2001)
Tobacco smoking	*N*-heterocycles (nicotine), iso and anteiso-alkanes, solanesol		Rogge et al. (1994), Daisey (1999)
Candles	wax esters		Fine et al. (1999), Wasson et al. (2002)
Incense			*Tracers not yet proposed*
Kerosene heaters	HONO, excess SO_4^{2-}		Leaderer et al. (1999)
Biogenic aerosols			Macher et al. (1999)
Pets		Allergens in dander, endotoxins	Song and Liu (2003), Erwin et al. (2003)
Microbes		Endotoxins	Song and Liu (2003), Erwin et al. (2003)
Mites		Allergens	Loan et al. (2003)
Human activities			
Pesticide use	chlorinated compounds		Rudel et al. (2003)
Air fresheners+O_3	organic aldehydes and acids		*Tracers not yet proposed*
Walking		Crustal elements, mite wastes	Abt et al. (2000), Ferro et al. (2002)
Cleaning activities	resuspended PM	Resuspended PM, crustal elements, mite wastes, dander	Abt et al. (2000), Ferro et al. (2002)
Cleaning prods+O_3	organic aldehydes and acids		Weschler et al. (1997)
Renovation		Cellulose, lignin, $CaSO_4^{2-}$	Harrison and Brooke (1999)

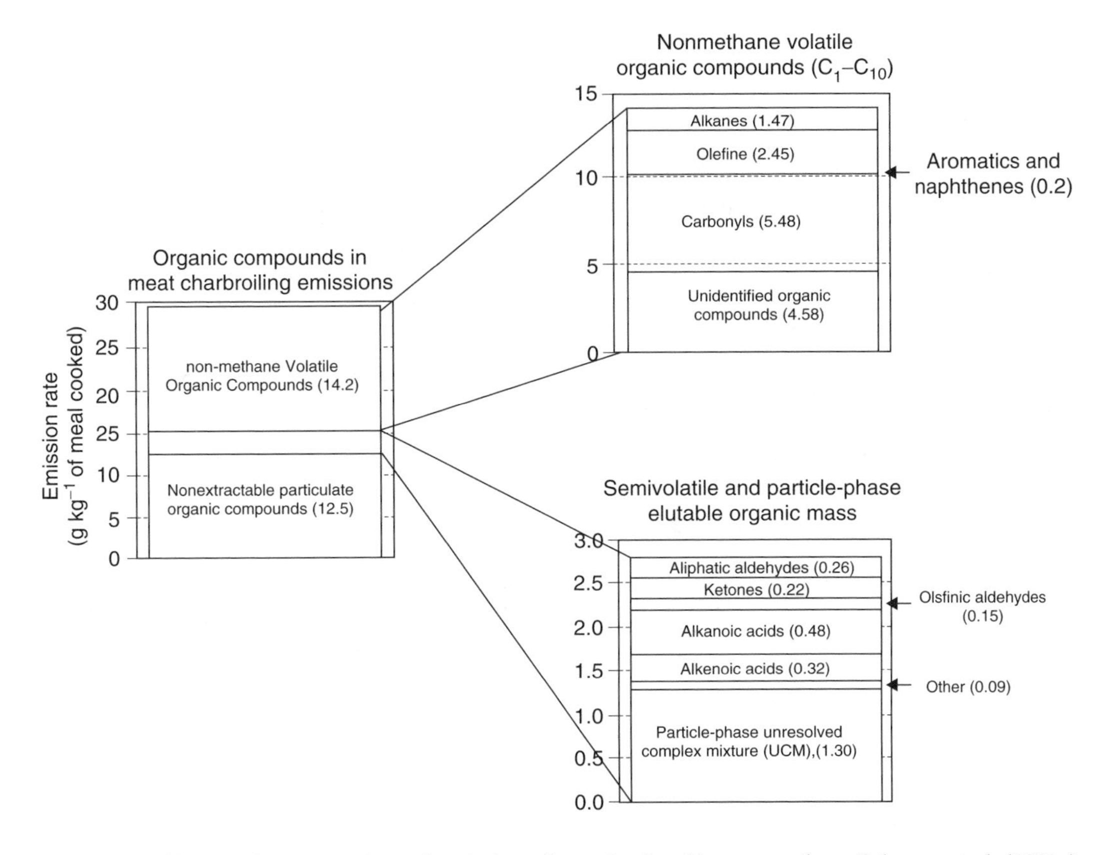

Figure 10.4 Chemical composition of emissions from charbroiling meat; from Schauer et al. (1999a) with permission.

suggested cholesterol as a tracer for the contribution of charbroiling to ambient PM. Neither Rogge nor Schauer reported concentrations of triglycerides.

10.2.1.2 Fireplaces

Schauer et al. (2001) found $PM_{2.5}$ emission factors of 5–10 g kg^{-1} for several types of wood burned in a fireplace, in general agreement with the results of other investigators (Hildemann et al., 1999; McDonald et al., 2000). Figure 10.5 illustrates the results of Schauer et al. for the composition of wood smoke from pine and oak burning, at the top and bottom, respectively. When wood burns, its oxygenated organic polymers (cellulose and lignins) decompose to distinctive fragments. Schauer et al. (2001) found that 18–31% of the fine particulate organic carbon mass from wood burning was levoglucosan, a pyrolysis product of cellulose. Hornig et al. (1985) suggested the use of this sugar anhydride as a woodsmoke marker. Methoxyphenols and resin acids are the next most abundant classes of identified components of fine woodsmoke particles. Methoxyphenols are pyrolysis products of lignin, a biopolymer that trees synthesize from aromatic vinyl alcohols to strengthen their cellulose frameworks.

Methoxyphenols from different types of wood yield characteristic composition patterns that can be used as tracers (syringols for oak; guaiacols for pine and oak; Schauer et al., 2001; McDonald et al., 2000; Rogge et al., 1998). Methoxyphenols are also found in the gas phase of woodsmoke, along with aldehydes, dicarbonyls, alkenes, and other compounds. Many woodsmoke components are irritants; some are genotoxic, and others (e.g., terpenes) can react with other indoor pollutants.

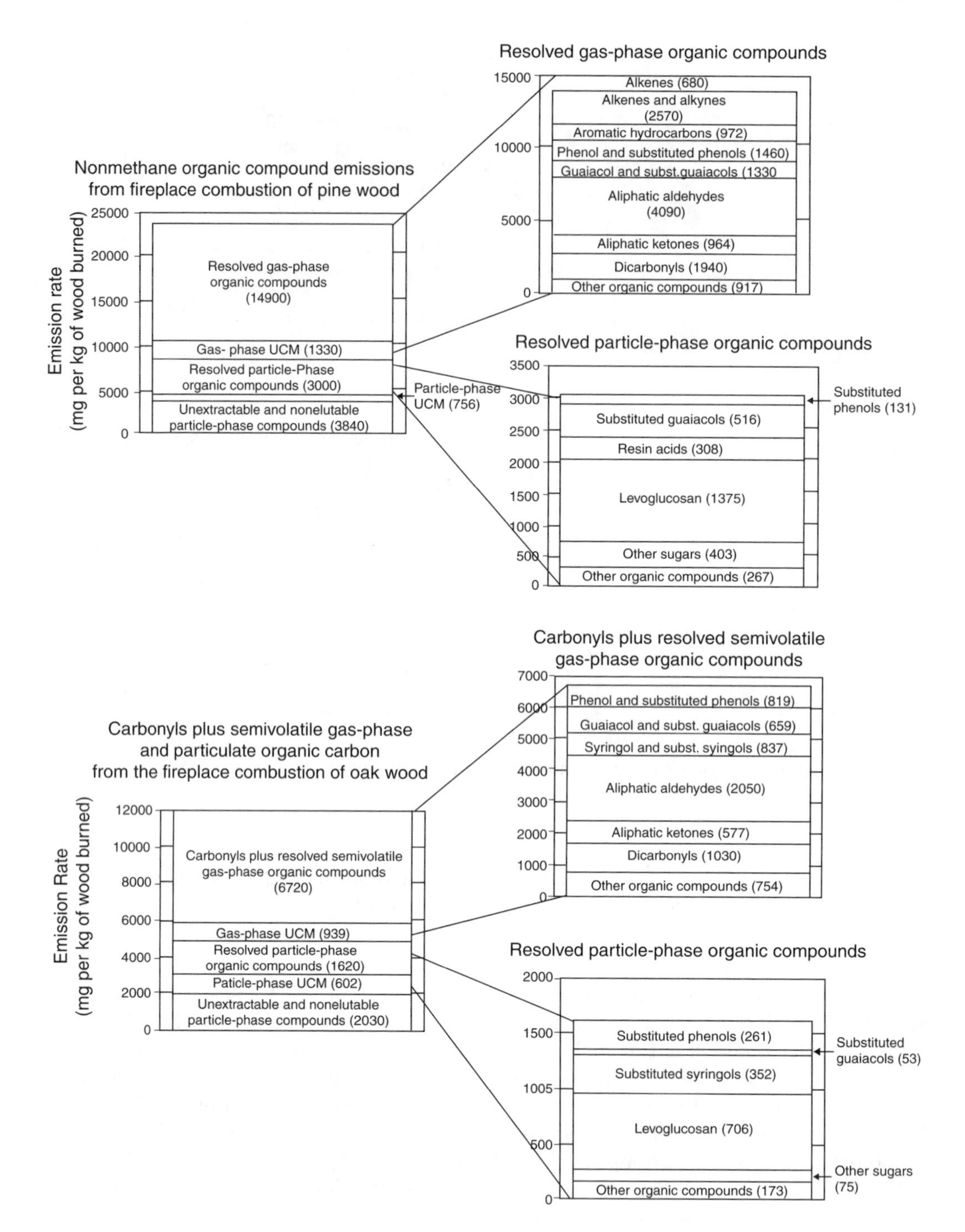

Figure 10.5 Chemical composition of smoke from burning pine and oak wood; from Schauer et al. (2000) with permission.

10.2.1.3 Tobacco

Although a wide range of tobacco smoke constituents have been identified (Jenkins et al., 2000), the most detailed characterization of environmental tobacco smoke (ETS) composition has been done by Rogge et al. (1994), who sampled exhaled mainstream smoke and sidestream smoke in a vertical dilution tunnel. $PM_{2.5}$ emission factors are roughly 10–20 mg cig^{-1} (Hildemann et al., 1991; Gundel et al., 1995 among others). Rogge et al. reported that nitrogen-containing heterocyclic compounds are the most abundant class, when gas and particulate phase concentrations of nicotine are included with other particulate *N*-compounds. Since nicotine is primarily in the gas phase in fresh ETS, and very soon after emission it sorbs to indoor surfaces, its suitability as a tracer for ETS exposure has been questioned (Daisey, 1999). When gas-phase nicotine is excluded from the reckoning of Rogge et al. (1994), particulate *N*-heterocycles are the third most abundant class in ETS particles, after alkanoic acids and alkanes. Phytosterols (primarily stigmasterol), phenols, alkanols, and branched (*iso*-[1-methyl] and *anteiso*-[2-methyl]) alkanes follow in order of decreasing abundance. Rogge et al. pointed out the potential usefulness of *iso*- and *anteiso*-alkanes as ETS tracers, and later Kouvaras et al. (1998) quantified them both indoors and outdoors in Greece. The carcinogenic components of ETS particles such as *N*-nitrosamines and polycyclic aromatic hydrocarbons are minor constituents. Aldehydes and terpenes are potential reactive gas-phase ETS constituents of ETS (Shaughnessey et al., 2001) and aerosol precursors.

10.2.1.4 Candles

Fine et al. (1999) characterized particles from burning candles at Caltech. They reported fine $PM_{1.8}$ emission factors of 0.5–4 and 2 mg g^{-1} for paraffin and beeswax, respectively. The highest emission factors and elemental carbon concentrations were found when paraffin candles were "sooting" (emitting visible smoke). For both types of wax, less than a third of the fine particulate organic carbon could be traced to individual compounds. Alkanes, aldehydes, and alkanoic acids were the most abundant classes of compounds in paraffin emissions, reflecting the condensation of unburned alkanes on combustion particles. Emissions from beeswax candles contained the same wax esters, alkanoic acids, and alkanes found in the unburned wax, and also decomposition (alkenes) and combustion products (aldehydes). The strong odd carbon number preference found in beeswax was reflected in its fine PM composition. The beeswax candles produced very little elemental carbon.

Lead has been found in emissions from some commercially available candles in the U.S. and elsewhere (van Alphen, 1999; Nriagu and Kim, 2000). Wasson et al. (2002) found lead in eight pairs among 100 purchased with metal cores or covers of wicks. They reported airborne lead emission rates of 100–1700 $\mu g\ h^{-1}$ from individual candles. At an air exchange rate of 0.3 h^{-1} in a 30 m^3 room, it would take only a few minutes for airborne lead concentrations from a single candle at the average emission rate of 550 $\mu g\ h^{-1}$ to exceed the EPA's ambient air Pb concentration level of 1.5 $\mu g\ m^{-3}$. Using this scenario, Wasson et al. (2002) estimated that the room concentration of Pb would be around 5 $\mu g\ m^{-3}$ for the last 3 h of a 4 h candle burn.

10.2.1.5 Incense

Incense burning can contribute an important source of indoor fine PM. Incense smoke has not yet been subjected to the type of comprehensive chemical characterization described above for particles from cooking, fireplaces, or cigarette smoke. Because of the variety of ingredients and configurations used to prepare incense around the world, acquiring representative data could be very challenging and expensive. Recently, Lung and Hu (2003) reported $PM_{2.5}$ and particulate polycyclic aromatic hydrocarbon (PAH) emission factors for two types of hand-made incense, averaging 32 mg g^{-1} $PM_{2.5}$ and 21 $\mu g\ g^{-1}$ PAH. Jetter et al. (2002) reviewed the literature and reported particulate emission factors for 23 different types of incense purchased in the U.S. They found large numbers of ultrafine particles and mass size distributions that typically peaked at around 0.5 μm. $PM_{2.5}$ and PM_{10} emission factors were statistically indistinguishable, with emission factors of 5–56 mg g^{-1} incense, and an average emission rate of 42 mg h^{-1}. They also found that the burn time ranged from 14 min for one type of incense cone to 3 h or more, for both a smudge bundle and an incense coil. Average burn time was 43 min. Using an indoor air quality model developed by one of the coauthors (Guo et al., 2000), they estimated that one unit of incense with an average emission rate and burn time would generate a peak concentration of 0.85 mg m^{-3} if used in a small room (30 m^3) with 0.5 air changes per hour.

10.2.1.6 Unvented kerosene heaters

Although the sale of portable unvented kerosene heaters (UKHs) has been banned in some U.S. states, over a million units are sold in the U.S.A. each year (Manuel, 1999). Carbon monoxide poisoning and fire are the greatest dangers from such heaters. Increased indoor moisture, kerosene odor, and fine particles are also associated with their use (Trayhor et al., 1983; 1986; 1990). Leaderer et al. (1999) found UKHs in about one third of over 200 homes in Virginia and Connecticut whose indoor air was monitored during the winters of 1995–1998. They reported that UKHs added about 40 μg m^{-3} of $PM_{2.5}$ and 15 μg m^{-3} particulate SO_4^{2-} to background indoor levels of 18 and 2 μg m^{-3}, respectively. They also found elevated levels of nitrous acid (HONO) in homes where UKHs or gas stoves were operated. In an earlier chamber study, Leaderer et al. (1990) measured emission factors (per gram of K-1 fuel with 0.04% S) of 33–392 and 15–227 μg g^{-1} for $PM_{2.5}$ and SO_4^{2-} $PM_{2.5}$, respectively, from the operation of four types of UKHs. The sulfate result suggests that UKHs are a major indoor source of fine acidic aerosol.

10.2.2 Human activities and consumer products

10.2.2.1 Pesticides

Organophosphates, chlorinated compounds, and permethrins have been found in indoor PM and dust (Rudel et al., 2003, and references therein; Rudel et al., 2002; Berger-Preiss et al., 2002; Becker et al., 2002). Foot traffic is thought to be a major vector of pesticide transport from outdoor soil particles (Nishioka et al., 1996). Indoor sources of particulate pesticides include flea powder and aerosolized insecticides, among others. Dry flea powder contributes to the coarse particle fraction (2.5–10 μm diameter, authors' comparison to fine powder of known mass size distribution), whereas sprays deliver semivolatile agents in liquid particles that settle on surfaces or evaporate within minutes (Bukowski and Meyer, 1995). Semivolatile pesticides like chlorpyrifos, a termiticide, continue to evaporate from indoor surfaces such as carpets for weeks or longer (Stout and Mason, 2003, and references therein).

10.2.2.2 Air fresheners

Most products sold as air fresheners include volatile fragrances that increase the concentrations of VOCs indoors. Frequently, the fragrant compounds include terpenes such as pinene from pine oil or limonene from citrus oils. These terpenes can react with ozone to produce fine PM (Hoffmann et al., 1997; Kamens et al., 1999). Since building envelopes scrub only about half of the infiltrating outdoor ozone, indoor secondary aerosol formation probably occurs to a larger extent than previously recognized (Weschler and Shields, 1997). This is exacerbated by the use of air cleaners that generate ozone.

10.2.2.3 Walking

In their study of real-time particle size distributions in four nonsmoking houses in Boston, MA, Abt et al. (2000) found that the movement of people increased the indoor concentrations of large particles (~4 μm median diameter) five times more than they increased the

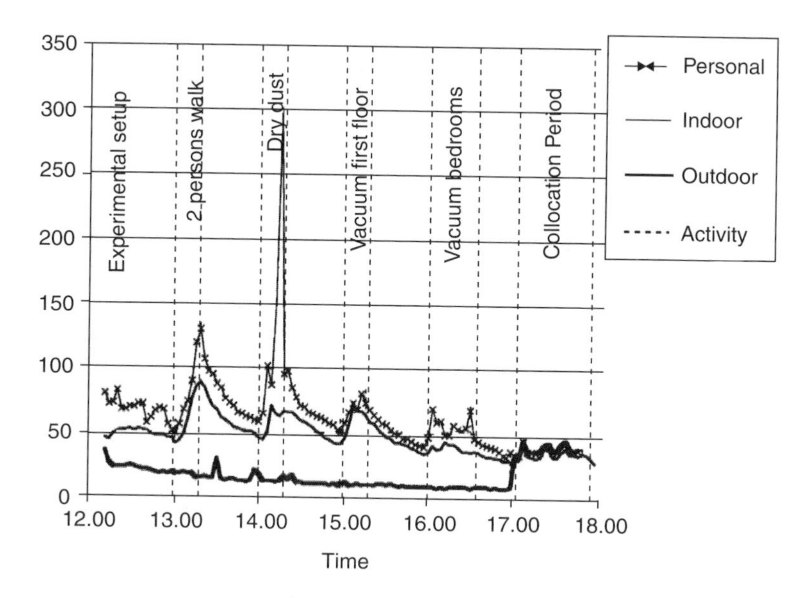

Figure 10.6 Concentrations of personal, indoor, and outdoor PM_5 while people walked around a house; from Ferro et al. (2002) with permission.

concentrations of fine particles (~0.2 μm). Ferro et al. (2002) found that simply walking around indoors increased personal exposure to both fine and coarse particles. Figure 10.6 shows real-time PM_5 profiles obtained by Ferro et al. The major contributor is resuspension of settled dust from floors and carpets (Thatcher and Layton, 1995).

10.2.2.4 Cleaning

Cleaning activities that involve only the movement of equipment and indoor furnishings, rather than the use of cleaning chemicals, have been found to elevate personal exposure to particles. For example, Ferro et al. (2002) reported that 15-min periods of bed making, folding clothes, and vacuuming increased (5-h average) $PM_{2.5}$ and PM_5 personal exposures to 1.4 and 1.6 times the indoor concentrations for the same time periods, as shown in Figure 10.7. Textiles like sheets, towels, and clothes shed fibers, but they can also collect and release airborne particles and indoor dust. Using vacuum cleaners without $PM_{2.5}$ filters, as well as dry dusting and sweeping, redisperse part of the settled dust and indoor PM (Abt et al., 2000).

10.2.2.5 Cleaning products

Some cleaning products and air freshners contain VOCs such as limonene and pinene that can participate in indoor aerosol chemistry (Weschsler and Shields, 1999). A recent comparison of personal, indoor, and outdoor exposures to VOCs in three Minnesota communities (Sexton et al., 2004) found that personal exposure to both the aerosol precursors pinene and limonene exceeded the indoor concentrations, and that indoor concentrations exceeded outdoor concentrations.

Much less is currently known about indoor concentrations of many semivolatile compounds that are used in cleaning products. Rudel et al. (2003) found the semivolatile disinfectant *o*-phenyl phenol in indoor air and dust, along with several alkylphenols and alkylphenol ethoxylates that are ingredients of detergents. Semivolatile compounds can adsorb onto and desorb from indoor surfaces and even sampling media, and they were mentioned in the introduction as participants in observed apparent violations of

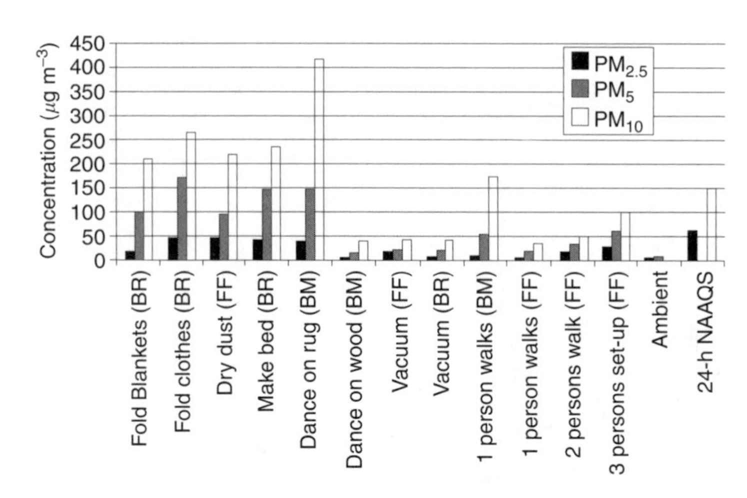

Figure 10.7 Personal exposure to three size fractions of PM during indoor activities; from Ferro et al. (2002) with permission.

conservation of mass by carbonaceous components of indoor fine mass. Semivolatile organic compound (SVOC) emissions from indoor materials and furnishings will be discussed later in this chapter.

10.2.2.6 Renovation

Building renovation includes many activities that can generate large quantities of indoor particles, including removing walls, carpets and flooring materials, sanding wood, stripping paint, installing gypsum board, and spray painting. Most of these activities generate coarse particles (Harrison and Brooke, 1999), but these can be ground to smaller respirable particles by the movement of people. Some materials used in renovation (adhesives, sealants) emit semivolatile species that can adsorb to indoor dust that eventually settles everywhere in the building (Hodgson et al., 2000, 2002; Morrison et al., 1998; Rudel et al., 2003). (The influence of building materials and furniture on indoor aerosol chemistry is discussed later.)

10.2.3 Indoor aerosol source apportionment

Indoor source apportionment is critical to understanding the links between PM exposure and health effects, and several approaches have been developed. Two of these will be mentioned only superficially, as this topic requires a fuller discussion than is possible here.

This review of indoor PM sources began by mentioning that the PTEAM found that smoking and cooking were the largest contributors to indoor particles, after outdoor air. To estimate the contribution of individual sources to indoor and personal PM concentrations by the PTEAM study, Ozkaynak et al. (1996) chemical mass balance used a (CMB) approach and a simple model of indoor pollutant behavior (Koutrakis and Briggs, 1992) to measured mass and elemental concentration data for outdoor, indoor, and personal $PM_{2.5}$ and PM_{10}. Indoor source profiles were derived from reviewing participants' activity reports and multiple linear regression.

Yakovleva et al. (1999) used positive matrix factorization (PMF) to identify and quantify the sources of the particles in the "personal cloud." This approach requires less prior knowledge of source profiles than earlier source–receptor models and allows different error estimates for each data point, with lower weighting as uncertainty increases. Yakovleva et al. identified five source profiles that influenced personal PM_{10} exposure: motor vehicles,

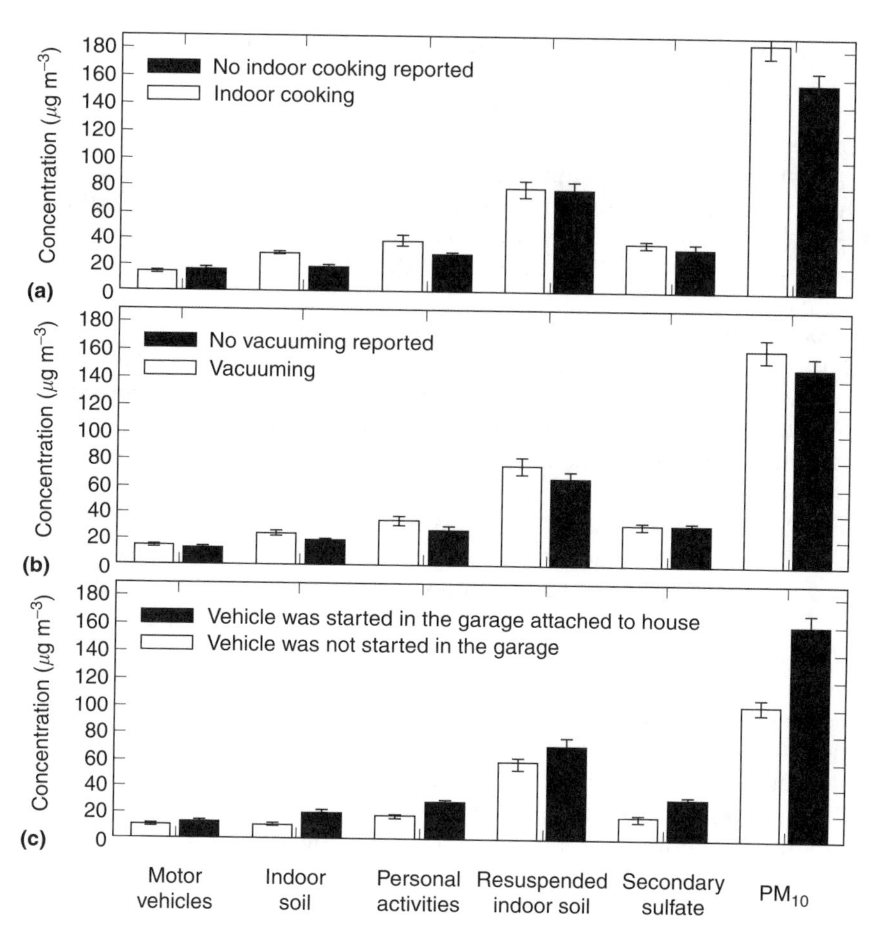

Figure 10.8 PMF source apportionment for PTEAM personal PM_{10}. The histograms show the contributions from each computed source factor compared to activity logs: (**a**) indoor cooking; (**b**) vacuuming; (**c**) starting a vehicle in an attached garage. From Yakovleva et al. (1999) with permission.

secondary sulfate, personal activities, resuspended indoor soil, and indoor soil. Figure 10.8 compares the contributions of these factors to the participants' reported activities.

10.3 Physical properties of indoor aerosols

This section presents a brief overview of a few key aspects of aerosol physics that influence the indoor behavior and fate of airborne particles. Hinds presents a clear and thorough discussion of aerosol physical properties in his classic text (Hinds, 1999), as well as in a chapter of this handbook. Seinfeld and Pandis (1998) also introduce aerosol physics with many examples from ambient air.

10.3.1 Size distributions

Particle size is a key physical parameter that influences aerosol behavior, for example, the penetration of aerosols through building envelopes (Thatcher et al., 2003), the deposition of fine particles in the human respiratory system and flow through sampling instruments (Hinds, 1999 (chapter 11)). Many studies indicate that fine particles <2.5 μm diameter appear to be more toxic to cells per unit mass than larger or coarse particles (with diameters between 2.5 and 10 μm). Because many sources contribute to airborne particles in an urban area, particle sizes vary in time and space. *Size distribution functions* are very useful characterization tools to describe the size dependence of particle number, surface area, and volume (or mass) concentrations.

Aerosol instrumentation design depends on understanding the size dependence of aerosol behavior, particularly aerosol flow dynamics. Number size distributions are measured with particle counters that depend on aerosol electrical mobility and optical properties. The size dependence of light scattering by airborne particles is used in optical particle counters, for example.

Nucleation, accumulation and coarse size modes. Particles that arise from nucleation are always small (~5–50 nm), and large numbers of ultrafine particles (<0.1 μm, nucleation mode) are typically found near combustion sources. After emission, they coagulate when they collide with each other and with larger particles, and land in the accumulation mode between about 0.1 and 1 μm. Atmospheric lifetimes of nucleation mode particles are of the order of minutes or less. Airborne particles between about 2 and 10 μm are part of the coarse mode. Resuspended soil particles, sea spray, plant debris, pollen, and spores contribute to the coarse mode. Outdoor lifetimes of accumulation and coarse mode particles are of the order of days to weeks and minutes to days (Seinfeld and Pandis, 1998).

10.3.1.1 Number

The top part of Figure 10.9 shows a typical urban aerosol *number size distribution*, as illustrated by Seinfeld and Pandis (1998, p. 431). In this example, the number size distribution peaks in the nucleation mode, even for this typical, rather than fresh, aerosol.

Normal or bell-shaped distributions of particle size are symmetrical about the average diameter. The standard deviation describes the spread around the mean. Most aerosol sizes are not normally distributed, but tail off with increasing size, and *log-normal* distributions are more appropriate for airborne particles. For log-normal distributions, the logarithm of particle diameter is normally distributed rather than the particle diameter, and the geometric standard deviation describes the spread of the log of particle diameter about the median. Many ambient aerosol number size distributions can be described as the sum of modes that are each log-normally distributed (Hinds, 1999; Seinfeld and Pandis, 1998).

Sampling applications. Particle number size distributions are measured with electrical mobility analyzers, condensation nuclei counters, differential mobility analyzers, electron microscopes and optical counters.

10.3.1.2 Surface area

The middle section of Figure 10.9 shows the *surface area size distribution* for the typical urban aerosol example. The peak in the surface area distribution occurs at larger particle size, within the accumulation mode. Interactions between particles and fluids take place at particle surfaces. Adsorption, absorption, condensation, and reactions all occur at surfaces or in liquid films on particle surfaces. The size dependence of concentrations of aerosol constituents can be used to infer mechanisms of aerosol processing. For example, if the size dependence of concentrations of semivolatile compounds such as PAHs or photochemical reaction products matches the size dependence of surface area, sorption or condensation is responsible rather than particle number or mass.

Sampling surface area. The epiphaniometer is an example of an aerosol instrument that measures surface area particle size distributions.

10.3.1.3 Volume and mass

The middle section of Figure 10.9 shows the *mass size distribution* of the typical urban aerosol example that is used above. There are two peaks, one in the accumulation mode and one in

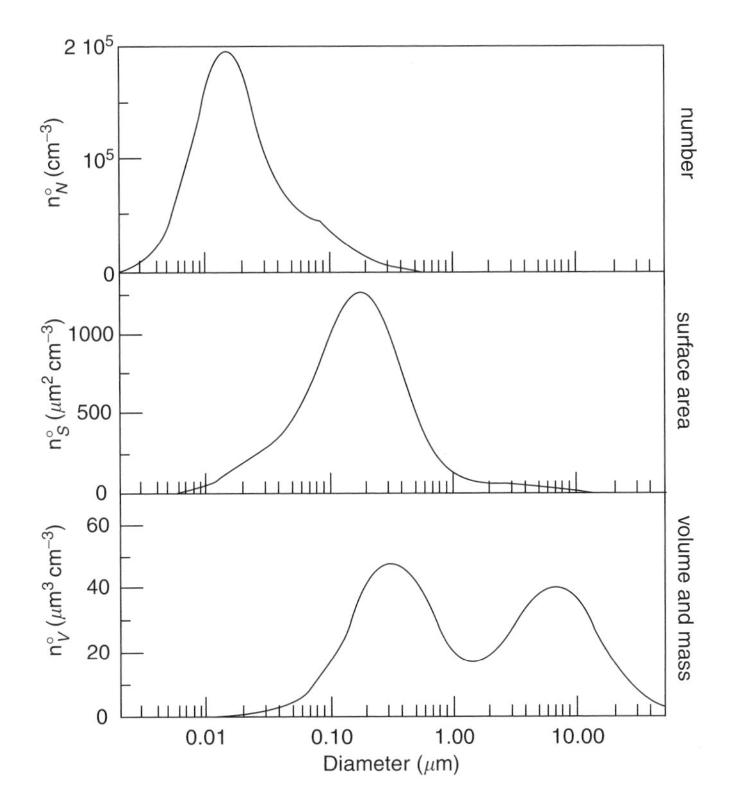

Figure 10.9 Three descriptions of the size distribution of "typical" urban airborne particles; from Seinfeld and Pandis (1998) with permission. The top panel shows number size distribution, the middle panel shows surface area size distribution, and the lower panel shows volume size distribution. The mass size distribution has the same shape as the volume distribution.

the coarse mode. This is a bimodal distribution. The minimum in mass concentration around 2 μm is usually quite pronounced in urban and rural air. Sometimes trimodal distributions are seen very near sources of fresh aerosol, when enough numbers of new tiny particles are present to add sufficient amounts of mass to contribute to the mass size distribution (Seinfeld and Pandis, 1998, p. 432).

Particle mass and size are a key physical parameters that influence the movement of particles in air. Diffusion rate depends on particle mass, as do the effects of inertia and fluid viscosity on particle behavior in moving air. During air sampling through tubes (respiratory system sampling tubing, for example), the heavier the particle, the more likely the particle is to be caught in a bend because it cannot keep up with the air flow. This property leads to particle deposition in bronchial tubes. Size-selective inlets, impactors, and virtual impactors take advantage of this property to separate particles based on their mass.

Instruments based on particle mass. Particle mass size distributions are measured directly by aerosol impactors. Mass size distributions can be integrated to yield total mass concentrations that can be compared directly to mass concentrations determined gravimetrically from filters. If particle number and mass size distributions are known, the size dependence of particle density can be calculated. Frequently, the particle density is estimated for spherical particles of the same aerodynamic diameter as the measured distribution, and the calculated density is used to convert from mass to number size distributions and *vice versa*. Geometric (smooth) surface area estimates depend only on the aerodynamic diameter and can be derived from either the mass or number size distributions. However, actual surface areas may be larger due to particle shape and porosity.

10.3.2 Particle dynamics

The behavior of indoor particles depends on their size, the forces they experience, and their interactions with surfaces in the indoor environment. These factors are embedded in the mass balance equations used in Section 10.5 to describe their fate and transport. The objective of this section is to illustrate how basic aerosol concepts apply in the indoor environment, starting with a room with dimensions 3.65 m × 3.65 m × 3 m. This room has a total volume of 40 m^3 and an assumed air exchange rate of 1 per hour.

10.3.2.1 Particle motion

The flow rate at one air exchange per hour through the example room is 11.1 $cm^3 s^{-1}$. A cube of air with this volume has a linear velocity of 2.23 $cm\,s^{-1}$. A quick way to get a sense of the behavior of particles in air at this velocity is to look at their Reynolds numbers (ratios of inertial to frictional forces), relaxation times (for adjustment of particle velocity to applied force), and terminal settling velocities in still air (Table 10.5). When particle Reynolds numbers are <1, the particles experience laminar flow. That is, they keep up with the streamlines, and gravity has little effect on them. In this example the particles move with the air flow unless they are large enough to be visible to the human eye. Respirable particles (diameter <2.5 μm) would remain entrained in room air movement even at much higher air exchange rates. The relaxation times show that they quickly adjust to the air exchange rate, and the settling velocities show that respirable particles do not deposit onto the floor quickly under the influence of gravity.

Deposition. Airborne particles deposit on building materials and other indoor surfaces after collision and adhesion. As outdoor air infiltrates through cracks in the building envelope, particles below 0.1 μm in diameter are lost because of diffusion much more efficiently than larger particles. However, larger particles deposit by impaction when they cannot keep up with the airflow around turns and obstacles. Once inside the building, both diffusion and gravitational settling contribute to particle loss. Table 10.5 shows that diffusion is responsible for more deposition to a horizontal surface than gravitational settling, for particles smaller than 0.2 μm, over the 100 sec period examined by Hinds (1999, p. 162).

Table 10.5 Properties of Particles with unit Density in a Room with Volume 40 m^3 and 1 ACH at Atmospheric Pressure and 20°C

Diameter (μm)	Re^a at 2.23 cm s^{-1}	Relaxation Time[b] (sec)	Terminal Settling Velocity[d] (cm s^{-1})	Diffusion coefficient[e] ($cm^2 s^{-1}$)	Ratio, Diffusion to Settling[f]
0.01	0.00015	6.8×10^{-9}	7.0×10^{-6}	5.2×10^{-4}	390
0.10	0.0015	8.8×10^{-8}	8.8×10^{-5}	6.7×10^{-6}	3.4
1.0	0.015	3.6×10^{-6}	3.5×10^{-3}	2.7×10^{-7}	1.7×10^{-2}
10	0.15	3.1×10^{-4}	0.29	2.4×10^{-8}	5.5×10^{-5}
100	1.5	3.2×10^{-2}	17	2.4×10^{-9}	2.2×10^{-7}

[a]Reynolds number $Re = 6.6\ V \times d$, where V is the particle velocity and d its diameter, in cgs units. (Hinds, 1999, p. 28). *Re* is shown for particle velocity and is equated to the linear flow velocity in the example room, 2.23 cm s^{-1}.

[b]Hinds (1999, p. 112, Table 5.1).

[d]In still air, Hinds (1999, p. 51, Table 3.1).

[e]Hinds (1999, p. 153, Table 7.1).

[f]Cumulative deposition of particles over 100 sec, Hinds (1999, p. 162, Table 7.5).

10.3.2.2 Particle formation and phase partitioning

10.3.2.2.1 Nucleation. Nucleation describes the process of particle formation from gas-phase precursor compounds. *Homogeneous nucleation* starts when the air becomes supersaturated with precursor gas molecules that collide with each other. No preexisting particles are necessary. Supersaturation means that the actual vapor pressure of a compound is higher than its equilibrium vapor pressure, as combustion exhaust expands and cools, for example. After multiple collisions, molecular clusters or 'embryos' are formed. They grow by further collisions with gas molecules, and when they reach a critical diameter that depends on the extent of supersaturation and gas molecular properties, some of them become stable enough to form nanoparticles. Homogeneous nucleation requires supersaturation ratios of 2–10 (ratios of actual vapor pressure to equilibrium vapor pressure; Hinds, 1982, p. 257). Particles continue to grow by condensation as long as supersaturation continues.

10.3.2.2.2 Condensation. *Heterogeneous nucleation* or *nucleated condensation* occurs at much lower supersaturation when enough gas clusters have formed from intramolecular collisions to provide some initial surface area for further growth. Above the oceans, water vapor condenses on soluble nuclei such as sodium chloride crystals at low supersaturation. When mixtures of gases are present, ultrafine particles may appear at supersaturation ratios somewhat <1.

10.3.2.2.3 Sorption/desorption. While supersaturation is necessary for particle formation entirely from gas molecules, the extent of gas *ad*sorption onto existing dry particles or *ab*sorption into liquid films on particles depends not only on the extent of saturation but also on the amount of available surface area. Sorption and desorption influence the size and mass of particles in both the nucleation and accumulation size modes.

The partition coefficient K_p is the most commonly used parameter for describing gas/particle partitioning at equilibrium, primarily because of its log-linear relationship to compound vapor pressure, $p^0{}_L$. Although a compound's vapor pressure at the temperature of interest has the greatest influence on partitioning, the interaction between compound structure and the sorptive medium plays an important role (e.g., compound size and polarity vs. *ad*sorptive affinity or *ab*sorptive capacity). Plots of log K_p vs. log $p^0{}_L$ can provide information on the nature of the partitioning.

Using Pankow's nomenclature (as presented in Chapter 3 of Lane, ed., 1999), the equilibrium partitioning of a semivolatile organic compound to an environmental surface S can be represented most simply by

$$G + S = P \tag{10.1}$$

where G and P represent the gas and particulate phases of the sorbing molecule. Since the gas and particulate phases of the molecule are usually collected on an adsorbent and filter, respectively, their concentrations have been conveniently represented by A and F in the literature (*ibid.*) If the sorbing surface is total suspended particulate matter, its concentration can be represented by TSP. (G/P theory takes the same form for size segregated particles, but TSP is used here to be consistent with Pankow's development.) At equilibrium, the gas/particle partitioning constant K_p for *ad*sorption of the SVOC compound i onto the solid surface of a particle can be expressed as

$$K_p = \frac{F_i}{A_i \, TSP} \quad \textit{ad}\text{sorption to a solid surface} \tag{10.2}$$

Pankow (1987) showed that Langmuir adsorption theory predicts that K_p (at constant temperature) is inversely proportional to the vapor pressure of *i*. If *i* is a solid, the subcooled liquid vapor pressure $p^0{}_L$ is used.

$$K_p = \frac{N_s a_{tsp} Te^{(Q_1 - Q_v)/RT}}{1600 p_L^0} \quad \text{(adsorption)} \tag{10.3}$$

N_s and a_{tsp} are terms for the number of adsorption sites per unit area and the surface area of the particles, respectively. For compounds of the same class, with similar enthalpies of desorption and vaporization Q among the members, plots of log K_p vs. log $p^0{}_L$ will have a linear slope of −1, as shown below:

$$\log K_p = -\log p_L^0 + \log \frac{N_s a_{tsp} Te^{(Q_1 - Q_v)/RT}}{1600} \quad \text{(adsorption)} \tag{10.4}$$

Semi-volatile compounds can also *ab*sorb into liquid particles such as environmental tobacco smoke or liquid (organic and/or water) films on particles with solid cores, as Pankow and colleagues have shown. Gas/particle partitioning of semivolatile organic compounds (SVOC) in urban areas is better explained as *ab*sorption than *ad*sorption. The *ab*sorptive partitioning of SVOC *i* into a liquid organic layer on a particle is like a gas dissolving in a liquid (Finlayson-Pitts and Pitts, 2000, p. 417), and the measured partitioning coefficient for *ab*sorption takes the same form as (10.2):

$$K_p = \frac{F_{i,om}}{A_i\, TSP} \quad \text{(absorption into a liquid film or droplet)} \tag{10.5}$$

$F_{i,om}$ represents the particle-associated concentration of *i* in air as measured from a filter, with explicit recognition that *i* has dissolved in liquid organic material, *om*, on the particle. For absorption into liquid films on particles, Pankow (1994) showed that K_p is proportional to the weight fraction of *om* to *TSP:*

$$K_p = \frac{f_{om} 760RT}{\mathrm{MW}_{om} p_L^0 \gamma \times 10^6} \quad \text{(adsorption)} \tag{10.6}$$

K_p is inversely proportional to the product of $p^0{}_L$ and the activity coefficient γ of *i* in the liquid phase. Vapor pressure is the most important factor influencing K_p, followed by activity coefficient and molecular weight. If the activity coefficient does not vary much across members of a class of SVOCs, plots of K_p vs. log $p^0{}_L$ will have a slope of −1 for both adsorption and absorption, as in Equations (10.4) and (10.7).

$$\log K_p = -\log p_L^0 + \log \frac{f_{om} 760RT}{MW_{om} \gamma \times 10^6} \tag{10.7}$$

Recent contributions to gas/surface partitioning theory address observed deviations from the predictions of Equations (10.4) and (10.7). Jang et al. (1997) applied a comprehensive thermodynamic approach to calculate group contributions to activity coefficients for adsorption of SVOC into nonideal organic films. This allows calculation of activity-normalized partitioning coefficients, $K_{p,g}$. Goss and Schwarzenbach (1998) argued that slope deviations from –1 in log K_p vs. log $p^0{}_L$ plots do not necessarily indicate nonequilibrium conditions, and they indicated how these deviations can be used to identify types of sorbate/sorbent interactions and thus characterize sorption processes. For example, they showed how acid/base interactions can influence gas/surface partitioning polar SVOC. Harner and Bidleman (1998) demonstrated that using laboratory-derived octanol/air partitioning coefficients circumvents the need to estimate activity coefficients for compounds absorbed in the organic films that coat urban particles. Mader and Pankow (2000, 2001a,b) and Mader et al. (2001) expanded partitioning theory to quartz and Teflon filter materials that are used to collect particles, thus tackling the sticky problem of SVOC adsorption artifacts in PM sampling on filters.

10.3.3 *Surface area in the indoor environment*

From the perspective of indoor air quality and personal exposure, understanding the influence of surface area on the behavior of airborne particles includes recognizing the importance of indoor surfaces on aerosol processing. The surface-to-volume ratio is two orders of magnitude greater indoors than outdoors. Weschler's instructive discussion (2003) of indoor surface area will be recounted briefly here. The example is based on a room with volume 40 m^3 and a carpet of 10 m^2.

10.3.3.1 *Particles*

In a now classic study of gas/particle partitioning, Liang and Pankow (1997) inferred the surface area of their airborne particles as 2 $m^2\ g^{-1}$. Using their data, Weschler (2003) calculated that surface loading of a semivolatile compound with subcooled liquid vapor pressure of 10^{-6} atm (hexadecane or phenanthrene) onto airborne particles (at their total suspended particle concentration of 20 $\mu g\ m^{-3}$) was $2.1 \times 10^3\ \mu g\ m^{-2}$. From the more recent data of Naumova et al., (2003) Weschler calculated the mass loading of the same compounds per unit of particulate surface area as $7.7 \times 10^3\ \mu g\ m^{-2}$ in a room of volume 40 m^3.

10.3.3.2 *Indoor materials*

Morrison and Nazaroff (2000) found that the measured the actual surface area of nylon carpet (with pad) was about 100 times its nominal surface area, and Weschler used their data to calculate the surface loading capacity for hexadecane or phenanthrene in the example room. He found that the surface loading ($7.1 \times 10^3\ \mu g\ m^{-2}$) for the carpet was similar to that found by Naumova et al. (2003) for fine particles. This means that indoor surfaces, with their much higher total surface area than indoor particles, are very likely to be important for any indoor aerosol process that depends on gas/surface interactions.

Weschler (*ibid*) estimated that the surface area of the particles would be $1.6 \times 10^{-3}\ m^2$, compared to the room's nominal surface area of 36 m^2, not counting surface porosity or furnishings. The carpet and pad contributed actual surface area of 1000 m^2. Although the surface area of the airborne particles is miniscule compared to the surface area of the room and its furnishings, it is the surface of the particles that makes contact with the respiratory systems of the occupants. Particle motion provides the vector for moving low-volatility pollutants from one indoor surface to another: Weschler showed that particle deposition is the main route for transport of the plasticizer diethylhexylphthalate (DEHP, vapor pressure 1.9×10^{-10} atm at 25°C) to indoor surfaces.

10.3.3.3 *Adsorption indoors*

Indoor aerosol behavior is strongly influenced by the presence of indoor surfaces because building materials and furnishings act as reservoirs of sorbed compounds. At equilibrium, sorption to indoor surfaces follows equation (1). Won et al. (2000, 2001) determined that partitioning coefficients of organic gases to indoor materials depend on p^0_L, the subcooled liquid vapor pressure. Weschler (2003) showed that the molecular weight and the negative log of vapor pressure are linearly related for a large number of organic compounds found indoors, as shown in Figure 10.10. Thus, for compounds whose vapor

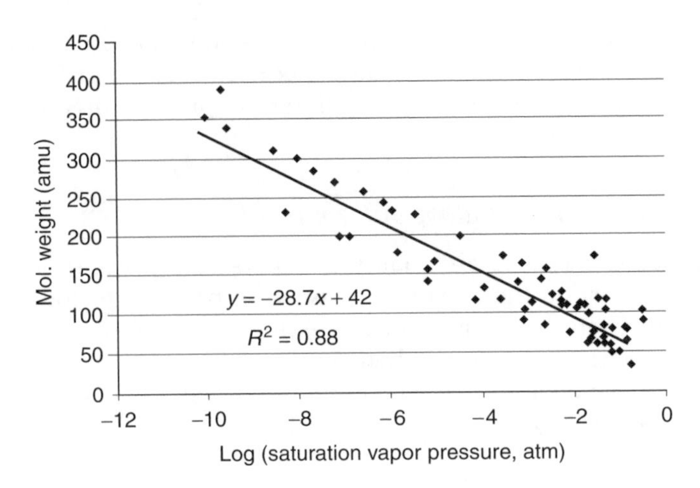

Figure 10.10 Molecular weight vs. the log of the saturation vapor pressure for nonpolar volatile and semivolatile compounds; from Won et al. (2000) as shown by Weschler (2003) with permission.

Table 10.6 Distribution of Selected Organic Compounds Between the Gas Phase and the Surfaces of Airborne Particles, A Carpet, and Walls Within a Typical Room

Compound	Mol. weight (amu)	Vapor Pressure at 25°C (atm)	Assumed Gas Phase Concentration (μg m^{-3})	Mass in Gas Phase (μg)	Mass on Particles (μg)	Mass on Carpet (μg)	Mass on Walls (μg)
MTBE	88	3.2E−01	10	400	2.3E−5	17	19
Toluene	92	3.7E−02	10	400	1.4E−4	100	70
Ethylbenzene	106	1.3E−02	10	400	3.6E−4	260	140
Propylbenzene	120	4.5E−03	10	400	8.9E−4	610	260
Naphthalene	128	1.0E−04	5	200	1.2E−2	7400	1390
Acenaphthene	154	5.9E−06	5	200	0.13	8.0E+4	8000
Hexadecane	226	9.1E−07	5	200	0.66	3.8E+5	2.6E+4
Phenanthrene	178	1.4E−06	1	40	0.093	5.4E+4	4000
Octadecane	254	2.5E−07	1	40	0.40	2.3E+5	1.1E+4
Pyrene	202	7.6E−08	1	40	1.1	6.2E+5	2.4E+4
Heneicosane	296	8.7E−09	0.5	20	3.6	1.9E+6	4.6E+4
Chrysene	228	5.0E−09	0.5	20	5.8	3.0E+6	6.4E+4
Tetracosane	338	2.8E−10	0.01	0.4	1.4	6.9E+5	7800
DEHP	390	1.9E−10	0.07	3.0	14	6.7E+6	6.9E+4
Pentacosane	352	8.7E−11	0.01	0.4	3.8	1.8E+6	1.6E+4

Values derived for a 3×3.65×3.65 m^3 room containing 20 μg m^{-3} of airborne particles (TSP), a 10 m^2 carpet with pad, and painted gypsum board walls. See text for further details.

pressures are unknown, partitioning behavior can be predicted from knowledge of only the molecular weight.

Table 10.6, taken from Weschler (2003) shows how a range of common organic pollutants partition in the example room. The lower the vapor pressure, the more significant is the particulate-bound fraction on the transport and fate of the compound, even if that fraction is very small. Inhalation effects then depend on aerosol dynamics. Accurate assessment of human exposure to indoor particulate organics is hindered by use of sampling methods that cannot exclude adsorption of gas phase semivolatile compounds by the sampling medium (Pang et al., 2002).

10.4 Fate and transport

The concentration of aerosols indoors is the result of several dynamic processes where the production of aerosols (source terms) is balanced by various removal or transformation mechanisms. These processes are shown schematically in Figure 10.3.

10.4.1 Model of indoor aerosol behavior

Definitions

Outdoor aerosol concentration, $C_o\ (d_p)$ (μg m^{-3}).

Interior volume of building, V (m^3).

Infiltration (as flow), Q_{in} (m^3 h^{-1}); $Q_{in}/V=\lambda_{in}$ (h^{-1}) (as rate).

Aerosol penetration, $P(d_p)$.

Ventilation (forced outside airflow via HVAC mechanical systems), Q_{oa} (m^3 h^{-1}).

Total mechanical system flow, $Q_m=Q_{oa}+Q_r$ (recirculation); in most residences with forced air systems, there is no explicit ventilation (outside air) flow so $Q_m=Q_r$.

Aerosol filtration efficiency, $\varepsilon(d_p)$; transport fraction through the filter is $[1-\varepsilon(d_p)]$.

Total Exfiltration/mechanical exhaust, $Q_{te} = Q_{ex} + Q_{eh} = Q_{in} + Q_{oa}$ (m^3 h^{-1}), where Q_{ex} and Q_{eh} are exfiltration and exhaust flows, respectively.

Indoor deposition rate, $k_j(d_p)$ (h^{-1}); the value of k depends upon the orientation of surface j; $k_j=v_j(d_p)A_j/V$, where v_j is the particle-size-dependent deposition velocity for surface j and A_j is the area of surface j.

Indoor resuspension, $R(d_p)$ (μg h^{-1}).

Indoor aerosol generation, $S_i\ (d_p)$ (μg h^{-1}); examples include cooking, smoking, etc.

Transformation processes T (μg h^{-1}): these include particle formation via gas–particle conversion, particle size change due to coagulation, particle size change from accumulation (hygroscopic growth) or loss (desiccation) of water, and particle formation or loss due to phase change.

Tracking processes: these are not a direct source of airborne particles; rather they contribute to the source term for indoor resuspension. Material may be transported from outdoors to indoors or from room to room — typically via the soles of shoes or attached to clothing or other objects. The uptake and release processes that affect transport and mass transfer are poorly defined or quantified at present.

Using these concepts and definitions the generalized mass-balance equation for aerosol concentrations in indoor air is given by

$$V\frac{dC_i(d_p)}{dt}=Q_{in}P(d_p)C_o(d_p)+Q_{oa}[1-\varepsilon(d_p)]C_o(d_p)+R(d_p)+S_i(d_p)$$
$$-Q_{te}C_i(d_p)-C_i(d_p)\sum_j v_j(d_p)A_j-Q_r[1-\varepsilon(d_p)]C_i(d_p)+T \qquad (10.8)$$

Note that in this description, we have explicitly incorporated particle size (d_p is the particle diameter) as an important variable for most of the parameters in the equation. As will be discussed later, some of these parameters are known to be highly dependent upon particle size. The first four terms in Equation (10.8) describe the indoor particle sources: the first term accounts for infiltrating outdoor air as an aerosol source, the second term provides for air brought into the building via a mechanical (HVAC) system, the third term represents resuspension of materials collected on the floor, and the fourth term describes the generation of particles indoors (e.g., combustion, cooking). Resuspension can, in principle, occur from materials collected on other surfaces, including other horizontal surfaces like a table top, but the dominant source of resuspension is that due to occupant activity (such as walking, vacuuming, etc.) on the floor surface. Note that all of the source terms are particle size dependent.

Three sink terms are shown here: the first term is removal of particles by air flow out of the building (either due to exfiltration or exhaust), the second term is deposition to surface j, and the third term accounts for filtration removal due to air recirculation in a heating/cooling system with a filter. Alternatively, this term can represent particle removal by a stand-alone air cleaner. Only the first of these terms is independent of particle size. The second term — particle deposition to various indoor surfaces — is dependent upon both particle size and surface orientation, while the third term incorporates filtration efficiency, which is inherently particle size dependent.

Other terms, represented here by T, can be added to the equation to account for particle transformation processes that can either add or remove particles or shift the particle size spectrum (Nazaroff and Cass, 1989). Coagulation, for example, removes small particles and creates (fewer) larger ones, although for this process to be significant, the number concentrations of small particles need to be elevated. Coagulation has been observed in studies of ETS where small particle concentrations exceed ~10^5 cm^{-3} (Klepeis et al., 2003). Studies of hygroscopic growth of various indoor aerosols have shown mixed effects; in some cases, very little change in particle size was observed, while in other situations, for example, wood smoke, the mass median diameter increased (Dua and Hopke, 1996; Dua et al., 1995).

As a final example of transformation processes, phase change has been observed to alter particle concentrations indoors. In particular, ammonium nitrate aerosol, a major component of ambient aerosols in the western U.S., is volatile and exists in equilibrium with its gas-phase constituents, ammonia and nitric acid. When these aerosols enter buildings via infiltration or ventilation, the temperature and relative humidity conditions may change, driving the equilibrium toward dissociation into the gas-phase species. These species in turn interact with indoor surfaces — especially nitric acid — leading to additional dissociation. Under these conditions, indoor concentrations of ammonium nitrate particles are significantly reduced (Lunden et al., 2003a).

In order to illustrate the main features of the mass balance equation and to keep the number of variables tractable, we drop the filtration and transformation terms from Equation (10.8) and simplify some of the variables to yield

$$\frac{dC_i(d_p)}{dt} = \lambda_{in}P(d_p)C_o(d_p) + [R(d_p) + S_i(d_p)]/V - \lambda_{in}C_i(d_p) - C_i(d_p)\sum_j k_j(d_p) \tag{10.9}$$

If we assume that the indoor space is well mixed and that the various terms vary slowly with time, the average indoor concentration can be approximated by the steady-state solution:

$$C_i(d_p) = \frac{\lambda_{in}P(d_p)C_o(d_p) + [R(d_p) + S_i(d_p)]/V}{\lambda_{in} + \sum_j k_j(d_p)} \tag{10.10}$$

Equation (10.10) is useful in "well-controlled" situations, such as laboratory-based experiments or where indoor sources, for example, are operated to produce high concentrations of aerosols so that variable contributions from outdoors, etc. can be neglected. This equation helps illustrate the balance between typical indoor aerosol sinks and sources.

However, in "real-world" situations, as would be the case in examining or estimating the transport and fate of aerosols in actual buildings, two important parameters in equation (10.9) are often time varying—outdoor aerosol concentrations and infiltration rates — especially in houses where mechanical systems are not used to supply ventilation air. If there is sufficient time-series information on the variability of these two parameters, then a "forward-marching" approach with a small time step, Δt, can be used, as has been recently demonstrated in the analysis of particle penetration data (Thatcher et al., 2003). The form of this equation is

$$C_i(d_p, t_2) = C_i(d_p, t_1) + P(d_p)C_o(d_p, t_1)\lambda_{in}(t_1)\Delta t + [R(d_p) + S(d_p)]/V$$
$$- C_i(d_p, t_1)[\lambda_{in}(t_1) + \sum_j k_j(d_p)]\Delta t \tag{10.11}$$

This example treats R and S as constant in time, but these could be restated to incorporate their time variability by using the form $R(d_p,t_1)\Delta t$. Use of this forward-marching approach requires that information/data are available for each parameter at each time step.

The key parameters determining aerosol transport and fate in buildings are discussed below.

10.4.2 *Airflow and aerosol transport through penetrations in building envelopes*

Airflow through building envelopes occurs as a result of temperature and wind-driven pressure differences between the inside and outside of the building. Houses are often kept "closed" during wintertime heating periods and, in some regions, during summertime cooling periods, thus limiting airflows to gaps, holes, or other inadvertent penetrations in the building shell, created as a result of (poor) construction practices, settling, and/or aging of building components, etc. Transport of particles into the building will occur via airflow through building leaks.

Few measurements have been made to determine the nature and significance of most building leaks. On average, the largest air leakage values are for leaks in the walls and floor (35%), in the ceiling (15%), and around windows and doors (15%) (Diamond and Grimsrud, 1984). Not all of these leaks will conduct aerosol into the building interior under most operating conditions. High leaks, such as those in the ceiling or openings such as fireplace or furnace chimneys, are usually locations for exfiltration. The physical dimensions of many building penetrations are poorly characterized and the flows across the building envelope may proceed through tortuous pathways, making *a priori* prediction of aerosol penetration efficiency difficult.

Many studies have used indoor and outdoor mass measurements to provide broad categorical penetration factors. In a summary paper (Wallace, 1996), penetration factors of ~1 are derived for both fine ($PM_{2.5}$) and coarse (PM_{10}) mode particles, based on a statistical analysis of data from the PTEAM study. However, these estimates do not take into account the details in the size dependence of the deposition rates nor can they account for any potential size-dependent effect of the penetration process itself. Because these values are based on mass, the results will heavily depend upon the underlying size distribution of the aerosols and, in particular, the populations of the largest particle size fractions.

Similar results have been reported by Thatcher and Layton (1995) based on experimental measurements of penetration; however, this study is limited to summertime measurements in one house and to particles larger than 1 μm. In contrast, results from a series of controlled room-size chamber experiments show a significant decline in the penetration factor as the test aerosol becomes larger than 2 μm in diameter (Lewis, 1995). For particles larger than ~6 μm, the penetration fraction was measured to be essentially zero.

A set of experiments was performed in two residential buildings to examine both aerosol penetration and deposition indoors (Thatcher et al., 2003). Continuous, size-resolved data were collected during a three-step experimental procedure: (1) artificially enhancing particle concentrations indoors and following the decay in concentrations, (2) rapid reduction in particle concentrations through induced exfiltration by pressurizing the dwellings with HEPA-filtered air, and (3) following the rebound in infiltrating particle concentrations when overpressurization was stopped. Penetration factors as a function of particle size are shown in Figure 10.11.

The dwelling labeled "Richmond" is an older structure with a much larger overall leakage area than the "Clovis" dwelling. As can be seen in Figure 10.11, the tighter structure had lower overall penetration factors and particles larger than ~3 μm had penetration factors as low as ~0.3. In contrast, penetration factors for the Richmond structure were close to 1 as a function particle size, with the lowest penetration factor = ~0.8.

10.4.3 Indoor aerosol deposition rates

Indoor aerosol deposition rates as a function of particle size have been developed empirically and the results are summarized by Thatcher et al. (2002) in Figure 10.12. An important feature of these curves is the strong dependence of the deposition decay rates on particle size. There is a minimum deposition rate for particle diameters between 0.1 and 0.2 μm ($k \sim 0.03\ h^{-1}$ under quiescent conditions), but the rates increase rapidly for particles both larger ($k \sim 0.3\ h^{-1}$ and $\sim 18\ h^{-1}$ for 1.5 and 5-μm-diameter particles, respectively) and smaller ($k \sim 0.15\ h^{-1}$ for 0.06-μm-diameter particles).

These deposition rates are a combination of deposition due to diffusion (more important for small particles) and that due to gravitational settling (more important for large particles). For a horizontal, upward-facing surface, the two deposition mechanisms are about equal for ~0.2-μm-diameter, unit-density spheres. In contrast, diffusion is about 400 times more important for 0.01 μm particles, while settling is about 60 times more important for 1 μm particles (Hinds, 1999, p. 162). There are also effects due to the turbulence of air within the enclosure, as reported by Lewis (1995), Xu et al. (1994), Lai and Nazaroff (2000), Thatcher et al. (2002), and as shown in Figure 10.12.

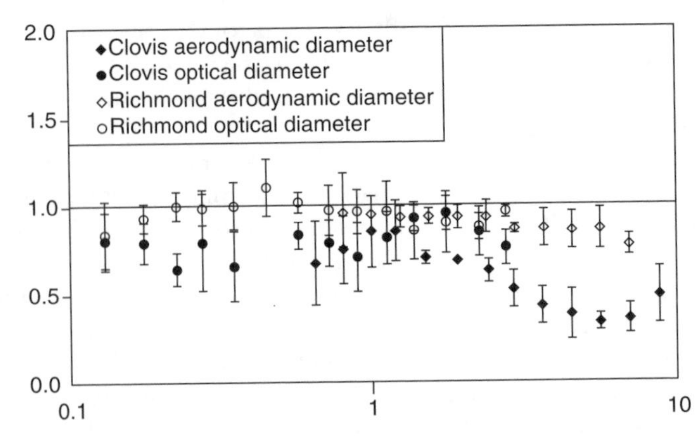

Figure 10.11 Particle penetration in two dwellings as measured by both optical and aerodynamic particle size instruments. From Thatcher et al., 2003.

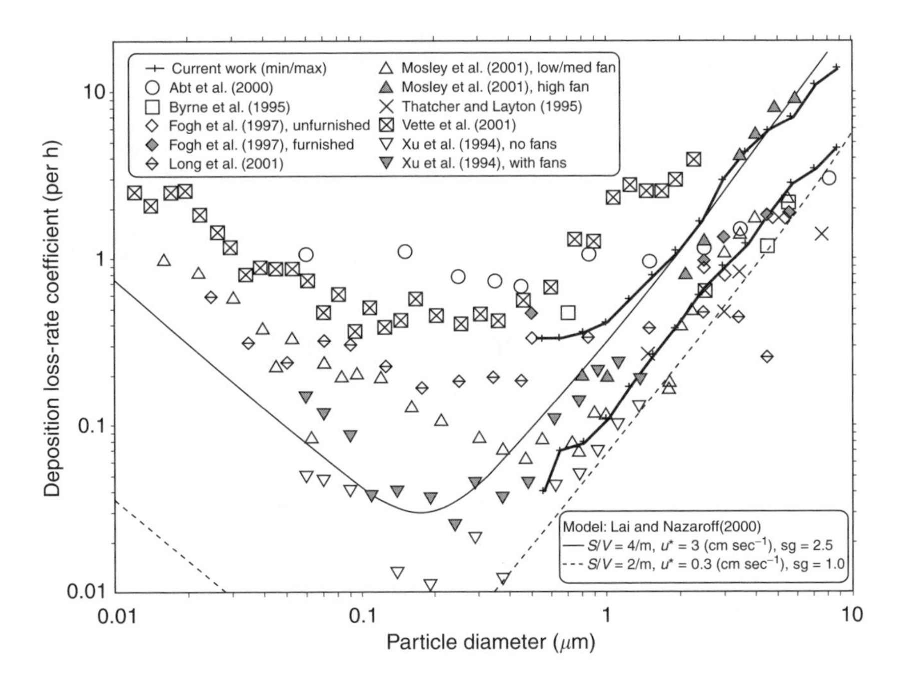

Figure 10.12 Particle deposition loss rates as reported in several studies. Current work refers to Thatcher et al. (2002). The model results cover a range of surface-to-volume ratios, internal air motion (friction velocities), and particle densities (from Thatcher et al. 2002).

Thus, indoor aerosol dynamics depends strongly on both particle size and surface orientation. These differences may also have important implications for depositional losses in building leaks, depending upon the flow rates and particle velocities.

10.4.4 Resuspension rate of particles on carpets/floors

Only limited data are available on the rates at which particles on floors are suspended into air (expressed as a fraction of particulate loading on floor surfaces suspended per unit time) by human activities. Early work on resuspension indoors focused on the movement of radionuclides from the floors to the air. Healy (1971) developed a time-weighted average resuspension rate of 5×10^{-4} h^{-1} for a house. This is comparable to the value of Murphy and Yocom (1986), who selected a value of 10^{-4} h^{-1}. Measurements of particles in indoor air using optical particle counters demonstrate that resuspension is a function of human activities as well as particle size. Kamens et al. (1991), for example, showed that the increase in suspended particles over the course of a day in a house corresponded to the activities of the residents. The apparent resuspension threshold of particles from floor surfaces is about 1–2 μm (Thatcher and Layton, 1995). The study of Clayton et al. (1993) supports this relationship. They found that the concentrations of fine PM (i.e., particles under 2.5 μm in diameter) in the main living area of a sample of houses were highly correlated with the outdoor levels recorded at fixed monitors; however, the PM_{10} concentrations had a correlation coefficient of only 0.37 with fixed-site monitors. This suggests that a significant portion of the PM_{10} particles collected were suspendable particles over 2 μm in diameter derived from human activities. The resuspension rates increase significantly with aerosols larger than ~2 μm, as illustrated in Figure 10.13.

The mass loadings of soil/dust on floors in the literature that we have examined vary from 0.136 to 0.870 g m^{-2}. The geometric mean of the mass loadings is 0.42 g m^{-2} with a

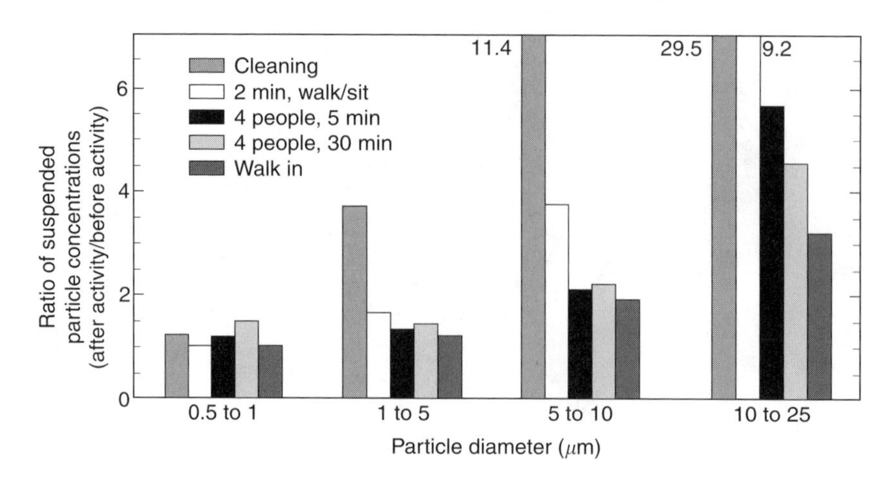

Figure 10.13 Ratios of suspended particle concentrations after human activities to concentrations before activity in a single-family residence (from Thatcher and Layton, 1995).

geometric standard deviation of 1.88. Most of the data on dust loadings are based on studies dealing with lead contamination of the indoor environment (see, e.g., Gulson et al., 1995) The mass loading of PM on carpets/floors represents the mass available for the resuspension of particles into indoor air. Floor dust also serves as a contact medium for infants/toddlers who crawl on floors and have hand-to-mouth behaviors that result in the ingestion of dusts. An unresolved issue is the relative suspendability of particles on carpets vs. bare floors.

References

Abt, E., Helen, H., Suh, H.H., Catalano, P., and Koutrakis, P., Relative contribution of outdoor and indoor particle sources to indoor concentrations, *Environ. Sci. Technol.*, 34, 3579–3587, 2000.

Andrejs, B., Fauss, J., Weigl, M., and Rietschel, P., Ventilation in kitchen — aerosol concentration and key components in the vapor, Proceedings of the 9th International Conference on Indoor Air Quality and Climate, Monterey, CA, June 30–July 5, 2002, 3, pp. 292–297, 2002.

Becker, K., Seiwert, M., Schulz, C., Kaus, S., Krause, C., and Seifert, B., German environmental survey 1998 (GerESIII); pesticides and other pollutants in house dust, Proceedings of the 9th International Conference on Indoor Air Quality and Climate, Monterey, CA, June 30–July 5, 2002, 2002, pp. 883–887.

Bennett, D.H. and Furtaw, E.J., Fugacity–based indoor residential pesticide fate model, *Environ. Sci. Technol.*, 38, 2142–2152, 2004.

Berger-Preiss, E., Levsen, K., Leng, G., Idel, H., and Ranft, U., Indoor monitoring of homes with wool carpets, treated with permethrin, Proceedings of the 9th International Conference on Indoor Air Quality and Climate, Monterey, CA, June 30–July 5, 2002, 1, 2002, pp. 1021–1025.

Bukowski, J.A. and Meyer, L.W., Simulated air levels of volatile organic compounds following different methods of indoor insecticide application, *Environ. Sci. Technol.*, 29, 673–676, 1995.

Butt, C.M., Diamond, M.L., and Truong, J., Spatial distribution of polybrominated diphenyl ethers in Southern Ontario as measured in indoor and outdoor window organic films, *Environ. Sci. Technol.*, 38, 724–731, 2004.

Byrne, M.A., Goddard, A.J.H., Lange, C., Roed, J., Stable tracer aerosol deposition measurements in a test chamber, *J. Aerosol Sci.*, 26, 645–653, 1995.

Carlsson, H., Nilsson, U., and Ostman, C., Video display units: an emission source of the contact allergenic flame retardant triphenyl phosphate in the indoor environment, *Environ. Sci. Technol.*, 34, 3885–3889, 2000.

Cheng, Y. S., Bechtold, W.E., Yu, C.C., and Hung I.F., Incense smoke — characterization and dynamics in indoor environments, *Aerosol Sci Technol.*, 23, 271–281, 1995.

Clayton, C.A., Perritt, R.L., Pellizzari, E.D., Wallace, L.A., Ozkaynak, H., and Spengler, J.D., Particle Total Exposure Assessment Methodology (PTEAM) Study: distributions of aerosols and elemental concentrations in personal, indoor, and outdoor samples in a southern California community, *J. Expos. Anal. Environ. Epidemiol.*, 3, 227–250, 1993.

Cowan, C.E., Mackay, D., Feijtel, T.C.J., van de Meent, D., Di Guardo, A., Davies, J., and Mackay, N., Eds., *The Multimedia Fate Model: A Vital Tool for Predicting the Fate of Chemicals*, SETAC Press, Penascola, FL, 1995, 78pp.

Daisey, J.M., Tracers for assessing exposure to environmental tobacco smoke: what are they tracing? *Environ. Health Perspect.*, 107 (Suppl. 2), 319–327, 1999.

Diamond, R.C. and Grimsrud, D.T., Manual on Indoor Air Quality, EPRI Report EM-3469, Electric Power Research Institute, 1984.

Dua, S.K. and Hopke, P.K., Hygroscopic growth of assorted indoor aerosols, *Aerosol. Sci. Technol.*, 24, 151–160, 1996.

Dua, S.K., Hopke, P.K., and Raunemaa, T., Hygroscopic growth of consumer spray products, *Aerosol. Sci. Technol.*, 23, 331–340, 1995.

Erwin, E.A., Woodfolk, J.A., Custis, N., and Platts-Mills, T.A.E., Animal danders, *Immunolo Allerg. Clin North Am.*, 23, 469–481, 2003.

Ferro, A.R., Kopperud, R.J., and Hildemann, L.M., Exposure to house dust from human activities, Proceedings of the 9th International Conference on Indoor Air Quality and Climate, Monterey, CA, June 30–July 2, 2002, 1, 2002, pp. 527–532.

Fine, P.M., Cass, G.R., and Simoneit, B.R.T., Characterization of fine particle emissions from burning church candles, *Environ. Sci. Technol.*, 33, 2352–2362, 1999.

Finlayson-Pitts, B.J. and Pitts, J.N., *Chemistry of the Upper and Lower Atmosphere: Theory, Experiments, and Applications*, Academic Press, New York, 1999, 969pp.

Fogh, C.L., Byrne, M.A., Roed, J., Goddard, A.J.H., Size-specific indoor aerosol deposition measurements and derived I/O concentration ratios, *Atmos. Environ.*, 31, 2193–2203, 1997.

Fujii, M., Shinohara, N., Lim, A., Otake, T., Kumagai, K., and Yanagisawa, Y., A study on emission of phthalate esters from plastic materials using a passive flux sampler, *Atmos. Environ.*, 37, 5495–5504, 2003.

Goss, K.-U. and Schwarzenbach, R.P., Gas/solid and gas/liquid partitioning of organic compounds: critical evaluation of the interpretation of equilibrium constants, *Environ. Sci. Technol.*, 32, 2025–2032, 1998.

Gulson, B.L., Davis, J.J., Mizon, J., Korsch, M.J., and Bawden-Smith, Source of lead in soil and dust and the use of dust fallout as a sampling medium, *Sci. Total Environ.*, 166, 245–262, 1995.

Harner, T. and Bidleman, T.F., Octanol–air partition coefficient for describing particle/gas partitioning of aromatic compounds in urban air, *Environ. Sci. Technol.*, 32, 1494–1502, 1998.

Harrison and Brooke, 1999.

Healy, J.W., *Surface Contamination: Decision Levels*, Los Alamos Scientific Laboratory, Los Alamos, NM, 1971.

Hildemann, L.M., Markowski, G.R., and Cass, G.R., Chemical composition of emissions from urban sources of fine organic aerosol, *Environ. Sci Technol.*, 25, 744–759, 1991.

Hinds, W.C., *Aerosol Technology*, Wiley, New York, 1982; 424pp; Second ed., 1999, 483pp.

Hodgson et al., 2000, 2002.

Hoffmann, T., Odum, J.R., Bowman, F., Colins, D., Klockow, D., Flagan, R.C., and Seinfeld, J.H., Formation of organic aerosols from the oxidation of biogenic hydrocarbons, *J. Atmos. Chem.*, 26, 189–222, 1997.

Hornig, J.F., Soderberg, R.H., Barefoot, A.C., III, and Galasyn, J.F., Woodsmoke analysis: vaporization losses of PAH from filters and levogucosan as a distinctive marker for woodsmoke, in Cooke, M. and Dennis, A.J., *Polynuclear Aromatic Hydrocarbons: Mechanism, Methods and Metabolism*, Eds., Batelle Press, Columbus, 1985, pp. 561–568.

Jang, M., Kamens, R.M., Leach, K.B., and Strommen, M.R., A thermodynamic approach to group contribution methods to model the partitioning of semi-volatile organic compounds on atmospheric particulate matter, *Environ. Sci. Technol.*, 31, 2805–2811, 1997.

Jenkins, R.A., Guerin, M.R., and Tompkins, B.A., *The Chemistry of Environmental Tobacco Smoke: Composition and Measurement*, 2nd ed., Lewis, Boca Raton, 2000, 467pp.

Jetter, J.J., Guo, Z.S., McBrian, J.A., and Flynn, M.R., Characterization of emissions from burning incense, *Sci. Total Environ.*, 295, 51–67, 2002.

Kamens, R., Lee, C.-T., Wiener, R., and Leith, D., A study to characterize indoor particles in three non-smoking homes, *Atmos. Environ.*, 25A, 939–948, 1991.

Kamens, R.M., Jang, M., Chien, C.J., and Leach, K., Aerosol formation from the reaction of α-pinene and ozone using a gas-phase kinetics and gas-particle partitioning theory, *Environ. Sci. Technol.*, 35, 1394–1405, 1999.

Kavouras, I.G., Stratigakis, N., and Stephanou, *Iso-* and *anteiso*-alkanes: specific tracers of environmental tobacco smoke in indoor and outdoor particle-size distributed urban aerosols, *Environ. Sci. Technol.*, 32, 1369–1377, 1998.

Kemmlein, S., Hahn, O., and Jann, O., Emissions of organophosphate and brominated flame retardants from selected consumer products and building materials, *Atmos. Environ.*, 37, 5485–5493, 2003.

Klepeis, N.E., Apte, M.G., Gundel, L.A., Sextro, R.G., and Nazaroff, W.W., Determining size-specific emission factors for environmental tobacco smoke particles, *Aerosol Sci. Technol.*, 37, 780–790, 2003.

Koutrakis, P. and Briggs, S.L.K., Source apportionment of indoor aerosols in Suffolk and Onondaga Counties, New York, *Environ. Sci. Technol.*, 26, 521–527, 1992.

Kumar, D. and Little, J.C., Characterizing the source/sink behavior of double-layer building materials, *Atmos. Environ.*, 37, 5529–5537, 2003.

Lai, A.C.K. and Nazaroff, W.W., Modeling indoor particle deposition from turbulent flow onto smooth surfaces, *J. Aerosol Sci.*, 31, 463–476, 2000.

Landis, M.S., Norris, G.A., Williams, R.W., and Weinstein, J.P., Personal exposures to $PM_{2.5}$ mass and trace elements in Baltimore, MD, USA, *Atmos. Environ.*, 35, 6511–6524, 2001.

Lane, D.A., Ed., *Gas and Particle Phase Measurements of Atmospheric Organic Compounds*, Vol. 2 of *Advances in Environmental, Industrial and Process Control Technologies*, Gordon and Breach, Amsterdam, 402pp.

Leaderer, B.P., Boone, P.M., and Hammond, S.K., Total particle, sulfate and acidic aerosol emissions from kerosene space heaters, *Environ. Sci. Technol.*, 24, 908–912, 1990.

Leaderer, B.P, Naeher, L., Jankun, T., Balenger, K., Holford, T.R., Toth, C., Sullivan, J., Wolfson, J.M., and Koutrakis, P., Indoor, outdoor, and regional summer and winter concentrations of PM_{10}, $PM_{2.5}$, SO_4^{2-}, H^+, NH_4^+, NO_3^-, NH_3, and nitrous acid in homes with and without kerosene space heaters, *Environ. Health Perspect.*, 107, 223–231, 1999.

Lewis, S., Solid particle penetration into enclosures, *J. Haz. Mats*, 43, 195–216, 1995.

Liang, C., Pankow, J.F., Odum, J.R., and Seinfeld, J.H., Gas/particle partitioning of semivolatile organic compounds to model inorganic, organic and ambient smog aerosols, *Environ. Sci. Technol.*, 31, 3086–3092, 1997.

Loan, R., Siebers, R., Fitzharris, P., and Crane, J., House dust-mite allergen and cat allergen variability within carpeted living room floors in domestic dwellings, *Indoor Air*, 13, 232–236, 2003.

Long, C.M., Suh, H.H., Catalano, P.J., Koutrakis, P., Using time- and size-resolved particulate data to quantify indoor penetration and deposition behavior, *Environ. Sci. Technol.*, 35, 2089–2099, 2001.

Lunden, M.M., Revzan, K.L., Fischer, M.L., Thatcher, T.L., Littlejohn, D., Hering, S.V., and Brown, N.J., The transformation of outdoor ammonium nitrate aerosols in the indoor environment, *Atmos. Environ.*, 37, 5633–5644, 2003a.

Lunden, M.M., Thatcher, T.L., Hering, S.V., and Brown, N.J., Use of time and chemically resolved particulate data to characterize the infiltration of outdoor $PM_{2.5}$ into a residence in the Sand Joaquin Valley, *Environ. Sci. Technol.*, 37, 4724–4732, 2003b.

Lung, S.-C.C. and Hu, S.-C., Generation rates and emission factors of particulate matter and particle-bound polycyclic aromatic hydrocarbons of incense sticks, *Chemosphere*, 50, 673–679, 2003.

Mader, B.T. and Pankow, J.F., Gas/solid partitioning of semivolatile organic compounds (SOCs) to air filters, 1. Partitioning of polychlorinated dibenzodioxins, polychlorinated dibenzofurans and polycyclic aromatic hydrocarbons to quartz fiber filters, *Atmos. Environ.*, 34, 4879–4887, 2000.

Mader, B.T. and Pankow, J.F., Gas/solid partitioning of semivolatile organic compounds (SOCs) to air filters, 2. Partitioning of polychlorinated dibenzodioxins, polychlorinated dibenzofurans and polycyclic aromatic hydrocarbons to teflon membrane filters, *Atmos. Environ.*, 35, 1217–1223, 2001a.

Mader, B.T. and Pankow, J.F., Gas/solid partitioning of semivolatile organic compounds (SOCs) to air filters, 3. An analysis of gas adsorption artifacts in measurements of atmospheric SOCs and organic carbon (OC) when using teflon membrane filters and quartz fiber filters, *Environ. Sci. Technol.*, 35, 3422–3432, 2001b.

Mader, B.T., Flagan, R.C. and Seinfeld, J.H., Sampling atmospheric carbonaceous aerosols using a particle trap impactor/denuder sampler, *Environ. Sci. Technol.*, 35, 4857–4867, 2001.

Manchester-Neesvig, J.B., Schauer, J.J., and Cass, G.R., The distribution of particle-phase organic compounds in the atmosphere and their use for source apportionment during the southern California children's health study, *J. Air Waste Manage. Assoc.*, 53, 1065–1079, 2003.

Macher, J.M., Ammann, H., Burge, H.A., Milton, D.K., and Morey, P.R., Eds., Bioaerosols: Assessment and Control, American Conference of Governmental Industrial Hygienists (ACGIH), Cincinnati, Ohio, 1999.

Mannix, R.C., Nguyen, K.P., Tan, E.W., Ho, E.E., and Phalen, R.F., Physical characterization of incense aerosols, *Sci. Total Environ.*, 193, 149–158, 1996.

Manuel, J., A healthy home environment? *Environ. Health Perspect.*, 107, A352–A357, 1999.

McDonald, J.D., Zielinska, B., Fujita, E.M., Sagebiel, J.C., Chow, J.C., and Watson, J.G., Fine particle and gaseous emission rates from residential wood combustion, *Environ. Sci. Technol.*, 34, 2080–2091, 2000.

Morawska, L. and Salthammer, T., Eds., *Indoor Environment: Airborne Particles and Settled Dust*, Wiley, New York, 2003, 350pp.

Morrison, G.C. and Nazaroff, W.W., The rate of ozone uptake on carpets: experimental studies, *Environ. Sci. Technol.*, 36, 4963–4968, 2000.

Mosley, R.B., Greenwell, D.J., Sparks, L.E., Guo, Z., Tucker, W.G., Fortmann, R., Whitfield, C., Penetration of ambient fine particles into the indoor environment, *Aerosol Sci. Technol.*, 34, 127–136, 2001.

Murphy, B.L. and Yocom, J.E., Migration factors for particulates entering the indoor environment, in Proceedings of the 79th Annual Meeting of the Air Pollution Control Association, Paper 86–7.2, APCA, Pittsburgh, PA, 1986.

Naumova, Y.Y., Offenberg, J.H., Eisenreich, S.J., Meng, Q., Polidori, A., Turpin, B.J., Weisel, C.P., Morandi, M.T., Colome, S.D., Stock, T.H., Winer, A.M., Alimokhtari, S., Kwon, J., Maberti, S., Shendell, D. Jones, J. Farrar, C., Gas/particle distribution of polycyclic aromatic hydrocarbons in coupled outdoor/indoor atmospheres, *Atmos. Environ.*, 37, 703–719, 2003.

Nazaroff, W.W. and Alvarez-Cohen, *Environmental Engineering Science*, Wiley, New York, 2000.

Nazaroff, W.W. and Cass, G.R., Mathematical modeling of indoor aerosol dynamics, *Environ. Sci. Technol.*, 23, 157–166, 1989.

Nazaroff, W.W., Weschler, C.J., and Corsi, R.L., Indoor air chemistry and physics, *Atmos. Environ.*, 37, 5431–5453, 2003.

Nishioka, M.G., Burkholder, H.M., Brinkman, M.C., and Lewis, R.G., Measuring transport of lawn-applied herbicide acids from turf to home, correlation of dis lodgeable 2, 4-D turf residues with carpet dust and carpet surface residues, *Environ. Sci. Technol.*, 30, 3313–3320, 1996.

Nriagu, J.O. and Kim, M.J., Emissions of lead and zinc from candles with metal-core wicks, *Sci. Total Environ.*, 250, 37–41, 2000.

Ozkaynak, H., Hue, J., Spengler, J., Wallace, L., Pellizari, E., and Jenkins, P., Personal exposure to airborne particles and metals; results from the particle TEAM study in Riverside CA. *J. Expos. Anal. Environ. Epidemiol.*, 6, 57–78, 1996.

Pang, Y., Gundel, L.A., Larson, T., Finn, D., Liu, S. (L.-J.), and Claiborn, C.S., Development and evaluation of a personal particulate organic and mass sampler (PPOMS), *Environ. Sci. Technol.*, 36, 5205–5210, 2002.

Pankow, J.F., Review and comparative analysis of the theories on partitioning between the gas and aerosol particulate phases in the atmosphere, *Atmos. Environ.*, 21, 2275–2283, 1987.

Pankow, J.F., An absorption model of gas/particle partitioning of organic compounds in the atmosphere, *Atmos. Environ.*, 28, 185–188, 1994.

Rogge, W.F., Hildemann, L.M., Mazurek, M.A., Cass, G.R., and Simoneit, B.R.T., Sources of fine organic aerosol. 1. Charbroilers and meat cooking operations, *Environ. Sci. Technol.*, 25, 1112–1125, 1991.

Rogge, W.F., Hildemann, L.M., Mazurek, M.A., Cass, G.R., and Simoneit, B.R.T., Sources of fine organic aerosol. 2. Non-catalyst and catalyst-equipped automobiles and heavy-duty diesel trucks, *Environ. Sci. Technol.*, 27, 636–651, 1993a.

Rogge, W.F., Hildemann, L.M., Mazurek, M.A., Cass, G.R., and Simoneit, B.R.T., Sources of fine organic aerosol. 3. Road dust, tire debris and organometallic brake lining dust: reads as sources and sinks, *Environ. Sci. Technol.*, 27, 1892–1904, 1993b.

Rogge, W.F., Hildemann, L.M., Mazurek, M.A., Cass, G.R., and Simoneit, B.R.T., Sources of fine organic aerosol. 4. Particulate abrasion products from leaf surfaces of urban plants, *Environ. Sci. Technol.*, 27, 2700–2711, 1993c.

Rogge, W.F., Hildemann, L.M., Mazurek, M.A., Cass, G.R., and Simoneit, B.R.T., Sources of fine organic aerosol. 5. Natural gas home appliances, *Environ. Sci. Technol.*, 27, 2736–2744, 1993d.

Rogge, W.F., Hildemann, L.M., Mazurek, M.A., and Cass, G.R., Sources of fine organic aerosol. 6. Cigarette smoke in the urban atmosphere, *Environ. Sci. Technol.*, 28, 1375–1388, 1994.

Rogge, W.F., Hildemann, L.M., Mazurek, M.A., Cass, G.R., and Simoneit, B.R.T., Sources of fine organic aerosol. 7. Hot asphalt roofing tar pot fumes, *Environ. Sci. Technol.*, 31, 2726–2730, 1997a.

Rogge, W.F., Hildemann, L.M., Mazurek, M.A., Cass, G.R., and Simoneit, B.R.T., Sources of fine organic aerosol. 8. Boilers burning No. 2 distillate fuel oil, *Environ. Sci. Technol.*, 31, 2731–2737, 1997b.

Rogge, W.F., Hildemann, L.M., Mazurek, M.A., Cass, G.R., and Simoneit, B.R.T., Sources of fine organic aerosol. 9. Pine, oak and synthetic log combustion in residential fireplaces, *Environ. Sci. Technol.*, 32, 13–22, 1998.

Rudel, R.A., Brody, J.G., Spengler, J.D., Vallarino, J., Geno, P.W., Sun, G., and Yau, A., Identification of selected hormonally active agents and animal mammary carcinogens in commercial and residential air and dust samples, *J. Air Waste Manage. Assoc.*, 51, 499–513, 2002.

Rudel, R.A., Camann, D.E., John D. Spengler, J.D., Leo R. Korn, L.R., Julia, G., and Brody, J.G., Alkylphenols, pesticides, phthalates, polybrominated diphenyl ethers, and other endocrine-disrupting compounds in indoor air and dust, *Environ. Sci. Technol.*, 37, 4543–4553, 2003.

Salthammer, T., Ed., *Organic Indoor Air Pollutants: Occurrence, Measurement, Evaluation*, Wiley, New York, 1999, 344pp.

Salthammer, T., Fuhrmann, F., and Uhde, E., Flame retardants in the indoor environment — Part II: release of VOCs (triethylphosphate and halogenated degradation products from polyurethane), *Indoor Air*, 13, 49–52, 2003.

Sarwar, G., Corsi, R., Allen, D., and Wechsler, C., The significance of secondary organic aerosol formation and growth in buildings: experimental and computational evidence, *Atmos. Environ.* 37, 1365–1381, 2003.

Schauer, J.J., Kleeman, M.J., Cass, G.R., and Simoneit, B.R.T., Measurement of emissions from air pollution sources. 1. C_1 through C_{29} organic compounds from meat charbroiling, *Environ. Sci. Technol.*, 33, 1566–1577, 1999ab.

Schauer, J.J., Kleeman, M.J., Cass, G.R., and Simoneit, B.R.T., Measurement of emissions from air pollution sources. 12. C_1 through C_{30} organic compounds from medium duty diesel trucks, *Environ. Sci. Technol.*, 33, 1578–1587, 1999b.

Schauer, J.J., Kleeman, M.J., Cass, G.R., and Simoneit, B.R.T., Measurement of emissions from air pollution sources. 3. C_1–C_{29} organic compounds from fireplace combustion of wood, *Environ. Sci. Technol.* 35, 1716–1728, 2001.

Schauer, J.J., Kleeman, M.J., Cass, G.R., and Simoneit, B.R.T., Measurement of emissions from air pollution sources. 4. C_1–C_{27} organic compounds from cooking with seed oils, *Environ. Sci. Technol.*, 36, 567–575, 2002a.

Schauer, J.J., Kleeman, M.J., Cass, G.R., and Simoneit, B.R.T., Measurement of emissions from air pollution sources. 5. C_1–C_{32} organic compounds from gasoline-powered motor vehicles, *Environ. Sci. Technol.*, 36, 1169–1180, 2002b.

Schauer, J.J., Fraser, M.P., Cass, G.R., and Simoneit, B.R.T., Source reconciliation of atmospheric gas-phase and particle-phase pollutants during a severe photochemical episode, *Environ. Sci. Technol*, 36, 3806–3814, 2002c.

Seinfeld, J.H. and Pandis, S.N., *Atmospheric Chemistry and Physics*, Wiley-Interscience, New York, 1998, 1326pp.

Sexton, K., Adgate, J.L., Ramachandran, G., Pratt, G.C., Mongin, S.J., Stock, T.H., and Morandi, M.T., Comparison of personal, indoor and outdoor exposures to hazardous air pollutants in three urban communities, *Environ. Sci. Technol.*, 38, 423–430, 2004.

Shaughnessy, R.J., McDaniels, T.J., and Weschler, C.J., Indoor chemistry: ozone and volatile organic compounds found in tobacco smoke, *Environ. Sci. Technol.*, 35, 2758–2764, 2001.

Shoeib, M., Harner, T., Ikonomou, M., and Kannan, K., Indoor and outdoor air concentrations and phase partitioning of perfluoroalkyl sulfonamides and polybrominated diphenyl ethers, *Environ. Sci. Technol.*, 38, 1313–1320, 2004.

Simoneit, B.R.T., Biomass burning — a review of organic tracers for smoke from incomplete combustion, *Appl. Geochem.*, 17, 129–162, 2002.

Sjödin, A., Carlsson, H., Thursesson, K., Sjolin, S., Bergman, A., and Ostman, C., *Environ. Sci. Technol.*, 35, 448–454, 2001.

Song, B.J. and Liu, A.H., Metropolitan endotoxin exposure, allergy and asthma, *Curr. Opin. Allerg. Clin. Immunol.*, 3, 331–335, 2003.

Stout, D.M., II. and Mason, M.A., The distribution of chlorpyrifos following a crack and crevice type application in the U.S. EPA Indoor Air Quality Research House, *Atmos. Environ.*, 37, 5539–5549, 2003.

Thatcher, T.L. and Layton, D.W., Deposition, resuspension and penetration of particles within a residence, *Atmos. Environ.*, 29, 1487–1497, 1995.

Thatcher, T.L., Alvin, C.K., Lai, A.C.K., Moreno-Jackson, R., Sextro, R.G., and Nazaroff, W.W., Effects of room furnishings and air speed on particle deposition rates indoors, *Atmos. Environ.*, 36, 1811–1819, 2002.

Thatcher, T.L., Lunden, M.M., Revzan, K.L., Sextro, R.G., and Brown, N.J., A concentration rebound method for measuring particle penetration and deposition in the indoor environment, *Aerosol Sci. Technol.*, 37, 847–864, 2003.

Traynor, G.W., Allen, J.R., Apte, M.G., Girman, J.R., and Hollowell, C.D., Pollutant emissions from portable kerosene-fired space heaters. *Environ. Sci. Technol.* 17, 369–371, 1983.

Traynor, G.W., Apte, M.G., Carruthers, A.R., Dillworth, J.F., Grimsrud, D.T., and Thompson, W.T., Indoor air pollution and inter-room pollutant transport due to unvented kerosene-fired space heaters. *Environment International*, 13, 159–166, 1986.

Traynor, G.W., Apte, M.G., Sokol, H.A., Chuang, J.C., Tucker, W.G., and Mumford, J.L., Selected Organic Pollutant Emissions from Unvented Kerosene Space Heaters. *Environ. Sci. and Technol.*, 24(8), 1265–1270, 1990.

U.S. Environmental Protection Agency, Latest Findings on National Air Quality: 2002 Status and Trends, Office of Air Quality and Standards, Air Quality Strategies and Standards Division, Research Triangle Park, NC, EPA Publication No. EPA 454/K-03-001, 2003, 31pp. (citation to data on p. 14).

Van Alphen, M., Emission testing and inhalation exposure-based risk assessment for candles having Pb metal wick cores, *Sci. Total Environ.*, 244, 53–65, 1999.

Vette, A.F., Rea, A.W., Lawless, P.A., Rodes, C.E., Evans, G., Highsmith, V.R., Sheldon, L., Characterization of indoor-outdoor aerosol concentration relationships during the Fresno PM exposure studies, *Aerosol. Sci. Technol.*, 34, 118–126, 2001.

Wallace, L., Indoor particles: a review, *J. Air Waste Manage. Assoc.*, 46, 98–126, 1996.

Wallace, L., Mitchell, H., O'Connor, G.T., Lucas Neas, L., Lippmann, M., Kattan, M., Koenig, J., Stout, J.W., Vaughn, B.J., Wallace, D., Walter, M., Ken Adams, K., and Liu, L.-J.S., *Environ. Health Perspect.*, 111, 1265–1272, 2003.

Wasson, S.J., Guo, Z.S., McBrian, J.A., and Beach, L.O., Lead in candle emissions, *Sci. Total Environ.*, 296, 159–174, 2002.

Weschler, C.J., Ozone in indoor environments: concentration and chemistry, *Indoor Air*, 10, 269–288, 2000.

Weschler, C.J., Indoor/outdoor connections exemplified by processes that depend on an organic compound's saturation vapor pressure, *Atmos. Environ*, 37, 5455–5465, 2003.

Weschler, C.J. and Shields, H.C., Potential reactions among indoor pollutants, *Environ. Sci. Technol.*, 31, 3487–4395, 1997.

Weschler, C.J. and Shields, H.C., Indoor ozone reactions as a source of indoor particles, *Atmos. Environ.*, 33, 2301–2312, 1999.

Weschler, C.J. and Shields, H.C., Experiments probing the influence of air exchange rates on secondary organic aerosols derived from indoor chemistry, *Atmos. Environ.*, 37, 5621–5631, 2003.

Williams, R., Suggs, J., Rea, A., Leovic, K., Vette, A., Croghan, C., Sheldon, L., Rodes, C., Thornburg, J., Ejire, A., Herbst, M., and Sanders, W., Jr., The Research Triangle Park particulate matter panel study: PM mass concentration relationships, *Atmos. Environ.* 37, 5349–5363, 2003a.

Williams, R., Suggs, J., Rea, A., Sheldon, L., Rodes, C., and Thornburg, J., The Research Triangle Park particulate matter panel study: modeling ambient source contribution to personal residential PM mass concentrations, *Atmos. Environ.*, 37, 5365–5378, 2003b.

Wilson, E.B., *Introduction to Scientific Research*, McGraw-Hill, New York, 1952, p.148.

Wirts, W., Grunwald, D., Schulze, D., Uhde, E., and Salthammer, T., Time course of isocyanate emission from curing polyurethane adhesives, *Atmos. Environ.*, 37, 5467–5475, 2003.

Wolkoff, P., Clausen, P.A., Wilkins, C.K., and Nielsen, G.D., Formation of strong airway irritants in terpene/ozone mixtures, *Indoor Air*, 10, 82–91, 2000.

Xu, M., Nematollahi, M., Sextro, R.G., and Gadgil, A.J., Deposition of tobacco smoke particles in a low ventilation room, *Aerosol. Sci. and Technol.*, 20, 194–206, 1994.

Yakovleva, E., Hopke, P.K., and Wallace, L., Receptor modeling assessment of particle total exposure assessment methodology data, *Environ. Sci. Technol.*, 33, 3645–3652, 1999.

Zheng, M., Cass, G.R., Schauer, J.J., and Edgerton, E.S., Source apportionment of $PM_{2.5}$ in the southeastern United States using solvent-extractable organic compounds as tracers, *Environ. Sci. Technol.*, 36, 2361–2371, 2002.

chapter eleven

Aerosols in the industrial environment

Andrew D. Maynard and Paul A. Baron
National Institute for Occupational Safety and Health

Contents

11.1 Introduction 225
11.2 Exposure metrics 228
11.3 Size-selective sampling....229
11.4 Exposure regulations 232
11.5 Measurement technologies....232
11.5.1 Samplers....232
11.5.1.1 General aerosol samplers 233
11.5.1.2 Inhalable samplers 236
11.5.1.3 Thoracic samplers....237
11.5.1.4 Respirable samplers 241
11.5.1.5 Multifraction samplers 242
11.5.1.6 Sample analysis 246
11.5.2 Direct reading instruments 248
11.5.2.1 Personal exposure measurements 248
11.5.2.2 Light scattering instruments....248
11.5.2.2.1 Optical particle counters 249
11.5.2.2.2 Photometers 251
11.5.2.3 Tapered element oscillating microbalance 252
11.5.2.4 Condensation particle counter 253
11.5.2.5 Pressure drop sensor 253
11.5.2.6 Aerosol surface area measurement 254
11.5.2.7 Specific applications of direct-reading instruments 255
11.5.2.7.1 Sampling cassette leakage testing....255
11.5.2.7.2 Respirator testing....255
11.5.2.7.3 Sampler testing 256
11.5.2.7.4 Combined aerosol and video monitoring 257
11.6 Summary....258
Disclaimer 258
References 258

11.1 Introduction

Aerosol exposure has been associated with occupational illness since the earliest times, and remains a major source of ill health within the workplace to this day. Recognition of

1-56670-611-4 / 05 / $0.00+$1.50

the hazards airborne particles present can be traced back to the ancient Greeks and Egyptians.[1] In the 4th century BC, Hippocrates (ca. 460–370 BCE) recorded details of occupational diseases associated with aerosols, including lead poisoning.[2,3] Plinius Secundus (Pliny the Elder, 23–79) is recorded as recognizing the harmfulness of inhaling dust in the 1st century AD, noting the use of loose bladders wrapped round refiners' faces to prevent inhalation of "fatal dust."[4] However, it was not until the 15th and 16th centuries that a clear understanding began to emerge on the relationship between aerosol exposure and occupational health. Around this time, technical and economic developments in Europe led to an increased demand for gold and lead. As mines became deeper, the injuries and poor health associated with mining became more obvious. The writings of the founders of modern occupational hygiene such as Paracelsus (1493–1541), Agricola (1495–1555), and Ramazzini (1633–1714) are clearly influenced by the incidence of ill health and death associated with mining at the time. However, they also extend to many other industries. Without exception, aerosols are acknowledged by these authors as presenting a major health hazard to workers in industrial environments. Ramazzini, in particular, documents many occupations where the inhalation of "...very fine particles inimical to human beings..." is a problem,[5] including the inhalation of metal particles, gypsum, flour, stone dust, and tobacco dust.

The industrial revoution of the late 19th and early 20th centuries introduced new and greater exposures to aerosols, and increasing awareness of the associated hazards. Mining was undertaken with increased intensity — particularly for coal — and exposure to soot, metal fumes, and aerosols such as cotton dust increased markedly. Alice Hamilton carried out seminal research in the early 1900s into the health of workers in America, and readily understood the close association between aerosol exposure and ill health. Her work laid the foundation for occupational hygiene in the United State.[3] At the same time, researchers such as Tyndall, Aitken, and Rayleigh were laying the foundations for modern aerosol science that would provide the means to understand and control occupational aerosols.

An understanding of aerosol toxicity and how to measure and control exposures developed rapidly over the 20th century, and much of the research from this period defines how we now approach occupational aerosols. Although the current understanding of occupational hygiene has expanded significantly from previous centuries, aerosols are still perceived as one of the highest profile health hazards. Numerous aerosols are widely understood to be harmful to health if inhaled, including lead particles, asbestos, diesel smoke, crystalline silica, a wide range of chemicals and metals, radon progeny, bacteria, viruses, fungal spores, and endotoxins. Unlike gases or vapors, aerosols pose a particularly complex hazard, as probable dose and toxicity are associated with particle size, shape, and chemical structure, as well as composition. Toxicity depends on the dose received, and the body's response to the deposited particles. Biological response will depend in turn on the chemical and physical nature of the particles, and on the deposition region.

Deposition region is primarily governed by particle size and shape (Chapters 6–8). Response may be a function of particle size, number, surface area, mass, morphology, chemical composition, and/or surface chemistry. Dose therefore needs to be measured in terms of the most appropriate standard of measurement or metric. As an example of the importance of particle size and chemical nature, SiO_2 presents a relatively low risk when present in its amorphous form, while the crystalline form is highly toxic. However, particles of crystalline silica larger than a few micrometers in diameter present a lower health risk, as the probability of entering the lower lungs when inhaled decreases. Dose depends on actual deposition in the respiratory system, interactions with the lung fluid, and removal mechanisms. In practice, it is more convenient to measure penetration to the relevant areas of the respiratory system rather than dose. Although not ideal, this approach results in a measure of potential dose, and provides an indicator of maximum potential risk.

Many chronic respiratory diseases are classified as pneumoconiosis — a broad term from the Greek meaning " dust in lungs." Severe forms of pneumoconiosis are associated with fibrotic lung change. The current definition of the disorder refers to an accumulation of dust in the lungs, and the tissue reaction to its presence — essentially associating it with solid, relatively insoluble particles that are respirable (i.e., are capable of penetrating to the terminal bronchioles and beyond, where gas exchange takes place). Most insoluble dusts are associated with pneumoconiosis at sufficiently sustained exposure levels. The least harmful of these dusts have little biological interaction with the lungs, and are generally classified as nuisance dusts or particles not otherwise classified (PNOCs). However, a number of dusts do interact with the respiratory system to a significant degree, leading to specific forms of pneumoconiosis, including asbestosis (asbestos inhalation), silicosis (crystalline silica inhalation), siderosis (iron particle or fume inhalation), and berylliosis (from inhaling beryllium compounds).

Silicosis and asbestosis are associated with a wide range of occupations, and deserve further mention. SiO_2 is the most abundant mineral on earth, with much of it present in a free (i.e., unattached to another mineral) crystalline form. Thus, any occupation that leads to dust being formed from rock, stone, or natural mineral products such as bricks and mortar has an associated risk of crystalline silica exposure. While amorphous silica is classified as a nuisance dust, the crystalline form is highly toxic in the lungs, leading to silicosis. The disease was known to the Greeks and Egyptians,[6] and in modern times it peaked in the late 19th and early 20th centuries, coinciding with the industrial revolution. Current work practices have lowered the incidence of silicosis significantly, although the prevalence of crystalline silica still results in exposures leading to the disease. Classic silicosis results from low to moderate exposure over 20 years or more, and primarily leads to an increased disposition to mycobacterial infections and progressive massive fibrosis.[7] Accelerated silicosis results from higher exposures over 5–10 years, and progressive respiratory illness is virtually certain, even following cessation of exposure. Very high exposures over as little as a year can lead to acute silicosis, resulting in progressive respiratory illness and death in a few years following exposure.

The term asbestos covers a group of commercially recognized fibrous hydrated silicates (fibrous polymorphs of chrysotile, amosite, crocidolite, fibrous tremolite, fibrous anthophyllite, and fibrous actinolite). All these materials are asbestiform, that is, they are capable of shedding increasingly fine fibers down to the fibrils that form the mineral, allowing a single inhaled fiber to divide into many thin fibers in the lungs. Fibers may be relatively straight, as in the case of crocidolite and other amphiboles, or curved, as is found with chrysotile. Asbestos can be used to form materials that are excellent insulators against heat, cold, and noise, have good dielectric properties, great tensile strength, are flexible, and resist corrosion by alkalis and most acids. Correspondingly, the material has been used since ancient times, and is currently associated with over 3000 commercial applications.[8] Recognition of the extreme toxicity of inhaled asbestos fibers in the latter part of the 20th century has led to a great reduction in its use. Today, most potential exposures in North America and Europe arise during asbestos abatement, although it is still mined, and used in applications such as lining brakes, asbestos cement, and roofing tiles. As well as being associated with asbestosis (fibrosis of the lungs), exposure can also lead to malignant mesothelioma, and all types of lung cancer. Toxicity is associated with the shape of the fibers, leading to exposure controls based on the number and shape of particles inhaled. Other minerals that show properties similar to asbestos exist and are in use, but are not formally classified as such. These include vermiculite (a group of silicate materials that expand on heating and can be contaminated with asbestiform minerals) and zeolite (hydrated aluminum silicates that may occur in a fibrous form). Both materials are associated with lung disease, lung cancer, and mesothelioma.[9]

Man-made vitreous fibers (MMVFs) provide a fabricated nonasbestiform substitute for asbestos, particularly for thermal and acoustic insulation. While MMVFs can have similar sizes and aspect ratios as asbestos fibers, they are often relatively soluble by comparison, leading to reduced lung residence times. By nature they are not crystalline and do not shed smaller fibers. Although animal studies have shown significant toxicity for some of these fibers, significant human toxicity has yet to be established.[10]

Unlike the dusts associated with pneumoconiosis, soluble particles and droplets are relatively short-lived in the lungs, and tend to lead to material-specific reactions that are less well associated with the physical nature of the particles. These may range from pulmonary irritation to systemic toxicity. Inhalation of isocyanates, for instance, may lead to a response ranging from transient irritation to chronic sensitization and reduced lung function. Lead, on the other hand, is a systemic poison, and toxicity is associated with the transport of material from the respiratory system to specific target organs.

A number of aerosols lead to short-term flu-like symptoms following inhalation, which are usually classified under the umbrella term of "inhalation fever." Symptoms are self-limiting, but can be temporarily debilitating. Agents include endotoxins, metal and metal oxide fumes (in particular, zinc oxide fumes, leading to metal fume fever), pyrolysis products of polytetrafluoroethylene (PTFE), and moldy grain dust. Workers usually build up tolerance to the exposure, although symptoms can recur following an absence of exposure, such as a weekend or short break away from work. Inhalation fever is usually distinguished from acute lung injury, although a number of agents associated with it can lead to acute injury following sufficiently high or prolonged exposures. Exposure to cotton dust is a case in point, with prolonged and excessive exposure leading to byssinosis. Symptoms occur a few hours following exposure, and are more acute after a period of nonexposure (leading to the colloquial term *Monday Morning Asthma*). Permanent dyspnea may develop following several years of exposure.

Inhalation of fungal spores can also lead to an acute response following high temporal exposures, and presents a particular hazard when the rapid spread of fungal growth in buildings occurs.[11] Biological aerosols are also widely associated with the airborne transmission of infections. The risk of infection following inhalation is associated with the number of viable organisms entering the respiratory system, leading to methods of measuring exposure geared to identifying inhaled organisms capable of reproducing.[12,13] However, the biological material associated with the organisms may also elicit a toxic response in its own right, requiring much broader classification of inhaled bioaerosols.

In recent years, interest has been shown in a group of aerosols characterized by low-solubility particles smaller than 100 nm in diameter (often termed ultrafine particles). Laboratory-based research has shown that on a mass-for-mass basis, chemically inert materials such as TiO_2 increase in toxicity with decreasing particle size.[14,15] There are indications that the toxicity of similar low-solubility materials is associated with the surface area of the particles.[16–18]

11.2 Exposure metrics

There is wide variation in the types of aerosols found in occupational settings, and in the mechanisms by which they interact with the body to elicit a biological response. As a result, sampling an aerosol and relating concentration to the health risk posed is not as clear cut as it may be for a gas or vapor. Ideally, aerosol sampling in the workplace should enable quantification of that aspect of the aerosol that leads to specific health effects. While there are many potential exposure metrics applicable to aerosols, mass concentration has been shown to correlate well with health effects in the past[19] and is now widely accepted as the predominant metric for characterizing occupational aerosol exposure. Notable exceptions are fibrous aerosols, including asbestos and MMVFs, and viable bioaerosols, where associations between respiratory disease and exposure are closely linked to particle number and (in the case of fibers) shape. In both cases, number concentration provides a better indicator of the hazard associated with inhalation exposure.

Recent research on response to low-solubility ultrafine and high specific surface area particles has challenged the general acceptance of mass as the predominant exposure metric. It has been suggested that response may be associated with the number of deposition sites within the lungs, leading to particle number concentration being an appropriate exposure metric.[20] However, there is increasing evidence that the surface area of low-solubility particles plays a key role in triggering biological responses in the lungs and elsewhere in the body. A number of studies indicate response vs. aerosol surface area to be independent of particle size for low-solubility materials, suggesting that surface area may be a more appropriate exposure metric for such aerosols.[16,17] Using surface area rather than mass concentration to measure exposure will be particularly important for aerosols with a high surface-area-to-mass ratio (specific surface area) if these results are found to be applicable to occupational aerosols. Examples within this category include aerosols resulting from combustion (such as diesel exhaust particulates), metal and metal oxide particles from hot processes such as smelting and welding, and particles formed during fine-powder production. Some fine powders, including ultrafine titanium dioxide, carbon blacks, and fumed silicas, have specific surface areas in excess of 2×10^5 m^2/kg. The use of ultrafine aerosols in the rapidly developing field of nanotechnology is also likely to pose new exposure problems requiring the consideration of alternative exposure metrics such as surface area.[21,22]

11.3 Size-selective sampling

Measuring aerosol exposure against an appropriate metric provides information on the hazard posed by an aerosol, but to estimate the risk associated with exposure, particle-size-dependent penetration within the respiratory system must be accounted for (Chapter 7). For instance, an aerosol of 50 μm diameter quartz particles may pose a significant health hazard on a mass concentration basis, but the low probability of 50 μm particles penetrating to the alveolar region of the lungs would lead to a low risk of silicosis following repeated exposure. Occupational exposures are conventionally expressed as the fraction of an aerosol capable of penetrating to a specific region of the respiratory system, thus giving a measure of the maximum potential risk the aerosol poses.[23] Weighting aerosol samples by penetration probability as a function of particle size is possible if measurements have sufficient size resolution, as may be the case with cascade impactor measurements. A simpler, more direct method is to use a sampling device such as a cyclone with an engineered bias toward specific particle sizes. If the particle size selectivity of an aerosol sampler matches the penetration probability of a given region of the respiratory system, the resulting samples will provide a good estimate of the health risk presented by an aerosol.

Aerosol penetration to specific regions of the respiratory tract forms a basis for standards against which size-selective occupational aerosol samplers can be designed and tested. Early estimates of penetration into the alveolar region — perhaps the most vulnerable part of the respiratory system — were proposed in the 1950s and 1960s, resulting in the British Medical Research Council (BMRC) and the American Conference of Governmental Industrial Hygienists (ACGIH) conventions describing respirable aerosols [24,25] (Figure 11.1). Over the following 30 years, a number of standards were proposed and used for size-selective sampling relevant to the upper, mid, and lower

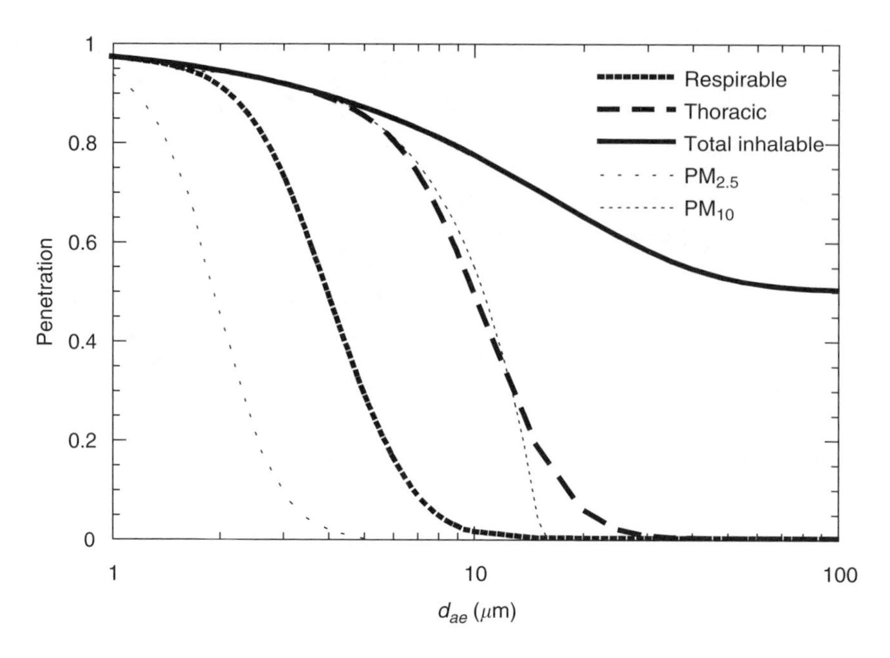

Figure 11.1 Comparison of different respirable sampling conventions.

airways, resulting in somewhat divergent standards in some cases. In the early 1990s, international consensus was reached on particle penetration standards between the International Standards Organization (ISO), ACGIH, and the European Committee for Standardization (CEN). The resulting conventions describe penetration as a function of particle aerodynamic diameter into the entire respiratory system (inhalable aerosol), into the tracheobronchial region (thoracic aerosol), and into the alveolar region (respirable aerosol), with thoracic and respirable aerosol defined as subfractions of the inhalable aerosol. These are now widely used as the standards to which industrial hygiene aerosol samplers should conform.[26] A review of the basis for these conventions was developed by the ACGIH.[27]

The inhalable convention is based on particle penetration through the mouth and nose of a breathing manikin over a range of wind speeds and orientations with respect to the wind, and is defined as

$$SI(d_{ae}) = 0.5 \times (1 + e^{-0.06 d_{ae}}) \tag{11.1}$$

for $0 < d_{ae} < 100$ μm. $SI(d_{ae})$ is the fraction of particle entering the system as a function of aerodynamic diameter d_{ae} (Figure 11.2).

Both the thoracic and respirable conventions are expressed as subfractions of the inhalable convention, and are based on lung penetration modeling and measurements (see Chapters 6 and 8). The thoracic convention is given as

$$ST(d_{ae}) = SI(d_{ae}) \times (1 - F(x))$$
$$x = \frac{\ln(d_{ae}/\Gamma)}{\ln(\Sigma)} \tag{11.2}$$

$ST(d_{ae})$ is the fraction of particles penetrating beyond the larynx as a function of aerodynamic diameter. $F(x)$ is a cumulative log-normal distribution, with a median aerodynamic diameter Γ of 11.64 μm and a geometric standard deviation Σ of 1.5 (Figure 11.2).

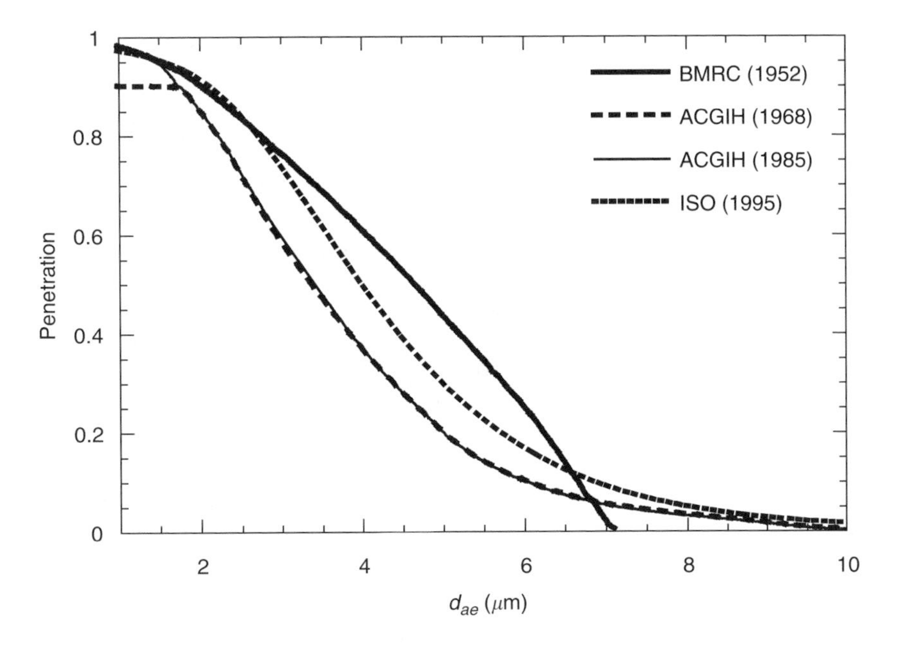

Figure 11.2 Occupational aerosol sampling conventions.[26] The PM_{10} and $PM_{2.5}$ ambient sampling size-selection curves are shown for comparison.

The respirable convention $SR(d_{ae})$ is similarly given as

$$SR(d_{ae}) = SI(d_{ae}) \times (1 - F(x))$$
$$x = \frac{\ln(d_{ae}/\Gamma)}{\ln(\Sigma)} \tag{11.3}$$

where the cumulative log-normal distribution has a median aerodynamic diameter Γ of 4.25 μm and a geometric standard deviation Σ of 1.5 (Figure 11.2). A respirable convention for susceptible groups is also defined, with Γ = 2.5 μm. Standards relating to penetration to the tracheobronchial and extrathoracic regions are defined by the difference between the respirable and thoracic conventions (tracheobronchial), and the thoracic and inhalable conventions (extrathoracic), respectively.

Of the three primary sampling conventions, the inhalable convention is least well supported by experimental data. The current convention is based on data collected at wind speeds between 0.5 and 4 m/s, with sampler performance averaged over all orientations to the prevailing wind.[28] Measurements have indicated that air movement in many workplaces lies below the range used to establish the inhalable convention.[29] Determination of aerosol inhalability at low air velocities has indicated that the current convention may not be a good indicator of particle penetration into the respiratory system in all cases.[30,31] In the study by Aitken et al.,[31] inhalability under low wind speeds was consistently found to be greater than the inhalable convention, and was dependent on breathing rate. The convention is also not defined above 100 μm, creating ambiguity over how larger particles should be treated. Aitken and Donaldson[32] have shown that very large particles may enter the nose or mouth as projectiles, and should be considered as being inhalable if airborne (although it has been suggested that few particles larger than 100 μm in diameter will reach the mouth[33]). Above 100 μm, inhalability reduces when averaged over all orientations, reaching zero around 300 μm.[32,34]

The thoracic and respirable samplers are influenced far less by external conditions, and are generally accepted as suitable approximations of aerosol penetration within the general population. The thoracic convention is very similar to the PM_{10} convention used for environmental sampling.[35]

11.4 Exposure regulations

Health-based aerosol exposure limits follow country-specific systems, but in the majority of cases they follow a similar philosophy.[36] In the United States, the Occupational Safety and Health, and Mines Safety and Health Administrations (OSHA and MSHA) enforce permissible exposure limits (PELs). PELs are based on health effects data, but economic and technological feasibility factors are also taken into consideration. The National Institute for Occupational Safety and Health (NIOSH) recommended exposure limits (RELs) are also used, as are the voluntary threshold limit values (TLVs®) published by the ACGIH. RELs and TLVs are primarily based on health effects data. NIOSH RELs are time-weighted average (TWA) concentrations for up to a 10 h workday during a 40 h workweek, while OSHA PELs are TWA concentrations that must not be exceeded during any 8 h work shift of a 40 h workweek. The ACGIH TLVs are 8 h TWA concentrations for a normal 8 h workday and a 40 h workweek, to which nearly all workers may be exposed continuously during their working lifetime, without adverse effects. In the U.K., a two-tier system of occupational exposure standards (OES) and maximum exposure limits (MELs) is employed.[37] Each represents an 8 h TWA exposure limit. An OES is set where a no-effect level can be identified for a substance, thus giving an exposure limit below which adverse effects are not expected (as for the ACGIH TLVs). MELs are employed where there is no clear no-effect level. As some health effects are manifested at whatever exposure limit is chosen (above zero), the choice of a limit is in essence a political decision. Reflecting the nature of substances having MELs, there is an obligation on U.K. industries to keep exposures as low as reasonably practicable, even when this results in a target exposure significantly below the limit. Similarly, the ACGIH defines a threshold limit value-ceiling (TLV-C) for some substances, which should not be exceeded in any part of the overall exposure. Other countries use exposure evaluation systems and limits similar to those used in the U.S. and U.K.

For substances that may potentially lead to health effects following short exposures or high peak exposures, short-term exposure limits (STELs) are set to complement the 8–10 h TWA limits. Samples are taken over shorter time periods — typically 15 min — and are collected during periods when the concentration of contaminant is likely to be highest.

Aerosol exposure in the workplace can be reduced by controlling emissions, or by using personal protective equipment (PPE). Controlling emissions is generally the preferred approach, either by removing the source, applying control measures such as containment or local exhaust ventilation (LEV), or substituting materials with less harmful ones. For instance, NIOSH recommends that respiratory protection be used only when engineering controls are not technically feasible, such as during the installation or repair of engineering controls, or when an emergency or other temporary situations arise.[38] Respirators are the least preferred method against worker protection to air contaminants because of the difficulties generally encountered during implementation of an effective respiratory protection program. Reliable protection depends on the cooperation of the workers to adhere to critical program guidelines.

11.5 Measurement technologies

11.5.1 Samplers

Aerosol samplers are generally defined as devices that collect aerosol particles for subsequent analysis. Historically, there have been a great number of different methods used to sample aerosols in the workplace,[39] and this is reflected in the broad range of devices currently in use. However, with a few exceptions most devices in use today fall into one of four main categories: size-selective samplers, size-differentiating samplers such as cascade impactors, general industrial aerosol samplers (with no clearly defined size selectivity), and biological aerosol samplers. Within these categories, devices may require placement at a static location (static or area samplers) or be worn on the person (personal samplers). Current thinking tends to favor personal samplers as providing a more representative estimate of aerosol exposure, although there are specific instances where static samplers are used. Personal sampler placement is recommended within the breathing zone, defined as a region within 30 cm from the mouth and nose (Chapter 4).[40] Most samplers are positioned on the left or right side of the chest, and occasionally in the center of the chest. Placing samplers to the side and above the head (for instance, on a helmet) has also been proposed for welding and mining.[41,42] OSHA states that for welding, the air sampler should be placed inside the helmet if either no respirator is used or a negative pressure respirator is used.[43]

11.5.1.1 General aerosol samplers

One of the early aerosol samplers in common use was the impinger. This was originally developed as an area sampler,[44] but miniaturized so that it could be worn by workers and operated at a flow rate of 2.8 l/min (Figure 11.3).[45] Air was pulled through a nozzle that was immersed in a liquid, usually water, and was aimed at the bottom surface of a glass chamber. Particles entering the sampler were impacted onto the bottom surface and washed into the liquid. The sampler had a lower particle size cutoff of about 1 μm so fumes and vapors may not have been efficiently collected.[46] The particles in the final liquid suspension were analyzed in a special cell under a light microscope at X100 magnification. At this magnification, only the larger dust particles could be counted and were reported in millions of particles per cubic foot. This device was widely used for various dusts, including coal mine dust and asbestos, until the development of membrane filter methods. It is currently used primarily for sampling bioaerosols and chemically unstable particles, for example, isocyanates. Impingers have been widely evaluated to determine their efficiency and internal losses for bioaerosol sampling,[47–49] and a tangential impinger has been developed specifically to increase the likelihood of bacterial viability (Figure 11.3, Biosampler, SKC Inc.[50]).

Historically, little attempt was made to carry out size-selective sampling other than for respirable aerosol, resulting in many samplers with poor or poorly defined size selectivity

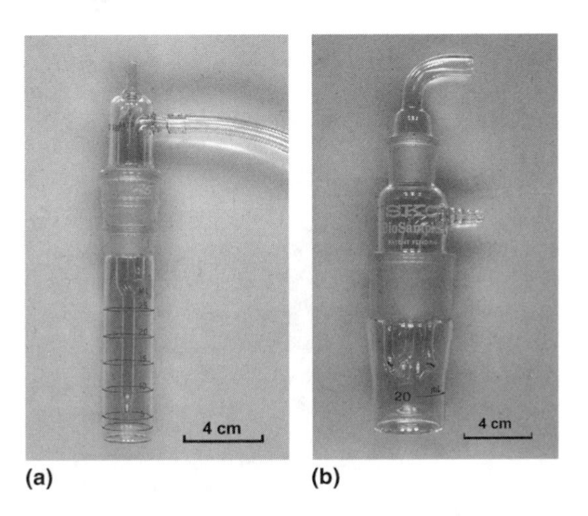

Figure 11.3 Glass impingers: (**a**) midget impinger and (**b**) biosampler (SKC Inc., U.S.A.).

being used. Generic filter holders are still widely used, both as "open-faced" samplers (filter face not enclosed) and "closed-faced" samplers (filter protected by a cover through which the aerosol is sampled) (Figure 11.4). The 37 mm filter holder is widely used in the closed form in the United States, and in the open form across Europe. Although the size selectivity of the sampler does not match any sampling convention,[51] its low cost and simplicity has led to its continued widespread use. These cassettes are also commonly used as the sample collection device on size-selective aerosol samplers. Both conductive and nonconductive filter cassettes are available, with the nonconductive variety being prone to electrostatic losses as collected material adheres to the cassette rather than being collected on the filter. Demange et al.[52] reported an average of 30% losses to the walls of nonconducting 37 mm cassettes. The push-fit filter cassettes are also prone to leakage unless assembled properly.[53,54]

The 25 mm cowled sampler (Figure 11.5) is a specialized general sampler used widely for asbestos and fibrous aerosol sampling. Sampling efficiency in calm air is not strongly dependent on particle size for aerodynamic diameters below 20 μm.[55,56] Uneven deposition within the sampler has been seen as a function of orientation[57] (the sampler is usually used facing down). Fiber samples are nominally taken at 2 l/min, although it is common for the sampler to be used at different flow rates to ensure optimal loading of the filter.[58]

Most aerosol samplers are used on the assumption that the collected aerosol will remain stable between collection and analysis. This is clearly not the case when sampling volatile particles, where deposited particles may evaporate during and following the sampling process. The conventional method of dealing with this is to place a sorption tube

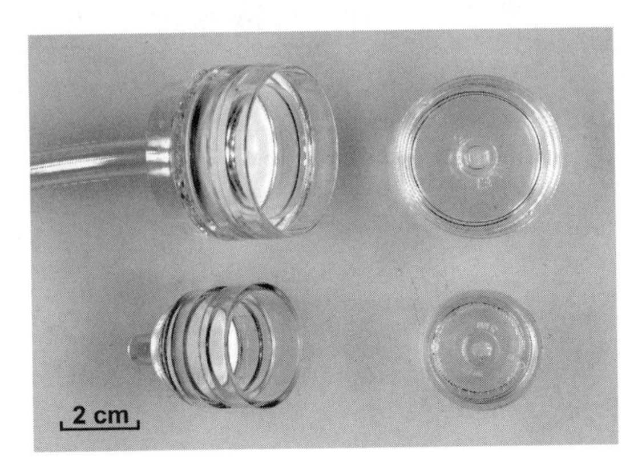

Figure 11.4 Filter holder-based personal aerosol samplers.

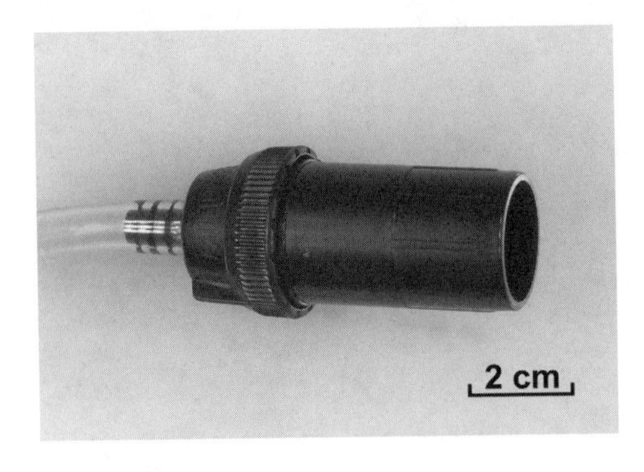

Figure 11.5 25 mm cowled sampler.

behind the sample filter, allowing total aerosol and vapor concentration of a particular substance to be measured. However, using this approach it is not possible to directly infer partitioning between the condensed and vapor phase in the workplace atmosphere. One solution proposed to collect volatile aerosols while minimizing evaporative losses has been to use a coaxial electrostatic precipitator[59–61] (Figure 11.6). The sampler consists of a central positive electrode and outer casing that is grounded. Particles entering the sampler are first charged, and then deposited on the outer electrode. Airflow close to the wall surface is low, reducing the rate of evaporation from deposited particles. After sampling, the sampler is sealed at both ends to prevent vapor release, and the component of interest is removed by washing out with a suitable solvent. The sampler has proved to be useful for measuring exposure to JP-8 aviation fuel, particularly in cold environments where the use of a conventional sampler can lead to substantial aerosol evaporation from the collection media.[60]

Most aerosol samplers rely on the use of a pump to move air through the sampling device. Several samplers that do not require a pump have been developed for personal sampling. These passive samplers are particularly attractive in that they can be made much smaller and lighter than conventional units. However, their size selectivity depends on local air movement and generally does not conform to any specific sampling convention.[62–66] Table 11.1 lists a selection of nonspecific personal aerosol samplers.

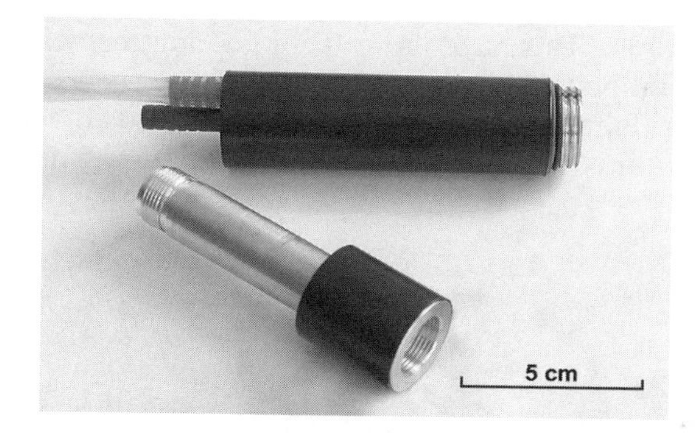

Figure 11.6 Coaxial electrostatic precipitator, designed for collecting volatile aerosols.

Table 11.1 A Selection of Nonspecific Personal Aerosol Samplers (the List is not Inclusive)

Sampler	Flow Rate (l/min)	Notes	References
37 mm cassette (open)	2	Standard filter cassette, worn facing down at 45° to the body. Conducting versions available. Figure 11.4.	51
37 mm cassette (closed)	2	Standard filter cassette with a cap containing a 2 mm diameter inlet. Figure 11.4.	51
Passive sampler	—	Electret-based sampler relying on aerosol charge and naturally occurring air movements. Correlation is good with some size-selective samplers.	65,153
Coaxial electrostatic precipitator		Designed to sample volatile aerosols. Figure 11.6.	60
Cowled sampler	2 (typical)	Used in the main for fiber sampling. Size selectivity not quantified. Figure 11.5	55,56

References provide information on sampler performance.

11.5.1.2 Inhalable samplers

Measurements of the aspiration efficiency of breathing mannikin established the ideas of inhalability and inhalable aerosol in the 1970s and 1980s.[67–69] As measurements were performed, efforts were made to develop a personal sampler that matched the mannikin's aspiration. The result was a sampler developed at the UK Institute of Occupational Medicine and referred to as the IOM inhalable sampler[70] (SKC Inc.; Figure 11.7). Air is sampled through a 15 mm diameter inlet at 2 l/min. The filter is held within a cartridge, and the whole filter cartridge assembly is weighed, allowing all particles entering the sampling inlet to be measured. Cartridges of either conductive plastic or stainless steel are available, with the latter being more weight stable.[71] The IOM inhalable sampler generally agrees with the inhalable convention at wind speeds between 0.5 and 2.6 m/s.[51,70] However, at low wind speeds there is a marked divergence, with the sampler oversampling.[72] The open inlet is susceptible to large particles entering through their own inertia (projectiles), and is some cause for concern when sampling near sources of such particles.[32]

Preventing projectiles from entering the IOM inhalable sampler has been addressed by considering placing screens in front of the inlet.[32] A similar approach has been used in the Button sampler (Figure 11.8; SKC Inc., U.S.A.[73]). This sampler consists of a 25 mm filter holder with a hemispherical perforated screen covering the sampling inlet. The screen serves a fourfold purpose — reduction of aspiration dependency on wind speed, formation of a uniform deposit on the filter, reduction of internal deposits, and exclusion of large projectiles from the sample. Tests have indicated a good agreement between the sampler and the inhalable convention.[74]

The CIP-10 sampler (Figure 11.9, ARELCO, France) avoids the aspiration of large particles by using a convoluted inlet design, which also serves as an inhalable fraction

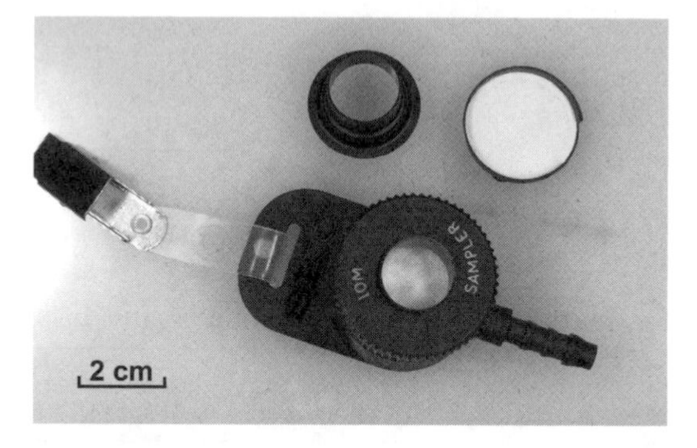

Figure 11.7 IOM personal inhalable sampler (SKC Inc., U.S.A.).

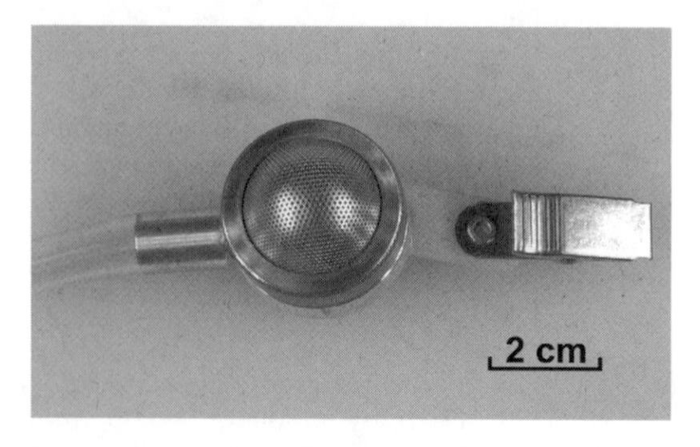

Figure 11.8 Button sampler, for inhalable aerosol.

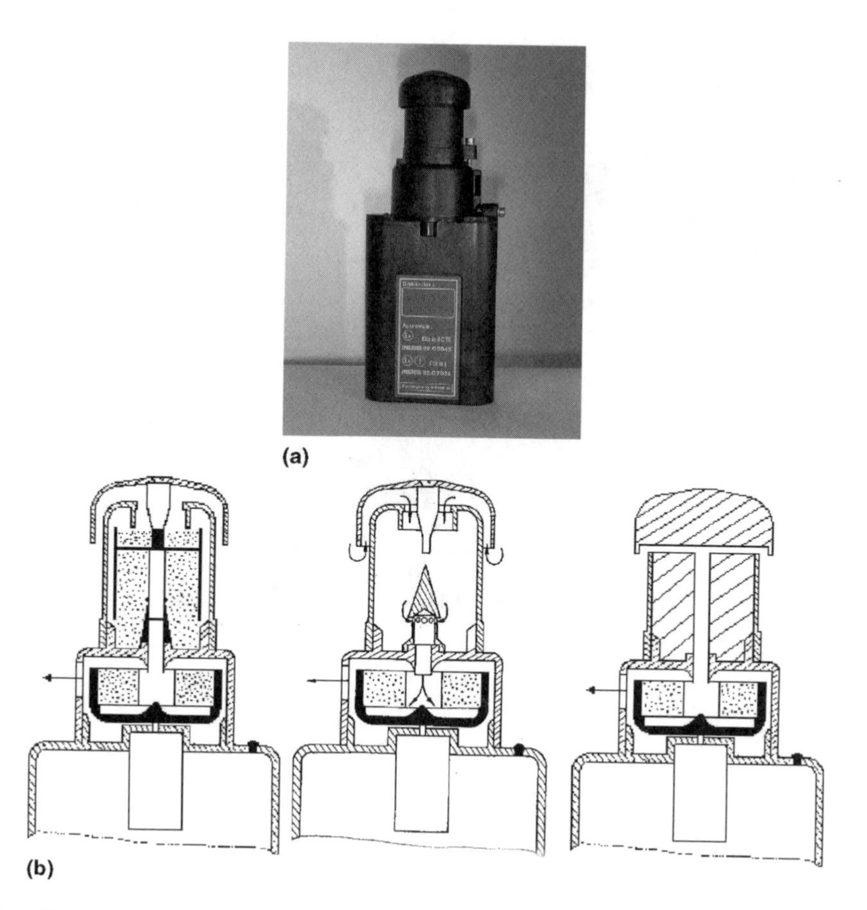

Figure 11.9 CIP-10 sampler (ARELCO, France), showing schematics of the three size-selective sampling heads: (**a**) respirable selector, (**b**) thoracic selector, and (**c**) inhalable selector.

preseparator. Air movement through the sampler is driven by a rapidly rotating foam ring held in a cup, acting as a centrifugal pump. With this arrangement, sampling rates up to 10 l/min are achievable. Sampling can only be achieved with a low pressure drop across the device, preventing the use of conventional filters to collect the sampled aerosol. Instead, the rotating foam is used to collect the particles. Reasonable agreement is seen between the sampler and the inhalable convention,[51,75,76] although submicrometer particles penetrate the foam and are not collected efficiently.

Various other inhalable samplers are in use, including the GSP inhalable sampler, the conical inhalable sampler (CIS, based on the GSP, JS Holdings, U.K.), and the PAS-6 inhalable sampler (Figure 11.10). These three samplers all use a conical inlet, and appear not to be as susceptible to external wind speed as other devices.[51] Table 11.2 lists a selection of inhalable personal samplers.

11.5.1.3 Thoracic samplers

Very few exposure limits are referenced to the thoracic fraction, and thus there are correspondingly few thoracic samplers available. One of the first to find widespread use was the vertical elutriator used to sample cotton dust. This is a static sampler operated at 7.4 l/min, and does not show good agreement with the thoracic sampling convention.[77]

The CIP-10 personal sampler was designed as a modular system, capable of being configured to match a range of sampling conventions.[75] A modified sampling head is

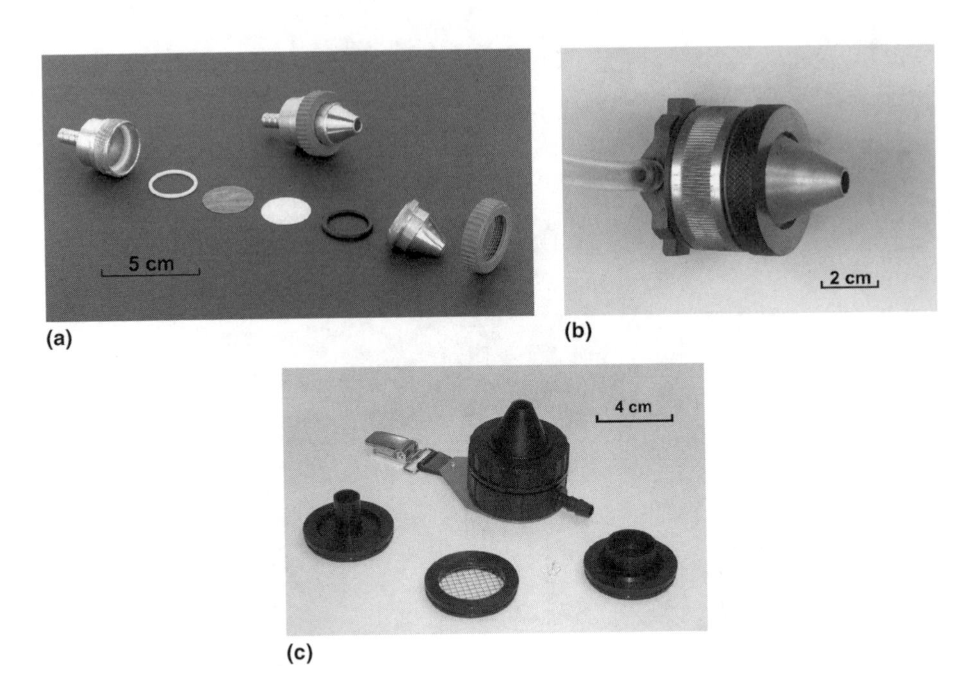

Figure 11.10 Conical inlet inhalable samplers: (a) PAS-6, (**b**) GSP sampler, and (**c**) conical inhalable sampler (CIS, JS Holdings, U.K.).

Table 11.2 A Selection of Inhalable Personal Aerosol Samplers

Sampler	Flow Rate (l/min)	Notes	References
IOM inhalable	2	Uses filter cassette. Susceptible to large projectiles. Wind speed dependent. Figure 11.7.	51,70,72
CIP-10I	10	Rotating porous foam acts as an air mover, and collection medium. Figure 10.9.	51,76,92
GSP inhalable	3.5	Conical inlet sampler. Figure 11.10b.	51
Conical inhalable	3.5	Based on the GSP sampler. Figure 11.10c.	51
Seven hole Sampler	2	Also known as the multiorifice, or UKAEA sampler.	51
Single hole	2	Used for lead aerosol sampling in the U.K.	51
PAS-6		Conical inlet sampler. Figure 11.10a.	51
Button sampler	4	Perforated inlet reduces wind speed dependence and intersampler variability, and leads to a uniform filter deposit. Figure 11.8.	73, 154

References provide information on sampler performance.

available for the device, allowing it to be used as a thoracic sampler at 7 l/min (in this configuration it is referred to as the CIP-10T).[78] A static version of the sampler — the CATHIA sampler — allows sampling onto a 25 mm filter with an external pump. The sampler shows reasonable agreement with the thoracic convention,[78] but a tendency to oversample in calm air.[55] The GK2.69 cyclone (Figure 11.11, BGI Inc., U.S.A.) is an alternative personal sampler designed to follow the thoracic convention at 1.6 l/min. It has been used for measuring metal working fluids.[79] It also doubles as a respirable sampler when operated at 4.2 l/min. Samples are collected on a 37 mm filter, and the sampling efficiency is close to the thoracic convention under calm air conditions.[55]

Figure 11.11 GK 2.69 cyclone (BGI Inc., U.S.A.). The sampler may be used for either respirable or thoracic aerosol sampling.

Research on aerosol separation in porous polyurethane foams (PUF) has indicated that penetration is predictable based on foam porosity and face velocity,[80,81] and PUF has subsequently been used in several personal aerosol samplers. The IOM thoracic sampler uses a 24 mm length of 30 pores per inch (ppi) PUF behind an inhalable inlet, collecting the penetrating aerosol onto a 37 mm filter. When operated at 2 l/min, the sampling efficiency is close to the thoracic convention, but the sampler tends to over-sample in calm air conditions.[55] The same idea has been used to modify the IOM inhalable sampler by placing an appropriate foam plug in the inlet.[82] Measurements using a 17.5 mm diameter, 10 mm deep, 45 ppi foam plug indicated that the sampler followed the thoracic convention reasonably well, although the sharpness of the penetration function led to a large predicted sampler bias for aerosols with mass median aerodynamic diameters much larger and smaller than 10 μm.[55] PUF inserts are available for the IOM inhalable sampler, and the GSP and CIS samplers, allowing them to be used as thoracic or respirable sampling devices.

Thoracic samplers can potentially be used to sample fibrous aerosols, allowing fibers to be sampled while preventing large compact particles and clumps of fibers from reaching the collection substrate.[83] It is thought that such an approach may increase the precision of fiber concentration measurements where large numbers of clumps and compact particles are present.[84–86] Maynard[87] has demonstrated that a number of currently available thoracic samplers may be suitable for fiber sampling, although deposition inhomogeneity on the collection substrate when using cyclones may contribute to measurement errors. Jones et al.[88] have shown that using thoracic samplers to sample for asbestos leads to results comparable with conventional techniques, although in this study there was little evidence for an increase in precision when using size-selective sampling.

PM_{10} samplers used for ambient aerosol sampling have similar size selection characteristics to thoracic samplers, and may be used in their stead with little reduction in sampling accuracy.[35] Versions of the personal environmental monitor (PEM) model 200 (MSP Corp., U.S.A.) allow personal PM_{10} sampling at either 4 l or 10 l/min (Figure 11.12). Table 11.3 lists a selection of personal thoracic samplers.

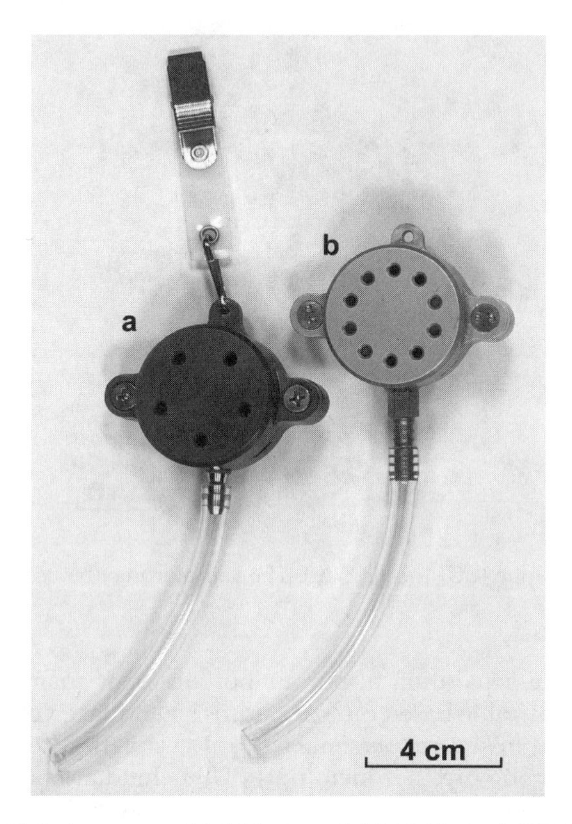

Figure 11.12 Personal PM_{10} impactors: (**a**) 4 l/min and (**b**) 10 l/min (MSP Corp., U.S.A.).

Table 11.3 A Selection of Thoracic Aerosol Samplers

Sampler	Flow Rate (l/min)	Notes	References
Elutriator	7.4	Static sampler. Specific to cotton dust.	77
CIP-10T	7	CIP-10I with a thoracic separation stage. Figure 11.9.	78
CATHIA	7	Static version of the CIP-10T.	78
IOM thoracic	2	Separation based on PUF	55
GK 2.69 cyclone	1.6	Can also be used as a respirable sampler (Table 11.4). Figure 11.11.	55
PEM Model 200	4, 10	PM_{10} personal sampler. Figure 11.12.	35
IOM inhalable + thoracic foam	2	IOM inhalable sampler with a size-selective PUF insert.	55

All samplers are personal except for the elutriator and the CATHIA. References provide information on sampler performance.

11.5.1.4 Respirable samplers

It has long been recognized that particles capable of reaching the alveolar region of the lungs are potentially more harmful than those depositing in the upper airways, and as a result many respirable aerosol samplers have been developed and used. One of the earliest was the U.K. Mines Research Establishment (MRE) MRE 113A horizontal elutriator, developed to monitor the concentration of respirable aerosol in mines (Figure 11.13). This sampler is still used as the main sampling method in U.K. mines and is the reference sampler in U.S. coal mines. As the BMRC respirable convention was developed from elutriation theory, this sampler shows close agreement with the old convention, but only fair agreement with the current international convention (see Figure 11.1). This device is only used as an area sampler as it is relatively large.

The SIMPEDS or Higgins and Dewell (U.K.) and Dorr-Oliver (U.S.A.) cyclones are both personal respirable samplers that have a long and continued history of use (Figure 11.14). Both were designed to follow older sampling conventions. However, both show good agreement with the current international respirable convention when operated at 2.2 and 1.7 l/min, respectively.[89,90] The Dorr-Oliver cyclone is constructed from nonconducting nylon, raising the possibility of electrostatic losses within the sampler. In addition,

Figure 11.13 MRE 113A respirable sampler.

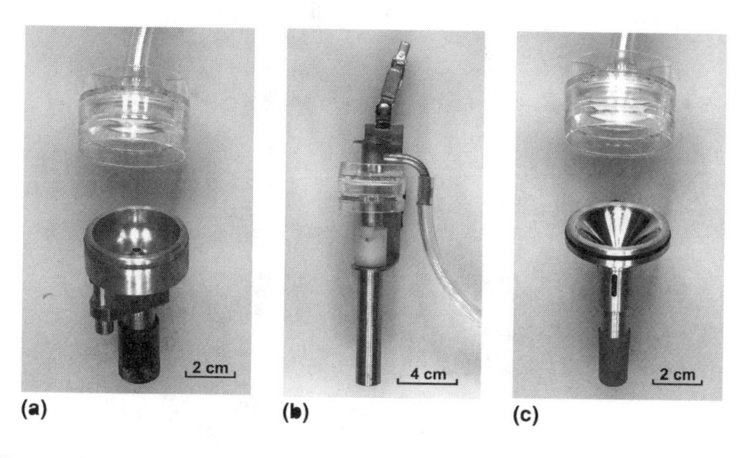

Figure 11.14 Respirable cyclone samplers: (**a**) Higgins and Dewell or SIMPEDS cyclone, (**b**) Dorr-Oliver (nylon) cyclone, and (**c**) SKC aluminum cyclone (SKC Inc., USA).

because the primary application for this cyclone does not require it, the manufacturer has little interest in high dimensional accuracy and so slight changes in cyclone dimension occasionally occur.

Cyclones lend themselves well to personal sampler design, and recent years have seen a number of improved samplers emerge. A novel extension of the cyclone principle is the virtual cyclone[91] (Figure 11.15, Omega Specialty, U.S.A.). Sampled air is passed round a 90° bend. Large particles are inertially separated from the flow, and collected in a large collection chamber. Respirable particles closely follow the airflow through the device, and are collected on a filter at the outlet. As the device does not rely on separated particles depositing on internal surfaces, it is potentially less prone to overload than other comparable samplers.

PUF preseparators are also used in respirable samplers. The CIP-10 sampler is available in a respirable form (as the CIP-10R) with a PUF preseparator in the sampling head. The sampling characteristics of the device are good,[92] although there is a possibility of nonrespirable particles migrating to the sample media under heavy agitation. It is also possible to convert the IOM inhalable sampler into a respirable sampler by inserting an appropriate PUF disk into the inlet.[93] Chen et al.[94] developed a sampler using two sections of PUF having different porosities in parallel to match the respirable curve more exactly (Figure 11.16). Table 11.4 lists a selection of personal respirable samplers.

11.5.1.5 *Multifraction samplers*

A small number of samplers are available that allow all three size fractions to be collected simultaneously. The PERSPEC[95,96] is a personal sampler designed to deposit different aerosol fractions in different regions on a 47 mm filter. Aerosol sampled at 2 l/min is winnowed in a highly divergent clean sheath flow, leading to deposition position on the filter being particle size dependent. When the filter is cut using an appropriate tool, the weight of material collected in different regions enables estimates to be made of the selected aerosol fractions.[97]

Sequential PUF sections of differing porosity and length are used in the IOM personal multifraction sampler to separate out the three different fractions.[80] The device uses an inhalable inlet, and thus the mass of all particles collected in the PUF sections and on the

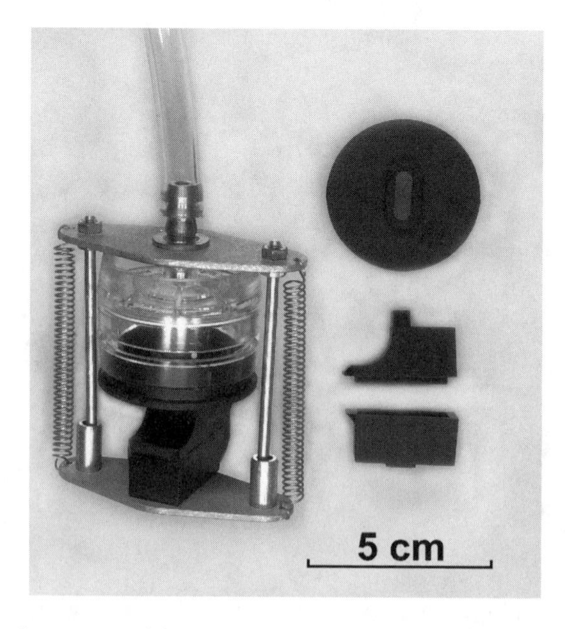

Figure 11.15 Virtual cyclone respirable aerosol sampler (Omega Specialty, U.S.A.).

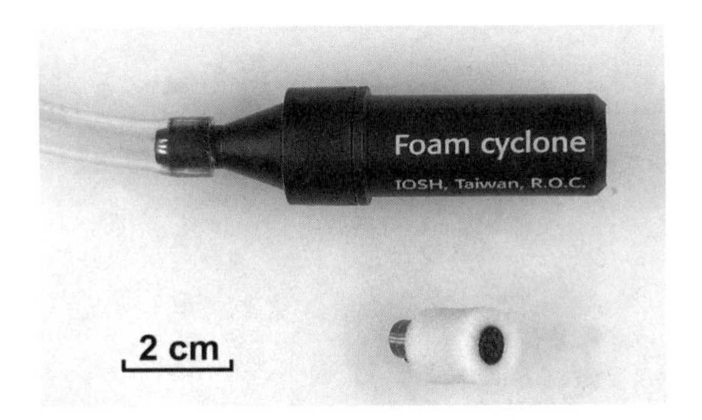

Figure 11.16 Foam cyclone respirable sampler (Omega Specialty, U.S.A.).

Table 11.4 A Selection of Respirable Personal Aerosol Samplers.

Sampler	Flow Rate (l/min)	Notes	References
CIP-10R	10	CIP-10I with a respirable separation stage. Figure 11.9.	92
SIMPEDS cyclone	2.2	Also known as the Higgins and Dewell (HD) cyclone. Figure 11.14.	89,90,155
SKC cyclone	1.9 – 2.75	Sampling flow rate depends on cyclone type and respirable convention used. Figure 11.14.	154
GK 2.69 cyclone	4.2	Can also be used as a thoracic sampler (Table 11.3). Figure 11.11.	55
Dorr-Oliver (10 mm) cyclone	1.7	Sampler constructed from nonconducting nylon. Figure 11.14.	89
MRE 113A (gravimetric dust sampler)	2.5	Static sampler. Use limited to sampling in mines. Figure 11.13.	156
IOM inhalable + respirable foam	2	IOM inhalable sampler with a size-selective PUF insert.	157
Foam cyclone	2	Cowled sampler with size-selective PUF Figure 11.16.	94
Virtual cyclone	3.3	Provides a good match with the respirable convention slope. Figure 11.15.	91

All samplers are personal except for the MRE113A. References provide information on sampler performance.

backing filter gives the inhalable fraction. The PUF sections are designed so that aerosol reaching the backing filter represents the respirable fraction, while that in the adjacent PUF section and filter combined gives the thoracic fraction. The same idea has been proposed for PUF inserts for the IOM personal inhalable sampler,[72] although it is not clear how the shallowness of the PUF disks will necessarily affect overload and measurement accuracy within the sampler.

The RESPICON sampler (Figure 11.17, TSI Inc., U.S.A.) presents a third approach to multifraction sampling.[98] Aerosol is passed through two virtual impactors in series. Particles of sufficient inertia are passed through to the next stage, while those that follow the major flow are collected on a filter. Thus, unlike a conventional cascade impactor, large particles collect on the final filter and small particles on the first stage of the impactor. All particles entering the device represent the inhalable fraction. Particles collecting on the first-stage filter represent the respirable fraction, while those on the first- and second-stage

Figure 11.17 Respicon® three-fraction aerosol sampler (TSI Inc., U.S.A.).

filters represent the thoracic fraction. About 10% of the smaller particles are deposited on each of the last two stages, and the results obtained have to be adjusted accordingly.

Cascade impactors provide somewhat more detailed information on particle size distribution (Chapter 3), but may be used to derive exposure to the three aerosol fractions. These are generally capable of giving the size distribution of an aerosol between around 0.1 and 15 μm aerodynamic diameter and above. Static cascade impactors such as the Anderson eight-stage impactor and the multiorifice uniform deposit impactor (MOUDI; MSP Corp., St. Paul, MN) have found relatively widespread use in the workplace. The Anderson impactor consists of eight multiorifice stages with cut points between 0.4 and 10 μm when operated at 28.3 l/min. Collection is usually onto aluminum foils, although other substrates are available. The use of multiorifices in the Anderson impactor allows deposits to be distributed with relative evenness onto substrates. This is taken further within the MOUDI, where many orifices per stage, together with rotating substrates, lead to highly uniform deposits. The MOUDI is available in an 8-stage or 10-stage version, and is capable of making aerosol size distribution measurements down to 0.056 μm at 30 l/min. An extension of the MOUDI — the nanoMOUDI — has recently become available from the manufacturer that adds stage cut points of 10, 18, and 32 nm.

Aerosol size distributions within the breathing zone are generally of greater relevance to health than static samples, and three cascade impactors have been developed to enable personal aerosol size distribution measurements to be made. The Marple personal cascade impactor (Anderson, U.S.A., Figure 11.18)[99] is configurable with up to eight stages, and will provide information on particle size distribution down to 0.5 μm at a flow rate of 2 l/min. The personal inhalable dust spectrometer (PIDS) is similar in concept to the Marple impactor, although the slot-shaped impactor jets of the Marple device are replaced by circular jets.[100] Cut points in the eight stages of the PIDS range from 0.9 to 19 μm at 2 l/min. Another personal cascade impactor designed for "home" use is the personal cascade impactor sampler (PCIS).[101] This device has cut points between 0.25 and 2.5 μm and operates at 10 l/min, making it particularly well suited to sampling low aerosol concentrations, and is commercially available as the Sioutas cascade impactor (SKC Inc., U.S.A.).

Cascade impactors are of limited use for measuring aerosol size distributions up to the limit of the inhalable convention (100 μm aerodynamic diameter), because the cut point of

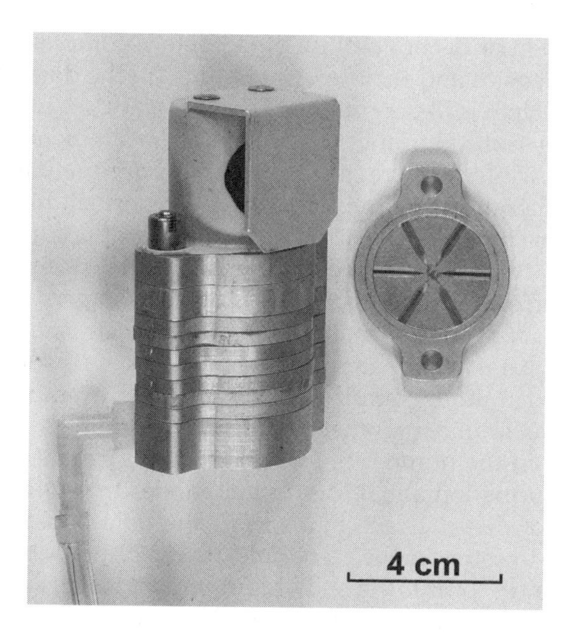

Figure 11.18 Marple personal cascade impactor (Anderson, U.S.A.). An impaction stage is shown to the right; each stage collects particles from the previous stage nozzle and has slot nozzles for the subsequent stage.

the upper stage of most of these devices is relatively low. Extrapolation of measured size distributions above this cut point is dependent on assumptions about the sampled aerosol and the aspiration efficiency of the device, and is generally not reliable. However, the PIDS was designed with an inlet designed to follow the inhalable convention.[100] It may be assumed that summing all deposits within the PIDS impactor gives a measure of the inhalable aerosol mass, and subsequent analysis of the deposits gives the size distribution as a function of inhalable aerosol. Such an approach is advantageous to industrial hygiene measurements, where ultimately measurements need to be related to the mass of particles inhaled. Vincent and co-lookers[102,103] have approached this by inserting a PUF plug with known separation characteristics into the inlet of an Anderson cascade impactor.

Analysis of cascade impactor data is a complex issue, to which there may not be a completely satisfactory solution. There is a long tradition in industrial hygiene of fitting the impactor data by assuming that a log-normal distribution is the true underlying distribution. This was done graphically for many years, but the availability of increased computer power has allowed more complex fitting of the data.[104] The simplest approach is to assume that the 50% cut point of each stage is an absolute value and that no particles smaller than that cut point deposit on subsequent stages. One can further assume that the size distribution of the dust measured can be represented by one or two log-normal size distribution modes. A simple spreadsheet program can be used to extract the mass median aerodynamic diameter (MMAD) and the geometric standard deviation (σ_g) for each mode.[105] However, the accuracy of this approach is questionable because of several confounding factors: the cut point of each stage is not an absolute value, that is, there is an overlap in the collection efficiency for adjacent stages; there are losses of particles within the cascade impactor; particle bounce can cause large particles to end up on stages downstream of their intended target stage; and there is variability associated with measurement of the particle mass on each stage. The first two factors can be accommodated by the accurate assessment of collection efficiency of each stage and of losses within the impactor. This information has been published for some cascade impactors.[99–101] Even with this information, the inversion or deconvolution of the data is complex because, with overlap in stage collection efficiencies and the presence of noise in the data, there is no single correct answer, i.e., the problem is ill-conditioned. There is an extensive literature on attempting to provide the best solution, and the technique recommended by Kandlikar and Ramachandran[106] in their literature review is the first-order regularization technique. An alternative to the regularization method for a single log-normal distribution is the nonlinear least-squares approach, which allows the calculation of accuracy estimates.[107] These relatively simple approaches take into account the stage efficiencies as well as the expected level of measurement error for the mass collected on each stage.

If there is clearly a single log-normal mode in the particle size distribution, either the regularization or nonlinear least-squares methods can give a reasonably accurate MMAD and σ_g. However, fitting a more complex distribution, for example, with two modes, is problematic because five or more parameters must be determined with only seven or eight degrees of freedom (from the number of stages). In many instances, it is probably better not to try to extract information about log-normal modes that might be present. There is usually a large range of distributions that can fit the data equally well. If the cascade impactor data are used to estimate the respirable, thoracic, and inhalable fractions, it may be more accurate not to try to fit the size distribution, but rather to use just the raw mass data and assume a sharp cut for each stage.

11.5.1.6 *Sample analysis*

Aerosol sample analysis methods are matched to the sampled material and the appropriate exposure metric. Samples are generally collected onto a filter, within a PUF, or onto an impenetrable impaction substrate such as aluminum or mylar (which is usually coated with a layer of grease or oil to prevent particle bounce).

Personal liquid impinger samplers are still occasionally used, allowing the aerosol to be directly collected into a liquid suspension. Because of the difficulties of handling and using liquid collection media in the field, the application of impingers for nonbiological aerosols is generally limited to isocyanate aerosols,[108,109] which are chemically unstable and must be immediately reacted with a complexing agent for accurate assessment. Impingers are also used for bioaerosols, but not usually as personal samplers.

Filters may be held in a cartridge within the sampler, as is the case with the IOM inhalable sampler, or may be mounted directly into the sampling head. Chemical analysis of aerosol components is usually component specific.[110] Analysis of metals is usually by spectroscopic analysis; atomic emission spectroscopy is frequently used, and mass spectrometry provides a more sensitive alternative. Crystalline silica is quantified using either x-ray powder diffraction or infrared spectroscopy. Radioactive aerosols are generally collected on filters, and characterized using off-line radiation detectors (see Chapter 13). Other chemical species are analyzed using a range of standard analytical techniques, including gas chromatography, high-performance liquid chromatography (HPLC), and so on. Viable biological organisms are generally characterized by the number of colony-forming units (CFUs) collected, following incubation on a suitable culture medium,[111,112] while the total concentration of specific organisms can be determined using techniques such as PCR.[113–116] Where samples are characterized on the collection substrate, care needs to be taken to ensure that the sample is presented appropriately for analysis, and background levels of the analyte (or any confounding components) are at suitably low levels. In many cases, collected particles are transferred to an appropriate medium for subsequent analysis, requiring the collection substrate to be matched to the preparation process.

Gravimetric analysis is used to characterize exposure to nuisance dusts, where specific chemical speciation is not required. The accuracy of gravimetric samples may be affected by water adsorption onto substrates and filter cartridges, and by losses or gains in material during transit.[117,118] In particular, cellulose ester membrane filters, PUFs, and conducting plastic filter cartridges are particularly prone to weight changes following water uptake.[71,119] Substrates such as mylar have been found to outgas substantially following removal from storage, and may take several days to reach a stable weight before being useable for gravimetric sampling. To combat bias from such sources, it is common practice to weigh a number of control or blank filters with each set of sample filters (typically one blank per ten samples, with a minimum of three blanks). Filters should be conditioned in the weighing area (preferably a temperature- and humidity-controlled environment) for up to 24 h before weighing to allow them to reach an equilibrium weight.[120,121] Desiccation is generally not advisable prior to filter weighing, as weight changes after removal of the filter can be sufficiently rapid to lead to significant weighing errors.[71] Where possible, blank filters should be transported with the sample substrates and exposed to the same conditions, to minimize bias resulting from handling, transport, and changes in environment.[122]

Other sources of bias include electrostatic attraction where substrates are highly charged, and buoyancy effects. Electrostatic charge buildup may be significant for substrate materials such as PVC and PTFE, particularly when working at low relative humidity. In all instances, samples should be neutralized using a source of bipolar ions. A common approach is to place samples close to a radioactive antistatic source prior to weighing. Buoyancy corrections only become necessary when the volume of the sample exceeds around 0.1 cm^3. For most substrates, this is not a problem, although it may be significant when using large integral filter cartridges or substrate supports.

Asbestos and other fibrous aerosols are characterized in terms of particle number concentration and aspect ratio. Fiber detection and analysis following collection is carried out using either phase contrast microscopy (PCM), polarized light microscopy (PLM), scanning electron microscopy (SEM), or transmission electron microscopy (TEM), depending on what information is required from the sample. PCM is most frequently used to measure airborne concentrations of fibers in occupational settings, as it is relatively quick and inexpensive. To image fibers in a light microscope, they are collected on cellulose ester filters that are chemically cleared on a glass slide, and to provide contrast when imaging the fibers (usually a liquid or resin with a refractive index close to that of the filter is used to impregnate the sample and fill the gap between the sample and the cover slip). The use of PCM allows fibers thicker than around 0.25 μm to be detected, and enables the number of fibers collected to be estimated. The exact definition of fibers in the context of exposure standards varies with the standard used. OSHA and NIOSH "A" asbestos counting rules [123] allow all particles longer than 5 μm with an aspect ratio equal to or greater than 3:1 to be counted, while the Environmental Protection Agency (EPA) uses the definition of particles longer than 0.5 μm with an aspect ratio greater than 5:1. The EPA method is used primarily for evaluating the cleanliness of locations being cleared of asbestos-containing materials and is not designed to provide concentrations that have health relevance. NIOSH "B" counting rules allow particles larger than 3 μm in diameter to be discounted when characterizing MMVF samples (which OSHA treats as a nuisance dust).

The measurement of asbestos by PCM involves the preliminary assumption that all fibers detected are asbestos, since asbestos fibers often cannot be accurately discriminated from other fibers. During the 1960s and 1970s, most occupational asbestos operations still created high levels of exposure and this assumption was valid. However, in recent years, the exposure levels have dropped significantly and other fibers may constitute a significant fraction of fibers sampled in workplaces and asbestos abatement or removal sites. NIOSH recommends the use of TEM to determine the fraction of asbestos fibers in a sample and apply that fraction to the PCM fiber count to obtain the final asbestos fiber count.[123]

There are a number of error sources encountered in fiber counting, including sampling error, nonuniformity on collection filters, and human counting error. As a result, the errors associated with measuring airborne fiber concentration can be relatively high. To minimize errors, several interlaboratory proficiency schemes exist to ensure that analytical laboratories are operating within acceptable accuracy limits (e.g., RICE (the UK Regular Inter-Laboratory Counting Exchange), AFRICA (the international Asbestos Fiber Regular Interchange Counting Arrangement), PAT (American Industrial Hygiene Association (AIHA) Proficiency Analytical Testing), and AAR (AIHA Asbestos Analysts Registry) programs). One of the limitations of PCM is that fibers around 0.25 μm in diameter are hard to detect with 100% efficiency. In principle, fiber counting using TEM or SEM overcomes this limitation, allowing fibers with diameters down to nanometer widths to be detected and counted.[123,124] However, somewhat inexplicably interlaboratory comparisons with TEM analysis have shown poorer accuracy over PCM.[125]

One of the advantages of SEM and TEM analysis is that analytical systems such as X-ray energy dispersive spectroscopy (EDS or EDX) and selected area electron diffraction (SAED) allow elemental analysis of fibers, and thus identification of fiber type/source. Optical PLM can also be used to identify asbestos fibers from nonasbestos fibers (asbestos fibers appear bright under cross-polarization), and goes some way toward identifying asbestos types. However, PLM can only be applied to fibers thicker than about 1 μm in diameter and cannot be used on the same sample used for PCM counting.

11.5.2 *Direct reading instruments*

11.5.2.1 *Personal exposure measurements*

Most occupational exposure measurements are made using traditional sampling and laboratory analysis techniques. This approach, while providing accurate measurements, often requires days or weeks to provide feedback on specific exposure situations. Quite often, the aerosol concentration in a workplace is highly variable and sources of the aerosol are close to the workers or associated with the workers' activities. Direct reading instruments are often used to aid in the development and optimization of dust control systems. Direct reading instruments have the potential to provide immediate feedback to the workers regarding their environment. If the workers can associate specific locations or actions with high aerosol concentration, then, by taking appropriate action, they may be able to quite significantly reduce their exposures.

Several approaches to direct reading instrumentation have been used to provide real-time readout. When sufficiently small instruments were not available, some researchers used larger instruments and placed them in the vicinity of the worker, extracted aerosol from the neighborhood of the worker and ducted it to the instrument, or put the instruments into a holster or backpack and placed them on the worker. A variety of physical mechanisms can be used to detect aerosol particles, and some of these have been incorporated into small portable commercial instruments. In many cases, these instruments are small enough to be worn by a worker.

11.5.2.2 *Light scattering instruments*

Light scattering detection of dust provides the advantage of designing instruments that have a quick response and are relatively inexpensive. The disadvantage of these instruments is that the response depends strongly on the dust particles' size, shape, and refractive index.[126] Most of these parameters are not only difficult to predict for a given occupational setting, but may also change with time. The effect of particle size can be accommodated to some extent by using an optical particle counter (OPC), which indicates particle optical diameter, a parameter approximately proportional to aerodynamic diameter. However, when the dust characteristics are not expected to change significantly, calibration of a light scattering instrument may result in good measurement accuracy.[127]

11.5.2.2.1 Optical particle counters The interaction between light and airborne particles is complex and except for some simple particle shapes, such as spheres and fibers, cannot be accurately predicted. However, there are some generalizations that can be made based on spherical particles and on experimental evidence. The amount of light scattered from a particle is a complex function of particle size, shape, and refractive index, as well as instrumental factors such as light beam size, shape, and intensity, and detector sensitivity and measurement angle. When the particle is smaller than the wavelength of light, this is called the Rayleigh scattering regime and the amount of light scattered increases proportional to the 6th power of the particle diameter. Thus, the lower limit of single-particle detection occurs because most light sources cannot scatter enough light from small particles to be detected by a detector. Inexpensive instruments that detect single particles (OPCs) typically do not detect particles smaller than 0.3 μm. By using a higher powered laser and more efficient detectors, more expensive instruments can detect down below 0.1 μm diameter. Another technique for increasing sensitivity is to decrease the wavelength of the light source. When shorter wavelength LEDs or solid-state lasers become less expensive, these may be incorporated into handheld instruments as well. A typical handheld OPC is shown in Figure 11.19.

When the particle is larger than the wavelength of light, this is called the Mie scattering regime and the amount of light scattered increases very roughly with the particle diameter. However, the interaction of light with a spherical or another regularly shaped particle, can produce resonances within the particle, and the amount of light scattered to

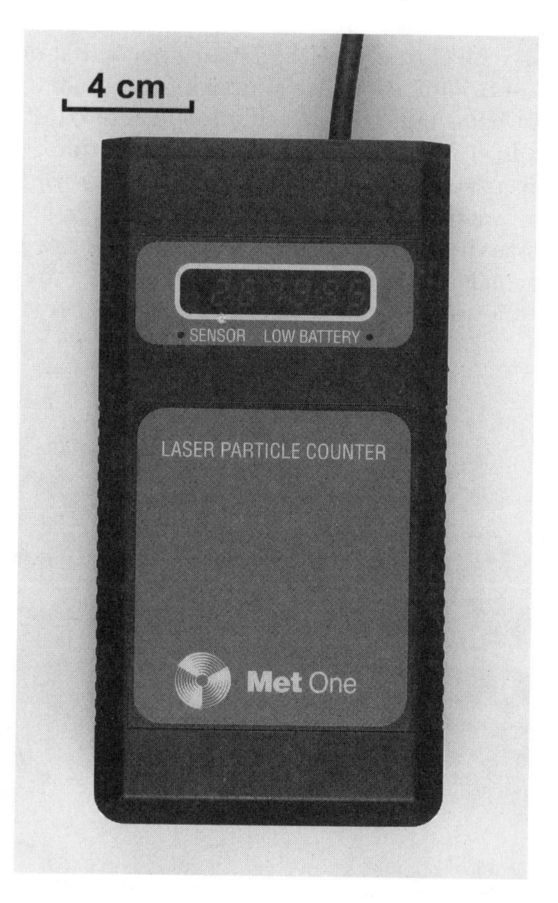

Figure 11.19 Met One handheld optical particle counter (Met One, U.S.A.).

a detector is a complex function of size. While the scattering of spherical particles can be predicted theoretically with great accuracy, irregularly shaped particles have a less well-defined scattering efficiency and cannot be predicted. The absorption component of the refractive index plays an important role in the amount of light scattered; for instance, a coal particle scatters less light and appears smaller optically than a glass particle when measured by an OPC. Thus, the amount of light scattered is only an approximate indicator of particle size. Although many OPCs indicate that they detect particles, for example, larger than 5 μm, this is not an accurate measure and is only approximately correct for the particles with which the instrument was calibrated.

An OPC detects the light scattered from a particle as the particle passes through a small detection volume. Several factors play a role in the detection efficiency of the OPC and its ability to handle a desired range of concentrations. If more than one particle passes through the detection volume at a time, these coincident particles are detected as a single larger particle, reducing the accuracy of the count. Thus, there is a limited range of concentrations for most particle counters. The upper limit is determined by coincidence, often represented by a 10% coincidence level, that is, 10% of the detected counts are of two or more particles. The lower concentration limit is determined by the acceptable statistical error of counting relatively few particles. Thus, an OPC that is used for clean rooms might have a high sampling rate through a relatively large detection volume to maximize the number of particles detected. On the other hand, for many workplace situations, the particle number concentration can be high, so an OPC for this application should have a small detection volume and low flow rate. Otherwise, a dilution system is needed to match the OPC to the concentration being measured.

Relatively inexpensive OPCs are available. Instruments that detect all particles above a cutoff size, such as 0.3 or 0.5 μm, provide a simple indication of total count. More sophisticated instruments may have several selectable lower size cutoffs (Figure 11.20). Some instruments provide up to six size bins so that a size distribution can be measured. More than six size bins are not useful for most applications because of the imprecision in light scattering response as a function of particle size. One example of a handheld instrument that measures particle size distribution is the GRIMM 1.10 series of OPCs (Figure 11.20a, Grimm Technologies Inc., Germany). These instruments not only measure size distributions, but the software allows the calculation of several dust fractions, such as thoracic,

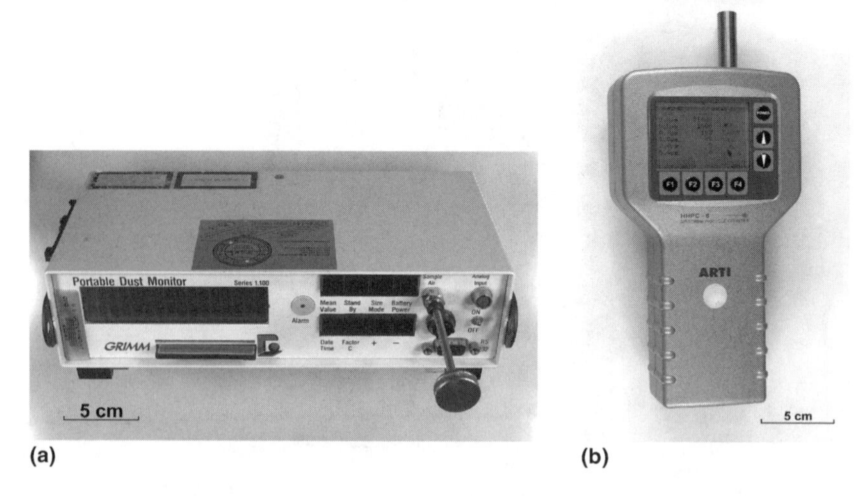

Figure 11.20 Handheld size-differentiating optical particle counters: (**a**) Grimm Series 1.10 dust monitor (Grimm Technologies, Germany) and (**b**) HHPC-6 optical particle sizer (ARTI, U.S.A.).

respirable, and $PM_{2.5}$. They also have a built-in filter sampler that can be used to calibrate the dust mass response. Another example of a similar instrument is the ART Instruments HHPC-6 handheld OPC (Figure 11.20b, ARTI, U.S.A.). The HHPC-6 is a six-channel handheld device that measures and displays size-resolved aerosol concentration in real time. Both of these instruments can download logged data to a computer for further analysis.

11.5.2.2.2 Photometers Photometers (also called nephelometers) rely on the same light scattering that OPCs measure, except that the detection volume is sufficiently large that the scattering from many particles is detected simultaneously. The integrated scattering signal provides an indication of concentration. The total light scattering detected in such an arrangement is a function of the number, sizes, and refractive indices of all the particles present in the detection volume. Thus, if an unknown airborne material is being detected, the mass concentration of that material cannot be predicted. However, if the material is from a source that is well characterized, or at least constant, the photometer can be calibrated to give a reasonably accurate estimate of mass concentration. Photometer response as a function of aerosol mass concentration is only weakly dependent on particle size in the transition region between Rayleigh and Mie scattering (roughly 0.3 to 10 μm). Beyond these limits, response depends on the size distribution of the aerosol, and the method is generally inappropriate for measuring aerosols dominated by particles outside this region.

The simplicity and rapid response of photometers make them attractive for detecting sources, evaluating controls, and evaluating time-dependent concentrations. Some instruments depend on local air motion to propel the aerosol to the sensor, but, generally, use of a pump to actively pull the aerosol through the sensor provides a more accurate and quicker response (as in the DustTrack (TSI Inc., U.S.A.) shown in Figure 11.21).[128] The size-dependent mass response of these instruments is somewhat similar to the response

Figure 11.21 DustTrak aerosol photometer (TSI Inc., U.S.A.).

required for respirable sampling, so, once calibrated for a specific dust, they can be used to obtain respirable mass concentrations with reasonable accuracy. This approach was described by Baron[129] and has been used for borate aerosols,[130] coal dust,[131] environmental and home aerosols,[127] and fire smoke.[132] Monitoring dust close to a source is more problematic because the size distribution and character of the dust is more likely to change with time, thus reducing the accuracy of calibration. This was noted for personal exposure correlations with filter samplers in home environments.[127]

It must be remembered that these direct reading instruments detect all aerosol particles with little difference in response due to chemical, biological, radiological, or morphological differences in composition. Therefore, when a specific component of the total aerosol present is being investigated, photometers or OPCs may provide a poor indication of the concentration of that component. This is often true in occupational settings where many potential sources of aerosol exist. It has also been noted in the environmental field, where personal exposure includes not only aerosol from the ambient environment but also aerosol from the personal environment, for example, clothing, home, office, and car. Thus, personal measurements may not be appropriate with a nonspecific direct-reading instrument in these cases.

11.5.2.3 Tapered element oscillating microbalance

The tapered element oscillating microbalance (TEOM®, Rupprecht and Patashnick, NY) has been used in the environmental field for many years and has been accepted as providing an accurate and rapid indication of airborne particle mass. This device collects particles on a filter attached to the end of a vibrating tube; the vibrational frequency is directly related to the mass of particles collected. Research is ongoing to refine and miniaturize the TEOM to develop an instrument that a person can wear.[42] The first attempt to produce such an instrument was under a contract funded by the Bureau of Mines and NIOSH.[133] In this development, the sensor was worn by a miner and, after the work shift, the sensor was placed in a readout unit. The technique provided accurate results, but was expensive for a device that was not direct-reading. More recently, NIOSH funded the development of a personal dust monitor (PDM) for use in coal mines. The PDM is integrated with the miner's lamp system, with the inlet mounted on the helmet and the detector and pump combined with the lamp battery pack (Figure 11.22). This minimizes the inconvenience to the miner and allows periodic readout of the instrument during the work shift; however, only the final measurement at the end of the work shift would be used for compliance

Figure 11.22 TEOM personal dust monitor (R&P Co., U.S.A.).

with the coal mine dust standard of 2 mg/m^3. Preliminary tests showed good agreement with conventional respirable dust measurements using a 10-mm nylon cyclone sampler.[42]

11.5.2.4 Condensation particle counter

The condensation particle counter was one of the original instruments developed to detect airborne particles in the late 1800s,[134] but has only recently become available in a handheld direct-reading instrument. A condensation particle counter contains a condensation section, in which particles enter a supersaturated vapor (e.g., butanol or isopropanol) region and the vapor condenses on the particles, allowing them to grow to a uniform size on the order of 1–3 μm in diameter. Once the particles have grown, they can be readily detected using an optical particle counter. This technique allows a count of all particles present from approximately 1 μm in diameter down to a lower diameter limit in the range of a few to 20 nm. Typically, the number concentration of particles in the submicrometer range is much larger than the concentration of larger particles, so this technique is used to monitor primarily smokes and fumes.

The handheld instruments currently available are similar in style and available from the same manufacturer. The P-TRAK (Model 8525, TSI Inc., St. Paul, MN) is a handheld device that counts particles continuously and gives a concentration in the range of 0–5 $\times$ 10^5 particles per ml with a lower particle diameter limit of 20 nm. A more sophisticated version of this instrument (Model 3007, TSI Inc.; Figure 11.23) has a flow-controlled pump, an upper concentration limit of 1$\times$10^5, and a detection limit of 10 nm.[135] These instruments can be useful for detecting and developing controls for sources of submicrometer particles, such as welding fumes, diesel fumes, and cigarette and other fire smoke.

11.5.2.5 Pressure drop sensor

When filters collect particles from an aerosol, the particles tend to clog the openings of the filter and increase the pressure drop across the filter. Volkwein et al.[136] used this principle to develop a small and inexpensive direct-reading respirable dust dosimeter (RDD), primarily for use in coal mines (Figure 11.24). Once the RDD is calibrated for a specific type of dust, it can provide accurate measurement of that dust. The accuracy of the measurement depends on the dust concentration and measurement period, since the product of these two factors determines the dust load on the sensor filter and, hence, the pressure drop. At concentrations below the 2 mg/m^3 standard for coal mine dust, several minutes may be required to obtain enough pressure drop change to produce an updated reading. However, this is still adequate for many situations. A major disadvantage of this method is that it is dependent on the size distribution of the dust. Within a single coal mine, good agreement with the conventional coal mine sampler was obtained. However, different calibration responses were obtained for some mine dusts.[137] In addition, when submicrometer particles such as cigarette smoke or

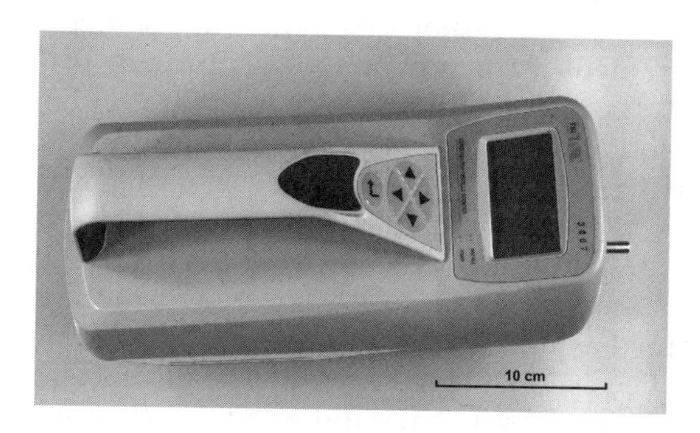

Figure 11.23 Model 3007 portable CPC (TSI Inc., U.S.A.).

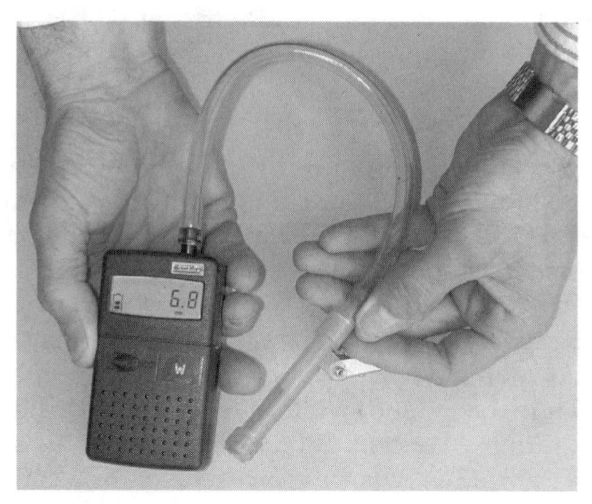

Figure 11.24 Respirable dust dosimeter.

welding fumes were sampled, the RDD produced a much larger pressure drop response. Therefore, if significant submicrometer aerosols are mixed with dust, the instrument is not capable of giving meaningful results.

11.5.2.6 Aerosol surface area measurement

Laboratory-derived associations between the surface area of inhaled insoluble particles and toxicity are leading to increased interest in the real-time measurements of aerosol surface area. Although off-line measurements of bulk material surface area have been possible for some time using the BET method,[138] instruments capable of measuring aerosol surface area in the field are not widely available at present. BET has been used with some success for measuring aerosol surface area. However, it requires the collection of relatively large amounts of material, internal surfaces are included in the measurement, and the collection/support substrate may contribute significantly to the measured surface area — particularly where the quantity of material analyzed is small. The first instrument designed specifically to measure aerosol surface area was the epiphaniometer[139,140] (Matter Engineering, Switzerland). This device measures the Fuchs or active surface area of the aerosols by measuring the attachment rate of radioactive ions. As yet it is unknown how relevant active surface area is to health effects following inhalation exposure. Below approximately 100 nm, active surface scales as the square of particle diameter, and thus is probably a good indicator of actual particle surface area for ultrafine particles. However, above approximately 1 μm it scales as particle diameter, and so the relationship with actual particle surface area is lost.[139] The epiphaniometer is not well suited to widespread use in the workplace due to the inclusion of a radioactive source.

The same measurement principle may be applied in the aerosol diffusion charger/electrometer.[141] The LQ1-DC diffusion charger (Matter Engineering, Switzerland) uses this combination to measure the attachment rate of unipolar ions to particles, and from this the aerosol active surface area is inferred.[141,142] This instrument is also available in a portable form as the DC2000CE, complete with rechargeable battery and built-in data logger (Figure 11.25). A similar instrument, the 3070a electrical aerosol detector (EAD), has also recently been developed by TSI Inc. (MN, U.S.A.). As in the case of the LQ1-DC, the sampled aerosol is charged using a unipolar ion source, and the mean charge is measured. However, the EAD has been shown to have a response much closer to particle diameter d to the power 1.16 ($d^{1.16}$). Recent research has indicated that the EAD's response may match the surface area of particles depositing in the lungs.[143]

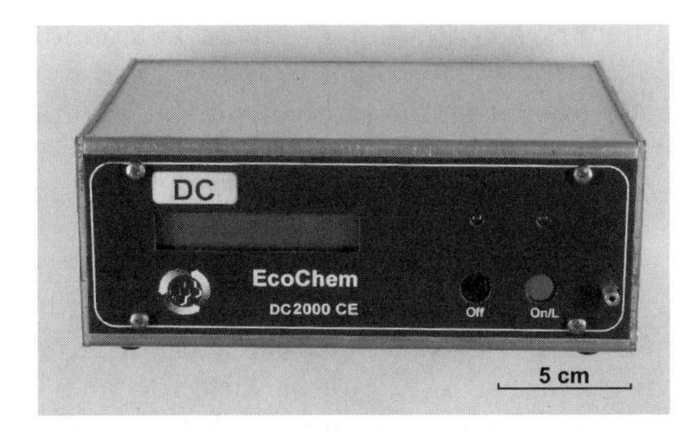

Figure 11.25 DC 2000 CE portable aerosol active surface area monitor (Matter Engineering, Switzerland).

For log-normal aerosol size distributions, it is possible in principle to estimate aerosol surface area by deriving the size distribution from three measurements of different aerosol parameters. This approach has been investigated by Woo et al.[144] using simultaneous measurements of aerosol mass concentration, number concentration, and charge. Maynard has proposed a simplified method of estimating surface area using measurements of aerosol number and mass concentration.[145] By assuming a geometric standard deviation of the aerosol size distribution, a reasonable estimate may be made of aerosol surface area using this approach. However, as the aerosol deviates from a log-normal distribution, errors rapidly become large. Theoretical modeling has suggested that in many cases, surface area concentration estimated using this approach will be within a factor of four of the actual value. Although relatively large, errors of this magnitude may still allow surface area exposures to be placed within broad categories.

11.5.2.7 Specific applications of direct-reading instruments

11.5.2.7.1 Sampling cassette leakage testing Sampling cassettes purchased from manufacturers are checked for proper compression and are generally leak free. However, cassette pieces are often purchased so that the sampler can be assembled by the user. It was found in one study that 15% of cassettes assembled by hand exhibited leakage and could have caused underestimation of worker exposure.[53] It was recommended that cassettes be assembled using a press and that bypass leakage around the collection filter be checked using either an optical particle counter or a condensation particle counter and ambient aerosol. The ratio of downstream to upstream particle concentration is an indication of the amount of filter bypass leakage. This leakage can occur because of incomplete compression of the seal at the edge of the cassette or because of cuts in the filter when cassette compression is too great. The amount of aerosol loss is a complex function of particle size, particle type (e.g., liquid, solid), and leak size.[54,146]

11.5.2.7.2 Respirator testing Respirators are devices used to prevent exposure to hazardous gases and aerosols. There are two basic types of respirators: (a) air-supplied respirators, in which clean air is provided to the breathing zone from an external source, and (b) air-purifying respirators, in which air passes through an air-cleaning device, for example, a filter or cartridge. Air-purifying respirators can be further classified as being powered or nonpowered. The nonpowered or negative pressure air-purifying respirators are the most common and use the person's lung suction to pull the air through the cleaning device. Respirators are classified by their assigned protection factor, a number assigned to a particular class of respirator for regulatory purposes.[147] It is a number that estimates by what factor an air contaminant concentration is reduced in a population of properly fitted and trained users. Actual protection factors for properly fitted respirators may be higher than the assigned protection factor or, alternatively, if the respirator is used incorrectly, lower. For aerosols, the efficiency with which the respirator prevents particles from being breathed in is determined by the efficiency of the filter and the sealing efficacy of the mask, that is, the prevention of particles bypassing the filter.

Filter technology allows the production of highly efficient filters, but often the high efficiency comes at the cost of higher pressure drop across the filter and greater production cost. A higher pressure drop across the filter can actually reduce the overall performance of a nonpowered air-purifying respirator because the suction caused by breathing reduces the air pressure inside the respirator and any leaks around the seal are increased. In addition, the higher pressure drop increases the physiological load on the wearer, making it less likely to be used properly.

Respirators are tested and certified by NIOSH by measuring the penetration of 0.3 μm particles through the respirator. Particles of approximately this size are the ones with the greatest likelihood of passing through the filter. Larger particles are collected by impaction and interception, while smaller ones are collected by diffusion. To test respirators for use by workers, either a qualitative or a quantitative fit test is performed. Fit testing is usually carried out by replacing the respirator filter with a high-efficiency (HEPA) filter so that particles detected inside the respirator are due to leakage. Qualitative fit tests are performed by exposing the respirator-equipped worker to an aerosol that the worker can smell or taste, for example, irritant fumes, sodium saccharin, or bitrex. A quantitative fit test is used to determine the factor by which aerosol is reduced inside the respirator. This requires measuring the efficiency of the respirator while it is worn and the worker moves about and speaks. This measurement can be done in several ways. In each case, the fit factor is determined from the ratio of the outside aerosol concentration to the inside concentration. One technique is to create an aerosol of an oil, such as corn oil, in a small tent enclosing the worker and measuring the aerosol concentration inside and outside the respirator using a photometer (e.g., from Air Techniques Inc.). The aerosol is generated using a nebulizer that produces submicrometer droplets.

Willeke and co-workers have investigated several alternative means for testing respirators. The particles that penetrate the leaks around the edge of a respirator are primarily submicrometer in size. Therefore, the most common particles available, namely ambient aerosol particles, were used to develop a test.[148] The condensation particle counter is capable of rapidly and accurately measuring the total number concentration of these particles. The PortaCount (Model 8020, TSI Inc.) can be used to fit test respirators. This device has two ports, one to detect inside a respirator and the other ambient aerosol. The proportion of ambient fine particles that penetrate the respirator provides an indication of the fraction of air leakage around the edge of the respirator. This test has the advantage of being less obtrusive to the worker and quicker to set up and use. Han et al.[149] discuss the details of quantitative fit testing techniques and the advantages and disadvantages of each approach.

11.5.2.7.3 Sampler testing Personal samplers for measuring respirable, thoracic, and inhalable dust require testing to determine their size-dependent characteristics. Two primary aspects of these devices are usually tested: the inlet or aspiration efficiency of the sampler and the internal loss characteristics of the sampler. The aspiration efficiency under most indoor conditions is close to 1 (or 100%) for particles smaller than about 5 μm. Thus, for respirable sampling, aspiration efficiency can be assumed to be 1 unless the sampler is to be used at high wind velocities (e.g., some mines and occasionally in outdoor air). Thoracic and especially inhalable samplers should be tested at a range of wind velocities to ensure accurate sampling. In the case of the inhalable sampler, the external wind velocity is critical to the sampling efficiency of the sampler so that internal loss characteristics are somewhat secondary.

Respirable and thoracic dust samplers are often tested in a stagnant or low-velocity chamber to determine the size-dependent efficiency of the classifier. There are two basic approaches to performing the testing. Traditionally, single-size (monodisperse) particles containing a tracer are generated, collected with the sampler, and then the tracer is measured in the various parts of the sampler, for example, the classifier, connecting pieces, filter, and filter cassette. This is done for several particle sizes over the range of interest near the cutoff of the classifier. This is a relatively tedious approach to testing, but provides useful information about the behavior of the classifier and internal losses in the system.

An alternative approach is to use a polydisperse aerosol and measure the aerosol upstream and downstream of the classifier with an aerodynamic sizing instrument, for example, the aerodynamic particle sizer (APS3321, TSI Inc., U.S.A.). Dilution air is introduced at the top of the chamber, aerosol is injected into a mixing region, the flow is straightened, and the samplers are placed in the test section. The APS is placed below the chamber and can be attached either to the sampler or to a reference inlet port. The aerosol is measured at the two ports and the concentration in each size bin of the classified aerosol distribution is divided by the corresponding size bin of the reference port. This provides a size-dependent penetration of the classifier. Penetration curves for a sampler can be obtained in a matter of minutes by this approach, although setup time and validation of the measurement can take longer.

There are no direct reading aerosol sizing instruments available for larger particles, so inhalable samplers are typically tested in a wind tunnel over a range of wind velocities using the traditional approach of monodisperse dusts with a tracer or gravimetric measurement of the collected material. Because sampler results have been shown to be affected by the human body, samplers are usually placed on a manikin torso to simulate placement on a worker. In addition, the manikin is rotated during sampling to average over wind direction.

11.5.2.7.4 Combined aerosol and video monitoring To reduce the exposure of workers to harmful aerosols, it is important to investigate the detailed interaction between workers and their environment. Although an industrial hygienist may observe a worker during the period that a personal measurement is being taken on that worker, capturing a video image of the worker during the sampling period provides the opportunity for a much more objective and quantitative measure of the exposure as a function of various worker activities and environmental factors. The technique involves using a video camera with time information imprinted on the image and a real-time monitoring system with the output data-logged, also with time information saved.[150,151] The video record can be analyzed to record various activities and occurrences, so that these can be correlated with changes in concentration noted in aerosol concentration data. Quite often, there is a time lag after an event before the concentration changes because of the time that the aerosol takes to reach the monitoring instrument and the instrumental response time. Mathematical analysis of the time lag allows a direct correlation of concentration and activity. Repetitive activities can be averaged to obtain a more accurate evaluation of the time course of exposure. Cooper et al.[152] were able to target specific locations and activities that produced high exposure levels of crystalline silica, and this information was successfully used to apply control measures that significantly lowered exposures. Using a video mixer to place the aerosol concentration data directly on the video image results in a video record of activities and correlated aerosol concentration. Such a video record can be used as a training or feedback tool for workers to demonstrate how their exposure is affected by various activities.

11.6 Summary

The observation of occupational exposure to aerosols has a long history, but great progress has been made in the past several decades in providing a range of measurement instruments and techniques. While dose/response relationships have been established for a wide range of workplace contaminants, there are still uncertainties as to which metric is the most appropriate to measure for specific contaminants. Current research emphasis on the ultrafine aerosol particle range may result in surface area being considered as an important exposure metric.

When attacking the problem of what instruments or approaches are to be used in making workplace measurements, a number of factors have to be considered. The size range of interest, the aerosol metric (size, mass, number surface area, radiological, chemical, or biological activity concentration), and purpose of measurement (compliance with standards, evaluation for source control, personal vs. environmental exposure) all have an impact on the selection of the most appropriate type of measurement technique or instrument. Because aerosol particles have a whole range of properties that can affect their measurability, one needs to have an adequate understanding of the aerosol to be measured and how its properties may affect the selected measurement technique.

The development of readily available, accurate direct-reading measurement tools (e.g., the SMPS and the APS) has spawned a range of classifiers and other aerosol measurement techniques. Many of these are innovative and have definite advantages over the traditional approaches to sampling and measurement. However, one has to be cautious in ensuring that adequate testing of these devices has been performed and that one understands the advantages and drawbacks of any instrument selected. The literature is replete with discoveries of subtle biases introduced by new devices that initially appeared to have clear advantages over traditional approaches. On the other hand, some traditional approaches were not carefully tested when they were originally selected and also have significant drawbacks. Hopefully, with the development of new instrumentation and appropriate testing, clearer choices and better characterized measurement tools will be available in the future.

Disclaimer

Mention of company names or products does not constitute endorsement by the Centers for Disease Control and Prevention.

References

1. Sigerist, H.E., Historical background of industrial and occupational diseases, *Bull. NY Acad. Med.*, 12, 597–609, 1936.
2. Rose, V.E., History and philosophy of industrial hygiene, in *The Occupational Environment — Its Evaluation and Control*, DiNardi, S.R., Ed., AIHA Press, Fairfax, Virginia, 1997, pp. 3–20.
3. Hunter, D., *The Diseases of Occupations*, 6th ed., Hodder and Stoughton, London, 1978.
4. Plogg, B.A. and Quinlan, P.J., *Fundamentals of Industrial Hygiene*, National Safety Council Press, U.S.A., 2002.
5. Ramazzini, B., *Diseases of Workers., De Morbis Artificuim*, Academy of Medicine, New York, 1964.
6. Corn, J., Historical aspects of industrial hygiene — silicosis, *Am. Ind. Hyg. Assoc. J.*, 41, 125–132, 1980.
7. Balaan, M.R. and Banks, D.E., Silicosis, in *Environmental and Occupational Medicine*, 3rd ed., Rom, W.E., Ed., Lippincot-Raven, Philadelphia, 1998, pp. 435–448.
8. Rom, W.N., Asbestos-related disease, in *Environmental and Occupational Medicine*, 3rd ed., Rom, W.N., Ed., Lippincott-Raven, Philadelphia, 1998, pp. 349–375.

9. Lockey, J.E. and Wiese, N.K., Man-made vitreous fiber, vermiculite and zeolite, in *Environmental and Occupational Medicine*, 3rd ed., Rom, W.N., Ed., Lippincott-Raven, Philadelphia, 1998, pp. 397–412.
10. IARC, *Man-Made Vitreous Fibers*, International Association for Research into Cancer, Lyon, France 2002.
11. Singh, J., Occupational exposure to moulds in buildings, *Indoor Built Environ.*, 10, 172–178, 2001.
12. Cox, C.S. and Wathes, C.M., *Bioaerosols Handbook*, Lewis Publishers, New York, 1995.
13. Cohen, B.S. and McCammon, C.S., *Air Sampling Instruments for Evaluation of Atmospheric Contaminants*, 9th ed., ACGIH, Cincinnati, 2001.
14. Oberdörster, G., Gelein, R.M., Ferin, J., and Weiss, B., Association of particulate air pollution and acute mortality: involvement of ultrafine particles? *Inhal. Toxicol.*, 7, 111–124, 1995.
15. Stöber, W., Morrow, P.E., and Oberdörster, G., Pulmonary retention of inhaled anatase (TiO_2) aerosols and lung-tumor induction in rats simulated by a physiology-oriented model, *Inhal. Toxicol.*, 7, 1059–1074, 1995.
16. Lison, D., Lardot, C., Huaux, F., Zanetti, G., and Fubini, B., Influence of particle surface area on the toxicity of insoluble manganese dioxide dusts, *Arch. Toxicol.*, 71, 725–729, 1997.
17. Brown, D.M., Wilson, M.R., MacNee, W., Stone, V., and Donaldson, K., Size-dependent proinflammatory effects of ultrafine polystyrene particles: a role for surface area and oxidative stress in the enhanced activity of ultrafines, *Toxicol. Appl. Pharmacol.*, 175, 191–199, 2001.
18. Tran, C.L., Buchanan, D., Cullen, R.T., Searl, A., Jones, A.D., and Donaldson, K., Inhalation of poorly soluble particles. II. Influence of particle surface area on inflammation and clearance, *Inhal. Toxicol.*, 12, 1113–1126, 2000.
19. Bedford, T. and Warner, C., Physical studies of the dust hazard and thermal environment in certain coalmines, in *Chronic Pulmonary Disease in South Wales Coalminers. II. Environmental Studies*, British Medical Research Council, HMSO, London, 1943, pp. 1–78.
20. Donaldson, K., Stone, V., Seaton, A., and MacNee, W., Ambient particle inhalation and the cardiovascular system: potential mechanisms, *Environ. Health Perspect.*, 109, 523–527, 2001.
21. Roco, M.C., Williams, S., and Alivisatos, P., *Nanotechnology Research Directions: IWGN Workshop Report*, Kluwer Academic Publishers, Dordrecht, Netherlands, 2000.
22. Borm, P.J.A., Particle toxicology: from coal mining to nanotechnology, *Inhal. Toxicol.*, 14, 311–324, 2002.
23. ACGIH, Particle Size-selective Sampling for Particulate Air Contaminants, American Conference of Government Industrial Hygienists, Cincinnati, OH, 1999.
24. BMRC, Recommendations of the BMRC panels relating to selective sampling, in From the Minutes of a Joint Meeting of Panels 1, 2 and 3, March 4, British Medical Research Council, 1952.
25. ACGIH, Threshold Limit Values of Airborne Contaminants, American Conference of Government Industrial Hygienists, Cincinnati, OH, 1968.
26. ISO, Air Quality — Particle Size Fraction Definitions for Health-related Sampling, Technical Report No. ISO 7708:1995, 1995.
27. Vincent, J.H., *Particle Size-Selective Sampling for Particulate Air Contaminants*, ACGIH, Cincinnati, 1999.
28. Vincent, J.H., Mark, D., Miller, B.G., Armbruster, L., and Ogden, T.L., Aerosol inhalability at higher windspeeds, *J. Aerosol Sci.*, 21, 577–586, 1990.
29. Baldwin, P.E.J. and Maynard, A.D., A survey of wind speeds in indoor workplaces, *Ann. Occup. Hygiene*, 42, 303–313, 1998.
30. Maynard, A.D., Aitken, R.J., Kenny, L.C., and Baldwin, P.E.J., Preliminary investigation of aerosol inhalability at very low wind speeds., *Ann. Occup. Hyg.*, 41 (Suppl. 1), 695–699, 1997.
31. Aitken, R.J., Baldwin, P.E.J., Beaumont, G.C., Kenny, L.C., and Maynard, A.D., Aerosol inhalability in low air movement environments, *J. Aerosol Sci.*, 30, 613–626, 1999.
32. Aitken, R.J. and Donaldson, R., HSE Contract Research Report No. 117/1996, 1996.
33. Liden, G. and Kenny, L.C., Errors in inhalable dust sampling for particles exceeding 100 micrometers, *Ann. Occup. Hyg.* 38, 373–384, 1994.
34. Aizenberg, V., Choe, K., Grinshpun, S.A., Willeke, K., and Baron, P.A., Evaluation of personal aerosol samplers challenged with large particles, *J. Aerosol. Sci.*, 32, 779–793, 2001.

35. Baron, P.A. and John, W., Sampling for thoracic aerosol, in *Particle Size-Selective Sampling for Particulate Air Contaminants*, Vincent, J.H., Ed., ACGIH, Cincinnati, OH, U.S.A., 1999.
36. Vincent, J.H., International occupational exposure standards: a review and commentary, *Am. Ind. Hyg. Assoc. J.*, 59, 729–742, 1998.
37. HSE, *Occupational Exposure Limits 2002*, HSE Books, London, 2002.
38. NIOSH, NIOSH Respirator Decision Logic., DHHS (NIOSH) Publication No. 87–108, 1987.
39. Walton, H.W. and Vincent, J.H., Aerosol instrumentation in occupational hygiene: an historical perspective, *Aerosol Sci. Technol.*, 28, 417–438, 1998.
40. Vincent, J.H., *Aerosol Science for Industrial Hygienists*, Pergamon, Oxford, 1995.
41. Chung, K.Y.K., Aitken, R.J., and Bradley, D.R., Development and testing of a new sampler for welding fume, *Ann. Occup. Hyg.*, 41, 355–372, 1997.
42. Volkwein, J.C., Tuchman, D.P., and Vinson, R.P., Performance of a prototype personal dust monitor for coal mine use, in *Mine Ventilation*, Souza, D., Ed., Ashgate Pub. Co., Kingston, Canada, 2002, pp. 633–639.
43. OSHA, Occupational Safety and Health Standards, Welding, Cutting and Brazing, 29CFR Part 1910.252, 1998.
44. Greenburg, L. and Smith, G.W., A New Instrument for Samping Aerial Dust, U.S. Bureau of Mines Report No. 2392, 1922.
45. Littlefield, J.B. and Schrenk, H.H., Bureau of Mines Midget Impinger for Dust Sampling, U.S. Bureau of Mines Report No. RI 3360, 1937.
46. Spanne, M., Grzybowski, P., and Bohgard, M., Collection efficiency for submicron particles of a commonly used impinger, *Am. Ind. Hyg. Assoc. J.*, 60, 540–544, 1999.
47. Juozaitis, A., Willeke, K., Grinshpun, S.A., and Donnelly, J., Impaction onto a glass slide or agar versus impingement into a liquid for the collection and recovery of airborne microorganisms, *Appl. Environ. Microbiol.*, 60, 861–870, 1994.
48. Terzieva, S., Donnelly, J., Ulevicius, V., Grinshpun, S.A., Willeke, K., Stelma, G.N., and Brenner, K.P., Comparison of methods for detection and enumeration of airborne microorganisms collected by liquid impingement, *Appl. Environ. Microbiol.*, 62, 2264–2272, 1996.
49. Grinshpun, S.A., Willeke, K., Ulevicius, V., Juozaitis, A., Terzieva, S., Donnelly, J., Stelma, G.N., and Brenner, K.P., Effect of impaction, bounce and reaerosolization on the collection efficiency of impingers, *Aerosol Sci. Technol.*, 26, 326–342, 1997.
50. Willeke, K., Lin, X.J., and Grinshpun, S.A., Improved aerosol collection by combined impaction and centrifugal motion, *Aerosol Sci. Technol.*, 28, 439–456, 1998.
51. Kenny, L.C., Aitken, R., Chalmers, C., Fabries, J.F., Gonzalez-Fernandez, E., Kromhout, H., Liden, G., Mark, D., Riediger, G., and Prodi, V., A collaborative European study of personal inhalable aerosol sampler performance, *Ann. Occup. Hyg.* 41, 135–153, 1997.
52. Demange, M., Gendre, J.C., Hevre-Bazin, B., Carton, B., and Peltier, A., Aerosol evaluation difficulties due to particle deposition on filter holder inner walls, *Ann. Occup. Hyg.*, 34, 399–403, 1990.
53. Baron, P.A., Khanina, A., Martinez, A.B., and Grinshpun, S.A., Investigation of filter bypass leakage and a test for aerosol sampling cassettes, *Aerosol Sci. Technol.*, 36, 857–865, 2002.
54. Baron, P.A. and Bennett, J.S., Calculation of leakage and particle loss in filter cassettes, *Aerosol Sci. Technol.*, 36, 632–641, 2002.
55. Maynard, A.D., Measurement of aerosol penetration through six personal thoracic samplers under calm air conditions, *J. Aerosol Sci.*, 30, 1227–1242, 1999.
56. Chen, C.C. and Baron, P.A., Aspiration efficiency and inlet wall deposition in the fiber sampling cassette, *Am. Ind. Hyg. Assoc. J.*, 57, 142–152, 1996.
57. Baron, P.A., Chen, C.C., Hemenway, D.R., and O'Shaughnessy, P., Nonuniform air-flow in inlets — the effect on filter deposits in the fiber sampling cassette, *Am. Ind. Hyg. Assoc. J.*, 55, 722–732, 1994.
58. Beckett, S.T., The effects of sampling practice on the measured concentration of airborne asbestos, *Ann. Occup. Hyg.* 23, 259–272, 1980.
59. Armendariz, A.J. and Leith, D., A personal sampler for aircraft engine cold start particles; Laboratory development and testing., *Am. Ind. Hyg. Assoc. J.* 64(6), 755–762, 2003.
60. Armendariz, A.J. Leith, D., Boundy, M., and Goodman, R., Sampling and analysis of aircraft engine cold start particles and development and field demonstration of an electrostatic particle sampler, *Am. Ind. Hyg. Assoc. J.* 64(6), 777–784, 2003.

61. Volckens, J. and Leith, D., Comparison of methods for measuring gas-particle partitioning of semivolatile compounds, *Atmos. Environ.*, 37, 3177–3188, 2003.
62. Wagner, J. and Leith, D., Passive aerosol sampler. Part I. Principle of operation, *Aerosol Sci. Technol.*, 34, 186–192, 2001.
63. Wagner, J. and Leith, D., Passive aerosol sampler. Part II. Wind tunnel experiments, *Aerosol Sci. Technol.*, 34, 193–201, 2001.
64. Wagner, J. and Leith, D., Field tests of a passive aerosol sampler, *J. Aerosol. Sci.*, 32, 33–48, 2001.
65. Brown, R.C., Hemingway, M.A., Wake, D., and Thompson, J., Field trials of an electret-based passive dust sampler in metal-processing industries, *Ann. Occup. Hyg.*, 39, 603–622, 1995.
66. Brown, L.M., Electron energy loss spectrometry in the electron microscope. Part 2 — EELS in the context of solid state spectroscopies, in *Impact of the Electron and Scanning Probe Microscopy on Materials Research*, Rickerby, D.G., Valdrè, G., and Valdrè, U., Eds., Kluwer Academic Press, Netherlands, 1999, pp. 231–249.
67. Ogden, T.L. and Birkett, J.L., The human head as a dust sampler, in *Inhaled Particles IV*, Walton, W.H., Ed., Pergamon Press, Oxford, 1977, pp. 91–105.
68. Vincent, J.H. and Mark, D., Applications of blunt sampler theory to the definition and measurement of inhalable dust., *Ann. Occup. Hyg.*, 26, 3–19, 1982.
69. Armbruster, L. and Breuer, H., Investigation into defining inhalable dust, in *Inhaled Particles*, Walton, W.H., Ed., Pergamon Press, Oxford, 1982, pp. 21–31.
70. Mark, D. and Vincent, J.H., A new personal sampler for airborne total dust in workplaces, *Ann. Occup. Hyg.*, 30, 89–102, 1986.
71. Smith, J.P., Bartley, D.L., and Kennedy, E.R., Laboratory investigation of the mass stability of sampling cassettes from inhalable aerosol samplers., *Am. Ind. Hyg. Assoc. J.*, 59, 582–585, 1998.
72. Kenny, L.C., Aitken, R.J., Baldwin, P.E.J., Beaumont, G.C., and Maynard, A.D., The sampling efficiency of personal inhalable aerosol samplers in low air movement environments, *J. Aerosol Sci.*, 30, 627–638, 1999.
73. Hauck, B.C., Grinshpun, S.A., Reponen, A., Reponen, T., Willeke, K., and Bornschein, R.L., Field testing of new aerosol sampling method with a porous curved surface as inlet, *Am. Ind. Hyg. Assoc. J.*, 58, 713–719, 1997.
74. Aizenberg, V., Grinshpun, S.A., Willeke, K., Smith, J., and Baron, P.A., Performance characteristics of the button personal inhalable sampler, *Am. Ind. Hyg. Assoc. J.*, 61, 398–404, 2000.
75. Görner, P., Witschger, O., and Fabries, J.F., Annular aspiration slot entry efficiency of the CIP-10 aerosol sampler, *Analyst*, 121, 1257–1260, 1996.
76. Gero, A. and Tomb, T., Laboratory evaluation of the CIP-10 personal dust sampler, *Am. Ind. Hyg. Assoc. J.*, 49, 286–292, 1988.
77. Robert, K.Q., Cotton dust sampling efficiency of the vertical elutriator, *Am. Ind. Hyg. Assoc. J.*, 40, 535–545, 1979.
78. Fabriès, J.F., Gorner, P., Kauffer, E., Wrobel, R., and Vigneron, J.C., Personal thoracic CIP10-T sampler and its static version CATHIA-T, *Ann. Occup. Hyg.*, 42, 453–465, 1998.
79. NIOSH, Occupational Exposure to Metal Working Fluids, NIOSH Report No. 98–102, 1998.
80. Vincent, J.H., Aitken, R.A., and Mark, D., Porous plastic foam filtration media: penetration characteristics and applications in particle size-selective sampling, *J. Aerosol Sci.*, 24, 929–944, 1993.
81. Kenny, L.C., Aitken, R.J., Beaumont, G., and Gorner, P., Investigation and application of a model for porous foam aerosol penetration, *J. Aerosol Sci.*, 32, 271–285, 2001.
82. Stancliffe, J.D. and Kenny, L.C., Sampling characteristics of modified personal inhalable sampler inlets, *J. Aerosol Sci.*, 25, S601, 1997.
83. Lippmann, M., Nature of exposure to chrysotile, *Ann. Occup. Hyg.*, 38, 459–467, 1994.
84. Baron, P.A., Application of the thoracic sampling definition to fiber measurement, *Am. Ind. Hyg. Assoc. J.*, 57, 820–824, 1996.
85. Lippmann, M., Workshop on the health risks associated with chrysotile asbestos: a brief review, *Ann. Occup. Hyg.*, 38, 638–642, 1994.
86. WHO, Determination of Airborne Fiber Number Concentrations: A Recommended Method, by Phase Contrast Optical Microscopy (membrane filter method), World Health Organization, 1997.

87. Maynard, A.D., Thoracic size-selection of fibers — dependence of penetration on fiber length for five thoracic sampler types, *Ann. Occup. Hyg.* 46, 511–522, 2002.
88. Jones, A.D., Aitken, R.J., Armbruster, L., Byrne, P., Fabriès, J.F., Kauffer, E., Lidén, G., Lumens, M., Maynard, A., Riediger, G., and Sahle, W., *Thoracic Sampling of Fibres*, HSE Books, Norwich, U.K., 2001.
89. Bartley, D.L., Chen, C.C., Song, R.G., and Fischbach, T.J., Respirable aerosol sampler performance testing, *Am. Ind. Hyg. Assoc. J.*, 55, 1036–1046, 1994.
90. Maynard, A.D. and Kenny, L.C., Performance assessment of three personal cyclone models, using an aerodynamic particle sizer, *J. Aerosol Sci.*, 26, 671–684, 1995.
91. Chen, C.-C., Huang, S.-H., LIin, W.-Y., Shih, T.-S., and Jeng, F.-T., The virtual cyclone as a personal respirable sampler, *Aerosol Sci. Technol.*, 31, 422–432, 1999.
92. Courbon, P., Wrober, R., and Fabriès, J.-F., A new individual respirable dust sampler: the CIP-10, *Ann. Occup. Hyg.*, 32, 129–143, 1988.
93. Teikari, M., Linnainmaa, M., Laitinen, J., Kalliokoski, P., Vincent, J., Tiita, P., and Raunemaa, T., Laboratory and field testing of particle size-selective sampling methods for mineral dusts, *Am. Ind. Hyg. Assoc. J.*, 64, 312–318, 2003.
94. Chen, C.-C., Lai, C.-Y., Shih, T.-S., and Yeh, W.-Y., Development of respirable aerosol samplers using porous foam, *Am. Ind. Hyg. Assoc. J.*, 59, 766–773, 1998.
95. Prodi, V., Belosi, F., Mularoni, A., and Lucialli, P., PERSPEC — a personal sampler with size characterization capabilities, *Am. Ind. Hyg. Assoc. J.*, 49, 75–80, 1988.
96. Prodi, V., Sala, C., Belosi, F., Agostini, S., Bettazzi, G., and Biliotti, A., PERSPEC, personal size separating sampler — operational experience and comparison with other field devices, *J. Aerosol Sci.*, 20, 1565–1568, 1989.
97. Kenny, L.C. and Bradley, D.R., Optimization of the PERSPEC multifraction aerosol sampler to new sampling conventions, *Ann. Occup. Hyg.*, 38, 23–35, 1994.
98. Koch, W., Dunkhorst, W., and Lodding, H., RESPICON TM-3 F: a new personal measuring system for size segregated dust measurement at workplaces, *Occup. Health Ind. Med.*, 38, 161, 1998.
99. Rubow, K.L., Marple, V.A., Olin, J., and McCawley, M.A., A personal cascade impactor: design, evaluation and calibration, *Am. Ind. Hyg. Assoc. J.*, 48, 532–538, 1987.
100. Gibson, H., Vincent, J.H., and Mark, D., A personal inspirable aerosol spectrometer for applications in occupational hygiene research, *Ann. Occup. Hyg.*, 31, 463–479, 1987.
101. Misra, C., Singh, M., Shen, S., Sioutas, C., and Hall, P.M., Development and evaluation of a personal cascade impactor sampler (PCIS), *J. Aerosol Sci.*, 33, 1027–1048, 2002.
102. Vincent, J.H., Ramachandran, G., and Kerr, S.M., Particle size and chemical species "fingerprinting" of aerosols in primary nickel production industry workplaces, *J. Environ. Monit.*, 3, 565–574, 2001.
103. Kerr, S.M., Vincent, J.H., and Ramachandran, G., A new approach to sampling for particle size and chemical species "fingerprinting" of workplace aerosols, *Ann. Occup. Hyg.*, 45, 555–568, 2001.
104. Baron, P.A. and Heitbrink, W.A., An approach to performing aerosol measurements, in *Aerosol Measurement: Principles, Techniques and Applications*, Baron, P.A. and Willeke, K., Eds., John Wiley & Sons, New York, 2001, pp. 117–139.
105. Hewett, P. and McCawley, M.A., A microcomputer spreadsheet technique for analyzing multimodal particle size distributions, *Appl. Ind. Hyg.*, 6, 865–873, 1991.
106. Kandlikar, M. and Ramachandran, G., Inverse methods for analysing aerosol spectrometer measurements: a critical review, *J. Aerosol Sci.*, 30, 413–437, 1999.
107. O'Shaughnessy, P.T. and Raabe, O.G., A comparison of cascade impactor data reduction methods, *Aerosol Sci. Technol.*, 37, 187–200, 2003.
108. NIOSH, Method 5522, Isocyanate, in *NIOSH Manual of Analytical Methods*, Eller, P.E., Ed., National Institute for Occupational Safety and Health, Cincinnati, 1998.
109. NIOSH, Method 5521 isocyanates, monomeric, in *NIOSH Manual of Analytical Methods*, Eller, P.E. Ed., NIOSH, Cincinnati, 1994.
110. NIOSH, *NIOSH Manual of Analytical Methods*, 4th ed., National Institute for Occupational Safety and Health, Cincinnati, 1994.

111. Reponen, T., Willeke, K., Grinshpun, S., and Nevalainen, A., Biological particle sampling, in *Aerosol Measurement: Principles, Techniques and Applications*, Baron, P.A. andWilleke, K., Eds., Wiley & Sons, New York, 2001.
112. Macher, J.M., *Bioaerosols: Assessment and Control*, ACGIH, Cincinnati, 1998.
113. Calderon, C., Ward, E., Freeman, J., and McCartney, A., Detection of airborne fungal spores sampled by rotating-arm and Hirst-type spore traps using polymerase chain reaction assays, *J. Aerosol. Sci.*, 33, 283–296, 2002.
114. Williams, R.H., Ward, E., and McCartney, H.A., Methods for integrated air sampling and DNA analysis for detection of airborne fungal spores, *Appl. Environ. Microbiol.*, 67, 2453–2459, 2001.
115. Alvarez, A.J., Buttner, M.P., and Stetzenbach, L.D., PCR for bioaerosol monitoring — Sensitivity and environmental interference, *Appl. Environ. Microbiol.*, 61, 3639–3644, 1995.
116. Schafer, M.P., Fernback, J.E., and Jensen, P.A., Sampling and analytical method development for qualitative assessment of airborne mycobacterial species of the mycobacterium tuberculosis complex, *Am. Ind. Hyg. Assoc. J.*, 59, 540–546, 1998.
117. van Tongeren, M.J.A., Gardiner, K., and Calvert, I.A., An assessment of the weight-loss in transit of filters loaded with carbon black., *Ann. Occup. Hyg.*, 38, 319–323, 1994.
118. Awan, S. and Burgess, G., The effect of storage, handling and transport traumas on filter-mounted dusts, *Ann. Occup. Hyg.*, 40, 525–530, 1996.
119. Vaughan, N.P., Milligan, B.D., and Ogden, T.L., Filter weighing reproducibility and the gravimetric detection limit, *Ann. Occup. Hyg.*, 33, 331–337, 1989.
120. American Society for Testing and Materials, Standard Practice for Controlling and Characterizing Errors in Weighing Collected Aerosols., ASTM, 2000.
121. Lawless, P.A. and Rodes, C.E., Maximizing data quality in the gravimetric analysis of personal exposure sample filters, *J. Am. Waste Manage. Assoc.*, 49, 1039–1049, 2001.
122. ISO, Workplace Atmospheres — Controlling and Characterizing Errors in Weighing Collected Aerosols, Report No. ISO 15767:2003, 2003.
123. NIOSH, Method 7402: Asbestos by TEM, in *NIOSH Manual of Analytical Methods*, Eller, P.E. Ed., NIOSH, Cincinnati, 1994.
124. EPA, Measuring Airborne Asbestos Following an Abatement Action, EPA 600/4-85–049, 1985.
125. Taylor, D.G., Baron, P.A., Shulman, S.A., and Carter, J.W., Identification and counting of asbestos fibers, *Am. Ind. Hyg. Assoc. J.*, 45, 84–88, 1984.
126. Gebhart, J., Optical direct-reading techniques: light intensity systems, in *Aerosol Measurement: Principles, Techniques, and Applications*, Baron, P. and Willeke, K., Eds., John Wiley & Sons, New York, 2001, pp. 419–454.
127. Liu, L.-J.S., Slaughter, J.C., and Larson, T.V., Comparison of light scattering devices and impactors for particulate measurements in indoor, outdoor, and personal environments, *Environ. Sci. Technol.*, 36, 2977–2986, 2002.
128. Willeke, K. and Degarmo, J., Passive versus active aerosol monitoring, *Appl. Ind. Hyg.*, 3, 263–266, 1988.
129. Baron, P.A., Aerosol photometers for respirable dust measurements, in *NIOSH Manual of Analytical Methods*, Eller, P.E., Ed., NIOSH, Cincinnati, 1985.
130. Woskie, S.R., Shen, P., Eisen, E.A., Finkel, M.H., Smith, T.J., Smith, R., and Wegman, D.H., The real-time dust exposure of sodium borate workers: examination of exposure variability, *Am. Ind. Hyg. Assoc. J.*, 55, 207–217, 1994.
131. Lehockey, A.H. and Williams, P.L., Comparison of respirable samplers to direct-reading real-time aerosol monitors for measuring coal dust, *Am. Ind. Hyg. Assoc. J.*, 57, 1013–1018, 1996.
132. Trent, A., Davies, M.A., Karsky, R., and Fisher, R., Report No. 0125-2832-MTDC, 2001.
133. Williams, K. and Vinson, R.P., Evaluation of the TEOM dust monitor, U.S. Bureau of Mines Information circular Report No. 9119, 1986.
134. Aitken, J., On the number of dust particles in the atmosphere, *Nature*, 37, 187–206, 1888.
135. Hämeri, K., Koponen, I.K., Aalto, P.P., and Kulmala, I., The particle detection efficiency of the TSI-3007 condensation particle counter, *J. Aerosol Sci.*, 33, 1463, 2002.
136. Volkwein, J.C., Schoeneman, A.L., and Page, S.J., Laboratory evaluation of pressure differential-based respirable dust detector tube, *Appl. Occup. Environ. Hyg.*, 15, 158–164, 2000.

137. Ramani, R.V., Mutmansky, J.M., He, H., Marple, V.A., Olson, B.A., Luna, P.C., and Volkwein, J.C., Evaluation of the respirable dust dosimeter for real-time assessment of airborne respirable coal mine dust exposures, in *Mine Ventilation: Proceedings of the North American Ninth US Mine Ventilation Symposium*, De Souza, E. Ashgate Pub. Co., Kingston, Canada, 2002, pp. 579–586.
138. Brunauer, S., Emmett, P.H., and Teller, E., Adsorption of gases in multimolecular layers, *J. Am. Chem. Soc.*, 60, 309, 1938.
139. Gäggeler, H.W., Baltensperger, U., Emmenegger, M., Jost, D.T., Schmidt-Orr, A., Haller, P., and Hofmann, M., The epiphaniometer, a new device for continuous aerosol monitoring, *J. Aerosol Sci.*, 20, 557–564, 1989.
140. Pandis, S.N., Baltensperger, U., Wolfenbarger, J.K., and Seinfeld, J.H., Inversion of aerosol data from the epiphaniometer, *J. Aerosol Sci.*, 22, 417–428, 1991.
141. Keller, A., Fierz, M., Siegmann, K., Siegmann, H.C., and Filippov, A., Surface science with nanosized particles in a carrier gas, *J. Vacuum Sci. Technol., — Vacuum Surf. Films*, 19, 1–8, 2001.
142. Baltensperger, U., Weingartner, E., Burtscher, H., and Keskinen, J., Dynamic mass and surface area measurements, in *Aerosol Measurement, Principles, Techniques and Applications*, 2nd ed., Baron, P.A. and Willeke, K., Eds., Wiley-Interscience, New York, 2001, pp. 387–418.
143. Wilson, W.E., Han, H.S., Stanek, J., Turner, J.H., and Pui, D.Y.H., The Fuchs surface area, as measured by charge acceptance of atmospheric particles, may be a useful indicator of the surface area of particles deposited in the Lungs, in *European Aerosol Conference*, Madrid, Spain, Pergamon, Oxford, 2003, p. S421.
144. Woo, K.-S., Chen, D.-R., Pui, D.Y.H., and Wilson, W.E., Use of continuous measurements of integral aerosol parameters to estimate particle surface area, *Aerosol Sci. Technol.*, 34, 57–65, 2001.
145. Maynard, A.D., Estimating aerosol surface area from number and mass concentration measurements, *Ann. Occup. Hyg.*, 47, 123–144, 2003.
146. Baron, P.A., Using a filter bypass leakage test for aerosol sampling cassettes, *Appl. Occup. Environ. Hyg.*, 17(9), 593–597, 2002.
147. OSHA, Respiratory Protection, 29CFR Part 1910.134, 1998.
148. Willeke, K., Ayer, H.E., and Blanchard, J.D., New methods for quantitative respirator fit testing with aerosols, *Am. Ind. Hyg. Assoc. J.*, 42, 121–125, 1981.
149. Han, D.-H., Willeke, K., and Colton, C.E., Quantitative fit testing techniques and regulations for tight-fitting respirators: current methods measuring aerosol or air leakage, and new developments, *Am. Ind. Hyg. Assoc. J.*, 58, 219–228, 1997.
150. Rosén, G., PIMEX^{TM}. Combined use of air sampling instruments and video filming: experience and results during six years of use, *Appl. Occup. Environ. Hyg.*, 8, 344–347, 1993.
151. Gressel, M.G. and Heitbrink, W.A., Eds., Report No. DHHS (NIOSH), Publication Number 92–104, 1992.
152. Cooper, T.C., Gressel, M.G., Froehlich, P.A., Caplan, P.E., Mickelson, R.L., Valiante, D., and Bost, P., Successful reduction of silica exposure at a sanitary ware pottery, *Am. Ind. Hyg. Assoc. J.*, 54, 600–606, 1993.
153. Brown, R.C., Wake, D., Thorpe, A., Hemingway, M.A., and Roff, M.W., Preliminary assessment of a device for passive sampling of airborne particulate, *Ann. Occup. Hyg.*, 38, 303–318, 1994.
154. Aizenberg, V., Grinshpun, S.A., Willeke, K., Smith, J., and Baron, P.A., Performance characteristics of the button personal inhalable aerosol sampler, *Am. Ind. Hyg. Assoc. J.*, 61, 398–404, 2000.
155. Lidén, G., Evaluation of the SKC personal respirable dust sampling cyclone, *Appl. Occup. Environ. Hyg.*, 8, 178–190, 1993.
156. Dunmore, J.H., Hamilton, R.J., and Smith, D.S.G., An instrument for the sampling of respirable dust for subsequent gravimetric assessment, *J. Sci. Instrum.*, 41, 669–672, 1964.
157. Kenny, L.C., Bowry, A., Crook, B., and Stancliffe, J.D., Field testing of a personal size-selective bioaerosol sampler, *Ann Occup Hyg.*, 43, 393–404, 1999.

chapter twelve

Medical and pharmaceutical aerosols

Hugh D.C. Smyth, Lucila Garcia-Contreras, Daniel J. Cooney, Robert J. Garmise, Latarsha D. Jones, and Anthony J. Hickey
University of North Carolina

Contents

12.1 Introduction 266
12.1.1 Historical perspective....................266
12.1.2 Future prospects 266
12.2 Therapeutic agents 266
12.2.1 Locally acting medical and pharmaceutical aerosols 266
12.2.1.1 β-Adrenergic agonists 266
12.2.1.2 Corticosteroids 267
12.2.1.3 Anticholinergics 268
12.2.1.4 Antiinflammatory agents 268
12.2.1.5 Antimicrobials....................268
12.2.1.6 Biotechnological agents, genes, and DNA aerosols 268
12.2.2 Systemically acting agents....................269
12.3 Classification 269
12.3.1 Metered dose inhalers 269
12.3.1.1 CFC systems 270
12.3.1.2 HFC systems 270
12.3.1.3 Alternative propellants 270
12.3.2 Dry powder inhalers 272
12.3.2.1 Passive 272
12.3.2.2 Active 273
12.3.3 Nebulizers 274
12.3.3.1 Solutions 274
12.3.3.2 Suspensions....................275
12.3.3.3 Macromolecules 275
12.3.4 Handheld aqueous systems 276
12.3.5 Topical drug delivery sprays 276
12.3.6 Specific measurement techniques and calibration....................277
12.3.6.1 Particle size measurement....................277
12.4 Specific measurement techniques and calibration 277
12.4.1 Inertial methods 277
12.4.1.1 Cascade impactor 277
12.4.1.2 Twin impinger....................279
12.4.2 Optical methods 280

1-56670-611-4 / 05 / $0.00+$1.50

12.4.3 Other *in vitro* measurement techniques commonly employed for pharmaceutical aerosols........280
12.5 Respiratory deposition, retention, and dosimetry........280
12.5.1 Deposition........280
12.5.1.1 Gamma scintigraphy, PET, and SPECT........280
12.5.2 Retention........281
12.5.3 Dosimetry........281
12.5.3.1 Bolus delivery........282
12.5.3.2 Continuous delivery........282
References........284

12.1 Introduction

12.1.1 Historical perspective

The use of aerosol therapy for the treatment of pulmonary disorders can be traced to India over 5000 yr ago.[1] These therapies were either palliative, in the form of smokes and mists, or therapeutic, containing pharmacologically active agents such as stramonium alkaloids.[2] There have been periodic improvements in our understanding of diseases, which give relevance to aerosol approaches.

Modern aerosol therapy with pure drugs was initiated in the 1950s with the development of the pressurized metered dose inhaler (pMDI) for the delivery of β-adrenergic agonists to facilitate bronchodilatation to relieve the bronchoconstriction occurring in asthmatic patients.[3] In the past 50 years, aerosol therapy has become a central element of asthma management, and its potential for the treatment of other pulmonary and systemic diseases has been explored.

12.1.2 Future prospects

The future of aerosol therapy seems assured. New developments occur at frequent intervals in the areas of drug discovery, formulation, and device development.[4,5] The major areas in which new therapies can be anticipated are treatment of lung diseases such as chronic obstructive pulmonary disease (COPD),[6] emphysema,[7] lung cancer,[8] and systemic diseases such as diabetes[9] and prostate cancer.[10]

New drugs have been developed for the treatment of asthma and COPD such as the anticholinergic agent, tiotropium.[11] Undoubtedly, new compounds will continue to be developed.

Formulation strategies have focused on methods of particle manufacture[12] and desirable physicochemical characteristics.[13,14] These particles can then be placed in one of three general categories of device: pMDIs, dry powder inhalers (DPIs), or nebulizers.[15] There continue to be new and exciting developments in each of the categories of the device.[5]

12.2 Therapeutic agents

12.2.1 Locally acting medical and pharmaceutical aerosols

12.2.1.1 β-Adrenergic agonists

Figure 12.1a shows the structure of a β-adrenergic agonist. This category of compounds was the first to be commercially available for the treatment of the symptom of bronchoconstriction in asthmatic individuals. Structurally, these agents fall into three groups: catechols, resorcinols, and saligenins.[16] Each compound was a structural analog to epinephrine (adrenaline). Indeed, epinephrine appeared in an early product, which is now available over-the-counter at pharmacies (Primatene, Whitehall-Robins, Richmond, VA,

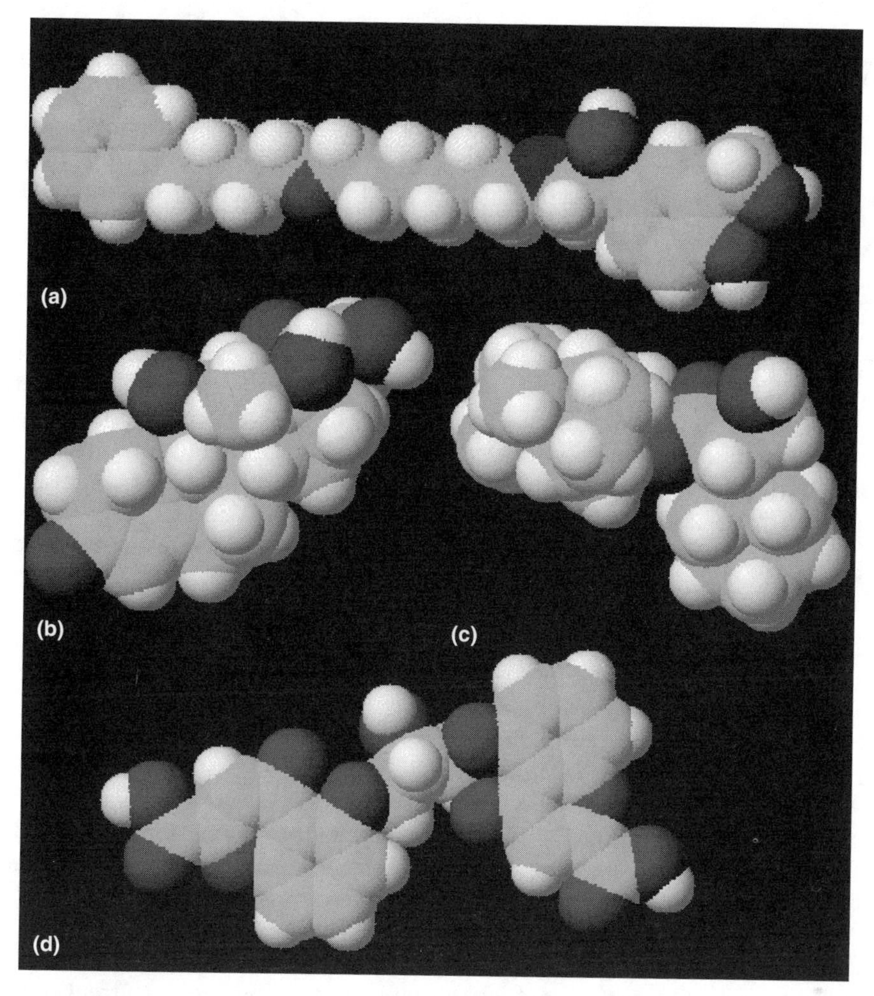

Figure 12.1 Examples of structures of common pharmacological agents used in medical and pharmaceutical aerosols for delivery to the lungs. (Space-filling models of **a** — salmeterol, **b** — fluticasone, **c** — ipratropium, and **d** — cromolyn.)

U.S.A). After some early toxicity issues with nonspecific β-adrenergic agonists, specific β_2-adrenergic agonists were produced and introduced with great success to manage the symptoms of asthma. The first specific product, albuterol (GSK, RTP, NC, U.S.A.), was a short-acting agent introduced in the 1960s. It was followed by a number of variants, including fenoterol (Boehringer Ingelheim, Ingelheim, Germany) and terbutaline (Astra-Zeneca, Lund, Sweden). In the late 1980s and early 1990s, these were replaced for routine maintenance therapy by long-acting β_2-adrenergic agonists, notably formoterol (Boehringer Ingelheim, Ingelheim, Germany) and salmeterol (GSK, RTP, NC, U.S.A.). The short-acting agents were retained as rescue medications for patients experiencing acute exacerbations.

12.2.1.2 Corticosteroids

The mechanism of action of β_2-adrenergic agonists is to act on the sympathetic system to cause muscle relaxation and, therefore, bronchodilation in the lungs.

Figure 12.1b shows the structure of a glucocorticosteroid molecule. Inhaled corticosteroids were introduced to treat the underlying cause of asthma: inflammation. Corticosteroids had been used as an oral therapy for severe, life-threatening conditions in the form of prednisone and prednisolone. The development of agents with some lung specificity gave rise to the concept of inhaled steroid therapy. The first agent to be developed was beclomethasone (GSK, RTP, NC, U.S.A. and Schering Plough, White Plains, NJ) followed closely by triamcinolone (Aventis, Collegeville, PA, U.S.A.), both in pMDIs. In the late 1980s and early 1990s, two steroids with a high degree of lung specificity, budesonide and fluticasone, were developed as DPI products. These molecules were shown to induce lower side effects, notably cortisol suppression, than their predecessors.

12.2.1.3 *Anticholinergics*

Figure 12.1c shows the structure of an anticholinergic agent. Anticholinergic agents are known parasympathetic antagonists that operate on the opposing arm, balancing bronchomotor tone to the sympathomimetic agents described in Section 12.2.1. The first anticholinergic agent was ipratropium, which was followed by oxitropium. Cholinergic receptors are known to be centrally located in the airways and consequently these agents are effective in dilating the bronchioles in this region. Tiotropium is a recent addition to this category of drugs and has been shown to be very effective in COPD, which requires not only bronchodilatation but also the loosening of mucus. All these agents are produced by Boehringer Ingelheim (Ingelheim, Germany).

12.2.1.4 *Antiinflammatory agents*

Figure 12.1d shows the structure of a unique antiinflammatory agent. Disodium cromoglycate (cromolyn sodium) (Aventis, Collegeville, PA, U.S.A.) was developed in the 1960s and 1970s and has been shown to have a number of effects in the lungs. The most prominent effect is mast cell stabilization and the prevention of release of inflammatory mediators. However, the nature of the dominant action that renders this molecule therapeutically effective is not clear. This molecule has the distinction of being the first in modern times to be administered as a dry powder aerosol. A follow-up molecule was developed by the same company in the 1980s: nedocromil sodium.

12.2.1.5 *Antimicrobials*

A number of antimicrobial agents have been delivered to the lungs to treat different diseases. Pentamidine (Fujisawa, North Chicago, IL, U.S.A. and Aventis, Collegeville, PA, U.S.A.) was delivered for the treatment of *Pneumocystis carinii* pneumonia (PCP). This organism was originally considered to be a parasite, but has recently been redesignated taxonomically as a fungus. Fungal therapy had already been attempted with amphotericin B for the treatment of aspergillosis. The occurrence of *Pseudomonas aeruginosa* infections as a corollary to cystic fibrosis has engendered considerable interest in antibiotic aerosol therapy. After initial work on amikacin products, most notably this has resulted in the development of tobramycin aerosol treatment (Chiron, San Francisco, CA, U.S.A.).

12.2.1.6 *Biotechnological agents, genes, and DNA aerosols*

The most sustained effort in the area of delivery of biotechnological agents has been for the treatment of cystic fibrosis. In the early 1990s, rDNase aerosols (Genetech, South San Francisco, CA, U.S.A.) were delivered to cleave leukocyte DNA, which was contributing to the viscocity of mucus in cystic fibrotic lungs. Delivery of this aerosol allowed the patient to expectorate readily and clear the lungs of mucin blockages. In combination with tobramycin to treat the bacterial infection, this appears to have been a successful approach. However, it has long been the objective of those involved in gene therapy to challenge their technology by expressing the gene for cystic fibrosis transport receptor (CFTR) in the epithelium of airway cells, thereby correcting the underlying chloride ion imbalance which gives rise to thickened mucus and poor mucociliary clearance. As yet this approach has met with limited success, but it remains the goal of a number of researchers.[17] Oligonucleotides have also been used to target the adenosine A(1) receptor, a G-protein-coupled-receptor (GPCR) that plays an important role in the etiology of asthma.[18]

12.2.2 Systemically acting agents

The lung has been considered a route of administration for systemically acting agents for decades. The first systemically acting agent delivered as an aerosol product in the 20th century was ergotamine tartrate for the treatment of migraine headaches.[19] More recently, the focus has been on proteins and peptides. Among these, the notable candidates to date have been insulin[9] for the treatment of diabetes, and leuprolide acetate, a leutinizing hormone releasing hormone analog,[10] for the treatment of prostate cancer.

Other agents have been evaluated, including calcitonin, human growth hormone, parathyroid hormone,[20] but these have progressed more slowly than the candidates mentioned in the previous paragraph.

12.3 Classification

12.3.1 Metered dose inhalers

Many devices used to deliver drugs to the respiratory tract do so by producing a metered dose of aerosolized drug that is inhaled by the patient. However, metered dose inhalers (MDIs) are specifically recognized as devices that contain a pressurized formulation that is aerosolized through an atomization nozzle. More precisely, these devices should be called pMDIs to avoid confusion with MDIs that incorporate and use dry powders and aqueous-based systems (discussed below).

The first MDI was commercialized in the mid-1950s to compete with glass nebulizers for the delivery of asthma medications. The acceptance and utility of these delivery systems was quickly established and they have now become the most common system for drug delivery to the respiratory tract. The basic design of pMDIs (Figure 12.2) has not changed greatly since their inception and they typically contain three basic components: the active

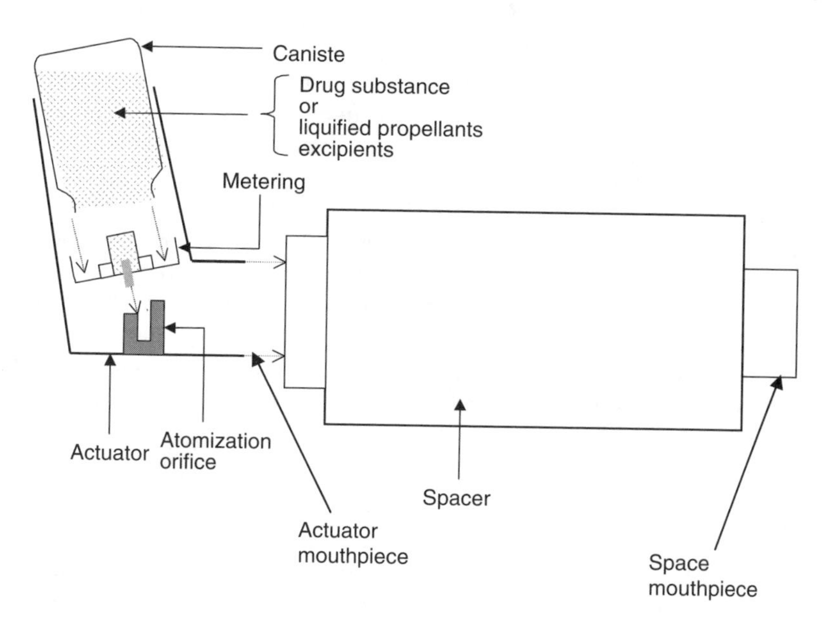

Figure 12.2 Basic components of a pMDI system.

substance, the propellant system, and other stabilizing excipients. These are enclosed within an aerosol container and a metering valve that connects to an actuator or aerosolization nozzle. An adaptor mouthpiece, which may include one of a variety of holding chambers, allows the patient to draw the aerosol into the lungs. The composition and design of pMDIs affects the performance of the drug delivery system.[21] Propellants serve as a source of energy for atomization of the liquid formulation as it exits the nozzle. They also function as a liquid phase for the dispersion of drug and other excipients that are often present in suspension. Surfactants aid dispersion or dissolution in addition to providing lubrication for valve components.[21–23] Solution formulations are typically attained by incorporation of cosolvents such as ethanol.[21,24] Currently marketed products and those in development can be divided into three classes based upon the propellant system used.

12.3.1.1 CFC systems

Chlorofluorocarbon (CFC) propellants have been the most common propellant type used in pMDIs.These propellants include CFC 12 (dichlorodifluoromethane), CFC 11 (trichlorofluoromethane), and CFC 114 (1,2-dichloro-1,1,2,2-tetrafluoroethane). The widespread use of these propellants was a consequence of their low pulmonary toxicity, high chemical stability, purity, and compatibility.[21] Also, mixtures of these propellants can be formulated to yield desirable vapor pressures, densities, and solvency properties for the successful formulation of a variety of drug substances. However, pMDIs containing CFCs are now being phased out and replaced by other propellant systems. This is a result of the linking of CFCs to the depletion of stratospheric ozone and the signing of the Montreal Protocol on substances that deplete the ozone layer.[25,26]

12.3.1.2 HFC systems

Several hydrofluorocarbon (HFC, also referred to as hydrofluoroalkanes or HFAs) compounds were identified as possible CFC propellant alternatives because of their non-ozone-depleting capacity and similar desirable characteristics in common with CFC pMDI propellants (nonflammability, chemical stability, and similar vapor pressures). These propellants currently include HFC 134a (1,1,1,2-tetrafluoroethane) and HFC 227ea (1,1,1,2,3,3,3-heptafluoropropane).[27] Extensive toxicological and safety testing demonstrated that these propellants were at least as safe as the CFC propellants.[27–29] Subsequently, inhalers have been approved by regulatory agencies for medical use (Table 12.1). Despite apparent similarities with CFCs, the HFC alternatives have required different formulation strategies and device designs.[25,30,31] In general, these challenges include overcoming different solvency properties,[31–33] designing different materials and coatings for compatibility issues,[34] and identifying different atomization behaviors.[35–37] An example of a reformulated product is an HFA inhaler containing beclomethasone dipropionate, which is now marketed as a solution formulation instead of a suspension formulation. [38] Regional lung deposition was shown to be different from the existing CFC product due to a smaller droplet size and throat deposition characteristics.[38] Thus, in addition to pharmaceutical issues, regulatory issues are also important in the transition to more "environmentally friendly" propellant systems. These issues have meant that 14 years since the signing of the Montreal Protocol, over half of the world's pMDIs are still CFC-based systems.[27]

12.3.1.3 Alternative propellants

This third category of propellant systems includes propellants such as dimethylether and low-molecular-weight hydrocarbons such as butane and propane.[39,40] Compressed gases have also been considered for incorporation into medical aerosol systems.[21,40] Although under development, the commercialization of these propellant systems has been restricted by

Table 12.1 Commonly Marketed pMDIs and their General Composition

Therapeutic Group	Drug	Surfactants/ Excipients	Propellant System	Formulation Type	Particle Size Estimate (μm)	Reference
			Bronchodilators			
Maxair	Pirbuterol acetate	Sorbitan trioleate	CFC 11, CFC 12	Suspension	3.3	127
Maxair Autohalor	Pirbuterol acetate	Sorbitan trioleate	CFC 11, CFC 12	Suspension	3.1	127
Proventil	Albuterol sulfate	Oleic acid	CFC 11, CFC 12	Suspension		
Proventil HFA	Albuterol sulfate	Oleic acid	HFA 134a, ethanol	Suspension	1.96	127
Tornalate	Bitolterol mesylate	Ascorbic acid, saccharin, menthol	38% w/w ethanol, CFC 11, CFC 12,	Solution	2.21	128
Ventolin	Albuterol sulfate	Oleic acid	CFC 11, CFC 12	Suspension	3.33	129
Ventolin HFA	Albuterol sulfate	None	HFA 134a	Suspension		
			Corticosteriods			
Aerobid	Flunisolide	Sorbitan trioleate, menthol	CFC 11, CFC 12, CFC 114	Suspension	4.14	127
Azmacort	Triamcinolone acetonide		CFC 12, 1% w/w ethanol	Suspension	4.33	130
Beclovent	Beclomethasone dipropionate	Oleic acid	CFC 11, CFC 12	Suspension	4	131
Becotide 100	Beclomethasone dipropionate	Oleic acid	CFC 11, CFC 12	Suspension		
Flovent	Fluticasone propionate		CFC 11, CFC 12	Suspension	2.5	131
QVAR	Beclomethasone dipropionate		HFA 134a, ethanol	Solution	1.0	131
QVAR Autohaler	Beclomethasone dipropionate		HFA 134a, ethanol	Solution	1.0	131
Vanceril	Beclomethasone dipropionate		CFC 11, CFC 12	Suspension		
			Other Anti-inflammatory			
Intal	Cromolyn sodium	Sorbitan trioleate	CFC 11, CFC 12	Suspension	4.65	127
Tilade	Nedocromil sodium	Sorbitan trioleate	CFC 11, CFC 12	Suspension		

formulation issues and/or safety issues. One concern has been the potential flammability of propellants such as butane, propane, and dimethylether.[21] Handheld aqueous systems are also under development and are similar in appearance and operation to pMDIs. However, these systems are more closely related to aqueous-based nebulizers and are discussed below.

12.3.2 *Dry powder inhalers*

DPIs have been available since the 1970s when the Rotahaler® (GlaxoSmithKline) and the Spinhaler® (Aventis) were introduced to the public.[41] DPIs provide an alternative to MDIs and offer the advantage of not using environmentally unfriendly propellants. However, without a propellant to create an aerosol cloud, other dispersion methods must be employed. Most DPIs employ the energy from patient inhalation to disperse the powder. The airflow is directed over or through the static powder bed in order to fluidize it and allow it to be entrained into the inspiratory airflow. This technique offers the advantage of automatic coordination with patient inhalation, a problem with MDIs, but does bring up a number of challenges. The energy from patient inhalation must be applied in a manner that is able to overcome the cohesive and adhesive forces in the powder to allow dispersion to the lung.[42] Some patients may not be able to generate a high enough flow to disperse particles and varying flow rates can lead to varying delivered doses.[43] These characteristics, along with hydration of the powders and surface electrical properties, are determined by a combination of drug formulation and dispersion device. For this reason, the dispersion problem must be viewed from two different angles: formulation and device design.

The aim of formulation optimization is to improve powder dispersion properties. The efficiency of an inhaler device is often measured by the amount of particles that enter the periphery of the lung during inhalation. These particles need to have an aerodynamic diameter smaller than 5–7 μm, and the portion of the aerosol falling into this category is generally referred to as the fine particle fraction (FPF).[44] The interparticulate adhesive forces become dominant in particles in this size range and tend to form agglomerates or aggregates.[5] The adhesive forces are influenced not only by particle size but also by shape, crystallinity, surface morphology, and surface chemistry.[42] A number of approaches have been taken to reduce these forces and increase the flowability and dispersibility of the powders.[45] The manufacture of small particles is accomplished by several means. Breaking methods such as milling are employed, but these particles tend to have irregular size, shape and surface characteristics, and high cohesiveness.[5] Constructive methods like spray drying, evaporation, extraction, and condensation have also been used and tend to produce more uniform and lower energy particles.[12] Even uniform particles in the 5 μm size range tend to adhere to one another and have poor dispersion and flow properties. For this reason, excipients are often included in dry powder formulations. Carrier particles such as lactose are blended with the active compound to control the dispersion.[42] The carrier particles are designed to keep the small drug particles from forming agglomerates by allowing adhesion to the carrier until the inhalation energy frees the particles. Similar to drug particles alone, the size, shape, and surface of the carrier particles affect the dispersion. Ternary components such as L-leucine and fine particle lactose have been added to blended formulation systems to decrease binding during inhalation, thus increasing FPF.[46,47] Another approach to increasing the FPF of the formulation is to produce large, low-density particles.[48] These particles have aerodynamic diameters in the appropriate range, but the adhesive forces are smaller in ratio because of the large size.

12.3.2.1 *Passive*

Formulation is only one side of the DPI. Even the best formulation needs a vehicle for dispersion, the inhaler device. The first DPIs developed used the patients' own inhalation airflow directed through or across a capsule that is broken prior to inhalation to aerosolize the powder.[5] Because the only energy source used to deaggregate the particles is the inhalation, these devices are deemed passive inhalers. To increase the deaggregation energy, baffles or deflected airflow are used to create turbulence around the powder bed during inhalation. [42] The first passive inhalers (Spinhaler and Rotahaler) required the loading of a new capsule for each use as single-dose inhalers. Since then multi-single-dose (device holds multiple capsules or blisters) and true multi-dose inhalers (dose taken from the reservoir for each inhalation) have become available. Table 12.2 lists some inhalers and their properties.

12.3.2.2 Active

Using the inspiratory force of the patient is no longer the only way in which powder is dispersed in inhaler design. There has been a recent push to develop active rather than passive inhalers. The inhalation flow rate and force developed varies from breath to breath and certainly from person to person. This makes it very difficult to insure both that a proper dose is delivered and that the inhaler is operating at maximum efficiency. Active DPIs use a source of energy other than inhalation to disperse the powder. Compressed air from a user-operated pump is used as the dispersion energy in the Inhance™ (Nektar Therapeutics, San Carlos, CA, U.S.A.) inhaler. Another device uses battery power to drive an impeller (Spiros, San Diego, CA, U.S.A.). Other patents have been filed for active DPI designs using energy sources such as vacuum pressure, an impaction hammer, and vibration.[5,49,50]

Table 12.2 Selected Dry Powder Inhalers and some of their Properties[a]

Inhaler	Company	Energy Source	Carrier	Powder Supply	Dosing	Doses
Rotahaler	GSK	Passive	Lactose	Capsule	Single-dose	1
Spinhaler	Fision/Aventis	Passive	None	Capsule	Single-dose	1
Inhalator	Boehringer Ingelheim	Passive	Glucose	Capsule	Multiple unit-dose	6
Diskus/ Accuhaler	GlaxoSmithKline	Passive	Lactose	Blister	Multiple unit-dose	60
Aerohaler	Boehringer Ingelheim	Passive		Capsule	Multiple unit-dose	6
Diskhaler	GlaxoSmithKline	Passive	Lactose	Blister	Multiple unit-dose	4, 8
Easyhaler	Orion	Passive	Lactose	Reservoir	Multidose	200
Airmax	IVAX	Passive		Reservoir	Multidose	
Novolizer	Sofotec	Passive	Lactose	Reservoir	Multidose	200
Twisthaler	Schering-Plough	Passive		Reservoir	Multidose	60
Turbuhaler	AstraZeneca	Passive	None	Reservoir	Multidose	200
Spiros	Elan Pharmaceuticals	Impeller	N/A	Blister, cassette	Single-dose, multiple unit-dose	1, 16, or 30
Inhance	Nektar Therapeutics	Compressed gas	Lactose	Blister	Single-dose	1
Dynamic Powder Disperser	Pfeiffer	Compressed gas	Lactose	Cartridge	Multiple unit-dose	12

[a]Adapted from. Dunbar et al.[42] and Newman and Busse.[49]

12.3.3 Nebulizers

Nebulization is probably the oldest means of administering drugs to the lungs as aerosols. Because coordination between breathing and aerosol generation is not necessary, nebulization of therapeutic agents is mostly used by children, elderly patients, in hospital settings, and for the treatment of lung diseases such as cystic fibrosis and asthma.[51–53] Nebulizers produce small polydisperse droplets capable of delivering therapeutic agents to the deep lung, in large doses, and can deliver a dose ten times larger than that of DPIs or MDIs.[51] Mass median diameters of droplets generated by nebulizers normally range from 2 to 5 μm, and have been used to deliver solutions and suspensions of a great variety of therapeutic agents including macromolecules and biotechnology products.[54] Miniaturization of the hardware and introduction of high-output nebulizers that shortens dosing times and increases drug delivery to the patient have increased the usage of nebulizers.[55,56] However, some disadvantages related to their use include lack of portability, use of other accessories such as tubing, mouthpiece, or facemasks, obstructive dosing, and cleaning requirements.[57] Others, depending on the nebulizer type, may include interdevice variability, greater costs of drug delivery as a result of the need for extensive assistance from health-care personnel, and the requirement of high doses to achieve therapeutic effect.[51]

Based on the mechanism of aerosol production, nebulizers can be classified into ultrasonic or air-jet. In jet nebulizers, the aerosol is produced by applying a high-velocity airstream from a pressurized source at the end of a capillary tube; liquid can be drawn up the tube from a reservoir in which it is immersed. When the liquid reaches the end of the capillary, it is drawn into the airstream and forms droplets that disperse to become an aerosol.[58] In ultrasonic nebulizers, the solution is aerosolized by the vibration of a piezoelectric crystal to induce waves in a reservoir of solution. Interference of these waves at the reservoir surface leads to the production of droplets in the atmosphere above the reservoir. An airstream is passed through this atmosphere to transport the droplets as an aerosol.[59] Ultrasonic nebulizers are operated electrically or by battery. Temperature increase in the nebulized solution occurs because of the production of heat due to frictional forces induced by the movement of the transducing crystal. This may be detrimental for proteins and thermolabile drugs. Ultrasonic nebulizers are less popular than jet nebulizers because they are more expensive and not disposable. A common problem with ultrasonic nebulizers is that continuous atomization does not occur if the volume in the chamber falls below 10 ml.[51] Detailed information on these mechanisms of aerosolization has been described elsewhere.[60,61]

Some factors that influence nebulizer performance and droplet size are density and velocity of the atomizing air, surface tension and viscosity of liquid, concentration, temperature, and nebulizer design.[60] The efficiency of nebulizers is often expressed in terms of energy use, function, or output. Output can be measured by simply weighing the nebulizer before and after operation. Variation in total output, particle size, and overall efficiency has been reported between different nebulizers. Several studies have evaluated the differences between nebulizers. Chan et al.[62] examined the aerosol characteristics of five handheld nebulizers in terms of aerosol output and droplet size. Smith et al.[63] tested the variability among 23 different nebulizer/compressor combinations in terms of MMD, while Weber et al.[64] optimized the nebulizer conditions of several nebulizers used to administer antibiotics for the treatment of cystic fibrosis.

12.3.3.1 Solutions

Nebulizers are commonly used with solutions of bronchodilators, such as albuterol or tertbutaline, and other drugs like sodium cromoglycate, corticosteroids, and pentamidine.[65,66]

Combinations of drugs could be administered at the same time; however, the stability and possible interactions of the components should be evaluated before administration.[67] A possible outcome can be an insoluble complex in certain conditions. Interactions during nebulization of amiloride hydrochloride and nucleotide UTP have been documented.[68] Other factors that should be accounted for are the effects that some excipients included in the solution to be nebulized may have on the patient. Inclusion of preservatives in the solutions of bronchodilators has been reported to cause bronchoconstriction instead, and the addition of osmotic agents may also cause side effects.[57,58,69–71] The effect of other additives, ionic strength, and contamination could also influence the effect of the nebulized solution.[72]

12.3.3.2 Suspensions

Insoluble or inert particles can be suspended in a solution and delivered by nebulization, provided the particle size is smaller than the droplet size and the density of the particle is relatively small. Most steroids are not soluble in water; therefore, they have to be administered as suspensions. It is important to note that when nebulized, suspensions of drugs behave differently than solutions. Cameron et al.[73] compared the performance of five different nebulizers with an amino-phylline aerosol solution and a suspension of budesonide. They found that even with the same nebulizer, different aerosol characteristics were obtained after nebulization of the solution and the suspension. Tiano and Dalby[74] determined how the differences in aerosolization mechanism (jet vs. ultrasonic) affected droplet and insoluble particle deposition of a nebulized model respiratory suspension. Both nebulizers produced droplets large enough to incorporate $<6\,\mu m$ insoluble latex spheres. However, droplets generated by the jet nebulizer contained spheres of all sizes, while with the ultrasonic nebulizer 99% of the spheres were not aerosolized and recovered from the nebulizer.[74]

Liposomal formulations have also been delivered by nebulization.[75–78] The effects of air pressure, temperature, buffer, osmotic strength, and pH on nebulized liposome dispersions were studied. Changes in air pressure produced large changes in the percentage of release of the encapsulated substance, and increasing the air pressure increased the percentage of release. Leakage of liposomes was increased in hypotonic solution but decreased in hypertonic solution; at low pH the leakage was increased compared to higher pH. Stability of liposomes was affected by the operating and environmental conditions of aerosolization, with air pressure having the greatest effect.[77]

Niosomes, which are considered to be more stable than liposomes, have been used to encapsulate all-*trans*-retinoic acid and have been effectively delivered by the PARI-LC STAR nebulizer with droplets in respiratory size and good entrapment.[79]

12.3.3.3 Macromolecules

Nebulization of macromolecules has been a delicate issue, since some factors inherent to nebulization such as shear stresses and volume have an effect on the stability and integrity of macromolecules.[60] Furthermore, the characteristics of macromolecule solutions are often different from those of regular solutions or suspensions, which in turn may affect nebulizer performance. Biotechnology products frequently form viscous solutions, with modified interfacial and surface tension.[72] The effects of pH, and additive and ionic strength on the delivery of recombinant consensus alpha interferon have been published.[80] It was found that the interferon molecule was destabilized by air-jet nebulizer aggregation of the plasmid. This was influenced by pH: the smaller the pH, the larger the aggregation. Ionic strength of the solution did not influence aggregation. Ultrasonic nebulization of the solution of plasmid also resulted in aggregation, but denaturation was dependent upon the type of nebulizer used and related to the heating of nebulizer solutions.

Contradictory statements are made regarding the nebulization of proteins by ultrasonic nebulizers. Some authors state that denaturation of proteins is unavoidable by ultrasonic nebulizers, while others have successfully aerosolized protein solutions using this type of nebulizer.[81–83] It has been observed that by preventing heating of the nebulizer fluid during operation, denaturation of the proteins was altered. In addition, by including 0.01% Tween 80 or 1% w/v PEG 8000, almost all activity of the proteins was retained. Therefore, cooling of the solution in conjunction with the addition of surfactant is one approach that could be used to stabilize proteins to ultrasonic nebulization. However, cooling may also reduce solute output from the nebulizer.[84,85]

Stribling et al.[17] were among the first to use a jet nebulizer to aerosolize plasmid/liposome vectors to mice for the treatment of cystic fibrosis, with encouraging results. Cipolla et al.[86] have also used this approach efficiently in monkeys. A review has been published by Niven[54] on the delivery of biotherapeutics by inhalation aerosols. It outlines the advantages of delivery of these molecules by inhalation stability problems when they are aerosolized, the problems with aqueous protein formulations, and also highlights a variety of biotherapeutics given by aerosols.

12.3.4 *Handheld aqueous systems*

A major disadvantage of nebulizer systems has been their lack of portability.[87] The next-generation aqueous systems include several smaller battery-powered devices that address this issue. Some examples of these are summarized in Table 12.3. These devices typically deliver small-volume doses equivalent to those emitted from pMDIs (10–50 μl).

12.3.5 *Topical drug delivery sprays*

Therapeutic topical aerosols have been used for centuries for various skin diseases. Topical sprays are used due to the ability of a spray to coat the target surface evenly with a minimum of excess drug that may lead to undesirable effects.[88] Modern topical sprays include antiinfectives (e.g., tolnaftate), local anesthetics (e.g., lidocaine, benzocaine), antiseptics (phenol), scabicides (piperonyl butoxide, pyrethrins), and sunscreens (octyl methoxycinnamate). These agents generally act locally or topically. Subsequently, the generation of the aerosol is generally not required to have the same stringent quality controls on particle

Table 12.3 Specific Challenges of Particle Size Measurement of Pharmaceutical Aerosols[88]

Device Manufacturer	Trade Name	Description of Atomization Mechanism	References
Aerogen	Aerodose®	Electrically induced vibrations to a concave surface generates an aerosol from aqueous solution or suspension formulation	134
Aradigm	AERx®	Computer-controlled device using a laser-machined nozzle through which the formulation is mechanically extruded to produce an aerosol	132 133
Battelle	Mystic™	Electrohydrodynamic atomization	8
Boehringer Ingelheim	Respimat	Mechanical propulsion	87
ODEM	TouchSpray™	Vibrating perforated membrane	135

size or dose as those aerosols used for delivery to the lungs. However, droplet sizes are generally larger, so that the probability of inertial impaction on the skin is increased while respirable particles are minimized.[88] In addition, systemically acting sprays have been developed (NitroLingual™). This sublingual spray is used to deliver nitroglycerin for the treatment of angina. A metered spray device is used. Topical sprays are also being investigated for other systemically acting agents such as hormones.[89] Related technologies also include high-speed powder transdermal delivery (Powderject®). This involves the acceleration of powder particles of a specific size range, density, and strength so that they penetrate the skin without the need for injections. Applications of this technology include traditional small-molecule pharmaceutical agents, peptides, proteins, and vaccines.

12.3.6 Specific measurement techniques and calibration

As with other aerosols, pharmaceutical and medical aerosols are described using properties such as particle size, electrostatics, hygroscopicity, and uniformity of drug dispersion.[88] Measurement is performed for product development and also for quality and regulatory purposes. Measurement techniques focus on characterizing the efficiency and reproducibility by which aerosols are generated and delivered to the respiratory tract of patients. *In vitro* and *in vivo* techniques are used and are summarized below.

12.3.6.1 Particle size measurement

Particle size analysis is extremely important in the characterization of medical and pharmaceutical aerosols as particle or droplet size and distribution are the most important physicochemical properties influencing lung deposition.[90] Hence, the measurement of particle size is important in several aspects of pharmaceutical aerosols, including particle manufacture, formulation optimization and stability, and quality control. Medical and pharmaceutical aerosols are generally sized using two general classes of particle size analysis: inertial methods and optical methods.

Particle size analysis in medical and pharmaceutical aerosols has unique issues relating to aerosol sampling. There are spatial restrictions between the generation site (device) and inhalation target site (lungs). The presence of inhalation flow rates and cyclical breathing patterns in vivo also complicates the interpretation of particle size observations. Measurement can be particularly challenging given the often unstable nature of generated particles due to evaporation, condensation, temperature and humidity changes.

12.4 Specific measurement techniques and calibration

12.4.1 Inertial methods

12.4.1.1 Cascade impactor

The cascade impactor has been a tool to determine the particle size of pharmaceutical aerosol dispersions for over 30 years. The impactor applies the principles of inertia discussed earlier in this text to separate particles by size. Particles are dispersed and travel through multiple stages with sequentially decreasing jets. Each stage plate should be thinly coated in oil to avoid particle bounce and blow-off.[91] There are multiple impactor systems currently in the market and their size specifications are given in Table 12.4.[42] The Next Generation Impactor (MSP Inc., Niwot, CO, U.S.A.) was developed to make particle size testing more efficient. It has low internal particle loss and requires minimal washing between tests.

Although cascade impactors are commonly used to characterize environmental materials such as pollutants, this chapter will focus on the use of the instrument for characterizing particles designed for delivery to the lung. There are some differences in the setup

Table 12.4 Cutoff Diameters (μm) for Instruments that Determine Particle Size by Inertial Impaction as Reported by the Manufacturer (modified from reference 42)

Apparatus	Twin Stage Liquid Impinger	USP B	Multistage Impinger				Andersen Viable Impactor	Andersen Nonviable Impactor			Next Generation Impactor			Marple-Miller Impactor			Delron
Flow rate (l/min)	60	60	30	60	80	100	28.3	28.3	60	90	30	60	100	30	60	90	12.5
Preseparator	—	—	—	—	—	—	—	—	—	—	15.0	13.0	10.0	—	—	—	—
stage −1	—	—	—	—	—	—	—	—	8.6	8.0	—	—	—	—	—	—	—
0	6.4	9.8	—	—	—	—	7.1	9.0	6.5	6.5	—	—	—	—	—	—	—
1	—	—	16.9	13.3	11.8	10.4	4.7	5.8	4.4	5.2	11.0	7.8	6.0	10.0	10.0	8.0	11.2
2	—	—	9.3	6.7	5.6	4.9	3.3	4.7	3.3	3.5	6.6	4.6	3.6	5.0	5.0	4.0	5.5
3	—	—	4.5	3.2	2.7	2.4	2.1	3.3	2.0	2.6	3.9	2.7	2.1	2.5	2.5	2.0	3.3
4	—	—	2.5	1.7	1.4	1.2	1.1	2.1	1.1	1.7	2.3	1.6	1.2	1.25	1.25	1.0	2.0
5	—	—	—	—	—	—	0.65	1.1	0.54	1.0	1.4	0.96	0.72	0.63	0.63	0.6	0.9
6	—	—	—	—	—	—	—	0.7	0.25	0.43	0.84	0.57	0.42	—	—	—	0.5
7	—	—	—	—	—	—	—	0.4	—	—	0.51	0.33	0.23	—	—	—	—
MOC	—	—	—	—	—	—	—	—	—	—	0.31	0.13	0.055	—	—	—	—

MOC= microorifice collector.

of the apparatus when sampling for pulmonary delivery as opposed to environmental testing. First, a throat is added to the impactor to model the anatomy of the human body.[92] Second, the flow rate of air through the impactor differs from that of environmental sampling.[93] When performing impactions of aerosols intended for pulmonary delivery, airflow rates of 28.3–90 l/min are employed.

12.4.1.2 *Twin impinger*

In the late 1980s, the twin impinger was devised to act as a suitable device for aerosol characterization while decreasing the analytical effort.[94] The twin impinger (Figure 12.3) uses two stages to collect dispersed particles, with the cut point being a median diameter of 6.4 μm. That is, all particles larger than 6.4 μm will remain in stage 1 and all particles smaller than 6.4 μm will remain in stage 2. The impaction surface is formed by the perpendicular between the airflow and the surface of the collection solution. Volumes of 7 and 30 ml are

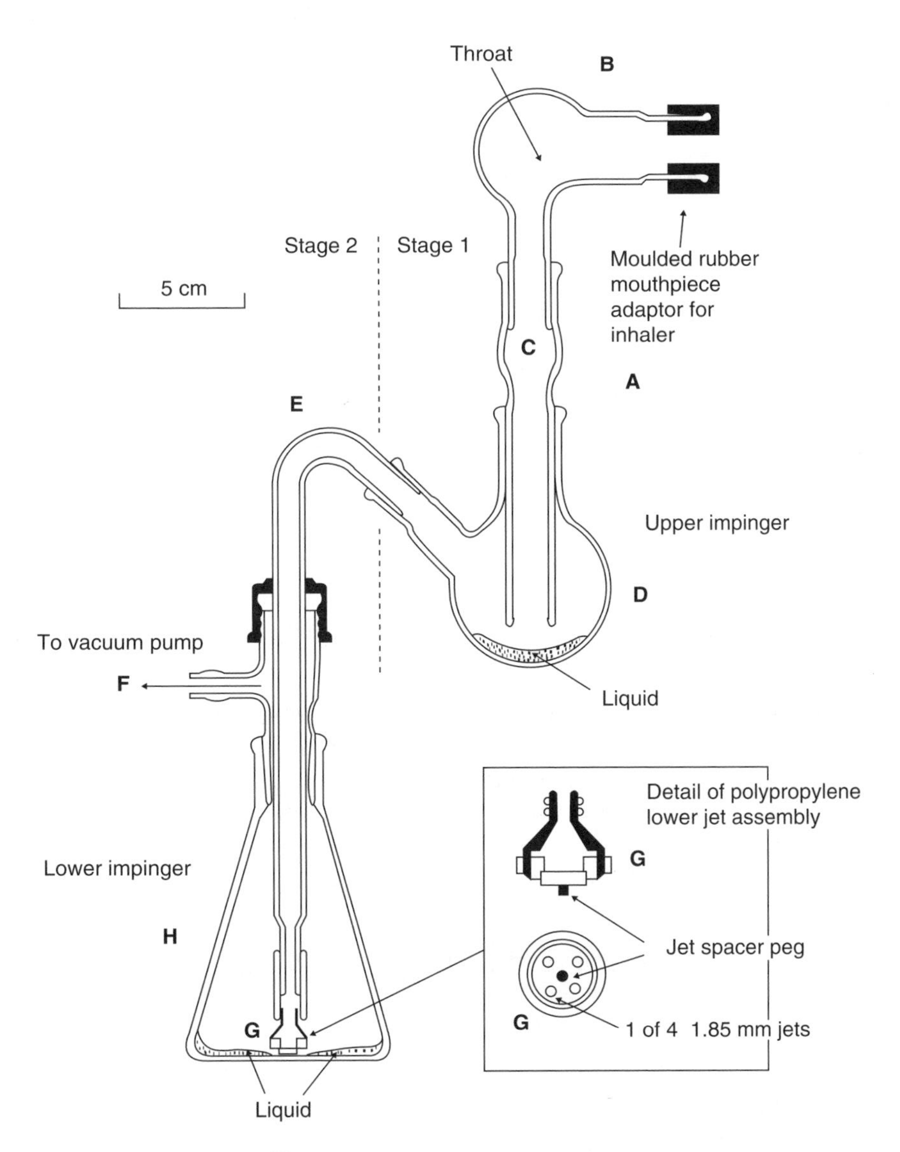

Figure 12.3 The twin impinger.[126]

placed in stages 1 and 2, respectively. The type of solution used does not have an effect on the cut point of the stage. The vacuum is applied at 60 l/min.

12.4.2 Optical methods

Optical particle size analysis techniques are commonly used in pharmaceutical aerosol development and testing. Due to the unstable and transient nature of emitted aerosols, these methods frequently perform dynamic measurements. Static measurements of particle size are still used however. In particular, microscopy is an essential component of the particle size characterization of pharmaceutical aerosols during formulation development.[95–97]

Phase Doppler anemometry (PDA) is an open laser system and can sample the aerosol from the point of generation to the extremities.[98,99] Accordingly, PDA has been most often used to characterize the development of nebulizer and MDI aerosols.[98] Time-of-flight (TOF) particle size analyzers are real-time optically based and are used frequently in medical and pharmaceutical aerosols.[100,101] Particles are separated on the basis of their inertia by accelerating the particles through a well-defined flow field and measuring the particle's TOF across a split laser beam. TOF analyzers measure single particles and collect distributions by collecting size data on a statistically valid number of particles. Powder dispersion, DPIs, MDIs, and nebulizers have all been investigated using this type of instrument.[101,102]

As a method of particle size analysis for pharmaceutical aerosols, laser diffraction has also been reviewed.[103–105] Laser diffraction techniques are widely used in the characterization of pharmaceutical aerosols due to the rapid nature of data collection. The principle of this technique is outlined elsewhere in this text. The laser light scattering patterns of an aerosol passing through the laser region allow instantaneous evaluation of multiple particles. Like PDA systems, various regions and time points of an aerosol produced from an inhaler can be sampled.[103]

Currently, optical particle sizing techniques are useful tools in evaluating pharmaceutical aerosols. In general, these techniques are not accepted as pharmacopeial methods because, from a regulatory point of view, they suffer from representative sampling issues and an absence of chemical analysis.[106]

12.4.3 Other in vitro measurement techniques commonly employed for pharmaceutical aerosols

From a development and regulatory viewpoint, particle size is a dominant analytical method used to characterize medical and pharmaceutical aerosols. However, regulatory agencies also recommend other characteristics of medical aerosols to be measured.[107] These include dose content uniformity of the emitted aerosol,[21] spray pattern and geometry measurements,[108] and how the aerosol changes with respect to different storage conditions, flow rates, and stability challenges.[21]

12.5 Respiratory deposition, retention, and dosimetry

12.5.1 Deposition

12.5.1.1 Gamma scintigraphy, PET, and SPECT

Assessment of lung deposition of pharmaceutical and medical drug delivery systems is a key step in the evaluation of pharmaceutical aerosol device performance. It has also been shown that with inhaled antiasthma drugs, lung deposition data can act as a surrogate for clinical response.[109] Thus, lung deposition studies can facilitate optimized drug delivery to the lungs during product development and can also be used in therapeutic equivalency studies. Gamma scintigraphy has been widely used in quantifying regional lung deposition of aerosolized drugs.[109] It is a planar imaging technique, where a gamma-ray-emitting nuclide of appropriate half-life (typically 99m Technetium for pulmonary systems) is included in the pharmaceutical formulation such that its aerodynamic particle size and deposition parallels that of the drug substance. The radionuclide distribution in the lungs is then analyzed using a gamma camera and can be quantified using computer-based software. Single-photon emission computed tomography (SPECT) is similar to gamma scintigraphy in that it involves similar radiolabeling procedures and measurement of deposition using a gamma camera.[110] However, in SPECT the emitted gamma rays are detected using a rotating gamma camera from various angles in order to obtain several views of lung deposition. Thus, it may be possible to define drug deposition more closely with respect to the three-dimensional anatomy of the lung. Nebulizers and MDIs have been investigated using this technique, but some significant drawbacks include the following: larger amounts of radionuclide are required for satisfactory scintigraphic data collection, and longer imaging periods result in drug clearance from the lung.[110] Positron emission tomography (PET) is a relatively new technique in analyzing lung deposition resulting from pharmaceutical systems.[110,111] The basis of the method is the use of positron-emitting nuclides (typically 11C, 15O, 13N, or 18F) that are detected after positron collision with electrons. This interaction results in two photons being emitted in opposite directions, which are detected using coincident counting of photons on ring-shaped arrays of detector elements. The advantages of this technique include the small radiation doses resulting from the relatively short half-lives of the radionuclides used, very short scanning times, detectable picomolar concentrations of nuclides, and nuclides often being chemically incorporated into the drug molecule rather than being physically attached.[110] However, these are balanced against the practical disadvantages of handling radionuclides with such short half-lives: transportation, formulation and labeling steps, and dosing volunteers. Apart from time limitations, the cost of PET can also be significant due to the need for a nearby cyclotron and detector equipment. These factors need to be weighed against the PET images that yield more precise relations of deposition data with the three-dimensional structural features of the respiratory tract.[110]

12.5.2 Retention

The retention of aerosols in the lungs is related to the physicochemical properties of the material, particularly those linked to dissolution rate,[112,113] and the mechanisms of clearance from the lungs, that is, absorption, mucociliary transport, and cell-mediated transport.[114]

Rapidly dissolving particles are subject to clearance from the lungs by absorption and mucociliary clearance, depending on the site of deposition. Slowly dissolving particles are cleared predominantly by mucociliary clearance in the upper airways and macrophage uptake in the periphery of the lungs.

Delaying dissolution has been used as a strategy for controlling drug delivery by increasing lung retention time.[115,116] However, there may be some concerns about the safety of such an approach.[112]

12.5.3 Dosimetry

Dosimetry is generally linked to the physicochemical properties of a drug; however, other factors that may influence the decision to formulate or administer a drug by a determined device or devices are technological implications, the health condition of the patient, target site in the airways, and therapeutic dose. Doses are drug specific and they are influenced by the potency of the drug itself. In addition, throughput of a device and respirable fraction should be considered. Throughput is used to describe the amount of drug deliverable from aerosol devices, while respirable fraction (RF) describes the fraction of aerosolized dose surviving filtration and impaction mechanisms of the nasopharynx.[117]

Although it was generally believed that dosimetry from MDIs was more efficient, convenient, and reproducible than DPIs and nebulizers, emerging technologies have changed this perspective. Gamma scintigraphy and pharmacokinetic (charcoal-block) methods have been used to quantify the amount and pattern of deposition in human lungs.[118–120] The influence of pulmonary physiology and pressure gradients on the dosimetry of inhaler devices has been studied.[121,122] Martonen et al.[123] have developed models to study human lung dosimetry and deposition. These models can simulate the effects of aerosol polydispersity and hygroscopicity, lung morphology and patient ventilation, age, and airway disease. Another model of mechanistic dosimetry developed by Lazardis et al. describes the dynamics of respirable particles. Model predictions of disposition and transport of aerosols are based on equations that describe changes in particle size and mass distribution as a consequence of nucleation, condensation, coagulation, and deposition processes.

12.5.3.1 Bolus delivery

Table 12.5 shows some common drugs administered by the inhalation device, the doses dispensed with each inhalation, and the therapeutic doses required per day.[124] In general, DPIs and MDIs are used for the single delivery of medications on a single breath; doses range from 4 to 500 μg and require administration twice to six times a day. MDIs are capable of delivering very accurate and reproducible doses. The valves in the device can deliver volumes of the formulation between 25 and 100 μl. Characteristics that can influence the amount of drug to be formulated, and therefore the amount of drug that can be administered, are drug characteristics such as solubility and concentration, and the addition of excipients such as surfactants (to minimize particle aggregation, improving physical stability and dose uniformity), solvents (to aid in drug solubility, carrier properties), and propellants. The use of accessories such as spacers with MDIs improves inhalant technique and drug delivery.[58] Some factors that influence the size of doses delivered by DPIs are drug characteristics, method used to produce fine particles, blend uniformity, and factors related to the device itself such as FPF and emitted dose of the formulation.

12.5.3.2 Continuous delivery

As mentioned in previous sections, the use of nebulizers is preferred in hospital settings, acute conditions, and treatment in elderly patients and children. This is mainly because the required doses of the prescribed drug are very large and have to be administered over long periods of time or because the health condition of the patient has limited his/her ability to breath adequately to use any other inhalation device. In most cases, once the clinical condition of an adult patient improves, the prescribed medications are administered by either MDI or DPI. Table 12.5 shows some commercial preparations to be delivered by nebulization. Doses range from a few milligrams, to be used at home, to several hundred milligrams or even grams when they are used in hospital settings. Nebulizers are more effective generators of small particles than MDIs and DPIs.[57] Recent nebulizer designs provide increased efficiency of delivery by gauging the patients' breathing and delivering aerosol only during inhalation. Other systems achieve increased delivery efficiency by collecting the aerosol generated during exhalation in a reservoir and administered during inhalation. Nebulizer reservoirs increase the delivery rate of nebulized

Table 12.5 Drugs Delivered by the Inhalation Route[124]

Product	Active	Dose/inhalation	Dose/day
	Metered Dose Inhaler[a]		
Aerobid	Flunisolide	250 μg	2 inhalations b.i.d. (1 mg)
Azmacort	Triamcinolone acetonide	200 μg	2 inhalations 3–4 times or 4 inhalations b.i.d. (1.6 mg)
Flovent	Fluticasone propionate	44 μg 110 μg 220 μg	2 inhalations b.i.d. (0.176 – 0.880 mg)
Maxair Autohaler	Pirbuterol acetate	200 μg	2 inhalations 4–6 times (1.6 – 2.4 mg)
Proventil HFA	Albuterol sulfate	90 μg	2 inhalations 4–6 times (0.72–1.08 mg)
Vanceril	Beclomethasone dipropionate	42 μg	2 inhalations 2–3 times (0.168–0.252 mg)
	Dry Powder Inhaler[a]		
Advair Diskus	Fluticasone propionate and salmeterol xinafoate	93/45 μg 233/45 μg 165/45 μg	1 inhalation b.i.d. (0.186/ 0.090 – 0.330/0.090 mg)
Bricanyl turbuhaler	Terbutaline sulfate	500 μg	1 inhalation 4 times (2 mg)
Foradil Aerolizer	Formoterol fumarate	10 μg	1 inhalation b.i.d. (0.02 mg)
Pulmicort turbuhaler	Budesonide	200 μg	1 inhalation 1–4 times (0.2 –0.8 mg)
Relenza	Zanamivir	4 mg	2 inhalations b.i.d. (8 mg)
Serevent diskus	Salmeterol xinofoate	50 μg	1 inhalation b.i.d. (0.1 mg)
	Nebulizer[b]		
Accuneb	Albuterol sulfate	1.25 mg/3 ml 0.63 mg/3 ml	1 nebulization 3–4 times as needed (3.75 mg)
Alupent solution	Metaproterenol sulfate USP	5%/10 ml (dropper) 0.4%/2.5 ml 0.6%/ 2.5 ml	1 nebulization 3-4 times (125 mg)
Atrovent solution	Ipratropium bromide	0.02% 500 μg/2.5 ml	1 nebulization 3–4 times (1.5 mg)
Duoneb solution	Ipratropium bromide/ albuterol sulfate	0.5/3.0 mg/ 3 ml 42%/46%/3 ml	1 nebulization 3 times (1.5/9.0 mg)
Pulmicort respules	Budesonide suspension	0.25 mg/2 ml 0.5 mg/2 ml	1 nebulization once or twice for a maximum of 0.5 mg
Pulmozyme solution	Dornase alfa	1.0 mg/ml (2.5 ml)	1 nebulization once (2.5 mg)
TOBI solution	Tobramycin	300 mg/5 ml	1 nebulization twice (600 mg)
Virazole solution	Rivabirin	6.0 g/300 ml	Continuous aerosol administration for 12 –18 h

[a] Doses in MDIs and DPIs are described from the mouthpiece.

[b] Nebulizations usually take between 15 and 30 min on average if used with the nebulizer and compressor recommended by the manufacturer observing the recommended parameters.

medications by conserving the aerosol generated during exhalation and making it available to the patient during the next inhalation.[125] The development of medications for nebulization should include the specification and testing of a nebulizer and compression source that is fixed by prescription along with the medication. Alterations of the nebulizer/compression source must be carefully considered at the physician level, as such alterations could significantly change the dose received by the patient even if the same nebulizer charge dose is used.[67]

References

1. Sciarra, J.J. and Cutie, A.J., Pharmaceutical aerosols, in *Modern Pharmaceutics*, Banker, G.S. and Rhodes, C.T., Eds., Marcel Dekker Inc., New York, 1990, pp. 605–634.
2. Duke, J.A., Ed., *Handbook of Edible Weeds*, CRC Press, Boca Raton, FL, 1992.
3. Thiel, C., From Susie's question to CFC free: an inventor's perspective on forty years of MDI development and regulation. in *Respiratory Drug Delivery, V*, Dalby, R.N., Byron, P.R., and Farr, S.J., Eds., Interpharm Press, Inc., Phoenix, AZ,1996, pp. 115–123.
4. Dunbar, C. and Hickey, A.J., A new millenium for inhaler technology, *Pharm. Technol.*, 21, 116–125, 1997.
5. Crowder, T.M. et al., An odyssey in inhaler formulations and design, *Pharm. Technol.*, 25, 99–113, 2001.
6. Barnes, P.J., New concepts in chronic obstructive pulmonary disease, *Ann. Rev. Med.*, 17, 17, 2002.
7. Hillerdal, G., New principles for the treatment of diffuse pulmonary emphysema, *J. Intern. Med.*, 242, 441–448, 1997.
8. Ding, J.Y. et al., Delivery of a chemotherapy agent for cancer treatment via nebulization, in *Respiratory Drug Delivery, VIII*, Dalby, R.N. et al., Eds., DHI, Tucson, AZ, 2002, pp. 359–362.
9. Patton, J.S., Bukar, J., and Nagarajan, S., Inhaled insulin, *Adv. Drug Delivery Rev.*, 35, 235–247, 1999.
10. Adjei, A. and Garren, J., Pulmonary delivery of peptide drugs: effect of particle size on bioavailability of leuprolide acetate in healthy male volunteers, *Pharm. Res.*, 7, 565–569, 1990.
11. Shukla, V.K., Tiotropium: a potential replacement for ipratropium in patients with COPD, *Issues Emerg Health Technol.*, 1–4, 2002.
12. Sacchetti, M. and Van Oort, M.M., Spray-drying and supercritical fluid particle generation techniques, in *Inhalation Aerosols: Physical and Biological Basis for Therapy*, Hickey, A.J., Ed., Marcel Dekker, New York, 1996, pp. 337–384.
13. Crowder, T.M. et al., Fundamental effects of particle morphology on lung delivery: predictions of Stokes' law and the particular relevance to dry powder inhaler formulation and development, *Pharm. Res.*, 19, 239–245, 2002.
14. Edwards, D.A. et al., Large porous particles for pulmonary drug delivery, *Science*, 276, 1868–1871, 1997.
15. Hickey, A.J., Delivery of drugs by the pulmonary route, in *Modern Pharmaceutics*, Rhodes, C. and Banker, G., Eds., Marcel Dekker Inc., New York, 2002.
16. Hickey, A.J., Summary of common approaches to pharmaceutical aerosol administration, in *Pharmaceutical Inhalation Aerosol Technology*, Hickey, A.J., Ed., Marcel Dekker Inc., New York, NY, 1992.
17. Stribling, R. et al., Aerosol gene delivery *in vivo*, *Proc. Natl. Acad. Sci. U.S.A.*, 89, 11277–11281, 1992.
18. Sandrasagra, A. et al., RASONs: a novel antisense oligonucleotide therapeutic approach for asthma, *Exp. Opin. Biol. Ther.*, 1, 979–983, 2001.
19. *British Pharmaceutical Codex*, Ed., The Pharmaceutical Press, London, 1973.
20. Byron, P.R. and Patton, J.S., Drug delivery via the respiratory tract, *J. Aerosol Med.*, 7, 49–75, 1994.
21. Purewal, T.S., Formulation of metered dose inhalers, in *Metered Dose Inhaler Technology*, Purewal, T.S. and Grant, D.J.W., Eds., Interpharm, Buffalo Grove, IL, 1998.
22. Clarke, J.G., Wicks, S.R., and Farr, S.J., Surfactant mediated effects in pressurized metered dose inhalers formulated as suspensions. I. Drug/surfactant interactions in a model propellant system, *Int. J. Pharm.*, 93, 221–231, 1993.
23. Blondino, F.E. and Byron, P.R., Surfactant dissolution and water solubilization in chlorine-free liquified gas propellants, *Drug Dev. Ind. Pharm.*, 24, 935–945, 1998.
24. Harnor, K.J. et al., Effect of vapor pressure on the deposition pattern from solution phase metered dose inhalers, *Int. J. Pharm.*, 95, 111–116, 1993.
25. McDonald, K.J. and Martin, G.P., Transition to CFC-free metered dose inhalers — into the new millennium, *Int. J. Pharm.*, 201, 89–107, 2000.

26. Forte, R., Jr. and Dibble, C., The role of international environmental agreements in metered-dose inhaler technology changes, *J. Allerg. Clin. Immunol.*, 104, S217–220, 1999.
27. Noakes, T., Medical aerosol propellants, *J. Fluorine Chem.*, 118, 35–45, 2002.
28. Alexander, D.J. and Libretto, S.E., An overview of the toxicology of HFA-134a (1,1,1,2-tetrafluoroethane), *Hum. Exp. Toxicol.*, 14, 715–720, 1995.
29. Emmen, H.H. et al., Human safety and pharmacokinetics of the CFC alternative propellants HFC 134a (1,1,1,2-tetrafluoroethane) and HFC 227 (1,1,1,2,3,3, 3-heptafluoropropane) following whole-body exposure, *Regul. Toxicol. Pharmacol.*, 32, 22–35, 2000.
30. Keller, M., Innovations and perspectives of metered dose inhalers in pulmonary drug delivery, *Int. J. Pharm.*, 186, 81–90, 1999.
31. Vervaet, C. and Byron, P.R., Drug surfactant propellant interactions in HFA formulations, *Int. J. Pharm.*, 186, 13–30, 1999.
32. Byron, P.R. et al., Some aspects of alternative propellant solvency, in *Respiratory Drug Delivery, IV*, Byron, P.R., Dalby, R.N., and Farr, S.J., Eds., Richmond, VA, Interpharm Press, Inc. Buffalo Grove, IL, 2002.
33. Dickinson, P.A. et al., An investigation of the solubility of various compounds in the hydrofluoroalkane propellants and possible model liquid propellants, *J. Aerosol Med.*, 13, 179–186, 2000.
34. Tiwari, D. et al., Compatibility evaluation of metered-dose inhaler valve elastomers with tetrafluoroethane (P134a), a non-CFC propellant, *Drug Dev. Ind. Pharm.*, 24, 345–352, 1998.
35. Brambilla, G. et al., Modulation of aerosol clouds produced by pressurised inhalation aerosols, *Int. J. Pharm.*, 186, 53–61, 1999.
36. Dunbar, C.A., Watkins, A.P., and Miller, J.F., Theoretical investigation of the spray from a pressurized metered-dose inhaler, *Atomization Sprays*, 7, 417–436, 1997.
37. Dunbar, C.A., Watkins, A.P., and Miller, J.F., An experimental investigation of the spray issued from a pMDI using laser diagnostic techniques, *J. Aerosol Med.*, 10, 351–368, 1997.
38. Leach, C., Effect of formulation parameters on hydrofluoroalkane-beclomethasone dipropionate drug deposition in humans, *J. Allerg. Clin. Immunol.*, 104, S250–S252, 1999.
39. Dalby, R.N., Possible replacements for CFC-propelled metered-dose inhalers, *Med. Device Technol.*, 2, 21–25, 1991.
40. Sommerville, M.L. et al., Lecithin inverse microemulsions for the pulmonary delivery of polar compounds utilizing dimethyl ether and propane as propellants, *Pharm. Dev. Technol.*, 5, 219–230, 2000.
41. Clark, A.R., Medical aerosol inhalers: past, present, and future, *Aerosol Sci. Technol.*, 22, 374–391, 1995.
42. Dunbar, C.A., Hickey, A.J., and Holzner, P., Dispersion and characterization of pharmaceutical dry powder aerosols, *KONA*, 16, 7–45, 1998.
43. Hindle, M. and Byron, P.R., Dose emissions from marketed dry powder inhalers, *Int. J. Pharm.*, 116, 169–177, 1995.
44. Hickey, A.J., Martonen, T.B., and Yang, Y., Theoretical relationship of lung deposition to the fine particle fraction of inhalation aerosols, *Pharm. Acta Helv.*, 71, 165–170, 1996.
45. Hickey, A.J. et al., Factors influencing the dispersion of dry powders as aerosols, *J. Pharm. Technol.*, 18, 58–64, 1994.
46. Lucas, P. et al., Enhancement of small particle size dry powder aerosol formulations using an ultra low density additive, *Pharm. Res.*, 16, 1643–1647, 1999.
47. Lucas, P., Anderson, K., and Staniforth, J.N., Protein deposition from dry powder inhalers: fine particle multiplets as performance modifiers, *Pharm. Res.*, 15, 562–569, 1998.
48. Musante, C.J. et al., Factors affecting the deposition of inhaled porous drug particles, *J. Pharm. Sci.*, 91, 1590–1600, 2002.
49. Newman, S.P. and Busse, W.W., Evolution of dry powder inhaler design, formulation, and performance, *Respir. Med.*, 96, 293–304, 2002.
50. Peart, J. and Clarke, M.J., New developments in dry powder inhaler technology, *Am. Pharm. Rev.*, 14, 37–45, 2001.
51. Smith, S.J. and Bernstein, J.A., Therapeutic uses of lung aerosols, in *Inhalation Aerosols, Physical and Biological Basis for Therapy*, Hickey, A.J., Ed., Marcel Dekker Inc., New York, 1996, pp. 233–269.

52. Taburet, A.M. and Schmit, B., Pharmacokinetic optimization of asthma treatment, *Clin. Pharmacokinetics*, 26, 396–411, 1994.
53. Ilowite, J., Asthma, corticosteroids, and growth, *N. Eng. J. Med.*, 344, 607, 2001.
54. Niven, R., Delivery of biotherapeutics by inhalation aerosol, *Crit. Rev. Therap. Drug Carrier Syst.*, 12, 151–231, 1995.
55. Newnham, D.M. and Lipworth, B.J., Nebulizer performance, pharmacokinetics, airways and systemic effects of salbutamol given via a novel nebulizer delivery system ("Venstream"), *Thorax*, 49, 762–770, 1994.
56. Baker, P.G. and Stimpson, P.G., Electronically controlled drug delivery systems based on the piezoelectric crystal, in *Proceedings of Respiratory Drug Delivery IV*, Byron, P.R., Dalby, R.N., and Farr, S.J., Eds., Interpharm Press, Buffalo Grove, IL, 1994, pp. 273–285.
57. Dalby, R.N., Tiano, S.L., and Hickey, A.J., Medical devices for the delivery of therapeutic aerosols to the lungs, in *Inhalation Aerosols, Physical and Biological Basis for Therapy*, Hickey, A.J., Ed., Marcel Dekker Inc., New York, 1996, pp. 441–473.
58. Hickey, A.J., Summary of common approaches to pharmaceutical aerosol administration, in *Pharmaceutical Inhalation Aerosol Technology*, Hickey, A.J., Ed., Marcel Dekker, New York, 1992, pp. 255–288.
59. Boucher, R. and Kreuter, J., The fundamentals of the ultrasonic atomization of medicated solutions, *Ann. Allerg.*, 26, 591–600, 1968.
60. Niven, R.W., Atomization and nebulizers, in *Inhalation Aerosols, Physical and Biological Basis for Therapy*, Hickey, A.J., Ed., Marcel Dekker Inc., New York, 1996, pp. 273–312.
61. Greenspan, B.J., Ultrasonic and electrohydrodynamic methods for aerosol generation, in *Inhalation Aerosols, Physical and Biological Basis for Therapy*, Hickey, A.J., Ed., Marcel Dekker Inc., New York, 1996, pp. 313–335.
62. Chan, K.N., Clay, M.M., and Silverman, M., Output characteristics of DeVilbiss No. 40 hand-held jet nebulizers, *Eur. Respir. J.*, 3, 1197–1201, 1990.
63. Smith, E.C., Denyer, J., and Kendrick, A.H., Comparison of twenty-three nebulizer/compressor combinations in domiciliary use, *Eur. Respir. J.*, 8, 1214–1221, 1995.
64. Weber, A. et al., Effect of nebulizer type and antibiotic concentration on device perfomance, *Pediatric Pulmonol.*, 23, 249–260, 1997.
65. Montgomery, A., Aerosolized pentamidine for treatment and prophylaxis of *Pneumocystis carinii* pneumonia in patients with acquired immunodeficiency syndrome, in *Pharmaceutical Inhalation Aerosol Technology*, Hickey, A.J., Ed., Marcel Dekker, New York, 1992, pp. 307–320.
66. Niven, R.W., Modulated drug therapy with inhalation aerosols, in *Pharmaceutical Inhalation Aerosol Technology*, Hickey, A.J., Ed., Marcel Dekker Inc., New York, 1992, pp. 321–359.
67. Garcia-Contreras, L. and Hickey, A.J., Pharmaceutical and biotechnological aerosols for cystic fibrosis therapy, *Adv. Drug Delivery Rev.*, 54, 1491–1504, 2002.
68. Jones, L.D. et al., Analysis and stability of pharmaceutical aerosols, *Pharm. Technol.*, 24, 40–54, 2000.
69. Summers, Q. et al., A non-bronchoconstrictor, bacteriostatic preservative for nebuliser solutions, *Br. J. Clin. Pharmacol.*, 31, 204–206, 1991.
70. Rocchiccioli, K. and Pickering, C., Airflow obstruction induced by ultrasonically nebulised water: the underlying mechanism, *Thorax*, 39, 710, 1984.
71. Borland, C. et al., The effect of ultrasonically nebulised distilled water on airflow obstruction, regional ventillation and lung epithelial permeability in asthma, *Thorax*, 39, 240, 1984.
72. Atkins, P.J., Barker, N.P., and Mathisen, D., The design and development of inhalation drug delivery systems, in *Pharmaceutical Inhalation Aerosol Technology*, Hickey, A.J., Ed., Marcel Dekker, New York, 1992, pp. 155–185.
73. Cameron, D., Clay, M., and Silverman, M., Evaluation of nebulizers for use in neonatal ventilator circuits, *Crit. Care Med.*, 18, 886–870, 1990.
74. Tiano, S. and Dalby, R., Comparison of a respiratory suspension aerosolized by an air-jet and an ultrasonic nebulizer, *Pharm. Dev. Technol.*, 1, 261–268, 1996.
75. Niven, R. and Schreier, H., Nebulization of liposomes. I. Effects of lipid composition, *Pharm. Res.*, 7, 1127–1133, 1990.

76. Niven, R., Speer, M., and Schreier, H., Nebulization of liposomes. II. The effects of size and modeling of solute release profiles, *Pharm. Res.*, 8, 217–221, 1991.
77. Niven, R., Carvajal, T., and Schreier, H., Nebulization of liposomes. III. The effects of operating conditions and local environment, *Pharm. Res.*, 9, 515–520, 1992.
78. Lange, C. et al., *In vitro* aerosol delivery and regional airway surface liquid concentration of a liposomal cationic peptide, *J. Pharm. Sci.*, 90, 1647–1657, 2001.
79. Desai, T. and Finlay, W., Nebulization of niosomal all-*trans*-retinoic acid: an inexpensive alternative to conventional liposomes, *Int. J. Pharm.*, 241, 311–317, 2002.
80. Ip, A. et al., Stability of recombinant consensus interferon to air-jet and ultrasonic nebulization, *J. Pharm. Sci.*, 84, 1210–1214, 1995.
81. Gale, A., Drug degeneration during ultrasonic nebulization, *J. Aerosol Sci.*, 16, 265, 1985.
82. Kosugi, T. et al., Effect of ultrasonic nebulization of Miraclid on the proteolytic activity in tracheobronchial secretions of rats, *Laryngoscope*, 99, 1281–1285, 1989.
83. Niven, R. et al., Some factors associated with the ultrasonic nebulization of proteins, *Pharm. Res.*, 12, 53–59, 1995.
84. Niven, R. et al., Systemic absorption and activity of recombinant consensus interferons after intratracheal instillation and aerosol administration, *Pharm. Res.*, 12, 1889–1895, 1995.
85. Niven, R. et al., The pulmonary absorption of aerosolized and intratracheally instilled rhG-CSF and monoPEGylated rhG-CSF, *Pharm. Res.*, 12, 1343–1349, 1995.
86. Cipolla, D.C. et al., Coarse spray delivery to a localized region of the pulmonary airways for gene therapy, *Hum. Gene Ther.*, 11, 361–371, 2000.
87. Dolovich, M., New propellant-free technologies under investigation, *J. Aerosol Med.*, 12 (Suppl. 1), S9–S17, 1999.
88. Hickey, A.J. and Swift, D., Characterization of pharmaceutical and diagnostic aerosols, in *Aerosol Measurement: Principles, Techniques, and Applications*, Baron, P.A. and Willeke, K., Eds., Wiley-InterScience, New York, 2001, pp. 1031–1052.
89. Finnin, B.C. and Morgan, T.M., Transdermal penetration enhancers: application, limitations and potential, *J. Pharm. Sci.*, 88, 955, 1999.
90. Gonda, I., Targeting by deposition, in *Pharmaceutical Aerosol Inhalation Technology*, Hickey, A.J., Ed., Marcel Dekker, New York, 1992, pp. 61–82.
91. Esmen, N.A. and Lee, T.C., Distortion of cascade impactor measured size distribution due to bounce and blow-off, *Am. Ind. Hygiene Assoc. J.*, 41, 410–419, 1980.
92. Hallworth, G.W. and Andrews, U.G., Size analysis of suspension inhalation aerosols by inertial separation methods, *J. Pharm. Pharmacol.*, 28, 898–907, 1976.
93. Rubow, K.A. et al., A personal cascade impactor: design, evaluation and calibration, *Am. Ind. Hygi. Assoc. J.*, 48, 532–538, 1987.
94. Hallworth, G.W. and Westmoreland, D.G., The twin impinger: a simple device for assessing the delivery of drugs from metered dose pressurized aerosol inhalers, *J. Pharm. Pharmacol.*, 39, 966–972, 1987.
95. Williams, R.O., 3rd and Hu, C., Moisture uptake and its influence on pressurized metered-dose inhalers, *Pharm. Dev. Technol.*, 5, 153–162, 2000.
96. Clarke, M.J., Tobyn, M.J., and Staniforth, J.N., The formulation of powder inhalation systems containing a high mass of nedocromil sodium trihydrate, *J. Pharm. Sci.*, 90, 213–223, 2001.
97. Evans, R., Determination of drug particle size and morphology using optical microscopy, *Pharm. Technol.*, 17, 146, 148, 150, 152, 1993.
98. Ranucci, J.A. and Chen, F.C., Phase Doppler anemometry: technique for determining aerosol plume-particle size and velocity, *Pharm. Technol.*, 17, 1993.
99. Dunbar, C.A. and Hickey, A.J., Selected parameters affecting characterization of nebulized aqueous solutions by inertial impaction and comparison with phase-Doppler analysis, *Eur. J. Pharm. Biopharm.*, 48, 171–177, 1999.
100. Mitchell, J.P., Nagel, M.W., and Cheng, Y.S., Use of the aerosizer aerodynamic particle size analyzer to characterize aerosols from pressurized metered-dose inhalers (pMDIs) for medication delivery, *J. Aerosol Sci.*, 30, 467–477, 1999.
101. Mitchell, J.P. and Nagel, M.W., Time-of-flight aerodynamic particle size analyzers: their use and limitations for the evaluation of medical aerosols, *J. Aerosol Med.*, 12, 217–240, 1999.

102. Mitchell, J.P., Nagel, M.W., and Archer, A.D., Size analysis of a pressurized metered dose inhaler-delivered suspension formulation by the API Aerosizer time-of-flight aerodynamic particle size analyzer, *J. Aerosol Med.*, 12, 255–264, 1999.
103. Ranucci, J., Dynamic plume-particle size analysis using laser diffraction, *Pharm. Technol.*, 16, 1992.
104. Annagragada, A. and Adjei, A., Analysis of the Fraunhofer diffraction method for particle size distribution analysis and its application to aerosolized sprays, *Int. J. Pharm.*, 127, 219–227, 1996.
105. de Boer, A.H. et al., Characterization of inhalation aerosols: a critical evaluation of cascade impactor analysis and laser diffraction technique, *Int. J. Pharm.*, 249, 219–231, 2002.
106. Hickey, A.J. and Jones, L.D., Particle-size analysis of pharmaceutical aerosols, *Pharm. Technol.*, 24, 48–58, 2000.
107. FDA., Guidance for industry metered dose inhaler (MDI) and dry powder inhaler (DPI) drug products, CDERGUID | 2180dft.wpd, November 1998, http://www.fda.gov/cder/guidance/2180dft.pdf.
108. Barry, P.W. and O'Callaghan, C., Video analysis of the aerosol cloud produced by metered-dose inhalers, *Pharm. Sci.*, 1, 119–121, 1995.
109. Newman, S.P. and Wilding, I.R., Gamma scintigraphy: *in vivo* technique for assessing the equivalence of inhaled products, *Int. J. Pharm.*, 170, 1–9, 1998.
110. Dolovich, M.B., Measuring total and regional lung deposition using inhaled radiotracers, *J. Aerosol Med.*, 14, S35–S44, 2001.
111. Lee, Z. et al., Mapping PET-measured triamcinolone acetonide (TAA) aerosol distribution into deposition by airway generation, *Int. J. Pharm.*, 199, 7–16, 2000.
112. Gonda, I., Drugs administered directly into the respiratory tract: modeling of the duration of effective drug levels, *J. Pharm. Sci.*, 77, 340–346, 1988.
113. Byron, P.R., Prediction of drug residence times in regions of the human respiratory tract following aerosol inhalation, *J. Pharm. Sci.*, 75, 433–438, 1986.
114. Hickey, A.J. and Thompson, D.C., Physiology of the airways, in *Pharmaceutical Inhalation Aerosol Technology*, Hickey, A.J., Ed., Marcel Dekker, New York, NY, 1992, pp. 1–27.
115. Pillai, R.S. et al., Controlled release from condensation coated respirable aerosol particles, *J. Aerosol Sci.*, 25, 461–477, 1994.
116. Pillai, R.S. et al., Controlled dissolution from wax-coated aerosol particles in canine lungs, *J. Appl. Physiol.*, 84, 717–725, 1998.
117. Adjei, A.L., Qiu, Y., and Gupta, P. Bioavailability and pharmacokinetics of inhaled drugs, in *Inhalation Aerosols, Physical and Biological Basis for Therapy*, Hickey, A.J., Ed., Marcel Dekker Inc., New York, 1996, pp. 197–231.
118. Borgstrom, L. et al., Pulmonary deposition of inhaled terbutaline: comparison of scanning gamma camera and urinary excretion methods, *J. Pharm. Sci.*, 81, 753–755, 1992.
119. Newman, S. et al., Comparison of gamma scintigraphy and a pharmacokinetic technique for assessing pulmonary deposition of terbutaline sulphate delivered by pressurized metered dose inhaler, *Pharm. Res.*, 12, 231–236, 1995.
120. Newman, S. et al., Deposition of fenoterol from pressurized metered dose inhalers containing hydrofluoroalkanes, *J. Allerg. Clin. Immunol.*, 104, S253–S257, 1999.
121. Altiere, R. and Thompson, D., Physiology and pharmacology of the airways, in *Inhalation Aerosols*, Hickey, A.J., Ed., Marcel Dekker, New York, 1996, pp. 85–137.
122. Clark, A.R. and Hollingworth, A.M., The relationship between powder inhaler resistance and peak inspiratory conditions in healthy volunteers — implications for *in vitro* testing, *J. Aerosol Med. — Deposition Clearance Effects Lung*, 6, 99–110, 1993.
123. Martonen, T.B. et al., Lung models: strengths and limitations, *Respir. Care*, 45, 712–736, 2000.
124. Commission, P. *Physicians' Desk Reference*, 56 ed., Medical Economics Company Inc., Montvale, NJ, 2002.
125. Corcoran, T.E. et al., Improving drug delivery from medical nebulizers: the effects of increased nebulizer flow rates and reservoirs, *J. Aerosol Med.*, 15, 271–282, 2001.
126. British Pharmacopoeia, Her Majesty's Stationary Office, London, U.K., 1998, pp. A194.

127. Tzou, T.Z., Aerodynamic particle size of metered-dose inhalers determined by the quartz crystal microbalance and the Andersen cascade impactor, *Int. J. Pharm.*, 186, 71–79, 1999.
128. Cheng, Y.S. et al., Respiratory deposition patterns of salbutamol pMDI with CFC and HFA-134a formulations in a human airway replica, *J. Aerosol Med. — Deposition Clearance Effects Lung*, 14, 255–266, 2001.
129. Wilkes, W., Fink, J., and Dhand, R., Selecting an accessory device with a metered-dose inhaler: variable influence of accessory devices on fine particle dose, throat deposition, and drug delivery with asynchronous actuation from a metered-dose inhaler, *J. Aerosol Med. — Deposition Clearance Effects Lung*, 14, 351–360, 2001.
130. Warren, S.J. and Farr, S.J., Formulation of solution metered dose inhalers and comparison with aerosols emitted from conventional suspension systems, *Int. J. Pharm.*, 124, 195–203, 1995.
131. Stein, S.W., Size distribution measurements of metered dose inhalers using Andersen Mark II cascade impactors, *Int. J. Pharm.*, 186, 43–52, 1999.
132. Deshpande, D. et al., Aerosolization of lipoplexes using AERx® pulmonary delivery system, *AAPS PharmSci*, 4, article 13, 2002.
133. Farr, S.J., Expanding applications for precision pulmonary delivery, *Drug Delivery Technol.*, 2, 42, 2002.
134. Fink, J. et al., Enabling aerosol delivery technology for critical care, in *Respiratory Drug Delivery, VIII*, Dalby, R.N., Byron, P.R., Peart, J., and Farr, S.J., Eds., DHI Publishing, Raleigh, NC, 2002, pp. 323–325.
135. Smart, J. et al., Touchspray technology: comparison of the droplet size measured with cascade impaction and laser diffraction, in *Respiratory Drug Delivery, VIII*, Dalby, R.N., Byron, P.R., Peart, J., and Farr, S.J., Eds., DHI Publishing, Raleigh, NC, 2002, pp. 525–527.

chapter thirteen

Bioaerosols

Maire S.A. Heikkinen
New York University

Mervi K. Hjelmroos-Koski
University of California at Berkeley

Max M. Häggblom
Rutgers University

Janet M. Macher
California Department of Health Services

Contents

13.1 Introduction 292
13.1.1 Health effects of bioaerosols in indoor and outdoor environments 293
13.1.1.1 Infectious diseases 293
13.1.1.2 Hypersensitivity diseases 294
13.1.1.3 Inflammatory and other diseases 295
13.1.2 Sources and transmission 304
13.1.2.1 Transmission of infectious agents 304
13.1.2.2 Bioaerosols from outdoor sources 304
13.1.2.3 Bioaerosols in manufacturing industries 305
13.2 Bioaerosol size distributions 306
13.2.1 Aerodynamic diameters of airborne biological agents 306
13.2.1.1 Allergens 307
13.2.1.2 Culturable bacteria and fungi 307
13.2.1.3 Fragments 308
13.2.2 Seasonal and diurnal variability in ambient bioaerosol concentrations 308
13.2.2.1 Time of day 313
13.2.2.2 Time of year 313
13.2.2.3 Geographic region 313
13.3 Respiratory dosimetry 314
13.3.1 Respiratory deposition and clearance 314
13.3.2 Dose–response relationships/exposure assessment 316
13.3.2.1 Infectious dose 316
13.3.2.2 Allergens 317
13.3.2.3 Glucan 318
13.3.2.4 Endotoxin 319

1-56670-611-4 / 05 / $0.00+$1.50

13.3.2.5 Other bioaerosol exposures 320
13.3.2.6 Interactions 321
13.4 Environmental measurement and analytical techniques 321
13.4.1 Measurement of surrogates and indicators 321
13.4.2 Bioaerosol measurement 323
13.4.3 Bioaerosol samplers 324
13.4.3.1 Inertial sampling 324
13.4.3.2 Filtration 325
13.4.3.3 Electrostatic precipitation 325
13.4.3.4 Future directions in bioaerosol sampling 326
13.4.4 Sample analysis 326
13.4.4.1 Microscopy 326
13.4.4.2 Cultivation-based methods for bacteria and fungi 327
13.4.4.3 Biological assays 328
13.4.4.4 Immunoassays 329
13.4.4.5 Chemical assays 329
13.4.4.6 Molecular genetic assays 330
13.5 Concluding remarks 330
References 331

13.1 Introduction

Bioaerosols are those airborne particles that originate from living organisms (e.g., bacteria, protists, plants, fungi, and animals) or that depend on living organisms (e.g., viruses). Bioaerosols may consist of entire microscopic structures, for example, viruses, intact bacterial cells and spores, protozoa and their cysts, fungal cells and spores, and plant pollen grains and spores. Cell fragments may be present in indoor and outdoor air and are also considered as bioaerosols; for example, airborne particles of decayed microbial, plant, and animal matter; wood and grain dusts; the droppings and dried body parts of arthropods; and particles of larger animal skin, saliva, feces, and urine. The term *biological agent* refers to any substance of biological origin that is capable of producing an effect on humans. Bioaerosols may elicit responses similar to those caused by nonbiogenic particles (e.g., a hypersensitivity, irritant, or inflammatory response) as well as unique reactions (e.g., infectious diseases and toxicoses). The respiratory tract responds to injury, including that caused by biological agents, in a limited number of ways.[1] Rhinosinusitis, pharyngitis, laryngitis, upper airway obstruction, bronchitis, alveolitis, and pulmonary edema are acute reactions to biological agents while asthma, bronchitis, and some pulmonary infections are chronic conditions (Figure 13.1; see also Figure 24.6).

Bioaerosols occur as airborne particles in a size range of approximately 0.02 to 100 μm; thus, different cells, spores, pollen grains, and biological fragments may deposit in all regions of the human airways. Microorganisms, plants, and animals are important components of surface waters and soil. Although not particulate in form, plants and animals release gases and vapors, for example, oxygen, carbon dioxide, methane, and water. Emissions from microorganisms (microbial volatile organic compounds, MVOCs) account for the earthy smell that follows a rain shower or comes from freshly turned garden soil. The distinctive flavors of certain foods and beverages as well as the less pleasant aromas of decay, body odor, and moldy buildings are also volatile microbial metabolites. The complex interactions of MVOCs and bioaerosols with other airborne particles are seldom studied but are likely important for a comprehensive understanding of the effects of biological agents on human health and comfort. It is impossible to cover this broad topic in great depth in a single chapter; therefore, we discuss bioaerosol measurement, dosimetry, and health effects with illustrations and examples from the vast literature on the subject.

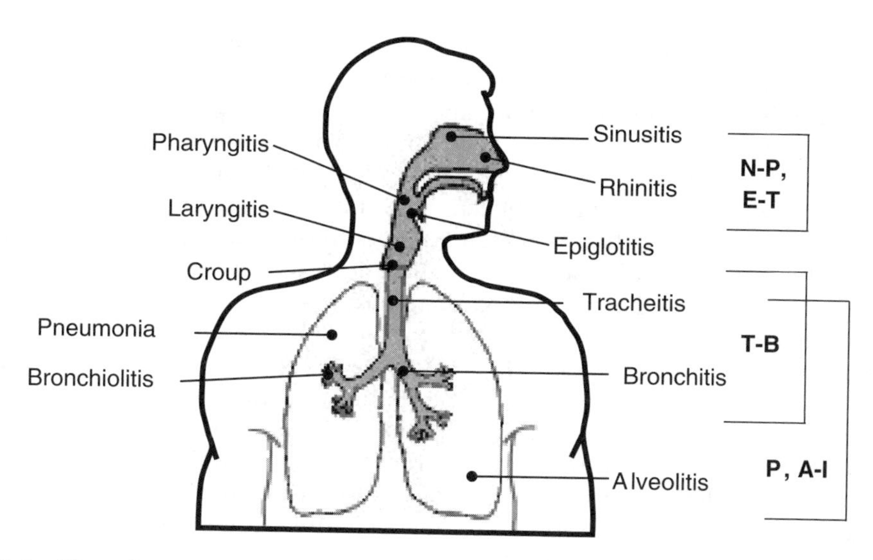

Figure 13.1 Clinical syndromes associated with infectious or hypersensitivity responses to bioaerosols at different locations within the respiratory tract (regions of the lung: N-P, nasopharyngeal; E-T, extrathoracic; T-B, tracheobronchial; P, pulmonary; A-I, alveolar-interstitial).

13.1.1 *Health effects of bioaerosols in indoor and outdoor environments*

Ambient air is comprised of suspended solid and liquid particles with relatively stable size distributions. Particles of biological origin are part of this mixture along with sea salt, salts formed from pollutant gases, soil fragments, water droplets, and combustion soot, and bioaerosols are inhaled along with other airborne particulate matter (PM) and gases. The fraction of the particle number concentration of outdoor PM originating from living sources has been estimated: 37% on a rooftop in Mainz, Germany[2] and 4 to 20% over the North Sea.[3,4] Carbon from fungi accounted for up to 10% of the coarse (2.5 to 10 μm) fraction of organic carbon in samples from the Austrian Alps.[6] Average concentrations of bacteria and fungi were 1.2×10^4 cells/m^3 and 7.3×10^2 spores/m^3, corresponding to 0.03 and 0.9% of organic carbon, respectively. The source strength of biogenic material (from plants and microorganisms) emitted as particles with aerodynamic diameters (d_a) smaller than 2.5 μm has been estimated at 6.5×10^{13} g/a.[7]

Some of the bioaerosols that we inhale are clearly unhealthy, but the vast majority of them are harmless or even essential for good health. For example, it has been postulated that early exposure to microorganisms may protect children from atopy and asthma (the hygiene hypothesis).[8] The evidence that exposure to cats and dogs in the first years of life increases a child's risk of developing hypersensitivity to these animals is conflicting.[9–11]

13.1.1.1 *Infectious diseases*

Infectious diseases are a fact of daily life, and acute respiratory infections are the most common of human illnesses. Most of the agents responsible for respiratory infections are spread through the air, primarily from person to person (anthroponoses) but also from living animals (zoonoses) and the abiotic environment, for example, soil, water, or decaying plant or animal matter (sapronoses).[12] Of the common, acute, respiratory infections, influenza produces the most severe illness and accounts for the greatest number of days of restricted activity in the United States.[13,14] The economic cost of lost productivity alone from the common cold has been estimated at over $25 billion ($17 billion due to on-the-job productivity losses, $8 billion to absence of ill persons, and $230 million for caregiver absenteeism).[15] Outbreaks of previously unrecognized airborne infectious diseases (e.g., Legionnaires' disease in the 1970s and severe acute respiratory syndrome (SARS) more recently) seriously affect the global economy.

Respiratory infections occur so often and are usually mild; therefore, many persons take them for granted. However, throughout human history, respiratory infections have imposed an immense burden on society.[16] A World Health Organization (WHO) study estimated that lower respiratory tract infections accounted for over 6% of the total, world-wide burden of disease and that tuberculosis alone represented 2.5% of the burden.[17,18] One third of the world's population is currently infected with *Mycobacterium tuberculosis*, which causes more deaths than any other microorganism.[19]

Among children under the age of 5 yr, approximately one third of all deaths (4.1 million in 1990) were due to acute respiratory tract infections, with over 95% of these deaths in children from developing countries.[17] The higher childhood mortality in developing countries probably reflects poorer nutrition and immunization practices and more frequent low birth weight, crowding, and exposure to indoor and outdoor air pollution.[20] Viruses are responsible for most childhood respiratory tract infections, although bacteria (e.g., *Mycoplasma* and *Chlamydia* spp.) also infect children at particular ages (Table 13.1). Respiratory tract infections in childhood may have long-term sequelae, including loss of lung function after severe episodes of lower respiratory tract infection, the development of asthma or bronchiectasis, and an increased risk of developing chronic obstructive pulmonary disease in adulthood.[21] Although disadvantaged population groups bear the greatest burden, communicable diseases occurring anywhere in the world are of concern to everyone, given the connectedness of all parts of the world via voluntary and involuntary human migrations and rapid international air travel.

13.1.1.2 *Hypersensitivity diseases*

An estimated 20% of Americans (>50 million persons) suffer from allergic rhinitis (hay fever) and other allergic diseases.[22,23] Hypersensitivity occurs in persons who (a) are genetically predisposed to mount an immunoglobulin E (IgE) response to a specific allergen; (b) have been exposed to a sensitizing dose of the allergen; and (c) are subsequently exposed to the allergen and manifest an allergic response. Hypersensitive persons may experience seasonal pollinosis or fungal allergy (if allergic to airborne pollen or spores that have a limited annual occurrence); episodic sneezing, rhinorrhea, or nasal congestion (following limited indoor or outdoor exposures); or perennial symptoms (if allergic to persistent indoor allergens). One of the consequences of nasal obstruction in hypersensitive persons is greater oral breathing, which bypasses some of the body's defenses and may lead to greater susceptibility to airborne infectious agents (see section 13.3.1).[24]

The WHO study mentioned above estimated that asthma accounts for 1% of the total burden of disease worldwide.[17] Approximately 8 to 17% of the United States population has asthma[22–27] and similar rates are seen around the world, increasing in most countries.[28] Biological agents may cause asthma without sensitization (e.g., endotoxin) or following immunological sensitization to high-molecular-weight agents (e.g., animal and plant proteins).[1] Persons with allergic rhinitis and asthma can control their illness to some extent by avoiding the agents to which they are sensitive. However, once asthma has developed, infections, exercise, and exposure to airborne irritants may also trigger acute or chronic symptoms.

Estimates suggest that approximately 80% of asthma in children and 30 to 50% in adults is allergic asthma.[29] At least 3% of asthma cases in the United States are occupational in origin, and work-related asthma (occupational and work-aggravated asthma) has been estimated to be 15% of all job-related disease.[1] For agricultural workers, rhinitis and asthma are the most important respiratory illnesses, which, although common, are seldom fatal.[30] Overall, respiratory hypersensitivity conditions cost billions of dollars a year in the

United States alone in terms of absences from work and school, lowered productivity, visits to doctors' offices and emergency rooms, hospitalizations, and the over 5000 annual deaths from asthma.[31,32]

13.1.1.3 *Inflammatory and other diseases*

Dusts of vegetable, animal, or microbial origin are often referred to as organic dusts. People are exposed to organic dust in a variety of environments, including agricultural operations, industrial or manufacturing processes, and residential, office, and commercial settings.[33,34] The composition of organic dust varies with the source material, but may contain bacteria, amebae, fungi, vegetable matter, arthropods, animal dander, and other proteins. Hypersensitivity pneumonitis (HP or extrinsic allergic alveolitis) refers to an immunologically mediated inflammatory disease of the lung parenchyma that is induced by exposure to certain low-molecular-weight chemicals or organic dusts (Table 13.1).[1,30]

Organic dust toxic syndrome (ODTS) is a poorly characterized, acute inflammatory response, in the airways and alveoli, to heavy inhalation exposures to organic material. However, unlike HP, ODTS is not an immune-mediated disease, does not require repeated exposures to occur, and does not appear to cause permanent lung damage.[35] The main features of the etiology and pathology of ODTS have been defined, but the precise mechanisms of the disease are unclear. It is likely that multiple agents may give rise to ODTS and that the syndrome represents a group of disorders.[30] Endotoxin is a common component of organic dust and may be involved in both asthma and ODTS.

β-(1–3)-D-glucans (glucose polymers in the cell walls of most fungi and plants and some bacteria) appear to play a role in bioaerosol-induced inflammatory responses and resulting respiratory symptoms (see Section 13.3.2.3).[36–39] However, only a limited number of field or experimental exposure studies are available on the health effects of inhaled glucans. Alwis et al.[40,41] studied woodworking sites and found exposure–response relationships between several symptoms and personal exposure to respirable glucan, inhalable glucan, and endotoxins. These researchers also observed a positive correlation between mean inhalable and respirable endotoxin exposures and Gram-negative bacteria and between mean inhalable and respirable glucan exposures and total fungal concentrations. Glucans have been associated with increased mucous membrane irritation and fatigue and may be a causative agent for the development of airway inflammation. In animal studies, glucan was found to be a powerful stimulator of humoral and cell-mediated immunity. Nevertheless, as for endotoxin, the relationship between glucan exposures and health complaints cannot yet be concluded to be causal. Exposure to glucan may simply indicate exposure to airborne fungal spores, and exposures to endotoxin and glucan may represent simultaneous exposures to other bioaerosols that are directly responsible for the observed respiratory and other effects.[42–45]

Endotoxins are a broad category of heat-stable, lipopolysaccharide (LPS)–protein complexes in the outer membranes of Gram-negative bacteria (a layer absent from Gram-positive bacteria). Endotoxin usually refers to the biologically active toxin present in the bacterial cell wall, while LPS is the chemically purified molecule with no other cell wall components present. Because Gram-negative bacteria comprise the natural microbiota on plant surfaces and are abundant in soil, endotoxin is ubiquitous in outdoor and indoor environments.[29] The polysaccharide portion of the molecule represents an antigenic surface and the lipid A portion is associated with toxicity.[46] The core polysaccharide and lipid A of endotoxin are conserved within bacterial species, but vary in structure and composition between species and, to a greater extent, between genera.

Inhalation of airborne endotoxin has been associated with workplace-related illnesses ranging from byssinosis, humidifier fever, mill fever, and grain fever, which characteristically recur on Mondays following return to work and dissipate later in the week because of tolerance on repeated exposure (Table 13.1, see Section 13.3.2.4).[45] Endotoxin is also

Table 13.1 Particle Sizes and Health Effects of Airborne Biological Agents

Biological Agent	Bioaerosol Particle Size Exposure Limit (TWA)	Location (Respiratory Tract)	Reservoir	Possible Health Effect	Susceptible Population
Infectious Agents					
Viruses					
Adenovirus	Droplets of respiratory discharges; 70–90 nm	Oropharynx	Humans	Acute febrile respiratory disease Acute viral rhinitis (common cold) Oropharyngitis (sore throat)	All persons Infants, young children, the elderly Nonimmune military recruits
Coronavirus	Droplets of nasal and throat secretions; 100–150 nm	Nasopharynx	Humans	Acute febrile respiratory disease Acute viral rhinitis (common cold)	All persons
Enteroviruses	Droplets of nasopharyngeal secretions; 30 nm	Oropharynx	Humans	Oropharyngitis (sore throat)	All persons
Hanta virus	Droplet nuclei of rodent excreta; 80–120 nm	Lung	Infected deer mice (*Peromyscus maniculatus*)	Hanta virus pulmonary syndrome	All persons, animal handlers, farmers
Influenza viruses type A, B, C	Droplet nuclei of respiratory secretions; 80–120 nm	Nasopharynx Larynx Bronchi	Humans, birds, swine	Influenza (A: pandemic; A, B: epidemic) Pneumonia	All persons, teachers, health care and public safety workers Most serious in elderly
Measles virus	Droplet nuclei of respiratory secretions; 120–250 nm	Bronchi	Humans	Measles (rubeola)	All persons, child care and health care workers, teachers
Mumps virus	Droplets of saliva; 150–200 nm	Salivary glands	Humans	Mumps (infectious parotitis)	Nonimmune children and adults
Parainfluenza virus	Droplets of respiratory secretions; 150–200 nm	Larynx-trachea Lung	Humans	Croup: obstruction of the upper airway in young children; hoarseness in adults	Young children Debilitated elderly persons

				Acute febrile respiratory disease Pneumonia	
Respiratory syncytial virus (RSV)	Droplets of respiratory secretions; 100–350 nm	Bronchioles Lung	Humans	Acute febrile respiratory disease Pneumonia	Infants and young children; Debilitated elderly persons
Rhinoviruses	Droplets, droplet nuclei of sputum, saliva, or nasal or throat secretions; 20–27 nm	Nasopharynx	Humans	Acute febrile respiratory disease Acute viral rhinitis (common cold)	All persons
Rubella virus	Droplets of nasopharyngeal secretions; 70 nm	Pharynx	Humans	Rubella (German measles)	All persons, child care and health care workers
Varicella-zoster virus	Droplet nuclei of respiratory secretions, vesicular fluid; 150–200 nm		Humans	Varicella (chicken pox)	All persons, child care and health care workers
Other respiratory viruses	Droplet nuclei of nasal and throat secretions	Nasopharynx Lung	Humans	Pneumonia	Infants and young children
Bacteria					
Bacillus anthracis	Aerosol of spores; cell diameter: 1.3 μm	Lung	Infected animals, pelts, and hides	Anthrax (woolsorter's disease); Pneumonia	Uncertain, especially for weaponized strains
Bordetella pertussis	Droplets of respiratory secretions; 0.2–1 μm	Ciliated epithelial cells in the upper respiratory tract	Humans	Pertussis (whooping cough)	All persons, health care workers
Burkholderia pseudomallei	Droplet nuclei; 0.5–1.0 $\times$ 1–5 μm	Lung	Water Soil	Melioidosis Pneumonia	Uncertain
Chlamydia pneumoniae	Uncertain, possibly droplet nuclei; ~0.3 μm	Lung	Humans	Pneumonia	All persons
Chlamydia psittaci	Droplet nuclei of dried feces, secretions, feathers, tissues; ~0.3 μm	Lung	Infected psittacine birds (parakeets, parrots)	Psittacosis Pneumonia	All persons, bird handlers

(Continued)

Table 13.1 *(Continued)*

Biological Agent	Bioaerosol Particle Size Exposure Limit (TWA)	Location (Respiratory Tract)	Reservoir	Possible Health Effect	Susceptible Population
Chlamydia trachomatis	Uncertain, possibly droplet nuclei; ~0.3 μm	Lung	Humans	Pneumonia	Infants and young children
Corynebacterium diphtheriae	Droplets of throat secretions; 0.3–0.8 × 1.5–8.0 μm	Throat (toxin passes into the bloodstream)	Humans	Diphtheria	Nonimmune children and adults
Coxiella burnettii	Droplet nuclei; 0.2 × 0.7 μm	Lung	Infected animals	Q fever Pneumonia	All persons, animal handlers
Francisella tularensis	Droplet nuclei; 1–2 μm	Lung	Infected animals	Tularemia Pneumonia	All persons
Haemophilus influenzae	Droplets of nasopharyngeal secretions; 0.3 × 1 μm	Middle ear and paranasal sinuses Epiglottis, Bronchi	Humans	Meningitis Pneumonia	Infants and young children
Legionella spp.	Droplet nuclei; 0.3–0.9 × 2 μm	Lung	Contaminated water	Legionnaires' pneumonia (Legionnaires' disease)	All persons, especially older persons, smokers, and immunocompromised persons
Atypical mycobacteria	Droplet nuclei of contaminated water or soil or respiratory secretions; 0.2–0.6 × 1–10 μm	Lung	Contaminated water or soil Humans Infected animals	Mycobacteriosis Pneumonia	Immunocompromised persons
Mycobacterium tuberculosis	Droplet nuclei of respiratory secretions; 1–5 μm	Lung	Humans	Tuberculosis Subacute or chronic pneumonia	All persons, especially very young and old, health care and public safety workers
Mycoplasma pneumoniae	Droplets of nasopharyngeal secretions; 0.1–0.2 × 1–2 μm	Bronchi	Humans	Pneumonia	All persons, especially older children and young adults
Neisseria meningitides	Droplets of nasopharyngeal secretions; 0.6–1.0 μm	Lung	Humans	Pneumonia	All persons, susceptibility decreases with age

Nocardia spp.	Airborne spores; 0.5–1.0 μm	Lung	Soil	Nocardiosis Pneumonia	Immunocompromised persons
Streptococcus pneumoniae (pneumococcus)	Droplets of nasopharyngeal secretions; <2 μm	Middle ear and paranasal sinuses Bronchi	Humans	Pneumonia Meningitis	All persons, especially <1 yr old, >60 yr old
Group A streptococcus (*Streptococcus pyogenes*)	Droplets of nasopharyngeal secretions; <2 μm	Oropharynx Middle ear and paranasal sinuses	Humans	Streptococcal sore throat Pneumonia	All persons
Yersinia pestis	Droplet nuclei of respiratory secretions; 0.5–0.8 × 1–3 μm	Lung	Infected animals Humans	Pneumonic plague Pneumonia	All persons
Fungi					
Aspergillus spp.	Airborne conidia; 2–5 μm	Lung	Soil, organic materials	Pneumonia	Immunocompromised persons
Blastomyces dermatitidis	Airborne conidia; 2–10 μm	Lung	Contaminated soil	Blastomycosis Subacute pneumonia	Workers with soil contact
Coccidioides immitis	Airborne arthroconidia; 2–4 × 3–6 μm	Lung	Contaminated soil	Coccidioidomycosis Subacute pneumonia	Workers with soil dust exposure
Cryptococcus neoformans	Yeast cells; 4–8 μm	Lung	Contaminated soil, pigeon droppings	Cryptococcosis Subacute pneumonia	Animal handlers, cleaners
Histoplasma capsulatum	Airborne microconidia; 2–5 μm	Lung	Infected birds, droppings	Histoplasmosis Subacute pneumonia	Bird handlers, cleaners
Zygomycetes (phycomycetes)	Spores	Lung	Soil, organic matter	Zygomycosis (phycomycosis) Pneumonia	Immunocompromised persons
Protists					
Pneumocystis carinii	Droplet nuclei of respiratory secretions; Trophozoite: 1–5 μm Precyst: 5–8 μm Cysts: 8 μm	Lung	Humans (uncertain)	Pneumonia	Immunocompromised persons, Malnourished children
Amebae *Acanthamoeba* spp. *Balamuthia* spp.	Trophozoite: 15–45 μm Cysts: 10–25 μm Trophozoite: 12–60 μm Cysts: 6–30 μm	Nasal passages	Contaminated water and soil	Granulomatous amebic encephalitis (GAE)	Unknown

(Continued)

Table 13.1 *(Continued)*

Biological Agent	Bioaerosol Particle Size Exposure Limit (TWA)	Location (Respiratory Tract)	Reservoir	Possible Health Effect	Susceptible Population
Naegleria fowleri	Trophozoite: 10–35 μm Cysts: 7–15 μm	Nasal passages	Contaminated water and soil	Primary amebic meningoencephalitis (PAM)	Unknown
Aeroallergens **Bacterial Antigens**	Cells and cell fragments	Lung Mucous membranes	Growth in water or on organic matter, settled dust	Building-related symptoms Hypersensitivity pneumonitis (HP)	Hypersensitive persons in contaminated buildings Repeatedly exposed, hypersensitive workers (REHW)
Bacillus subtilis	0.8 μm diameter	Lung	Detergent enzyme	HP: Detergent worker's lung	REHW
Mycobacteria *Mycobacterium immunogenum*	Bacterial cells or cell fragments; 0.2–0.6 × 1–10 μm	Lung	Contaminated water-based metalworking fluid	HP	Machinists and other workers
Saccharopolyspora (Faenia) rectivirgula	Spores	Lung	Moldy hay	HP: Farmer's lung	REHW (agricultural)
Thermoactinomyces candidus	Spores	Lung	Heated water reservoirs	HP: Humidifier lung	REHW
Thermoactinomyces sacchari	Spores	Lung	Moldy sugar cane fiber	HP: Bagassosis	REHW
Thermoactinomyces vulgaris	Spores	Lung	Moldy grain, compost	HP: Grain worker's lung, Mushroom worker's lung	REHW
Fungal Allergens	Spores, cell fragments	Lung Mucous membranes	Growth in water or on organic matter, settled dust	Allergic conjunctivitis, rhinitis, sinusitis, asthma; Building-related symptoms	Hypersensitive persons in contaminated buildings
Aspergillus clavatus	3.0–4.5 × 2.5–4.5 μm	Lung	Moldy malt	HP: Malt worker's lung	REHW
Aureobasidium pullulans	8–12 × 2–5 μm	Lung	Moldy redwood dust	HP: Sequoiosis	REHW
Graphium spp.	4–7 μm				
Cryptostroma corticale	4–6.5 × 3.5–4 μm	Lung	Moldy maple bark	HP: Maple bark stripper's lung	REHW

Penicillium commune (*P. casei, P. camembertii*)	3.0–4.5 × 4.0–5.0 μm	Lung	Moldy cheese	HP: Cheese worker's lung	REHW
Penicillium glabrum (*P. frequentans*)	3.0–3.5 μm	Lung	Moldy cork dust	HP: Suberosis	REHW
Aspergillus flavus; *Aspergillus fumigatus;* *Aspergillus terreus*	3–8 μm 2–3.5 μm 1.5–2.5 μm	Lung	Contaminated materials	Allergic bronchopulmonary aspergillosis	Exposed persons with asthma
Penicillium spp.	2–3.5 μm	Lung	Contaminated materials	Allergic bronchopulmonary mycosis	Exposed persons with asthma
Amebic Allergens *Acanthamoeba castellani*	Trophozoite: 15–45 μm long Cysts: 10–25 μm	Lung	Contaminated water and soil	HP: Humidifier lung	All exposed persons
Naegleria gruberi	Trophozoite: 10–35 μm long Cysts: 7–15 μm	Lung	Contaminated water and soil	HP: Humidifier lung	All exposed persons
Plant Allergens	Pollen grains, plant fragments	Lung, uncertain	Vegetation, settled dust	Allergic conjunctivitis, rhinitis, sinusitis, asthma	Exposed hypersensitive persons
Ambrosia spp., ragweed	17–19 × 16–20 μm				
Artemisia spp., mugwort	20–23 × 19–21 μm				
Betulaceae, birch family	20–29 × 18–29 μm				
Cryptomeria japonica, Japanese cedar, Sugi	36–35 μm				
Cupressaceae, cypress family	23–31 × 21–29 μm				
Fagaceae, beech family (*Quercus* spp., oaks)	32 × 29 μm				
Juglandaceae, walnut family	35–41 × 32–38 μm				
Mimosoidae, mimosa subfamily (*Acacia* spp.) *Prosopis* spp., mesquites	*Acacia* spp.: 40 × 60 μm *Prosopis* spp.: 30–27 μm				
Morus spp., mulberry	18–17 μm				
Oleaceae, olive family	*Olea* spp.: 21–19 μm				
Poaceae, grass family	Wild grass: 28–36 × 26–33 μm Cultivated grass: >42 μm				

(*Continued*)

Table 13.1 *(Continued)*

Biological Agent	Bioaerosol Particle Size Exposure Limit (TWA)	Location (Respiratory Tract)	Reservoir	Possible Health Effect	Susceptible Population
Parietaria spp., pellitory Orbicules, plant fragments	12–15 μm 1.5–2 μm				
Arthropod Allergens	Cockroach or house dust mite excreta and body fragments	Lung, uncertain	Settled dust	Allergic conjunctivitis, rhinitis, sinusitis, asthma Building-related symptoms	Exposed hypersensitive persons
*Blatella germanica, Bla g*1, cockroach allergen	Roach: > 5 μm				
*Dermatophagoides farinae, Der f*1, *Dermatofagoides pteronyssinus, Der p*1, dust mite allergens	Mites: 10–35 μm				
Avian Proteins	Particles of bird droppings, feathers	Lung	Birds	HP: Bird breeder's lung Asthma	Exposed hypersensitive persons, bird handlers
Mammalian Allergens *Felis domesticus, Fel d*1, cat allergen *Canis familiaris, Can f*1 allergen *Rattus norvegicus, Rat n*1, rat allergen *Muscus musculus, Mus m*1,2, house mouse allergen *Cavia porcellus, Cav p*1, guinea pig allergen	Particles of cat, dog, or rodent skin, saliva, or urine Cat and dog allergens: <5 μm	Lung, uncertain	Pets, pests, settled dust	Allergic conjunctivitis, rhinitis, sinusitis, asthma Building-related symptoms HP: Animal handler's lung	Exposed hypersensitive persons, animal handlers
Biological Toxins and Inflammatory Agents					
Bacterial endotoxin	Gram-negative bacterial cells and cell fragments	Lung	Growth in water or on organic matter	Respiratory inflammation Humidifier fever Cottonworker's lung (byssinosis)	Highly exposed workers and other persons

				Mill fever Grain fever Building-related symptoms	
Legionella spp.	Bacterial cells or cell fragments	Lung	Contaminated water	Pontiac fever	Highly exposed workers and other persons
Fungal glucan	Spores, cell fragments	Lung	Growth on organic matter	Airway inflammation Mucous membrane irritation, Fatigue, Antitumor activity	Highly exposed workers and other persons
Fungal toxins	Spores, cell fragments	Lung, uncertain	Growth on organic matter	Toxic and irritant effects	Highly exposed workers and other persons
Organic dust	Fungal spores, plant, animal, and microbial cell fragments Total dust: 15 mg/m^3 Respirable dust: 5 mg/m^3	Airways and alveoli	Moldy silage, compost, wood chips Sewage sludge	Organic dust toxic syndrome	Highly exposed workers and other persons
Cotton, hemp, jute, and flax dust	Microorganisms, cell fragments TWA (raw cotton dust): 0.2 mg/m^3	Lung	Plants	Mill fever Acute airway obstruction (byssinosis) Chronic bronchitis	Highly exposed textile workers (especially cigarette smokers) and other persons
Grain (oat, wheat, barley) dust Flour dust	Microorganisms, cell fragments Grain dust: 4 mg/m^3 Flour dust: 0.5 mg/m^3	Lung	Plants	Irritation Grain fever Asthma Chronic bronchitis Obstructive lung disease	Highly exposed grain workers (especially cigarette smokers) and other persons
Wood (hard woods: beech and oak; soft wood) dust	Dust particles Allergenic species: 0.5 mg/m^3 Nonallergenic species: 1 mg/m^3	Lung Nose	Trees	Irritation Lung function Asthma Cancer (nasal)	Highly exposed workers and other persons
Vegetable oil mists (castor bean, sesame, acacia, cashew nut)	Sprays and mists 10 mg/m^3 Total dust: 15 mg/m^3 Respirable dust: 5 mg/m^3	Lung	Plants	Nuisance particles Respiratory irritation Asthma	Highly exposed workers

Sources: APHA,[250], Salyers and Whitt,[165], Hjelmroos et al.,[85]; Samson et al.,[171] Larone,[251] Watanabe,[252] ASM.[253]
TWA: time-weighted average.

suspected of contributing to the severity of asthma and ODTS and of playing a role in nonspecific building-related symptoms.[47] In the late 1990s, some researchers considered the possibility that endotoxin may play a beneficial role by promoting the type 1 helper T (Th1) cell phenotype and, therefore, protecting exposed persons from atopic disease and asthma, but conclusive evidence of this effect is not yet available.[44, 45]

13.1.2 Sources and transmission

13.1.2.1 Transmission of infectious agents

Inhalation is the most important and efficient route by which infectious agents enter the human body, and infections contracted by this route are the most difficult to control.[48] The great pandemics in human history (e.g., plague and influenza) as well as many common childhood infections (e.g., measles, rubella, varicella, and mumps) are acquired from the respiratory secretions of others (Table 13.1). Transmission by air allows an infectious agent to reach a larger number of potential hosts than would be possible if the victims had to come into direct contact to transfer microorganisms from person to person. Humans release particles from the larynx, mouth, and throat during normal breathing, speaking, singing, and expectorating, but sneezing and coughing expel droplets with greater efficiency. Most of the droplets from a sneeze originate in the mouth and do not carry infectious agents. However, an increase of nasal secretions due to an upper respiratory tract infection can trigger sneezing and explosive expulsion of infectious particles from that region. Coughing is an efficient means of transmitting viruses and bacteria if an infection causes an increase of mucus secretions that, in turn, induce the cough reflex. A sneeze has been estimated to generate millions of aerosol particles and a cough many thousands. The fate of these particles depends on ambient environmental conditions (e.g., humidity and air movement) and particle size.

Droplet contact involves large particles ($>5\ \mu m$) that humans release from the nasopharyngeal region or that contaminated bodies of water generate through sprays or splashes.[48] Laboratory, surgical, and dental procedures can also produce aerosols of blood, saliva, enamel, dentin, bone, and tissue, some of which contain infectious agents. Particles ranging from 0.3 to 20 μm have been measured from surgical procedures, oral operations, laser smoke, and dermabrations, their concentration and size depending on the method of aerosol generation.[49] After water evaporates, the dried residue of a droplet is called a droplet nucleus. The important differences between droplets and droplet nuclei are, respectively, (a) their sizes (greater or less than $\sim 5\ \mu m$), (b) the distances they can travel (less or more than a few meters), and (c) their deposition sites in the respiratory tract (the airways of the head and upper respiratory system or the lower lungs) (Figure 13.1, Table 13.1).

13.1.2.2 Bioaerosols from outdoor sources

A variety of environments, including soil, surface water, and plants, serve as natural reservoirs of biological agents and sources of bioaerosols. In number, bacteria are the most abundant organisms in surface soils, where their concentration can exceed 10^{10} cells/g, of which typically 10^7 to 10^8 cells are culturable on standard media.[50,51] Torsvik et al.[52] estimated that as many as 4000 different bacterial species may be present in a gram of soil. Protozoa and algae are also abundant in soil (10^3 to 10^6 cells/g) along with fungi (typically 10^5 to 10^6 cells/g), but the overall biomass of fungi is greater because of their larger cell sizes. Soil microorganisms can be aerosolized during mechanical disturbance, such as the tilling of agricultural fields or dust storms.[53] Many plant pathogens, in particular fungi, are transmitted by air (e.g., during crop harvesting), but these agents do not pose a hazard for people. However, some human infections result from exposure to soil dust contaminated with infectious bacteria or fungi. Some of these agents are opportunistic pathogens, which can grow either as saprophytes living on decaying organic matter or as parasites infecting the human lung, for example, *Burkholderia pseudomallei*: melioidosis; *Histoplasma capsulatum*: histoplasmosis; *Coccidioides immitis*: coccidioidomycosis; and *Cryptococcus neoformans*: cryptococcosis (Table 13.1). Bioaerosols are also generated during the mechanical turning of compost and from large landfill sites.[54] The activated sludge systems in wastewater treatment plants can generate large amounts of bioaerosols.[55] Irrigation with reclaimed water and the application of solid waste materials from water treatment processes to soil can also aerosolize microorganisms and particles of organic matter.

13.1.2.3 Bioaerosols in manufacturing industries

Microbial contamination is a continuing problem in a variety of manufacturing processes.[56,57] One setting that has been investigated extensively is the metalworking industry, for example, facilities that manufacture automobile components. Metalworking fluids (MWFs) are widely used in the machining of metal parts to prolong the life of the cutting tool and protect product surfaces by cooling and lubricating the parts and carrying away debris. Water-based MWFs provide an excellent medium for the growth of a variety of bacteria and fungi to concentrations exceeding 10^6/ml despite the expanded use of biocides and the implementation of other control measures.[58–62] Rapidly moving metal parts can generate droplets of contaminated cutting fluid, and occupational MWF exposures have been linked with a number of respiratory diseases.[62–66] Microorganisms or their components in MWFs are likely causes for several of the observed adverse pulmonary effects in machinists and metal workers. Investigations of HP cases at several metalworking plants have shown that *Mycobacterium* species and other Gram-positive bacteria may be pervasive contaminants of MWFs.[62,67]

Exposure to organic dusts that contain microbes or microbial cell fragments is of concern in industrial and agricultural environments.[49,68,69] Gram-negative bacteria and associated endotoxin are among the primary biological contaminants of concern, with inhalation of endotoxin resulting in inflammatory responses, for example, cotton workers' lung. Workers are exposed to airborne bacteria and endotoxins as well as fungi in large animal confinement buildings and sheds[70,71] and industrial food-processing facilities.[72]

Anthrax, although currently of concern as a biological weapon, may be one of the first infectious, respiratory diseases to be associated with certain occupations. Inhalation anthrax has been associated with exposure to *Bacillus anthracis* spores in contaminated animal products, such as during the industrial processing of wool (woolsorters' disease).[73] The intentional distribution of *B. anthracis* spores through mail established a new and unusual route of occupational exposure to aerosolized spores from contaminated letter-sorting machines.[74] In October 2001, four cases of inhalation anthrax occurred in employees at a Washington, DC postal processing and distribution center. These cases were part of a multistate outbreak of inhalation and cutaneous anthrax associated with intentional distribution of envelopes containing *B. anthracis* spores to media and federal government offices. Twenty-two cases of anthrax were identified: 11 inhalation infections, of which five were fatal, and 11 cutaneous infections.[75] Twenty (91%) of the infections were either mail handlers or were exposed at worksites where contaminated mail was processed or received. Together, these represent the first reported cases of inhalation anthrax in postal workers and the first outbreak of inhalation anthrax caused by occupational exposure in the United States since 1957. One of the key concerns is "weaponization" of infectious particles through the addition of surface-active materials that cause them to separate from each other and to maintain a uniform and inhalable particle size. Threats of bioterrorism and biowarfare have heightened the need for accurate and sensitive methods for the measurement of airborne infectious agents and have encouraged the development of rapid detection methods (see Section 13.4.3.4).

13.2 Bioaerosol size distributions

Bioaerosols vary greatly in particle diameter from nanometer-sized viruses and cell fragments to single cells, cell agglomerates, and aeroallergen particles in the micrometer range. Viruses are generally 20 to 300 nm in diameter, intact bacteria 0.3 to 10 μm, fungal spores 1 to 100 μm, and pollen grains 5 to 200 μm (Table 13.1). However, information is incomplete on the size distribution of airborne viruses, bacteria, pollen, fungi, and animal allergens, and very little information is available on the size distributions of cell fragments (see Section 13.2.1.3). The prevalence and distribution of bioaerosols vary considerably with atmospheric and indoor conditions and between and within sites. The dynamic properties of all airborne particles depend on their physical characteristics such as size, shape, surface features, density, hygroscopicity, and electrostatic charge (see Section 13.3). Electrostatic charges on particles may enhance their deposition on oppositely charged surfaces or due to image charging (induction of an opposite charge on a surface by a charged particle), but few studies have measured naturally occurring charges on biological particles. Mainelis et al.[76,77] observed a net negative charge on laboratory-generated bacterial aerosols. The amount of charge that particles carry depends on the generation mechanism, environmental conditions, and the time that has elapsed between aerosol generation and measurement.

Most airborne particles have a thin layer of water molecules on their surfaces and can absorb or lose water. Substantial changes in the particle size of hygroscopic cells can alter their deposition behavior. Hygroscopic particle growth that is rapid enough to occur during passage through the respiratory tract will increase deposition in the regions where settling and impaction are important and decrease deposition in regions where diffusion dominates. Although fungal spores are assumed to be hydrophobic, Reponen et al.[78] found an almost 30% growth in the size of some spores at relative humidities (RHs) above 90% and estimated that particle sizes from 0.5 to 2 μm would be affected most by hygroscopic growth. They calculated that a 30% increase in spore size would cause a 20% increase in respiratory deposition, primarily in the bronchi. Madelin and Johnson[79] observed an immediate increase in particle diameter and separation of fungal and actinomycete spores as they passed briefly through a test chamber of warm, humid air (38°C, 95% RH) compared with exposure to cooler, dryer air (20°C, 40% RH).

13.2.1 Aerodynamic diameters of airborne biological agents

The most important parameter for describing the behavior of particles with diameters larger than 500 nm is d_a (see Chapter 2). The density of common airborne fungal spores varies from 0.56 to 1.44 g/cm^3 and that for pollen grains from 0.39 to 1.1 g/cm^3.[80] Because the density of most microbial spores and pollen grains is near unity, their aerodynamic diameters depend primarily on particle size and shape.

A fraction (~10%) of the pollen from angiosperms (flowering plants) and all of that from gymnosperms (conifers and other plants that produce seeds within cones) is dispersed by wind,[81,82] but not all of these pollen are allergenic. Pollen and spores that rely on wind transport (anemophilous species) are generally smaller (<50 μm) than those transported by other means, for example, by insects (entomophilous species). However, there is considerable overlap, and some strictly wind-pollinated plants (e.g., the pine family) produce larger pollen.[83] The texture or ornamentation of the outer pollen wall differs between windborne and insectborne grains. Anemophilous pollen are often nearly spherical and have relatively smooth surfaces. Conversely, entomophilous pollen tend to have more elaborate surface features, which help them adhere[83] but also affect their dynamic shape factor and thus their d_a when airborne.

For a given genus or species of microorganism or plant, the variability (geometric standard deviation, GSD) in spore or pollen grain diameter can be small. For example, using a six-stage impactor, Reponen[84] observed the following GSDs for culturable fungi: *Penicillium* species 1.1 to 1.4; yeasts 1.1 to 1.5; *Cladosporium* species 1.1 to 1.9; and *Aspergillus* species 1.4 to 1.9. However, measurement of the dimensions of individual spores by microscope has shown much higher variability.[85] Part of the variability in measurements of d_a may be due to the occurrence of clusters or chains of cells or the attachment of biological matter to other particles. Comparisons of measurements made with a multistage viable impactor and an aerodynamic particle sizer have revealed that fungal particles in chains behave aerodynamically like single fungal particles.[79,86] Reponen[84] calculated the effect of several forms of particle aggregation on the average d_a of a cluster of *Penicillium chrysogenum* spores (physical size: 2.8 × 3 μm), respective d_a at two particle densities (0.5 and 1 g/cm^3): (a) single particle — 2.0 and 2.9 μm; (b) four-particle chain — 2.8 and 4.0 μm; (c) compact four-particle cluster — 3.0 and 4.2 μm; (d) eight-particle chain — 3.1 and 4.4 μm; and (e) eight-particle cluster — 3.7 and 5.2 μm. These calculations illustrate the effect of particle density and cell arrangement on d_a. For example, the average diameter of a unit-density, single spore was 45% larger than one of 0.5 g/cm^3 (2.9 vs. 2.0 μm). As compared to a unit-density, eight-cell chain, the average diameter of eight cells in a cluster was 18% larger (5.2 vs. 4.4 μm).

13.2.1.1 Allergens

Air sampling has provided information on the size of indoor allergens that are of health significance (cat and dog: <5 μm; cockroach: >5 μm; house dust mite: >10 μm). [29,87–90] In a cat challenge facility, Wood *et al.*[91] found the majority of airborne allergen (*Fel d*1) on particles larger than 17 μm in diameter. While less than 15% of cat allergen was detected on particles smaller than 4 μm, allergen-containing particles of that size were present whether cats were caged or free and under all ventilation conditions. Using a ten-stage sampler, Custovic *et al.*[89] collected dog allergen (*Can f*1) in a dog-handling facility as well as in homes with and without dogs. Almost half of the total allergen was found on particles larger than 9 μm and approximately 20% on particles smaller than 5 μm. Another study compared the concentrations of cat, dog, and mite allergens in reservoir dust and indoor air.[92] The investigators detected airborne cat and dog allergens in all 65 homes with pets and in 26 and 64%, respectively, of 62 homes without cats or dogs. However, in these homes, dust mite allergen was not detected in any air samples whether or not it was found in settled dust because of the large size and rapid settling of the antigenic particles. Fecal pellets from mites, which contain digestive enzymes and partially digested food (primarily human skin flakes), are estimated to range from approximately 10 to 35 μm in diameter.[22] Only a small fraction of allergen particles may be smaller than 5 μm and able to enter the lungs. However, this variable fraction of particles is assumed to produce inflammatory responses in hypersensitive persons. [29]

13.2.1.2 Culturable bacteria and fungi

Bacterial spores resist environmental stress much better than vegetative cells and are considerably smaller.[93] Many fungal spores are well suited for dispersal through the air with special launching mechanisms that are triggered by climatic changes, for example, an increase or decrease in humidity or the impaction of raindrops.[94] A study of the release of spores from cultures of *Aspergillus fumigatus, Penicillium* species, and *Cladosporium* species at different air velocities and humidities showed that the conidiophores of the first two groups released spores at air velocities as low as 0.5 m/s, whereas *Cladosporium* species required a velocity of at least 1.0 m/s.[95] Górny et al.[96] observed that the release of fungal propagules from cultures of *Aspergillus versicolor, Penicillium melinii,* and *Cladosporium cladosporioides* varied, depending on fungal species, air velocity across the contaminated surface, vibration, and the texture of the contaminated material.[96] An air velocity of 0.3 m/s (still indoor air) released more spores from smooth surfaces than air moving at 29.1 m/s (typical flow rate through ventilation ducts) but did not affect the removal of fragments, indicating a different release mechanism for the latter.

The size distributions of viable, airborne bacteria and fungi (reported as colony-forming units (CFU)/m^3) have been studied in many environments. The highest air concentrations of culturable bacteria and fungi are typically found on the stages of size-segregating samplers that correspond to the diameters of intact cells or cell agglomerates. For example, the maximum concentration of culturable fungal particles in indoor and outdoor air has been measured between 1.1 and 2.1 μm using the Andersen six-stage impactor (Cascade impactors, Table 13.2).[53,97,98] Reponen[84] found the highest concentrations of the major fungal genera in the size range of 2.1 to 3.3 μm in residences with mold problems and control homes, respectively. Her calculations predicted that 30 to 50% of these particles would deposit in the nose and 30 to 40% in the alveoli during nasal breathing, whereas 70% would deposit in the alveoli during oral breathing (see Section 13.3.1).

13.2.1.3 Fragments

All the particle sizes reported in Table 13.1 are for intact forms, but fragments of biological materials may be found as particles much smaller than their original structures. Górny et al.[96] measured whole fungal spores and fragments entrained in air that had passed over contaminated ceiling tiles or agar surfaces placed in a test chamber. The fragments and spores that they collected carried common antigens suggesting their shared origin and potential importance to human health. Fungal fragments can remain airborne longer, travel further, and are found in finer particle size fractions than would be expected given the size of their original spores and hyphae.

High concentrations of birch pollen and antigen are occasionally found before and after the pollination period, and birch antigens may also be detected in particles smaller than intact pollen grains, which may explain the occurrence of allergy symptoms outside the pollination period.[99–101] In Japan, Takahashi et al.[102] found plant antigens in two particle size ranges (0.5 to 1.4 and 29 to 40 μm, the latter is the size of intact pollen grains), as had earlier researchers who studied ragweed and grass pollen.[103–105] The antigenic activity of the smaller particle size fraction was eight times greater than that of the larger fraction,[102] suggesting a greater allergenicity for the smaller particles due to their origin on the plant and higher numbers relative to intact pollen grains. Zureik et al.[106] hypothesized that the size of fungal spores allows them to reach the lower airways and trigger asthma, but that a different exposure mechanism must be involved for larger pollen grains.[106] There is evidence that during thunderstorms, pollen is concentrated by changes in airflow, the grains are ruptured by osmotic shock, and each grain releases hundreds of allergen-containing starch granules that are small enough to be respired.[107–110] Lewis et al.[111] saw a significant interaction between the effects of grass pollen and weather conditions on emergency room visits for asthma in England, with the increase most marked on days of light rainfall. Dales et al.[112] also saw a relationship between thunderstorms and asthma admissions in Canada. However, while changes in pollen and air pollutants on thunderstorm days were relatively small, fungal air concentrations increased twofold.

13.2.2 Seasonal and diurnal variability in ambient bioaerosol concentrations

Bioaerosol prevalence and fluctuations in concentration are influenced by climate and weather (resulting in diurnal and seasonal cycles) and by local sources (resulting in regional variation).

Table 13.2 Bioaerosol Samplers

Sampler	Principle of Operation	Sampling Rate (l/min)	Manufacturer/Supplier	Commercial Name	Application[a]
Slit Agar Impactors					
Rotating slit or slit-to-agar impactors	Impaction onto agar on rotating surfaces	28	Barramundi Corp.	Mattson-Garvin Air Sampler	C
		175–700	Casella Ltd.	Casella Airborne Bacteria Sampler	C
		15–55	New Brunswick Scientific Co. Inc.	New Brunswick Slit-to-Agar Air Samplers	C
Impactors					
Single-stage impactors	Impaction onto agar or other surface	10, 20	Burkard Manufacturing Co. Ltd.	Burkard Portable Air Sampler for Agar Plates	C
		28	ThermoAndersen Inc.	Andersen Single Stage Viable Particle Sampler; N6 Single Stage Viable Impactor	C
		28	Aerotech Laboratories Inc.	Aerotech 6 Bioaerosol Sampler	C
		28, 71, 142	Veltek Associates Inc.	Sterilizable Microbiological Atrium (SMA): MicroSampler and MicroPortable Viable Air Sampler	C
		100 or 180	International PBI; Scientific Products Corp.; Bioscience International	Surface-Air-Sampler (SAS): Super 100 Sampler, HiVac Sampler, HiVac Impact, HiVac Petri, and Super 180 Sampler	C
		100	F.W. Parrett Ltd.	MicroBio Air Sampler: MB1, MB2	C
		100, 200	Microbiology International	Sampl'Air Air Sampler	C
		100	Merck KGaA/VWR Scientific Products Corp.	Merck Air Sampler MAS 100	C
		140, 180	Millipore Corp.	M Air T, Millipore Air Tester	C

(Continued)

Table 13.2 (*Continued*)

Sampler	Principle of Operation	Sampling Rate (l/min)	Manufacturer/Supplier	Commercial Name	Application[a]
Cascade impactors	Impaction onto agar or other surface	28	ThermoAndersen Inc.	Andersen Two- and Six-Stage Viable Sampler/Cascade Impactor	C
		2	ThermoAndersen Inc.	Andersen Personal Cascade Impactor, Series 290 Marple Personal Cascade Impactor	O
Pollen, Spore, and Particle Impactors					
One- to seven-day tape/slide impactors	Impaction onto rotating drum with tape strip or glass slide	10	Burkard Manufacturing Co. Ltd.	Burkard Recording Volumetric Spore Trap	M
		10	Lanzoni, S.R.L.	Lanzoni Volumetric Pollen and Particle Sampler	M
			G-R Manufacturing Co.	Kramer-Collins Suction Trap	M
Moving slide impactors	Impaction onto moving glass slides	15	Allergen LLC; McCrone Microscopes and Accessories	Allergenco Air Sampler (MK-3)	M
		10	Burkard Manufacturing Co. Ltd.	Burkard Continuous Recording Air Sampler	M
		10	Lanzoni, S.R.L.	Lanzoni Volumetric Pollen and Particle Sampler	M
Stationary slide impactor	Impaction onto stationary glass slide	10	Burkard Manufacturing Co. Ltd.	Burkard Personal Volumetric Air Sampler	M
Cassette slide impactor	Impaction onto stationary glass slide	15	Zefon Analytical Associates; Aerotech Laboratories Inc.; McCrone Microscopes and Accessories; SKC Inc.	Air-O-Cell Sampling Cassette	M

Cassette tape sampler	Deposition onto stationary tape	10	*Metha*pharm Inc.	Partrap FA52	M
Rotating rod impactor	Impaction onto rotating rods	48	Sampling Technologies Inc.	Rotorod	M
Personal aeroallergen sampler	Impaction on a protein-binding membrane held in a nasal insert	Dependent on breathing rate	Available for research only	Inhalix	O
Liquid Impingers					
All-glass impingers	Impingement into liquid	12.5	Ace Glass Inc.; Hampshire Glassware; Millipore Corp.	All-Glass Impingers (AGI): AGI-4, AGI-30	C, M, O
Three-stage impingers		10, 20, 50	Burkard Manufacturing Co. Ltd.	Burkard Multiple-Stage Liquid Impinger	C, M, O
		10, 20, 50	Hampshire Glassware	Hampshire Glass Three-Stage Impinger	C, M, O
Centrifugal Samplers					
Centrifugal agar impactors	Impaction onto agar in plastic strips	40, 50, 100	Biotest Diagnostics Corp.	Reuter Centrifugal Samplers (RCS): Standard RCS, RCS Plus, and RCS High Flow Microbial Air Sampler	C
Wetted cyclone samplers	Tangential impingement into liquid	50–55 167, 500	Hampshire Glassware	AEA Technology PLC Aerojet Cyclones	C, M, O
		167	F.W. Parrett Ltd.	MicroBio MB3 Portable Cyclone	C, M, O
		700–1000	Life's Resources Inc.	Aerojet-General Liquid-Scrubber	C, M, O
		100–800	MidWest Research Institute	SpinCon High-Volume Cyclonic Liquid Sampler	C, M, O
		300	InnovaTek Inc.	Mini-Cyclone Aerosol Collector	C, M, O

(Continued)

Table 13.2 (*Continued*)

Sampler	Principle of Operation	Sampling Rate (l/min)	Manufacturer/Supplier	Commercial Name	Application[a]
Dry cyclone sampler	Reverse flow cyclone	20	Burkard Manufacturing Co. Ltd.	Burkard Cyclone Sampler	H, M, O
Three-jet, tangential sampler	Tangential dry impaction or impingement into liquid	12.5	SKC Inc.	BioSampler	C, M, O
Rotating arm impactor	Impaction onto rotating arm with liquid rinse	125	MesoSystems Technology Inc.	BioCapture Air Sampler	C, H, M, O
Filter Sampling					
Filter holder	Inhalable sampler for filter collection	4	SKC Inc.	Button Sampler	H, M, O
Membrane filter	0.025–8 μm pore size	0.5–50	Millipore Corp.	MF-Millipore Membrane	H, M, O
Nucleopore filter	0.05–12 μm pore size 0.01–20 μm pore size	0.5–50	Millipore Corp.; Osmonics	Isopore Membrane Poretics Polycarbonate Membrane	H, M, O
Real-Time Sampler					
Time-of-flight spectrometer	Aerodynamic size 0.5–15 μm, fluorescence characteristics of individual particles	1	TSI Inc.	UV-APS	O

[a]C, culture of sensitive and hardy microorganisms, for example, vegetative bacterial and fungal cells and spores; H, culture of hardy microorganisms only, for example, spore-forming bacteria and fungi; M, microscopic examination of collected particles; O, other assays, for example, immunoassays, bioassays, chemical assays, or molecualr detection methods.

13.2.2.1 Time of day

Diurnal patterns in bioaerosol concentrations are caused by changes in air and surface temperatures and RH as well as fluctuations in wind speed and turbulence, all of which affect the emission, suspension, and removal of pollen grains, fungal spores, and other bioaerosols from the atmosphere.[83,94,113] Each biological species responds differently to external conditions. Many plants (e.g., mugwort and sorrel) have their peak hour of pollen release just before noon whereas other plant species (e.g., grasses) peak in the early afternoon.[113] Diurnal cycles have also been measured for bacterial and fungal air concentrations (maxima in the early morning and before nightfall), but patterns are species and microclimate specific. For example, in an urban area of subtropical Taipei, Lin and Li[114] saw diurnal patterns for *Cladosporium* and *Alternaria* species but not for *Aspergillus* and *Penicillium* species. Contrary to the pattern typically seen elsewhere in summer, yeasts peaked during the daytime, but no peak was observed in winter. The authors recorded the highest fungal colony counts under conditions of 25 to 30°C, 60 to 70% RH, and wind speed less than 1 m/s.

13.2.2.2 Time of year

Seasonal patterns in bioaerosol concentrations are caused by temperature, moisture availability, and hours of daylight. Several studies have found that culturable bacteria are more prevalent in summer than winter in some regions due to dry, dusty conditions and associated agricultural or human activities in summer in contrast to wet conditions with snow cover in winter.[115–118] Tong and Lighthart[119] found a maximum for the concentration of ambient bacteria in mid- to late summer in an agricultural area. Particle-associated culturable bacteria were collected with a six-stage agar impactor and total and culturable bacteria with a wetted cyclone. Larger particles were found in summer samples, smaller particles in fall and winter samples, and single bacterium particles during rain or storm conditions. Particle-associated and total bacterial concentrations had several peaks while culturable bacteria had a single maximum in summer. Concentrations of total and culturable bacteria had positive linear correlations with temperature and solar radiation, negative correlations with RH, and no correlation with wind speed. Particle-associated bacteria were not correlated with any meteorological parameters.

Outdoor fungi are dominated by local sources and are often found year round with maxima in spring, summer, or fall.[120] Particularly in temperate northern regions, outdoor concentrations of fungi decrease in winter due to subfreezing temperatures and snow cover. Usually outdoor and indoor spora are similar, although indoor concentrations of fungi are slightly lower in buildings without interior sources, whereas bacterial concentrations are higher throughout the year and associated primarily with human shedding.[121] Water vapor can condense on cold indoor surfaces and this, along with releases of liquid water, may create conditions that allow microbial growth with both qualitative and quantitative changes in the bioaerosol composition inside buildings.[122–124]

Airborne pollen are found predominantly during their respective pollination periods from early spring to fall when plants are producing pollen and releasing whole grains. Spring typically is the main pollen season for those trees and shrubs that release airborne pollen grains. Pollen dehiscence (opening of mature anthers) occurs before foliage is fully grown, which otherwise would impair the release and distribution of pollen grains. Normally, the primary pollen season for trees is no longer than 2 to 3 weeks. Herbs and grasses can flower several times during each growing season; thus, airborne pollen grains from these plants can be found throughout the year in snow-free parts of the world.

13.2.2.3 Geographic region

Bioaerosol concentrations, in particular bacterial concentrations, are generally higher in urban than rural atmospheres in the absence of local sources such as animal houses, agricultural operations, waste treatment plants, or composting facilities. The distribution of pollen-producing plants is a result of natural floristic patterns, but landscaping has significantly changed the air biota in many parts of the world.[94] Northern boreal forests produce large amounts of pine, spruce, hemlock, and birch pollen. Sugi (Japanese cedar, *Cryptomeria japonica*) is very important in Japan and other parts of East Asia. Oak-hickory forests cover large areas of the United States, maples are abundant in the east, and the Southwestern mountains support mountain cedar. Olive trees cannot tolerate cold winters and grow only in regions with Mediterranean climates where they contribute significant amounts of pollen to the air.[125] In Southern Europe, North Africa, and parts of Australia and the American Southwest, olive pollen is a major cause of respiratory disorders. Ragweed (*Ambrosia* species), a dominant pollen allergen in the Mid- and South-Western United States, is increasing in importance in Central Europe and Australia. Ragweed does not reproduce in colder climates, but mugwort (*Artemisia vulgaris*, St. John's plant) often causes problems in these regions. *Parietaria* (pellitory) is an important allergen in areas of Southern Europe and Australia with moderate-to-warm maritime climates. Grass is an important pollen source throughout the world.

13.3 Respiratory dosimetry

Exposures to bioaerosols may be chronic if the episodes occur repeatedly over extended time periods, for example, as experienced by the occupants of microbially contaminated residences or workplaces. Acute exposures, on the other hand, take place within relatively short time intervals, often during periods of elevated air concentrations, for example, in agricultural and manufacturing workplaces and outdoors during peak pollen seasons. More accurate estimations of bioaerosol exposures can be made if both particle size and concentration are known to determine the dose of an agent that is delivered to a particular area of the respiratory tract (see Section 13.2.1).

13.3.1 Respiratory deposition and clearance

The size of an airborne particle determines the time it remains suspended and, when inhaled, the site at which it deposits in the respiratory tract. The most important parameter describing the behavior of particles with diameters larger than 500 nm is particle aerodynamic diameter, d_a. Diffusive or thermodynamic diameter, d_{th}, characterizes the behavior of ultrafine particles, that is, those smaller than 100 nm in diameter. As particle size decreases, the diffusion coefficient and diffusivity increase as does particle deposition. Particle deposition and dose of a biological agent in the respiratory tract also depend on pulmonary anatomy, breathing rate and pattern (nasal vs. oral breathing), health of the lung tissue, and particle characteristics (see Chapters 6 and 8).

The respiratory tract has traditionally been divided into three anatomical regions: (a) nasopharyngeal (N-P) or extrathoracic (E-T), which extends from the nose and mouth to the epiglottis and larynx; (b) tracheobronchial (T-B), which consists of the conducting airways from the trachea to the terminal bronchioles; and (c) pulmonary (P) or alveolar-interstitial (A-I), which includes the respiratory bronchioles and alveolar ducts and sacs (Figure 13.1; Chapter 24). The total regional deposition for a given particle size distribution cannot be predicted accurately because there are considerable differences in respiratory parameters among the individuals in a population determined by their health status, age, gender, and ethnic origin. Several theoretical models (i.e., mathematical predictions) of regional deposition and particle fate as functions of particle size have been presented. One of the most commonly used models is from the International Commission on Radiological Protection (Chapters 8 and 24).[126] This model uses representative values for normal respiratory parameters, such as an inhalation rate of 15 breaths per minute, ventilation rates ranging from 7.5 l/min during sleep to 50 l/min during heavy exercise, and daily, inhaled air volumes of 23 m^3 for light work and 27 m^3 for heavy work.[126]

The different regions of the lung have distinct functions, for example, cleaning of inhaled air (by the removal of large particles in the N-P region) and adjustment of air temperature and humidity (in the N-P region and bronchi of the T-B region) as well as gas exchange (uptake of oxygen and removal of carbon dioxide), pulmonary clearance, and immunological defense (in the A-I region). Airway passages are lined with small hairs and ciliated mucous membranes. The bronchi are lined with beating ciliated epithelial cells that move the respiratory mucous layer toward the trachea. All regions of the respiratory tract contain lymphatic tissues. Fluid that accumulates in interstitial connective tissue collects in lymph capillaries, flows into lymph vessels, and passes through lymph nodes.

Inhaled particles are retained if they contact the lung surface. Relatively soluble material that is deposited in the lung can enter the bloodstream. Less soluble material can directly influence the airway epithelium, cause cellular damage, and affect cell responses through various airway receptors. Not all particles that are inhaled deposit in the respiratory tract. Deposition is minimal (approximately 35%) for 0.3 to 0.5 μm particles and increases rapidly for larger and smaller particles. For particles larger than 0.5 μm, the most important deposition mechanisms are impaction and gravitational settling, while smaller particles are collected by diffusion. Interception may increase the deposition of elongated and fibrous particles. Electrostatic effects may also increase the deposition of charged particles by image charging. Gravitational settling occurs throughout the respiratory tract, from the bronchi to the A-I region. Inertial impaction is the dominant collection mechanism in the E-T and T-B regions and increases with increasing particle size. Diffusion is dominant in the A-I region and in the E-T region for particles 10 nm in diameter and smaller.

Major differences in deposition patterns occur with nasal and oral breathing. The nose is an efficient filter for particles larger than 10 μm, thus limiting the pulmonary deposition of 2 to 10 μm particles. Mouth breathing allows particles to bypass the nose at least partially and enhances aerosol deposition in the deep lung. Nearly all inhaled particles larger than 10 μm are deposited in the N-P region during nasal breathing, but only 65% collect there with oronasal breathing.

Deposited particles may either be cleared from the respiratory tract or retained for variable periods of time. Deposited particles are cleared by absorption into the blood, transport to the gastrointestinal tract via the pharynx, or movement to the regional lymph nodes via the lymphatic channels. Particles deposited on the anterior nasal passages are usually removed by extrinsic means, for example, nose blowing or wiping. Some material is carried further by mucus flow and sniffling. Some surfaces in the N-P region are covered by a fluid layer and some areas contain ciliated cells, which assist in the movement of deposited material to the pharynx, where it is swallowed.

In the ciliated T-B airways of the lung, relatively insoluble material moves with the mucus flow toward the epiglottis and is swallowed. In the nonciliated bronchioli and alveoli, particles are engulfed by pulmonary alveolar macrophages. Some of these macrophages enter the mucus flow and others the pulmonary lymph vessels, from which they are transported to the T-B lymph nodes. Clearance rates from one region to another depend on the deposition location, physicochemical form of the material, and time since deposition. Clearance rates vary greatly and can change over time. The half-time to clear particles from the N-P region to the outside of the body is estimated to be 17 h, whereas removal of particles to the gastrointestinal tract may take only 10 min.[126]

It has been shown that to establish disease, infectious particles must be deposited as virulent organisms at a susceptible site, the critical location depending on the organism. Infection produces an immune response that plays a role in recovery and prevents reinfection with the same or immunologically similar agents. For certain substances, such as allergens, deposition at a specific site in the respiratory system may not be necessary to cause an adverse reaction, although response mechanisms may vary according to region (see Section 13.3.2.2). Small protein molecules from microorganisms, plants, and animals have been assumed to cause symptoms in humans through induction of IgE antibody responses that trigger release of the mediators (e.g., histamine and leukotrines) that generate the allergic signs and symptoms. However, research also suggests that, for instance, mast cells and respiratory epithelial cells may respond to an allergen challenge in a non-IgE-dependent manner.[127]

Injury to the respiratory tract manifests itself in several ways. Exposure to cytotoxic agents may cause lung cells to increase in size or number or may kill them. Many biological materials induce inflammation with or without an immune response. Inflammation is characterized by changes in the types of cells that are present in the region and recruitment of other inflammatory cells, such as macrophages, monocytes, and neutrophils, into the region. Endocytosis of a particle may induce cell activation that results in the production of different proinflammatory cytokines, such as interleukins and tumor necrosis factor-alpha (TNF-α), regulatory mediators, and penetration of cytokines into the blood. Development of fibrosis often occurs following chronic inflammation. Because the responses of the respiratory tract are not unique to specific agents, attributing disease to a single biological, chemical, or physical agent based solely on epidemiological associations is difficult (see Section 13.1.1.3).

13.3.2 Dose–response relationships/exposure assessment

Experimental inhalation studies of bioaerosols in animal models have been used to estimate infectivity in humans, define the pathogenesis of airborne diseases, study the mechanisms by which biological components exert their effects, and understand the action of anti-inflammatory agents.[128] The number of inhalation tests on humans or animals as well as field exposure studies of bioaerosols is increasing, especially for endotoxin and glucans. Animal studies can provide exposure–response data under well-controlled and defined conditions when a biological agent is delivered at an environmentally relevant concentration and route of exposure. However, extrapolation from animal models to humans is not straightforward due to interspecies differences. There is relative certainty about potential risks for some bioaerosols, such as infectious agents, whereas for other bioaerosols, such as endotoxin and other bacterial and fungal toxins, the uncertainty is much greater. Furthermore, some effects that have been inferred from epidemiological population studies have not been proven. The contribution of biological agents to the adverse health effects observed for PM has yet to be evaluated.

A dose-dependent relationship between an exposure to a biological agent and a response can be understood most completely if information is available on all steps in the pathway: the origin of the agent and means by which it becomes airborne as well as objective signs or symptoms of a host response following receipt of a measured dose of the agent.[129] Research tools suitable for the collection of data on the first parts of the chain are reasonably widely available. Therefore, investigators have gathered a great deal of information on the sources of biological agents, methods of biogenic (natural) and anthropogenic (human) bioaerosol generation, and indoor and outdoor air concentrations. Documentation of received dose and measurement of dose–response reactions that identify threshold values for different human populations has been more difficult.

13.3.2.1 Infectious dose

Respiratory tract infections are not equally communicable.[130] The outcome of a human exposure to an infectious agent depends on the number of microorganisms encountered, the virulence of the agent, and the strength of the body's defenses. Infectivity and virulence can be expressed in quantitative terms. The 50% infectious dose (ID_{50}) is estimated from the number of microorganisms or amount of microbial toxin needed to initiate infection in half of the exposed subjects. The virulence of an infectious agent can be measured as the 50% lethal dose (LD_{50}) (the number of microorganisms required to kill half of the infected animals). In an outbreak investigation, the infectivity of a particular strain of an agent can be determined from the observed attack rate (i.e., the number of cases of clinically apparent disease divided by the number of susceptible persons). Host defenses that determine the severity of an encounter with an infectious agent include mucosal immune factors, which may block adherence and local proliferation, and humoral or cellular immune responses, which may contain proliferation and cell invasion, resulting in an asymptomatic infection rather than a more serious illness.[130]

An early study by Wells et al.[31] demonstrated the relationship between infectious dose, particle size, and pulmonary deposition site. This group showed that the number of 2 to 3 μm droplet nuclei of *M. tuberculosis* to which rabbits were exposed approximately equaled the number of infectious foci (tubercles) that developed in their lungs. In contrast, only 6% of 13 μm particles caused tubercles because few particles of this size reach the deep lung and the bacteria are innocuous if they implant in the upper respiratory tract. It is difficult to measure the number of infectious particles to which people are exposed and determine the number of viable organisms in each infectious particle; therefore, Wells[132] introduced the idea of quantal infection. He noted that the number of persons who become infected bears a Poisson relationship to the number of infective particles that they breathe. Approximately 63% of persons homogeneously exposed to an airborne agent will be infected when, on average, each of them has breathed one infective particle or "quantum." The infectivity of the organism in combination with the strength of the host's defenses determines the number of organisms required to induce a quantal response.

Some of the information on infectious doses for inhalation exposure has been modeled from outbreak situations or calculated from experimental exposures. For example, the dynamics of measles and tuberculosis transmission have been analyzed using the concept of infectious quanta in mass-balance equations similar to those applied to the study of other environmental contaminants.[133–137] Investigations of outbreaks in which the number of infected persons, their duration of exposure, and the indoor ventilation rate were known have allowed estimation of the average air concentrations of infectious agents and their generation rates. Infectious doses may be as low as ten airborne adenoviruses, *Coxiella burnettii, Franciscella tularensis,* or *M. tuberculosis* cells, and as high as 10^3 to 10^4 *B. anthracis* spores. [48,138,139]

13.3.2.2 Allergens

A committee of the Institute of Medicine applied a rigorous system to the evaluation of the evidence for a biological gradient (dose–response relationship) between exposures in indoor air and associations with asthma. [29] However, our understanding of the relevant sites and nature of the interactions between inhaled agents and the human body that lead to asthma or exacerbate the condition remains uncertain. The committee recognized that it could not expect to observe the pattern usually seen for chemical exposures, that is, a greater response to a higher or more frequent exposure to an agent. For hypersensitivity, infections, and some inflammatory diseases, genetic susceptibility and prior encounters with a biological agent govern the human body's response, in some instances increasing but in other cases decreasing reactions. Thus, the absence of a dose–response effect for a biological agent does not always constitute evidence against a causal relationship between a bioaerosol and an outcome.

House dust mites are among the most thoroughly studied allergens and the most important factors linked with asthma causation and symptom exacerbation. An example of a study of the exposure–response relationship between mite allergens and asthma was that conducted by Arshad *et al.*[140] These researchers sought to mimic natural allergen exposure for nine mildly asthmatic adults who were highly sensitive to dust mites (i.e., skin prick responses to *Der f* or *Der p* at least 5 mm greater than a saline control). The subjects were exposed in three weekly sessions for 4 weeks to low doses of mite extract estimated to be equivalent to daily exposures in a typical bedroom (~400 pg of Group 1 allergen). The exposures adversely affected pulmonary function and caused bronchial hyperreactivity, but the changes were rapidly reversible upon cessation of the exposure. Schelegle *et al.*[141] developed an animal model to study whether allergic asthma could be induced in a nonhuman primate (rhesus monkeys) by aerosol challenge following sensitization by subcutaneous injection. For both these studies, the test aerosols were generated by nebulizing saline solutions of the allergens. Such studies most often use a pure agent and nebulizers typically generate particles of 1 to 3 μm in diameter, which may not reproduce natural environmental exposures.[142]

Recognizing the difficulties of human experiments, risk guidelines for indoor allergens in settled dust have been derived from epidemiological surveys and are based on allergen concentrations in dust rather than air (Table 13.3). Assessment of viable fungi in house dust has also been proposed as a useful measure of longer term and cumulative exposure to indoor fungi because settled dust is influenced less by indoor activities than air samples.[143,144] However, many of the particles in floor dust were tracked inside on the feet and coverings of the occupants and do not represent previously airborne particles. Karlsson *et al.*[145] found no correlation among four methods to sample allergens in dust and air in classrooms. Chew *et al.*[120] found a weak but significant ($r = 0.13$, $p < 0.05$) correlation between total culturable fungi in indoor air and house dust.[120] Outdoor fungal concentrations were often predictive of indoor levels, and *Cladosporium* and *Penicillium* species in dust were positively associated with their indoor air concentrations.

The National Bureau of the American Academy of Allergy, Asthma and Immunology bases its risk guidelines for ambient pollen and fungal spores on geographical, phenological (seasonal and climatic), and clinical information; thus, the severity scales differ throughout the country (Table 13.4).[146] The concentrations of tree, grass, and weed pollen as well as fungal spores are reported in daily and annual databases as absent, low, moderate, high, or very high to alert persons with different levels of sensitivity and their physicians to take appropriate precautions throughout the year. Similar guidelines have been in use in Sweden and Finland since 1973, and several other European countries began aeroallergen forecast services during the 1980s.[147]

13.3.2.3 Glucan

Fewer inhalation studies have been performed with β-(1→3)-D-glucans than endotoxin (see Section 13.3.2.4) and no exposure limits have been proposed. Glucan has been found

Table 13.3 Allergen Concentrations Associated With Sensitization and Symptom Provocation [22,29,170,254,255]

Allergen	Sensitization Level	Symptom Provocation
House dust mite allergen	2 μg/g	>10 μg/g
(*Dermatophagoides farinae, Der f*1,)	100 mites/g	500 mites/g
and *Dermatophagoides pteronyssinus, Der p*1	0.6 mg guanine/g	
Cat allergen (*Felis domesticus, Fel d*1)	8 μg/g	≥8 μg/g
Dog allergen (*Canis familiaris, Can f*1)	10 μg/g	≥10 μg/g
Cockroach allergen (*Blatella germanica, Bla g*2)	10 units/g	≥10 units/g

Table 13.4 National Concentration Guidelines for Ambient Airborne Plant Pollen and Fungal Spores[146]

Concentration Category (Percentile of Total Distribution)	Allergic Persons Who May Experience Allergy or Asthma Symptoms	Concentration (count/m³)			
		Weed Pollen	Grass Pollen	Tree Pollen	Fungal Spores
Absent (not detected)	None	Below detection	Below detection	Below detection	Below detection
Low (<50th)	Extremely sensitive persons only	>0–9	>0–4	>0–14	>0–6499
Moderate (50th–90th)	Many sensitive persons	10–49	5–19	15–89	6500–12,999
High (90th–99th)	Most sensitive persons	50–499	20–199	90–1499	13,000–49,999
Very high (>99th)	Almost all sensitive persons (extremely sensitive persons may experience severe symptoms)	>500	>200	>1500	>50,000

to initiate the secretion of different inflammatory cytokines *in vitro*, but in inhalation tests no such effect has been found.[148] Macrophages have specific receptors for glucan, but after endocytosis there is no enzymatic pathway for glucan degradation, which is broken down oxidatively. Acute inhalation exposures to glucan do not elicit an inflammatory response but seem to modulate the response to a simultaneous or subsequent exposure to another agent. Simultaneous exposure to glucan and endotoxin increases the severity of symptoms and airway responsiveness. Persons sensitive to allergens without apparent clinical disease seem to be more sensitive to glucan, but there is no evidence that inhaled glucan itself can elicit an immune response.[148]

13.3.2.4 Endotoxin

Animal and human toxicity tests with endotoxin and LPS have shown that inhaled endotoxin causes inflammation with the release of different cytokines and increased production of oxygen metabolites. Macrophages and epithelial cells are the primary targets, and endothelial cell damage has also been observed. The concentrations of cytokines peak a few hours after inhalation. Thorn[149] and Rylander[150] have reviewed studies of human inhalation challenges and environmental measurements of endotoxin concentrations. Inhalation challenges demonstrate local and systemic inflammatory responses to endotoxin at lower doses, while higher doses cause clinical responses such as chest tightness, fever, airway irritation, headache, chills, tiredness, and lung function changes. Studies of asthmatic subjects suggest that endotoxin could be an important factor contributing to the severity of asthma and other chronic pulmonary diseases. Hyperresponsiveness of the airways and evidence of airway obstruction in asthmatics was found to increase with exposure to endotoxin.

Inhalation tests with endotoxin that have been performed to date are most relevant to occupational environments. The threshold dose of pure LPS for induction of symptoms in controlled trials has been found to be higher than was estimated based on field studies of different dusts, for example, cotton, soybean, and corn dust. Exposures in the field and their measurement depend on several factors, for example, weather conditions, work practices, particle size, sampling system, and analytical method. Differences in the methods used in field studies and experimental inhalation tests may affect the comparability of results across studies and relevance to natural exposures (see Section 13.3.2.2). The measured indicators include cytokines and leucocytes in blood, bronchoalveolar lavage fluid, urine, and sputum in addition to observations of lung function changes. Gradations in response may be due to differences in the bioavailability of the LPS or the presence of other agents in the dust or dust extract. In addition, the question of differences in the potency of endotoxin from different Gram-negative bacteria is only beginning to be addressed.

Clear correlations have been observed between endotoxin exposures and decreased pulmonary function in humans, and consensus recommendations have been published for the measurement of environmental endotoxin in workplaces.[151–153] In 1998, the Dutch Expert Committee on Occupational Standards proposed a health-based occupational exposure limit of 50 endotoxin units/m^3 or approximately 5 ng/m^3 based on personal inhalable dust measured as an 8-h time-weighted average.[154] The American Conference of Governmental Industrial Hygienists (ACGIH) has proposed alternative relative limit values (RLVs) for endotoxin based on the presence or absence of symptoms (e.g., fatigue, malaise, cough, chest tightness, and acute airflow obstruction) and the endotoxin concentration at an appropriately chosen control location.[155] In the presence of respiratory symptoms, ACGIH proposed an RLV action level of 10 times background; and in the absence of symptoms, a maximum of 30 times background. The RLV approach accommodates differences among laboratories and *Limulus* amebocyte lysate (LAL) reagent lots.

13.3.2.5 Other bioaerosol exposures

Occupational risk assessment studies involving bioaerosols have been limited by the lack of accepted reference values for specific agents other than those for which occupational exposure limits have been determined, for example, cellulose, wood dust, cotton dust, grain dust, nicotine, pyrethrum, starch, subtilisins, sucrose, turpentine, and vegetable oil mist.[156] For indoor air quality assessments, concentrations measured in test environments are typically compared to baseline data from reference areas or to data reported in the literature.[157] Reponen et al.[121] recommended using the extremes (>95th percentile) of the data distributions of indoor, culturable bacteria and fungi (5000 CFU/m^3 and 500 CFU/m^3, respectively) as indicators of the presence of abnormal indoor sources or insufficient ventilation (but not health risk) in urban and suburban residences in a subarctic climate (Finland).[121] However, the general consensus is that it is not possible to set numeric concentration limits for bacteria or fungi in indoor environments.[156,158–163] In place of permissible exposure limits, investigators use various means to identify signs of dampness and microbial growth.[124] For example, Dillon et al.[164] explained that the ratio of the sum of the concentrations of soil fungi (e.g., *Aspergillus*, *Penicillium*, and *Eurotium* species) to the sum of the concentrations of phylloplane fungi (*Alternaria*, *Cladosporium*, and *Epicoccum* species) should be near one if fungi are not multiplying indoors, because primarily soil fungi grow on building materials rather than the fungi found on leaf surfaces.

A quantitative understanding is needed of the relationships among bacterial and fungal concentrations, doses delivered to respiratory tissues, and specific pulmonary responses in addition to information on the deposition and ultimate fate of microorganisms and their products in the respiratory tract. Chemical agents can alter physiological responses, particularly during exercise, by changing pulmonary function. It is not yet clear to what extent microorganisms need to be identified (e.g., class, family, genus, species) or if microbial markers are suitable for risk assessment (e.g., glucan or ergosterol for fungi and muramic acid or fatty acids for bacteria). Exposure–response data are limited and many different microorganisms and microbial components may cause similar health effects.

13.3.2.6 Interactions

Complicating studies of human responses to bioaerosol exposures are simultaneous exposures to gaseous and particulate air pollutants that may irritate or damage the respiratory tract, altering people's sensitivity to bioaerosols. Any practice or condition that impairs the natural defenses of the lung increases a person's susceptibility to infection.[165] For example, influenza is often a precursor to secondary bacterial pneumonia because the virus temporarily destroys ciliated cells. Smoking also depresses the effectiveness of the ciliated cell defense, which explains why persons with a long history of smoking have a higher risk of bacterial and fungal infections than nonsmokers. Intubation of hospitalized patients makes it easier for microorganisms to cause pneumonia because the air bypasses the defenses of the upper airway and is neither filtered nor humidified.[166]

Chemical agents can modify physiological responses by changing pulmonary function, particularly during exercise. Few studies have measured the complex interactions that occur among bioaerosols, air pollutants, and meteorological factors. Failure to measure environmental aeroallergens in studies of the health effects of ambient air pollutants can lead to confounding. For example, a clinical study found that following ozone exposure, sensitized asthmatics responded to lower concentrations of allergen than subjects exposed to clean air.[167]

It has been postulated that interactions between gaseous pollutants and pollen particles can alter their allergenic potency.[168] The contribution of biological agents to the adverse health effects that are seen with exposure to airborne PM is only beginning to be evaluated. Airborne particles provide adsorptive and absorptive surfaces for inorganic and organic gases and vapors. Particles of vehicle exhaust emissions are generally 50 to 200 nm in diameter and are the most abundant ambient particles by number. Diesel exhaust particles of respirable size have been shown to bind to grass pollen allergen,[105] and these particles may act as carriers, which could result in the periods of the highest exhaust emissions being the most allergenic.[169] Lewis et al.[111] found no statistically significant interactions between the effects of any individual aeroallergen and outdoor air pollutants on emergency room visits or asthma admissions, but did observe increased asthma admissions with higher counts of *Cladosporium* species. Lin and Li[114] noted that (i) hydrocarbons were correlated positively with *Cladosporium* species, (ii) PM_{10} mass was correlated positively with *Penicillium* species and yeasts, and (iii) ozone was correlated negatively with total fungal concentration.

13.4 Environmental measurement and analytical techniques

13.4.1 Measurement of surrogates and indicators

Assessments of exposure to environmental agents in indoor and outdoor air play a central role in epidemiological studies seeking to characterize population risks, in screening studies aimed at identifying individuals at risk, and in interventions designed to reduce risk.[29] However, the assessment of exposure to many biological agents remains difficult, and there is little agreement on standard methods for the analysis of bioaerosol components. Information on indoor air concentrations of infectious agents is limited, because it is seldom possible or practical to collect airborne infectious agents except for certain environmental pathogens, for example, opportunistic fungi. Wider availability of molecular methods may change the situation. However, methods based on detection of nucleic acid sequences are of limited value because they do not distinguish between viable (potentially infective) and inactive microorganisms. Instead, exposures to infectious agents are recognized when persons develop signs of disease and seek medical care or through health surveillance programs. Environmental sampling of suspected sources of infectious agents (e.g., contaminated water or equipment) and collection of clinical specimens (e.g., throat swabs and sputum samples) are far more common than air sampling to detect the presence of infectious agents. Exposure to common indoor allergens is most often measured indirectly by sampling settled dust rather than air. Indicators of the presence of pets (cats, dogs, birds, or other small animals), pests (cockroaches or rodents), or microbial contamination (visible fungal growth) are also used to classify home and work environments into risk categories in the absence of air or source samples.[170]

Even when air samples are collected to determine bioaerosol concentrations, exposure typically is estimated from a readily measured surrogate of an active agent, for example, fungal spores, ergosterol, or glucans to represent fungal allergens or toxins, whole pollen grains for plant allergens, guanine for house dust mite allergens, or CFUs of Gram-negative bacteria for endotoxin. Certain bacteria and fungi that have been observed in moisture-damaged buildings have been suggested as "indicator species" (Table 13.5).[171,172] Several "indicator" fatty acids have also been identified for the detection of specific groups of microorganisms (see Section 13.4.4.5). This approach risks over- or underestimating exposure, but is the only means available in some cases. Indoor microbial growth may produce various Volatile Organic Compounds (VOCs), many with known irritant and other health effects, for example, ethanol and other alcohols, acetone, benzene, and hexane. However, no single volatile compound has been found to be a reliable indicator of biological contamination of building materials, that is, to have no source other than bacterial or fungal metabolism.[173]

Researchers often substitute safe and convenient surrogate organisms for ones that are difficult or hazardous to use in experimental studies of bioaerosol generation or dispersion, evaluations of collection and analytical methods, or tests of respirators and air cleaners. For example, Schafer et al.[174] aerosolized a nonpathogenic strain of *M. tuberculosis* (i.e., H37Ra) to evaluate two size-segregating, air sampling methods. The investigators reported that the bacterium was collected predominantly in the 0.6 to 2.1 μm size range of the six-stage Andersen sampler and the 0.3 to 1.8 μm range of an eight-stage cascade impactor. Henningson and Ahlberg[175] summarized the many test aerosols that investigators have

Table 13.5 Guidance on recognition of moisture-indicating microorganisms

Materials with a High Water activity (a_w >0.90 – 0.95)	Materials with a Moderately High Water Activity (0.90 > a_w >0.85)	Materials with a Lower Water Activity(a_w ≤0.85)
Indicator organisms[171]		
Aspergillus fumigatus	*A. versicolor**	*Aspergillus versicolor**
Trichoderma		*Eurotium*
Exophiala		*Wallemia*
*Stachybotrys**		Penicillia (e.g., *Penicillium chrysogenum*, *Penicillium aurantiogriseum**)
Phialophora		
*Fusarium**		
Ulocladium		
Yeasts (*Rhodotorula*)		
Actinomycetes and Gram-negative bacteria (e.g., *Pseudomonas*)		

Where moisture and/or health problems are signaled, the above fungi can be regarded as indicator organisms if above a baseline level (to be established) in air samples, or if isolated from surfaces. Names indicated with an * are important toxigenic taxa. Stachybotrys is always a sign of serious moisture problems (e.g., leakage, flooding, or extreme condensation) and has serious health implications for occupants of affected buildings.

a_w = available water (water requirement for growth).

chosen for studies of bioaerosol sampler performance. In an effort to standardize performance evaluations, Griffiths *et al.*[176] surveyed researchers and concluded that for experimental aerosol work *Escherichia coli, Saccharomyces cerevisiae,* and *Penicillium expansum*, were reasonable representatives of Gram-negative bacteria, fungal yeasts, and filamentous, spore-forming fungi, respectively.

13.4.2 Bioaerosol measurement

Bioaerosols are collected for a wide variety of purposes, for example, to measure inhalation exposure, characterize indoor and outdoor environments, identify emissions from work activities, and evaluate the effectiveness of control methods. Some commercially available bioaerosol instruments are listed in Table 13.2. The collection of particles of biological origin follows the same principles and mechanisms used to collect other aerosols: gravitational settling, inertial impaction, interception, diffusion, and electrostatic forces. Particle size and air velocity determine an instrument's collection efficiency, and all samplers should be characterized for particle losses, for example, inlet, wall, and line losses. In addition to ensuring collection of a representative sample unbiased by the sampling process, biological agents often have special requirements related to the maintenance of cell integrity or biological activity. Currently, there is only one commercially available, real-time instrument for the measurement of bioaerosols (UV-APS, time-of-flight spectrometer, Table 13.2); all other methods concentrate particles for later analysis, for example, in a liquid, on an agar or adhesive surface, or on a filter.

Some air sampling instruments were developed specifically for bioaerosol collection (e.g., ones that incorporate agar plates), but many widely used devices (e.g., liquid impingers and membrane filters) have been adapted from other applications. The impactors, impingers, and cyclones that are used most often collect bioaerosols by inertial impaction; thus, they capture larger, micron, and submicron particles down to approximately 500 nm in diameter. Liquid impingers and wetted cyclones collect particles in fluids from which cells and biological fragments can be examined microscopically (Section 13.4.4.1), culturable microorganisms can be grown for enumeration and identification (Section 13.4.4.2), and other molecules can be analyzed using biological, immunological, chemical, or molecular methods (Sections 13.4.4.3–13.4.4.6). Biological particles can be collected with high efficiency on filters, although desiccation of bacteria and fungi limits subsequent identification by cultivation-based methods. Biological particles concentrated on filters can be extracted for analysis with noncultivation-based methods, for example, staining and direct microscopic examination, chemical assays, immunoassays, and molecular analyses.[177]

The collection and measurement of airborne microorganisms has focused mainly on viable and culturable particles. Measuring living bacteria and fungi that have maintained their ability to reproduce is essential when studying infectious agents and the transmission of communicable diseases, but is not always necessary for investigations of the relationships between other biological agents and respiratory disease. Measuring the concentration of only the viable fraction of airborne particles seriously underestimates the total bioaerosol concentration because most bacteria as well as many fungi in ambient air are not in a culturable state and will not multiply on laboratory growth media.[178,179] Sampling without particle size discrimination or information on the cut point of an instrument may also over- or underestimate exposure to the site in the respiratory tract that is associated with the initiation of infection or symptom provocation.

No single collection method has yet been developed that can be coupled with the entire range of analytical procedures that are needed to provide a complete picture of the bioaerosol composition of indoor or outdoor air. However, optimal collection methods can be chosen for individual biological agents or groups of agents that were identified in advance as the focus of an environmental assessment. Investigators designing a study must consider the limitations of the measurement method (such as sampler performance), the temporal and spatial variability of bioaerosol composition and concentration in the study environment, indoor and outdoor activities near the sampling site, and bioaerosol deposition on and resuspension from surfaces. For reviews of bioaerosol samplers and study design, see Cox and Wathes,[180] AIHA,[159,181] ACGIH,[129,156] Hurst,[182] and Macher and Burge.[183]

13.4.3 Bioaerosol samplers

Gravitational settling is the simplest way to collect airborne particles. Exposed filters, adhesive-coated microscope slides, or open dishes of agar (i.e., settle plates) placed horizontally or vertically have been used to study indoor surface contamination and particle resuspension. This method has also been used to collect aeroallergens in outdoor air and can be as simple as an open glass jar covered with a rain shield. Passive, nonvolumetric sampling depends highly on particle size and airflow patterns around a sampler. Gravitational deposition is biased toward large particles, for example, pollen grains, larger fungal spores, and cell clusters over cell fragments, smaller spores, and single bacteria. In still air, the deposition rate of a particle is proportional to its settling velocity, which is proportional to the square of the diameter of the particle (Chapter 2). Air concentration cannot be calculated from gravitational samples; thus, this method provides only qualitative information.

13.4.3.1 Inertial sampling

Impactors operate on the principle of drawing air through a nozzle and forcing the jet to turn sharply after exiting the inlet (see Chapter 2). Particles will deposit on the collection surface beneath an inlet if they cannot travel with the air stream due to their inertia. The largest particles collect in the center of the impaction zone and the smallest ones along the edges. Thus, an impactor divides the entering particles into two size classes: (a) particles larger than a median aerodynamic diameter (d_{50}) for which collection efficiency increases with increasing particle size; and (b) particles smaller than the median diameter for which fewer and fewer are collected until all of the smallest particles pass through with the exiting air stream.

Many single- and multiple-jet impactors have been developed to collect bioaerosols (single-stage and cascade impactors, Table 13.2). The nozzles may be round holes or rectangular slits and vary in number from one to several hundred. A positive-hole correction may be applied to the colony counts obtained with multiple-hole agar impactors to account for coincidental impaction at high particle concentrations.[184,185] The impaction stage typically is removable to allow easy replacement of the collection medium, which may be an agar plate, microscope slide, filter, or tape. A slit inlet or the collection surface beneath it may be moved laterally or rotated during the sampling period to allow examination of the variation of air concentration with time.

A cascade impactor is formed by operating several impactors in series, proceeding from the stage with the largest cutoff diameter at the inlet to that with the smallest cut point at the exit (see Chapter 11). Cutoff diameter decreases with each stage because jet velocity increases through a reduction in nozzle diameter. A final filter may be added to collect the finest particles. The most popular device for measuring the aerodynamic size distribution of bioaerosols has been a six-stage, 400-hole impactor in which particles collect on a semisolid culture medium, although other collection surfaces, such as filters, have also been used.[186] The sampler is available in one-, two-, and six-stage versions for use with agar plates as well as an eight-stage version for other collection substrates.

Liquid impingers collect particles by impaction followed by suspension in a collection fluid. Air is drawn through a narrow, straight, or curved tube, usually smaller than 5 mm in diameter, which is positioned above a liquid or immersed in it. One impinger uses three curved jets that induce a swirling motion in the capture liquid, which increases collection efficiency and decreases particle re-entrainment (three-jet, tangential sampler, Table 13.2).[187] Liquid volumes from 5 to 50 ml may be used, and fluids as diverse as sterile distilled water and nonevaporating mineral oil can be chosen to collect bioaerosols.

Centrifugal samplers and cyclones collect particles by impaction onto a solid medium or into a liquid. For example, particles may be rotated at high speed by impeller blades, which force large particles to deposit on an outer wall covered with a removable agar tape (centrifugal agar impactors, Table 13.2). Air tangentially enters the cylindrical or inverted conical chamber of a cyclone sampler, spins down along the chamber walls, flows up through the center, and exits at the top (wetted and dry cyclone samplers, Table 13.2). Large particles deposit on the walls of a cyclone and very large particles fall to the bottom. In commercial cyclone samplers used to collect bioaerosols, liquid is often pumped into the inlet to wash the particles into a collection bottle at the bottom. Whirling-arm impactors with flat or round bars that are rotated at high speeds (e.g., 2000 to 3000 rpm) have been used to collect ambient plant pollen and fungal spores (rotating rod impactor, Table 13.2).

13.4.3.2 Filtration

Filter samplers collect particles of all sizes, the upper limit depending on the filter holder's inlet characteristics and addition of precollectors. Most filters have high collection efficiency for large ($d_p > 1\ \mu m$) and small ($d_p < 0.05\ \mu m$) particles. Between these sizes, the minimum collection efficiency depends on face velocity (see Chapter 17).[185,188] Generally, two kinds of membrane filters are used for bioaerosols: porous and capillary pore filters (filter sampling, Table 13.2). The former filters have complex pore structures, low porosity, high collection efficiency, and relatively high pressure drop. The pores in capillary pore filters are approximately uniform in size and perpendicular to the filter surface.

Filters are available in sizes from 13 to 47 mm diameter for use at flow rates from 0.5 to 50 l/min; higher sampling rates have been used with larger filters. A small, metal filter holder with a perforated spherical inlet (developed for the collection of inhalable samples with uniform particle deposition) is suitable for area and personal sampling of biological agents (Button sampler, Table 13.2).[189]

After sampling, a filter may be analyzed microscopically, placed on an agar surface for culture, or the collected material may be washed into solution or suspension for further processing. Porous membrane filters are often chosen for culturing and immunostaining of bioaerosols. Dehydration of vegetative bacterial cells during filtration sampling significantly reduces their viability, but hardy bacterial and fungal spores are not as susceptible to damage. For instance, Li *et al.*[190] and Lin and Li[191,192] found filtration better than liquid impingement for fungal spores in terms of culturability. Capillary pore filters have smooth, flat surfaces suitable for examining particles with an optical or scanning electron microscope (see Section 13.4.4.1). Investigators are increasingly using filters to collect bioaerosols in conjunction with analyses not based on culture (see Sections 13.4.4.3–13.4.4.6) because of the ease of use and high collection efficiency of membrane filters.

13.4.3.3 Electrostatic precipitation

The electrostatic precipitator is an example of an instrument long used for particle collection that has been adapted for laboratory sampling of bioaerosols.[76,77] Particles first pass through a charging section because, although ambient particles may carry one or more elementary charges, all particles need to be charged in order for an electrostatic precipitator to have a predictable collection efficiency. After charging, particles travel through a narrow space between two plates, of which the upper one is maintained at either a positive or negative voltage — the same as the charge that was applied to the particles. In this electrical field, charged particles move toward collectors on the lower plate. The collectors in the bioaerosol precipitator were rectangular agar dishes, and collection efficiency varied from 50 to 90% depending on air flow rate and applied voltage.

13.4.3.4 Future directions in bioaerosol sampling

Several real-time instruments for bioaerosol measurement are under development. The only direct-reading sampler currently available commercially, the ultraviolet aerodynamic particle size spectrometer (UV-APS, time-of-flight spectrometer, Table 13.2), measures particle size by light scattering and detects ultraviolet irradiation-induced fluorescence. All microorganisms contain fluorophores (molecules that fluoresce when excited by certain wavelengths of electromagnetic radiation) such as riboflavin, nicotinamide adenine dinucleotide, and tryptophan. Brosseau *et al.*[193] presented a method that used this natural fluorescence to distinguish between biological and nonbiological particles. Although specific organisms could not be identifed, variations in fluorescence were distinguishable between species. Both of these rapid systems rely on optical detection of particles; thus, their use is not feasible for particles smaller than 0.1 μm. An experimental analytical method, consisting of a fiber optic biosensor that ran four simultaneous immunoassays, was developed for remote sensing of bacteria when coupled with a sampling system for airplanes.[194] A personal sampling device that fits into the nostrils of a test subject and is reported to collect 95% of 10-μm and 50% of 5-μm particles by impaction onto a protein binding membrane is among the more imaginative of bioaerosol samplers.[195]

13.4.4 Sample analysis

Before collecting samples, investigators should identify the biological agents in which they are interested and then select appropriate analytical methods, depending on the information sought. Analytical methods can be divided into direct counts (microscopic examination of stained and unstained particles) (Section 13.4.4.1), cultivation-based methods (*in vitro* or *in vivo* growth) (Section 13.4.4.2), biological assays (infectivity assays and tests of cytotoxicity) (Section 13.4.4.3), immunoassays (based on antigen–antibody recognition) (Section 13.4.4.4), chemical and biochemical analyses (for carbohydrates, fatty acids, lipids, proteins, and metabolic products) (Section 13.4.4.5), and molecular methods (using RNA or DNA probes or sequence analysis) (Section 13.4.4.6). No single method measures all bioaerosols, and the methods vary widely in their sensitivity and specificity. Microscopy and cultivation assays measure intact and viable microorganisms, respectively, while bioassays, immunoassays, and chemical analyses measure different compounds or biologically active molecules in whole cells, cell fragments, or extracellular metabolites. Methods that detect selected genetic sequences are becoming increasingly available and their utility relative to the traditional methods of environmental bioaerosol sampling is under study.

13.4.4.1 Microscopy

Light microscopy is the most powerful method available for the qualitative and quantitative analysis of those pollen grains and fungal spores (both viable and nonviable) that have unique morphological features. The lower limit of resolution for this method is 1 μm. Thus, viruses are outside the recognition limits of this procedure, and stained bacteria are only recognizable at high magnification ($\times$1000). Particles are collected on transparent tapes or glass slides that have been coated with an adhesive, most often a silicon-based substance. In the laboratory, the particle trace is overlaid with a cover slip, which is held in place by a mounting medium, and the deposit is viewed under the microscope at magnifications up to ×1000. Particles can also be collected on membrane filters for microscopic examination, but the filter must be made transparent before viewing, for example, by adding immersion oil or a solvent. Other adhesive surfaces (e.g., pressure-sensitive adhesive tapes) are sometimes used for simultaneous immunodetection and microscopic examination of aeroallergen fragments as well as intact pollen and spores with their associated allergens.[196,197]

Identification, to the genus or species level, is not possible with light microscopy for microorganisms and pollen that lack distinctive morphological details, for example, many fungal spores and conidia and grass pollen. Complementary analysis with cultivation-based methods is informative for some fungal spores (see Section 13.4.4.2). Microscopic identification of pollen grains from wild grass species is seldom possible without a scanning electron microscope because of the small size of their distinguishing structures.

There have been many attempts to automate the identification and quantification of individual pollen grains and fungal spores using pattern recognition and gray-scale features[198–201] or translucent, autofluorescence, confocal techniques.[202,203] However, no one has yet automated the differentiation and counting of pollen grains or fungal spores to the point that a machine can replace a human for routine analyses.

Schumacher et al.[104] developed a method to distinguish pollen grains of *Cynodon dactylon* (Bermuda grass) from morphologically identical grass-pollen types by fluorescence microscopy and immunoblotting. Takahashi et al.[204] modified the procedure for *Cryptomeria japonica* (Japanese cedar). This analysis can be used to distinguish whole pollen, fragments of pollen grains, and antigen-containing particles with known immunological uniqueness. Emilson et al.[205] used confocal, laser scanning microscopy to locate the major allergen in birch pollen. Grote[206] and Grote et al.[207–210] used transmission electron microscopy, field emission scanning electron microscopy, and scanning electron microscopy combined with immuno-gold labeling to locate grass allergens in intact pollen grains, the walls and cytoplasm of grains, and respirable allergen-bearing fragments. Scanning electron microscopy has also been combined with aeroallergen immunoblotting[211] and used to investigate orbicules as possible allergen carriers.[211,212] Orbicules are granules (1.5 to 2 μm in diameter) that can separate from pollen surfaces and the lobes of anthers at anthesis (when flowers open).

13.4.4.2 Cultivation-based methods for bacteria and fungi

In cultivation-based (culture- or growth-based) methods, airborne microorganisms are collected in a liquid, on a filter, or directly on a semisolid growth medium. After an appropriate incubation period, the resulting visible colonies can be counted and the isolates identified. Cultivation-based systems are widely used in many fields of environmental and medical microbiology. Investigators can identify selective growth media and optimal incubation conditions when the microorganisms under study are known, based on their habitats in the natural environment or in human or animal hosts. Otherwise, cultivation-based methods are limited for the study of airborne microorganisms because even broad-spectrum growth media that support many microorganisms are somewhat selective, and therefore unsuitable for a portion of the collected agents. This limitation can be overcome, in part, by replication of samples on multiple culture media in combination with incubation under several growth conditions (e.g., carbon and nitrogen sources, trace elements, dyes, antibiotics, pH, water activities, redox conditions, incubation temperatures, and oxygen concentrations) to meet the specific requirements of different organisms and identify selected ones.[182,213–215] Except for constraints of time and resources, there is no limit to the combinations of media and conditions an investigator can choose. Following the initial cultivation of samples, individual colonies can be isolated further and identified through standard microbiological methods.[182,216,217]

In addition to the problem of anticipating the right growth medium and incubation condition for environmental samples, physical trauma and desiccation can render bacterial or fungal cells nonviable, and thus undetectable by cultivation.[218–221] Stresses that microorganisms encounter before becoming airborne, while in the air, and during collection prevent some of them from multiplying, which underestimates the true air concentration. Dehydration of airborne microorganisms can result in lethal cellular water loss, and exposure to oxygen radicals and radiation can further damage airborne cells.[218,219] Gram-positive bacteria are generally more resistant to drying than Gram-negative bacteria because the outer membranes of Gram-negative bacteria are very susceptible to stress-induced damage.[218] The peptidoglycan layer makes up 80% of the cell walls of Gram-positive bacteria but accounts for only 5% to 10% in Gram-negative bacteria.[222] Gram-positive endo- and exospores (e.g., *Bacillus* spp. and actinomycetes) maintain their viability longer than vegetative bacterial cells.[223] Likewise, many fungal spores tolerate desiccation well, yielding closer agreement between viable and total spore counts.

Various types of resuscitation and recovery media have been formulated to facilitate the repair of sublethal cell injuries and allow slow hydration, which substantially increases viable counts.[222] Examples of modifications to standard media for the enrichment of airborne microorganisms are the addition of protective agents such as pyruvate or betaine or the utilization of diluted growth media. Resuscitation in either a liquid or on an agar medium is typically done in a preincubation step before inoculation of the final growth medium. The degree of osmotic stress that the cells have experienced and the rate of cell rehydration following collection are the most important factors affecting recovery and culturability.[218]

One of the primary limitations of cultivation-based methods is that they provide a measure of only those organisms that are able to grow in the laboratory. Differences in viability can lead to the under- or overrepresentation of particular bacteria or fungi in the analysis step. Furthermore, even when the growth medium and incubation conditions are appropriate, aggregation of bacterial or fungal cells may lead to errors in the enumeration of the total number of culturable organisms, especially with direct agar impactors as opposed to samplers that collect particles into liquid, that is, a CFU may result from the growth of one or a cluster of several cells.

13.4.4.3 Biological assays

The effects of bioaerosols on other biological systems can be measured by different methods, among them whole animal exposure, infectivity assays for viruses, bioassays utilizing prokaryotic or eukaryotic cell lines, and nonspecific toxicity, cytotoxicity, and genotoxicity assays. One of the most commonly used biological assays, the LAL assay, measures bacterial endotoxin and fungal glucans. The LAL assay is an *in vitro* biological test that uses a lysate of blood cells from the horseshoe crab (*Limulus polyphemus*). The lysate contains a serine protease that triggers an enzyme cascade when activated by endotoxin. This amplification makes the assay sensitive, and the reaction can be observed as coagulation, an increase in turbidity, or a chromogenic reaction. Kits for kinetic chromogenic and turbidimetric assays are available commercially. The enzyme activation occurs also with β-glucans. Therefore, this pathway either should be inhibited to measure endotoxin alone or LAL preparations should be fractionated to obtain lysates that are sensitive only to endotoxin or β-glucan.

Variations have been found between LAL preparations from different manufacturers and within reagent lots from single producers. This variability severely limits the assay for comparisons of measurements made in different environments or by different researchers. LAL is a comparative dilution assay and the procedure must be standardized strictly. The response of the LAL assay varies for LPS molecules from different microorganisms. It has been assumed that the responses in this assay correspond to the toxic potencies of different endotoxins in humans, but this assumption has not been proven.

A variety of other assays (many of which use continuous cell lines) have been developed to detect overall toxicity, cytotoxicity, or mutagenicity and to screen for the presence of particular toxins, for example, aflatoxin from *Aspergillus flavus*. Brera *et al.*[224] reported the detection of aflatoxin in air samples collected in food-processing plants. A boar sperm cell, motility inhibition assay has also been used to detect bacterial depsipeptide and other toxins in foods and indoor environments.[225–227] The assay is sensitive to mitochondrial toxins that inhibit sperm motility (e.g., valinomycin) but is relatively insensitive to toxins that affect protein or nucleic acid synthesis (e.g., many mycotoxins). Mitochondriotoxin-producing species of bacteria and fungi have been found in building materials and air samples from water-damaged houses even when the toxins themselves were not detected in air samples.[225,227]

13.4.4.4 *Immunoassays*

Immunoassays rely on the binding of an antigen to an antibody (e.g., in human serum to identify prior exposure to an infectious agent or an allergen) or binding of an antibody to a target antigen (e.g., in an environmental sample to detect a virus, bacterium, fungus, protist, plant allergen, or animal allergen). Critical for this technique is the availability of well-characterized antigens and sufficiently specific antibodies, the latter typically obtained through the production of monoclonal antibodies, which recognize single antigens. The specificity and sensitivity of immunoassays are their main advantages and they are used widely on clinical specimens, to detect human hypersensitivity to environmental allergens and current and past infections, and on environmental samples, to measure allergens in air and dust samples.[228–230] However, specific antibodies for environmental bacteria and fungi, other than some infectious agents, have been difficult to develop.[182,231–233]

Immunological products can be coupled with fluorescent, enzymatic, and radiological assays. Gazenko *et al.*[234] studied reactions between fluorogenic substrates and different groups of enzymes to detect actinomycete spores. Speight *et al.*[235] developed polyclonal antibodies for an enzyme-linked immunosorbent assay (ELISA) to recognize a range of different epitopes on microbial cell surfaces. While application of the method to measure bioaerosols is feasible, it is not very sensitive. However, airborne bacteria have been collected with liquid impingers and their concentration measured with fluorescent *in situ* hybridization and flow cytometry.[220]

13.4.4.5 *Chemical assays*

Chemical analyses for bioaerosols have gained expanded utility with the development of increasingly sensitive instruments that allow analysis of smaller and smaller amounts of material. Because preservation of cell viability is not an issue in chemical assays, a wider variety of methods can be used to collect bioaerosols. The cell walls and membranes of fungi and bacteria contain unique chemical components by which they can be identified and quantified.[236,237] These compounds include lipids and various proteins and peptides, peptidoglycan in bacterial cell walls, teichoic acid in the thick peptidoglycan layer of Gram-positive bacteria, LPS in the outer membranes of Gram-negative bacteria, and ergosterol in the cytoplasmic membranes of fungi. Aerosolized toxins and volatile metabolites can also be analyzed by standard chemical methods.

Cellular fatty acids are commonly used to identify microorganisms after they have been grown in culture. For example, tuberculostearic acid, a component of lipoarabinomannan in the cell walls of coryneform bacteria (e.g., *Mycobacterium, Rhodococcus,* and *Corynebacterium* species), can be analyzed readily by gas chromatography (GC).[236] Chemical analysis of LPS is generally based on the detection of 3-hydroxy fatty acids by high-performance liquid chromatography (HPLC) or gas chromatography–mass spectrometry (GC–MS).[238] Chemical analysis of endotoxin has the advantage of being insensitive to variations in biological activity, but these techniques are two to three orders of magnitudes less sensitive than the LAL assay (see Section 13.4.4.3).[155]

Ergosterol is a principal sterol in the membranes of fungal hyphae and spores and has been used to measure fungal biomass, but it is not species specific. Ergosterol is stable in air-dried conditions and can be extracted in basic aqueous methanol followed by microwave heating and analysis by HPLC, GC, or GC–MS.[164,239,240]

When studying the production of mycotoxins, other secondary metabolites, and MVOCs, investigators should be aware that the substrate on which a bacterium or fungus has grown determines the types and quantities of metabolites it produces. Profiles of volatile compounds show the potential for identification of environmental microorganisms. Korpi et al.[241] collected air samples on a sorbent material and used thermal desorption-GC and HPLC to detect metabolic activity of fungal and actinomycete species on building materials placed in a test chamber. Activity occurred at 90 to 99% RH, and the main VOCs were 3-methyl-1-butanol, 1-pentanol, 1-hexanol, and 1-octen-3-ol. Using GC–MS, Ahearn et al.[242] found that ventilation system filters colonized with fungi released acetone, hexane, and organic compounds that uncolonized air filters did not emit.

13.4.4.6 Molecular genetic assays

Molecular genetic techniques have received increased attention as diagnostic tools in the study of airborne microorganisms. Most importantly, molecular techniques can provide genus-, species-, or strain-specific identification without the need for culturing of a sample. These procedures allow the detection of specific organisms or genes with precision but do not yet allow easy quantification. Polymerase chain reaction (PCR) assays amplify target nucleic acid sequences. Genus- and species-specific nucleic acid probes have been designed that can be used for microscopic visualization.[243] PCR-based methods have been used to detect bacteria on contaminated surfaces,[244] *Legionella pneumophila* in filter and impinger air samples,[245] fungal spores in a miniature cyclone-type sampler,[246] *M. tuberculosis* in samples of the air in patient isolation rooms,[247] and the aerodynamic size range of airborne mycobacteria associated with whirlpools.[248] Solid-phase PCR from filtered impinger samples was found to be more sensitive than a cultivation-based method for aerosolized *E. coli*.[249]

13.5 Concluding remarks

Particles of microbial, plant, or animal origin cover a wide range of sizes and deposit in all regions of the airways. Some of the effects that bioaerosols cause in the respiratory tract are similar to reactions to other allergenic or inflammatory particles, but certain responses are unique, for example, infection as a result of microbial multiplication. The economic burden of bioaerosol-related diseases in the community and workplaces is great. Public health management of new infectious diseases as well as reemergent ones requires national and international cooperation. Increased support, in response to terrorist threats, of research on the airborne transmission of infectious agents and biological toxins, engineering controls, personal protective equipment, and clean-up procedures are expected to benefit other areas of public health preparedness as well.

Continued research is needed to gain a clearer understanding of the positive and negative effects of exposure to biological agents at different ages and stages of lung development. We can look forward to a wider availability of methods that measure exposure to biological agents themselves rather than surrogates, for example, all allergen-bearing particles rather than only intact pollen and spores. The fields of aerobiology and bioaerosol research would advance more rapidly if it were convenient to measure personal exposure and inhaled dose over extended time periods rather than estimate exposures from the concentration of biological agents in settled dust or small-volume, grab air samples collected with stationary samplers. The continued development and wider availability of rapid detection procedures and methods not based on microbial cell multiplication or visual recognition will provide better exposure information.

Particles of biological origin account for substantial fractions of airborne PM, and the interactions of bioaerosol and other air pollutants deserve greater attention. More effort should be made to understand the mechanisms of action and fate in the respiratory tract of particles of biological origin. Better human exposure–response data and information from animal models may lead to the establishment of exposure limits for more biological agents. However, for many bioaerosols, reliance on approaches not based on environmental measurements will continue to be the best means to ensure public safety and health. For example, avoidance of dampness and prompt removal of contaminated materials provide better protection from indoor mold than would adherence to numeric concentration limits. Selection of less allergenic plants for landscaping projects would protect hypersensitive persons and is preferable to drug treatment, exposure avoidance by remaining indoors, and use of energy-consuming air cleaners.

References

1. Balmes, J.R. and Scannell, D.H., Occupational lung diseases, in *Occupational and Environmental Medicine*, 2nd ed., LaDou, J., Ed., Appleton & Lange, Stamford, 1997, chap. 20.
2. Matthias-Maser, S. and Jaenicke, R., Examination of atmospheric bioaerosol particles with radii >0.2 μm, *J. Aerosol Sci.*, 25, 1605, 1994.
3. Gruber, S., Matthias-Maser, S., Brinkmann, J., and Jaenicke, R., Vertical distribution of biological aerosol particles above the North Sea, *J. Aerosol Sci.*, 29, S771, 1998.
4. Gruber, S., Matthias-Maser, S., and Jaenicke, R., Concentration and chemical composition of aerosol particles in marine and continental air, *J. Aerosol Sci.*, 30, S9, 1999.
5. Ebert, M., Weinbruch, S., Hoffmann, P., and Ortner, H. M., Chemical characterization of North Sea aerosol particles, *J. Aerosol Sci.*, 31, 613, 2000.
6. Bauer, H., Kasper-Giebl, A., Loflund, M., Giebl, H., Hitzenberger. R., Zibuschka, F., and Puxbaum, H., The contribution of bacteria and fungal spores to the organic content of cloud water, precipitation and aerosols, *Atmos. Res.*, 64, 109, 305, 2002.
7. Penner, J.E., Carbonaceous aerosols influencing atmospheric radiation black and organic carbon, in *Aerosol Forcing of Climate,* Charlson, R.J. and Heintzenberg, J., Eds., John Wiley & Sons, New York, 1995, 91.
8. Douwes, J. and Pearce, N., Asthma and the westernization "package," *Int. J. Epidemiol.*, 31, 1098, 2002.
9. Ahlbom, A., Backman, A., Bakke, J., et al., "NORDPET" Pets indoors — a risk factor for or protection against sensitization/allergy, *Indoor Air*, 8, 219, 1998.
10. Pearce, N., Douwes, J., and Beasley, R., Is allergen exposure the major primary cause of asthma? *Thorax*, 55, 424, 2000.
11. Murray, C.S., Woodcock, A., and Custovic, A., The role of indoor allergen exposure in the development of sensitization and asthma, *Curr. Opin. Allergy Clin. Immunol.*, 1, 407, 2001.
12. Hubálek, Z., Emerging human infectious diseases: anthroponoses, zoonoses, and sapronoses, *Emerg. Infect. Dis.*, 9, 403, 2003.
13. Akazawa, M., Sindelar, J.L., and Paltiel, A.D., Economic costs of influenza-related work absenteeism, *Value Health*, 6, 107, 2003.
14. Thompson, W.W., Shay, D.K., Weintraub, E., Brammer, L., Cox, N., Anderson, L.J., and Fukuda, K., Mortality associated with influenza and respiratory syncytial virus in the United States, *J. Am. Med. Assoc.*, 289, 179, 2003.
15. Bramley, T.J., Lerner, D., and Sarnes, M., Productivity losses related to the common cold, *J. Occup. Environ. Med.*, 44, 822, 2002.

16. Storch, G.A., Respiratory system, in *Mechanisms of Microbial Disease*, 2nd ed., Schaechter, M., Medoff, G., and Eisenstein, B.I., Eds., Williams & Wilkins, Baltimore, 1993, p. 675.
17. WHO, *The World Health Report 2002, Reducing Risks, Promoting Healthy Life,* World Health Organization, Geneva, 2002.
18. Ezzati, M., Lopez, A.D., Rodgers, A., vander Hoorn, S., Murray, C.J.L., and the Comparative Risk Assessment Collaborating Group, Selecting major risk factors and global and regional burden of disease, *Lancet*, 360, 1347, 2002.
19. Snider, D.E., Raviglione, M., and Kochi, A., Global burden of tuberculosis, in *Tuberculosis: Pathogenesis, Protection, and Control,* Bloom, B.R., Ed., American Society for Microbiology, Washington, 1994, p. 3.
20. Graham, N.M.H., The epidemiology of acute respiratory infections in children and adults: a global perspective, *Epidemiol. Rev.*, 12, 149, 1990.
21. Coultas, D.B. and Samet, J.M., Respiratory disease prevention, in *Public Health and Preventive Medicine*, 14th ed., Wallace, R.B., Ed., Appleton & Lange, Stamford, 1998, p. 981.
22. IOM (Institute of Medicine), Magnitude and dimensions of sensitization and disease caused by indoor allergens, in *Indoor Allergens, Assessing and Controlling Adverse Health Effects*, Pope, A. M., Patterson, R., and Burge, H., Eds., National Academy Press, Washington, 1993, p. 44.
23. Bellanti, J.A. and Wallerstedt, D.B., Allergic rhinitis update: epidemiology and natural history, *Allergy Asthma Proc.*, 21, 367, 2000.
24. Shusterman, D., Upper respiratory tract disorders, in *Occupational and Environmental Medicine,* 2nd ed., LaDou, J., Ed., Appleton & Lange, Stamford, 1997, chap. 19.
25. Montealegre, F. and Bayona, M., An estimate of the prevalence, severity and seasonality of asthma in visitors to a Ponce shopping center, *P. R. Health Sci. J.*, 15, 113, 1996.
26. CDC, Forecasted State-Specific Estimates of Self-Reported Asthma Prevalence — United States, 1998, Centers for Disease Control and Prevention, MMWR, 47, 1022, 1998.
27. CDC, Surveillance for asthma — United States, 1980–1999, Centers for Disease Control and Prevention, MMWR, 51/SS-1, 1, 2002.
28. WHO, *Bronchial Asthma,* Fact Sheet No. 206, World Health Organization, Geneva, 2000.
29. IOM (Institute of Medicine), Major issues in understanding asthma, Methodological considerations in evaluating the evidence, and Indoor biologic exposures, in *Clearing the Air: Asthma and Indoor Air Exposures*, Committee on the Assessment of Asthma and Indoor Air, National Academy Press, Washington, 2000, chaps. 1, 2, and 5.
30. Linaker, C. and Smedley, J., Respiratory illness in agricultural workers, *Occup. Med.*, 52, 451, 2002.
31. Ray, N., Barniuk, J., Thamer, M., Rinehart, C., Gergen, P., Kaliner, M., Josephs, S., and Pung, Y., Healthcare expenditures for sinusitis in 1996: contributions of asthma, rhinitis, and other airway disorders, *J. Allergy Clin. Immunol.*, 103, 408, 1999.
32. ALA, *Epidemiology and Statistics Unit*, Best Practices and Program Services, Trends in Asthma Morbidity and Mortality, American Lung Association, 2002.
33. Rylander, R. and Jacobs, R.R., Eds. *Organic Dusts: Exposure, Effects, and Prevention*, Lewis Publishers, Chicago, 1994.
34. Harding, A.L., Fleming, D.O., and Macher, J.M., Biological hazards, in *Fundamentals of Industrial Hygiene*, 5th ed., Plog, B. A., and Quinlan, P. J, Eds., National Safety Council, Itasca, 2002, chap. 14.
35. NIOSH, Preventing Organic Dust Toxic Syndrome, DHHS (National Institute of Occupational Safety and Health) Publication No 94–102, Cincinnati, 1994.
36. Rylander, R., Persson, K., Goto, H., Yuasa, K., and Tanaka, S., Airborne beta-1–3 glucan may be related to symptoms in sick buildings, *Indoor Environ.*, 1, 263, 1992.
37. Williams, D.L., (1→3)-β-D-glucans, in *Organic Dusts: Exposure, Effects, and Prevention*, Rylander, R. and Jacobs, R.R., Eds., Lewis Publishers, Chicago, 1994, p. 83.
38. Fogelmark, B., Sjöstrand, M., and Rylander, R., Pulmonary inflammation induced by repeated inhalations of beta(1,3)-D-glucan and endotoxin, *Int. J. Exp. Pathol.*, 75, 85, 1994.
39. Gehring, U., Douwes, J., Doekes, G., Koch, A., Bischof, W., Fahlbusch, B., Richter, K., Wichmann, H., and Heinrich, J., β (1 3)-Glucan in house dust of German homes: housing

characteristics, occupant behavior, and relations with endotoxins, allergens, and molds, *Environ. Health Perspect.*, 109, 139, 2001.

40. Alwis, U., Mandryk, J., Hocking, A.D., Lee, J., Mayhew, T., and Baker, W., Dust exposures in the wood processing industry, *Am. Ind. Hyg. Assoc. J.*, 60, 641, 1999.
41. Alwis, K.U., Mandryk, J., and Hocking, A.D., Exposure to biohazards in wood dust: bacteria, fungi, endotoxins, and (1→3)-beta-D-glucans, *Appl. Occup. Environ. Hyg.*, 14, 598, 1999.
42. Flannigan, B., Air sampling for fungi in indoor environments, *J. Aerosol Sci.*, 28, 381, 1997.
43. Rylander, R., Airborne (1→3)-beta-D-glucan and airway disease in a day-care center before and after renovation, *Arch. Environ. Health*, 52, 281, 1997.
44. Douwes, J., Pearce, N., and Heederik, D., Does environmental endotoxin exposure prevent asthma? *Thorax*, 57, 86, 2002.
45. Niven, R., The endotoxin paradigm: a note of caution, *Clin. Exp. Allergy*, 33, 273, 2003.
46. Sonesson, H.R.A., Zähringer, U., Grimmecke, H.D., Westphal, O., and Rietschel, E.T., Bacterial endotoxin: chemical structure and biological activity, in *Endotoxin and the Lungs*, Brigham, K., Ed., Marcel Dekker Inc., New York, 1994, Vol. 77.
47. Myatt, T.A. and Milton, D.K., Endotoxins, in *Indoor Air Quality Handbook*, Spengler, J.D., Samet, J.M., and McCarthy, J.F., Eds., McGraw-Hill, New York, 2001, chap. 42.
48. Evans, D., Epidemiology and etiology of occupational infectious diseases, in *Occupational and Environmental Infectious Diseases*, Couturier, A. J., Ed., OEM Press, Beverly Farms, 2000, p. 37.
49. Mauderly, J.L., Cheng, Y.S., Johnson, N.F., Hoover, M.D., and Yeh, H.C., Particles inhaled in the occupational setting, in *Particle–Lung Interactions*, Gehr, P. and Heyder, J., Eds., Marcel Dekker Inc., New York, 2000, chap. 3.
50. Maier, R.M., Pepper, I.L., and Gerba, C.P., *Environmental Microbiology*, Academic Press, San Diego, 2000.
51. Sylvia, D.M., Fuhrmann, J.J., Hartel, P.G., and Zuberer, D.A., *Principles and Applications of Soil Microbiology*, Prentice-Hall, Upper Saddle River, NJ, 1998.
52. Torsvik, V., Sorheim, R., and Goksoyr, J., Total bacterial diversity in soil and sediment communities — a review, *J. Ind. Microbiol.*, 17, 170, 1996.
53. Yeo, H.G. and Kim, J.H., SPM and fungal spores in the ambient air of west Korea during the Asian dust (Yellow sand) period, *Atmos. Environ.*, 36, 5437, 2002.
54. Huang, C.Y., Lee, C.C., Li, F.C., Ma, Y.P., and Su, H.J.J., The seasonal distribution of bioaerosols in municipal landfill sites: a 3-yr study. *Atmos. Environ.*, 36, 4385, 2002.
55. Rylander, R., Health effects among workers in sewage treatment plants, *Occup. Environ. Med.*, 56, 354, 1999.
56. Wald, P.H. and Stave, G.M., Biological hazards in the workplace, in *Physical and Biological Hazards of the Workplace*, Van Nostrand Reinhold, New York, 1994, Part II.
57. Couturier, A.J., *Occupational and Environmental Infectious Diseases*, OEM Press, Beverly Farms, 2000.
58. Rossmore, H.W., Antimicrobial agents for water-based metalworking fluids, *J. Occup. Med.*, 23, 247, 1981.
59. Elsmore, R. and Hill, E.C., The ecology of pasteurized metalworking fluids, *Int. Biodeterior.*, 22, 101, 1986.
60. Foxall-van Aken, S., Brown, J.A. Jr., Young, W., Salmeen, I., McClure, T., Napier, S. Jr., and Olsen, R.H., Common components of industrial metal-working fluids as sources of carbon for bacterial growth, *Appl. Environ. Microbiol.*, 51, 1165, 1986.
61. Mattsby-Baltzer, I., Sandin, M., Ahlström, B., Allenmark, S., Edebo, M., Falsen, E., Pedersen, K., Rodin, N., Thompson, R.A., and Edebo, L., Microbial growth and accumulation in industrial metal-working fluids, *Appl. Environ. Microbiol.*, 55, 2681, 1989.
62. Kreiss, K. and Cox-Ganser, J., Metalworking fluid-associated hypersensitivity pneumonitis: a workshop summary, *Am. J. Ind. Med.*, 32, 423, 1997.
63. Kennedy, S.M., Greaves, I.A., Kriebel, D., Eisen, E.A., Smith, T.J., and Woskie, S.R., Acute pulmonary responses among automobile workers exposed to aerosols of machining fluids, *Am. J. Ind. Med.*, 15, 627, 1989.
64. Mackerer, C.R., Health effects of oil mists: a brief review, *Toxicol. Ind. Health*, 5, 429, 1989.

65. Zacharisen, M.C., Kadambi, D.P., Schlueter, D.P., Kurup, V.P., Schack, J.B., Fox, J.L., Anderson, H.A., and Fink, J.N., The spectrum of respiratory disease associated with exposure to metal working fluids, *J. Occup. Environ. Med.*, 40, 640, 1998.
66. Fox, J., Anderson, H., Moen, T., Gruetzmacher, G., Hanrahan, L., and Fink, J., Metal working fluid-associated hypersensitivity pneumonitis: an outbreak investigation and case–control study, *Am. J. Ind. Med.*, 35, 58, 1999.
67. Wallace, R.J., Jr., Zhang, Y., Wilson, R.W., Mann, L., and Rossmore, H., Presence of a single genotype of the newly described species *Mycobacterium immunogenum* in industrial metalworking fluids associated with hypersenstivity pneumonitis, *Appl. Environ. Microbiol.* 68, 5880, 2002.
68. Rylander, R., Bake, B., Fischer, J.J., and Helander, I.M., Pulmonary function and symptoms after inhalation of endotoxin, *Am, Rev. Respir. Dis.*, 140, 981, 1989.
69. Pillai, S.D. and Ricke, S.C., Bioaerosols from municipal and animal wastes: background and contemporary issues, *Can. J. Microbiol.*, 48, 681, 2002.
70. Anderson, H.R., Ponce de Leon, A., Bland, J.M., Bower, J.S., Emberlin, J., and Strachan, D.P., Air pollution, pollens, and daily admissions for asthma in London 1987–92, *Thorax*, 53, 842, 1998.
71. Chang, C.W., Chung, H., Huang, C.F., and Su, J.J., Exposure of workers to airborne microorganisms in open-air swine houses, *Appl. Environ. Microbiol.*, 67, 155, 2001.
72. Dutkiewicz, J., Krysinska-Traczyk, E., Skorksa, C., Cholewa, G., and Sitkowska, J., Exposure to airborne microorganisms and endotoxin in a potato processing plant, *Ann. Agric. Environ. Med.*, 9, 225, 2002.
73. Friedlander, A.M., Anthrax, in *Textbook of Military Medical Medicine*, Walter Reed Army Medical Center, On U.S. Army Medical NBC Online Information Server, 1997, chap. 22 (http://www.nbc-med.org/SiteContent/HomePage/WhatsNew/MedAspects/Ch-22electrv699.pdf; website accessed April 2003).
74. Dull, P.M., Wilson, K.E., Kournikakis, B., Whitney, E.A.S., Boulet, C.A., Ho, J.Y.W., Ogston, J., Spence, M.R., Mckenzie, M.M., Phelan, M.A., Popovic, T., and Ashford, D., *Bacillus anthracis* aerosolization associated with a contaminated mail sorting machine, *Emerg. Infect. Dis.*, 8, 1044, 2002.
75. Jernigan, D.B., Raghunathan, P.L., Bell, B.P., Brechner, R., Bresnitz, E.A., Butler, J.C., Cetron, M., Cohen, M., Doyle, T., Fischer, M., Greene, C., Griffith, K.S., Guarner, J., Hadler, J.L., *et al.*, Investigation of bioterrorism-related anthrax, United States, 2001: epidemiologic findings, *Emerg. Infect. Dis.*, 8, 1019, 2002.
76. Mainelis, G., Willeke, K., Baron, P., Grinshpun, S.A., and Reponen, T., Induction charging and electrostatic classification of micrometer-size particles for investigating the electrobiological properties of airborne microorganisms, *Aerosol Sci. Technol.*, 36, 479, 2002.
77. Mainelis, G., Willeke, K., Adhikari, A., Reponen, T., and Grinshpun, S.A., Design and collection efficiency of a new electrostatic precipitator for bioaerosol collection, *Aerosol Sci. Technol.*, 36, 1073, 2002.
78. Reponen, T., Willeke, K., Ulevicius, V., Reponen, A., and Grinshpun, S.A., Effect of relative humidity on the aerodynamic diameter and respiratory deposition of fungal spores, *Atmos. Environ.*, 30, 3967, 1996.
79. Madelin, T.M. and Johnson, H.E., Fungal and actinomycete spore aerosols measured at different humidities with an aerodynamic particle sizer, *J. Appl. Bacteriol.*, 72, 400, 1992.
80. Gregory, P.H., Spores: their properties and sedimentation in still air, in *The Microbiology of the Atmosphere*, Aylesbury, Leonard Hill Books, U.K., 1973, chap. 2.
81. Linder, H.P., Morphology and the evolution of wind pollination, in *Reproductive Biology*, Owens, S.T. and Rudall, P.J., Eds., Royal Botanic Gardens, Kew, 1998, p. 123.
82. Linder, H.P., Pollen morphology and wind pollination in Angiosperms, in *Pollen and Spores: Morphology and Biology*, Harley, M.M., Morton, C.M., and Blackmore, S., Eds., Royal Botanic Gardens, Kew, 2000, p. 73.
83. Muilenberg, M.L., Pollen in indoor air: sources, exposures, and health effects, in *Indoor Air Quality Handbook*, Spengler, J.D., Samet, J.M., and McCarthy, J.F., Eds., McGraw-Hill, New York, 2001, chap. 44.

84. Reponen, T., Aerodynamic diameters and respiratory deposition estimates of viable fungal particles in problem buildings, *Aerosol Sci. Technol.*, 22, 11, 1995.
85. Hjelmroos, M., Benyon, F.H.L., Culliver, S., and Jones, A.S., *Airborne Allergens. Interactive Identification of Allergenic Pollen Grains and Fungal Spores*, Institute of Respiratory Medicine, University of Sydney, Australia, 1999.
86. Lacey, J., Aggregation of spores and its effect on aerodynamic behaviour, *Grana*, 30, 437, 1991.
87. Anderson, M.C. and Baer, H., Allergenically active components of cat allergen extracts, *J. Immunol.*, 127, 972, 1981.
88. Platts-Mills, T.A. and Carter, M.C., Asthma and indoor exposure to allergens, *N. Engl. J. Med.*, 336, 1382, 1997.
89. Custovic, A., Green, R., Fletcher, A., Smith, A., Pickering, C.A., Chapman, M.D., and Woodcock, A., Aerodynamic properties of the major dog allergen *Can f* 1: distribution in homes, concentration, and particle size of allergen in the air, *Am. J. Respir. Crit. Care. Med.*, 155, 94, 1997.
90. Custovic, A., Simpson, A., Pahdi, H., Green, R.M., Chapman, M.D., and Woodcock, A., Distribution, aerodynamic characteristics, and removal of the major cat allergen *Fel d* I in British homes, *Thorax*, 53, 33, 1998.
91. Wood, R.A., Laheri, A.N., and Eglleston, P.A., The aerodynamic characteristics of cat allergen, *Clin. Exp. Allergy*, 23, 733, 1993.
92. Custovic, A., Simpson, B., Simpson, A., Hallam, C., Craven, M., and Woodcock, A., Relationship between mite, cat, and dog allergens in reservoir dust and ambient air, *Allergy*, 54, 612, 1999.
93. Reponen, T.A., Gazenko, S.V., Grinshpun, S.A., Willeke, K., and Cole, E.C., Characteristics of airborne Actinomycete spores, *Appl. Environ. Microbiol.*, 64, 3807, 1998.
94. Burge, H.A. and Rogers, C.A., Outdoor allergens, *Environ. Health Perspect.*, 108 (Suppl.), 653, 2000.
95. Pasanen, A.-L., Pasenen, P., Jantunen, M.J., and Kalliokoski, P., Significance of air humidity and air velocity for fungal spore release into the air, *Atmos. Environ.*, 25A, 459, 1991.
96. Górny, R.L., Reponen, T., Willeke, K., Schmechel, D., Robine, E., Boissier, M., and Grinshpun, S.A., Fungal fragments as indoor air biocontaminants, *Appl. Environ. Microbiol.*, 68, 3522, 2002.
97. Hyvarinen, A., Vahteristo, M., Meklin, T., Jantunen, M., Nevalainen, A., and Moschandreas, D., Temporal and spatial variation of fungal concentrations in indoor air, *Aerosol Sci. Technol.*, 35, 688, 2001.
98. Meklin, T., Reponen, T., Toivola, M., Koponen, T., Husman, T., Hyvarinen, A., and Nevalainen, A., Size distributions of airborne microbes in moisture-damaged and reference school buildings of two construction types, *Atmos. Environ.*, 36, 6031, 2002.
99. Hjelmroos, M., Evidence of long-distance transport of *Betula* pollen, *Grana*, 30, 215, 1991.
100. Rantio-Lehtimaki, A., Viander, M., and Koivikko, A., Airborne birch pollen antigens in different particle sizes, *Clin. Exp. Allergy*, 24, 23, 1994.
101. Matikainen, E. and Rantio-Lehtimaki, A., Semiquantitative and qualitative analysis of preseasonal birch pollen allergens in different particle sizes — background information for allergen reports, *Grana*, 37, 293, 1998.
102. Takahashi, Y., Sasaki, K., Nakamura, S., Mikihirosige, H., and Nitta, H., Aerodynamic size distribution of the particle emitted from the flowers of allergologically important plants, *Grana*, 34, 45, 1995.
103. Agarwal, M.K., Swanson, M.C., Reed, C.E., and Yunginger, J.W., Airborne ragweed allergens: association with various particle sizes and short ragweed plant parts, *J. Allergy Clin. Immunol.*, 74, 687, 1984.
104. Schumacher, M.J., Griffith, R.D., and O'Rourke, M.K., Recognition of pollen and other particulate aeroantigens by immunoblot microscopy, *J. Allergy Clin. Immunol.*, 82, 608, 1988.
105. Knox, R.B., Suphioglu, C., Taylor, P., Desai, R., Watson, H.C., Peng, J.L., and Bursill, L.A., Major grass pollen allergen *Lol p* 1 binds to diesel exhaust particles: implications for asthma and air pollution, *Clin. Exp. Allergy*, 27, 246, 1997.
106. Zureik, M., Neukirch, C., Leynaert, B., Liard, R., Bousquet, J., and Neukirch, F., Sensitization to airborne moulds and severity of asthma: cross sectional study from European Community respiratory health survey, *Br. Med. J.*, 325, 325, 2002.

107. Packe, G.E. and Ayres, J.G., Asthma outbreak during a thunderstorm, *Lancet*, 2, 199, 1985.
108. Knox, R.B., Grass pollen, thunderstorms and asthma, *Clin. Exp. Allergy*, 23, 354, 1993.
109. Newson, R., Strachan, D., Archibald, E., Emberlin, J., Hardaker, P., and Collier, C., Effect of thunderstorms and airborne grass pollen on the incidence of acute asthma in England, 1990–94, *Thorax*, 52, 680, 1997.
110. Marks, G.B., Colquhoun, J.R., Girgis, S.T., Hjelmroos-Koski, M., Treloar, A.B.A., Hansen, P., Downs, S.H., and Car, N.G., Thunderstorm outflows preceding epidemics of asthma during spring and summer, *Thorax*, 56, 468, 2001.
111. Lewis, S.A., Corden, J.M., Forster, G.E., and Newlands, M., Combined effects of aerobiological pollutants, chemical pollutants and meteorological conditions on asthma admissions and A & E attendances in Derbyshire UK, 1993–96, *Clin. Exp. Allergy*, 30, 1724, 2000.
112. Dales, R.E., Cakmak, S., Judek, S., Dann, T., Coates, F., Brook, J.R., and Burnett, R.T., The role of fungal spores in thunderstorm asthma, *Chest*, 123, 745, 2003.
113. Käpylä, M., Diurnal variations of non-arboreal pollen in the air in Finland, *Grana*, 20, 55, 1981.
114. Lin, W.H. and Li, C.S., Associations of fungal aerosols, air pollutants, and meteorological factors, *Aerosol Sci. Technol.*, 32, 359, 2000.
115. Di Giorgio, C., Krempff, A., Guiraud, H., Binder, P., Tiret, C., and Dumenil G., Atmospheric pollution by airborne microorganisms in the City of Marseilles, *Atmos. Environ.*, 30, 155, 1996.
116. Jones, B.L. and Cookson, J.T., Natural atmospheric microbial conditions in a typical suburban area, *Appl. Environ. Microbiol.*, 45, 919, 1983.
117. Bovallius, A., Bucht, B., Roffey, R., and Ånäs, P., Three-year investigation of the natural airborne bacterial flora at four localities in Sweden, *Appl. Environ. Microbiol.*, 35, 847, 1978.
118. Kelly, C.D. and Pady, S.M., Microbiological studies of air masses over Montreal during 1950 and 1951, *Can. J. Bot.*, 32, 591, 1954.
119. Tong, Y. and Lighthart, B., The annual bacterial particle concentration and size distribution in the ambient atmosphere in a rural area of the Willamette Valley, Oregon, *Aerosol Sci. Technol.*, 32, 393, 2000.
120. Chew, G.L., Rogers, C., Burge, H.A., Muilenberg, M.L., and Gold, D.R., Dustborne and airborne fungal propagules represent a different spectrum of fungi with differing relations to home characteristics, *Allergy*, 58, 13, 2003.
121. Reponen, T., Nevalainen, A., Jantunen, M., Pellikka, M., and Kalliokoski, P., Normal range criteria for indoor air bacteria and fungal spores in a subarctic climate, *Indoor Air*, 2, 6, 1992.
122. Levetin, E., Fungi, in *Bioaerosols*, Burge, H.A., Ed., Lewis Publ., Boca Raton, FL, 1995, chap. 5.
123. Muilenberg, M.L., The outdoor aerosol, in *Bioaerosols*, Burge, H.A., Ed., Lewis Publishers, Boca Raton, FL, 1995, chap. 9.
124. Bornehag, C.G., Blomquist, G., Gyntelberg, F., Jarvholm, B., Malmberg, P., Nordvall, L., Nielsen, A., Pershagen, G., and Sundell, J., Dampness in buildings and health. Nordic interdisciplinary review of the scientific evidence on associations between exposure to "dampness" in buildings and health effects (NORDDAMP), *Indoor Air*, 11, 72, 2001.
125. Dallman, P.R., Plant life in the world's Meriterranean climates, University of California Press, Berkeley and Los Angeles, California, 1998.
126. ICRP, International Commission on Radiological Protection, Human Respiratory Tract Model for Radiological Protection, ICRP Publication 66, *Ann. ICRP*, 24, 1, Oxford, Pergamon Press, Elmsford, 1994.
127. Stewart, G.A., McWilliam, A.S., Thompson, P.J., King, C.M., and Robinson, C., Potential consequences of interactions between aeroallergens and cells within the respiratory tree, in *Particle–Lung Interactions*, Gehr, P. and Heyder, J., Eds., Marcel Dekker Inc., New York, 2000, chap. 12.
128. Thorne, P.S., Inhalation toxicology models of endotoxin- and bioaerosol-induced inflammation, *Toxicology*, 152, 13, 2000.
129. ACGIH, Data interpretation, in *Bioaerosols: Assessment and Control*, Macher, J.M., Ammann, H.M., Burge, H.A., Milton, D.K., and Morey, P.R., Eds., American Conference of Government Industrial Hygienists, Cincinnati, 1999, chap. 7.
130. Musher, D.M., How contagious are common respiratory tract infections?, *N. Engl. J. Med.*, 348, 1256, 2003.

131. Wells, W.F., Ratcliffe, H.L., and Crumb, C., On the mechanics of droplet nuclei infection. II. Quantitative experimental air-borne tuberculosis in rabbits, *Am. J. Hyg.*, 47, 11, 1948.
132. Wells, W.F., Response and reaction to inhaled droplet nuclei contagium, in *Airborne Contagion and Air Hygiene,* Harvard University Press, Cambridge, MA, 1955, chap. 11.
133. Riley, E.C., Murphy, G., and Riley, R.L., Airborne spread of measles in a suburban elementary school, *Am. J. Epidemiol.*, 107, 421, 1978.
134. Catanzaro, A., Nosocomial tuberculosis, *Am. Rev. Respir. Dis.*, 125, 559, 1982.
135. Remington, P.L., Hall, W.N., Davis, I.H., Herald, A., and Gunn, R.A., Airborne transmission of measles in a physician's office, *J. Am. Med. Assoc.*, 253, 1574, 1985.
136. Nardell, E.A., Keegan J., Cheney, S.A., and Etkind, S.C., Airborne Infection. Theoretical limits of protection achievable by building ventilation, *Am. Rev. Respir. Dis.*, 144, 302, 1991.
137. Gammaitoni, L. and Nucci, M.C., Using a mathematical model to evaluate the efficacy of TB control measures, *Emerg. Infect. Dis.*, 3, 335, 1997.
138. U.S. Department of Health and Human Services, *Biosafety in Microbiological and Biomedical Laboratories*, 4th ed., U.S. Government Printing Office, Washington, DC, 1999.
139. Health Canada, *Material Safety Data Sheets for Infectious Agents*, Office of Laboratory Security, Population and Public Health Branch, Ottawa, Canada (http://www.hc-sc.gc.ca/pphb-dgspsp/msds-ftss/index.html; website accessed March 2003).
140. Arshad, S.H., Hamilton, R.G., and Adkinson, N.F., Repeated aerosol exposure to small doses of allergen: a model for chronic allergic asthma, *Am. J. Respir. Crit. Care Med.*, 157, 1900–1906, 1998.
141. Schelegle, E.S., Gershwin, L.J., Miller, L.A., Fanucchi, M.V., van Winkle, L.S., Gerriets, J.P., Walby, W.F., Omlor, A.M., Buckpitt, A.R., Tarkington, B.K., Wong, V.J., Joad, J.P., Pinkerton, K.B., Wu, R., Evans, M.J., Hyde, D.M., and Plopper, C.G., Allergic asthma induced in rhesus monkeys by house dust mite (*Dermatophagoides farinae*), *Am. J. Pathol.*, 158, 333, 2001.
142. Eggleston, P.A., Biological pollutants, in *Air Pollutants and the Respiratory Tract*, Swift, D.L. and Foster, W.M., Eds., Marcel Dekker Inc., New York, 1999, chap 9.
143. Jacob, B., Ritz, B., Gehring, U., *et al.*, Indoor exposure to molds and allergic sensitization, *Environ. Health Perspect.*, 110, 647, 2002.
144. Lioy, P.J., Freeman, N.C.G., and Millette, J.R., Dust: a metric for use in residential and building exposure assessment and source characterization, *Environ. Health Perspect.*, 110, 969, 2002.
145. Karlsson, A.S., Renstrom, A., Hedren, M., and Larsson, K., Comparison of four allergen-sampling methods in conventional and allergy prevention classrooms, *Clin. Exp. Allergy*, 32, 1776, 2002.
146. AAAAI. American Academy of Allergy, Asthma, and Immunology, National Allergy Bureau (NAB), Reading the Charts (www.aaaai.org/nab/index.cfm?p=reading_charts; website accessed April 2003).
147. Nilsson, S. and Spieksma F.Th.M., Eds., *Allergy Service Guide in Europe,* Palynological Laboratory, Swedish Museum of History, Stockholm, Sweden, 1994.
148. Rylander, R. and Lin, R.H., (1–3)-beta-D-glucan — relationship to indoor air-related symptoms, allergy and asthma, *Toxicology.*, 152, 47, 2000.
149. Thorn, J., The inflammatory response in humans after inhalation of bacterial endotoxin: a review, *Inflamm. Res.*, 50, 254, 2001.
150. Rylander, R., Endotoxin in the environment — exposure and effects, *J. Endotoxin Res.*, 8, 241, 2002.
151. ASTM, Standard Practice for Personal Sampling and Analysis of Endotoxin in Metalworking Fluid Aerosols in Workplace Atmospheres, Practice E2144-01, American Society for Testing and Materials, W. Conshohocken, 2001.
152. CEN, European Committee for Standardization, Workplace Atmospheres — Determination of Airborne Endotoxin, CEN/TC 137 Work Programme, Project Reference EN 14031, Brussels, 2001.
153. White, E. M., Environmental endotoxin measurement methods: standardization issues, *Appl. Occup. Environ. Hyg.*, 17, 606, 2002.
154. DECOS, Endotoxins, publication 1998/03WGD, Dutch Expert Committee on Occupational Standards (DECOS), Health Council of the Netherlands, Rijswijk, 1998.

155. Milton, D.K., Endotoxin and other bacterial cell-wall components, in *Bioaerosols: Assessment and Control*, Macher, J.M., Ammann, H.M., Burge, H.A., Milton, D.K., and Morey, P.R., Eds., ACGIH, Cincinnati, 1999, chap. 23.
156. ACGIH, Biologically derived airborne contaminants and agents under study, in *2002 TLVs and BEIs*, American Conference of Governmental Industrial Hygienists, Cincinnati, 2003, p. 181.
157. Rao, C.Y., Burge, H.A., and Chang, J.C.S., Review of quantitative standards and guidelines for fungi in indoor air, *J. Air Waste Manage. Assoc.*, 46, 899, 1996.
158. Verhoeff, A.P. and Burge, H.A., Health risk assessment of fungi in home environments, *Ann. Allergy Asthma Immunol.*, 78, 544, 1997.
159. AIHA, Report of Microbial Growth Task Force, American Industrial Hygiene Association, Fairfax, 2001.
160. USEPA, Mold Remediation in Schools and Commercial Buildings, United States Environmental Protection Agency, Washington, DC, 2001.
161. CDC, State of the Science on Molds and Human Health, Statement for the record before the Subcommittees on Oversight and Investigations and Housing and Community Opportunity, Committee on Financial Services, United States House of Representatives, 2002.
162. Institute Of Medicine (IOM), *Damp Indoor Spaces and Health*, National Academy Press, Washington, 2004.
163. WHO, *WHO Guidelines for Biological Agents in the Indoor Environment*, World Health Organization, Geneva, 2003, *in press*.
164. Dillon, H.K., Miller, J.D., Sorenson, W.G., Douwes, J., and Jacobs, R.R., Review of methods applicable to the assessment of mold exposure to children, *Environ. Health Perspect.*, 107(Suppl. 3), 473, 1999.
165. Salyers, A.A. and Whitt, D.D., Introduction to infectious diseases, and the lung, a vital but vulnerable organ, in *Microbiology: Diversity, Disease, and the Environment*, Fitzgerald Science Press, Bethesda, 2001, pp. 289, 315.
166. Mims, C.A., Nash, A., and Stephen, J., Attachment to and entry of microorganisms into the body, in *Mim's Pathogenesis of Infectious Disease*, 5th ed., Academic Press, New York, 2001, chap. 2.
167. Peden, D.B., Setzer, R.W., Jr., and Devlin, R. B., Ozone exposure has both a priming effect on allergen-induced responses and an intrinsic inflammatory action in the nasal airways of perennially allergic asthmatics, *Am. J. Respir. Crit. Care Med.*, 151, 1336, 1995.
168. Lebowitz, M.D. and O'Rourke, M.K., The significance of air pollution in aerobiology, *Grana*, 30, 31, 1991.
169. Mastalerz, M., Glikson, M., and Simpson, R.W., Analysis of atmospheric particulate matter; application of optical and selected geochemical techniques, *Int. J. Coal Geol.*, 37, 143–153, 1998.
170. Chew, G.L., Burge, H.A., Dockery, D.W., Muilenberg, M.L., Weiss, S.T., and Gold, D.R., Limitations of a home characteristics questionnaire as a predictor of indoor allergen levels, *Am. J. Respir. Crit. Care Med.*, 157, 1536, 1998.
171. Samson, R.A., Flannigan, B., Flannigan, M.E., Verhoeff, A.P., Adan, O.C.G., and Hoekstra, E.C., Eds., Recommendations, and Media, in *Health Implications of Fungi in Indoor Environments*, Elsevier, New York, 1994, chap. 7, Appendix II.
172. Hoekstra, E.S., Samson, R.A., and Summerbell, R.C., Methods for the detection and isolation of fungi in the indoor environment, in *Introduction to Food- and Airborne Fungi*, 6th ed., Samson, R.A., Hoekstra, E.S., Frisvad, J.C., and Filtenborg, O., Eds., Centraalbureau voor Schimmelsultures, Utrecht, the Netherlands, 2000, chap. 3.
173. Ammann, H.M., IAQ and human toxicosis: empirical evidence and theory, in *Bioaerosols, Fungi and Mycotoxins: Health Effects, Assessment, Prevention and Control*, Johanning, E., Ed., Eastern New York Occupational and Environmental Health Center, Albany, NY, 1999, p. 84.
174. Schafer, M.P., Fernback, J.E., and Ernst, M.K., Detection and characterization of airborne *Mycobacterium tuberculosis* H37Ra particles, a surrogate for airborne pathogenic *M. tuberculosis*, *Aerosol Sci. Technol.*, 30, 161, 1999.
175. Henningson, E.W. and Ahlberg, M.S., Evaluation of microbiological aerosol samplers: a review, *J. Aerosol Sci.*, 25, 1459, 1994.

176. Griffiths, W.D., Stewart, I.W., Reading, A.R., and Futter, S.J., Effect of aerosolization, growth phase and residence time in spray and collection fluids on the culturability of cells and spores, *J. Aerosol Sci.*, 27, 803, 1996.
177. Eduard, W. and Heederik, D., Methods for quantitative assessments of airborne levels of noninfectious microorganisms in highly contaminated work environments, *Am. Ind. Hyg. Assoc. J.*, 59, 113, 1998.
178. Lighthart, B. and Tong, Y.Y., Measurements of total and culturable bacteria in the alfresco atmosphere using a wet-cyclone sampler, *Aerobiologia*, 14, 325, 1998.
179. Tong, Y.Y. and Lighthart, B., Diurnal distribution of total and culturable atmospheric bacteria at a rural site, *Aerosol Sci. Technol.*, 30, 246, 1999.
180. Cox, C. S. and Wathes, C. M., Eds., *Bioaerosols Handbook*, Lewis Publishers, Boca Raton, FL, 1995.
181. AIHA, *Field Guide for the Determination of Biological Contaminants in Environmental Samples*, American Industrial Hygiene Association, Fairfax, 1996.
182. Hurst, C.J., Ed., *Manual of Environmental Microbiology*, 2nd ed., American Society for Microbiology, Washington, 2001.
183. Macher, J.M. and Burge, H.A., Sampling biological aerosols, in *Air Sampling Instruments for Evaluation of Atmospheric Contaminants*, 9th ed., American Conference of Government Industrial Hygienists, Cincinnati, 2001, chap. 22.
184. Willeke, K. and Macher, J.M., Air sampling, in *Bioaerosols, Assessment and Control*, Macher, J.M., Ammann, H.M., Burge, H.A., Milton, D.K., and Morey, P.R., Eds., American Conference of Government Industrial Hygienists, Cincinnati, 1999, chap. 11.
185. Hinds, W.C., Filtration, Bioaerosols, in *Aerosol Technology: Properties, Behavior, and Measurement of Airborne Particles*, 2nd ed., John Wiley and Sons Inc., New York, 1999, chap. 9, p. 19.
186. Andersen, A.A., New sampler for the collection, sizing and enumeration of viable airborne particles, *J. Bacteriol.*, 76, 471, 1958.
187. Willeke, K., Lin, X., and Grinshpun, S.A., Improved aerosol collection by combined impaction and centrifugal motion, *Aerosol Sci. Technol.*, 28, 439, 1998.
188. Liu, B.Y.H. and Lee, K.W., Efficiency of membrane and nucleopore filters for submicrometer aerosols, *Environ. Sci. Technol.*, 10, 345, 1976.
189. Aizenberg, V., Bidinger, E., Grinshpun, S.A., *et al.*, Airflow and particle velocities near a personal aerosol sampler with a curved, porous aerosol sampling surface, *Aerosol Sci. Technol.*, 18, 247, 1998.
190. Li, C.S., Hao, M.L., Lin, W.H., Chang, C.W., and Wang, C.S., Evaluation of microbial samplers for bacterial microorganisms, *Aerosol Sci. Technol.*, 30, 100, 1999.
191. Lin, W.H. and Li, C.S., The effect of sampling time and flow rates on the bioefficiency of three fungal spore sampling methods, *Aerosol Sci. Technol.*, 28, 511, 1998.
192. Lin, W.H. and Li, C.S., Evaluation of impingement and filtration for yeast bioaerosol sampling, *Aerosol Sci. Technol.*, 30, 119, 1999.
193. Brosseau, L.M., Vesley, D., Rice, N., Goodell, K., Nellis, M., and Hairston, P., Differences in detected fluorescence among several bacterial species measured with a direct-reading particle sizer and fluorescence detector, *Aerosol Sci. Technol.*, 32, 545, 2000.
194. Ligler, F.S., Anderson, G.P., Davidson, P.T., *et al.*, Remote sensing using an airborne biosensor, *Environ. Sci. Technol*, 32, 2461, 1998.
195. Tovey, E.R., Taylor, D.J.M., Graham, A.H., O'Meara, T.J., Lovborg, U., Jones, A.S., and Sporik, R., New immunodiagnostic system, *Aerobiologica*, 16, 113, 2000.
196. Razmovski, V., O'Meara, T.J., Hjelmroos, M., Marks, G., and Tovey, E.R., Adhesive tapes as capture surfaces in Burkard sampling, *Grana*, 37, 305, 1998.
197. Razmovski, V., O'Meara, T.J., Taylor, D.J.M., and Tovey, E.R., A new method for simultaneous immunodetection and morphologic identification of individual sources of pollen allergens, *J. Allergy Clin. Immunol.*, 105, 725, 2000.
198. Benyon, F.H.L., Jones, A.S., Tovey, E.R., and Stone, G., Differentiation of allergenic fungal spores by image analysis, with application to aerobiological analysis, *Aerobiology*, 15, 211, 1999.

199. Jones, A., Hjelmroos-Koski, M., and Tovey, E., Image analysis can be used to identify airborne allergenic pollen, *J. Allergy Clin. Immunol.*, 103, S188, 1999.
200. Hjelmroos, M.K., Jones, A., and Tovey, E.R., Image analysis differentiates the airborne pollen grains of *Urtica* and *Parietaria*, *J. Allergy Clin. Immunol.*, 101, S132, 1998.
201. Bonton, P., Boucher, A., Thonnat, M., Tomczak, R., Hidalgo, P., Belmonte, J., and Galan, C., Colour image in 2D and 3D microscopy for the automation of pollen rate measurement. Proc. 8th European Congress for Stereology and Image Analysis (ECSIA), *Image Anal. Stereol.*, 20(Suppl. 1), 527, 2001.
202. Aronne, G., Cavuoto, D., and Eduardo, P., Classification and counting of fluorescent pollen using an image analysis system, *Biotech. Histochem.*, 76, 35, 2001.
203. Ronneberger, O., Heimann, U., Schultz, E., Dietze, V., Burkhardt, H., and Gehrig, R., Automated pollen recognition using gray scale invariants on 3D volume image data, Second European Symposium on Aerobiology, Vienna/Austria, Sept. 5 – 9, 2000.
204. Takahashi, Y., Nagoya, T., Watanabe, M., Inouye, S., Sakaguchi, M., and Katagiri, S., A new method of counting airborne Japanese cedar (*Cryptomeria japonica*) pollen allergens by immunoblotting, *Allergy*, 48, 94, 1993.
205. Emilson, A., Takahashi, Y., Svensson, A., Berggren, B., and Scheynius, A. Localization of the major allergen *Bet v* I in birch pollen by confocal laser scanning microscopy, *Grana*, 35, 199, 1996.
206. Grote, M., *In situ* localization of pollen allergens by immunogold electron microscopy: allergens at unexpected sites, *Int. Arch. Allergy Immunol.*, 118, 1, 1999.
207. Grote, M., Dolecek, C., Van Ree, R., and Valenta, R., Immunogold electron microscopic localization of timothy grass (*Phleum pratense*) pollen major allergens *Phl p* I and *Phl p* V after anhydrous fixation in acrolein vapor, *J. Histochem. Cytochem.*, 42, 427, 1994.
208. Grote, M., Vrtala, S., Niederberger, V., Valenta, R., and Reichelt, R., Expulsion of allergen-containing materials from hydrated rye grass (*Lolium perenne*) pollen revealed by using immunogold field emission scanning and transmission electron microscopy, *J. Allergy Clin. Immunol.*, 105, 1140, 2000.
209. Grote, M., Vrtala, S., Niederberger, V., Wiermann, R., Valenta, R., and Reichelt, R., Release of allergen bearing cytoplasm from hydrated pollen: a mechanism common to a variety of grass (Poaceae) species revealed by electron microscopy, *J. Allergy Clin. Immunol.*, 108, 109, 2001.
210. Grote, M., Stumvoll, S., Reichelt, R., Lindholm, J., and Valenta, R., Identification of an allergen related to *Phl p* 4, a major timothy grass pollen allergen, in pollens, vegetables, and fruits by immunogold electron microscopy, *Biol. Chem.*, 383, 1441, 2002.
211. El-Ghazaly, G., Takahashi, Y., Nilsson, S., Grafström, E., and Berggren, B., Orbicules in *Betula pendula* and their possible role in allergy, *Grana*, 34, 300, 1995.
212. Vinckier, S. and Smets, E., The potential role of orbicules as a vector of allergens, *Allergy*, 56, 1129, 2001.
213. Gerhardt, P., Ed., *Methods for General and Molecular Microbiology*, American Society for Microbiology, Washington, 1993.
214. Atlas, R.M., *Handbook of Microbiological Media*, 2nd ed., CRC Press, Boca Raton, FL, 1996.
215. Samson, R.A., Hoekstra, E.S., Frisvad, J.C., and Filtenborg, O., *Introduction to Food- and Airborne Fungi*, 6th ed., Centraalbureau voor Schimmelcultures, Utrecht, 2000.
216. Buttner, M.P., Willeke, K., and Gringshpun, S.A., Sampling and analysis of airborne microorganisms, in *Manual of Environmental Microbiology*, 2nd ed., Hurst, C. J., Ed., American Society for Microbiology, Washington, 2001, chap. 73.
217. NIOSH, *Aerobic Bacteria by GC-FAME*, Method 0801, DHHS (National Institute of Occupational Safety and Health) Publication 94–102, Cincinnati, 1998.
218. Marthi, B. and Lighthart, B., Effects of betaine on enumeration of airborne bacteria, *Appl. Environ. Microbiol.*, 56, 1286, 1990.
219. Marthi, B., Schaffer, B., Lighthart, B., and Ganio, L., Resuscitation effects of catalase on airborne bacteria, *Appl. Environ. Microbiol.*, 57, 2775, 1991.
220. Lange, J.L., Thorne, P.S., and Lynch, N., Application of flow cytometry and fluorescent *in situ* hybridization for assessment of exposures to airborne bacteria, *Appl. Environ. Microbiol.*, 63, 1557, 1997.

221. Mohr, A.J., Fate and transport of microorganisms in air, in *Manual of Environmental Microbiology*, 2nd ed., Hurst, C.J., Ed., American Society for Microbiology, Washington, 2001, chap. 74.
222. Marthi, B., Resuscitation of microbial aerosols, in *Atmospheric Microbial Aerosols — Theory and Applications*, Lighthart, B. and Mohr, A.J., Eds., Chapman & Hall, New York, 1994, chap. 7.
223. Setlow, P., Mechanisms which contribute to the long-term survival of spores of *Bacillus* species, *J. Appl. Bacteriol.*, 76, S49, 1994.
224. Brera, C., Caputi, R., Miraglia, M., Iavicoli, I., Salerno, A., and Carelli, G., Exposure assessment to mycotoxins in workplaces: aflatoxins and ochratoxin A occurrence in airborne dusts and human sera, *Microchem. J.*, 73, 167, 2002.
225. Andersson, M.A., Nikulin, M., Köljalg, U., Andersson, M.C., Rainey, F., Reijula, K., Hintikka, E.-L., and Salkinoja-Salonen, M., Bacteria, molds, and toxins in water-damaged building materials, *Appl. Environ. Microbiol.*, 63, 387, 1997.
226. Andersson, M.A., Mikkola, R., Kroppenstdt, R., Rainey, F.A., Peltola, J., Helin, J., Sivonen. K., and Salkinoja-Salonen, M.S., Mitochondrial toxin produced by *Streptomyces griseus* strains isolated from indoor environments is valinomycin, *Appl. Environ. Microbiol.*, 64, 4764, 1998.
227. Peltola, J., Andersson, M.A., Haahtela, T., Mussalo-Rauhamaa, H., Rainey, F.A., Kroppenstedt, R.M., Samson, R.A., and Salkinoja-Salonen, M.S., Toxic-metabolite-producing bacteria and fungus in an indoor environment, *Appl. Environ. Microbiol.*, 67, 3269, 2001.
228. ASM. *Manual of Clinical Laboratory Immunology*, 6th ed., Rose, N. R., Ed., ASM Press, Washington, 2002.
229. Constantine, N.T. and Lana, D.P., Immunoassays for the diagnosis of infectious diseases, in *Manual of Clinical Microbiology*, 8th ed., P.R. Murray, Ed., ASM Press, Washington, 2003, p. 218.
230. Chapin, K.C. and Murray, P.R., Principles of stains and media, in *Manual of Clinical Microbiology*, 8th ed., P. R. Murray, Ed., ASM Press, Washington, 2003, p. 257.
231. Hensel, A. and Petzold, K., Biological and biochemical analysis of bacteria and viruses, in *Bioaerosols Handbook*, Cox, C.S. and Wathes, C.M., Eds., Lewis Publishers Inc., 1995, p. 335.
232. Horner, W.E, Hebling, A., Salvaggio, J.E., and Lehrer, S.B., Fungal allergens, *Clin. Microbiol. Rev.*, 8, 161, 1995.
233. Horner, W.E. and Lehrer, S.B., Why are there still problems with fungal allergen extracts? in *Bioaerosols, Fungi and Mycotoxins: Health Effects, Assessment, Prevention and Control*, Johanning, E., Ed., Mount Sinai School of Medicine, New York, 1999, p. 313.
234. Gazenko, S.V., Reponen, T.A., Grinshpun, S.A., and Willeke, K., Analysis of airborne actinomycete spores with fluorogenic substrates, *Appl. Environ. Microbiol.*, 64, 4410, 1998.
235. Speight, S.E., Hallis, B.A., Bennett, A.M., and Benbough, J.E., Enzyme-linked immunosorbent assay for the detection of airborne microorganisms used in biotechnology, *J. Aerosol Sci.*, 28, 483, 1997.
236. Goodfellow, M. and O'Donnell, A.G., Eds., *Chemical Methods in Prokaryotic Systematics*, John Wiley & Sons, Chichester, U.K., 1994.
237. Spurny, K.R., Chemical analysis of bioaerosols, in *Bioaerosols Handbook*, Cox, C.S. and Wathes, C.M., Eds., Lewis Publishers Inc., 1995, chap. 12.
238. Saraf, A., Park, J.-H., Milton, D.K., and Larsson, L., Use of quadrupole GC–MS and ion-trap GC–MSMS for determining 3-hydroxy fatty acids in settled house dust: relation to endotoxin activity, *J. Environ. Monit.*, 2, 163, 1999.
239. Young, J.C., Microwave-assisted extraction method of the fungal metabolite ergosterol and total fatty acids, *J. Agric. Food Chem.*, 43, 2904, 1995.
240. Miller, J.D. and Young, J.C., The use of ergosterol to measure exposure to fungal propagules in indoor air, *Am. Ind. Hyg. Assoc. J.*, 58, 39, 1997.
241. Korpi, A., Pasanen, A.L., and Pasanen, P., Volatile compounds originating from mixed microbial cultures on building materials under various humidity conditions, *Appl. Environ. Microbiol.*, 64, 2914, 1998.
242. Ahearn, D.G., Crow, S.A., Simmons, R.B., Price, D.L., Mishra, S.K., and Pierson, D.L., Fungal colonization of air filters and insulation in a multi-story office building: production of volatile organics, *Curr. Microbiol.*, 35, 305, 1997.

243. Liu, W.-T. and Stahl, D.A., Molecular approaches for measurement of density, diversity, and phylogeny, in *Manual of Environmental Microbiology*, 2nd ed., Hurst, C.J., Ed., American Society for Microbiology, Washington, 2001, chap. 11.
244. Buttner, M.P., Cruz-Perez, P., and Stetzenbach, L.D., Enhanced detection of surface-associated bacteria in indoor environments by quantitative PCR, *Appl. Environ. Microbiol.*, 67, 2564, 2001.
245. Mukoda, T.J., Todd, L.A., and Sobsey, M.D., PCR and gene probes for detecting bioaerosols, *J. Aerosol Sci.*, 25, 1523, 1994.
246. Williams, R.H., Ward, E., and McCartney, H.A., Methods for integrated air sampling and DNA analysis for detection of airborne fungal spores, *Appl. Environ. Microbiol.*, 67, 2453, 2001.
247. Mastorides, S.M., Oehler, R.L., Greene, J.N., Sinnott, J.T., Kranik, M., and Sandin, R.L., The detection of airborne *Mycobacterium tuberculosis* using micropore membrane air sampling and polymerase chain reaction, *Chest*, 115, 19–25, 1999.
248. Schafer, M.P., Martinez, K.F., and Mathews, E.S., Rapid detection and determination of the aerodynamic size range of airborne mycobacteria associated with whirlpools, *Appl. Occup. Environ. Hyg.*, 18, 41, 2003.
249. Alvarez, A.J., Buttner, M.P., Toranzos, G.A., Dvorsky, E.A., Toro, A., Heikes, T.B., Mertikas-Pifer, L.E., and Stetzenbach, L.D., Use of solid-phase PCR for enhanced detection of airborne microorganisms, *Appl. Environ. Microbiol.*, 60, 374, 1994.
250. APHA, *Control of Communicable Diseases Manual*, 17th ed., Chin, J., Ed., American Public Health Association, Washington, 2000.
251. Larone, D.H., *Medically Important Fungi: A Guide to Identification*, 3rd ed., ASM Press, Washington, 1995.
252. Watanabe, T., *Pictorial Atlas of Soil and Seed Fungi, Morphologies of Cultured Fungi and Key to Species*, 2nd ed., CRC Press, Boca Raton, FL, 2002.
253. ASM. *Manual of Clinical Microbiology*, 8th ed., Murray, P. R., Ed., ASM Press, Washington, 2003.
254. Kitch, B. T., Chew, G., Burge, H. A. *et al.*, Socioeconomic predictors of high allergen levels in homes in the Greater Boston area, *Environ. Health Perspect.*, 108, 301, 2000.
255. Leaderer, B. P., Belanger, K., Triche, E. *et al.*, Dust mite, cockroach, cat, and dog allergen concentrations in homes of asthmatic children in the Northeastern United States: impact of socioeconomic factors and population density, *Environ. Health Perspect.*, 110, 419, 2002.

chapter fourteen

Radioactive aerosols

Lev S. Ruzer
Lawrence Berkeley National Laboratory

Contents

14.1 Historical overview 345
14.2 The Aerosol Laboratory of the All-Union Institute of Physico-Technical and Radiotechnical Institute (VNIIFTRI) in Moscow (former U.S.S.R.) 346
14.3 Reported uncertainties in the exposure of miners (some BEIR VI remarks) 346
14.4 Direct measurement of activity in the lungs: Problems with practical application 348
14.5 Personal experience 348
14.6 Geography and underground conditions of mine regions 350
14.7 Diversity of mining and working conditions 351
14.8 Direct method: The Tadjikistan study as an opportunity to reduce lung dosimetric uncertainty 352
14.9 Radioactive aerosols and lung irradiation 354
14.10 Aerosol concentration measurement 358
14.10.1 Radon and its contribution to absorbed dose 358
14.10.1.1 Experimental study on animals 358
14.10.1.2 Methods and measurement techniques for air radon concentration monitoring 361
14.10.1.3 Radon concentration distribution measurement 362
14.10.1.4 Radon and lung cancer 363
14.10.2 Measurement of the concentration of decay products of radon, thoran, and actinon 364
14.10.2.1 Characteristics of radon progeny 364
14.10.2.2 Basic equations for radon decay product series 366
14.10.2.3 General activity methods of measuring the concentration of radon decay products 370
14.10.2.4 Measurement of radon decay products in air by alpha- and beta-spectrometry 373
14.10.2.4.1 Measurement procedure and experimental results 374
14.10.2.5 Absorption of alpha-radiation in the sample 377
14.10.2.6 Measurement procedures for the determination of the activity of RaA, RaB, RaC, and RaC′ on the filter by alpha- and beta-spectrometry 381
14.10.2.6.1 ^{218}Po (RaA) activity measurement 381
14.10.2.6.2 ^{214}Po (RaC′) activity measurement 382

1-56670-611-4 / 05 / $0.00+$1.50

14.10.2.6.3 ^{214}Pb (RaC) activity measurement 382
14.10.2.6.4 ^{214}Bi (RaB) activity measurement 382
14.10.2.7 Other methods of determination of the radon decay products concentration in air 383
14.10.2.7.1 Radon progeny concentration measurement 385
14.10.2.7.2 Equilibrium factor and unattached fraction of radon progeny 386
14.10.2.8 Methodical errors in the RaA, RaB, and RaC concentrations measurement 387
14.10.2.9 Characteristics of thoron and actinon decay products 390
14.10.2.10 Basic equations for the thoron and actinon series 390
14.10.2.10.1 Actinon series 391
14.10.11 Unattached fraction measurements 395
14.10.11.1 Correlation between the unattached activity of radon decay products and aerosol concentration 395
14.10.11.2 Measurements of other radon decay product unattached activity concentrations 401
14.10.11.3 Effect of recoil nuclei being knocked off aerosol particle unattached concentrations of radon-decay products 402
14.10.12 Measurement of artificial radioactive aerosol concentration 406
14.10.12.1 Measurement technique for artificial aerosol concentration measurements 409
14.10.12.2 Artificial radioactive aerosol concentration measurement 416
14.10.13 Aerosol particle size measurements 417
14.10.13.1 Concept of the scale of particle size of aerosols....417
14.10.13.2 Ultrafine aerosols 420
14.10.13.3 Portable instrument for measuring UFA in mines 422
14.10.13.4 Installation for generating and investigating aerosols in the range of 2×10^{-3} to 1 μm 422
14.10.13.4.1 Method for investigating aerosols 427
14.10.13.4.2 Experimental results of the study of UFA aerosol generator 428
14.10.13.4.3 Measurement errors 430
14.10.13.5 Diffusive particle deposition in the inlet segment of a tube....431
14.10.13.6 Errors in determination of the parameters of the logarithmically normal size distribution of aerosol particles by the diffusion method 434
14.10.13.7 Fine aerosols 436
14.10.13.7.1 Determining the composition of aerosols by means of two mean radii 438
14.10.13.8 Standard for generating and measuring the electrical properties of aerosols 439
14.10.13.8.1 Generators of aeroions and electroaerosols....439
14.10.13.8.2 Method and instrumentation for the measurement of electrical parameters of radioactive UFA 440
14.11 Dosimetry 441
14.11.1 Intake vs. exposure: Propagation of the uncertainties in dose assessment in mining studies 441
14.11.1.1 Discussion on miner radiation dosimetry: Quantitative approach 442

14.11.2 Measurements of the dosimetric parameters 445
14.11.2.1 Radon measurements....................................445
14.11.2.2 Radon decay progeny measurements 447
14.11.2.3 Distribution of radon decay product concentration 448
14.11.2.4 Unattached fraction measurements 450
14.11.2.5 Breathing zone concentration measurements 450
14.11.2.6 Breathing volume rate and deposition coefficient measurements for miners 453
14.11.2.7 Assessment of the deposition and the upper-bound average breathing rates for miners 454
14.11.2.8 Assessment of the breathing rates of miners 456
14.11.2.9 Assessment of the dose (activity, intake) in the lungs of miners 459
14.11.2.10 Results 464
14.11.3 The method of direct measurement of activity (dose) in the lungs of miners 464
14.11.3.1 Theory of the method....................................465
14.11.3.2 Assessment of uncertainties in the evaluation of dose....................................467
14.11.3.3 Correction for the shift of equilibrium of radon progeny in the air and in the lungs 468
14.11.3.4 Accounting for parametric variations: Variations of concentrations, breathing rate, and deposition coefficients in real working conditions 472
14.11.3.5 Model measurement....................................474
14.11.3.6 Phantom measurements and geometric corrections 475
14.11.3.7 Assessment of the errors of the direct method 476
14.11.3.8 Portable instrument for direct measurement of the activity of radon decay products in the lungs of miners 480
14.11.3.9 Radon decay products as a radioactive marker in studying the deposition and dosimetry of nonradioactive aerosols 481
14.12 Radioactive aerosols epidemiology: Miners studies 483
14.12.1 Lung cancer mortality and lung sickness among nonuranium miners in Tadjikistan 483
14.12.1.1 Lung cancer mortality data....................................483
14.12.1.2 Lung sickness data 487
14.12.1.3 Comparison of Tadjikistan data with data from other epidemiological studies 490
14.12.2 Quality of dosimetry and the risk assessment for miners some aspects of the comparison of a "Joint Analysis of 11 Underground Miners Studies" and a study of nonuranium miners in Tadjikistan 490
References 493

14.1 Historical overview

The existing lung cancer risk estimates and guidelines for exposure to radon, both occupationally and in the home, are based on the cancer experience of miners in underground mines. Here, we explain how and why we arrived at the conclusion that it is very useful to discuss radioactive aerosols, as our experience in mines in Tadjikistan relates to the validity of the existing values of risk that are adopted by many countries.

The details of the work presented show the progress made to better evaluate the actual exposure of miners to radon and its decay products. Originally, much effort was made to measure and characterize the air concentrations in mines. Many of the characterization techniques were developed at the Aerosol Laboratory in Moscow. These are described in later sections. A method was developed to measure the radioactivity content of the miners' lungs directly, and it was shown that measured air concentrations alone did not accurately describe a miner's exposure. Ultimately, the direct measurements in the lungs could be correlated with health effects, and this is the fundamental reason for this study.

14.2 The Aerosol Laboratory of the All-Union Institute of Physico-Technical and Radiotechnical Institute (VNIIFTRI) in Moscow (former U.S.S.R.)

I am a nuclear physicist by training. For most of my professional life in the former Soviet Union, nearly 20 years, I worked as the Chairman of the Laboratory of Aerosols in one of the main Soviet Metrological Centers in Moscow, the VNIIFTRI.

The primary goal of the laboratory was to develop the metrological basis for the measurement of aerosol parameters, both radioactive and nonradioactive. As a result of this work, lasting approximately 15 years, the State Standard on Aerosols, which consisted of a facility for the generation of stable aerosols and precision measurement techniques for measuring its parameters, was developed and officially approved by the Committee of Standards of the U.S.S.R. This facility allowed highly accurate measurements of

- The concentration of radioactive aerosols, both natural and artificial
- The particle size distribution in all particle size ranges
- The electrical parameters of aerosol particles

A secondary goal of this laboratory was to test the methods developed in the laboratory for measurement of the concentration of radon decay products in mine atmospheres and to develop a direct measurement of the activity (dose) levels in the lungs of miners under actual mining conditions.

The metrological part of the work demanded a careful consideration and quantitative assessment of every source of error (uncertainty) and methods to calculate the total error. By definition, the error (uncertainty) should be a small fraction of the value of interest itself.

14.3 Reported uncertainties in the exposure of miners (some BEIR VI remarks)

If we follow the typical path of metrology, that is, carefully assess every uncertainty in each step from the air concentration measured using standard procedures to the activity (dose) of individuals (or groups) of miners, we find that the total uncertainty in this case will be in hundreds of percent. This fact makes the data on dosimetry, and consequently on the risk assessment, open to question. Because the error cannot exceed 100% in the negative direction, it is better to say that the true value (of concentration, and consequently exposure) can be an order (or even more) of magnitude greater or smaller than the current estimated value. A broad discussion of some (not all) sources of uncertainties in dosimetry is presented in BEIR VI (NAS 1999).

We were not surprised that the accuracy in the measurement of radon decay product concentrations in mines was in tens of percent (40–50%), because in mines the instrumentation must be portable and easy to operate. Such poor accuracy is necessarily acceptable under some field conditions.

What was not acceptable from a dosimetric point of view was the fact that the air concentration of radon decay products (or exposure) measured by a standard procedure, especially in real mining conditions, was not the physical parameter that is directly responsible for the biological effect, that is, lung cancer.

Moreover, no serious attempt was made to study the correlation between radon and its decay product measurement in real mining conditions, and the damaging biological factor, the activity or dose to the lungs of miners.

According to standard metrological terminology, the uncertainty estimate in such cases is called the propagation of the uncertainties. From a metrological point of view, the worst scenario in the assessment of the physical value is not when the error (uncertainty) is large, because very often by analyzing the sources of the errors we can find a way of introducing the corresponding corrections. The worst scenario is when the analysis shows that the uncertainty is itself uncertain. This is exactly what takes place in the dosimetry of miners.

The (NRC 1999) report "Health Effect of Exposure to Radon" (BEIR VI, 1999) discussed this problem briefly as a heterogeneity of the exposure–response trends among the various miner studies. That is, there were very different results from the 11 underground mining cohorts studied to determine the relationship between radon decay product exposure and lung cancer.

Still, it seems that the assessment of miner dosimetry appears more qualitative than quantitative because no serious attempt to simply follow the rules of metrology was practiced.

Another question arises. Are there different approaches, that is, methods in improving this situation, to diminish partial and consequently total uncertainty? It seemed to us that such an opportunity to at least diminish the uncertainty existed in using a direct measurement on miners' lungs, and this technique was subsequently proved in measurements on miners in Tadjikistan. Most of the results of these studies were not published, and exist only in a few dissertations and my *Radioactive Aerosols* (2001, in Russian).

We had access to the health effects data in these same miners where lung measurements were made. We decided it could clarify the risk in mines from radon and its decay product if we carried out a study on the health effects as a function of radon and especially the detailed decay product exposure.

Some (if not most) of the studies on exposure–effect results are presented without error bars on both vertical and horizontal axes, that is, without any serious attempt to assess the reliability of the dosimetric and consequently risk data, as well as the death data.

Decreasing the statistical uncertainty in the biological effect (lung cancer mortality in the case of miners) is usually achieved by choosing a substantial number of miners with reported exposure histories. The uncertainties in exposure in most of the studies are not mentioned at all. This is the irony: because the uncertainty is so large and undetermined, we simply ignore it.

If someone suggested that in grouping many mines, averaging of the various exposure factors takes place, and that there is therefore no problem in risk assessment, we can respond: maybe. But such a point of view should be validated by statistical modeling or on-site experimentation in the mines. We should also remember that miner data are the only source of quantitative risk assessment in the epidemiology of radon and its decay product.

Our measurements in mines, and especially the direct measurement of radon decay product activity in the lungs of miners, suggest that there is no correlation between the data on measured concentration as is measured routinely and the activity (dose) in the lungs of miners.

The significance of the uncertainty in risk estimates in this case pertains not only to the abstract metrological factors but also to the economic factors. Risk estimates are directly related to the permissible concentration in mines and therefore to the atmospheric controls, ventilation rate, predicted health effects, etc.

On the other hand, it will be shown through these studies that the radon decay product concentration, together with the work load, directly correlated with the lung sickness of miners, which in turn correlated with the economic factors.

14.4 Direct measurement of activity in the lungs: Problems with practical application

When I studied dosimetry, the idea and the resulting analyses to correlate the absorbed dose from alpha-emitting radon decay products with the external body counting of gamma-emitting decay products arose.

The importance of such a correlation was obvious. Alpha-radioactivity and the absorbed dose from radon progeny in the lungs, the physical values directly responsible for the health effect due to alpha-particles, cannot be measured directly. Alpha-particles have relatively high energy and a very short range in tissue and thus can cause significant damage to cellular DNA in the lungs because of this high linear energy transfer (LET).

On the other hand, the gamma-activity from two of the radon decay products in this decay chain is measurable, and therefore this correlation presented an opportunity to assess the absorbed dose *in vivo*, at least in principle.

Model measurements were initiated using a human phantom and human body spectrometer in order for quantitative external measurement of the radon progeny gamma-activity to be performed. The amounts of radioactivity used in the phantom approximated the level that corresponds to that in the lung from breathing the average air concentration in the mine atmosphere. Preliminary measurements confirmed that such measurements are possible.

After the direct method was officially approved by the Committee of Radiology of the Ministry of Health of the U.S.S.R., the following statements were made:

1. The proposed method of estimating radiation effects from radon decay products contained in the lung is in contrast to the generally applied method of indirect estimation of the effect based on the measurement of concentration of radioactive aerosol in the atmosphere.
2. The proposed method is original.
3. There is no doubt that it is possible to measure the intensity of the gamma-radiation of radioactive aerosols deposited in the lungs of miners using high-sensitivity equipment designed to estimate low concentrations of radioactive substances in human organisms.

The next logical step was testing the method under actual working conditions, that is, making direct measurements on uranium miners. However, for me personally, this was to be a very difficult task.

14.5 Personal experience

In 1948, I graduated from the University of Moscow with a degree in nuclear physics. It was a time of rapid development of work on the Atomic Weapons Programs and an extensive number of scientific institutes were involved in this program. At the same time, there was the start of anti-Western, especially anti-American and anti-Semitic, propaganda in science, literature, and art in all walks of Soviet life.

Looking back 50 years, we can say that it was natural for these circumstances to be connected. The Soviet Union was a dictatorial state by definition and therefore had zero tolerance for people connected in some way to other countries, especially beyond the "Iron Curtain." It was well known that the majority of Soviet Jews, in some way or another, had relatives and close friends in the United States and only recently had established the new state of Israel. I also had two cousins in the United States with whom I was in correspondence.

It was not a surprise for Soviet Citizens that none other than KGB chief Beriya was appointed as Chairman of all Atomic Programs. Suspicion and spying were rampant everywhere. It was not a "Gulag", as described by Solzhenitzin, that is, a prison or concentration camp surrounded by walls in Siberia or the Far East. It was just everyday life, without material walls but under constant surveillance. Even when I was a student of the fourth and fifth year at the faculty of nuclear physics, I felt it. We lived in the university dormitory. There were six of us in one room and we knew who was the spy. We called him "stukach," which translates as a "knocker" or "informer."

There were many Jews among the older generation of nuclear physicists working on the Atomic Program. But for the young, recent graduates it was extremely difficult to enter the atomic circle. I even passed a special mathematical test in the Department of Theoretical Physics at the Institute of Soviet Academy of Sciences, which had only recently started work on the Atomic Weapons Program. Without any explanation and despite a plea from the chairman of the department, academician Zeldovich, I was not accepted at the institute. It became clear that I would not find a job as a nuclear physicist.

After many months of searching, I was finally enlisted in an institute that was impoverished and quite far from the atomic circle. I worked quietly for 1 year. But, again, I was unlucky. This institute was included in the Soviet Space Program and all employees were placed on the "secret list." Consequently, I was called into the local KGB headquarters and interrogated. Because I tried to conceal the fact that my father was shot in the 1930s without any trial on official explanation, the KGB chief shouted and threatened me saying that they knew all about my father. The chief said that only if I would collaborate with the "organs" (KGB) could I continue work as an institute employee. I pretended that I did not understand what the chief was talking about, indirectly refusing to work as "stukach" among my co-workers. The next day I was discharged.

It now became clear that I could not work as a nuclear physicist — more importantly, I could not work as a scientist at all. I was a Jew, my father was shot, and I had relatives in the U.S.A. With these three strikes, I was out. With great difficulty, I finally found a job as a teacher in a Moscow middle school.

I worked as a teacher for 8 years until Stalin died and Khrustchev's "thaw" began. However, the memory and fear of the KGB interrogation were alive inside of me throughout those years.

In the Medical Institute, I begin scientific work on the irradiation of animals using radon and its decay products. I defended for candidate dissertation (equivalent to a Ph.D.) in 1961 and then worked for 20 years as the Chairman of the Aerosol Laboratory at VNIIFTRI. There I defended my doctor's dissertation (no analogy in the U.S.A.) and was the scientific supervisor of the aspirants (postgraduate students).

In 1968, I published my book *Radioactive Aerosols* (in Russian), where new ideas about the measurement of radioactive aerosols in the air and in the lungs of miners were presented.

It was a time when the importance of the development of the State Standard for different physicotechnical parameters was very well understood by government officials. In order to put this scientific and technical problem in a political perspective, the Chairman of the Committee of Standards proclaimed the slogan:

Without State Standards we cannot achieve Communism!

It was not an analogy of the Aerosol Standard of the world, so from a scientific point of view the development of such a standard was very interesting and a challenging task.

On the other hand, it was even dangerous not to work on this problem due to political implications. So, we worked very hard in order to achieve this goal and, finally, the Aerosol State Standard was officially approved by the Special Government Commission in 1978.

Unfortunately, for me personally, the next year I was discharged from my position as Chairman of the Aerosol Laboratory, spent 8 years without a job, and finally arrived in the United States: in the midst of capitalism.

In terms of the second part of the problem, the dose measurement in the lung of the miners' situation was also not so simple at that time.

Now, fifteen years later (in the 1960s), I was afraid to have contacts with the KGB in order to get clearance and permission to make measurements in uranium mines. Fortunately, this time I was luckier. I received a call from a young scientist who worked for the sanitary–epidemiological station in the capital of Tadjikistan, Dushanbe. This scientist worked as a dosimetrist in the measurement of radon and its decay products in nonuranium (not secret) mines in South and North Tadjikistan and he asked me for a topic for his dissertation. He wanted to be a postgraduate student by correspondence. As a topic, I proposed a test of the method of direct measurement of gamma-activity from radon decay products in the lungs of miners. During conversations with him it became clear that radon decay product concentrations in these nonuranium mines were, in most cases, close to the annual permissible concentration (APC) of 3.9×10^4 MeV/l (0.3 WL), and in some cases even higher.

When I first proposed the dissertation topic, I saw disappointment in the eyes of this young man, who felt that it was too far-fetched and not achievable in mining conditions. But the young man said, "OK, I will try." After a while, I received a telegram from him saying: "The method is working now. I will make a portable instrument in order to make these measurements systematically." A little later, a radiologist epidemiologist from the same station joined us in this study, which resulted in a more than 10+ year dosimetric–epidemiological study of 2400 miners from 98 nonuranium mines in Tadjikistan. A by-product of the results of these first measurements is that they became known to colleagues from other institutes, who worked on the problem of radon measurements in uranium mines.

The representative (official) from the secret ministry ("Ministry of the Middle Machinery") visited the laboratory and expressed interest in the preliminary results of the direct measurements. I immediately obtained clearance and the opportunity for me and other members of the laboratory to make similar measurements in uranium mines.

14.6 Geography and underground conditions of mine regions

The mining departments in which the study was conducted were located in the territory of the former Soviet Republic of Tadjikistan and also partly in the former Soviet Republic of Uzbekistan (Tashkent province).

Tadjikistan is a mountainous country, with 93% of its territory being occupied by mountains with narrow valleys in between. The main part of Tadjikistan belongs to the Tjan-Shan mountain system. The southeastern part of the territory is located in the Pamir mountain system. All mining departments of the Republic were located on the southern spurs of Tjan-Shan and so were the mines of Uzbekistan. Only a few mines (six prospecting mines) were located in the Pamir mountain system.

Deposits of nonferrous and rare metals (lead, zinc, bismuth, tungsten, molybdenum, etc.) are concentrated mainly in North Tadjikistan. Almost all of them belonged to the

Karamazar mining region. In Central Tadjikistan, deposits of Hg, W, and fluorspar were located (Takob region) and some of the uranium mines were located in North Tadjikistan.

14.7 *Diversity of mining and working conditions*

The complexity of the terrain causes great diversity in the climate of underground working sites. North of the Republic, mines were located at an altitude of 1–2 km above sea level. The summer weather is hot and dry with temperatures between 25 and 28°C. Mines in Central Tadjikistan had average temperatures of +12 and −12 in July and January, respectively (altitude 2100–2600 m). Some mines, primarily involved in geology and prospecting, are located above 3 km and activity in this region takes place only in the summer because daily temperatures are less than 15 during the day and less than 0 at night. In this zone of permafrost, the walls and roofs of the pit were covered with ice, which formed a solid monolith with the rock.

The mining departments in this area represented either the industrial complex, consisting of the mining section with a concentration factory or simply a mining section, consisting of the mines, which were independent in the sense of production and developed separate mining fields. Thus, a mine represents a distinctive "organism" with distinctive characteristics of ventilation, geological conditions, and production functions.

The mining industry, especially gold and silver mining, was primitive in this region. We were told that these were open pit mines made by the soldiers of Alexander the Great. Skeletons in the vertical position still exist in some mines.

The results of the primary and secondary studies in mines, together with measures for improving the working condition, were presented in the dissertations of A.D. Alterman and S.A. Urusov. In all mines, only one type of technological working plan (exploitation) was used: boring (drilling) of the blast hole (bore hole) with a pneumatic boring instrument (types PR-30k, TR-45) followed by blasting, ventilation of the ore face, loading the ore mass in the trunk, and exporting (carting) by electric locomotive to the vertical shaft or mouth (estuary) of the mining gallery. The most labor-intensive activities were preparation and timbering (fastening) work, which required substantial physical pressure, especially in a drift (a rise in the mine shaft).

All work is done by the "brigade" method. All members of a "brigade" (team) take part in the management of any emergency situation (derailing of the trucks, water breaks in the main line, etc.) and other hard work.

The important parameters for working conditions are:

- Time distribution of work with known atmospheric decay product characteristics
- Work load, which determines the breathing rate

For many job categories, it is difficult to determine the specific work site (locksmiths, electricians, mine surveyors, samplers, electric locomotive engineers, mining engineers, etc.). For them a "working site" can be all of the working sites, including "nonworking sites." A much more constant working place exists for the drillers, despite the fact that in specific situations their time present at the pit face can vary within broad limits. According to the time sheets provided on six shaft sinkers (drifters) during two shifts, the working time varied from 8 to 92% of the shift time. Some of the miners were constantly situated at the same working site.

Most of the mining industry in this region used artificial ventilation. In the period of the primary study, only three mines did not have artificial ventilation. Blind (cul-de-sac) mine shafts were ventilated by ventilators (types VAM-450, BM-200). In most cases, suction-type ventilation was used together with a dispersed set of ventilators in the pipelines. A substantial disadvantage to this method is the possibility of contamination (pollution) of the clean air entering the mine working with radon and also the possibility of suctioning polluted air from nonworking sites.

In general, numerous nonworking sites connected with working sites created a negative influence on ventilation and consequently on radon and its decay product concentration in the working sites, increasing it. Due to the noise, often miners turn off ventilation during the shift and mine working takes place without air replacement. The worst air conditions, therefore, took place in blind shafts, particularly in shaft sinker (drifter) working sites. In cleaning sites, due to trough ventilation, air exchange was constant.

In the nonworking sites, air usually stagnates. These nonworking sites are visited from time to time by geologists, mine surveyors, and others, who carry out different types of work, such as dismantling, taking down the rail, pipes, cables, etc.

The main tool in the fight with dust, besides ventilation, was wet drilling. Even in mines where the average dust concentration did not exceed the APC standard level in mine shafts, the dust concentration was substantially higher and therefore, in terms of dust, the worst situation was for ore face-workers (getters).

Especially difficult working conditions arose for the geology-prospecting groups due to:

- Remoteness (distance) of this type of work from main transport arterial roads and industrial centers
- Difficulties in energy supply
- Difficulties in alpine (mountain), arid (waterless), deserted, permafrost conditions
- Difficulties with maintenance equipment
- Absence of ventilation and mountain life-saving services

All these situations existed in Tadjikistan. The prospecting mine shaft is blind, that is, it has only one exit to the surface, which makes the ventilation of such working sites very difficult. In addition, a cross-section of such sites is usually less than 5 m^2, making it impossible to use ventilation pipes with large diameters. A low cross-section also makes it impossible to use loading mechanisms (gearing), which resulted in most work being manual with a very high physical work load. All these circumstances came to our attention after we became closely acquainted with mine working conditions.

It became clear that mining conditions are very diverse and, therefore, too complicated to use one or two measurements per month of radon and its decay product concentration (or exposure) as a characteristic of the dose of individuals or a group of miners.

14.8 Direct method: The Tadjikistan study as an opportunity to reduce lung dosimetric uncertainty

Our first underground experience was a very dramatic one. We arrived in the capital of Tadjikistan, Dushanbe, after an 8-h flight from Moscow, late in the evening. The next day, early in the morning, we went to the mine located outside the city.

In order to obtain the best results, we asked the mining administration to show us places with the highest concentrations. At first, our guide led us to the wrong site and it was only by chance that we did not perish. We arrived at the site, where the concentrations were 10^4–10^5 times higher than permissible. We wore respirators, but I was so tired after the flight and the almost sleepless night that I had difficulty breathing, and so took off the respirator. Radon decay products were not only in our lungs, they were on our faces, clothes, everywhere.

We immediately went to another place where the concentration was closer to the average levels, and continued measuring the radon decay products in our lungs. After this experience, we continued to make direct lung measurements in metal mines in North Tadjikistan (Leninabad, now the Khodgennt region). The measurements in uranium mines were performed in the mines of the Uranium Industrial Complex, located in the city of Chkalovsk near the Sir-Darjinsky reservoir. The mines of this complex were located on the huge territory of three former Soviet Republics (all now independent states) — Tadjikistan, Uzbekistan, and Kirgizstan. For many years, together with the highly qualified and enthusiastic personnel of the local dosimetric laboratory, we made measurements in the mines of the Yangiabad region. Then, similar measurements were performed in North Kazakhstan.

It was rumored among the Soviet population that prisoners were working as miners in uranium mines. People called them "smertniki" (prisoners sentenced to death). However, we did not see any prisoners underground. Most of the uranium miners were recruited from other parts of the former Soviet Union in order to make money. They became worried when we measured the radioactivity in their lungs instead of routine air measurements. To hide their emotions, they jokingly said that the goal of our measurements was to check if they were faithful to their wives. From this, we understood that these types of measurements have a special psychological impact on both miners and the mining administration.

At one meeting at the Industrial Mining Complex, the minister of this former secret ministry ("Middle Machinery") told an audience of miners: "You should not worry about your health, because radon is going in and out of your body." But suddenly a voice from the hall loudly said: "But the decay products remain in the lung."

Our measurements directly proved that radon decay products are in miners' lungs in measurable amounts, and this made an impression on the mining administration. As will be shown later, the practical application of this method inaugurated some improvements in working conditions, which resulted in declines in the lung sickness of miners.

As far as prisoners in the Soviet mines were concerned, we saw thousands of prisoners in one mine of northern Kazakhstan; they worked as builders and decorators on the surface, but not underground. Because most of them were sentenced for 15–25 years, they became good professionals and earned money. We met one couple working as freelancers at this prison camp: he worked as a hair dresser and she as a prostitute among the prisoners. They made a lot of money. They told us that they planned to stop their business soon, buy a house in a good place, and begin a new life. It was a wild life for us — the Wild East!

It should also be mentioned that when the dosimetric data in miner studies are used, the problem of objectivity (lack of bias) should be taken into account. We will discuss this problem later in connection with the Polish study. From our own experience, we know that each time we asked for official approval of our measurements (which was mandatory), the chief engineer would very politely and cunningly exclude the results that exceeded permissible levels. With a smile, he hinted that it would be difficult for us to come again if such "incorrect" data were reported.

Those years until the middle of the 1970s were very productive, especially in measurements in Tadjikistan. Together with the measurements of the distribution of radon and radon decay products for different working sites, groups of workers, and the direct measurement of activity in the lungs of miners, a study of lung cancer mortality and lung sickness was provided. Unfortunately, after the conclusion of the study in Tadjikistan, it was impossible to continue similar measurements in other mines, including in the Far East.

After the middle of the 1970s, direct measurements were not provided. The first reason was because the laboratory was in preparation for the official approval of the Aerosol Standard.

The second reason was, again, political for me personally. It was a time of strong dissident activity, in which my children took part. I was again interrogated by the institute's KGB authority and finally discharged in 1979 without any official explanation. However, obviously this time it was because of my son, not my father. I call this story: For Father and for Son.

After arriving in the United States, I received an invitation from a publisher in Washington, DC to prepare a monograph on Aerosol Research in the U.S.S.R.

When I left Moscow, I was afraid to take some of my papers, dissertations, and other materials. No one knew what the KGB would demand from me at the airport.

Based on what was available for me, I prepared a monograph in 1989 entitled "Aerosol R&D in the Soviet Union." The book was published mostly for government organizations. After a couple of years, I decided to try and find this book on the Internet. The exactly same title could not be found, but there were two monographs with the closest topic. One was "The rise and the fall of the Soviet Union" and the other was "closest" "Why the Soviet Union collapsed." When I had told this story to my American colleagues, they said: "In the past we thought that the Soviet Union collapsed because of the policy of President Reagan and because Gorbachev was such a great democrat. Now we know that it happened because of your aerosols."

I attended the Aerosol Conference in Reno in 1989 and asked the speaker at the plenary session Dr. Naomi Harley about the quality of mining dosimetric data. From her response, I understood that the data of our Tadjikistan miners' study could present useful information to this problem.

My last book *Radioactive Aerosols* was published in Russia in 2001. This book is different from previous publications, because it includes materials on developing Aerosol Standards, dose measurements in the lungs of miners, and other materials on radioactive aerosols published together with American colleagues.

14.9 *Radioactive aerosols and lung irradiation*

The incidence of elevated rates of lung cancer among miners exposed to radon decay products has led to programs that control worker exposure. Furthermore, such evidence is the basis for concern about lung cancer risk among the general public due to radon in homes, which can lead to a substantial number of putative lung cancer deaths.

In conducting epidemiological studies or in controlling worker or general-public risks, human exposures to the decay products must represent or estimate, in some way, time-integrated airborne concentrations of the relevant radionuclides ("exposure") or calculated energy deposition in the lung tissues affected ("absorbed dose"). Intermediate quantities that may also be used are the amount of activity deposited in the whole lung or relevant part of the lung ("deposited activity") or intake, which was officially included together with concentration in the Norms of Radiation safety (ICRP, 1979). The use of any of these quantities in either epidemiological studies or control programs involves, either explicitly or implicitly, presumptions about their relationships with one another. However, although it is the dose that is the cause of changes in tissue that lead to cancer, it is airborne concentration that is measured and corresponding estimates of exposure that are used as correlates or predictors of risk.

The absorbed dose or the activity (intake) in lung tissue is calculated to understand the actual insult leading to lung cancer, or to permit the quantitative comparison of doses arising from atmospheres having different characteristics such as in mines and homes (BEIR VI, 1998). If exposure is used as an index of dose (activity), for example, in epidemiological studies, presumptions are made implicitly about a comparability or constancy of breathing rates and of the retention of decay products in the lung.

Since both dosimetric and epidemiological data for miners are the main source of a risk cancer mortality assessment for miners and the general population, it is important to understand to what degree correlation between the measured radon decay products concentration in the air and dose to the lung takes place in the real mining environment.

In this connection, it should be noted that, especially in mining environments, inhalation rates depend on the groups involved and on working conditions, specifically the physical load, which can vary substantially. Retention of the decay products depends on the properties of the aerosols and also on the load of work and is typically not known accurately.

It should be pointed out that measurements of the airborne concentrations are not complete enough to provide directly the value of exposure, because the concentrations to which individuals are exposed vary substantially in time and space. For example, variation of the ventilation rate even for a short period of time can lead to a substantial change in concentration. Moreover, the concentrations of radon decay products and other nuclides in the breathing zone may differ substantially from the value measured by the actually implied standard instrument (Schulte, 1967; Domanski et al (1989). Finally, the very concept of "workplace," associated concentration, and load of work are indefinite, since miners are typically at a number of places during their working day, with variable concentrations (as well as nature of work). Since the measurements of concentrations in mines may be performed only once or twice a month (not to mention estimated retrospectively), we cannot expect reliable correspondence between exposures estimated from measurement results and actual personal exposures, that is, time-integrated concentrations. Thus, the use of measurements of airborne concentration as a basis for estimating or comparing dose can lead to substantial errors, both because of lack of correspondence to concentration measurements and personal exposures and because of uncertainty and variability in breathing rate and radionuclide retention. These errors, in aggregate, may be as much as an order of magnitude, and therefore make dosimetric and consequently risk assessment data unreliable (Ruzer, 1970; Lekhtmakher and Ruzer, 1975).

It should also be mentioned that the dose to the lung from radon itself is negligible in comparison with that from its decay products. However, until now assessment of the risk in many studies is based on the measurement of radon with an assumption on the degree of equilibrium between radon and radon progeny in mines. The degree of equilibrium varies substantially in time and space, both for underground workers and the general population.

Unfortunately, even after concentration (exposure) of decay products (not of radon) was established as a measure of the irradiation of the lung more than 40 yr ago, no discussion and experimental study in mines were conducted to determine the degree of correlation between the measured concentration and the actual dose (intake) for individuals and a group of miners.

To assess the reliability of dosimetric data for miners, we should carefully consider every factor related to dose (activity) calculated or measured together with assessment of the uncertainties for every partial factor and the total error (uncertainty).

It seems that a lack of such careful consideration of the metrological aspects in the assessment of the dose to the lungs of miners from radon decay products in the past resulted in large uncertainty in the dose to the lung, and correspondingly unreliable data in estimating the risk of lung cancer.

It is obvious that because irradiation of the lung by radioactive aerosols, and particularly radon decay products, depends on physical activity, there are some biological factors that will contribute to the total uncertainty in the dose assessment. In order to simplify the task, we will put aside, at least for now, the biological factors and focus only on the physical factors, especially because the uncertainty in physical factors is very large in itself.

Let us call all these factors that contribute to the release of energy to the lung tissue as "dosimetric factors." There are two groups of dosimetric factors. The first group — radioactive dosimetric factors — determines the total energy of radiation in the air or in the lung. The second — nonradioactive group — is mainly responsible for the portion of this energy deposited in the lung.

The following are the factors in the first group — radioactive parameters:

1. Concentration measured by standard procedure.
2. Breathing zone concentration.
3. Concentration of unattached radon decay products.
4. Exposure.
5. Activity in the lung.
6. Energy of the alpha and beta particles.
7. Integral absorbed dose.
8. Absorbed dose.

The parameters of the second nonradioactive group are:

1. Counting, surface area, and weighted aerosol concentration.
2. Aerosol particle size distribution.
3. Breathing rate (minute volume).
4. Deposition coefficient.
5. Filtration ability of the lung (FAL) — combination of the volume breathing rate and deposition coefficient.
6. Efficiency in using respirators (when applicable).
7. Parameters of biokinetic processes.

Besides all these factors, mention of the "work itinerary" ("scenario of exposure") is also a necessary element in the correct dose assessment.

In the NRC Report (BEIR VI, 1998), the structure of the lung inner surface layers from the point of view of alpha-dosimetry was studied.

A diagrammatic representation of the respiratory tract region in humans is shown in Figure 14.1.

It is well known that the dose to lung tissue from the inhalation of radon progeny cannot be measured. It must be calculated by modeling the sequence of events involved in inhalation, deposition, clearance, and decay of radon progeny within the respiratory airways.

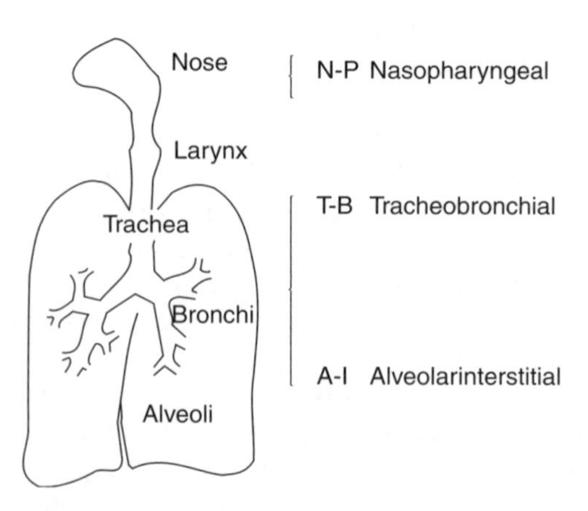

Figure 14.1 Regions of the respiratory tract.

We must discuss this statement because of its great importance to the dosimetry of radon decay products.

In the first step of this consideration, that is, the correlation of the concentration in the breathing zone and that in the place of sampling, it is very difficult to predict how some of the parameters of the first and second group will interact. It is difficult, or even impossible, to model the concentration in the breathing zone, especially for hard-working miners. Even the sparse data available showed that the ratio between the breathing zone concentration and that measured at some distance varies within very broad limits. This is especially true for aerosol particle size distribution and activity particle size distribution, which are responsible for the deposition in different parts of the lungs.

The second step of the distortion of particle size distribution spectra is the change of particle size distribution inside the lung pathways due to change in humidity and temperature. The modeling of this step can lead to very high uncertainty in the calculation of particle activity deposition.

It is true that the physical effect of alpha-radiation of the lung tissue cannot be measured directly. But very often, when the direct measurements of the physical factors are not possible, a different approach is used. In such cases, we often try to find other measurable physical parameters, and at the same time use a correlation (analytically or in another way) between this measurable and nonmeasurable physical value. One method is external gamma-ray counting of the lung.

Such analytical correlation of the gamma-activity of radon progeny (a measurable factor) and the alpha-dose to the lung tissue (nonmeasurable factor) was derived in (Ruzer, 1958, 1964). The possibilities of practical application of this "direct method," including an introduction of the necessary corrections and assessment of the accuracy of measurement, were presented in Ruzer (1962, 1968), Vasin et al. (1975); Ruzer and Urusov (1969) and Urusov (1972).

The practical applications of the direct measurement of the activity in the lungs of miners were carried out in uranium and nonuranium mines of the republics of Kazakhstan, Uzbekistan, and Tadjikistan on more than 500 miners. The results of these studies were published in three dissertations (Ruzer, 1970; Urusov, 1972; Alterman, 1974)(in Russian) and in my book *Radioactive Aerosols* published in Russia in 2001.

The radon decay products of gamma-radiation of ^{214}Pb and ^{214}Bi are natural markers, and the relatively high historic radon progeny concentration in mines makes it possible to obtain information from direct measurements of the gamma-activity of radon decay products deposited in the lungs of miners.

In principle, using more sophisticated gamma-detectors together with collimators, it is possible to measure the distribution of radon progeny in different parts of the lungs under real mining conditions.

A detailed analysis of all corrections in such measurements suggested that the accuracy in the activity assessment is satisfactory from the point of view of practical dosimetry.

In all calculations of the dose to the lung tissue provided in NRC Report (1998), it is assumed that activity of the radon decay products in the lung or in part of it is known, without taking into account uncertainties in the activity assessment.

However, here lies the main problem. There are no real data on the activity in the lung. From a practical point of view, it makes no sense to use data concerning the activity of radon progeny without mention of the errors of the calculation of the dose distribution through lung tissue, especially because the errors in the activity assessment are so large.

The purpose of practical dosimetry is to present a set of rules (algorithm) for determining the quantity responsible for a biological effect (in this case, the absorbed dose due to alpha-radiation of radon progeny) using the value — in the case of miners — measured by the standard procedure radon progeny concentration. This should be done with the assessment of associated errors.

The method of direct measurement of activity in the lungs of miners allowed the study of the transition coefficient between activity in the air and in the lungs. It was demonstrated in Ruzer (1970), Ruzer and Urusov (1969), Urusov (1972), Alterman (1974) and Ruzer et al. (1995) that this coefficient — filtration ability of lungs — is different for different physical activities, which resulted in different doses (intake) in different groups of miners.

As a result of this nonuniformity in dose, the variability in lung sickness and lung cancer mortality among nonuranium miners in Tadjikistan was established.

This aspect of nonuniformity both for dosimetric and epidemiological data is very important. In previous epidemiological studies, a uniformity approach to dosimetry was used and this can lead to a substantial error in risk determination.

The nonuniformity problem together with other aspects mentioned below suggest that there is good cause for reevaluating the previous epidemiological data.

14.10 Aerosol concentration measurement

14.10.1 Radon and its contribution to absorbed dose

In a real atmospheric environment, we have a combination of radon (inert gas) and its decay products (natural radioactive aerosols). For the assessment of the biological effect of the inhalation of radon and its decay products, it is important to estimate the contribution to the dose of radon itself and its decay products separately. Such assessment is important because of different physicochemical properties of radon and radon progeny, and subsequently their different behaviors in the lung. Such studies related to the separate assessment of radon and its decay products were described in Aurund and Schraub (1954), Cohen et al. (1953); Bail (1955); Jacobi et al. (1956) and Kushneva (1957).

14.10.1.1 Experimental study on animals

In Leites and Ruzer (1959), results of the experiments on white rats inhaling radon with its decay products in the small chamber ("emanatorium") are presented. Doses were determined separately for radon and radon progeny. Besides dosimetry, pathology–anatomical study and weighting were provided.

The animals inhaled radon for 2 h every day with radon concentrations of 4.8×10^8 Bq/m^3(1st series), 5.0×10^7 Bq/m^3 (2nd series), and 1.3×10^6 Bq/m^3 (3rd series).

As there were no available methods for the measurement of the concentration of radon decay products for the time of this study, especially for the small-volume chamber, the simplified variant of the method proposed in Ruzer (1960) was used.

Let us denote the following: e, the average shift of equilibrium of radon decay products in the air; N_{Rn}, the current rate in the scintillation chamber due to radon itself at $t=0$; N_p, the current rate due to decay products at $t=0$; $N=N_{Rn}+N_p$, the sum of the current rate at $t=0$; and N_{max}, the maximum value of the current rate achievable 3 h after sampling. Taking $N_{Rn}=0.46N_{max}$ (Nesmejanov, 1956), we obtain

$$\varepsilon=N_p/(N_{max}-N_{Rn})=(N-N_{Rn})/(N_{max}-N_{Rn})$$

$$=(N-0.460N_{max})/0.540N_{max}=1.85\ (N/N_{max}-0.460) \qquad (14.1)$$

Besides radon and radon decay product concentration measurements, measurements of the gamma-activity of radon decay products in the organs of animals, particularly in the lungs, were measured.

In Figure 14.2, the curve of a buildup and clearance in time of radon and decay of radon progeny was shown according to the average data from all the three series of experiments. On the vertical axis, the activity of radon decay products (radon) in the organism of animals (or in the lungs) per unit air concentration (A/q) was shown. The area under the curve is proportional to the absorbed energy due to the decay of nuclides (integral absorbed dose).

From these data, it is clear that the general absorbed energy due to the decay of radon in the organism is much smaller than that due to the decay of radon progeny. The main reason for this is that the deposition of radon decay products in the lung is very high and at the same time biological clearance of the inert gas radon from the organism is high. In other words, the behavior of decay products in the organism are determined completely by radioactive decay in the lungs with no observed trace of the clearance. At the same time, radon, due to its chemical nature, does not deposit in the organism for a long period and therefore does not produce a substantial amount of decay products and a subsequently absorbed dose. A difference in terms of absorbed dose will be much more substantial if we take into account the mass of irradiation tissue — lungs for decay products and whole body for radon — because with some approximations we can assume that the distribution of radon in the body is uniform. According to our calculations, the absorbed dose to the lung from decay products of radon is 2–3 times higher in magnitude than from radon itself.

The results of these measurements were confirmed by biological data (Leites and Ruzer, 1959). In Figure 14.3, data on the changes in the weights of animals were presented.

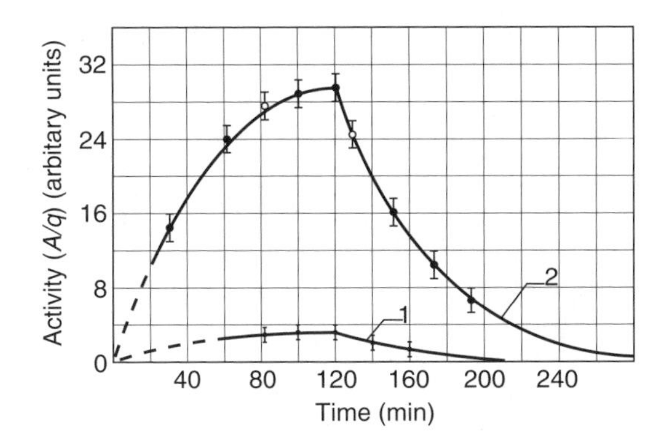

Figure 14.2 Buildup and clearance of radon (1), and buildup and decay of radon progeny (2) in rats.

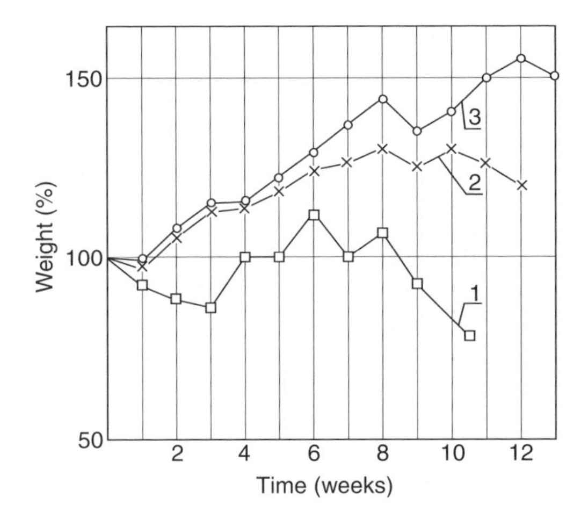

Figure 14.3 Weight of animals exposed to radon; 1 — 1st series; 2 — 2nd series; 3 — 3rd series and control.

In the 1st series of experiments, animals beginning from the second week of exposure were inert and sluggish, with decreasing appetite and decreasing weight in comparison with the control group. They died after completion of the irradiation within 7–72 days in the 1st series of experiments and within 69–94 days in the 2nd series. The main macro- and microscopic changes took place in the lungs, especially in the front and median parts of it. In the 3rd series, no changes in the weight and behavior of animals were detected during the period of exposure and even 10 months later.

These biological results show that in this case typical radiation sickness of the lung took place, which was in agreement with the dosimetric data.

For directly checking the fact that the biological effect in this case was determined mainly by the radon decay products, and not from radon itself, animals took a bath in radon water in such a way as to practically exclude the entry of radon and its decay products through the lungs. The amount of radon that entered the body from the skin was close to that in the 1st series of experiments, that is, the highest concentration. Contrary to the case of breathing, no difference in the biological effect from the control group was found.

From the curves in Figure 14.2, which represent the accumulation of radon in rats due to radon inhalation, it is obvious that the number of radon atoms in the body, n, which is numerically equal to the area enclosed by the curve, can be represented in the form of three integrals. The first integral takes into account the accumulation of radon in the organism (this portion can be considered as linear up to an instant of time (t_1). During the next interval of the time, from t_1 to t_2 (removal of the subject from the "emanatorium"), the radon content in the organism remains almost constant (a is the maximum amount of radon in the organism for unit radon concentration in the air to be inhaled). Finally, the third integral takes into account the number of radon decay events after the subject is removed from the emanatorium up to the instant of time t_3, when the amount of radon in the organism is almost equal to zero. For this portion of the curve, the radon content in the organism is proportional to $e^{-\lambda_b t}$, where λ_b is the biological constant of the clearance.

By using the expression from Shtukkenberg (1956), which relates the isotope activity in the organism to the isotope elimination rate, we obtain

$$(\mathrm{d}a/\mathrm{d}t)_{t_2} = \lambda_b a \tag{14.2}$$

where $(\mathrm{d}a/\mathrm{d}t)_{t_2}$ is the rate of elimination of radon from the organism after the end of exposure and l_b is the biological constant of the clearance.

Finally, we obtain (Ruzer, 1960)

$$D_{\mathrm{Rn}} = (E_{\mathrm{Rn}} + E_a + E_b + E_c + E_c')\, q \times 1/\lambda_b \times (\mathrm{d}a/\mathrm{d}t)_{t_2} \{t_2 - (t_1/2) - (1/\lambda_b)\,[e^{-\lambda_b(t^3 - t^2)} - 1]\} \tag{14.3}$$

where $D_{\mathrm{Rn}} = D_{\mathrm{Rn},1} + D_{\mathrm{Rn},2}$ is the integral of absorbed dose due to radon and its decay products that accumulated during the decay of radon in the organism.

In Equation (14.3), the integral absorbed dose of radon is expressed directly in terms of the radon exhaling rate at the moment of the end of exposure.

In the experiments on rats (Leites and Ruzer, 1959) for

$$q = 1.35 \times 10^{-5}\ \mathrm{Ci/l}\ (4.8 \times 10^8\ \mathrm{Bq/m^3})$$

$$(\mathrm{d}a/\mathrm{d}t)_{t_2} = 30.5 \times 10^8\ \mathrm{decay} \times 1/\mathrm{Ci\,min^2}$$

$$\lambda_b = 0.023\ \mathrm{min^{-1}},\ t = 2\ \mathrm{h}$$

the absorbed energy due to radon was approximately equal to 6250 erg.

If we consider that the value of $(da/dt)_{t_2}$ for humans is approximately 30–40 times as large as for rats (the breathing rate of rats is ~0.5 l/min, and the breathing rate of humans is ~20 l/min) and if we assume that λ_b for humans has the value indicated above, the absorbed dose (D_{Rn}/m) will be equal to ~0.25 mrad/yr for 3.7×10^3 Bq/m^3, t=6 h, and m=70 kg. Equation (14.3) can be used without modifications for thoron and actinon. The total energy of alpha-emitters for thoron per single decay differs little from the corresponding value for the radon family, and $(da/dt)_{t_2}$ and λ_b should be equal for all emanation isotopes.

The comparative contribution of radon itself and its decay products is important in terms of understanding what accuracy of measuring concentration of radon itself is important for dosimetric purposes.

The measurement of radon concentration itself is practically important because it shows the potential upper bound of danger associated with radon decay products, that is, with the equilibrium concentration of radon progeny.

However, it does not represent the real qualitative assessment of the dose, because depending on the shift of equilibrium the ranges of the dose can be of 2–3 orders of magnitude. The contribution of radon will be substantial only in cases when the shift of equilibrium is small.

Still, the results of radon concentration measurements are practically important in cases when preliminary assessment of the danger associated with natural radioactivity, especially in houses, should be made.

14.10.1.2 Methods and measurement techniques for air radon concentration monitoring

There are two types of radon monitoring methods:

1. Active sampling, in which radon is brought into the vicinity of a detector by pumping
2. Passive sampling, in which radon is collected through natural diffusion or permeation close to the detector

In passive sampling, one of the most often used techniques is sampling based on radon adsorption in activated charcoal, which is analyzed by gamma-spectrometry at the end of the exposure. Other techniques include sampling based on radon solubility or absorption, diffusion-based, permeation-based, and polyethylene permeation-based sampling (Tomassino, 1998). Some advantages and disadvantages of measuring radon and its decay products concentration based on long-lived radon progeny ^{210}Po trapped in the glass or alpha-track detectors are described in Risica et al. (2001). The problem of the background of passive radon monitors, such as alpha-track detectors, is discussed in Fujimoto (1994). A tentatively comprehensive summary of the measurement results and review of the most innovative measurement technique for thoron (^{220}radon) is presented in Nuccetelli et al. (1998).

In Tokonamii et al. (1996) the continuous radon monitor using the two-filter method is described. The short-term radon concentration was measured using a silicon semiconductor detector by measuring the alpha-activity of ^{218}Po, which was deposited on the internal surface of the cylindrical vessel. The minimum detectable concentration of the monitor was 2.7 Bq/m^3 for a 30 min continuous interval. This instrument showed great stability in counting, and is suitable for determining very low radon concentrations in the natural environment.

In Kotrappa et al. (1994), the application of the NIST ^{222}radon standard for radon monitoring is discussed. The certified parameters include ^{222}radon strength and the emanation coefficient. The paper describes a study involving 34 randomly chosen commercially available electret-passive environment radon monitors (E-PERMs) and 17 NIST sources in 17 different calibration jars. The study indicated that E-PERMs give results within about 5% of the predicted results. The paper also included a study of continuous radon monitors. The availability of NIST sources with precisely known ^{222}radon emanation characteristics is a major advantage in radon metrology.

14.10.1.3 *Radon concentration distribution measurement*

Radon concentration distribution measurement is studied in Cohen (1994), Alexander et al. (1994), White et al. (1992), Price et al. (1994), Miller (1998), Giannardi et al. (2001), Doi et al. (1994), Giovan et al. (2001), Vaupotic et al. (2001), Bem et al. (1999) and Gaidolfi et al. (1998). In Cohen et al. (1994), the compilation of average radon concentrations in 1729 U.S. counties with a standard error of ~20% is presented together with a correlation over multicount areas. Previous maps of indoor radon in U.S. counties were based on far less data and do not distinguish between measurements in basements and living areas of homes.

Results of the study of areas in 38 of 48 states of the United States with elevated levels of ^{222}radon are presented in [18]. The surveys produced short-term screening measurements in 55,000 randomly selected houses. The 38 states were divided into 225 geographic regions and summary statistics of ^{222}radon concentration were calculated for each region. Of the 225 regions, 24 had arithmetic means exceeding 222 Bq/m^3 (6 pCi/l). The U.S. EPA has assisted 30 of the 48 having common boundary states in completing surveys of indoor ^{222}radon (White et al., 1992). In 43,054 randomly selected houses, the lowest level was determined using charcoal canisters exposed for 48 h.

As expected, ^{222}radon concentration varies widely from one state to another and, in every state, the basement houses showed higher concentrations than nonbasements. The lognormal distribution is shown to be a good approximation to the distribution of radon concentration over 30 state areas.

Measurements in outdoor air in Nevada (Price et al., 1994) indicate that the state-wide median is the same as the national median (15 Bq/m^3), with a considerable range from 2.6 to 52 Bq/m^3. Variations in these data can generally be correlated with different concentrations of radon in the soil, and uranium and its progeny in the rocks. Silica-rich rocks appear to be the main source of high levels of radon in outdoor air in Nevada. Towns for which >20% of the houses have indoor air radon concentrations of 48 Bq/m^{33} (4 pCi/l) generally have relatively high soil-gas radon, relatively high outdoor air radon, or both.

In Miller (1998), the property of the data set of radon measurements made in U.K. houses with etched-track detectors is examined. Methods to use the lognormal model to estimate the geometric means radon concentration and geometric standard deviation for data groups by area have been developed. These data are then used to map radon-prone areas of England and Wales.

Identification of the two radon prone-areas in Italy with the maps of the percentages of dwellings with more than 200 Bq/m^3 are presented in Giannardi et al. (2001) Areas with magmatic rocks clearly emerge with more than 10% of dwellings above 200 Bq/m^3.

In Doi et al. (1994), the spatial distribution of thoron and radon concentration in the indoor air of a traditional Japanese wooden house is presented. The study was provided by means of a radon–thoron discriminative passive "dosimeter," which can measure both radon and thoron concentrations at the same time by means of two polycarbonate films used as a solid-state nuclear track detector. The lower detection limit was 2.9 Bq/m^3 for radon concentration and 9.0 Bq/m^3 for thoron. Examples of the study of the distribution of radon concentration in schools and kindergartens are presented in Giovani et al. (2001), Vaupotic et al. (2001), Bem et al. (1999) and Giodolfi et al. (1998). In Risica (1998), the choices made by European countries, the European Union (EU), and International Agencies of the origin, development, and state of the art of legislation on radon indoors in both domestic and working environments are presented and discussed. Based on EU and IAEA guidelines, the national authorities of many European countries have issued recommendations and regulations on radon concentrations in dwellings, balancing the dimension of the radon problem in their country with the social and economical feasibility of choice.

14.10.1.4 Radon and lung cancer

An extensive number of studies have been devoted to the association between residential radon concentration and lung cancer (Lubin et al. 1997; Neuberger, 1992; Field, 2001; Keirim-Markus, 2000; Largarde, 2001; Field et al., 2001; Estimations for——2001; Tomasek et al., 2001; Krienbrock et al., 2001; Meloni et al., 1999; Sobue et al., 2000; Gerken et al., 2000; Krewski et al., 1999; Alvanja et al., 1999; Goldsmith, 1999, Field, 1998; Darby et al., 1998; Cohen, 1997; Lagarde et al., 1997; Lubin et al., 1997; Bochiccio et al., 1996; Lubin, 1994; 1995; Letourneau et al., 1994; Ennemoser et al., 1994; Pisa et al., 2001; Vonstille et al., 1990; Samet et al., 1989; Murihead, 1994; Bowie, 1991; Neuberger et al., 1990). Studies (Melonni et al., 1999; Goldsmith, 1999; Lubin, 1995; Vonstille et al., 1990; Murihead, 1994; Bowie, 1991; Neuberger, 1990) pointed out inconclusive results in terms of correlation between residential radon concentration and lung cancer mortality for the general population. There can be many reasons for such inconclusiveness and contradictory results due to variations in both biological and physical parameters, especially when radon concentration is very low. But we have to focus on one very important source of uncertainty — not reliable and even poor dosimetry. At the beginning, we have already proved that the contribution of radon itself to the dose is very small.

The difference between radon and its decay products in terms of their contribution to the dose, and consequently to health effects, is due to their different physicochemical nature. Radon — the inert gas — after entering the lung, distributes through the blood vessels uniformly in the body, whereas decay products — aerosols — deposit only in the lungs, where the process of irradiation of the lung tissue and the corresponding biological effect take place. The role of decay products in the different parts of the body due to decay of radon is negligible.

So, the question arises as to how and why in the studies of correlation between residential radon concentration and lung cancer mortality, radon concentration, a surrogate for the causative factor dose, can be used without even a discussion of their direct correlation.

As already pointed out in Chapter 7, in order to get reliable results from the exposure–effect study, three main conditions should be met:

1. The dose, the physical parameter responsible for the health effect, should be defined correctly.
2. If, as in most practical cases, the dose cannot be measured directly, the correlation should be established between measurable surrogate value and the dose.
3. The uncertainty related to the replacement (substitution) of the dose by a practically measurable surrogate value should be assessed correctly.

The last is especially important in the case of aerosols where uncertainties are very high due to spatial and temporary variability, and also uncertainties associated with interaction between aerosols and humans.

It may be suggested that all factors determining concentration–dose correlation, such as variations of the equilibrium factor between radon and decay products, including in the breathing zone, lung deposition, breathing rate, and other factors, are averaged spatially and temporally; therefore, we have reason to believe that radon concentration is proportional to the dose. But this should be proved theoretically by modeling or experimentally. On the contrary, measurements on miners [Ruzer, Z.S., 1970; Urusov, S.A., 1972; Altezman, A.D. 1974] showed no correlation between measured radon concentration and dose to the lung.

Thus, it is very easy to measure radon concentration. But interpretation of such results in terms of correlation with the causative factor–dose and health effect—is very complicated. *It is not only not so simple. It is simply not so.*

In terms of health effects, radon concentration measurements present a qualitative (not quantitative) picture. These data on the surrogate of the dose should not be used without a careful consideration of the dose–effect studies. This is especially true of the correlation between residential radon and lung cancer mortality because of very low radon levels.

14.10.2 *Measurement of the concentration of decay products of radon, thoran, and actinon*

14.10.2.1 *Characteristics of radon progeny*

The presence of radon and its decay products is due to the abundance of heavy metals — radioactive elements at the end of the periodic table — in the earth. One of them, uranium, undergoes a long series of transformations to yield radium. The chain of radioactive decay continues further; however, nature dictates that the member of the chain after radium, radon, is a radioactive noble gas. Due to its inert chemical properties, radon does not remain in the earth or in water, but enters the atmosphere.

The links that follow in the radioactive chain — isotopes of polonium, bismuth, and lead — attach to aerosol particles to become radioactive aerosols or exist in the unattached form in the air. Eventually, they may be deposited in the lung and cause irradiation to the lung tissue.

Specific biological consequences not only depend upon the amount (concentration) of radioactive aerosols but also on physiological characteristics of humans, especially physical activity.

The concentration depends not only on the amount of radium in the soil and in the air but also on atmospheric conditions in the open air, dwellings, and underground environment.

Table 14.1 presents the basic characteristics of radon, thoron, actinon, and its decay products.

Figure 14.4 presents diagrams of the radioactive families corresponding to ^{226}Ra, ^{232}Th, and ^{227}Ac, respectively. Clearly, the decay products represent a very complicated system, consisting of a series of radioactive elements with various types of decay (alpha, beta, and gamma).

In terms of radiation danger, the most important radioisotopes are alpha-emitters, because the alpha-particles have the greatest ionization density (linear energy transfer (LET)). Given identical absorbed energy, the biological effect of alpha-particles is supposed to be 20 times greater than the corresponding effect of beta-particles and gamma-radiation (the "quality coefficient" for alpha-particles is 20). However, it is impossible on a practical basis to measure the alpha-activity of aerosols deposited in the lung of a living subject. As a result, this alpha-radioactivity should be measured in the air, and the absorbed dose to the lungs is calculated according to known concentration, breathing rate, and the coefficient of deposition in the lungs. The concentration of radon in open air depends to a great extent on atmospheric conditions and air movement. The average concentration of radon decay products in the atmosphere fluctuates between 4 and 40 Bq/m^3. However, there are instances in which local or temporary concentrations are 2 orders of magnitude higher. In closed premises, whether buildings, residences, or mines, the concentration depends as much on the rate of ventilation as it does on the amount of radon in the soil (construction materials) and water. The measurement of the

Table 14.1 Basic Characteristics of Rn, Tn, An, and their Progeny

Nuclide	Symbol	τ	λ (s^{-1})	Radiation	E_α, E_β	E_γ
Radon	^{222}Rn	3.8 days	2×10^{-6}	α	5.486	
RaA	^{218}Po	3.05 min	3.788×10^{-3}	α	5.998	
RaB	^{214}Pb	26.8 min	4.310×10^{-4}	β^-, γ	0.7	0.35(0.43) 0.29(0.24) 0.61(0.36)
RaC	^{214}Bi	19.7 min	5.864×10^{-4}	β^-, γ	1.65(23%) 3.17(77%)	1.76(0.22) 1.12(0.27)
RaC′	^{214}Po	1.6×10^{-4} sec	4.23×10^{3}	α	7.68	
RaD	^{210}Pb	22 yr	$9{\cdot}98\times10^{-10}$	β^-, γ	0.027	
RaE	^{210}Bi	4.99 days	1.608×10^{-6}	β	1.17	
RaF	^{210}Po	138.4 days	$5{\cdot}8\times10^{-8}$	α	5.298	
Thoron	^{220}Rn	54.5 sec	1.27×10^{-2}	α	6.282	
ThA	^{216}Po	0.16 sec	4.387	α	6.774	
ThB	^{212}Pb	10.67 h	1.816×10^{-5}	β^-, γ	0.37	0.30(034) 0.24(0.33)
ThC	^{212}Bi	1.09 h	1.766×10^{-4}	α(33.7%) β(66.3%)	6.055 2.25	0.81(0.10) 1.81(0.05)
ThC′	^{212}Po	2.9×10^{-7} sec	2.3×10^{6}	α	8.476	
ThC″	^{208}Tl	3.1 min	3.73×10^{-3}	β^-, γ	1.792	2.62(0.34) 0.58(0.26)
An	^{219}Rn	3.92 sec	0.1767	α	6.41	
AcA	^{215}Po	1.83×10^{-3} sec	3.787×10^{2}	α	7.365	
AcB	^{211}Pb	36.1 min	3×10^{-4}	β^-, γ	1.23	0.83(0.13)
AcC	^{211}Bi	2.16 min	5.348×10^{-3}	α	6.56	
AcC′	^{207}Tl	4.79 min	2.412×10^{-3}	β	1.50	

τ — half-time of decay, λ — decay constant, E_α, E_β — energy of alpha- and beta-particles, respectively (MeV), E_γ — energy of gamma-radiation (MeV), exit per decay shown in parentheses.

concentration of the natural radioactive aerosol — decay products of radon, thoron, and actinon — is important due to its own irradiation of the lungs. With the advent of nuclear power and nuclear weapons systems, artificial radioactive aerosols began to enter the atmosphere due to nuclear experiments, leakages of radioactive substances, and nuclear accidents.

In this sense, it should be pointed out that the background natural radioactive aerosols are serious obstacles to the correct measurement of artificial radioactive aerosol concentration.

The maximum permissible concentrations (MPCs) for artificial radioactive aerosols, especially for alpha-emitters, are 2–3 orders of magnitude lower than the concentrations of radon decay products in the atmosphere.

Thus, in measuring the concentration of artificial radioactive aerosols, which are mostly long-lived isotopes, two different approaches can be used:

1. The "waiting method," that is, making measurements only after the short-lived decay products of radon and thoron decay substantially enough (sometimes for 3–4 days after sampling). Such an approach is not always acceptable in practical situations.
2. Using spectroscopical methods. Even with this type of technique, the natural radioactive aerosol background together with aerosol concentration itself will play a major role and should be taken into account.

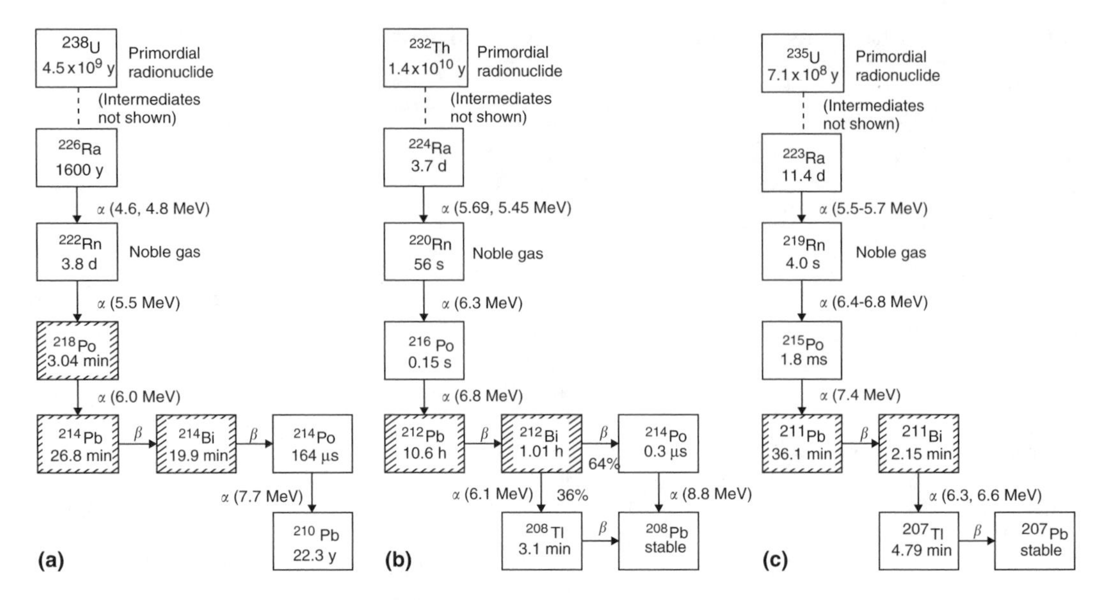

Figure 14.4 Radioactive decay chains: (a) ^{238}U to ^{210}Pb, (b) ^{232}Th to ^{208}Pb, and (c) ^{235}U to ^{207}Pb. For each isotope the radioactive half-life and primary mode of decay are shown. For each decay chain, inhalation of the shaded isotopes constitutes the primary health concern. If inhaled and retained in the respiratory tract, the decay from the shaded isotopes to the long-lived or stable lead isotope produces one or two alpha particles that cause damage to tissue adjacent to the decay site (Browne and Firestone, 1986; Martz et al., 1989).

14.10.2.2 *Basic equations for radon decay product series*

The basic equations in the most generalized form for the radon series was derived in (Ruzer, 1958, 1968).

The derivation is based on equations described in Bateman (1910) with some transformations. The following is a chart for deriving these equations, where v is the volume rate of inhalation, and k is the retention coefficient in the lungs or part of it (in deriving the equations, we supposed that this value is equal for all decay products)

	Concentration	Breathing rate	Rate of Intake	
^{218}Po (RaA)	q_a	v	$Q_a = q_a vk/\lambda_a$	$\text{RaA} \xrightarrow{\alpha} \text{RaB} \xrightarrow{\beta,\gamma} \text{RaC} \xrightarrow{\beta,\gamma} \text{RaC}' \xrightarrow{\alpha}$
^{214}Pb (RaB)	q_b		$Q_b = q_a vk/\lambda_b$	$\text{RaB} \rightarrow \text{RaC} \rightarrow \text{RaC}'$
^{214}Bi (RaC)	q_c		$Q_c = q_c vk/\lambda_c$	$\text{RaC} \rightarrow \text{RAC}'$

For determination of the correlation between measured concentration and activity of each decay product on the filter, the Bateman equation for the chain of radioactive transformations was used.

In order to insert the mechanism of the buildup activity on the filter due to filtration (or physiological breathing) in these equations, we assumed that the first member of this chain for each decay product are the decay products in the air, which supply the decay products to the lungs with a constant rate $Q_i = q_i vk/\lambda_i$. From a mathematical point of view, the constant rate of supply is equal to the equilibrium between the first and the second member of the chain of radioactive transformation. In this case, the number of atoms N of each decay product in the lungs can be found as follows:

$$N_i = c_1 e^{-\lambda_1 t} + c_2 e^{-\lambda_2 t} + \cdots + c_i e^{-\lambda_i t} \tag{14.4}$$

where

$$c_i = N_{1,0}\, \lambda_1\, \lambda_1\lambda_2 \cdots \lambda_{\iota-1}/(\lambda_1-\lambda_i)\,(\lambda_2-\lambda_i)\cdots(\lambda_{i-1}-\lambda_i) \tag{14.5}$$

Let us examine the quantitative correlation between the activity $A_i(\theta, t)$ of a given isotope on a filter, corresponding to a given duration of filtration (breathing) θ, time after the conclusion of filtration (exposure) t, and concentration of each decay product q_i.

We will assume first that the concentrations for the duration of filtration do not change, the radioactivity on the filter is affected only by the short-lived radon decay products, and the coefficients of retention are identical for each decay product.

It is also presupposed that during the time of sampling, the basic parameters of the aerosol system are constant.

The activity of ^{214}Bi (RaC) on the filter, $A_c(\theta, t)$, is made up of ^{214}Bi (RaC) that has deposited during filtration, taking into account its decay $A_{c,c}(\theta, t)$, and also of ^{214}Bi (RaC) formed as a result of the decay of ^{218}Po (RaA) — $A_{c,a}(\theta, t)$ and of ^{214}Pb (RaB) — $A_{c,b}(\theta, t)$.

The formulas for each of these activities can be obtained by applying expressions (14.4) and (14.5).

We will provide the equations in two steps. First, Equations (14.4) and (14.5) will be used for describing the changes taking place on the filter after sampling. At the moment t after completion of the sampling, activity is

$$A_{c,a}(\theta, t) = A_a(\theta, 0)\lambda_b\lambda_c\{[(e^{-\lambda_b t} - e^{-\lambda_a t})/(\lambda_a-\lambda_b)\,(\lambda_c-\lambda_c b)]$$

$$+[(e^{-\lambda_c t} - e^{-\lambda_a t})/(\lambda_a-\lambda_c)\,(\lambda_b-\lambda_c)]\} \tag{14.6}$$

$$A_{c,b}(\theta, t) = A_b(\theta, 0)\,[\lambda_c/\lambda_b-\lambda_c)]\,(e^{-\lambda_c t} - e^{-\lambda_b t}) \tag{14.7}$$

$$A_{c,c}(\theta, \mathrm{t}) = A_c(\theta, 0)\, e^{-\lambda_c t} \tag{14.8}$$

In the second step of derivation, the same formulas (14.4) and (14.5) were used in order to find the correlation between activity on the filter after sampling and concentration in the air.

Expressions for the activities of RaA, RaB, and RaC at the moment of the end of filtration ($t=0$)—$A_a(\theta, 0)$, $A_b(\theta, 0)$, $A_c(\theta, 0)$—were obtained by applying the solution for the chain of radioactive transformation when the maternal isotope is long-lived. Each of the decay products is deposited at the constant rate of $q_i vk/\lambda_\iota$, which corresponds to the decay rate of the long-lived maternal nuclide.

$$A_a(\theta, 0) = aq_a vk/\lambda_a\,(1-e^{-\lambda_a\theta}) \tag{14.9}$$

$$A_b(\theta, 0) = avk(q_a/\lambda_a)\,[(1+(\lambda_b e^{-\lambda_a\theta} - \lambda_a e^{-\lambda_b\theta})/(\lambda_a-\lambda_b)) + (q_b/\lambda_b)\,(1-e^{-\lambda_b\theta})] \tag{14.10}$$

$$A_c(\theta, 0) = avk\,[\,q_a\,\xi_a(\theta) +_b \xi_b(\theta) + q_c\,\xi_c(\theta)] \tag{14.11}$$

where

$$\xi_a(\theta) = (1/\lambda_a)\,[1-\lambda_b\lambda_c\, e^{-\lambda_b\theta}/(\lambda_b-\lambda_a)\,(\lambda_c-\lambda_a)$$

$$-\lambda_a\lambda_c\, e^{-\lambda_b\theta}/[(\lambda_a-\lambda_b)\,(\lambda_c-\lambda_b)] - \lambda_a\lambda_b\, e^{-\lambda_c\theta}/[(\lambda_a-\lambda_c)\,(\lambda_b-\lambda_c)] \tag{14.12}$$

$$\xi_b(\theta) = 1/\lambda_b\,[1+(\lambda_c\, e^{-\lambda_b\theta} - \lambda_b \mathrm{e}^{-\lambda_c\theta}\,\lambda_b)/(\lambda_b-\lambda_c)] \tag{14.13}$$

$$\xi_c(\theta) = 1/\lambda_c\,(1-e^{-\lambda_c\theta}) \tag{14.14}$$

where $a = 2.22\times10^{12}$ decay/min for 1 Ci.

Finally,

$$A_c(\theta, t) = A_{c,a}(\theta, t) + A_{c,b}(\theta, t) + A_{c,c}(\theta, t)$$

$$= avk\,[q_a\,\Phi_{a,c}(\theta, t) + q_b\,\Phi_{b,c}(\theta, t) + q_c\,\Phi_{c,c}(\theta, t)] \tag{14.15}$$

Similarly, the expression for $A_b(\theta, t)$ is

$$A_b(\theta, t) = avk\,[q_a\,\Phi_{a,b}(\theta, t) + q_b\,\Phi_{b,b}(\theta, t)] \tag{14.16}$$

and for $A_a(\theta, t)$ it is

$$A_a(\theta, t) = avkq_a\,\Phi_a(\theta, t) \tag{14.17}$$

The final expression for the alpha-activity of ^{218}Po (RaA) and ^{218}Po (RaC′) on the filter (i.e., lungs) is

$$A_\alpha = avk\,\{q_a\,[\Phi_a(\theta, t) + \Phi_{a,c}(\theta, t)] + q_b\,\Phi_{b,c}(\theta, \mathrm{t}) + q_c\,\Phi_{c,c}(\theta, \mathrm{t})\} \tag{14.18}$$

Correspondingly, for the beta- and gamma-activity of RaB and RaC, it is

$$A_{\beta,\gamma} = avk\,\{q_a\,[\Phi_{a,b}(\theta, t) + \Phi_{a,c}(\theta, t)] + q_b[\Phi_{b,b}(\theta, t) + \Phi_{b,c}(\theta, t)] + q_c\,\Phi_{c,c}(\theta,t)\} \tag{14.19}$$

The expressions for the functions $\Phi_{i,j}(\theta, t)$, where $i,j = a,b,c$, are presented in Ruzer (1968).

The graphs of the functions and values of $\Phi_{i,j}(\theta, t)$ for fixed θ and t are presented correspondingly in Figures 14.5–14.8 and for different θ and t in Table 14.2.

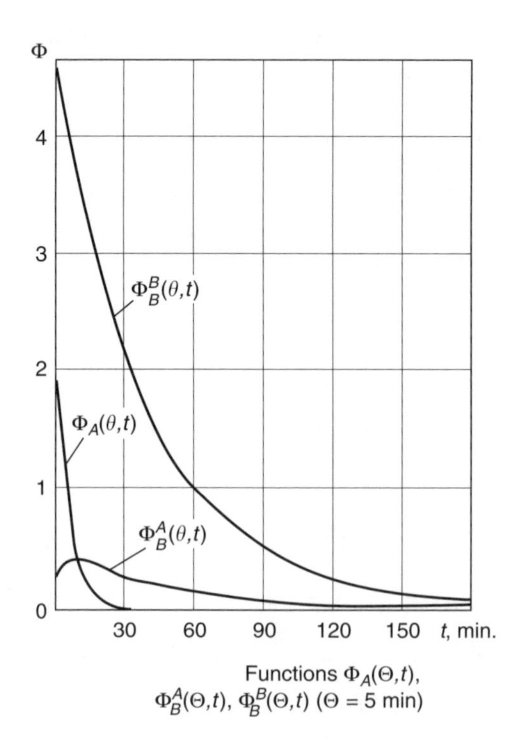

Figure 14.5 Functions $\Phi_A\,(\theta, t)$, $\Phi_B^A\,(\theta, t)$, $\Phi_B^B\,(\theta, t)$ ($\theta = 5$ min).

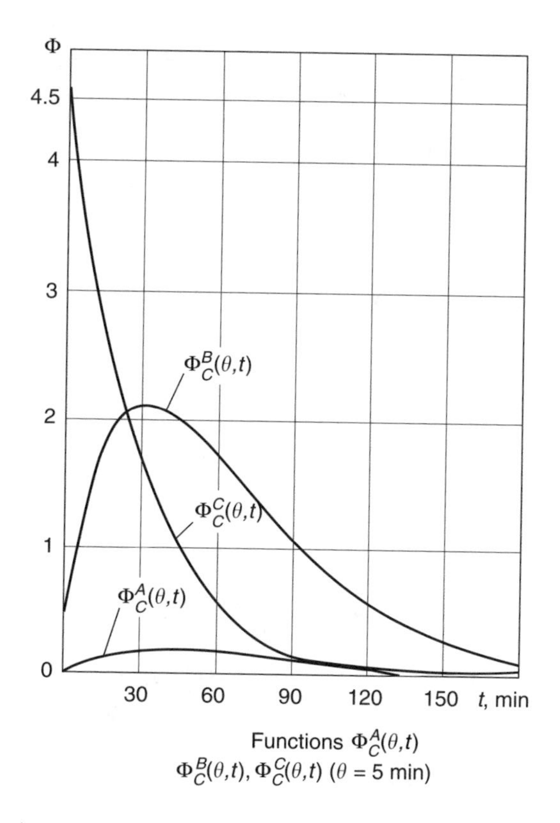

Figure 14.6 Functions $\Phi_C^A(\theta, t)$, $\Phi_C^B(\theta, t)$, $\Phi_C^C(\theta, t)$ ($\theta = 5$ min).

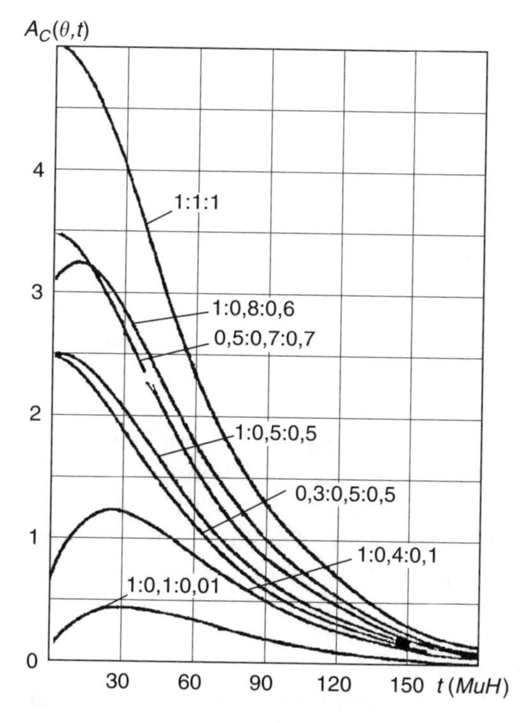

Figure 14.7 $A_C(\theta, t)$ in relative units for different ratios of q_a:q_b:q_c ($\theta = 5$ min).

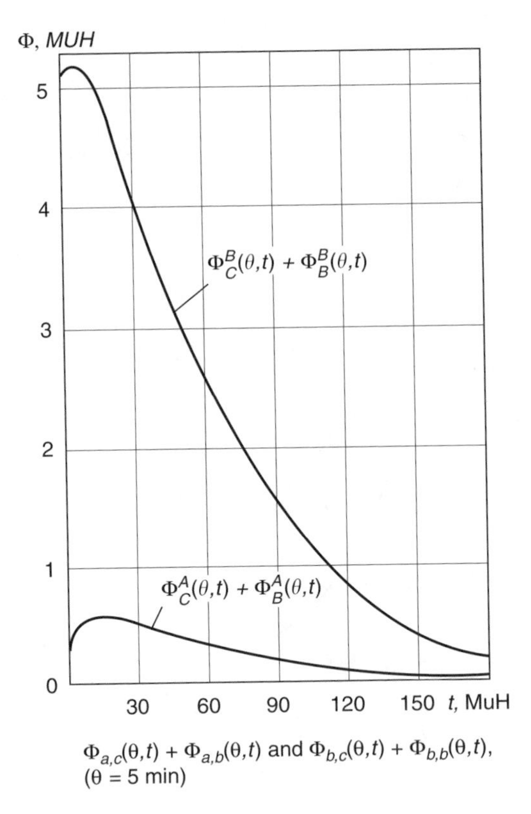

Figure 14.8 $\Phi_{a,c}(\theta, t) + \Phi_{a,b}(\theta, t)$ and $\Phi_{b,c}(\theta, t) + \Phi_{b,b}(\theta, t)$, $(\theta = 5$ min)

Graphs for A_c in relative units for different shifts of equilibrium are presented in Figure 14.7. Graphs for $A_\alpha = A_a + A_c$, are presented in Figures 14.9 and 14.10.

14.10.2.3 *General activity methods of measuring the concentration of radon decay products*

The first method for the concentration of radon decay products measurement was published in Tsivoglou and Ayer (1953). The premise of this method was in analysis of the curve of radioactive decay after sampling was concluded. To determine the three unknown quantities q_a, q_b, and q_c, the values of the total radioactivity were measured at three different times, specifically according to (Tsivoglou and Ayer, 1953) 5, 15, and 30 min.

Different variants of this method were used, but all of them presented difficulties in practical situations. This is true in mines because they are not accurate enough and because they cannot yield results in very short periods of time, which is important in practice.

In Kuznets (1956) and Holeidi et al. (1961) another approach was presented which does not require individual determination of the atmospheric concentration of ^{218}Po (RaA), ^{214}Pb (RaB), and ^{214}Bi (RaC). As a characteristic of the contamination of the atmosphere with radon decay products, the concept of the potential alpha energy concentration (PAEC) was proposed, which is the energy produced in complete alpha-decay, resulting in the production of ^{210}Pb (RaD), of all the short-lived radon decay products contained in a unit volume of air. The maximum permissible value of PAEC may be taken to be 3.8×10^4 MeV/l, corresponding to the equilibrium percentage of the radon decay products at a concentration of 1100 Bq/m^3.

Table 14.2 Values of Functions $\Phi_{i,j}(\theta, t)$

θ (min)	$\Phi_{i,j}$	*t (min)*									
		0	1	2	3	5	10	15	30	60	180
2	Φ_a	1.61	1.28	1.02	0.81	0.52	0.17	0.05	0	0	0
	$\Phi_{a,c}$	0	0.01	0.01	0.02	0.04	0.08	0.10	0.12	0.08	0.01
	$\Phi_{a,b}$	0.04	0.08	0.11	0.13	0.15	0.17	0.16	0.11	0.05	0
	$\Phi_{b,b}$	1.95	1.90	1.85	1.80	1.71	1.50	1.32	0.90	0.41	0.18
	$\Phi_{b,c}$	0.08	0.15	0.21	0.26	0.37	0.56	0.70	0.86	0.68	0.06
	$\Phi_{c,c}$	11.6	11.3	10.9	10.5	9.77	8.20	6,87	4.05	1.41	0.02
5	Φ_a	3.00	2.38	1.90	1.51	0.96	0.31	0.10	0	0	0
	$\Phi_{a,c}$	0.01	0.02	0.04	0.06	0.10	0.18	0.23	0.22	0.17	0.01
	$\Phi_{a,b}$	0.22	0.28	0.33	0.36	0.41	0.43	0.40	0.18	0.13	0.01
	$\Phi_{b,b}$	4.69	4.57	4.45	4.34	4.12	3.62	3.18	2.16	0.99	0.04
	$\Phi_{b,c}$	0.42	0.48	0.70	0.82	1.06	1.51	1.81	2.13	1.66	0.14
	$\Phi_{c,c}$	4.59	4.43	4.27	4.13	3.85	3.22	2.70	1.60	0.56	0.01
10	Φ_a	3.60	2.87	2.27	1.82	1.16	0.37	012	0	0	0
	$\Phi_{a,c}$	0.08	0.10	0.13	0.16	0.21	0.33	0.40	0.41	0.24	0.01
	$\Phi_{a,b}$	0.63	0.66	0.74	0.77	0.81	0.79	0.72	0.50	0.23	0.01
	$\Phi_{b,b}$	8.81	8.59	8.37	8.15	7.74	6.80	5.97	4.05	1.86	0.08
	$\Phi_{b,c}$	1.41	1.52	1.90	2.12	2.52	3.28	3.78	4.22	3.19	0.26
	$\Phi_{c,c}$	8.43	8.14	7.86	7.59	7.07	5.93	4.97	2.93	1.02	0.01
15	Φ_a	3.88	3.09	2.46	1.96	1.25	0.40	013	0	0	0
	$\Phi_{a,c}$	0.20	0.23	0.27	0.31	0.38	0.51	0.58	0.56	0.30	0.01
	$\Phi_{a,b}$	1.05	1.11	1.16	1.18	1.20	1.15	1.03	0.71	0.33	0.01
	$\Phi_{b,b}$	12.4	12.1	11.8	11.5	10.9	9.59	8.43	5.72	2.63	1.17
	$\Phi_{b,c}$	2.95	3.06	3.57	3.84	4.35	5.30	5.90	6.29	4.61	0.37
	$\Phi_{c,c}$	11.6	11.3	10.9	10.5	9.77	8.20	6.87	4.05	1.41	0.02
30	Φ_a	4.40	3.50	2.79	2.22	1.41	0.45	015	0	0	0
	$\Phi_{a,c}$	0.80	0.85	0.90	0.95	1.03	1.11	1.21	1.01	0.47	0.02
	$\Phi_{a,b}$	2.12	2.16	2.19	2.20	2.18	2.01	1.80	1.23	0.57	0.03
	$\Phi_{b,b}$	20.9	20.3	19.8	19.2	18.3	16.1	14.1	9.60	4.41	0.20
	$\Phi_{b,c}$	8.84	8.90	9.62	9.95	10.6	11.6	12.2	11.9	8.21	0.62
	$\Phi_{c,c}$	18.5	17.9	17.3	16.7	15.5	13.0	11.1	6.45	2.42	0.03
60	Φ_a	4.40	3.51	2.79	2.26	1.41	0.45	015	0	0	0
	$\Phi_{a,c}$	2.37	2.40	2.43	2.46	2.51	3.52	2.42	1.79	0.75	0.02
	$\Phi_{a,b}$	3.35	3.37	3.36	3.34	3.26	2.97	2.64	1.80	0.83	0.04
	$\Phi_{b,b}$	30.4	29.7	28.9	28.2	26.7	23.5	20.6	14.0	6.44	0.29
	$\Phi_{b,c}$	20.7	20.5	21.3	21.6	22.0	22.5	22.4	20.1	12.9	0.92
	$\Phi_{c,c}$	25.0	24.1	23.3	22.4	20.9	17.6	14.7	8.69	3.02	0.04
180	Φ_a	4.40	3.51	2.79	2.23	1.41	0.45	015	0	0	0
	$\Phi_{a,c}$	4.48	4.48	4.47	4.46	4.42	4.21	3.88	2.72	1.08	0.03
	$\Phi_{a,b}$	4.35	4.31	4.31	4.27	4.14	3.74	3.32	2.26	1.04	0.05
	$\Phi_{b,b}$	38.2	37.3	36.3	35.4	33.6	29.5	26.0	17.6	8.09	0.36
	$\Phi_{b,c}$	37.4	37.0	37.4	37.3	37.2	36.2	34.8	29.2	17.6	1.18
	$\Phi_{c,c}$	28.4	27.4	26.4	25.5	23.8	19.9	16.7	9.87	3.43	0.05

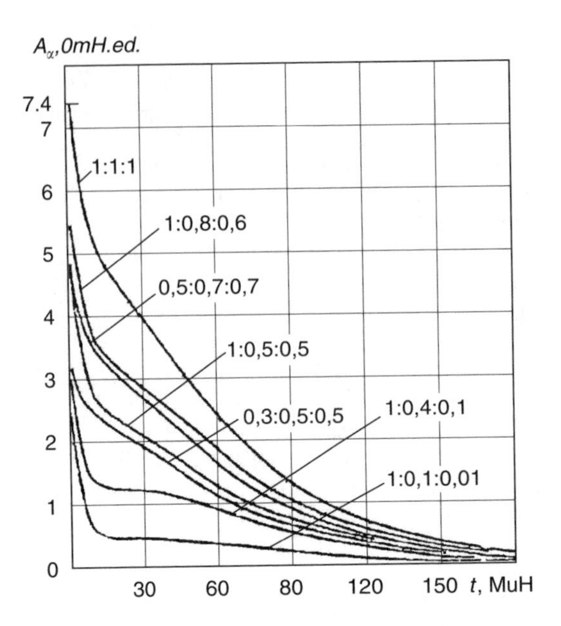

Figure 14.9 $A_\alpha(\theta, t) = A_a(\theta, t) + A_c(\theta, t)$ in relative units for different ratios of $q_a{:}q_b{:}q_c$ ($\theta = 5$ min).

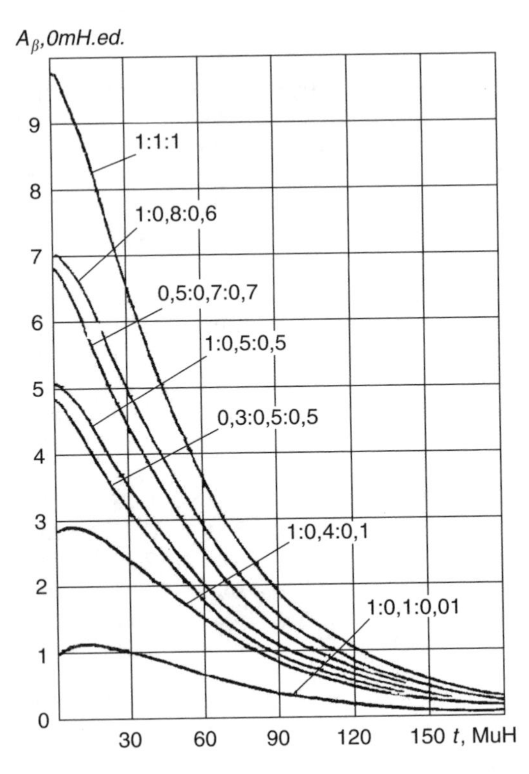

Figure 14.10 $A_\beta(\theta, t) = A_b(\theta, t) + A_c(\theta, t)$ in relative units for different ratios of $q_a{:}q_b{:}q_c$ ($\theta = 5$ min).

To satisfy the conditions in mines where the rapid assessment of results is sometimes more important than the accuracy of the measurement, the authors in Markov and Ryabov (1965, 1962) presented different variants of the method of measurement of general alpha-activity, called the "express method."

The "express method" allows the rapid (within 10 min) determination of the concentration of each radon decay product and PAEC with an accuracy of 20–40%, which is completely acceptable for dosimetric purposes.

In this method, the 5-min circulation time and the measurement of the alpha-activity twice, once from the first to the fourth minute, and again from the seventh to the tenth minute after conclusion of the sampling, was proposed.

The formulas for determining each q and PAEC are

$$q_a = 1.18\,(n_1 - n_2)/\text{ekv}\ 10^{-10}\ \text{Ci/m}^3$$

$$q_b = 3n_2/\text{ekv}\ 10^{-11}\ \text{Ci/m}^3$$

$$q_c = (5.97n_1 - 2.44n_2)/\text{ekv}\ 10^{-11}\ \text{Ci/m}^3$$

$$\text{PAEC} = 40n_2/\text{ekv MeV/l}$$

In Markov and Ryabov (1962), the authors also pointed out that the 5-min sampling time is not mandatory for this method. In the case when the slightly lower accuracy of the PAEC measurements is acceptable, the time of the PAEC determination can be decreased using the formula

$$\text{PAEC} = mn/\text{ekv}$$

in which the coefficient m and the corresponding methodical error in the determination of PAEC can be found from the nomogram.

14.10.2.4 *Measurement of radon decay products in air by alpha- and beta-spectrometry*

The methods described before were used, for example, for daily monitoring of underground mining operations in the atmosphere. In such situations, the monitoring equipment must be portable, light, simple to operate, and capable of functioning in considerable gamma-fields. High accuracy is often not required for this equipment.

All methods of determining the concentrations of each radon decay product, based on the measurement of total alpha-radioactivity collected on a filter, can have substantial errors (Nazaroff, 1988, 1984; Labushkin and Ruzer, 1965; Ruzer and Sextro 1997; Thiessen, 1994). If, for example, the equilibrium ratios are 1:0.5:0.5, and the total alpha-radioactivity on the filter is analyzed using the three-count method of Tsivoglou and Ayer (1953), the ratios of the methodical and statistical errors in the inferred concentrations of ^{218}Po, ^{214}Bi, and ^{214}Pb will be 16, 5.7, and 8, respectively, for measurements made at 5, 15, and 30 min after sample collection. For count timings of 1, 15, and 60 min, the ratios are 5.8, 2.7, and 4, respectively.

Although the health effects associated with radon decay products exposures do not critically depend upon the ratio of the ^{218}Po and ^{214}Po concentrations, in some cases, high inherent accuracy in the measurement of radon decay products is important, such as in

1. the study of the behavior of aerosols, including the attached and unattached fraction of radon decay products,
2. examination of the correlation between the ventilation rate and the shift of equilibrium, and
3. standardization and calibration of decay product measurements.

As a practical matter, in many cases determining the dose due to the unattached radon decay products will require that airborne concentrations of ^{214}Pb and ^{214}Bi be accurately measured. Even though their concentrations may be lower than that of ^{218}Po, their indirect effect on the dose, as estimated by the PAEC, can be substantial due to ^{218}Po alone. Thus, the accuracy afforded by alpha-spectrometry may make it the preferred measurement method.

In some practical situations such as high aerosol (dust) concentrations or high humidity, the results of the measurements can be distorted due to absorption of alpha-radiation by the aerosol or water on the filter.

The method, which will be presented below, of determining ^{218}Po, ^{214}Pb, and ^{214}Bi (^{214}Po) in air based on the measurement of the activity on the filter using both alpha- and beta-spectrometry permits the assessment of the effect of absorption of alpha-radiation with high accuracy (Ruzer and Labushkin, 1965).

14.10.2.4.1 Measurement procedure and experimental results

The short-lived alpha-emitters in the ^{222}Rn series are ^{218}Po and ^{214}Po with alpha-particle energies of 6.00 and 7.69 MeV, respectively; the short-lived beta-emitters are ^{214}Pb and ^{214}Bi with maximum (end point) beta-energies of 1.024 and 3.27 MeV, respectively. In the former case, the strongest beta-decay transitions have end point energies of 0.67 (48%) and 0.73 (42%) (Brown and Firestone, 1986). As a result of the considerable difference between the energies for either alpha- or beta-emissions from these nuclides, high-energy resolution detection methods are not needed and the activity of the progeny on the filter can be relatively easily measured separately by means of scintillation detectors. The equipment used for these measurements is shown in Figure 14.11.

Alpha-particle counting was carried out with either a 0.1 mm thick Cs(Tl) scintillator coupled to a multiplier tube or a Si semiconductor detector, with alpha-particle energy resolutions of 9 and 1%, respectively. For measuring the beta-spectra, a stilbene (1,2-diphenylethene) beta-scintillator with a thickness of 13 mm was used in conjunction with a photomultiplier tube.

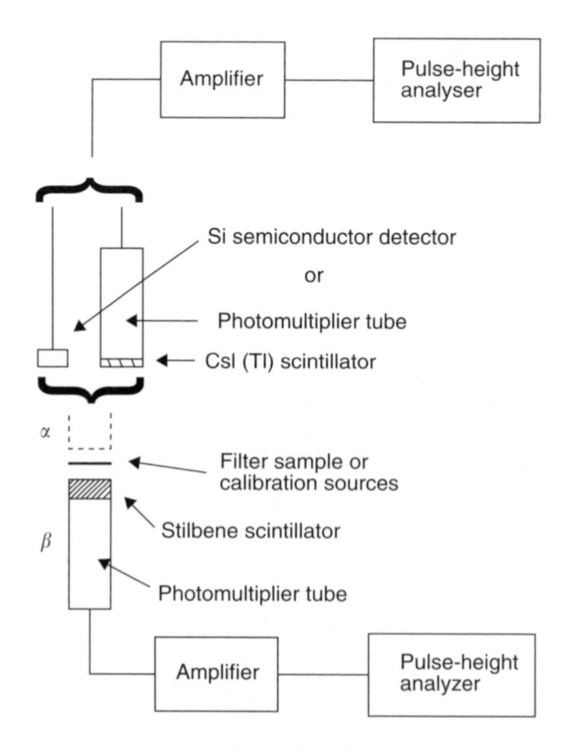

Figure 14.11 Block diagram of the detectors used for the measurement of alpha and beta particles.

In order to determine the absolute concentration of each decay product, it is necessary to calibrate the detection efficiencies of the measurement equipment with the standard sources of short-lived radon decay products. Such sources were described in (Volkova et al., 1996; Ruzer, 1993). An alpha-decay energy spectrum from a nonemanating standard sample is shown in Figure 14.12; as can be seen from this spectrum (acquired with a Si semiconductor detector), the peaks from the decays of ^{226}Ra, ^{222}Rn, ^{218}Po, and ^{214}Po are well resolved.

In this nonemanating standard sample, all the radon decay products (both alpha- and beta-emitters) are in radioactive equilibrium and therefore have the same activity. Because these sources were mounted on the thin foil backing, the alpha- and beta-spectra can be acquired at the same time, as suggested by Figure 14.13. The resulting beta-spectrum from this source is shown in Figure 14.13; superimposed on this figure are data collected from a ^{137}Cs source, which produces internal conversion electrons. The main emission peak is at 0.624 MeV, which accounts for the peak of beta-spectrum between channels 20 and 25.

Because this energy is close to the endpoint energies for the most intense beta-decays of ^{214}Pb, the ^{137}Cs internal conversion peak provides a convenient marker for dividing the energy spectrum for the beta-decays from the ^{222}Rn decay products ^{214}Pb and ^{214}Bi.

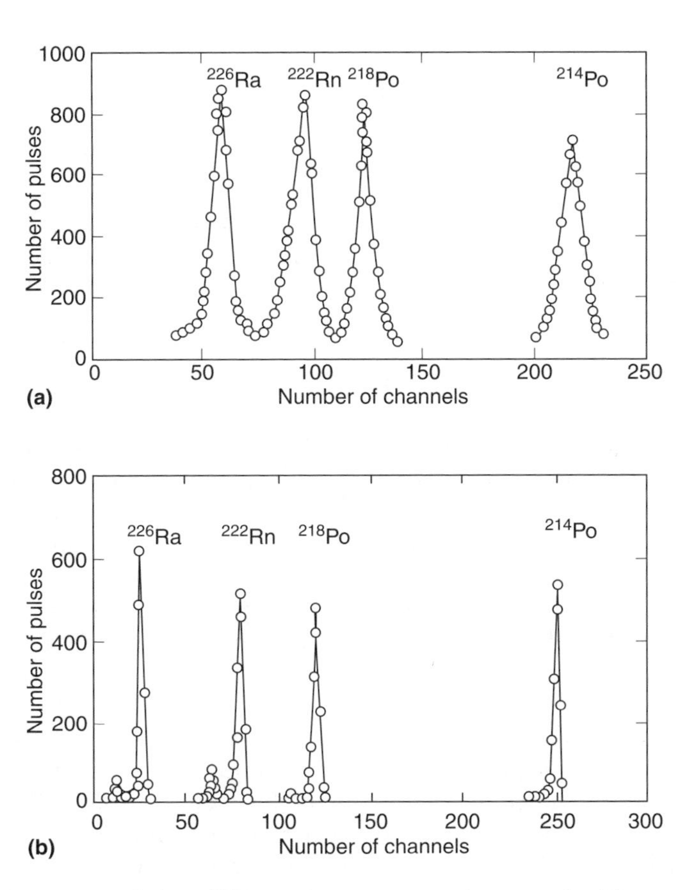

Figure 14.12 Alpha spectra from a ^{222}Rn series nonemanating sample. The upper spectrum was obtained using a CsI(TI) scintillator, the lower a semiconductor detector. The small unidentified peak to the left of the ^{222}Rn peak (lower spectrum) is due to the 5.3 MeV α particle from ^{210}Po, which has grown into the sample over a period of more than a year.

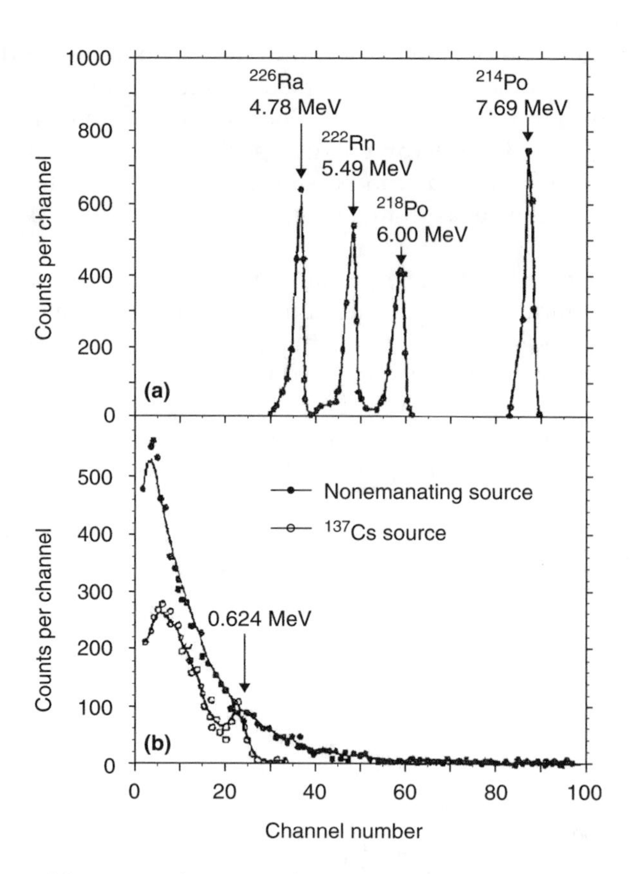

Figure 14.13 Alpha and beta particle spectra from the nonemanating standard source. (a) Alpha spectra from the nonemanating source acquired with a Si semiconductor detector. Although the peak heights vary, the integrated counts under each peak are essentially the same. (b) Corresponding beta decay spectrum from the nonemanating reference source, along with the beta spectrum from ^{137}Cs. The 0.624 MeV peak in the ^{137}Cs beta spectra is mainly due to internal conversion electrons from the K shell in ^{137}Ba produced by decay of the isometric 0.6617 MeV state in ^{137}Ba.

Using the nonemanating standard sample to provide detection efficiencies and measuring the air flow rate through the sampling filter, all activity measurements can be converted directly into airborne activity concentrations (Bq per volume of air).

For these experiments, airborne radon decay products were collected by pumping a specified volume of air at a rate of 20 l/min from the test chamber through a fiber filter especially designed for alpha-spectrometry (filter-type LFS). Measurements of the alpha-activity from the front (collection) side of the filter and beta-activity from the back of the filter were done simultaneously. The resulting alpha- and beta-spectra are shown in Figure 14.14.

The activities of ^{218}Po and ^{214}Po, indicated in Figure 14.14(a), on the filter are determined by comparing the integrated counts for each nuclide on the filter with that for the standard nonemanating source. As before, the beta-spectrum from ^{137}Cs is superimposed on the beta-spectrum from the filter; these spectra are shown in Figure 14.14(b).

In order to determine the ^{214}Pb and ^{214}Bi concentrations, the ^{137}Cs source of monoenergetic inner conversion electrons noted earlier was used to find the edge of the ^{214}Pb beta-spectrum. Since all but 10% of the ^{214}Pb decays have maximum beta-decay energies <0.73 MeV, it is assumed that all electrons with energies higher than this 0.73 MeV edge are due to the beta-decay of ^{214}Bi, while that portion of the energy spectrum with energies below this value consists of beta-decays from both ^{214}Pb and ^{214}Bi. Based on the usual beta-decays

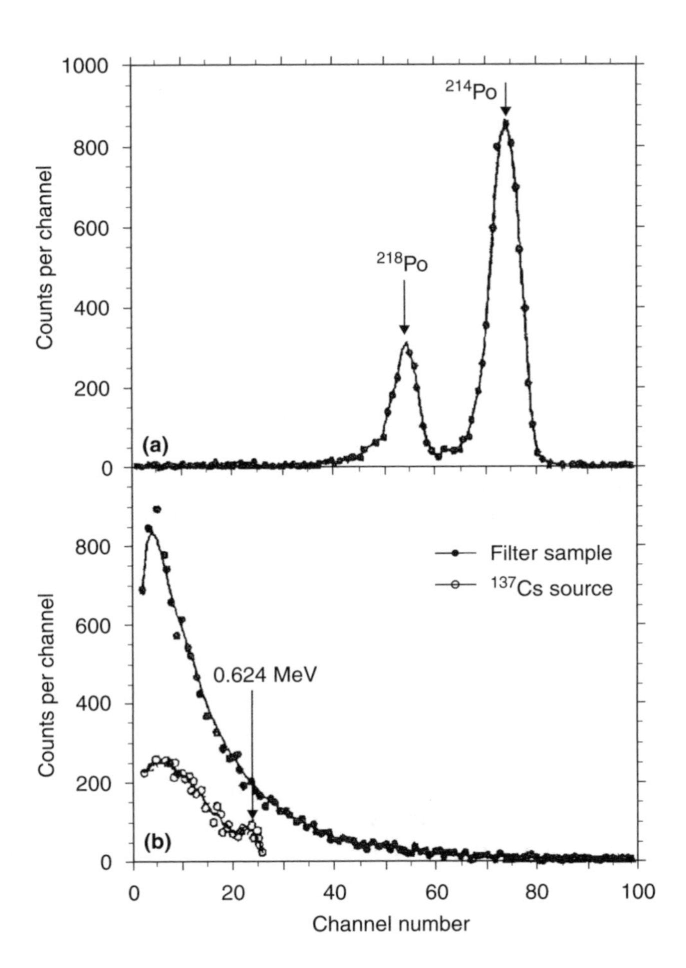

Figure 14.14 Spectra acquired from a filter sample of ^{222}Rn decay products from the chamber are similar to Figure 14.10. Alpha spectra data, using a CsI(TI) detector, from ^{222}Rn decay products collected on the filter are shown in (a). The beta spectra are shown in (b), again with the ^{137}Cs beta spectra added to help establish the edge of the ^{214}Pb beta energies.

from energy distribution, it is reasonable to neglect the small number of beta-particles from ^{214}Pb with energies greater than 0.73 MeV. A comparison of the integrated beta-decays with energies above 0.73 MeV obtained from the nonemanating standard source with that from the filter sample permits the direct determination of ^{214}Bi activity. A similar procedure for the total beta-decays with energies lower than 0.73 MeV provides the combined ^{214}Pb and ^{214}Bi activities on the filter sample. The ^{214}Pb activity can be obtained by simple subtraction of the ^{214}Bi activity from the combined activity.

14.10.2.5 Absorption of alpha-radiation in the sample

Using the technique described above to provide an independent measurement of both ^{214}Bi and ^{214}Po on the filter (Lubushkin et al., 1965), one can then evaluate the apparent loss of alpha-activity due to self-absorption in thick filter samples.

The coefficient of absorption is defined as the fractional difference between the actual ^{214}Po alpha-activity on the filter and the measured ^{214}Po activity, corrected for detector efficiencies:

$$k(m)=1-A_2/A_1$$

where $k(m)$ is the absorption coefficient as a function of collected aerosol mass (mg/cm^2), A_2 is the measured ^{214}Po alpha-activity on the filter, and A_1 is the actual ^{214}Po alpha-activity collected on the filter. Due to the very short half-life of ^{214}Po, the alpha-activity of this nuclide is identical to the beta-activity of its precursor, ^{214}Bi. Thus, assuming that absorption of beta-particles can be neglected, measurement of the betas from ^{214}Bi can be used to determine A_1:

$$A_1 = N_1(\beta)/e_\beta$$

and

$$A_2 = N_2(\alpha)/e_\alpha$$

where $N_1(\beta)$ is the beta-count rate for ^{214}Bi and $N_2(\alpha)$ is the alpha-count rate for ^{214}Po. The quantities e_β and e_α are the detection efficiencies for alpha- and beta-radiation, respectively. These can be determined directly from measurements using the nonemanating standard source:

$$A_0 = N_0(\alpha)/e_\alpha = N_0(\beta)/e_\beta$$

where A_0 is the activity of ^{214}Po in the nonemanating standard source (which is also the ^{214}Bi activity of that source) and $N_0(\alpha)$ and $N_0(\beta)$ are the count rates for the alpha-decay from ^{214}Po and the beta-decay from ^{214}Bi, respectively. In principle, the detection efficiencies for both alpha- and beta-particles are energy dependent. Any energy dependence can be evaluated by using the nonemanating source.
Substituting in Equation (1), we have the following formula for the absorption coefficient in terms of measured quantities:

$$k(m) = 1 - N_2(\alpha)/e_\alpha / N_1(\beta)/e_\beta = 1 - N_2 e_\beta / N_1 e_\alpha$$

In this work, the nonemanating standard source had a ^{226}Ra activity of 44 Bq and had the same geometry as the filter.

Experiments were performed to study the correlation between the mass of aerosol collected and the self-absorption of the alpha-radiation of the activity deposited on the filter. The distortion of the alpha-particle spectrum was studied and, using the measured sampling air flow rates, a method for the assessment of aerosol concentration in air was proposed based on the measurement activity of ^{214}Bi and ^{214}Po (Ruzer and Lubushkin, 1965). The coefficient of absorption of the alpha-activity of ^{214}Po in the filter samples was measured for aerosol mass collected on the filter ranging in thickness from 0 to 18 mg/cm^2.

For these experiments, NH_4Cl aerosols were generated by gas-phase condensation from the gas-phase interaction $HCl + NH_3 = NH_4Cl$. The aerosols were generated in a 2 m^3 chamber. Radon was then injected into the chamber. Following the establishment of radioactive equilibrium between radon and its decay products, filter samples were collected using the fiber filter developed for alpha-spectrometry noted previously. The increasing distortion of the alpha-spectra due to absorption as the amount of aerosoldeposited on the filter increases can be seen for both ^{218}Po and ^{214}Po in Figure 14.15 and, after an elapsed time period of several ^{218}Po half-lives, for only ^{214}Po in Figure 14.16.

The absorption results are summarized in Figure 14.17, where curve 1 shows $k(m)$ plotted against the thickness of the aerosol loading on the filter. Curve 1 can also be used for the assessment of aerosol concentration in air, but because the slope of the curve is not very large the uncertainty in this estimation procedure will be high.

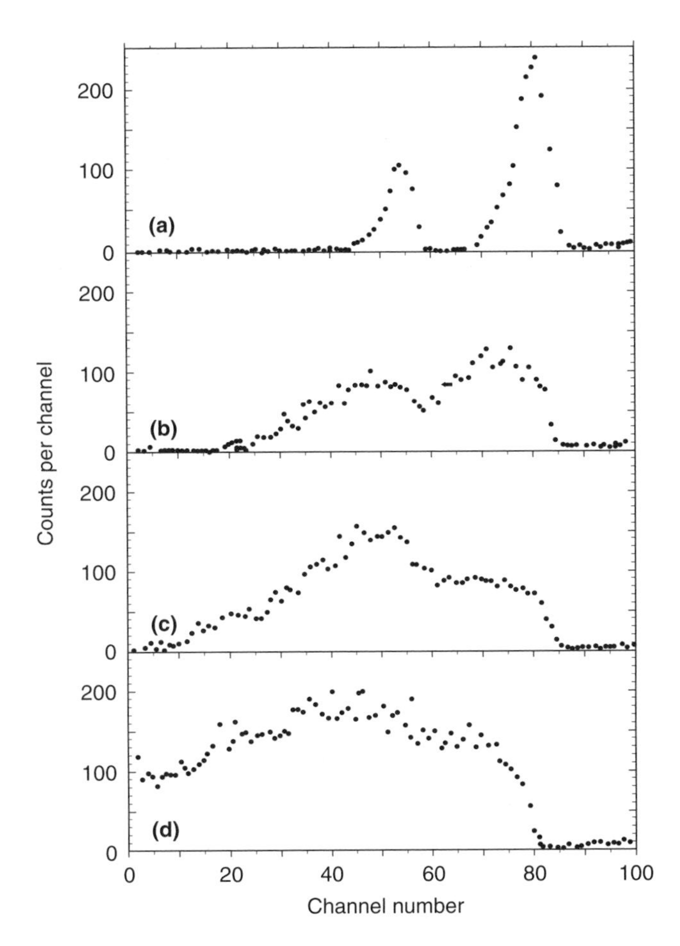

Figure 14.15 Energy spectra for alpha particles from both ^{218}Po and ^{214}Po collected on a filter for four different aerosol conditions. The aerosol loading on a filter is (a) 0 mg cm^{-2}, (b) 2.1 mg cm^{-2}, (c) 4.47 mg cm^{-2}, and (d) 9.0 mg cm^{-2}.

In order to increase the slope, thus reducing the uncertainty in these aerosol concentration estimates, particularly for small values of m, the discrimination threshold for the alpha-spectrometer was increased. These alpha-particles are counted and thus increase the proportion of the high-energy alpha-particles counted.

Curves 2, 3, and 4 in Figure 14.17 show the effect of increasing discrimination thresholds. While the collected aerosol mass can be determined directly by weighing the filter, using absorption curves like these to estimate the mass loading offers the advantage of not requiring any additional equipment and will, in many cases, provide sufficient accuracy.

The problem of accuracy in measuring the concentration of radon decay products — especially at the low concentrations typical of indoor environments — has been discussed in many papers (Nazaroff, 1984; Labushkin and Ruzer, 1965; Ruzer and Sextro, 1997; Ruzer and Labushkin, 1965).

Modifications of one of the standard three-point methods can substantially reduce the uncertainties (Nazaroff, 1984); however, even with this improvement, the uncertainties can still be 5–10 times higher than the counting statistical errors alone. These uncertainties are also dependent upon the shift in the radon decay product equilibrium, as discussed above. When these uncertainties are coupled with the need to make measurements in low concentration environments, such as those typically found indoors, the resulting uncertainties in

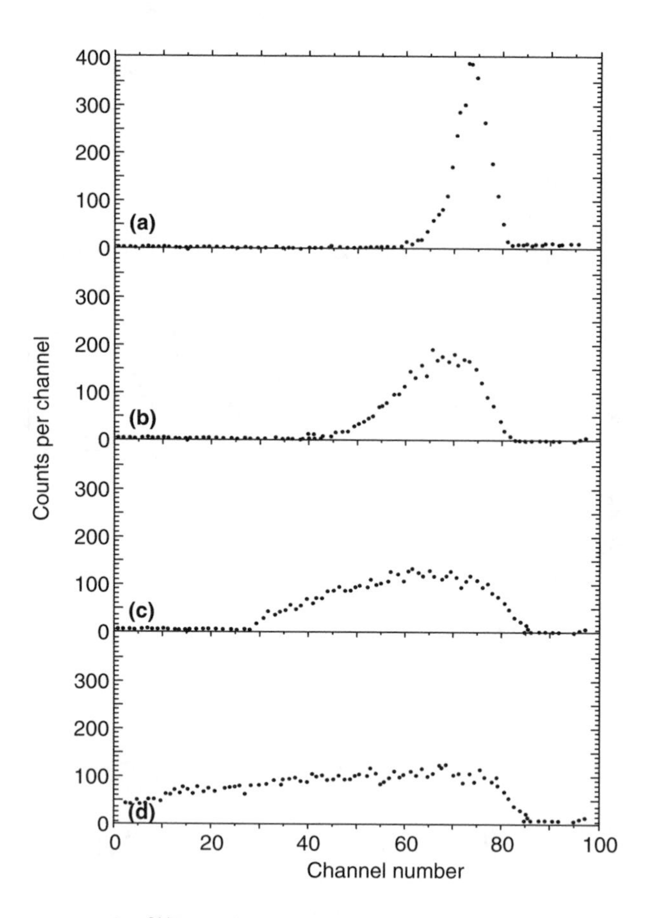

Figure 14.16 Energy spectra for ^{214}Po only.

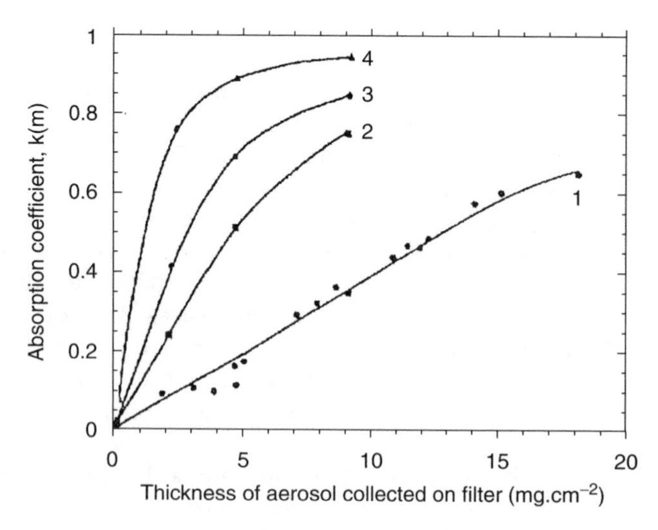

Figure 14.17 Absorption coefficient plotted against aerosol mass collected on filter samples. The four curves are based on data obtained with different discriminator thresholds for alpha particle detection: curve 1, 1V; curve 2, 50 V; curve 3, 60 V; curve 4, 70 V.

the determination of the radon decay product concentrations using a general three-point measurement procedure are often too large. These may be unacceptable when the purpose of the investigation is to examine the effects of changes in aerosol or ventilation conditions.

This work has also shown that the problem of collecting an adequate sample of airborne activity can compound the analytical difficulties due to the need to sample large quantities of air, which may contain significant concentrations of inactive (nonradioactive) aerosol. This material, when collected on the filter along with the radon decay products, may distort the alpha-spectra due to absorption. The method described here can be used to correct the measurements.

The use of a nonemanating reference standard as part of this process has several advantages. First, it provides a source of both alpha- and beta-particles with exactly similar energies as are observed in the measurement of radon decay product samples, so that detector efficiencies as a function of energy can be precisely obtained.

Second, the source provides both alpha- and beta-particles with essentially the same geometry as the filter samples, so no further correction is needed.

Third, the nonemanating source provides a tool for checking the linearity of the spectrometer across the wide range of energies needed for alpha-spectrometric measurements of Rn decay products. Even higher alpha-energies can be obtained by using a nonemanating sample of ^{228}Th, where one of the ^{212}Bi alpha-peaks has an energy of 8.78 MeV.

14.10.2.6 *Measurement procedures for the determination of the activity of RaA, RaB, RaC, and RaC′ on the filter by alpha- and beta-spectrometry (Labushkin and Ruzer, 1965; Ruzer and Sextro, 1997; Ruzer and Labushkin, 1965)*

14.10.2.6.1 ^{218}Po (RaA) activity measurement As seen from Figure 14.15, the division of RaA and RaC′ present no difficulties even if we use the scintillation spectrometer.

Let A_0 denote the activity of ^{226}Ra in the standard nonemanating source. Due to equilibrium in this source, the activity of ^{222}Rn, ^{218}Po, and ^{214}Po will also be A_0. When taking the measurements, the width of the window of the differential discriminator should be such that it can count only the pulses from RaA. The width and exact location of the window should be chosen by means of the nonemanating sample.

Let us denote $N_{a,0}$ as the count rate of RaA corresponding to the width and location of the window of the alpha-spectrometer in measuring the nonemanating source, N_a the count rate of RaA corresponding to the width and location in measuring the filter, and n_a the number of pulses for the time of measurement of the filter of the activity of RaA at the time t after the completion of sampling.

Then,

$$N_0 = e_\alpha A_0 \text{ and } N_a = e_\alpha A_a(\theta, t)$$

and the number of pulses from RaA on the filter n_a is

$$n_{a=a}\, e^{-\lambda_a t} dt = N_a(1 - e^{-\lambda_a t})/_{\lambda a}$$

and, consequently, from the two previous equations

$$A_a(\theta, t) = A_0\, N_a / N_0 = A_0\, n_a\, {}_{\lambda a} / N_{a,0}(1 - e^{-\lambda_a t})$$

In this case, it is not important at what exact moment the determination of A_a takes place, because by means of the exponential factor we can calculate the activity each time and also for the completion of sampling:

$$A_a(\theta, 0) = A_a(\theta, t) e_a^{\lambda t}$$

Of course, in these measurements as in other measurements, all geometric differences should be minimal.

14.10.2.6.2 ²¹⁴Po (RaC′) activity measurement The activity of RaC′ on the filter is measured in a similar manner. In this case, it is not necessary to use the differential discriminator; the integral discriminator can be used to discriminate all pulses with energies a little higher than the energy of alpha-particles of the RaA. Due to the fact that contribution in the count rate of RaC′ on the filter is not only by RaC but also by RaA and RaB, the correlation between the activity of RaC on the filter for different times cannot be established in the same simple way as that for RaA. Therefore, in the case of RaC′, it is better to decrease the duration of the measurement as much as possible.

As is clear from Figures 14.5–14.8, in cases when the shift of equilibrium is not so large, the activity of RaC on the filter in the first few minutes after sampling completion does not change substantially, which permits an increase in the duration of the measurement to achieve the lower statistical error. In the case of a substantial shift of equilibrium, the curves in Figures 14.5–14.8 allow one to assess the maximum error and to choose the necessary duration of the measurement for achieving the appropriate accuracy of measurement. The activity of RaC′ on the filter can be determined according to the formula

$$A_{c'}(\theta, t)=A_0\, N_{c'}/N_0$$

14.10.2.6.3 ²¹⁴Pb (RaC) activity measurement The division of RaB and RaC in the beta-spectra is not so obvious as for alpha-spectra of RaA and RaC′ because beta-spectra are indiscrete. If was possible to overcome this difficulty for two reasons: (i) the maximum energies of beta-spectra of RaB and RaC differ substantially from each other; and (ii) the monoenergetic line of the electrons of the inner conversion of ^{137}Cs corresponds to the edge between RaB and RaC. Therefore, for RaC activity on the filter determination, it is first necessary to establish the integral discriminator threshold in such a way that the detector will count only pulses with energies larger than that for the source of the monoenergetic electrons of ^{137}Cs, covered with very thin foil.

If $N_{c,0}$ is the count rate of RaC from the nonemanating source with the background with chosen discriminator threshold, N_c the count rate of RaC on the filter with the same background and discriminator threshold, and N_{bg} the count rate for the same discriminator threshold, the activity of RaC on the filter is

$$A_c(\theta, t)=A_0\,(N_c-N_{bg})/(N_{c,0}-N_{bg})$$

What was mentioned about the correlation of the activity of RaC′ on the filter in different moments of time is true for RaC.

14.10.2.6.4 ²¹⁴Bi (RaB) activity measurement RaB activity on the filter is measured by subtraction of the count rate of RaC from the summary count rate of RaB+RaC. Let us denote $N_{b+c,0}$ as the count rate from the standard nonemanating source with the background, N_{b+c} the count rate from RaB+RaC on the filter with the background, and N_{bg} the background count rate.

Then the count rates of RaB in the standard source $N_{b,0}$ and on the filter N_b will be

$$N_{b,0}=(N_{b+c,0}-N_{bg})-(N_{c,0}-N_{bg})$$

$$N_{b,0}=(N_{b+c}-N_{bg})-(N_c-N_{bg})$$

and the activity of RaB on the filter will be

$$A_b(\theta, t)=A_0\, N_b/N_{b,0}=A_0[(N_{b+c}-N_{bg})-(N_c-N_{bg})]/[(N_{b+c,0}-N_{bg})-(N_{c,0}-N_{bg})]$$

Due to the fact that the expression for $A_b(\theta, t)$ consists of two factors, one of which is determined by the activity of RaA, it is possible to establish a correlation between the activities of RaB in different moments of time.

Transfer from the measured activities of each decay product on the filter to the corresponding concentrations should be made according to the formulas

$$q_a=[N_a(\theta, t)/N_{a,0}]\, A_0/vk\Phi_a(\theta, t)$$

$$q_b=A_0/vk\Phi_{b,b}(\theta, t)\{[N_b(\theta, t)/N_{b,0}]-[N_a(\theta, t)/N_{a,0}]-[\Phi_{a,b}(\theta, t)/\Phi_a(\theta, t)]\}$$

$$q_c=A_0/vk\Phi_{c,c}(\theta, t)\{\{[N_c(\theta, t)/N_{c,0}]-[N_b(\theta, t)/N_{b,0}][\Phi_{b,c}(\theta, t)/\Phi_{b,b}(\theta, t)]\}$$

$$-[N_a(\theta, t)/N_{a,0}]\{[\Phi_{a,c}(\theta, t)/\Phi_a(\theta, t)]+[\Phi_{b,c}(\theta, t)/\Phi_{b,b}(\theta, t)][\Phi_{a,b}(\theta, t)/\Phi_a(\theta, t)]\}\}$$

The values for the functions $\Phi_{i,j}(\theta, t)$ for different θ and t are presented in Table 14.2. For example, for θ=5 min and t=1 min and the duration of the measurement 1 min,

$$q_a=0.032A_0N_a/N_{a,0}$$

$$q_b=A_0[(0.0115N_b/N_{b,0})-0.0028N_a/N_{a,0}]$$

$$q_c=A_0[(0.0115N_c/N_{c,0})-0.0023N_b/N_{b,0}-0.0009N_a/N_{a,0}]$$

Alpha- and beta-spectrometry can be used for research purposes or as a standard measurement technique. It is impossible to use this instrumentation in mining environments or even in measurements in dwellings. For these purposes, the simplified spectrometric method and instrumentation were developed based only on the alpha-spectrometer [18, 19]. In this case $N_a(\theta, t)$ should be measured once in a moment t_1 and $N_{c'}(\theta, t)$ twice in the moments t_2 and t_3 to obtain the concentration of RaA, RaB, and RaC according to the formula

$$q_a=2N_a(\theta, t_1)\, C_1/vk$$

$$q_b=2/vk[N_a(\theta, t_3)\, C_2-N_c(\theta, t_2)\, C_3-N_a(\theta, t_1)\, C_4]$$

$$q_c=2/vk[N_c(\theta, t_2)\, C_5-N_c(\theta, t_3)\, C_6-N_a(\theta, t_1)\, C_7]$$

where C_i are the coefficients that take into account the radioactive transformations of short-lived radon decay products on the filter as a function of the pumping time and times of measurements t_1, t_2, and t_3.

14.10.2.7 Other methods of determination of the radon decay products concentration in air

Besides different variants of measurement of the general activity on the filter and also alpha- and beta-spectrometry of the sample, in some situations another approach can be used based on the count rate measurement in the scintillation chamber or correspondingly current in the ionizing chamber [Ruzer, L.S., 1960].

This idea is based on the buildup curve of count rate in the scintillation chamber after the outside air with radon and its decay products enter the chamber. If pure radon (filtered completely from decay products) enters the chamber, the buildup curve will be according to the radioactive transformation of radon to RaA, then to RaB to RaC, and finally to RaC′. On the other hand, if the air entering the chamber consists of radon in equilibrium with decay products, the count rate will be constant within at least a 3 h period.

In cases when the equilibrium is between 0 and 1, the buildup curve will be between these two extreme situations.

Let us denote q as the concentration of radon itself, and s_a, s_b, and s_c the degrees of equilibrium of RaA, RaB, and RaC, respectively.

The activity due to RaA alone, that is, RaA present in a unit volume of the chamber, is $qs_a\, e^{-\lambda_a t}$ and the RaA activity formed from radon in a unit volume of the chamber is $q(1-e^{-\lambda_a t})$. The total activity in the volume v due to RaA is

$$A_a = qv(s_a e^{-\lambda_a t} + 1 - e^{-\lambda_a t})$$

The activity due to RaC′ in the chamber at any instant is $A_{c'}(t) = A_c(t)$, and will be a sum of the activities of RaC($A_{c,1}$), RaB($A_{c,2}$), RaA($A_{c,3}$), and Rn($A_{c,4}$). Using the solution for a chain of radioactive transformations [3], we obtain

$$A_{c,1} = qvs_c\, e^{-\lambda_c t}$$

$$A_{c,2} = qvs_b[\lambda_c/(\lambda_c-\lambda_b)]\,(e^{-\lambda_b t} - e^{-\lambda_c t})$$

$$\begin{aligned} A_{c,3} = qvs_a\lambda_b\lambda_c\, \{&[e^{-\lambda_a t}/(\lambda_b-\lambda_a)\,(\lambda_c-\lambda_a)] \\ &+[e^{-\lambda_b t}/(\lambda_a-\lambda_b)\,(\lambda_c-\lambda_b)\,] \\ &+[e^{-\lambda_c t}/(\lambda_a-\lambda_c)\,(\lambda_b-\lambda_c)\,]\} \end{aligned}$$

$$\begin{aligned} A_{c,3} = qv\,\{1&-[\lambda_b\lambda_c e^{-\lambda_a t}/(\lambda_b-\lambda_a)\,(\lambda_c-\lambda_a)] \\ &-[\lambda_a\lambda_c e^{-\lambda_b t}/(\lambda_a-\lambda_b)\,(\lambda_c-\lambda_b)] \\ &-[\lambda_a\lambda_b e^{-\lambda_c t}/(\lambda_a-\lambda_c)\,(\lambda_b-\lambda_c)]\} \end{aligned}$$

Finally, we have

$$A_\alpha = A_{Rn} + A_a + A_{c'} = A_{Rn} + qv[f_{Rn}(t) + s_a f_a(t) + s_b f_b(t) + s_c f_c(t)] = A_{Rn} + qvF(t) \quad (14.20)$$

where A_{Rn} is the background count rate of radon itself.

The graphs plotted for functions $f_i(t)$ give some idea of the contribution of radon and its decay products to the summary count rate and may be seen in Figure 14.18.

The ratios $s_i = q_i/q$ can be found if three values of the count rate for three different instants are used. The values of $f_{Rn}(t)$, $f_a(t)$, $f_b(t)$, and $f_c(t)$ for any time t are read off from the graph in Figure 14.18.

Figure 14.19 shows a graph of the function $F(t)$, which gives us some idea of the nature of the increase in the count rate in the scintillation chamber at different values of s_a, s_b, and s_c. As seen from Figure 14.19, the count rate values in the chamber corresponding to different ratios of the degree of equilibrium s_a, s_b, and s_c differ markedly from each other in the first 60–80 min following sampling of the air.

The low curve corresponds to the count rate increase in the case of radon without decay products. In the case when the decay products are in equilibrium with radon

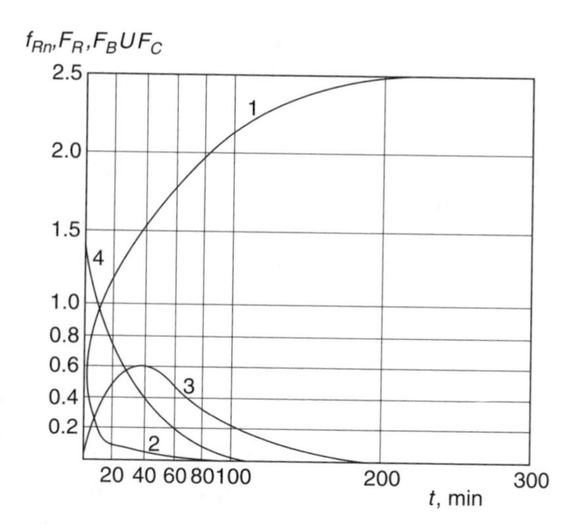

Figure 14.18 Graphs of functions f_{Rn} (t), f_A (t), f_B (t), and f_C (t) (curves 1–4, respectively).

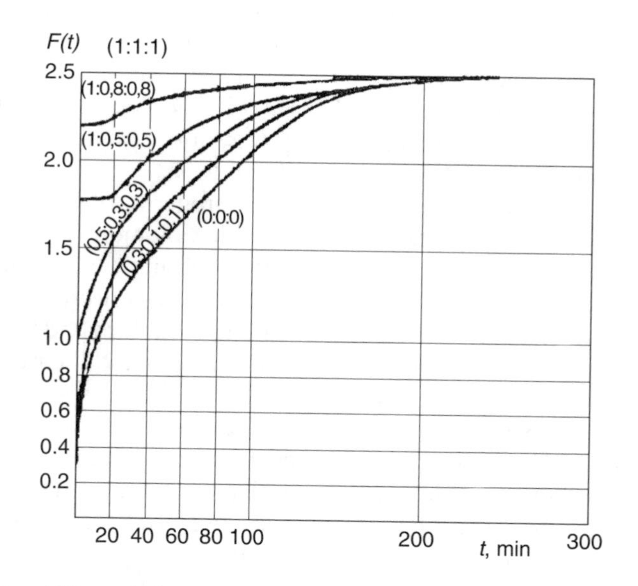

Figure 14.19 Function $F(t)$ at different values of η_A, η_B, η_C. Parentheses enclose the ratios of the degrees of equilibrium as $\eta_A : \eta_B : \eta_C$.

$(s_a = s_b = s_c = 1)$, $F(t) =$ constant. The method described here is not very precise but it does not need filtration, and therefore can be used for radon decay product measurement in small volumes.

The same idea can be used in the case when radon concentration is measured by the charcoal technique. In this case, however, the curve of buildup of the gamma-emitters RaB and RaC should be used according to the corresponding equation for gamma-activity.

14.10.2.7.1 Radon progeny concentration measurement In Furuta et al. (2000), a continuous radon progeny monitor with a silicon barrier detector (CAM PIPS) coated with a thin layer is described. The membrane filter 60 mm wide and 36 m long with a 0.8 μm pore size was used. It was printed with black lines at intervals of 5 m each for the skip movement by optical sensors. The energy resolution for ^{241}Am standard source (full-width at half-maximum (FWHM)) was 81.2 keV under a pressure of 25 mmHg, which was enough for discrimination of ^{218}Po and ^{214}Po with energies of 6.00 and 7.69 MeV, respectively.

In Hadler et al. (1994), the absolute efficiency of the supergrade PM 355 CR-39 track detector was studied, and it was shown that it could be used as a detector in ^{222}Rn and radon progeny monitoring by employing it as an alpha-spectrometer. With this device, the separate measurement of ^{222}Rn activity, radon daughter activity in the air, and plateout on the detector surface was studied. The relationship between track density and radon and progeny alpha-activity was studied, and possible application of the experimental results on dose calculation is discussed.

A "bronchial dosimeter" for radon decay products was presented in Yu et al. (1998). It consists of a 400 mesh wire screen and one filter. A velocity of 12.0 cm/s^1 was used during the sampling and measurements of the screen and the filter were provided at the same time by gross alpha-counting. The "dosimeter" can be used in conjunction with an activated charcoal canister for the radon concentration measurement to yield the effective dose conversion factor (DCF in mSv/y^1 per Bq/m^3) and the validity of the "dosimeter" by measuring the annual effective dose and the effective dose conversion factor in different environments. A comparison between total alpha-count data and alpha-spectrometry for obtaining radon and thoron progeny concentration from data collecting from air sampling filters is presented in Thiessen (1994). Unfortunately, due to imperfect resolution, mathematical compensation is needed in order to obtain accurate results. A two-stage ^{222}Rn progeny sampler (HE-sampler) is described . The collection efficiency of the sampler was optimized to match the particle size dependency of radon decay products conversion factor derived from the latest respiratory tract model of ICRP.

A method of measurement of radon and thoron progeny concentration in air by beta-counting using a Geiger–Müller counter is presented in Solomon (1997). The efficiencies of the progenies were evaluated one by one. The detector limits were about 0.1, 0.2, and 0.01 Bq/m^3 for ^{214}Pb, ^{214}Bi, and ^{212}Pb, respectively. Indoor activity concentrations were measured in 86 buildings in Hungary, where industrial waste rich in uranium had been used as building materials. Radon gas concentration was measured in 26 cases and the minimum, maximum, and average values of the equilibrium factors were 0.17, 0.73, and 0.40, respectively.

In Papp et al. (1997), a stochastic approach was used to determine the posterior probability density function, which described the uncertainty about radon progeny concentration measurement. Poisson distribution was used to constitute the likelihood given the observed number of counts. Using Bayes' theorem, posterior densities were obtained for the number of atoms of ^{218}Po, ^{214}Pb, and ^{214}Bi per unit volume.

14.10.2.7.2 Equilibrium factor and unattached fraction of radon progeny In Groer et al. (1997), the authors pointed out that there were a few reports available describing the behavior of ^{222}Rn and the properties of the equilibrium factor F and unattached fraction f. It was observed in Cavallo et al. (1999) that the diurnal variations of ^{222}Rn concentration in two office rooms depend on the operation of the air conditioners. ^{222}Rn concentration was considerably higher during holidays than during weekdays. No clear differences between holidays and weekdays were observed in the equilibrium factor. Correlation coefficients of 0.78 and 0.72 were estimated in two rooms to ^{222}Rn concentration and equilibrium equivalent ^{222}Rn concentration (EERC) and unattached fraction, f_p, between 0.02 and 0.03.

The equilibrium factor and unattached fraction of radon progeny in nuclear power plants are studied in [Hattori et al. (1995). The frequency distributions of the equilibrium factor, F, and the unattached fraction of the PAEC, f_p, in two power plants were a normal distribution with an arithmetic mean of about 0.3, and a lognormal distribution with a geometric mean of about 0.06–0.07, respectively.

Hattori et al. (1994) pointed out that the equilibrium factor, a measure of the degree of radioactive equilibrium between radon and its short-lived decay products, is often assumed to be 0.4 for the risk assessment. But in some cases, in active mines where the ventilation rate is high, the equilibrium factor can be much less than 0.4. In such cases, the lung dose rate will be substantially different. The critical problem was ignored in a comparative section of the BEIR VI Report. Thus, the ratio of home lung dose rate to diesel miner lung dose rate is actually 0.44 and not 1 as stated in BEIR VI. This may have major public policy implication.

In the paper "Radon progeny unattached fraction in an atmosphere far from radioactive equilibrium," the author demonstrated that an equation that establishes a correlation between aerosol particle concentration and unattached fraction of PAEC, f_p, will underestimate f_p by a large factor in some important situations, in particular, in atmospheres far from radioactive equilibrium, such as areas that are ventilated at a high rate. The lower portion of the particle size distribution ($d<5$ nm), often termed the unattached or ultrafine aerosols, may be great importance in the assessment of dose to the lung.

14.10.2.8 Methodical errors in the RaA, RaB, and RaC concentrations measurement

The methodical errors in radon decay product concentration measurements, that is, errors that depend only on the chosen method, are determined by

- Ratio $q_a : q_b : q_c$
- Duration of the sampling θ
- Chosen time of measurement after sampling t

Due to the strict time correlation between short-lived radon decay products, it is possible in principle to measure concentration by measurement of the activity on the filter by each of the alpha-, beta-, or gamma-isotopes. There can be different types of measurements:

- Measurements of the general alpha-activity of the sample in different moments after sample completion
- Measurements of the general beta- or gamma-activity of the sample. In some of the cases it makes sense to avoid the absorption of alpha-radiation due to dust or water in the sample
- Measurement of each radon decay separately on the filter by means of alpha- and beta-, or alpha-spectrometry

The correlation between the general activity on the filter, concentration of each decay product, duration of filtration, and time of measurement is given by formulas (14.18) and (14.19).

For q_a, q_b, and q_c determination, three values of the general activity on the curve of the alpha-decay should be chosen and the solution for the system of three equations should be found.

Expressions for relative errors in the measurements of q_a, q_b, and q_c by general alpha-activity measurement for $t_1=5$, $t_2=15$, and $t_3=30$ min are

$$\delta q_a/q_a=[5.94\delta N_5+13.45\delta N_{15}+8.49\delta N_{30}]/[5.94N_5-13.45N_{15}+8.49N_{30}]$$

$$\delta q_b/q_b=[0.45\delta N_5+1.48\delta N_{15}+3.59\delta N_{30}]/[-0.45N_5-1.48N_{15}+3.59N_{30}]$$

$$\delta q_c/q_c=[0.29\delta N_5+4.02\delta N_{15}+3.62\delta N_{30}]/[-0.29N_5+4.02N_{15}-3.62N_{30}]$$

Table 14.3 Methodical Errors in the Determination of Concentrations of radon Decay Products

	Shift of Equilibrium				
T=5,15,30 min	1:1:1;	1:0.8:0.6;	1:0,5:0.5;	0.5:0.7:0.7;	1:0.1:0.01
Total α					
$\Delta q_a/q_a$	30.0	21.0	16.0	44.0	3.5
$\Delta q_b/q_b$	5.3	4.5	5.7	5.4	5.7
$\Delta q_c/q_c$	7.5	8.9	8.0	3.5	46.0
Total α					
t=1,15,60 min					
$\Delta q_a/q_a$	12.0	8.0	5.8	15.0	1.9
$\Delta q_b/q_b$	2.3	2.2	2.7	2.2	5.6
$\Delta q_c/q_c$	3.7	4.4	4.0	3.7	20.0
Total β					
t=5,15,30 min					
$\Delta q_a/q_a$	643	298	407	880	69
$\Delta q_b/q_b$	68	69	70	68	90
$\Delta q_c/q_c$	18.0	19.5	19.0	18.0	42.0
Express method					
$\Delta q_a/q_a$	8.6	7.6	5.9	10.5	2.3
$\Delta q_b/q_b$	1.0	1.0	1.0	1.0	1.0
$\Delta q_c/q_c$	3.1	3.3	3.7	2.8	55.0
α-, β-spectrometry					
$\Delta q_a/q_a$	1.0	1.0	1.0	1.0	1.0
$\Delta q_b/q_b$	1.1	1.1	1.2	1.1	2.2
$\Delta q_c/q_c$	1.3	1.3	1.3	1.2	4.9

The values of the ratios of the methodical to statistical errors for the chosen method in the determination of q_a, q_b, and q_c are presented in Table 14.3.

For general beta-activity measurement, expressions for the calculation of the errors are

$$\delta q_a/q_a=[12.84\delta N_5+29.08\delta N_{15}+18.35\delta N_{30}]/[-12.84N_5+29.08N_{15}-18.35N_{30}]$$

$$\delta q_b/q_b=[1.71\delta N_5+4.30\delta N_{15}+3.18\delta N_{30}]/[1.71N_5-4.30N_{15}+3.18N_{30}]$$

$$\delta q_c/q_c=[1.18\delta N_5+1.97\delta N_{15}+1.10\delta N_{30}]/[1.18N_5-1.97N_{15}+1.10N_{30}]$$

For the "express" method (Markov, Ryabor and Stas (1965) and (1962)),

$$\delta q_a/q_a=(\delta n_1+\delta n_2)/(n_1-n_2)$$

$$\delta q_b/q_b=\delta n_2/n_2$$

$$\delta q_c/q_c=(0.244\delta n_1+0.597\delta n_2)/(0.597n_2-0.244n_1)$$

where n_1 and and n_2 are the number of impulses measured from the first to the fourth and from the seventh to the tenth minutes after sampling, respectively.

For the method of alpha- and beta-spectrometry, the expressions will be

$$\delta q_a/q_a=\delta n_a/n_a$$

$$\delta q_b/q_b=(3.79\delta n_b+0.75\delta n_a)/(3.79n_b-0.75n_a)$$

$$\delta q_c/q_c=[3.96\delta n_c+0.64\delta n_b+0,05\delta n_a]/[3.96n_c-0.64n_b+0.05n_a]$$

where n_a, n_b, and n_c are the number of pulses registered from the first to the fourth minute.

The methodical errors in radon decay product concentration measurements by means of different methods are shown in Table 14.3, expressed as a ratio of the methodical to statistical errors (Labushkin and Ruzer, (1965)).

The selection of the time of the measurement of the activity should be made according to:

- Achieving the lowest statistical error
- Achieving the lowest methodical error, which depends on the time chosen for the measurement

The analysis shows that in terms of satisfying both these demands, the time of measurement $t=0$ is the best, that is, the closer the beginning of the measurement to the end of the sampling, the lower the errors.

The selection of the duration of the sample θ is determined by the necessity to achieve the lowest statistical error and depends on the concentrations of RaA, RaB, and RaC.

The time of achieving equilibrium on the filter of each decay product depends on the decay half-time. The equilibrium for RaA will be established in $\theta=20$ min, for RaB in 150 min, and for RaC in 160 min, which results in a great difference in the equilibrium activity of every radon decay product.

In Table 14.4 the ratios of alpha $(A_a/A_{c'})$ - 1 and beta (A_b/A_c) — 2 of radon decay products for different durations of filtration θ are shown.

Table 14.4 Ratio of the Activities of the Alpha-Nuclides $A_a/A_{c'}$ — 1 and Beta (A_b/A_c) — 2 of Radon Decay Products for Different Durations of Filtration θ

1:1:1	$\theta=2$	$\theta=5$	$\theta=15$	$\theta=30$	$\theta=60$	$\theta=90$	$\theta=180$	$\theta=300$
1	0.80	0.60	0.29	0.16	0.09	0.07	0.06	0.06
2	1.02	0.98	0.93	0.82	0.71	0.65	0.6	0.6
1:0.8:0.6								
1	1.33	0.97	0.45	0.23	0.13	0.10	0.08	0.08
2	1.35	1.30	1.18	1.01	0.82	0.75	0.67	0.67
1:0.5:0.5								
1	1.61	1.20	0.57	0.31	0.18	0.14	0.12	0.12
2	1.04	1.03	1.01	0.88	0.74	0.68	0.62	0.62
0.5:0.7:0.7								
1	0.57	0.43	0.21	0.11	0.07	0.05	0.04	0.04
2	1.01	0.97	0.91	0.81	0.69	0.65	0.59	0.59
1:0.1:0.01								
1	59.52	31.9	7.40	2.56	0.98	0.69	0.53	0.52
2	9.00	7.35	4.46	2.47	1.42	1.17	0.98	0.97

It is obvious that for the long duration of the filtration θ the relative amount of the activity of RaA will be low, and therefore the measurement of RaA activity will have substantial errors. Even in the case of alpha-spectrometry, the peak of RaA will be measured on the background of the "tail" of RaC′ comparable with the peak of RaA.

In the case of equilibrium, the activity of RaA on the filter after 3 h of sampling will be no more than 10% of RaC′ activity. Of course, such differences in activities will result in additional error in the determination of RaA concentration by every above-described method.

This, particularly, led to the conclusion that the methods, based on the long duration of sampling, have large errors, especially in the determination of RaA. For example, for the method where the duration of the sample is 3 h, $A_a/A_{c'}=0.06$, which does not make it useful for RaA concentration measurement.

As can be seen from Table 14.4, when the duration of filtration is low, we can ignore the error related to the duration of filtration in the determination of q_a, q_b, and q_c, because the relative quantities of activities on the filter are close to each other in the wide range of the ratios $q_a{:}q_b{:}q_c$.

It should also be pointed out that the determination of q_a, q_b, and q_c only by alpha-measurement can result in additional errors as a result of the self-absorption effect due to the dust or water in the sample in cases of high aerosol concentration or humidity. Methods of alpha- and beta-spectrometry allow one to assess the errors associated with this problem.

14.10.2.9 Characteristics of thoron and actinon decay products

As is seen from Figure 14.2, the difference between the chain of thoron from radon is in the fact that the half-lives of Tn and ^{216}Po (ThA) are small. In fact, the whole thorium series contains no long-lived isotopes.

For dosimetry purposes, the four isotopes ^{212}Pb (ThB), ^{212}Bi (ThC), ^{212}Po (ThC′), and ^{208}Tl (ThC″) are of primary interest. ThA is not usually taken into consideration because of its short half-life.

A "thorium branching decay" is typical for this series. The half-lives of ThC′ and ThC″ are considerably smaller than that of ThC. Thus, the activities of ThC′ and ThC″ can be deduced. It is also important to note the presence of the strong alpha-emitter ThC′ with an energy of 8.8 MeV in this series, maximal for natural emitters. Despite the fact that ThB and ThC are beta-emitters, they particularly make the greatest contribution through ThC′ in the activity and, consequently, absorbed dose of alpha-emitters, deposited in the lungs.

The actinium chain is relatively simple. In this series, the alpha-emitters are very short-lived: AcA (Po^{215}) and AcB(Po^{211}). The beta-emitters are AcB(Pb^{211}) and AcC″ (Tl^{207}). The half-life of the most long-lived isotope in this series, AcB, is 36.1 min and its contribution to the total alpha-radioactivity and absorbed dose will be the greatest.

14.10.2.10 Basic equations for the thoron and actinon series

Let us denote the following for thoron progeny as in the case of radon decay products: λ_a, λ_b, λ_c, $\lambda_{c'}$, and $\lambda_{c''}$ are decay constants for ThA, ThB, ThC, ThC′, and ThC″, respectively; $A_a(\theta, t)$, $A_b(\theta, t)$, $A_c(\theta, t)$, $A_{c'}(\theta, t)$, and $A_{c''}(\theta, t)$ are the activities in the samples.

Applying the same relationships as for radon, we obtain the following expressions for activity on the filter (in the lungs):

$$A_a(\theta, t)=avkq_a\, \Phi_a(\theta, t)$$

$$A_b(\theta, t)=vk\,[q_a\, \Phi_{a,b}(\theta, t)+q_b\, \Phi_{b,b}(\theta, t)]$$

$$A_c(\theta, t)=vk\,[q_a\,\Phi_{a,c}(\theta, t)+q_b\,\Phi_{b,c}(\theta, t)+q_c\,\Phi_{c,c}(\theta, t)]$$

$$A_{c'}(\theta, t)=0.663A_c(\theta, t)$$

Because the contribution of ThA in the activity of ThB and ThC is very small ($\Phi_{a,b}(\theta, t)=0$ and $\Phi_{b,b}(\theta, t)=0$), the equations can be rewritten as

$$A_b(\theta, t)=vk[q_b\,\Phi_{b,b}(\theta, t)]$$

$$A_c(\theta, t)==vk\,[q_b\,\Phi_{b,c}(\theta, t)+q_c\,\Phi_{c,c}(\theta, t)]$$

If the activities of ThC′ (ThC) are measured for the two different moments of time t_1 and t_2 after sampling, q_b and q_c can be calculated from the equations

$$q_b=1/kv\{[A_c(\theta, t_1)\Phi_{c,c}(\theta, t_2)-A_c(\theta, t_2)\Phi_{c,c}(\theta, t_1)]$$

$$/[\Phi_{b,c}(\theta, t_1)\Phi_{c,c}(\theta, t_2)-\Phi_{b,c}(\theta, t_2)\Phi_{c,c}(\theta, t_1)]\}$$

$$q_c=1/kv\{[A_c(\theta, t_2)\Phi_{b,c}(\theta, t_2)-A_c(\theta, t_1)\Phi_{b,c}(\theta, t_1)]$$

$$/[\Phi_{b,c}(\theta, t_2)\Phi_{c,c}(\theta, t_1)-\Phi_{b,c}(\theta, t_1)\Phi_{c,c}(\theta, t_2)]\}$$

Graphs of the functions of $\Phi_{i,j}(\theta, t)$ are presented in Figure 14.20.

The values of the functions $\Phi_{i,j}(\theta, t)$ for different θ and t are presented in Table 14.5, which allow one to calculate the concentrations of ThB and ThC. In Figure 14.21, the graphs of the functions $A_b(\theta, t)/vk$ and $A_c(\theta, t)/vk$ are presented for different q_b:q_c for $\theta=$ 30 min, which allow one, under known θ values, to choose the more appropriate time of measurement. The correlation between $A_c(\theta)/vk$ for different q_b, q_c in Figure 14.22 allows one to choose the appropriate time of filtration.

14.10.2.10.1 Actinon series Let us denote the following for actinon progeny as in the case of radon decay products: λ_a, λ_b, λ_c are decay constants and $A_a(\theta, t)$, $A_b(\theta, t)$, $A_c(\theta, t)$ are

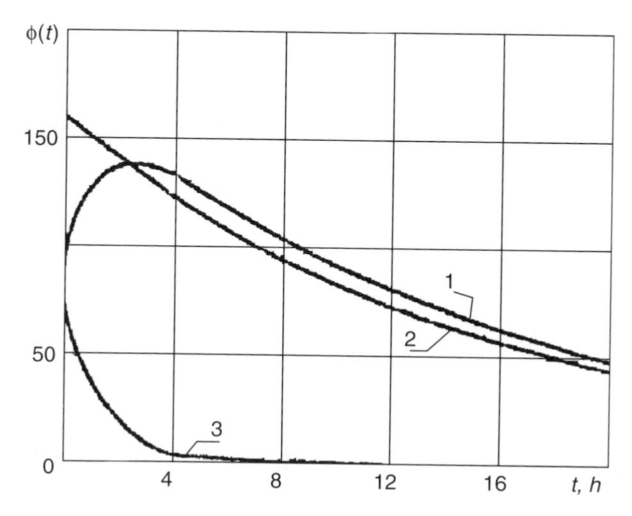

Figure 14.20 Graph functions $\Phi_{b,c}$ (1), $\Phi_{b,b}$ (2), and $\Phi_{c,c}$ (3) for thoron decay products (θ = 180 min).

Table 14.5 Values of $\Phi_{b,b}(\theta, t)$, $\Phi_{b,c}(\theta, t)$ and $\Phi_{c,c}(\theta, t)$ for Thoron Progeny

θ, $\Phi_{i,j}$	$\Phi_{i,j}$ for t (min)						
	0	5	10	15	30	60	180
5 $\Phi_{b,b}$	5.0	5.0	4.93	4.91	4.83	4.67	4.10
$\Phi_{b,c}$	0.14	0.41	0.66	0.90	1.53	2.46	3.85
$\Phi_{c,c}$	4.86	4.59	4.33	4.09	3.45	2.44	0.62
10 $\Phi_{b,b}$	9.95	9.90	9.84	9.79	9.63	9.32	8.18
$\Phi_{b,c}$	0.55	1.07	1.56	2.02	3.23	5.05	7.71
$\Phi_{c,c}$	9.45	8.92	8.38	7.92	6.67	4.73	1.20
15 $\Phi_{b,b}$	14.9	14.8	14.7	14.64	14.4	13.9	12.2
$\Phi_{b,c}$	1.20	1.96	2.68	3.34	5.11	7.74	11.55
$\Phi_{c,c}$	13.8	13.0	12.3	11.6	9.77	6.93	1.76
$\Phi_{b,b}$	29.5	29.4	29.2	29.0	28.6	27.65	24.3
$\Phi_{b,c}$	4.56	5.95	6.55	8.46	11.7	16.4	23.2
$\Phi_{c,c}$	25.4	24.0	22.6	21.4	18.0	12.8	3.23
60 $\Phi_{b,b}$	58.1	57.8	57.5	57.1	56.2	54.4	47.7
$\Phi_{b,c}$	16.2	18.6	20.7	22.8	28.1	36.0	46.7
$\Phi_{c,c}$	43.4	41.0	38.7	36.55	30.8	21.8	5.52
180 $\Phi_{b,b}$	163.4	162.6	161.7	160.8	158.2	153.1	134.3
$\Phi_{b,c}$	96.4	100.1	103.6	106.8	115.1	126.8	137.7
$\Phi_{c,c}$	76.2	72.0	67.9	64.2	54.0	38.3	9.69

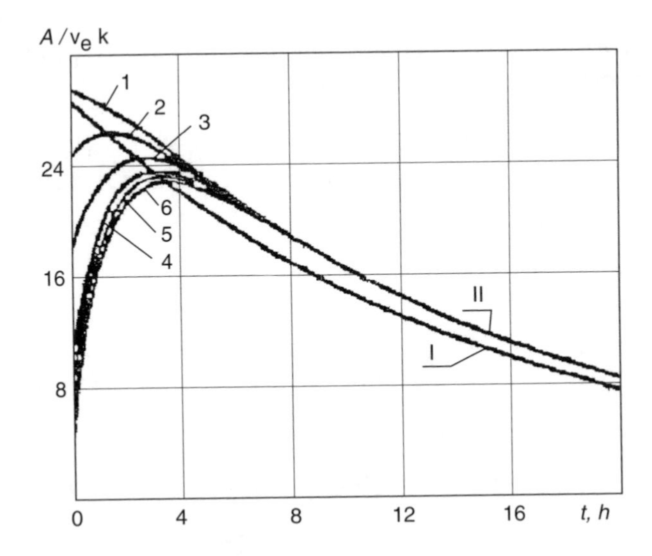

Figure 14.21 Graph functions $A_b(\theta, t)/vk$ (i), $A_c(\theta, t)/vk$ (ii) for different ratios q_b:q_c (θ = 30 min): 1 — 1:1; 2 — 1:0.8; 3 — 1:0.5; 4 — 1:0.2; 5 — 1:0.1; 6 — 1:0.01.

the activities of isotopes AcA, AcB, AcC on the filter, respectively. Taking into account the very small half-life of RaA, we have the following equations:

$$A_b(\theta, t)=vk[q_b\, \Phi_{b,b}(\theta, t)]$$

$$A_c(\theta, t)==vk\, [q_b\, \Phi_{b,c}(\theta, t)+q_c\, \Phi_{c,c}(\theta, t)]$$

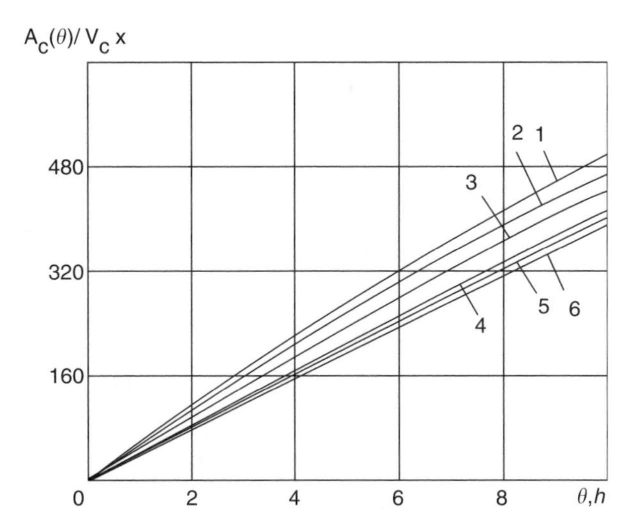

Figure 14.22 Graph function $A_c(\theta)/vk$ for different ratios: 1 — 1:1; 2 — 1:0.8; 3 — 1:0.5; 4 — 1:0.2; 5 — 1:0.1; 6 — 1:0.01.

where

$$\Phi_{b,b}(\theta, t)=(1/\lambda_b)(1-e^{-\lambda_b\theta})\, e^{-\lambda_b\theta}$$

$$\Phi_{b,c}(\theta, t)=(1/\lambda_b)\{[1+(\lambda_c e^{-\lambda_b\theta}-\lambda_b e^{-\lambda_c\theta})(\lambda_b-\lambda_c)]e^{-\lambda_c t}$$
$$+(1/\lambda_b)(1-e^{-\lambda_b\theta})\lambda_c(e^{-\lambda_c\theta}-e^{-\lambda_b\theta})/(\lambda_b-\lambda_c)\}$$

$$\Phi_{c,c}(\theta, t)=(1/\lambda_c)(1-e^{-\lambda_c\theta})\, e^{-\lambda_c\theta}$$

We can obtain similar equations for AcC. For the alpha-emitter AcC with alpha-particle energies of 6.56 MeV, it is better to determine the concentration of each decay product. As in the case of the radon decay products series, the most accurate method of the measurements of decay products in air is spectrometry. This method was described in [32] for the thoron and actinon decay products series Zhivet'nev et al. (1966).

For the alpha-spectrometry of thoron and radon decay products measurements, similar nonemanating samples consisting of radiothorium (^{228}Th) mixed with actinium (^{227}Ac), and consequently with all their decay products, were developed. All these nonemanating sources were developed as a cooperative effort between the All-Union Institute of Physico-Technical and Radiotechnical Measurements in Moscow and the Radium Institute in Leningrad (U.S.S.R.). Each sample consists of a backing, either a foil or a thick metal blank, onto which is deposited the appropriate long-lived parent radionuclide (^{226}Ra, ^{228}Th, ^{227}Ac) in a solution that is then evaporated. This is then sealed by deposition of a 0.2 μm layer, which retains Rn with an efficiency nominally greater than 99.9% (i.e., after more than 3 y only 0.1% Rn emanated from the source). The peak width (FWHM) due to the loss in the source is approximately 50 keV. Samples with activities in the range of 40–4000 Bq have been developed with areas from 5 to 20 cm^2. The alpha-spectrum for such nonemanating samples of thoron and actinon and their decay products is presented in Figure 14.23.

Figure 14.24 presents an alpha-spectrum of decay products of thoron and actinon from the sample, collected from the air on a fine-fiber filter. The alpha-radiation detector is a Cs(Tl) crystal of thickness ~0.1 mm.

The peaks on the alpha-spectrum are: (1) ThC with the energy of alpha-particles E_a= 6.04 MeV, (2) AcC (E_a=6.62 MeV), (3) RaC′ (E_a=7.68 MeV), and (4) ThC′ (E_a=8.87 MeV).

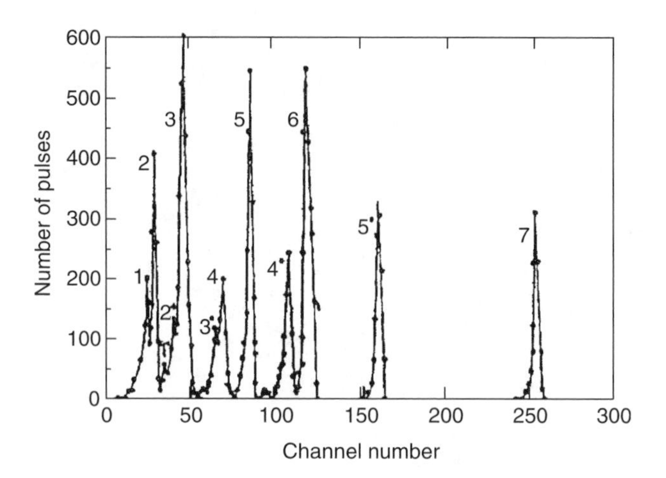

Figure 14.23 Alpha spectrum from a ^{220}Rn and ^{219}Rn series sample obtained using a semiconductor detector. The numbered peaks correspond to the following α particles. In the ^{228}Th (^{220}Rn) series, peaks 1 and 2, 5.34 and 5.42 MeV (from ^{228}Th); peak 3, 5.45 and 5.69 MeV (^{224}Ra); peak 3, 5.45 and 5.69 MeV (^{224}Ra); peak 4, 6.05 MeV (^{212}Bi); peak 5, 6.29 MeV (^{220}Rn); peak 6, 6.78 MeV (^{216}Po), and peak 7, 8.78 MeV (^{212}Po). In the ^{227}Ac (^{219}Rn) series, peaks 1* and 2*, 5.61 and 5.71 MeV (^{223}Ra); peak 3*, 5.90 MeV (^{227}Th); peak 4*, 6.55 MeV (^{211}Bi), and peak 5*, 7.39 MeV (^{215}Po). (Alpha energies here and in text are taken from Browne and Firestone,1986.)

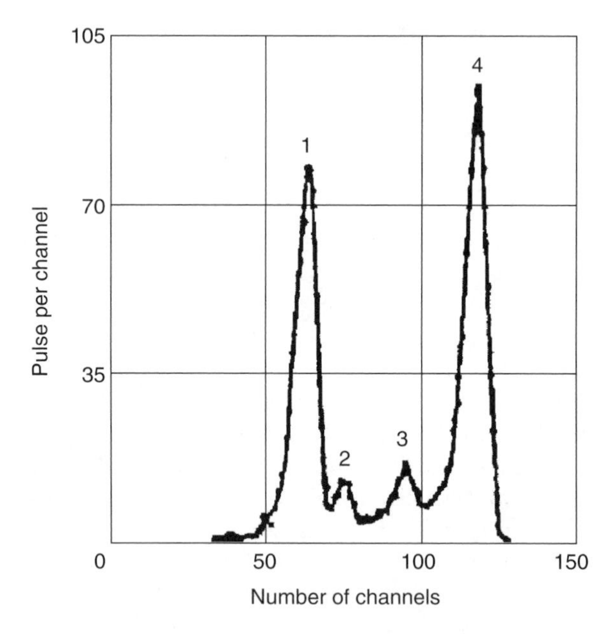

Figure 14.24 Spectrum of alpha-emission for the dispersion phase. Sample thorium/radium/actinium mix. 1 = ThC (E_α = 6.04 MeV); 2 = AcC (E_α = 6.62 MeV); 3 = RaC^1(E_α = 7.68 MeV); 4 = ThC^1 (E_α = 8.77 MeV).

For thoron decay products, as a result of self-absorption in the dust or water on the sample, there can be a substantial amount of alpha-particles with energies lower than for ThC′ (E_a=8.87 MeV). But because the closest peak of RaC′ is very far from ThC′, the "tail" from ThC′ can be easily detected.

Another advantage of the nonemanating samples and spectrometric methods is the opportunity for using it in the environment when all three or two of the decay product families are present in the air, for example, in rare metal facilities.

14.10.11 Unattached fraction measurements

Several studies published in the 1950s and 1960s noted the particular role of the unattached radioactivity of radon decay products (Knutson, 1998; Phillips et al., 1998; Yamasaki et al., 1982; Pistikopoulos et al., 1990; Chamberlain and Dyson, 1956; George, 1972; Holub and Knutson, 1987; Hopke, 1989). It was postulated that upon inhalation, the high mobility (large diffusion coefficient) of this radioactivity should cause significant diffusion in the lung airways and hence the main biological effect of the radiation was in the trachea. This was in fact confirmed when medical researchers observed the localization of lung cancer seen in the respiratory systems of miners. Although the effects of unattached fractions were documented long ago, they are still not fully understood today, especially in connection with the danger of radon in homes. To date, there is hardly any equipment that has been developed for commercial production to measure this type of radiation.

One such type of instrument was used for measuring the unattached activity in mines (Kartashev, 1966, 1967).

Reviews of the reliability of different methods for measuring the unattached activity, including some theoretical considerations, were provided in Polev (1967), Polev and Ruzer (1968), Dokukina et al. (1974), Knutsons et al. (1988).

The most commonly used methods for measuring unattached fractions are based on their small size and therefore the high diffusion coefficient. It should be mentioned, however, that the real deposition coefficient of the unattached activity on the walls of a cylindrical tube, the wires of the screen, and the channels of walls of diffusion batteries can have different values from 0 to 100%.

For the assessment of the reliability of instrumentation in this case as for other measurement techniques, a standard and calibration technique should be developed. This technique should provide stability and accurate measurement of the unattached fraction of radon decay products in the complete range of its values, including close to 0 and 100%. Such a standard and calibration technique was developed in the framework of the Special State Standard for the Volumetric Activity of Radioactive Aerosols in the former U.S.S.R. (Antipin et al., 1980).

14.10.11.1 Correlation between the unattached activity of radon decay products and aerosol concentration

Aerosol concentration can be expressed in several ways, such as particle number, particle number surface area, or particle volume (mass), depending upon on the aerosol properties of interest. The particle surface area concentration (expressed in units of cm^2/cm^3 or cm^{-1}) is, for example, an important parameter in the study of the interaction of radon decay products with available aerosols (Knutson, 1998; Phillips et al., 1988). The aerosol surface also plays a role in the behavior of gas-phase species indoors, such as polycyclic aromatic hydrocarbons, as it is a site for sorption and desorption of these species (Yamasaki et al., 1982; Pistikopoulos et al., 1990).

Light scattering is one of the main physical effects on which instruments for measuring small aerosol concentrations are based. It is widely used, for example, in the measurement of particle number concentrations through the use of a condensation nuclei counter or in measuring the size distribution of aerosols by means of an optical particle counter. In some cases, such as for optical particle counters, there are lower limits to the sizes of particles detected, typically in the vicinity of 0.1 μm.

Because the distribution of the radiation dose in the lung is not uniform but is highly dependent upon the proportion of unattached activity compared with that associated with the coexisting aerosol, many studies have been devoted to developing both the theory and techniques for measuring the unattached fraction and to correlating, both theoretically and experimentally, unattached fraction and aerosol concentration. In the last few years, the technique based on the collection of airborne activity on wire mesh screens became the most commonly used method for estimating the unattached fraction of radon decay products. Measurements using this technique have been made in both underground mining and residential environments.

The unattached fraction of ^{218}Po as a function of aerosol concentration in New Mexico uranium mines was studied in George and Hinchliffe (1972, 1975). The results of similar measurements were presented in Cooper et al. (1973) and Raghavayya and Jones (1974). The unattached fraction of potential alpha-energy concentration indoors has been reported in Postendorfer (1987), Reineking (1985, 1990).

The results of experiments in laboratory conditions will be presented below [23]. These conditions allow one to change the aerosol concentration in broad limits by simultaneously measuring both the radon decay products unattached concentration and aerosol concentration.

In Dokukina et al. (1974), the correlation between unattached fraction of ^{218}Po and aerosols was studied in a 2 m^3 calibration chamber. The chamber has an inlet for introducing Rn and aerosols, and an outlet for collecting samples of radon decay products and measuring aerosol concentration. In separate experiments for each aerosol size, seven different sizes of monodispersed latex aerosol, ranging from 0.3 to 2.1 μm, were generated and injected into the chamber. Their concentrations were measured using an aerosol counter, type AZ-5, which operates on the basis of light scattering. This instrument was a standard aerosol counter in widespread use in the former Soviet Union.

Measurement of the ^{218}Po unattached activity was carried out using a rectangular diffusion battery followed by a backing filter. An open-face filter operating in parallel to the diffusion battery sampled the total airborne ^{218}Po activity.

A single-channel alpha-particle spectrometer was used to count ^{218}Po alpha-particles. This device consisted of a Cs(Tl) scintillator coupled with a photomultiplier tube which, along with the appropriate electronics, identified the alpha-particles emitted from the decay of ^{218}Po (Antipin et al., 1976).

The diffusion battery consisted of a flat brass disc 2.3 mm thick with 117 rectangular apertures, each 0.18 mm wide and 5 mm high (Figure 14.25).

The 3.5 cm diameter diffusion battery disc was the same as the diameter of the standard filter used to collect the aerosol. The filter samplers were specially designed filters (filter-type LFS) used for alpha-spectroscopy measurements in the former Soviet Union.

A nonemanating sample of ^{226}Ra (Ruzer, 1993), also deposited on a 3.5 cm disc, was used to calibrate the efficiency of the detector for ^{218}Po deposited on or in the channels of the diffusion battery and for activity deposited on the filter.

In the former case, a mask was built for the nonemanating sample to provide a counting geometry similar to that of the diffusion battery. The nonemanating sample was also used to set up the discriminator windows for the alpha-spectrometer.

As mentioned before, it is critical, in the measurement provided with the diffusion battery, to determine the efficiency of unattached activity deposition on the walls of the battery. The following formulas were used to determine the unattached fraction of ^{218}Po based on the measured alpha-activities. The same formulas should be used in studies based on wire mesh screen or other types of measurement techniques.

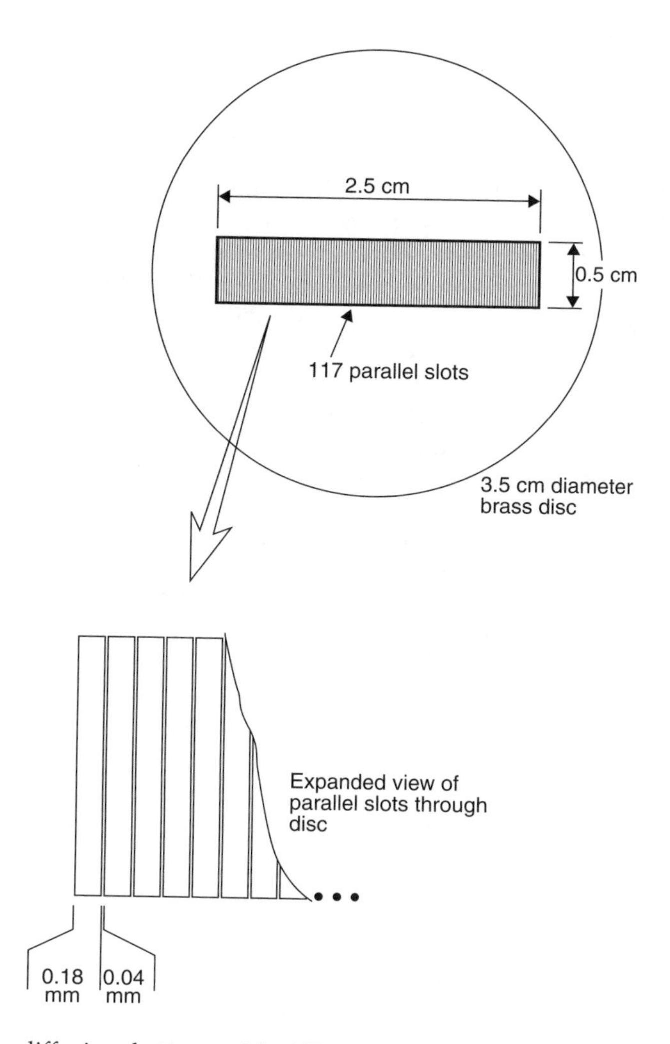

Figure 14.25 Disc diffusion battery with 117 rectangular channels. Each channel measures 0.18 × 5 mm. The disc has a thickness of 2.3 mm. The expanded view shows a portion of the parallel slots in more detail.

If N_i is the detected ^{218}Po activity for the three collection surfaces — $i=db$ for the diffusion battery, $i=bf$ for the backing filter, and $i=of$ for the (parallel) open-face filter—then

$$N_{db}=e_1e_3A_u+e_2e_4A_a \tag{14.21}$$

$$N_{bf}=e[(1-e_1)A_u+(1-e_2)A_a] \tag{14.22}$$

$$N_{of}=e(A_u+A_a)=eA_t \tag{14.23}$$

where e_1 and e_2 are the collection efficiencies of the diffusion battery for the unattached and attached ^{218}Po, respectively, e_3 and e_4 are the detection efficiencies for the unattached and attached ^{218}Po deposited on/in the diffusion battery, respectively, e is the detection efficiency for ^{218}Po alpha-activity collected on the filter, and A_u, A_a, and A_t are the airborne activity concentrations of the unattached, attached, and total ^{218}Po activity, respectively.

The unattached fraction, f, of ^{218}Po is defined as

$$f = A_u / A_t \tag{14.24}$$

The detection efficiency for the ^{218}Po activity collected on the filters, denoted as e in Equations (14.22) and (14.23), was determined using the nonemanating radium source noted earlier, which had the same source–detector geometry as activity on the filters. Because this calibration source is nonemanating, the ^{222}Rn and its short-lived decay products are in complete radioactive equilibrium with the ^{226}Ra parent; thus, the detection efficiency for the ^{218}Po alpha-particles can be easily measured. In the case of the diffusion battery, the parameters of the battery were originally calculated using formulas for diffusion deposition in rectangular channels (Fuchs, 1964). The size of the channels was selected to achieve a collection efficiency of nearly 100% for the unattached fraction ^{218}Po. The battery and the measurement technique itself were then calibrated experimentally using the chamber with 100% of the ^{218}Po unattached (achieved by near-zero aerosol concentration) and a zero level of unattached ^{218}Po, corresponding to an aerosol concentration of $>10^6$ particles cm^{-3}. These two cases provided the means for determining the collection and detection efficiencies based on Equations (14.21)–(14.23), as follows.

At 100% unattached fraction, $A_t \sim A_u$, so $N_{of} = eA_u$ from Equation (14.23), or $A_u = N_{of}/e$. Then from Equation (14.22),

$$N_{bf} = e(1 - e_1)A_u = (1 - e_1)N_{of}$$

and

$$e_1 = 1 - N_{bf}/N_{of} \tag{14.25}$$

The detection efficiency for the unattached ^{218}Po deposited on the diffusion battery can be found from Equations (14.21) and (14.23):

$$e_3 = eN_{db}/e_1N_{of} \tag{14.26}$$

Similarly, at 100% attached fraction, $A_t = A_a$, so $A_a = N_{of}/e$. Then from Equation (2),

$$N_{bf} = e(1 - e_2)A_a = (1 - e_2)N_{of}$$

and

$$e_2 = 1 - N_{bf}/N_{of} \tag{14.27}$$

The detection efficiency for unattached ^{218}Po deposited on the diffusion battery is

$$e_4 = eN_{db}/e_2N_{of} \tag{14.28}$$

Chamber calibration experiments were conducted using both monodisperse aerosols and known mixtures of different aerosol sizes to determine the operating parameters of the diffusion battery. Based on these experiments, the deposition and detection efficiency parameters of the diffusion battery are $e_1 = 0.95$, $e_2 = 0.24$ (this depends on the particle size distribution; the value used here is an average of all calibration aerosol sizes), $e_3 = 0.55$, and $e_4 = 0.24$. The difference in detection efficiency is due to the difference in deposition location of the unattached and attached ^{218}Po within the diffusion battery aperture.

Based on Equation (14.24) and combining Equations (14.21) and (14.22), a formula for the unattached fraction of ^{218}Po in terms of measured quantities can be derived:

$$f = A_u e/N_{of} = [eN_{db}(1-e_2)-e_2e_4N_{bf}]/N_{of}[e_1e_3(1-e_2)-e_2e_4(1-e_1)] \tag{14.29}$$

This formula applies generally to any similar system. This approach can be used, for example, with methods designed to increase the collection of low concentrations of radioactive aerosols, by using higher flow rates and/or modified diffusion batteries, as long as appropriate calibration procedures can be used to provide the efficiency parameters in Equation (14.29).

A summary of the results of this approach is presented in Figure 14.26, which shows the unattached concentration of ^{218}Po relative to particle surface area concentration, based on calibration measurements made with monodispersed latex aerosols of different sizes and concentrations (Dokukina and Ruzer, 1976).

In each case, the measured aerosol concentration was converted to the aerosol surface area concentration. These results suggest that for aerosols in the size range covered by this calibration, from 0.3 to 2.1 μm in diameter, the particle surface area concentration, s, in the corresponding range from 10^{-5} to 0.3 cm^{-1} is related to the unattached fraction of ^{218}Po, f. This is the basis for the approach proposed here for measuring small concentrations of aerosols (Ruzer, 1964).

The calibration procedure used relies upon monodisperse spherical particles, from which surface area can be directly calculated. In practice, an "equivalent surface area" should be used, which is the surface area of a spherical aerosol having the same diffusion deposition property as the real aerosol. Under actual measurement conditions (e.g., in a building or in a mine), the only measurements necessary to determine the unattached fraction and to infer the average aerosol surface area are those described above — a diffusion battery followed by a backing filter and an open-face filter operated in parallel. The alpha-activity from ^{218}Po on each of these three collectors is then measured and used with appropriate calibration factors to yield the unattached fraction.

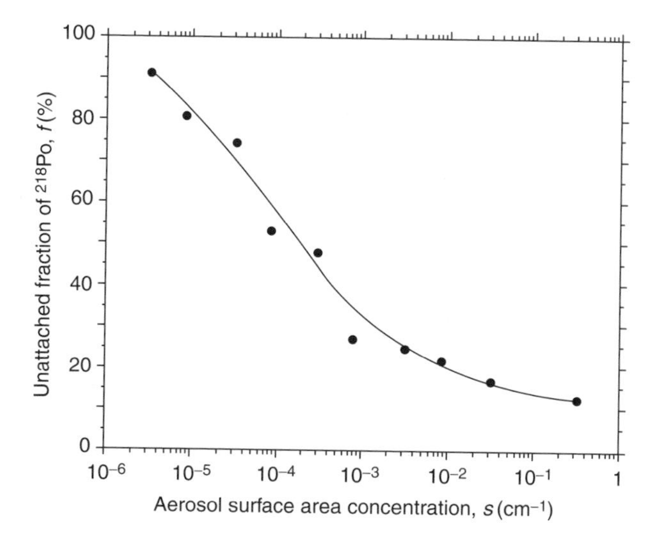

Figure 14.26 Measured ^{212}Po unattached fraction plotted against particle surface area concentration based on calibration experiments with the rectangular diffusion battery. The line is to guide the eye.

The curves in Figure 14.26 can be used for calibration purposes. For example, for spherical particles with diameters of 0.1 and 1.0 μm and concentrations of 10^5 cm^{-3}, the particle surface area concentration is 3×10^{-5} and 3×10^{-3} cm^{-1}, respectively. The corresponding unattached fractions of ^{218}Po are 70 and 25%, respectively.

Figure 14.27 shows calculated errors in the measurements of both *s* and *f*. Errors for *f* were calculated for three different total activity concentrations of ^{218}Po.

The estimated error for the area of particles of any size, *i*, is shown as a function of $(1/N_i)$, where N_i is the concentration of particles of the *i*th size range. Combining the results shown in Figures 14.26 and 14.27 permits one to estimate the lower detection limit on aerosol concentration afforded by this approach. From Figure 14.26, the largest unattached fraction measurement achieved in the calibration chamber, 92%, corresponds to an aerosol concentration of 3×10^{-6} cm^{-1}. In terms of particle number concentration, the limits, which depend upon particle size, are from 20 (at 2.1 μm particle diameter) to 1000 particles cm^{-3} (at 0.3 μm particle diameter). For these concentrations, relative errors range from 30 to 40% as seen from Figure 14.27.

A quantitative comparison of results of the experiments described here with those previously reported (George and Hinchliffe, 1972, 1972; Cooper et al., 1973; Raghavayya and Jones, 1974; Postendorfer, 1987), Reineking (1985) presents some difficulties because most of the previously reported measurements were conducted under laboratory conditions that yielded a wide range of unattached fractions for similar aerosol conditions. These studies did not directly investigate the problem of measurement and calibration across a broad range of aerosol and radon decay product concentrations, particularly in the case of low aerosol concentration and the concomitant high fraction of unattached radon progeny.

Note also that the surface area of aerosol particles has also been used in studies of aerosol concentrations using an instrument called an epiphanometer (Gaggler, 1989;

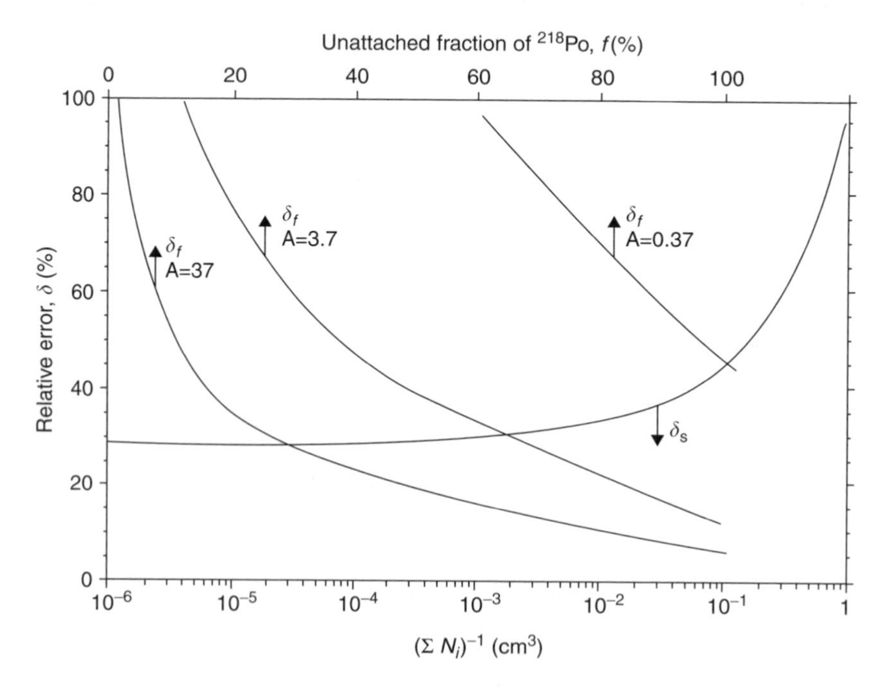

Figure 14.27 Relative errors (standard deviation/absolute value) in estimating ^{218}Po activity, as shown. The curve for δ_s is plotted against the inverse of the total particle concentration. *A* is the total ^{218}Po activity in kBq m^{-3}.

Baltensperger, 1991). The physical principle on which these measurements were based is Fuchs' coagulation theory for the attachment coefficient of radon decay product atoms to particles with small aerodynamic diameters. The epiphanometer signal is roughly proportional to the surface area of aerosol particles. However, in contrast to the method described here in which the unattached activity is measured, the epiphanometer technique measures only the attached radon decay product activities because a capillary tube acts as a diffusion barrier for the unattached lead atoms. Thus, when the aerosol concentration is small, the corresponding unattached activity concentration will be high. Under these conditions, the statistical uncertainty in determining the aerosol area based on measurement of the unattached fraction of ^{218}Po should be smaller than the measurements that are based on the corresponding attached fractions.

An examination of Figures 14.26 and 14.27 indicates the applicability of this proposed method. Two factors are important: (1) the unattached fraction of ^{218}Po should be at least 15–20% or higher to provide sensitivity to changes in particle surface area concentration and (2) the amount of ^{218}Po collected should be enough to achieve the desired statistical accuracy. As a practical matter, these conditions exist in many underground mining situations and in some residual environments.

14.10.11.2 *Measurements of other radon decay product unattached activity concentrations*

In most of the studies, only ^{218}Po unattached fraction concentration measurements were provided. Both from theoretical considerations and some experimental results, the concentrations of unattached ^{214}Bi and ^{214}Pb are much smaller than that for unattached ^{218}Po. This is true in terms of concentration.

However, from the point of view of the contribution to the dose, the unattached activity of ^{214}Bi and ^{214}Pb should not be ignored according to the formula for potential alpha-energy concentration (PAEC), which is equally true both for attached and unattached activity:

$$\text{PAEC} = 1.05q_a + 5.16q_b + 3.80q_c$$

This formula shows that even when the ratio between unattached fraction concentrations of radon decay products is in the range of 1:0.3:0.1, their contribution to the dose will be comparable. That means that from a dosimetric point of view in providing unattached fraction measurements for dosimetric purposes, the unattached fraction of all three short-lived decay product activities should be measured.

One example of such a study is presented in Kojima and Abe (1988). Continual and simultaneous measurements of unattached radon daughters (RaA, RaB, RaC) and the total radon daughters were performed in a house under normal living conditions, which included ordinary living habits, the style of house construction, and the normal natural ventilation rate in Japan. The hourly concentrations of three radon daughters were obtained separately and simultaneously with a radon daughter monitor, which had air filtration and gross alpha-counting systems. The instrument was automatically operated (and maintenance free) over a period of 2 months.

The unattached atoms of radon daughters were measured with another radon daughter monitor with wire screens substituting for the filter. From the measurements for 7 months (1986–1987), it can be stated that:

1. The concentration of RaA is the most predominant of these daughters, and unattached RaB is always detectable.
2. The median value of the unattached fraction, indicated by potential alpha-energy f_p, is around 0.04; RaB makes a contribution of more than 30% to the overall f_p.

3. Data for the unattached fraction show a typical diurnal variation, having a low level in intervals of 7–9 and 20–22 local hours. It corresponds to aerosol generation from the use of gas ranges in the house.

The collection of unattached radon daughters was performed by a wire screen of 500 mesh with a roll shape 10 m in length and 60 mm width. Using this apparatus, the hourly data for a continual 5 days were obtained. The collection efficiency was calculated by the equation of Thomas and Hinchliffe (1972). The efficiency at a linear air velocity of 29.7 cm s^{-1} was found to be 0.99.

Because the counting efficiency of the wire screen is not the same as that of the membrane filter due to attachment of some of the unattached daughters to the back of the screen, the authors used a special procedure to obtain the correction factor, that is, using two types of technique — for unattached and attached measurements. This correction factor was 0.79 ± 0.21.

Because some of the unattached atoms may adhere to the monitor's sampling head during the sampling, which will decrease unattached radon daughters collected on the wire screen, another correction factor was introduced using other types of experiments. This factor was 0.820×10 (normalized to the result of the filter sampler) at a flow rate of 35 l m^{-1}. The concentration of RaA and RaB was estimated to be about 2 and 0.3 Bq m^{-3}, respectively, with a standard deviation of 50%.

Field measurements were conducted in a detached two-storied concrete house under ordinary living conditions in Japan. In the house, the mechanical ventilation system was not set. The frequency of door and window opening increased during the hot summer months. In cold months, the rooms in the house were heated by oil heaters. The ventilation rate was found to be 0.4–0.8 h^{-1} with closed windows. The apparatus were set in a living room at the ground floor of the house. The total radon daughters were measured continuously without a pause, and unattached radon daughters were measured for about 10 days in a month. Other parameters, like aerosol particles, were measured only occasionally.

In these measurements, unattached RaA has the highest concentration and RaB the second highest. The concentration of unattached RaC was extremely low, and therefore can be ignored if aerosol concentration is in the range between 3×10^3 and 2×10^4 cm^{-3}. However, unattached RaB cannot be ignored because its concentration is 10–20% of unattached RaA. It is concluded that the difference in concentration found between unattached RaB and unattached RaC is due to the recoil effect of the alpha-decay of attached RaA, which will produce some unattached RaB.

14.10.11.3 *Effect of recoil nuclei being knocked off aerosol particle unattached concentrations of radon-decay products (Kolerskii et al., 1973)*

At the present time, the common opinion is that radon atoms do not deposit on aerosol particles and, consequently, unattached daughters of the first decay product are developed only by decay of the atoms of Rn. However, atoms of the second, third, and further daughter products can be developed in two ways: from the unattached fraction of the previous daughter products by its decay; and from the attached fraction to aerosol particle atoms of the previous product, which by its decay recoiled from the aerosol particles due to the "knocking off" (recoil) effect.

Taking into account recoil under radioactive decay, the equation can be written for the concentration of unattached activity of radon progeny for the change in time:

$$dn_i/dt = \lambda_{i-1} n_{i-1} + \alpha_{i-1} \lambda_{i-1} N_{i-1} - \lambda_{pi} n_i - \lambda_i n_i \tag{14.30}$$

where i is the number of daughter products, n the unattached concentration of the corresponding product, λ the decay constant, α the recoil coefficient (the ratio of the numbers of attached atoms of the ith products, which will recoil under radioactive decay), N_i the concentration of aerosol particles with ith decay product, and λ_{pi} the constant of deposition of the ith product to the aerosol particles. When aerosol concentration is not changed or changes insignificantly, it is possible to say that λ_{pi} is proportional to aerosol concentration, that is,

$$\lambda_{pi}=kN \tag{14.31}$$

For the equilibrium situation, that is, for $dn_i/dt=0$ and $dN_i/dt=0$, the expressions for n_i and N_i will be

$$n_i=(\lambda_{i-1}\, n_{i-1}+\alpha_{i-1}\, \lambda_{i-1}\, N_{i-1})/(\lambda_i+\lambda_{pi}) \tag{14.32}$$

$$N_i=(\lambda_{pi}\, n_i+(1-\alpha_{i-1})\lambda_{i-1}\, N_{i-1})/\lambda_i \tag{14.33}$$

The fraction f_i of i-product unattached atoms is

$$f_i=n_i/(n_i+N_i) \tag{14.34}$$

By substituting the values for n_i and N_i from (14.32) and (14.33) in (14.34), we obtain

$$f_i=[\lambda_i/(\lambda_i+\lambda_{pi})]\,[f_{i-1}+\alpha_{i-1}(1-f_{i-1})] \tag{14.35}$$

Radon is not attached to aerosol particles and, therefore, its unattached atom fraction $f_1=1$ and its recoil factor is $\alpha_1=0$.

By substituting f_2 for f_3, f_3 for f_4, etc., we obtain the following general expression for the fraction of i-product of the unattached activity:

$$f_i=f_1[\lambda_k(1-\alpha_{k-1})/(\lambda_k+\lambda_{pk})]+[(\alpha_{j-1}\lambda_j)/(\lambda_j+\lambda_{pj}) \\ \times(1-\alpha_{k-1})\lambda_k/(\lambda_k+\lambda_{pk}))]+\alpha_{i-1}\,\lambda_i/\lambda_i+\lambda_{pi} \tag{14.36}$$

The ejection factor of daughter emanation-product atom aerosol particles, which is produced by recoil in radioactive decay, has not been measured to date. It is obvious that the value of this factor is in the range of 0–1. Within the limits, it assumes the form

$$f_i=[\lambda_k/(\lambda_k+\lambda_{pk})] \quad \text{for all } \alpha_i=0 \tag{14.37}$$

and

$$f_i=\lambda_i/(\lambda_i+\lambda_{pi}) \quad \text{for all } \alpha_i=1 \tag{14.38}$$

The relationship of the fraction of unattached atoms of radon daughter products from λ_p was calculated from (14.37) and (14.38) and represented in the form of curves. In plotting these curves, it was assumed that the permanent attachments of all radon decay products to the aerosol particles are equal to each other. Since $0<\alpha_i<1$, the true curves of $f_i=f_i(\lambda_p)$ should be located between the curves of the relationships for $\alpha_i=0$ and 1. Their position, in the main, is determined by the type and energy of radiations which produce the ejection of radioactive atoms from the aerosol particles owing to the recoil effect.

It was found that the relative value of the recoil effect contribution to the fraction of the radon daughters unattached activity increases with an increase in λ_p, that is, with an increase in aerosol concentration.

On the basis of an analysis of (14.36), it is possible to suggest the following method for determining experimentally the ejection factor of the unattached decay radon products.

According to (14.36) and bearing in mind that $f_1=1$ and $\alpha_i=0$, the fraction of the unattached first daughter product by radioactive decay can be written as

$$f_2=\lambda_2/(\lambda_2+\lambda_p) \tag{14.39}$$

In a similar manner, we can find the unattached atoms for the second product:

$$f_3=[\lambda_3(\lambda_2+\lambda_p\alpha_2)]/(\lambda_2+\lambda_p)\,(\lambda_3+\lambda_p) \tag{14.40}$$

By eliminating λ_p from (14.31) and (14.32), we obtain

$$\alpha_2=(\lambda_2/\lambda_3)+f_2/(1-f_2)\,f_3/f_2-f_2/(1-f_2) \tag{14.41}$$

This expression can be used for the experimental determination of the ejection factor of the radon decay product atoms from the aerosol particles produced by the recoil effect in the course of the radioactive decay. For this purpose, it is necessary to measure the unattached activity fractions f_2 and f_3 of the first and second decay products, respectively.

In preliminary measurements of the unattached fraction in a chamber where the air was enriched with radon, the following mean values were obtained:

$$f_2(\text{RaA})=0.33 \text{ and } f_3=0.03$$

By using these values in (14.41), we obtain

$$\alpha_2(\text{RaA})=0.4$$

The expression for α_2(RaA) can be used for the evaluation of the contribution to the unattached activity of RaB by the attached RaA atoms, which in the course of the decay were knocked off the aerosol particles. For this purpose, let us rewrite (14.40) in the following manner:

$$f_3=\lambda_2\lambda_3/(\lambda_2+\lambda_p)(\lambda_3+\lambda_p)+\alpha_2\lambda_3\lambda_p/(\lambda_2+\lambda_p)(\lambda_3+\lambda_p) \tag{14.42}$$

$$f_3=f_{31}+f_{32} \tag{14.43}$$

where f_{31} is the fraction of the unattached activity of RaB atoms, which are produced in the course of unattached RaA decay, and f_{32} is the fraction of unattached RaB produced by attached RaA atoms being knocked off the aerosol particles due to the recoil effect of radioactive decay (Figure 14.28).

Figure 14.28 shows the relationship of f_3, f_{31}, and f_{32} to λ_p, obtained from Knuston, Solomon and Strong, 1988 and Antipin, et al. 1980 for $\alpha_2(\text{RaA})=0.4$.

The average aerosol concentration in the continental surface layer of the atmosphere can be considered for calculating purposes to be $N=10^4$ cm^{-3} (Junge, 1965).

Jacoby (1964) obtained the following results for this value of N: $\lambda_p=0.85\times10^{-2}$ sec^{-1} for a diffusion mechanism of radioactive atoms precipitation on aerosol particles and 1.5×10^{-2} sec^{-1} for the case of superposition and diffusion of electrostatic precipitation.

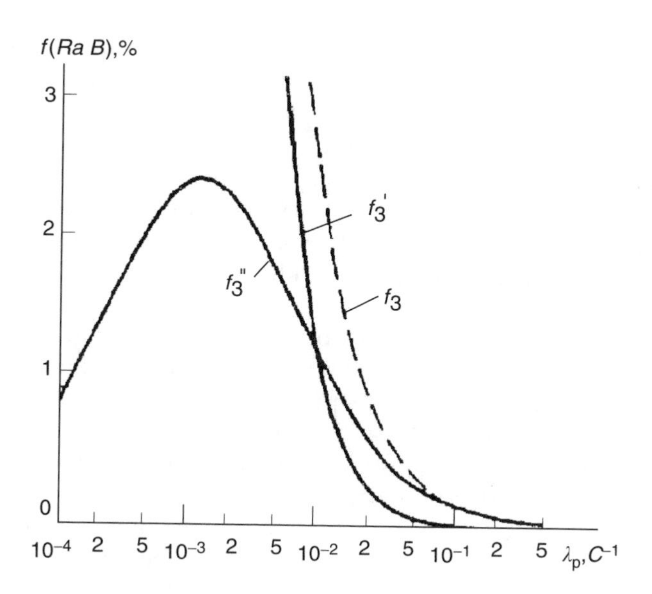

Figure 14.28 Relationship of f_3, f_3', and f_3'' to λ.

From these values of λ_p, it is possible to evaluate the mean fraction of unattached radon decay products for the continental surface area layer of the atmosphere.

It is seen from the graph that the contribution of ejected attached RaA atoms to the fraction of unattached RaB is small for small values of λ_p. For $\lambda_p=9.5\times10^{-3}$ sec^{-1}, this amounts to half of all the unattached atoms and as it follows from (14.34) that $\lambda_p>\lambda_2/\alpha_2$ $=9.5\times10^{-3}$ sec^{-1}, it becomes predominant; however, the absolute value of unattached RaB becomes very small. It is interesting to note that the fraction of unattached RaB atoms due to the ejection of attached RaA has a clearly expressed maximum on the curve representing the relationship of the former to λ_p. The exact position of this maximum can be found by differentiating it with respect to λ_p and equating it to zero:

$$df_{32}/d\lambda_p=\alpha_2\lambda_3(\lambda_2\lambda_3-\lambda_\pi^2)/[(\lambda_2+\lambda_p)\ (\lambda_3+\lambda_p)]^2=0 \tag{14.44}$$

The position of the maximum is therefore determined from the condition

$$\lambda_2\lambda_3-\lambda_\pi^2=0 \tag{14.45}$$

or

$$\lambda_p(\text{max})=1.3\times10^{-3}\ \text{sec}^{-1}$$

If (14.31) holds and the value of $\lambda_p(max)$ for the continental surface layer of the atmosphere is taken into account, then the value of $\lambda_p(max)$ will correspond to the approximate aerosol concentration:

$$N\ (max)=\lambda_p(max)10^4/1.5\times10^{-2}=10^3\ \text{cm}^{-3}$$

It is seen from (14.45) that the position of the maximum does not depend on the ejection factor α_2(RaA) and is determined only by the ratio of the decay constants of RaA and RaB.

The physical meaning of this maximum is simple; when the aerosol concentration starting from low values (small value of λ_p) grows, the number of unattached RaA atoms increases as well as that of the attached RaA atoms subject to ejection.

For $N=N(max)$, the ejection of decayed RaA atoms and their secondary attachment to aerosol particles balance each other.

For $N > N(max)$ [$\lambda_p > \lambda_p(max)$], the secondary attachment of ejected atoms to aerosol particles exceeds their ejection and, with a further increase in aerosol concentration, the part of the unattached RaB fraction due to knocking off of the RaA atoms attached to the aerosol particles decreases.

In conclusion, an extensive review of the history of development and utilization of diffusion batteries as a main tool in the measurement of nanometer-sized particles, including the unattached fraction of radon progeny, is presented in Knutson (1999).

In Hopke (1992) it was mentioned that the term "unattached fraction" was originally developed for lung dosimetry, but now some ambiguity exists in its definition between a physical size basis and a dosimetric definition. So, according to Hopke (1992), we can use the term "unattached activity" to refer to activity with thermodynamic diameters <2 nm, where the atoms of radon progeny are associated with a small number of other molecules. The rest of the radon progeny activity with sizes >2 nm can also deposit on the tracheobronchial tree.

It should be pointed out, however, that this deposition, according to recent studies Martonen (1987), will be different from uniform because of differences between real structure and unevenness of the inner surface of this part of the lung and that used in modeling.

In reality, as shown in Nemmar et al. (2000), a substantial part of the particles in the range of sizes between 5 and 12 nm penetrate the tracheobronchial tree and deposit on different organs of the body through the bloodstream.

The only correct understanding of the role of unattached fractions in lung dosimetry can be achieved by using them in experiments on humans, as was provided, after approval from a special medical committee, in Butterweck et al. (2001).

14.10.12 Measurement of artificial radioactive aerosol concentration

Among all the radioactive aerosols that come into contact with respiratory organs, artificial aerosols are the most dangerous. Especially harmful are such nuclides as the alpha-emitters ^{210}Po, ^{226}Ra, ^{227}Ac, ^{228}Th, and ^{239}Pu, nuclides of other transuranium elements, and also the beta-emitter ^{90}Sr. The Norms of Radiation Safety includes these nuclides in a group of elements noted for their especially high toxicity (NRS-96, 1999). For this group of radioactive aerosols, the Norms sets the maximum permissible concentration (MPC) in the workplace at less than 3.7 Bq/m^3, more than two orders of magnitude lower than for the decay products of radon. The low MPC is explained by the fact that these nuclides are present in the air as soluble components, which allow them to migrate from the lungs into the bloodstream and to eventually precipitate in bone, where they emit radiation into the bone tissue for a long period of time. According to the International Commission of Radiologiocal Protection (ICRP), the half-life of soluble compounds of ^{239}Pu in bone is about 200 yr. Low MPC values significantly complicate the problem of measuring and monitoring concentrations of this type of radioactive aerosol and in determining the accuracy of measurements for the following reasons:

- To obtain the optimal statistical error, a substantial quantity of air must be circulated and filtered.
- When large quantities of air are filtered, a similarly large quantity of nonradioactive aerosols is collected on the filter. This in turn distorts the results of the measurements, especially for the alpha-emitters.

- It was found that artificial radioactive aerosols are often large-particle aerosols; thus, alpha-emission can be absorbed by the aerosol particles.
- The concentration of artificial radioactive aerosols must be measured against a background of the short-lived nuclide decay products of radon and thoron. The concentrations of these short-lived isotopes are two to three orders of magnitude higher than the MPC for long-lived isotopes and, no less important under actual conditions, are subject to significant variations.

The fourth condition mentioned above, in particular, complicates the process of measurement and makes interpretation of the results especially difficult. It is obvious that the accumulation of alpha-emitters during air filtration proceeds differently for natural and artificial radioactive aerosols. When the air containing short-lived radon decay products is filtered, equilibrium is reached after a certain time between the number of atoms of a given nuclide precipitating on the filter and the numbers decaying per unit of time. The shorter the half-life, the more quickly this saturation point is reached; Figures 14.29–14.31 illustrate this.

In contrast, the accumulation of primary artificial radioactive aerosols during filtration occurs linearly because of their long half-lives. The decay products of ^{212}Pb (ThB) and ^{212}Bi (ThC) will accumulate nearly linearly, since their half-lives are sufficiently long.

After filtration, the activity of the long-lived artificial aerosols does not change, whereas the activity of natural radioactive aerosols — radon decay products — decreases with an average half-life of 35 to 40 min. This means that 3/h after filtration, about 2% of the activity

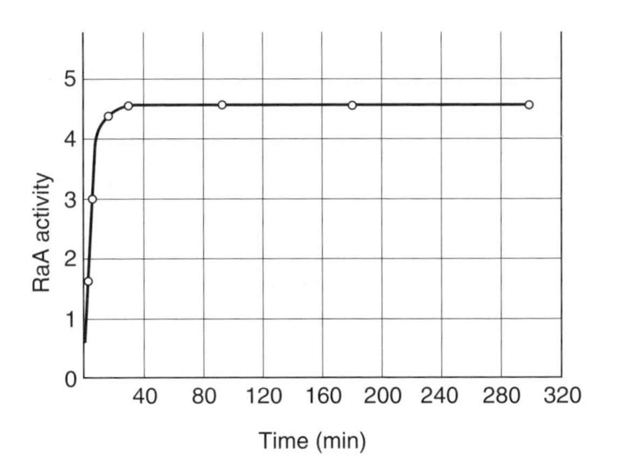

Figure 14.29 Dependence of RaA activity on duration of filtration (q_A=1).

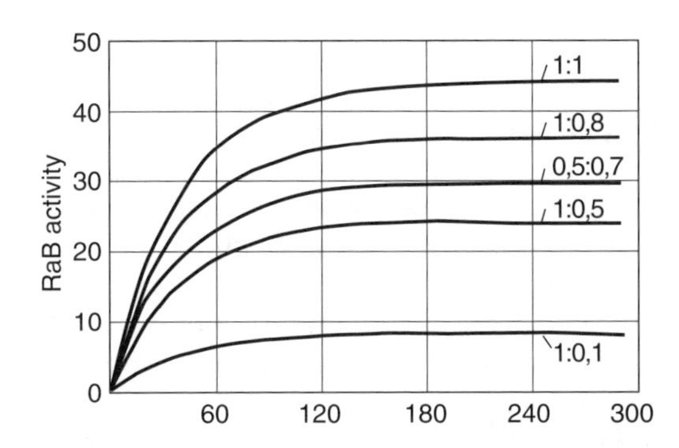

Figure 14.30 Dependence of RaB activity on duration of filtration for various values of q_A:q_B.

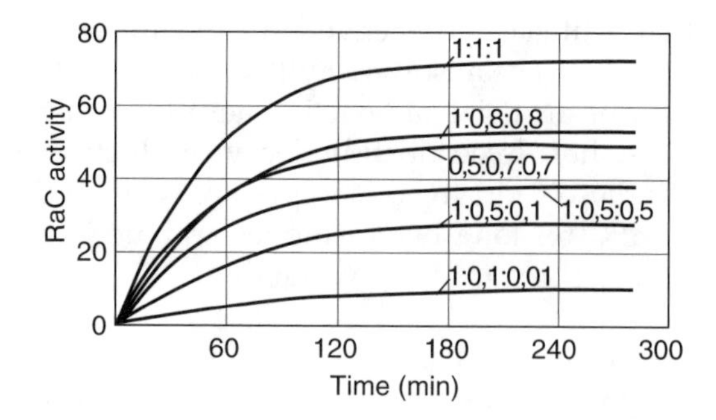

Figure 14.31 Dependence of RaC activity on duration of filtration for various values of q_A:q_B:q_C.

of radon decay products remain, which were present at the moment filtration was concluded. If we make the average value of atmospheric radioactivity for radon decay products 3.7·Bq/m^3, then, even with measurements of the concentration of the radioactive aerosols of ^{239}Pu at the level of MPC and of the fraction of MPC, the natural and artificial activity on the filter after a 3-h exposure will be of the same order; then, distinguishing them using general activity measurements methods and devices will be impossible. In addition, radon decay product concentrations can be, and often are, higher than indicated above, which strengthens the "masking effect" caused by short-lived radon decay products even more.

The activity of thoron decay products decreases with an average half-life of approximately 10.5 h. However, because of the peculiarity in the thorium decay chain, the activity can even increase 3 h after filtration. If it decreases, it does so quite insignificantly. According to Belousova (1961), the average concentration of thoron decay products in the atmosphere is 1.5 Bq/m^3. This means that even after an exposure of 40 h after sampling, the activity of thoron decay products on a filter will be equal to the artificial nuclide ^{239}Pu when its presence in the air is at MPC level. Of course, waiting for more than 40 h to obtain information on concentration values is unsatisfactory in practical dosimetry.

To determine the accuracy of any measurements of artificial radioactive aerosol concentration, it is necessary to know the amount of radon and thoron decay product concentration in the air or on the filter. Therefore, only measurement methods based on the separation of nuclides in the sample, that is, primarily methods of alpha-spectrometry, can avoid systematic errors associated with the presence of natural aerosols in the sample. The general expression for the alpha-activity of natural and artificial radioactive aerosols on a filter is

$$A(\theta,t)=A^{azt}+A^{nat}_{\mathrm{Rn},(}+A^{nat}_{\mathrm{Th},}$$

$$=qvk\theta+vk\,\{q_{a,\mathrm{Rn}}\,[\Phi_a(\theta,t)+\Phi_{a,c}(\theta,t)]+q_{b,\mathrm{Rn}}\,\Phi_{b,c}(\theta,t)+q_{c,\mathrm{Rn}}\,\Phi_{c,c}(\theta,t)\}+$$

$$+1.663\,kv\,[q_{b,\mathrm{Tn}}\,\Phi_{b,c}(\theta,t)\,+q_{c,\mathrm{Tn}}\,\Phi_{c,c}(\theta,t)],$$

where $A(\theta,t)$, A_{art}, $A_{\mathrm{Rn},nat}$, and $A_{\mathrm{Th},nat}$ are the activities of artificial and natural nuclides of the filter together, the activity of only artificial nuclides, and the activity of radon decay products and thoron decay products, respectively; q_i, Rn, q_i, Tn are the concentrations of the corresponding decay product of each family; functions $F_{i,j}(\theta,t)$ represent the buildup and corresponding decay of each decay product within time θ of the sampling and time t after the completion of sampling.

The first term on the right-hand side in this equation corresponds to the contribution of artificial, long-lived radioactivity. The next terms are in the following order: ^{218}Po (RaA),

^{214}Pb (RaB), and ^{214}Bi (RaC) from the radon series, and ^{212}Pb (ThB) and ^{212}Bi (ThC) from the thoron series. In the textbook on dosimetry [3] data are given for the relationship of the radioactivity of artificial and natural isotopes on a filter for different durations of filtration θ and a time interval after its conclusion t.

According to Ivanov (1964), the formula for radon decay products is

$$q^{art}/q_{\mathrm{Rn}}=(A_{art}/A_{\mathrm{Rn}})\cdot(\theta/\tau)\cdot(e^{t/\tau})/(1-e^{-\theta/\tau}),$$

where q_{art}, qRn, A_{art}, and ARn are the corresponding concentrations and activities of artificial and natural nuclides at time t after the conclusion of filtration, and τ is the "average" half-life of the short-lived radon decay products, 35 min. The results of the calculations confirm that when measuring artificial aerosols, the portion of natural aerosols in the sample should be carefully evaluated.

14.10.12.1 *Measurement technique for artificial aerosol concentration measurements*

One of the first pieces of spectrometric equipment developed in the former U.S.S.R., dating from the late 1950s to the early 1960s, was described in Bolotin et al. (1961). The dispersion phase of artificial aerosols was precipitated onto a filter thickness of 0.3 g/m^2. A Cs (Tl) scintillator, 0.2–0.4 mm thick, combined with a photomultiplier serves as the transducer. Using this equipment, with a 40 mm diameter filter, the energy resolution for alpha-particles of ^{239}Pu (E=5.15 MeV) is 13–15%, while for an electroplated source of ^{239}Pu of the same dimensions the resolution was 10–13%.

Figure 14.32 shows the results of measurements with this equipment. As seen from the figure, such energy resolution causes spectra of natural and artificial nuclides to overlap.

Therefore, it is necessary to resort to measuring activity in two channels. One channel is used to count alpha-particles of natural and artificial aerosols (N_1 — corresponding count rate) and the other is used for natural aerosols alone (N_2 — count rate). The concentration of the artificial radioactive aerosols can be determined by the formula

$$q=(N_1-kN_2)/e_d v\theta t_{mes}$$

where k is the ratio of the counting rate from natural nuclides in the first to the second channel, determined experimentally, e_d is the efficiency of detection of the alpha-particles from artificial radioactive aerosols in the first channel, and t_{mes} is the duration of the measurement.

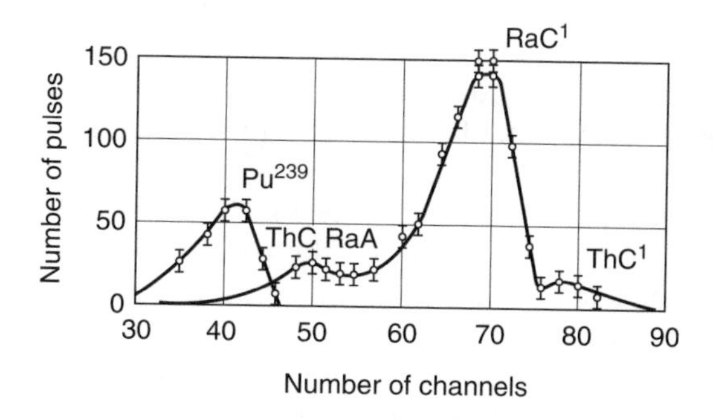

Figure 14.32 Spectrum of alpha-radioactive aerosols (scintillation spectrometer).

In Grigorov (1966), Bolotin et al. (1961) and Ruzer (1993), the further development of this method is provided. In Grigorov (1966), a method is proposed for calculating the spectra of alpha-particles from thick sources with the help of matrixes for calculating what is termed the "compensation coefficient" *k*. In Bolotin et al. (1961) and Ruzer (1993), a study was conducted by both calculation and experiment on the dependence of coefficient *k* on:

- Variations in the concentration ratios of (^{218}Po)RaA, (^{212}Bi)ThC, (^{214}Po) RaC′, and (^{212}Po) ThC′ in air
- Variations in dispersion, chemical composition, and dust concentration
- Variations in the performance of transducers over time and temperature range
- The stability of the discrimination level

For such types of measurements, it is critically important to carefully select the discrimination threshold and a collimator and to prove that coefficient *k* will be equal for all decay products of radon and thoron. The results also show that for these measurements, it is better to focus on alpha-spectrometry with a significantly better energy resolution, including scintillation spectrometers. The nonemanating sources from ^{226}Ra, ^{228}Th, and ^{227}Ac [7] have become a very important tool in spectrometry, because they

1. all have short-lived decay products of Rn, Tn, and Ac in equilibrium, that is, with the constant activity in time,
2. establish the discrimination threshold for the alpha-spectrometer, and
3. check the linearity of the spectrometer.

In some of the studies, the concentration of artificial radioactive aerosols was measured by means of ionization chambers (Osborne and Hill, 1964) with a resolution of 1–3%. The pulse ionization chamber had a high sensitivity, but it also had its deficiencies. The most important is the complicated and cumbersome equipment entailing special systems.

Some works for improving the resolution and detecting the efficiency of alpha-spectrometers are presented below.

Figure 14.33 shows how it is possible to find the optimal voltage on semiconductor detectors by using a nonemanating source. It is clear from Figure 14.33 that when U=6.0 V, the energy resolution of alpha-particles with energy 7.68 MeV ^{214}Po (RaC′) is worse than for the lowest energies.

By increasing the voltage correspondingly to 10.5 V (b), 15 V (c), and 19.5 V (d), the energy resolution for the alpha-spectrum of ^{214}Po (RaC′) will improve substantially; at the same time, the spectrum of the ^{226}Ra, ^{222}Rn, and ^{218}Po (RaA) will change insignificantly. Similar experimental results for the scintillation detector of Cs (Tl) with a surface area of 3 cm^2 are presented in Table 14.6.

It should be pointed out that in the experiments described in Labushkin et al. (1973), an energy resolution of 0.6% for the alpha-particles of ^{239}Pu was achieved with the semiconductor detector with a high vacuum in the chamber and a detector surface area of 0.5 cm^2. At the same time, a resolution of 3.5% was achieved on the scintillation detector with a detector surface area of 3 cm^2.

From the spectrogram presented in Figure 14.34, it is clear that by means of the scintillation spectrometer it is possible to divide the peaks of ^{239}Pu (E=5.15 MeV), ^{210}Po (E= 5.3 MeV), and ^{226}Ra in the background of radon and thoron decay products, and first of all RaA and Thc with an energy of 6.05 MeV.

In Gladkova et al. (1967), the results of the experiments were provided using a CsI (Tl) crystal in conjunction with a photomultiplier utilizing a multialkali photocathode. This photocathode yielded a light output twice that of an SbCs cathode. After analyzing many

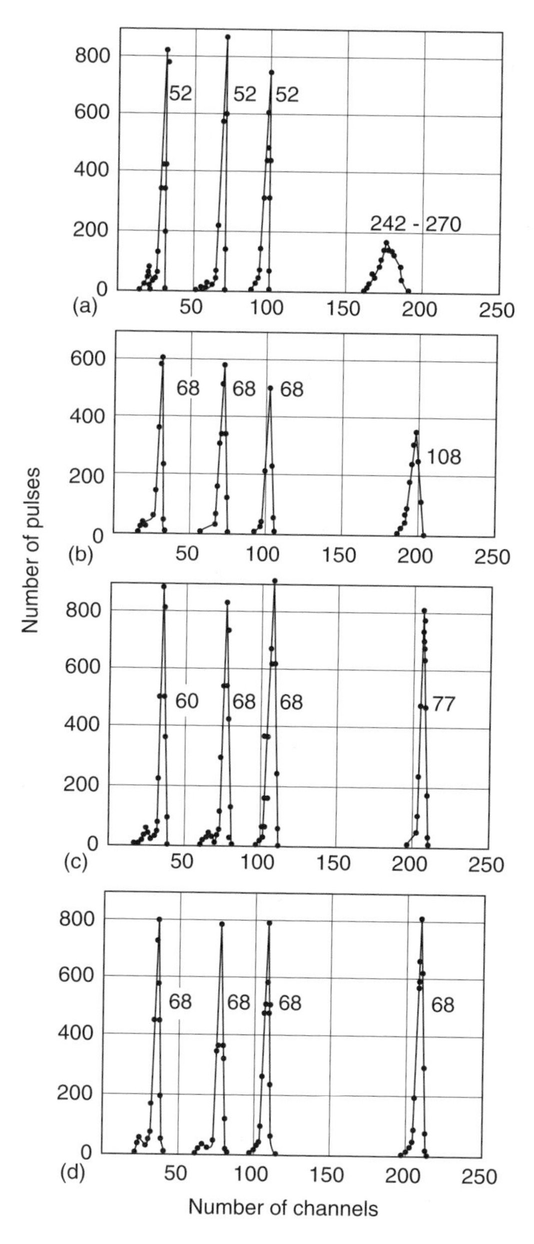

Figure 14.33 Energy resolution for nonemanating source measured by semiconductor detector vs. voltage.

crystals of different sizes and several multipliers with various photocathode sensitivities, it was established that the best resolution was obtained with a crystal thickness of 0.35–0.55 mm and the integral photocathode sensitivity greater than 150 mA/lm (Labushkin, 1967). Using semitransparent films for greater light dispersion between the crystal and the multiplier photocathode also gave positive results. The experimental results in Table 14.6 show how the resolution of a scintillation spectrometer depends on the area of the crystal Cs (Tl).

Figure 14.35 shows the alpha-spectrum of ^{239}Pu taken against a background of radon daughter products, obtained by using a semiconductor detector with an area of 1.8 cm^2, 10

Table 14.6 Crystal Size vs. Resolution

Area of the crystal (cm^2)	Line resolution for ^{226}Ra E_α=4.78 MeV		Line resolution for ^{214}Po E_α=7.68 MeV	
	keV	%	keV	%
0.8	220	4.6	240	3.1
2.0	228	4.8	262	3.4
4.0	264	5.5	284	3.7
7.0	270	5.6	287	3.7
12.0	292	6.1	292	3.8

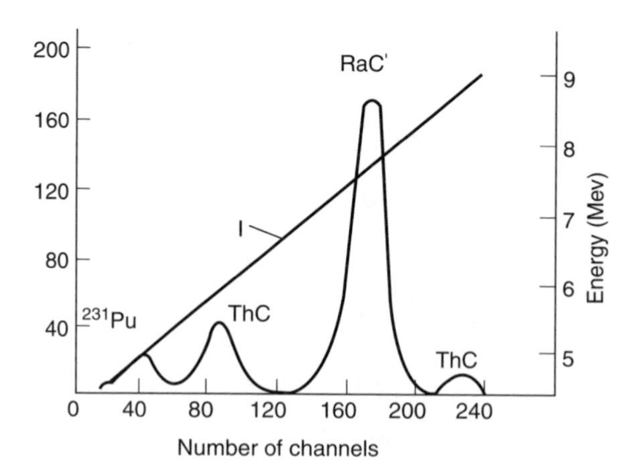

Figure 14.34 Spectrum of alpha-radioactive aerosols by means of a scintillation crystal of Cs(TI) with a 3 cm^2 surface.

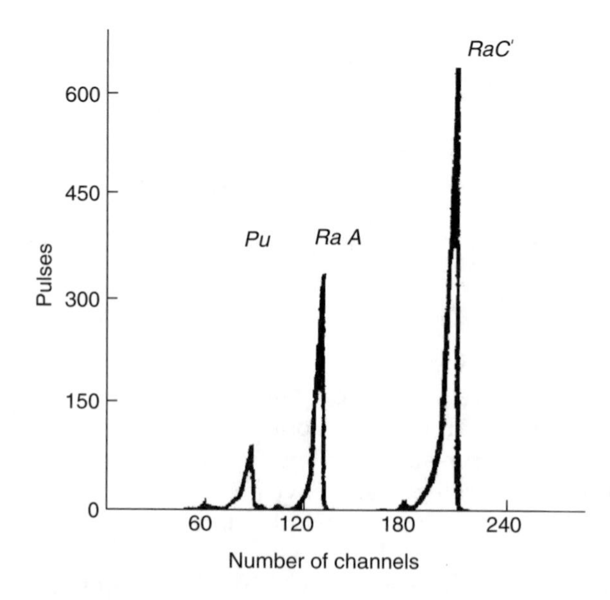

Figure 14.35 Spectrum of alpha-radioactive aerosols obtained by a 1.8 cm^2 semiconductor detector.

min after the conclusion of sampling with an exposure time of 10 min. The resolution is 1.4%, and the effectiveness of the registration is in the range of 2–16%.

Silicon surface-barrier detectors are used when the concentration of ^{239}Pu is at the MPC level, and the duration of the measurement is 30 min. Methodological error, associated with the radon and thoron decay products in the sample, will be negligible under these conditions. To account for various levels of radon and thoron decay product concentrations, computer-generated deviations of the concentration measured indirectly, q_{mes}, from the true concentration q_{art} were introduced (Kutnezov, 1978). This correction factor is $k=A^{nat}_{ect}/A^{art}_{nck}$:

$$k=1+(\gamma_{Rn}/\theta\cdot n_{Rn})+(\gamma_{Tn}/\theta\cdot n_{Tn})$$

where n is the ratio of the concentration of the artificial radioactive aerosols to each of the natural radioactive aerosols of the radon and thoron series and γ are a combination of functions that take into account buildup and decay of each of the radon and thoron decay products and, for the alpha artificial radioactive aerosols, have the following appearance:

$$\gamma_{Rn}=\Phi_a(\theta,t)+\Phi_{a,c}(\theta,t)+(q_b/q_a)\ \Phi_{b,c}(\theta,t)+(q_c/q_a)\ \Phi_{c,c}(\theta,t);$$

$$\gamma_{Tn}=1.36\ [\Phi_{b,c}(\theta,t)+(q_c/q_b)\ \Phi_{c,c}(\theta,t)].$$

Analogous expressions are cited in Kuznetzov (1978) for the beta-emitters.

Tables 14.7 and 14.8 show the results of these calculations.

To eliminate natural background error it is necessary to multiply q by a correction coefficient k. The establishment of the Standard on Radioactive Aerosols measurement in

Table 14.7 Values of γ_{Rn} for Alpha- and Beta-Emitters (θ=40 min)

Deviation from Equilibrium	Values of γ_{Rn} for Alpha-Emitters, t (min)				Values of γ_{Rn} for Beta-Emitters, t (min)			
	40	70	130	190	40	70	130	190
1:1:1	21	11	3.0	0.7	31	16	4.0	0.9
1:0.9:0.7	18	10	2.7	0.6	27	14	3.6	0.8
1:0.8:0.6	16	9.0	2.4	0.6	24	13	3.2	0.8
1:0.6:0.3	11	6.7	1.8	0.4	18	9.7	2.5	0.6
1:0.4:0.2	8.1	4.8	1.3	0.3	13	7.0	1.8	0.4
1:0.2:0.05	4.6	2.8	0.8	0.2	7.5	4.2	1.1	0.3
1:0.1:0.01	3.0	1.9	0.6	0.1	5.0	2.8	0.7	0.2

Table 14.8 Values of γ_{Tn} for Alpha- and Beta-Emitters (θ=40 min)

Deviation from Equilibrium	Values of γ_{Tn} for Alpha-Emitters, t (min)				Values of γ_{Tn} for Beta-Emitters, t (min)			
	40	190	720	4320	40	190	720	4320
1:1	53	47	27	0.5	63	54	31	0.6
1:0.8	48	46	27	0.5	60	54	31	0.6
1:0.6	42	45	27	0.5	57	53	31	0.6
1:0.4	37	44	27	0.5	55	53	31	0.6
1:0.2	31	43	27	0.5	52	52	31	0.6
1:0.05	27	43	27	0.5	50	51	31	0.6

1973 Antipin et al., (1973) in the former U.S.S.R. allows one to experimentally determine the corresponding corrections. The further upgradation of the Standard (Antipin et al., 1980), the use of new thoron and actinon decay products generators, and also the experience of government testing accumulated over the years gave much greater precision to the new equipment.

The thoron and actinon decay products have some advantages in terms of their use as radioactive markers, due to the energies of the alpha-particles of ^{212}Bi (ThC) and ^{212}Po (ThC′) differing substantially (8.78 and 6.05 MeV) from that of ^{218}Po (RaC′) and (^{214}Po) RaA (7.69 and 6.00 MeV), which can greatly simplify the measurement of (^{212}Po) ThC′ in a sample by means of alpha-spectrometry (Ruzer, 1993; Labushkin and Ruzer, 1965; Ruzer and Sextro, 1997). As far as actinon daughter products are concerned, the measurement is simple to perform by the general account method, which gives advantage in terms of efficiency of registration of the alpha-particles. It is possible to use this general account method because ^{207}Tl (AcC′) is the only practical alpha-emitter in the sample, and also because the half-time of the second alpha-emitter in the sample AcA is very small (1.778×10^{-3} sec) and therefore it does not have enough time for a buildup on the filter.

Developing the radon and thoron decay product generators presents some difficulties. For example, due to the short half-life (55.3 for thoron and 4.00 for actinon), Tn and An do not have enough time to go from the vessel to the air in the chamber. On the other hand, bubbling through the solution of thorium and actinium is impossible because it does not produce the necessary reproducible concentration. The volume rate of the bubbling cannot be large enough in order to not pulverize the droplets of solution and consequently not pollute the chamber.

Therefore, the special solid nongermetic sources were developed with thin layers of substance based on ^{227}Ac and ^{228}Th with a substantial emanation capability. The sources were developed by neutron bombardment of ^{226}Ra. Such sources had the minimum of other admixtures; for example, for ^{227}Ac with an activity of $(39.2 \pm 1.1) \times 10^4$ Bq, the additional content of ^{226}Ra was not more than 0.9%, and for ^{228}Th it was 0.04%. Correspondingly, for ^{228}Th with an activity of $(5.55 \pm 0.30) \times 10^4$ Bq, the content was not more than 0.03% of ^{226}Ra and ^{227}Ac. The sources were produced by a "droplet method" on the special glass backing diameter 23 to 25 mm. The sources were placed in boxes type 2BP2-OS with a volume of 1 m^3. The sampling was provided on spectrometric filters of AFA-RSP type. Measurements were provided with the installation, which consisted of two alpha-spectrometer: one based on the semiconductor detector with and without vacuuming, and the second based on a scintillation crystal of CsJ (Tl) with the photomultiplier. The third instrument was used for the beta-activity measurements. The calculations of the decay product concentrations were provided according to the count rate of each decay product on the filter for two measurement intervals of t_1 and t_2 by the formulas

$$q_{\text{ThB}} = (1/av\varepsilon\eta)[c_1 N_c(\theta,t_2) - c_2 N_c(\theta,t_1)];$$

$$q_{\text{ThC}} = (1/av\varepsilon\eta)[c_3 N_c(\theta,t_1) - c_4 N_c(\theta,t_2)];$$

$$q_{\text{AcB}} = (1/av\varepsilon\eta)[c_5 N_c(\theta,t_2) - c_6 N_c(\theta,t_1)];$$

$$q_{\text{AcC}} = (1/av\varepsilon\eta)[c_7 N_c(\theta,t_1) - c_8 N_c(\theta,t_2)],$$

where v is the pumping volume rate through the filter, v is registration efficiency of alpha-radiation, ε is filtration efficiency, $a = 2.22 \times 10^{12}$ decays /min, and θ and t_i are the times of duration of sampling and measurement after sampling.

For the thoron series,

$$c_1=\Phi_{c,c}(\theta,t_2)/[\Phi_{b,c}(\theta,t_1)\Phi_{c,c}(\theta,t_2)-\Phi_{b,c}(\theta,t_2)\Phi_{c,c}(\theta,t_1)];$$

$$c_2=\Phi_{c,c}(\theta,t_1)/[\Phi_{b,c}(\theta,t_1)\Phi_{c,c}(\theta,t_2)-\Phi_{b,c}(\theta,t_2)\Phi_{c,c}(\theta,t_1)];$$

$$c_3=\Phi_{b,c}(\theta,t_2)/[\Phi_{b,c}(\theta,t_2)\Phi_{c,c}(\theta,t_1)-\Phi_{c,c}(\theta,t_2)\Phi_{b,c}(\theta,t_1)];$$

$$c_4=\Phi_{b,c}(\theta,t_1)/[\Phi_{c,c}(\theta,t_1)\Phi_{b,c}(\theta,t_2)-\Phi_{c,c}(\theta,t_2)\Phi_{b,c}(\theta,t_1)].$$

For the actinon series,

$$c_5=\Phi_{c,c}(\theta,t_2)/[\Phi_{b,c}(\theta,t_1)\Phi_{c,c}(\theta,t_2)-\Phi_{b,c}(\theta,t_2)\Phi_{c,c}(\theta,t_1)];$$

$$c_6=\Phi_{c,c}(\theta,t_1)/[\Phi_{b,c}(\theta,t_1)\Phi_{c,c}(\theta,t_2)-\Phi_{b,c}(\theta,t_2)\Phi_{c,c}(\theta,t_1)];$$

$$c_7=\Phi_{b,c}(\theta,t_2)/[\Phi_{b,c}(\theta,t_2)\Phi_{c,c}(\theta,t_1)-\Phi_{c,c}(\theta,t_2)\Phi_{b,c}(\theta,t_1)];$$

$$c_8=\Phi_{b,c}(\theta,t_1)/[\Phi_{c,c}(\theta,t_1)\Phi_{b,c}(\theta,t_2)-\Phi_{c,c}(\theta,t_2)\Phi_{b,c}(\theta,t_1)].$$

Calculated values of the functions $\Phi_{i,j}$ for the typical measurement conditions are presented in Tables 14.9 and 14.10.

An established procedure for calibrating each type of aerosol radiometer includes the determination of the transitional coefficient k_t. This value is the ratio of the concentration

Table 14.9 Values of the Functions $\Phi_{i,j}(\theta,t)$ for the Thoron Decay Products

θ (min)	$\Phi_{i,j}$	t (min)							
		3	5	7	20	25	30	35	40
3	$\Phi_{c,b}$	0.15	0.21	0.28	0.65	0.77	0.89	1.00	1.11
	$\Phi_{c,c}$	2.85	2.79	2.72	2.35	2.22	2.09	1.98	1.89
5	$\Phi_{c,b}$	0.30	0.41	0.51	1.12	1.33	1.53	1.71	1.88
	$\Phi_{c,c}$	4.70	4.59	4.49	3.87	3.65	3.45	3.26	3.08
8	$\Phi_{c,b}$	0.61	0.78	0.94	1.89	2.22	2.53	2.81	3.08
	$\Phi_{c,c}$	7.39	7.22	7.06	6.08	5.74	5.42	5.12	4.84
10	$\Phi_{c,b}$	0.87	1.07	1.27	2.45	2.85	3.23	3.58	3.92
	$\Phi_{c,c}$	9.13	8.92	8.72	7.52	7.10	6.70	6.33	5.98

Table 14.10 Values of the Functions $\Phi_{i,j}(\theta,t)$ for the Actinon Decay Products

θ (min)	$\Phi_{i,j}$	t (min)							
		3	5	7	20	25	30	35	40
3	$\Phi_{c,b}$	2.16	2.41	2.50	2.31	2.11	1.92	1.74	1.58
	$\Phi_{c,c}$	0.72	0.38	0.20	0.01	0	0	0	0
5	$\Phi_{c,b}$	3.79	4.09	4.16	3.78	3.45	3.13	2.85	2.59
	$\Phi_{c,c}$	0.93	0.49	0.25	0.02	0	0	0	0
8	$\Phi_{c,b}$	6.30	6.57	6.58	5.89	5.36	4.88	4.43	4.02
	$\Phi_{c,c}$	1.07	0.56	0.29	0.02	9	0	0	0
10	$\Phi_{c,b}$	7.95	8.16	8.13	7.23	6.58	5.98	5.44	4.94
	$\Phi_{c,c}$	1.11	0.58	0.30	0.02	0	0	0	0

at the entrance of the apparatus (the actual value) ($q1$) to the concentration measured by the equipment ($q2$). If deviation of k_t from 1 occurs during calibration, the source of this deviation is determined and possible methods of elimination are explored. If the source of any additional existing error cannot be eliminated, the value of k_t is entered on the registration certificate for the equipment.

14.10.12.2 Artificial radioactive aerosol concentration measurement

In Ruzer and Sextro (1997) the alpha- and beta-spectra were measured by using a GM-50 Si surface-barrier detector system in order to determine the volume activity of alpha/beta artificial aerosols and radon/thoron progeny background compensation. The measurement of the alpha-peak of ^{218}Po and ^{214}Po in two time intervals and the activities of ^{218}Po, ^{214}Po, and ^{214}Bi were determined, and the total beta-count of ^{214}Pb and ^{214}Bi was calculated from their decay scheme. The volume activities of thoron progeny were approximately calculated by ^{212}Po alpha peak counts. The alpha/beta-count of transuranic aerosols was determined by subtracting radon/thoron progeny aerosol concentration. In these experiments, if the radon progeny concentration was less than 15 Bq/m^3, the lower limit of detection (LLD) of transuranics concentration is less than 0.1 Bq/m^3. If the radon progeny concentration is as high as 75 Bq/m^3 LLD of the total beta-activity, the concentration of artificial aerosols is less than 1 Bq/m^3.

In order to discriminate the most interfering nuclides — naturally occurring radon and thoron progeny — Liu et al. (1997) proposed an approach based on the beta:alpha ratio instead of gross beta- or alpha-counting. A theoretical model was applied (Chen et al., 1994) in the simulation of radon and thoron progeny behavior in the environment and the filter. It was found that:

- The beta:alpha ratio obtained from an air sample filter can converge to a constant value.
- No matter what the change in concentration of radon and thoron progeny, the beta:alpha ratio is rather stable with a small change under certain conditions.
- The beta:alpha activity ratio for thoron progeny is more stable than that for radon progeny.
- If artificial radioactive aerosols exist, the beta:alpha ratio will dramatically change. The presence of artificial beta-emitters will cause the ratio to rise, and the presence of alpha-emitters will cause the ratio to fall.

In the nuclear power industry, high specific activity particles can arise from two main sources: from corrosion of the irradiated fuel with defective clodding and from neutron activation of corrosion product particles originating within the coolant circuit. The particle activity found in the first type tends to be slightly more active with isolated examples up to 40 GBq. It should be pointed out that even a single 5 kBq point source of ^{60}Co is potentially hazardous. The characteristics and origin of beta emitting hot particles were studied, and the method of dose calculation and corresponding difficulties have been reviewed. Parallel measurements and calculations have been carried out for well-defined model hot particle sources in order to evaluate measurement and calculation [(NRS-96, 1999). The published data are presented and reviewed based on physicochemical characteristics and the behavior of radionuclides, which are responsible for the dose calculation following release of these nuclides into the environment after a nuclear accident or incident, including those at Chernobyl, Maralinga, Palomares, Kozloduy, as well as from nuclear fallout. Four main parameters can be measured: the particle size distribution of aerosols, shape, isotope composition, and dissolution parameters such as *in vitro* distribution rates or *in vivo* absorption parameters to the blood, which are more relevant in terms of inhalation. Some examples of *in vivo* experiments have been chosen as a methodological guide for future experiments.

14.10.13 Aerosol particle size measurements

The radiation dose to the lung is determined by both radioactive and nonradioactive aerosol parameters. Radioactive parameters are responsible for the quantity of activity that they deliver to the lung tissue. Nonradioactive parameters such as particle size distribution, counting, surface area, and mass concentration, together with physiological factors, are responsible for aerosol deposition and their distribution in the breathing organs. Because the activity in the lung (and corresponding absorbed dose) is a combination of these factors — activity in the air and deposition coefficient — the accuracy in dose assessment depends equally on the accuracy of determination of these factors. In most of the studies, the assessment of dose from radon decay products is provided by the calculation of deposition coefficient and distribution in the lung according to the ICRP model.

It should be noted that particle size distribution and activity particle size distribution are usually measured only at a distance from the breathing zone, especially for hard-working personnel. Particle size distribution will be different in the breathing zone and inside the lung. It is practically very difficult or even impossible in this case to assess distortion in size distribution and, therefore, determine the real particle distribution inside the lung.

Still, it is important to know the degree of accuracy really achievable in measurements of dosimetric aerosol parameters such as particle size distribution, particle surface area, and mass concentrations. In other words, in aerosols, as in all types of measurements of physical parameters, it is important to establish the uniformity and correctness of measurements.

The goal of this section is to present the theoretical and experimental data that are necessary for improving the quality of aerosol measurement. To achieve this goal, the metrological provision or the scale of aerosol particles should be established. This metrological assessment can be provided by:

- *Hardware.* Aerosol generators covering all ranges of particle size with high stability and measuring equipment for measuring aerosol parameters with high accuracy.
- *Software.* System of procedures for calibration, convolution of measurements and corrections together with an assessment of the accuracy of measurement of aerosol parameters.

14.10.13.1 Concept of the scale of particle size of aerosols

For the purpose of the uniformity and correctness of the measurements of physicotechnical parameters such as electrical current, temperature, pressure, electromagnetic frequency, etc., the scale of these values is established in countries where metrological service is available. The concept of the scale of particle sizes of aerosols was at first presented in Kravchenko and Ruzer (1978).

The wide application of aerosols in the national economy requires a further investigation of their parameters.

Marketed instruments for aerosol particle size distribution measurements and those under development cover virtually all particle size ranges: ultrafine aerosols (UFA) with particle sizes from 10^{-3} to 10^{-1} μm, fine aerosols (FA) with sizes of 10^{-1} to 1 μm, and coarse aerosols (CA) with sizes of 1 to 100 μm. Therefore, it is necessary to produce a set of methods and equipment for obtaining aerosols of a given size in the entire aerosol particle dimension range and measure their characteristics with the highest accuracy. The set should ensure these metrological provisions for marketed measuring equipment and for that being prepared for mass production, and it should serve to carry out research work.

The set should consist of basic and auxiliary equipment. The basic equipment is intended for transmitting units of dispersed compositions and aerosol counted, surface area and mass concentrations, whereas the auxiliary set is intended for detecting and evaluating factors that affect the basic aerosol characteristics measurements results.

The basic equipment should ensure:

- The obtaining of stable, virtually monodispersed aerosols for the entire particle range; and the possibility of varying the produced-aerosol counted concentration over a wide range
- Stability and reproducibility of the generated aerosol parameters
- Obtaining particles of a spherical shape
- Measuring aerosol parameters with the highest precision over the entire range of aerosol particle sizes and counted concentrations

The concept of aerosol stability applies in this case only to particle evaporation and condensation growth, without taking into account their sedimentation, coagulation, and diffusion, which occur on the vessel walls and are determined mainly by aerosol parameters, namely, by their dispersed composition and number concentration. Aerosol stability is determined by the absence, in the gas phase, of high-pressure vapors, which are close to saturation, saturated, or oversaturated and prevent condensation growth or even the formation of aerosol particles, and by low vapor pressure of the aerosol particle material, which prevents their substantial evaporation.

Instruments for analyzing the aerosol dispersed composition are designed on different principles. Therefore, different instruments provide different mean diameters of aerosol particles in measurements, for instance, the mean geometrical, mean square, Stokes or sedimentary, and other diameter values. In the case of polydispersed aerosols, the mean particle diameters can differ substantially from one to another. For monodispersed aerosols, they coincide. Thus, in order to obtain comparable results in measuring the aerosol particle dimensions by means of different methods and instruments, it is necessary to use monodispersed or almost monodispersed aerosols. A measure of the distribution width is the monodispersity factor, which is defined as the standard deviation in the particle diameters referred to as the arithmetic diameter (Fuks and Sutugin, 1964). An aerosol with this factor <0.2 is taken as monodisperse, whereas one >0.2 is polydisperse. However, this criterion is not always applicable in calibration and apparatus testing. The problems of monodispersity of aerosols, especially for calibration purposes, are discussed in Kravchenko (1983).

Aerosol equipment is produced for various purposes. Some of it is intended for the measurement of aerosols in clean and very clean premises, whereas other equipment has been developed for measuring technological aerosol flows. This makes it necessary to obtain aerosols with number concentrations from tens of particles per cubic meter to the maximum possible number for which, however, it is still possible to neglect the aerosol particle coagulation.

Since this equipment is intended to meet the metrological requirements of aerosol instruments, special requirements are set for the stability and reproducibility of the generated aerosol parameters. Evaluation of the tested measuring-equipment characteristics may become distorted if these parameters are inadequate. Stability requirements of the aerosol generators are determined by the tested instruments' characteristics, their errors in evaluating aerosol parameters, and the time spent on sampling.

The shape of aerosol particles is important in evaluating the composition's mass concentrations, and other aerosol diameters.

The concept of the equivalent particle parameter is usually adopted for particles of irregular shape. Thus, the sedimentary diameter of an irregularly shaped particle is understood to be the diameter of a spherical particle that sinks at the same rate as the irregular. For spherical particles, all the equivalent diameters are equal, and for precisely this reason it is necessary to generate spherical particles. On this basis, it is easy to obtain, for a known dispersed composition and counted concentration of aerosols, the mass concentration, specific surface concentration, and other characteristics.

Based on analogy with the different types of standards, achievable accuracy, and economical factors, the aerosol standard equipment should have a precision 3–10 times higher than that of the working equipment.

The auxiliary equipment is intended for detecting and evaluating:

- The shape of aerosol particles
- The properties of their constituent material
- The aerosol particle losses that occur in communication channels and are due to sedimentation, inertial precipitation, diffusion, etc.
- The type of particle distribution according to the size of aerosols
- Electrical charges of aerosol particles
- Volume of the aerosol pumped samples, and several other factors

Deviation of the actual aerosol particles from a spherical shape can lead not only to a distortion of the results of measuring aerosol parameters, but also to changes in the pattern of particle behavior. Thus, flat particles slide at an angle to the vertical for a given orientation of the force of gravity in its field.

Properties of the aerosol particle material, such as refractive index, density, permittivity (dielectric constant), etc., play an important role in evaluating errors of the equipment under development. Correction for aerosol particle photoelectric counters for the refractive index of the tested particle material can serve as an example of the above phenomenon. This correction for the photoelectric counter type AZ widely used in the former U.S.S.R. may attain ~70% (Kravchenko, 1974).

The aerosol particle losses in pipelines depend on aerosol-dispersed composition, the specific geometry of pipes, aerosol propagation speed, and a number of other factors. Therefore, losses can be evaluated only for each specific case.

Elucidation of errors due to the type of aerosol distribution according to size is important in testing the equipment intended for measuring the aerosol mean parameters.

Electrical charges on aerosol particles can lead to changes in aerosol parameters both in the case of their transportation and during measurements.

It is very important to measure the pumped sample volume in evaluating such aerosol parameters as the number and weighted concentration, specific surface, etc. In connection with the development of photoelectric counters with an aerosol jet being blasted by clean air, it becomes necessary to measure small flow through thin capillary tubes.

It is as yet impossible to obtain the required aerosols from any arbitrary shaped particles and any type of distribution. Therefore, in evaluating reading corrections of any specific instrument for variation of any factor, it is possible to use two methods. The first one consists in an attempt to obtain aerosols whose prevailing characteristic differs from that of aerosol generated by the basic equipment of the standard set. Thus, in order to evaluate the effect of the particles' shape, it is possible, for instance, to attempt obtaining particles of cubic or any other regular shape by using crystallization properties of certain substances. The second method consists in finding corrections by computation as was done previously, for instance, for the type AZ instrument, when the material refractive-index correction was calculated according to Mie theory (Kravchenko, 1974).

By taking this into consideration, it is possible to outline the following composition for the auxiliary equipment:

- Generators for obtaining aerosols from various materials with very diverse properties
- Equipment for generating aerosol particles with shapes widely different from a spherical one
- Generators of aerosols with the most common distributions
- Equipment for investigating the precipitation of aerosols in pipelines, and a number of other instruments

Based on the principles presented above, the set of installation was developed at the All-Union Scientific-Research Institute of Physico-Technical and Radiotechnical Measurements (Moscow, former U.S.S.R.) in the framework of the Special State Standard for the Concentration Unit of Artificial and Natural Radioactive Aerosols. The installation—called GERA-06—intended to measure the dispersed composition and the counted concentration of aerosols, generate aerosols with a narrow distribution of sizes in the range of 5×10^{-3} to 2×10^{2} μm, and investigate their parameters.

Aerosols with $\sim5\times10^{-3}$ μm sized particles were investigated by means of the diffusion method, with a set of nine diffusion batteries with cylindrical channels and three zero batteries. The diffusion batteries have the parameter ml (m is the number of tubes in a battery and l is its length) equal to 1.1×10^{2}, 2.6×10^{2}, 7.5×10^{2}, 2.4×10^{3}, 7.2×10^{3}, 2.2×10^{4}, 6.6×10^{4}, 1.9×10^{5}, 5.6×10^{5} cm. The technique was developed for the determination of measured factors from their passage through the batteries, comparing them with their plotted curves (Fuks and Skechkina, 1962) and evaluating errors in determining the aerosol parameters (Kravchenko et al., 1971; Lekhtmakher and Ruzer, 1972).

The FA composition starting with a size of 0.4 m is measured by means of reference of a photoelectric counter type AZ, which was specially certified for these purposes (Kravchenko, 1974). The dispersed composition of particles in suspensions and that of roughly dispersed aerosols was investigated by means of an electron and an optical microscope, together with equipment for microphotography and for processing results. The UFA counted concentration and that of aerosols with a mean dispersion was measured by means of a photoelectric counter type AZ in combination with diluting filters (Kravchenko, 1974). Before measuring the counted concentration, the aerosol sizes can be enlarged in a type KUST (Kogan and Burnashova, 1959) instrument or in a generator developed by Fuks and Sutugin (1969). These installations were suitable for covering almost the entire scale of aerosol particle sizes.

14.10.13.2 *Ultrafine aerosols*

Highly dispersed aerosols play an important role due to their high mobility in the deposition of radioactive aerosols in the upper lung. When the short-lived decay products of ^{222}Rn — ^{218}Po, ^{214}Pb, and ^{214}Bi(^{214}Po) — are inhaled, a small fraction, that is, 0.02–0.05, deposits in the bronchial airways, while a large fraction, up to half, deposits in the lower lung or pulmonary region. Nevertheless, the dose from the alpha-emitters ^{218}Po and ^{214}Po deposited in the bronchial region is responsible for the observed increase in lung carcinoma among miners following ^{222}Rn exposure, presumably because different geometries result in a higher proportion of the alpha-energy being deposited in the epithelial cells at risk in the bronchi, or because of the great sensitivity there.

On the other hand, the practical application of UFA measurements presents some difficulties in terms of theory in the introduction of corresponding corrections on some of the factors, which affect the results of the measurements. In the range of UFA particle sizes, in principle three methods can be used for determining dispersion composition: electron microscopy, the aspirator condenser method, and the diffusion method.

The application of electron microscopy is restricted because of small sizes; besides, this is difficult in terms of practical usage. The aspiration condenser method demands the preliminary charge of the aerosol particles and, consequently, a knowledge of the conditions of the charge deposition on aerosol particles.

The theoretical basis for using the high mobility of UFA in the measurement of diffusion deposition was done in Fuks and Stechkina (1962), Kravchenko (1974), Fuks and Sutugin (1969), Fuks (1964), Lekhtmakher (1977). The main expression for "passage" (the ratio of the average concentrations of monodispersed aerosols at the exit and entry of the cylindrical tube) due to diffusion deposition was presented in Gormley and Kennedy (1949):

$$k=\begin{cases} 0.8191e^{-7.314h}+0.0975e^{-44.6h}+0.0325e^{-114h}, & h>0.0156 \\ 1-4.07h^{2/3}+2.4h+0.446h^{2/3}, & h<0.0156 \end{cases} \qquad (14.46)$$

where k is a passage and $h=Dl/2Q$ (l is the length of the tube, Q is the volume aerosol passing rate, and D is the diffusion coefficient of aerosol particles).

For practical purposes, it is necessary to determine conditions for applying Equation (14.46) in all ranges of UFA sizes, because it was derived on the basis of simplified assumptions. It is also necessary to study the possibility of using the diffusion method for the determination of polydispersed aerosol parameters.

Because of the sharp decrease in diffusion deposition with particle size, a set of parallel channels (diffusion batteries) are practically used for UFA aerosol measurement, which automatically allows a decrease in the volume rate to the necessary degree. The largest battery in Kravchenko (1974) and Ivanova et al. (1974) had 3700 cylindrical tubes (m=3700 and length l=20 cm); that is, the summary length of the channels (ml), which is a parameter of the deposition capability, was 74,000. This was not enough for the study of UFA in all size ranges of the aerosol particles. In addition, due to 20 cm diameter batteries, it was practically impossible to provide for laminar flow in the diffusor, which connects the battery with communications.

The expression for the "passage" through the flat and conic diffusor and confusor (Ivanova et al., 1974) is

$$1-2.3\,\{[\pi Dl(\tau_0)^{1/2}(1+\tau_0)]/Q(1+2\tau_0)^{1/2}\}^{2/3}$$

where l is the length of the confusor generatrix and $\tau_0=\cos\,\theta_0$ (θ_0 is a half of the conic angle).

During the passage of radioactive aerosols of RaA through the diffusion battery, the decay of atoms of RaA on aerosol particles to a lesser degree are compensated by the deposition of RaA atoms on particles due to deposition of the aerosol particles in the walls. This effect of decreasing the "passage" depends on the parameter $\alpha=b\sqrt{\lambda_a/D}$, where λ_a is a constant of decay of RaA. Distortions of the "passage" were calculated for values of passages from 0.1 to 0.9 in Lekhtmakher (1977).

In the case of polydispersed aerosols, the expression for the "passage" through the diffusion battery was expressed by Equation (1) — the integral equation of Fredholm 1st type

— the solution for which represents an incorrect problem. Therefore, the method of approximation of experimental data by calculated data (Fuks and Stechkina, 1962; Kravchenko, 1971) (reverse method) is used.

14.10.13.3 Portable instrument for measuring UFA in mines

For the purpose of measurement of radioactive UFA (ultrafine aerosols) in mines, a portable instrument was developed: diffusion batteries had lengths of 47, 20, and 15 cm and the tubes had wall thicknesses of 0.005 cm and outer diameters of 0.04, 0.05, 0.1, and 0.3 cm (Lekhtmakher, 1977). The working diameter of the battery was equal to the diameter of the standard radiometric filter. The parameters *ml* of the batteries were 2100, 20,000, 110,000, and 410,000 cm. The measurement of the activities of the filters was performed by a single-channel alpha-spectrometer with a peak resolution of RaC′ 5% and detection efficiency 8%. This instrument was used in measurements in mines with weight aerosol concentrations from 0.1 to 1.0 mg/m^3, the level of PAEC from background to 10^6 MeV/l, unattached RaA concentrations from 370 to 18,500 Bq/m^3, and aerosol concentrations of RaA from 5920 to 85,100 Bq/m^3. The count aerosol concentration of particles with diameters >0.4 μm changes in the range from 30 to 300 cm^{-3} measured by means of a standard aerosol counter AZ-5.

For the determination of the "passage" of aerosol particles, air was pumped simultaneously through a battery and control filter during the time required to achieve the necessary statistical accuracy. Because the measurements were provided with ^{218}Po, the increase in the activity of a filter will take place only till 30 min of pumping.

The passage for each battery *k* was determined by the formula

$$k = A_b Q_b / A_c Q_c$$

where A_b, Q_b, A_c, and Q_c are activity on the exit filter of the diffusion battery, velocity through the battery, activity of the front control filter, and corresponding velocity, respectively. Then the experimental passage curve was compared with the calculated curve for cylindrical diffusion batteries (Figure 14.39) and the errors in the determination of a counted median radius according to procedure are described later in this section.

It should be noted that the diffusion batteries of this instrument had no enter diffuser. The air was pumping directly from the air through the whole cross-section of the battery. Therefore, air velocity distribution was homogeneous and the loss of particles in the enter diffuser did not occur.

Data for the passage $t = \log(3.657\ ml/Q)$ through each of the four instrument batteries are presented in Table 14.11.

The results of the measurements are presented in Tables 14.12 and 14.13.

In the case of high humidity and temperature, which usually exist in mines, at first the condensation of water on battery walls took place. But after one or two pumps, it disappeared and high humidity and temperature did not disturb the experiments. A similar situation was observed with the relatively high dust concentration.

The counting median radius of radioactive UFA of RaA was in the range of 0.02–0.05 μm, depending on the chosen working site. The results of measurements confirm the practical usage of such instruments in mining conditions.

14.10.13.4 Installation for generating and investigating aerosols in the range of 2×10^{-3} to 1 μm

UFA were generated by means of two methods. The interaction of hydrochloric acid vapor flow is used in the first method. As a result of this reaction, aerosols are produced. The size of the particles is adjusted by varying the concentration of ammonia vapors in an excessive

Table 14.11 Values of Parameter t for Different Pumping Velocities Q

Q (l/min)	ml (cm[a])			
	2100	20,000	110,000	410,000
0.25	3.762	4.741	5.482	6.053
0.5	3.461	4.440	5.181	5.752
0.75	3.285	4.264	5.005	5.576
1.0	3.160	4.139	4.880	5.451
1.25	3.063	4.046	4.783	5.354
1.5	2.984	3.963	4.703	5.275
2.0	2.859	3.838	4.579	5.150
2.5	2.762	3.741	4.482	5.053
3.0	2.683	3.662	4.403	4.974
3.5	2.616	3.595	4.336	4.907
4.0	2.558	3.537	4.278	4.849
4.5	2.507	3.486	4.226	4.798

[a]Values of ml rounded

Table 14.12 Measurement of the Concentration of Radioactive Aerosols

	1	2	3	4	5	6
Chamber near exit ventilation	0.5	1.7	0.12	2.9	0.02 ± 007,	4×10[4]
30 m from working site	0.8	1.1	—	1.6	0.04 ± 02,	3×10[5]
Nonworking dead-end site	0.2	9.6	1.8	5.8	0.050 ± 02,	2.4×10[5]
All mine exit ventilation	1.1	2.0	5.2	18	0.050 ± 01,	1.5×10[5]
All mine exit ventilation	1.0	1.7	5.0	4.8	0.020 ± 01,	1.5×10[5]

1 — mass concentration (mg/m^3); 2 — PAEC 10^5 (MeV/l); 3 — unattached concentration of RaA × 3700 Bq/m^3: 4 — attached concentration of RaA×3700 Bq/m^3; 5 — median radius UFA (μm); 6 — concentration of particles with diameters more than 0.4 μm.

Table 14.13 Particle Size Distribution in the Range 0.4–10 μm (particles/l)

	Site 1	Site 2	Site 3	Site 4	Site 5
0.4	8000	—	60,000	0	0
0.5	2000	60,000	50,000	0	0
0.6	2000	50,000	40,000	0	0
0.7	1000	40,000	20,000	0	0
0.8	3000	50,000	15,000	0	0
0.9	4000	40,000	15,000	0	0
1.0	2500	34,000	5000	0	50,000
1.5	2500	13,000	7560	0	25,000
2.0	10,400	12,000	10,800	110,000	85,000
4.0	3100	840	1600	0	89,000
7.0	350	75	100	0	1000
10	0	0	0	0	0

concentration of hydrochloric acid vapors, or vice versa. This method is suitable for obtaining aerosols of size 0.002–0.2 μm.

Figure 14.36 shows a block diagram for generating and measuring parameters of the ammonium chloride aerosol. After entering from the main line, compressed air is cleansed in the filter, 1, and its pressure is stabilized in the stabilizer, 2. It then forks into two streams. One stream enters the device itself, 3, which generates the steam–vapor mixture. At the exit, 3, is the choke, 6, which creates the pressure necessary for the normal function of unit 3. The other stream enters the buffer space, 4, which also chokes at its exit. The pressure in the buffer space is measured by a manometer, 6, and regulated by a needle cock, 5. The volumetric consumption of air at the exit of the buffer space is adjusted with a rotameter, 7. The ammonium chloride aerosol, formed in device 3 by a combination of ammonia vapor with hydrochloric acid vapor, passes through the choke, 8, and into the mixer, 9, where it is diluted with pure air fed in from the buffer space, 4. Dilution is necessary to reduce the concentration of unreached vapors and stop or otherwise inhibit aerosol formation in the next line. Dilution also increases the volumetric consumption of the aerosol. Upon exiting the mixer, 9, the aerosol passes through a block of filter-diluters, 10, which dilute the aerosols's calculated concentration by a predetermined factor (10, 100, or 1000). It then enters into a block of diffusion batteries. The block is made up of four batteries with the parameters ml=50, 500, 6000, and 74,000 cm (l is the length of the battery and m is the number of channels). The coefficient of the "passage" through the diffusion battery is measured by a nephelometer, 18. To enhance the light dispersion of the aerosol for easier measurement, it is first enlarged in the KUST apparatus. This is accomplished by pure air being brought in through a purifying system consisting of a filter, 12, an antigas box, 14, and a stabilizer, 13. The enlarged aerosol can be fed either into the nephelometer, 18 (position "B"), to measure light dispersion, or into the photoelectric meter, 16 (position "A"), to the exit of which a conversion apparatus is attached to measure calculated concentration n. The true value of the concentration of ammonium chloride (N) is found by the formula

$$N=nk(q_a+q_b)/q_a$$

where k is the coefficient of dilution for the filter-diluter, q_a is the volumetric consumption of aerosol at the exit, 3, and q_b is the volumetric consumption of pure air fed into the mixer, 9.

It was shown in Kravchenko (1974) that when the aerosol formation reaction occurs without irradiation of the mixing chamber by ionizing radiation, the parameters of generated

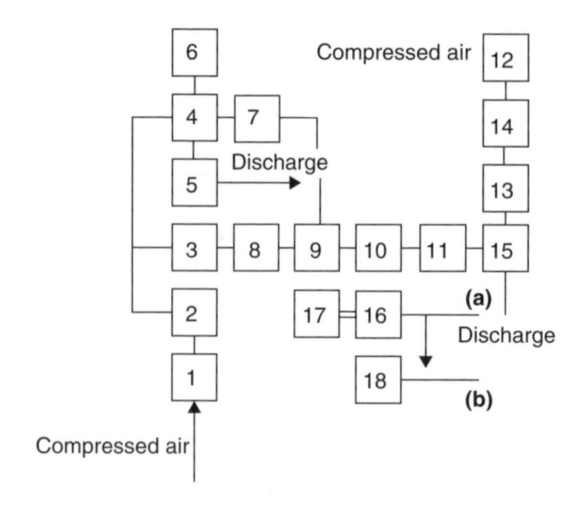

Figure 14.36 Device for NH_4Cl UFA generation.

aerosol are the most irregular. In some cases, light dispersion of the aerosol over the course of several minutes fell to zero. The study of aerosol formation was provided both with and without irradiation of the mixing chamber. Irradiation was performed with a BIS-M-type source of beta-emission ($^{90}Sr+^{90}Y$) having an activity of 555×10^4 kBq. The results of the measurements are shown in Table 14.14, which indicate the time of measurement from the moment the device is turned on and the amount of light dispersion in relative units. The first series of measurements was conducted with irradiation of the stream-mixing chamber, and the second without irradiation. The third series of measurements was made with irradiation of only the ammonia vapor stream, and the fourth with irradiation of only the hydrochloric acid vapor stream.

As Table 14.14 shows, the greatest particle concentration was obtained when the stream-mixing chamber was irradiated. It is also shown that irradiation of the chamber substantially increases the stability of the aerosol formation process. The study shows that mixing streams with different initial reagent concentrations cause r_g and ln β_g to vary within a very wide range. However, in all cases the aerosol remains highly dispersed and its dispersion composition can be determined with good accuracy by the diffusion method.

The second type of (UFA) generator was based on another principle. In a number of studies (Fuks and Sutugin, 1969; Espenshkeid et al., 1964; Sinclair and La Mer, 1949; Gormley and Kennedy, 1949), generator designs that are suitable for obtaining highly dispersed aerosols (UFA) have been described, with narrow-dispersion (close to monodispersed) aerosols. Their disadvantages are based on the fact that at the time of establishing the scale of aerosol particles dimension, most of them were suitable for obtaining aerosols only over a narrow range of sizes. Among all these generators, there was not a single one that could cover the entire range from 10^{-3} to 1 μm. Moreover, most of the existing generators were used only for research.

The production of permanently operating installations for generating monodispersed aerosols over a wide range of sizes and for the determination of their parameters, in combination with adequate means for generating and measuring particles in the range larger than 0.1 μm, will substantially extend the possibility of carrying out different work of scientific and production value and permit the calibration of aerosol instrumentation.

The schematic of the equipment for generating and testing (UFA) aerosols (Kravchenko, 1974) is shown in Figure 14.37.

The equipment consists of two units: the aerosol generating unit I and the aerosol measurement unit II. Compressed air is supplied through the purifying filters F1 and F2 and the monostat with pressure reducers Rd1 and Rd2, and is finally purified in gas-discriminating compartments (GDCs). The volumetric speed of airflow through the equipment is regulated

Table 14.14 Results of Aerosoal Formation Measurements Both with and without Irradiation

1	2	3	4	5	6	7	8
9	1.43	6	410	7	5.9	6	31
15	1.18	15	52.5	14	8.2	12	13.1
20	0.97	23	26.2	20	5.6	20	9.4
25	0.89	32	16.6	25	4.1	25	6.5
31	0.87	37	5.9	30	3.1	30	7.3
37	0.77	54	5.8	35	2.6	—	—
45	0.79	—	—	—	—	—	—

1, 3, 5, 7 — experimental time in the first, second, third, and fourth series, respectively (min); 2, 4, 6, 8 — corresponding light dispersion (relative units).

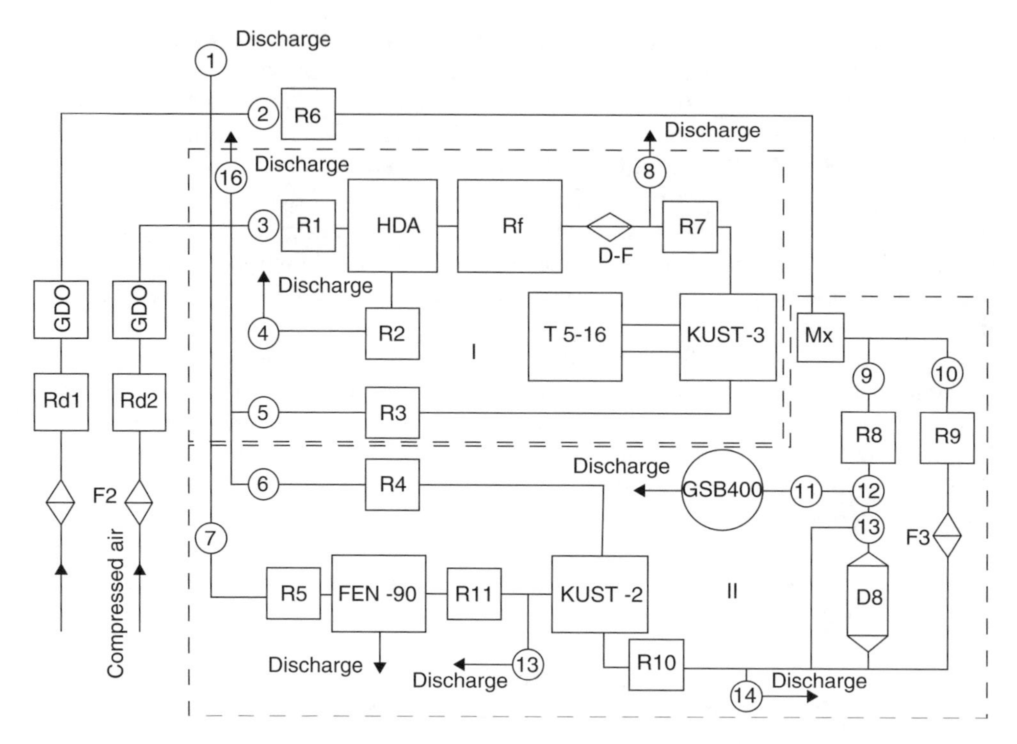

Figure 14.37 Device for NaCl UFA generation.

by means of valves 2 to 7 and checked with the rheometers R1 to R6. Valves 1, 4, 8, and 14 to 16 serve to discharge the surplus air and aerosols into the atmosphere.

The generating unit consists of the NaCl UFA generator, the refrigerator Rf, the set of diluting devices, and the KUST-3 instrument (Kogan and Burnashova, 1959). The principle of operation and design of the UFA generator described in Kravchenko (1974) is similar to those described in Fuks and Sutugin (1969) and differ from them only in the fact that its ceramic components are coated with fused NaCl. This coating increases considerably the duration of the generator's operation without replacing the gas adsorber.

The main components of the aerosol measurement unit consist of a set of cylindrical diffusion batteries, two of which have 64 channels each and ml=947.2 cm and one battery has 3700 channels and ml=74,111 cm (m is the number of channels in a battery and l is its length); the mixer Mx; the instrument KUST-2; and the nephelometer type FEN-90. The longest diffusion battery used in experiments has ml=5600 m.

Primary NaCl aerosol is formed in the UFA generator and then fed through refrigerator Rf to the diluting filter D-F, where it is diluted to the required calculated concentration. A part of the aerosol flow controlled by the thermistor flowmeter R7 is deviated, for increasing the size of its particles, to the KUST-3 instrument, which is heated by means of thermostat type TS-16. The aerosol is taken from the output of KUST-3 for measurements. In order to feed the NaCl primary aerosol to the measurement unit, it is sufficient to shut valve 5, which disconnects the KUST-3 instrument.

The aerosol taken for investigation is at first combined in the mixer with pure air in order to obtain a larger volumetric speed of pumping in the measurement unit, and it is then separated into two flows. One of them is fed through valve 9, thermistor flowmeter R8, and valves 12 and 13 to the output of the diffusion battery DB. The other flow passes through valve 10, rheometer R9, and filter F3, and it is then combined with the aerosol, which is delivered from the diffusion battery, in order to ensure feeding to the

KUST-2 instrument's input of the aerosol with an adequate volumetric speed. The part of the aerosol controlled by the thermistor flowmeter R10 is then fed to the KUST-2 instrument to increase the size of its particles. The impregnating liquid used in the KUST-2 instrument consists of dibutylphthalate. The aerosol is fed from the KUST-2 instrument's output through the thermistor flowmeter R11, where its speed is controlled, to the FEM-990 instrument for light dispersion measurement. Valve 13 serves to measure the diffusion of light in the aerosol at the output of the diffusion battery. Valves 11 and 12 are used for checking the thermistor flowmeter R8 readings by means of the gas counter GSB-400.

The ratio of light dispersion in the aerosol that leaves and enters the DB measures the penetration of the aerosol through the battery, n/n_0 (ratio of the flow of particles at the output to that at the input of the battery).

14.10.13.4.1 Method for investigating aerosols The parameters of particle distribution, according to their sizes in the generated aerosol, are determined by the diffusion method. The penetration of a polydispersed aerosol $(n/n_0)_p$ through the diffusion battery is represented by the formula

$$(n/n_0)_p = f(Q) = \int_{r,min}^{r,max} (n/n_0)_m \Phi(r) dr \tag{14.47}$$

where Q is the volumetric speed of the aerosol's pumping through the battery, r is the distribution function of particles according to sizes, and $(n/n_0)_m$ is the penetration of a monodispersed aerosol determined by the well-known formula (Gormley and Kennedy, 1949).

Several methods that were analyzed in Fuks (1964) are suggested in the literature for the solution of Gormley and Kennedy (1949). For the case of a logarithmically normal distribution of particles according to sizes, the expression can be rewritten in the form:

$$(n/n_0)_p \approx (\sqrt{2\pi} \ln\beta_g)^{-1} \int_{\ln R_1}^{\ln R_2} \exp(-\ln^2 r/r_g/2\ln^2 \beta_g)(n/n_0)_m \, d\ln r \tag{14.48}$$

where R_1 and R_2 are the lower and upper boundaries of the integrated interval with respect to radius r, r_g is the geometric mean radius, and g is the standard geometric deviation.

In this work, it was assumed that the aerosol at the output of the generator is characterized by the logarithmically normal size distribution.

The calculation for plane-parallel channels has been described by Fuchs et. al. (1962). The curves in that publication proved to be very convenient for investigations of particle size compositions of aerosols. In Kravchenko and Lekhtmakher (1971), calculations were provided for the deposition of aerosols in cylindrical channels. Formulas for the diffusion deposition of monodisperse aerosol particles were taken from Gormley and Kennedy (1949).

The "passage" (penetration) curves were calculated for different parameters of batteries t:

$$t = \log(3.657ml/Q)$$

where m is the number of channels, l their length, and Q the volume flow rate. Calculation was provided for the following parameters of logarithmic-normal distribution: $r_g = 10^{-7}$, 2×10^{-7}, 4×10^{-7}, 8×10^{-7}, 2×10^{-6}, 4×10^{-6}, 8×10^{-6} cm; $\ln \beta_g = 0.1, 0.25, 0.5, 0.7, 1$. In Figure 14.41, for every value of r_g there is a family of five curves that are different only in $\ln \beta_g$. The neighboring families of curves virtually coincide when shifted along the t axis; it is therefore possible, with the aid of the diagram, to obtain curves for a certain intermediate r_g by moving the nearest family, corresponding to r_g, through the distance

$$\Delta t = \log_{10} [D(r_g - \Delta)/D(r_g)] \quad (14.49)$$

where D is the diffusion coefficient of particles in cm^2/s.

In Kravchenko (1974), the study of the profile of the flow rate on the results of measurement by diffusion batteries was provided in the cross-section by different configurations of diffusors and confusors and for different rates of pumping. Measurements were provided by means of specially prepared small thermistor detectors, with the diameter of the working body of the detectors being ~0.25 mm. These data allowed one to choose the most rational from the studied configurations of diffusors and confusors and was used for the calculation of correction on the measured values of the passages.

Analysis of the data (Kravchenko, 1974) showed that by determination of the parameters of aerosols by the diffusion method, nonuniformity of the flow rate in the battery cross-section leads to substantial errors in the determination of $\ln_g$, at the same time not changing the accuracy of the determination of r_g. For the volume flow rate of 20 l/min, the correction for $\ln \beta_g$ reaches ~40%.

For plotting experimental results of passage vs. parameter t, it is necessary to make measurements at three points of the experimental penetration coefficient $k=0.8$, 0.4, 0.1. It is clear from formula (14.48) that such measurements can be provided by different aerosol flow rates through one battery, or with the same flow rate through batteries with different parameters ml. Studies suggest that the method of the fixed flow rate has advantages because, on the one hand, it allows the use of diffusion batteries in complicated measurements, in which the changes in flow rates are difficult or even impossible to obtain. On the other hand, by this approach the correction on the flow rate profile will be minimal.

It follows from the calculated curves of Figure 14.39 that the experimental curves may be obtained by means of three diffusion batteries with parameters ml differing one from another by approximately three times. Thus, for overlapping the whole range of dimensions of HDA with a flow rate of 1 l/min, it is necessary to have a set consisting of 14 diffusion batteries with parameters ml from 2.3 to 5×10^6 cm. The preparation of batteries with $ml > 5\times10^6$ cm is a very difficult task; therefore, for measurements of the distribution of aerosol particles larger than 10^{-6} cm, it is better to use very low flow rates. This leads to a greater duration of measurements because batteries with substantial parameters ml will have large volumes of diffusor and confusor. Thus, diffusion batteries with large parameters ml should be made from tubes with the smallest possible diameters and at the same time the smallest achievable thicknesses of the walls.

14.10.13.4.2 Experimental results of the study of UFA aerosol generator The primary NaCl aerosol was tested by means of two diffusion batteries with $ml=947.2$ cm and corresponding null batteries (Ivanova et al., 1974). The reason for using null batteries, which consist only of diffusors and confusors, is to exclude deposition by introducing a corresponding correction.

The generated aerosol's parameters (r_g and β_g) were determined for the HDA generator. For a generator gas absorber temperature of 550°C and a heater temperature of 680°C, the aerosol is generated with $r_g=51\times10^{-7}$ cm and $\beta_g=1.5$. Raising the generator and heater temperature to 700 and 800°C, respectively, leads to the following results: r_g $(10\pm1)\times10^{-7}$ cm and $\beta_g=1.1$. Thus, the mean size of the generated aerosols increases with rising temperature of the generator. The standard geometric deviation then changes within the limits of 1.1–1.5; it was noticed that for a gas absorber temperature exceeding the melting point of NaCl, the generator produces an aerosol whose parameters are unstable with time.

The relationship between aerosol parameters and the volumetric speed of air pumping through the gas absorber of the UFA generator was studied. For a speed of 150 cm^3/min, the NaCl aerosol is generated with $r_g=(10\pm1)\times10^{-7}$ cm and $\beta_g=1.65$, and for a speed of 250 cm^3/min with $r_g=(5\pm1)\times10^{-7}$ cm and $\beta_g=1.3$. Air was provided at a speed of 15 l/min. These results indicated that by raising the speed of air pumping through the gas absorber, the generator will produce particles with smaller size.

Using a flow-through ultramicroscope type VDK-4, it was shown that particle concentration in the primary NaCl aerosol was at least 2×10^6 $1/cm^3$. For decreasing the counted concentration, a set of calibrated diluting devices with dilution coefficients of 10, 100, and 1000 were used.

The reproducibility of the generator's operation was evaluated by comparing the parameters of the generated aerosol immediately after loading the generator with the gas absorber to the results obtained after a lapse of 50–100 h of the generator operation during which it was frequently turned off. The results have shown that the mean size of aerosol particles is in the range of $5\times10^{-7}<r_g<10^{-6}$ cm. It was difficult to come to any conclusion about the limits of the β_g variations. It is possible to conclude that the stability of the NaCl generator's operation was satisfactory, since the process of obtaining each characteristic of the generator lasted several hours, and during this time no substantial variation in the parameters of the generated aerosol was observed.

Aerosols with particle sizes exceeding 5×10^{-7} cm were obtained by increasing the size of NaCl primary aerosol particles in the enlarger KUST-3 instrument. The impregnating liquid then consisted of dioctylphthalate.

The final size of aerosols was adjusted by changing the temperature of the gas absorber in the enlarger KUST-3 instrument. Figure 14.38 shows the curves of penetration of an aerosol with enlarged particles through the diffusion battery, with ml=74,111 cm taken at different temperatures of the gas absorber in the KUST-3 instrument.

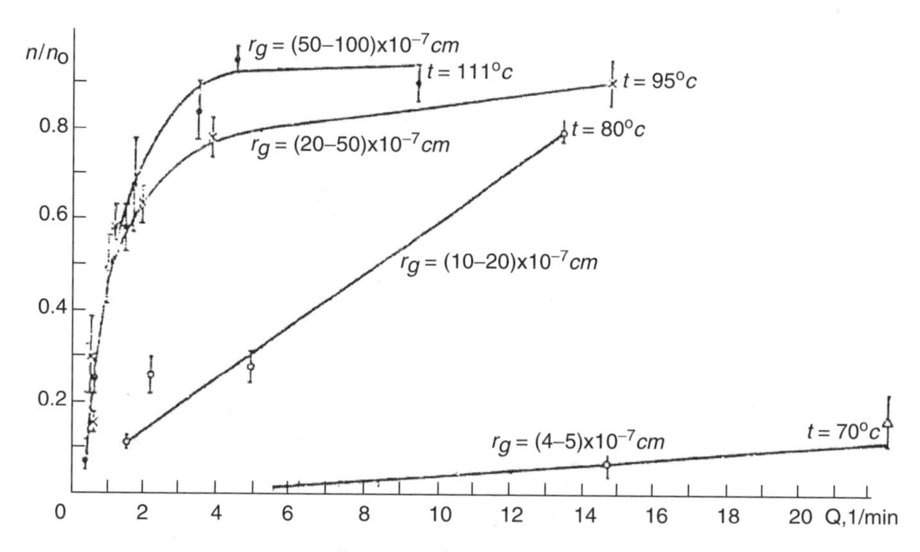

Figure 14.38 Curves for deposition of aerosol particles in a circular cylindrical channel. Penetration $t = \log(3.657y)$, $y = \pi Z/Q$; Z, channel length; Q, volume flow rate; values of r_g (cm): (i) 10^{-7}; (ii) 2×10^{-7}; (iii) 4×10^{-7}; (iv) 8×10^{-7}; (v) 2×10^{-7}; (vi) 4×10^{-6}; (vii) 8×10^{-6}. Values of In β_g: (a) 0.01; (b) 0.25; (c) 0.5; (d) 0.7; (e) 1.

14.10.13.4.3 Measurement errors (a) Error in measuring penetration by means of a nephelometric method. The penetration of aerosols through the diffusion battery can be represented as

$$k=n/n_0=\zeta_2/\zeta_1=(j_1-j_0)/(j_2-j_0)$$

where ζ_2 and ζ_1 are the nephelometer sensitivities in measuring n and n_0, j_1 and j_2 are the instrument's output current in measuring n and n_0, and j_0 is the instrument's output current produced by the inherent light diffusion of the nephelometer chamber.

In this case, a relative measurement of root-mean-square error of the penetration through the battery $\Delta k/k$ is

$$\Delta k/k=\sqrt{(\Delta\xi_2/\xi_2)^2+(\Delta\xi_1/\xi_1)^2+(\Delta j_1^2+\Delta j_0^2)/(j_1-j_0)^2+(\Delta j_2^2+\Delta j_0^2)/(j_2-j_0)^2}$$

Measurements were made in such a manner that $\xi_1=\xi_2=\xi$ and $\Delta\xi_1=\Delta\xi_2$. As a result of measurements, it was also found that instrument sensitivity over the entire range was $\Delta\xi/\xi=0.02$. As a result of testing of the nephelometer, it was found that $\Delta j=\Delta j_1=\Delta j_2=\Delta j_0=3$ mA. The values of j_1-j_0 and j_2-j_0 are determined on the basis of the consideration that, in the majority of cases, measurements were made for the condition that 25 mA $<(j_1-j_0)<(j_2-j_0)$. In the limiting case when $j_1-j_0=25$ mA and $n/n_0=1$, we have

$$\Delta k/k=\{2(\Delta\xi/\xi)^2+4[\Delta j/(j_1-j_0)]^2\}^{1/2}=0.25 \quad (14.50)$$

Thus, the maximum root-mean-square error of a single penetration measurement did not exceed 25%. In practice, the penetration for each speed is measured several times and the root-mean-square error is calculated for each point.

(b) The assessment of the error associated with measuring the volumetric speed of aerosol pumping through a diffusion battery based on the following. When the speed Q is measured by means of a thermistor flowmeter, its value is determined from the formula $U=f(Q)$, where U is the reading of the galvanometer and f is a function with unknown form.

The relative root-mean-square error in measuring Q in this case is $Q/Q=U/Qf'(Q)$.

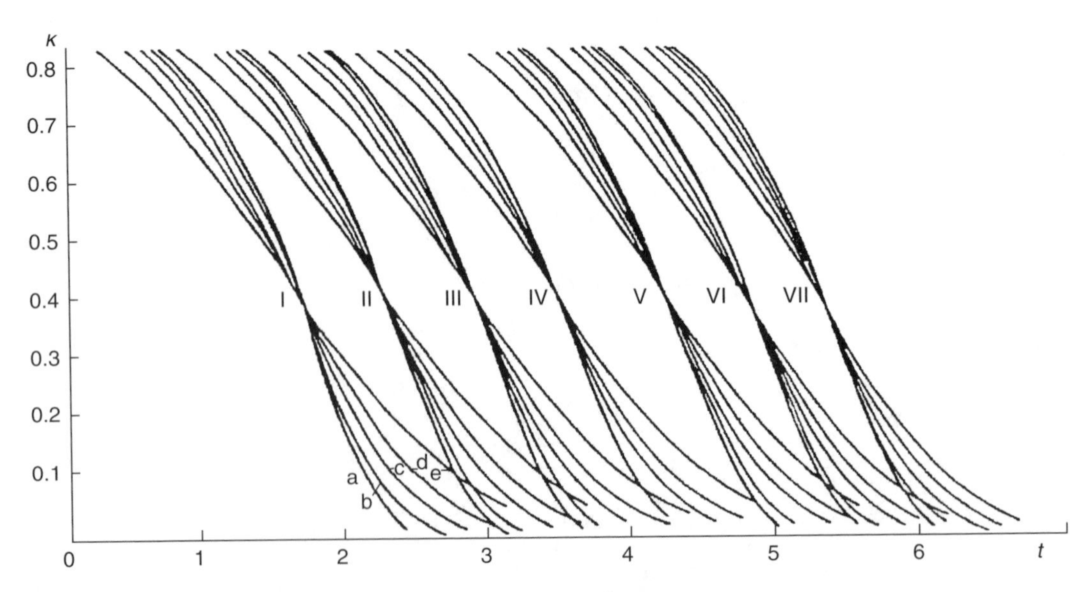

Figure 14.39 Penetration of aerosol through diffusion battery.

In these measurements, the voltage error was $U=5\times10^{-4}$ V. The value of $f'(Q)$ was found with the flowmeter being calibrated by means of the gas meter:

$$f'(Q)=(U_2-U_1)/(Q_2-Q_1)$$

where U_1 and U_2 are galvanometer readings for the pumping speeds of Q_1 and Q_2. It was found that the error in measuring over the entire range of velocities does not exceed 2%.

(c) Error in determining the parameters of the particle size distribution function can be assessed taking into account such circumstances. Unfortunately, owing to the complicated nature of relationship (14.48), it is impossible to calculate errors in determining r_g and β_g from known errors in n/n_0 and, therefore, the errors in determining r_g are found approximately from graphs. It should be stated that the penetration depends mainly on r_g and to a much smaller extent on ln β_g. On the basis of the results obtained in Kravchenko (1974), it can be asserted that errors in β_g are in the range of 0–0.5. It should also be noted that in measuring the penetration of aerosols through diffusion batteries, it is necessary to bear in mind the precipitation of aerosols in the diffusor and confusor of the battery. In order to make corrections on this effect, zero-type batteries were developed and used in the measurements.

14.10.13.5 *Diffusive particle deposition in the inlet segment of a tube*

The following symbols are introduced:

Q	volumetric speed
U_{av}	$(=Q/a^2)$ speed at the entrance to the pipe (average speed)
a	radius of the pipe
r	radial coordinate
φ	particle concentration
U	component vector of speed in direction x
V	component vector of speed in direction r
v	viscosity
δ	thickness of boundary layer
Re	Reynolds number
l_e	length of the entrance section
m	root of the equation $I_2=0$
I_0, I_2	Bessel functions

The formulas developed by Gormley and Kennedy (1949) for the diffusion deposition of aerosol particles in the pipes were obtained for a Poiseuille flow.

$$U=2U_{av}(1-r^2/a^2) \tag{14.51}$$

Actually, this expression yields an asymptotic approach for the actual speed at great distances from the entrance. Therefore, it was possible to assume that this effect can play a substantial role in the case of aerosols with a large diffusion coefficient, since the precipitation in this case is already considerable in the area of hydrodynamic flow stability. Analogous questions have been addressed in research associated with heat exchange problems (Petukhov, 1967).

In Lekhtmakher et al. (1972), the calculation of the passage k of particles in their dependence on the "special length" of the tube $\zeta=x/aRe$ was provided. The results of the calculation show that for the purpose of practical measurements, the "entrance effect" may be disregarded for Schmidt numbers $Sc\sim1$ and greater.

Under usual approximations (Petukhov, 1967), the equation of diffusion in the entrance area looks like

$$D\Delta\varphi - U(\partial\varphi/\partial x) - V(\partial\varphi/\partial r) = 0 \quad (14.52)$$

where U and V are the parts of the vector of velocity in the direction of x and r, respectively. For simplification, the approximated solution was used for U and V from Sadikov (1967). For the location close to the entrance, the more accurate solution was obtained by Schiller's method (Campbell and Slatery, 1963).

$$U = 6/(m^2 - 4m + 6) \text{ if } r/a < 1 - m;$$

and

$$U = 6/(m^3 - 4m^2 + 6m)\,[2(a-r)/a - (a-r)^2/ma^2] \text{ if } r/a > 1 - m \quad (14.53)$$

while

$$420 = \{336m - 26\ln(1-m) + 318\ln(2-m) + 148\ln(m^2 - 4m + 6)$$
$$+ 27(52m + 3)/(m^2 - 4m + 6) - 2084/2^{1/2}[\arctan(m-2)2^{1/2}]\} \quad (14.54)$$

At the distance of the entrance, the more accurate solution becomes that obtained by integrating the Ozyeyen equation Sadikov (1967):

$$U = 2\overline{U}(1 - r^2/a^2)$$
$$-4\overline{U}\sum_{m=1}^{\infty}\gamma_m^{-2}\{1 - [I_0(\gamma_m r/a)/I_0(\gamma_m)]\}\exp(-\gamma_m^2\, vx/\overline{U}a^2) \quad (14.55)$$

Figure 14.41 shows the curves obtained by two of these formulas — (14.53) and (14.55). The parameter here is the distance from the axis of the pipe. Formula (14.53) was used for $\zeta < 0.02$ and formula (14.55) was used for $\zeta > 0.02$ (Figure 14.44).

The expression for the radial component of the speed vector is found from the equation

$$V = 1/r$$

If at the entrance the particle concentration changes with respect to the cross-section of the pipe, the variation in the profile of the stream speed itself leads to equalization of the particle concentrations. The homogeneous distribution of particles stabilizes as the stream stabilizes. This reflects the resilience of the medium and the medium's ability to completely carry the particles.

Introducing the nondimensional variables with respect to Equation (14.55),

$$U' = U/\bar{U},\ V' = (v/\bar{U})\,Re$$

$$\zeta = x/aRe;\ \varphi = \varphi/\varphi_0(\varphi_0 = \varphi(x=0));\ \rho = r/a$$

we get for a round cylindrical pipe

$$\partial^2\varphi/\partial\rho^2 + (1/\rho)\partial\varphi/\partial\rho - Sc\, u\partial\varphi/\partial\zeta - Sc\, v\partial\varphi/\partial\rho = 0$$

where $Sc = \nu/D$ is the Schmidt number.

The limiting factors are $\varphi|_{\xi=0} = 1$ and $\varphi|_{\rho=1} = 0$.

The equation was solved numerically by the matrix method.

The passage of the particles through the pipe is equal to

$$k(\zeta)=1/\pi a^2\varphi_0 U_{av}\int_0^a 2\pi r u \psi \mathrm{d}r=2\int_0^1 2\psi\rho u \mathrm{d}\rho$$

from which, using the trapezoid formula, we obtain

$$k(\zeta_k)=\sum_{n=1}^{N=1} u_{k,n}\varphi_{k,n} n2/N^2$$

The results of the calculations are shown in Figure 14.40 (the lower curve) in comparison with the curves derived by Gormley and Kennedy (1949). It is evident that as *Sc* decreases,

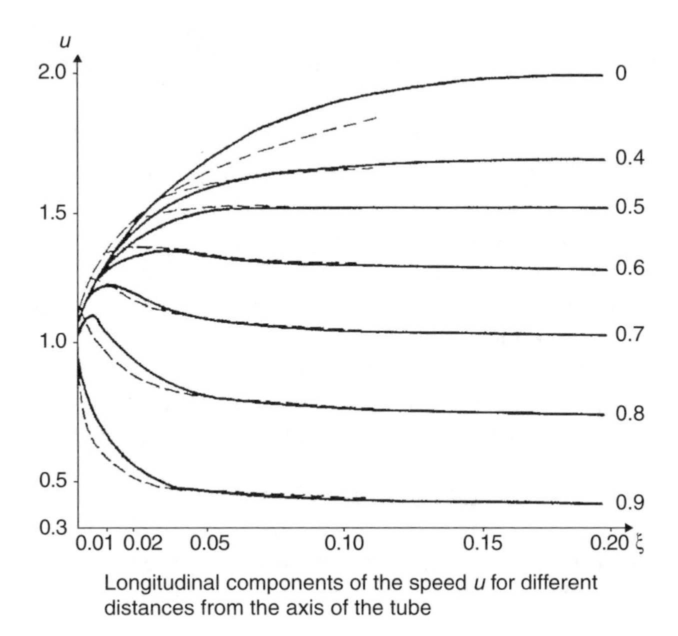

Figure 14.40 Longitudinal components of the speed u for different distances from the axis of the tube.

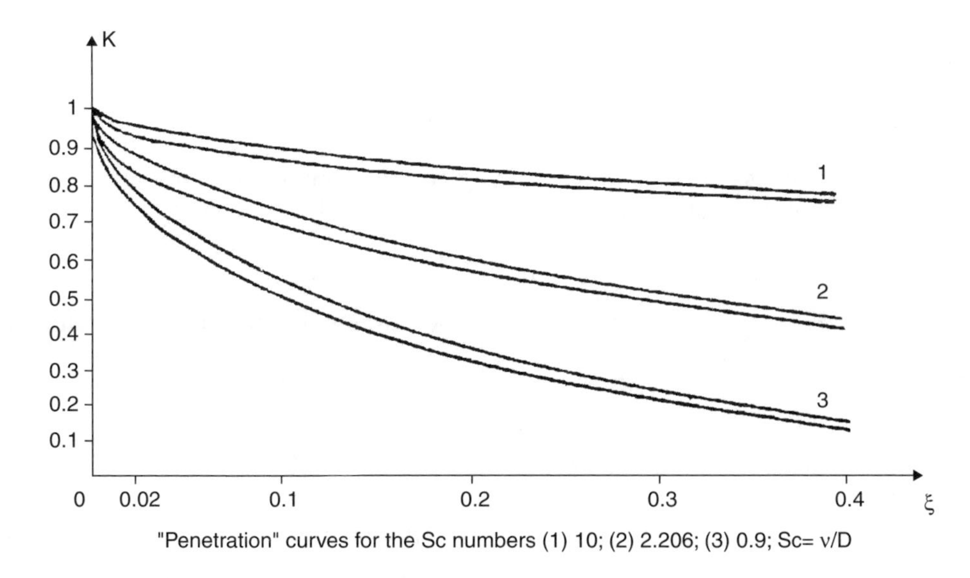

Figure 14.41 "Penetration" curves for *Sc* numbers: (1) 10; (2) 2.206; (3) 0.9; $Sc = v/D$.

the divergence between the particles suspended in the air, the "entrance effect," grows, but even for $Sc=0.9$ the asymptote is already sufficient when $Sc=0.4$; however, at $Sc<0.4$, the difference does not exceed 10%. Applying these results to the aerosol can be disregarded. It is necessary, however, to note that this applies only to the case most frequently encountered in practice, that is, when the distribution of speed at the entrance is homogeneous. Obviously, this conclusion is justified when the speed distribution at the entrance is characterized by a certain intermediate profile arising when a "pivoted" flow is converted into Poiseuille flow. In the general case, the speed distribution at the entrance to the pipe can exert a notable influence on the precipitation of particles. Some aspects of the "entrance effect" are due to the homogeneous speed distribution at the entrance. The asymptotic profile is reached fairly quickly: already at $\zeta=0.2$ (i.e., at 0.1 of the length of the entrance section), the thickness of the boundary layer is 0.527 of the pipe radius and precipitation occurs near the wall, precisely as would be expected from the fundamentals of flow dynamics.

14.10.13.6 *Errors in determination of the parameters of the logarithmically normal size distribution of aerosol particles by the diffusion method (Lekhtmakher and Ruzer, 1972)*

Fuchs and Stechkina (1962) and Kravchenko et al. (1971) gave calculated curves for the passage of aerosols with a logarithmically normal particle size distribution through plane-parallel and cylindrical channels. By comparison with experimental curves, they make it possible to find approximate values of r_{g0} and β_g.

We present formulas below that make it possible to refine the values of r_g and β_g and to evaluate the error in the determination of sought quantities. Using the nomenclature adopted in Kravchenko et al. (1971), the expression for the passage of particles is written as

$$k(t,r_g,a)=1/\sqrt{2\pi a}\int_{r1}^{r2} F(t,\, D)\exp[-(\ln r/r_g/2a^2)^2]\mathrm{d}\ln r$$

where $t=\log_{10}(3.657l/Q)$ (l is the channel length and Q is the volume flow rate).

We expand this function in a series at the point $(r_{g0},\, a_0)$:

$$k(t,r_g,a)=k(t,r_{g0},a_0)+(\mathrm{d}k/\mathrm{d}a)a+(\mathrm{d}k/\mathrm{d}r_g)r_g+\cdots$$

$$=k_0+(t,r_{g0},a_0)x+(t,r_{g0},a_0)y+\cdots$$

where

$$x=a/a_0,\; y=r_g/r_{g0},\; a=\ln\beta_g \tag{14.56}$$

In view of the smallness of x and y, we can neglect subsequent terms in the expansion. If $N>2$, we have the system of relative equations from Kravchenko et al. (1971):

$$l_i=\alpha_i x+\beta_i y,\; i=1,2,\ldots,N \tag{14.57}$$

where $l_i=k(t,r_g,a)-k(t,r_{g0},a_0)$, N is the number of measurements, and k is the experimentally obtained value of the passage at $t=t_i$.

$$\alpha_i=\alpha\,(t_i,r_g,a),\; \beta_i=\beta(t_i,r_{g0},a_0) \tag{14.58}$$

In accordance with the method of least squares, we obtain (Malikov, 1949)

$$x=(\Sigma\alpha_i\, l_ip_i\Sigma\beta^2{}_ip_i-\Sigma\beta_il_ip_i\Sigma\alpha_i\beta_ip_i)\theta \tag{14.59}$$

$$y=(\Sigma\beta_i\, l_ip_i\Sigma\alpha^2{}_ip_i-\Sigma\beta_il_ip_i\Sigma\alpha_i\beta_ip_i)\theta$$

where

$$\theta = \Sigma\alpha^2_i p_i \Sigma\beta^2_i p_i - (\Sigma\alpha_i\beta_i p_i)^2$$

For the mean-square errors, we have the formulas

$$\sigma_x = \sigma/\sqrt{P_x},\ \sigma_y = \sigma\sqrt{P_y},\ \sigma = \left(\sum^{N} \theta_i^2 p_i\right)/N-2 \tag{14.60}$$

where $P_x = \theta_i/\Sigma\beta^2_i p_i$, $P_y = \theta_i/\Sigma\alpha^2_i p_i$

$\theta_i^2 = (l_i - \alpha_i x - \beta_i y)^2$ (the square of the residual error of the ith relative equation), P_i is the weight of the ith relative equation $P_i = A/(\sigma_i)^2$, A is an arbitrary constant, and σ_i is the relative error in measurement of the passage at a given point, t_i. Thus,

$$a = a_0(1 + x \pm \sigma_x),\ r_g = r_{g0}(1 + y \pm \sigma_y)$$

For intermediate values of r_{g0}, the data can be taken from the same Figure 14.42, since the form of the curves depends only slightly on r_g. This fact leads to a situation in which the errors in the determination of a and r_g are practically independent of r_g.

The procedure for finding the corrections x and y of the errors ε_x and ε_y consist in the following. Let the values of the passage be

$$k_1 = 0.7 \text{ (at } t_1 = 0.7),\ k_2 = 0.55 \text{ (at } t_2 = 1.5),\ k_3 = 0.7 \text{ (at } t_3 = 2.3)$$

Comparing these data with the calculated curves given in Kravchenko et al. (1971) we find

$$r_{g0} = 10^{-7} \text{ cm},\ \ln \beta_g = 1$$

For the differences between the experimental values of the passage and the corresponding values of the passage on the calculated curve, we have

$$l_1 = -0.05,\ l_2 = +\ 0.06,\ l_3 = +0.03$$

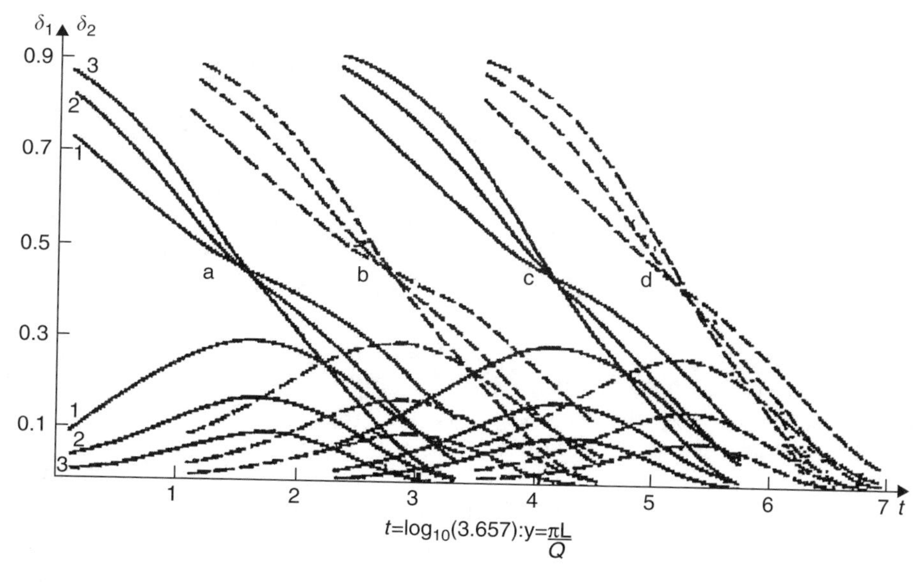

Figure 14.42 Functions S_1 and S_2 as a function of t, $t = \log(3.657\ \pi L/Q)$: L, length of diffusion battery (cm); Q, volumetric pumping rate (cm^3/sec); a_0 equal to 1 (1), 0.7 (2), 0.5 (3); r_g equal to 10^{-7} (a), 4×10^{-7} (b), 2×10^{-6} (c) and 8×10^{-7} cm (d).

From Figure 14.42, we find S_1 and S_2 for a given value of t_i:

$$S_1(t_1)=0.22,\ S_1(t_2)=0.31,\ S_1(t_3)=0.25$$

$$S_2(t_1)=0.62,\ S_2(t_2)=0.47,\ S_2(t_3)=0.36$$

We determine $_i$ and, using (14.58), we seek the correction x and y, assuming that all $p_i=1$, $x=0.31$, $y=0.07$. We substitute the values found for x and y into the relative equations for l_i and seek residual errors:

$$\sigma=\sqrt{\sum_{j=1}^{3}\vartheta^2/(n-2)}=0.05$$

$$\sigma_x=0.27,\ \sigma_y=0.11$$

Thus,

$$r_g=10^{-7}(1+0.07\pm0.11)\ \text{cm},\ \ln\beta_g=1+0.31\pm0.27$$

14.10.13.7 Fine aerosols

The generation of liquid MDA with a narrow size distribution was provided in a condensation-type generator (Kravchenko, 1974). The experiments showed that the generator produced dioctylphthalate aerosols in the size range 0.15–0.45 μm, with a counted concentration of $(5–14)\times10^6$ particles/cm^3. Changing the parameters of the studied aerosol was possible by means of changing the flow rate through the generator. The results of the study of the stability and reproducibility of the generator showed that r and n and their average square root errors are

$$r=0.25\ \mu\text{m},\ n=7.8\times10^6\ \text{cm}^{-3},\ \sigma_r=4\%,\ \sigma_n=10\%$$

For generating aerosols in the intermediate ranges of 5×10^{-7} to 2.5×10^{-5} and 2.5×10^{-5} to 5×10^{-4} μm, aerosol from the HDA generator or the MDA generator was enlarged in the KUST-3 instrument with an impregnating liquid of dioctylphthalate. The degree of enlargement of particle size was achieved by changing the temperature of the thermosetting liquid of the KUST-3 instrument.

Besides this type of generator, the generator of the latex was used with particle diameters of 0.08, 0.163, 0.35, 0.47, 0.74, 1.49, and 3.49 μm. The study shows that in this generator, aerosols were dry (without liquid "coat") with a good reproducibility of particle concentration; after 1 h of work, the counted aerosol concentration on the outlet of a generator changes no more than 5.4%.

The majority of studies of MDA was provided by means of the photoelectric aerosol counter AZ type. The main source of systematic errors in this case are errors associated with the difference between the refraction coefficient of the substance of the studied aerosol and the aerosol used for calibration.

Calculations of the amplitudes of signals U on the integral discriminator of AZ-type instrument from particle radius r with different refraction coefficient m, registered by the detector of the instrument, were provided in Kravchenko (1974). Taking into account the working principle of the instrument, we obtain

$$U(r,m)=A\int_{\theta}\int_{\lambda}a(\lambda)b(\lambda)I(\text{r},\theta,\lambda,m)\,d\theta\,d\lambda \tag{14.61}$$

where A is a constant, the wavelength of light, the refraction angle, $a(\lambda)$ the spectral characteristic of the light source (lamp OP6-15), and $b(\lambda)$ the spectral characteristic of the detector (FEM-51). Function I is determined by the Mie formula:

$$I=\lambda^2/8\pi^2R^2(i_1+i_2) \tag{14.62}$$

where R is the distance from the aerosol particle to the photocathode of a photomultiplier.

Functions i_1 and i_2, which represent the intensities of parallel and perpendicular components of refraction light, are

$$i_1=\left\{\sum_{v=1}^{\infty}(2v+1)/v(v+1)[a_v\prod{}_v+p_v(\prod{}_v\cos\theta-\prod{}'_v\sin^2\theta)]\right\}^2 \tag{14.63}$$

$$i_2=\left\{\sum_{v=1}^{\infty}(2v+1)/v(v+1)[a_v(\prod{}_v\cos\theta-\prod{}'_v\sin^2\theta)+p_v\prod{}_v]\right\}^2 \tag{14.64}$$

Coefficients a_v and p_v can be written as

$$a_v=(-1)^v i[m\Psi'_v(\alpha)\psi_v(\beta)-\Psi'_v(\beta)\Psi_v(\alpha)]$$

$$/[m\zeta'_v(\alpha)\Psi_v(\beta)-\Psi'_v(\beta)\zeta_v(\alpha)] \tag{14.65}$$

$$p_v=(-1)^{v+1} i[m\Psi_v(\alpha)\Psi'_v(\beta)-\Psi_v(\beta)\Psi'_v(\alpha)]$$

$$/[m\zeta_v(\alpha)\Psi'_v(\beta)-\Psi_v(\beta)\zeta'_v(\alpha)] \tag{14.66}$$

where a_v and p_v are expressed through cylinder functions of half–whole order;

$$\alpha=2\pi r/\lambda,\ \beta=m\alpha \tag{14.67}$$

Functions are expressed through Legendre polynomials. Calculations were provided by changing integrals by sums. The values U were calculated for 14 radii of the particles in the range of 0.01–0.5 μm and for six values of the refraction factor: m_1=1.59 (polystyrol); m_2=1.333 (water);

m_3=2.06 (sulfur); m_4=1.485 (dioctylphthalate);

m_5=1.1–2.02·i (gold for=0.5 μm); m_6=4.0 (limit refraction coefficient for the solid substance).

From these data, the correction of measurement results on the refraction coefficients was made under conditions that the calibration aerosol was polystyrol latex.

Aerosols with particle sizes of 5–200 μm were generated by means of a disk pulverizer whose principle of operation consists of the following. The original liquid is fed to the center of a rapidly rotating disk. Under the effect of centrifugal force, the liquid is displaced toward the disk edge and is ejected from it in the form of monodispersed drops. The particle size was adjusted by changing the disk rotation speed. The disk pulverizer serves to yield aerosols with a liquid dispersed phase from type VM-4 oil and with solid dispersed phase from a rosin solution in alcohol from steric acid melts. These types of coarse aerosol (CA) generators are suitable for obtaining virtually monodispersed aerosols with particle sizes of 0.002–200 μm.

The stability and reproducibility of aerosol generators in all ranges of aerosol particle size is especially important in terms of using these generators in testing the measurement technique, both newly developed and currently used in practice. From the experience in using the installations of the Aerosol Standard, it was clear that developers of some aerosol measurement technique sometimes ignored the precipitation of aerosol particles in transportation channels both outside and inside the instrument due to diffusion, sedimentation, and inertia, which lead to the distortion of measurement results.

In order to estimate losses in the connecting tubes, special measurements were made for both inactive and radioactive aerosols. Experiments were carried out to determine inertial precipitation of the dispersed phase of aerosols in metallic pipes and its dependence on volumetric speed. The polydispersed radioactive aerosol phosphorus-32 (a salt of the monosubstituted KH_2PO_4, with a density of 2.34 g/cm^3) was used. The pipes, bent to an angle of 90° with a radius of curvature of 50 mm, had a length of 250 mm and internal diameters of 14 and 22 mm. Precipitation was measured as a function of the difference in activities of the filters at the entrance and exit of the pipe. The results of the measurements also showed that precipitation depends on the bend angle of the pipe. Thus, with bend angles of 45 and 90° at a speed of 200 l/min, the inertial precipitation is 0.15 and 0.23 for a pipe with an inside diameter of 14 mm, and 0.13 and 0.18 for an inside diameter of 22 mm.

For monodispersed aerosols of polystyrol latex having a particle size of 1.4 μm and a density of 1.05 g/cm^3, precipitation was measured in a pipe of diameter 8 mm and length 250 mm, with a bend angle of 90°. Volumetric speeds were 25 and 50 l/min, and precipitations were 0.06 and 0.12, respectively.

These data are given only to show that the inertial precipitation is essential to calculate the loss of aerosol particles in the main lines approaching the transducer of the apparatus. Special attention should be paid to CA, both inactive and radioactive. Substantial precipitation of CA was discovered in two places in the test equipment. Precipitation occurred in the horizontally placed channel at the entrance of the equipment, and also in the conical portion. Both cases necessitated reconfiguring the equipment.

These experiments clearly demonstrate that both these types of precipitation (inertial and sedimentation) must be taken into account when measuring equipment in main lines. This, in particular, also applies to artificial radioactive aerosols, which frequently coarse aerosols, and have high-density particles (e.g., ^{239}Pu).

14.10.13.7.1 Determining the composition of aerosols by means of two mean radii In Novozhilova (1963), the method of measurement of aerosol parameters by means of measuring two mean radii was proposed. It has been shown that in many cases the particle size is distributed according to a logarithmically normal law. Two parameters, log r_g and log β_g, must be known for calculation of the particle size distribution of an aerosol.

In Novozhilova (1963), formulas were derived for the calculation of these parameters from the values of two mean particle radii, determined experimentally by two methods based on different averaging laws. The experimental methods for determining mean particle radii considered for this purpose were, for example, the following:

1. The count-weight method, averaging by the $\overline{r^3}$ law.
2. The count-nephelometric method, averaging for $\overline{r^6}$ in the region of $r < 0.1$ μm, and for $\overline{r^2}$ in the region of $r > 0.5$ μm.
3. The diffusion method, averaging for $\bar{r}$ at $r >> \lambda$ and for $\overline{r^2}$ at $r << \lambda$ (λ — free path length of gas molecules).

The metrological aspects of this method were discussed in Kolerskii and Ruzer (1968). It was shown that:

1. In measuring the dispersed composition of aerosols by the method of two mean values, the problem of determining the parameters of aerosol particle distribution can be completely solved, if the errors in measuring the mean values of the tested quantity are known.
2. The diffusion and nephelometric measuring methods are the most promising, and their combination provides the minimum error in determining log β_g and log r_g as well as the possibility of measuring the most monodispersed aerosols.

Problems of the calibration of spectrometers and determination of the apparatus function were discussed in Lekhtmakher (1985, 1987).

14.10.13.8 Standard for generating and measuring the electrical properties of aerosols
The electrical properties of aerosols are important in terms of their behavior in air, and deposition on surfaces and in the lung. They play a substantial role in determining particle size distribution, especially by the diffusion method, and in calibration of the aerosol measurement technique. A standard of electroaerosols and aeroions was developed as part of the National Aerosol Standard of U.S.S.R. (Antipin et al., 1980).

14.10.13.8.1 Generators of aeroions and electroaerosols The generators of the Standard should have:

- Stability of the polar (+and−) density charge of aeroions with a definite mobility distribution during a long period of time.
- Opportunity for the regulation of polar volume density charge in broad limits. In Kolerskii (1975, 1974) and Kolerskii et al. (1969) the generator of the bipolar–unipolar ions was described, which satisfied both demands essentially in terms of calibration and testing aeroion and electroaerosol counters. It consists of the radioactive aeroionizer, diluter, exposure chamber, and cylindrical electrostatic precipitator (ESP).

The aeroionizer is analogous to the Steinbok radioactive ionizer with the advanced electronic system. One of the electrodes of the aeroionizer is the metal casing with the radioactive source.

The aeroion unipolarity coefficient is controlled by changing the voltage between electrodes. For the purpose of the broadening of ranges for regulating unipolarity coefficient, the size of the aeroionizer and type of radioactive alpha- or beta-source is chosen in such way that the distance from the source to the axis of the channel, through which air is passed, is approximately equal to the distance of the run of particles of the used nuclide.

Air, purified from admixtures by means of aerosol and vapor filters, passes with a definite flow volume rate through the aeroionizer, where it is enriched with ions and mixed in a diluter with nondusted deionized air. Next it is passed through the exposure chamber during the time necessary for establishing the stationary mobility of ions distribution and then through ESP on the generator outlet. The generator regulates the unipolarity coefficient, mark of the charge, and the polar volume density of charge of predominantly light ions. The polar volume density of charge is controlled from the zero up to the maximum value, which is determined by the activity and type of radioactive source, unipolarity coefficient from bipolar to completely unipolar ions. Instability of the polar volume density of the charge does not exceed 1% during the 8 h of uninterrupted work of the generator.

The generator of liquid electroaerosols consists of an atomizer (pulverizer) of the Bergson–Barkovsky type, a diluter, a light ions precipitator tank with a liquid, and a heater with autoregulation of temperature. Charging of the atomized liquid takes place under the electrical field of the induction electrode. The generator allows one to regulate the polar volume density of the charge in the range of 10^5–10^7 elemental electrical charges per cm^3, and unipolarity coefficient from 50 to 90%. Aerosol dispersion was regulated by using a sprayer with different diameters of the aperture and changing the distance between the atomizer and reflector. The instability of the polar volume density of the charge of electroaerosols did not exceed 10% during the 8 h of uninterrupted work of the generator. For monodispersed polystyrol latex generation, a generator of liquid aerosols is used in the mode of thin dispersed atomization latex suspension in distillation water and substantial dilution by the flow of dry and dust-free air.

14.10.13.8.2 Method and instrumentation for the measurement of electrical parameters of radioactive UFA In Kolerskii (1975), a study of the aspirating method for measuring electrical parameters of radioactive aerosols was provided — concentration and portion of charged aerosols, differential concentrations charged radioactive UFA, and their electrical mobility $k(k_1, k_2)$. The formulas were derived for the assessment of errors in the determination of aerosol electric parameters depending on the errors of measurements of the concentration of radioactive aerosols, type of ESP, and the mode of its work.

Radioactive aerosols integral ESP were produced with an outer electrode diameter of 3.2 cm, and inner electrodes had different diameters from 0.015 to 1.78 cm and lengths of 5, 10, and 30 cm.

Every precipitator has maximum sensitivity in a special part of the spectra, and a set of precipitators allows one to study particles with mobilities from some units up to 10^{-3} (cm^2 s)V^{-1}.

By means of integral ESP, concentration of the uncharged UFA is measured by the activity of aerosols deposited on the filter, which is installed at the entrance of ESP. Concentrations of charged aerosols are determined by the difference in the activity at the entrance and outlet filter.

For spectra characteristics of radioactive UFA measurements with mobilities of some units to 10^{-3} (cm^2/s)V^{-1}, differential first-type (cylindrical and plate-parallel) ESP were developed for the measurement of activity through the length of electrodes.

For the mobility distribution of radioactive aerosols in the range of 10^{-1} to 10^{-5} (cm^2/s)V^{-1}, differential second-type ESP were produced with a radius and length of the inner electrode of 1.5 and 100 cm, respectively, and the radius of the outer electrode was 6.5 cm.

Measurement of the radon decay products was provided by means of an alpha-spectrometry instrument with four alpha spectrometric detectors, which allowed one to measure the activity of aerosols, precipitated on the filters and on ESP electrodes.

Unexcluded systematic error of the measurement of ^{218}Po concentration did not exceed 5%. The error in measuring the concentration of charged radioactive aerosols was 30% if their concentration was more than 1850 Bq/m^3.

UFA aerosols of ^{218}Po were studied in a chamber of 0.6 m^3 volume with an open source of ^{226}Ra inside. The sampling took place on diffusion batteries, ESP, and spectrometric filters AFA-RSP type, which provided the overall measurement of aerosol parameters.

It was shown that in "dust-free" air, UFA aerosols of ^{218}Po are mainly neutral (if concentration is in the range of 1850–370,000 Bq/m^3) and the portion of charged aerosols does not exceed 10%. With increasing dust concentration, the portion of charged HDA increases. The charged UFA aerosols of ^{218}Po consisted of some groups with mobilities of 2.1 ± 0.2, 1.4 ± 0.1, 0.9 ± 0.1, 0.6 ± 0.1, 0.45 ± 0.06, and 0.23 ± 0.04 (cm^2/s)V^{-1}.

In a different group of experiments, these or other groups prevailed, but in the majority of cases maximum distribution was associated with ions of mobilities 1.4 and 0.9 (cm^2/s)V^{-1}. For example, in the case of ^{218}Po concentration of 18,500–185,000 Bq/m^3, it was shown that in the range of mobilities of 4–0.1 (cm^2/s)V^{-1} the average mobility was

$$k^+=0.99\pm0.19 \text{ (cm}^2\text{/s)V}^{-1}; k^-=1.58\pm0.22 \text{ (cm}^2\text{/s)V}^{-1}$$

The diffusion coefficient of the "unattached" ^{218}Po was in the range of $(1–5)\times10^{-3}$ to 6×10^{-2} cm^2/s and the average diffusion coefficient was $(3.7\pm1.4)\times10^{-2}$ cm^2/s. The average number of the elemental charges on the particles was one.

It was shown in Kolerskii (1975) that the UFA aerosols of ^{218}Po consisted of some groups of ions and neutral particles. Depending on the aerosol concentration and of the time of life of ^{218}Po HDA, their properties change. This leads to changes in the portion of charged aerosols, average diffusion coefficient, and electrical mobility.

The study of the portion of charged ^{131}I was provided on a ^{131}I vapor generator (Antipin et al., 1980) based on the principle of oxidation of water by potassium bichromate. Reproducibility was not worse than 10% in the range of concentrations from 37 to 185,000 Bq/m^3 and the portion of charged particles was no more than 15%.

14.11 Dosimetry

14.11.1 Intake vs. exposure: Propagation of the uncertainties in dose assessment in mining studies

Epidemiological studies of underground miners is the basis for estimating the risk of indoor radon (NCRP, 1984; National Research Council, 1988; U.S. EPA, 1992). Although there have been a number of studies investigating the health effects of exposure to radon decay products in mines (the most recent compilation has been carried out by NCI [1994], Lubin et al. [1994], and NRC 1998 [BEIR VI]), there are several unresolved issues in the assessment of the actual radiation dose to the miners' lungs:

- Lack of detailed spatial and temporal data on radon and radon decay product concentrations
- Variability in the ratio between concentrations as measured by the standard inspection procedure and as measured in the breathing zone
- Variability in breathing rates and deposition coefficients for radon decay products in the lungs for different types of work and among different groups of miners
- Information not known about the use of respirators by different groups of miners
- Very little data on the work itinerary (scenario of exposure) for individuals or groups of miners

The presence of such errors in the exposure estimates for miners has been widely recognized and discussed in NCRP (1984), DOE/ER (1991), and Radon and Lung Cancer Risk (1994), NRC 1998 (BEIR VI). In this connection, it has been noted that concentrations of radon and its decay products vary spatially and temporally within mines, although little data have been published in this regard. In New Mexico mines, for example, information presented on dosimetry documented extensive variation in the concentrations of radon progeny across various locations within mines in Ambrosia Lake, New Mexico (BEIR VI).

It has also been pointed out in BEIR VI that: exposure estimates for individual miners would be ideally based on either a personal dosimeter, as used for low-LET occupational exposure, or on detailed information on concentrations at all locations where participants in the studies received significant exposure (SENES, 1989). For miners, information would be needed on the location where time was spent, the duration of time spent in the location, and the concentration in the location when miners were present. Personal dosimeters for radon and its progeny have not been developed until recently and their usage has been limited; hence, detailed information on concentrations within mines and time spent in various locations has not been available to most epidemiological studies.

In the epidemiological studies, these ideal approaches have been replaced by various pragmatically determined strategies for exposure estimates that draw on measurements made for regulatory and research purposes and extend the measurements using interpolation and extrapolation to complete gaps for miners in particular years. Additionally, missing information for mines in the earliest years of some of the studies was completed by either expert judgment or by recreation of operating conditions. It should be noted that very little has been done to quantitatively assess uncertainties related to the replacement of this ideal approach with the pragmatic approach.

This is a very important issue because, in reality, uncertainties in the assessment of the dose and even exposure are many times greater than statistical errors in the assessment of lung cancer mortality, which make risk assessment very uncertain. In Ruzer (1970), Alterman (1974), Urusov (1972), Ruzer et al. (1995) and measurements of radon decay product concentrations in different working sites, the direct measurement of the activity in the lungs of individual miners and lung cancer mortality and lung sickness for different groups of miners was studied in nonuranium mines of Tadjikistan. It was shown that variations in the radon and radon decay product concentrations vary by a factor of 2–10 and that calculated radon progeny intake varies by a factor of 3–14. A similar nonuniformity was found in lung cancer mortality and lung sickness. The uniformity approach can also be the source of uncertainties much greater than the statistical errors in the assessment of mortality.

From the point of view of dose (alpha particle energy deposited in the lung) assessment of all the factors, that is, concentration of radon decay products measured by standard procedure, the ratio between this measurement and the concentration in the breathing zone, volume breathing rate, and deposition coefficient are equally important, including efficiency in using respirators, because all these factors affect the amount of radioactivity deposited in the lung.

In this section, based on the quantitative analysis of accuracy of dosimetric factors, we will show that the conception of uniformity is incorrect in terms of both dosimetry and epidemiology.

As a result, we suggest that intake, which takes into account not only radon progeny concentration in the working site but also the physical load of work for different groups of miners and the scenario of exposure, is a better measure of radon and its decay products dosimetry than exposure.

14.11.1.1 Discussion on miner radiation dosimetry: Quantitative approach

In every miner epidemiological study, two important parameters should be determined together with their errors (uncertainties):

1. Lung cancer mortality
2. Radon decay products dosimetric characteristics: concentration, exposure, activity in the lung, intake, dose of alpha-radiation

In a majority of published studies, only the assessment of concentration and exposure takes place, despite the fact that both are characteristics of air and not of tissue irradiation. It is not even a characteristic of air really breathed in by miners, as was shown in studies (Breslin et al., 1969; Schulte, 1967; Domanski et al., 1989).

It should also be mentioned that to achieve reliable data on risk assessment, uncertainties (errors) for both mortality and dosimetry must be comparable.

One of the most comprehensive studies of this kind is BEIR VI, which summarizes dosimetric and epidemiological data from 11 local studies in uranium and nonuranium mines in different countries. Similar data from the former Soviet uranium industry were not included in this report, because at the time of preparation data were yet to be classified. In the report BEIR VI, two objective criteria were established for inclusion in the study:

1. A minimum of 40 lung cancer deaths. This criterion established the level of uncertainty in lung cancer mortality in the range of 15 to 20%.
2. Estimates of Rn progeny exposure in units of WLM for each member of the cohort based on historical measurements of either Rn or Rn progeny.

The authors presented the study of tin miners from Cornwall and Devon in the U.K. as an example of an omitted cohort, which failed to meet the second criterion for inclusion, that is, no individual estimates of Rn progeny exposure were available at the time the analysis began. Still this report noted in general that "incomplete measurements data for years and for work areas were common, resulting in uncertainty in WLM estimation." Despite this quantitative assessment of the uncertainty in the assessment even for the exposure was not done in this or other studies. The situation with the uncertainties of the real doses for miners is even more complicated than for exposure.

In Lekhtmakher and Ruzer (1975), an attempt was made to assess the uncertainties in the dose (or activity, or intake) of radon decay products in the lungs based on standard measurement procedure. Using uncertainties in concentrations of radon progeny, breathing rate, and deposition coefficient in the lungs, the ratio of the real activity A to the calculated activity A_c based on measured concentration and standard breathing rate and deposition will be

$$0.18 < A/A_c < 5.5 \text{ with } 68\% \text{ probability}$$

and

$$0.03 < A/A_c < 30 \text{ with } 95\% \text{ probability}$$

We will try to assess the uncertainties in every step of the calculation from the air concentration to the activity (intake, dose) of radon decay products in the lung, assuming that activity (intake) is the main physical value responsible for the biological effect in this case.

1. In some of the studies mentioned in Radon and Lung Cancer risk (1994), assessment of radon progeny was made based on reconstruction and some assumptions (China, New Foundland, Sweden, Beaverlodge). In such cases, it is impossible even to determine the uncertainty (errors). The use of these data should be called into question. It seems that criterion 2 was not applied in this case because it was impossible to estimate exposure for each member of the cohort. In the Workshop on Uncertainty in Estimating Exposure to Radon Progeny Studies of Underground Miners (BEIR VI, E — Annex 2), questions were raised about removing certain cohorts, ranking of cohorts, and obtaining additional information.
2. In many of the earliest studies, only radon concentration was measured. It is well known that the contribution of radon itself to the dose is negligible in comparison to its decay products. It was shown that the equilibrium factor varies in mines and has a wide range from 0.2 to 0.9 (Radon and Lung Cancer risk [1994], China; Domanski et al. [1989], Poland). Therefore, assuming that the average equilibrium factor is 0.5 for all situations, we can have uncertainties of the order of 100%.
3. Measurements of radon decay products are directly related to irradiation of lung tissue. Errors in calibration and measurements should be taken into account. Assessment of the errors related to radon progeny measurements was the topic of discussion in the special workshop (BEIR VI, E — Annex 2). It was estimated that uncertainty in this case was about 50%. This is consistent with the standard adopted in the former Soviet Union for radon decay products measurements in mines — the errors should be of the order of 30–40% (Antipin et al., 1980).
4. Another source of error is the correlation between concentration measured by a standard procedure (or area monitor) and that in the breathing zone. In (BEIR VI, 1998), the suggestion was made that in some cases there are no substantial differences in these two concentrations. But in Schulte (1967), results showed that the ratio of concentrations measured by a personal aerosol sampler (PAS) and a standard aerosol sampler (SAS) depends substantially on the strategy of measurements and sampler location. If the source of contamination (activity) is evenly distributed in space, the ratio PAS/SAS is from 2 to 3. However, local contamination may be

influenced by material deposited on the miners' clothing, in which case the ratio may increase to 10.

In Domanski et al. (1989), results were also presented on the correlation between these two factors. The one measured by the air sampling system (ASS) was based on the field monitoring of radon progeny in air; the second one, called the individual dosimetric system (IDS), was based on the individual dosimeter worn by miners. The ratio ASS/IDS varies from 11.0 to 0.14.

In short, this means that if the concentration was measured only by standard procedure and the ratio PAS/SAS is not known, additional uncertainty in the assessment of the concentration related to lung irradiation can be in the order of hundreds of percent.

5. By definition, irradiation of the lung by radon decay products should depend on physical activity. The data on the breathing rate for different types of physical activity are presented in Alterman (1974), Ruzer et al. (1995) and Layton (1993). The problem for miners is that this parameter changes substantially within the shift from 10 to 30–40 l/min, and by using the average value for the breathing rate an error of 100% can be made. It should be mentioned that the measurement of the actual breathing rate was usually made only for low physical activity because the measurement method itself disturbs real breathing conditions, especially in the case of hard work.
6. In George and Breslin (1969), the deposition coefficient for radon decay products in mines (for nonminers) was measured using a special mask and the difference in concentrations in inhaled and exhaled air only for the "rest" situation. A similar technique was used in Holleman et al. (1969), and the range of deposition for miners was found from 0.25 to 0.55.
7. For deposited activity in lungs, the combination of breathing rate and deposition coefficient is important, and not these values separately.
In Ruzer (1970), a new value, filtration ability of lungs (FAL), was introduced as a combination of breathing rate v and deposition coefficient k (FAL=vk) as a bridge between concentration in the breathing zone and activity in the lungs. FAL can be estimated by measurement of the gamma-activity in the lungs of miners and radon progeny concentration in the breathing zone. This method permits measurements on miners without disturbing their real working conditions (noninvasion approach) and was used for the measurement of different groups of miners. The error in estimating FAL was 20%, which is many times smaller than the combined error in the assessment of breathing rate and deposition coefficient (Alterman, 1974; Urusov, 1972; Ruzer et al., 1995).
8. In Ruzer (1958, 1964) and Ruzer and Urusov (1969), the method of direct measurement of the activity in the lungs of miners was developed and used in miner measurements both for the assessment of activity in the lung and for the estimation of deposition of radon decay products. The accuracy of activity assessment in the lungs of miners, including corrections on the shift equilibrium in the air and geometric factors, was 30–40%.

No correlation was found between direct measured activity in the lungs of miners and calculated activity based on radon progeny concentrations measured by the standard procedure. On the other hand, a relationship was found between the concentration in the breathing zone, the parameter of physical activity (FAL), and direct measured activity in the lungs.

As a result of this, conclusions should be made that the total uncertainty of the activity in the lungs or the dose for miners calculated on concentration measurements can be of the order of hundreds of percent. But because errors close to even 100% make no sense, it is better to express this uncertainty in such a way that the true value of the dose or the activity in the lung can be at least of the order of magnitude lower or higher than calculated.

Only direct measurements of activity can produce an accuracy acceptable from the dosimetric point of view and close to the statistical error that is used in the assessment of lung cancer mortality.

14.11.2 Measurements of the dosimetric parameters

14.11.2.1 Radon measurements

Measurements of radon concentration were used both in mines and in homes in order to assess potential danger from natural radioactivity. There is a diversity of methods and instruments, based on different physical events: ionization and scintillation chambers, charcoal absorption technique, track detectors, etc. For every type of instrumentation, a calibration technique was developed, and an accuracy of measurements no more than 10% in both homes and mines is achievable.

Radon concentration measurement itself can determine the upper limit of the potential alpha-activity in air, that is, in the case when radon progeny are in equilibrium with radon itself. However, the contribution of radon to all the energy of alpha-particles is very small, and taking into account the accuracy of the measurements can be even negligible.

Thus, the accuracy of 10% of radon measurements is more than enough in terms of dosimetry. Still, in the case of a very low shift of equilibrium, the contribution of radon should be taken into account, and therefore, in principle, radon measurements for dosimetric purposes are necessary.

One opportunity for measuring radon progeny together with radon itself should be pointed out (Ruzer, 1960). This technique is based on using a buildup curve of radon and radon progeny after sampling. The problem of using radon concentration as a surrogate for dose is discussed in Section 14.9.

Here are some examples of radon concentration measurement techniques together with the results of measurements. It was observed that surface activity of the long-lived radon progeny ^{210}Po on window glass in a basement and on the surface of rocks in a cave were proportional to the mean radon concentration. In Fleishert et al., 2001 the retrospective radon exposure assessment procedures based on the measurement of alpha-recoilimplanted long-lived radon progeny ^{210}Po are critically described. The results are presented in a series of radon retrospective measurements made in a former uranium mining district in Germany.

Lenses for eyeglasses in U.S.A. are most commonly made of CR-39 (allyl diglycol carbonate), an alpha-particle-sensitive plastic of usefulness and reliability for radon monitoring. For regular wearers of glasses, the accumulated tracks of alpha-particles from ^{222}Rn, ^{218}Po, and ^{214}Po typically give a multiyear record of the exposure of wearers. In J.A. Mahaffey et al., 1993 the procedure of calibration and measurement is described. The average radon concentrations to which the wearers were exposed were in the range of 14–130 Bq m^{-3} (0.4–3.5 pCi l^{-1}). A simple protocol has been recommended for measuring individual lenses.

In Hadler, J.C. 1994, a study is described in which the CR-39 plastic was attached to household glass objects to learn if residual alpha-radioactivity from radon progeny could be measured and correlated with cumulative radon exposure. It was found that CR-39 measurements of alpha-activity in the surface of selected objects, correlated with ambient radon measurements (R^2=0.48), provided that reliable information. Conceivably, data collected from CR-39 may be more representative than data collected using ambient tracketch detectors, because the track-etch detectors may inadequately reflect subject lifestyle, home room arrangement, and heating/cooling mode.

Radon concentrations in natural gas are the highest at points closest to the gas reservoir. According to Dixon D.W., radon concentration in U.K. natural gas can reach 500 Bq m^{-3}, but then the average concentration decreases in gas entering the distribution system to close to 200 Bq m^{-3}.

Radon decay product concentration in plants may be significantly higher, and many employers undertake routine surveillance of exposure conditions in accordance with health and safety regulations. The average annual dose of natural gas containing radon is estimated at about 4 mSv in the public, which is less than 1% of natural radon background. Exposure from cooking with gas on commercial premises is around 19 mSv and the critical group of users can receive annual doses of a few tens of mSv.

A model proposed in A.M. Marenny et al., 2001 for interfering with the radon annual average collective and personal doses, as well as the dose distribution of the population of Russia (148,000,000 as of 1996), suggested that it is very important to assess adequately the annual role of radon in the exposure of populations. From the Chernobyl accident in the contaminated area of Russia (about 2.7 million people are living in the area where the ^{137}Cs level exceeds 1 Ci km^{-2}), the collective whole-body dose from external exposures for that population was about 20,000 man Sv for 10 yr (1986–1995), while the radon-induced collective dose was about 26,000 man Sv. Even in the most contaminated areas of the Bryansk region, about 45% of the current annual average individual dose induced by radon and as little as about 36% is due to external and internal exposures to the Chernobyl accident.

A survey Nikl J. of indoor radiation in Hungary shows that the 129 Bq m^{-3} average radon concentration lies in the upper part of the interval cited in the worldwide literature. The resulting 107 Bq m^{-3} weighted mean is very close to the results for Sweden. A significant difference in radon concentration was found between flats in the upper floor and dwellings with ground contact. The mean value of absorbed dose rate, including terrestrial and cosmic rays, is 127 nGy h^{-1}. The annual effective doses to the Hungarian population based on the weighted means are 1.3–1.8 from radon inhalation and 0.4–0.6 mSv from external sources, according to the occupancy of 5008–7008 h per year.

In Göran Pershagen et al., 1992 the results of the study on residential radon exposure and lung cancer in Swedish women were provided. The study included 210 women with lung cancer and radon concentration measurement in the county of Stockholm with arithmetic and geometric means of 127 and 96 Bq m^{-3}, respectively. Risk estimates obtained from the study appear within the same range as those calculated from miner studies. The authors underlined the great need for reducing the uncertainty in present risk assessment following residential exposure, both for assessment of the public health impact and as a basis for control strategies.

In the study provided in Slovenian spas Vaupotic et al., 2001 it was shown that in the majority of the rooms of all the spas investigated, radon concentration was lower than that recommended by the ICRP level of 200–600 Bq m^{-3} for nonoccupational exposure. Such low radon concentrations are believed to result from the highly effective ventilation of all the therapetic rooms, reflected also in low equilibrium factors. The conclusion was drawn that under present operational conditions, there is no basis for concern about elevated radon exposure of the personnel of the spas under consideration.

In Lubin Jay H., 1998 discrepancies between epidemiological studies of lung cancer and residential radon concentration in individuals and as proposed by Cohen ecological regression were discussed. The paper demonstrated that ecological analysis by Cohen cannot produce valid conclusions unless other lung cancer factors (smoking, age, occupation, etc.) for individuals are statistically not correlated with radon concentration within counties or unless risk effects for radon and other factors are additive.

Unfortunately, the fact that radon concentration does not correlate with the dose to the lung directly has not been discussed at all.

14.11.2.2 Radon decay progeny measurements

The short-lived radon decay products are the main factors in irradiation of the lungs. Different types of methods and instruments were used for the measurements in mines. For mine conditions, such measurement techniques should be portable and simple to operate. Even now in many of the studies, different types of total alpha-activity measurement (three points measurement) were used. These methods can have substantial errors (Labushkin, V.G., Ruzer, L.S. 1965).

In most of the Western countries, methods described in Dixon (2001) were used. In most of the studies of the former U.S.S.R. and Eastern block countries, the standard method for measuring radon progeny concentration, called "express method," was used (Markov, K.P. et al, 1962). According to this technique, the total alpha-activity should be measured from the first to the fourth minute, and then from the seventh to the tenth minute after the conclusion of the sampling collection, with a total duration of measurement of 15 min and accuracy in each radon decay product and potential alpha-energy concentration (PAEC) measurements of 20 and 30%, respectively, which is satisfactory from the point of view of dosimetry.

The next generation of techniques based on alpha-spectrometry and nonemanating sources from radium-226 was developed in Labushkin, V.G., Ruzer, L.S. 1965; Ruzer, L.S., 2001. It should be mentioned that although an accuracy of 30–40% is satisfactory, for practical purposes, for radon progeny concentration measurements in mines, it is still higher than errors in the assessment of lung cancer mortality in epidemiological studies (Ruzer, L.S., 2001).

And finally, in terms of dose assessment, uncertainty in the radon progeny measurements is not so important, because concentration data used for the calculation of the dose to the lungs of miners depend very substantially on the strategy of the measurements. Even if we want to use only concentration as a measure of the irradiation of the lung, we should use the concentration in the breathing zone of the miners.

The correlation between ^{210}Po counted in the skull and exposure to radon progeny in the lung was studied in Scheler, R. et al. 1998. The following conclusions were drawn:

- Measurement of ^{210}Pb in the skulls of individuals could be a source of information on the estimation of previous exposure to ^{222}Rn decay products.
- In the present stage of development, the method is practicable for the *in vivo* measurement of miners and inhabitants with high radon concentration.
- The detection limit could be improved by reducing the background in the measurement chamber.
- This method has some uncertainty related to the insufficient data on the accumulated ^{210}Pb by other sources than inhalation of the short-lived radon progeny.
- The detection limit of ^{210}Pb for the total skeleton in a counting time of 7200 sec was estimated to be 17 Bq (250 WLM).

In Obersted, S. et al. 1995, a measurement system called bronchial dosimeter was developed to assess the lung dose directly. The design consisted of multiple wire screens based on the work of Hopke, Philip, K, 1992, The first test measurement was performed under laboratory conditions, including efficiency determination. The results show that this device is a suitable instrument to survey the deposition characteristics of the short-lived radon decay products in the nasal cavity and bronchial tree. Fraction deposition as a function of radon progeny concentration should be studied, that is, their dependence on the equilibrium factor *F*, in order to obtain more reliable results.

In Skowronek, J. 1999, the assessment of exposure due to natural radioactivity in Polish coal mines is presented. It was shown that in 1997 the radiation exposure slightly decreased due to several years of the existence and application of the radioprotection system, as well as better qualifications of the supervision staff and mine services. Recently, new methods of control of the radiation exposure were developed and implemented, such as radium removal from the mine waters, and new methods of control of the radon progeny at every stage of mining work. Also, computerized software for the forecast of radon progeny exposure was developed, which now serves to assist ventilation engineers.

In El-Hady, M.A. et al. 2001, it was shown that not only workers in uranium mines, but also the staff of other underground mines, such as workers in underground phosphate mines, can be exposed to radon progeny. From the results of radon progeny measurements in three mines, it was concluded that radon decay products concentration increased with increasing depth and reached a minimum at the entrance of the air shaft. All mean values of radon progeny significantly exceeded the action level for working places recommended by ICRP 65. In this case, the main reason for such high radon progeny concentration and corresponding levels of exposure was poor ventilation conditions in mines. Corresponding measures for improving the situation were recommended.

The authors of Wasiolek, P.T., et al. 1995 found that by taking 240 samples of radon gas concentration, together with unattached and attached-to-ambient-aerosol radon progeny concentration at 16 outdoor sites in four U.S. states, the correlation coefficient of effective dose with respect to radon gas concentration between these 16 sites was 0.98, at least when the dose calculation was based on a known PAEC and the assumption of a discrete unattached and attached particle size model. The dose conversion coefficient at 15 sites in the southwestern U.S. averaged 67 nSv h^{-1} per Bq m^{-3} radon gas concentration. The dose conversion coefficient with respect to radon gas for outdoor air also appears to be several fold higher than the currently recommended value for indoor air (2.3 nSv h^{-1} per Bq m^{-3}).

For aerosol-type monitors developed in France Grivaud, L. et al., 1998, two different tests were used:

1. A static test with solid standard radioactive sources in order to measure their efficiency.
2. A dynamic test provided on the special bench, which continuously generated natural and artificial radioactive aerosols, calibrated for size and activity. This enables the true performances of radioactive aerosol monitors to be defined under normal operation conditions. The measurement efficiency was obtained by sampling and measuring, in real time, the concentration of aerosols labeled with ^{239}Pu and ^{137}Cs. The influence of natural radioactivity upon artificial activity monitoring was determined by aerosols of radon decay products. Minimum volume detective activities for both types of aerosols and different monitors were measured.

14.11.2.3 Distribution of radon decay product concentration

In Urusov, S.A., 1972, Alterman A.D., 1974, Ruzer, L.S., 1970, the results of the distribution of concentration of radon progeny in nonuranium mines of Tadjikistan are presented:

- In all mines (Table 14.15)
- In 56 mines and 15 geology prospecting teams (Table 14.16)
- In the Karamazar lead–zinc–bismuth–fluorite mine (Table 14.17)

Measurements were provided in 1521 working locations of 98 mines. For each and every mine and group of underground workers, parameters of different types of distribution were calculated:

$$y = a + b\,x \quad \text{(linear)}$$

Table 14.15 Distribution of the Concentrations of Radon Progeny in All 98 Nonuranium Mines in Tadjikistan

Concentration (WL)	Mines					
	Median Number	%	Minimum Number	%	Maximum Number	%
<0.1	43	44	78	78	32	32
0.10–0.3	23	24	11	13	13	13
0.31–0.9	18	18	7	7	13	14
1.0–2.0	2	2	1	1	4	4
2.1–3.0	5	5	1	1	4	4
3.1–4.0	4	4	0	0	4	4
4.1–5.0	1	1	0	0	1	1
5.1–9.0	2	2	0	0	9	9
> 9	0	0	0	0	10	10
Total	98	100	98	100	98	100

Table 14.16 Radon Decay Product Concentration in Working Locations in 56 Mines and 15 Geology Prospecting Teams (1963–1972)

1	2	3	4						
			5	6	7	8	9	10	11
All	827	658	145 (22)	112 (17)	151 (23)	197 (30)	53 (8)	0.68	57.3
Ore removal	510	341	65 (19)	39 (11)	81 (24)	122 (36)	34 (10)	0.16	50.6
Nonworking	268	222	16 (7)	18 (8)	36 (16)	60 (28)	92 (41)	5.12	132.3

1 — site; 2 — number of samples; 3 — number of sites; 4 — number of sites with conc., MeV/l×10^5 (%); 5 — <0.13; 6 — 0.14–0.4; 7 — 0.41–1.2; 8 — 1.2–12; 9 >12; 10 — median; 11 — maximum.

Table 14.17 Radon Decay Product Concentration in Working Locations in the Karamazar Lead–Zinc–Bismuth–Fluorite mine (1963–1964)

1	2	3	4						
			5	6	7	8	9	10	11
All	199	101	30 (30)	15 (15)	23 (23)	31 (30)	2 (2)	0.70	31.2
Faces	85	38	9 (24)	4 (10)	9 (24)	15 (39)	1 (3)	1.06	31.2
Cleaning	38	29	6 (33)	2 (10)	7 (35)	5 (22)	-	0.67	6.0
Drifting	47	18	3 (17)	2 (11)	2 (11)	10 (55)	1 (6)	1.61	31.2
Nonworking	72	37	4 (11)	4 (11)	3 (7)	8 (22)	18 (49)	5.6	156

1 — site; 2 — number of samples; 3 — number of sites; 4 — number of sites with conc., MeV/l×10^5, (%); 5 — <0.13; 6 — 0.14–0.4; 7 — 0.41–1.2; 8 — 1.2–12; 9 >12; 10 — median; 11 — maximum.

$$y=a+b\ln(x) \quad \text{(logarithmic)}$$

$$y=a+b^x \quad \text{(exponential)}$$

$$y=a+x^b \quad \text{(power)}$$

where y is the cumulative percentage and x is the upper limit of the interval of measured concentrations. The correlation coefficient r represents the goodness-of-fit of the equation with the measured data. In general, the closer r is to 1 or -1, the better the fit.

In the majority of cases, the distribution of concentration was logarithmic, with the correlation coefficient being greater than 0.97:

- The distribution of radon progeny concentrations in all of the 98 observed mines is logarithmic, with the correlation coefficient being close to 0.97 and more.
- The distribution of concentrations in local groups of mines is also logarithmic, with the correlation coefficient being close to 1 both for all working sites in general and for every type of work such as faces, cleaning, drifting sites with the exception for data on nonworking sites, where the correlation coefficient is slightly higher for power than for log distribution.
- There is a great percentage of nonworking sites with very high concentrations (Tables 14.15, 14.16).
- Concentrations in the working sites of prospecting teams are very high.
- The results suggested a great diversity in radioactive air pollution in mine environments.
- From the dosimetric point of view, it is incorrect to use the average in time and space measured concentration for the assessment of miner exposure.

Epidemiological data on miners in the same mines also showed great nonuniformity.

14.11.2.4 *Unattached fraction measurements*

In most practical cases, especially in mines with high aerosol concentration, the unattached fraction is very small and subsequently the uncertainty in the assessment of unattached activity is very high as shown in Markov et al. (1962), Ruzer and Sextro (1997) and Antipin et al. (1976).

The absence of systematic studies related to the unattached fraction measurements in mines should also be mentioned. Thus, the role of unattached fractions in the assessment of dose to the lungs of miners is another uncertainty. At least we can assume that in hard working conditions such as drilling, the contribution of unattached activity is small.

14.11.2.5 *Breathing zone concentration measurements*

Unfortunately, not many studies were devoted to breathing zone concentration measurements, especially to the correlation between measurements based on the standard procedure and breathing zone measurements. To determine the ratio between concentration in the breathing zone and the fixed sampler, the personal sampler was developed (Sherwood, R.J., et al., 1960). It consisted of a small, compact, battery and a powered suction source (pump) connected by a flexible tube to a small filter paper holder. The pump was worn on the belt or in the pocket, and the filter paper holder was clipped to the clothing like a film badge. In all, 20% of the maximum of a permissible concentration (MPC) of 1.5 Bq m^{-3} of ^{239}Pu can be detected with this device. The personal sampler was used (Domanski et al., 1989) for the study of the ratio between sampling in the breathing zone and by the fixed sampler.

In most cases, concentrations measured by the personal sampler were considerably higher than those measured by the fixed sampler. The ratio between these results generally varies from 2 to 10, but some of the results are as high as a factor of 30 or even more. Such results caused the authors to reexamine the location of the fixed samplers (area monitor) and to conduct a careful study of individual operations. For the 40 pairs of samples, the median value of the ratio was 4, but it varies from 1.5 to 50; 85% of the ratios were less than 10. In no case was the personal sampler concentration less than 1.5 times that of the fixed sampler.

Since the personal sampler is attached to clothing, it may be contaminated by clothing material that can reach the worker's nose. Also, since the personal sampler is in the dust cloud created by the worker's operations, the average particle size of the material being sampled may be considerably larger than that collected on the fixed sampler a short distance away. Since the smaller particles are more apt to reach the depth of the lungs, the ratio of the concentration measured by the personal sampler to that of the fixed sampler underestimates the hazard more than it actually does. This points to the need for more studies of particle size distribution of airborne contaminants. Like a film badge, a personal sampler is still only an estimate of the individual's dose, because its sampling efficiency is not the same as the human respiratory tract with respect to various particle sizes and it does not sample exactly the same air as that breathed by the wearer.

Schulte, H.F., 1968, a multistage sampler was developed in an attempt to better the data on the quantity of contaminant actually reaching the pulmonary region of the lungs. While a specific deposition pattern is assumed by ICRP in calculating permissible concentration levels, the actual distribution of deposition in the respiratory tract is determined by the particle size distribution of the contaminant. The multistage sampler is supposed to make corrections in a simple manner; however, its limitations are not always evaluated.

Domanski, T., 1989, the results of a long-term study were described by comparing two independent systems, that is, the "air sampling system" (ASS) and the "individual dosimetry system" (IDS), which were implemented and tested for 6 yr simultaneously in Polish underground metal-ore mines.

Each of these systems has certain different inherent advantages and critical weak points. The main feature of the ASS is usually the relatively high precision of each single measurement; however, the strategy of monitoring and the selection of the proper frequency of monitoring and the site of the system on the area of the mine still remains the weak point of the ASS.

On the other hand, the critical point of the IDS lies in the cost of the measuring devices or in the doubtful precision of the measurement technique. The term "dosimetry system" should also not be used, because such types of instruments measure only concentration, not the dose, which in this case depends on physical activity.

The ASS system was implemented in all Polish underground copper and zinc–lead ore mines between 1981 and 1983, and the crucial point, that is, strategy of monitoring and sampling, was thoroughly considered, discussed, and finally approved by the Institute of Occupational Medicine. It was recommended that from tens to several hundreds of potential alpha-energy concentration measurements of radon progeny in the year would be done at the sites where miners actually work and in the local air stream outlets.

The IDS, based on "individual dosimeters" worn by miners on the backs of their helmets, was introduced in all the mines under consideration in the period between 1977 and 1979. This system is based on the use of small cassettes containing a tracketch detector foil sensitive to alpha-particles emitted by radon decay products.

Thousands of measurements conducted simultaneously in 11 metal-ore mines under these two technically compatible, but entirely independent, long-term systems of radiation exposure assessment have brought results that have led to important conclusions concerning the reliability and validity of these systems.

The summary of the results of these measurements is presented in Figure 14.43. The first general conclusion is that these systems yield inconsistent results. It can be seen from Figure 14.43 that the correlation coefficient is very low — 0.16. Thus, the general question arises as to which system is more reliable and practicable.

The ratio of annual concentration, that is, mean ASS/mean IDS, varies from 11 to 0.14, depending on the mine and the year of implementation of the ASS. The results show that the policy of air sampling is very important, because soon after ASS implementation the effect which the authors called "hunting for results" takes place, that is, radiation officers or dosimetrists look for places where concentration will be higher, or in other words they

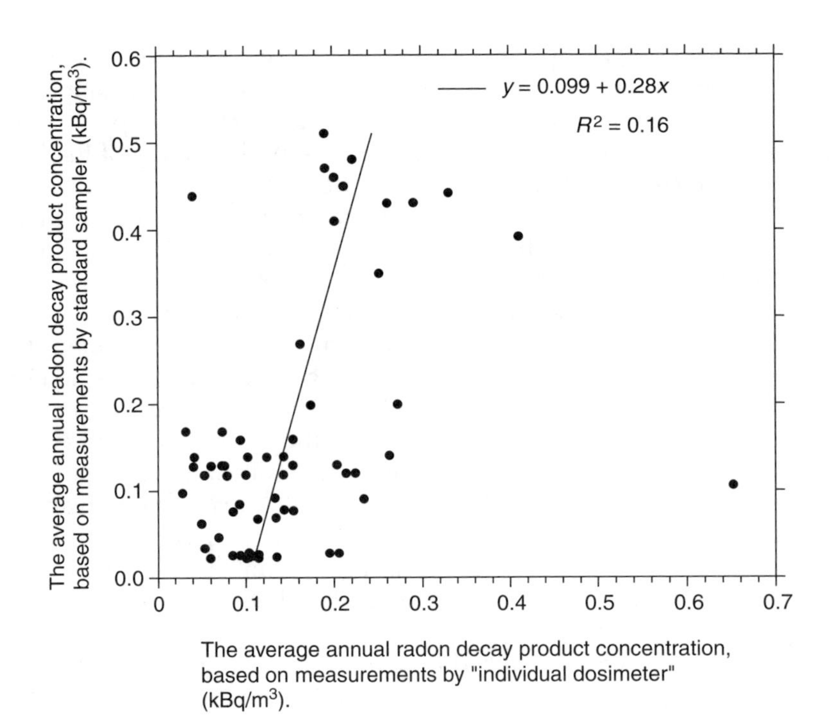

Figure 14.43 Average annual radon decay product concentration, based on measurements by individual dosimeter (kBq m^{-3}).

"hunt" for results, neglecting the strict instruction given to them during the training courses before implementing ASS. The opposite effect, revealed by results collected in two other mines, can be called the "avoidance of inconvenient results." The authors suspected that in these cases, the radiation safety officers probably tried to hide the results that would increase the mean value, or simply avoid places in which miners work at relatively higher concentration.

Both the above-mentioned effects distort the real picture of both concentrations and miners' exposure. The "hunting" effect appears to be disappearing over the years, but it sometimes turns into the "avoidance" effect, and the "avoidance" effect appears to be more serious.

The conclusion was drawn that IDS is more reliable in miners' exposure assessment than the currently implemented ASS. Nevertheless, ASS should not be abandoned, because it is obviously better than nothing and plays a useful role as a tool for technical preventive actions.

Still, as a measure of real irradiation of the lungs of all miners, these standard types of measurements should be called into question, because the calculation of the activity (dose) to the lungs of individuals or groups of miners based on the results of such measurements will yield results with such uncertainty, which is unacceptable in the dose and, correspondingly, the risk assessment.

The authors mention that the results presented in Ruzer and Sextro (1997) can have a possible impact and contribution to the discussion on the value of radiation mortality risk. It was shown in a study of American miners that the mortality risk was several times lower than in an analogous study of Czechoslovakian mines. Such a discrepancy can at least be partially explained by different strategies of measurement. This would lead to an overestimation of miners' expected exposure and, consequently, to the obtaining of a much lower mortality risk for American miners than for Czechoslovakians.

Summing up, Domanski, T. et al. 1989 declared the following opinion:

> The postirradiation lung cancer risk factor due to exposure to radon progeny should be thoroughly reconsidered in the light of discovered effects. The correctness of the miners' exposure can be provided only by the IDS.

14.11.2.6 Breathing volume rate and deposition coefficient measurements for miners
For the correct assessment of the activity (dose) to the lungs of miners in real working conditions, data on the breathing volume rate and deposition in the lungs are critically important. Standard methods of breathing rate measurements (such as the gas meter) are usually used only in cases of low physical activity, which is not typical for underground operations.

Calculated data on the breathing volume rate for different types of physical activity are presented in Layton, D.W. 1993 (Table 14.18).

There is a shortage of systematic data on the deposition coefficient of radon progeny in the lungs of miners. In Obersted et al. (1995), the deposition of radon progeny in the respiratory system of humans in radium mines as well as in the laboratory was studied and found to be from 14 to 51% in the mines. In Skowronek (1999), the study yielded results of 40–50% in two human subjects in the laboratory.

In Harley, N.H., et al. 1951, the fractional deposition of natural radioactive particles in human subjects was determined by using apparatus consisting of a modified filter respirator, a rubber bag, a sample filter holder, an air flow meter, and an air pump. Preliminary measurements of four human subjects breathing New York City air are in the range 17.9–50.7%.

In George, A.C., et al. 1967, uranium mines and laboratory experiments were described for the determination of the total deposition of radon daughters together with the effect of particle size, tidal volume, respiratory frequency, nasal deposition, and growth of radon daughter particle size inside the lungs. The total respiratory deposition of radon daughters measured in humans exposed in uranium mines was found to be in the range of 23–45%.

From the mine and laboratory experiments, both particle size and tidal volume were found to influence deposition, fractional deposition increasing with decreasing particle

Table 14.18 Short-term (i.e., minutes or hours) Ventilation Rate Estimates[a]

Gender and Age (yr)	Weight (kg)	Inhalation Rate (l min^{-1})				
		Rest	Sedentary	Light	Moderate	Heavy
Male						
0.5 to <3	14	3.2	3.8	6.4	13	32
3 to <10	23	4.0	4.8	8.1	16	40
10 to <18	53	6.3	7.5	13	25	63
18 to <30	76	7.2	8.7	14	29	72
30 to <60	80	7.0	8.4	14	28	70
60+	75	5.7	6.9	11	23	57
Female						
0.5 to <3	11	2.4	2.9	4.9	10	24
3 to <10	23	3.8	4.5	7.5	15	38
10 to <18	50	5.3	6.4	11	21	53
18 to <30	62	5.5	6.6	11	22	55
30 to <60	68	5.4	6.5	11	22	54
60+	67	5.0	6.0	9.9	20	50

[a]*Source:* Layton, David W., 1993

size, and increasing tidal volume. Particle size was found to be different with location within a mine.

In George, A.C., et al. 1969, the total deposition in the human respiratory tract of the short-lived radon progeny present in a "typical" mine atmosphere was determined in 30 separate experiments. Respiration rates and tidal volumes were also measured in each experiment.

The total percent working level (WL) depositions, defined by the relationship

$$100\ [1 - \text{WL in exhaled air}/\text{WL in inspired air}]$$

were in the range of 30–65%. Percent depositions for individual radon daughters were also measured, with mean values 50.4, 44.7, and 40.6% determined for RaA, RaB, RaC, respectively, for an average respiration rate of 18 min^{-1} and an average tidal volume of 1.21 l.

14.11.2.7 *Assessment of the deposition and the upper-bound average breathing rates for miners*

Lung deposition fractions and breathing rates, or more specifically minute volumes, are important quantities in the dosimetry of radon decay products, but there have been few measurements of these quantities in the occupational setting. Some results have been reported for the total decay product deposition fraction in the lungs by measuring the difference in concentration in inhaled and exhaled air in laboratory chambers and in mines. Breathing rates, particularly in underground mines, cannot usually be determined accurately because the usual measurement techniques are cumbersome and disturb normal conditions. New measurements are reported here, using an alternative approach, yielding deposited activity per unit exposure and minute volumes for miners working underground in lead, zinc, and bismuth mines in Tadjikistan in the late 1960s.

When the short-lived decay products of ^{222}Rn — ^{218}Po, ^{214}Pb, and ^{214}Bi(^{214}Po) — are inhaled, a small fraction, that is, 0.02–0.05, deposits on the bronchial airways, while a large fraction, up to half, deposits in the lower lung or pulmonary region. Nevertheless, the dose from the alpha-emitters ^{218}Po and ^{214}Po deposited in the bronchial region is responsible for the observed increase in lung carcinoma among miners following ^{222}Rn exposure (NCRP, 1984), presumably because of different geometries causing a higher proportion of the alpha-energy to be deposited in the epithelial cells at risk in the bronchi or because of the great sensitivity there. The dose from the beta-and gamma-emitters from ^{214}Pb and ^{214}Bi, by comparison, is relatively insignificant because a smaller fraction of emitted energy is deposited in the relevant cells, and because it has a lower biological effectiveness.

Nonetheless, the gamma-radiation from these radionuclides provides a means of monitoring deposited decay product activity. The study reported here takes advantage of high airborne concentrations occurring previously in mines, and the large pulmonary deposition fraction, to measure directly ^{214}Pb and ^{214}Bi activity deposited in the lungs of approximately 100 miners in three job categories, and to calculate deposited activity per unit exposure, proportional to the product of breathing rate (minute volume) and deposition fraction (the fraction of inhaled activity deposited in the lung). Deposited gamma-activity was measured using portable gamma scintillation detectors against which miners could press their chests, usually in the mine near the site of the mine operations. Together with measurements of breathing rate and deposited activity in a small group of miners engaged in light activity, yielding an estimate of deposition fraction, these data permit the estimation of upper-bound average breathing rates for miners engaged in three levels of activity.

Substantial work has been devoted to estimating deposited decay product activities, and the resulting dose to cells in bronchial airways, based on airborne decay product concentrations and presumed lung physiology (Ruzer, L.S. 1958, 1964). Of more interest for present purposes are equations derived by Ruzer, L.S. 1958, 1964 that relate deposited activity directly to airborne concentrations and that also provide a basis for estimating dose. Gamma-ray activity in the lung (in Bq) from decay product deposition was described in a general mathematical form by equation in 14.10.22.

Only a small fraction of lung gamma-activity is from bronchial deposition, so that this may be taken to represent primarily pulmonary gamma-activity, which is dominated by the portion of decay products attached to the particles. The much smaller "unattached" fraction (particularly in mines) is largely included in the small percentage of activity that deposited in the bronchi (George and Breslin, 1969).

The gamma-activity in the equation from Part I, Section 2 can be given in a simplified form after attaining steady state, assuming an average breathing rate and deposition fraction:

$$A_\gamma=(vk)\,(8.8q_a+77.3q_b+28.4q_c) \tag{14.68}$$

The data available from this study are the lung gamma-ray activity A_g and the total ^{214}Pb and ^{214}Bi activity concentration in air. Over a wide range of decay product equilibrium (ratios of decay product concentrations from 1:1:1 to 1:0.1:0.1), the ratio of these measurements can be reduced to

$$A_\gamma/(q_b+q_c)=60(vk) \tag{14.69}$$

This simplification is analogous to that associated with the simple fast decay product monitoring technique.

The lung gamma-activity measurements were used historically as a control over the permissible pulmonary dose for miners in several types of underground mines in Tadjikistan (Ruzer, L.S., 1970, Urusov, S.A., 1972). Equations (14.68) and (14.69) for the gamma-activity in the lung led to the idea of using gamma-activity measurements as a source of information on the breathing characteristics of the working personnel.

Although arising primarily from pulmonary gamma measurements, this information may in turn yield information on the variability in bronchial deposition. The fractional aerosol deposition in the bronchial tree has been measured in hollow casts of human airways (NCRP, 1984) and was shown to vary approximately with the inverse square root of flow rate, $v^{-1/2}$, over a wide range of particle size. Total deposition (and thus alpha dose) in an airway, proportional to the flow rate times the deposition fraction, may therefore be given as

$$\text{Total airway deposition} \sim vv^{-1/2}=v^{1/2}$$

Thus, the alpha-dose in the airways is related crudely to $v^{1/2}$. The variability of the bronchial dose due to breathing rate differences can then be assessed if the latter are known. (Doses to particular airways could also depend on geometric factors.)

The measurements were carried out in a nonuranium mine in Tadjikistan (former U.S.S.R.) with a special instrument having two probes. The measurement of the concentration of Rn decay products in the breathing zone was performed by alpha-counting of 1.8 or 3.6 cm diameter filtered air samples with a ZnS scintillation probe, using a two-count analytical procedure (Ruzer, L.S., 1970; Urusov, S.A., 1972; Alterman, A.D., 1974).

The measurement of gamma-ray activity deposited in the lung was performed using a low background gamma-probe consisting of a collimated 80×40 mm NaJ(Tl) crystal in 50 mm of lead shielding. The procedure of the direct measurements of gamma-activity in the lungs of miners was presented in Section 14.10.11.

With the high Rn decay product concentrations observed in these mines, both the time-average activity in the lung and the air concentration in the breathing zone could be measured accurately. Based on Equation 14.68, these results yield values for the product vk, called filtration ability of lungs (FAL) (Nazaroff, 1984), for example,

$$FAL = vk = A_\gamma / F(\theta, t)(60q)$$

where $q = q_b + q_c$; $F(\theta, t)$ is the theoretical function accounting for the duration of exposure q and decay in the lung after time t, if the measurements are not immediate.

$F(q, t)$ is given approximately by the expression $\lambda_{eff}/(1-e^{-\lambda_{eff}})$, $\lambda_{eff}=\ln 2/T \sim \ln 2/40$, where T (in min) is the observed effective half-life for decay in the lung.

Measurements were performed for three groups — drillers, auxiliary drillers, and inspection personnel — totalling approximately 100 workers, without disturbing the working conditions. The average, standard error, and the median values for a total of 297 air samples and 391 lung measurements are shown in Table 14.19.

From the average (arithmetic mean) and median values, the air concentrations and the chest gamma-activity are estimated to be distributed lognormally with a geometric standard deviation of 2.0–2.5.

From the air concentration measurements in the breathing zone and lung activity measured on the same day for individual miners, individual FALs were calculated. The average FALs calculated for the three groups — drillers, auxiliary drillers, and inspection personnel — are 0.0079, 0.0067, and 0.0052 $m^3\ min^{-1}$, respectively (with standard errors on each of approximately 20%). Thus, FAL varies only moderately among the three groups, consistent with the observation mentioned below indicating that, even for drillers, most time is spent at light to moderate work. Note also that the average and median FALs are similar for each group, suggesting a normal rather than lognormal distribution.

The study also included a comparison of the putative absorbed dose in the lung as indicated by the direct monitoring of lung activity and associated airborne activity and by inference using standard lung values. The indirect calculation was based on standard values for the breathing rate, 0.020 $m^3\ min^{-1}$, and deposition coefficient. 0.25, in use at the time. The average and median ratios of FAL for the measured and calculated values are shown in the last column of Table 14.5.

The average ratio is substantially greater than 1, indicating a tendency for the actual FAL (and hence dose) to exceed the standard value based on assumed breathing rates and deposition fractions. The calculation of FAL using standard values for the minute volume showed that in 76% of the measurements (from 168 man-shifts), individual activities of decay products in the lungs of miners could be higher than predicted from assumed average values of breathing rate and pulmonary deposition, by up to a factor of 8. There was no direct correlation between average concentration in the air and calculated activity in the lung. The reason for this difference was clearly that the actual breathing rate and deposition coefficient are substantially different from the standard values used in the calculation.

The FAL is a characteristic of each individual, and group averages may be taken to be typical for the type of work considered. The FAL is a result of various factors such as physical effort (nature of the job activity), physical characteristics such as lung morphometry, aerosol particle size distribution, etc.

14.11.2.8 *Assessment of the breathing rates of miners*

The minute volume of working miners may be estimated from a measurement of $FAL = vk$ if we have an independent measurement of deposition fraction, k.

Available techniques for the measurement of breathing rate change the actual breathing conditions in most cases, especially for hard-working personnel. Furthermore, meas-

Table 14.19 Measured Air Concentration of ^{214}Pb and ^{214}Bi, Gamma Ray Activity in Miners' Lungs and Calculated Filtration Ability of Lungs (FAL) for Different Groups of Miners in a Metal Mine in Tadjikistan

Job Category	Number of Air Samples	Concentration of $^{214}Pb+^{214}Bi$ ($kBq.m^{-3}$)			Activity in Lungs (kBq)[a]				FAL=vk ($m^3.min^{-1}$)			FAL_{means}/FAL_{std}[b]	
		Average	Std error	Median	No of lung measurements	Average	Std error	Median	Average	Std error	Median	Average	Median
Drilling	92	5.92	0.55	3.7	219	2.66	0.22	2.59	0.0079	0.0014	0.0090	1.6	1.8
Auxiliary	76	6.88	0.81	5.2	104	2.63	0.30	1.92	0.0067	0.0015	0.0062	1.3	1.25
Inspection Personnel	129	11.1	1.11	7.4	68	3.26	0.33	2.11	0.0052	0.0011	0.0055	1.4	1.1
Average		8.9	0.9	5.2		2.85	0.26	2.22	0.0066	0.0011	0.0069	1.4	1.4

[a] Actual gamma activities at time of measurement. In calculating FAL a correction was made on A_{γ} to account for decay between end of shift and time of measurement, in all cases less than 1 h.

[b] FAL_{means} is the value based on measured data, FAL_{std} is the calculated value based on assumed standard breathing rate of 0.020 $m^3.min^{-1}$ and standard fractional lung deposition of 0.25.

urements of the deposition coefficient, based on the difference in inhaled and exhaled air, are probably not as reliable as a direct measurement on lung deposition. For this reason, the measurement of FAL in different groups of workers may yield the best assessment of breathing rate when deposition is known and *vice versa*.

In Tadjikistan mines, the breathing rate for 14 inspection personnel doing light work was measured. Measurements could be made using the conventional minute volume measurement technique with Douglas air bags and a gas meter. This did not disturb their normal light working conditions markedly (Ruzer, L.S., 1970; Urusov, S.A., 1972; Alterman, A.D., 1974).

Breathing rate was measured approximately every hour during 2 to 3 shifts, and the average for each shift was calculated from 3 to 8 measurements. The FAL was calculated for each individual using Equation 14.68 from an average of 10 to 12 measurements of Rn decay products in the breathing zone of miners and from an average of a series of 10 to 12 gamma-ray activity measurements, made through and at the end of a shift. The results of these measurements and the calculated deposition coefficient are presented in Table 14.20 for 14 technical inspection personnel. The average fractional pulmonary deposition for this group was 0.340.03.

It is argued that the average deposition coefficient for workers other than inspection personnel (especially drilling miners) should not be less than 0.34, since total lung deposition should increase with the level of physical activity, as seen in measurements of deposition fraction, discussed above. Using a value of 0.34 for lung deposition, nominal breathing rates for drillers, auxiliary drillers, and supervisory personnel were calculated based on the group average FALs given in Table 14.20. These values are shown in Table 14.21, indicating an average breathing rate of 0.023 $m^3\ min^{-1}$ for the group with the highest level of activity (drillers).

This average minute volume deposition fraction during drilling may be somewhat higher than the value of 0.34 for inspection personnel used in the calculation.

Table 14.20 Measured Minute Volume, FAL, and Calculated Lung Deposition Fraction of Technical Inspection Personnel in an Underground Metal Mine in Tadjikistan

Ind. Job Categ.	Date	FAL=vk ($m^3\ min^{-1}$)	Min. Vol. ($m^3\ min^{-1}$)	Pulmonary Deposition Coeff. (k)
Dosimetrist A	7/27/1968	0.003	0.012	0.25
Dosimetrist A	9/12/1968	0.0028	0.012	0.23
Dosimetrist A	9/15/1968	0.0029	0.012	0.25
Dosimetrist B	7/27/1968	0.0039	0.013	0.30
Dosimetrist B	9/12/1968	0.0022	0.011	0.2
Dosimetrist C	7/27/1968	0.0072	0.015	0.48
Dosimetrist C	9/12/1968	0.0068	0.014	0.5
Dosimetrist C	9/12/1968	0.008	0.014	0.57
Signalist A	7/27/1968	0.0021	0.010	0.21
Signalist A	9/18/1968	0.0037	0.012	0.32
Signalist B	9/12/1968	0.0041	0.011	0.37
Signalist B	9/15/1968	0.0034	0.010	0.34
Sample man	9/12/1968	0.0041	0.014	0.3
Sample man	9/15/1968	0.0056	0.014	0.4
Average				
Median		0.0038	0.012	0.31
Minimum		0.0021	0.010	0.20
Maximum		0.0080	0.015	0.57

The correct assessment of the average breathing rate for an 8 h work shift must take into account the time distribution of different types of work (especially heavy work, such as drilling) during the shift. Observation of the time distribution for six miners with the job category of driller and drilling supervisor (bore master) in this metal mine was conducted (Ruzer, L.S., 1958, 1970). The results of these observations are presented in Table 14.22.

14.11.2.9 Assessment of the dose (activity, intake) in the lungs of miners

As we have shown below in real mining conditions between measured concentration and activity of radon progeny in the lungs, there are so many uncertainties (errors) in hundreds of percent, which make it practically impossible to calculate correctly the activity (dose) for an individual or a group of miners based on concentrations measured by standard procedure.

In Ruzer, L.S., 1958, 1970, the direct correlation between dose from alpha-radiation D_a from ^{218}Po and ^{214}Po and gamma-activity A_g from ^{214}Pb and ^{214}Bi was established:

$$D_a = kA_\gamma$$

which, in principle, permits one to measure the alpha-dose. Later this idea was used in hundreds of measurements on miners in the Middle Asia mines of the former U.S.S.R.

Table 14.21 Upper-bound Estimates of Average Minute Volume for Three Different Job Categories in an Underground Metal Mine in Tadjikistan

Group of Workers	Upper Bound Estimated Group Average, Minute Volume ($m^3.min^{-1}$)	
	Arithmetic Mean ±Standard Error	Median
Drilling	0.023±0.004	0.0264
Auxiliary	0.019±0.004	0.018
Inspection personnel	0.015±0.003	0.016

Table 14.22 Time distribution of the work status of miners in an underground metal mine in Tadjikistan

Job category	Number of shifts	Duration in specificed work status (min)*				Average nominal breathing rate for the shift (m^3 min^{-1})**
		Resting	Easy	Medium	Hard	
Shaft sinker A	2	86	180	86	3	0.019
Shaft sinker B	2	14	100	217	29	0.022
Shaft sinker C	1	34	116	163	47	0.0224
Shaft sinker D	1	25	180	94	61	0.0236
	Average±SD	45±16	142±29	144±40	35±11	0.0213±0.0044
Bore master-supervisor A	1	220	46	90	4	0.0168
Bore master-supervisor B	1	162	18	148	32	0.0195

* Total observation time 480 min.

** Average calculated assuming standard breathing rates, ranging from 10 m^3 min^{-1} for resting to approximately 40 m^3 min^{-1} for hard work, based on results from a coal mine[25].

Note: According to this study, the average time distribution for the shift was: resting 13%; easy work 39%; medium work 40%; and hard work 8%. These data show that even if the breathing rate is high for a short period of time, the average breathing rate for the shift is similar to that for medium to light work. Taking into account the time distribution of work type, the average breathing rates for each shift were calculated, using nominal values from other work, and these are in shown in the table.

The accuracy of these measurements, including the introduction of all corresponding corrections, was in the range of 30–40%, which is fully acceptable from a dosimetric point of view.

A detailed description of the method is presented in Section 14.10.11 and in Ruzer, L.S., 2001, published in Russia (2001). This is a completely different approach based on direct measurement of the activity instead of calculation and is based on particle size distribution and the ICRP model of deposition.

This direct approach yields results of measurements of the breathing characteristics of miners without disturbing their real working conditions, including hard work.

It should be mentioned that from the point of view of assessment of the dose to the lungs, the combination of the volume rate v and deposition coefficient k is important, and not each of these values separately.

In Ruzer, L.S., 1970, the new value FAL=vk was introduced as a measure of physical activity. In underground measurement, this value was found to be substantially different for different groups of miners (Ruzer, L.S., 1970; Urusov, S.A., 1972; Alterman, A.D., 1974).

Taking into account the large diversity of mining conditions, that is, distribution of concentration in the working sites, different types of physical activity and working itinerary, we propose that intake, which takes into account all these differences, is a more adequate measure than exposure in the dosimetric assessment of miners. The value of intake instead of exposure should also be used in comparison between miners and the general population.

Data of breathing rate, deposition in the lungs, and FAL are presented in Figure 14.44 for different types of physical activity. Unfortunately, the data on deposition coefficient for light physical activity are not available now, and we have to use extrapolation.

In W. Hoffman, 1998, different aspects of radioactive aerosol dosimetry are presented. An overview of different modeling used in radon lung dosimetry, that is, the sequence of events beginning from the inhalation of radon decay products, their deposition at the bronchial airway surface, clearance from the initial deposition sites, simultaneous radioactive decay, and irradiation of the target cells, is presented. The difference in modeling based on the ICRP respiratory tract model with a conversion factor of approximately 15 mSv WLM^{-1} and ICRP recommendations derived from epidemiological studies of 5 mSv WLM^{-1} for workers based on an equality of detriment but not of dose is studied.

The average concentration of the airborne radionuclides in the workplace is based on the annual limit of intake (ALI) according to the current guidance of the International Commission on Radiological Protection. In Boecker, B.B. et al., 1994 the possible approaches based on the use of a personal sampler, a fixed air sampler breathing zone, a fixed air sampler area, and a continuous monitor were studied. Based on this analysis, a strategic approach will be recommended for designing and implementing an optimized monitoring program. The use of optimized programs should maximize the health protection value obtained from the total resources expended.

The assessment of the dose and intake due to the inhalation of dust particles bearing ^{239}Pu and ^{241}Am from the aviation accident at Palomares in 1966 shows that it is still a potential health hazard for populations living or working in this area Espinosa, et al., 1998. It was concluded that for an aerosol 1 μm activity median aerodynamic diameter, the most appropriate dose coefficients for adults, 10-, and 1-yr-old children are 2.3×10^{-5}, 2.6×10^{-5}, and 4.8×10^{-5} Sv Bq^{-1}, respectively. For the same purpose in Espinosa et al., 1998 site and material specific data on airborne contamination and absorption biokinetics of ^{239}Pu in animals was used in conjunction with ICRP biokinetic models. The calculated effective doses over 30 yr are 37 and 210 μSv for urban and agricultural workers, respectively, below the ICRP value of 1 mSv yr^{-1} for members of the public. The predicted urinary excretion value for ^{239}Pu is similar to the measured value.

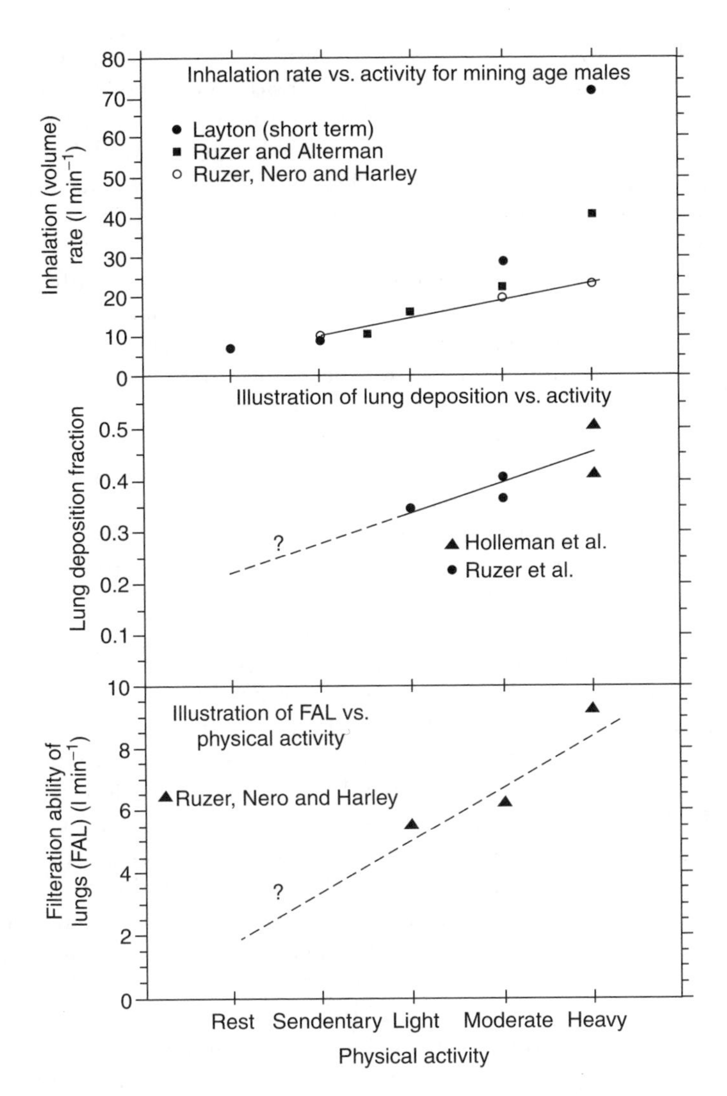

Figure 14.44 Inhalation volume rate, deposition fraction, and filtration ability of lungs (FAL).

In Lipztein, J.L., et al., 1998, the results of bioassay and cytogenetic monitoring during emergency phase and follow-up of the radiological accident in Brazil involved a large number of people who became exposed to a stolen ^{137}Cs source. A Prussia Blue (PB) drug given to persons in order to enhance the elimination of ^{137}Cs from the body reduced the committed absorbed dose by a factor between 1.7 and 3.4 for children and adolescents and between 2.1 and 6.3 for adults. In terms of dose reconstruction, the model suggested by the authors for ^{137}Cs retention in the body may produce substantial differences when calculating intake based on bioassay monitoring results.

Paper Lipztein, J.L., et al., 1998, described the problem of reconstruction of the internal doses from exposure of a population of 30,000 people living on the banks of the Techa River (South Ural, Russia) contaminated by fission products in the early 1950s. Monitoring, which started in 1951 for this cohort, has provided an objective basis for analyzing different approaches to individual dose reconstruction. The data on ^{90}Sr for this human cohort could be used for the validation and improvement of biokinetic models of bone-seeking radionuclides to individual doses for risk assessment purposes.

According to Degteva, M.O., et al., 1998, in Germany a wide range of industrially manufactured products contain thorium compounds. Thorium has the lowest value for the ALI in the natural radiation protection guidelines. In Eisemenger, et al., 1998, a method for the assessment of inhaled thorium intake based on the measurement of ^{220}Rn in exhaled air has been developed. Decay products of thoron (^{216}Po and ^{212P}Po) were collected electrostatically and measured by alpha-spectrometry. This makes the method capable of detecting thorium in lungs at an amount of 3% of ALI.

In Hattory, T. et al., 1994 the correlation between indoor and outdoor radon and its decay product concentration (equilibrium equivalent ^{222}Rn concentration) at two different operating Japanese nuclear plants was measured. In both plants, the indoor equilibrium equivalent ^{222}Rn concentration, $EERC_{in}$, of the standard measured sites was strongly related to outdoor concentration, $EERC_{out}$. Outdoor radon decay products can be collected by the prefilter in the air supply system, but outdoor Rn gas passes through the prefilter. The correlation between EERC and ^{222}Rn concentration is usually strong. Therefore, it should be suggested that the primary source of indoor radon gas was outdoor Rn gas. The annual dose was estimated for workers and the public living around the plants.

The long-term median radon gas concentration was measured and radon progeny concentration, EEC_{Rn}, was calculated based on the assumed equilibrium factor 0.4 by means of passive radon cups in 12 residential areas in Korea Chung Woo Ha, et al., 1992. The annual average effective dose equivalent from chronic inhalation exposure to the mean EEC_{Rn} of 20 Bq m^{-3} was estimated to be 1.2 mSv y^{-1} for Korean adults, according to the conversion factor proposed in ICRP (1987); therefore, ^{222}Rn and its decay products are the greatest natural source of radiation exposure in the Korean population. The expanded nationwide program of outdoor radon surveillance is recommended.

In Berkowski, V., et al., 1998 the authors pointed out the complex role of the size of aerosol particles on lung dose assessment. For example, in UO_2 particles of 20 and 100 activity median aerodynamic diameter (AMAD) containing uniformly distributed ^{239}Pu, the committed equivalent dose in extrathoracic tissues is reduced by a factor of 1.1 and 3.9, respectively. These corrections should be introduced because of alpha-absorption factors for large particles and the size distribution of deposited aerosols. The ICRP 66 Respiratory Tract Model takes no account of energy self-absorption within aerosol particles.

A study of the clearance from extrathoracic (ET) airways measured in nine volunteers for up to 5 days following inhalation of 3 μm aerodynamic diameter particles has enabled a more realistic model of ET clearance. The model is provisional, since further work will be conducted with different particle sizes and subject exercise rates. Measurements of the nose blows suggest that under appropriate conditions nose blow sampling may be used as a quantitative intake assessment technique.

In Chazel, V., et al., 1998, correlation between the specific area (SSA) of U_3O_8 samples from different processes on its *in vivo* dissolution, its biokinetics in rats, and dose coefficient are studied. Eight preparations of U_3O_8 with different SSA have shown an increase in solubility with SSA, that is, the amount of uranium absorbed in the blood increases three fold. Correspondingly, dose coefficients calculated from *in vivo* absorption parameters vary from 5.5 to 4.2 μSv Bq^{-1} when SSA increases from 0.7 to 3.2 m^2 g^{-1} In V.A. Kutkov, 1998, the attempt was made based on measured data on ^{137}Cs in the body in 15 subjects in the period from 40 to 600 days and $^{238+239+240}$Pu in 22 subjects from 30 to 90 days, and in 20 subjects from 1600 to 1800 days after exposure in the Chernobyl accident to reconstruct the sizes of inhaled aerosols. Three different biokinetic models for these radionuclides were used, including that on ICRP publication 66. Practically all different models yielded the same AMAD value of the nuclear fuel. Aerosol particle size was 11 ± 4 μm.

In S.K. Dua, et al., 1996, the hygroscopicity of indoor aerosols, including radon progeny, and its influence on deposition in different compartments of the respiratory system and correspondent dose was studied. The study of hygroscopic growth was performed using a tandem differential mobility analyzer (DMA). Dry monodispersed particles were selected by the first DMA, and then exposed to high relative humidity (99–99.5%) in a growth chamber. The sizes of these particles were measured by the second DMA. The particles that do not grow have low solubility in water. The equivalent dose rates, due to inhalation of radon progeny, for the grown Potpourri particles are higher than for the dry ones. For grown wood particles, the dose rate is lower than that for dry particles.

A survey of the data on the AMAD in the working sites was conducted in M.-D. Dorrian et al., 1995. The value of AMAD from all studies ranged from 0.12 to 25 μm and was fitted by a lognormal distribution with a median value of 4.4 μm. Both the nuclear power and nuclear fuel industries gave a median diameter of 4 μm, and uranium mills gave a medium value of 6.8 μm with AMAD frequently greater than 10 μm. It was apparent that particle size distribution should be measured for individual work practices to provide realistic information for use in radiation dose assessment, and the existing large amount of unpublished information on AMAD should be published.

In Yu, K.N. et al., 1995, the deposition of charged particles in hollow-cast modes was studied. The size of single-charged, charge-neutralized, and zero-charge 20 and 125 nm particles was chosen, because they are about where modal peaks occur for the activity of short-lived radon progeny in indoor air. For single-charged 20 nm particles, the deposition in casts was 3.4 $\pm$ 0.3 times that for charge-neutralized aerosols and 5.3 $\pm$ 0.3 times the amount deposited for zero-charged aerosols. Corresponding ratios for 125 nm particles were 2.3 $\pm$ 0.3 and 6.2 $\pm$ 0.7. This effect should be taken into account when the dose assessment for ultrafine particles is calculated.

In Yu, K.N. radon concentration, total PAEC, and aerosol distribution characteristics, such as the fraction of unattached radon decay products, f_p , have been tested in a typical aerosol environment in Hong Kong. The data on the effect of positive and negative air and the values for untreated air were compared. Then, according to the regional lung dose model and the mean lung dose model of ICRP, the annual equivalent lung dose for unit radon concentration was calculated for both untreated and treated air with positive and negative ions.

In Strong, J.C. et al., 2002, the regional lung deposition of aerosols of aged and diluted sidestream tobacco smoke *in situ* was measured by labeling the particulate phase of aerosols with gamma-emitting ^{212}Pb on volunteers. The mean deposition value for nine males varied from 22 to 59% under different breathing conditions, including nasal as against mouth breathing. When these data are used for the deposition assessment of the attached fraction, it is calculated that the annual effective dose is reduced by a factor of 3. These preliminary results confirm that radon dosimetry can be affected significantly by the physical environment in which exposure occurs, and that ambient particles from whatever source can influence dosimetry by changing the deposition pattern.

In Strong, J.C. the effect of sources other than ^{222}Rn and its decay products contributing to the dose of miners is studied. According to the author, the contribution of other sources in the mine environment may be 25–75%, and neglecting these sources results in a systematic overestimation of the risk assessment and dose miscalculation. The magnitude and relative contribution of each source can be extremely variable between facilities. This makes different populations of uranium mines noncomparable in metaanalysis without taking into account-neglected dose and the corresponding uncertainty. From this point of view, the extrapolation of radon risk from occupational to indoor exposures should also take into account the effect of doses neglected in uranium mines studies.

The uncertainty in the effective dose per unit exposure from radon progeny was studied in Bavhall, et al., 1994, from the point of view of ICRP risk-weighting factors. In order to calculate the effective dose per unit exposure in mines and homes, the recently (1994) published model of the human respiratory tract was used; results showed that according to the ICRP model the ratio is 15 mSv WLM^{-1}, and around 5 mSv WLM^{-1} determined from the uranium miner epidemiological data. The source of uncertainties in [ ] has been divided into uncertainty on aerosol conditions, ICRP respiratory tract model parameters, and other ICRP assumptions.

14.11.2.10 Results

Based on the results the following conclusions presented, were drawn:

- Since the beginning of the study of the health effects of radon and its decay products, concentration and exposure were chosen as an adequate measure of the irradiation of the lungs of miners. No serious and systematic attempt in establishing the individual dosimetric control for underground workers from alpha-emitters of radon series were made despite the fact that for low-LET radiation in occupational exposure such individual control is mandatory.
- In the majority of dosimetric and epidemiological studies on miners, no quantitative considerations were made according to the rules of metrology for accuracy in the evaluation of exposure, intake, and dose to the lungs for every dosimetric parameter, that is, concentration in the breathing zone of miners, volume breathing rate, and deposition for different degrees of the load of work.
- Routine methods of assessment of exposure (intake, dose) in the lungs of miners are very sensitive to the strategy of the measurements and consist of errors (uncertainties) in hundreds of percent, which make the correct risk assessment of miners, and consequently of the general population, very uncertain.
- Direct measurements of the activity (dose) in the lungs of miners permit one to assess dosimetric factors with an accuracy close to the statistical error in the assessment of lung cancer mortality.
- Because practically it is impossible to provide individual measurements for every miner, an intake for the group of miners is proposed, which takes into account diversities of mine environment as an alternative to exposure as a characteristic of miners' irradiation.

Nomenclature

D	integral absorbed dose (erg; g rad)
m	mass of the lung (g)
D/m	absorbed dose (rad)
q_i	radon decay products concentration, $i=a, b, c$ (Bq/m^3)
v	rate of inhalation (m^3/s)
k	retention coefficient (k)
A_i	activity in the lung, $i=\alpha, \beta, \gamma$ (Bq)
I	intake (rate of intake) (Bq/s)
η_i	degree of equilibrium of radon progeny, $i=a, b, c$
e	detection efficiency registration for the decay products

14.11.3 The method of direct measurement of activity (dose) in the lungs of miners

The determination of absorbed radiation dose in the lungs of miners due to the inhalation of radon and its decay products can, in principle, be estimated from the air concentration of radon progeny at the work sites, the rate of inhalation, and the retention in lung airways. However, the value of the rate of inhalation is indefinite, since it depends on the physical load (the nature of the work) and varies within broad limits from 10 l/min for rest, to 30–40 l/min for hard work. The air concentration varies in time and the value of the lung deposition fraction depends on the particle size of inhaled aerosols and the inhalation rate.

The accurate measurement of air concentration during exposure is complex. First, the concentration at the same working site is subject to substantial variations. For example, a variation of the intensity of ventilation even for a short period of time leads to a change in the radon and its progeny concentration of several-fold. Moreover, the content of aerosols directly in the breathing zone may differ substantially from the value measured by an instrument, located in another part of the working area.

Finally, the very concept of the "workplace" and "concentration at the workplace" are indefinite since miners are at several workplaces during their work shift, each with different possible concentrations of radon decay products and physical work loads.

For a correct evaluation of the exposure under actual working conditions, we have also taken into account that in some cases miners used respirators for protection of the lung from aerosol conditions in the mine. All these factors are important for the assessment of the irradiation of the lungs, and we will call them "the exposure scenario" or "the working itinerary."

Since practically all measurements of the air concentration of radon progeny at work sites in mines are performed only once or twice a month, there cannot be a precise correspondence between the actual and measured individual (or for groups of miners) breathing zone concentrations.

No systematic studies have been made for the assessment of the breathing rate and deposition in lungs for individuals or a group of miners in real underground conditions. Therefore, it is impossible to calculate individual exposures correctly.

We present here a method for making direct measurements of the radioactivity in miners' lungs, including the experimental development, the assessment of error, the necessary corrections, and the results of direct measurements of the activity (dose) of miners using portable instruments. The research, that is, the model and phantom measurements, was performed in Moscow (in the former Soviet Union) at the All-Union Institute of Physico-Technical and Radio-Technical Measurements (VNIIFTRI) and the Institute of Biophysics (Ministry of Health). The practical application took place in uranium and nonuranium mines in Uzbekistan and Tadjikistan (former U.S.S.R.). Some aspects of the work have been published, for example (Ruzer et al., 1995). The complete experimental and theoretical details of the method along with the compilation of the results and observed health effects in these mines have not been published before in English.

14.11.3.1 Theory of the method

The basic equations in the most generalized form for the radon series were derived in Ruzer (1958, 1960, 1968). The derivations of the equations are based on the equations described in Bateman, (1910) with some appropriate transformations. The graphs of the applicable equations are presented in 14.10.2.2. Bateman's equations for the decay chain transformations were used for the determination of the correlation between the measured air concentration and activity of each decay product on the filter (or lungs).

In order to use these equations for the buildup of activity in the lungs due to filtration, that is, breathing, we assumed that each member of this chain of decay products supplies the decay products to the lungs at a constant rate $Q_i = q_i vk/\lambda_i$. From a mathematical point of view, the constant rate of supply is equal to the equilibrium between the first and the second member of the chain of radioactive transformation. In this case, the number of atoms N of each decay product in the lungs can be found according to

$$N_i = c_1 e^{-\lambda_1 t} + c_2 e^{-\lambda_2 t} + \cdots + c_i e^{-\lambda_i t}$$

where $c_i = N_{1,0}\ \lambda_1\lambda_2\cdots\lambda_{i-1}/(\lambda_1-\lambda_i)\ (\lambda_2-\lambda_i)\cdots(\lambda_{i-1}-\lambda_i)$ and λ_i is the decay constant of the "i" progeny.

The idea in Ruzer (1958) was to find a correlation between the dose from the alpha-radiation of ^{218}Po and ^{214}Po, which cannot be measured directly in the lung, and the gamma-radiation from ^{214}Pb and ^{214}Bi, which at least theoretically can be measured by external counting. The carcinogenic dose is delivered by the alpha-radiation due to its high-energy transfer to the irradiated cells.

Due to the gamma-emission from the decay products ^{214}Pb and ^{214}Bi and also the relatively high maximum permissible air concentration in comparison with other radioactive aerosols, radon decay products present a unique opportunity for the direct measurement of activity in the lungs.

The final expression for the alpha-activity of ^{218}Po and ^{214}Po on the filter (i.e., lungs) is

$$A_\alpha = vk\ \{q_a\ [F_a(\theta,t) + F_{a,c}(\theta,t)] + q_b\ F_{b,c}(\theta,t) + q_c\ F_{c,c}(\theta,t)\} \tag{14.70}$$

Correspondingly, the beta- and gamma-activity of ^{214}Pb and ^{214}Bi is

$$A_{\beta,\gamma} = vk\ \{q_a\ [F_{a,b}(\theta,t) + F_{a,c}(\theta,t)] + q_b[F_{b,b}(\theta,t) + F_{b,c}(\theta,t)] + q_c\ F_{c,c}(\theta,t)\} \tag{14.71}$$

The integral absorbed dose to the lung will be

$$D = E_A \int A_a(t)\ dt + E_{c'} \int A_{c'}\ (t)dt \tag{14.72}$$

where E_A and $E_{c'}$ are the energies of the alpha-particles of ^{218}Po and ^{214}Po.

After a series of transformations, Equation (14.72) can be present in the form

$$D_a = vk[q_a\ X_a(\theta) + q_b\ X_b(\theta) + q_c\ X_c(\theta)] \tag{14.73}$$

where X_i represents the contribution of each decay product in the absorbed dose. Expressions (14.72) and (14.73) are true for the dose to the whole lung. But practically measurements were made over the pulmonary region of the lung. By comparing (14.73) and (14.71), the correlation between the integral absorbed alpha-dose to the lung and the gamma-activity of ^{214}Pb and ^{214}Bi A_γ can be presented in the form

$$D_a = A_\gamma F(\theta,t;\ \eta_{ba}\ ;\ \eta_{ca}) \tag{14.74}$$

where F is the ratio of the right-hand sides of Equations (14.73) and (14.71), θ is the duration of exposure, and t is the time after exposure; $\eta_{ba} = q_b/q_a$ and $\eta_{ca} = q_c/q_a$.

The direct correlation between the absorbed dose from alpha-emitters ^{218}Po and ^{214}Po and beta- and gamma-emitters ^{214}Pb and ^{214}Bi in the lungs was at first introduced in Ruzer (1958, 1960, 1968).

As a first step, it is important to know to what extent function F depends on shift of equilibrium between ^{218}Po, ^{214}Pb, and ^{214}Bi. The results of the calculations are presented in Table 14.23.

The results in Table 14.23 suggest that by measuring directly the gamma-activity in the lungs, we could avoid the radon decay products concentration measurements. In this case, the maximum deviation from the mean value will be not more than 15%.

The final expression for the integral absorbed dose of alpha-radiation for an exposure time of more than 3 h can be written as

$$D_a = 2.1\times10^8\ A_\gamma\ \theta\ \text{erg} \tag{14.75}$$

Table 14.23 Relationship Between the Function $F(\theta, \eta_{ba}, \eta_{ca})$ and the Shift of Equilibrium of Radon Decay Products for Various Times of Inhalation

η_{ba}, η_{ba}	$F(\theta, \eta_{ba}, \eta_{ba})$ θ=3 h	$\times 10^4$ erg min θ=6 h
$\eta_{ba}=\eta_{ba}=1$	23.8	47.8
$\eta_{ba}=\eta_{ba}=0$	40.8	81.4
$\eta_{ba}=\eta_{ba}=0.8$	24.2	48.2
$\eta_{ba}=\eta_{ba}=0.6$	24.5	49.0
$\eta_{ba}=\eta_{ba}=0.5$	24.9	49.8
$\eta_{ba}=\eta_{ba}=0.4$	25.3	50.7
$\eta_{ba}=\eta_{ba}=0.2$	27.2	54.5
$\eta_{ba}=\eta_{ba}=0.1$	30.0	60.0
$\eta_{ba}=\eta_{ba}=0.01$	38.2	75.8
$\eta_{ba}==0.8$; $\eta_{ba}=0.4$	24.5	48.8
$\eta_{ba}==0.6$; $\eta_{ba}=0.2$	25.3	50.5
$\eta_{ba}==0.2$; $\eta_{ba}=0.1$	28.2	56.3
$\eta_{ba}==0.1$; $\eta_{ba}=0.01$	32.2	64.4

where θ is the duration of inhalation in hours and A_γ is the activity of gamma-emitters in Ci.

For the maximum permissible concentration (APC) of 30 pCi/l (1100 Bq/m^3) (or in terms of [PAEC] 3.8×10^4 MeV/l), a breathing rate of 20 l/min, and a standard lung deposition coefficient of 0.25 for a duration of 3 h or more, the activity in the lungs will be about 0.02 μCi (740 Bq). This activity can be measured by external counting with good statistical accuracy. To prove this, model measurements were taken.

14.11.3.2 *Assessment of uncertainties in the evaluation of dose*

According to Ruzer (1970), the propagated errors in the calculation of the absorbed dose (exposure), based on measured air concentrations of decay products, are estimated to be at approximately 5 to 10-fold in comparison with the actual value of internal irradiation. Such large errors demand a special statistical treatment in the assessment of the dose. The assessment of uncertainties in dose measurement was derived in Lekhtmakher et al. (1975).

Strictly speaking, a Gaussian function cannot serve to describe physical quantities whose domain is limited on the left. The contradiction is insignificant, however, when distribution dispersion remains low, as it also takes place in classical error theory. In the instance examined here, the conditions of low dispersion are violated.

The expression for the absorbed dose D in the lungs over exposure time q can be presented by formula (14.73).

When we estimate the total error, it is necessary to take into account all partial errors such as:

- Correlation between the concentration in the respiratory zone and the site where the sample was taken.
- Variation in the concentration of radon progeny in the respiratory zone (up to 200%).
- Variation in the volumetric breathing rate and the lung deposition fraction during the work shift (up to 300%).
- All these uncertainties will be even greater in calculating the dose to the different parts of the lungs, that is, the trachea, bronchi, etc., as other factors are involved.

Therefore in this case, when the error of the estimated activity in the lungs is so large, it is natural to suggest that the quantities found for q, v, and k can be, with equal probability, several times larger or smaller than the true values. A logarithmic normal distribution in this case is warranted. The magnitude of the activity will be lognormally distributed according to a logarithmic normal law:

$$f(A)\ dA = 1/(2\pi)^{1/2}\ \sigma_{g,A} \times 1/A\ e^{-\lg^2 \times A/A_g}\ /2\sigma^2_{g,A}\ dA$$

where

$$A_g = q_g v_g k_g X(q),\ (\sigma_{g,A})^2 = (\sigma_{g,q})^2 + (\sigma_{g,v})^2 + (\sigma_{g,k})^2$$

Assuming that $\sigma_{g,q} = \ln 2$, $\sigma_{g,v} = \ln 3$, $\sigma_{g,k} = \ln 3$, we obtain

$$\sigma_{g,A} = 1.7$$

This signifies that with a probability of 95%, 68%, the true value of Ag is between the limits of

$$0.03A < A_\gamma < 30\ A \text{ and } 0.18\ A < A_\gamma < 5.5\ A$$

where A is the calculated value of the activity. The same magnitude of uncertainty will naturally be in the absorbed dose assessment.

The results of this calculation suggest that the uncertainties in the dose assessment of miners are very high.

14.11.3.3 *Correction for the shift of equilibrium of radon progeny in the air and in the lungs*

The laboratory measurement of radon decay product gamma-activity in simulated lungs was conducted using a calibrated sample of ^{226}Ra. The ^{226}Ra gamma-spectrum is identical to the entire gamma-spectrum of radon progeny. However, there is a difference in the measurement of miners. In the ^{226}Ra calibration source, all radon decay products are in equilibrium, that is, their activities are equal. The shift in equilibrium in the lungs will depend on the shift in equilibrium in the air, and this will change with time in the lungs. This is especially important because the gamma-spectra of RaB and RaC are different, which will result in the change of gamma-detection efficiency with the change of equilibrium both in the air and in the lungs.

Let us denote e_0 as the detection efficiency of RaB and RaC in equilibrium in the lungs or in the standard source of ^{226}Ra, e_b the detection efficiency of RaB, e_c the detection efficiency of RaC, e the detection efficiency in the case of the shift of equilibrium between RaB and RaC, and $k_2(\theta,t) = e_0/e$ the correction factor on inequality of detection.

$$e_0 = (N_{b,0} + N_{c,0})/2A_0$$

where $N_{b,0}$, $N_{c,0}$ are the count rates of RaB, RaC, respectively, from equilibrium RaB and RaC in the lungs or in the standard source of ^{226}Ra in the phantom, and A_0 is the activity in the equilibrium source.

$$e = [N_b(\theta,t) + N_c(\theta,t)]/[A_b(\theta,t) + A_c(\theta,t)]$$

where N_b, N_c, A_b, A_c are the count rates and activities of RaB and RaC, respectively, in the lungs at time t after exposure time θ. Then

$$k_2 = e_0/e = (N_{b,0}+N_{c,0})/2A_0/(N_b+N_c)/(A_b+A_c)$$

$$N_{b,0} = e_b A_0,\ N_{c,0} = e_c A_c$$

$$N_b = e_b A_b,\ N_c = e_c A_c$$

$$k_2 = (e_b A_0 + e_c A_0)\ (Ac + A_b)/2A_0(e_b A_b + e_c A_c)$$

and finally

$$k_2(q,t) = (1 + e_c/e_b)\ (1 + A_c/A_b)/2[1 + (e_c/e_b)\ (A_c/A_b)]$$

When the shift of equilibrium between RaA, RaB, and RAC in the air changes from 1:1:1 to 1:0.03:0.01, the ratio of $A_c(t)/A_b(t)$ in the lungs changes from 0.31 to 2.93. But due to the mode of the function k_2 (the ratios e_c/e_b and A_c/A_b are both in the numerator and denominator), the difference in k_2 is not significant. For example, for a NaI(Tl) crystal 80×40 mm, the change in k_2 is not more than 20% (Ruzer and Urusov, 1969).

In calculating the detection efficiency for RaB and RaC, we have to take into account the number of gamma quanta per decay for RaB (0.823) and RaC (1.397) with their average energies of 0.316 and 0.717 MeV, respectively.

Relative efficiency e_c/e_b was calculated for the 80×40 mm NaI(Tl) crystal and the model of the lungs (Ruzer and Urusov, 1969) according to the formula

$$e_c/e_b = \Sigma h_i k_{l,i} k_{b,i} k_{c,i} / \Sigma h_j k_{l,j} k_{b,j} k_{c,j}$$

where h is the yield of the correspondent gamma-line, k_l, k_b are the absorption coefficients of the lungs and the body, respectively, and k_c is the detection efficiency of the crystal. Indexes i and j belong to RaC and RaB, respectively.

Simplifying the expression for the average absorbed dose, assuming that the mass m of the lung is 1000 g,

$$D/m = 8.05\times10^{-7}\ k_1 k_2 k_3\ N_{0,\gamma}\ \theta \text{ mrad} \qquad (14.76)$$

where θ is the duration of the inhalation, $N_{0,\gamma}$ is the count rate above the "background" (the "background" is the count rate measured for the person before entering the workplace), k_1 is the correction coefficient for decay of radon progeny in the lungs after the end of exposure, and k_3 is the reciprocal of detection efficiency of the gamma-radiation of radon progeny in equilibrium.

Results of the calculations of k_1, k_2, k_3, function F, ratio A_c/A_b, etc. are shown in Figures 14.46–14.49.

Values of the function F for different θ and t are presented in Figure 14.46. Each point on the graph is an average for the shift of concentrations from 1:1:1 to 1:0.1:0.03. Values of the coefficients k are presented in Figure 14.47. After exit from the work site, the ratio of the activities of RaC and RaB in the lungs of miners will increase. The correction for this factor is substantial, when direct measurement is not provided immediately after exit from the working site. The ratio of the activities of RaC (^{214}Bi) and ^{214}Pb (RaB) at the time after exposure is presented in Figure 14.48.

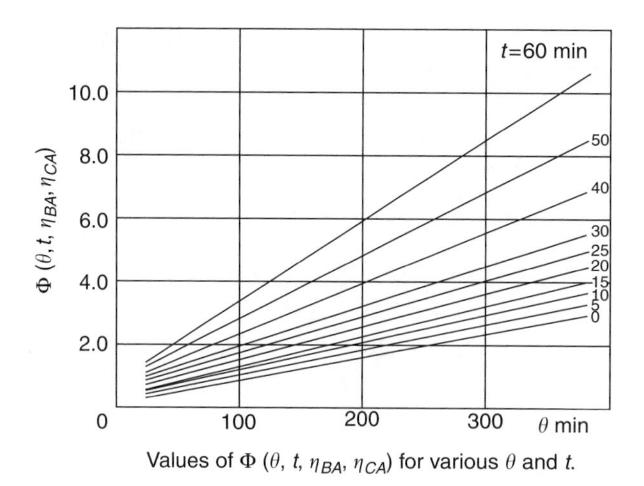

Figure 14.45 Values of Φ (θ, t, η_{ba}, η_{ca}) for various θ and t (Ruzer, L.S., 2001).

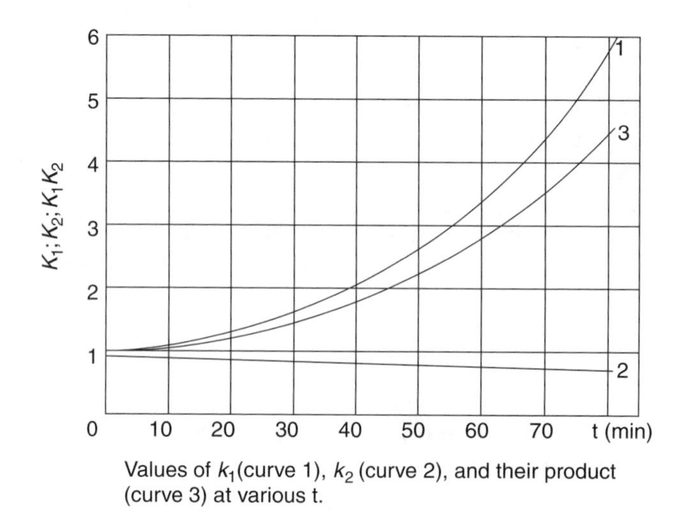

Figure 14.46 Values of k_1 (curve 1), k_2 (curve 2), and their product (curve 3) at various t.

The efficiency of registration of the gamma-ray from ^{214}Pb depends on the size of the crystal and the geometrical (position) factor. The dependency of efficiency e_b of the gamma-radiation of ^{214}Pb from ^{226}Ra for different equilibrium ratios, e_c/e_b, is shown in Figure 14.49.

The ratio e_c/e_b as a function of crystal thickness (mm) (1 — ^{226}Ra source and 2 — registration from the chest) is presented in Figure 14.49.

It should be pointed out that measurement of the gamma-activity in the lungs of miners was provided in the environment with a relatively high gamma-background, especially when measurements took place near the working sites. In such cases, the parameter "quality of measurement" $q=n^2/n_b$ (n — count rate above the background, n_b — background count rate) was chosen to achieve the best measurement conditions. The results of this study are shown in Figures 14.50 and 14.51.

Figure 14.50 shows the statistical error (1) and the "quality factor" (2) vs. the voltage on the photomultiplier (FEA). Figure 14.51 shows the statistical error (1) and the "quality factor" (2) vs. amplification coefficient.

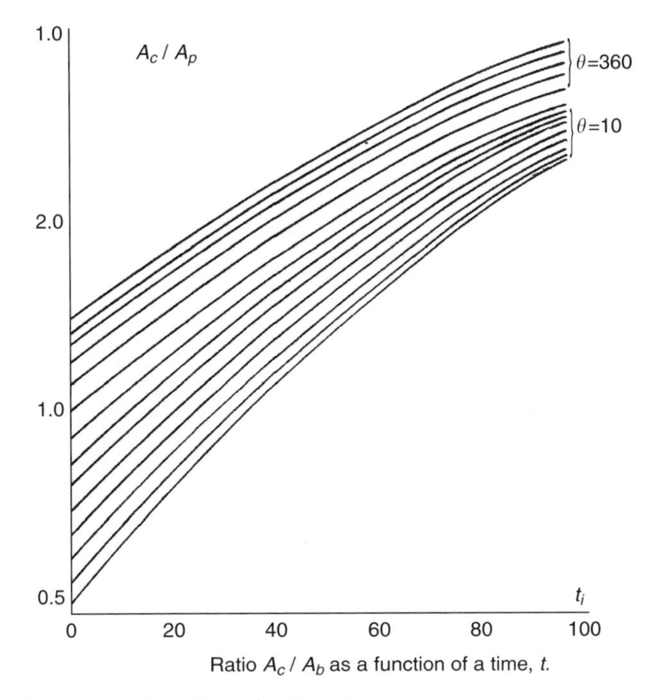

Figure 14.47 Ratio A_c / A_b as a function of a time, t.

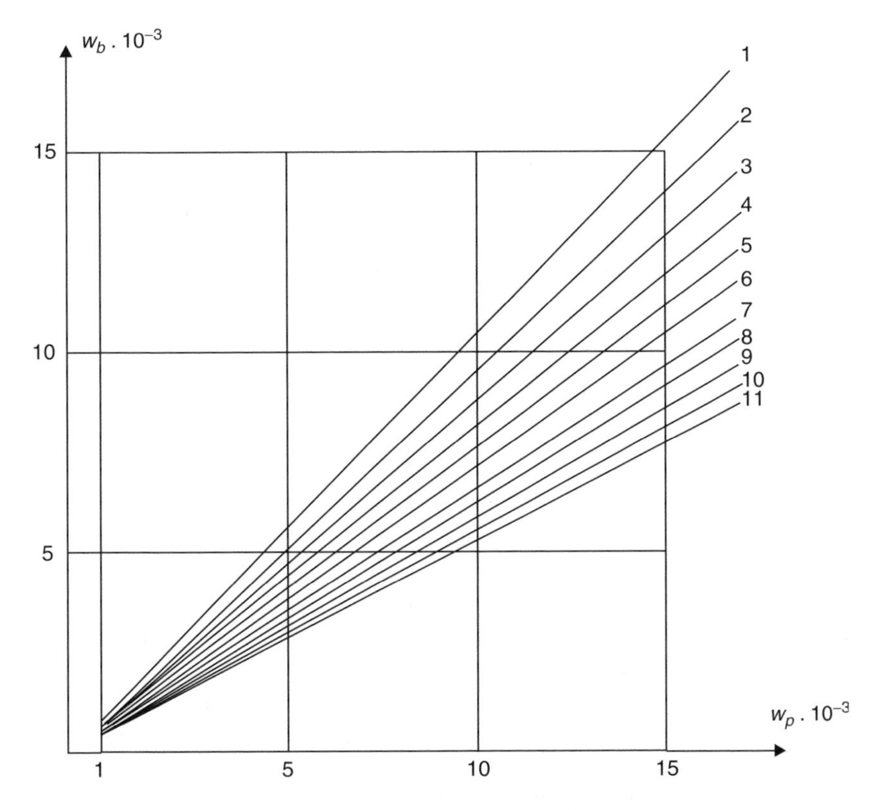

Figure 14.48 Efficiency registration of gamma-radiation of ^{214}Pb e_b vs. efficiency registration of the equilibrium source of ^{226}Ra for different e_c / e_b

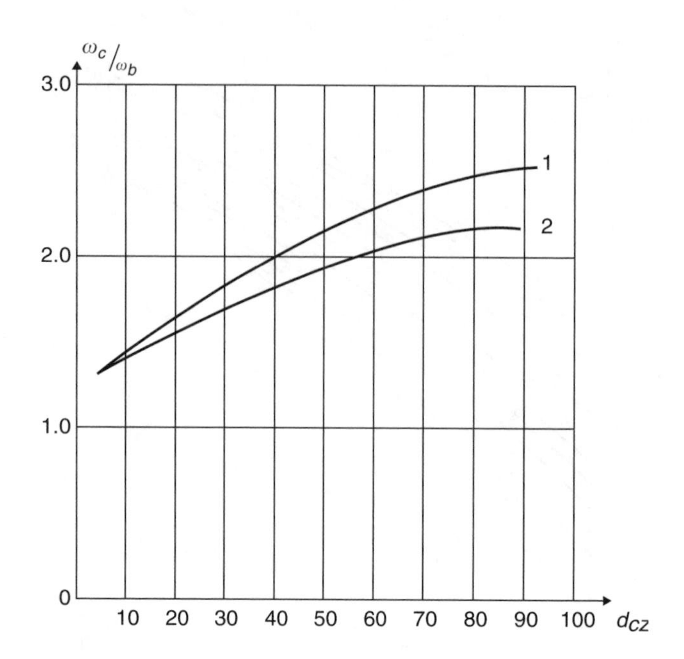

Figure 14.49 Ratio e_c / e_b as a function of crystal thickness (mm): 1 — source of ^{226}Ra; 2 — registration from the chest.

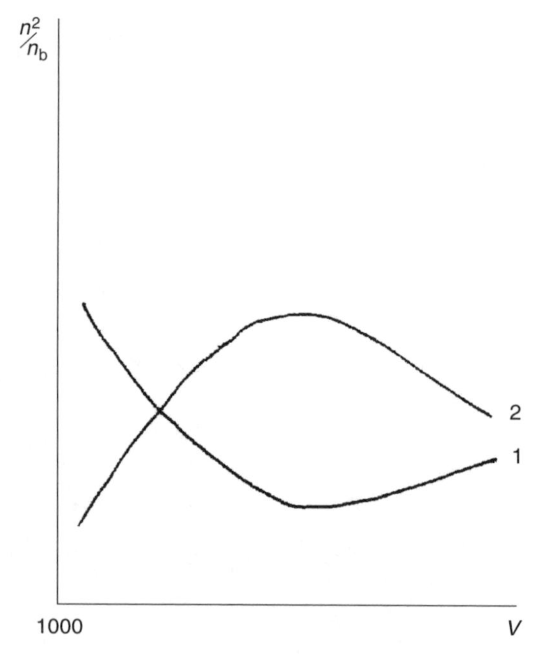

Figure 14.50 Statistical error (1) and the quality factor (2) vs. voltage on the photomultiplier (Ruzer, L.S., 2001).

14.11.3.4 *Accounting for parametric variations: Variations of concentrations, breathing rate, and deposition coefficients in real working conditions*

All dosimetric parameters under actual working conditions are not uniform. This is important both in the case of the estimated or indirect calculation of the dose (lung activity) and in the case of direct measurement.

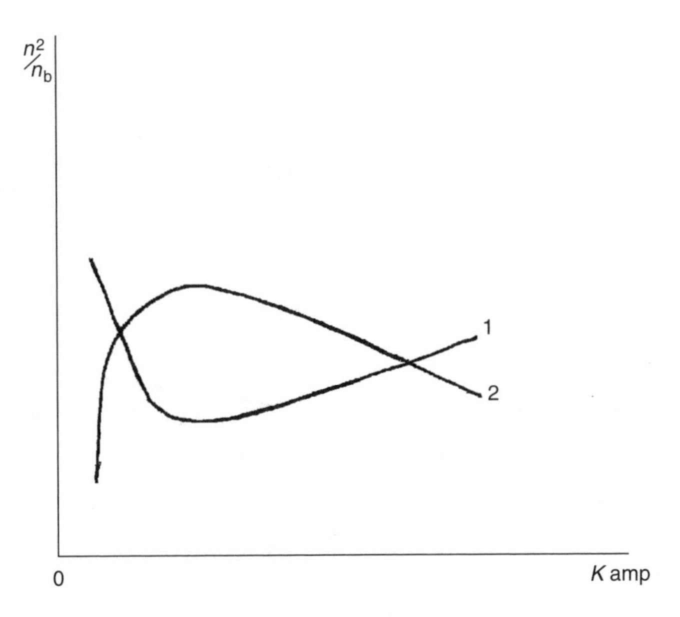

Figure 14.51 Statistical error (1) and the quality factor (2) vs. amplification coefficient.

Let us denote Q (the activity in the lung) as a combination of all parameters (air concentration, breathing rate, lung deposition, respirator efficiency):

$$Q = qvk\,(1 - f)$$

where f is the coefficient of effectiveness of a respirator and Q is the rate of the intake.
To take into account the variation of intake, two options were used:

1. Experimental, by placing a portable gamma-counter near the work site and, for parts of the shift, with special attention to the situation when Q is high.
2. Theoretical, by calculation of corresponding corrections on the inequality of Q during the time of exposure for different types of working itinerary (scenario of exposure).

To perform such a calculation, the whole working shift time, θ, should be divided by n periods in which every dosimetric parameter will be constant. Different typical scenarios of exposure (work itinerary) in terms of the rate of intake, Q, changing during the shift were calculated in Gerasimov and Ruzer (1973) and presented in Figure 14.52.

Variant 1 can represent the situation on Monday morning, since the ventilation was usually turned off over the weekend. This resulted in increasing q, and consequently of Q. Variants 7 and 8 present situations when during the work shift:

- Ventilation was turned off (on), which resulted in increasing (decreasing) q.
- Change in physicochemical properties of aerosols that result in an increase (decrease) of the coefficient k.
- Change in the work load resulting in changed breathing rate and deposition coefficient.
- Use of respirators by Miners resulting in a change of the factor $1-f$.

For the introduction of a correction factor for Q, we have to obtain additional information on each concrete working situation (variation of Q during the shift). The degree of

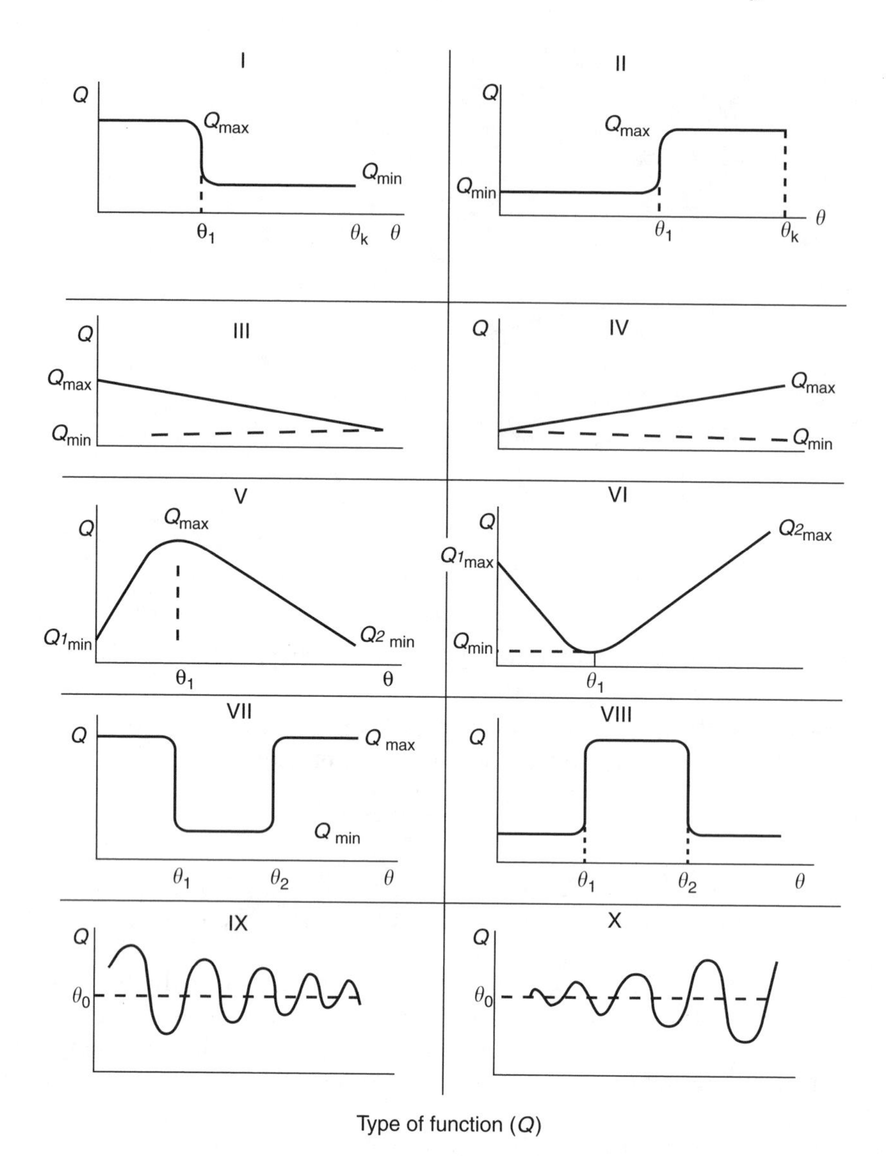

Figure 14.52 Type of function (Q).

error in determining the correction factor depends on the proximity of the chosen variant of the function Q to the variant under actual conditions.

According to the calculations, (Urusov, 1972) even if the magnitude of Q_{max}/Q_{min} is close to 30 the total error in the measurement of activity in the lungs of miners by the direct method will not exceed 40%, which is acceptable for the purposes of individual dosimetry.

14.11.3.5 Model measurement

Experiments to determine the detection limit of the gamma-activity measurement of radon progeny in the lungs of miners were provided using a Human Body Spectrometer at the Biophysics Institute in Moscow. It consisted of a measurement chamber protected by a 20 cm layer of iron. The detector was a NaI(Tl) crystal, and samples of ^{226}Ra containing $RaCl_2$ solution with activity 590 Bq (a) and 175.4 Bq (b) were placed in different positions in the chest of the phantom, and an 80×50 mm NaI(Tl) detection crystal and photomultiplier was placed on the back. The activity of the sample corresponded (a) to approximately 0.75 of APC and (b) 0.25 APC in air.

Tables 14.24 and 14.25 show the results of these measurements concerning the assessment of the detection limit and the statistical errors with one and two detectors (the minimal distance from the sample to the detector was 23 cm).

The results of these model experiments suggest that at the level of concentration of radon progeny in the air of 0.75 APC, 4.3×10^4 MeV/l (or correspondingly 30 pCi/l), the corresponding gamma-activity in the lungs of miners can be measured by one detector within 3 min with an accuracy of 8% if the measurement of the background is 10 min. Even if we take into account that in real underground conditions measurements take place 30–40 min after completing the work, during which time the activity will decrease by a factor of two, it will still be possible to measure the activity in the lungs with an accuracy of 15% for a measurement of 5 min.

14.11.3.6 Phantom measurements and geometric corrections

For calibration purposes in laboratory conditions, a phantom of the human torso was used. The measurements were made with a NaI(Tl) crystal with a window diameter of 77 mm, a lead shield, and a cylindrical sample of ^{226}Ra activity of 3.9×10^{-5} Ci size 160×16 mm placed in different positions in the phantom (Zalmanson et al., 1973; Vasin et al., 1975) (Figure 14.53).

The results of phantom measurements, including positions of the sample and detector, are shown in Table 14.26. Measurements were provided by a scintillation detector of NaI(Tl) with a lead shield (diameter of the window 77 mm).

For the field measurements, a more primitive phantom was used, that is, a long-sized vessel filled with water size close to the size of lungs, in which the sources of ^{226}Ra were placed in different positions.

The systematic errors associated with the contamination of body and work clothes in real working conditions, for example, the influence of radon accumulated in the adipose tissue of the abdominal cavity, were eliminated by the introduction of corresponding

Table 14.24 Human Body Spectrometer Measurements

Mode of Measurement	Impulses per 100 sec	
	Single Detection	Dual Detection
Background (open door)	914	1540
Background (closed door)	427	843
Background (closed door) with a person	500	978
Person with a ^{226}Ra sample, 590 Bq	—	1244
Person with a ^{226}Ra sample, 175.4 Bq	557	1088
Person with a ^{226}Ra sample, 0.016 μCi	704	—

Table 14.25 Relative Errors (%) in Human Body Spectrometer Measurements

Activity of ^{226}Ra		One Detector		Two Detectors			
t_b		10 min		100 Sec		10 min	
t_b+p+s	100 s.	3 min	5 min	10 min	100 s.	5 min	10 min
590 Bq	9.5	7.4	6.4	—	12	—	—
175 Bq	46	36	30	24	57	15	12

t_b — time of the background measurement, t_b+p+s — time of the measurement of the person+$^{+226}$Ra source+background.

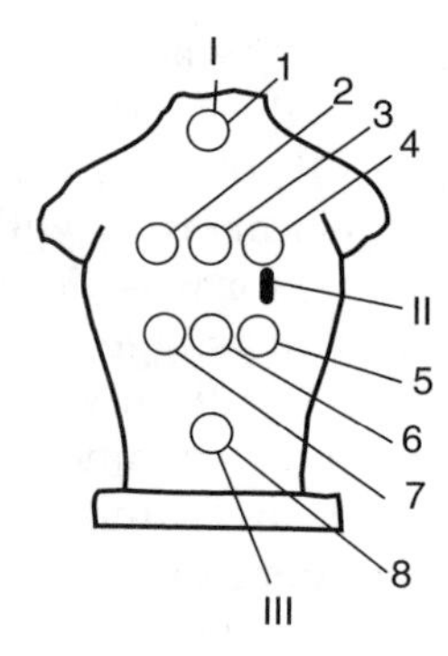

Figure 14.53 Phantom measurement: 1–8, detector positions; i, ii, iii, positions of the source.

Table 14.26 Phantom Measurements[a]

i	*j*=1	*j*=2	*j*=3
1	3500	450	50
2	1600	600	130
3	2200	1050	130
4	1550	1350	110
5	620	800	460
6	750	1400	470
7	600	1900	400
8	100	650	3650

[a]i — Detector position; j — source of ^{226}Ra position

corrections. It has been shown that the contribution of gamma-radiation from the abdomen to the total count rate 30 min after leaving the workplace does not exceed 5%.

Errors associated with the contamination of clothes and the bodies of miners dropped to 15 and 1% of the measured value, respectively, after work clothes were removed and a shower was taken.

The variability of the background in underground measurements with the same person was no more than ±3% (Figure 14.54).

Dependency of gamma-background on body thickness is shown in Figure 14.55.

The decay of radon progeny in the lungs of miners is presented in Figure 14.56. Figure 14.56 shows that activity in the lungs of miners decreases according to the law of radioactive decay of radon progeny, that is, no substantial clearance was found during a 3 h period.

The dependence of the count rate on the distance between the detector and the chest is shown in Figure 14.57, where *a* is the efficiency of registration, *b* is the statistical error, 1 represents one detector with a single window, 2 represents one detector with a double window, and 3 represents two detectors with a double window.

The dependency of the count rate from the horizontal (x) and vertical (z) shift of the body for one (I) and two (II) detectors is shown in Figure 14.63.

14.11.3.7 *Assessment of the errors of the direct method*

As it is obvious from formula (7) that the errors associated with direct measurement of the activity of radon progeny in the lungs depends on many factors, each should be studied carefully:

- Error associated with the measurement of activity in the lungs, including geometrical factors, quality of measurement factor, etc. (δ_1)

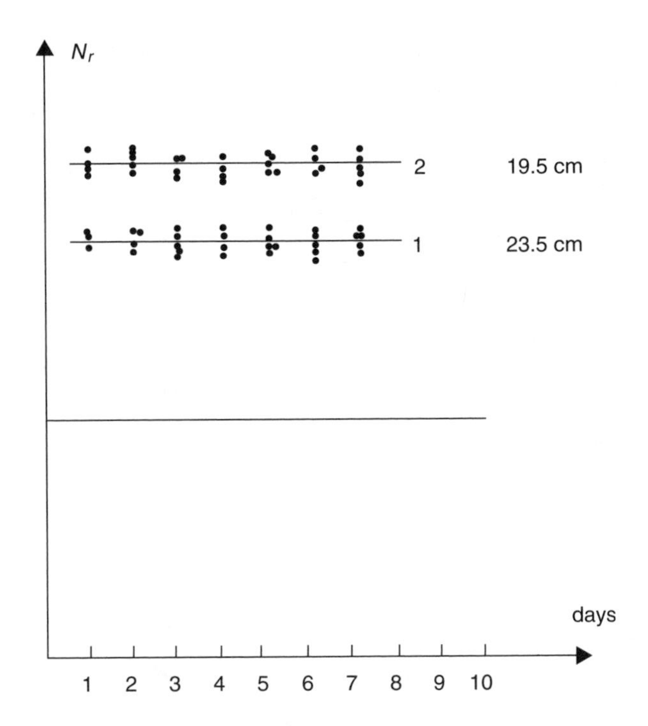

Figure 14.54 Gamma-background of miners (arbitrary units) for two different thicknesses of the body (d) within 7 days of the week (upper curve $d = 19.5$ cm; lower curve $d = 21$ cm).

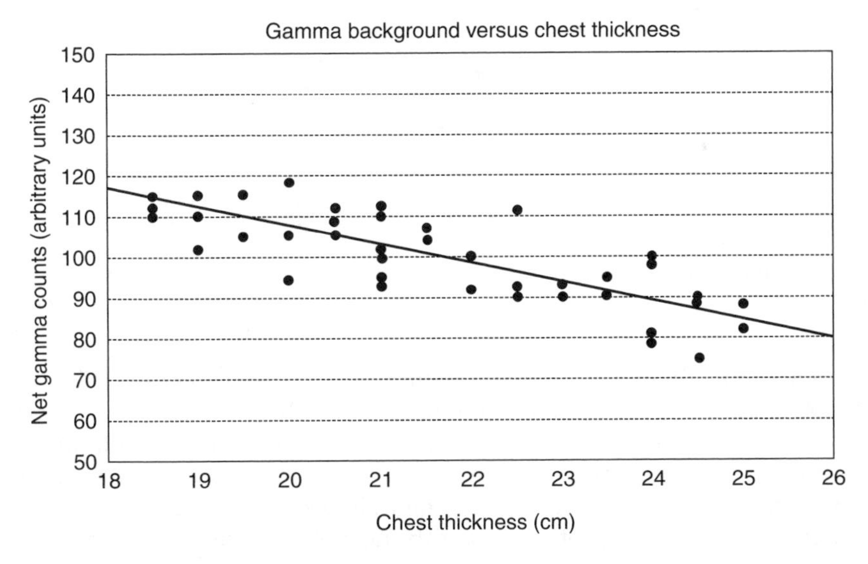

Figure 14.55 Gamma-background vs. chest thickness.

- Error associated with contamination of the body and clothes of miners (δ_2)
- Errors associated with calculation of the coefficients k_2 and k_3 (respectively δ_3, δ_4)
- Errors associated with the variation of Q during the time of exposure (δ_5)
- Errors associated with the averaging of the function F (δ_6)
- Errors associated with the uncertainty of the times of exposure and the time t (δ_7, δ_8)

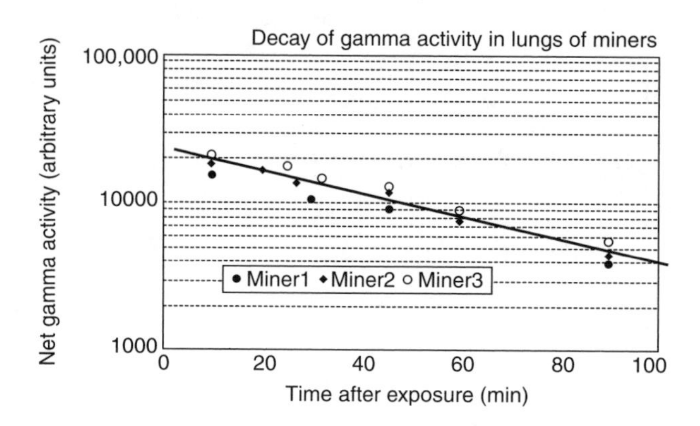

Figure 14.56 Decay of gamma activity in the lungs of miners.

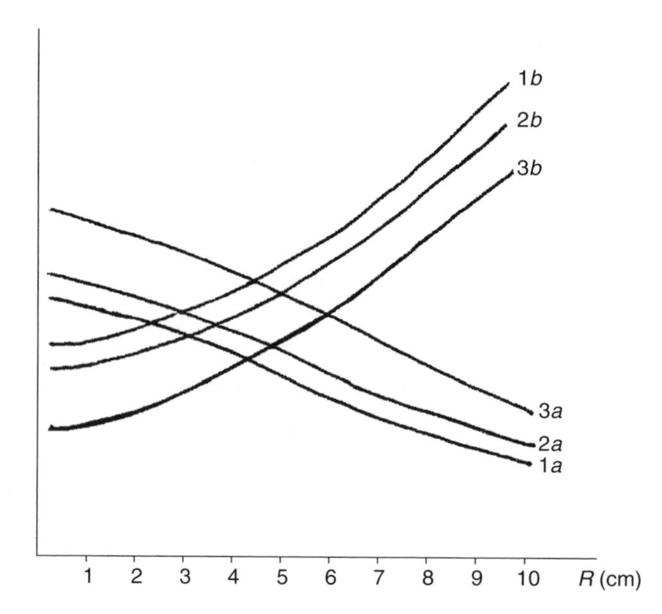

Figure 14.57 Dependence of the efficiency (a) and statistical error (b) from the distance of the body from the detector.

Calculations show that for the measurement using a crystal size of 80×40 mm with the average radon progeny concentration equal to 3.9×10^4 MeV/l will not exceed 26%.

Under mining conditions, it is necessary to take into account the contamination of the body and clothing of miners by radon decay products, otherwise the systematic error of measurement activity in the lung can be very high. It was shown that in radon decay products measurements in the lungs of miners with the clothes taken off, δ_2 decreases to 16%; after a shower was taken, δ_2 decreases to 1%. The error in the determination of k_2 — δ_3 — does not exceed 2%.

The error δ_4, which is a characteristic of the degree of difference between the real situation and measurement in the lungs, includes: error in the measurement of the activity in the standard source of −226 Ra — $\delta_{4,1}$, error related to the difference between the size of the real chest and the average that was taken into account in calculation — $\delta_{4,2}$, errors $\delta_{4,3}$, $\delta_{4,4}$ associated with the position of the body to the detector and in nonuniformity of the distribution of activity in the lungs. The error $\delta_{4,1}$ was not more than 3%. The assessment

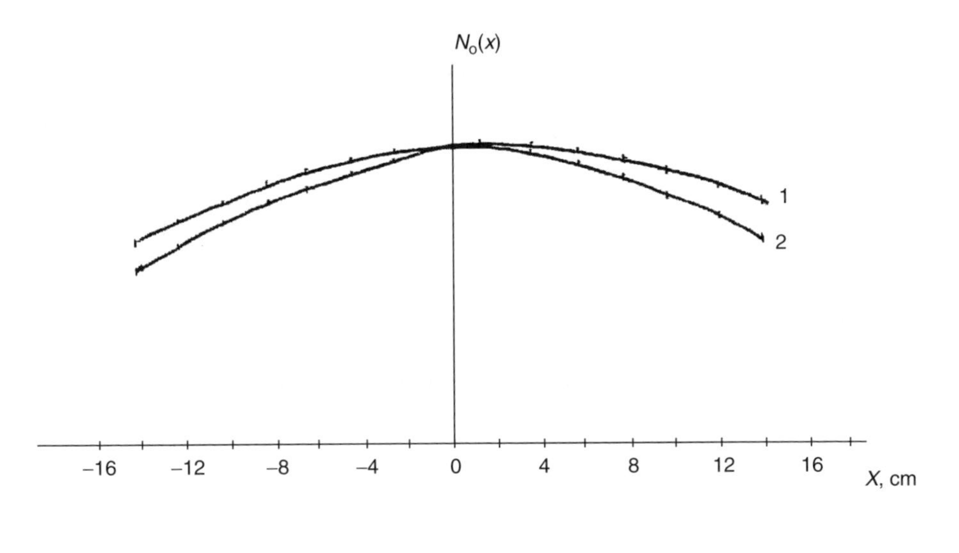

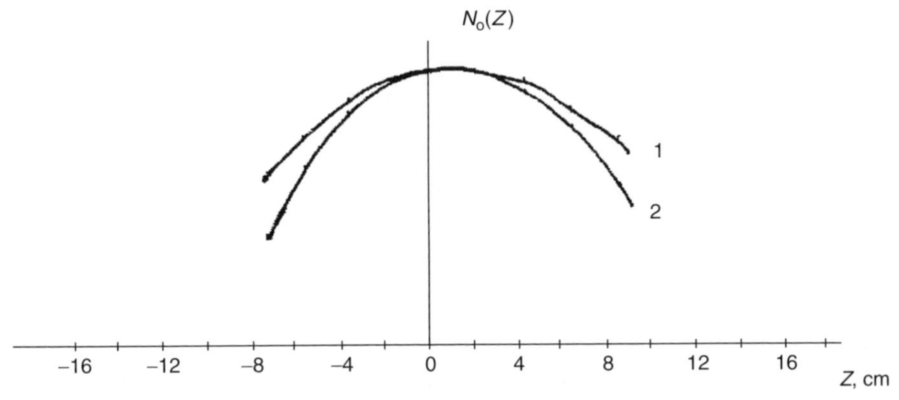

Figure 14.58 Dependence of the count rate from the horizontal (upper curve) and vertical (lower curve) change of position of the body (i–one detector; ii–two detectors) (Ruzer, L.S., 2001).

of $\delta_{4,2}$ was $\delta_{4,3}$ on chest size measurements of miners and also on the study of the correlation between the vertical and horizontal changes of the position of the body and the count rate. The results show that

$$\delta_{4,2}=5\% \text{ and } \delta_{4,3}=10\%$$

The assessment of $\delta_{4,4}$ was based on the fact [Urusov, S.A., 1972] that nonuniformity in the activity distribution in the lungs does not exceed 30%. The calculation of this error was determined theoretically by using the lung model, represented by two cylinders in a case when the activity in one of the cylinders was changed by 30%. In this case, the error $\delta_{4,4}$ was 5%. Therefore, δ_4 was 12%.

Evaluation of δ_5 presents special difficulties because it requires additional information associated with the work itinerary (scenario of exposure), that is, type of the function Q.

For this assessment, measurements of radon decay product concentrations were made at 208 working sites; in 98% of the cases, q_{max}/q_{min} does not exceed 5 with a maximum of 9.

Taking into account the ratio of Q_{max}/Q_{min}, the difference in the breathing rate was 15. If we suppose that the changes in Q within the shift take place not more than 10 times, δ_5 will be no more than 30% for Q_{max}/Q_{min} =15. The error δ_6 was no more than 3% with known θ and t. An error associated with known θ — δ_7 — in the real working situation was no more than 4%. The uncertainty in time (t), if the instruction of workers was made with periodic control of the time of leaving the working site, should not be more than 5 min in terms of time and consequently 9% as an error.

Thus, the resulting error in the measurement of radon decay products activity in the lungs of miners in a real mining environment will be

$$\delta = \sqrt{(\delta_1)^2 + (\delta_2)^2 + (\delta_3)^2 + (\delta_4)^2 + (\delta_5)^2 + (\delta_6)^2 + (\delta_7)^2 + (\delta_8)^2}$$

For the average PAEC of 1.3×10^5 MeV/l (1 working level (WL)) δ will be near 38%, and for 3.9×10^4 MeV/l it will be 50%. These results were confirmed by the measurement in real underground conditions.

Similar calculations were provided for thoron decay products and for the mixture of radon and thoron.

According to Urusov (1972), the level of thoron decay products in nonuranium mines in Tadjikistan was on average 10 times lower than for radon progeny.

All coefficients and formulas in the case of the thoron and radon decay products mixture will be similar to what it was in the case of radon progeny itself.

In the case of thoron and radon decay products, the measurement of the activity can be provided by sequential gamma-ray activity due to differences in the half-time decay of each family. Ten hours after miners exit the working site, the activity of radon decay products will be 0.1% that at the moment of the end of work, while thoron decay products represent the original activity.

The background measurement should be provided for each miner using the maximum interval between two shifts, that is, for example, on Monday, when the activity of thoron daughters will be 1% that at the moment of leaving the workplace on Friday.

14.11.3.8 *Portable instrument for direct measurement of the activity of radon decay products in the lungs of miners*

Some variants of portable instruments were developed for measuring miners both near the working sites and in the sanitary building, where miners took showers and changed their working clothes. Instruments with one and two detectors (for the front and back of the chest) were developed. For field measurement, NaI(Tl) crystals were used with standard lead blocks as a shield. The main demand on the instrument was to measure the concentration of radon progeny at a concentration of 1.3×10^5 MeV/l (1 WL), taking into account the general tendency of decreasing the maximum permissible concentration even at the level of 3.9×10^4 MeV/l (0.3 WL). The radiometric quality of the instrumentation can be improved by increasing the counting efficiency by using a crystal of larger size, higher-density CsI(Tl), semiconductor detectors, or by decreasing the background. A study of the correlation between background and shield thickness has shown that after increasing the thickness above 5 cm, no decrease of the background took place. Therefore, a 5 cm thickness of the lead shield was chosen as optimal (Urusov, 1972).

The measurement of activity in the lungs can be provided by means of general gamma-activity measurement or spectrometry. The calculation shows that the quality factor of the first group of measurement is higher.

One type of instrument for direct measurement is a portable instrument allowed transport very close to the working site and performed measurements within 1–2 min, without interruption of work.

More than 500 measurements of the activity (dose) in the lungs of miners were provided in uranium and nonuranium mines of the former Middle Asia Republic of the U.S.S.R. over an 11-yr period. Some of the results are presented in Table 14.27. Measurements took place in June 1969 in Tadjikistan.

Table 14.27 Measurements of the Activity (Dose) in the Lungs of Miners

Names	1	2	3 (N_0)	4 (N)	5	6
ä-v	420	20	574	671	17.0	55
V-n	420	25	574	667	17.0	57
S-v	420	25	574	638	16.0	39
ä-v	420	25	574	664	16.0	55
O-v″	420	25	574	609	3.0	21
ï-v″	420	25	574	614	3.0	24
ë-Í	420	30	574	614	4.0	26
Ñ-Ó″	420	30	574	608	4.0	22
ë-n″	420	30	574	604	13.0	20
F-v	420	30	574	613	13.0	26
ë-n	420	40	574	562	14.0	0
L-i″	420	30	574	550	14.0	0
ë-i″	420	30	574	609	13.0	23
I-Ó″	420	30	574	615	4.0	26
V-n	420	25	574	586	3.0	0
K-n″	420	20	574	680	17.0	60
S-j	420	20	574	681	17.0	61
S-n	420	25	574	671	16.0	59

N_0 — background count rate, that is, before entering radon atmosphere; N — count rate after leaving the working site. 1 — time of exposure, min; 2 — time of measurement, min; 3, 4 — count rates (multiplied by 16); 5 — PAEC, MeV/l; 6 — absorbed dose, mrad.

14.11.3.9 Radon decay products as a radioactive marker in studying the deposition and dosimetry of nonradioactive aerosols

As presented above, the theory, model measurements, corrections, assessment of the errors, and results of measurements on miners suggest that even by using portable instruments it is possible to measure the activity of radon decay products in the lungs of miners at a level even lower than the APC.

By using such a technique, it was possible to introduce actual individual dosimetry for hundreds of thousands of underground workers in mines where levels of radon were similar to those found in nonuranium mines in Tadjikistan. Besides this, direct measurement presents an opportunity for noninvasion measurement of the FAL value responsible for aerosol deposition in the lungs. This value, which is a "bridge" between concentration in the air and deposited aerosol activity (or mass) in the lungs, plays an important role in the inhaled particles dosimetry for all kinds of aerosols. The uncertainty in FAL measurement is much lower than in the determination of breathing volume rate and deposition coefficient separately.

As is known, the biological effect of aerosols depends on local dose, because sensitivity of the lung tissue to different substances, deposition, clearance, and transport of deposited aerosols inside the lung is not uniform. In this connection, the local FAL (LFAL) is an important parameter to correlate dose with the concentration for both radioactive and nonradioactive aerosols.

It should also be mentioned that radon decay products are good natural markers due to their short-lived nature and relatively high permissible concentration in comparison with most artificial aerosols. In mines, the primitive portable technique was used due to special underground conditions.

For the purpose of studying the deposition of aerosols in lungs with radon progeny as a marker in laboratory conditions and by using spectrometric detectors, a scanner, and a three-dimensional technique, it is possible to study experimentally, and not by modeling, the deposition of aerosols in different parts of the lung for different particle size distributions. As a marker, radon decay products have some advantages:

- Due to its nature, radon, after entering a volume, distributes uniformly in the air and produces the marker itself with a known constant rate — "unattached fraction," consisting of an atom of nuclide, surrounded by a group of 8–10 molecules.
- Due to its extremely small size, such a marker does not change particle size distribution of the aerosols of interest.
- Measurements in mines showed what kinds of concentrations are appropriate for the study.
- Radon and its decay products are with natural radioactive background. The risk of these nuclides was studied intensively. Based on the data provided in Leonard (1963) and U.S. EPA (1993), the risk of using it in humans will be hundreds of times smaller than from radon and its decay products in open atmosphere.

Some examples of the studies related to direct measurements of the radionuclides in the lungs are given below.

In Webb et al. (2000), the construction, procurement, and evaluation of parameters of the *in vivo* lung counting monitoring system in the Carlsbad Environment Monitoring and Research Center, New Mexico State University (CEMRC) was described. The goal of the *in vivo* radioassay program is to

- Provide measurement for low- and high-energy photons of radionuclides deposited in the lung and the whole body
- Provide the highest measurement sensitivity

The evaluation of the parameters of the system included evaluation of the detector thickness for the whole body and lung counting, electronic modification to reduce low-energy noise from cosmic rays, the shield, and the cryostat alignment.

The objective of the European whole-body counter measurement intercomparison (Thime et al., 1998) was to compare the performance of selected European whole-body counters. Forty-four whole-body counting facilities from 42 organizations in 19 countries plus Hungary, the Czech Republic, Switzerland, and Norway took part in the intercomparison. For the study, a 79 kg tissue equivalent "St. Petersburg phantom" was used with rods containing ^{40}K, ^{57}Co, ^{60}Co, and ^{137}Cs. The work was provided in the frames of article 25 of the Basic Safety Standard Directive, which requires that measuring instruments that are used for radiation protection purposes be regularly checked and tested to ensure their effectiveness and correct use.

The goal of the study described in Butterweck et al. (2001) was to carry out nose breathing and mouth breathing measurements to determine the deposition and retention of unattached radon decay products in the respiratory tract. After ethical approval by a special commission, volunteers were exposed for 30 min in the chamber with a volume of 10 m^3 to atmosphere enriched with unattached radon progeny with a radon concentration of 20 kBq/m^3. The concentration and the size distribution of each unattached short-lived decay product was measured by an eight-channel tube diffusion battery in combination with on-line alpha-spectrometry. The total results of these measurements are in agreement with the ICRP Publication 66 Human Respiratory Tract Model. The results showed no significant transport of particle diameters close to 1 nm to the gastrointestinal tract as generally reported for large aerosol particles.

The high-purity germanium detector was studied in Boschung (1998) for whole-body monitoring in Paul Scherer Institute (PSI) in Switzerland. The advantages of this detector compared to the presently used NaI(Tl) monitor are high-energy resolution, improved radionuclide identification capability, and improved minimum detectable activity (MDA) for the same counting time. However, when used as a future routine measurement system, this detector has some disadvantages, such as higher operation cost, more demanding maintenance, and increased operator qualifications.

In Ishigure et al. (1998), the validity of using ^{241}Am as a tracer of inhaled Pu in external chest counting was presented. Pu isotopes are the most difficult radionuclides in terms of direct chest measurement. They can be detected through weak emission of low-energy L x-rays. Another opportunity is to measure ^{241}Am produced from ^{241}Pu by beta-decay, and gamma-rays of 60 keV with an emission rate of 0.36. The ratio of ^{241}Am and ^{241}Pu was studied for 15 months *in vivo* counting using rats that inhaled plutonium dioxide with 4% of ^{241}Am activity. The result indicates that ^{241}Am is cleared from the lung at the same rate as Pu, so ^{241}Am can be used as a tracer of inhaled plutonium in extended chest counting.

It was demonstrated in Chen et al. (1998) that, among different methods of the long-term monitoring of thorium inhaled by workers observed for 20 yr in China, the method of exhaled thoron decay products using the electrostatic collecting system is the best in terms of monitoring and assessing the body burden of thorium (ThO_2). This is particularly useful for workers having thorium activities lower than the detection limit for whole-body counting. Among other methods used in the study were thorium concentration in urine samples and exhaled thoron using a ZnS detector.

The results of *in vivo* monitoring about 250 persons from the test group of witnesses of the Chernobyl accident who inhaled spent nuclear fuel (SNF) particles are presented in Kutkov (2000). The intakes of ^{144}Ce, ^{131}I, ^{106}Ru, ^{103}Ru, ^{137}Cs, and ^{134}Cs were calculated. It should be mentioned that detailed information on the properties of the aerosol for the first post-accident period is absent. The biokinetic model used in this work is based on the ICRP 66 model of the Human Respiratory Tract.

14.12 Radioactive aerosols epidemiology: Miners studies

14.12.1 Lung cancer mortality and lung sickness among nonuranium miners in Tadjikistan

14.12.1.1 Lung cancer mortality data

In order to study the lung cancer mortality of miners, data were collected in Alterman (1974) about mortality among all male populations older than 19 yr in the cities of Tadjikistan (200,000), a similar contingent of mining settlements (11,000), and different groups of underground workers (2400) from 1960 to 1970. In this region, 30 mines and geology-prospecting teams were located.

The primary diagnosis of cancer took place in the health department of the industrial complex. All such complexes had x-ray diagnostic equipment with special medical personnel trained in the diagnosis of the sickness of breathing organs. Patients with suspected cancer were directed to the oncology department (hospitals) in the cities of Leninabad (now Hodjent), Dushanbe, Tashkent, and Alma-Ata for final diagnosis.

The mortality of miners was studied according to registry office records. For all men older than 18 yr (the minimum age for underground work) from mining settlements, who died from breathing organs illness, special files were set up. Then according to the data from the personnel department, duration of underground work, professional itinerary, and number of underground workers per year, including miners in the drilling group, were established. Data on male populations of the mining settlement were taken from local Soviet authorities according to the 1959 and 1970 census.

A study of lung cancer mortality was provided by means of long-time observations of miners in mines, where radiation–hygienic conditions were studied. As a control, the average of three very close mortality rate data of men older than 18 yr was used: the population of all cities of Tadjikistan, mining settlements, and miners who worked in mines with a very low radon progeny concentration.

Measurements of the radon progeny concentrations were provided by means of the procedure described in Markov et al. (1962); in some of the cases, direct measurements of gamma-activity (dose) in the lungs of miners were provided.

Besides this, the gamma-ray intensity, thoron decay products concentration, and ^{210}Po in the air were measured, but it is known that from a dosimetric point of view the contribution of these factors is very small.

For every group of miners, the annual intake was calculated based on the average radon decay products concentration and the filtration ability of lungs (FAL) (see Tables 14.28 and 14.29). In order to diminish the influence of other factors such as dust concentration (and correspondingly, silicosis) and smoking habits of the miners as much as possible, data on these factors were collected.

According to Alterman (1974), the sickness and mortality of miners in this region from 1949 to 1959 was registered for miners with a duration of work for 3 yr. In 1959–1961, when the dust concentration was diminished to the level of some mg/m^3, only eight workers

Table 14.28 Intake (*I*) of Radon Progeny of Miners in all 98 Observed Mines in the Numbers of the Limit Permissible Intake (LPI), Calculated According to the Average Concentration in the Numbers of the Annual Permissible Concentration (APC) and the Average (for the Group of Professionals) FAL in the Numbers of the Standard Value (5 l/min)

Group of Mines	Number of Mines (miners %)	Groups of Miners											
					Drilling Group								
		All Miners			All Drillers			Shaft Workers			Other Sinkers		
		q	*FAL*	*I*	*q*	*FAL*	*I*	*q*	*FAL*	*I*	*q*	*FAL*	*I*
All	98 (100%)	0.5	1.4	0.7	0.7	1.8	1.3	1.2	1.8	2.2	0.5	1.2	0.6
Ore	72 (84%)	0.7	1.4	1.0	1.0	1.8	1.8	1.7	1.8	3.1	0.7	1.2	0.7
Output	34 (60%)	0.5	1.4	0.7	0.7	1.8	1.3	1.2	1.8	2.2	0.5	1.2	0.6
Prospect.	38 (24%)	1.7	1.4	2.4	2.9	1.5	4.3	2.9	1.8	5.2	1.7	1.2	2.0
Other	26 (16%)	0.1	1.4	0.14	0.14	1.8	0.25	0.2	1.8	0.4	0.1	1.2	0.1

Table 14.29 Average Radon Decay Product Concentration (*q*) in Numbers of APC and Calculated Values of the Annual Intake (*I*) in the Numbers of LPI for Two Groups of Mines

Groups of Mines		Groups of Miners							
	All Miners	All Drilling Workers			Including Passing Work			Other	
		q	*I*	*q*	*I*	*q*	*I*	*q*	*I*
All mines	(30)	0.7	1.0	1.0	1.8	1.8	3.3	0.7	0.3
Including mines with $q <$ APC	(17)	0.15	0.2	0.21	0.38	0.4	0.7	0.15	0.18
Including mines with $q >$ APC	(13)	3.1	4.2	4.3	7.8	8.1	14.5	3.1	3.7

Table 14.30 Statistical Data on Lung Cancer Mortality of Nonuranium Miners in Tadjikistan (1960–1970)

Cohort of Observation	Number of Men Older than 19 years (in 10^3)	Number of Deaths on 10^4 Older than 19 years ($P \pm m$)	Mortality Index	
			Ordinary (P_{or})	Standardized (P_{or})
All cities of Tadjikistan	200	860	43±1.5	43±1.5
Mining settlements	11	34	31±5	52±7
Nonminers	8.6	20	23±5	41±7
Miners	2.4	14	58±16	95±20
Miners, working in mines with q <APC				
Miners of the drilling group	0.6	1	17±16	35±24
Others	0.4	2	50±35	19±22
Miner, working in mines with q >APC				
Miners of the drilling group	0.72	9	125±41	265±60
Others	0.68	2	29±21	9±11

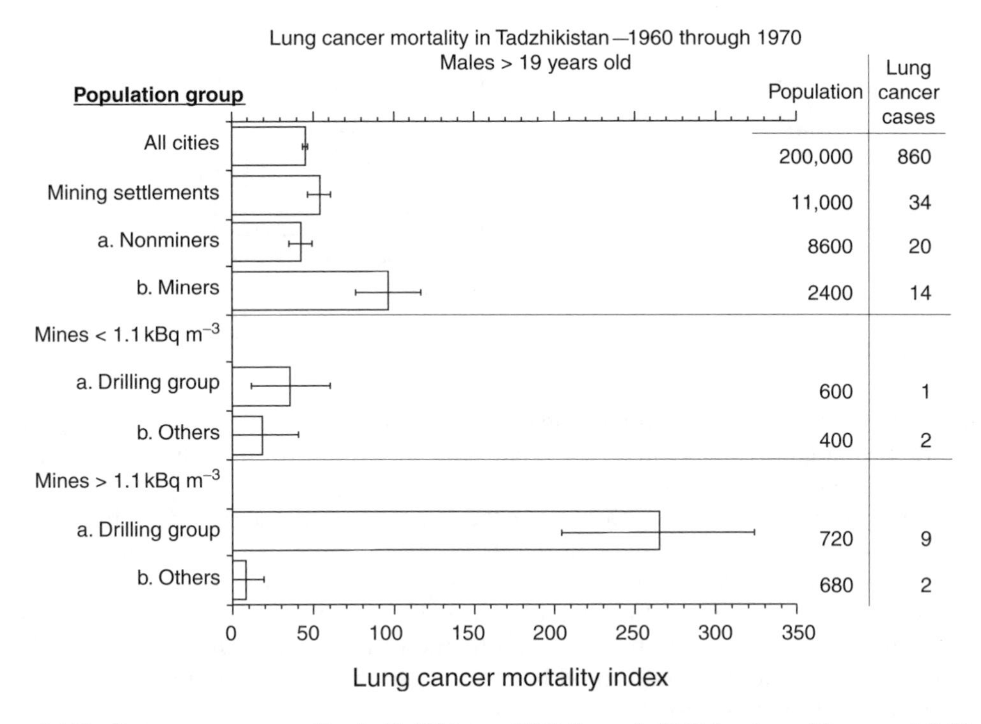

Figure 14.59 Lung cancer mortality in Tadjikistan: 1960 through 1970 (males > 19 years old) (Ruzer, L.S., 2001).

were diagnosed with silicosis from miners with a duration of work for more than 10 yr. At the end of the 60th year, only a single occurrence of silicosis took place.

As far as the problem of smoking is concerned, there is a high probability that the distribution of smokers among underground workers is very similar as for the population of

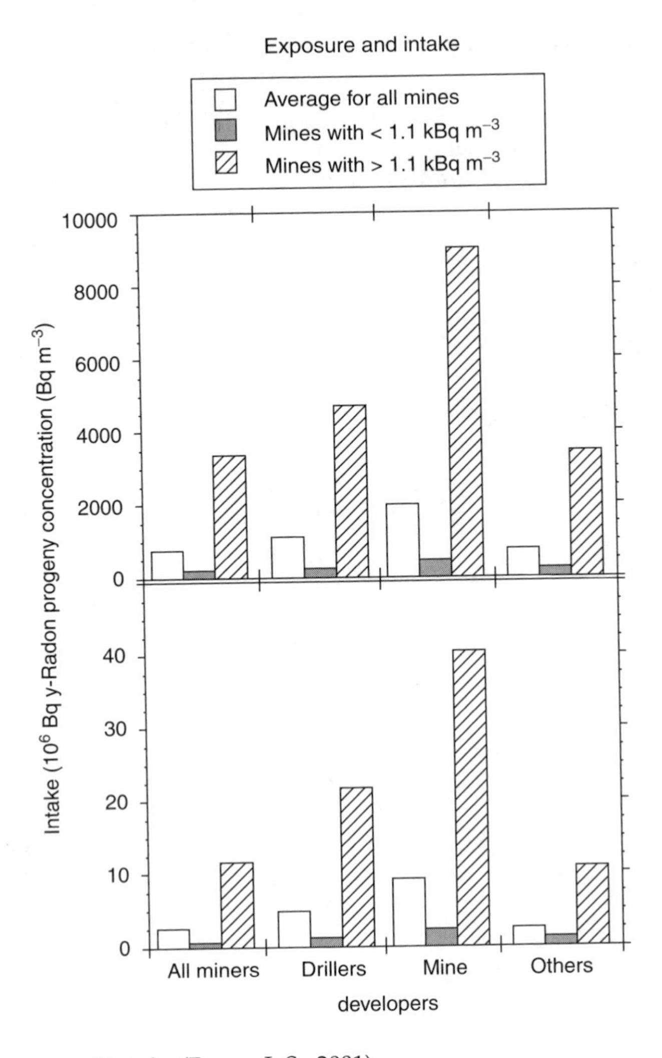

Figure 14.60 Exposure and intake (Ruzer, L.S., 2001).

mining settlements (control group) of the same age. Moreover, it is difficult to imagine that the distribution of smokers should be different for different groups of miners, working in sites with different levels of radon decay products concentrations.

However, it was shown that the lung cancer mortality of nonuranium miners in this study [Alterman, A.D., 1974] differs substantially for different groups of miners, with the highest level for the drilling group. Thus, we can say with great probability that, in this case, the excessive lung cancer mortality took place due to irradiation by radon decay products. Information on the personnel of the mining settlements is presented in Table 14.30.

Lung cancer mortality was studied in connection with the level of exposure. For this purpose, all mines were divided into two groups:

1. Mines with the median concentration of radon decay products in the working sites (q) lower than APC
2. Mines with q larger than APC

These data suggest that the mortality of miners exceeded the mortality of the control group only in the second group of mines, with concentrations larger than APC according

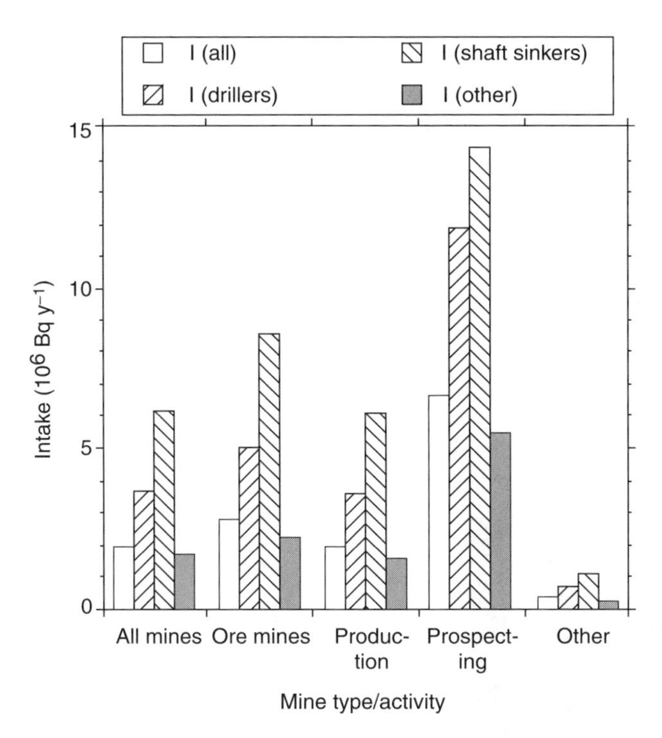

Figure 14.61 Intake (~ dose) by type of miner and activity (Ruzer, L.S., 2001).

to the ordinary index by 3.5 times and according to the standard index by 3.1 times. At the same time, the mortality of drilling workers was greater than for the control group by 5.4 and 6.5 times, respectively.

The mortality of the drilling workers of the first group of mines, and the mortality of other miners in both groups did not differ from the mortality of the control group.

Thus, the higher lung cancer mortality of miners in this mining region was caused mainly by the high level of mortality of drilling workers of the second group. Apparently, it is difficult to consider it accidental that this group also had the highest intake of radon decay products.

14.12.1.2 Lung sickness data

Lung cancer mortality represents a long-time effect of the irradiation of miners. The study of this effect presents many difficulties in terms of collecting information during a long period of time, when professional itinerary, working conditions and other factors can change substantially.

On the other hand, both from the medical and economical point of view, it is important to assess the effect of radon progeny on sickness rate disability as a short-time effect.

For this purpose, a comparative study was provided in the mining department of Karamazar, Tadjikistan (Alterman, 1974), by studying the radon decay products concentration with a parallel observation of the sickness rate disability of miners. The study was carried out in two steps. In the first period (1963–1965) and in the second (1967–1971), measures were taken to reduce the concentration of radon decay products mainly by improving the efficiency of the ventilation system. The results in decreasing concentrations for these two periods are shown in Table 14.32.

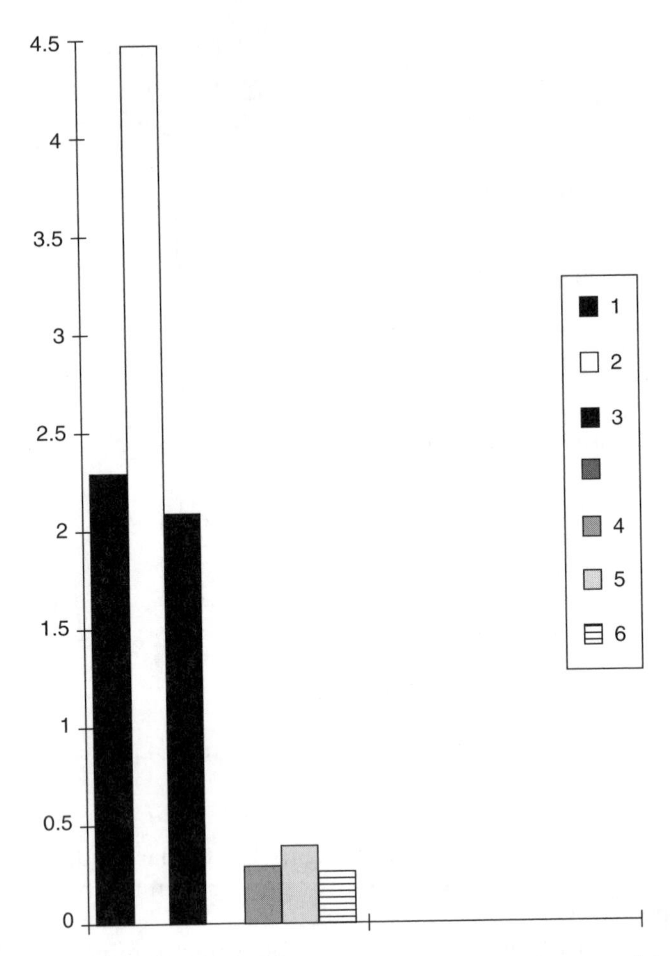

Figure 14.62 Intake of radon decay product in the number of LPI for period A (63–66) – 1,2,3; and period B (67–70) — 4, 5, 6; 1,4 for all miners; 2, 5 for drillers; 3,6 for others (Ruzer, L.S., 2001).

These results suggest that due to measures provided at the end of the first period, the radon decay product concentration was reduced:

- In all working sites, 4 to 9 times
- In drilling sites, 6.6 to 13 times
- Including 4 to 12 times in passing sites

Results with the maximum concentration decreased even more sharply: in all working sites 8–32, in drilling 14–43, and in passing 10–60 times. Corresponding differences in intake are shown in Figure 14.63.

Similar data on aerosol concentration suggested that from period 1 to 2, a reduction in aerosol concentration took place on an average of 2.3 and 5 times in terms of the maximum and was very close to the level of a limit dust permissible concentration of 2 mg/m^3.

Comparative results on the sickness of breathing organs in terms of the number of days of disability for periods 1 and 2 are presented in Figure 14.63. It should be mentioned that during the period of observation no substantial changes took place in terms of technology of the output and mining–geological conditions.

Table 14.31 Radon Decay Products Concentration in Numbers of APC in 28 Mines of Four Mine Departments and Three Geology-Prospecting Teams for Two Periods of Observation[a] (Ruzer, L.S., 2001)

Mine Department	Number of Measurements	Concentration in Numbers of APC (%)			Number of APC	
		<0.3	0.3 to 1.0>	1.0	Median	Maximum
Gigikurt	61/65	51/59	25/29	24/12	0.3/0.25	25/5.8
Takob	110/111	29/51	27/31	44/18	0.9/0.3	61/12.5
Karamazar	101/80	30/71	15/15	55/14	1.7/0.2	78/4.8
Chorukh-Dajron	103/70	5/40	14/43	81/17	4.2/0.4	29/6
Geology-prospecting teams	132/150	9/45	17/37	74/18	3.5/0.4	143/6
Balance	507/476	22/52	19/32	59/18	2.1/0.25	143/12

[a]1. (1963–1966) — nominator; 2. (1967–1971) — denominator

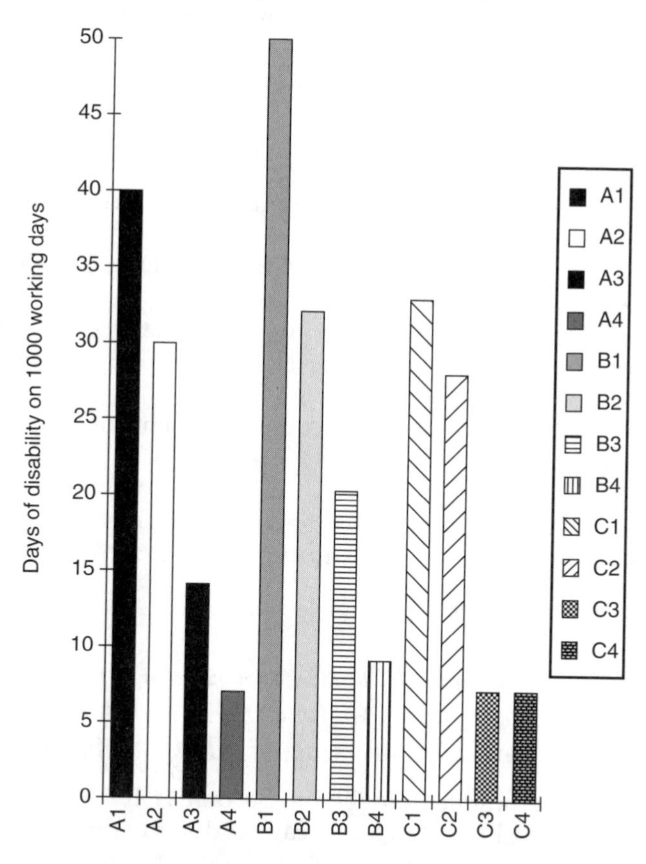

Figure 14.63 Sickness of different groups of miners in terms of the days of disability for two periods of observation with the higher and lower concentration 1 — high exposure (radon decay product concentration), general number of days of sickness 2 — low exposure (radon decay product concentration), general number of days of sickness 3 — high exposure, days in a hospital, A — for all miners, B — for drillers, C — for others.

Figure 14.63 Sickness of breathing organs of miners for two periods of observation: 1 — higher, 2 — lower concentration. A — for all miners; B — for drilling groups; C — for others. 1,2 — for general days of disability; 3,4 — for days in hospital.

Data in Figure 14.63 suggest that sickness of breathing organs in terms of days of disability for all workers has decreased by 23%, mainly due to the reduction of average duration of the case by 28% and in terms of numbers of days in hospital by 44%. In the first period, sickness of drilling workers was 44% higher than among others; in the second period, it was not a reliable difference in this index.

In general, these results suggest that it is a great possibility that correlation between concentration of radon decay products in the working sites and sickness of breathing organs in terms of days of disability of miners took place especially for drilling workers, that is, for the group with the higher intake.

14.12.1.3 *Comparison of Tadjikistan data with data from other epidemiological studies*

Analyses of many epidemiological studies are presented in Lubin (1994) and BEIR VI (1999) together with advantages and disadvantages in every study. It was also mentioned, especially in Ruzer, L.S., 2001, that the uncertainties in dose assessment are very high and no data are available on the individual dose of miners or at least for different groups of miners.

In this sense, data on nonuranium miners in Tadjikistan present a new approach to the problem:

- A large diversity in terms of radon progeny concentration for different groups of miners, with especially high concentrations for the drilling group, was pointed out.
- The parameter of physical activity, FAL, differs substantially within groups of miners, with the highest for the drilling group.
- As a result, the drilling group had the highest intake among other miners.
- Both lung cancer mortality and lung sickness of miners were highest for the drilling group.
- For the first time in such studies, direct measurements of the activity (dose) in the lungs of miners were used.

From these results, the conclusion should be drawn that it is incorrect to consider miners as a uniform cohort in the risk assessment study, both from a dosimetric and epidemiological point of view.

14.12.2 *Quality of dosimetry and the risk assessment for miners some aspects of the comparison of a "Joint Analysis of 11 Underground Miners Studies" and a study of nonuranium miners in Tadjikistan*

The direct measurement technique described in Section 14.10.11 presents an opportunity to measure directly the deposition of radon decay products in the lung and, due to the correlation between alpha-dose and gamma-activity, to determine the absorbed dose to the lung tissue.

Direct measurement also provides important information about the correlation between concentration in the air and activity in the lung. This transition coefficient called filtration ability of the lungs (FAL) is a product of minute volume rate and deposition coefficient. It also plays an important role in dosimetry of nonradioactive inhaled particles because it does not depend on radioactivity but depends on physical activity and particle size distribution.

The accuracy of dose assessment by direct measurement is the highest achievable (35 to 40% uncertainty), and for this reason it presents an opportunity to assess the optimal uncertainty in the dose under real conditions and, consequently, the uncertainty in risk assessment.

In simplest terms, the risk (R) in studies of miners is determined by the ratio of lung cancer mortality (M) and dose (D), or its usual substitute, exposure(E). This means that the reliability in the risk assessment, expressed in the fractional (relative) standard deviation $\delta=\sigma/\text{mean}$, depends on the standard deviation of mortality σ_M and dose σ_D. For noncorrelated errors,

$$\delta_R=\sqrt{(\delta_M)^2+(\delta_D)^2} \tag{14.77}$$

It follows that it makes no sense to improve the accuracy of one value (M or D) if the uncertainty in the other is much larger, because in this case δ_R is determined by the larger standard deviation.

In real underground conditions, dose D is a product of the concentration (exposure, E), breathing volume rate, V, and deposition coefficient, K, or their product FAL$=VK$. Therefore,

$$\delta_D=\sqrt{(\delta_E)^2+(\delta_V)^2+(\delta_K)^2}=\sqrt{(\delta_E)^2+(\delta_{FAL})^2} \tag{14.78}$$

where δ_E, δ_V, δ_K, and δ_{FAL} are the fractional standard deviations of the exposure, minute volume rate, deposition coefficient, and FAL, respectively.

As was shown in previous sections, the results of the Tadjikistan study suggested that the absorbed dose D depends on physical activity and, correspondingly, on FAL. For the purpose of comparison with Lubin (1994) and BEIR VI (1998), we simplify the problem and assume that dose directly correlated with exposure ($\delta_{FAL}=0$):

$$\delta_R=\sqrt{(\delta_E)^2+(\delta_M)^2} \tag{14.79}$$

In Lubin (1994) and BEIR VI (1998), a joint analysis of underground miners in China, Czechoslovakia, Colorado, Ontario, Newfoundland, Sweden, New Mexico, Beaverlodge, Port Radium, Radium Hill, and France was presented. The objective criteria for inclusion in the study of joint analysis were:

- A minimum of 40 lung cancer deaths
- An estimate of Rn progeny exposure in WLM for each member of the cohort based on historical measurements of either Rn or Rn progeny

Let us look at both these criteria from the point of view of reliability (uncertainty) in the risk assessment.

Criterion (1) suggests a willingness to tolerate 15% in mortality.

From Table D-12 of BEIR VI (1998), we can see that from the highest number of deaths (980) in the China study to the minimum by established criteria (40), the fractional statistical error in mortality σ_M will change only from 3 to 15%. And even when the number of deaths is in the range 20–30, σ_M will be in the range of 20%.

In contrast to the criteria that mortality must be known with an accuracy of about 15%, the criteria on dose (or exposure) are much more lenient.

Assessment of radon progeny concentration in mines is subject to an uncertainty of about ±50%. Furthermore, in most of the studies, exposure was reconstructed from work histories and historical radon measurements rather than being directly measured.

Taking into account that the contribution to the dose from radon itself is negligible in comparison with its progeny and that, according to Lubin (1994) and BEIR VI (1998), the range of the shift of equilibrium in mines varies from 0.2 to 0.9, we should assume that uncertainty in exposure assessment is in hundreds of percent.

It should also be pointed out that personal dosimetry even in terms of so-called "personal dosimeters," that is, devices that measure concentration in the breathing zone of miners, except in a French study with a very small number of deaths, were not provided.

From this point of view, data on WLM for all studies presented in Table D-12 of BEIR VI (1998) look completely unrealistic (accuracy up to tenths of WLM instead of at least ±50%).

Because statistical error in mortality is already much smaller than uncertainty in exposure, it makes no sense to decrease δ_M by including it in the risk assessment data from the early years when uncertainty in the exposure δ_E was very high.

One way of improving the accuracy of dose (exposure) assessment is by using the weighted average exposure where the weights are inversely proportional to the square of the variance:

$$w \sim 1/(\delta)^2$$

If these data are excluded, the number of deaths will decrease, but statistical error in mortality will still be much lower than the error in dose or exposure assessment.

There is a trade-off: by choosing the largest cohort and correspondingly the greatest number of deaths, to try to increase the statistical accuracy of mortality, we include cases with great dosimetric (exposure) uncertainty, which results in decreasing the reliability of risk assessment.

A professor of epidemiology said at a conference: "Unfortunately, not so many people died from this epidemic." Paradoxically, in the case of miner studies, by improving the mortality statistic we can make the risk assessment less reliable.

In Lubin (1994), a summary of the strengths and weaknesses of the various studies is presented. Among the strengths, the authors mostly mentioned the large cohort and the long follow-up of the studies. Among the weaknesses, limited exposure data, no or limited smoking data, and in six studies limited numbers of lung cancer were cited.

In the study of nonuranium miners in Tadjikistan, the number of deaths was even more limited: 34 (14 miners and 20 men from mining settlements). We can say, following the professor of epidemiology: "Unfortunately, only 14 miners died from lung cancer in our study." In this case, the fractional statistical error in mortality will be in the range of 40% for miners.

On the other hand, the assessment of dosimetric factors for all 11 yr of the Tadjikistan study was much better than in most of the studies:

- All instruments for radon and radon progeny measurements were properly calibrated with the accuracy of PAEC measurement better than 25 to 30%.
- Radon decay product concentration measurements were provided in different working sites and for different groups of workers.
- On hundreds of miners, direct measurements of activity (dose) in lungs were provided.
- By making around 300 measurements of the activity of radon progeny in the lungs of miners and a similar number of measurements of the concentration in the breathing zone, FAL was established for different physical activities (group of miners).

Thus, for the Tadjikistan study,

$$\delta_M = 40\% \text{ and } \delta_E = 30\%$$

implying δ_R=50%, which is less uncertainty than is present in the larger studies with poor dosimetry.

In the Tadjikistan study by dividing all mines into two categories, with an average annual concentration q<APC (annual permissible concentration) and q>APC, another goal was achieved.

In the control group, three categories of miners were included :

1. Men older than 19 yr from all cities of Tadjikistan
2. The same age nonminers from miner settlements
3. Miners who worked in mines with concentrations of the first group

All these three groups had very close mortality rates. One can suggest that smoking habits of groups 1 and 2 are different from miners. But it is likely that smoking habits for miners who work in mines with high radon concentrations are similar to those of miners who work in mines with lower concentrations.

Comparison of lung cancer for miners in high and low radon concentration mines thus allows the estimation of mortality as a function of cumulative exposure.

This chapter focuses on the importance of correct dosimetry, that is, measurement or calculation of dose factors with a critical assessment of uncertainty in measurement or calculation.

It focuses on dosimetry even in a broader sense, particularly on how values measurable in real mining conditions (in the case of miners, concentration of Rn and its decay product, or calculated exposure) are related to the main physical factor that determines the radiation damage to the lung-absorbed dose. It especially focuses on what kind of propagation errors occur, and what kinds of measures and methods can be used to diminish uncertainty in the dose, which makes data of risk assessment unreliable.

It seems to us that, based on the data and ideas presented in this chapter, some review of the data in epidemiological studies, including Lubin (1994) and BEIR VI (1998), should be made:

- Try to find some additional data on concentration and mortality among different groups of miners.
- Try to achieve more reliable data in the risk assessment (optimal value of σ_R) by choosing groups of miners with more reliable dosimetry.

References

Abelenzev, V.V., Sarsemdinoff, S.S., and Sevostyanov, V.N., Instrument for measuring the Volume Activity of Radon, in Proceedings of the National Academy of Sciences of Kazakhstan, Vol. 1, 1994, pp. 18–22 (in Russian).

Alavanja, M.C. et al., Residential radon exposure and risk of lung cancer in Missouri, *Am. J. Public Health*, 89, 1042–1048, 1999.

Alfred Cavallo et al., Radon progeny unattached fraction in an atmosphere far from radioactive equilibrium, *Health Phys.*, 76, 532–536, 1999.

Alterman, A.D., The Problems of Radiation Hygiene in Underground NonUranium Mines, Candidate's dissertation, Institute of Hygiene and Professional Diseases, Medical Academy of Science, Moscow, USSR, 1974, (in Russian).

Altshuler, B., Nelson, N., and Kuschner, M., *Health Phys.*, 10, 1137–1161, 1964.

Ansoborlo, E. et al., Review of the characterization of hot particles released into the environment and pathways for intake of particles, *Radiat. Prot. Dosim.*, 92, 139–143, 2000.

Antipin, N.I. et al., *Izmer. Tekh.*, Moscow, USSR No. 1, 5, 1980.

Antipin, N.I. et al., *Izmer. Tekh.*, Moscow, USSR No. 12, 1973.

Antipin, N.I. et al. Special state standard for the volumetric activity of radioactive aerosols, Translated from *Izmer. Tekh.* January, Moscow, USSR No. 1, 5–7, 1980.

Antipin, N.I., Kuznetzov, Y.V., and Ruzer, L.S., Single-channel Alpha-spectrometer for radon decay products concentration measurement, *At. Energ.*, 40, 14, 1976.

Aurand, K. et al., *Naturwissenchaften*, 42, 398, 1955.

Aurand, K., *Schraub. Stralen Therapie*, 94, 272, 1954.

Bail, Shapiro, Proceedings of the International Conference in Geneva on Peaceful Use of Atomic Energy, Vol. 13, 1955, p. 283.

Baltensperger, U. et al., *Atmos. Environ.*, 25A, 629–634, 1991.

Barbara Alexander, et al., Areas of the United States with elevated screening levels of ^{222}Rn, *Health Phys.*, 66, 55–54, 1994.

Bateman, H., *Proc. Cambridge Philos. Soc.*, 15, 423, 1910.

Belousova, I.M. and Shtukkenberg, Yu.M., *The Natural Radioactivity*, Medgiz, Moscow, USSR 1961.

Belyaev, S.P., Nikiforova, N.K., Smirnov, V.V., and Schchelchkov, G.I., *Optoelectronic method for the study of aerosols*, Energoizdat, Moscow, 1981 p. 189 (in Russian).

Bem, H. et al., Radon concentrations in kindergartens and schools in the Lodz Region of Poland, *Radiat. Prot. Dosim.*, 82, 147–149, 1999.

Berkovski, V. et al., Time-spatial peculiarities of dose formation after inhalation of alpha emitters, *Radiat. Prot. Dosim.*, 79, 387–390, 1998.

Bochiccio, F., Results of the representative Italian National survey on radon indoors, *Health Phys.*, 71, 741–748, 1996.

Boecker, B.B. et al., Evaluation of strategies for monitoring and sampling airborne radionuclides in the workplace, *Radiat. Prot. Dosim.*, 53, 69–71, 1994.

Bolotin, V.F., Grigorov, V.P., and Chutkin, O.A., *At. Energ.* Moscow, USSR 21, 1967.

Bolotin, V.F., Grigorov, V.P., and Chutkin, O.A., *Dosimetry and Radiometry of Ionizing Radiation*, Gosatomizdat, Moscow, USSR 1961, p. 157.

Boschung, M., The high purity germanium detector whole-body monitor at PSI, *Radiat. Prot. Dosim.*, 79, 481–484, 1998.

Bowie, Cameron et al., Radon and health, *Lancet*, 337, 1991: 409 (5 pages).

Boxue Liu et al., Determination of the volume activity concentration of alpha artificial radionuclides with alpha spectrometer, *Health Phys.*, 73, 938–931, 1997.

Breslin, A.J., George, A.C., and Weinstein, M.S., HASL-220, 1969.

Brown, E. and Firestone, R.B., in *Table of Radioactive Isotopes*, Shirley, V.S., Ed., Wiley-Interscience, New York, 1986.

Butterweck, G. et al., *Radiat. Prot. Dosim.*, 94, 247–250, 2001.

Butterweck, G. et al., *In vivo* measurement of unattached radon progeny deposited in human respiratory tract, *Radiat. Prot. Dosim.*, 94, 247–250, 2001.

Buvhall and James, A.C., Uncertainty analysis of the effective dose per unit exposure from radon progeny and implications for ICRP risk-weighting factors, *Radiat. Prot. Dosim.*, 53, 133–140, 1994.

Campbell, D. and Slatery, Ts., *Tech. Mech.*, 85, 1963.

Cavallo, A., The radon equilibrium factor and comparative dosimetry in homes and mines, *Radiat. Prot. Dosim.*, 92, 295–298, 2000.

Chamberlain, A.C. and Dyson, E.D., The dose to the trachea and bronchi from the decay products of radon and thoron, *Br. J. Radiol.*, 29, 317–325, 1956.

Chazel, V. et al., Effect of U3O8 specific surface area on *in vivo* dissolution, biokinetics, and dose coefficients, *Radiat. Prot. Dosim.*, 79, 39–42, 1998.

Chen Cing-an et al., Long-term monitoring of thorium inhaled by workers and assessment of thorium lung burden in China, *Radiat. Prot. Dosim.*, 79, 91–93, 1998.

Ching-Jang Chen et al., Discrimination of airborne radioactivity from radon progeny, *Health Phys.*, 66, 557–564, 1994.

Chung-Woo Ha et al., Dose assessment to inhalation exposure of indoor 222Rn in Korea, *Health Phys.*, 63, 453–456, 1992.

Cohen, B.L., Questionnaire study of the lung cancer risk from radon in homes, *Health Phys.*, 72, 615–622, 1997.

Cohen, Beverly S. et al., Deposition of charged particles on lung airways, *Health Phys.*, 74, 554–560, 1998.

Cohen, B. et al., Indoor radon maps in the United States, *Health Phys.*, 66, 201–205, 1994.
Cohn, H., Skow, R.K., and Gong, J.K., *Arch. Ind. Hyg. Occup. Med.*, 7, 508, 1953.
Cooper, L.A., Jackson, P.O., Langford, L.C., Petersen, M. R., and Stuart, B.O., *Characteristics of Attached Radon-222 Daughters Under Both Laboratory and Field Conditions ...*, Battelle Pacific Northwest Laboratories, Richland, WA, 1973.
Cumulative Contents and Index 1981–1993, *Radiat. Prot. Dosim.*, 45, 62–65, 1992; The Natural Environment, Proceedings of the Fifth International Symposium on the Natural Radiation Environment held in Saltzburg, Austria, September. 22–28, 1991.
Darby, S. et al., Risk of lung cancer associated with residential aerosol exposure in south-west England: a case-control study, *Br. J. Cancer*, 78, 394–408, 1998.
Degteva, M.O. et al., Retrospective dosimetry related to chronic environmental exposure, *Radiat. Prot. Dosim.*, 79, 155–160, 1998.
Dixon, D.W., Radon exposures from the use of natural gas in buildings, *Radiat. Prot. Dosim.*, 97, 259–264, 2001.
DOE/ER - 0488P, U.S. Department of Energy, 1991.
Dokukina, V.L. and Ruzer, L.S., All-Union Institute of Physico-Technical and Radiotechnical Measurements (VNIIFTRI), Moscow, 30(60), 1976 (in Russian).
Dokukina, V.L., Polev, N.M., and Ruzer, L.S., *Izmer. Tekh.*, 7, 60–61, 1974.
Domanski, T., Kluszczynski, D., Olszewski, J., and Chrusciewski, W., Field monitoring vs individual miner dosimetry of radon daughter products in mines, *Polish J. Occup. Med.*, 2, 147–160, 1989.
Dorrian, M.-D. and Bailey, M.R., Particle size distributions of radioactive aerosols measured in workplaces, *Radiat. Prot. Dosim.*, 60, 119–133, 1995.
Dua, S.K. and Hopke, P.K., Hygroscopicity of indoor aerosols and its influence on the deposition of inhaled radon decay products, *Environ. Int.*, 22 (Suppl. 1), 5941–5947, 1996.
Duport, P., Is the radon risk overestimated? Negected doses in the estimation of the risk of lung cancer in uranium underground miners, *Radiat. Prot. Dosim.*, 98, 329–338, 2002.
Eisemenger, et al., Monitoring of thorium incorporation by thoron in breath measurement: technical design of a routine method, *Radiat. Prot. Dosim.*, 79, 491–494, 1998.
El-Hady, M.A. et al., Radon progeny in Egyptian underground phosphate mines, *Radiat. Prot. Dosim.*, 95, 63–68, 2001.
Ennemoser, O. et al., High indoor radon concentrations in an Alpine region of western Tyrol, *Health Phys.*, 67, 151–154, 1994.
EPA Air Quality Criteria for Particulate Matter, Vols. II and III, 1996.
Espenskheid, W.F., Matyevic, E., and Kerker, M., *J. Phys. Chem.*, 68, 1964.
Espinosa et al., Assessment of doses to adult members of the public in Palomares from inhalation of plutonium and americium, *Radiat. Prot. Dosim.*, 79, 161–164, 1998.
Field, R.W. et al., The Iowa Radon Lung Cancer Study — Phase I: residential radon gas exposure and lung cancer, *Sci. Total Environ.*, 272, 67–72, 2001.
Field, R.W. et al., Retrospective temporal and spatial mobility of adult Iowa women, *Risk Anal.*, 18, 575–584, 1998.
Field, R.W. et al., Does exposure to residential radon increase the risk of lung cancer? Topic under debate, *Radiat. Prot. Dosim.*, 95, 75–81, 2001.
Fleishert, R.I. et al., Personal radon dosimetry from eyeglass lenses, *Radiat. Prot. Dosim.*, 97, 251–258, 2001.
Fuchs, N.A., *The Mechanics of Aerosols*, Pergamon, New York, 1964.
Fujimoto, K., Background of passive radon monitors, *Radiat. Prot. Dosim.*, 55, 273–277, 1994.
Fuchs, N.A. and Sutugin, A.G., *Kolloidn. Zh.*, 26, 110–115, 1964.
Fuchs N.A. and Sutugin, A.G., Highly dispersed aerosols, in *Scientific Results, Physical Chemistry Series*, VINITI, Moscow, 1969, (in Russian).
Fuchs, N.A., The Mechanics of Aerosols, Pergamon, New York, 1964.
Fuchs, N.A. et al., Stechkina, I.B., and Starosel'ski, *Inzh. Fiz. Zh.*, No. 5, 100, 1962.
Furuta, S. et al., A continuous radon progeny monitor with a vacuum vessel by alpha spectrometry, *Radiat. Prot. Dosim.*, 90, 429–435, 2000.
Gaggler, H.W. et al., *J. Aerosol Sci.*, 20, 557–564, 1989.

Gaidolfi, L. et al., Radon measurements in kindergartens and schools of six Italian regions, *Radiat. Prot. Dosim.*, 78, 73–76, 1998.
George, A. and Breslin, A.J., Deposition of radon daughters in humans exposed to uranium mine atmospheres, *Health Phys.*, 17, 115–124, 1969.
George, A.C. and Breslin, A.J., *Health Phys.*, 13, 375–378, 1967.
George, A.C., Hinchliffe, L., and Sladowski, R., *Am. Ind. Hyg. Assoc. J.*, 36, 484–490, 1975.
George, A.C. and Hinchliffe, L., *Health Phys.*, 23, 791–803, 1972.
George, A.C., Measurement of the uncombined fraction of radon daughters with wire screens, *Health Phys.*, 23, 390–392, 1972.
Gerasimov, Yu. S., Ruzer, L.S., Methods of Determining the amount of Radioactive Isotopes in Human Organism. Symposium Materials. Leningrad NII Radiation Hygiene, Leningrad, 1973, p. 58.
Gerken, M. et al., Models for retrospective quantification of indoor radon exposure in case-control studies. *Health Phys.*, 78, 268–278, 2000.
Giannardi, C. et al., In progress identification of radon prone areas: Toscana and Veneto, *Radiat. Prot. Dosim.*, 97, 349–354, 2001.
Gilbert, B. et al., Uranium dust concentration measured in a conversion plant by aerosol sampling and application for dose calculation, *Radiat. Prot. Dosim.*, 79, 77–81, 1998.
Giovani, C. et al., Radon survey in schools in North-East Italy, *Radiat. Prot. Dosim.*, 97, 341–344, 2001.
Gladkova, I.V. et al., Abstracts of Proceedings of the 17th Annual Conference on Nuclear Spectroscopy, Moscow, 1967.
Goldsmith, J.B., The residential radon-lung cancer association I U.S. counties: a commentary, *Health Phys.*, 76, 553–557, 1999.
Göran Pershagen et al., Residential radon exposure and lung cancer in Swedish women, *Health Phys.*, 63, 179–186, 1992.
Gormley, P.G. and Kennedy, M., *Proc. R. Ir. Acad.*, 52a, 163, 1949.
Grigorov, V.P., *At. Energ.*, 20 N(6), pp. 517–518, 1966.
Grivaud, L. et al., Measurement of performances of aerosol type radioactive contamination monitors, *Radiat. Prot. Dosim.*, 59, 495–497, 1998.
Groer, P.G. et al., Measurement of airborne radon daughters — Bayesian approach, *Radiat. Prot. Dosim.*, 69, 281–288, 1997.
Hadler, J.C. et al., Indoor radon daughter contamination monitoring: the absolute efficiency of CR-39 taking into account the plateout effect and environmental conditions, *Radiat. Prot. Dosim.*, 51, 283–296, 1994.
Harley, J.H. and Fresco, J.M., Retention of Radon Daughter Products in the Respiratory System, U.S. Atomic Energy Comission, Health and Safety Laboratory, Report No. 22, May 22, 1951.
Harley, N.H. and Cohen, B.S., in *American Chemical Society Symposium Series*, Vol. 331, Hopke, P.K. Ed., American Chemical Society, Washington, DC, 1986, p. 419.
Harley, N.H. and Pasternack, B.S., *Health Phys.*, 23, 771–782, 1972.
Harley, N.H. and Pasternack, B.S., *Health Phys.*, 42, 789–799, 1982.
Harley, N.H., *Radiat. Prot. Dosim.*, 7, 371–375, 1984.
Harley, N.H., Meyers, O.A., and Robbins, E.S., Biomodelling of Rn-222 decay product dose. *Health Phys.* (submitted) in 1998.
Hattori, T. et al., Equilibrium factor and unattached fraction of radon progeny in nuclear power *plants, Radiat. Prot. Dosim.*, 55, 191–197, 1994.
Hattori, T. et al., Behaviour of radon and its progeny in a Japanese office, *Radiat. Prot. Dosim.*, 62, 151–155, 1995.
Hattory, T. et al., Dose due to inhalation of radon progeny in Japanese nuclear power plants, *Radiat. Prot. Dosim.*, 58, 53–59, 1995.
Health Effect of Exposure to Radon. BEIR VI. National Academy Press, Washington, D.C. 1999.
Health Effect of Exposure to Radon. BEIR VI. National Academy Press, Washington, D.C. 1998.
Health Effect of Exposure to Radon, BEIR VI. National Research Council (1998), National Academy Press, Washington, D.C. 1998.
Hoffman, W., Overview of radon lung dosimetry, *Radiat. Prot. Dosim.*, 79, 229–236, 1998.

Holeidi, D.A. et al., *The Problem of Radon in Uranium Mines*, Gosatomizdat, Moscow, 1961 (in Russian).

Holleman, D.E., Martz, D.E., and Schiager, K.J., Total respiratory deposition of radon daughters from inhalation of uranium mine atmosphere, *Health Phys.*, 17, 187–192, 1969.

Holub, R.F. and Knutson, E.O., Measuring polonium-218 diffusion coefficient spectra using multiple wire screens, in *Radon and its Decay Products; Occurrence, Properties and Health Effects*, Hopke, P.K., Ed., ACS Symposium Series, Vol. 331. American Chemical Society, Washington, DC, 1987, pp. 340–356.

Hopke, P.K., Ed., *Symposium Series*, Vol. 331, American Chemical Society, Washington, DC, 1986, pp. 400–18.

Hopke, P.K., The initial behavior of ^{218}Po in indoor air, *Environ. Int.*, 15, 288–308, 1989.

Hopke, Philip K., Some thoughts on the unattached fraction of radon decay products, *Health Phys.*, 63, 209–212, 1992.

Kravchenko, I.I., Method and Instrumentation for the Measurement of Dispersion Composition of Aerosols in the Range of $5{\cdot}10^{-7}$–$5{\cdot}10^{-4}$ cm, Candidate's dissertation, VNIIFTRI, Moscow, USSR, 1974.

International Comission on Radiological Protection (ICRP), *Limits of Intakes of Radionuclides by Workers*, Pergamon Press, Oxford, ICRP Publication 30, Part I, Ann ICRP 2 (3/4), 1979.

Ishigure, N. et al., Validity of ^{241}Am a tracer of inhaled Pu in external chest counting, *Radiat. Prot. Dosim.*, 79, 133–136, 1998.

Ivanov, V.I., *Dosimetry of Ionizing Radiation*, Atomizdat, Moscow, 1964.

Ivanova, A.P. et al., *Izmer. Tech.*, April, No. 4, 74–76, 1974.

Jacobi, W., Aurand, K., and Schraub, A., Proceedings held in Stockholm, August 1956.

Jacobi, W., *Health Phys.*, 10, 1163–1175, 1964.

Jacoby, V., *Atmospheric Aerosols and Radioactive Pollution of Air*, 1964. Gidrometizdat (in Russia) Leningrad.

James, A.C., *A Reconsideration of Cells at Risk and other Key Factors in Radon Daughter Dosimetry*, American Chemical Society, Washington, DC.

John, S., Neuberger residential radon exposure and lung cancer: an overview of ongoing studies, *Health Phys.*, 63, 503–509, 1992.

Jon Miller, Mapping radon-prone areas by lognormal modeling of house radon data, *Health Phys.*, 74, 370–378, 1998.

Jonathan G. Price et al., Radon in outdoor in Nevada, *Health Phys.*, 66, 433–438, 1994.

Junge, H., *Chemical Composition and Atmospheric Radioactivity*, 1965. (Russian translation) Miz, Moscow.

Kartashev, N.P., *At. Energ.*, 24, 144, 1966.

Kartashev, N.P., Geophysical Collection, Nuclear Physical Research, No. 6, Ural Branch of the Academy of Sciences of the USSR, Sverdlovsk, 1967.

Keirim-Marku, I.B.S., A discussion around ecological studies of correlation between indoor radon concentration and lung cancer mortality by Cohen, *Radiat. Biol. Radioecol.*, 40, 465–470, 2000.

Knutson, E.O., Modeling indoor concentrations of radon's decay products, in *Radon and its Decay Products in Indoor Air*, Nazaroff, W.W. and Nero, A.V., Eds., John Wiley, New York, 1988, pp. 161–202.

Knutson, E.O., Tu, K.W., Solomon, S.B., and Strong, L., *Radiat. Prot. Dosim.*, 24, 261–264, 1988.

Knutson, Earl O., History of diffusion batteries in aerosol measurement, *Aerosol Sci. Technol.* 31, 83–128, 1999.

Kogan, Ya.I. and Burnashova, A.V., *Zh. Fiz. Khimii*, Moscow, USSR 39, 1959.

Kojima, H. and Abe, S., *Radiat. Prot. Dosim.*, 24, 241–244, 1988.

Kolerskii, S.V. and Ruzer, L.S., *Izmer. Tekh.*, October, Moscow, USSR No. 10, 37–40, 1968.

Kolerskii, S.V., Methods and Instrumentation for the Measurement of the Electric Parameters of Aeroions and High Dispersed Aerosols, Candidate's dissertation, VNIIFTRI, Moscow, USSR, 1975.

Kolerskii, S.V., *Izmer. Tech.*, Moscow USSR, No. 2, 1974.

Kolerskii, S.V., Lekhtmakher, S.O., Polev, N.M., and Ruzer, L.S., Proceedings of the 9th Republican Conference on Gas Dynamics of the Dispersed Systems, Odessa, USSR, 1969.

Kolerskii, S.V., Kuznetzov, Yu.V., Polev, N.M., and Ruzer, L.S., *Izmer. Tekh.*, October, Moscow USSR No. 10, 57–58, 1973.

Kotrappa, P. et al., Application of NIST 222-Rn, Emanation standards for calibrating ^{222}-Rn monitors, *Prot. Dosim.*, 55, 211–218, 1994.

Kravchenko, I.I. and Ruzer, L.S., Reproducing the scale of aerosol-particle dimensions, *Izmer. Tekh.*, April, No. 4, 65–67, 1978.

Kravchenko, I.I., *Izmer. Tech.*, February, Moscow USSR No. 2, 52–53, 1983.

Kravchenko, I.I., Lekhtmakher, S.O., and Ruzer, L.S., *Kolloidn. Zh.*, 33, 923, 1971.

Krewski, D. et al., Characterization of uncertainty and variability in residential radon cancer risks, *Ann. N.Y. Acad. Sci.*, 895, 245–272, 1999.

Krienbrock, L. et al., Case–control study on lung cancer and residential radon on Eastern Germany, *Am. J. Epidemiol.*, 153, 42–52, 2001.

Kushneva, V.S., *Study of Toxicology of Radioactive Substances*, Moscow, USSR 1957, pp. 130–148 (in Russian).

Kutkov, V.A., Application of human respiratory tract models for reconstruction of the size of aerosol particles through the investigation of radionuclide behavior in the human body, *Radiat. Prot. Dosim.*, 79, 265–268, 1998.

Kutkov, V.A., Results of *in vivo* monitoring of the witnesses of the Chernobyl accident, *Radiat. Prot. Dosim.*, 89, 193–197, 2000.

Kuznets, H.H., *Am. Ind. Hyg. Assoc. Q.*, N(5) 17, 1956.

Kuznetzov, Yu.V., Proceedings of VNIIFTRI, No. 36 (66), Moscow, 1978, p. 81.

Labushkin, V.G. and Ruzer, L.S., A method of determining the concentration of the short-lived daughter products of radon in the air by the alpha- and beta-radiations, *At. Energ.*, 19, 24–28, 1965.

Labushkin, V.G., Candidate's dissertation, VNIIFTRI, Moscow, 1967.

Labushkin, V.G. and Ruzer, L.S., *At. Energ.*, 19, 24–28, 1965.

Labushkin, V.G., Polev, N.M., and Ruzer, L.S., Determining the self-absorption of alpha-radiation in a sample during air filtration., *At. Energ.*, 19, 39, 1965.

Labushkin, V.G., Popov, V.I., and Ruzer, L.S., *Izmer. Tekh.*, July, Moscow USSR No. 7, 59–61, 1973.

Lagarde, F. et al., Residential radon and lung cancer among never-smoker in Sweden, *Epidemiology*, 12, 396–404, 2001.

Lagarde, F. et al., Residential radon and lung cancer in Sweden: risk analysis accounting for random error in the exposure assessment, *Health Phys.*, 72, 269–276, 1997.

Layton, D.W., Methodically consistent breathing rate for use in dose assessments, *Health Phys.*, 64, 23–26, 1993.

Leites, F.L. and Ruzer, L.S., *Arkiv Patologii*, Moscow USSR No. 1, 20–27, 1959 (in Russian).

Lekhtmakher, S.O. and Ruzer, L.S., Estimation of the errors in determining the uptake of radioactive isotopes in the respiratory organs, *Izmer. Tekh.*, May, Moscow USSR No. 5, 75–76, 1975.

Lekhtmakher, S.O., Candidate dissertation, VNIIFTRI, Moscow, 1977.

Lekhtmakher, S.O., *Kolloidn. Zh.*, 49, 349–351, 1987.

Lekhtmakher, S.O., *Kolloidn. Zh.*, 47, 48–53, 1985.

Lekhtmakher S.O. and Ruzer, L.S., *Kolloidn. Zh.*, 34, 805–807, 1972.

Lekhtmakher, S.O., Polev, N.M., Ruzer, L.S., and Stoyanova, *Ing. Physic. J.*, XXII, 155, 1972.

Lekhtmakher, S.O. and Ruzer, L.S., *Izmer. Tekh.*, Moscow USSR No. 5, 1975.

Leonard A. Cole, *Element of Risk, The Politics of Radon*, USA 1993.

Letourneau, E.G. et al., Case–control study of residential radon and lung cancer in Winnipeg, Manitoba, Canada, *Am. J. Epidemiol.*, 140, 310–322, 1994.

Lipztein, J.L., et al., The Goiânia 137Cs accident — a review of the internal and cytogenetic dosimetry, *Radiat. Prot. Dosim.*, 79, 149–154, 1998.

Lubin, J.B., Errors in exposure assessment, statistical power and the interpretation of residential radon studies. *Radiat. Res.*, 144, 329–341, 1995.

Lubin, J.H. et al., Estimating lung cancer mortality from residential radon using data for low exposures of miners, *Radiat. Res.*, 147, 126–134, 1997.

Lubin, J.H. et al., Radon exposure in residences and lung cancer among women: combined analysis of three studies, *Cancer Causes and Control*, USA 1994.

Lubin, J.H. et al., Lung cancer risk from residential radon: meta-analysis of eight epidemiological studies, *J. Nat. Cancer Inst.*, 89, USA 1997.

Lubin, Jay H., On the discrepancy between epidemiological studies in individuals of lung cancer and residential radon and cohen's ecological regression, *Health Phys.*, 75, 4–10, 1998.

Mahaffey, J.A. et al., Estimating past exposure to indoor radon from household glass radon, *Health Phys.*, 64, 381–391, 1993.

Malikov, M.F., *Fundamentals of metrology*, Izd. Trudrezervizdat, Moscow USSR 1949, p. 185 (in Russian).

Marenny, A.M. et al., Estimation of the radon-induced dose for Russian population: methods and results, *Radiat. Prot. Dosim.*, 90, 403–408, 2000.

Markov, K.P., Ryabov, N.V., and Stas', K.N., Proceedings of the SNIIP, Vol. 2, Gosatomizdat, Moscow, 1965, p. 315.

Markov, K.P., Ryabov, N.V., and Stas', K.N., A Rapid method for estimating the radiation hazard associated with the presence of radon daughter products in air, *At. Energ.*, 12, 315, 1962.

Martonen et al., Cigarette smoke and lung cancer, *Health Phys.*, 52, 213–217.

Masahiro Doi et al., Vertical distribution of outdoor radon and thoron in Japan using a new discriminative dosimeter, *Health Phys.*, 67, 385–392, 1994.

McLaughlin, J.P., The application of techniques to assess radon exposure retrospectively, *Radiat. Prot. Dosim.*, 78, 1–6, 1998.

Melonni, B. et al., Radon and domestic exposure, *Rev. Mal. Resperratoires*, 17, 1061–1071, 1999, (in French).

Muirhead, Colin R., Radon risks, *Lancet*, USA 344, 1994, 143.

National Council of Radiation Protection and Measurements, Exposures from the Uranium Series with Emphasis on Radon and its Daughters, NCRP Report 77, Washington, DC, 1984.

National Research Council, Committee on Biological Effects of Ionizing Radiations. Health risks of radon and other internally deposited alpha-emitters; BEIR 4, National Academy Press, Washington, DC, 1988.

Nazaroff, W.W., Appendics: measurement techniques, in *Radon and its Decay Products in Indoor Air*, Nazaroff, W.W. and Nero, A.V., Eds., John Wiley, New York, 1988, pp. 491–504.

Nazaroff, W.W., Optimizing the total-alpha three-count technique for measuring concentrations of radon progeny in residences, *Health Phys.*, 46, 395–405, 1984.

NCRP, Evaluation of Occupational and Environmental Exposures to Radon and Radon Daughters in the United States, National Council on Radiation Protection and Measurements Report No. 78, NCRP Publications, Bethesda, MD, 1984.

NCRP, Exposures from the Uranium Series with Emphasis on Radon and its Daughters, National Council on Radiation Protection and Measurement Report No. 77, NCRP Publications, Bethesda, MD, 1984.

Nemmar, A. et al., Brief rapid communication, 2002.

Nesmejanov, An. N., et al., *Practical Handbook on Radiochemistry*, Goskhimizdat, Moscow, 1956, p. 389.

Neuberger, John S. et al., Residential radon exposure and lung cancer: evidence of an inverse association in the Washington State, *J. Environ. Health*, 53, 25, 1990.

Nicholas P. Thiessen, Alpha particle spectroscopy in radon/thoron progeny measurements, *Health Phys.*, 67, 632–640, 1994.

Nikl, I., The radon concentration and absorbed dose rate in Hungarian dwellings, *Radiat. Prot. Dosim.*, 67, 225–228, 1996.

Novozhilova, D.V., *Kollolidn. Zh.*, 25, 175–177, 1963.

NRC, Health Risks of Radon and Other Internally Deposited Alpha-Emitters, National Research Council Report, BEIR 4, National Academy Press, Washington, DC, 1988.

NRC, Comparative Dosimetry of Radon in Mines and Homes, National Research Council Panel on Dosimetric Assumptions Affecting the Applications of Radon Risk Estimates, National Academy Press, Washington, DC, 1991.

Nuccetelli, C. et al., The thoron issue: monitoring activities, measuring techniques and dose conversion factors, *Radiat. Prot. Dosim.*, 78, 59–64, 1998.

Obersted, S. et al., The bronchial dosimeter, *Radiat. Prot. Dosim.*, 59, 285–290, 1995.

Osborne, P.V. and Hill, C.R., *Nucl. Instr. Methods*, 29, 1964.

Papp, Z. et al., Measurement of radon decay products and thoron decay products in air by beta counting using end-window Geiger–Muller counter, *Health Phys.*, 72, 601–610, 1997.

Petukhov, B.S., *Heat Exchange and Resistance to the Laminar Flow in Tubes*, Energia, Moscow, USSR, 1967, p. 219, (in Russian).

Phillips, C.R., Khan, A., and Leung, H.Y., The nature and determination of the unattached fraction of radon and thoron progeny, in *Radon and its Decay Products in Indoor Air*, Nazaroff, W.W. and Nero, A.V. Eds., John Wiley, New York, 1988, pp. 203–256.

Pisa, Federica E. et al., Residential radon and risk of lung cancer in an Alpine area, *Arch. Environ. Health*, 56, 208, 2001.

Pistikopoulos, P., Wortham, H.M., Gomes, L., Masclet-Beyne, S., Bon Nguyen, E., Masclet, P.A., and Mouvier, G., Mechanisms of formation of particulate polycyclic aromatic hydrocarbons in relation to the particle size distribution; effect on meso-scale transport, *Atmos. Environ.*, 24A, 2573–2584, 1990.

Polev, N.M., Candidate's dissertation, VNIIFTRI, Moscow, 1967.

Polev, N.M. and Ruzer, L.S., *At. Energ.*, 25, 227–228, 1968.

Porstendorfer, J., *ACS Symposium Series*, Vol. 331, American Chemical Society, Washington, DC, 1987, pp. 285–300.

Lubin, J.H., Radon and Lung Cancer Risk; A Joint Analysis of 11 Underground Miners Studies, National Institute of Health, Bethesda, MD, U.S.A., Publication No. 94–3644, 1994.

Radiat. Prot. Dosim., 79, 1998.

Raghavayya, M. and Jones, J.H., *Health Phys.*, 26, 417–430, 1974.

Reineking, A. et al., *Health Phys.*, 58, 717–727, 1990.

Reineking, A. et al., *Sci. Total Environ.*, 45, 261–270, 1985.

Risica, S., Legislation on radon concentration at home and at work, *Radiat. Prot. Dosim.*, 78, 15–21, 1998.

Risica, S. et al., Experimental and measurement issues in natural radioactivity, *Radiat. Prot. Dosim.*, 97, 345–348, 2001.

Ruzer, L.S., Nero, A.V., and Harley, N.H., Assessment of lung deposition and breathing rate of underground miners in Tadjikistan, *Radiat. Prot. Dosim.*, 58, 261–268, 1995.

Ruzer, L.S. and Labushkin, V.G., The Method of Determining the Concentrations of the Short-lived Daughter Products of Radon, Patent No. 171478, Bulletin of Inventions, Vol. 11, Moscow, USSR, 1966 (in Russian).

Ruzer, L.S. and Sextro, R.G., *Radiat. Prot. Dosim.*, 71, 135–140, 1997.

Ruzer, L.S., and Urusov, S.A., *At. Energ.*, 26, 301–303, 1969.

Ruzer, L.S., Determination of degrees of equilibrium of short-lived radon in the air, *At. Energ.*, 8, 557–559, 1960.

Ruzer, L.S., Radioactive Aerosols, Determination of the Absorbed Dose, Doctoral dissertation, Moscow, USSR, 1970 (in Russian).

Ruzer, L.S., Radioactive Aerosols, Energoatomizdat, Moscow, Russia, 2001, pp. 230 (in Russian).

Ruzer, L.S., Estimate of dose in the inhalation of radon. *At. Energ.*, February, 189–194, 1958, (in English).

Ruzer, L.S., Gamma-control in the inhalation of radon, *At. Energ.*, 13, 384–385, 1962.

Ruzer, L.S., *Radiat. Prot. Dosim.*, 46, 127–128, 1993.

Ruzer, L.S., Radioactive Aerosols, Determination of the Absorbed Dose, Doctoral dissertation, Moscow, USSR, 1970 (in Russian).

Ruzer, L.S., *Radioactive Aerosols*, Moscow, 1968, 191pp. (in Russian).

Ruzer, L.S., The Manner of Determination of the Absorbed Dose upon the Inhalation of Radon, USSR Patent No. 165250, Bulletin of Inventions No. 18, 1964, (in Russian).

Ruzer, L.S., The Method of Determining the Concentration of the Gas, Patent No. 234746, *Bulletin of Inventions* No. 28, 1964 (in Russian).

Ruzer, L.S., Determination of absorbed doses in organisms exposed to emanations and their daughter products, *At. Energ.*, 8, 542–548, 1960.

Ruzer, L.S. and Sextro, R.G., Measurement of radon decay products in air by alpha and beta spectrometry, *Radiat. Prot. Dosim.*, 72, 43–48, 1997.

Ruzer, L.S. and Sextro, R.G., ANRI, No. 3, 1999, pp. 18–20 (in Russian).

Ruzer, L.S. and Sextro, R.G., *New Information Technologies in Medicine and Ecology*, Gurzuf, Ukraine, May 1997.
Ruzer, L.S. and Urusov, S.A., Determining the lung irradiation dose from radon disintegration products from the chest gamma radiation, *At. Energ.*, 26, 301–303, 1969 (in English).
Sadikov, I.N., *Engl. Phys. J.*, Moscow, USSR 12, 1967.
Samet, Jonathan M. et al., Indoor radon and cancer, *New Engl. J. Med.*, 320, 591, 1989.
Scheler, R. et al., Retrospective estimation of exposure to short-lived 222Rn progeny by measurements of 210Po in the skull, *Radiat. Prot. Dosim.*, 79, 129–132, 1998.
Schulte, H.F., *Personal air sampling and multiple stage sampling*, Los Alamos Scientific Laboratory, Los Alamos, New Mexico, U.S.A., Symposium Radiation Dose Measurements, Paris, 1967.
Schulte, H.F., *Personal Air Sampling and Multiple Stage Sampling*, Interpretation of Results from Personal and Static Air Samplers, Los Alamos Scientific Laboratory, Los Alamos, New Mexico, U.S.A., Proceedings of the Conference, Paris, 1968.
SENES Consultants, Limited, Uncertainty in exposure of underground miners to radon daughters and the effect of uncertainty on risk estimates, Report to Atomic Energy Control Board, Ottawa, 1989.
Sherwood, R.J. and Greenhalgh, D.M.S., A personal air sampler, *Ann. Occup. Hyg.*, 2, 127-132, 1960.
Shtukkenberg, Yu.M., *At. Energ.*, 5, 124, 1956.
Sinclair, D. and La Mer, V., *Chem. Rev.*, 44, 1949.
Skowronek, J., Radiation exposures to miners in Polish coal mines, *Radiat. Prot. Dosim.*, 82, 293–300, 1999.
Sobue, T. et al., Residential radon exposure and lung cancer risk in Misasa, Japan: a case-control study, *J. Radiat. Res.*, 41, 81–92, 2000.
Solomon, S.B., A radon progeny sampler for the determination of effective dose, *Radiat. Prot. Dosim.*, 72, 31–42, 1997.
Strong, J.C. et al., The regional lung deposition of thoron progeny attached to the particulate phase of environmental tobacco smoke, *Radiat. Prot. Dosim.*, 54, 43–56, 1994.
Takonamii, S. et al., Continuous radon monitor using a two-filter method, *Radiat. Prot. Dosim.*, 63, 123–126, 1996.
The Norms of Radiation Safety (NRS-96), Moscow, Russia, 1999.
Thime, M. et al., European whole body counter measurement intercomparison, *Health Phys.*, 74, 465–471, 1998.
Thomas, J.W. and Hinchliffe, J., *Aerosol Sci.* 3, 387–393, 1972.
Tomasek, L. et al., Radon exposure and lung cancer risk Czech cohort study on residential radon, *Sci. Total Environ.*, 272, 43–51, 2001.
Tomassino, L., Passive sampling and monitoring of radon and other gases, *Radiat. Prot. Dosim.*, 78, 55–58, 1998.
Tsivoglou, E.C. and Ayer, H.E., *Arch. Ind. Hyg. Occup. Med.*, 8, 125, 1953.
United State Environment Protection Agency, Home Buyer's and Seller's to Radon, 1993.
Urusov, S.A., Method and Measurement Technique for Determination of the Intake of Radon Decay Products in the Lungs of Underground Workers, Candidate's dissertation, Biophysics Institute of the USSR Ministry of Health, Moscow, 1972, (in Russian).
U.S. Environmental Protection Agency, Technical Support Document for the 1992 Citizen's Guide to Radon, EPA-400-R-92-011, U.S. Government Printing Office, Washington, DC, May 1992.
Vasin, V.A., Zalmanson, Yu.E., Kuznetzov, Yu.V., Lekhtmakher, S.O., Ruzer, L.S., and Sidorov, V.V., Phantom measurements for the determination of the intake of the radon decay products in the lungs by direct method, *Proc. VNIIFTRI*, 22, 46–47, 1975.
Vaupotic, J. et al., Radon exposure in Slovene kindergartens based on continuous radon measurements, *Radiat. Prot. Dosim.*, 95, 359–364, 2001.
Vaupotic, J. et al., Radon exposure in Slovenian spas, *Radiat. Prot. Dosim.*, 97, 265–270, 2001.
Volkova, E.A., Ziv, D.U., Labushkin, V.G., Ruzer, L.S., Stepanov, E.K., and Tyutikov, N.V., *Prib. Tek. Eksp.*, 4, 36–39, 1966.
Vonstille, W.T. et al., *J. Environ. Health*, 53, 25, 1990. Radon and cancer: Florida study? Finds no evidence of increased risk.

Wasiolek, P.T. et al., Outdoor radon dose conversion coefficient in South-western and South-eastern United States, *Radiat. Prot. Dosim.*, 79, 269–278, 1995.

Webb, J.L. et al., An evaluation of recent lung counting technology, *Radiat. Prot. Dosim.*, 89, 325–332, 2000.

White, S.B. et al., Indoor ^{222}radon concentration in a probability sample of 43,000 houses across 30 states, *Health Phys.*, 62, 41–50, 1992.

Yamasaki, H., Kuwate, K., and Miyamoto, H., Effect of ambient temperature on aspects of airborne polycyclic aromatic hydrocarbons, *Environ. Sci. Technol.*, 16, 189–194, 1982.

Yu, K.N., et al., A portable bronchial dosimeter for radon progenies, *Health Phys.*, 75, 147–152, 1998.

Yu, K.N. et al., The effects of positive and negative ions on the lung dose from environmental radon, *Radiat. Prot. Dosim.*, 58, 65–68, 1995.

Zalmanson, Yu.E. et al., Geterogene Phantom of the Human Torso, Patent No. 402070, Bulletin of Inventions No. 41, 1973.

Zhivet'ev, V.M., Labushkin, V.G., and Ruzer, L.S., *At. Energ.*, 20, 511, 1966.

chapter fifteen

Dosimetry and epidemiology of Russian uranium mines

I.V. Pavlov
VNIPI PT, Moscow, Russia

Contents

15.1 Individual radon decay products concentration distribution in the 1090 dwellings of the village near Krasnokamensk from 1990 to 1991. The maximum level of radon progeny concentration — 20,000 Bq m^{-3} 503
15.2 Method and instrumentation for the integral volume activity of radon progeny and long-lived nuclides for a long period of time (from 8 h to 3 months) 504
15.2.1 A sampling device for breathing zone measurements of the integral volume activity of radon and thoron progeny, long-lived nuclides, and also nonradioactive aerosols 504
15.2.1.1 General thesis 504
15.2.1.2 Theoretical basis of the method 505
15.3 Epidemiological data on miners in the city of Lermontov, Russia 509
15.4 Individual dose distribution of uranium mines personnel (2500 workers) to radon progeny, long-lived nuclides, and external gamma-radiation from 1990 to 1994 in the city of Krasnokamensk, Chita region, Siberia 513
Acknowledgment 515
References 515

15.1 Individual radon decay products concentration distribution in the 1090 dwellings of the village near Krasnokamensk from 1990 to 1991. The maximum level of radon progeny concentration — 20,000 Bq m^{-3}

Settlement Oktjabr'skij is located at a distance of 13 km from the city of Krasnokamensk, Chita province in Siberia, Russia, in direct proximity (<1000 m) to two uranium mines. The area of the settlement is 1×1.5 km, and it has been built over, mainly with one-story wooden two-room buildings. In the past, it was a base for geology-prospecting expeditions. Measurements of radon progeny concentration indoors have been provided from 1987. The maximum level of radon decay product concentration was around 20,000 Bq m^{-3}. The data of Table 15.1 (Pavlov et al., 2001) are related to 1990–1991, when the levels

1-56670-611-4 / 05 / $0.00+$1.50

Table 15.1 Distribution of the Annual Average Equivalent Equilibrium Volume Activity (EEVA) and Effective Dose

EEVA (Bq m^{-3})	Number of Apartments	Calcuated Average Dose (mSv/year)
<200	951	4
201–300	47	10
301–500	26	17
501–700	20	25
701–960	7	34
961–1400	12	46
1401–1930	7	70
1931–2760	2	90
2761–3860	7	140
3861–4960	3	180
4961–6100	8	230
All	1090	9

of radon progeny concentration extended to 6000 Bq m^{-3}. After that high level was discovered, all habitants of the houses were evicted.

In the northern part of the settlement at a depth of 120–180 m, the mining pits of one of the mines are located. Here, in the dwellings, the variations of high radon progeny concentration were discovered after the system of forced ventilation was stopped. At the same time, the highest levels of radon progeny concentration occur in the southern part of the settlement, where there are no mine-working sites. Here, in calm weather, levels of radon progeny concentration outdoors higher than 200 Bq m^{-3} were observed.

15.2 *Method and instrumentation for the integral volume activity of radon progeny and long-lived nuclides for a long period of time (from 8 h to 3 months)*

15.2.1 *A sampling device for breathing zone measurements of the integral volume activity of radon and thoron progeny, long-lived nuclides, and also nonradioactive aerosols*

15.2.1.1 *General thesis*

For the assessment of doses caused by airborne short-lived radon decay products (DPR), it is necessary to have data on the average value of the equivalent equilibrium volume activity (EEVA) of radon for the long period of observation. In cases when values of the EEVA of radon are very variable in time and space (e.g., in uranium mines), the most representative results can be obtained by measurements in the breathing zone by wearing devices.

In order to practically solve this problem, serious difficulties should be overcome. If we try to use the portable aerosol radiometer, it is necessary to collect a large number of air samples. Automatic stationary monitors of EEVA of radon with bend drown-out mechanisms are too expensive. The most acceptable measurement technique, which uses the method of continuous air sampling on the filter with simultaneous registration of the filter alpha-activity, is that of the trek or semiconductor detectors. Among the disadvantages of these methods are the self-absorption in the dust deposit and the short range of the permissible density of treks in detectors from cellulose nitrate.

This method of measurement of the integral EEVA of radon in air for a very long period of time [1] does not require activity of the filter registration during sampling, and, moreover, has a much broader range of measurement.

This method is based on the correlation between the value of EEVA of radon decay products in the air and the number of Pb-210 atoms (RaD) on the filter, while the value of the proportional coefficient is approximately the same in practically all cases of "equilibrium coefficients." As a result of the decay of RaD on the filter, Po-210 (RaF) is formed, for which alpha-radiation can be measured by the trek detector.

The increase in RaF activity on the filter will continue after air sampling. The last circumstance allowed one to measure not only EEVA DPR but also the volume activity of the long-lived alpha-radionuclides in air: U-238, Th-232, and also plutonium-239. For this purpose, it is necessary to make the second measurement not later than 180 days after air sampling.

The main problem of the long-lived radionuclide activity measurement, deposited on the filter, is the restriction in air volume due to the possibility of alpha-particles in the dust being deposited on the filter. So, in order to obtain the integral values of the volume activities of radionuclides in air for a long period of time, the sampling devices should work in a repeated short-time manner, in which the value of the "porosity" (ratio of work/work+pause) should be established according to the expected mass concentration at the point of control.

The method of sampling and measurement in the general case includes the following operations:

1. The short-time (1–2 h) sampling of the large-volume (1–2 m^3) air sample at the point of control, with subsequent determination of EEVA DPR, mass concentration, and the summary volume activity of the long-lived radionuclides on the sample (after exposure of the filter for 6 days in the non-radon camera) by known methods. The optimal volume of the sample, and also the duration (if necessary) of the second measurement activity of the filter by the trek detector, is calculated.
2. Setup of the calculated regime of "porosity," taking into account the neccesary duration of the sampling.
3. Air sampling:
   (a) Filter exposure in the non-radon camera for 6 days.
   (b) Filter together with trek detector exposure during a chosen time for the first measurement.
   (c) Calculation of the number of treks measured in the first measurement, and correction of the calculated duration of the second measurement.
   (d) Providing the second measurement and treatment of the measurement results.

15.2.1.2 *Theoretical basis of the method*

In order to choose optimal technical characteristics of the sampling device, the following were chosen: dependence of the number of atoms of Pb-210 (RaD), Bi-210 (RaE), and Po-210 (RaF) in the filter from the "age" of radon in the air sample ("equilibrium factor"), radon volume activity (VA), duration and speed of filtration, and also time after sampling completion.

In column 8 of Table 15.2, the number of atoms of RaD on the filter is shown after filtration of 1 Bq EEVA of radon and complete decay deposits of RaA, RaB, and RaC. The age of radon and corresponding "equilibrium factor" are shown in columns 1 and 2 of Table 15.2. Calculation is made according to known formulas, which describe the buildup and decay of radon decay products in air with the radon decay constant and without deposition of radon decay products. The decay of RaD in the filter is not taken into account due

Table 15.2 Number of Atoms of RaD on the Filter per 1 Bq of Radon Depending on the "Age" of Radon

Radon "Age"	Equilibrium Factor	Number of Atoms of RaD on the Filter per 1 Bq of Radon				Number of Atoms of RaD on the Filter per 1 Bq	
		RaA	RaB	RaC	RaD	Rh	EEVA
1	2	3	4	5	6	7	8
5	0.098	179	116	5	0,5	301	3070
10	0.174	237	329	31	3,7	601	3452
15	0.242	255	554	78	13	902	3725
20	0.305	261	763	145	33	1202	3942
25	0.365	263	949	223	65	1501	4112
30	0.432	264	1116	310	112	1802	4172
35	0.474	264	1262	401	175	2102	4434
40	0.524	264	1390	495	254	2402	4584
50	0.611	264	1601	675	460	3000	4909
60	0.685	264	1766	844	723	3597	5251
70	0.746	264	1891	994	1047	4196	5624
80	0.797	264	1988	1123	1419	4795	6016
90	0.838	264	2065	1233	1834	5396	6439
100	0.871	264	2123	1325	2284	5996	6884
120	0.920	264	2202	1461	3265	7192	7817

to its small influence on the result of the assessment, if all the cycles of air sampling and measurement of the activity do not exceed the 1 yr period.

Data in Table 15.2 show that the value of the proportional coefficient between numbers of RaD atoms in the filter and composition of EEVA of radon decay products and volume of the sample can be constant and equal to 4400 at/Bq EEVA. Associated uncertainty in the range 0.2–0.7 for the "equilibrium factor" will not exceed 10%.

For the calculation of the radioactive chain transformation of RaD → RaE → RaF, in the filter, the following parameters were used: E — average EEVA of radon in the sample (Bq/m^3); ω — sample speed (m^3/days); t — time from the beginning of the sampling (days); T_s — sampling duration (days); τ — time from the end of the sampling (days); N_D, N_E, N_F — number of atoms of RaD, RaE, and RaF; λ_D, λ_E, λ_F — decay constants of RaD, RaE, and RaF, 8.516×10^{-5}, 0.1384, and 0.005 days^{-1}, respectively; n_d — number of RaD atoms per 1 Bq EEVA of radon in the air sample.

During air sampling, the chain transformation RaD → RaE → RaF on the filter is described by the system of equations

$$\frac{dN_D}{dt} = N_d E \omega t \quad (15.1)$$

$$\frac{dN_E}{dt} = N_D \lambda_D - N_E \lambda_E \quad (15.2)$$

$$\frac{dN_F}{dt} = N_E \lambda_E - N_F \lambda_F \quad (15.3)$$

A solution of these equations for the moment of the sampling completion ($t=T_S$, $\tau=0$) after substitution of the numerical values of, λ_D, λ_E, λ_F, and $N_d=4400$ will be as follows:

$$N_D^0 = 4400 E \omega T_S \quad (15.4)$$

$$N_E^0 = 2.71 E\omega T_S\left[1 - \frac{7.23}{T_S}(1 - e^{-0.1384T_S})\right] \quad (15.5)$$

$$N_F^0 = 75 E\omega T_S\left[1 + \frac{0.271}{T_S}(e^{-0.005T_S} - e^{-0.1384T_S}) - \frac{207}{T_S}(1 - e^{-0.005T_S})\right] \quad (15.6)$$

During the time after air sampling, transformations in the chain of RaD → RaE → RaF on the filter are described by the system of equations

$$\frac{dN_E}{d\tau} = N_D^0\lambda_D - N_E\lambda_E \quad (15.7)$$

$$\frac{dN_F}{d\tau} = N_E\lambda_E - N_F\lambda_F \quad (15.8)$$

solutions of which have the form

$$N_E = N_E^0 e^{-\lambda_E\tau} + N_D^0\frac{\lambda_D}{\lambda_E}(1 - e^{-\lambda_E\tau}) \quad (15.9)$$

$$N_F = N_F^0 e^{-\lambda_F\tau} + N_D^0\frac{\lambda_D}{\lambda_F}(1 - e^{-\lambda_F\tau}) + \frac{\lambda_E N_E^0 - \lambda_D N_D^0}{\lambda_E - \lambda_F}(e^{-\lambda_F\tau} - e^{-\lambda_E\tau}) \quad (15.10)$$

After substituting the numerical values of λ_D, λ_E, λ_F, $n_D^{УД} = 4400$ in (15.10), we obtain (taking into account (15.4)–(15.6) and $A_F = N_F\lambda_F$) Equation (15.11), which links the activity of RaD on the filter (A_F, Bq) at the time (τ, days) after air sampling with an integral value of EEVA of radon decay products in air (E, Bq/m^{-3}):

$$A_F = 4.36 \times 10^{-6} E\omega T_S[1 - \Phi_1 e^{-0.005\tau} - \Phi_2(e^{-0.005\tau} - e^{-0.1384\tau})] \quad (15.11)$$

where

$$\Phi_1 = \frac{207}{T_S}(1 - e^{-0.005T_S}) - \frac{0.271}{T_S}(e^{-0.005T_S} - e^{-0.1384T_S}) \quad (15.12)$$

$$\Phi_2 = \frac{0.262}{T_S}(1 - e^{-0.1384T_S}) \quad (15.13)$$

If the measurement of the activity of RaD on the filter was provided by means of the trek detector, with the efficiency of registration for alpha-particles having an energy of 5 MeV equal to ε (relative units), the number of treks developed in the detector during common exposure detector and filter from $\tau = T_H$ to $\tau = T_K$ will be described according to the equation (here M is the number of seconds in 1 day — $8.64 \cdot 10^4$)

$$N_{TP} = \varepsilon M \int_{T_H}^{T_K} A_F \, d\tau \quad (15.14)$$

a solution of which is

$$N_{T,m} = 0.38\varepsilon E\omega T_S\Phi_3 \quad (15.15)$$

where

$$\Phi_2 = T_K - T_H - 200(\Phi_1 + \Phi_2)(e^{-0.005T_H} - e^{-0.005T_K}) + 7.23\Phi_2(e^{-0.1384T_H} - e^{-0.1384T_K}) \quad (15.16)$$

All variables are the same as above.

Table 15.3 Number of Treks in Detector Exposed from T_n Till T_k (days)

T_s (days)	T_n=5T_k=60	T_n=60 T_k=180	T_n=5 T_k=90	T_n=90 T_k=180
30	38	212	76	173
90	60	244	109	195
180	87	282	148	222
360	123	334	200	258

Calculating according to formula (15) correlation $N_{TP}=f(\tau, T_S)$ when E = 100 Bq/m^3, ε = 0.1, and the value ωT = 1 m^3 are shown in Table 15.3.

If in the sample there is a substantial quantity of long-lived aerosols of alpha-active radionuclides of U-238 and Th-232 families (U-238, U-234, Th-230, Ra-226, Po-210 [which was not developed from radon in the air], Th-232, Th-228, and Ra-224), in order to exclude their association with uncertainty in the EEVA of radon assessment, it is necessary to make two measurements of the activity of the filter: first during the interval $(T_k - T_n)_1$ and the second during $(T_k - T_n)_2$.

We introduce the following parameters: C — integral value of the summary volume activity of alpha-activity of the long-lived radionuclides in the air sample (Bq/m^3); ε — activity average weighted efficiency of registration of long-lived radionuclides on the filter (relative units); (Φ_{3_1}), (Φ_{3_2}) — value of functions Φ_3 for the first and second measurements; N_{TP_1}, N_{TP_2} — number of treks for the first and second measurements.

Finally,

$$N_{TP_1}=\varepsilon_{ДРН}C_{ДРН}\omega T_{ОТБ}(T_K-T_H)_1+0.38\varepsilon E\omega T_S\Phi_{3_1} \qquad (15.17)$$

$$N_{TP_2}=\varepsilon_{ДРН}C_{ДРН}\omega T_{ОТБ}(T_K-T_H)_2+0.38\varepsilon E\omega T_S\Phi_{3_2} \qquad (15.18)$$

After transformations,

$$\mathrm{E}=\frac{N_{TP_1}-N_{TP_2}(T_K-T_H)_1/(T_K-T_H)_2}{0.38\varepsilon\omega T_S[(\Phi_3)_1-(T_K-T_H)_1/(T_K-T_H)_2(\Phi_3)_2]} \qquad (15.19)$$

and

$$C_S=\frac{N_{TP_2}-0.38\varepsilon E\omega T_S\Phi_{3_2}}{\varepsilon_{ДРН}\omega T_S(T_K-T_H)_2} \qquad (15.20)$$

The method described above allowed one to also measure the integral value of the summary volume activity of the long-lived radionuclides in air. The method was developed in two versions of the sampling devices:

1. Sampling device for stationary sampling (SSA-1).
2. Wearing sampling device for sampling in the breathing zone (ISA-1).

Both devices are automatic.

15.3 *Epidemiological data on miners in the city of Lermontov, Russia*

Results of the epidemiological study "dose–effect" in the city of Lermontov (North Caucasus, Russia) — 1500 miners, duration of work 1948–1973, period of observation 1958–1978. Cumulative exposures to radon progeny are as follows: for 750 miners, more than 240 WLM; for another group of miners (400), more than 600 WLM. Lung cancer mortality is 11-fold higher than for the control group (population of the region). There are additional data on the high radon concentrations in dwellings.

A retrospective epidemiological study of the dependency of the lung cancer mortality of miners in the city of Lermontov from the cumulative dose of radon decay products was conducted by Shalaev et al. (1992) in 1963–1986. The cohort consisted of 1500 miners, who lived constantly in the relatively restricted area of the uranium mine (North Caucasus, Russia). The main characteristics of the population of the observed group are shown in Table 15.4.

In this group of miners, during the period of observation from 1958 (the first observed lung cancer mortality) to December 1978 (the year of the end of the observation), 80 cases of mortality were observed with the diagnosis of "lung cancer." The general number of those who died for controversial reasons was near 15%. That is, in the whole structure, the lung cancer mortality was around 36%; at the same time, in the nearest city of Pjatigorsk for the same period, the corresponding number was only 55 of all deaths. The miners of the retrospective observed group died on an average of 11.7 yr earlier than the men of the city of Pjatigorsk (Table 15.5). The value of the induction-latent period (time of the beginning of the work in the mine until death from lung cancer) was increased during the time of observation by more than 2 times, and for the last years reached 25 yr.

A comparison of the lung cancer mortality factors between the observed and control group (Table 15.6) shows substantial excess observed mortality in comparison with the expected one.

Restoration of the working conditions for the observed group of miners for the whole period and calculation of the radon decay product cumulative exposure represent

Table 15.4 Characteristics of the Observed 1500 Miners

Beginning of the work	Years	1947–1949	1950–1954	1955–1959	1960–1965	Not included after 1965
	% of group	6	46	37	11	
End of the work	Years	1955–1959	1960–1964	1965–1969	1970–1974	1975–1978
	% of group	21	28	30	19	2
Duration of the work time	Years	2–5	6–10	11–15	16–20	>20
	% of group	22	32	28	13	5

Table 15.5 Lung Cancer Mortality Data on the Observed Group of Miners

Years	Number of Lung Cancer Mortality Cases in the Observed Group	Average Age of Deceased Miners with Lung Cancer (yr)			Average Induction-Latent Period (yr)
		Observed Group	Control Group	Age Difference Between Control and Observed Group	
1958–1963	5	52.6	62.0	9.4	11.1
1964–1969	30	47.2	63.0	15.8	14.8
1970–1975	24	53.7	63.7	10.0	20.0
1976–1978	21	55.4	64.7	9.3	24.6
— ——— …	80	51.6	63.3	11.7	18.6

Table 15.6 Lung Cancer Mortality for the Period 1958–1978 Among Men in the Observed and Control Group of Miners

Age (yr)	Lung Cancer Mortality of Men, Control Group (Number of Cases per 10,000 Men-Years)	Observed Group			Observed Lung Cancer Mortality, Number of Cases on 10^4 Men-Years	Ratio of the Observed and Expected Mortality	Surplus Lung Cancer Mortality, Number of Cases on 10^4 Men-Years
		Number of men-years of the risk, $n \times 10^4$	Number of lung cancer cases				
			Expected	Observed			
18–29	0.08	0.225	0.02	—	—	—	—
30–39	0.12	1.008	0.12	3	3.0	25	2.8
40–49	2.6	1.015	2.6	34	33.5	13	31
50–59	5.7	0.396	2.2	26	65.6	12	60
60	15.5	0.147	2.3	17	115	7.4	100
____	4.7	2.79	7.2	80	28.7	6.0[a]	26

[a] The ratio nonstandardized by age. After standardization of the control group, the mortality in the observed group decreases to 2.85 per 10^4 men-years and the ratio of the observed and expected factor becomes equal to 11.1.

extremely complicated tasks. There was a very small amount of data on concrete working sites of individual miners in underground conditions, and also data on individual control of the levels of the equivalent equilibrium volume activity (EEVA) of radon in air. Preserved radiation control data were extremely small. In using these data, we should be very careful because in the first years of the uranium industry, radiation safety was focused mainly on unfavorable working sites, at the same time ignoring measurement in the relatively favorable sites.

In the framework of this study, in order to obtain information on the characteristics of working conditions, not only the annual reports of the radiation safety system were used, but also the primary materials (dosimetric measurement register, memos to the administration and high-level organizations). Documents from the mine archives and reports of measurements provided by scientific research organizations were also analyzed.

Based on these data, the values of the annual average and mine levels of calculated radon EEVA (Table 15.7) were used for the calculation of individual radon decay product exposures for miners of the observed group. During the study, it was found that mine average levels of EEVA are the most informative data under the retrospective study, because at the present time it is practically impossible to restore detailed information on the working routes for every miner. Moreover, the information available in the mine does not represent in reality the character of work for each miner. Not in one of the saved documents was the individual working itinerary taken into account, including change of the working site during the shift related to the character of the work. Whereas exactly such constant dislocations of miners in the limits of one lot during the shift, transfer from one working site to another during the year period, the character itself of the many professional groups of miners (electric locomotive machinists, explosives, railway engineers, managing personnel, etc.) allowed one to use average mine EEVA radon levels for cumulative exposures calculation. This was supported by the results of special studies carried out on 16 uranium mines of the former U.S.S.R. in the middle of the 1960s that is, after individual radiation control and registration were introduced. During such control and registration, the assessment of the real working itinerary of miners in concrete working sites took place, the net of the control points are determined as more purposeful, and control periodic chose correspondingly variability of the levels of EEVA.

Table 15.7 Mine-Average Levels of EEVA(1 WL=3700 Bq/m³) and Dust Concentration of the Atmosphere During the Period of Observations

Year	EEVA of Radon in the Air (WL)	Dust Concentration (mg/m³)	Year	EEVA in the Air (WL)	Dust Concentration (mg/m³)
1948	20	200	1961	3.1	1.2
1949	19	160	1962	1.9	1.0
1950	17	140	1963	1.3	0.8
1951	14	110	1964	1.0	0.8
1952	12	80	1965	0.7	0.8
1953	10	20	1966	0.6	0.8
1954	8.6	4.1	1967	0.5	0.8
1955	7.0	2.8	1968	0.4	0.8
1956	5.1	2.1	1969	0.4	0.8
1957	3.8	1.7	1970	0.3	0.8
1958	2.7	1.3	1971	0.3	0.8
1959	2.6	1.3	1972	0.3	0.8
1960	4.6	1.4	1973	0.3	0.8

Table 15.7 shows that the mine-average annual-average EEVA levels of radon in the majority of cases are in good agreement with an average value of the annual exposure of the underground contingent, increasing it on an average by 20–30%. The real individual exposures are more variable (up to 2–5 times), but in the majority of cases differences are no more than 1.5 times from the value calculated according to the mine-average EEVA of radon. Thus, there is great probability that the result will be higher, not lower, than the actual. Owing to mutual compensation of the annual variabilities, calculated cumulative exposure for the whole length of service substantially accurately describes the actual exposure, than under annual observation, and the value of possible relative uncertainty of the cumulative dose based on annual-average EEVA of radon decreases with the time of increasing exposure.

In order to determine the dependence between lung cancer mortality and radon decay product cumulative exposures, the observed group was divided into three categories: less than 240 WLM — category A; 240–600 WLM — category B; and more than 600 WLM — category C (Table 15.8). In each of these categories during 20 years of observation, according to ages (for the expected cases of lung cancer prediction) the summary numbers of men-years risk were calculated. By this, the contribution of every miner, who finished his job in the category with highest exposure, into the men-years of corresponding ages in the category with smallest exposure, was taken into account. This took place at the beginning of the working period.

Lung cancer mortality indexes were shown according to exposure in Table 15.9.

As expected, the index of surplus lung cancer mortality increased with an increase in the cumulative exposure, and the line of regression, representing surplus lung cancer mortality vs. cumulative exposure, represents the straight line passing through the coordinate beginning, which corresponds to the nonthreshold effect concept of ionizing radiation.

Differences in the indexes of the risk (number of surplus lung cancer cases on 10^6 and on 1 WLM) between different exposure categories can be explained by the high EEVA of radon in the dwellings of the city of Lermontov, where the majority of miners lived. Unfortunately, EEVA radon measurements in this city began only from 1993 (Table 15.10), and therefore these results could not be used in the calculation of the cumulative exposures of the observed group.

Table 15.8 Main Characteristics of the Observed Group of Miners Depending on their Radon Decay Product Cumulative Exposures

Index		Cumulative Exposure (WLM)		
		<240	240–600	>600
Category of Miners		A	B	C
Part of the men-years risk in this category from the whole group (%)		45	27	28
Number of lung cancer mortality cases during 20 yr		16	24	40
Average age (yr)	All miners in 1975	45	49	53
	Died from lung cancer	50	50	54
Average working period in mines (yr)	All miners	7.8	15.0	13.8
	Died from lung cancer	9.8	9.9	12.0
Average radon decay product cumulative exposure (WLM)	All miners	140	400	800
	Died from lung cancer	180	420	820

Table 15.9 Lung Cancer Mortality Indexes in the Observed Group of Miners for 20 years

Exposure, Category	Number of Men-Years Risk, $n \times 10^4$	Lung Cancer Numbers			Lung Cancer Surplus Index, Number of Cases on 10^4 Men-Years Risk (Average, 95% Confidence Interval)	Average Value of the Risk Index, Number of Surplus Lung Cancer Cases on 10^6 Men-Years Risk and on 1 WLM
		Observed	Expected	Ratio of Observed to Expected		
A	1.29	16	1.9	8.4	10.9 (6.0–18.3)	7.8
B	0.74	24	2.2	10.9	29.4 (18.4–44.2)	7.0
C	0.76	40	3.1	12.9	48.6 (34.2–66.4)	6.0
All	2.79	80	7.2	11.1	26.1 (20.9–32.2)	7.2

The city of Lermontov existed from 1948. Before 1948, there were only two-story many-apartments and one-apartment private houses. As can be seen from Table 15.10, the average EEVA of radon in two-story buildings was 240 Bq/m^3, and in the private houses it was 520 Bq/m^3. If we assume that all miners of the observed group lived constantly in the city from 1948 to 1978 (30 yr) and that on an average they were located in the dwellings for 5000 hours per year (taking into account that 1 WLM=6.3×10^5 Bq h m^3), they received radon decay product cumulative exposures in the range of 60–120 WLM (average 90 WLM). These exposure values are entirely are comparable with the calculated average cumulative exposures received by workers in the uranium mine, which according to separate categories of miners, deceased from lung cancer, are equal to 180, 420, and 820 WLM (Table 15.8). If we add to these numbers, for example, 45 WLM (one half of the cumulative exposure in dwellings), the risk index (Table 15.9) will be equal to 6.2 (category A), 6.3 (category B), and 5.7 (category C). That is, the tendency of decreasing this index with an increase in the cumulative exposure practically disappeared.

The later observations were provided by Glushinskij and others in 1987–1990. The observed group was increased up to 1721 men mainly due to the inclusion of miners from

Table 15.10 EEVA Radon in the Dwellings of the City of Lermontov (Bq/m^3)[a] (1993–1997)

Floor	Multifloor Houses				Private Houses	
	Upper Part of the City		Lower part of the city			
	Number of Measurements	EEVA of Radon	Number of Measurements	EEVA of Radon	Number of Measurements	EEVA of Radon
	26	170	59	460	4	3900
1	463	180	1121	310	456	520
2	113	90	394	170	4	230
3	128	60	100	110	—	—
4	58	40	21	110	5	12
5 and higher	49	32	4	90	—	—
All	837 (12)[b]	130	1699 (37)[b]	260	469 (51)[b]	540

[a] Calculated average EEVA is shown according to the results of measurements of the integral for 4–5 days volume activity of radon (gas) using the equilibrium factor 0.5.

[b] Values in parentheses represent portions of measurements with EEVA higher than 200 Bq/m^3 (%).

other closely located uranium mines. The general number of miners who died from lung cancer from 1950 to 1990 was 161 men (9.4% of the population of the observed group), and was 6.5 times higher than the expected number. The index of surplus lung cancer mortality was on average 7.3×10^{-6}/ (men-years) /WLM.

As a result of the retrospective analysis, in general, all the obtained data in the study of the quantitative lung cancer mortality risk characteristics of miners are in good agreement with the data of analogous studies provided by other countries.

15.4 Individual dose distribution of uranium mines personnel (2500 workers) to radon progeny, long-lived nuclides, and external gamma-radiation from 1990 to 1994 in the city of Krasnokamensk, Chita region, Siberia

In Tables 15.11 and 15.12, the assessments of individual effective doses of irradiation of 2500 miners, who worked in 1990–1997 in the uranium mines near the city of Krasnokamensk (Chita region, Siberia), are shown (Pavlov et al., 1999). Radon decay product doses were calculated based on the results of the measurement of the radon progeny concentration in air (1 Bq/m^3=2.7×10^{-4} WL of the "potential alpha-energy concentration") *in situ* by using aerosol radiometer IZV-3 type. The activity of RaA and RaC′ on the filter was measured at the moment of air sampling with a velocity of 10 l/min during 1–4 min. Points of observation were located in the working mine-pits (2–3 places for pits) and in the transporting working sites. In every place, more than 50 measurements per year were provided. Exposure for every miner was calculated according to the actual number of working hours in every working site. For the transfer from exposure to effective dose, the ratio 1 Bq h m^{-3}=0.95×10^{-5} mSv was used.

The values of the volume activity of the long-lived radionuclides of the U-238 series were calculated on the basis of the measurement of the alpha-activity of the filters. Air sampling (volume 800 l) and measurement of the dust concentration of the atmosphere in the working sites under different technological operations and in transporting working sites were taken. Activity measurements were provided by means of the low-background radiometers after filters were exposed in non-radon boxes for 6 days. For the exposure–effective dose transfer, the ratio 1 Bq h m^{-3}=1.5×10^{-2} mSv was used.

For the assessment of the possible decreasing of the dose from radon decay products and long-lived radionuclides by using respirators, the actual time in using respirators and

Table 15.11 Average Individual Effective doses of Irradiation of Underground Personnel (1990–1997) (mSv yr^{-1})

Year	Number of Miners	Contribution to the Dose			Total Dose	
		Radon Progeny	Long-Lived Radionuclides	Outer Gamma-Radiation	Without Accounting for Respirators	With Accounting for Respirators
1990	2771	8.6	1.5	2.7	12.8	8.5
1991	2504	7.6	1.8	2.9	12.3	8.5
1992	2651	9.6	1.4	2.2	13.2	9.4
1993	2458	9.0	1.5	1.6	12.1	9.0
1994	2111	11.5	1.3	1.7	14.5	10.1
1995	2590	8.8	1.2	3.0	13.0	9.9
1996	2910	8.1	1.1	3.5	12.7	9.5
1997	2554	9.3	1.5	3.7	14.5	11.6

Table 15.12 Underground Personnel of Mines Distribution by Annual Average Individual Doses 1990–1994[a]

Dose Range (mSv yr^{-1})	Number of Miners	Personnel Quota (%)
0–5	617	24.6
6–10	1008	40.3
11–15	635	25.4
16–20	176	7.03
21–25	40	1.60
26–30	5	0.20
31–35	6	0.24
36–40	4	0.16
41–45	3	0.12
46–50	2	0.08
51–55	1	0.04
56–60	2	0.08
61–65	3	0.12
66–70	1	0.04
71–75	1	0.04
Including		
0–20	2436	97.3
21–50	60	2.4
51–75	8	0.3
All	2504	100

[a] Accounting for the actual use of respirators.

its actual efficiency were taken into account during the shift, and the degree of complete fitting respirators contoured to the skin face was studied. According to special studies, the actual efficiency of respirators was around 85%. The actual use of respirators, "Lepestok" type, by mine personnel is on average, for different parts of the mines, from 25 to 60% working time, which makes it possible to decrease the dose of irradiation by 15–40%. Doses from gamma-radiation were measured by means of wearing dosimeters. For transfer from the absorbed to the effective dose, the ratio 1 Gy=0.7 Sv was used.

Acknowledgment

Chapter translated from original Russian by Lev S. Ruzer.

References

1. Glushinskij, M.V., Beljaev, A.V., Nagaev, A.I., and Zuevich, F.I., The New Data for the Assessment of Professional Risk of Lung Cancer Mortality of the Uranium Mines Personnel, in Materials of the 3rd Symposium of the Hygiene of the Work, Radiation Safety, Environment Protection, and Professional Pathology of Mining and Primary Re-making of Radioactive Ores, 4–9 September 1991, Saint-Petersburg, 1992, pp. 250–253.
2. Pavlov, I.V., Kamnev, E.N., Panfilov, A.P., Shevchenko, O.A., and Sidorov, Yu.M., Security of the radiation safety of the uranium mines personnel, *Mining J.*, No.12, 62–64, 1999.
3. Pavlov, I.V., Beljaev, B.M., and Tsapalov, A.A., Automatic sampling device for integral measurements of the EEVA of radon for the long period of time, in "Development of the normative-methodical, metrological and informative service for decreasing the level of irradiation of population from natural sources," Report (First stage of governement contract 1.02.30.01.011) (part-program 11 of the federal program "Nuclear safety of Russia for 2000-2006), Research-Scientific Test Center of Radiation Safety of Cosmical Objects M: 2001, pp. 41–45.

chapter sixteen

Radioactive aerosols of the Chernobyl accident

A.K. Budyka and B.I. Ogorodnikov
Karpov Physico-Chemical Institute, Moscow

Contents

16.1 Introduction 517
16.2 Dynamics of the ejection of radionuclides from the reactor 518
16.3 Global transfer of the accident products 519
16.4 Sampling devices 520
16.5 Aerosol characteristics in the first half year after the accident 520
16.6 Gaseous components I, Te, and Ru in the atmosphere 526
16.7 Characteristics of radioactive aerosols near the earth's surface 529
16.8 Forest fires in the exclusive zone 531
16.9 Aerosols of the "Shelter" 531
16.9.1 Types of aerosols 532
16.9.2 Aerosol concentration inside the "Shelter" 532
16.9.3 Aerosol transport from the "Shelter" into the atmosphere 534
16.9.4 Aerosol dispersity inside the "Shelter" 535
16.10 Radioactive aerosols close to the surface layer of the atmosphere near "Shelter" 536
16.11 Conclusion 538
Acknowledgment 538
References 538

16.1 Introduction

The accident in the 4th block of the Chernobyl Nuclear Power Plant (NPP) resulted in radioactive contamination not only in a territory of the European part of the former U.S.S.R., but also in the entire northern hemisphere, including the U.S.A. and Japan. This accident should be classified as a global disaster.

It is very clear that contamination of territories occurred due to atmospheric transfer and fallout of radioactive aerosol particles on the earth's surface. Therefore, understanding the physicochemical properties of radioactive aerosols, formed as a result of reactor explosion, allows not only the correct assessment of the scale of the disaster, but also the choice of a more effective means of defense for more than 300,000 people, who participated in the liquidation of the consequences of the Chernobyl accident.

Despite the fact that radioactive aerosol concentrations after some months on the majority of territory surrounding Chernobyl (30 km around Chernobyl NPP) did not exceed permissible concentrations, still today there are places with a dangerously high

1-56670-611-4 / 05 / $0.00+$1.50

level of air contamination. For example, there are some places inside the object "Shelter" (previously called "Sacrophagus") with high levels. "Shelter" was built in 1986 in order to localize and subsequently control the situation of nuclear fuel from the destroyed reactor of the 4th block of Chernobyl NPP.

The fallout on the earth's surface slowly migrates deep into the soil. For 17 yr after the accident, the property of the primary fallout changed. Secondary airborne aerosol particles were formed due to winds, forest fires, and also technogene activity (moving transport, building, and agriculture activity).

The data of this chapter are unique, because it is impossible to repeat such a situation due to the exclusiveness of the accident. Even now, for example, when we measure the radioactive aerosol concentration inside the "Shelter," the smallest change in meteorological conditions (temperature, pressure, humidity, speed and direction of the wind) and some technogene activity results in changes in the physicochemical characteristics of aerosols in the same sampling point.

It should be understood under what circumstances the measurements in the first weeks and months after the accident were provided. Dosimetric control of inner irradiation air sampling was provided in extreme conditions. In some sites, the radiation levels were so high that for dosimetric personnel to spend only a couple of minutes or hours endangered their health. Often it was difficult to study the samples carefully, and dosimetrists were restricted only to measure the summary of gamma- and beta-activity of the sample.

Still we hope that these materials, the majority of which was obtained by scientists and technicians of the aerosol laboratory of Karpov Institute of Physical Chemistry, will inform readers of the main properties of radioactive aerosols and the gaseous products of the accident in Chernobyl NPP.

16.2 Dynamics of the ejection of radionuclides from the reactor

On April 26, 1986 as a result of the explosion in the 4th block of Chernobyl NPP, the core and upper part of the reactor were completely destroyed, as were all defense barriers and systems of security. It was the largest accident in the atomic industry. Most of the radionuclides were ejected from the 4th block from April 26 up to May 6, 1986. The dynamics of the ejection are shown in Figure 16.1.

In the first day after the accident, aerosol sampling and analysis of sampling composition above the destroyed reactor already began. Due to many circumstances (nonstationary

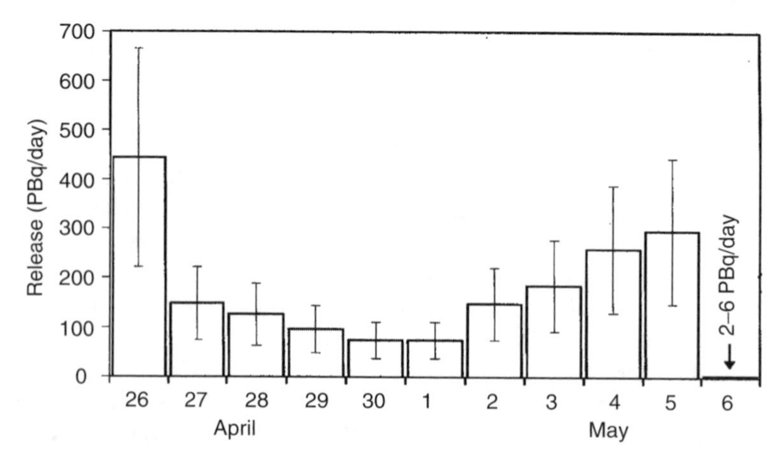

Figure 16.1 Daily integral ejection of radioactive products (without inert gases) from destroyed reactor of 4th block of Chernobyl NPP in April–May, 1986.

character of the ejection, meteorological conditions, intensive work on covering the destroyed reactor by different materials), the accurate determination of the dynamics of radioactive substances entering the atmosphere was difficult to assess. Two assessments of the composition of the emission rejection are shown in Table 16.1.

In the initial period after the accident, most radionuclides were ejected from the destroyed reactor in the form of dispersed fuel (mainly with a UO_2 matrix). During lava formation on fuel materials, which took place at a temperature of about 2000°C, only volatile and light fusible substances, like Te and alkaline metals, evaporated from the fuel. More than 95% of nuclear fuel (more than 180 t) remained inside the "Shelter," which was built above the destroyed block at the end of November 1986.

The integral ejection of radionuclides with a half-time of $T_{1/2}$>20 h, without taking into account inert gases from the destroyed reactor, was around 3×10^{18} Bq.

16.3 Global transfer of the accident products

At the moment of the accident, winds near the earth's surface were weak and without special direction. Still, at altitudes of more than 1500 m the wind was mostly in the south-east direction, with a speed around 8–10 m/s. Part of the radioactivity was raised to this level and moved throughout the western regions of the U.S.S.R. to Finland and Sweden. There, on April 27, radioactive aerosols from the Chernobyl accident were discovered [3] for the first time outside the U.S.S.R. At the end of April and the beginning of May, radioactive aerosols, including I and Cs, were detected in the upper troposphere and lower stratosphere (up to 15 km) above the territory of Poland [4]. The fallout of refractory elements such as Ce, Zr, Np, and Sr was detected mainly on U.S.S.R. territory.

Due to changing meteorological conditions, especially wind direction, and also the continuous ejection of a large mass of radioactive aerosols and gases for 10 days resulted in a very complicated picture of the distribution of radionuclides in the atmosphere. Part of the nuclides was shifted on the low highs to Poland and Germany. By April 29 and 30, radioactive clouds reached other countries of Eastern and Central Europe. Radioactive substances were in the north of Italy by April 30, and in Central and South Italy the next day. In France, Belgium, and Holland, radioactive contamination was detected on May 1, and in Great Britain the next day. Gases and aerosol products from the Chernobyl accident reached the north of Greece on May 2, and the south of Greece by May 3. At the beginning of May, radioactive substances were detected in Israel, Kuwait, and Turkey.

The observation of aerosol and gaseous components of radioactive iodine was provided in 19 stations in Europe, with sampling and analysis of 171 samples. Only aerosol components of radionuclides were studied on 1892 samples from 85 sites [5].

Table 16.1 Emission (in %) from the Total Activity of Radioactive Substances Saved up in the Core During the Operating Period of the 4th Block of Chernobyl NPP.

Radionuclide	According to Data from the U.S.S.R. for the International Atomic Agency (August 1986) [1]	According to Results of the Study at the End of 2000 [2]
^{133}Xe, ^{85}Kr	~100	~100
^{131}I	20±10	55±5
^{134}Cs	10±5	33±10
^{137}Cs	13±7	33±10
Uranium and transuranium elements[a]	3±1.5	3±1

[a] Taking into account the fragments of the reactor active zone, ejected around the 4th block.

The transfer of radioactive products in the eastern direction took place very fast. Already by May 2, the first aerosol and gaseous samples of ^{131}I were found in Japan, May 4 in China, May 5 in India, and May 5–6 in Canada and the U.S. On the North American continent, products of the Chernobyl accident came practically simultaneously from the west and east. All this confirms that the intermix of aerosols in the horizontal and vertical direction was substantial, and also that transfer of activity took place mostly on submicrometer-sized particles. No substantial amount of radionuclides was detected in the southern hemisphere. Thus, the distribution of aerosols after the Chernobyl accident took place in the earth's atmosphere according to ideas known before on examples of volcano eruption and tests of nuclear weapons.

16.4 Sampling devices

In the initial period after the accident and during the building of the "Shelter," measurements of aerosol characteristics (mainly for radiation reconnaissance and dosimetry) were provided practically continuously at different sampling points. After the "Shelter" was built, the main goal became the study of the temporary and spatial evolution of radionuclides and the disperse composition of aerosols, and also the study of secondary sources of aerosols (wind raising of dust, forest fires, etc.). Sampling became more focused and related to meteorological conditions. For sampling, usually filter materials AFA-RMP and also packages from different filter materials were used.

Three-layer packages, consisting of filter materials FPA-70-0.15, FPA-70-0.25, and FPA-15-1.5, were used for the simultaneous measurement of the concentration and disperse composition of aerosols (with the help of the so-called method of multilayer filters [MMF], see Chapter seventeen, p. 555).

For simultaneously sampling aerosol and gaseous substances with I, Te, Ru, and Cs, sorption-filter material SFM-I was used [6]. This material consists of two sorption layers, covered from above and below with FPP-70-0.2, and the frontal layer of FPP-15-1.5 is only for aerosol deposition. Each sorption layer was developed from FPP-70-0.3 material, on the fiber of which was fine-grained carbon with $AgNO_3$. The quantity of sorbent was some mg/cm^2, and the efficiency of molecular iodine catching was not less than 90%. All used filter materials and analytical filters were developed in the Karpov Institute of Physical Chemistry.

The areas of filter materials, depending on sampler devices, were in the range of 3 cm^2 to 3 m^2. Such variety was needed for operational assessment of radiation contamination of the air, taking a great number of air samples at different sites and times.

All used sampler devices, both ground and aircraft, were variations of the measurement technique widely used in the system of routine and research aerosol monitoring in Russia. All measurements were provided by means of standard methods of spectrometry and radiometry of ionizing radiation. In some cases, radiochemical methods were used for the preliminary separation of studied substances from the samples.

16.5 Aerosol characteristics in the first half year after the accident

The first aerosol sample of "Chernobyl origin" was taken on April 26 above the western portion of the European part of the Soviet Union. These data were used for the assessment of the ejection of radioactive substances from the destroyed 4th block of Chernobyl NPP in the first day after the accident. Regular sampling above the breakdown reactor began at night from 27 and 28 April from aircraft An-24, which belonged to the Ministry of Defense of the U.S.S.R. In addition, for the same purpose, a helicopter was used [7].

Aerosol sampling from the aircraft above the 4th block continued until the beginning of August 1986 [8]. In Figure 16.2, the values of concentrations of beta-radioactive aerosols

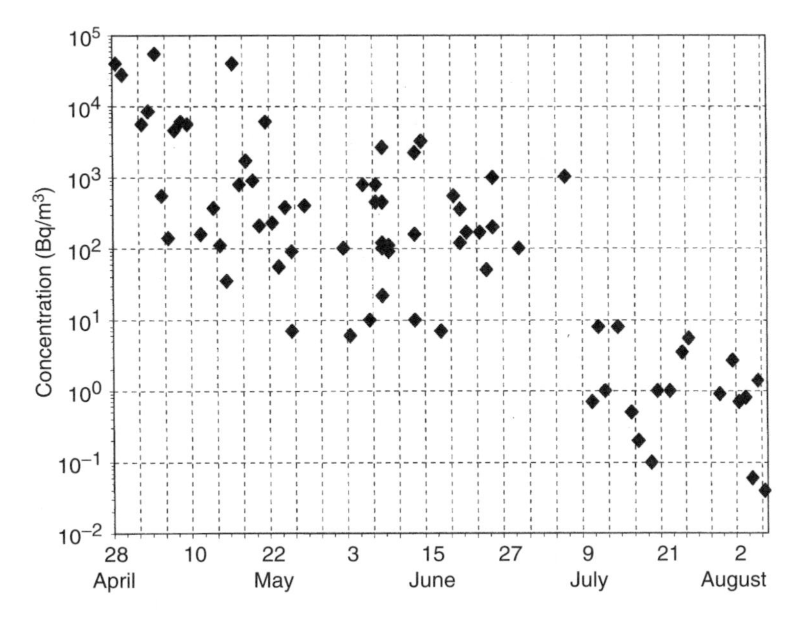

Figure 16.2 Concentration of the sum of beta-radioactive aerosols over Chernobyl NPP in April–August, 1986. Sampling from aircraft An-24.

for this period are shown. As we can see from these data, the concentration of radioactive isotopes in the atmosphere, despite some variations, decreased by 5–6 orders of magnitude already after 100 days.

This took place due to decreasing temperature in the destroyed reactor, radioactive decay of short-lived radionuclides, deactivation of the territory, and measures for decreasing dust concentration, etc. Table 16.2 presents the distribution of radionuclides in aerosols that were sampled above the destroyed reactor of the 4th block of Chernobyl NPP in April–May 1986 from helicopter Mi-8 [7], and aircraft An-30 [1] and An-24 [9].

As we can see from the Table 16.2 data, the composition of the ejection changed strongly not only day by day, but also during the same day (see the results of measurements on May 8, 14, 15, and 16).

Figure 16.3 presents the dynamics of changes in aerosol concentration of ^{95}Zr, ^{95}Nb, ^{103}Ru, ^{106}Ru, ^{131}I, ^{134}Cs, ^{137}Cs, ^{140}La, ^{140}Ba, and ^{141}Ce sampling from An-24 aircraft 200–300 m above the destroyed reactor on May 8–19, 1986. This period can be described as the beginning of systematical decreasing of ejection of radioactive substances from the destroyed reactor. At this time, the burning of graphite was stopped, and temperature in breakdown began to gradually decrease. Still, concentrations measured on May 8 and 16 (some thousands of Bq/m^3) were comparable with concentrations measured on May 4–5 (see Figure 16.2).

Usually, the radioactive composition of aerosols did not correspond to the composition of irradiated fuel at the moment of reactor explosion. For example, on May 8 and 19, volatile radioactive nuclides I, Ru, and Te were measured. On other days, the main contribution to the summary samples of gamma-radiation were delivered from refractory radionuclides Zr, Nb, and Ce.

Such variations of the ratio of the activities of nuclides in every sample were similar for the aerosols sampled above the territory of Belarus and Russia, which were close to Chernobyl NPP from the north and northeast, and from the research ship in the Atlantic Ocean at the beginning of May [9].

Table 16.2 Relative Content of Radionuclides in Aerosol Particles above the Reactor of the 4th Block of the Chernobyl NPP in April–May 1986

	April			May													
	26	28	29	1	2	3	4	8		14		15		16		17	18	19
	An-30	Mi-8	Mi-8	Mi-8	Mi-8	Mi-8	Mi-8	Mi-8	An-24	Mi-8	An-24	Mi-8	An-24	Mi-8	An-24	An-24	Mi-8	An-24
^{131}I	39	31	11.6	81	68.4	100	17.6	19.6	35.4	-	4.2	-	8.8	1.8	1.7	3.2	-	3.7
^{132}Te	34.7								6.9		0.92							6.2
^{103}Ru	5.2	12.7	8.3				13.4	58.7	37.2		8.4		14	10.3	11	8	48.7	36.9
^{106}Ru	1.7	4.6	2.5				5.1	20.6	1.3		1.1		7	4	3	2	13.5	11.4
^{134}Cs	1.3						1.3		0.65	31.8	0.37	3.4	1	2.7	0.35	0.46	13.5	0.62
^{137}Cs	2.6		2.5				1.9		3.3	68.2	1.3	8.2	2.5	5.8	0.87	0.93	24.3	1.5
^{95}Zr	3.9	12.7	34.7		15.8		12.3		5.3		18.3	18.4	12.6	16.4	16.8	23.5		8.3
^{95}Nb		16.3	28.1	19	15.8		17.6		2.4		21.6	24.2	22.3	21.2	30.2	33.2		13.5
^{141}Ce	3.5	13.6	7.4				18.4		2.6		16.8	27.6	11.6	22.7	13.8	12.3		7.1
^{144}Ce	3.9	9.1	5				12.3		1.5		6.2	18.4	9.1	15.2	14.2	12.3		7.4
^{140}Ba	4.3								2.6		8.4		8.3		6.6	3.1		2.5
^{140}La									0.9		12.5		2.8		1.5	1.3		1.2

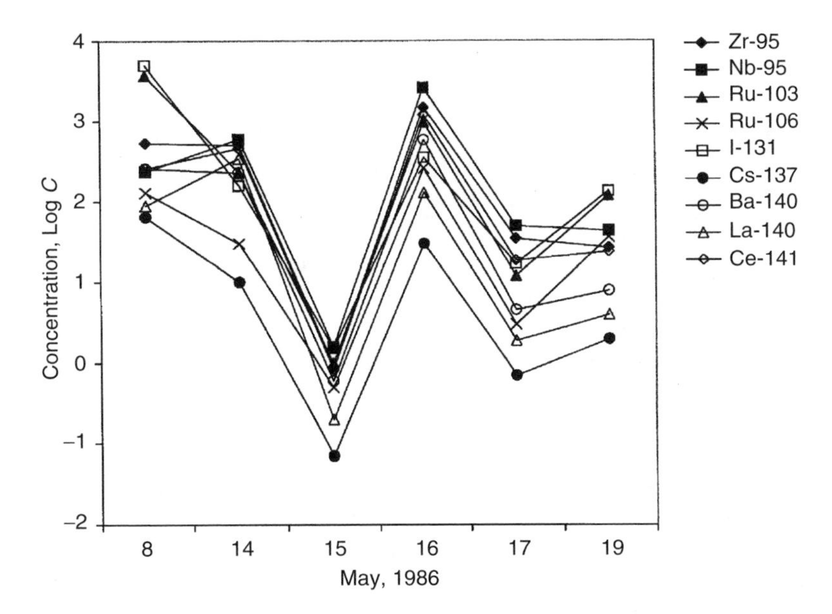

Figure 16.3 Concentration of the aerosols contained gamma-radioactivity above 4th block of the Chernobyl NPP in May, 1986.

It is also possible to form an opinion about the composition of primary aerosols based on the fallout, which were sampled in a 30-km zone around the atomic station. Almost 90% of the particles were fragments of irradiated nuclear fuel. It consisted of two groups of fuel particles. The composition of one of them was identical to the composition of fuel elements at the moment of the accident. These particles were formed under heat explosion of the reactor. 85% of the activity of the second group consists of refractory ^{144}Ce. The quantity of radionuclides of ^{95}Zr and ^{95}Nb was larger than in irradiation fuel at the moment of the accident, but Cs and Ru were substantially lower. Among the particles, there were some radioactive microparticles that were displaced on a nonradioactive carrier [10]. Radioactive particles of condensation origin have, mainly, nuclides of Cs and Ru [11]. At a distance of 15 km from the reactor, the portion of fuel particles depending on azimuth were from 50 to 100%. In the far zone, including beyond borders of the U.S.S.R., to the contrary, particles of fuel origin were measured only in a small number of cases (see, e.g., [12]).

Radioactive aerosol particle sizes directly above the destroyed reactor were measured for the first time on May 14, 1986. For this purpose, the gondola of aircraft An-24 was supplied with a package of filters for disperse analysis composition.

It was shown that the smallest size had particle carriers of ^{131}I, ^{132}Te, and ^{103,106}Ru. The activity median aerodynamic diameter (AMAD) of aerosols was 0.3–0.4 μm with geometric standard deviation σ=2.3–2.5. The sizes of carriers of refractory radionuclides Zr, Nb, La, and Ce were larger: AMAD=0.7 (σ =1.6–1.8). On the same carriers, $^{134,\,137}$Cs were discovered. In some countries of the northern hemisphere, at the beginning of May [13–19] practically similar values of AMAD of radionuclides carriers of chernobyl origin were measured. In these measurements, different types of impactors were used.

In the summer of 1986, the surface of the soil, buildings, vegetable layer, rods, etc. polluted with primary fallout became comparable to or even more important sources of contamination closest to the surface layer of the atmosphere. In this period, it was impossible even quantitatively to describe all mechanisms of forming aerosols, because in the atomic station area and around a 30-km zone very intense deactivation took place, often resulting in aerosol formation. An additional source was intense transport movement near the Chernobyl NPP.

Figure 16.4 presents the dynamics of aerosol concentration in the summer of 1986.

This correlation was based on measurements of the total gamma-activity of samples provided by scientists of Khlopin Radium Institute from military vehicles, which traveled over Chernobyl NPP by perimeter [20]. As we can see, average aerosol concentrations near the station in the background of substantial variations were comparable with measured concentrations above the destroyed reactor in the second half of May (Figure 16.2).

Measurements provided in July–September 1986 demonstrated that practically all radionuclides were disposed on the same particle carriers. AMAD values of aerosols, averaging for the period of observation, were 2.95±0.60 μm for ^{95}Zr, 2.89±0.37 μm for ^{95}Nb, 2.93±0.62 μm for ^{134}Cs, ^{137}Cs, 2.83±0.26 μm for ^{144}Ce, and 3.00±0.28 μm for isotopes of Pu. At the same time, isotopes of ^{103}Ru, ^{106}Ru were discovered on smaller particle carriers (AMAD=2.57±1.01 μm) with the greatest variation from average size. This can be explained by the desorption of volatile compounds Ru from particle carriers, and subsequently their deposition on aerosol particles of submicron range. On some days, the portions of small disperse fractions of Ru were near 40%.

Observation in the 1986 summer–autumnal period showed that AMAD values of aerosols near the earth's surface, began to increase, which was confirmed by later studies.

Table 16.3 shows typical aerosol concentrations, based on samples that were taken in the middle of June in the machine room of the 3rd block (neighboring the destroyed 4th block) by means of packets of fiber and sorption filter materials [22].

It is clear from these data that day concentrations were substantially higher than night concentrations due to extensive activity in the daytime, resulting in the formation of higher dust concentrations.

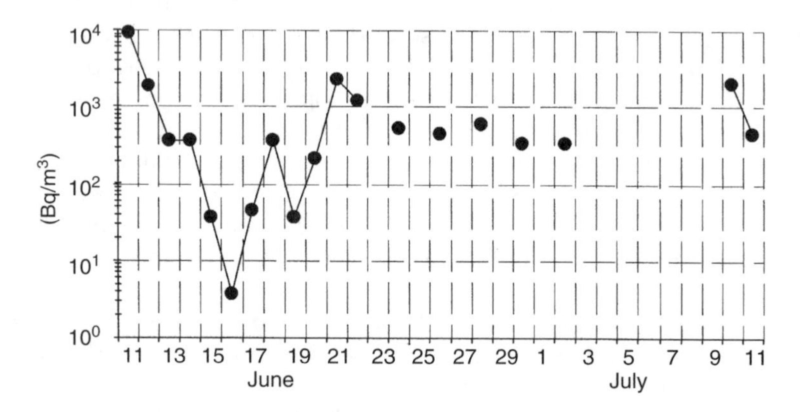

Figure 16.4 Concentration of the sum of gamma-radioactive nuclides in the aerosols sampled in June–July, 1986 on territory of Chernobyl NPP.

Table 16.3 Concentrations of Radionuclides in the 3rd Block of the Chernobyl NPP (Bq/m^3)

Sample	Sampling Date and Time	^{144}Ce	^{141}Ce	^{131}I	^{103}Ru	^{106}Ru	^{95}Zr	^{95}Nb	^{137}Cs	^{134}Cs
A	13.07, 18^{15}–19^{07}	437	156	1	210	—	—	—	40.7	25.9
B	13.07, 18^{15}–19^{07}	465	175	—	124	79	334	516	46.7	24.2
C	13–14.07, 19^{10}–10^{10}	51.8	18.5	8.5	15.8	12.2	36	54	5.9	2.9
D	13–14.07, 19^{10}–10^{10}	54.1	23.3	—	17.4	14.1	46.3	70.4	6.4	2.5

The sizes of aerosol particles at night (AMAD=0.75 μm) were smaller than in the daytime (AMAD=1 μm), and all radionuclides were on the same carriers. In the "day" sample A, Ru was found in both aerosol (70%) and gaseous form (30%). But in the "night" sample C, Ru in gas form did not exceed 2%, possibly due to differences in air temperature. At the same time, in samples A and C, the portion of gaseous ^{131}I was 98%.

In July–September of 1986, in some of the rooms of the 3rd block, aerosol concentration (400–1000 Bq/m^3) was not lower than at the same period near the destroyed reactor.

Before building the roof of the "Shelter," it was necessary to ensure the necessity of reconstruction and connecting up the filtration station, which was common for the 3rd and 4th blocks. For this purpose, in August and September of 1986 11 air samples were taken from a height of 10—30 m above the reactor [23].

Analysis of samples shows that aerosol concentrations (C) consisted of gamma-radiation nuclides, were stable and their values usually did not exceed limit permissible numbers. Short-lived radionuclides were not found in the samples, which proved that no chain reaction was in the fragments of the fuel; the values of C for alpha-active aerosols were in good correlation with C for gamma-radiation (C of the last was 500±200 times higher). By measuring ^{242}Cm activity on aerosol particles, assessments were provided for C ^{239}Pu. It was shown that the aerosol concentration of ^{239}Pu was in the range of 0.1–0.9 Bq/m^3 and was usually higher than the limit permissible value. Despite the fact that at the end of August ^{131}I had to be decayed, in one of the samples it was found completely in the gaseous form (0.6 Bq/m^3). Small amounts of Ru (2–5%) were also in the gaseous form; AMAD values of aerosol consisting of isotopes of Ce, Cs, Zr, and Nb were the same (1.06±0.08 μm (September. 10, 1986). There were slightly smaller sizes of particles for Ru (0.84 ±0.08 μm), but according to one measurement it was difficult to judge the difference. The absence of diversity of isotopes by particle sizes showed that the temperature in the higher layer of the destroyed reactor was close to the temperature of outdoor air. The next day, particle sizes were higher (AMAD=1.7±0.1 μm) and sizes of Ru were the same as for other isotopes.

During the same period, samples were taken 1 m above the earth's surface around 300 m from the destroyed reactor. The aerosol concentrations were on average an order of magnitude lower than near the reactor. Assessment showed that the contribution of aerosols from the reactor to aerosol concentration near the earth's surface was not more than some percents.

Important information about the physicochemical forms of radioactive aerosols, sampled near the surface not far from the destroyed reactor, was obtained by analysis of radionuclide composition. As seen from Figure 16.5, radionuclides that made the larger contribution in the summary gamma-radiation of samples can be divided into two groups.

In the first group are refractory nuclides Zr, Nb, La, and Ce, which despite the high temperature during the accident remained in the destroyed 4th block. The second group — ^{103}Ru, ^{106}Ru, ^{134}Cs, ^{137}Cs — includes volatile compounds, which condensed on aerosol particles in the air.

From the data in Figure 16.5, we can see that very high concentration refractory elements were in samples of 20–22, 25–26, and 30–31 of August. Their contribution of Ru and Cs did not exceed 20%; therefore on these days, the contribution of fuel aerosol was highest.

The sample of September 2 was different. Here, the contribution of Ru and Cs was near 65%. It has been mentioned very often in the literature [24–26] that the radiation of some highly radioactive particles sampled even at a great distance from Chernobyl consisted of a large portion of Ru. On other days there was a mixture of fuel and condensation particles in the atmosphere.

Only in October–November 1986, when a substantial portion near the destroyed reactor was cleaned up from radioactive fallout, sealed by gravel, sand, and concrete, aerosol concentrations decreased by an order of magnitude and reached relative levels, which were observed in July–September of 1986.

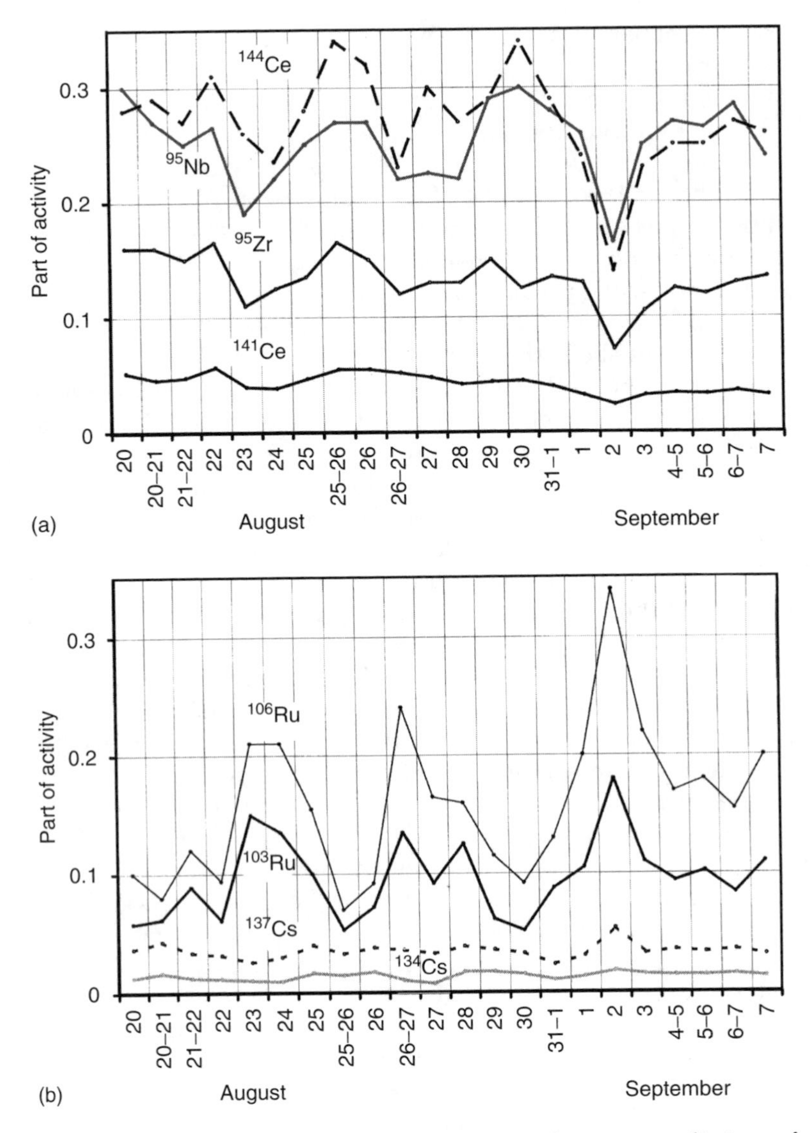

Figure 16.5 Relative contribution of radionuclides in total gamma-radiation of the aerosols sampled from ground level of atmosphere (1 m above a surface of ground) near destroyed reactor in 1986: (a) refractory nuclides, (b) nuclides with volatile components.

16.6 Gaseous components I, Te, and Ru in the atmosphere

The products of the Chernobyl accident in the atmosphere were not only included in aerosols but also in the form of gases. During the sampling in May 1986, both aerosol and gas co pounds such as I, Te, and Ru were measured. For the detection of gases, multilayer filter material SFM-I, consisting of an aerosol layer, upper facing filter material, and two sorption layers, was used. The ratio between aerosol and gaseous components of radionuclides was determined according to the distribution in layers of the package. It was assumed that the sorption deposition of I, Te, and Ru was near 100%.

The layer-by-layer distribution of volatile nuclides in SFM-I, which was used for air sampling above the destroyed reactor, is present in Table 16.4.

Table 16.4 Distribution (%) by Layers of Material SFM-I of I, Te, and Ru in Samples

Sampling Date (1986)	Component	SFM-I Material Layers			
		Aerosol	Upper Facing	First Sorption Layer	Second Sorption Layer
08.05		—	?	?	?
	^{131}I	69.4	19.4	11.2	—
	^{103}Ru	88.7	4.2	7.1	—
	^{106}Ru	84.0	5.2	10.8	—
	^{132}Te	91.7	3.3	5.5	—
14.05		—	0.25*	5.1*	2.5*
	^{131}I	72.5	19.8	7.7	—
	^{103}Ru	97.4	1.0	1.6	—
	^{106}Ru	96.0	2.0	2.0	—
	^{132}Te	99.0	—	1.0	—
15.05		—	0.29*	4.7*	2.7*
	^{131}I	42.4	19.2	18.7	19.7
	^{103}Ru	99.4	—	0.6	—
	^{106}Ru	97.3	—	2.7	—
16.05		—	0.26*	2.4*	2.7*
	^{131}I	41.4	34.5	20.4	3.7
	^{103}Ru	96.1	1.9	2.0	—
	^{106}Ru	97.1	—	2.9	—
17.05		—	0.28*	5.0*	4.2*
	^{131}I	28.0	41.0	28.0	3.0
	^{103}Ru	97.7	—	2.3	—
	^{106}Ru	100	—	—	—
19.05		—	0.24*	5.0*	3.5*
	^{131}I	8.9	64.5	22.2	4.4
	^{103}Ru	87.0	7.0	6.0	—
	^{106}Ru	87.8	5.7	6.5	—

* quantity of Sorbent in layer (mg/cm^2)

In middle of May, 1986 in the gaseous form, it was 0.6–13% ^{103}Ru, 2.7–16% ^{106}Ru, and 1–8% ^{132}Te.

The similarity of distribution for both isotopes of Ru showed that they are present in air in the same gaseous compounds, and their absence in the second layer shows that sorption was good. Even with high filtration velocity, the efficiency of gaseous compounds of Te was also good.

Radioactive iodine was found in the second sorption filter. So the gaseous compound in which it was, sorbed worse than that with Ru and Te. Only in samples taken above the reactor on May 8 and 14 (Table 16.4) was iodine found in the upper facing and in the first sorption layer. This can be explained by the fact that iodine was in the form of vapors of I_2. These measurements point out that from May 15 to 19, new gaseous iodine compounds were in the atmosphere with smaller coefficients of dynamic sorption, that is, possible organic forms (CH_3I, etc).

Measurements of radioactive components of I and Ru were provided also at a great distance from Chernobyl NPP [27] from aircraft at a distance of 100–300 north of Chernobyl above Ukraine, Belarus, and Russia at a height of 1200 m, and from a research ship 4000 km in the Atlantic ocean. The results of measurements of concentrations of ^{131}I and the ratio of aerosol and gaseous components of I and Ru are shown in Table 16.5.

As we can see from Table 16.5, gaseous ^{131}I was present in all samples, and gaseous Ru in the majority of samples. Based on date and site of sampling (near 4000 km from Chernobyl), samples from May 4–5 belong to the first ejection of radioactive materials from the exploded reactor. Measurements provided in May 1986 showed a weak dependence of gaseous ^{131}I from the distance of the Chernobyl NPP — the lowest near the reactor and the highest in the Atlantic, North, and Baltic sea.

Unfortunately, in the majority of countries around the world, atmosphere monitoring was provided only by aerosol filters [5]. It is, however, necessary to pay attention to the measurements of gaseous components of the Chernobyl accident provided in Europe, U.S.A., and Japan. Summary data of the global atmosphere monitoring phase composition of ^{131}I at different periods after the Chernobyl accident [28] are presented in Figure 16.6.

From the majority of measurements performed in the first 29 days after the accident, it is clear that gaseous I concentration was higher than aerosol concentration, and 1 month after the accident (data from Japan and U.S.A.) it was still higher than 0.7.

After deposition on the particle I compounds can leave its surface due to desorption. If desorption probability is close to the duration of sampling, the second desorption can take place between different parts of the sampler. Therefore, experimental data on aerosol-gas ratio can be higher than the actual value [29].

Taking into account the correction on desorption, it was found that the real portion of gaseous iodine was in the range of 0.33–0.63 with a mode of 0.48 [29]. In [30, 31] it was found that ^{133}I was in gaseous form in the same proportion with aerosols as ^{131}I, and in the

Table 16.5 Volume Activity of ^{131}I, Phase Composition I and Ru in the Atmosphere above Ukraine, Belarus, and Russia and from a Research Ship in May 1986

Sampling Date, 1986	Sampling Site	Sampling Velocity (m/sec)	Time of Sampling (hour)	Concentration ^{131}I (Bq/m^3)	Portion of Gaseous Component (%)		
					^{131}I	^{103}Ru	^{106}Ru
8.05	Above the reactor	0.9–1.2	1	5000	30	11.3	16
14.05				160	28	2.6	4
15.05				1.5	58	0.6	2.7
16.05				350	59	3.9	2.9
17.05				17	72	2.3	
19.05				136	91	13	12.2
14.05	Ukraine, Belarus, Russia, north of Chernobyl	0.9–1.2	1–2	0.34	57	0.7	
16.05				1.5	46	2.2	2.7
17.05				1.7	77	4	3.5
19.05				0.81	90	8	12
4–5.05	Atlantic, North, and Baltic sea	1.0	31	0.008	79		
5–6.05			25	0.015	84	23	
6–7.05			18	0.034	90	18	
7–9.05			44	0.031	73	2	7
10–11.05			27	0.001	79		

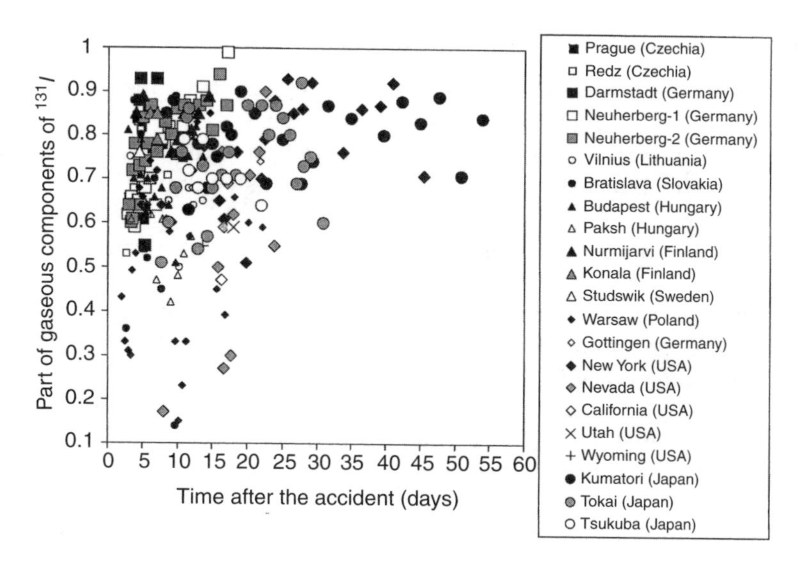

Figure 16.6 Results of global monitoring of the gaseous component of ^{131}I of Chernobyl origin [48].

period from April 29–May 2 the gaseous fraction was 60–80%. From this, the important conclusion can be drawn that the behavior of ^{129}I and ^{135}I in the atmosphere was similar. It was also found that ^{132}Te was in the air in gaseous form with 10% of its general concentration, which was the same as the results of [27]. After the first days of the accident, both aerosol and gaseous forms of ^{137}Cs were also found [32].

So in the process of monitoring the products of the Chernobyl accident and the new previously unknown data on physicochemical forms of existence in the atmosphere, some radionuclides such as Te, Ru, Cs, and I isotopes were found.

16.7 Characteristics of radioactive aerosols near the earth's surface

Extensive territories were polluted as a result of fallout on the earth's surface of gaseous–aerosol products of the accident at Chernobyl NPP. The most polluted territories were close to the Chernobyl station, where the density of pollution by ^{137}Cs was more than 3.7 MBq/m^2 (100 Ci/km^2).

After fallout was stopped from the reactor, the main mechanism of producing aerosol was the secondary raising (resuspension) from polluted surfaces of the soil, trees, buildings, etc. Concentration of these secondary aerosols depends on many factors. Among them are meteorological situation, characteristics of the surfaces (presence of vegetable layer, buildings, surface humidity and type of soil, intensity of mechanical activity on the surface, presence of the snow layer), physicochemical properties of pollution, time after forming of primary pollution, etc.

The value that characterizes the danger of radioactive fallout, as a source of aerosols, is the resuspension coefficient

$$K_r = C/S.$$

where C is aerosol concentration (Bq/m^3) and S is the density of pollution (Bq/m^2).

Because S depends only on the decay constant of radionuclides, the value of C is in the same way as informative as K_r (in the case of absence of aerosol transport from other regions).

As a result of extensive studies on secondary aerosols of Chernobyl origin, it was found that

- The concentration of secondary aerosols decreases 3–4 times approximately linearly at a height of up to 15 m above the soil [33].
- ^{137}Cs and ^{144}Ce concentrations change in the region by more than an order of magnitude, and concentration decrease takes place faster than it should according to radioactive decay.
- The intensity of fallout decreases with increasing distance from roads and sites of agricultural activity [34].
- Aerosol concentration increases linearly with the speed of wind [35].
- Values of K_r decrease with time after primary fallout according to exponential law with a time constant of 0.02–0.12 month^{-1}[36].
- Strong winds resulted in aerosol transport from the zone of pollution by hundreds of kilometers, which resulted in an increase in aerosol concentration on remote territories by tens–hundreds of times [37].
- The values of K_r measured in the same periods, but at different points of Europe, differ from Chernobyl data in the range of order of magnitude, and variations are accidental in nature. In 1994, K_r was around 2.3×10^{-10} month^{-1} [36]

After completion of the "Shelter" building, measurement of aerosol disperse composition around the Chernobyl NPP continued at some points in a radius of 0.5–5 km around the reactor. All samples were taken at a height of 1.5 m from the surface. The method of multilayer filters was used for sampling [21].

More extensive data on disperse composition of aerosol carriers of radionuclides until April 2002 were obtained by analysis of ^{137}Cs. ^{144}Ce disperse composition was measured only until the end of 1992, when this radionuclide decayed practically completely.

Figure 16.7 shows the value of AMAD of ^{137}Cs and ^{144}Ce for all periods of monitoring, where AMAD is in the range of 3–8 μm. These data are comparable with the data presented in [38]. It should be noted that in the majority of cases, the sizes of particle carriers of ^{90}Sr and plutonium isotopes are the same as for ^{137}Cs.

From the presented data, it is clear that average AMAD increased from the summer–autumn of 1986 until 1991–1992, after which it became constant until now. It is difficult to find a reason for this. The sizes cannot be different from the sizes of nonactive

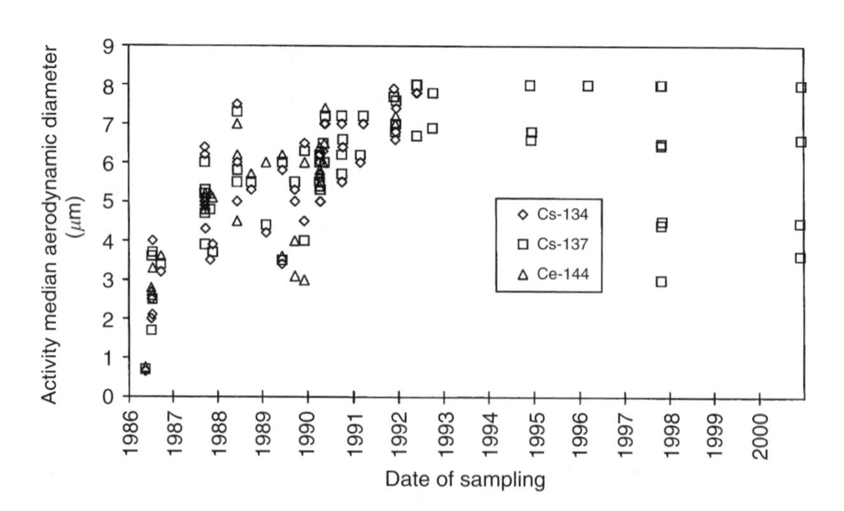

Figure 16.7 AMAD (in μm) of aerosols of ^{134}Cs, ^{137}Cs, ^{144}Ce in 5-km zone around the Chernobyl NPP in 1986–2000.

aerosol, which were developed from resuspension from surfaces of soil, vegetable, roads, buildings, etc. because of natural causes. Such was the situation in 1992. However, the continuation of activity near the Chernobyl NPP (transportation of soil, transport movement, etc.) led to changes in aerosol concentration and disperse composition, which was observed frequently [39].

16.8 Forest fires in the exclusive zone

Forest fires taking place on territories polluted with radionuclides are an additional factor of radiation danger, because under these circumstances aerosol particles that consist of radioactive substances are formed. Their concentration can substantially exceed what was typical for this site before the fire. Physicochemical characteristics of aerosols of fire genesis can be different from "background" aerosols in the atmosphere. Smoke trains are dangerous for people who take care of the fire, as for people located at great distances from the fire. Transportation of radioactive smoke leads to a redistribution of radionuclides between "dirty" and "clean" territories.

Nearly 600 different fires were observed from 1993 to 2002 in the 30-km zone of the Chernobyl NPP. For the first time, attention was paid to the radiation effects of fires after the hot and dry summer of 1992 [40]. Then radioactive substances, moving with the smoke trains from Chernobyl, were observed as far as 500 km in Lithuania [41] and Sweden [42]. Monitoring of the radiation situation at 30 points in the 30-km zone of Chernobyl showed that even at a distance of 5–10 km from forest fires on relatively "clean" territories, the concentration of radioactive aerosols increased up to 10–100 times.

Radioactive aerosol concentration growth was caused by a sublimation of light volatile ^{137}Cs from the firing zone. It was found experimentally that the growth of concentration at a distance of some kilometers from the fire was more than 2 orders of magnitude. Assessments show that if the density of pollution of the surface is 7 Ci/km^2 (2.6 $\times 10^5$ Bq/m^2) in the case of a fire close to the earth's surface, the concentration of radioactive Cs can exceed the limit permissible values [40].

A study that was provided in the framework of the experimental fire in 1993 suggested that ^{137}Cs concentration increased 1000 times, particle size distribution was bimodal (75% AMAD>10 μm, 25% AMAD=0.4 μm), and the water-soluble form of Cs increased by 38% [41]. It should be noted that the primary radioactive fallout of the accident contained 99% of Cs in nonsoluble chemical form. This means that fires are responsible not just for concentration growth. We have a new submicrometer compound of spectra size of particles much more dangerous from the point of view of dosimetry of internal irradiation. The amount of water-soluble Cs becomes more substantial. Ultrafine components of aerosol make a substantial contribution in remote transfer. For example, ^{137}Cs concentration in Vilnjus (Lithuania) during the fire of 1992 increased a 100 times [41].

16.9 Aerosols of the "Shelter"

Object "Shelter" is the aggregate of structures that sealed radioactive sources outside the active zone of the reactor of the 4th block after the accident. It was completed by November 30, 1986. As a result, a unique system was created, comprising already destroyed and rebuilt constructions, which make it generally stable without any guarantee for the destruction of some of its elements. "Shelter" is a temporary construction for the localization of nuclear fuel and radioactive materials.

"Shelter" was connected with engineer communications and was equipped with devices for fuel diagnostics, neutron sorption solutions, dust suppression, contrafire lines, etc., which present a complex of security of "Shelter" [43].

Years of operation of "Shelter" showed that the goal of the "Shelter" building was correct. Still, the difficult conditions under which it was built did not allow the construction of a truly hermetic structure. There are cracks, apertures, and technological openings for cables and pipe-lines in the structure. Natural ventilation of "Shelter" takes place through ventilation tube of the 3rd and 4th blocks.

For characterization of "Shelter" as a source of radioactive aerosols and assessment of its influence on environment, the following information is required: (a) mechanism of formation of aerosols inside "Shelter"; (b) physicochemical characteristics of aerosol particles in different parts of the object; and (c) values and dynamics of ejection of aerosols in the free atmosphere.

16.9.1 *Types of aerosols*

At the present time, radioactive aerosols of "Shelter" are formed from dust, present in its compartments (the concentration of uranium in dust is assessed as 5–10 t [2]), and from fuel-containing materials (FCM). FCM is destroyed due to radioactive decay, strong fields of radiation, variations of humidity and air temperature, and also building activity and drilling work.

During the observation period, two types of radioactive aerosols were found in the "Shelter". First, aerosols of disperse origin consist of particles of irradiated fuel and, usually, with only small amounts of volatile isotopes of Cs, Ru, Te, I. Second, condensation aerosols were formed due to the absorption of these and other radionuclides on the particle carriers.

The fuel particles also consist of two groups. To the first group belong large particles (average diameter near 30 μm), consisting of grains of UO_2. To the second group belong smaller (diameter some micrometers) particles, formed after graphite burning, fuel oxidation, and its partial melting. These particles consist of small amounts of fragmental radionuclides, especially Cs.

16.9.2 *Aerosol concentration inside the "Shelter"*

In the first years of work of the "Shelter," the main cause of aerosol forming was drilling work inside the structure. Small radioactive particles were present in large amounts in boring liquids, on the floor, which led to high aerosol concentration.

During the existence of "Shelter," a large number of aerosol samples were taken in room 207 — the corridor was divided into three sections. After the completion of drilling activity, some of the boreholes remained open and became channels of intake for air and water from sites with high concentrations of radionuclides products of the accident. According to the measurements provided in 2000, the dose rate of gamma-radiation in 207 changed in the range 0.1–10 mSv/h.

The results of systematic aerosol sampling in room 207/5, taken every 1 h in 1988–1991, were very significant. Figure 16.8 shows variations of summary aerosol alpha-activity. After cut out the bore equipment, their concentrations were usually at the level of 0.03 Bq/m^3, and when the boring work started it increased by 1–2 or even 3–4 orders of magnitude.

A large amount of dust formed during boring activity, and its deposition on all surfaces was very substantial even for other types of activity. For example, when the metal platform was cut, the concentration of alpha-active aerosol increased by 2 orders of magnitude, up to 7 Bq/m^3. The concentration decreased to normal level only 8–10 h after completion of the work.

The secondary long monitoring of radioactive aerosols in room 207/5 took place from November 28 to December 26, 2000 after Chernobyl NPP was closed. During those days, 20 samples were taken with 2–3 h exposure [44]. As can be seen from Figure 16.9,

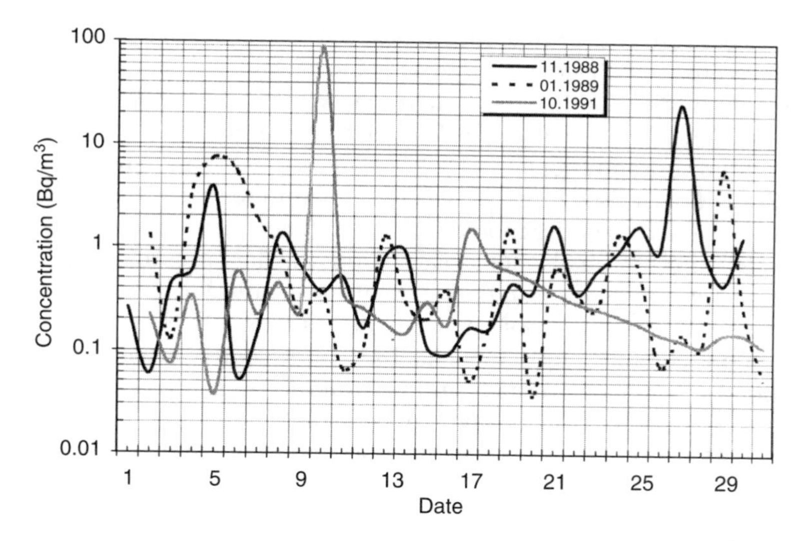

Figure 16.8 Concentration of alpha-active aerosols during drilling works in room 207/5 of "Shelter" (1988–1991).

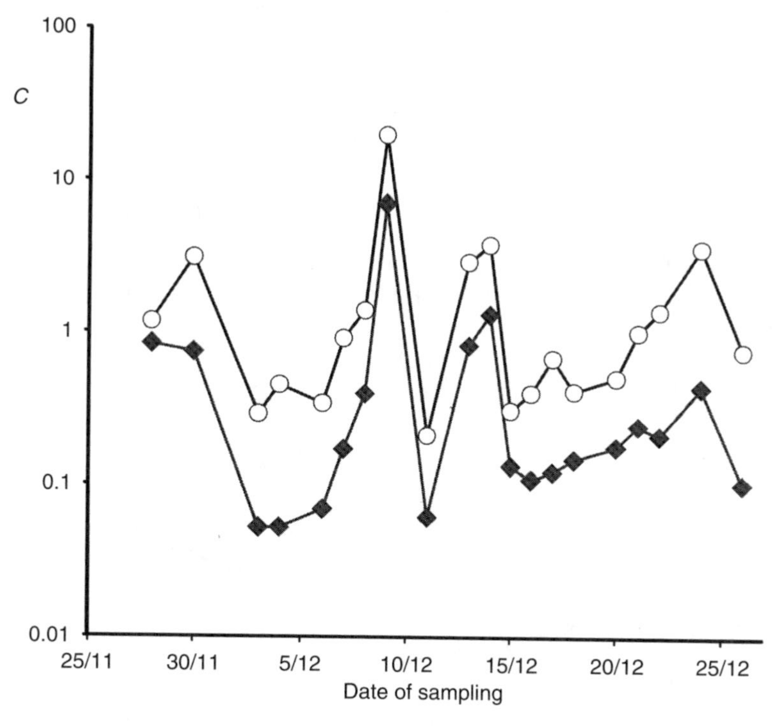

Figure 16.9 Contents of aerosols — carriers of ^{137}Cs (in Bq/m^3) and long-lived beta-active radionuclides (in pulses per second per cubic meter) in room 207/5 of "Shelter " in 2000.

^{137}Cs concentration varies from 0.2 to 20 Bq/m^3. There was a positive correlation between ^{137}Cs concentration and long-lived beta-emitters. During monitoring, some splashes of radioactive aerosol concentration were measured, but all of them were lower than during the drilling activity.

As can be seen from Figure 16.9, the highest concentration of radioactive aerosols was measured on December 9. Besides ^{137}Cs, also ^{134}Cs, ^{154}Eu, ^{155}Eu, ^{90}Sr, ^{241}Am, and ^{60}Co were found in aerosol samples. Taking radioactive decay into account, the ratio of activities of $^{134}Cs/^{137}Cs$ was 0.48, which is close to the number at the time of the accident [45].

^{241}Am and ^{154}Eu concentrations were 0.05 Bq/m^3, and for ^{60}Co it was 3 times higher. The ratio $^{137}Cs/^{90}Sr=1.15$ was also close to that calculated for the fuel of the 4th block.

Besides aerosol characteristics monitoring, simultaneous measurements of the daughter products of radon and thoron also took place. From the data presented in Figure 16.10, it is clear that, unlike accident aerosols, the concentration of natural radioactive aerosols in room 207/5 continuously declined in the second decade of the month and increased in the third.

From a comparison of Figures 16.9 and 16.10, it is clear that the dynamics of changes of concentrations of accident products and natural radionuclides are different. The cause of the high concentrations of the first type of aerosols is connected to working operations, and for the second it is due to gas emissions from building materials and constructions of the destroyed block, air exchange, and atmospheric pressure.

One important cause of sharp increasing aerosol concentrations in "Shelter" is fire inside the building. During 16 years, in the room of "Shelter" seven fires were recorded. During the fire: (1) the formation of radioactive aerosols takes place, and these aerosols are transported with smoke and hot air; and (2) inside the "Shelter" a very intensive movement of air mass takes place, which overwhelms the usual ventilation flow. This results in additional ejection of radioactive aerosols into the free atmosphere.

During the 4 hours fire on January 14, 1993, around 33 MBq of radioactive material was ejected additionally into the atmosphere. The normal ejection for that period was approximately 1.4 MBq. The average concentration of gamma-radioactive nuclides through the ventilation stack increased 140 times.

16.9.3 Aerosol transport from the "Shelter" into the atmosphere

Assessment of the influence of the "Shelter" on the environment is complicated and many factors are problematic. Radiation, including the aerosol situation, both inside and outside

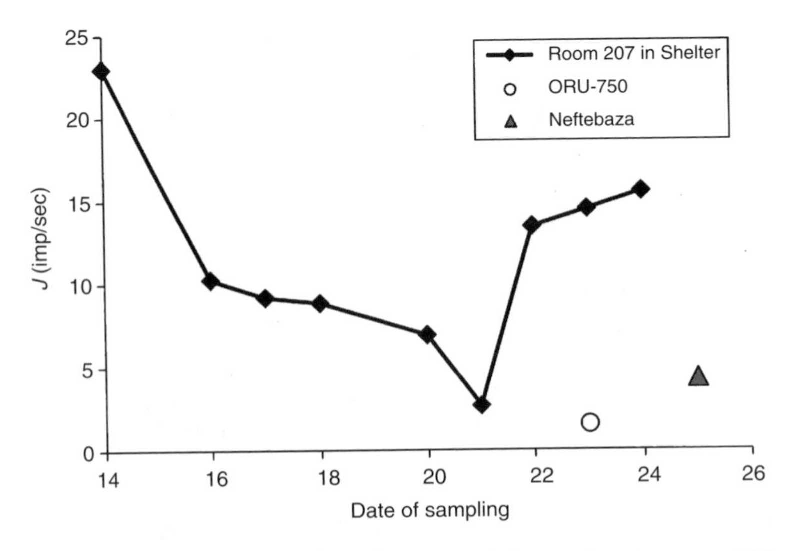

Figure 16.10 Rate of count of beta-particles of radon and thoron daughters on 50th min after the finish of aerosol sampling in room 207/5 of "Shelter", and in two points near Chernobyl NPP in December, 2000.

of the "Shelter" is stable. Contribution of the "Shelter" as a source of radioactive aerosols is only some percents from the level permissible for the normal working of a nuclear block of 10^3 MWt.

During the experimental study provided in 1996–2000 [46], parameters of air flows in the apertures in the walls and characteristics of radioactive aerosols transported into the atmosphere were determined by means of aerosol filters.

It was found that the maximum aerosol concentrations of ^{137}Cs, ^{90}Sr, and $^{239+240}$Pu in the exits of apertures are close to permissible values. In the cold time of the year, the transportation of radioactive aerosol from the "Shelter" was substantially higher than during the warm period. In summary, the rate ejection of 290 Bq/sec (7×10^9 Bq/yr), part of ^{137}Cs was 220 Bq/sec (76%), for ^{90}Sr — 58 Bq/sec (20%), and for $^{239+240}$Pu — 1.1 Bq/sec (0.38%). The rest of the activity was determined by ^{60}Co, ^{106}Ru, ^{125}Sb, ^{134}Cs, ^{144}Ce, ^{154}Eu, ^{155}Eu, and ^{241}Am. Data presented in [47] are smaller, but in this study a plane table and not a filter was used for sampling.

In August 2000, the concentration of aerosols of ^{137}Cs was in the range of 0.01–0.06 Bq/m^3 and exceeded the concentration of radiocesium near the earth's surface on the territory only slightly [48].

16.9.4 Aerosol dispersity inside the "Shelter"

The first study of disperse composition of aerosols inside the "Shelter" was provided in 1988 [49]. Analysis of 180 particles with about 80% of aerosol activity, collected in filters, showed that aerosol particle size distribution can satisfactorily be described by a logarithmic normal law with a median diameter of 5 μm. The results of the measurements of disperse composition of aerosols provided by the method of multilayer filters in room 207/5 [44] is presented in Figure 16.11.

The dispersity of carriers of accident products was in the range of AMAD 0.5–6 μm. Such a wide range of sizes was probably because aerosols were produced from different

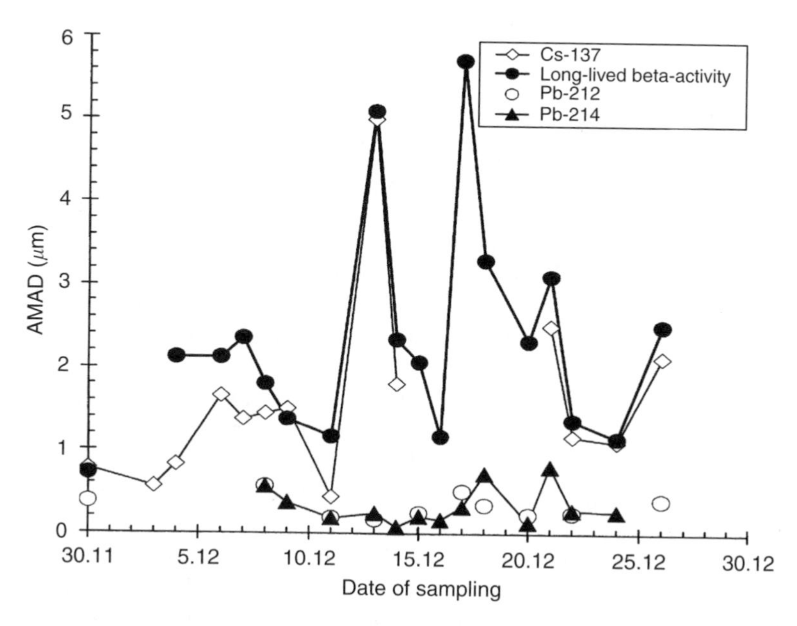

Figure 16.11 AMAD of aerosols — carriers of ^{137}Cs, long-lived beta-active nuclides, ^{212}Pb, and ^{214}Pb in room 207/5 of "Shelter" in December, 2000.

sources, with different mechanisms of formation. Large particles were probably of disperse origin, that is, they become small, erode, and disperse some materials. Submicrometer-sized particles can be produced in the "Shelter" by dispersion of liquids, including radioactive solutions formed by condensation of moisture or sediments. After evaporation of moisture from micrometer-sized or even larger particles, submicrometer particles with small amounts of even dry particles can be formed. It should be noted that during welding work, sizes of radiocesium were AMAD=1.02 μm.

Aerosols of daughter products of radon and thoron belong to condensation aerosols. They form due to deposition of atoms of metals (Po, Pb, Bi) that arise during radioactive decay of gaseous maternal substances on very small aerosol particles, called condensation nuclei. These aerosols can be formed both inside the "Shelter" due to emanation of radon and thoron from building constructions and from the free atmosphere. The AMAD values of such particles are in the range of 0.08–0.8 μm.

Important data were received on December 9 in room 207/5, when aerosol concentration was the highest. Three methods were used: beta-radiation radiometry, gamma-spectrometry, and beta-spectrometry. It was found that all radionuclides (^{90}Sr, ^{134}Cs, ^{137}Cs, ^{241}Am) were disposed on the same particles with AMAD values near 1.5 μm. The ratio of ^{137}Cs/^{90}Sr was near 1.15 and was practically equal to data presented in [45] for average nuclear fuel of the 4th block at the moment of the accident.

During the study of ejection from the "Shelter" from 1996 to 2000, a five-cascade impactor of the Institute of Biophysics (Russia) and an Andersen impactor PM-10 were used.

A majority of the measurements were provided for ^{137}Cs (22 samples) and ^{241}Am (16 samples). It was found that the AMAD for aerosols of ^{137}Cs was 2.9$\pm$1.5 μm, and for ^{241}Am it was 1.4$\pm$0.4 μm. No systematically measured AMAD values of aerosols of ^{90}Sr and $^{239+240}$Pu were in the range of 3.1–4.5 and 6.0–9.3 μm, respectively. It should be noted that these data are similar to the data obtained by the method of multilayer filters. Unfortunately, in [46] there were no comments on the substantial difference in the size of Pu and Am, which is a daughter product of ^{241}Pu.

16.10 Radioactive aerosols close to the surface layer of the atmosphere near "Shelter"

Aerosol monitoring on territory close to the "Shelter" began after "Shelter" was built. For many years, sampling was provided by means of aspirators, disposed at a distance of 50–80 m from the "Shelter" in the south, north, and northwest of it. Filter materials FPP-15-1.5 were used and the exposure time was 2 weeks. Figure 16.12 presents the results of monitoring during 2000 close to the earth's layer of air.

A convincing illustration of the influence of "Shelter" on the composition and concentration of radioactive aerosols near Chernobyl NPP was the fire of January 14, 1993 [43]. During the fire, concentration exceeded the usually observed level by approximately 30 times.

In 2000, aspiration in the north and northwest showed the same concentration values: for ^{137}Cs — 1–3 mBq/m^3, and for ^{241}Am — 0.02–0.06 mBq/m^3. During wintertime, the values were a little lower than in the summer. In the south filter installation they are usually higher than for the north and northwest, which can be explained only by the higher density of pollution of the soil in this site.

A substantial concentration growth took place in the spring of 2000, when at a distance of 300–500 m from the "Shelter" building activity took place. Figure 16.12 reveals that in this period aerosol concentration increased substantially. It should also be mentioned that the ^{137}Cs/^{241}Am ratio increased at the beginning and end of 2000 (Figure 16.13).

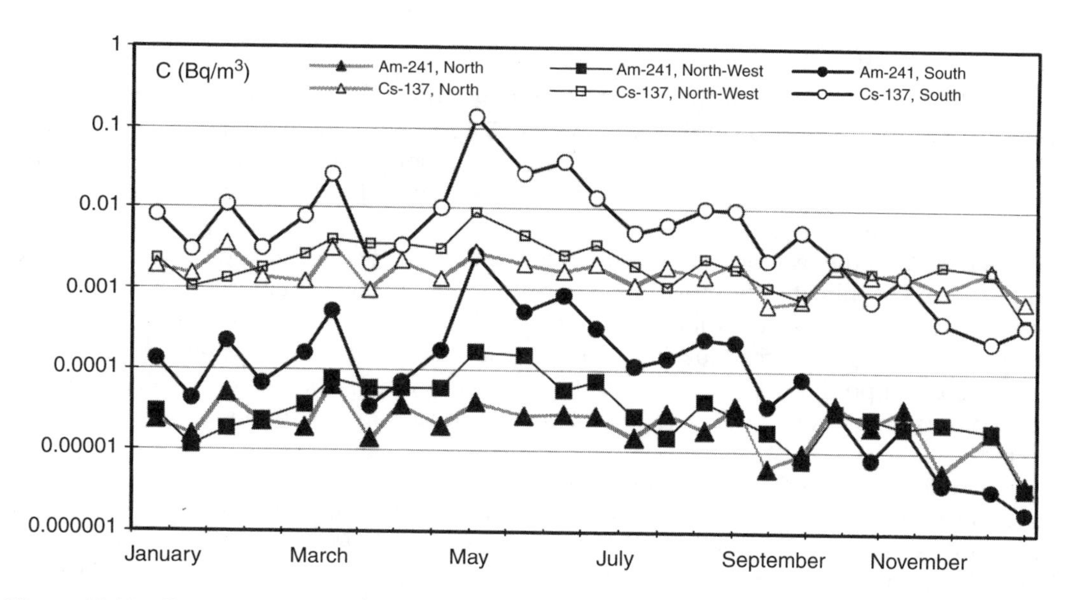

Figure 16.12 Concentration of radioactive aerosols of ^{137}Cs and ^{241}Am in ground level of atmosphere near "Shelter" in 2000. Sampling was executed in three points.

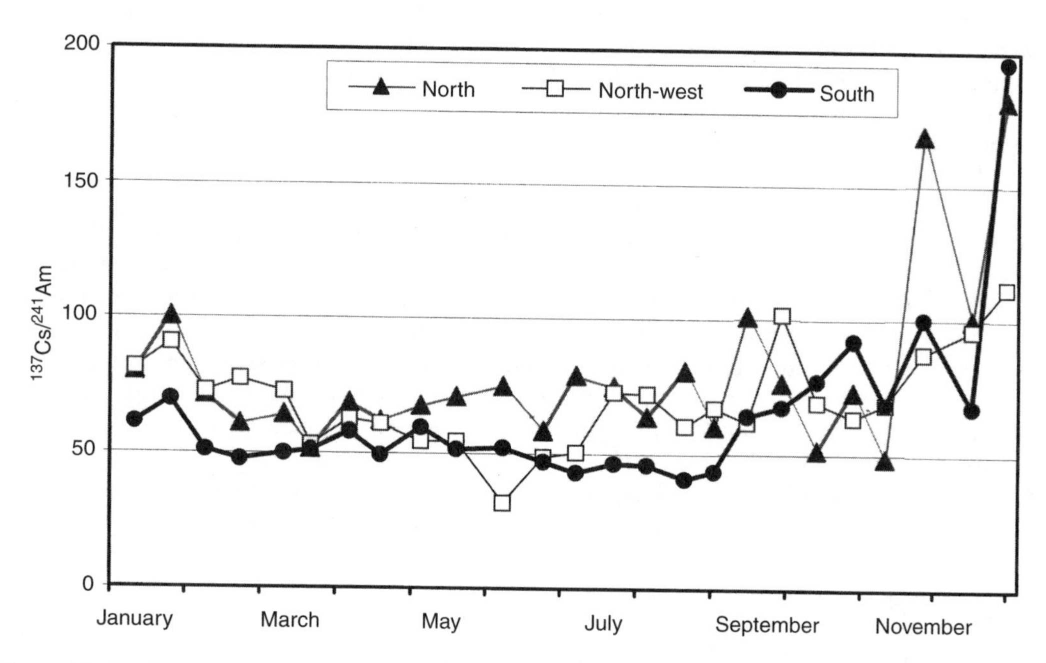

Figure 16.13 Ratio of activity ^{137}Ce/^{241}Am in aerosols of ground level of atmosphere near "Shelter" in 2000.

This can be explained as a result of changing the ratio between contributions of aerosols due to dust raising in the territory and transportation from the "Shelter." In the limit situation, when dust raising is minimum, for example, in the case of a stable snow layer, the ^{137}Cs/^{241}Am ratio can be close to 200, which is typical for aerosols ejected from cracks in the walls [46].

16.11 Conclusion

The accident in the reactor of the 4th block of Chernobyl NPP resulted in the ejection of radioactive materials with a summary activity of 3×10^{18} Bq without taking inert gases into account. As a result of fallout of radioactive gas–aerosol products of the accident on the earth's surface, extensive territories not only of the former U.S.S.R. but also of other countries were polluted.

Ejection from the reactor was not stable both in terms of concentration and radioactive composition. Usually, the composition of aerosol particles did not correspond to the composition of irradiated nuclear fuel at the moment of the accident. Transport of activity from the reactor at long distances took place on particle carriers of submicrometer range.

Volatile compounds of I, Te, Ru, and Cs for a very long period of time were found in both aerosol and gaseous form. The largest portion of gaseous component was for radioactive iodine.

After stopping ejection from the destroyed reactor, radioactivity at the ground level was determined by secondary raising of aerosol particles from polluted surfaces. Aerosol concentration depends on many natural and technogene factors. Variations of concentration under typical conditions were in the range of an order of magnitude. A decrease in concentration took place substantially faster than according to the law of radioactive decay. Radionuclides were disposed on the same carriers, and sizes of the particles were in the micrometer range.

Fires on the territory, polluted by radionuclides of Chernobyl origin, resulted in the sharp growth of aerosol concentrations of radiocesium both in the zone close to the source and in the far zone at a distance of hundreds of kilometers. Its transport from the fire zone takes place on submicrometer carriers and the water-soluble forms of Cs in aerosols of fire origin are increased substantially.

"Shelter" was built above the reactor in order to prevent pollution of the environment from radioactive materials. Only some percents of permissible levels for 1 GWt capacity nuclear object ejection go through cracks in the walls of the "Shelter". Inside this construction, aerosol concentration depends on the type and intensity of provided work and changes in the range of some orders of magnitude. The composition of particles usually corresponds to the composition of nuclear fuel, and long-lived α- and β-nuclides are disposed on particles of the same sizes of micrometer and sometimes submicrometer range.

The physicochemical characteristics of radioactive aerosols of Chernobyl origin are still being studied very extensively. At the present time, the 30-km zone around Chernobyl NPP and "Shelter" are unique experimental testing areas, where it is possible to provide a wide spectrum of studies, particularly studies of the behavior of disperse systems, and the testing of new methods of measurements and analysis of radioactive aerosols.

Acknowledgment

Chapter translated from original Russian by Lev S. Ruzer.

References

1. U.S.S.R. State Committee on the Utilization of Atomic Energy, The Accident at the Chernobyl NPP and its Consequences, IAEA Post-Accident Review Meeting, Vienna, 25–29 August, 1986.
2. Borovoi, A.A. and Gagarinski, A.Yu., Chernobyl 15 years after: radioactive release, *Nucl. Eur. Worldscan*, No. 1–2, 34–35, 2001.
3. Persson, C. et al., The Chernobyl accident — a meteorological analysis of how radionuclides reached and were deposited in Sweden, *Ambio*, 16, 1, 1987.

4. Kownacka, L. et al., Vertical distribution of ^{131}I and radiocesium in the atmosphere over Poland after Chernobyl accident, *Acta Geophys. Polonica* 34, 405–412, 1986.
5. Raes, F. et al., Radioactive measurements in air over Europe after the Chernobyl accident, *Atmos. Environ.*, 24A, 900–916, 1990.
6. Borisov, N.B., Isotopy v SSSR, 52/53, 1978 pp. 66–67 (in Russian).
7. Dobrynin, Yu.L. and Khramtsov, P.B., Date verification methodology and new data for Chernobyl source term, *Radiat. Prot. Dosim.*, 50, 307–310, 1993.
8. Gavrilin, Yu.I., Bjulleten po atomnoi energii, 8, 20–28, 2001 (in Russian).
9. Borisov, N.B. et al., Composition and Concentration of Gaseous and Aerosols Fractions Above the 4-th Unit of Chernobyl NPP and Far from Reactor in May 1986, Science Technical Report Series, Environment Protection, Questions of Ecology and Food Quality Control, Moscow, No. 1, 1992, pp. 11–17 (in Russian).
10. Ter-Saakov, A.A., Glebov, M.V., and Gordeev, S.K., in Chernobyl–90, Reports of All-Union Conference, Chernobyl, Vol. 1, part 2, 1990, p. 3 (in Russian).
11. Ter-Saakov, A.A., Kurinny, V.D., and Michaelyan, A.I., in Chernobyl–90, Reports of All-Union Conference, Chernobyl, Vol. 1, part 2, 1990, p. 9 (in Russian).
12. Dewell, L., Nuclide composition of Chernobyl hot particles, in *Hot Particles from the Chernobyl Fallout*, von Philisborn, H. and Steinhausler, F., Eds., Bergbau, Theuern, 1988, p. 16.
13. Ooe, H., Sirinuntavid, S., Ootsuji, M., Seki, R., and Ikeda, N., Size distribution of radionuclides in airborne dust (April 1986), *J. Radiat. Res.*, 28, 68, 1987.
14. Yanase, N., Parpyatipsakul, Y., Matsunaga, T., and Kasai, A. Concentration and particle size distribution of airborne dust at Chiba from reactor accident at Chernobyl, *J. Radiat. Res.*, 28, 67, 1987.
15. Ooe H., Seki R., and Ikeda N., Particle size distribution of fission products in airbornedust collected at Tsukuba from April to June 1986, *J. Environ. Radioactivity*, 6, 219, 1988.
16. Maqua, M. and Bonka, H -G., Deposition velocity and washout coefficients of radionuclides bound to aerosol particles and elemental radioiodine, *Radiat. Prot. Dosim.*, 21, 43, 1987.
17. Georgi, B., Helmeke, H.-J., Hietel, B., and Tschiersch, J., Particle size distribution measurements after the Chernobyl accident, in *Hot Particles from the Chernobyl Fallout*, von Philisborn, H. and Steinhausler, F., Eds., Bergbau, Theuern, 1988, p. 16.
18. Kauppinen, E.I., Hillemo, R.E., Aaltonen, S.H., and Sinkko, K.T.S., Radioactivity size distributions of ambient aerosols in Helsinki, Finland, during May, 1986 after the Chernobyl accident: premilinary report, *Environ. Sci. Technol.*, 20, 1257, 1986.
19. Erlandsson, B., Askind, L., and Swietlicki, E., Detailed early measurements of the fallout in Sweden from the Chernobyl accident, *Water, Air, Soil Pollut.*, 35, 335, 1987.
20. Belovodsky, L.F. and Panfilov, A.P. Ensuring radiation safety during construction of the facility "Ukrytie" and restoration of unit 3 of the Chernobyl nuclear power station, One Decade After Chernobyl: Summing up the consequences of the Accident, Poster Presentations, Vol. 2, International Conference, Vienna, April 8–12, 1996, Vienna, IAEA, 1997, pp. 574–597.
21. Skitovich, V.I., Budyka, A.K., and Ogorodnikov, B.I., Method and results of aerosol size definition in 30-km zone of Chernobyl nuclear power plant in 1986–1987, in Rogers, J.T., Ed., Proceedings of the International Seminar on Fission Products Transport Processes in Reactor Accidents, Dubrovnik, Yugoslavia, May 22–26, 1989, Hemisphere Publishing Corporation, New York, 1990, pp. 779–787.
22. Ogorodnikov, B.I., Pavluchenko, N.I., and Pazukhin, E.M., Radioactive Aerosols of "Shelter" Object (a review), Part 1. Aerosol Statement in Industrial Zone of the Chernobyl NPP under Building of the "Shelter" Object, Preprint No. 02-10, National Academy of Sciences of Ukraine, Interdisciplinary Scientific and Technical Centre "Shelter," Chernobyl, 2002, 48 pp. (in Russian).
23. Ogorodnikov, B.I., Radioactive products over the damaged block-4 of Chernobyl nuclear power plant before the completion of "Sarcophagus," in Rogers, J.T., Ed., Proceedings of the International Seminar on Fission Products Transport Processes in Reactor Accidents, Dubrovnik, Yugoslavia, May 22–26, 1989, Hemisphere Publishing Corporation, New York, 1990, pp. 799–806.

24. Devell, L., Tovedal, H., Bergstrom, U. et al., Initial observations of fallout from the reactor accident in Chernobyl, *Nature*, 321, 192–193, 1986.
25. Devell, L., Tovedal, H., Bergstrom, U. et al., Initial observations of fallout at Studsvik from the reactor accident at Chernobyl, Studsvik Energiteknik AB, Sweden, Report No. NP-86/56, 1986, 23 pp.
26. Interim Report on Fallout Situation in Finland from April 26 to May 4 1986, Finnish Centre for Radiation and Nuclear Safety, Finland, STUK-B-YALO 44, May 1986, 38 pp.
27. Borisov, N.B. et al., Observation of Gaseous-aerosols Components of Radioiodine and Radioruthenium at First Weeks after the Chernobyl NPP Accident, Science Technical Report Series, Environment Protection, Questions of Ecology and Food Quality Control, Moscow, No.1, 1992, pp. 17–24 (in Russian).
28. Ogorodnikov, B.I., Problems of environment and natural resources, *Rev. Inform.*, 53, 1998 (in Russian).
29. Budyka, A.K., *J. Aerosol Sci.*, 31, Suppl. 1, S478, 2000.
30. Hotzl, et al., Ground depositions and air concentrations of Chernobyl fallout radionuclides at Munich-Neuherberg, *Radiochim. Acta*, 41, 181–190, 1987.
31. Radioactivity Measurements in Europe after the Chernobyl Accident Part 1, Air. EUR-12269, 1989, pp. 35, 229.
32. Styro, B.I., Filistovich, V.I., and Nedvetskaite, T.A., Isotopes of iodine and radiation safety, St. Petersburg, Gidrometeoizdat, 1992 (in Russian).
33. Garger, E.K., *J. Aerosol Sci.*, 25, 745, 1994.
34. Kashparov, V.A., Protsak, V.P., Ivanov, Y.A., and Nicholson, K.W., *J. Aerosol Sci.*, 25, 755, 1994.
35. Hollander, W., *J. Aerosol Sci.*, 25, 789, 1994.
36. Garland, J.A. and Pomeroy, I.R., *J. Aerosol Sci.*, 25, 793, 1994.
37. Budyka, A.K. and Ogorodnikov, B.I., Radioactive aerosols generated by Chernobyl, *Russ J. Phys. Chem.*, 73, 310–319, 1999.
38. Dorrian, M.D. and Bailey, M.R. Particle size distributions of radioactive aerosols measured in workplaces, *Radiat. Prot. Dosim.*, 60, 119–133, 1995.
39. Pasukhin, E.M. and Ogorodnikov, B.I., Radionuclide Conduct upon Forest Fire, International Scientific Workshop Radioecology of Chernobyl Zone, September 18–19, 2002, Slavutych, Ukraine, Abstracts of Poster Display Presentations, Slavuyich, 2002, pp. 132–133.
40. Budyka, A.K. and Ogorodnikov, B.I., *Radiazionnaya biologia, Radioekologia*, 35, 102–112, 1995 (in Russian).
41. Lujaniene, G., Ogorodnikov, B.I., Budyka, A.K. et al., An investigation of changes in radionuclide carrier properties, *J. Environ. Radioactivity*, 35, 71–90, 1997.
42. Hollander, W. and Garger, E.K., Eds., Contamination of Surfaces by Resuspended Material, Final Report EUR 16527 EN, 1996.
43. Gerasko, V.N., Klyuchnikov, A.A., Korneev, A.A. et al., Unit "Shelter": history, state and perspektive, Slavutych, 1998.
44. Ogorodnikov, B.I. and Budyka, A.K., *At. Energ.*, 91, 470–475, 2001 (in Russian).
45. Begichev, S.N. et al., Radioactive release due to the Chernobyl accident, in Rogers, J.T., Ed., Proceedings of the International Seminar Fission Products Transport Processes in Reactor Accidents, Dubrovnik, Yugoslavia, May 22–26, 1989, Hemisphere Publishing Corporation, New York, 1990.
46. Garger, E.K., et al., Problems of Chernobyl, #10, Part 2, Slavutitch, 2002, pp. 60–71.
47. Borovoy, A.A., Problems of Chernobyl, #10, Part 2, Slavutitch, 2002, pp. 192–198.
48. Ogorodnikov, B.I., *At. Energ.*, 93, 42–46, 2002.
49. Bogatov, S.A., *At. Energ.*, 69, 36–40, 1990.

chapter seventeen

Aerosol filtration (aerosol sampling by fibrous filters)

A.K. Budyka and B.I. Ogorodnikov
Karpov Physico-Chemical Institute, Moscow

Contents

17.1 Introduction 541
17.2 Terminology and definitions 542
17.3 Particle capture mechanisms 544
 17.3.1 Diffusion 544
 17.3.2 Interception 545
 17.3.3 Inertia 545
 17.3.4 Electrostatic 545
 17.3.5 Gravitational effect 546
 17.3.6 Combined actions of the filtration mechanisms 546
17.4 The most (more) penetrating size 546
17.5 Pressure drop 547
17.6 Effect of nonstationary filtration 547
17.7 Fiber filters FP (Petryanov's filters) 548
17.8 Using the fiber filters for particle size measurement 549
 17.8.1 Size range 551
 17.8.2 Sampling 551
 17.8.3 Comparison with impactor 551
 17.8.4 MMF application 552
17.9 Filters for detecting gaseous compounds 552
17.10 Correction on desorption of volatile substances 553
17.11 Filter material composition for atmospheric monitoring 553
17.12 Conclusion 554
Acknowledgment 554
References 554

17.1 Introduction

Aerosol characteristics should be determined for the description of aerosol particles behavior in the free atmosphere, industrial conditions, dwellings, and in the lungs. These characteristics are determined by different methods. Nonetheless, the most accurate analysis of aerosol parameters can be carried out in the dispersed phase, where particles will be isolated from the gaseous medium. The most effective and prevalent method of capturing of aerosol particles is filtration.

1-56670-611-4 / 05 / $0.00+$1.50

Certainly, filtration is used not just for aerosol sampling. The main goal of filtration is the refinement of air and technological gases from dispersed admixtures of very different origin. Filter materials used in units for air refinement, or in devices for individual protection of the breathing organs should have a high capturing efficiency of aerosol particles. For example, the efficiency of ULPA filters [1] is 99.9999% for particles of 0.3 μm.

Filter materials intended for aerosol sampling should meet different requirements. On the analytical filter, the largest amount of the substance should be sampled during the least time. The efficiency of capturing should not be so high as in the case of cleaning filters, because the uncertainty of particle characteristic measurements is usually not less than several percents. Such analytical filters should have relatively low aerodynamical resistance in order to provide high productivity for aerosol sampling. They should have high durability and plasticity. And finally, the analytical filters should consist of substances that will not hamper but instead facilitate the process of sediment analysis. For aerosol capture, aerosol filters consisting of fibers (fiber filters) or representing other pore structures are used [2].

Sediments of the particles are then analyzed by means of physicochemical or nuclear-physical methods. In particular, radioactive aerosols are investigated by alpha-, beta-, and gamma-spectrometry methods, often in combination with autoradiography. For the isolation of alpha- and beta-activity, radiochemical methods are used.

Different types of filters are produced for aerosol analysis in developed countries, information of which can be found in the firms' reference books and on the Internet.

In Russia, in the majority of studies fiber filter materials of FP type, consisting of ultrafine polymer fibers, are used [3]. For the first time in 1938, it was obtained by I.V. Petryanov and N.D. Rosenblyum under the direction of N.A. Fuchs at the aerosol laboratory of the Karpov Institute of Physical Chemistry.

17.2 Terminology and definitions

Let us assume that N_0 aerosol particles enter the filter and N particles penetrate it (Figure 17.1).

The ratio $K=N/N_0$ is defined as penetration through the filter, and $E=1-K$ is efficiency of the filter. Because the filter renders resistance to gas flow, the pressure before and

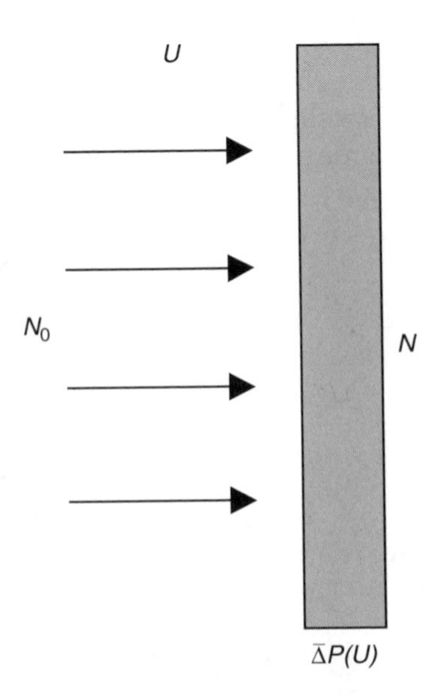

Figure 17.1 Parameters for the definition of main filter characteristics.

after it is unequal. The pressure after the filter is smaller than before it by a value ΔP. This value is called pressure drop. The value of the pressure drop depends, besides other parameters, on the velocity of the gas flow U. Pressure drop measured when the flow velocity is 1 cm/sec is called standard filter resistance and marked as $[\Delta P]$.

Fibrous filters consist of thin fibers, randomly oriented in planes perpendicular to the flow. The volume portion of the fibers, or the packing density α is very small, and usually no more than 5%. The porosity of the filter is $1-\alpha$.

By increasing the thickness of the filter material indefinitely, it is possible to achieve as much high efficiency as required. But in this case, the more effective the filter material, the higher the aerodynamic resistance.

The main filtration law, formulated by Langmuir [4], established that the number of monodispersed particles, deposited by a layer of thickness dx, is proportional to the thickness of the layer and the number of particles before the filter N_0. From this, we obtain

$$K=e^{-\gamma\delta x} \tag{17.1}$$

where γ is the filtration coefficient, which depends on the properties of the particles, medium, and the filter. It is obvious that under fixed flow velocity through the filter, the pressure drop on it will be directly proportional to the thickness of the layer: $\Delta P = k\delta x$. The ratio of the logarithm of the penetration to the pressure drop, taken with the opposite sign, does not depend on the thickness of the filter, and characterizes its capability in capturing aerosol particles:

$$q_F=\log(K)/\Delta P \tag{17.2}$$

The higher the value of q_F ("filter quality") [2] or the "coefficient of the filter operation" [3], the better the filter. In other words, a smaller thickness of filter material is needed to provide the necessary efficiency of particle capture. The value q_F depends on the properties of the particles and filter.

The filter characteristics mentioned above are macroscopic values, that is, they determine the filter as a whole, irrespective of its structure. The following question could be asked: how is it possible to obtain such high filter efficiency if the porosity is so high? Many are mistaken in suggesting that aerosol filters work like a sieve, catching all particles larger than a definite size and letting much smaller particles pass.

For theoretical calculation of the filter properties and an understanding of the mechanisms of particle capture, the concept of the "single fiber capture coefficient" is used. It is determined as the number of particles (N_1) colliding with a fiber of unit length, divided by the number of particles (N) in the flow with a width equal to fiber diameter d_f. For a value N calculated for a substantial distance from the fiber, that is, in undistorted flow,

$$\eta = N_1/N \tag{17.3}$$

After transformation, we obtain correlation between the efficiency of the filter and penetration and single fiber capture coefficient:

$$E=1-K=1-e^{4\alpha H\eta/\pi d_f} \tag{17.4}$$

where H is filter thickness.

The experimentally measured value of the single fiber capture coefficient is the average characteristic. At the present time, an accurate calculation of the "local" capture coefficient, based on peculiarities of the field of flow near the concrete fiber and the hydrodynamic influence of neighboring fibers, is impossible to make. Therefore, in the calculation of the capture coefficient, regulated models of the fiber filters are used, with the correction data obtained by different empirical and half-empirical coefficients [5].

17.3 Particle capture mechanisms

What kinds of mechanisms lead to the capture of aerosol particles by filter fibers? There are several such mechanisms. First, it should be mentioned that the so-called "sieved" mechanism is not the main mechanism for fiber filters.

17.3.1 Diffusion

1. For low filtration velocity (some cm/sec), diffusive capture by fiber filters plays an important role. Aerosol particles not moving along streamlines flow round the obstacle (in our case, the fiber) and become displaced from them due to constant collisions with gas molecules (Figure 17.2).
2. The smaller the size of the particle and its velocity in the fiber direction, the higher the probability of collision.

Kirsch and Stechkina [5] found that the diffusion capture coefficient is equal to

$$\eta_D = 2.7 Pe^{-2/3} \tag{17.5}$$

where Pe is the Peclet number, which is the characteristic of the predominance of convection transfer at diffusion:

$$Pe = d_f U_0 / D \tag{17.6}$$

where U_0 is the velocity of flow and D is the diffusion coefficient of aerosol particles. The diffusion coefficient is equal to:

$$D = \frac{k_B T C_{Kn}}{6\pi\mu r},$$

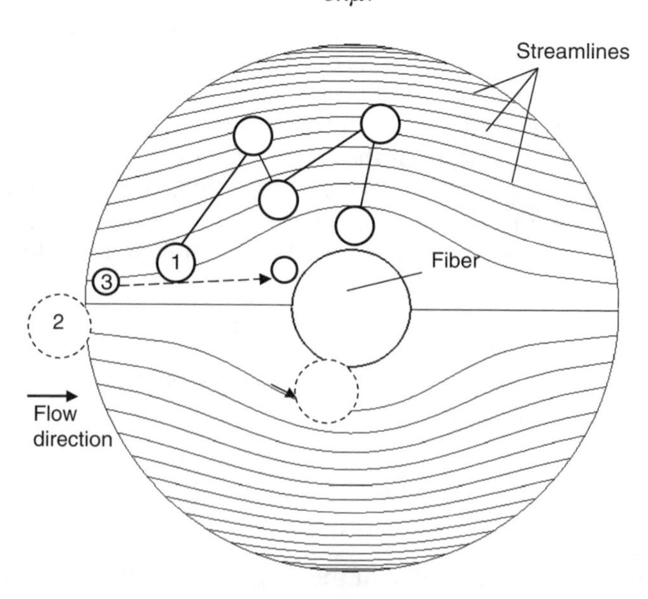

Figure 17.2 Main mechanisms of aerosol capture by fibers: (1) diffusion; (2) interception; (3) inertia.

Table 17.1 Diffusion Coefficients in Air [a]

d (μm)	0.01	0.1	0.5	1
D (cm^2/sec)	5.31E −04	6.84E −06	6.31E −07	2.76E −07

[a] Pressure 1 atm, Temperature 20°C [6]

where C_{Kn} is the Cunningham correction factor, k_B is the Boltzmann constant, and T is temperature, (μ is gas viscosity). In Table 17.1, the values of the diffusion coefficient smaller than 1 (m) are presented. It should be mentioned that the diffusion capture coefficient can be higher than 1.

17.3.2 Interception

If particles move with a streamline in the direct proximity of the cylindrical fiber, then by increasing the particle size the probability of its interception increases. This mechanism is called "interception effect" and the interception coefficient can be calculated according to the formula [5]

$$\eta_R = \frac{1}{2k}\left[\frac{1}{1+R} - (1+R) + 2(1+R)\ln(1+R)\right], \quad R = \frac{r}{a} \tag{17.7}$$

where r represent particle radii, a represent fiber radii, and k is the Kuwabara hydrodynamic factor equal to [7]

$$k = -0.5\ln\,\alpha - 0.75 + \alpha - o(\alpha) \tag{17.8}$$

17.3.3 Inertia

Large and (or) heavy particles can be displaced from the flow due to inertia and can collide with a fiber. The probability of collision depends on the Stokes number (St) and the Reynolds (Re) number. Particularly for $Re < 0.5$ [8], and for $0.8 < St < 5$

$$\eta_{St} = 0.45(\lg\,St + 0.4) \tag{17.9}$$

The critical value of St, below which inertia deposition does not take place [9], for the cylindrical fiber is $St_c = 0.25$ [10, 11].

Inertial capture of particles is a predominant mechanism for the deposition of submicrometer and micrometer aerosols with high filtration velocities (1 cm/sec and higher). As a result, the filter efficiency sharply increases with an increase in particle size (proportionally to the square radii and velocity). Later we will show that this rule has been successfully used for the separation of aerosol particles by size fraction.

17.3.4 Electrostatic

If the particle and fiber are charged, the probability of capture due to electrical charge is increased. This effect is widely used for dust deposition. In some filtering installations, particles are intentionally charged in the corona field in order to increase filtration efficiency in the atmosphere and charges on the particle are distributed according to the Boltzmann law.

Fibers of FP materials are charged enough extensively; therefore, they find wide use as a filter layer as a means of individual safety of breathing organs [12].

At the beginning of filtration, the efficiency of deposition of aerosol particles by the charged filter can be of an order of magnitude much higher than the efficiency of the uncharged material.

Still, during use, especially in the case of humid air, the charges flow down from the fibers, and then efficiency decreases, as mentioned below, because now efficiency is determined mainly by diffusion, inertia, and interception.

It should also be mentioned that the contribution of the electrostatic mechanism decreases rapidly with an increase in filtration velocity; therefore, for velocities of 0.5 m/sec and higher, we can often ignore the contribution of the electrostatic mechanism.

17.3.5 *Gravitational effect*

Gravity, which acts on aerosol particles, contributes to the particle capture coefficient. This effect is substantial for heavy particles and mostly takes place under filtration from "below to up" or from "up to bottom." In the first case the summary effect is increasing, and in the second it is diminishing. It is on the value of order, $\eta_G \propto G(1 + R)$, where G is the ratio of the velocity of gravitational deposition to filtration velocity [2].

17.3.6 *Combined actions of the filtration mechanisms*

All capture mechanisms of aerosol particles by fibrous filters work simultaneously. The most important task when we try to calculate the filter efficiency is the assessment of the contribution of each mechanism in the final catching coefficient. By studying the movement of particles in direct proximity to the fiber, it is difficult to separate catching due to diffusion, interception, and inertia. It turns out that the summary catching coefficient is larger than the sum of coefficients for each mechanism. Therefore, the expression for the entire coefficient is

$$\eta = \eta = \sum_{i} \eta_i + \sum_{i \neq j} \eta_{ij}$$

where i, j are indexes of the mechanisms of catching [5]. For example,

$$\eta_{DR} = 1.24 k^{-1/2} Pe^{-1/2} R^{-2/3} \tag{17.10}$$

By increasing the size of particles, contribution of the diffusion deposition declined but hooking increased, and there is an optimal size of particle by which the filter obtained minimum efficiency. This size depends upon filtration velocity and filter parameters. Figure 17.3 shows the dependence of the penetration through fiber filters from particle sizes [14] calculated according to formulas (5), (7), and (10).

17.4 *The most (more) penetrating size*

The size of particles, through which filter efficiency is minimum is called the most (more) penetrating size. For filters with fiber diameters of order 1–2 μm and filtration velocity of some cm/sec, the diameter of the more penetrating size equals 0.3–0.4 μm. This means that particle diameters of 0.2 and 0.5 capture better than particle diameters of 0.3–0.4 μm.

The test of the filters and respirators should be provided exactly with such particles. For this, generators of monodisperse aerosols were developed. Therefore, the nominal value of the efficiency of the means of defense of breathing organs is minimal. The real efficiency as for larger and smaller particles is always higher.

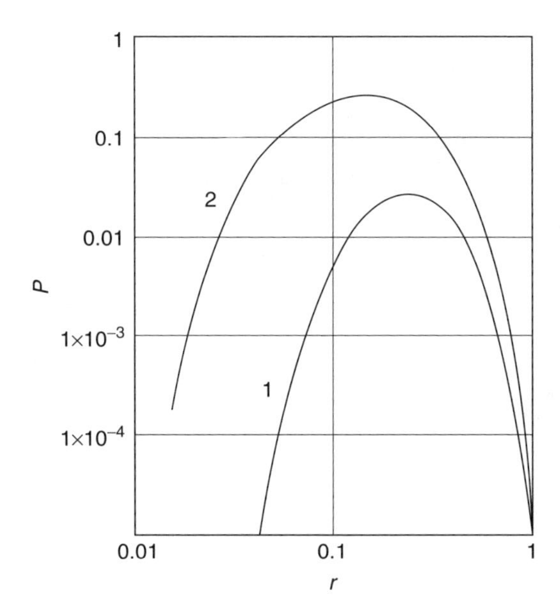

Figure 17.3 Particle penetration vs. particle radius (in μm): fiber radius a=2 μm, air velocity U=1 cm/s (curve 1) and U=10 cm/s (curve 2), packing density α=0.05, and filter thickness H=1 mm. Calculation according to formulae (5), (7), and (10).

17.5 Pressure drop

As already mentioned, for the quality of filter assessment it is necessary to know not only its efficiency but also the pressure drop. It is obvious that if material is uniform, the pressure drop will be proportional to material thickness. The value of ΔP also depends on fiber diameter, flow velocity, viscosity, and packing density. The dependence of pressure drop on filtration velocity when *Re* numbers are less than 0.5 can be expressed by the formula [15]

$$\Delta P = \frac{4\mu U_0 \alpha H}{a^2(-0.5\ln\alpha - 0.5)} \tag{17.11}$$

that is, the pressure drop is proportional to the filtration velocity. If $Re > 1$, the dependence becomes square. The calculation of pressure drop in such cases can be performed based on empirical correlation (see, e.g., [16]).

17.6 Effect of nonstationary filtration

Classical filtration theory is based upon the proposition that particles colliding with the fiber will be kept by it in the future. Still, a situation can exist when the particle will rebound from the fiber surface. This effect, which takes place when the filtration velocity is high, can result in decreasing the filter efficiency (Figure 17.4). The problem of rebound was studied in [10, 13, 17].

Another assumption of the classical theory of aerosol filtration is that the filter efficiency is not changed during the process of filtration. In reality, however, gradual cover with dust change its filtration characteristics. With an increase in dust load, the pressure drop on the filter increases. The front layer of the material can obstruct it to such a degree

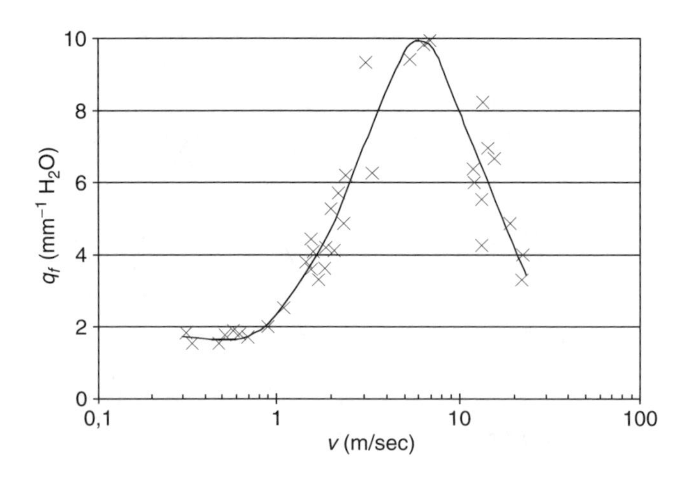

Figure 17.4 Effect of rebound of aerosol particles on filtration characteristics of fibrous filter: quality factor (q_f) vs. velocity of particles (v) (mean fiber diameter = 7 μm, packing density = 0.03, particle diameter = 0.4 μm).

that filtration changes from the surface to the deep end of the filter. This leads to a sharper increase in pressure drop.

Every filter is characterized by its resource of work, determined by the limit permissible dust load. This value depends not only on aerosol concentration but also on filtration velocity, the nature of particles, and their size. The general theory of nonstationary filtration does not exist at the present time; still in this direction some results have already been achieved [18, 19].

17.7 Fiber filters FP (Petryanov's filters)

Fiber filter materials FP are statistically uniform structure [20], which consists of charged ultrafine polymer fibers (Figure 17.5).

The fibers build up by the stretching of polymer solutions under the influence of an electrostatic field (electrospinning technology) [21]. At the present time in the Karpov Institute

Figure 17.5 SEM photograph of typical structure of FP filter (mean fiber diameter = 1 μm, Teflon fibers, packing density = 0.03).

of Physical Chemistry, more than 30 filter materials have been developed. From different polymers, the Russian industry produced some millions of m^2 per year of filter materials.

Filter materials FP were produced in the form of standard linen size of 0.7×1.3 m. These pieces of linen were later used for making respirators, analytical filters, and aerosol filters type HEPA and ULPA. Radioactive aerosol sampling in outdoor environment used pieces of linen according to the size of the filters. As usual, they were used in high output aspirators, counted for hundred thousand m^3 air per hour. Airplane mean sampling of atmospheric aerosols (filter gondolas) has an output up to 20,000 m^3/h [22]. It is possible to accommodate up to 6 m^2 filter material. After sampling filter material FP, it is easy to compress in compact pills. This is especially useful in the measurement of gamma-radioactive aerosols. In the industrial environment, stationary or portable sampling devices are used, where analytical aerosol filters of AFA type are displaced by an area of 3–20 cm^2. The output of such devices is relatively low (tens to hundreds of l/min), which is enough for dosimetric purposes.

Analytical filters AFA represent disks of filter material. Table 2 shows a list of the filters AFA that are manufactured by Russian industry [23].

17.8 *Using the fiber filters for particle size measurement*

For particle size measurement, different methods are used. These represent well-known accurate and time-consuming methods analysis by means of optical and electronic microscopy. Still, if aerosol was built from some of the sources, it is difficult to estimate what kinds of chemical compositions or elements are in the particles of different sizes. The same problem exists when using the measurement technique for disperse analysis based on light scattering [24]. Impactors [25], very widely used for disperse analysis of aerosols, have relatively small outputs of sampling. Therefore, despite the relatively high accuracy of the division on size fraction, they cannot be used in cases when the aerosol concentration is very small. The task of restoration of the spectra of aerosol particles belongs to the class of "incorrect" if we do not make special propositions about unknown functions. In the case of filtration of aerosols by means of a package, which consists of consequently displaced fiber materials, this problem can be written as

$$N_i(\vec{u}) = \int_0^\infty E_i(r) f(\mathrm{r}, \vec{u}) \mathrm{d}r, \; i = 1 \ldots n \tag{17.12}$$

Table 17.2 Russian Filters for Aerosol Analysis

Type AFA Filter	Purpose	Type of Filter According to the Method of Analysis
AFA XP AFA XA	Determination of the concentration of chemical and radiochemical aerosol composition	Chemical
AFA RMP AFA RMA	Determination of volume aerosol activity	Radiometric
AFA RSP	Determination of the volume aerosol activity of isotopic aerosol composition	Radiospectrometric
AFA RGP	Determination of the size and activity of aerosol particles	Radiography
AFA VP	Determination of mass aerosol concentration	Mass
AFA DP	Determination of the size of aerosol particles under a microscope	Dispersion
AFA BA	Determination of aerosol concentration	Bacteriological

where r is particle radius, N_i the portion of substance detained by, i filter, E_i the efficiency of the i-filter in the package taking into account detained substance of the previous $i - 1$ filter, the totality of the distribution parameters, f the unknown function (density function), and n the number of filters.

Efficiency of the i-rfilter E_i in the package is equal to

$$E_i = E_i^* \prod_{j=1}^{i-1} (1 - E_j^*) \tag{17.13}$$

where E_j^* is the filter efficiency of the j filter.

In the system of n integral equations of Fredholm's I type (17.12), the core of the equation is function $E_i(r)$. When we change the integral equations by sums, a system of ill-condition is built. Lockhart et al. [26,27] analyzed dispersed composition submicrometer radioactive aerosols by means of the set of some fiber filters, trying to solve the reciprocal problem (17.12) without additional preconditions. Naturally, such an approach became practically inconvenient.

Bogolapov et al. [28], due to introduction of prior information on aerosol particle size distribution, managed to simplify problem (17.12) to determine the dispersion of alpha-active submicron aerosols close to the earth's layer of air.

It is known that the sizes of the majority of the natural and artificial aerosols can be approximated by lognormal distribution (LND), the density of which is equal to

$$f(r) = \frac{1}{\sqrt{2\pi} \ln\sigma r} \exp\left\{ -\frac{(\ln r - \ln r_0)^2}{2(\ln\sigma)^2} \right\} \tag{17.14}$$

where r is particle radius, and r_0 and σ are parameters of LND (median radii and standard geometric deviation). So if we assume that the size of aerosol particles is distributed according to LND, the problem of determination of the dispersion composition will be to find parameters of LND from the system of equations

$$N_i = \int_{R_1}^{R_2} \frac{E_i(r)}{\sqrt{2\pi} \ln\sigma} \exp\left\{ -\frac{(\ln r - \ln r_0)^2}{2\ln^2\sigma} \right\} \frac{dr}{r} \tag{17.15}$$

For finding the LND parameters, it is enough to have the package from three filters, and the third should be highly effective (catching practically all particles).

The problem (17.15) is solved by minimization of the functional of the discrepancy:

$$Q = \sqrt{\sum_{i=1}^{2} (N_i - \int_{R_1}^{R_2} E_i(r) f(r) dr)^2}$$

As already mentioned under the inertia mechanism of deposition, the filter efficiency is proportional to the square of the size of the particle. For solving (17.15), the quantitative dependence filtration efficiency of filter materials FP from Stokes number was found in the range of 0.2–5 [29], and algorithms were developed for searching LND parameters [30, 31]. Extending the range of the method [32], the optimal composition of the set of filters and their characteristics [29] were defined. The technology of making fiber filters FP was improved, and especially the packages of analytical filters for the disperse analysis of aerosols, which allow one to make very thin layers of homogeneous and plastic filter materals. These advantages of the technology of FP filters allowed the practical realization of this method, known in Russia as the method of multilayer filters (MMF).

17.8.1 Size range

The range of the size of particles determined with the MMF method depends on filtration velocity and combination of filters. All restrictions are determined at the beginning and end of working of the inertia mechanism of aerosol deposition. If the Stokes number is lower than 0.2, the inertia mechanism does not take place; therefore, under low filtration velocities of very small particles, this method is not effective. When velocities are high (more than 2 m/sec) the catching coefficient is not increased according to (17.9), but lowered, and by increasing the velocity we will decrease the catching coefficient due to the effect of rebound from fibers.

It is obvious that determination of the disperse composition by means of MMF is impossible if all particles are deposited on the first layer of the package. So, the larger the particles, the less effective the first and second layers. For this purpose, we can choose thinner filters with the same fiber diameter, or filters with thicker fibers.

If the velocity of sampling is around 1.5 m/sec and [ΔP] of the first layer is around 1.2 Pa/cm/sec, analysis of the particle sizes in the range of 0.1–2 μm is possible. By decreasing the velocity to 0.5 m/sec and decreasing the resistance of the first layer to 0.8 Pa/cm/sec, it is possible to determine the size of the particle with median diameters up to 7–8 μm [33].

17.8.2 Sampling

The filter package for the dispersed analysis is displaced in the filter holder. Filter layers are displaced in the direction of the airflow in order of increasing of [ΔP]. In our measurements, the regime of the inertia catching of aerosols is in three-layer packages, in which the first and second filters were made from polymer fibers around 7 μm. The standard resistance of the first filter was usually 1 Pa/cm/sec. To decrease the errors of analysis, [ΔP] of the second filter should be 1.73 times higher [29]. In order to catch all particles by the third filter, penetrating through the first two, it should consist of thin fibers and have [ΔP] no less than 20 Pa/cm/sec. Usually the third filter consists of filter material FPP-15-1.5. The areas of filter materials of the package can be different, but in our studies we used areas in the range of 0.3–6$\times$10^4 cm^2.

The linear filtration velocity should be the same during sampling. Depending upon the supposed size range of particles, its value should be chosen in the range 0.5–2 m/sec, and the volume velocity should be up to 20,000 m^3/h. Because of this, dispersed analysis of aerosols with a very small concentration will be possible. The time of sampling is determined by instrumentation sensitivity for the measurement of deposited material on each filter layer of the package.

17.8.3 Comparison with impactor

- *Distribution function*. In MMF the density distribution is given a *priori*, and then its parameters are restored. In the case of an impactor, according to the obtained gystogram, the distribution function can be restored. The use of MMF for the analysis of bimodal distribution is ineffective, but it is possible to change the last cascade of the impactor on the filter package [34].
- *Number of measurements*. In the case of MMF, three filters are used, impactors have no less than five stages of division, and some (Andersen Ambient, Andersen MK III) consist of eight, and even ten (Sierra/Marple Model 210) cascades. So in using MMF, we need less measurements, which is important from the point of view of efficiency and cost of the study.

- *Sampling output*. Theoretically there is no restriction on the MMF method output. It is restricted only by linear sampling velocity and but not the filter area. Under real conditions, by using the standard sampling technique it was possible to achieve out put up to 20,000 m^3/h. The outputs of the most powerful impactors are substantially lower (around 300 times).
- *Size range*. MMF developed on the basis of 7 μm fiber allows one to measure the size of particles in the range of -0.04–7 μm. The error is higher at the edge of this range than in the middle, and we obtain the best results in the range of 0.1–5 μm. No one existing model of impactor allows one to make measurements in such a broad range, but with impactors we can get better results only when the diameter of the particle is higher than 5 μm.
- *Auxiliary technique and sampling procedure*. For MMF, it is possible to use the sampling devices for sampling on fiber aerosol filters. An impactor, which is a complicated construction with high aerodynamic resistance, needs special powerful ventilators for the high velocity of flow. In this sense, MMF is more simple and cheaper in operation.

17.8.4 MMF application

The use of MMF for the determination of aerosol dispersity allowed one to obtain many interesting results. For example, beginning in 1986, this method is the main experimental tool for studying radioactive aerosols — products of the Chernobyl accident (see chapter 16).

In the process of the wide range of study of aerosol transfer from the Aral region [35], the same atmospheric aerosol was studied by means of an impactor, a laser aerosol spectrometer, and MMF. The difference in the value of mass median aerodynamic diameter (MMAD) was no more than 20%. The method was also used for the study of the size of natural radioactive aerosols ^{7}Be, ^{212}Pb [36], and atmospheric aerosols near Moscow [37] for the assessment of the dispersity of some special aerosols in technological communications of industrial installations [38], nuclear reactors [39], and for the complex study of other aerodynamic systems [40].

17.9 Filters for detecting gaseous compounds

Some chemical compounds (I, Ru, Te, Cs, Hg) exist in the atmosphere simultaneously in aerosol and gaseous form. It is known that fiber filters can retain aerosols, but cannot retain gases. Therefore, two-component sorption filter materials FP (SFM) were developed, consisting of fibers and particles of sorbent, introduced between the fibers [41]. Due to their small sorption capacity, such materials cannot often be used in technological clearance, but as analytical means they have some advantages in comparison with other types of detection (coal cartridge, sorption columns, etc.). First, they have a relatively high efficiency and low flow resistance. Second, such analytical filters can be placed in standard sampling devices for fiber filters. And finally, they are simple to use. On the basis of SFM sorption, analytical filters AFAS are produced for the detection of gaseous compounds of some dangerous substances. Table 17.3 shows the characteristics of some of these analytical filters AFAS.

It should be pointed out that the sorption of gaseous compounds of the same chemical elements takes place in a different manner. For example, radioactive iodine was detected in the light (I_2) and difficult sorption (CH_3 I, etc.) compounds, which are characterized by different dynamic coefficients of sorption and efficiency of deposition. For the separation of different compounds of iodine after the aerosol filter several analytical filters

Table 17.3 Characteristics of Some Analytical Filters for the Detection of Gaseous Compounds [23]

Analytical Filter	Sorbent	Efficiency of Detection of Gaseous Compounds under Velocities up to 5 cm/sec	Resistance to Airflow, Pa, Velocity 1 cm/sec	Application
AFAS-I	Carbon OU-A type with nitrogen-oxide silver	90–99	50	I
AFAS-P	Carbon OU-A type	99	60	Po
AFAS-R	Carbon OU-A type	99	50	Hg
AFAS-U	Carbon OU-A type	99	60	Low volatile compounds

AFAS are displaced. According to the distribution of iodine in this package not only the phase composition of iodine, but also the ratio between the light and heavy sorption component was determined [42].

Layers of SFM are displaced after aerosol filters. Depending on the substance, SFM with different sorbents are used. The use of SFM substantially simplified the monitoring. Plasticity of the matrix SFM allowed, as in the case of filters used in MMF, one to prepare the device in the form of compact pills.

17.10 *Correction on desorption of volatile substances*

After deposition on the aerosol particle, volatile substances can desorb from its surface. The time on the particle is determined by desorption probability (λ_{des}). If this time is comparable with sampling time, the secondary redistribution of the substance between stages of the sampler takes place. In this case, the experimental data of the gaseous substance can be substantially enhanced in comparison with the real data. This will lead to incorrect dose assessment.

Unfortunately, it is impossible to calculate λ_{des}, but it can be assessed based on experimental data. The majority of data are available for radioactive I^{131} — a product of the Chernobyl accident. Using the simplest model of transportation of iodine from the aerosol particle, by means of nonlinear regressive analysis it was possible to determine $\lambda_{des}=0.13\ h^{-1}$ [43].

From this value of desorption probability, it is possible to conclude that during the characterized time $t_{1/2des}=\ln 2/\lambda_{des}$ equal approximately to 5.3 h, around half of the radioactive iodine from the aerosol filter will be deposited on the second stage of the sampler.

17.11 *Filter material composition for atmospheric monitoring*

For deposition and subsequent analysis of aerosol characteristics and substances simultaneously present in the atmosphere in the form of aerosols and gases, sampling should be provided on a row of consequently displaced filters and SFM (Figure 17.6).

The first three filters are necessary for the determination of the aerosol dispersed compositions by means of multilayer filters. The next layer SFM is used for the determination of gas components. The number of filters in the package are variable. The use of such packages allows one, after measurement of the deposited material on every filter, to determine

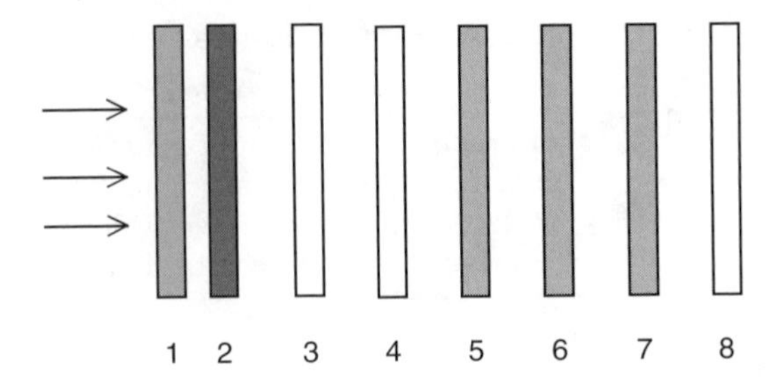

Figure 17.6 Filter pack for air monitoring: 1–3, fibrous flters for the analysis of dispersity; 4, filter for preventing the penetration of aerosols to sorption-filtering material; 5–7, SFM; 8, protective fibrous filter.

all the main characteristics of the gas–aerosol system, including aerosol dispersity and the ratio between aerosol and gas components.

17.12 Conclusion

Fiber filter materials represent a powerful tool of sampling. An understanding of the nature of filtration allowed one to choose analytical filters (the most appropriate in each concrete case), determine the duration and velocity of the sampling, and the optimal condition of deposition. The use of composition from different filter materials makes it possible not only to measure the concentration and chemical (radionuclide) composition, but also dispersed composition, the ratio between aerosol and gaseous fraction, including correction for desorption.

Acknowledgment

Chapter translated from original Russian by Lev S. Ruzer.

References

1. Hajakawa, I., Ed., *Clean Rooms*, Mir, Moscow, 1990, 456 pp. (Transl. from Japanese).
2. Brown, R.C., *Air Filtration*, Pergamon, Oxford, 1992, 272 pp.
3. Petryanov, I.V., Kozlov, V.I., Basmanov, P.I., and Ogorodnikov, B.I., *Fibrous Filter Materials FP*, Znanie, Moscow, 1968 (in Russian).
4. Davies, C.N., *Air Filtration*, Academic Press, London, 1973, 173 pp.
5. Kirsch, A.A. and Stechkina, I.B., in *Fundamentals of Aerosol Science*, Ed., Show, D.T., Wiley, New York, 1968, pp. 165–256.
6. Hinds, W.C., *Aerosol Technology*, Wiley, New York, 1982, 424 pp.
7. Kuwabara, S., *J. Phys. Soc., Jpn.*, 14, 527, 1959.
8. Budyka, A.K., Ogorodnikov, B.I., and Petryanov, I.V., *Nucl. Physic Methods of Analysis for Environmental Control*, Gidrometeoizdat, Leningrad, 1985, p. 128 (in Russian).
9. Levin, L.M., *Dokl. AN SSSR*, 91, 1329, 1953 (in Russian).
10. Budyka, A.K., Ogorodnikov, B.I., Skitovich, V.I., and Petryanov, I.V., *Dokl. AN SSSR*, 284, 1161, 1985 (in Russian).
11. Uchakova, E.N., Kozlov, V.I., and Petryanov, I.V., *Dokl. AN SSSR*, 206, 916, 1972 (in Russian).
12. Basmanov, P.I., Kaminsky, S.L., Korobeinikova A.V., and Trubizyna, M.E., *Means of Individual Protection of Breath Organs, Handbook*, Sankt-Peterburg, GIPP Iskusstvo Rossii, 399c, 2002 (in Russian).
13. Loffler, F., *Separation Sci. Techn.*, 15, 297, 1980.

14. Kirsch, A.A., "CESAT-6," Sixth Session, Karpov Institute (publisher), Moscow, 2001, p. 20 (in Russian).
15. Ogorodnikov, B.I., *Trudy IPG, 21, Atmospheric Aerosols*, Gidrometeoizdat, Moscow, 1976, p. 50 (in Russian).
16. Budyka, A.K., Ogorodnikov, B.I., and Skitovich, V.I., *Electron. Ind.*, 10, 20, 1988 (in Russian).
17. Skitovich, V.I., Efimenko, V.S., and Ogorodnikov, B.I., *Trudy IPG, Vol. 21, Atmospheric Aerosols*, Gidrometeoizdat, Moscow, 1976, p. 8 (in Russian).
18. Kirsch, V.A., *Russ. Colloid. J.*, 58, 786, 1996.
19. Kirsch, V.A., *Russ. Colloid. J.*, 60, 480, 1998.
20. Budyka, A.K., *Russ. Colloid. J.*, 54, 54, 1992.
21. Filatov, Yu.N., *Electrospinning of Fibrous Materials*, Ed., Kiritchenko, V.N., Neft I Gaz, Moscow, 1997, 298 pp. (in Russian).
22. Matushenko, A.M., *Chernobyl: Catastrophe. Exploit., Lessons and Conclusions*, Inter-Vesy, Moscow, 1996, p. 436, (in Russian).
23. Basmanov, P.I. and Borisov, N.B., and Filters AFA, Catalogue-Handbook, Atomizdat, Moscow, 1970, 44 pp. (in Russian).
24. Belyaev, S.P., Nikiforova, N.K., Smirnov, V.V., and Scheltchkov, G.I., *Optical-Electronic Methods of Aerosol Studying*, Energoizdat, Moscow, 1981, 232 pp. (in Russian).
25. Hering, S.V., Tut. No. 17, presented at 1995 Annual Meeting AAAR, Pittsburg, Pennsylvania, October 9, 1995, 69 pp.
26. Lockhart, L.B., Patterson, R.L., and Sanders, A.W., *J. Geophys. Res.*, 24, 6033–6041, 1965.
27. Lockhart, L.B., Patterson, R.L., and Sanders, A.W., NRL Report 6164, UN Docum. A/AC. 82/g/l 1138, 1964.
28. Bogolapov, N.V., Kashenko, N.I., and Konstantinov, I.E., *Problems of Dosimetry and Health Physics*, Vol. 14, Atomizdat, Moscow, 1975, p. 136 (in Russian).
29. Budyka, A.K., *Ph.D. thesis*, Moscow Engineering Physical Institute, 1986, 127 pp. (in Russian).
30. Sankov, Yu.A. and Budyka, A.K., *Problems of Environmental Protection, Ecology and Quality Control*, Vol. 9, NIITEChim, Moscow, 1992, p. 35 (in Russian).
31. Budyka, A.K., Konstantinov, I.E., and Maksimov, V.Yu., *Radiation Safety and Protection of NPP*, Vol. 11, Energoatomizdat, Moscow, 1986, p. 106 (in Russian).
32. Skitovich, V.I., Budyka, A.K., and Ogorodnikov, B.I., in *Fission Product Transport Processes in Reactor Accidents*, Hemisphere Publ. Corp., New York, 1990, p. 779.
33. Budyka, A.K. and Ogorodnikov, B.I., *Russ. J. Phys. Chem.*, 73, 310, 1999.
34. Bad'in, V.I., Budyka, A.K., Ogorodnikov, B.I., and Skitovich, V.I., *Problems of Environmental Protection, Ecology and Quality Control*, Vol. 3, NIITEChim, Moscow, 1992, p. 21 (in Russian).
35. Andronova, A.V., Private communication, 1996.
36. Lujaniene, G., Ogorodnikov, B.I., Budyka, A.K., Skitovich, V.I., and Lujanas, V., *J. Environ. Radioactivity*, 35, 71, 1997.
37. Ogorodnikov, B.I., Budyka, A.K., Skitovich, V.I., and Brodovoi, A.V., *Izv. Russ. Acad. Sci. Atmos. and Ocean Phys.*, 32, 163, 1996.
38. Skitovich, V.I., Sharapov, A.G., and Ogorodnikov, B.I., in Materials of Science — Technical Seminar of Russian Nuclear Society, Sergiev Posad, 1995, p. 23 (in Russian).
39. Budyka, A.K., and Fedorov, G.A., *Isotopes in the USSR*, Energoatomizdat, Moscow, #1(72), 1987, p. 113 (in Russian).
40. Budyka, A.K., Dr. Sc. thesis, Karpov Institute of Physical Chemistry, 2001, 219 pp. (in Russian).
41. Borisov, N.B., *Isotopes in the USSR*, Energoatomizdat, Moscow, #52/53, 1978, p. 66 (in Russian).
42. Borisov, N.B., Budyka, A.K., Verbov, V.V., and Ogorodnikov, B.I., *J. Aerosol Sci.*, 25(Suppl. 1), S271, 1994.
43. Budyka, A.K., *J. Aerosol Sci.*, 31(Suppl.1), S480, 2000.

chapter eighteen

Radioactive aerosol standards

L.S. Ruzer
Lawrence Berkeley National Laboratory (USA)

Yu.V. Kuznetzov and V.L. Kustova
All-Russian Scientific Research Institute to Physico-Technical and Radiotechnical Measurements ("VNIIFTRI"), Moscow, Russia

D.E. Fertman and A.I. Rizin
Scientific Engineering Centre, "SNIIP" Moscow, Russia

Contents

18.1 U.S.S.R. special state standard for the volumetric activity of radioactive aerosols 557
18.2 Currently applicable radioactive aerosol standards 562
 18.2.1 Radon and its decay products 562
 18.2.2 Artificial radioactive aerosols 563
 18.2.3 Developing model aerosol sources SAS in order to test aerosol radiometers directly on the consumer place 565
References 566

18.1 U.S.S.R. special state standard for the volumetric activity of radioactive aerosols

Particular attention has been paid in recent years to the pollution of ambient air and production premises by radioactive and other substances owing to rapid development rates of atomic power generation, the electronic industry, use of aerosols in medicine, the pharmaceutical industry, etc. A number of organizational measures have been adopted in many countries to solve this problem; namely, work is in hand for developing and mass producing instruments for testing aerosols of all types and for raising the measurement precision. The most important problem consists of improving the existing metrological-provision means and, above all, those of highest precision.

In the previous chapter, different methods and measuring techniques were presented for measuring the aerosol concentration of different types of radioactive and nonradioactive aerosols together with both theoretical and experimental studies focused on improving the accuracy and quality of the measurements of aerosol parameters.

All this information was used in the development of the Aerosol Standard in the former U.S.S.R. The set of installation [1] approved in 1973 as a special state standard for the unit of artificial (man-made) and natural radioactive aerosol concentration had no equivalent in metrological practice of the U.S.S.R. or other countries. Experience gained in its

1-56670-611-4 / 05 / $0.00+$1.50

operation indicates that its composition and design were, on the whole, selected correctly and that it was reliable in determining radioactive pollution. This confirms the validity of principles used as a basis for determining the precision levels and number of steps in the transition of units, which were established by the scheme for testing the radioactive aerosol concentration-measuring equipment.

At the same time, it was found necessary to reduce the physical standard random error S by a factor of two, that is, to the value of 5×10^{-2}. Moreover, the considerable amount of work carried out in the state testing, certification, and checking of aerosol radiometers by means of physical-standard installations revealed the possibility of reducing the maximum errors of the references and working measuring equipment (ME) to 15–30 and 30–60%.

This was promoted to a considerable extent by the new aerosol-radiometer design. It also became evident that the physical-standard installations should be supplemented by generators of thoron and actinon daughter products, and that it is necessary to modernize the electronic equipment used in measuring the activity of aerosol samples and to extend considerably the potentialities of installations for measuring the aerosols, disperse composition, and electrical parameters.

The improved special state standard for the volumetric activity (VA) unit of radioactive aerosols incorporates generators type GERA-1 of artificial aerosols; generators GERA-2 of natural radioactive aerosols; generators GERA-3 of vapors ^{131}I; spectrometric equipment type GERA-4 for measuring the activity of alpha-, beta-, and gamma-radiating aerosol samples; installation GERA-5 for generating and measuring the disperse composition and counted concentration of inactive aerosols; and installation GERA-6 for generating and measuring the parameters of aeroions and electroaerosols.

The generators type GERA-1 are based on the method of bubbling pure air through a radioactive solution of the appropriate radionuclide salt: (^{90}Sr + ^{90}Y) and ^{239}P to produce beta-radiation, and ^{239}Pu and ^{210}Po to produce alpha-radiation. The aerosols are formed in the following way. Under the effect of surface tension, the films of bubbles that are formed on the surface of the solution are disintegrated into drops. These drops are subsequently dried at temperatures of 140–150°C and converted into solid crystals. The artificial radioactive aerosols with a solid disperse phase thus obtained are diluted with pure air. The VA of artificial radioactive aerosols can be varied by changing the activity of the solution, which is poured into the generation unit, as well as the rate of bubbling and pure air dilution. The generators have a high reproducibility of radioactive aerosol VA in the range of $7\times10^{-2} - 10^{-3}$ s^{-1} m^{-3} and a mean-square deviation (MSD) for measurement results not exceeding 5×10^{-2}.

The GERA-2 set comprises generators of radon, thoron, and actinon daughter products. The radon daughter generator is based on the method of accumulating these products in an enclosed volume and is suitable for reproducing their VA in the range of $(2\times10^2 - 4\times10^5)$ s^{-1} m^{-3} with a measurement-results MSD not exceeding 5×10^{-2}. Investigations of the generator's radon atmosphere carried out by means of the GERA-6 installation have shown that the fraction of the RaA charged aerosols amounts to 89±8% for a radon daughter products VA of 259 Bq/l^{-1} and a median aerosol particle radius of R_m=0.11 μm±15%.

The thoron and actinon daughter products generators are also based on the method of cumulating emanation daughter products in an enclosed volume. Since their half-lives represent insufficient time for penetrating from the bubbler to the atmosphere of the container, the bubblers are replaced by solid unsealed sources with thin layers of a substance based on ^{227}Ac and ^{228}Th and possessing an adequate emanation capacity. The basic raw material for obtaining ^{227}Ac and ^{228}Th preparations consisted of ^{226}Ra irradiated with neutrons which form from it the required preparations as a result of the following reaction:

$$^{226}\text{Ra}+\text{n}_\text{r}\rightarrow{}^{227}\text{Ra}\xrightarrow{\beta}{}^{227}\text{Ac}+\text{n}_\text{r}\rightarrow{}^{228}\text{Ac}\xrightarrow{\beta}{}^{228}\text{Th}$$

The ^{227}Ac and ^{228}Th sources were prepared at the V.G. Khlopin Radium Institute and placed in 1 m^3 containers type 2BP2-OS.

The thoron and actinon daughter-products generator's reproducibility expressed in MSD terms does not exceed 35%. This value is due to the small generator volume, but for practical cases it is completely satisfactory.

The natural radioactive aerosol VA can be varied by means of diluting filters. Aerosol samples are obtained with spectrometric filter types AFA-RSP-3 and AFA-RSP-10.

The GERA-3 generator of vaporous ^{131}I is based on the method of distilling by oxidized KI potassium dichromate solution. It possesses a high VA reproducibility in the range of 7×10^{-4}– 10^{56} s^{-1} m^{-3} with a measurement-results MSD not exceeding 5×10^{-2}.

The GERA-4 spectrometric equipment for measuring the activity of aerosol samples includes a spectrometric installation for measuring the activity of the disperse phase deposited on the filter and of artificial and natural alpha- and beta-radiating aerosols, and an instrument for measuring ^{131}I activity in the sample by means of alpha- and beta-radiation.

The spectrometric installation comprises three detection units (DUs), a pulse amplifier, one integral and one differential discriminator, a switching unit with three inputs, two count rate meters, a conversion device, a stable-amplitude generator, and a power pack. The samples' activity measurements by means of the alpha-spectrometric method are carried out with two DUs: the first one includes a semiconductor detector with a working area of 2 cm^2 whereas the second one is a scintillation detector consisting of a 63×0.35 mm CsI(Tl) crystal and a spectrometric photoelectron amplifier type FEU-82. The first DU is suitable for measuring, under atmospheric pressure, an atmosphere rarefied down to ~0.7 Pa, and at different distances between the detector and filter. The alpha particles' registration efficiency can attain 0.7 for a resolution of ~150 keV with respect to ^{238}Pu. The second DU comprises a collimator with a 2 mm aperture diameter and height, thus making it possible to record only the alpha-particles that leave the filter almost perpendicularly to the detector. Its registration efficiency is substantially lower than that of the first DU and amounts to 0.10 for a resolution of ~250 keV with respect to RaC′. Both DUs are suitable for measuring either by means of the alpha-spectrometric or counting method. The discriminator levels are set in the first DU by means of a spectrometric alpha-source (RSAS) with ^{226}Ra radionuclides.

The activity of the samples with respect to beta-radiation is measured with a DU based on a halogen counter type SBT-11. The registration efficiency of the radionuclide ^{90}Sr + ^{90}Y beta-particles amounts to 0.22.

The instrument for measuring the VA of ^{131}I vapors is simultaneously a beta- and gamma-radiator and consists of two DUs, a conversion device, and a power pack. One of the DUs, made on the basis of a 40×40 mm NaI(Tl) monocrystal and a type FEU-82 photomultiplier, is intended for recording gamma-count; the second DU, based on a halogen counter type SBT-11, is intended for recording beta-particles. The recording efficiency of gamma-quanta amounts to 0.25, and for beta-particles to 0.18. Measurements are made by means of the counting method.

For computing the VA from the number of recorded pulses N, the following formula are used:

For the counting method:

$$q = \frac{N}{\varepsilon \upsilon \eta \theta};$$

For the alpha-spectrometric method of radon decay products:

$$q_{RaA} = \frac{N_A(\theta, t_1)C_1}{\varepsilon \upsilon \eta};$$

$$q_{\mathrm{RaB}}=\frac{1}{\varepsilon\upsilon\eta}\,[C_2N_c(\theta,t_2)-C_3N_c(\theta,t_2)-C_4N_A(\theta,t_1)],$$

$$q_{\mathrm{RaC}}=\frac{1}{\varepsilon\upsilon\eta}\,[C_5N_c(\theta,t_1)-C_6N_c(\theta,t_2)+C_7N_A(\theta,t_1)],$$

For thoron decay products:

$$q_{\mathrm{ThB}}=\frac{1}{\varepsilon\upsilon\eta}\,[C_1'N_c'(\theta,t_2)-C_2'N_c'(\theta,t_1)],$$

$$q_{\mathrm{ThC}}=\frac{1}{\varepsilon\upsilon\eta}\,[C_3'N_c'(\theta,t_1)-C_4'N_c(\theta,t_2)],$$

$$q_{\mathrm{AcB}}=\frac{1}{\varepsilon\upsilon\eta}\,[C_1''N_c''(\theta,t_2)-C_2''N_c''(\theta,t_1)],$$

$$q_{\mathrm{AcC}}=\frac{1}{\varepsilon\upsilon\eta}\,[C_3'N_c''(\theta,t_1)-C_4''N_c''(\theta,t_2)],$$

where θ is the pumping through time, Fi,j are the coefficients that take into account the radioactive transformation of short-lived daughter products in the filter, ε is the radiation-recording efficiency, υ is the speed of the aerosol pumping through the filter, and η is the trapping efficiency of aerosol particles on the filter. The nonexcluded systematic error in measuring the activity of the sample does not exceed 5×10^{-2}. The random-error MSD does not exceed 5×10^{-2}, and it is due to the unstable operation of the radioactive aerosol and the vaporous ^{131}I generators.

The GERA-5 installation is intended for generating and investigating the disperse composition of aerosols virtually in the entire range of their particle dimensions, that is, for reproducing the "aerosol-particle dimension scale" [2], which includes highly dispersed aerosols with particle sizes smaller than 0.1 μm, medium-dispersed aerosols with particle sizes of ~0.1 –1, and coarsely dispersed aerosols with particle sizes exceeding 1 μm.

Highly dispersed aerosols are generated by means of two methods. The first one consists of obtaining a primary NaCl aerosol with particle sizes of ~0.005 μm by means of spontaneous condensation and subsequent consolidation of particles up to the required size in the instrument, whose principle of operation is based on condensing the vapors of given substances on nuclei. The aerosol-particle sizes are adjusted in this method by varying the instrument mixture temperature or changing the ratio of air flows with the NaCl particles and the consolidating-substance vapors.

The interaction of hydrochloric acid vapors with ammonia vapors flow is used in the second method. As a result of this reaction, ammonium chloride aerosol particles are produced. The size of generated particles is adjusted by varying the concentration of ammonia vapors in an excessive concentration of hydrochloric acid vapors, or *vice versa*. This method is suitable for obtaining aerosols with ~0.002–0.2 μm particles.

Aerosols with 0.2–10 μm particles are generated by the method of diffusing monodisperse suspensions. Coarsely dispersed aerosols with particle sizes of 5–200 μm are generated by means of a disk pulverizer whose principle of operation consists of the following. The original liquid is fed to the center of a rapidly rotating disk. Under the effect of centrifugal force, the liquid is displaced toward the disk edge and is ejected from it in the form of monodispersed drops. The particle size is adjusted by changing the disk rotation speed. The disk pulverizer serves to obtain aerosols with a liquid dispersed phase from type VM-4 oil and with a solid dispersed phase from a rosin solution in alcohol from steric acid melts. The described generators are suitable for obtaining virtually monodispersed aerosols with particle sizes of 0.002–200 μm.

The disperse composition of highly dispersed aerosols as well as medium-dispersed aerosols up to 0.5 μm is determined by means of the diffusion method. For this purpose, 13 diffusion batteries are used with a total channel length of 0.5 – 5600 m in each battery. The error in measuring the average size of particles by means of the diffusion method does not exceed 20%.

For certifying suspensions with particles smaller than 1 μm and also for studying the dispersed composition of ammonium chloride aerosols, the electron microscope type EM-9 was used. Suspensions with particles exceeding 1 μm are certified and the dispersed composition of coarsely dispersed aerosols is studied by means of optical microscope type MBI-11. Medium-dispersed aerosol particles in the range of 0.25–10 μm are also determined by means of photoelectric aerosol particle counters type AZ-4. The counted concentration of aerosols is measured after preliminary consolidation is diluted (if required) with the AZ-4 instrument, whose measurement-result MSD does not exceed 20%. Counted concentrations of particles with diameter 2–4 μm are measured by means of the sedimentation method with an error not exceeding 7%. Aerosols with particle sizes exceeding 10 μm are measured (owing to their rapid sedimentation) by a newly developed technique for determining the number of particles entrapped in the instrument sampler with an error not exceeding 10%.

The GERA-5 installation was used for studying the dispersed composition of aerosols generated by the GERA-1 and GERA-2 installations and also for metrological investigation of instruments used in measuring the parameters of inactive aerosols.

The GERA-6 installation serves to generate aerosols and electroaerosols and to measure electrical characteristics of aerosols. Its generating unit consists of light-ion and electroaerosol generators. The light-ion generator is provided with air ionization by means of ^{239}Pu radionuclides.

The volumetric density of the charge is adjusted by varying the voltage across the generator electrodes. Pneumatic pulverization of liquid forced through a nozzle and the charging of particles in the field of the indexing electrode is used in the electroaerosol generator. The volumetric density of the charge is measured by means of the aspiration condenser method based on depositing charged aerosol particles in the electric field of the condenser.

The GERA-6 installation serves to reproduce the volumetric density units of positive and negative aerosol and aeroion electric charges with a random error characterized by the measurement-results MSD not exceeding 10% and a nonexcluded systematic error also not exceeding 10%. The range of reproduced charge volumetric densities amounts to 2×10^{-11} to 2×10^{-7} C m^{-3} for light ions and 1×10^{-9} to 1×10^{-6} C m^{-3} for liquid electroaerosols. This installation was used to certify and test ion counters, to measure the electric characteristics of aerosol generated on other physical-standard installations, and also in research work.

It was found in practice that the transmission of the VA unit of radioactive aerosols in three stages — from the physical standard to the reference equipment and then to the working measuring equipment — is optimal and can be used as a basis for developing test schemes for equipment used in measuring the counted and mass concentrations of non-radioactive aerosols, as well as for equipment used in measuring the aerosol's electrical parameters.

Thus, as a result of improving the special state standard for the volumetric activity unit of radioactive aerosols, its metrological characteristics were raised and the reference and working VA measuring-equipment errors were reduced; the nomenclature of natural radioactive aerosols measured with standard was extended and the range of generated monodisperse inactive aerosols was increased.

18.2 Currently applicable radioactive aerosol standards

At the present time, state standards [3], in a form that includes a technique for generating and measuring parameters for both radioactive and nonradioactive aerosols in a wide range of sizes and activities, do not exist. Instead, in some countries there are local standards for different groups of aerosols and radioactive gases.

18.2.1 Radon and its decay products

In [4], the application of National Institute of Standards and Technology (NIST) of U.S.A. ^{222}Rn emanation standards for calibration of ^{222}Rn is described. NIST certified parameters include ^{222}Rn strength and emanation coefficient. When a source of ^{222}Rn is loaded into a leak-tight jar of known volume, ^{222}Rn will accumulate over time. It is possible to calculate the time-integrated average radon concentration after any given accumulation time. The radon detector in the jar should be non-radon absorbing and a true integrator. In this case, the radon detector must yield the theoretically predicted results. In case of differences, the NIST traceable correction can be derived.

The study [4] involves 34 randomly chosen electret ion chamber system (E-PERM) detectors and 17 NIST sources. The procedures of calibration are not possible with simple equipment. E-PERM detectors were found to give predicted measurement results with an accuracy of about 5%. Commercially available continuous radon monitors also gave satisfactory performance. With the availability of this technology, ^{222}Rn measurement instruments can be made NIST traceable — a great step forward in radon metrology.

In [5], within the framework of the International Atomic Energy Agency (IAEA) and the European Union (EU) International Radon Metrology Program (IRMP), the results of international intercomparison were presented in order to evaluate radon and radon decay product measurement techniques. The work was organized jointly with the U.S. Environment Protection Agency Radiation and Indoor Environment National Laboratory (EPA) at Las Vegas, Nevada, and the former U.S. Bureau of Mines (BOM). The primary goal of this project was to compare the performance of radon and radon decay product measurement instruments from around the world under both laboratory and field exposure conditions.

Nineteen organizations from seven countries participated in this project with 32 types of radon and radon decay product measurement instruments. Laboratory exposures were conducted within an environmental radon chamber at EPA's Radon Laboratory in Las Vegas under very stable, controlled environmental conditions at relatively low concentrations of radon and radon decay products. This part of the study was provided in order to compare the instruments under such controlled environmental conditions. The field exposure was provided at a former underground uranium mine in Colorado maintained by BOM in order to compare the instruments under fluctuating and uncontrolled conditions.

NIST provided the EPA Radon Laboratory with multiple spherical glass sample ampoules, each containing an activity of ^{222}Rn gas known only to NIST. The primary radon measurement system used by the EPA Radon Laboratory to determine radon concentration within the chambers comprise 0.36 l scintillation cells (also termed Lucas cells).

The BOM facility is a previously operated uranium–vanadium mine, which was used for the purpose of conducting research. Radon concentrations were in the range from 5.110 to 15.535 Bq m^{-3} and radon decay product concentrations varied from about 3.54 to 74.32 μJ m^{-3}. The mean equilibrium factor was 14.2%.

Of the six participants who measured radon concentrations in the EPA laboratory with charcoal collectors, four produced results within 8% of the unity for performance ratio. Only one of the four participants who used alpha-track detectors in the EPA laboratory produced results within 11% of unity, and two participants who used the electret ion chamber in the EPA laboratory gave results of 10% of unity. Of the four organizations that measured radon decay product concentrations in the EPA laboratory using grab methods, two produced 10% of unity, and the remaining two within 30%.

The overall conclusion drawn is that more international intercomparisons are needed as there are differences even within the same organizations. This is especially important when making interpretations about health effects, which is the main goal of radon and its decay product measurements.

In [6], the intercomparison measurement of soil-gas radon concentration is presented.

The soil-gas ^{222}Rn concentration is defined as an average radon concentration in the air-filled part of soil pores in a given volume of soil. This value has a wide range of practical applications:

1. when soil-gas radon is used as an indicator for uranium, indoor radon, seismic activity, location of subservice faults, etc., and
2. when studies focus on radon itself.

From a metrological point of view, there are many problems with organizing field intercomparison measurement in the natural geological environment. Previous such intercomparisons were organized in 1991 and 1995. This project was organized during the Third International Workshop on the Geological Aspect of Radon Risk Mapping (Prague, Czech Republic, September 1996) measurements. Participants representing ten organizations from eight countries took part in the intercomparison. The ratio of the standard deviation to the arithmetic mean (SD/mean) was used as a measure of spread on intercomparison. For the soil-gas radon concentration, the agreement among participants was very good. If all single values that were obtained over the whole area of the test site were taken into account, the intercomparison differences, expressed as the ratio SD/mean, were 24%. A more detailed assessment shows an even smaller number — about 20%. This result has been previously considered as a reliable target for intercomparison measurements of soil-gas radon concentration.

18.2.2 Artificial radioactive aerosols

In [7], the installation for testing radioactive aerosol measurement instruments is described. The EPICEA laboratory (Laboratoire d'Essais Physiques des Instruments de Mesure de la Contamination de l'Eau de del'Air), which belongs to the Institute de Protection et de Surete Nucleaire (IPSN), France was established to carry out tests on atmospheric contamination monitors under the conditions recommended by the International Electrotechnical Commission (IEC).

These tests are carried out at the request of users, scientific or industrial manufacturers, either French or foreign, to define the performance of a given aerosol radioactive contamination monitor in order to obtain type approval for the monitor by the IPSN Centre Technique d'Homologation de l'Instrumentation de Radioprotection (CTHIR). Tests can also be used for defining prototypes manufactured by industry. There are two types of tests:

1. Static tests are performed with solid standard radioactive sources.
2. Dynamic tests are performed on the ICARE bench. This bench, continuously generating natural and artificial radioactive aerosols, calibrated for size and activity,

enables the true performances of radioactive aerosol monitors to be defined under normal operating conditions. The true measurement efficiency is obtained by sampling and measuring, in real time, the activity of aerosols labeled with ^{239}Pu and/or ^{137}Sr. The influence of natural activity upon artificial activity measurement channels is determined by aerosols bearing radon decay products whose concentration and attached fraction can be adjusted. A knowledge of the factor of influence of the natural activity and the type of treatment (algorithm) used on this monitor makes it possible to calculate the monitor detection threshold under normal operating conditions. The dynamic test procedure described in the document has been adopted as an international standard by IEC in 1996.

Installation for performing the testing of continuous air monitors (CAMs) for alpha-radioactive nuclides is described in [7]. These instruments must have adequate sensitivity to alert potentially exposed individuals that their immediate action is necessary to minimize or terminate an inhalation exposure.

The air monitor test facility at the Lovelace Respiratory Research Institute (formerly the Inhalation Toxicology Research Institute) has been developed for the U.S. Department of Energy and used to test the performance of prototype and commercially available alpha-CAMs, personal air samplers, and fixed-area filter samplers. Test conditions for these instruments are consistent with the 1995 recommendations of IEC [8].

The facility includes a station for instrument receipt and inspection, a test bench for determining detection efficiency and energy response for alpha-radioactive radionuclides using point-type and area-type electroplated sources and ambient radon progeny; an in-line aerosol delivery for testing the internal collection efficiency of sampling heads with fluorescent and other inert aerosols; an aerosol wind tunnel in which inert aerosols can be used to evaluate the inlet and transport efficiency of sampling probes and aerosol collection devices; and systems for testing the normal response of monitors to ambient radon progeny aerosols or providing aerosols of plutonium or uranium with or without radon progeny aerosols and interfering dusts to air monitors under different conditions of concentration and time. The Lovelace Air Monitor Test Facility is similar, but not identical, to the EPICEA Laboratory Institute, which belongs to ISPN.

Many different types of tests on air monitors were provided at this facility.

Static tests are performed using clean collection substrates with no sampling flow to verify proper reports of background in the absence of radioactive source. For this purpose, the standard radioactive source with traceability to the National Institute of Standard and Technology was used to determine the overall detection efficiency for uniformly distributed sources.

Dynamic tests of collection efficiency of in-line sampling heads were provided by connecting them to an aerosol generation system with parallel sampling ports for head and in-line reference filter.

Dynamic tests with radon progeny were performed to evaluate the influence of natural radioactivity on the ability of CAMs to report correctly the absence of plutonium in the presence of a low concentration of radon decay products when no plutonium is present.

Dynamic tests with radon progeny and artificial radioactivity were provided with higher concentrations of radon progeny (up to 370 Bq m^{-3} in the current system) in the special radioactive aerosol generation system located in the Lovelace plutonium test facility.

Tests for system performance in the presence of interfering dusts were provided because of special concern from the influence of salt dusts with the proper detection of actinide aerosols that may be released from the storage of transuranium wastes in underground salt formation. To evaluate these interferences, the aerosol generation system also provides dust aerosols to CAMs, either as a homogeneous mixture with actinide aerosols and radon progeny, or as separate aerosols.

Tests with computer-generated pulses were provided in order to understand which aspect of CAM performance can be evaluated with sources and ambient radon progeny and which responses must be evaluated with actual aerosols, including the effect of alpha-energy degradation from the deposition of actinide particles in a filter substrate. The alpha-energy spectrum simulator has been used to provide realistic electrical pulses to the instrument preamplifier to simulate different temporal concentration and combinations of actinide aerosols and radon progeny.

The minimum detectable activity and false alarm rates were also studied in the dynamic tests for a range of radon progeny concentrations, and for conditions involving both the absence and presence of plutonium. The performance of different types of filters was also studied.

It should be pointed out that the comparison results of facilities such as ISPN and the Lovelace system in the future should provide confidence that results in the two systems do not contain unexpected bias.

The results of development and practical application of new radioactive source types (special aerosol sources (SAS)) are presented in [9]. SAS are manufactured for certain types of radiometers using model aerosols of plutonium-239, strontium-yttrium-90, or uranium of natural isotopes composition. The original technology for source production allows one to take into account the features of sampling, as well as the geometry and conditions of measurement of the activity in the sample.

The highest accuracy in calibration, certification, and verification of the radiometers can be provided by using national standard precision equipment [3]. But its direct usage for metrological provision of the measurement technique is too expensive. Each stage of developing and using the measurement technique for metrological purposes should be performed with optimal cost-effective ratio.

The proposed method of manufacturing SAS includes radioactive substances depositing onto a substrate, its fixation, and substrate mounting into the frame-holder intended for the measurement device. The check of the radioactive substance fixation quality showed that the relative activity variation of sources with ^{239}Pu after 1–4 days by the dry cotton tampon is within the limits of registration instability (less than 5%). The authors have developed and tested the procedure of certifying and verifying the nuclear power plant radiation security monitoring system aerosol channels by using ^{90}Sr+^{90}Y SAS.

In [10], the problem of the Russian secondary standard for the volume activity of long-lived radionuclide aerosols and the system of international comparison was discussed. The national standard design was practically reproduced as a Russian secondary standard for the volume activity of long-lived radioactive aerosols (RUSSVALRA) in the metrological center of the Ministry of Atomic Energy in SEC "SNIIP," Moscow. It is used for the calibration, certification, and verification of the radiometers, and other metrological provision. In addition, the important problems that arose in radioactive aerosols metrology in the last decades were resolved: to this was added the generation of natural uranium aerosols.

18.2.3 Developing model aerosol sources SAS in order to test aerosol radiometers directly on the consumer place

Model aerosols are prepared in special generators mounted in the box. Parameters of model radioactive aerosol particle sizes were estimated by a six-cascade impactor and radiometer type MS-01P. Measurement showed that model radioactive aerosols are polydisperse with a maximum diameter close to 1 μm. Thus, radionuclides and characteristics of model aerosols are close to similar parameters of national standard [3]. The international comparison was provided between SEC "SNIIP" (Russia) and IPSN-CEA (France). As a result of cooperation between these two installations, a number of aerosol samples with certain radionuclide ingredients were prepared. In Russia, aerosol samples were prepared acccording to the technology of SAS manufacturing, and in France according to sources of ^{239}Pu and ^{137}Cs.

The review of air monitoring standards not as installations for testing and verification of air monitors, but as requirements, instructions for measurement, and calibration is presented in *An International Review of Currently Applicable Standards for Measuring Airborne Activity (International Electrotechnical Commission (IEC), TC News, 2003*. The report can be found electronically at: http://www.iec.ch/support/tcnews.

The following is a list of some standards related directly to radioactive aerosol measurements:

IEC Standards

1. IEC 6071-1 (2002-01): specific requirements for radioactive aerosol monitors including transuranics.
2. IEC61172 (1992-09): Radiation protection instrumentation — monitoring equipment — radioactive aerosols environment.

International Organization for Standardization (ISO) Standards
ISO 2889 (19750): general principles for sampling airborne radioactive materials.

European Standards
EN 481 (1993): workplace atmospheres — size fraction definitions for the measurement of airborne particles.

The American National Standards Institute (ANSI)
ANSI N323C (in preparation): Radiation Protection Instrumentation Test and Calibration Air Monitoring Instruments.

Applicable French National standards
NF X 43022 (1985): air quality — ambient air — concepts relating to the sampling of particular matter.
NF X 43257 (1998): air quality — air in workplaces — individual sampling of inspirable fractions of particulate pollution.
NF M 60-763 (1998): Nuclear Energy — measurement of radioactivity in the environment-air-radon and short-lived decay products in the atmospheric environment.
NF M 60-767 (1999): nuclear Energy — measurement of environmental radioactivity air-radon-222 measurement methods of the volumic activity of radon in the atmospheric environment.

References

1. Antipin, N.A. et al., *Izmer. Techn.*, No. 2, 24–26, 1973.
2. Kravchenko, I. and Ruzer, L.S., *Izmer. Tekhn.*, No. 4, 67–69, 1978.
3. Antipin, N.A. et al., Special State Standard for the volumetric activity of radioactive aerosols, *Izmer. Techn.*, No. 1, 5–7, J 1980.
4. Cotrappa, P. et al., Application of NIST ^{222}Rn emanation standards for calibration ^{222}Rn monitors, *Radiat. Prot. Dosim.*, 55, 211–218, 1994.
5. Budd, G. et al., Intercomparison of radon and decay product measurements in an underground mine and EPA radon laboratory: a study organized by the IAEA Radon Metrology Programme, *Health Phys.*, 75, 465–474, 1998.

6. Neznal M. et al., Intercomparison measurement of soil-gas radon concentration, *Radiat. Prot. Dosim.*, 72, 139–144, 1997.
7. Grivaud, L. et al., Measurement of performances of aerosol type radioactive contamination monitors, *Radiat. Prot. Dosim.*, 79, 495–497, 1998.
8. Hoover, M.D. et al., Performance testing of continuous air monitors for alpha-emitting radionuclides, *Radiat. Prot. Dosim.*, 79, 499–504, 1998.
9. Belkina, S.K., Zalmanson, Y.E., Kuznetsov, Y.V., Rizin, A.I., and Fertman, D.E., Special aerosol sources for certification and test of aerosol radiometers, *J. Aerosol Sci.*, 22 (Suppl. 1), 801, 1991.
10. Fertman, D. and Rizin, A.I., Using Polydisperse Aerosols and Special Aerosol Sources for Calibration of Aerosol Radiometers, Materials of International Congress for Particle Technology (PARTEDC 98), Nurnberg, Germany, 1998.

chapter nineteen

Radon and thoron in the environment: Concentrations and lung cancer risk

Naomi H. Harley
New York University School of Medicine

Contents

19.1 Introduction 569
19.2 Environmental concentrations 570
19.3 Indoor concentrations....570
19.4 Outdoor concentrations 570
19.5 Stratospheric concentrations 571
19.6 Radon in drinking water 573
19.7 Bronchial lung dose 574
19.8 Dose to the fetus from radon in drinking water 577
19.9 Confounding of ^{222}Rn and ^{220}Rn in the measurement of "radon"....577
19.10 Guidelines for indoor ^{222}Rn 579
19.11 Lung cancer risk projections 579
19.12 Summary 581
References 582

19.1 Introduction

Planet earth is mainly rock and metal orbiting the sun, and its rock is remarkably radioactive. The sun supplies most of our energy but the radioactivity in the earth contributes perhaps 20% of the total heat balance of the earth. The primordial decay series beginning with ^{238}U (4.5×10^9 yr half-life), present in all terrestrial materials, supports a chain of 13 alpha-, beta-, and gamma-emitting radionuclides, which includes the gas radon (^{222}Rn, 3.8 day half-life). The primordial decay series beginning with ^{232}Th (1.4×10^{10} year half-life) is also present in all basic earth materials, and supports a chain of 11 radionuclides, which includes another isotope of radon, commonly called thoron (^{220}Rn, 55 sec half-life). Both radon isotopes are produced in all soil or rock from their parent radium (^{226}Ra or ^{224}Ra). A fraction of the gas is then released from all terrestrial substances, and can be measured in any dwelling, outdoors, and in the case of ^{222}Rn, even at stratospheric altitudes. Thoron gas is an emerging issue because there are only sparse measurements of the gas itself. Thoron is also found both indoors and outdoors. Thoron, however, should not exist in the stratosphere because of its short half-life and the time required for transit to altitude.

Epidemiologic follow-up studies of 13 mining groups show a lung cancer dose response that increases with increasing radon concentration. Combined with this fact and

1-56670-611-4 / 05 / $0.00+$1.50

that some homes can attain the concentrations that exist in mines, interest was spurred in environmental measurements, and their risk estimates. The global environmental measurements to date are described, showing that the current indoor home environment averages about 40 Bq m^{-3} (about 1 pCi l^{-1}). An emerging problem in radon gas measurement is the presence of ^{220}Rn (thoron), often unaccounted for, and this undoubtedly affects the existing calculated risk estimates for homes. There are many domestic follow-up studies that try to link home exposure with lung cancer risk from radon, but so far no study has clearly defined a precise risk that could be used in domestic risk assessment.

19.2 Environmental concentrations

The concentration of radon depends on the dilution and mixing properties of the space into which it is released. Underground miners work in a relatively enclosed space close to the soil and rock source; therefore, radon concentrations can be very high. In the case of uranium mines, the rock (ore) concentration of uranium and ^{226}Ra is much higher than typical soil concentrations and so radon levels in mines can be extraordinarily high, but depends upon the local ventilation. Radon gas is quite soluble in water, and even in non-uranium mines water bearing ^{222}Rn can produce high concentrations. An example of this are the Canadian fluorspar mines (de Villiers, 1966; de Villiers et al., 1971; Morrison et al., 1988, 1998).

Radon concentrations in homes or offices are not generally high because of a lesser source and higher ventilation rates. Average indoor environments are generally about a factor of 2 higher than outdoor concentrations.

Most countries with a public health program have conducted comprehensive indoor radon surveys in homes. Results of the surveys have been used to estimate lung cancer risk for the populations involved. Again, a problem with these risk estimates is that the radon measurements do not account for the potential to measure two radon isotopes, ^{222}Rn (radon) and ^{220}Rn (thoron). Thoron is a decay product in the naturally occurring thorium series and its presence is emerging in essentially all measurements unless effort is made to exclude it. The effect of this omission on the results of the actual radon concentrations and subsequent risk evaluation is discussed in more detail in Section 19.9.

19.3 Indoor concentrations

More measurements have been made of indoor radon than perhaps for any other natural radionuclide. The reason for this is that some measurements of radon in homes in the late 1970s were consistent with underground mining concentrations. The emergence of statistically significant lung cancer risks associated with radon in these underground miners initiated a broad measurement regime, and countries rapidly began indoor radon surveys to assess potential domestic risk (NCRP, 1984a,b; NAS, 1988, 1999).

The results of 50 country indoor radon surveys are reported in NCRP (2005), with data derived mainly from UNSCEAR (2000), and are reproduced in Figure 19.1. From Figure 19.1, the global indoor radon average from the data reported so far is 50 to 30 Bq m^{-3} (1.4 ± 0.8 pCi l^{-1}).

19.4 Outdoor concentrations

Many countries, especially those in tropical climates, have indoor concentrations that are essentially outdoor concentrations. For example, indoor and outdoor measurements over a 4-yr period in Bangkok and Chiang Mai, Thailand, averaged 14 ± 1, and 15 ± 1 Bq m^{-3},

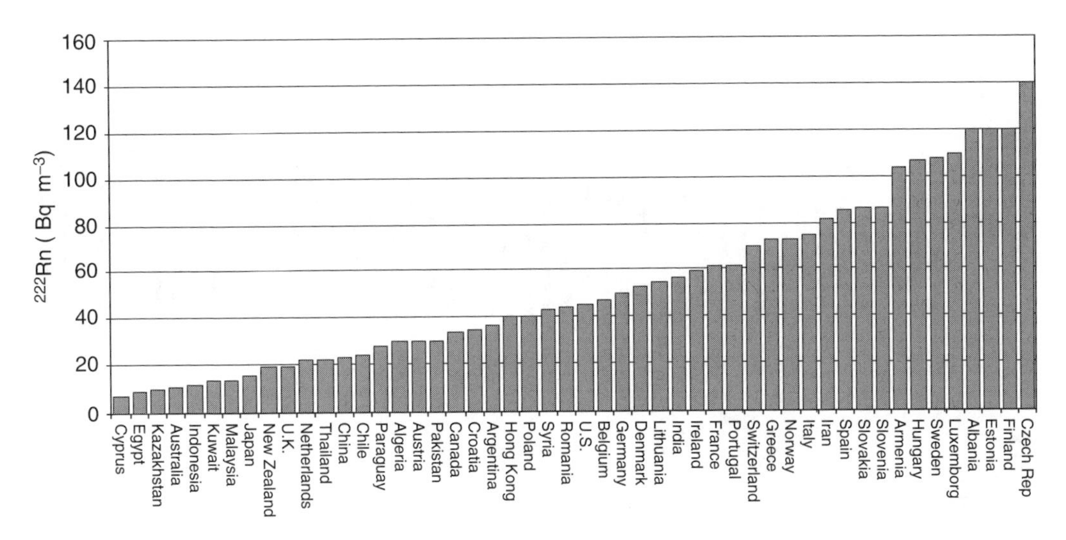

Figure 19.1 Global indoor radon measurements in 50 countries.

and 44±1 and 39±1 Bq m^{-3}, respectively (Chittaporn and Harley, 2000). The U.S. Environmental Protection Agency (EPA) performed an outdoor radon survey that included each major city in all 50 U.S. states. Given that the ventilation rate is from 5 to 10 air exchanges with outdoor air per day in most homes, the measurement of outdoor radon becomes of interest. The baseline for indoor radon is clearly outdoor radon. Even the indoor concentration of long-lived ^{210}Pb, a decay product late in the radon chain, is derived mainly from outdoor concentrations of ^{210}Pb (Harley et al., 2000).

Outdoor radon concentrations have not been reported from many countries, but the U.S. average appears to represent the sparse published results. A summary of published outdoor concentrations is given in Harley (1990). Outdoor radon concentration can also be a factor in remediation of home drinking water. Any risk from radon in drinking water is associated with inhalation exposure from radon released during water use. EPA has suggested target values for radon in drinking water using a multimedia approach (see the section on Radon in drinking water) that depends on a knowledge of outdoor air concentrations. In a report on the risk from radon in drinking water (NAS, 1999), this committee obtained all original outdoor radon data from the EPA survey and reevaluated the combined data. The graph of all U.S. state data is shown in Figure 19.2. The average outdoor concentration in the U.S. is 14.8 Bq m^{-3} (median 14.6 Bq m^{-3}, maximum 35.9 Bq m^{-3}).

19.5 Stratospheric concentrations

Radon is transported to the stratosphere by eddy turbulent diffusion and provides information on vertical transport rates. Stratospheric air samples were collected by the U.S. Weather Bureau (now NOAA) in 1962 to explore the possibility of using radon profiles as an atmospheric tracer (Fisenne et al., 2004). These data were available only in an internal report until reevaluated by Fisenne et al. for publication in 2004. In the spring of 1962, WB-57 aircraft collected tropospheric and stratospheric air samples by pressurizing steel gas collection spheres to a pressure of about 21 MPa at each sampling altitude.

Following sampling at predetermined locations and altitudes, the steel tanks were shipped to the U.S.D.O.E. Environmental Measurements Laboratory (formerly the Health and Safety Laboratory, HASL). Each collection sphere contained approximately 2 m^3 of air

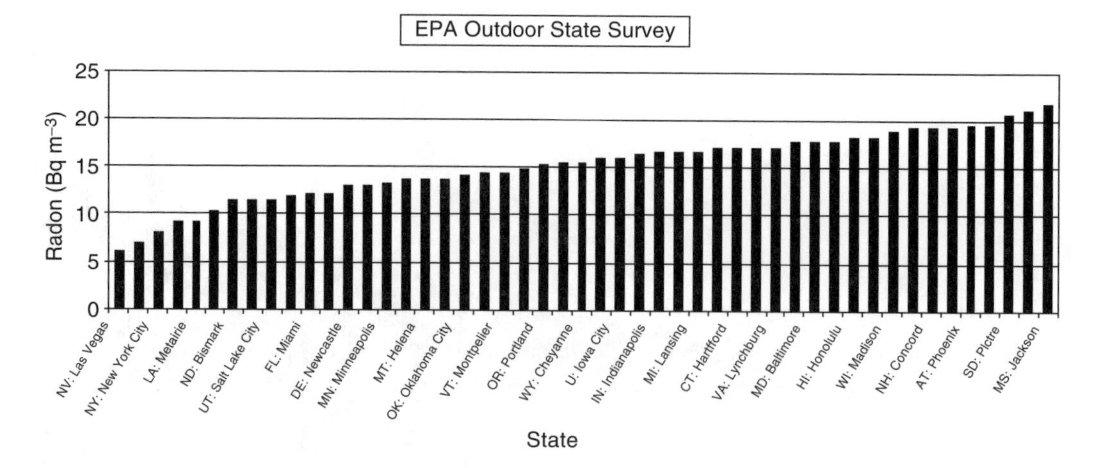

Figure 19.2 Outdoor radon measurements made by U.S. EPA in all 50 states.

at standard temperature and pressure. The sample was transferred through a sample train consisting of two gas washers to remove carbon dioxide and water vapor and then into a specially produced low-background activated charcoal trap. The adsorbed radon on the charcoal was transferred to low-background (6 counts per hour) 2 l ionization chambers for alpha-counting. The calibration of these chambers was carried out with the National Bureau of Standards (now NIST) standard ^{226}Ra sources, emanating radon gas into the chambers from the radium-bearing solution. The measurement error for the tropospheric samples was, in general, about 10% and up to 100% for the stratospheric collections, where the activity of some samples was below the detection limit.

The locations and trajectories of the flights were selected in order to investigate the influences of tropospheric height, the underlying land mass, and thermal gradients on radon concentrations in the atmosphere. The steel tanks were shipped to HASL and radon measurements were performed within 4 days of collection

The sampling locations were Alaska, southwest U.S.A., and the Panama Canal Zone at 8, 32, and 70° North, respectively. The samples were collected at different altitudes ranging from 4 to 20 km, both below and above the tropopause. The tropopause varied from 8 to 17 km. The ^{222}Rn concentrations ranged from 1 mBq m^{-3} at 20 km to 1000 mBq m^{-3} at 4 km, compared with an average ground-level concentration of 15,000 mBq m^{-3}.

The lowest tropospheric radon concentration might be expected in the Canal Zone samples because of the decay of radon over the oceans and the tropopause height. However, the highest concentrations of radon in the troposphere were measured in these samples. The stratospheric concentration at the Canal Zone was the lowest measured. These data are supported by similar tropospheric concentrations reported by Machta and Lucas (1962) for air samples collected over Hawaii. The tropospheric radon concentrations in the southwest U.S.A., samples are similar to the Canal Zone profiles and are supported by the later measurements of Moore et al. (1973). The lowest tropospheric radon concentrations measured were from Alaska, again supported by the data of Machta and Lucas (1962).

The ^{222}Rn concentration profiles for the Canal Zone and southwest U.S.A. were similar with height, while the Alaskan profiles were lower by a factor of 5 to 10. The tropopause appears to be a more effective barrier to ^{222}Rn transport in the Canal Zone and Alaska than over southwest U.S.A. The reason for this effect is not known but may reflect the bearing of the jet stream on lower altitude turbulence. An effective vertical transport rate was calculated at six different altitudes and the average was 0.5 to 0.1 cm sec^{-1}.

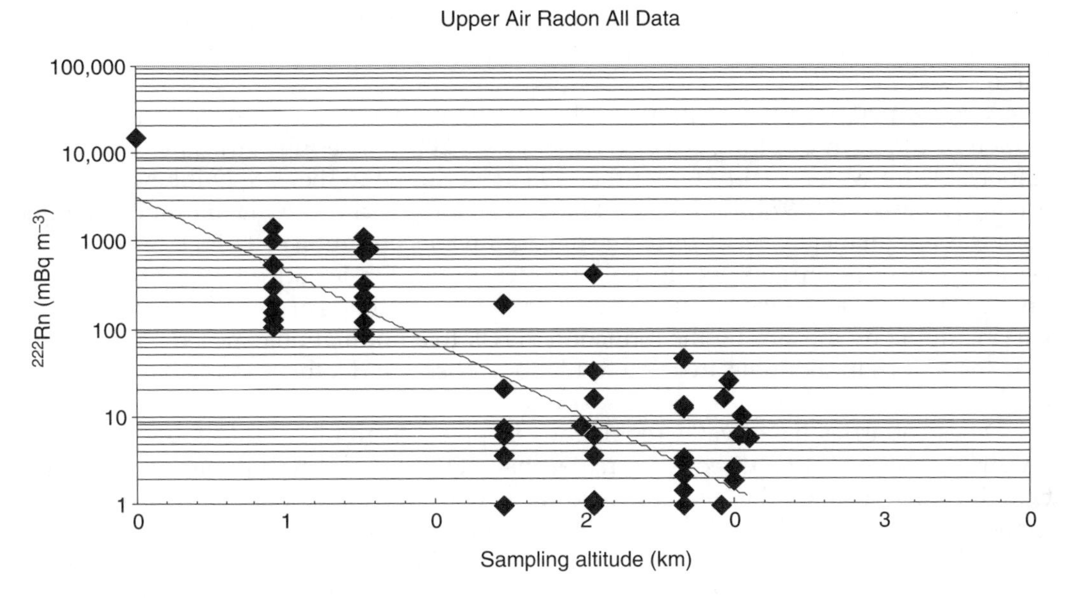

Figure 19.3 Radon measurements in the troposphere and stratosphere made over Alaska, the Canal Zone, and southwest U.S. by WB-57 high-flying aircraft.

The entire data set of 54 samples is shown in Figure 19.3. A full description of the sampling and measurement, and the individual data sets for the four locations can be found in Fisenne et al. (2004).

19.6 Radon in drinking water

The major determinant of internal dose and thus risk from radon in drinking water is not from the ingestion of water but the lung dose from radon decay products in air following the release of radon from water during use (NAS, 1999). The highest internal dose is to the stomach and this is very small compared with the lung dose from inhaled decay products resulting from the release of radon into the atmosphere. Radon in water is basically a radon in air issue. For a given radon concentration in drinking water, the risk ratio from total exposure (stomach cancer to lung cancer) is calculated to be from 1% (Harley and Robbins, 1994) to 11% (NAS, 1999), and depends upon the model used to transport the gas through the stomach wall to the cells identified as targets for stomach cancer.

Radon released from water during showering and other uses combines with and is indistinguishable from normal soil and outdoor sources. The best estimate for the transfer factor from water to air is 10,000 l^{-1}, and is based on an analysis of all published data (NAS, 1999). That is, 10,000 Bq m^{-3} of ^{222}Rn in water will, on average, add 1 Bq m^{-3} of ^{222}Rn to indoor air.

Outdoor radon is important in that the decisions for setting limits for remediation of radon in drinking water from public supplies can be affected by outdoor radon concentrations (NAS, 1999). EPA has suggested a value of 11 Bq l^{-1} ^{222}Rn (300 pCi radon l^{-1} water) for water. This value can be technically difficult to obtain. A multimedia approach suggested by EPA is to obtain an equivalent risk by reducing the water concentration to a value that will yield the standard outdoor concentration upon release, and reducing the overall risk by other means. Thus, the average U.S. outdoor concentration of 15 Bq m^{-3} would permit a water concentration of 150,000 Bq m^{-3} (150 Bq l^{-1}) to be delivered if alternative means of risk reduction are put in place. The risk reduction then needed to reduce the total risk could be, for example, home radon remediation that attained the same calculated risk reduction in a population affected, similar to reducing the water to the 11 Bq l^{-1} value.

A policy perspective was prepared to investigate the various issues regarding remediation of radon in drinking water (Hopke et al., 2000). The EPA and state agencies are responsible for water quality under the 1996 amendment to the Safe Drinking Water Act, originally passed in 1974. Radon is a known carcinogen and the maximum contaminant level was automatically set at zero. The maximum concentration of 11 Bq l^{-1} is considered because zero concentration could only be measured with an uncertainty of 30%. These involve technical and social decisions concerning the methods of risk reduction. There are varying opinions concerning the reduction in risk to a population exposed vs. risk reduction for a few individuals in the population. These and other issues have yet to be resolved.

Other countries have water regulations in place. Beginning on October 1, 1998, the Swedish government required that greater than 100 Bq l^{-1} ^{222}Rn in public water supplies be reduced, and stated that water exceeding a concentration of 1000 Bq l^{-1} is unsafe and cannot be supplied. Finland has set a recommended maximum limit of 300 Bq l^{-1}. The U.S. EPA is required to set a regulation for ^{222}Rn in drinking water in the United States. The standard they have proposed is 11 Bq l^{-1}, but to date (2004) no regulation has been set officially.

19.7 Bronchial lung dose

Saccomanno et al. (1996) evaluated lung cancer histology in 467 Colorado Plateau uranium miners and 311 nonminers to determine the localization of lung tumors. They showed that 84% of the lung cancer in uranium miners was located in the bronchial airways and 77% in the bronchial airways of nonminers who were mostly smokers. The tumors in the pulmonary or gas exchange region were 16 and 23% in miners and nonminers, respectively. Thus, lung tumors arise mainly in the bronchial airways whether from ^{222}Rn exposure or tobacco smoke. The fractional localization in the central and middle airways was somewhat different, indicating that the bronchial deposition of carcinogens in tobacco smoke and radon decay products differs slightly. Most of the radon decay products, that is, about 20% of what is inhaled, deposit in the lower lung while only a few percent deposit in the bronchial region. However, the very large surface area (square meters) for deposition in the lower or pulmonary lung compared with a much smaller area (a few hundred square centimeters) in the bronchial region accounts for the much larger bronchial dose, in spite of the lower actual mass deposition of the decay products.

The relevant bronchial dose from radon is actually from the particulate short-lived (30 min effective half-life for the chain) alpha-emitting decay products (^{218}Po, ^{214}Po). As the decay product atoms form in air from ^{222}Rn decay, they attach rapidly to the ambient aerosol particles. Bronchial and pulmonary deposition is determined by the ambient particle size distribution. The dose as a function of particle size is described later. There are some published short-term data on the size distribution of radon decay products and the median diameter indoors is about 150 to 200 nm (Tu and Knudson, 1988a, b; Li and Hopke, 1993; Reineking and Porstendorfer, 1986; NAS, 1999).

Harley and Chittaporn (1999, 2000) and Harley et al. (2004) developed an integrating particle size sampler that operates for up to 6 months indoors or outdoors. The short-lived radon decay products ultimately form long-lived ^{210}Pb/^{210}Po (22 yr, 138 days) and these are measured easily in the filtration stages in this instrument. Figures 19.4 and 19.5 show the size distribution indoors and outdoors over a 6- and 2-month period, respectively, at a suburban New Jersey home. The median particle diameter outdoors (0.15 μm) is somewhat

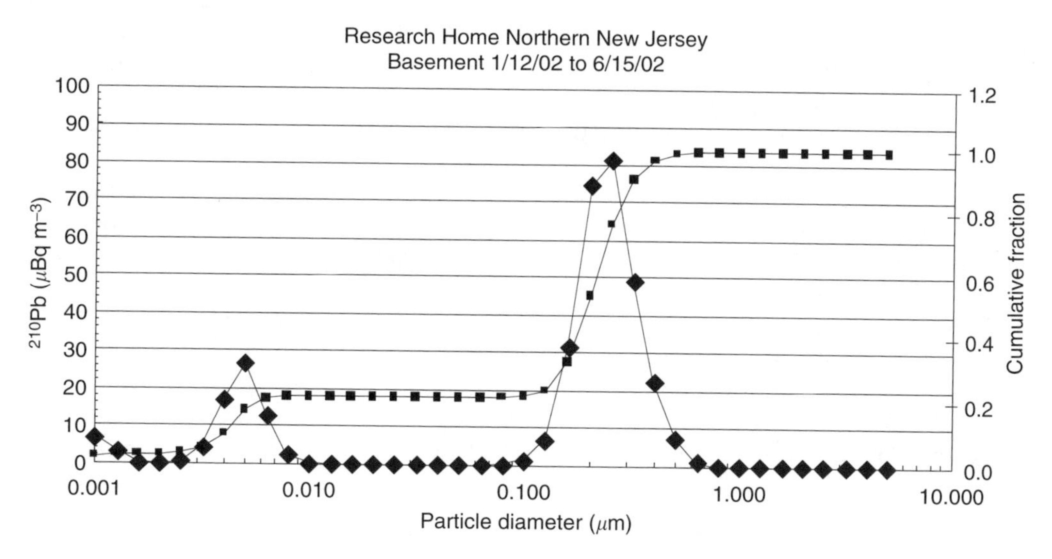

Figure 19.4 Measured particle size distribution indoors in the basement of a suburban New Jersey home.

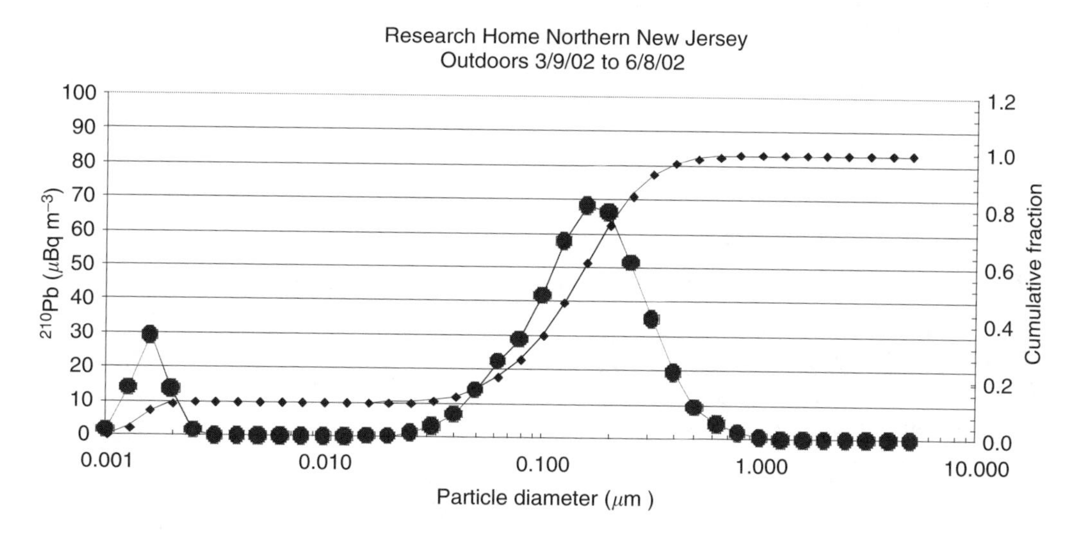

Figure 19.5 Measured particle size distribution outdoors at a suburban New Jersey home.

smaller than in the basement (0.25 μm). The larger size mode in the basement is most likely due to the oil-fired heating system that dominates the size distribution.

A small fraction of the ^{218}Po formed from radon decay associates with water vapor or other gases and remains as a small cluster of atoms with a diameter of a few nanometers (nm). All atmospheres contain this nanometer or unattached fraction. Because of their small diameter and rapid Brownian motion of small molecules, the unattached fraction deposits efficiently in the bronchial airways. Therefore, a small percentage of the total air activity inhaled accounts for a disproportionate fraction of the bronchial dose, with up to about 25% due to the unattached fraction.

Ruzer et al. (1995) measured the deposited radon decay product gamma-ray activity in the chests of metal miners in Tadjikistan by external counting. Measurements on about 100 miners were performed along with filtered air samples to assess the decay product exposure. These two measurements are proportional to the average minute volume. In this case, actual measured breathing rates were obtained for drillers, assistant drillers, and inspection personnel. The group average breathing rates were 0.0079, 00067, and 0.0052 m^3 min^{-1} respectively, with upper bound values of 0.023 to 0.004, 0.020 to 0.004, and 0.015 to 0.003 m^3 min^{-1}.

The calculated bronchial dose for radon decay products deposited on the bronchial airways is derived from five factors. These are (1) ratio of the decay products in the atmosphere, (2) breathing rate, (3) the particle size of inhaled aerosol particles, (4) unattached or ultrafine fraction of the aerosol, and (5) bronchial clearance rate.

The bronchial dose coefficients for a range of median particle diameters from 1 to 1000 nm (0.001 to 1 μm) is shown in Figures 19.6 and 19.7 (UNSCEAR, 2000). Figure 19.6 gives the dose coefficient as a function of breathing rate, and Figure 19.7 the dose coefficient as a function of the unattached (or nanometer size) decay products present in the decay product mixture. The radon concentration for the dose factors is given in terms of equilibrium

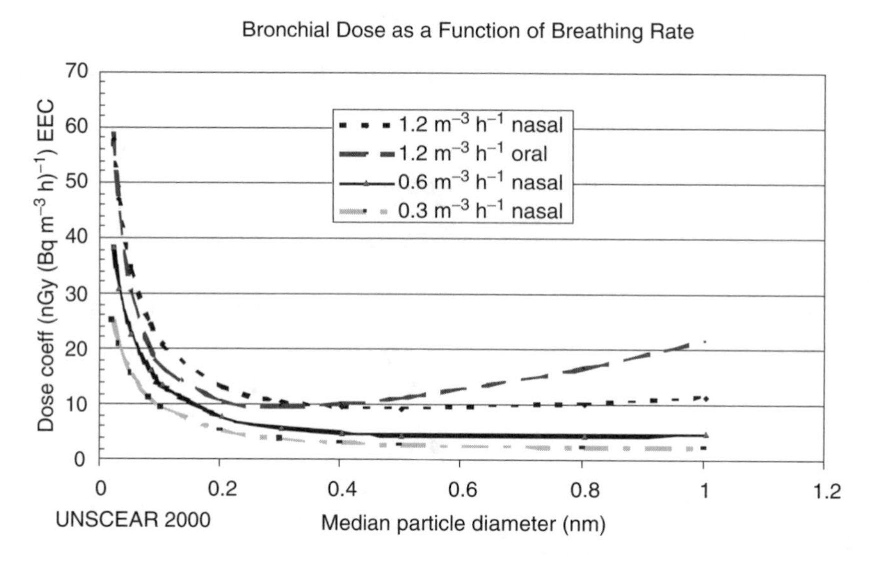

Figure 19.6 ^{222}Rn dose coefficients for the bronchial alpha-dose as a function of breathing rate.

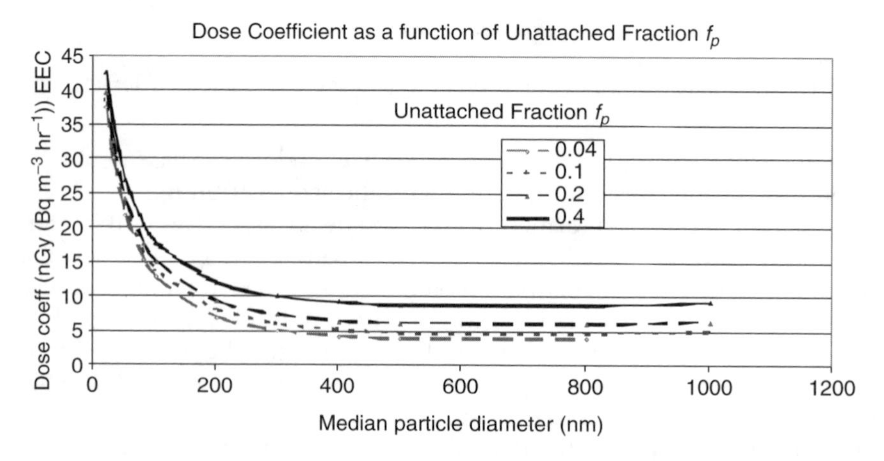

Figure 19.7 ^{222}Rn dose coefficients for the bronchial alpha-dose as a function of unattached or nanometer fraction.

equivalent concentration (EEC). The EEC is equal to the radon concentration if it were in equilibrium with its decay products. UNSCEAR adopts 40% as an equilibrium factor for indoor exposure and 60% for outdoor exposure. Thus, the radon gas concentration times 0.4 or 0.6 is accepted as the numerical value of the EEC for indoor and outdoor environments. The equilibrium factor for thoron and the calculation of EEC for thoron is discussed in the section on Confounding of ^{222}Rn and ^{220}Rn in the measurement of "Radon."

Clearly, from Figures 19.6 and 19.7, as the inhaled particle size changes with the particular environment the dose can change by factors of 2 to 3. The median particle size of inhaled particulates present in various indoor and outdoor environments does change significantly, and the unattached fraction changes significantly with the total particle loading.

Radon is also quite soluble in body tissues and there is a smaller dose to organs other than the lung. The annual bronchial dose from ^{222}Rn and that to other organs such as soft tissues, female breast, and skin (from atmospheric plate out of decay products) is small compared with the bronchial dose. For an average radon exposure of 40 Bq m^{-3} and a decay product equilibrium of 40% (UNSCEAR, 2000), the dose factor from Figure 19.6 for a normal breathing rate of 1.2 m^3 h^{-1} and a median particle size of 0.15 μm is 10 nGy per Bq m^{-3} h. Assuming this exposure is full time each year, the calculated dose for $40 \times 0.4 \times 10 \times 8760$ h is about 1.4 mGy yr^{-1}. The dose to female breast and soft tissues would be 0.003 and 0.001 mGy yr^{-1}, respectively (Harley and Robbins, 1992), or less than 1% of the bronchial dose. UNSCEAR (2000) has adopted an annual effective dose of 1.2 mSv for the inhalation of ^{222}Rn for the global population.

19.8 Dose to the fetus from radon in drinking water

Robbins and Harley (2004) calculated the dose to the developing fetus from ingested water by the mother. The model indicates an increase from 9 weeks to about 14 weeks and then a decrease. This is due to the assumed changing blood flow rates. The dose at 1 week is zero. This is a consequence of the very small ^{222}Rn concentration in maternal blood in which the embryo floats. Not even one alpha-particle traversal of the embryo is calculated for an initial ingestion of 100 Bq at that time. The highest calculated equivalent dose occurs between weeks 6 and 16. This is due mainly to the small body mass during that time interval.

The equivalent dose values can be compared with the average dose to all developing fetuses from the natural external gamma-ray and cosmic-ray radiation of approximately 1000 μSv (1 mSv) during pregnancy (NCRP, 1984). The maximum equivalent dose to the fetus is about 0.4% of the fetal-life external gamma-ray and cosmic-ray dose for each 100 Bq ingested by the mother.

If we assume an average consumption per day of 0.6 l of raw tap water at a concentration of 100 Bq l^{-1}, the calculated total dose to the fetus over the term of the pregnancy is 250 μSv or 25% of the normal background radiation dose.

19.9 Confounding of ^{222}Rn and ^{220}Rn in the measurement of "radon"

Presently, no other detector measures both radon and thoron gas simultaneously as a personal or area monitor, with the exception of the miniature monitor developed at New York University School of Medicine (Chittaporn and Harley, 1999). In most cases, thoron bronchial dose is assessed through the measurement of the decay products of thoron ^{212}Pb and ^{212}Po (Schery, 1985, 1990; Schery and Zarcony, 1985; Harley and Chittaporn, 1999), and ^{220}Rn concentrations are also estimated from these decay products. Thoron gas itself is rarely measured, because of the difficulty in measuring an alpha-particle-emitting gas with a very short half-life ($t_{1/2}$ = 55 sec). The measurement of the two gases normally requires real-time instrumentation with various types of decay chambers to permit a difference in signal with and without the ^{220}Rn (Israel, 1964; NCRP, 1988).

The NYU passive alpha-track detector measures the integrated signal from the alpha-decay of thoron gas, over any time period, depending upon the exposure assessment needs.

The ^{222}Rn and ^{220}Rn personal monitor is designed to use alpha-track detection film in four separate chambers using different entry diffusion barriers for signal differentiation between the two gases. Two chambers for each measurement permit the precision to be calculated. An electrically conducting foam, directly beneath the gas entrance holes to the monitor, eliminates solid decay products and permits only radon or thoron gas passage. An additional barrier in two chambers permits only passage of the long-lived radon. The monitor shape is four-lobed (similar to a cloverleaf) with a face dimension of 5 cm across chambers and a thickness of 1 cm. It is molded from conducting ABS (CNi) plastic, that is, plastic with embedded nickel-coated carbon fibers to prevent charge artifacts from disturbing the calibration. Nuclear track film (9×9 mm CR-39) is used for the detection of alpha-particles emitted within each chamber. The pristine film background is 5 – 15 tracks depending upon the batch, and the entire film area is counted.

Measurements at the National Weather Service station in Manhattan at Central Park have been made for about 4 yr. The radon and thoron concentration was similar at both heights, averaging 12 and 17 Bq/m^{-3} for radon and thoron, respectively, over a measurement interval from March to December 2002. Outdoor radon and thoron concentrations at the quality control home in suburban New Jersey averaged 5 and 18 Bq m^{-3} radon and thoron, respectively, from March to December 2002. Most measurements made either indoors or outdoors for ^{222}Rn and ^{220}Rn have identified a ^{220}Rn component. This is an emerging area of research that should provide information on the actual ^{222}Rn exposures. Many of the existing measurements and risk assessments are undoubtedly hindered by the presence of thoron in the measured signal unless measures have been applied to exclude its presence. UNSCEAR (2000) provides central dose factors for radon and thoron EEC; these are

$$\text{radon(EEC)} = 9 \text{ nSv per (Bq m}^{-3}\text{ h)}$$

$$\text{thoron(EEC)} = 40 \text{ nSv per (Bq m}^{-3}\text{ h)}$$

The EEC for radon or thoron is = F (equilibrium ratio) × (gas concentration).

As stated above, the accepted value of F for radon (UNSCEAR, 2000) is 0.4 for indoor environments and 0.6 for outdoor environments, that is, 40 or 60% equilibrium with the decay products. From long-term measurements of thoron gas and the thoron decay product ^{212}Pb, Chittaporn et al. (2003) have shown that the average, F, for thoron is 0.02 indoors and 0.003 outdoors, that is, 2 and 0.3% equilibrium with the decay products. Thus, the thoron bronchial dose can be derived from its gas measurement similar to radon dose estimates.

Although the dose factor per unit gas concentration for thoron is larger than that for radon, this is offset by the much smaller thoron equilibrium factor, F. Therefore, the dose from thoron decay products is usually less than that for radon decay products. Because the measurement of total gas has been used to identify radon, the historic dose and risk assessments should be revisited in the future.

It is unlikely that the historic measurements in uranium and other underground mines are compromised by thoron, because the ore seems to contain primarily ^{238}U, the parent of ^{226}Ra and ^{222}Rn.

19.10 Guidelines for indoor ^{222}Rn

Based on the domestic radon surveys conducted in many countries, both NCRP and ICRP have set guidelines for indoor radon concentrations (NCRP, 1984, 2005; ICRP, 1993). NCRP recommends that a prudent concentration guideline for home remediation should be above an annual average concentration of 400 Bq m^{-3} (10 pCi l^{-1}) in the living areas. ICRP recommends a guideline of 200 Bq m^{-3} for existing dwellings with 150 Bq m^{-3} for new construction, the U.S. Environmental Protection Agency (EPA) 150 Bq m^{-3} (4 pCi l^{-1}), and Canada 750 Bq m^{-3} (20 pCi l^{-1}). These guidelines for acceptable average annual concentrations are based on calculations that extrapolate the lung cancer risk derived from underground miner exposures at high concentration to lower home concentrations. None of the domestic studies have the precision to estimate lung cancer risk with any certainty and the extrapolations must be necessarily based on the underground miner risk calculations.

19.11 Lung cancer risk projections

More published information exists concerning the lung cancer risk from radon than for any other internal radioactive emitter. This is due to the fact that, beginning in 1984, numerical risk estimates were emerging from the underground miner population, and homes with extraordinarily high radon concentrations were discovered. It appeared to be a serious public health concern for the population, and this section will attempt to lay out the existing evidence.

Evidence of radon concentrations in homes being equal to those in mines, along with the evidence of lung cancer associated with radon in underground miners, stimulated many countries to conduct large-scale, comprehensive surveys of indoor radon. Many countries set guidelines for radon concentrations in homes, and these guidelines were based on the calculation of the lung cancer risk extrapolated from the higher exposures of underground miners. New home construction had to be sensitive to the possible radon reduction measures that could be called upon if the home showed concentrations above the guidelines.

A biological model for the number of cells hit in order to form lung cancer was derived by Harley et al. (1996). The calculations were based on the modeled radon decay product activity on the airways, the measured number of cycling stem cells in the bronchial epithelium, and the known lung cancer response in underground miners for a given radon decay product exposure. The estimate is that 3×10^6 cycling basal cell nuclei are hit in order to yield a high probability of tumor formation. Thus, development of lung cancer from radon decay product alpha-particles targeting cells in the bronchial mucosa is a relatively rare event.

Lubin and Boice (1997) performed a meta-analysis from eight of the published domestic radon risk estimates. Their analysis is shown in Figure 19.8. The line is said to indicate the calculated risk from the 13 underground miner risk estimates. This line is an underestimate from the miner risk because the line is calculated for a 40 yr exposure, and not whole-life exposure. The actual number of deaths recorded in the meta-analysis, and attributed to radon, are obviously for whole-life exposure. Lubin and Boice (1997) estimate from the meta-analysis that the lung cancer risk for lifetime domestic exposure to 150 Bq m^{-3} (4 pCi l^{-1}) results in an increase in lung cancer mortality of $14 \pm 14\%$.

An epidemiologic follow-up study of women in Saxony shows results that are consistent with the meta-analysis considering the uncertainty in the meta-analysis. The core study areas are the two closely located towns of Schneeberg and Schlema in Saxony, with

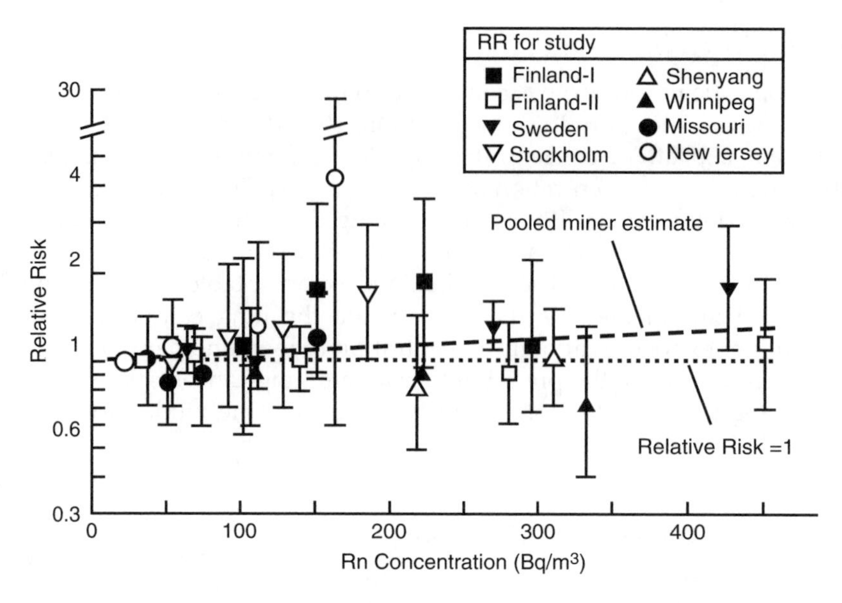

Figure 19.8 Lung cancer risk estimates as a function of exposure from a meta-analysis of eight domestic indoor ^{222}Rn studies.

about 20,000 and 3000 inhabitants, respectively. The study is unique in that the study power is high for the following reasons:

- There is a wide range of ^{222}Rn exposure from 50 Bq m^{-3} to >3000 Bq m^{-3}.
- The exposed fraction of the population is high.
- The majority of the study population of women are nonsmokers.
- The population has a very low residential mobility.
- The study region is included in a cancer registry from 1952 to the present day.

The high exposure level in the core study area is caused by mining and geologically induced factors. The towns of Schneeberg and Schlema are partially undermined by medieval mining activities mostly for silver and after World War II by present-day mining activities for uranium. The recent mining activities beneath the two towns started in 1945 and ended in Schneeberg and Schlema earlier than 1960. After this, a natural ventilation existed in the shafts and galleries, keeping the radon levels in the average of the year quite stable, despite seasonal fluctuations. This resulted in equally stable indoor radon levels in houses, especially those affected by former mining activities. To measure the possible influence of mining activities on current radon measurements, studies concerning the reconstruction of exposure are introduced into the study, limited to special areas of Schneeberg and Schlema, where such influences were suspected.

The median exposure level for cases was 209 Bq m^{-3}, 160 Bq m^{-3} for register controls, and 104 Bq m^{-3} for hospital controls. An increased and significant odds ratio (OR) could be established with the Schneeberg study by two forms of analysis in the higher exposure categories only. Below a radon concentration of 48×10^6 Bq h m^{-3}, accordingly 1000 Bq m^{-3} and a residential duration of 20 yr, the OR is not elevated. Significantly elevated OR after both forms of the analysis are detectable at an exposure level of >1500 Bq m^{-3}.

The results of the Schneeberg study are shown in Figure 19.9. Although these authors state that the OR is not elevated below 1000 Bq m^{-3} (25 pCi l^{-1}), the overall uncertainty in the point estimates from this and the other studies does not permit an exact numerical

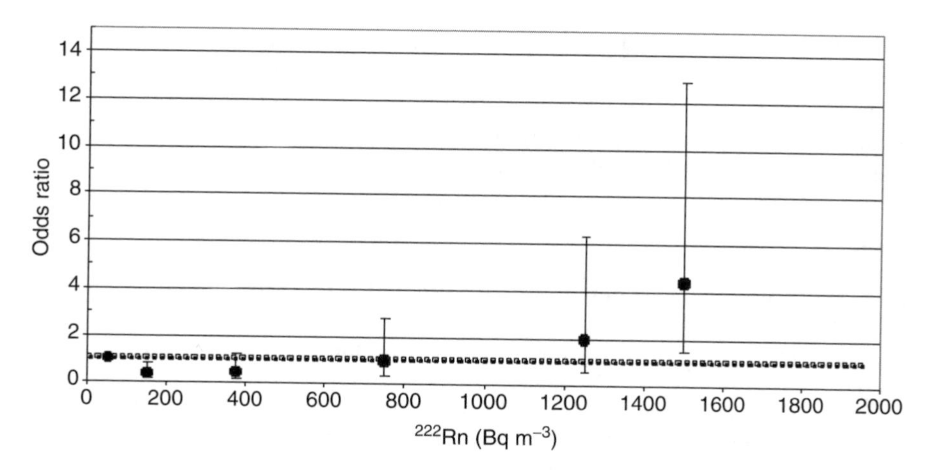

Figure 19.9 Lung cancer risk estimates as a function of exposure from indoor ^{222}Rn studies in Saxony.

value to be placed on the apparent threshold. It is certain that radon is a carcinogen, but the lung cancer risk for long-term exposure to concentrations of ^{222}Rn below 1000 Bq m^{-3} is too low to define with accuracy.

Pavia et al. (2003) performed a meta-analysis of 17 case control residential studies and found an odds ratio (OR) of 1.24 (C.I. 1.11 to 1.38) for life time exposure to 148 Bq m^{-3}. A lifetime lung cancer risk of 1% (C.I. 0.06 to 1.9%) can be estimated from these data. This may be the most definitive risk estimate possible from domestic radon exposure studies.

19.12 Summary

Radon is present in all atmospheres, whether indoors or outdoors, with a global indoor average based on 50 countries of 50 to 30 Bq m^{-3} (1.4 to 0.8 pCi l^{-1}). Outdoor radon concentration is not reported frequently, but an EPA survey of outdoor radon in all 50 U.S. states determined an average of 15 Bq m^{-3} at ground level. Even at 10 km into the troposphere, near stratospheric heights, the concentration has been measured at 0.1 Bq m^{-3}.

The alpha-emitting short-lived decay products of ^{222}Rn are always present in any atmosphere in an approximate equilibrium ratio of 40% with the gas in indoor environments and 60% outdoors. A fraction of these short-lived decay products deposit on the airway surfaces and deliver the lung dose of concern. The majority of lung tumors associated with high ^{222}Rn exposure are located in the upper airways.

Radon in water is an emerging issue. There are some locations where the ^{222}Rn gas concentration in water is quite high. The inevitable release of the gas during water use adds to the indoor inventory, with 10,000 units in water adding on average 1 unit to air. Air concentrations derived from water can in some cases require remediation of the water. A few countries have set concentration limits for ^{222}Rn in drinking water. Also, a small dose to a fetus from ^{222}Rn in maternal drinking water must be considered because ^{222}Rn is soluble in blood and body tissues.

Although there is a clear dose response for lung cancer in underground miners exposed to high concentrations of ^{222}Rn, the risk at home from domestic exposure has not been demonstrated. The exception to this is for long-term exposures to indoor concentrations that exceed about 1000 Bq m^{-3}, but the error associated with these risks does not permit precise risk estimates. The lack of clear risk data for home exposure has led to a disparity in the setting of indoor guidelines, with a range from 150 to 750 Bq m^{-3} (4 to 20 pCi l^{-1}). These values are the results of prudent judgment of what is considered an acceptable risk. The possible presence of ^{220}Rn (thoron) in the historic ^{222}Rn (radon) measurements poses an interesting problem in exposure, dose, and risk reconstruction in domestic studies. This is an issue that is worthy of continued evaluation.

References

Chittaporn, P. and Harley, N.H., A new personal ^{222}Rn and ^{220}Rn (RnTn) monitor, *Health Phys.*, 76, S163, 1999.

Chittaporn, P. and Harley, N.H., Indoor and outdoor ^{222}Rn measurements in Bangkok and Chiang Mai, Thailand *Technology*, 7, 491–495, 2000.

Chittaporn, P., Harley, N.H., Medora, R., and Merrill, R., Indoor and outdoor thoron decay products equilibrium at Fernald, OH, New York City, and New Jersey. *Health Phys.*, 84, S199, 2003.

de Villiers, A.J., Cancer of the Lung in a Group of Fluorspar Miners, Proceedings/Canadian Cancer Conference, Vol. 6, 1966, pp. 460–474.

de Villiers, A.J., Windish, J.P., Brent, F.de. N., Hollywood, B., Walsh, C., Fisher, J.W., and Parsons, W.D., Mortality experience of the community and of the fluorspar mining employees at St. Lawrence, Newfoundland, *Occup., Health Rev.*, 22, 1–15, 1971.

Fisenne, I.M., Machta, L., and Harley, N.H., *Stratospheric Radon, Radioactivity in the Environment*, Elsevier, Amsterdam, 2004.

Harley, J.H., Radon is out, in Indoor Radon and Lung Cancer: Reality or Myth, Twenty Ninth Hanford Symposium on Health and the Environment, Battelle Press, Richland, 1990.

Harley, N.H. and Chittaporn, P., An aerosol particle size sampler using ^{222R}Rn decay products as tracers, *Health Phys.*, 76, S163, 1999.

Harley, N.H. and Chittaporn, P., Long term measurement of indoor and outdoor ^{212}Pb decay products, with estimates of aerosol particle size, *Technology*, 7, 407–413, 2000.

Harley, N.H. and Robbins, E.S., ^{222}Rn alpha dose to organs other than the lung, *Radia. Prot. Dosim.*, 45, 619–622, 1992.

Harley, N.H. and Robbins, E.S., A biokinetic model for ^{222}Rn gas distribution and alpha dose in humans following ingestion, *Environ. Int.*, 20, 605–610, 1994.

Harley, N.H., Meyers, O.A., Chittaporn, P., and Robbins, E.S., A biological model for lung cancer risk from ^{222}Rn exposure, *Environ. Int.*, 22, S977–S989, 1996.

Harley, N.H., Chittaporn, P., Fisenne, I.M., and Perry, P., ^{222}Rn decay products as tracers of indoor and outdoor aerosol particle size, *J. Environ. Radioactivity*, 51, 27–35, 2000.

Harley, N.H., Chittaporn, P., Heikkinen, M.S.A., Medora, R., and Merrill, R., Airborne Particle Size Distribution Measurements at USDOE Fernald, American Chemical Society Monograph Series, 2004.

Hopke, P.K., Borak, T.B., Doull, J., Cleaver, J.E., Eckerman, K.F., Gunderson, L.C.S., Harley, N.H., Hess, C.T., Kinner, N.E., Kopecky, K.J., McCone, T.E., Sextro, R.G., and Simon, S.L., Health risks due to radon in drinking water. *Environ. Sci Technol.*, 34, 921–926, 2000.

ICRP, Protection Against Radon-222 at Home and at Work, ICRP Publication 65, Volume 23, No. 2, Elsevier Science, Inc., Tarrytown, NY, 1993.

Israel, H., The radon-220 content of the atmosphere, in '*The Natural Radiation Environment*', Adams, J.A.S. and Lowder, W.M., Eds., University of Chicago Press, Chicago, 1964, p. 313.

Li, C.S. and Hopke, P.K., Initial size and distributions and hygroscopicity of indoor combustion aerosol particles, *Aerosol Sci Technol.*, 19, 305–316, 1993.

Lubin, J.H., and Boice, J.D., Lung cancer risk from residential radon: meta-analysis of eight epidemiologic studies, *J. Natl. Cancer Inst.*, 89, 49–57, 1997.

Machta, L. and Lucas, H.F., Jr., *Science*, Washington, DC, 35, 296, 1962.

Moore, H.E., Poet, S.E., and Martell, E.A., *J. Geophys. Res.*, 78, 7065, 1973.

Morrison, H.I., Semenciw, R.M., Mao, Y., and Wigle, D.T., Cancer mortality among a group of fluorspar miners exposed to radon progeny, *Am. J. Epidemiol.*, 128, 1266–1275, 1988.

Morrison, H.I., Villeneuve, P.J., Lubin, J.H., and Schaubel, D.E., Radon-progeny exposure and lung cancer risk in a cohort of Newfoundland fluorspar miners, *Radiat. Res.*, 150, 58–65, 1998.

NAS, *Health Effects of Radon and other Internally Deposited Alpha Emitters (BEIR IV)*, National Academy Press, Washington, DC, 1988.

NAS, *Health Effects of Exposure to Radon (BEIR VI)*, National Academy Press, Washington, DC, 1999.

NCRP, Exposures from the Uranium Series with Emphasis on Radon and its Daughters, Report 77, National Council on Radiation Protection and Measurements, Bethesda, MD, 1984a.

NCRP, Evaluation of Occupational and Environmental Exposures to Radon and Radon Daughters in the United States, Report 78, National Council on Radiation Protection and Measurements, Bethesda, MD, 1984b.

NCRP, Measurement of Radon and Radon Daughters in Air, Report 97, National Council on Radiation Protection and Measurements, Bethesda, MD, 1988.

NCRP, Evaluation of Occupational and Environmental Radon Risk. National Council on Radiation Protection and Measurements, Bethesda, MD, 2005.

Pavia, M., Bianco, A., Pileggi, C., and Angellilo, I., Meta-analysis of residential exposure to radon gas and lung cancer. *Bulletin of the World Health Organization*, 81, 732–738, 2003.

Reineking, A. and Porstendorfer, J., High volume screen diffusion batteries and alpha spectroscopy for measurement of the radon daughter activity size distributions in the environment, *J. Aerosol Sci.*, 17, 873–879, 1986.

Robbins, E.S. and Harley, N.H., *The Fetal Dose from ^{222}Rn in Drinking Water, Radioactivity in the Environment*, Elsevier, Amsterdam, 2004.

Ruzer, L., Nero, A.V., and Harley, N.H., Assessment of lung deposition and breathing rate of underground miners in Tadjikistan, *Rad. Prot. Dosim.*, 58, 261–268, 1995.

Saccomanno, G., Auerbach, O., Kuschner, M., Harley, N.H., Michaels, R.Y., Anderson, M.W., and Bechtel, J.J., A comparison between the localization of lung tumors in uranium miners and in nonminers from 1947 to 1991, *Cancer* 77, 1278–1283, 1996.

Schery, S., Measurement of airborne ^{212}Pb and ^{220}Rn at various indoor locations within the United States, *Health Phys.*, 49, 1061, 1985.

Schery, S. and Zarcony, M.J., Thoron and Thoron Daughters in the Indoor Environment, Proceedings of the 18th Midyear Topical Symposium, Health Physics Society, 1985.

Schery, S., Thoron in the environment, *J. Air Waste Manage.*, 40, 493, 1990.

Tu, K.W. and Knudson, E.O., Indoor outdoor aerosol measuremetns for two residential buildings, *Aerosol Sci. Technol.*, 9, 71–82, 1988a.

Tu, K.W. and Knudson, E.O., Indoor radon progeny particle size measurements made with two different methods, *Radiat. Prot. Dosim.*, 24, 251–255, 1988b.

UNSCEAR, Ionizing Radiation: Sources and Biological Effects, United Nations Scientific Committee on the Effects of Radiation, New York, NY, 2000.

chapter twenty

Risk from inhalation of the long-lived radionuclides uranium, thorium, and fallout plutonium in the atmosphere

Isabel M. Fisenne
USDHS Environmental Measurements Laboratory

Contents

20.1 Introduction 585
20.2 Similarities and differences 586
20.3 Sampling 586
20.4 Uranium in TSP 587
20.5 Th in TSP 587
20.6 Pu in TSP 588
20.7 Resuspension 590
20.8 Respirable concentrations of U, Th, and Pu 591
20.9 Value of long-term monitoring 592
20.10 Exposure estimation from the inhalation of U, Th, and Pu 592
References 593

20.1 Introduction

The global population is chronically exposed to naturally occurring and man-made radionuclides by both inhalation and ingestion. The inhalation pathway is the principle focus of this chapter and the ingestion pathway will only be touched upon as necessary.

Soils and rocks contain the Uranium and Thorium Series, headed by ^{238}U ($t_{1/2}=4.468\times 10^9$ yr) and ^{232}Th ($t_{1/2}=1.405\times10^{10}$ yr), respectively. The natural forces of erosion and weathering reduce the particle size of the host rock or soil and surface winds suspend the small particles. These are removed from the atmosphere by the usual scavenging processes. It is known that "dust storms" resulting from desertification carry material for thousands of kilometers and in some instances around the world. The distribution of uranium and thorium in the terrestrial environment is relatively constant at 2–3 μg U g^{-1} (25–36 mBq ^{238}U g^{-1}) of soil and 10 μg Th g^{-1} (4 mBq ^{232}Th g^{-1}) of soil.[1,2] There are geographical areas that do have higher concentrations and have had the deposits exploited for commercial purposes. The bulk of the atmospheric inventory of uranium is soil derived from the earth's surface. Additional sources of uranium are emissions from energy-generating plants (coal, oil, nuclear), fallout resuspension from atmospheric nuclear weapons

1-56670-611-4 / 05 / $0.00+$1.50

tests, satellite failures, and nuclear-related accidents.[3] Atmospheric thorium is also derived from soil resuspension and, to a lesser degree, energy-related emissions.

Plutonium is a man-made element and like U and Th is an actinide element. The principal source of terrestrial plutonium was atmospheric nuclear weapons testing. Plutonium is produced by an (n, γ) reaction on ^{238}U and is separated for use in nuclear weapons. Fallout plutonium consists of ^{239}Pu ($t_{1/2}$=2.411×10^4 yr) and ^{240}Pu ($t_{1/2}$=6.563×10^3 yr) in a 240/239 atom ratio of 0.18. For convenience, the pair will be denoted as Pu. A third Pu isotope, ^{238}Pu ($t_{1/2}$=87.7 yr), was introduced into the atmosphere in the Southern Hemisphere stratosphere when a satellite that included a Systems for Nuclear Auxiliary Power Generator (SNAP-9A) failed and reentered the atmosphere. Harley estimated that in 1970 in the Northern Hemisphere, <4% of atmospheric Pu was due to ^{238}Pu from weapons plus SNAP-9A.[4] His ^{238}Pu estimate for the Southern Hemisphere was 18% of the total Pu in that hemisphere. The total Pu injection into the atmosphere was about 400 kCi (1.5×10^{16} Bq), some of which remained on or close to the test sites. Because the majority of the weapons tests were conducted in the Northern Hemisphere, ~80% of the Pu fallout occurred there.

20.2 Similarities and differences

As noted above, the principal source of U and Th in the atmosphere is derived from soil and rock. The base chemical composition is usually silicate or carbonate. The U in atmospheric aerosols exists primarily in the +6 valence state, while Th resides in the +4 state. Fallout Pu deposits on the earth's surface as the oxide and is principally in the +4 valence state.

Once aerosolized, these actinides become of interest as part of the total human exposure to radiation. Unlike U, the principal exposure route for Th and Pu is inhalation rather than ingestion. The International Commission on Radiological Protection (ICRP)[5] suggests that for insoluble compounds of Th and Pu, the gastrointestinal uptake (f_1) is of the order of 5×10^{-4}. ICRP[5] has adopted an f_1 value of 2×10^{-2} for U, assuming equal absorption from diet and water. However, Spencer et al.,[6] in the only controlled study of uptake of U in humans, showed that the principal source of uptake of U was from drinking water to the extent of about 5%.

The air concentrations of U and Th are influenced by climate and land mass while fallout Pu is dependent on the latitude of the initial injection, primarily the Northern Hemisphere.

20.3 Sampling

The collection of atmospheric aerosols is almost exclusively that of total suspended particulates (TSP), that is, particles <500 μm in diameter. More important are inhalable particulates, <10 μm in diameter, which may enter the nose and throat. Samplers that are designed to collect this class of particulates are referred to as PM_{10} samplers. Respirable particulates, <2.5 μm in diameter ($PM_{2.5}$), reach the functional areas of the lungs.

For fallout radionuclides, high-volume pumps were and are still used to draw air through filters made of material such as polystyrene. The large air volumes, usually in excess of 25,000 m^3, were necessary to collect sufficient sample for wet radiochemical analyses. Naturally occurring radionuclides were also present in the total aerosol, but from terrestrial sources.

There are remarkably few radionuclide measurements of environmental respirable particulates. Two such studies will be described later. Resuspension of deposited material has attracted little attention except for studies in the environs of nuclear weapons production facilities. There is one study of the resuspension of uranium from environmental sources. This will also be discussed later.

The sites for air collections to determine the concentration of these long-lived α-emitters are almost exclusively land based in the mid-latitudes of the Northern Hemisphere. A few measurements of U have been made from collections at remote sites, including the Atlantic Ocean and Antarctica. Measurements of the Th concentration in air are rare. The U.S. Department of Energy Environmental Measurements Laboratory's (EML) Surface Air Sampling Program (SASP) was the most geographically comprehensive monitoring program for fallout Pu. Air samples were collected and analyzed from stations at Nord, Greenland (80°N, 17°W), to South Pole Station, Antarctica (90°S,0°W), roughly along the 80th meridian. The EML database for SASP can be accessed at their website (http://www.eml.doe.gov).

20.4 Uranium in TSP

The largest number of measurements of U in air has been performed under a U.S. Environmental Protection Agency (EPA) program named Environmental Radiation Ambient Monitoring System (ERAMS). The ERAMS air program consists of 50 stations collecting air particulates on filters. The ERAMS filters were analyzed for U starting in the late 1970s. Presently, annual composites of the air particulate filters are analyzed for U and Pu.

The longest running program for the collection and measurement of air particulates is conducted at the Argonne National Laboratory (ANL), Argonne, IL. Begun in 1973, their air sampling program includes an off-site station to assess the background concentrations of nuclides of interest, including U, Th, and Pu.

Published data from the EPA and ANL sampling programs were evaluated to assess the U air concentration in the continental United States.[7] The U concentrations from 25 ERAMS sites were evaluated and an anomaly was found for stations in the northern and mid-section of the U.S. This was confirmed by the independent air filter collections and measurements of U performed at ANL. The 40% decrease in the air concentrations of U at these sites was attributed to "regulatory compliance in reducing emissions from fossil burning facilities."

The data from 22 ERAMS locations averaged 2.1 ± 0.7 μBq ^{238}U m^{-3} (170 pg U m^{-3}). Only the ANL data were useful to investigate seasonal variations as the ERAMS composite samples overlapped seasons. At ANL, they showed a slight rise in April (spring rise), a low point in August, and another small rise in November. It was thought that the November rise was due to emissions from a local coal burning electrical generating plant. The noncarbonaceous residues from the TSP collections were analyzed for U and the concentrations were found to reflect the U concentration in soil. The ^{234}U/^{238}U ratio for 22 ERAMS sites was 1.14 ± 0.24, indicative of the U source, soil.

The U concentrations in surface air are shown in Table 20.1. The impact of land mass, climate, and industrialization is apparent. Basically there is no soil at the remote sites, the ground being frozen year round or at oceanic locations. Even the remote sites in Antarctica show a difference between the pristine environment and human encampments. The global average concentration of U in air as adopted by UNSCEAR[1] of 1 μBq m^{-3} is based almost exclusively on data obtained in the mid-latitudes of the Northern Hemisphere.

20.5 Th in TSP

Measurements of Th concentration in air are sparse. This was partially due to the difficulties encountered with radiochemical separation, specifically unacceptable reagent, and material blanks. Again, ANL'S Environmental Monitoring Program has the longest running record of Th in air measurements. As stated earlier for their U in air measurements, the impact of regional sources causes a marked decrease with time. The ANL data are included in Table 20.2 as they are unremarkable when compared with other measurements. Again, measurements of the ashed filter residues indicated that the Th was derived from soil. The activity concentrations of Th in air are in the same general range as those for U.

Table 20.1 Uranium Concentrations in Total Suspended Particulates

	pg U m^{-3}		μBq ^{238}U m^{-3}		
Remote Sites	Mean	SD	Mean	SD	Reference
N. Atlantic Ocean	4.1	1.2	0.05	0.015	8
S. Atlantic Ocean	2.2	0.5	0.027	0.006	8
Antarctic Ice Pack	1.2	0.3	0.015	0.004	8
Antarctic Base Camp	3.0	1.0	0.037	0.012	8
Norway — Skibotn	3.0	1.3	0.037	0.016	9
Vardo	5.5	0.9	0.068	0.012	10
Asia					
Japan — Tokyo	25	15	0.31	0.19	11
Tsukba Science City	14	10	0.17	0.12	11
Kamisaibama	234	98	2.9	1.2	12
Europe					
Belgium — Mol	115	95	1.4	1.2	13
Olen	114	36	1.4	0.5	13
Germany — Berlin	133	20	1.6	0.2	9
Braunschweig	85	40	1.1	0.5	9
U.K. — Sutton	62	78	0.77	0.97	8
North America					
Canada — Ontario	100		1.2		14
U.S.A. — 23 states	170	54	2.1	0.7	15
Global average	80		1		1

Table 20.2 Thorium Concentrations in Total Suspended Particulates

	pg Th m^{-3}		μBq ^{232}Th m^{-3}		
Remote Sites	Mean	SD	Mean	SD	Reference
Norway — Skibotn	100	50	0.4	0.2	9
Vardo	15	7	0.065	0.003	10
Asia					
India — Bombay	250 to 2500		1 to 10		16
Europe					
Germany — Berlin	270	20	1.1	0.1	9
Braunschweig	150	70	0.6	0.3	9
North America					
United States — Argonne, IL	50	20	0.2	0.1	17
New York, NY	100	50	0.4	0.2	18
Global average	250		1		1

20.6 Pu in TSP

Measurements of a long-lived alpha-emitter Pu were necessitated by widespread public concern over its potential hazard. The measurements of Pu in TSP on a global basis began in 1966. By use of the known Pu to ^{90}Sr ratio (0.017), estimations of atmospheric Pu in TSP were extended back to 1963 from SASP samples. The U.S. DOE EML SASP network results for Pu were part of an extensive review by Harley.[4] He estimated that some 15 PBq (400 kCi) had been produced and dispersed globally. Hardy et al.[19] accounted for over 12 PBq (330 kCi) without estimating the deposition around test sites. The measurements of Pu on a global scale continued until mid-1985 when the Pu concentrations in monthly composited samples approached the detection limit for the measurement. The U.S. EPA ERAMS measurements included Pu beginning in 1978. The collections and measurements continue on an annual basis as the Pu concentrations in the TSP are at the detection limit for the measurement. The radionuclide concentrations from the ERAMS sites are available on their website (http://www.epa.gov/enviro/html/erams). The off-site aerosol collections at ANL begun in 1973 and continuing to the present time are still analyzed for Pu on a monthly basis, but the concentrations are nearly at the detection limit for the measurement.

In order to summarize this vast database, 11 sites were selected and the average annual concentrations of Pu for the years the site was operational were summed. The site, latitude, years in operation, and the sum of the Pu concentrations measured in the TSP samples are shown in Table 20.3. The distribution pattern of fallout Pu shows the highest concentrations in the Northern Hemisphere mid-latitudes, a minimum in the equatorial region, and an increase in the Southern Hemisphere mid-latitudes. Most of the atomic weapons testing was carried out in the Northern Hemisphere, but a few test sites were located in the Southern Hemisphere. A summary of atmospheric weapons testing has been prepared by UNSCEAR.[20]

Harley[4] stated that the Pu inventory for the Northern Hemisphere decreased with a half-time of 10 months until 1968 when the People's Republic of China began a series of atmospheric weapons tests. The concentrations of Pu in TSP collected in New York City are depicted in Figure 20.1. With no substantial atmospheric injection since 1981, the TSP

Table 20.3 Plutonium in TSP

Location	Latitude	Years	Total μBq Pu m^{-3}
Thule, Greenland	76°N	1963 to 1974	65.1
Moosonee, Ontario	51°N	1963 to 1985	71.0
New York, NY	40°N	1963 to 1985	98.9
Mauna Loa, HI	29°N	1963 to 1985	95.0
Miami, FL	25°N	1963 to 1985	82.5
Guayaquil, Eucador	2°S	1963 to 1976	10.3
Lima, Peru	12°S	1968 to 1985	11.6
Chalcaltaya, Chile	16°S	1963 to 1985	22.6
Santiago, Chile	33°S	1963 to 1985	30.3
Punta Arenas, Chile	53°S	1963 to 1985	6.5
Antarctica	64°S	1966 to 1975	3.4

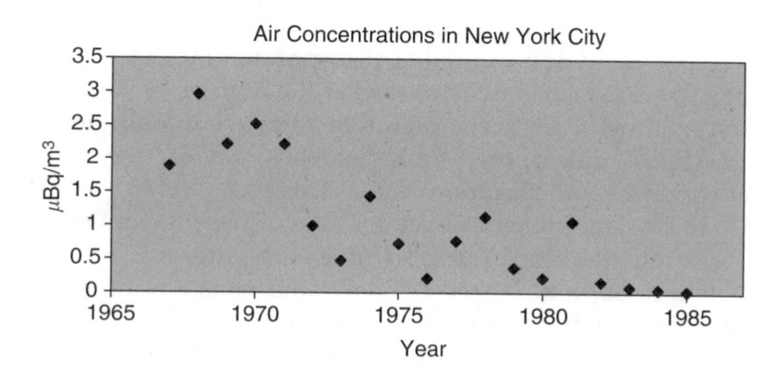

Figure 20.1 Plutonium in New York City Aerosol Samples

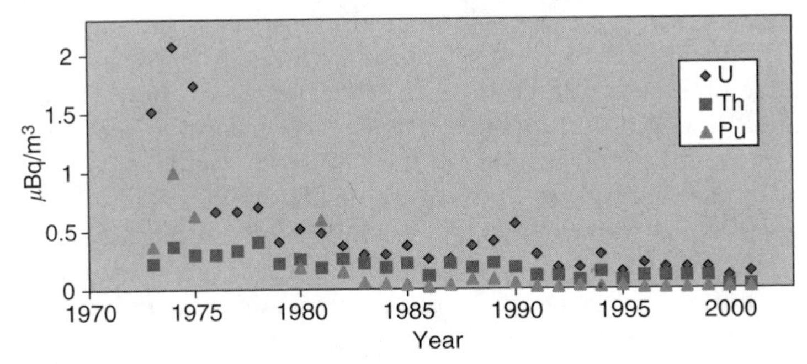

Figure 20.2 U, Th and Pu in ANL Aerosol Samples. Golchert et al.[17]

Pu concentrations have decreased to <0.05 μBq m^{-3} in 1985. The annual average Pu in air concentration at ANL was <0.01 μBq m^{-3} in 2001.

Lee et al.[21] estimated the stratospheric mean residence time of 1.2 yr, while Holloway and Hayes[22] estimated a 71 d mean residence time for Pu aerosol fallout in the troposphere. Both these pieces of information suggest that all the fallout Pu was deposited on the surface of the earth by 1985. The fact that Pu is still measurable in TSP, albeit with error terms of 100% or more, leads to the conclusion that resuspension of material from the surface is now the controlling mechanism for dispersion, as it is with U and Th.

20.7 *Resuspension*

Particle resuspension, while of importance, is an elusive process to quantify. Sehmel[23] published a definitive review of the subject that gives an appreciation for the uncertainties associated with this topic. Resuspension is a form of large-scale erosion describing the continual movement of particles as a function of surface stresses. The stresses are saltation, surface creep, and suspension, while the transport means depend on particle diameter, wind speed, and turbulence. Newman et al.[24] defined saltation as a process by which particles with diameters of 100 to 500 μm rise or bounce in a layer close to the surface–air interface. Surface creep particles with diameters of 500–1000 μm slide or roll, pushed along the surface by wind stresses and the impact of saltation particles. The smallest particles, <100 μm in diameter, move by suspension, following air motion. The interplay of the stresses causes suspension particles to leave a surface when saltation particles impact the surface. Although particles <50 μm and particularly <10 μm in diameter are almost impervious to wind erosion, when mixed with saltation particles, they become transportable by suspension.

Modelers have not been able to predict resuspension factors for general situations. Langham[25] defined the resuspension factor (RF) as the ratio of the airborne pollutant concentration per unit volume of air to the pollutant surface concentration per unit area on the surface. Thus, RF has units of m^{-1}. From Sehmel's[23] review, the resuspension factors developed under experimental conditions range from 10^{-12} to 10^{-2} m^{-1} for wind resuspension and 10^{-10} to 10^{-2} m^{-1} from human activities. These factors were developed not only from wind erosion situations but also from agricultural practices, vehicular and pedestrian traffic, and household chores. Similarly, the resuspension half-life ranges from days to years and is dependent on the situational parameters.[23]

Several resuspension studies were conducted at nuclear facilities and the Nevada test site, but there appears to be only one study of the resuspension of U, Th, and Pu. Golchert

Table 20.4 Resuspension Factors for U, Th Isotopes and Fallout Pu at Argonne, IL[a]

Nuclide	Air Concentration (μBq m^{-3})	Ground Deposition (Bq m^{-2})	Resuspension Factor (m^{-1})
^{228}Th	0.32	629	5.1×10^{-10}
^{230}Th	0.54	1036	5.3×10^{-10}
^{232}Th	0.30	555	5.5×10^{-10}
U	1.55	2516	6.1×10^{-10}
Pu	0.026	37	7.0×10^{-10}

[a] Adapted from Golchert and Sedlet.[26]

and Sedlet[26] collected and analyzed TSP air filters and soil samples at the off-site ANL monitoring station. They assumed that the top 1 cm of soil was available for resuspension and contained 925 μBq g^{-1} of soil. The weight of the noncarbonaceous material remaining after dry ashing the TSP filter was taken to be resuspended soil. Resuspension factors for U, Th isotopes and fallout Pu were calculated and are shown in Table 20.4. The RFs estimated in this study are internally consistent, despite the different sources, fallout and naturally occurring radionuclides. The Pu was "aged" deposition, assuming most of the Pu was deposited in the 1960s. This would suggest that the Pu in soil was in a relatively steady-state condition. Anspaugh et al.[27] estimated RFs for 20-yr-old Pu deposition at the Nevada test site to be 3×10^{-10} and 2×10^{-9} m^{-1}. Golchert and Sedlet[26] considered the U and Th representative of the "ultimate aged source and their resuspension factors, the equilibrium condition." However, these RF estimates are site specific and cannot be taken as the general case. It does suggest that resuspension will be the source of fallout Pu (in the absence of any atmospheric weapons testing or nuclear accidents) for a long period of time.

20.8 Respirable concentrations of U, Th, and Pu

The term TSP refers to particles <500 μm in diameter. The next smaller fraction is called inhalable particulates with sizes of <10 μm in diameter, sometimes designated as PM_{10}. Their size permits penetration into the nose and throat. The respirable particulates have diameters of <2.5 μm ($PM_{2.5}$) and penetrate into the lung. The NCRP has defined the respirable fraction as "the fraction of airborne material that can be inhaled and possibly deposited in the lung."[28] "Respirable dust" was defined as the portion of inhaled dust that is deposited in the nonciliated portions of the lung.[28]

Golchert and Sedlet[26] determined the particle size distribution of U and Pu at the ANL off-site location. A commercial high-volume cascade impactor[29] was operated for 1 month to collect a total air volume of 2.28×10^4 m^3. The summary of their results is shown in Table 20.5. It is worth noting that the total particulate concentration of 55 μg m^{-3} of air is in excellent agreement with the UNSCEAR[1] adopted value of 55 μg m^{-3} of air. As expected, the Pu-bearing particulates are very small, 87% of the total activity associated with particle sizes of $\leq$2 μm. The soil-derived U particles tend to be much larger with <20% of the total activity in the $\leq$2 μm fractions. Volchok et al.[30] collected air samples with a horizontal elutriator to determine the respirable fraction of Pu in an urban and a rural setting. The results were the same for the two locations (84% respirable) and in excellent agreement with the ANL data. With the uncertainties associated with the collections and measurements of these samples, fallout Pu will be considered 100% respirable. Although no measurements were reported for the respirable fraction of Th, it is assumed that it is the same as for U, that is, $\leq$20%.

Table 20.5 Particle Size Distribution of U and Pu in air at ANL[a]

Particle Size Range (μm)	Particulate Weight (g)	Particulate Concentration (μg m^{-3})	Air Concentration (μBq m^{-3})		Mass Concentration (mBq g^{-1})	
			U	Pu	U	Pu
>7	0.440	19.3	1.07 (37.2%)	0.036 (2.5%)	55.5 (14.5%)	1.81 (1.3%)
3.3–7	0.169	7.4	0.67 (23.3%)	0.044 (3.1%)	89.9 (23.5%)	5.92 (4.1%)
2.0–3.3	0.097	4.3	0.59 (36.4%)	0.11 (7.8%)	139.1 (36.4%)	25.9 (18.1%)
1.1–2.0	0.102	4.5	0.41 (14.3%)	0.28 (19.7%)	91.0 (23.8%)	61.4 (42.9%)
<1.1	0.445	19.5	0.13 (4.5%)	0.96 (67.0%)	6.7 (1.8%)	48.1 (33.6%)
Total	1.253	55.0	2.87	1.43		

[a] Adapted from Golchert and Sedlet.[26]

20.9 Value of long-term monitoring

The work performed and published by Golchert et al.[17] demonstrates the scientific significance of long-term monitoring efforts. The measurements of two naturally occurring actinide elements, U and Th, and the fallout actinide, Pu, at the same location on a monthly basis is a rare database. Figure 20.2 displays the annual average TSP in air concentrations for U, Th, and Pu at the ANL off-site location. Even with this gross depiction, it is evident that, in the absence of atmospheric weapons testing, fallout Pu rapidly decreased below the level of the natural emitters. The graph also shows the impact of local sources (coal-fired power plants) on the U air concentration and the subsequent decline due to the enforcement of clean air act requirements. The annual means mask the features of the monthly and seasonal variations measured at this site. Fallout-derived Pu concentrations followed the well-known spring rise pattern, while the U and Th air concentrations rose in the winter, a direct result of the heating pattern in the area. The data also revealed that the combined air concentrations of U and Th were greater than Pu from 1973 to 2002. The record is incomplete at ANL, but Pu in air measurements in New York City, NY, the same latitude band as Argonne, IL, from 1963 through 1973 show that >92% of the Pu was deposited during this period. The data obtained for the New York City site and the ANL off-site location were in good agreement for the years of mutual collections. Another interesting fact concerning the ANL off-site data and the U.S. EPA ERAMS data is that the Chicago (IL) U in air concentration of 2.4 μBq m^{-3} is close to the national average of 2.1 μBq m^{-3} of air, but a factor of 5 to 10 greater than the ANL off-site data. This comparison of a single site with measurements of a naturally occurring and fallout radionuclide with other studies points out the difficulties in deriving reasonable estimates of air concentrations based on sampling site location.

20.10 Exposure estimation from the inhalation of U, Th, and Pu

The measurements of U, Th, and Pu in TSP show the vast differences in concentrations with geographical location. To generalize the inhalation estimate of these radionuclides, the UNSCEAR global averages for U and Th (1 μBq m^{-3} of air) are adopted, and for Pu, a single location, New York, NY, is selected for a 23 yr period of 1963 to 1985. The estimations are for the adult male with a daily breathing rate of 22.2 m^{-3} of air. The total exposure to U and Th for a 23 yr period is 186 mBq for each or 372 mBq for U and Th. The exposure estimate based on New York, NY, Pu measurements from samples collected in the latitude with the highest fallout air concentration is 802 mBq or a factor of 2.2 greater than the sum of the U and Th exposures.

These estimates are made only for illustrative purposes and are not representative of specific geographical areas or particular time periods, except for Pu. The ANL TSP measurements for calendar year 2001 yield air concentrations of 0.15 μBq U m^{-3}, 0.04 μBq Th m^{-3}, and 0.01 μBq Pu m^{-3}. The sum of U and Th annual exposures at site for 2001 is a factor of 15 greater than the Pu exposure.

As noted earlier, all three environmental aerosols are refractory and, thus, only very slowly transferred from the alveolar region of the lung to the bloodstream. Some of the deposited aerosol is sequestered for long periods of time in the pulmonary lymph nodes, in effect reducing the systemic body burden.

The alpha-dosimetric consequences from the inhalation of these environmental aerosols are minor compared to the dose from radon progeny.

References

1. United Nations Scientific Committee on the Effects of Atomic Radiation (UNSCEAR), Annex B, Exposures from Natural Radiation Sources, Vienna, 2000.
2. National Council on Radiation Protection and Measurements Report No. 94, Exposure of the Population in the United States and Canada from Natural Background Radiation, Bethesda, MD, 1987.
3. United Nations Scientific Committee on the Effects of Atomic Radiation, Annex B, Exposures from Natural Radiation Sources, Vienna, 1988.
4. Harley, J.H., Plutonium in the environment — a review, *Jpn J. Radiat. Res.*, 23, 83, 1980.
5. International Commission on Radiological Protection (ICRP), Age-Dependent Doses to Members of the Public from Intake of Radionuclides, ICRP Publication 67, Part 2 Ingestion Dose Coefficients, Annals of the ICRP 23 (3/4), Elsevier Science Ltd., Oxford, 1993.
6. Spencer, H., Osis, D., Fisenne, I.M., Perry, P.M., and Harley, N.H., Measured intake and excretion patterns of naturally occurring ^{234}U, ^{238}U, and calcium in humans, *Radiat. Res.*, 124, 90–95, 1990.
7. Stevenson, K.A. and Pan, V., An Assessment of uranium in surface air within the continental US. *J. Environ. Radioactivity*, 31, 223, 1996.
8. Hamilton, E.I., The concentration of uranium in air from contrasted natural environments, *Health Phys.*, 19, 511–520, 1970.
9. Kolb, W., Seasonal fluctuations of the uranium and thorium contents of aerosols in ground-level air, *J. Environ. Radioactivity*, 9, 61–75, 1989.
10. Kolb, W., Thorium, uranium and plutonium in surface air at Vardo. *J. Environ. Radioactivity*, 31, 1–6, 1995.
11. Hirose, K. and Sugimura, Y., Concentration of uranium and the activity ratio of ^{234}U/^{238}U in surface air: effect of atmospheric burn-up of Cosmos-954. *Meterol. Geophys.*, 32, 317–322, 1981.
12. Yunoki, E., Kataoka, T., Michihiro, K., Sugiyama, H., Shimizu, M., and Mori, T. Background levels of ^{238}U and ^{226}Ra in atmospheric aerosols, *J. Radioanal. Nucl. Chem.*, 189, 157–164, 1995.
13. Janssens, M., Desmet, B., Dams, R., and Hoste., J., Determination of uranium, antimony, indium, bromine and cobalt in atmospheric aerosols using epithermal neutron activation and a low-energy photon detector, *J. Radioanal. Chem.*, 26, 305–315, 1975.
14. Tracy, B.L. and Prantl, F.A., Radiological impact of coal-fired power generation, *J. Environ. Radioactivity*, 2, 145–160, 1985.
15. U.S. EPA, Environmental Radiation Data Report Series 402-R-93, National Air and Radiation Environmental Laboratory, Montgomery, AL, 1978–1993.
16. Sunta, C.M., Dang, H.S., and Jaiswal, D.D., Thorium in man and environment: uptake and clearance, *J. Radionanal. Nucl. Chem.*, 115, 149–158, 1987.
17. Golchert, N.W. et al., Argonne National Laboratory-East Site Environmental Report for Calendar Years 1973–2001, Argonne National Laboratory, Argonne, IL.
18. Fisenne, I.M., Perry, P.M., Decker, K.M., and Keller, H.K., The daily intake of ^{234,235,238}U, ^{228,230,232}Th, and ^{226,228}Ra by New York City residents, *Health Phys.*, 53, 357–363, 1987.

19. Hardy, E.P., Krey, P.W., and Volchok, H.L., Global inventory and distribution of fallout plutonium, *Nature* 241, 444–445, 1973.
20. United Nations Scientific Committee on the Effects of Atomic Radiation (UNSCEAR), Sources and Effects of Ionizing Radiation: United Nations Scientific Committee on the Effects of Atomic Radiation 2000 Report to the General Assembly, with Annexes, United Nations, New York, 2000.
21. Lee, S.C., Rao, H.S.C., Sakuragi, Y., Bakhtiar, N., Jiang, F.S., and Kuroda, P.K., The origin of plutonium in the atmosphere, *Geochem. J.*, 19, 283–288, 1986.
22. Holloway, R.W. and Hayes, D.W., Mean residence time of plutonium in the troposphere. *Environ. Sci. Technol.*, 16, 127–129 1982.
23. Sehmel, G.A., Particle resuspension: a review, *Environ. Int.*, 4, 107–127, 1980.
24. Newman, J.E., Abel, M.D., Harrison, P.R., and Yost, K.J., Wind as related to critical flushing speed versus reflotation speed by high-volume sampler particulate loading, in Proceedings of the Atmosphere-Surface Exchange of Particulate and Gaseous Pollutants — 1974 Symposium, Richland, WA, September 4–6, 1974, U.S. Energy Research and Development Administration Symposium Series, CONF-740921, National Technical Information Service, Springfield, VA, 1976, pp. 466–296.
25. Langham, W.H., Plutonium distribution as a problem in environmental science, in Proceedings of Environmental Plutonium Symposium, Report LA-4756, National Technical Information Service, Springfield, VA, 1971, p. 9.
26. Golchert, N.A. and Sedlet, J., Resuspension studies on fallout level plutonium, in Selected Environmental Plutonium Research Reports of the Nevada Applied Ecology Group, Report NVO-192, 1978.
27. Anspaugh, L.R., Shinn, J.H., Phelps, P.L., and Kennedy, N.C., Resuspension and redistribution of plutonium in soils. *Health Phys.*, 29, 571–582, 1975.
28. National Council on Radiation Protection and Measurements Report No. 97, (NCRP), Measurement of radon and radon daughters in air, Bethesda, MD, 1988.
29. Lippmann, M. and Harris, W.B., Size-selective samplers for estimating "respirable" dust concentrations, *Health Phys.*, 8, 155–163, 1962.
30. Volchok, H.L., Knuth, R., and Kleinman, M.T., Respirable Fraction of Sr-90, Pu-239 and Pb in Surface Air, U.S. Atomic Energy Commission Report HASL-278, I-36-39, National Technical Information Service, Springfield, VA, 1974.

chapter twenty-one

Health physics considerations of aerosols in radiosynthesis laboratories

Mark L. Maiello
Wyeth Research, R&D Environmental Health & Safety

Contents

21.1 Introduction 595
21.2 Airborne radioactivity 596
  21.2.1 Laboratory air effluents 597
  21.2.2 Laboratory surface contamination 597
21.3 Control of contamination in the radiosynthesis lab 598
References 599

21.1 *Introduction*

Radiosynthesis laboratories are vital parts of pharmaceutical research facilities. As part of the drug development regime, an experimental compound is submitted to radiosynthesis chemists who then consider methods for incorporation of a radioactive isotope into the molecule. The purpose is to produce a compound that can be tested for metabolic transformation, biokinetic characteristics, and other *in vivo* parameters. The radioactivity provides a signal that can be measured by standard methods from plasma, blood, and other tissues. These data elucidate the organs affected by the compound, its metabolic profile, and whether the compound remains long enough in the target organ to be efficacious.

Radiosynthesis chemists are tasked with a formidable job. They must construct (synthesize) a molecule that incorporates a radioactive atom, for example, a carbon-14 (C-14) atom, at just the proper place within the molecule. Placement is crucial because some metabolic breakdown of the molecule is expected. The radioactive tag must be attached to the unmetabolized molecular component with the desired efficacious effect. The compound must be produced in a form that can be stored for long periods without molecular disintegration. The specific radioactivity (μCi per unit weight) must be well characterized. Other researchers will require this information to produce quantifiable dilutions for injections into experimental animals. The purity of the radioactive compound must be such that these subsequent *in vivo* experiments meet strict Food and Drug Administration guidelines. Application of "good lab practice" guidelines are required in the radiochemical laboratory to assure that calibrated equipment, reagents of certain ages and grades, and work practices meet rigorous quality standards.

1-56670-611-4 / 05 / $0.00+$1.50

Radioactive work practices must also follow the federal radiation safety guidelines or those of the local radiological health bureau. These are enforceable by the institutional radiation safety officer (RSO) with support from management.

To tag the candidate drug entities, chemists use various techniques that often involve the evolution of vapors and gases. Some release of particulate matter can also be expected. Solvent extraction is almost always involved in the process with concomitant releases of the gas phase to an effluent air stream. Thus, most chemical reaction work is performed in fume hoods that are vented to roof-top exhausts. Typical chemical manipulations can be expected involving radioactive precursors of the final product such as heating, pouring, shaking, sonication, and weighing. Modern analytical devices such as liquid chromatographic mass spectrometers (LCMS) and high-performance liquid chromatography (HPLC) units also handle radioactive compounds. Finally, there is production of the final form of the compound. This is achieved by lyophilizing or "freeze-drying." This produces a low-mass physical form that must be handled with great care. Experience indicates that this form can become electrostatically charged, making manipulation for weighing difficult and the ability to meet radiological contamination limits in laboratories equally troublesome.

The synthesis of drug compounds with C-14 often requires hundreds of mCi of activity. Synthesis chemistry involving tritium (H-3) involves Ci amounts of that isotope. In this case, the tritium is applied to the compound as a pressurized gas in order to facilitate incorporation of the H-3 atom at the desired molecular site ("tritiation").

21.2 *Airborne radioactivity*

The health physics of radiosynthesis labs can be implemented by determining the sources of radioactive contamination and attempting to moderate the problems they cause. As implied above, contamination in these facilities is partly due to chemical operations that produce gases, vapors, and airborne particulates. Some surface radioactivity on equipment and bench-tops will also arise from contaminated gloves that have not been regularly changed out.

Clearly, the major sources of radioactivity are the starting (stock) solutions obtained by radiochemists from commercial laboratories. The chemical form will be a compound that can be used to literally reconstruct the experimental drug entity. In place of a carbon atom, the reconstructed molecule will contain a C-14 atom at a predetermined location. These starting solutions often contain hundreds of mCi to accommodate the poor chemical yields encountered in synthesis work. The starting chemical form may itself be volatile or the synthesis methodology, often hundreds of steps, may call for some volatility to occur.

The tritiation process is a different synthesis track. Tritium must be manipulated as a gas. A commercial device is often used by research pharmaceutical facilities as both an H-3 storage bed and a delivery system (Rapkin et al., 1995). This unit can store hundreds of Ci of H-3 and can supply at least 4 Ci to the reaction vessel in order to incorporate H-3 into the target molecule. The unit houses several uranium "beds" to trap the H-3 when unneeded and a heating system to drive the H-3 gas into a manifold at the proper pressure for tritiation work. Although tightly sealed, some release of H-3 to effluent air is inevitable, especially since there is manipulation of the reaction vessel. Particulate matter is released as well, but tends to be a small fraction of the total radioactivity in the effluent from this source.

Evaporators (sometimes referred to as "roto-vaps") use vacuum pressure and temperature gradients to separate solvents from other chemical components. They are used throughout the synthesis procedure. The vacuum pumps associated with these devices may discharge radioactive mists unless properly filtered or otherwise exhausted.

In the final stages of the synthesis, lyophilization appears to be a further source of airborne contamination. These devices can release some particulate radioactivity into the surrounding environment during the freeze-drying process.

21.2.1 Laboratory air effluents

To restrict airborne releases to the work environment, chemical fume hoods with at least 80 feet per minute (FPM) face velocity are a necessity (Saunders, 1993). The chemist will often place as much equipment subject to high levels of radioactivity into the hoods as possible. Because of this potential to interfere with the work-surface airflow, it is imperative that hood face velocities be measured periodically to maintain the minimum safe level. Low-flow alarms are also required to warn the chemist of exhaust-fan malfunction or other conditions that may decrease the face velocity to levels that may create an inhalation hazard.

Regulatory agencies often restrict airborne releases to the environment via air permits. Due to these legal limits, monitoring of fume hood effluent radioactivity is a necessity. Effluent monitoring techniques for C-14 and H-3 are available and include integrating trapping techniques and real-time ion chambers. A typical method was discussed by Maiello and Linsalata (2000a, b) using sodium hydroxide trapping of C-14 as CO_2 (and H-3) in a gas dispersion bottle during radiochemical operations. They also describe a means to measure and record in live time the H-3 emissions from tritiation work using an ion chamber, a desktop computer, and software to quantify any measured H-3 emissions.

These sampling devices indicate that the majority of the H-3 and C-14 effluent is in nonparticulate form. Particle filters (glass fiber types of 1 μm pore size) indicate that particulate matter associated with H-3 effluents comprises only a fraction of 1% of the total tritium radioactivity.

21.2.2 Laboratory surface contamination

Within the work environment of the radiosynthesis laboratory, a few sources of surface contamination can be identified, inferred mainly from the sampling and scanning of surfaces by wipe-testing and Geiger counter, respectively. Wipe-testing is particularly useful for this task because it employs sensitive liquid scintillation counting (LSC) technology. Due to the extremely low energies of the H-3 beta-particle emissions, it is the only survey method that can be used for this radioisotope. Wipe-testing can reach surfaces that are inaccessible to Geiger probes. Other areas like bench-tops, sills, equipment surfaces, and lab cabinetry can be surveyed with this technique with relative ease.

Geiger counter instruments are suitable for the detection of C-14 contamination. Because they are easy to use, they can be employed during the work regime to periodically survey equipment, lab surfaces, and personnel. They lack high sensitivity for C-14 beta-emissions; hence, they must be used carefully. Geiger probes must also be maintained free of contamination. For increased sensitivity of detection of C-14, the RSO may use a gas proportional counter in place of the Geiger counter. Gas proportional probes cover much larger areas than Geiger probes, making laboratory surveys less burdensome.

The sources of contamination can be inferred from the surface tested. For example, a contaminated sill, shelf edge, or other flat surface not regularly disturbed was most likely contaminated from airborne radioactivity. The cause of contamination of a doorknob, bottle top, or faucet handle can be inferred to be manual handling.

Since most of the heavy use of radioactive substances is relegated to the fume hood, the laboratory surface contamination due to airborne particulates is most likely from equipment like lyophilizers and vacuum pumps. The process of weighing freeze-dried material is also a contributing factor.

21.3 Control of contamination in the radiosynthesis lab

There are radiation safety techniques that can mitigate radioactive contamination problems caused by aerosols in a radiosynthesis facility. These fall into the categories of engineering controls and administrative controls. Engineering controls employ a device to reduce the source of contamination or otherwise make feasible the control or cleanup of the contamination.

For example, maintaining the radiosynthesis lab at a slight negative pressure relative to adjacent rooms may reduce the transport of radioactive particulate matter under door spaces.

Floor surfaces and bench-tops can be covered with a durable, adhesive-backed Teflon-coated paper similar to old-fashioned "contact paper." The Teflon-coated version can withstand heavy foot and equipment traffic. Because it is smooth and nonstick on the Teflon side, it makes cleaning radioactive contamination easier. However, some chemicals, particularly those not soluble in water such as organic substances, are not readily removable from most surfaces, including Teflon. In that case, the health physicist can remove the affected area of Teflon or, if documented, carefully cover it with more Teflon until a full decontamination or decommissioning of the radiosynthesis lab is warranted.

For devices not normally housed in fume hoods like vacuum pumps, filters can be installed on the exhausts to reduce the radioactive aerosol emission. These must be periodically replaced, depending on frequency of use.

Another engineering technique is to enclose certain devices inside customized exhaust-boxes. Such boxes do not seal off the instrument from the laboratory environment as much as a glove-box can. The boxes are equipped with doors to permit access to device controls and to the sample input area. When the instrument is operated, the doors are closed to prevent mists or other aerosols from escaping to the work area. The box exhaust can be ganged to the radioactive fume hood effluent line, assuming that the fan can sustain the needed pressure. The exhaust will also dissipate some of the heat generated from instrument operation.

As indicated earlier, the freeze-dried compound will have to be weighed for solubilization in dosing solutions. Carefully weighing out potential drug product can be difficult. Static electric charges can build up on the experimental material because it is in granulate or powder form. Charge may also build up on containers since many of these are made of insulating materials that dissipate charge poorly. The difficulty is exacerbated because usually less than 20 mg of compound is weighed at any one time. Weigh balance results can be spurious due to the effects of static charge. Balances often become contaminated with fly-away material that is either blown by air currents or subjected to electrostatic forces of attraction or repulsion.

Ionizing devices can be employed to dissipate static charges. The devices include Po-210 alpha-emitting ionizers available from many laboratory suppliers or similar devices that use electrical power to produce ionization (both positive and negative ions) from small electrodes arranged on a small stand in a "U"-shaped configuration. Other means to mitigate the static charge problem include the maintenance of adequate air humidity in the balance room (perhaps 50%). Use of metal or metallized containers is also recommended.

Other contamination control techniques fall partly under the administrative category of contamination control. The enforcement of scheduled laboratory cleaning supplemented by frequent and documented radiological surveys (Geiger counter and wipe-testing) is useful for preventing or at least controlling buildup of radioactive contamination of all types. This falls under the purview of the institutional health physicist with cooperation from the radiochemistry staff.

For personnel protection, protective gear may include half-face respirators equipped with standard cartridge filters for protection against particulate matter. Ordinarily, these would be employed when volatile compounds are used even in fume hoods, which admittedly may provide imperfect protective airflow. Other personal protective equipment (PPE) includes disposable garments such as laboratory coats, sleeve protectors, gloves, and shoe covers. A change-out room can be supplied, which serves as a control point between the "hot" radiosynthesis lab and another room, lab, or access-way where radioactivity is not permitted or permitted at much lower levels. The change room can be used to store and don PPE and provide a sink for personnel cleaning purposes. Requiring documented surveys of PPE in the change room, where contaminated items can be discarded to radioactive waste containers, is another form of administrative control.

Since it is recognized that gases, vapors, and aerosols may be evolved from radiochemical operations, inhalation must be considered as a route to radiation exposure of internal organs and tissues. A powerful tool for determining internalization of radioactivity, particularly H-3 and C-14, is bioassay with LSC analysis. Periodic submission of urine samples to the RSO is needed, depending on the levels of radioactivity in use and the radiation exposure regulations imposed by agencies and/or the laboratory RSO. The critical counting level of a typical LS counter for H-3 and C-14 using 1 ml of urine counted for 30 min is about 1 pCi/ml. To translate urine concentration levels to radiation doses, a table of dose conversion factors such as that compiled by Labone (1983) for acute intakes of radionuclides like H-3 and C-14 can be employed.

References

LaBone, T.R. and Smith L.R., Evaluation of Acute Intakes of Radioactive Materials with an ICRP 30 Based Method: A Progress Report, Presented at the Health Physics Society Meeting, Baltimore, MD, June 20 – 24, 1983.

Maiello, M.L. and Linsalata, P., Real Time ^{3}H Monitoring of Effluent Air from a Radiosynthesis Laboratory, Proceedings of the 33rd Midyear Topical Meeting of the Health Physics Society: Instrumentation, Measurements and Dosimetry, Virginia Beach, VA, January 30 – February 2, Medical Physics Publishing, 2000a.

Maiello, M.L. and Linsalata, P., An Air Sampler for ^{3}H and ^{14}C Effluent Monitoring from a Radiopharmaceutical Laboratory, Presented at the Health Physics Society Mid-Year Topical Meeting, Virginia Beach, VA, January 30 – February 3, 2000b.

Rapkin, E., Steele, G., and Schavey, R., Modern tritium handling in the synthesis laboratory, *Am. Lab.*, October 1995. Vol. 27, No. 15, pp. 31–38.

Saunders, G.T., *Laboratory Fume Hoods*, John Wiley & Sons, Inc., New York, 1993.

chapter twenty-two

Diesel exhaust

Jonathan M. Samet
Johns Hopkins Bloomberg School of Public Health

Contents

22.1 Introduction 601
22.2 Diesel engines and diesel fuel 602
22.3 Characteristics of diesel emissions and particles 602
22.4 Health risks of diesel exposure 603
22.4.1 Lung cancer 603
22.4.2 Asthma and other allergic diseases 603
References 604

22.1 Introduction

Diesel engines are widely used in transportation and industry, particularly for heavy-duty applications including trucks, buses, construction equipment, locomotives, and ships [1]. Additionally, diesel engines are used to power passenger vehicles with extensive use in some countries. While diesel passenger vehicles are presently uncommon in the United States, in Europe more than 50% of new passenger vehicles are powered by diesel engines. Because of the reliability of diesel engines and the cheaper cost of diesel fuel in many countries, the use of diesel engines is projected to increase.

Given the widespread use of diesel engines, there has been concern for several decades as to the health consequences of exposure to diesel exhaust, both for more highly exposed occupational groups and for the general public [2]. Diesel particles are relatively small, with average aerodynamic diameters of less than 1 μm, so that they have the potential to penetrate to the small airways and alveoli of the lung. Diesel exhaust also contains nitrogen oxides and aldehydes, at relatively high concentrations, and lower concentrations of carbon monoxide and hydrocarbons [1]. Diesel emissions thus contribute not only primary particles but also precursor pollutants for secondary particle formation and generation of oxidant pollution. While the respiratory effects of diesel exhaust were assessed in specific worker groups as early as the 1960s, the first major concern as to the consequences of diesel emissions for public health came during the early 1970s. At that time, diesel engine use rapidly increased because of gasoline shortages. Diesel exhaust was found to be mutagenic in the Ames assay and this finding, along with the rising exposure to the population, led to research on cancer risks, which is still ongoing. More recently, laboratory and epidemiological findings have led to the hypothesis that diesel particles may enhance immune responses and contribute to the burden of allergic diseases, including asthma, in the population [3]. With the rise of asthma in inner cities, some neighborhoods

1-56670-611-4 / 05 / $0.00+$1.50

are concerned that heavy bus and truck traffic may be contributing to the high rates of asthma in their communities.

This brief review provides an introduction to the physical and chemical characteristics of diesel exhaust, key toxicological evidence, principal epidemiological findings, and the overall status of the evidence on the health effects of diesel exhaust. These topics have been covered in numerous reviews, including reports of the Health Effects Institute [4–6], agency reports [7–9], and other reviews [1,3,10].

22.2 Diesel engines and diesel fuel

There are several types of diesel engines that are manufactured, and diesel fuel is quite variable in its characteristics. The diesel engine operates by compression ignition, rather than spark ignition as with gasoline-powered engines; in a diesel engine, air is compressed and then fuel injected as the piston reaches maximum compression. The fuel ignites from the rise in temperature caused by the compression. Light-duty diesel engines typically have indirect injection with fuel injected into a small prechamber, ignited, and then mixed and burned with air in the main cylinder. By contrast, heavy-duty engines are usually direct-injected. Also, there are both two-stroke and four-stroke engines, with two-stroke engines increasingly being less used.

A variety of diesel fuels are available, which differ in ignitability, aromatic hydrocarbon content, sulfur content, and distillation characteristics. Particle emissions are increased by higher levels of hydrocarbons and also of sulfur. Future fuels will have lower sulfur content, and a variety of newer types of diesel fuels, such as biodiesel derived from plant oils, are in development.

22.3 Characteristics of diesel emissions and particles

Without controls, diesel engines emit high concentrations of particles because combustion in the engines is fuel-rich and at high temperatures. The high temperatures also favor the formation of nitric oxide (NO), some being converted to NO_2 in the exhaust. Aldehydes are also formed, but emissions of carbon monoxide and hydrocarbons tend to be low compared with gasoline engines because of the excess air in the cylinders of diesel engines. Diesel emissions also contain numerous organic compounds, including some that are volatile (VOCs) and some that are semivolatile (SVOCs).

In considering the health risks of diesel engines, emphasis has been placed on diesel particles, in part because they can be more readily investigated than the full complex mixture of diesel exhaust. Much of the mass of particles emitted by diesel engines lies in the accumulation mode, with mass median and aerodynamic diameters around 0.3 μm. The particles in this size range tend to consist of smaller particles that have agglomerated. These particles may contain absorbed hydrocarbons and condensed materials, including SVOCs and sulfates from sulfur in the diesel fuel. Particles in the nanometer size range may also be present; although these particles may be present in large numbers, they account for only a small fraction of the total mass of diesel particles and they agglomerate into larger particles. Particle emissions are affected quantitatively and qualitatively by the use of emission control technologies, engine design, and fuel characteristics.

With regard to chemical composition, diesel particles consist primarily of elemental carbon (EC), organic carbon (OC), and sulfate [1]. The composition is quite variable and EC, a proposed marker for diesel particles, can account for 30 to 90% of mass. Much of the mass consists of unburned or partially burned materials. These materials can include a variety of polycyclic aromatic hydrocarbons (PAHs), sulfate, and metals. These PAHs do include known carcinogens.

22.4 Health risks of diesel exposure

22.4.1 Lung cancer

For several decades, lung cancer risk has been the principal focus of research on the health consequences of diesel exposure [1,2,10]. The initial finding of mutagenicity of diesel particles was followed by a series of inhalation studies in various animal species as well as epidemiological investigations to directly assess lung cancer risk and to quantify risk in relation to exposure. Inhalation studies in rodents showed species sensitivity to the development of lung cancer, with rats showing a dose-related increase in lung tumor frequency. The animals were exposed at high concentrations that overwhelm clearance mechanisms, a phenomenon referred to as "particle overload." In fact, overloading the lungs of rats with particles thought to be noncarcinogenic, that is, titanium dioxide, produced a similar frequency of lung tumors to that observed with diesel exhaust exposure. The relevance of the rat model to human lung cancer is uncertain, given that the underlying mechanism appears to be nonspecific, rather than a specific response to carcinogens in diesel particles.

The epidemiological literature on diesel exhaust and lung cancer is now extensive [1,2,6]. Two general approaches have been used: case–control studies of lung cancer in the general population with questionnaire-based estimates of diesel exposure, primarily from work exposures, and cohort studies of diesel-exposed worker groups. In the case–control studies, exposure has been inferred primarily based on job or industry or on qualitative descriptions, while some cohort studies have incorporated semiquantitative estimates of exposure. For both designs, a principal limitation is the inevitable measurement error in assessing prior exposures; generally, measurement error would tend to reduce the magnitude of associations, but in case–control studies there is a potential for positive bias as well. These studies have been reviewed extensively [2,4,7,9,11], particularly in regard to the utility of findings for quantitative risk assessment.

For the purpose of risk assessment to support policy development, emphasis has been placed on two studies that have exposure estimates for participants and the potential to characterize the exposure–response relationship. These studies are the study of United States railroad workers carried out by Garshick et al. [12,13] from Harvard and the study of United States Teamsters in the trucking industry carried out by Steenland et al. [14,15] at the National Institute for Occupational Safety and Health. While risk estimates have been developed from these studies, they are subject to substantial uncertainty because past exposures were not directly measured as a result. These exposures are inferred based on more recent measurements and knowledge of past working conditions. Nonetheless, risk assessments have been carried out using risk coefficients from these studies (see Lloyd and Cackette [1] for a summary). Several new studies now in progress (U.S. trucking industry and underground miners) will provide relevant findings over the next decade.

Although the literature has limitations, particularly in regard to support of quantitative risk estimation, diesel exhaust has been designated as a probable carcinogen by a number of reviewing authorities. These include the U.S. National Toxicology Program and Environmental Protection Agency and the California Environmental Protection Agency.

22.4.2 Asthma and other allergic diseases

The prevalence of childhood asthma, and possibly other allergic diseases, has increased dramatically over the last several decades [16]. Some recent epidemiological studies show prevalence rates as high as 20% for asthma in school children; the increase does not seem to reflect only changing patterns of diagnosis and disease labeling. While asthma clearly has a genetic basis, environmental factors play a key role in triggering exacerbations and possibly in causing the disease. Dating back to the 1980s, a series of experimental studies have shown that diesel particles enhance immune responses to allergens [3]. A variety of experimental systems have been used, with a relatively consistent evidence of enhancement of immune responses by diesel particles. This evidence comes not only from animal investigations, but also from human exposure studies in which nasal immune responses have been investigated in volunteers. The animal and human studies provide evidence that exposure to diesel particles triggers cytokine release and other responses [3].

Recent epidemiological studies indicate that children living closer to roadways appear to have higher rates of many respiratory diseases, including indicators of allergic diseases, in comparison with children living farther from roadways [17–22]. While these studies do not specifically provide effects of diesel exhaust exposure, they are suggestive that exposure to the complex pollutant mixtures emitted by vehicles on roadways may have a role in the causation or exacerbation of allergic respiratory diseases. The specific contribution of diesel emissions to these adverse effects has not been characterized, and more research, both epidemiological and basic, is needed to determine public health implications of the finding that diesel particles can enhance immune responses.

References

1. Lloyd, A.C. and Cackette, T.A., Diesel engines: environmental impact and control, *J. Air Waste Manag. Assoc.*, 51, 809, 2001.
2. Mauderly, J.L., Diesel exhaust, in *Environmental Toxicants: Human Exposures and Their Health Effects*, Lippmann, M., Ed., Van Nostrand Reinhold Company, Inc., New York, 2000, p. 193.
3. Sydbom, A., Blomberg, A., Parnia, S., Stenfors, N., Sandstrom, T., and Dahlen, S.E., Health effects of diesel exhaust emissions, *Eur. Respir. J.*, 17, 733, 2001.
4. Health Effects Institute, Diesel Exhaust: A Critical Analysis of Emissions, Exposure, and Health Effects, A Special Report of the Institute's Diesel Working Group, Health Effects Institute, Cambridge, MA, 1995.
5. Health Effects Institute, Request for Applications, Spring 1998 Research Agenda, RFA 98-3: Epidemiologic Investigations of Human Populations Exposed to Diesel Engine Emissions: Feasibility Studies, Health Effects Institute, Cambridge, MA, 1998.
6. Health Effects Institute and Diesel Epidemiology Working Group, Research Directions to Improve Estimates of Human Exposure and Risk from Diesel Exhaust, Health Effects Institute, Boston, 2002.
7. California Environmental Protection Agency (Cal EPA), Health Risk Assessment for Diesel Exhaust: Proposed Identification of Diesel Exhaust as a Toxic Air Contaminant, Office of Environmental Health Hazard Assessment, California Environmental Protection Agency, Sacramento, CA, 1998.
8. World Health Organization, Diesel Fuel and Exhaust Emissions, Environmental Health Criteria 171, International Programme on Chemical Safety, WHO, Geneva, 1996.
9. U.S. Environmental Protection Agency (EPA), Health Assessment Document for Diesel Exhaust, National Center for Environmental Assessment, Washington, DC, 2000.
10. Mauderly, J.L., Diesel emissions: is more health research still needed? *Toxicol. Sci.*, 62, 6, 2001.
11. Health Effects Institute, Diesel Emissions and Lung Cancer: Epidemiology and Quantitative Risk Assessment, Health Effects Institute, Cambridge, MA, 1999.
12. Garshick, E., Schenker, M.B., Muñoz, A., Segal, M., Smith, T.J., Woskie, S.R., Hammond, S.K., and Speizer, F.E., A case–control study of lung cancer and diesel exhaust exposure in railroad workers, *Am. Rev. Respir. Dis.*, 135, 1242, 1987.
13. Garshick, E., Schenker, M. B., Muñoz, A., Segal, M., Smith, T.J., Woskie, S.R., Hammond, S.K., and Speizer, F.E., A retrospective cohort study of lung cancer and diesel exhaust exposure in railroad workers, *Am. Rev. Respir. Dis.*, 137, 820, 1988.
14. Steenland, N.K., Silverman, D.T., and Hornung, R.W., Case–control study of lung cancer and truck driving in the Teamsters Union, *Am. J. Public Health.*, 80, 670, 1990.
15. Steenland, K., Deddens, J., and Stayner, L., Diesel exhaust and lung cancer in the trucking industry: exposure–response analyses and risk assessment, *Am. J. Ind. Med.*, 34, 220, 1998.

16. Tang, E.A., Wiesch, D.G., and Samet, J.M., Epidemiology of asthma and allergic diseases, in *Allergy: Principles and Practice*, Middleton, E., Reed, C.E., Ellis, E.F., Adkinson, N.F., Yunginger, J.W., and Busse, W.W., Eds., Mosby-Year Book, St. Louis, MO, 2003.
17. Wyler, C., Braun-Fahrlander, C., Kunzli, N., Schindler, C., Ackermann-Liebrich, U., Perruchoud, A.P., Leuenberger, P., and Wuthrich, B., Exposure to motor vehicle traffic and allergic sensitization, The Swiss Study on Air Pollution and Lung Diseases in Adults (SAPALDIA) Team, *Epidemiology*, 11, 450, 2000.
18. Yang, C.Y., Yu, S.T., and Chang, C.C., Respiratory symptoms in primary schoolchildren living near a freeway in Taiwan, *J. Toxicol. Environ. Health A*, 65, 747, 2002.
19. Zmirou, D., Gauvin, S., Pin, I., Momas, I., Just, J., Sahraoui, F., Le Moullec, Y., Bremont, F., Cassadou, S., Albertini, M., Lauvergne, N., Chiron, M., and Labbe, A., Five epidemiological studies on transport and asthma: objectives, design and descriptive results, *J. Expos. Anal. Environ. Epidemiol.*, 12, 186, 2002.
20. Gehring, U., Cyrys, J., Sedlmeir, G., Brunekreef, B., Bellander, T., Fischer, P., Bauer, C.P., Reinhardt, D., Wichmann, H.E., and Heinrich, J., Traffic-related air pollution and respiratory health during the first 2 yrs of life, *Eur. Respir. J.*, 19, 690, 2002.
21. Perera, F.P., Illman, S.M., Kinney, P.L., Whyatt, R.M., Kelvin, E.A., Shepard, P., Evans, D., Fullilove, M., Ford, J., Miller, R.L., Meyer, I.H., and Rauh, V.A., The challenge of preventing environmentally related disease in young children: community-based research in New York City, *Environ. Health Perspect.*, 110, 197, 2002.
22. Yu, O., Sheppard, L., Lumley, T., Koenig, J.Q., and Shapiro, G.G., Effects of ambient air pollution on symptoms of asthma in Seattle-area children enrolled in the CAMP study, *Environ. Health Perspect.*, 108, 1209, 2000.

chapter twenty-three

Health effects of ambient ultrafine particles

Beverly S. Cohen
New York University School of Medicine

Contents

23.1 Introduction 607
23.2 Formation 609
23.3 Composition 610
23.4 Lung deposition 612
23.5 Toxicology 612
23.6 Summary 616
References 616

23.1 Introduction

More than 90% of all airborne particles are generally found in nuclei less than 150 nm in diameter [1–3]. Number concentrations in an urban environment vary, depending on local sources of particles and gaseous precursors, season, and weather. In a boreal forest in Finland, the measured concentration of ultrafine particles was about 1000 cm^{-3} [4]. When there is substantial vehicular traffic, the number concentration is frequently of the order of 10^4 to 10^5 particles cm^{-3}. In a quiet rural environment, or indoors in an undisturbed clean room, the concentration more typically ranges from a few 100 to a few 1000 particles cm^{-3}. These particles, on average, represent only about one half of 1% of the total airborne particulate volume and mass. Thus, they have not generally been regarded as important contributors to the toxic effects of inhaled ambient air. However, recent evidence suggests that they may have an important role in health decrements associated with ambient particulate matter (PM). Other evidence suggests that particle number may also be a better metric than mass on which to base risk estimates for certain occupational diseases [2,5].

Various boundaries are in current use for both the upper and lower limits for the diameter of ultrafine particles. Ultrafine particles include those in the atmospheric nuclei mode, plus some of the smaller particles classified as "accumulation" mode particles. The lower limit is generally regarded as about 1 nm, but some reports refer to those up to 10 nm as "nanometer" particles, reserving ultrafine for particles with diameters larger than 10 nm. A convenient upper limit of 100 nm is frequently used. This is because a 100 nm diameter particle is an acceptable lower size limit to the accumulation mode, and an upper limit to the nucleation mode (modes are defined in Chapter 2). In this size range, particle aerodynamic behavior is dominated by Brownian diffusion, and particle size is adequately described by a thermodynamic diameter. The thermodynamic diameter is the

1-56670-611-4 / 05 / $0.00+$1.50

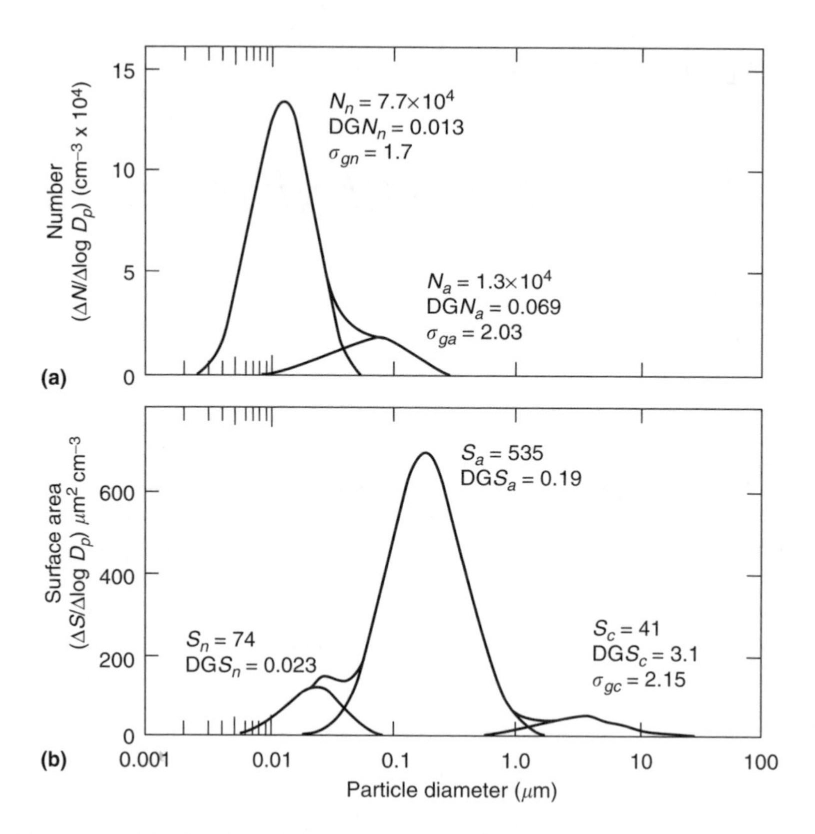

Figure 23.1 Frequency distribution of particles averaged over 1000 measured ambient particle size distribution in the United States: (**a**) by number, (**b**) by surface area. DGN and DGS: geometric mean diameter by number and surface, respectively; D_p: geometric diameter. Source EPA (1996).

diameter of a sphere of unit density that would have the same diffusion coefficient in air as the particle of interest. An upper limit value of 150 nm is chosen in this chapter, because it represents the particle size at which gravitational and inertial effects are of little importance when particles are inhaled. Thus, ultrafine particle behavior in human airways is dominated by diffusion.

The grand average concentration of over 1000 particle size distributions of ambient particles measured in the U.S. [6] is shown in Figure 23.1. The size distributions are shown as a number, or surface concentration. It is clear that ultrafine particles dominate the atmosphere when particles are counted and provide a dominant fraction of particle surface. The number in the nuclei mode is clearly an order of magnitude greater than the number in the larger size ranges. Taking an upper boundary diameter at 150 nm, almost all observed particles are in this ultrafine size range.

In the 1997 revision of the U.S. National Ambient Air Quality Standards, the U.S. EPA set 15 μg m^3 as an average annual mass concentration that should not be exceeded for particles with diameters less than 2.5 μm in aerodynamic diameter. The number of airborne unit density particles per cm^3 of a specific diameter that would result in this mass concentration is shown in Table 23.1. The number of 0.1 μm particles would be three orders of magnitude greater than the number of 1 μm diameter particles.

In recent years, very detailed data have become available on the ambient number concentration segregated by particle size. This resulted from the development of the scanning mobility particle size analyzer (SMPS). The data are acquired into narrow size classes by automatically selecting particles in a specific size range that penetrate a differential

Table 23.1 Number Concentration of Unit Density Monodisperse Particles at a Mass Concentration of 15 $\mu g\ m^{-3}$

Particle Diameter (μm)	Number Concentration (number cm^{-3})
0.01	28,600,000
0.05	229,000
0.1	28,600
0.15	8490
0.2	3580
0.5	229
1	28.6
1.5	8.5
2.5	1.8

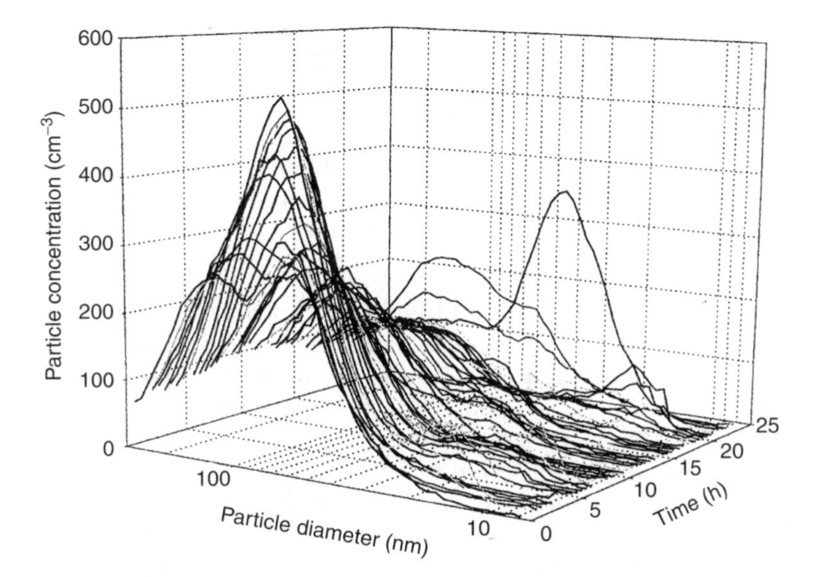

Figure 23.2 Particle size spectra collected over the course of a 24 h period at a rural location in Tuxedo, NY.

mobility analyzer (DMA) and counting them with a condensation nucleus counter (CNC). Automation of this combination of instruments has permitted semicontinuous acquisition of particle size spectra.

A few examples of such data are shown in Figures 23.1–23.3. These size distribution measurements were made with an SMPS system (TSI, St. Paul, MN) averaging three scans every 30 min for the duration of the measurements. The particle size range was from 7 to 300 nm in 32 size bins. Each clearly demonstrates both the temporal variability and the dominance of ultrafine particles when ambient particles are counted.

23.2 *Formation*

Ultrafine particles are formed by condensation reactions of precursor atmospheric gases and by nucleation of gas-phase species. The boundaries of the ultrafine region, as previously noted, are not exact, but the upper boundary of the nuclei mode is between 50 and 100 nm, whereas the lower boundary of the accumulation mode is roughly 100–200 nm. Nuclei mode particles are formed by nucleation, condensation, and coagulation. The

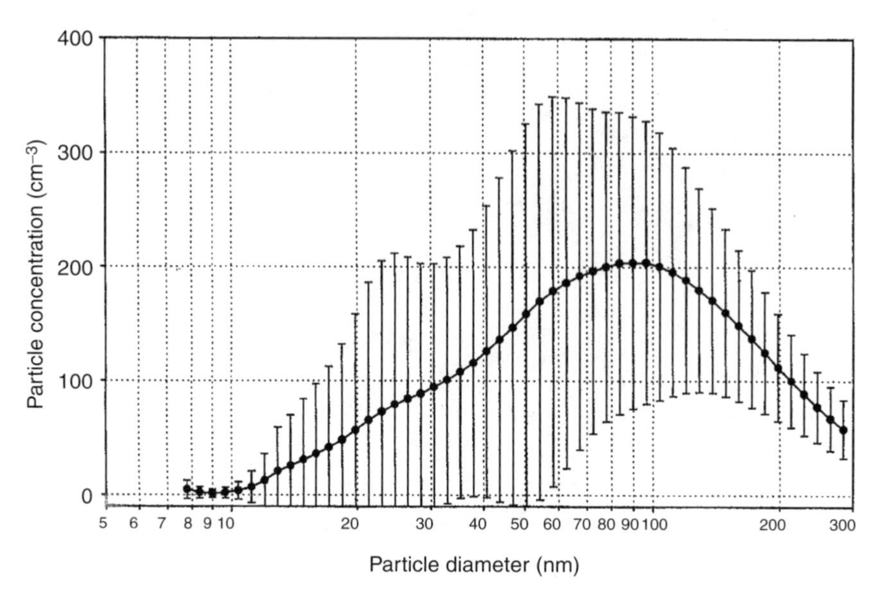

Figure 23.3 Average particle size distribution (mean and standard deviation) collected for 7 d (July 14–21, 1999) in Tuxedo, NY. SD represents the variability of the number detected by sampling every 30 min (NIEM outdoor, SMPS: sampling every 30 min, 3 scans/sample).

nuclei mode is sometimes further split into the nucleation (diameter <10 nm) and Aitkin modes to indicate that only the smallest particles are formed directly by nucleation. The particles are created when gas-phase species form condensed-phase species with very low equilibrium vapor pressure. Those ultrafine particles that are in the lower size range of the accumulation mode are primarily formed by condensation and coagulation, and also from evaporation of fog and cloud droplets in which dissolved gases have reacted [2,4].

Nucleation events in the atmosphere in which large concentrations of nanometer-sized particles form and grow over a period of hours occur about one quarter of the days per year. Concentrations of particles smaller than 50 nm can be increased by a factor of 10 or more when nucleation occurs. A nucleation event is shown in Figure 23.4, which displays the development of the particle number concentration over 24 h at a rural location in New York.

23.3 Composition

There are few data on the composition of ambient ultrafine particles. Because they are a very small fraction of the total ambient particle mass, specialized methods are required to isolate and analyze them. Reported studies indicate that they are composed, in large part, of organic and elemental carbon together with trace metals, ammonium, and sulfates [7]. Hughes et al. [8] reported the chemical composition of particles between 56 and 97 nm in diameter measured at Pasadena, CA. They reported that the largest fraction is organic compounds, followed by elemental carbon, then trace metals and sulfates. This is not unexpected since combustion sources are major contributors to ambient ultrafine particles. Detailed size distributions of ambient sulfates, ammonium, trace metals, and sulfuric acid in ambient particles smaller than 100 nm were measured by Hazi [9]. Although most of the acid and sulfate mass was measured in the 0.38 μm midpoint diameter fraction, the ultrafine fraction (<0.1 μm) was found to have lower pH. Iron, zinc, and sulfur were the dominant trace elements of the nine measured in the ultrafine fraction. Iron is the most abundant metal measured in ultrafine particles [7,9,10].

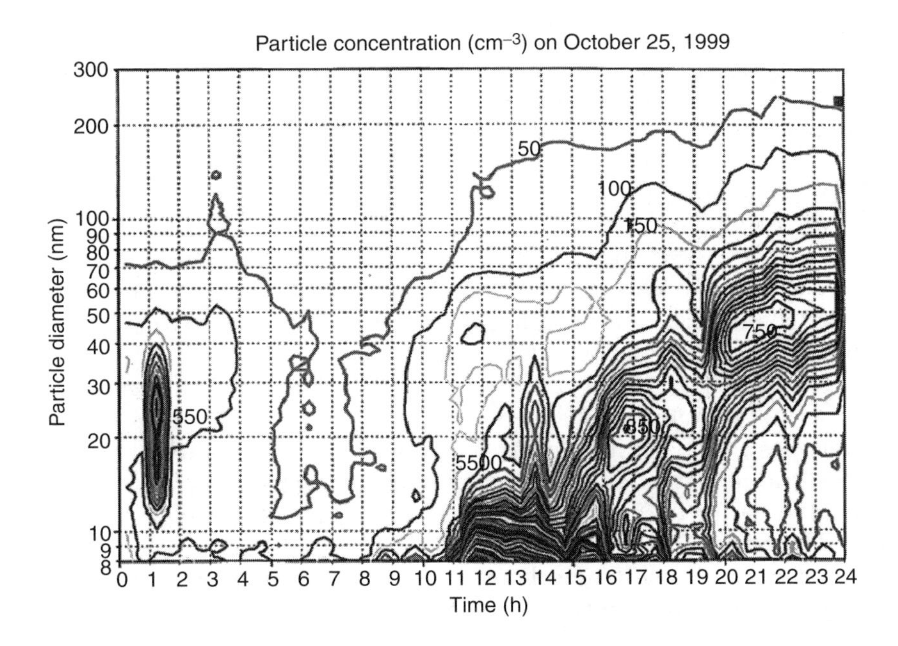

Figure 23.4 A nucleation event showing the development of a very high concentration of particles less than 10 nm. The contours are lines of equal particle concentration in particles cm^{-3}. The interval between contours is 50 particles cm^{-3}.

Measurements of the number of strong-acid ultrafine particles in New York City and Tuxedo, NY, have been reported by Cohen et al. [11]. The number was measured by the deposition of ambient particles less than 100 nm in diameter onto detectors that are coated with a 20 nm thick layer of iron applied by vapor deposition. The interaction of acid droplets and iron is detected by scanning the detector surface topography with an atomic force microscope. They report that the fraction of ultrafine particles that were acidic varied from 10 to 88% for the different seasons and sites. The average concentrations of acidic particles ranged from about 100 to 1500 particles cm^{-3} over the sampling periods.

Both vehicular and stationary combustion sources contribute ultrafine particles to the atmosphere. Gasoline and diesel vehicles are major sources. Emission testing indicates that organic carbon comprises roughly 70 to 90% of the carbon emitted from gasoline vehicles and 40–50% from diesel vehicles [12]. Sulfur has also been reported as a major component of emissions from combustion vehicles [13].

High concentrations of ultrafines are found on roadways and people are exposed to them in vehicles and alongside the roadways. However, because of particle growth, they have very short residence times, and number concentrations diminish very rapidly with distance from the road. In some cases, particle number concentrations at 50 m from a busy roadway were 5 to 8 times those measured at 100 and 500 m [14].

Recent technological advances provide excellent single-particle data via aerosol time-of-flight mass spectrometry (ATOFMS) instruments [15]. However, data reduction is arduous because a vast amount of data is collected during a sampling session. Methods for analysis of the data continue to be developed. These spectrometers are used for continuous single-particle measurements of size and composition simultaneously [16]. Particles are size selected as they enter the inlet of the spectrometer. They are then disintegrated by a laser beam, and the resulting ions are driven down separate positive and negative ion channels for identification of the charged fragment masses. In some systems, particle concentrators are being used to increase the concentration at the inlet. The concentrators have been developed for use in testing the effects of inhalation exposure to ambient particles. Measurements with these time-of-flight instruments have been undertaken in cities located in different geological areas of the U.S. to characterize fine and ultrafine urban PM.

23.4 Lung deposition

Ultrafine particles deposit in the respiratory tract primarily by diffusion, with deposition increasing as particle size decreases. The total lung deposition of ultrafine particles measured in human volunteers confirms diffusion as the mechanism [17,18]. When the smallest ultrafine particles are inhaled, they deposit very efficiently in the nasal and oral passages. The high deposition in the extrathoracic region was determined in human nasal/oral casts and in human volunteers [19,20].

Those particles that penetrate to the thorax are also very efficiently deposited. The deposition of ultrafine particles in the human tracheobronchial airways was measured in a hollow airway cast for particle sizes between 40 and 200 nm by Cohen et al. [21]. The data indicated higher deposition for the smaller diameter particles, again confirming diffusion as the primary deposition mechanism. Additionally, this work demonstrated that deposition in the tracheobronchial airways was higher than predicted by diffusional deposition in a tube, assuming a fully developed or parabolic flow profile. Subsequent studies showed that the deposition of iodine vapors with a diffusion coefficient of 0.08 $cm^2\ s^{-1}$, used as a surrogate for an approximately 1.8 nm particle, agreed with the theoretical prediction of diffusional deposition [22]. The same research group also showed that the charge on ultrafine particles enhances deposition in the airway replica [23,24], that is, charged particles and particles in charge equilibrium have higher deposition efficiency in the tracheal region as compared to neutral particles of the same size. The ratio of deposition efficiency for charge-equilibrium and neutral particles was 1.6 and 2.7 for 20 and 125 nm particles, respectively [25]. This is important because most ultrafine ambient particles carry one, or a few charges.

The deposition in airway casts for 1.75-, 10-, and 40-nm particles was measured at flow rates corresponding to respiratory minute volumes at rest and during moderate exercise [26]. Replicate casts of the upper tracheobronchial airways of 3-, 16-, and 23-yr-old humans were used, including the larynx, trachea, and bronchial airways down to generations 5–8. The deposition of the 1.75 nm particle was substantially higher than that of the 10 and 40 nm particles. The dependence of particle deposition on the flow rate was relatively weak, and deposition efficiencies were only slightly higher at the lower flow rates. The deposition models for diffusion from parabolic flow underestimated aerosol deposition, whereas the diffusion deposition predicted for plug flow overestimated the tracheobronchial deposition. This is in agreement with the earlier studies [21].

23.5 Toxicology

Submicrometer particles with demonstrated health effects include diesel exhaust, radon progeny, cigarette smoke, metal fumes, acidic aerosols, and trace metals. When generated, these primary aerosol particles are ultrafine. Additionally, biofragments such as endotoxin extend into this size range. Table 23.2 shows the toxic effects demonstrated when ultrafine particles are inhaled.

A substantial body of experimental data on the toxicity of ambient particles has developed in recent years as an indirect result of The Clean Air Act enacted by the United States in 1970. This Act established the Environmental Protection Agency and mandated the setting of Primary Ambient Air Quality Standards that would protect the public against

Table 23.2 Toxic Effects of Ultrafine Particles

Inhalation Exposure	Result
High concentration metal or polymer fume (occupational exposures)	Fever, diffusion impairment, respiratory symptoms
Aggregated ultrafines (TiO_2, carbon black, diesel soot)	Epithelial cell proliferation, occlusion of inter-alveolar pores, impairment of alveolar macrophages, chronic pulmonary inflammation, pulmonary fibrosis, induction of lung tumors
11 nm CuO at $10^9/cm^3$ 60 min (hamsters) [43]	Fourfold increase in pulmonary resistance. Particles dispersed throughout lung (interstitium, alveolar capillaries, pulmonary lymphatics)
Teflon (PTFE) fume (26 nm) $10^6/cm^3$, 10–30 min [31]	Highly inflammatory/mortality
TiO_2 1000 $\mu g/m^3$, 7 h [44]	Oxidative stress in lung
PTFE fumes, whole body inhalation, 1, 2.5, or 5×10^5 particles/cm^3, 18 nm, rat, 15 min, analysis 4 h postexposure [45]	Increased PMN, mRNA of MnSOD, and MT, IL-1α, IL-1β, IL-6, MIP-2, TNF-α mRNA of MT and IL-6 expressed around all airways and interstitial regions; PMN expressed IL-6, MT, TNF-α; AM and epithelial cells were actively involved
PTFE fumes, whole body inhalation, 1, 2.5, or 5×10^5 particles/cm^3, 18 nm, mice, C57BL/6J, 8 weeks and 8 months old, mice 30 min exposure analysis 6 h following exposure [46]	Increased PMN, lymphocytes, and protein levels in old mice over young mice; increased TNF-α mRNA in old mice over young mice; no difference in LDH and β-glucuronidase
CdO fumes, 8 nm, rats and mice, 1×3 h [47]	Mice created more metallothionein than rats, which may be protective of tumor formation

Adapted from: U.S. EPA (1996).

adverse health effects of ubiquitous pollutants (such as ambient PM) with an adequate margin of safety. The Clean Air Act Amendments of 1977 then required that the air quality standards be reviewed at 5-yr intervals and revised as necessary.

Evidence has accumulated that implicates ultrafine particles as a cause of the adverse effects of exposure to ambient PM, but the mechanisms are as yet unclear. Mechanisms that have been proposed for the induction of lung injury include irritant signaling [27], acid effects [28], and inflammation [29]. It is clear that ultrafine particles generally exhibit greater toxic potency than larger particles of the same material [30]. Table 23.3 provides a summary of experiments that compared responses to ultrafine particles with response to larger diameter particles of the same material.

Other suggestions for the increased potency are that (1) they are biologically more reactive, (2) there is a much higher number and surface area for the same total particle mass, and (3) they deposit with very high efficiency in lungs. Additionally, it has been demonstrated that ultrafine particles are more rapidly transferred to the interstitium than are fine particles of the same composition, and exhibit a greater accumulation in the regional lymph nodes and a greater retention in the lung [31,32]. Thus, ultrafine particles may penetrate the epithelium better and they may be less effectively cleared by macrophages, and the nonphagocytosed particles may penetrate the interstitium in a few

Table 23.3 Studies Comparing Ultrafine and Fine Particles Composed of the Same Material

Particle Type	Size	Exposure	Endpoint	Reference
TiO_2 anatase	Ultrafine 20 nm Fine 250 nm	Inhalation 23 mg/m^3 7 h/day 5 days/12 weeks	Inflammation in BAL Lung lymph node burdens Slowed clearance	[48]
TiO_2 rutile	Utrafine 12 nm Fine 230 nm	Instillation of 500 μg	Interstitialization measured as unlavageable fraction	[48]
Al_2O_3	Ultrafine 20 nm Fine 500 nm	Instillation 500 μg	Lung lymph node burden	[49]
Carbon black	Ultrafine 20 nm Fine 200–250 nm PM_{10}	Instillation of 50–125 μg in 0.2 ml	Increased PMN, protein, and LDH following PM_{10}, greater response with ultrafine CB but not CB; decreased GSH level in BAL; free radical activity (deplete supercoil DNA); Leukocytes from treated animals produced greater NO and TNF	[50,51,52]
TiO_2	Ultrafine 20 nm Fine 300 nm	Instillation of 10,000 μg particles/kg BW	Increased inflammatory indicators in BAL Pathology changes in alveolar ducts Increased pulmonary retention of ultrafine	[53]
TiO_2	Ultrafine 20 nm Fine 250 nm	Instillation 500 μg	Increased acute inflammation indicators in BAL at 24 h with ultrafines	[31]
MnO_2	Surface area of 0.16, 0.5; 17, 62 mg^2/g	Intratracheal instillation; *In vitro* 0.037, 0.12, 0.75, 2.5 mg/animal	LDH, protein and cellular recruitment increased with increasing surface area; Freshly ground particles had enhanced cytotoxicity	[54]
TiO_2	Ultrafine 21 nm Fine 250 nm	Intratracheal inhalation and intratracheal instillation	Inflammation produced by intratracheal inhalation (both severity and persistence) was less than that produced by instillation; ultrafine particles produced greater inflammatory response than fine particles for both dosing methods	[31]

(Continued)

Table 23.3 *(Continued)*

Particle type	Size	Exposure	Endpoint	Reference
TiO_2	Ultrafine 21 nm Fine 250 nm	Intratracheal inhalation and Intratracheal instillation (rat) Inhalation at 125 $\mu g/m^3$ Instillation at 500 μg for fine 750 μg for ultrafine Inhalation exposure 2 h; sacrificed at 0, 1, 3, and 7 days postexposure for both techniques	MIP-2 increased in lavage cells but not in supernatant in those groups with increased PMN (more in instillation than in inhalation; more in ultrafine than in fine); TNF-α levels had no correlation with either particle size or dosing methods	[56]
H_2SO_4 and O_3	Ultrafine 60 nm Fine 300 nm	Inhalation, nose-only 500 $\mu g/m^3$ H_2SO_4 aerosol (two diffferent particle sizes) with or without 0.6 ppm O_3 Rats, Sprague–Dawley, male, 250–300 g 4 h/day for 2 days	The volume percentage of injured alveolar septae was increased only in the combined ultrafine acid/O_3 animals BrdU labelling in the periacinar region was increased in a synergistic manner in the combined fine acid/O_3 animals	[57]

Expanded from: Donaldson, Li, and MacNee, *J. Aerosol Sci.*, 29, 553-560, 1998.

hours. Those with low solubility appear to be significantly more inflammatory in the lung than are larger sized particles of the same composition [33].

It is unlikely that all components of PM are equally toxic. Cohen et al. [11] have recently noted that candidates for especially active components of ambient PM are H^+, ultrafine number, and soluble transition metals. While all three have some supporting toxicological evidence consistent with known mechanisms of toxicity, only H^+ and ultrafine particles have produced effects at exposure levels that could occur in ambient air [2]. Perhaps the most likely candidate is a hybrid of H^+ and ultrafine particles, that is, acid-coated ultrafine particles.

Chen et al. [34] exposed guinea pigs to varying amounts of sulfuric acid layered onto 10^8 ultrafine (90 nm) carbon core particles cm^{-3}, and to a constant (300 $\mu g\ m^{-3}$) concentration of acid layered onto 10^6, 10^7, or 10^8 particles cm^{-3}. Indicators of irritant potency on macrophages harvested from the lungs of exposed animals clearly showed an increased response to a constant dose of acid when it was divided into an increased number of particles, as well as a response to an increased dose of acid at a constant number concentration. Recently, Oberdörster et al. [35] reported that nonreactive ultrafine particles do not appear to cause inflammation in young healthy rats.

Evidence, some of which was noted above, supports the hypothesis that the number of ultrafine particles that deposit per unit surface of the epithelial lining of the human respiratory system is an important determining factor affecting lung injury [27, 29, 34, 36–41], and that the resulting alveolar inflammation is able to provoke attacks of acute respiratory illness in susceptible individuals [29]. In particular, the work of Peters et al. [39] and Wichmann et al. [42] suggest that at ambient levels, the number concentration of inhaled particles may be a significantly more important determinant of the risk than inhaled mass measures. Very recent evidence has also shown that ultrafine particles are associated with human mortality [42].

23.6 Summary

Ultrafine particles constitute the largest share of particles in the atmosphere by both number and surface area. These particles, which are formed by condensation reactions of precursor atmospheric gases and by nucleation of gas-phase species, range in size from approximately 1 to 150 nm. Studies indicate that they are composed, in large part, of organic and elemental carbon, together with trace metals, ammonium, and sulfates, but there are few data on the temporal or spatial variations in the composition.

High concentrations of ultrafines are emitted from vehicles and other combustion sources and people are exposed to them in vehicles and alongside roadways. These small particles deposit very efficiently in the respiratory system, and when inhaled, the evidence derived from both experimental and occupational exposures indicates that they are more toxic than larger particles of the same composition. Some epidemiological studies associate ambient ultrafine particles with morbidity and mortality [12]. No mechanism has as yet been demonstrated by which these very small particles exert their toxicity on people or in animals at exposure levels that could occur in ambient air.

References

1. National Research Council (NRC), *Airborne Particles*, National Academy of Sciences, 1977.
2. U.S. EPA, Air quality for particulate matter, Vols. I, II, III, EPA/600/P-95/001aF, EPA/600/P-95/001bF, EPA/600/P-95/001cF, 1996.
3. Whitby, K.T., Husar, R.B., and Liu, B.Y.H., The aerosol size distribution of Los Angeles smog, *J. Colloid Interface Sci.*, 39, 177, 1972.
4. Makela, J.M. et al., Observations of ultrafine aerosol particle formation and growth in boreal forest, *Geophys. Res. Lett.*, 24, 1219, 1997.
5. McCawley, M.A., Kent, M.S., and Berakis, M.T., Ultrafine beryllium number concentration as a possible metric for chronic beryllium disease risk, *Appl. Occup. Environ. Hyg.*, 16, 631, 2001.
6. Whitby, K.T., The physical characteristics of sulfur aerosols, *Atmos. Environ.*, 12, 135, 1978.
7. Cass, G.R. et al., The chemical composition of atmospheric ultrafine particles, *Philos. Trans. R. Soc. London Ser. A*, 358, 2581, 2000.
8. Hughes, L.S. et al., Physical and chemical characterization of atmospheric ultrafine particles in the Los Angeles area, *Environ. Sci. Technol.*, 32, 1153, 1998.
9. Hazi, Y., Measurements of Acidic Sulfates and Trace Metals in Fine and Ultrafine Ambient Particulate Matter: Size Distribution, Number Concentration and Source Region, Ph.D. thesis, New York University School of Medicine, NY, 2001.
10. Gone, J.K., Olmez, I., and Ames, M.R., Size distribution and probable sources of trace elements in submicron atmospheric particulate material, *J. Radioanal. Nucl. Chem.*, 244, 133, 2000.
11. Cohen, B.S. et al., HEI Report, Field Validation of Nanofilm Acid Detectors for Assessment of H^+ in Indoor and Outdoor Air and Measured Ambient Concentrations of Ultrafine Acid Particles, Health Effects Institute, Boston, MA, 2004.
12. U.S. EPA, Particulate Matter Air Quality Criteria Document, Vols. 1 and 2, June 2003.
13. Gertler, A.W. et al., Preliminary Results of a Tunnel Study to Characterize Mobile Source Particulate Emissions, Presented at PM 2000; Particulate Matter and Health — The Scientific Basis for Regulatory Decision-Making, Specialty Conference and Exhibition, January, Charleston, SC, Air and Waste Management Association, Pittsburgh, PA, 2000.
14. Kleinman, M.T. et al., Exposure to Concentrated Fine and Ultrafine Ambient Particles Near Heavily Trafficked Roads Induces Allergic Reactions in Mice, AAAR PM Meeting, March 31–April 4, 2003, Pittsburgh, PA, 2003.
15. Prather, K.A., Chemical characteristics of ambient PM2.5, in *AAAR '97* (Abstracts of the Sixteenth Annual Conference, October 13–17, 1997, Denver, CO), American Association for Aerosol Research, Cincinnati, OH, Abstract No. 9SE3, 1997, p. 302.
16. Zhao, Y. et al., Using Ultrafine Concentrators to Increase the Hit Rate of Single Particle Mass Spectrometers, AAAR PM Meeting, March 31–April 4, 2003, Pittsburgh, PA, 2003.

17. Schiller, C. et al., Deposition of monodisperse insoluble aerosol particles in the 0.005 to 0.2 μm size range within the human respiratory tract, *Ann. Occup. Hyg.*, 32, 41, 1988.
18. Jaques, P.A. and Kim, C.S., Measurement of total lung deposition of inhaled ultrafine particles in healthy men and women, *Inhal. Toxicol.*, 12, 715, 2000.
19. Cheng, Y.S. et al., Deposition of thoron progeny in human head airways, *Aerosol Sci. Technol.*, 18, 359, 1993.
20. Cheng, Y.S. et al., Nasal deposition of ultrafine particles in human volunteers and its relationship to airway geometry, *Aerosol Sci. Technol.*, 25, 274, 1996.
21. Cohen, B.S., Sussman, R.G., and Lippmann, M., Ultrafine particle deposition in a human tracheobronchial cast, *Aerosol Sci. Technol.*, 12, 1082, 1990.
22. Li, W., Xiong, J.Q., and Cohen, B.S., The deposition of unattached radon progeny in a tracheobronchial cast as measured with iodine vapor, *Aerosol Sci. Technol.*, 28, 502, 1998.
23. Cohen, B.S. et al., Deposition of inhaled charged ultrafine particles in a simple tracheal model, *J. Aerosol Sci.*, 26, 1149, 1995.
24. Cohen, B.S., Xiong, J.Q., and Li, W., The influence of charge on the deposition behavior of aerosol particles with emphasis on singly charged nanometer sized particles, in *Aerosol Inhalation, Lung Transport and Deposition and the Relation to the Environment*, Marijnissen, J., and Gradon, L., Eds., Recent Research and Frontiers, Kluwer Scientific Press, Dordrecht, 1996, p. 153.
25. Cohen, B.S. et al., Deposition of charged particles on lung airways, *Health Phys.*, 74, 554, 1998.
26. Smith, S., Cheng, Y.-S., and Yeh, H.C., Deposition of ultrafine particles in human tracheobronchial airways of adults and children, *Aerosol Sci. Technol.*, 35, 697, 2001.
27. Hattis, D. et al., Acid particles and the tracheobronchial region of the respiratory system — an "irritation-signaling" model for possible health effects, *JACPA*, 37, 1060, 1987.
28. Lippmann, M., Background on health effects of acid aerosols, *Environ. Health Perspect.*, 79, 3, 1989.
29. Seaton, A.W. et al., Particulate air pollution acute health effects, *Lancet*, 345, 176, 1995.
30. Donaldson, K., Li, X.Y., and MacNee, W., Ultrafine (nanometer) particle mediated lung injury, *J. Aerosol Sci.*, 29, 553, 1998.
31. Oberdörster, G. et al., Role of the alveolar macrophage in lung injury: studies with ultrafine particles, *Environ. Health Perspect.*, 97, 193, 1992.
32. Oberdörster, G. et al., Association of particulate air pollution and acute mortality: involvement of ultrafine particles, *Inhal. Toxicol.*, 7, 111, 1995.
33. Driscoll, K.E., Macrophage inflammatory proteins: biology and role in pulmonary inflammation, *Exp. Lung Res.*, 20, 473, 1994.
34. Chen, L.C. et al., Number concentration and mass concentration as determinants of biological response to inhaled irritant particles, *Inhal. Toxicol.*, 7, 577, 1995.
35. Oberdörster, G. et al., Acute Pulmonary Effects of Ultrafine Particles in Rats and Mice, HEI Report 96, Health Effects Institute, Boston, MA, 2000.
36. Amdur, M.O. and Chen, L.C., Furnace-generated acid aerosols: speciation and pulmonary effects, in Symposium on the Health Effects of Acid Aerosols, Research Triangle Park, NC, October, 1987; *Environ. Health Perspect.*, 79, 147, 1989.
37. Oberdörster, G. et al., Particulate air pollution and acute health effects, *Lancet*, 345, 176, 1995.
38. Hattis, D.S., Abdollahzadeh, S., and Franklin, C.A., Strategies for testing the "irritation-signaling" model for chronic lung effects of fine acid particles, *J. Air Waste Manage. Assoc.*, 40, 322, 1990.
39. Peters, A. et al., Respiratory effects are associated with the number of ultrafine particles, *Am. J. Respir. Crit. Care Med.*, 155, 1376, 1997.
40. Wichmann, H.-E. and Peters, A., Epidemiological evidence of the effects of ultrafine particle exposure, *Philos. Trans. R. Soc. Ser. A*, 358, 2751, 2000.
41. Penttinen, P. et al., Number concentration and size of particles in urban air: effects on spirometric lung function in adult asthmatic subjects, *Environ. Health Perspect.*, 109, 319, 2001.
42. Wichmann, H.-E. et al., Daily Mortality and Fine and Ultrafine Particles in Erfurt, Germany, Part I. Role of Particle Number and Particle Mass, HEI Report 98, Health Effects Institute, Boston, MA, 2000.

43. Stearns, R.C. et al., Detection of Ultrafine Copper Oxide Particles in the Lungs of Hamsters by Electron Spectroscopic Imaging, in Proceeding of International Conference of Electron Microscopy, ICEM 13, July, Paris, France, 1994, p. 763.
44. MacNee, W. et al., Pro-inflammatory effect of particulate air pollution (PM10) *in vivo* and *in vitro*, *Ann. Occup. Hyg.*, 41(Suppl. 1), 7, 1997.
45. Johnston, C.J. et al., Characterization of the early pulmonary inflammatory response associated with PTFE fume exposure, *Toxicol. Appl. Pharmacol.*, 140, 154, 1996.
46. Johnston, C.J. et al., Pulmonary inflammatory responses and cytokine and antioxidant mRNA levels in the lungs of young and old C57BL/6 mice after exposure to Teflon fumes, *Inhal. Toxicol.*, 10, 931, 1998.
47. McKenna, I.M. et al., Expression of metallothionein protein in the lungs of Wistar rats and C57 and DBA mice exposed to cadmium oxide fumes, *Toxicol. Appl. Pharmacol.*, 153, 169, 1998.
48. Ferin, J., Oberdorster, G., and Penney, D.P., Pulmonary retention of ultra-fine and fine particles in rats, *Am. J. Respir. Cell Mol. Biol.*, 6, 535, 1992.
49. Ferin, J. et al., Increased pulmonary toxicity of ultra-fine particles? I. Particle clearance, translocation, morphology, *J. Aerosol Sci.*, 21, 381, 1990.
50. Li, X.Y. et al., Free radical activity and pro-inflammatory effects of particulate air pollution (PM_{10}) *in vivo* and *in vitro*, *Thorax*, 51, 1216, 1996.
51. Li, X.Y. et al., *In vivo* and *in vitro* proinflammatory effects of particulate air pollution (PM_{10}), in Proceedings of the Sixth International Meeting on the Toxicology of Natural and Man-Made Fibrous and Non-Fibrous Particles, Driscoll, K.E. and Oberdorster, G., Eds., *Environ. Health Perspect.* Suppl., 105 (5), Lake Placid, NY, September, 1996, p. 1279.
52. Li, X.Y. et al., Short term inflammatory responses following intratracheal instillation of fine and ultrafine carbon black in rats, *Inhal. Toxicol.*, 11, 709, 1999.
53. Driscoll, K. and Maurer, Cytokine and growth factor release by alveolar macrophages: potential biomarkers of pulmonary toxicity, *Toxicol. Pathol.*, 19 (4 Part 1), 398, 1991.
54. Lison, D. et al., Influence of particle surface area on the toxicity of insoluble manganese dioxide dusts, *Arch. Toxicol.*, 71, 725, 1997.
55. Osier, M. and Oberdorster, G., Intratracheal inhalation vs. intratracheal instillation: differences in particle effects, *Fundam. Appl. Toxicol.*, 40, 220, 1997.
56. Osier, M., Baggs, R.B., and Oberdorster, G., Intratracheal instillation versus intratracheal inhalation: influence of cytokines on inflammatory response, in Proceeding of the Sixth International Meeting on the Toxicology of Natural and Man-Made Fibrous and Non-Fibrous Particles, Driscoll, K.E., and Oberdorster, G., Eds., *Environ. Health Perspect.* Suppl., 105 (5), Lake Placid, NY, September, 1996, p. 1265.

chapter twenty-four

Health effects of aerosols: Mechanisms and epidemiology

Ira B. Tager
University of California at Berkeley

Contents

24.1 Introduction 619
24.2 Characteristics of outdoor (ambient) and indoor aerosol sources of importance to human health 622
24.2.1 Outdoor aerosol (Table 24.1, Figure 24.1) 622
24.2.2 Indoor aerosol (Table 24.1 and Figure 24.3) 624
24.3 Deposition and clearance of aerosols from the human respiratory tract 626
24.3.1 Deposition and retention 626
24.3.2 Clearance 629
24.4 Mechanisms of toxicity of aerosol components 633
24.4.1 Particle aerosol induction of oxidative damage 634
24.4.2 Particle aerosol induction of inflammation 635
24.5 Human health effects associated with ambient PM 640
24.5.1 Health effects associated with chronic exposure to PM 649
24.5.1.1 Effects of exposure to ambient PM and birth outcomes (Table 24.9) ..649
24.5.1.2 Increased mortality related to long-term exposure to PM aerosol ..651
24.5.1.3 Atopic allergy and asthma 656
24.5.1.4 Cancer 659
24.5.2 Health effects associated with acute and subacute exposures to PM 660
24.5.2.1 Mortality 660
24.5.2.2 Physiological mechanisms related to the association of daily changes in PM and daily mortality 676
24.6 Conclusions 679
References 679

24.1 Introduction

Health effects have been associated with a variety of aerosol components commonly encountered in the environments of people. The body of published data on the characterization (sources, composition, environmental transformations, and fate) of health-relevant aerosols is extensive as is the literature related to aerosol-specific health effects. Therefore, this chapter will discuss only selected, broad categories of health-relevant aerosols:

1-56670-611-4 / 05 / $0.00+$1.50

Table 24.1 Aerosols Discussed in this Chapter

	Formation (F)/Primary Sources(S)	Major Components[a] [265]
Outdoor Aerosol (Figure 24.1)		
Combustion-Related PM		
Ultrafine (nuclei) mode (<100 nm)[b] (includes nanoparticles (<50 nm))	F. Gas-to-particle conversion or primary during incomplete fuel combustion S. Motor vehicles, power plants, industrial sources	Organic and elemental carbon metals (especially transition metals)/elements, sulfates, nitrates, sulfuric acid
Accumulation (fine) mode (≤1 μm) (includes ultrafine PM)	F. Primary particles, gas-to-particle conversion during fuel combustion: condensation and coagulation of ultrafines; secondary transformation in atmosphere S. Motor vehicles, power plants, industrial sources	Sulfate, nitrate, ammonium Elemental carbon Organic compounds Metals (especially transition metals)
Coarse PM (>1 μm)[c]	F. Grinding, crushing and abrasion of surfaces with suspension by wind or anthropogenic activities S. Suspended dust from: industry, soil tracked on to roads and streets, mining, construction, farming Ocean spray Biological materials	Crustal metals Tire fragments Sulfates (ocean) Pollens[d] Fungal spores Pesticides Bacteria and bacterial products (e.g., endotoxin, bacterial spores)
Indoor Aerosol[e] **(Figure 24.2)**	F. See Outdoor Aerosol; heterogeneous chemical reactions [14]	See Outdoor Aerosol Tobacco products (sample of tar): nicotine

Combustion-Related PM	S. Tobacco smoke; fireplaces/wood stoves kerosene heaters, tobacco smoke, outdoor aerosol, cooking	fluoroanthrenes, acrolein, benzene, pyrenes, *N*-nitrosoamines, 1,3 butadiene, arsenic, chromium IV, lead, cadmium
Noncombustion PM		
Dust/lint	F. Disturbance of settled dust during cleaning and other human/pet activities S. Dirt, human	Soil and road dust tracked indoors Skin scales
Biological	F. Detritus composed of insect excrement and body parts, animal shedding, water damage; penetration of outdoor aerosol into buildings S. (1) Mites, (2) cockroaches, (3) pets, (4) water-damaged building materials/high humidity, (5) bacteria/bacterial products, and (6) outdoor aerosol	1. Fecal allergen 2. Multiple allergen-specific sources not known 3. Dander (cat, dog, etc.) 4. *Bacteria*: whole bacteria spores, peptidoglycans, endotoxin, *Molds and fungal* spores, toxins, $1 \rightarrow 3$-β-D glucans 5. Endotoxins, peptidoglycans 6. Bacteria, bacterial products, fungal spores, toxins, pollens

[a] Components of specific aerosols are spatially heterogeneous. Entries represent components found typically.

[b] Definitions taken from Reference [266]. These terms are used because most important combustion-related components of $PM_{2.5}$ are located in the fine fraction.

[c] For regulatory purposes, coarse-mode particles are usually referred to particles between 2.5 and 10 μm [2]. However, since most particles larger than 1 μm are formed by mechanical processes (see Figure 24.1), this definition seems more useful in terms of possible health effects. Some combustion-related PM may be found at the lower end of the distribution of the coarse mode (see Figure 24.1).

[d] See Chapter 20 for a more detailed presentation of the composition of bioaerosols.

[e] References [267–269].

aerosols derived from combustion processes, and aerosols composed of biological materials that are encountered in ambient and/or indoor environments (Table 24.1).[1] Specifically omitted are radioactive aerosols, aerosols encountered in industrial environments, and medical and pharmaceutical aerosols.

Broadly, the health-relevant aerosols to be considered can be classified into those derived from the activities of man (e.g., combustion of fuels) and those that occur in natural environments (e.g., bioaerosols) or as the result of natural processes (e.g., windblown dusts that carry man-made products such as pesticides or bioaerosols such as soil-resident fungal spores).

24.2 Characteristics of outdoor (ambient) and indoor aerosol sources of importance to human health

Clearly, the activities of people can modify the distribution and the concentrations of aerosol components that would exist in nature even in the absence of human activity. Humans come into contact with these aerosols both indoors and outdoors. Components of outdoor aerosols infiltrate into indoor environments to varying degrees, but, with the exception of products of indoor combustion (e.g., fireplaces, biomass burning stoves), indoor sources are not usually important contributors to the outdoor aerosol.

24.2.1 Outdoor aerosol (Table 24.1, Figure 24.1)

The typical outdoor aerosol can be described conveniently in terms of three components. Combustion-generated particulate matter (PM) is the source for the vast majority of PM ≤ 1 μm (aerodynamic diameter), and these particles are the target of Federal and State clean air criteria. With regard to human health, the particle size distribution observed for PM ≤ 1 μm has implications for the deposition distribution within the respiratory tract and toxicological properties that are related to surface area characteristics independent of the toxicology of the specific composition of the particle (e.g., see Reference [1]). The relationship between particle size, deposition, and surface area is shown in Figure 24.2, and the implications for toxicology will be discussed in a later section. The chemical composition of the combustion-related aerosol is complex, and its exact composition depends upon the mix of sources that contribute to it (Table 24.1; primary source contributions and long-range transport), meteorological conditions, and large-scale regional variations.

The so-called coarse component of the outdoor aerosol is defined in various ways. In the context of U.S. National Ambient Air Quality Standards (NAAQS), the classification of "coarse" is reserved for PM between 2.5 and 10 μm aerodynamic diameter (see Vol. I of Reference [2]). However, given the substantial differences in sources and deposition properties and contribution to overall surface area (Figure 24.2), the classification based on a cut point of 1 μm seems more useful for health assessment. As noted in Figure 24.2, there is a small overlap of sources and particle modes, which is related to cut-point characteristics of the sampling devices and conditions during sampling particle agglomeration and to meteorological factors such as high winds (see Vol. I of Reference [2]). Among the components of coarse PM, the bioaerosol has considerable implications for human health. Plant pollens have long been known to be important triggers of allergic reactions. However, more recently, the importance of fungal spores and toxins (e.g., see References

[1]Chapter 20 presents detailed information about the relevant aspects of the bioaerosol and some data on health effects. This chapter will focus on disease-specific health effects based on epidemiological and clinical data.

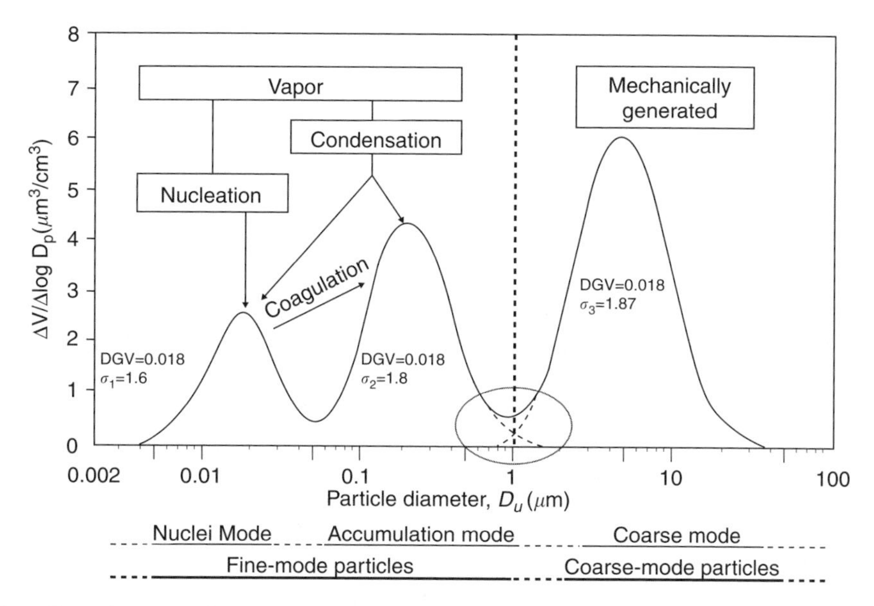

Figure 24.1 Distribution of particle mean diameter by particle volume for typical outdoor PM aerosol. DGV: geometric mean diameter by volume (volume mean diameter); σ_g: geometric standard deviation. The circle points to overlap area between particle diameter and source contribution. *Source*: U.S. Environmental Protection Agency, [2].

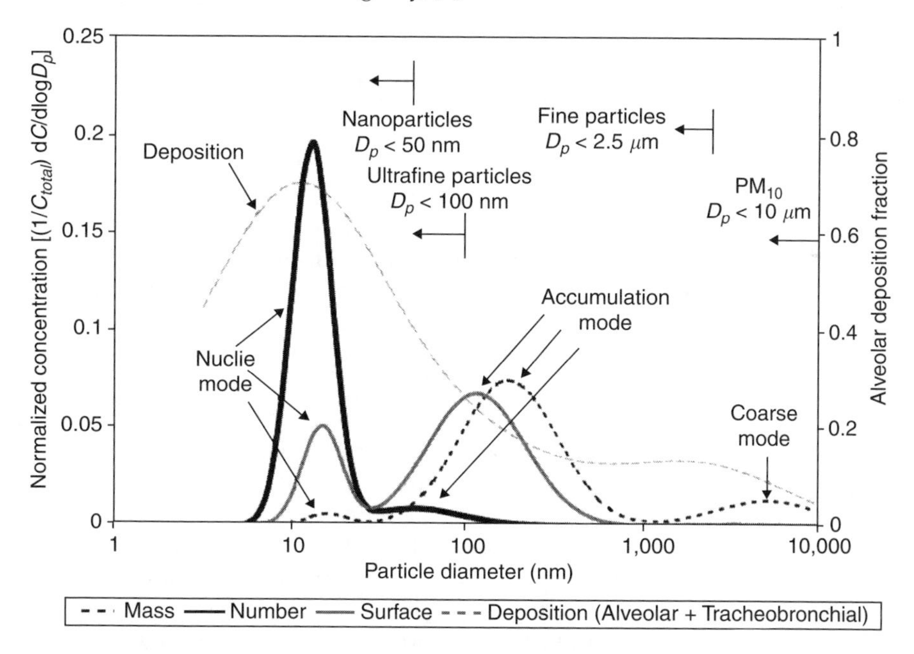

Figure 24.2 Normalized, mass-weighted and number-weighted particle size distributions and alveolar deposition from typical diesel exhaust. Note that in this figure "coarse mode" is defined as $PM_{10-2.5}$. *Source*: Kittelson D, Watts W, Ramachandran G, Paulsen D, Kreager C. Measurement of diesel aerosol exposure: A feasibility study. In: Health Effects Institute Special Report-Research Directions to Improve Estimates of Human Exposure and Risk from Diesel Exhaust; April, 2002, p. 153–79.

[3–5]) and bacterial products, especially endotoxin (e.g., see References [6, 7]), has been recognized. Recent studies have suggested that a major contributor of the toxicological and inflammatory potential of PM is related to bacterial endotoxins, and this activity is found almost exclusively in the $PM_{10-2.5}$ of mass distribution [6–9].

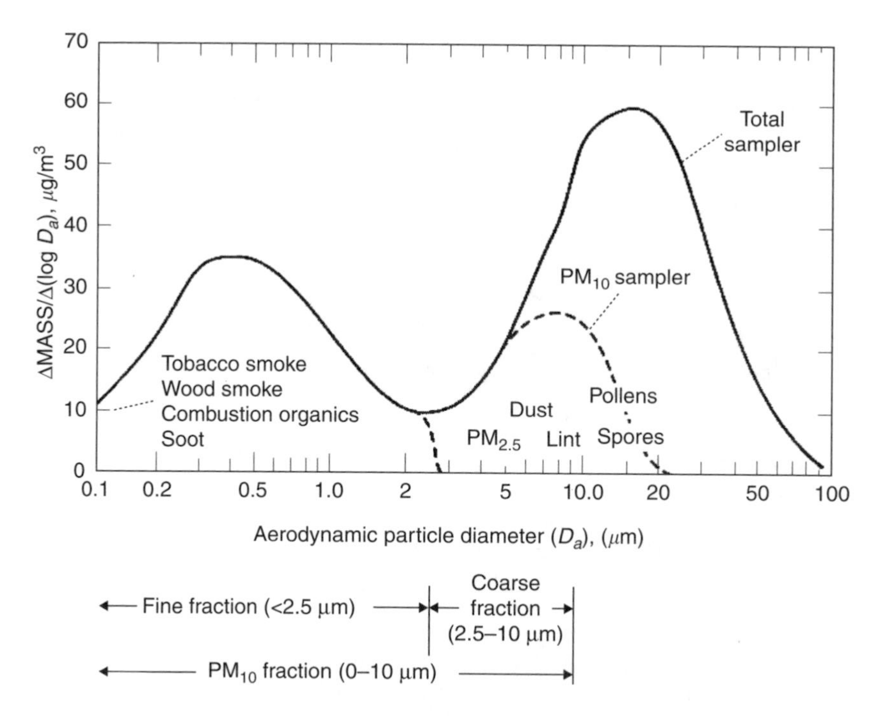

Figure 24.3 Distribution of particle mean diameter by particle volume for typical indoor PM aerosol. *Source:* Mc Donald and Ouyang [267].

24.2.2 *Indoor aerosol (Table 24.1 and Figure 24.3)*

Dependent on the characteristics of a given structure and the mode of ventilation, the outdoor aerosol can be an important contributor to the indoor aerosol. Indoor penetration of fine and coarse ambient particulate has been estimated to be close to one [10]. Source apportionment studies have indicated that the contribution of outdoor PM to the total indoor aerosol ranges from 60% or more for PM ≤ 1 μm and declines with increasing PM to about 20% for PM between 6 and 10 μm [10,11]. Estimates based on studies in the eastern U.S. indicate that 75% of the fine indoor aerosol during summer months is derived from the outdoors [12]. In contrast, studies carried out in the western U.S. indicate a much lower penetration of fine PM (~27% for sulfate and 12% for $PM_{2.5}$ ([13]). In addition, chamber studies indicate that ozone, which in most residences has no indoor sources, can react with volatile organic compounds in indoor environments to produce fine particulates (<0.5 μm) that are not generated originally in the outdoor environment [14].

In homes with smokers, cigarette smoke is, by far, the most important contributor to indoor aerosols. Figure 24.4 shows the effect of the presence of a smoker in an indoor environment on the concentration of PM in the air [15]. Nicotine is found primarily in the tar (particulate) fraction of tobacco smoke [16,17] and indoor residential concentrations range from 2 to 14 μg/m^3 in homes with a single smoker [17]. Concentrations over 1000 μg/m^3 have been recorded in vehicles with ventilation systems turned off [16,17].

Cooking is an important source of indoor PM, which is often not considered in terms of potential contribution to the total indoor particle burden [18]. Frying or use of an oven transiently can be the dominant source of indoor aerosol, and a single episode of cooking on a stove can lead to increases in indoor PM that are between 1 and 2 orders of magnitude of the background PM concentrations (Figure 24.5).

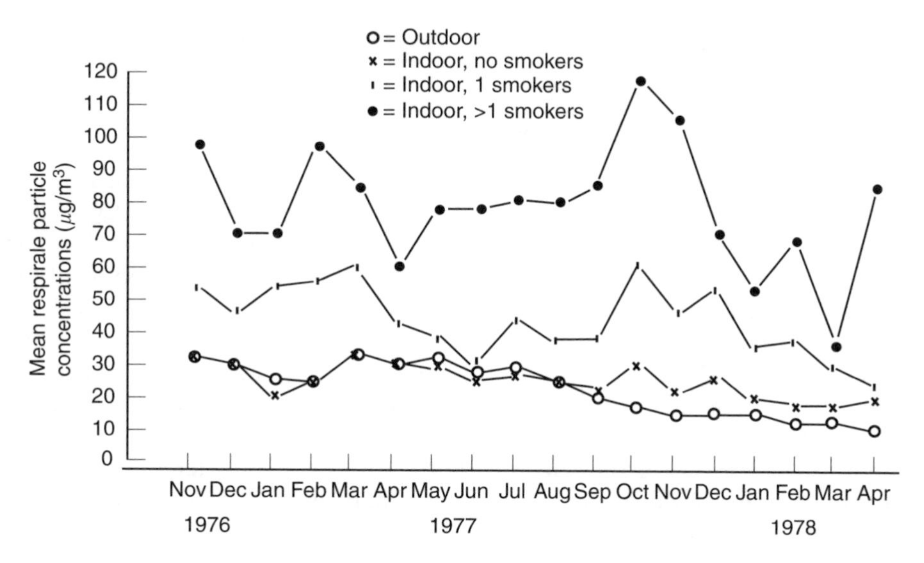

Figure 24.4 Effect on cigarette smoking on indoor concentrations of respirable particles (approximately PM_{10}). *Source*: Spengler [15].

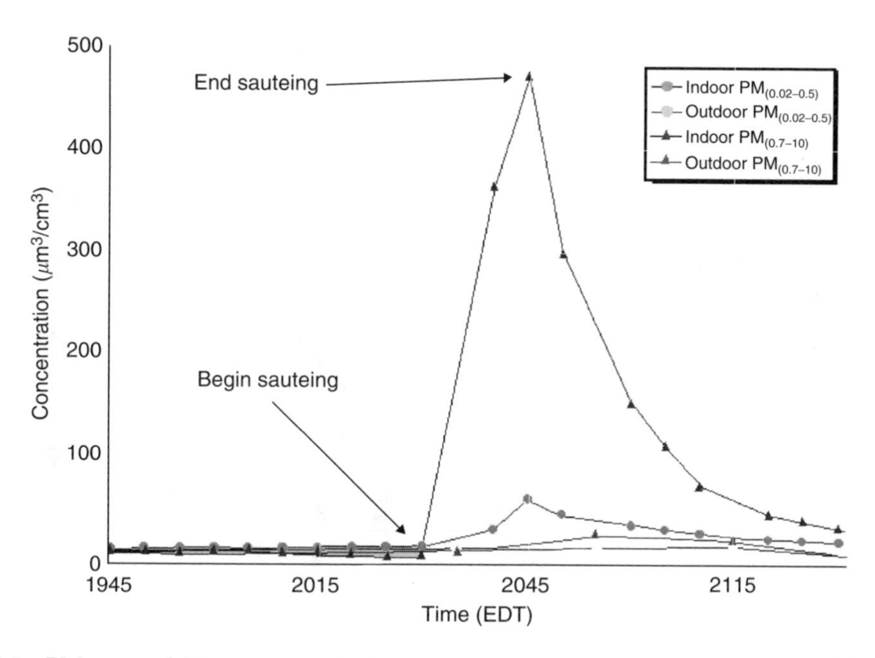

Figure 24.5 $PM_{10\text{–}2.5}$ and $PM_{0.7\text{–}1}$ concentrations from a sautéing event in one home. EDT=Eastern daylight savings time. *Source*: Abt et al. [18].

The noncombustion component of indoor aerosol makes the largest contribution to total mass (Figure 24.3). Of greatest interest is the biologically active components of this aerosol. The details related to the components of this aerosol are covered in Chapter 10.

24.3 Deposition and clearance of aerosols from the human respiratory tract[2]

24.3.1 Deposition and retention

For most people during quiet breathing, air enters the respiratory tract primarily through the nose at high velocity (Figure 24.6). Particles >5 μm are removed efficiently at this level. Particles <100 nm are also removed at this level, largely by impaction. Inertial impaction is proportional to velocity, the square of the particle diameter, and the sharpness of the angle of the airway [21,22]. Therefore, impaction occurs at places where there is a sudden change in the direction of the air stream, such as airway branch points, and involves a similar size range as that for gravitational impaction [23]. As air moves through the trachea and bronchi, impaction and sedimentation both play a role in the deposition of particles. Sedimentation due to gravity occurs in airways, except those that are vertical and for particles >0.5 μm aerodynamic diameter (vol. 1 of Reference [19]). As airway diameters narrow, particles whose distance to the airway wall surface is less than the particle's size are removed by interception [21]. As the small particles (<1 μm down to the nanometer range) move down the airways in an ever-slowing air stream, diffusion plays a greater role in deposition. Diffusion is an important mechanism for particles <0.5 μm [22]. Small particles that carry a surface charge (e.g., freshly generated combustion particles) may also be deposited as a result of electrostatic forces. Electrostatic forces are not important for particles >4 μm [22].

The nasal passages are effective in removing particles larger than 5 μm and remove virtually all particles larger than 10 μm [22]. In addition, condensation nuclei are effectively

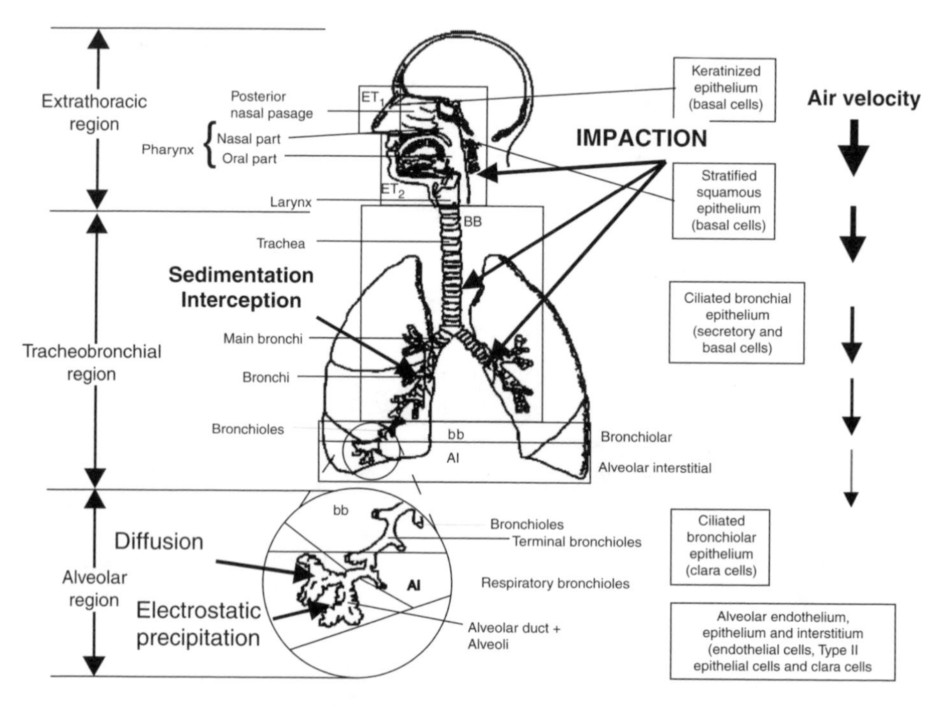

Figure 24.6 Schematic of human respiratory tract. Adapted from U.S. Environmental Protection Agency [2], which was adapted from Casarett [319] and Lippmann and Schlesinger [320].

[2]An exhaustive discussion of issues related to deposition of particles in the human respiratory tract can be found in Reference [19].

removed in the nasal passages by diffusion. Figure 24.7 and Table 24.2 summarize the estimated deposition fractions for particles of various mass median aerodynamic diameters (MMAD). Particles <10 μm MMAD are deposited at all levels of the respiratory tract, with particles <1 μm being deposited primarily in the tracheobronchial tree and the alveoli ("Pulmonary" in Figure 24.7). A number of factors affect deposition (Table 24.3). Patterns of ventilation and underlying lung disease are of particular importance for health considerations. Exercise is associated with a switch from primarily nasal breath to oral breathing, with a resultant increase in the alveolar deposition of particles >1 μm (see Section 10.5.1.4 in Reference [19]).

Retention of PM in human lungs has been studied by a number of investigators. The number of particles retained is related to the ambient concentrations (Figure 24.8; [24]). Particle retention is also a function of the level of the airways in which the particle is deposited, the type of particle, and the functional integrity clearance mechanisms (for an extensive discussion of specific mechanisms, see Reference [25]). In 42 left lungs obtained from the Coroner's Office in Fresno, CA (19 from cigarette smokers; all Hispanic males),

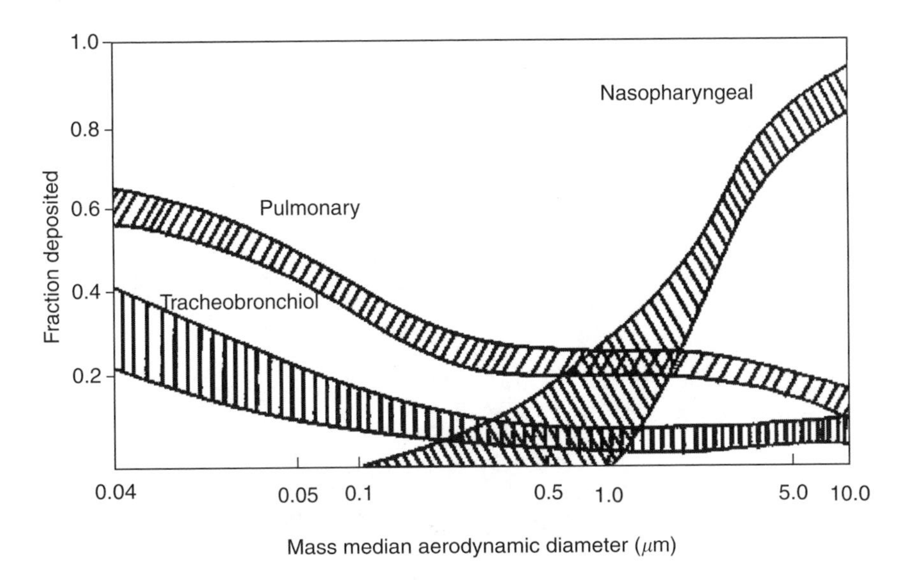

Figure 24.7 Fractional deposition of particles in the respiratory tract of humans. *Source*: Morrow [321] as cited by Brain and Valberg [21]. The *x*-axis is plotted on a logarithmic scale.

Table 24.2 Respiratory Tract Penetration of Particles of Various Sizes[a]

Particle Size Range (μm)	Level of Penetration (Generation Number)[b]
≥11	Do not penetrate
7–11	Nasal passages
4.7–7	Pharynx
3.3–4.7	Trachea and primary bronchi (1st)
2.1–3.3	Secondary bronchi (2nd–7th)
1.1–2.1	Terminal bronchi (8th)
0.65–1.1	Bronchioles (9th–23rd)
<0.65	Alveolar ducts (24th–27th) and alveoli

[a] Adapted from Reference 270.

[b] Generation numbers are from Reference [271].

Table 24.3 Factors that Affect Particle Deposition Exclusive of Particle Characteristics[a]

Respiratory Tract Geometry
Airway caliber
Airway branching pattern
Airway path length to terminal airways and alveoli
Ventilation
Pattern of breathing: oral, nasal, oronasal
Ventilation rate ([272])
Duration of pauses between breaths
Tidal volume (volume of each breathing during quiet breathing)
Ventilation distribution
Other Factors
Respiratory tract disease [273]
Altered airway geometry, branching patterns, and path lengths
Altered ventilation patterns
Altered distribution of ventilation
Changing patterns of pattern of breathing with age
Infants are preferential nasal breathers [274]
Sex
Females with greater deposition of nanoparticles ([272])

[a] Adapted from Reference [25].

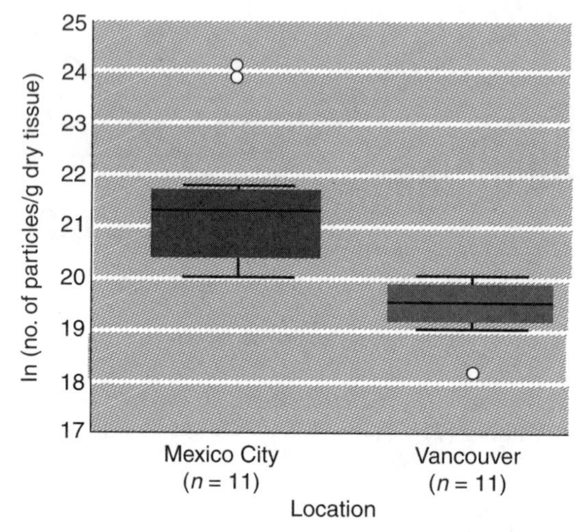

Figure 24.8 Number of particles per gram of lung tissue from a high PM area (Mexico City: mean $PM_{2.5}$=29.5 μm; Vancouver, Canada: mean $PM_{2.5}$=10.5 μm). The *y*-axis is plotted on a natural logarithm scale. Lungs from females were obtained from a general autopsy population from a referral hospital in Mexico City and a general hospital in Vancouver. *Source*: Braver et al. [24].

carbonaceous and birefringent silica particles were rarely found in the walls of larger airways [26]. In contrast, at the level of respiratory bronchioles (beyond the 12th generation of airways), such particles were frequently found in the airway walls [26]. Similar observations were made by Churg and Vidal [23] in the lungs of nonsmokers from Vancouver, Canada (Figure 24.9). These investigators also noted that ultrafine PM constituted <15% of the retained PM (expressed as millions of particles/gram of tissue), virtually all of

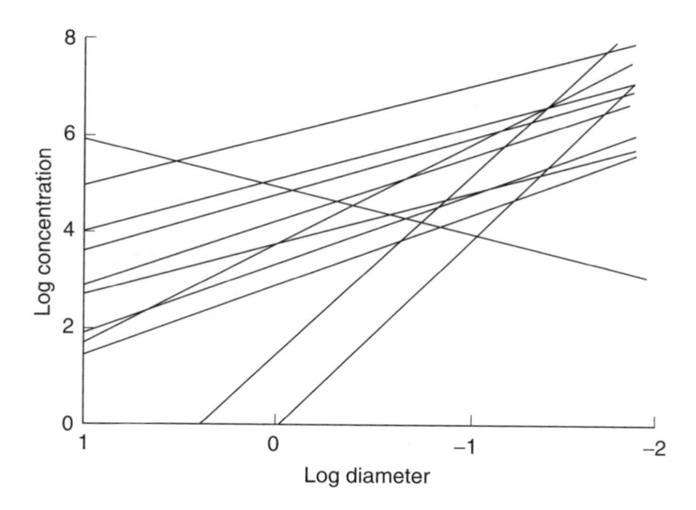

Figure 24.9 Relationship between airway concentration of particles and airway diameter both plotted on logarithmic scales. Lungs are from 11 nonsmoking residents of Vancouver, Canada, whose lungs were obtained from a general hospital. Vancouver is an area of low PM concentration (see text). The figure shows wide interindividual variation but fairly similar retention patterns across individuals. In these specimens, ultrafine particles constituted $<15\%$ of the retained particles at any airway site. *Source*: Churg and Vidal [23].

which were crustal minerals. This was in contrast to their findings in lungs from Mexico City, where lungs contained an average of 25% of chained aggregates of carbonaceous particles [24]. As noted previously, airway branch points are also sites of increased deposition [23]. For example, in the 4th generation of airways, Churg and Vidal [23] observed that the geometric mean particle number concentrations (mean of ten lungs) were two logs greater (per 10^{-6}/g dry weight) at bifurcation points (4.0 ± 1.2) compared to tubular segments (6.1 ± 1.4) both in the upper and lower lobes.

Deposition of particles is influenced by sex (greater deposition in females in the extrathoracic and tracheobronchial tree), age (increased deposition in the tracheobronchial region in children and young adults based on modeling), and underlying respiratory tract disease [27, 28]. Figure 24.10 presents simulated deposition data presented by U.S. EPA [28] and illustrates the variability in mass deposition (μg/d) as a function of age and sex. The largest predicted mass depositions for the tracheobronchial and alveolar regions are seen for ultrafine particles and particularly for children aged 14–18 yr, a group that is likely to be physically active in outdoor activities. Figure 24.11 shows simulated data for the fraction of the total number of particles inhaled that is deposited in various regions of the respiratory tract as a function of particle diameter. The predicted deposition fraction mode falls between particles in the range of 0.1–1.0 μm for all lung regions and drops as particles increase and decrease in size.

24.3.2 *Clearance*[3]

Clearances of nonbiological and biological aerosol particles have many common elements, and these are discussed in this section. These general pathways are depicted in Figure 24.12. The one major difference between the clearance of insoluble nonbiological particles and particles that are intact bacterial or fungal spores relates to the latter's

[3]This section is a synthesis of extensive discussions presented in Reference [25,28].

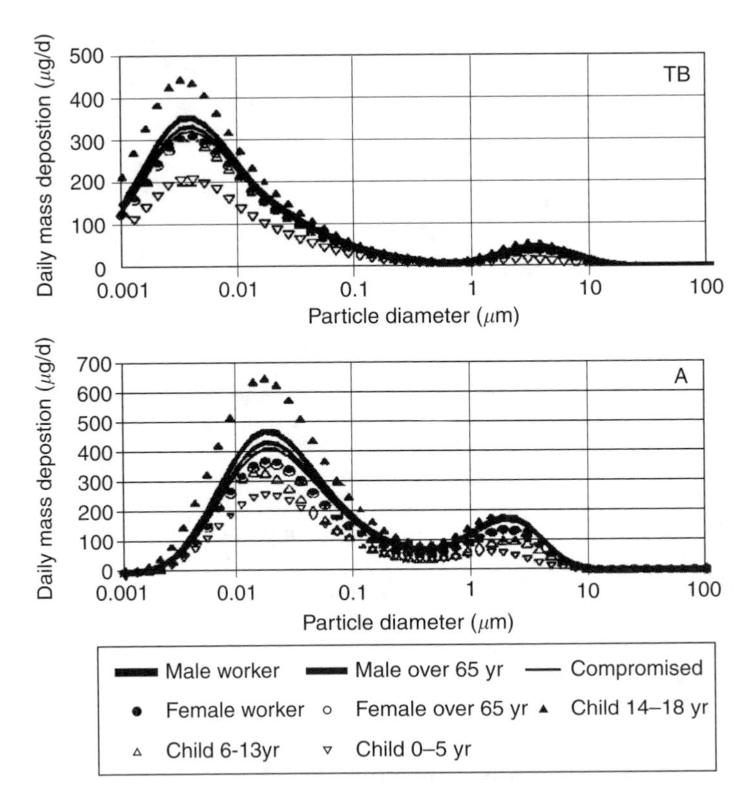

Figure 24.10 Daily mass particle deposition rate for 24-h exposure at 50 $\mu g/m^3$ in tracheobronchial and alveolar regions as predicted by the Commission on Radiological Protection Publication 66 model [322]. Simulations used daily minute volume patterns for different demographic groups. *Source*: U.S. Environmental Protection Agency [28], adapted from Figure 10-43).

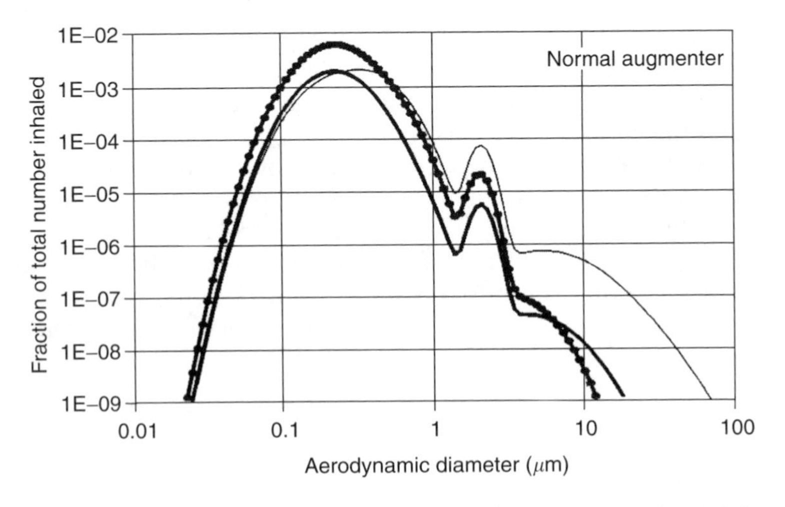

Figure 24.11 Fractional number deposition for a male with normal nose/mouth breathing with a general population activity pattern as predicted by the Commission on Radiological Protection Publication 66 model [322] for exposure to Philadelphia aerosol. *Source*: U.S. Environmental Protection Agency [28], adapted from Figure 10-50).

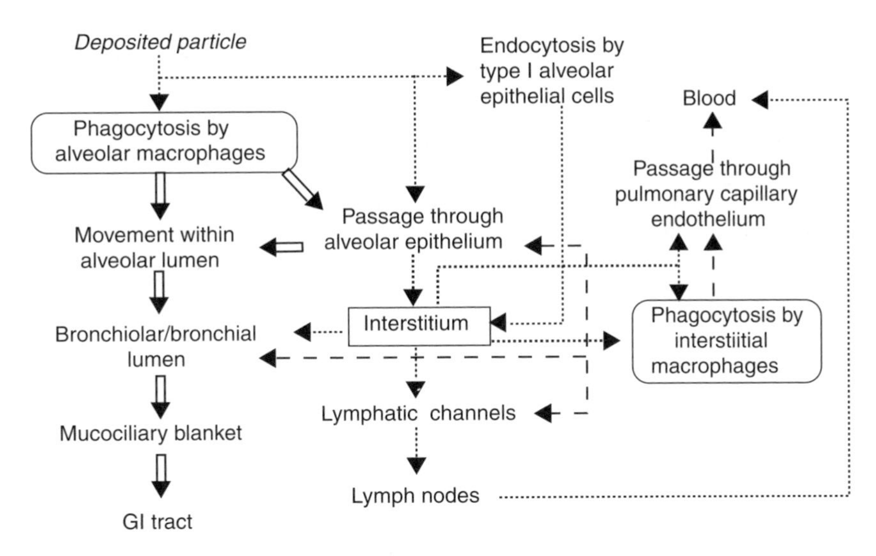

Figure 24.12 Known and suspected pathways for clearance of insoluble particles. *Source*: Schlesinger [25]. This diagram does not include all of the pathways for the clearance of bioaerosol particles such as fungal spores and viable bacteria which may involve primary phagocytosis by polymorphonuclear leukocytes as well as macrophages and other immunologically mediated effector mechanisms. Mucociliary blanket (transport) refers to the combined effects of ciliary action of epithelial cells to the level of terminal bronchioles and mucus secretion by mucus glands (only found in airways with cartilage [seven generations] and goblet cells [found to the level of respiratory bronchioles] [29].

activation of the complex immunological pathways responsible for killing these agents. These latter processes are summarized very briefly in Table 24.4 and are not discussed further.

Particles that enter the nasal cavity are cleared largely by muciliary clearance (Figure 24.11 and Reference [29]), sneezing, nose blowing, or dissolution for soluble particles. In the tracheobronchial tree, poorly soluble particles are cleared by mucociliary transport (the net movement of which is toward the oropharynx) and are swallowed or removed by coughing. More soluble particles may be absorbed through the mucosa and enter the bloodstream.

Insoluble particles that reach the alveolar region of the lung are ingested by alveolar macrophages. Increasing particle burden results in increased numbers of cells, the maximum accumulation of which appears to be more a function of particle number than of particle mass. These particle-filled cells are cleared by the mucociliary apparatus, by migration into the connective tissue that separates the alveoli (air sacs) and surrounding airways, and into the bloodstream via lymphatic channels. Some of these latter particles may then migrate to lymph nodes that are located along the tracheobronchial tree and then enter the bloodstream. Uningested, ultrafine particles may enter the bloodstream directly across the alveolar and capillary epithelium, and this clearance into the blood can be very rapid. Within 1 min after human inhalation of ^{99m}Technetium-labeled carbon particles (<100 nm), labeled particles can be detected in the blood, with peak concentrations achieved between 10 and 20 min [30]. Once in the bloodstream, free particles and particles contained in macrophages can be deposited in any extrapulmonary organ. Particles that are ingested by macrophages may dissolve in certain environments within the cells (e.g., the acidic environment of phagolysosomes).

Table 24.4 Summary of Immunological Mechanisms for the Clearance of Intact Biological Particles[a]

Innate Immune (Nonspecific) Response[b]
Cytokines/chemokines (proinflammatory, anti-inflammatory, activating)
Specific ligands (e.g., CD14 receptor for endotoxin)
Natural antibodies
  Microbe-specific antibodies found in healthy people in the absence of overt infection
Opsonin-independent phagocytosis
  Engulfment of biological particles in the absence of specific antibodies by macrophages, and polymorphonuclear leukocytes
Acute-phase proteins

Specific Immune Responses[c]
Pathogen-specific antibodies
Cell-mediated immunity

Complement System
System of more than 30 proteins that acts with or without antibody to initiate inflammatory reactions and to kill viable biological particles

[a] This table is created from material in Chapters 4–9 of Reference [275].

[b] The innate immune system is that part of the immune response that does not require specific response to pathogens and is the first response to pathogens not previously encountered by a host. The innate immune response is rapid compared to the delayed response of specific immunity.

[c] Specific immune responses are classified broadly as those related to antibodies and those related to direct cellular effects. This system is activated by a complex process of recognition, processing, and presentation of foreign antigens, which leads to activation of specific limbs of the specific immune response system.

Soluble particles are removed by absorption. Absorption is described as a two-stage process: dissolution (dissociation of particles into material that can be absorbed) and uptake of the material. Each of these steps is time-dependent, for which surface properties, chemical structure, and surface-to-volume ratio are important determinants.

Whether a particle is cleared or retained for some period of time is dependent on the physicochemical properties of the particle, site of deposition, presence of underlying diseases, and occurrence of tobacco smoking [22]. Explanations for the increased deposition and possible retention in persons with underlying lung disease are not well understood but, in part, may be related both to structural alterations and functional changes in the lung (e.g., altered epithelial permeability) [31,32]. In healthy individuals, poorly soluble particles are estimated to be cleared over a 2.5–20 h period. Persons with chronic lung disease and certain infections (e.g., influenza) may have impaired clearance of particles. The effect of physical activity on deposition and clearance is unresolved. However, a recent study that used a mouthpiece exposure system showed that deposition of ultrafine carbon particles (count median diameter=26 nm, GSD=1.6) was increased in 12 healthy subjects who exercised moderately, with a 4.5-fold increase in the total number of deposited particles [33].

Figure 24.13 illustrates the effects of duration of exposure on retention of particles for three particle modes and total lung burden as a function of age for a specific aerosol composition. Of interest for health considerations is the different time course and magnitude of lung burden for the three different modes. Given the importance of exposures at critical points of human development, this type of heterogeneity of accumulation may have important implications for health effects related to chronic exposures to indoor and outdoor aerosols.

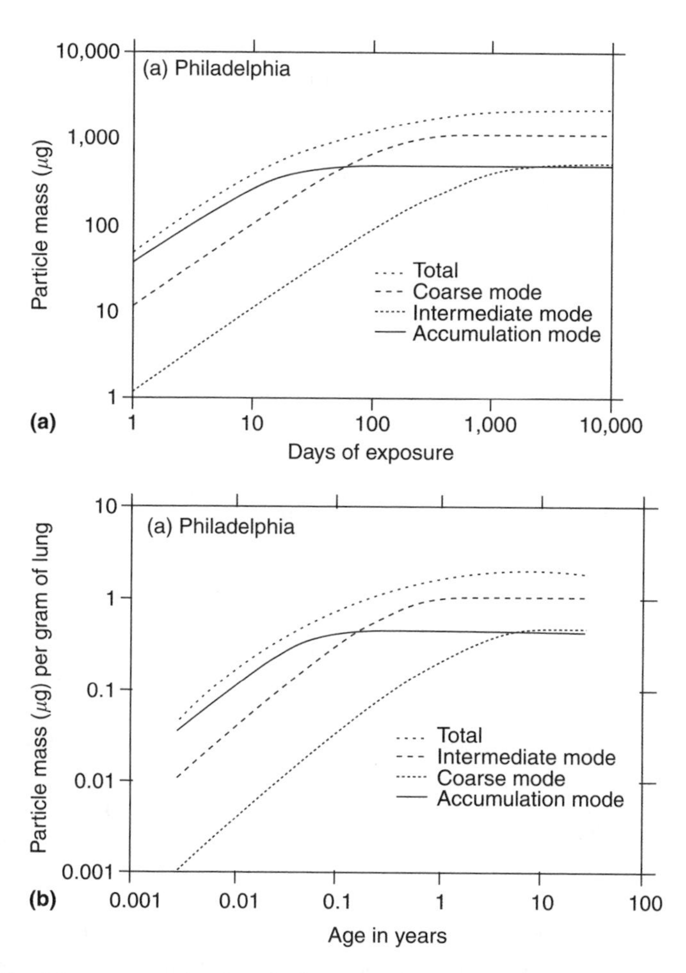

Figure 24.13 (**a**) Particle mass retained in the lung vs. days of exposure, predicted by [322] for exposure to Philadelphia aerosol, under the assumption of dissolution absorption half-times of 10, 100, 1000 days for three modes of particles. (**b**) Same data converted to specific lung burden. Based on continuous exposure to Philadelphia aerosol at 50 $\mu g/m^3$. *Source*: U.S. Environmental Protection Agency [28, adapted from Figures 10-55, 10-56].

24.4 *Mechanisms of toxicity of aerosol components*

The mechanisms that underlie the effects of the nonbiological ambient and indoor aerosols and the biological components of the aerosols have a number of features in common. Both components directly trigger inflammatory responses and/or trigger (directly or indirectly) immunological responses that lead to specific inflammatory responses that characterize various allergic phenotypes (e.g., asthma allergic rhinitis/conjunctivitis). Reactive oxygen and nitrogen species are major products of inflammatory reactions [34,35] and can be generated by neutrophils, macrophages, and eosinophils in response to contact (ingestion) with biological materials [36] or ambient particles [37]. Moreover, the organic carbon and transition metal components of ambient and indoor aerosol components can lead directly to the production of reactive oxygen species (metabolites — ROM) [38,39]. ROM (superoxide radical [$O_{2-}{}^{\bullet}$]), hydrogen peroxide [H_2O_2], hydroxyl radical [$OH^{\bullet}$]; sometimes included are singlet O_2 and hypochlorous acid), in turn, are potent inducers of inflammation [40], and initiators of lipid peroxidation and DNA strand breaks [41], which may be important in carcinogensis (summarized in Figure 12 of Reference [42]). Since a large body of research suggests that oxidant stress may be one of the principal (if not the principal) final common pathways for the effects of ambient aerosols on human health, this section begins with a consideration of the evidence for this view. The section then considers pathways less proximate: inflammation, immunological modulation, and particle overloading effects.

24.4.1 Particle aerosol induction of oxidative damage

As epidemiological evidence of health effects related to particulate air pollution mounted, animal studies were undertaken to try to identify the components of PM and mechanisms that might explain the epidemiological observations. Some of the most important of these were animal exposure studies undertaken with residual oil fly ash (ROFA), which focused on the role of transition metals. ROFA is generated from the burning of fossil fuels and has a high concentration of metals, particularly transition metals ([43]; reviewed in Reference [44]). These metals, which can exist in more than one valence state, can participate in electron cycling, which can lead to the production of highly reactive OH through the Haber–Weiss reaction (Table 24.5) [34]. This generation can take place in acellular and cellular systems [44,45]. First-row transition metals (titanium, vanadium, chromium, manganese, iron, cobalt, nickel, copper) are found in highest concentrations in crustal material and in the atmosphere as a consequence of human-generated pollutants [45]. A series of studies by U.S. EPA investigators and collaborators (selected References include [45–49]) established the role of the oxidative potential of the transition metal content of ROFA to generate ROM and the specificity of these metals as causes of lung injury in rodents. The metal contents of dusts from a variety of sources other than ROFA were also shown to produce lung injury that was specific to the generation of ROM [46,47,50]. A recent, comprehensive evaluation of the role of metals in the toxicology of PM generated from coals from three different locations in the United States and from PM derived from several formulations of gasoline and diesel fuel (Utah, Illinois, North Dakota) provides strong support for the toxicological importance of the oxidant properties of transition metals [51]. These investigators documented that bioavailable iron (nmol/mg of coal fly ash as measured after ferric ammonium citrate extraction) was greatest for $PM \leq 1\ \mu m$, but bioavailable iron could also be found in $PM_{10-2.5}$. Chelation of iron significantly reduced evidence of oxidative damage as measured by malondialdehyde [51].

The ability of PM_{10} to generate ROM unrelated to transition metals has been well documented [52,53]. Investigators at the University of California, Los Angeles, have conducted an extensive series of studies that establish that organic components of PM lead to the production of ROM (reviewed in Reference [54]). In particular, polycyclic aromatic hydrocarbons (PAH) and PAH-derived quinones found in diesel exhaust particles have been shown to produce *in vitro* evidence of oxidant stress [55]. These investigators have developed a "stratified oxidative stress model" to characterize both the production of ROM by diesel exhaust particles and the mechanisms through which these ROM exert their effects on inflammation and allergic response ([56] — the reader is referred to the extensive bibliography in this review — selected references relevant to this section include

Table 24.5 Haber–Weiss Reaction: Ferrous Iron-Catalyzed Generation of Hydroxyl Radicals

$O_2^{\bullet -} + Fe^{3+} \rightarrow O_2 + Fe^{2+}$
$Fe^{2+} + H_2O_2 \rightarrow Fe^{3+} + HO^{\bullet} + OH^-$
$O_2^{\bullet -} + H_2O_2 \rightarrow O_2 + HO^{\bullet} + OH^-$

All first-row transition metals can participate in this reaction.

[38,52,55,57–61]). The model is summarized in Figure 24.14. These investigators demonstrated that low concentrations of redox active PAHs and their oxy-derivatives initially activate genes that lead to the induction of hemoxygenase-1 (HO-1) and superoxide dismutases. "Low-level" oxidative stress is defined as the normal ratio of reduced glutathione to oxidized glutathione (GSH:GSSG). The oxidation of GSH is mediated by several enzymes that convert reactive oxygen species precursors to less reactive species and is one of the major cellular pathways for the control of reactive oxygen species [62]. HO-1 is involved in the metabolism of hemoglobin and leads to the production of antioxidants (bilirubin) and anti-inflammatory (carbon monoxide) end products [62]. Superoxide dismutases catalyze the conversion of $O_2^{\bullet-}$ to H_2O_2, which is converted to water through a catalase-mediated reaction. A decreasing GSH:GSSG ratio indicates a failure of these and other host defenses against oxidant stress (e.g., antioxidants in lung lining fluid and blood [vitamin C, urea] or in cell membranes [vitamin E]) to "neutralize" the oxidative stress [62,63]. Persistence of ROM leads to the activation of cell signaling cascades that ultimately lead to activation of nonallergic and allergic inflammation, peroxidation of cell membranes, programmed cell death, and cell necrosis [57,62,64]. The inflammatory process itself generates additional ROM that are part of normal cellular defense mechanisms ([36,65] and Section V in Reference [66]) and downregulates enzymes involved in the metabolism of ROM [67], thereby creating an ongoing cycle of cellular injury (see the discussion in Reference [68]). Ultrafine particles derived from urban formation and receptor sites have been found to be important sources of the oxidative stress induced by PM *in vitro* [38] and in animal toxicology studies [69,70]. Components of fine and coarse PM also lead to oxidant stress either directly [38] or indirectly through bioaerosol components such as endotoxin that are part of the coarse fraction of PM and are potent inducers of inflammation [71] (see below). Several studies of ambient PM have indicated that endotoxin may be as important a source for inflammation as are metals and organic components [6,9,72]. Through interactions with specific lymphocyte receptors, endotoxin activates many of the same signaling pathways that are activated by organic components of PM created by the activities of man [73]. This activation is the source of the potent inflammatory potential of endotoxin.

Oxidative stress provides a link between the immunomodulatory effects of diesel exhaust particles with regard to the immunological mechanisms that underlie asthma and allergic conjunctivitis/rhinitis (so-called "hay fever") [57, 74–79]. The ability of the metal content of ROFA to enhance sensitization to house dust mite in rats indicates the likelihood that a wide variety of redox-active PM sources also contribute to effects of allergen response enhancement through the generation of ROM and the pathways summarized in Figure 24.14 [80,81]. Oxidative stress plays an important role in the carcinogenic potential attributed to diesel exhaust through its capacity to induce mutagenesis, the production of DNA strand breaks and DNA adducts, and the induction or inhibition of lung metabolic enzymes [67, 82–84].

24.4.2 *Particle aerosol induction of inflammation*

The production of ROM may be the final common pathway through which anthropogenic and biological components of particle aerosols lead to the production of inflammation. Nonetheless, the ability of PM aerosol to produce inflammation is of sufficient importance to justify some expansion of this component of PM aerosol effects.

Exposure of pulmonary macrophages [6,9,72,85,86], lung epithelial cells [50, 87–89], and peripheral blood monocytes [8] leads to the activation of various molecular signaling pathways for the induction of inflammation (see Figure 24.14). Transition metals, endotoxin, and organic carbon fractions of PM have all been found to trigger inflammation

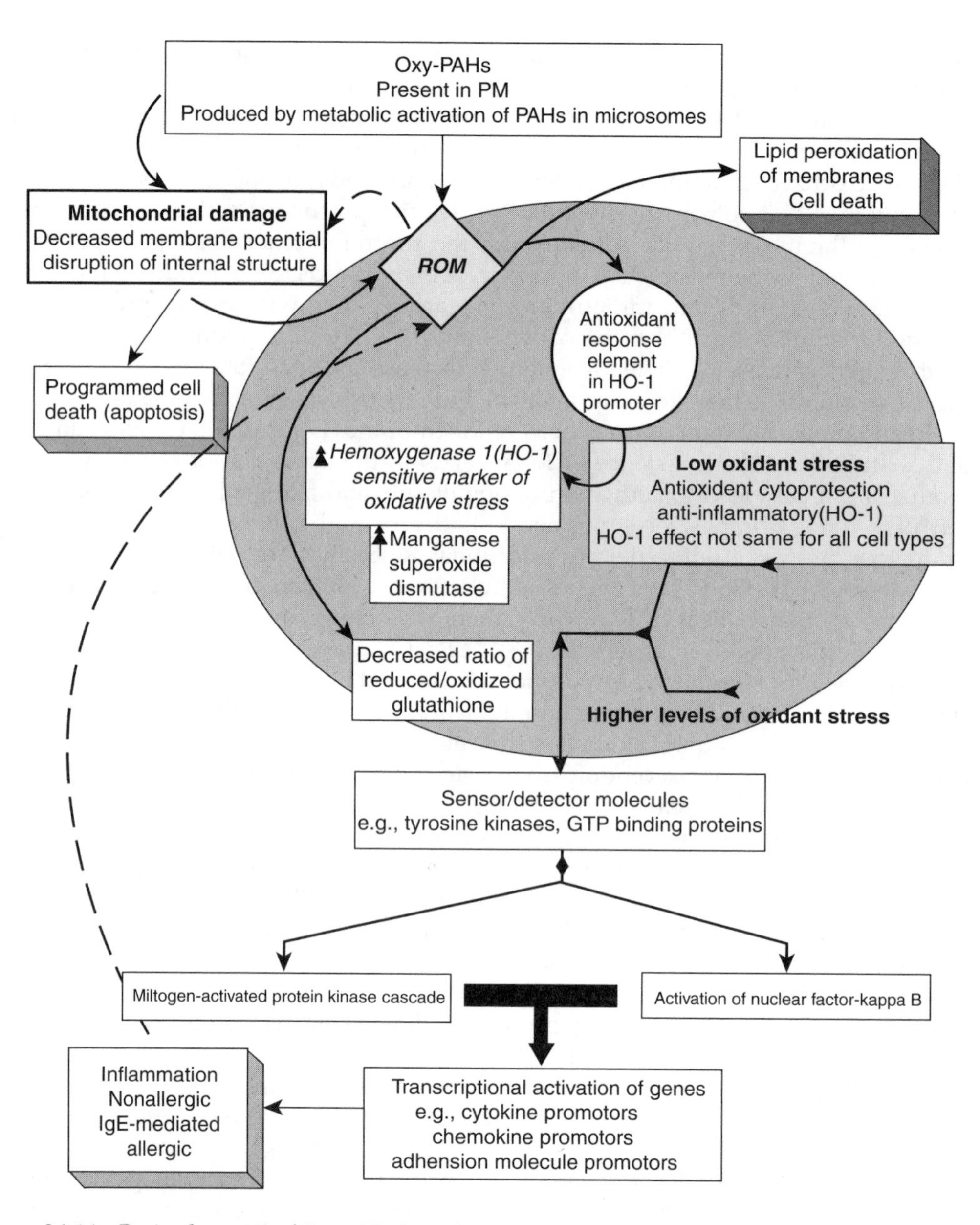

Figure 24.14 Basic elements of "stratified oxidative stress model" of Li et al. [56] (area enclosed in the gray circle) and consequences of ROM that are not controlled by antioxidant defenses. Solid arrows represent the pathways; dashed arrows identify feedback loops that perpetuate the adverse outcomes. ROM: reactive oxygen metabolites. Adapted from the following references: Li et al. [38,52,55,56], Net et al. [57], Hiura [58,59], and Kumagai [82].

in vitro (see references in previous sections and [88]). The relevance of these *in vitro* findings to human health is supported by findings of increased levels of the proinflammatory cytokine interleukin 8 (IL-8) in the airways [90] and increased neutrophils and myeloperoxidase in the sputum [91] of healthy volunteers exposed to diesel exhaust in an exposure chamber under controlled conditions. Inhaled PM of anthropogenic origin has systemic effects evidenced by the stimulation of bone marrow (increased production and release of polymorphonuclear leukocytes by supernatants from human alveolar macrophages stimulated with PM_{10} [85,92,93]). Intratracheal exposure of rats to ROFA has been shown to increase the blood levels of fibrinogen, a procoagulant protein that is associated with acute inflammation and blood clotting [94]. Humans exposed to the forest fires in Southeast Asia in 1997 [85,95] showed evidence of increased proinflammatory cytokines and circulating neutrophils that were related to the level of exposure. A recent study of controlled exposure of healthy and asthmatic volunteers to concentrated ambient PM (CAP) showed increases in markers of systemic inflammation [96]. Other data suggest that PM stimulation of sensory nerves in the epithelium of lung airways (nonadrenergic/noncholinergic fibers and C-fibers) leads to airway inflammation through peptidenergic transmitters (substance P, neurokinin A, and calcitonin-related gene product) and that neural sensitivity is greater than that of airway epithelium to the effects of PM on inflammation [97] (see Table 24.6).

A considerable body of data has accumulated on the mechanism of the effects of diesel PM on allergic inflammation. These have been summarized in detail by Nel et al. [57]. Table 24.7 summarizes specific findings that are related to the enhancement of IgE-mediated immune responses in humans by diesel PM. The driving of the immune system

Table 24.6 Summary of Effects of PM-Associated Transition Metals Derived from Combustion of Coals from Three Different Sources in the United States and from Gasoline and Diesel Exhaust [51]

Oxidant Generating Capacity
Cu(II)>Fe(II)>Va(III)>Ni(II)>Co(II)≅Zn(II)
Each with different time kinetics and dose–response curves
Distribution of Bioavailable Fe
Greatest in $PM_{2.5}$
$PM_{10-2.5}$ about 50% of $PM_{2.5}$
Oxidant Potential
Coals: PM_1>$PM_{2.5}$>$PM_{10-2.5}$
$PM_{10-2.5}$ approximately 20–100% as potent, dependent on source
Gasoline and diesel exhaust PM
Most oxidant production associated with transition metals (inhibited by desoxferamine (45–97%)
Biological Activity
Capacity of PM to produce IL-8 directly related to Fe content
Endotoxin content nondetectable in PM from three coal sources

Table 24.7 Summary of Adjuvant Effects of Diesel Exhaust Particles on *In Vivo* Allergic Responses in Humans[a]

Nasal Allergic Responses
Increased production of IgE and IgE-secreting cells [76]
Qualitative difference in IgE isoforms [76,276]
Increased production of allergen-induced antigen-specific IgE [77]
Induction of broad cytokine profile in the absence of antigen [277]
Induction of Th-2-like cytokine profile with coadministration of intranasal allergen and diesel PM [77]
Allergen-specific isotype switching of B cells to IgE synthesis
Increased production of chemokines [278]

[a] Modified Table III of Nel et al. [57]. References cited are those cited by Nel et al.

toward IgE-mediated responses could be reproduced with PAHs [98]. Of particular interest for the pathogenesis of new onset allergic respiratory disease is the observation that diesel PM can induce sensitization to a neoantigen in persons already sensitized to an allergen that triggers IgE-induced allergic response (Figure 24.15) [75]. In this study, the levels of the cytokine IL-4, which is a potent inducer of IgE synthesis, were increased in the presence of antigen and diesel PM above those seen with antigen alone. In contrast, the levels of INF-γ (interferon-γ), a Th-1 cytokine that reflects the part of the T-helper lymphocyte system that downregulates IL-4 production and stimulates cell-mediated immunity, were not increased above those observed with antigen alone. While most of the studies related to PM immune enhancement have focused on diesel PM, other sources of PM, such as ROFA, are also capable of the same type of immune enhancement [80,81,99]. These findings are not surprising, if the underlying mechanism for this enhancement ultimately relates to the production of ROM as discussed above.

In contrast to PM enhancement of IgE-mediated immune pathways, a number of studies have demonstrated that PM can inhibit cell-mediated immune responses that represent important host defenses against a variety of microbial pathogens. Studies with *Listeria monocytogenes*, a bacterium whose clearance is dependent on an intact cell-mediated immune system, have demonstrated that inhalation and intratracheal instillation of diesel particles lead to decreased alveolar macrophage clearance of the bacteria and decreased production of cytokines (IL-1β, IL-12, tumor necrosis factor-α [TNF-α]), which are essential for initiation and maintenance of cellular immune responses and the production of nitric oxide (NO) that is part of the oxidant antibacterial defense response [100–102]. Responses to endotoxin, a potent stimulator of IL-12 and TNF-α, were also inhibited [101,103]. Carbon black particles did not demonstrate this inhibitory effect, which suggests that the organic carbon and other components carried on the carbon core (e.g., metals) are responsible for these effects. The suppression of IL-12 may play a role in the adjuvant effect of diesel PM on IgE-mediated immune responses, since IL-12 is a potent inhibitor of

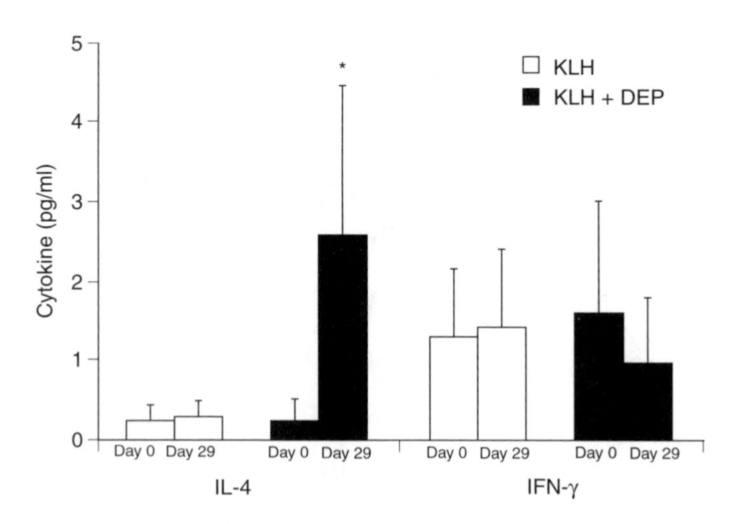

Figure 24.15 Concentration of IL-4 following nasal challenge of atopic subjects with keyhole limpet hemocyanin (KHO-antigen to which humans are not exposed) alone and in combination with diesel exhaust particles (DEP). IL-4 is a Th-2 lymphocyte cytokine and a potent inducer of IgE antibody whose levels are increased in persons with "hay fever" and with asthma. IFN-γ is a Th-lymphocyte cytokine that is not thought to participate in IgE-mediated immune response. The figure shows a lack of IL-4 response to KLH in the absence of DEP and a lack of any IFN-γ response above that seen with KLH alone. Day 29 is one day after the last of three challenge days [75].

Th-2 cytokine production [104]. *In vitro* studies have demonstrated that human alveolar macrophages exposed to urban air PM have impaired production of cytokines important for the control of respiratory syncytial virus and reduced ability to phagocytose the virus [105]. This virus is an important cause of lower respiratory illness in infants and upper respiratory infection in people of all ages. These same investigators also observed inhibition of phagocytosis and the oxidative burst to yeast particles in human alveolar macrophages exposed *in vitro* to ambient PM [6]. This effect was limited to the insoluble fraction (high concentration of organics) of PM_{10} and was not inhibited by chelation of iron or inhibition of endotoxin.

Products derived from bacteria, fungi, and plants may be carried by anthropogenic PM or, as is the case for plant allergens, may exist as intact particles in the air. In this section, discussion is limited to endotoxin and plant allergens. The mechanisms that related allergic responses to indoor allergens (see Chapter 14) such as dust mite and cat allergy are similar to those described for plant allergens.

Endotoxins are lipopolysaccharides (LPS) that make up the outer cell wall of Gram-negative bacteria, which are ubiquitous in the environment (see the review in Reference [106]). Endotoxins are bound by an LPS-binding protein, which then binds to specific cell receptors (CD14 [107] and Toll-like receptor [108]) and ultimately activate many of the same pathways noted for ROM (Figure 24.14) that lead to inflammation [73]. *In vitro* exposure of rat alveolar macrophages to urban air PM demonstrated that stimulation of inflammatory cytokines could be blocked by specific inhibitors of endotoxin but not by chelation of iron [72]. Further work by the same investigators showed that endotoxin, in the $PM_{2.5-10}$ fraction of ambient PM, was largely responsible for the production of proinflammatory cytokines (IL-6, IL-8) after *in vitro* exposure of human peripheral blood monocytes [8]. In these later experiments, chelation of iron inhibited cytotoxicity but not cytokine production. Similar results were observed with human alveolar macrophages; however, endotoxin did not appear to be responsible for the increased programmed cell death (apoptosis, which does not lead to an inflammatory stimulus) observed in the insoluble fraction of PM_{10} in which the endotoxin-mediated effects were observed [6]. Finally, exposure of rat alveolar macrophages to PM_{10} from four sites in Switzerland demonstrated that cytotoxicity, generation of ROM, and production of proinflammatory cytokines could be inhibited when extracts of PM were treated with an endotoxin neutralizing protein [9]. In aggregate, these studies point to endotoxins as important contributors to the inflammatory potential of PM aerosol and the involvement of ROM in this effect.

Aeroallergens derived from plants, animals, and insects produce their effects through the stimulation of the IgE-mediated pathways (Figure 24.16). As noted above, organic components of PM are thought to enhance response mechanisms along this pathway and inhibit negative feedback controls by specific cytokines. Due to the current concerns related to the contribution of PM aerosol to the rise in asthma prevalence over the last several decades [109], a brief discussion of these pathways is useful. Although it undoubtedly represents an oversimplification [110,111], the immunobiology of asthma is thought to be due to a skewing of the immune response to certain agents toward what is described as the Th-2 phenotype of helper (CD4+) T-lymphocytes [104]. Th-2 helper lymphocytes are defined by the secretion of IL-4, IL-5, IL-13, and granulocyte macrophage colony stimulating factor (GM-CSF), in particular and in contrast to Th-1 helper cells, which secrete IL-2, IL-2, and INF-γ. Each phenotype results in negative feedback of the alternate phenotype [112,113]. Similar phenotypes have been described for CD8+ (suppressor) T lymphocytes. The skewing of the immune response begins early in life [112] in response to as yet incompletely defined genetic predisposition and encounters with appropriate environmental antigens [112]. The details of the molecular pathways that regulate Th-1/Th-2 balance have been summarized in detail by Nel [114,115].

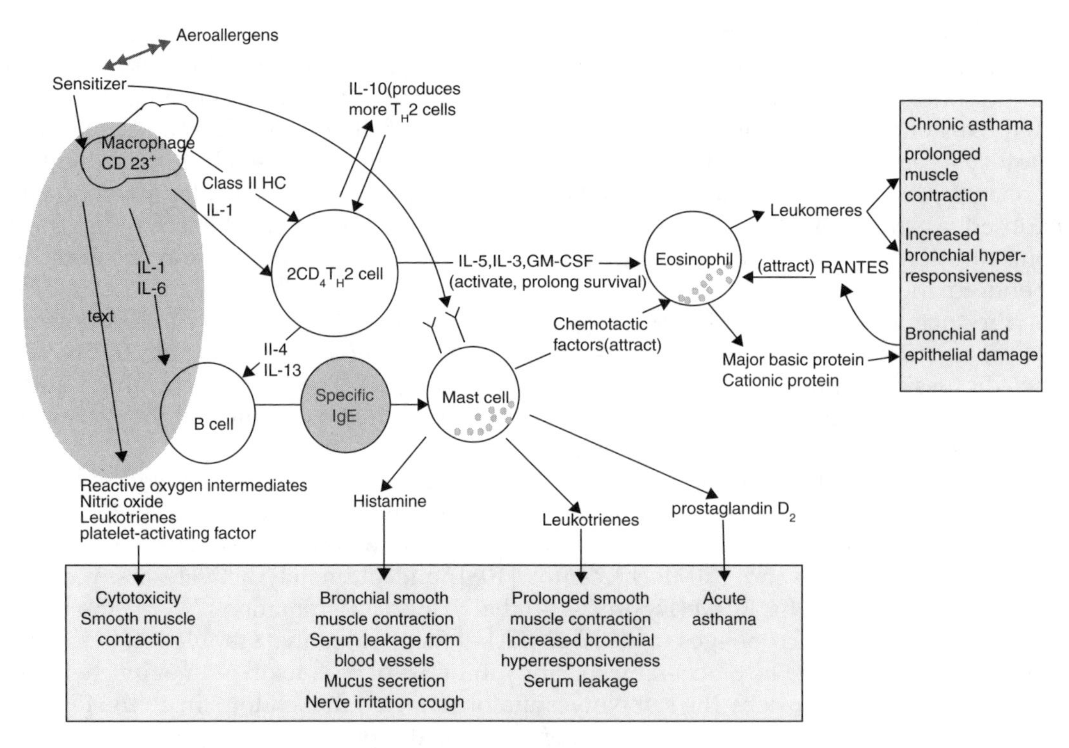

Figure 24.16 Major immunologic pathways related to IgE-mediated immune responses. Same pathways apply to allergens deposited in the nose. Pink-shaded areas define (approximately) pathways influenced by diesel PM. (Adapted from Pandya et al. [79]). See reference for detailed explanation and references (323,324) for an additional summary related to details of cytokine network and allergic effects.

24.5 Human health effects associated with ambient PM

This section summarizes an enormous body of data on the health effects that have been associated with ambient aerosols. Of necessity, the section is selective and limits itself to health effects related to PM that results from the activities of man. The emphasis is on data derived from epidemiological studies, but clinical or controlled exposures studies are cited when they serve to buttress or refute data from larger epidemiological studies. A few issues that transcend a specific consideration of specific health effects are presented prior to a more detailed discussion of health effects.

General introduction

Relationship between studies of mechanisms and studies of human health effects: Although many of the studies on the molecular and cellular mechanisms relate specific components of the ambient aerosol to specific mechanistic pathways, such clarity is not possible in observational studies of humans that are the principal sources of data on human health effects. In these studies, subjects are exposed to a "chemical soup" that arises from multiple indoor and outdoor sources of anthropogenic PM aerosol that have changing source contributions over short and long time scales and is made more complex by the aging of the aerosol through atmospheric chemical processes as it moves away from its sources. In addition, the PM aerosol is experienced in the presence of other oxidant pollutants (oxides of nitrogen, ozone) and stress (e.g., infections, diet) that may contribute to and/or "account for" effects that are attributable to the PM aerosol. Animal studies provide direct evidence for the occurrence of such complex interactions. Intranasal instillation of ambient urban PM exposed to 100 ppb of ozone (O_3) prior to instillation in rats produced greater inflammation and biochemical evidence of cellular injury than a similar concentration of PM not preexposed to O_3 [116]. Inhalation exposure of rats to urban PM and ozone demonstrated synergistic effects of O_3 and PM on lung inflammation, an effect that was not seen for stimulation of endothelin (potent vasoconstrictor produced by vascular endothelial cells and thought to contribute to ambient PM aerosol cardiovascular events) that was associated only with PM exposure [117]. This latter study points to the additional complexity that ambient PM likely has differential effects on different organ systems and metabolic processes that depend upon the chemical context in which the aerosol is presented to the host. A further complexity relates the qualitative and quantitative spatial heterogeneity of the aerosol constituents between areas in close geographical proximity. The effect of heterogeneity of the biological effects of PM aerosol collected from different parts of the same city is illustrated in a study conducted in Mexico City in which PM_{10} was collected from parts of the city with different sources of PM and different concentrations of O_3 [118]. Significant differences were observed in the effects of PM_{10} from the different locations on cell viability, occurrence of apoptosis, evidence of DNA damage (comet assay), and secretion of the proinflammatory cytokines IL-6 (Figure 24.17) and TNF-α. While controlled exposure studies of human subjects do permit the assessment of biological responses to PM aerosol exposure, they cannot provide direct evidence for acute health effects, since PM aerosol is presented in isolation or in higher than ambient

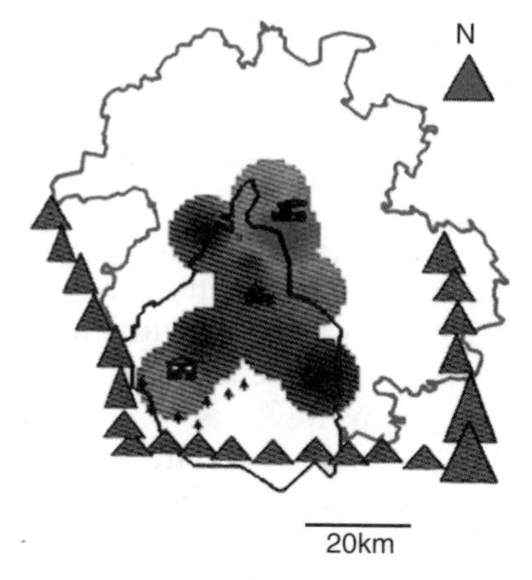

Map of Mexico City, The metropolitan area is composed of two regions: the Federal District (black outline) and the Estade de Mexico (■ outline). Air quality monitoring stations cover the area shown in ■. The most polluted areas for particles and ozone are in the northeast (■) and the southwest (■), respectively. We collected particles for this study in northern (industrial), central (business), and southern (residential) zones, sites of collection are marked with ■ squares. Mountains (■ triangles) surround the city, mainly in the south, Adapted from Gobierno del Distrito Federal [15]

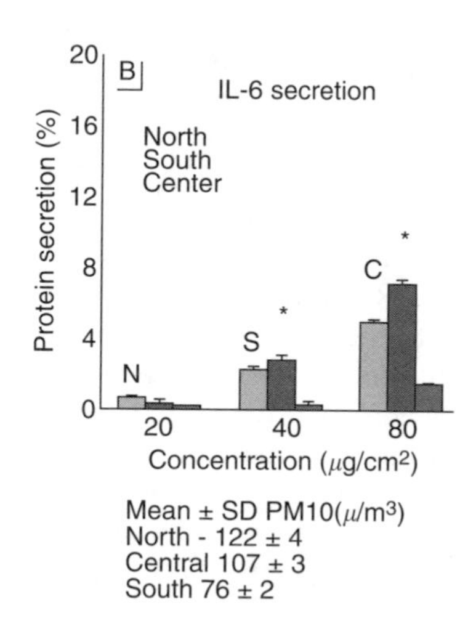

Mean ± SD PM10(μ/m³)
North - 122 ± 4
Central 107 ± 3
South 76 ± 2

Figure 24.17 Effects of PM_{10} collected from three different locations in Mexico City with different levels of PM_{10} and different concentrations of ozone. IL-6 secretion was measured in mouse monocytes. *Source*: Alfaro-Moreno [118].

concentrations and the exposure is finite. Moreover, such controlled studies cannot offer even indirect evidence for health effects related to exposure over long periods of time.

Exposure assessment in studies of human health: Since the actual distribution of source-specific dosages of aerosol received by a population cannot be known, epidemiological studies have had to focus on measures of exposure to estimate health effects related to exposure to the toxic components of the ambient PM aerosol. Two strategies have been used to try to ascribe health effects either to the PM aerosol itself or specific components of it. By far the most common approach is the comparison of the effect estimates from statistical models that include only PM aerosol or the mass concentration of a specific component with results from the same models with other elements of the aerosol or ambient gas concentrations included. This approach is hampered by the relatively high correlations that often exist between components of the aerosol or other ambient gases and can result in unstable or biased effect estimates (for examples of the approach, see References [119,120]). A less common approach is to group components of the aerosol by sources without attempting to attribute health effects to any specific component from the source [121]. A variation of this approach is to identify marker components of the aggregate PM mix to narrow the source contribution to health effects. The use of black smoke (correlated with elemental carbon) and nitrogen dioxide (NO_2) as a marker for exposure to diesel exhaust PM [122,123] and traffic-related pollution [124], respectively, are examples.

Most studies of health effects of PM aerosol have based their assessment of health effects on mass concentrations of total suspended particulates (TSP, older studies) PM_{10}, $PM_{2.5}$, and $PM_{10-2.5}$ or components such as element carbon (often measured as black smoke) (e.g., see Reference [2]). Attempts to attribute health effects to one mass fraction or another remain somewhat clouded. Some investigations, in areas in which $PM_{2.5}$ dominates the PM_{10} mass, have suggested that most health effects related to the anthropogenic component of the PM aerosol are confined to the $PM_{2.5}$ component [125–128]. Studies that have focused on the sulfate and acid component of PM_{10} (e.g., References [129–131]), by implication, emphasize the contribution of fine PM (see Figures 1 and 2), in which most of the sulfate and acid sulfates (and salts of acids from nitric acid) are found [132]. However, studies from environments in which $PM_{10-2.5}$ makes up a larger fraction of the PM_{10} mass have found that a variety of health effects are more closely associated with this coarse fraction [133,134]. Similar findings have been reported from environments whose PM aerosol mass is not dominated by $PM_{10-2.5}$ [135,136]. Further, claims have been made for a unique role of the ultrafine component of the PM aerosol in causing health effects [137–139]. The specificity of the results has been questioned in a discussion of one study (see the panel discussion in Reference [138]) and questioned by the findings in other studies [140,141]. Thus, despite the toxicological properties attributed to ultrafine particles (see above), it has been difficult to quantify what proportion of health effects attributed to the anthropogenic PM aerosol are related specifically to ultrafine particles or to the number of particles rather than their mass.

Sources of PM aerosol and human health effects

Mobile sources: Much of the early epidemiological research on the human health effects of ambient aerosol focused on fixed industrial sources and use home heating fuels (e.g., soft coal) that made major contributions to the ambient aerosol. The rapid growth of motor vehicle use has led to a research focus on mobile sources of the anthropogenic PM aerosol and its health effects (e.g., see Reference [142]). This focus has fostered the extensive research on the immunotoxicology of diesel exhaust particles discussed previously. This emphasis derives from a large number of epidemiological studies that have observed increased risks of a variety of health outcomes related to exposure to traffic and traffic-related PM aerosol. Table 24.8 summarizes a representative sample of these studies. In general, most of these studies support an association between one of several different

Table 24.8 Findings from Selected Studies on the Relationship Between Exposure to Traffic Sources of PM Aerosol and Health Outcomes

Location of Study/Year Subjects (Reference)	Marker for Exposure to Traffic Source	Health Outcomes	Results	Comments
Munich/1993/4th grade children [279]	Traffic counts per 24 h in school districts/ distance ≤ 2km	Lung function Respiratory symptoms	(<1% decrease in peak flow and mid-expiratory flow)/25,000 vehicles per 24 h	Semiindividual study[280]; effect size very small—Questionable meaning for health
Birmingham, U.K./ 1994/children < 5 yr [281]	Nearness (200 m cutoff) to roads with >24,000 vehicles/24 h	Hospitalization for asthma	4–13% increase in odds of hospitalization	Case–control study with hospital and community controls; effects most closely related to density than distance from roadway
Netherlands/1996/adults and children [282]	Air pollution model based on type-specific number of vehicles, fuel source, mean traffic density, emission rates for specific engine types, local topography (e.g., street canyons), regional meteorology	Respiratory symptoms by mailed questionnaire	4–15-fold increased odds of wheeze, shortness of breath only for girls 0–15 yr Weak association with shortness of breath for adults	Comparison of subjects living along busy roads vs. quiet roads Adjusted for confounders that included SHS[1], indoor heating, moisture Data consistent with studies of SHS that show a female predominant effect best traffic-related exposure data of any study
Austria/1997/1st and 2nd grade children [124]	Eight communities without industrial sources of PM aerosol and differing levels of diesel traffic	Respiratory symptoms; NO_2 from fixed monitor as marker	Respiratory symptoms associated only with community level NO_2 (no O_3 or SO_2) Association of "ever asthma" with community-level diesel traffic and tire dust (tons/year)	Semiindividual study Effects adjusted for indoor sources and SHS[a] Correlation between ambient NO_2 and diesel

(*Continued*)

Table 24.8 (*Continued*)

Location of Study/Year Subjects (Reference)	Marker for Exposure to Traffic Source	Health Outcomes	Results	Comments
Netherlands/1997/schools and children 7–12 yr [122][b]	Six areas with different traffic density Distance from roadway Density of truck/auto traffic Indoor black smoke; NO_2; wind direction at schools	Lung function	1–6% decrease in flows at low lung volumes Effects largest for truck density and black smoke effects larger for children living <300 m from roadway	Effects adjusted for indoor sources, parental respiratory history, SHS, pets, home moisture effects: girls>boys effects on small airways similar to that seen for SHS
San Diego County/1999/childhood (<14 yr) asthma from Medical[c] database [283]	Average daily traffic flow near home by GIS	Number of claims for asthma-related medical visit	Increased occurrence of >2 visits for care for children with asthma for those living near high traffic flow (41,000 cars/day)	Case (asthma)–control (any other diagnosis) study Health implications of outcome not clear
London/1999/children 5–14 yr [284]	Traffic at centroid of post code of residence (GIS) Simple Euclidean distance to main road distance to main road with modeled peak/hour of >1000 vehicles computed (vehicle-meters/hour) on road within radius of 150 m	Emergency hospital visits for asthma and respiratory illness	No association between residence within 150 m of main road or traffic volume and respiratory admission	Case–control (nonrespiratory, noninjury acute admission All studies based on hospital admission have potential bias related to criteria for admission
Harvard Six Cities/2000/all ages [121]	Factor analysis of 15 daily ambient PM metals concentrations; selenium-coal marker; lead-mobile source marker	Mortality, 1979–1988	"Mobile source" associated with 3.4% increased mortality; "coal" 1.1%; no increase with crustal factor	Time series analysis with exposure expressed as mobile source, crustal, coal factors

Nottingham, U.K./2000/ primary and secondary schools [285][d]	1 km^2 grid around each school; traffic activity index (TAI — vehicle meters/d/km^2)	Current wheezing by questionnaire	No association in ecological or cross-sectional analyses Weak association (5% increased odds) for persistence of wheeze 7–8 yr later	Ecological, cross-section and longitudinal data Results inconclusive
Amsterdam/2001/all ages [287]	Residence near road with > 10^4 vehicles per day Day NO_2, NO, black smoke, O_3, CO, SO_2	Mortality 1987–1998	Black smoke > NO_2 associated with increased risk (BS 25% increase; NO_2 5%) for residences with >10^4 vehicles/day	Daily time series Black smoke, CO, and NO highest at "traffic-influenced" fixed monitors
Netherlands/2002/adults 55–69 yr [197]	Long-term average exposure to ambient pollutants based on regional background and pollutants from local sources (nearby streets) Distance of homes from streets (50, 100 m) with estimated exposure to black smoke and NO_2	Mortality 1986–1994	Black smoke and NO_2 associated with all-cause and cardiopulmonary mortality Larger effect seen with indicator of distance lived from major road	Cohort study started in 1986; imbedded case-cohort (with no loss to follow-up in sub-cohort) Detailed exposure assessment with source apportionment Individual covariate data (smoking, diet, etc.) Analyses raises point that measurement error in exposure assignment offsets benefits of specificity of exposure assignment; simple distance marker more efficient[e]

(*Continued*)

Table 24.8 (*Continued*)

Location of Study/Year Subjects (Reference)	Marker for Exposure to Traffic Source	Health Outcomes	Results	Comments
Netherlands/2002/birth cohort [289]	Pollutant concentrations at residences modeled based on GIS traffic data and 40 local monitoring sites Black smoke, NO_2, $PM_{2.5}$	Respiratory symptoms by questionnaires at age 2 yr	Weak associations with all measures of local ambient PM aerosol	Birth cohort study of asthma and mite allergy
California/2003/births [290]	1994–1996 traffic counts to calculate distance–weight traffic density (DWTD) based on dispersion of motor vehicle exhaust from roadways	Birth weight; preterm delivery	Association of preterm births with quintile of DWTD (8% increase in odds from lowest to highest) Largest effect seen for exposure during the 3rd trimester (39% increase); no effect for spring/summer 3rd trimester exposure Same results for zip code with highest CO, NO_2, and PM_{10}	Control for multiple factors that affect birth outcomes 3rd trimester effect most consistent with effects on somatic growth rather than organogenesis Fall/winter increased risk matches "PM season" for California Major limitation is lack of data on maternal smoking during pregnancy and other indoor sources; adjustment for previous low-birth-weight delivery could have biased results

[a] Second-hand (passive) smoke exposure.

[b] Results reported for respiratory symptoms from this study in Reference [123]. Truck density and black smoke associated with respiratory symptoms.

[c] California's Medicaid program.

[d] A later study of the same population by the same authors [286], showed association of increased wheezing and living 150 m from the roadway for children aged 4–16 yr. Most of the increase was observed for children who lived within 90 m of a roadway.

[e] Study in Erie County, NY [288], showed association with hospitalization for asthma in children and living within 200 m of a state highway.

metrics of exposure to mobile sources (residential or school distance from roadway and markers of vehicle combustion [black smoke as a surrogate for elemental carbon, NO_2 in the absence of fixed sources]) and the increased occurrence of respiratory symptoms in children, decreased lung function in children, and increased mortality across all ages. In many of the studies, the methods of exposure assessment are somewhat crude (many do not even take location of school or residence in relation to predominant wind direction) and the choice of health outcomes is of variable quality; nonetheless, the fact that distance from roadways has the most consistent association with health outcomes adds some coherence to a causal association [143]. Figure 24.18 shows the decay of black carbon as a marker of diesel exhaust from two freeways with different diesel vehicle loads in southern California [144]. At a distance between 80 and 150 m from the roadway (dependent on vehicle road), the levels of black carbon return to background. Given the diversity of methods of exposure assignment and types of health outcomes, only a qualitative conclusion is reasonable: that is, there is a strong association between exposure to traffic sources of ambient PM and a variety of health outcomes.

Indoor combustion sources: As noted above, indoor environments of a variety of types (home, work, vehicles) are important sources of anthropogenic PM aerosol in both developed [11,18,145] and developing countries [146]. No further comment is made with regard to human health effects that are attributable to indoor combustion sources, since in general they do not differ qualitatively from those that would be related to outdoor sources. The major issue in this regard is the relative contribution of indoor and outdoor sources to the overall burden of PM aerosol experienced by humans. A full discussion of this is beyond the scope of this chapter (see References [10,146,147]. The health effects related to second-hand exposure to cigarette smoke have been reviewed recently by the National Cancer Institute [148] and an update will appear in a new Report of the Surgeon General of the United States that has not yet been released.

Bioaerosols: Human health effects related to bioaerosols have been discussed briefly in Chapter 14 and the mechanisms for the noninfectious components of PM aerosols have been discussed above.

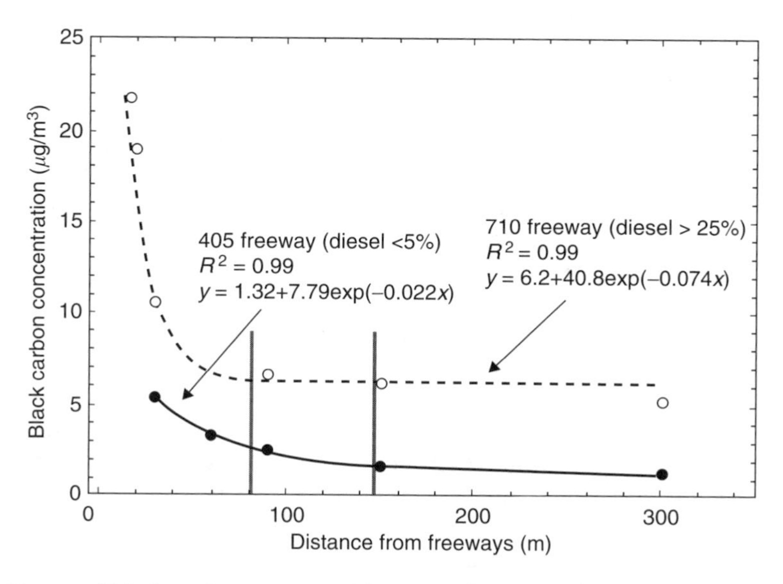

Figure 24.18 Decay of black carbon concentrations as a function of distance from two freeways in southern California with different loads of diesel vehicle traffic. *Source*: Zhu et al. [144].

Endotoxins are of particular interest, since they are ubiquitous in the environment either adsorbed onto the surfaces of particles generated through combustion processes (see above) or as part of indoor dust created by human activity and tracking of soil into homes and the presence of animal pets [149–151]. Inhalation of endotoxin leads to activation of proinflammatory cytokines, which leads to marked inflammation, increases in epithelial permeability, and evidence of activation of systemic inflammation [71]. Although aerosolized endotoxin has long been known as a source of lung disease in cotton (textile) workers [152] and swine handlers [153], more recent interest has focused on the complex role of endotoxin in the occurrence of IgE-mediated allergy and asthma [107]. As noted previously, the biological responses to endotoxin, in theory, could lead both to suppression of IgE-mediated responses as well as the worsening of the lung airway inflammation, which is a hallmark of asthma.

A number of studies have associated levels of house dust endotoxin with increased respiratory symptoms in infants [154] and the worsening of asthma that is independent of the levels of other common indoor allergens (e.g., house dust mite) [155,156]. In contrast, studies have shown that house dust levels of endotoxin are inversely related to the occurrence of IgE-mediated skin test positivity in infants [151] and to the occurrence of hay fever and skin test positivity in children 6–13 yr of age [157]. However, the reported cytokine responses related to other exposures appear to conflict. A study of asthma-prone infants [151] reported increased INF-γ production by CD4 and CD8 T cells with increased house dust endotoxin concentrations (expressed as EU/ml of extract), which is consistent with the finding of decreased prick skin test positivity in relation to house concentrations. The study of school-aged children [157] found decreased IL-12 and INF-γ with increased house dust (expressed as EU/m^2 of surface area sampled). These investigators invoked immune tolerance to endotoxin, which would explain the cytokine results but not necessarily the decreased prevalence of atopic disease. Several investigations have suggested that a specific polymorphism in the promotor region of the CD14 gene (see above for the role of CD14 in endotoxin biology) may increase the severity of allergy through decreased expression of CD14 and subsequent increase in IgE concentrations [107,158], although a more recent study did not confirm this observation [159]. Given the complex associations between CD14 levels and allergic manifestations at different developmental periods [160,161], it is difficult to reconcile these studies and provide a clearer picture of how the combination of genetic predisposition and developmental timing and intensity of exposure to endotoxin that is part of the indoor and ambient PM aerosol affects the risk of atopic allergy and asthma. The picture is further complicated by the findings from studies that have evaluated the effects of other microbial surface chemicals that can be found in PM aerosol. One study that evaluated the levels of endotoxin, (1→3)-β-*d*-glucan (extracellular polysaccharide of fungal origin), and allergens from mites and pets found that variability in peak expiratory flow (a measure of lung function) in children aged 7–11 yr was no longer associated with endotoxin after adjustment for mite allergy and the presence of dogs/cats, but the association for (1→3)-β-*d*-glucan persisted [162]. This result may have been influenced partly by overadjustment of pets in the endotoxin analysis, but it nonetheless points to the complexity of the study of endotoxin-related health effects in human populations.

Pollens are well-known sources of respiratory allergy (see Chapter 14) that exert their effects via IgE-mediated mechanisms similar to those described for asthma (Figure 24.16). Pollens exist in the air as pollen grains, which can penetrate into the respiratory tract or contact mucous membranes of the eye. However, it is the protein content of the grains that is responsible for the allergic responses, and pollen exists in the air as pollen grains and allergenic aerosols [163]. In addition to any direct effects of ambient PM on modulation of immunological responses, there is good evidence that pollen and allergenic aerosol agglomerate on to ambient PM [164], and the degree of agglomeration is proportional to the aromatic component of PM [165,166]. Pollen aerosol that is carried on PM may have enhanced access to immunologically competent cells as a consequence of the inflammation that is triggered by the ambient PM itself [166]. To the extent that aromatic compounds also upregulate IgE-mediated immune responses, it is possible that the allergenic potential of pollens could be enhanced. In addition, air pollutants can alter the allergenic content of plant pollen granules [167–169] as well as the antigenicity of proteins [165]. There are limited human data that have evaluated this possibility and they fail to find evidence of interactions with pollen counts and particles [170–173]. However, all of these studies assessed pollen concentrations in terms of standard counts (and not in terms of allergenic aerosol), which the studies cited above indicate may not be the relevant measurement. Moreover, the presence of other pollutants may modify these effects by decreasing the release of antigens [163], which further complicates studies of this problem in human populations. Probably the strongest evidence to suggest that there are human health consequences to interactions between pollen and PM comes from studies of cedar pollinosis in Japan, which suggest that the marked increase in allergy to cedar trees in Japan after 1964 can be traced to increases in mobile sources of air pollution [174]. A study reported by Dutch investigators observed that daily all-cause mortality and cardiovascular mortality were associated with average weekly concentrations of common pollens, an effect that was not confounded by air pollutants (low correlation between pollens and ambient air pollutants, which included black smoke), day of the week, or meteorological factors [175]. The explanation for these findings is not clear, as there was no evidence of interaction with ambient pollutants. The authors offer several speculations, some of which are supported by other data [175]. In contrast to these studies, a cross-sectional study in Switzerland found an association between traffic density near residences and sensitization to pollen allergen (Timothy grass) but not with symptoms of hay fever or the number of allergens to which residents (ages 18–60 yr) were sensitized [176]. The association was strongest for those who lived at their residences for 10 or more years.

24.5.1 Health effects associated with chronic exposure to PM

There is no universally agreed-upon duration that separates chronic from acute or subacute exposures to PM aerosol. This section defines chronic exposures as those to which humans are exposed over months to years or during defined periods of life events such as pregnancy. There is a vast literature based on cross-sectional, population-based studies that have associated chronic exposures to elevated pollutants and the occurrence of human disease. Due to the difficulty of the evaluation of such data with respect to inferences about causality, this section is restricted to more recent cohort studies of mortality and studies of effects on birth outcomes (these latter studies can be considered as retrospective cohort studies in which the cohort is defined by onset of pregnancy and birth and exposure histories are reconstructed after the fact). The focus is on data related to outdoor PM aerosol. Indoor sources are not discussed specifically — a complete discussion of health effects related to second-hand tobacco smoke is beyond the scope of this chapter. This latter subject is reviewed comprehensively in two recent publications from the National Cancer Institute [17,148] and is the subject of a soon-to-be-released Report of the Surgeon General.

24.5.1.1 Effects of exposure to ambient PM and birth outcomes (Table 24.9)

In 1992, investigators from the Czech Republic reported that annual ambient levels of TSP, measured in 46 of 85 administrative districts in the Czech Republic, were associated with an increased risk of infant mortality [177]. Studies in the United States [178] and a subsequent study in the Czech Republic confirmed these findings [179]. Two studies of the relationship between daily changes in TSP [180] and $PM_{2.5}$ [181] also observed

Table 24.9 Summary of Effects of Ambient PM Aerosol on Birth and Neonatal Outcomes

Outcome	PM species measured[a]	Location	References
NON-ACUTE EXPOSURE			
Neonatal (within 1st year) mortality	1. Total suspended particulates (TSP) (SO_2, NO_x)—cross-sectional, no personal exposure estimates	Czech Republic 1986–1988	[177]
	2. PM_{10} first 2 months of life	U.S. (NCHS 1989–1991)	[178]
	3. TSP (SO_2, NO_2) — monthly exposure for each month of pregnancy	Czech Republic 1989–1991	[179]
	4. Distance from a coke works within 7.5 km (no effect observed)	England, Wales, Scotland 1981–1992	[291]
Birth outcomes[b]			
1. LBW (+)	1. TSP and SO_2 exposure during each trimester and weekly prenatal moving averages	Beijing, China 1988–1991	[184]
	2. TSP (SO_2, NO_x) — ecological study (SO_2 effect only)	Czech Republic 1986–1988	[292]
2. SB (−), LBW (+)	3. PM_{10}, $PM_{2.5}$ (monthly exposure estimates)		
	4. PM_{10}, CO, NO_2, O_3 (monthly exposure estimates)		
3. IUGR (+)	5. Modeled TSP (SO_2)	Czech Republic 1994–1996	[293]
4. PTD (+)	6. PM_{10}, $PM_{2.5}$, $PM_{2.5}$ PAH (monthly average exposures)	Southern California 1989–1993	[183]
5. VLBW (+)	7. TSP (SO_2, NO_x) (trimester-specific exposure)	Georgia, U.S. 1988	[294]
6. IUGR (+)	8. TSP (SO_2, NO_2, CO, O_3) 1st and 3rd trimester-specific exposure)	Czech Republic 1994–1998	[182]
7. LBW (+), PTD (+) IUGR (−)	9. PM_{10}, (CO, NO_2, O_3) month-specific exposure estimate — association only with CO	Czech Republic 1990–1991	[295]
8. LBW (+)	10. Distance-weighted traffic density in relation to residence (based on dispersion of vehicle exhaust within 750 m radius of residence)	Seoul, Korea 1996–1997	[296]
9. Birth defects: Cardiac (+), Orofacial (±)	11. Personal PAH during 2 days of the 3rd trimester (nonsmoking subjects)	Southern California 1987–1993	[297]
10. PTD (+), LBW (±)		Southern California, 1994–1996	[290]
11. LBW (+)		New York City, U.S.	[186]
Daily (Acute) Exposure3			
Neonatal mortality	1. $PM_{2.5}$ (O_3, NO_2, NO, SO_2)	Mexico City 1993–1995	(181)
Intrauterine mortality	1. Daily PM_{10} (NO_2, SO_2, CO, O_3) last 28 weeks of pregnancy	São Palo, Brazil 1991–1992	(180)
Birth outcomes			
1. GA (+), PTD (+)	1. TSP, SO_2	Beijing, China 1988	(298)

[a] Pollutants in parentheses are the non-PM pollutants also evaluated.

[b] Studies often assess more than one outcome. A "+" after the outcome indicate that an adverse association was seen. A "−" indicates that the outcome was evaluated but no or an imprecisely estimated association was observed. IUGR=intrauterine growth retardation, LBW=low birth weight, VLBW=very low evaluated but no or an imprecisely estimated association was observed. IUGR=intrauterine growth retardation, LBW=low birth weight, VLBW=very low birth weight, SGA=small for gestational age, PTD=preterm delivery, GA=gestational age, DP=duration of pregnancy, SB=stillbirth.

increased mortality risk with increased concentrations. The extent to which PM aerosol contributed uniquely to the increased risks could not be determined by any of these studies, since associations were also found for gaseous pollutants (Table 24.9). Studies from many parts of the world have identified associations between various birth outcomes (see footnote b of Table 24.9) and exposure to PM aerosol, measured most frequently as TSP but also based on PM_{10} and $PM_{2.5}$ and personal exposure to airborne PAHs (Table 24.9). While there is consistency with regard to the fact that PM aerosol affects birth outcomes, there is less consistency with regard to the specific parameters and the critical developmental periods. Figure 24.19 presents data from three representative studies from three different countries. The studies from the Czech Republic (Figure 24.19A) [182] and southern California [183] found effects of exposure during the first trimester, but the former study failed to find effects of exposure later in pregnancy. A study from Bejing, China [184], observed that exposure to TSP during the third trimester increased the risk for low-birth-weight babies without a major shift in the overall population mean. Since the third trimester is the time of maximum somatic growth, this latter study would suggest that there is a susceptible subset of pregnancies for which exposure to ambient PM may be particularly harmful. A study from the Czech Republic evaluated the effects of ambient levels of SO_2 on fecundity [185] and observed that exposure to increasing levels of SO_2 in the 2 months prior to attempts at getting pregnant was associated with decreased conception the first unprotected menstrual cycle. Given the role of SO_2 in the formation of particle sulfates, these findings are relevant in the context of health effects of ambient PM aerosol. A recent study of nonsmoking pregnant women, which included personal monitoring of PAHs, observed associations between personal PAH exposure and decreased birth weight, birth weight and head circumference [186]. Analyses were adjusted for dietary PAHs and continine as a measure of exposure to second-hand tobacco smoke. A very recent study evaluated the effects on the sharp fall in TSP levels during the 1981–1982 recession in the United States on birth outcomes [187]. These investigators estimated that over 70% of the overall reduction in infant mortality during the first year of life could be attributed to the average 15 $\mu g/m^3$ decline in TSP levels that occurred during the recession. This estimate was robust to many different model specifications. Reductions were observed for deaths in the first 24 h, first 28 d, and first year of life [187].

The fact that studies show inconsistencies in the findings related to specific outcomes does not detract from the striking extent to which the findings agree on the general effects. Differences in PM metrics, use of central monitors for exposure monitoring, inherent population differences, and individual differences in susceptibility likely all contribute to the differences observed. The associations observed with PAHs are consistent with oxidative stress as well as direct toxic effects on pregnancies. The potential implications for these findings are highlighted by the observations that asthma appears to be more common in children of low birth weight and the observation of one study that found that asthmatic children who were born either before 37 weeks gestation or of low birth weight (<2500 g) had a substantially increased risk of symptoms and reduced lung function in responses to increased levels of summertime air pollutants in the eastern half of the United States [188]. Summertime air pollution in this part of the United States is characterized by high PM (especially sulfates) and ozone.

24.5.1.2 *Increased mortality related to long-term exposure to PM aerosol*

While there had been many cross-sectional studies that suggested that air pollution might be associated with increased mortality, the seminal study in 1993 by Dockery et al. [189] colleagues gave new credence to such an association. This prospective cohort study, based on 14–16 yr of follow-up in six U.S. cities, reported increased risk of death from a

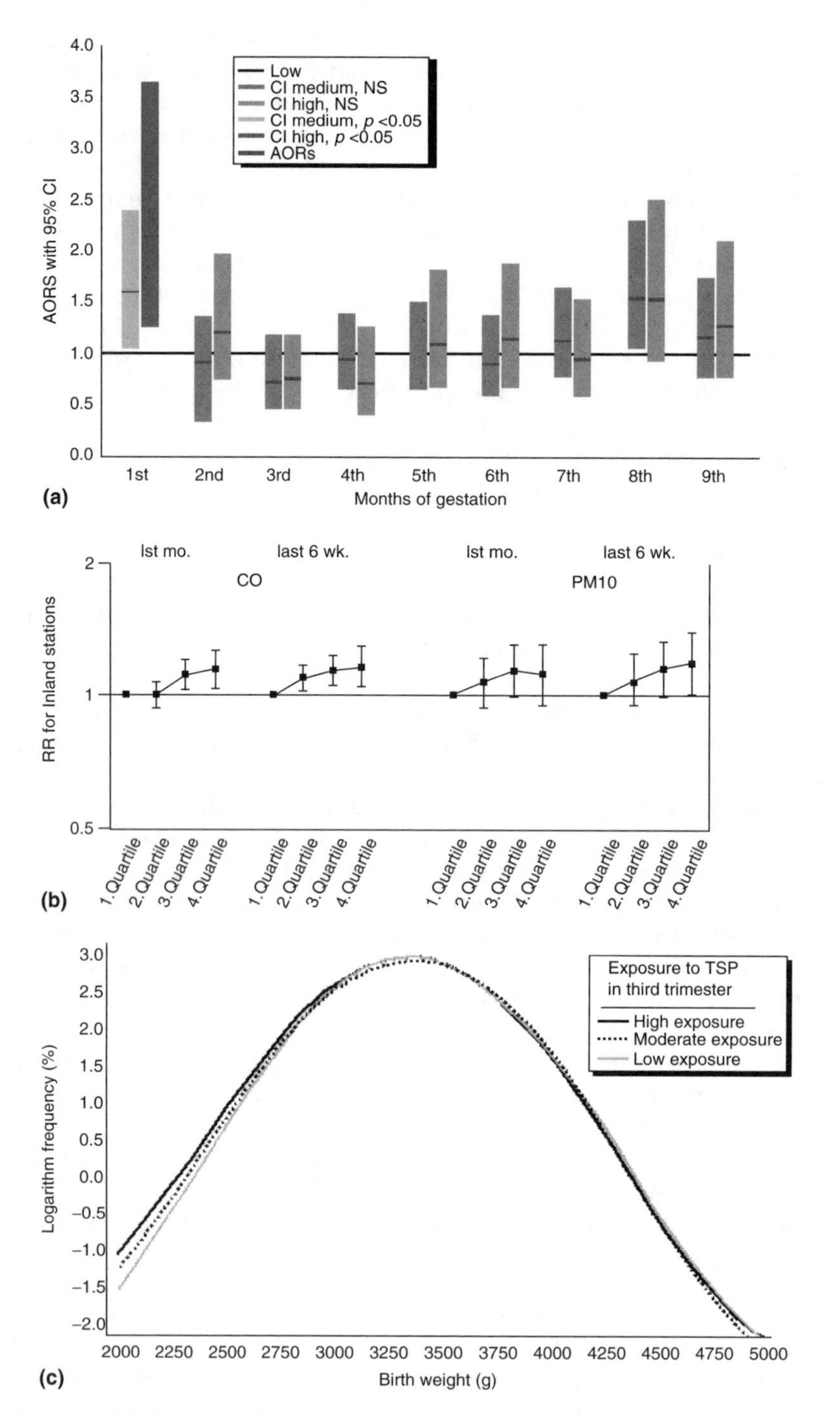

Figure 24.19 Timing of exposure to ambient PM aerosol and effects on birth outcomes. (**A**) Adjusted odds ratios (AOR) for associations of trimester-specific estimates of exposure to TSP on IUGR in the Czech Republic [182]. (**B**) RRs for PTD based on quintiles of exposure to CO and PM_{10} during the first and last 6 weeks of pregnancy by quintile of exposure in southern California, U.S. [183]. (**C**) Effect of exposure to TSP in the third trimester on the distribution of BW in Beijing, China [184].

variety of causes that were associated with mean levels of PM_{10} and $PM_{2.5}$ (Table 24.10, rows labeled "original"). The results were not confounded by smoking history, body mass index, or education (see Table 24.2; Reference [189]). In 1995, Pope et al. [190] report on the mortality experience of the American Cancer Society Study (ACS) cohort in relation to air pollution. This study used 151 metropolitan statistical areas and observed results broadly similar to those of the Harvard Six Cities study reported by Dockery (see Table 24.11, rows labeled "original"). Both studies pointed to $PM_{2.5}$ as the relevant mass fraction. In response to skepticism that the effect sizes were so large relative to results

Table 24.10[a] RRs (95% CI) of Mortality by Cause of Death Associated with an Increase in Fine Particles in Risk Models with Alternative Time Axes in the Reanalysis of the Six Cities Study[b]

Risk Model	Calendar Year	Age
All Causes (100%)[c]		
Base[d]	1.33 (1.14–1.54)	1.33 (1.15–1.55)
Original	1.29 (1.11–1.50)	1.29 (1.11–1.50)
Full	1.27 (1.09–1.49)	1.27 (1.09–1.48)
Extended	1.28 (1.09–1.49)	1.27 (1.09–1.48)
Cardiopulmonary Disease (54%)		
Base	1.39 (1.13–1.70)	1.39 (1.14–1.71)
Original	1.35 (1.10–1.66)	1.34 (1.09–1.65)
Full	1.31 (1.06–1.62)	1.30 (1.05–1.60)
Extended	1.32 (1.07–1.63)	1.31 (1.06–1.61)
Cardiovascular Disease (47%)		
Base	1.43 (1.15–1.78)	1.44 (1.16–1.79)
Original	1.41 (1.13–1.76)	1.40 (1.12–1.74)
Full	1.38 (1.10–1.72)	1.35 (1.08–1.69)
Extended	1.39 (1.11–1.73)	1.37 (1.09–1.70)
Respiratory Disease (7%)		
Base	1.11 (0.62–1.97)	1.10 (0.63–1.95)
Original	0.93 (0.51–1.71)	0.95 (0.53–1.72)
Full	0.89 (0.47–1.67)	0.94 (0.51–1.73)
Extended	0.88 (0.47–1.64)	0.93 (0.51–1.69)
Lung Cancer (8%)		
Base	1.53 (0.91–2.55)	1.64 (0.99–2.72)
Original	1.31 (0.76–2.25)	1.53 (0.90–2.60)
Full	1.30 (0.76–2.23)	1.42 (0.84–2.42)
Extended	1.29 (0.75–2.22)	1.45 (0.85–2.47)
Other Cancers (20%)		
Base	1.05 (0.74–1.48)	1.04 (0.73–1.47)
Original	1.04 (0.73–1.47)	1.02 (0.72–1.45)
Full	1.11 (0.78–1.59)	1.09 (0.77–1.55)
Extended	1.10 (0.77–1.57)	1.08 (0.76–1.54)

[a] Adapted from [191].

[b] RRs were calculated for a change in the pollutant of interest equal to the difference in mean concentrations between the most polluted city and the least polluted city; in the Six Cities Study, this difference for fine particles was 18.6 $\mu g/m.^3$

[c] Percentages are percentages of all deaths

[d] Base: air pollution, no covariates; original: used by original investigators; full: many covariates; extended: most parsimonious model.

from studies of the effects of daily changes in PM aerosol on daily mortality (see the section below), an extensive reanalysis of these data was undertaken by investigators from Health Canada and the University of Ottawa [191]. Extensive additional modeling (Tables 24.10 and 24.11) and extensive spatial analyses confirmed the results reported by the original investigators. These investigators also evaluated effect modification and observed a consistent increased risk associated with lower levels of education in both studies. A subsequent analysis of the ACS data that included a doubling of the follow-up time to more than 16 yr, threefold more deaths, and more extensive $PM_{2.5}$ data confirmed the earlier findings (Figure 24.20) [192]. Effect modification by education again was

Table 24.11[a] RRs (95% CI) of Mortality by Cause of Death Associated with an Increase in Fine Particles or Sulfate in Risk Models with Alternative Time Axes in the Reanalysis of the ACS Study[b]

	Time Axis			
	Calendar Year		Age	
Alternative Risk Model	Fine Particles	Sulfate	Fine Particles	Sulfate
All Causes (100%)				
Base	1.27 (1.18–1.37)	1.26 (1.19–1.33)	1.26 (1.17–1.35)	1.25 (1.18–1.32)
Original	1.18 (1.10–1.27)	1.16 (1.10–1.23)	1.18 (1.10–1.27)	1.16 (1.10–1.22)
Full	1.17 (1.09–1.26)	1.15 (1.08–1.21)	1.16 (1.08–1.25)	1.14 (1.07–1.20)
Extended	1.18 (1.09–1.26)	1.15 (1.09–1.21)	1.17 (1.09–1.25)	1.14 (1.07–1.20)
Cardiopulmonary Disease (50%)				
Base	1.41 (1.27–1.56)	1.39 (1.28–1.50)	1.41 (1.27–1.56)	1.38 (1.27–1.49)
Original	1.30 (1.18–1.45)	1.27 (1.17–1.38)	1.30 (1.18–1.45)	1.27 (1.17–1.37)
Full	1.28 (1.15–1.42)	1.25 (1.15–1.35)	1.28 (1.15–1.42)	1.24 (1.14–1.34)
Extended	1.30 (1.17–1.44)	1.25 (1.16–1.36)	1.29 (1.17–1.43)	1.25 (1.15–1.35)
Cardiovascular Disease (43%)				
Base	1.47 (1.32–1.65)	1.47 (1.35–1.60)	1.46 (1.31–1.63)	1.46 (1.34–1.59)
Original	1.36 (1.22–1.52)	1.36 (1.25–1.48)	1.36 (1.22–1.52)	1.35 (1.24–1.47)
Full	1.34 (1.20–1.49)	1.33 (1.22–1.45)	1.33 (1.19–1.48)	1.32 (1.21–1.43)
Extended	1.35 (1.21–1.51)	1.34 (1.23–1.46)	1.34 (1.20–1.50)	1.33 (1.22–1.44)
Respiratory Disease (7%)				
Base	1.07 (0.80–1.42)	0.94 (0.76–1.17)	1.09 (0.82–1.45)	0.95 (0.76–1.18)
Original	1.00 (0.76–1.33)	0.83 (0.67–1.04)	1.01 (0.76–1.34)	0.85 (0.68–1.05)
Full	0.96 (0.72–1.27)	0.81 (0.65–1.01)	0.99 (0.74–1.31)	0.82 (0.66–1.03)
Extended	0.98 (0.74–1.30)	0.82 (0.65–1.02)	1.00 (0.76–1.33)	0.83 (0.66–1.03)
Lung Cancer (8%)				
Base	1.23 (0.96–1.57)	1.63 (1.35–1.97)	1.21 (0.95–1.54)	1.62 (1.34–1.95)
Original	1.02 (0.80–1.29)	1.36 (1.13–1.65)	1.02 (0.80–1.30)	1.36 (1.12–1.64)
Full	0.99 (0.78–1.26)	1.32 (1.09–1.60)	0.98 (0.77–1.25)	1.31 (1.09–1.59)
Extended	1.00 (0.79–1.28)	1.33 (1.10–1.61)	0.99 (0.78–1.26)	1.32 (1.09–1.60)
Other Cancers (27%)				
Base	1.18 (1.03–1.36)	1.15 (1.03–1.28)	1.17 (1.02–1.34)	1.14 (1.02–1.26)
Original	1.14 (0.99–1.30)	1.10 (0.99–1.23)	1.13 (0.98–1.29)	1.10 (0.99–1.22)
Full	1.14 (1.00–1.31)	1.10 (0.99–1.23)	1.13 (0.98–1.29)	1.09 (0.98–1.21)
Extended	1.14 (0.99–1.31)	1.10 (0.99–1.22)	1.12 (0.98–1.29)	1.08 (0.97–1.21)

[a] Adapted from Reference [191].

[b] RRs were calculated for a change in the pollutant of interest equal to the difference in mean concentrations between the most polluted city and the least polluted city; in the ACS study, this difference for fine particles was 24.5 $\mu g/m^3$, and for sulfate it was 19.9 $\mu g/m^3$.

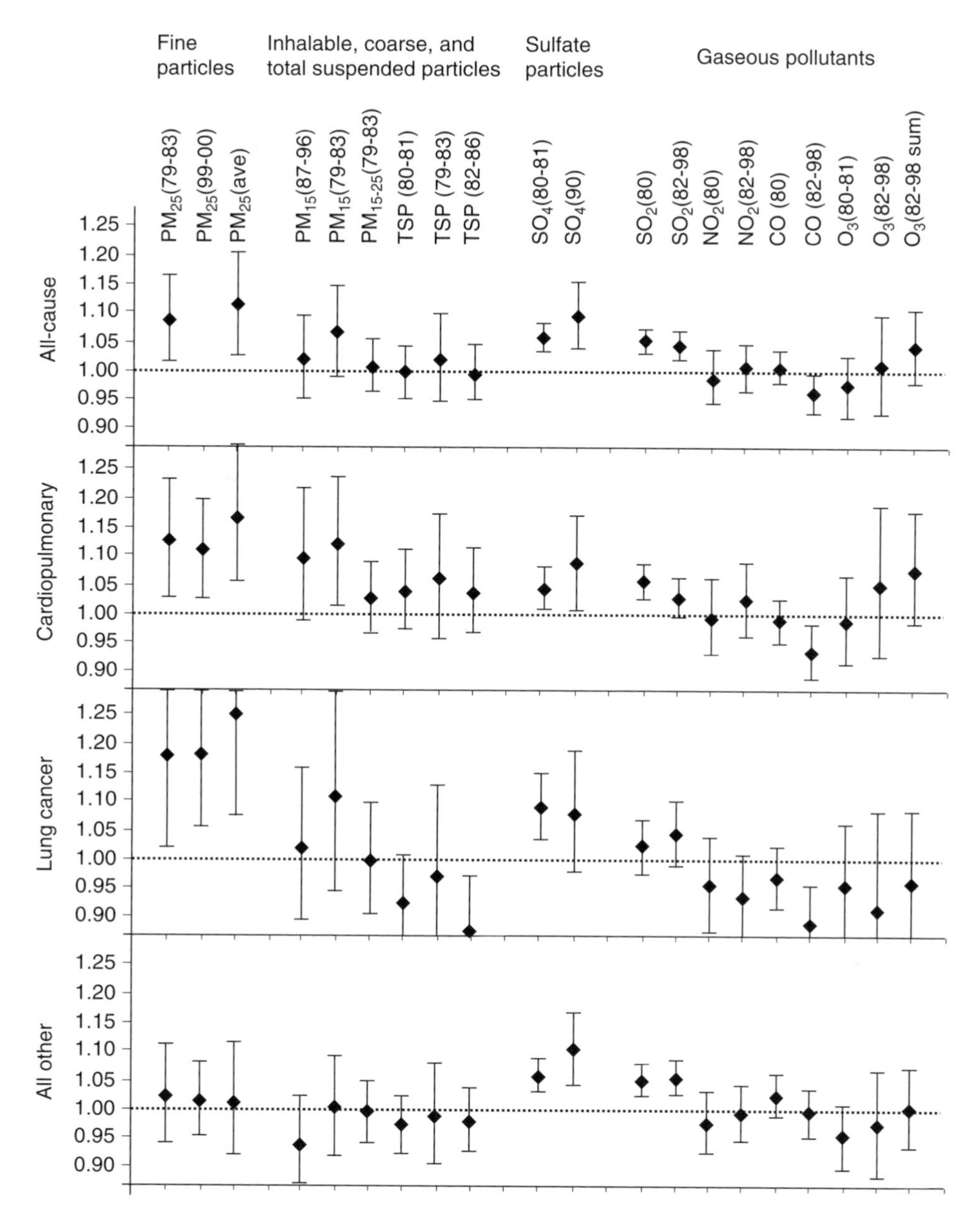

Figure 24.20 Summary of associations between various indicators of PM aerosol and gaseous pollutants for the ACS study over different averaging periods. From Pope et al. [192] as presented by U.S. EPA (U.S. Environmental Protection Agency, 2002, 1074 [264]).

observed, and the effects were largest and most consistent for $PM_{2.5}$ (similar when expressed as sulfate). Over the 21 yr of follow-up each, 10 $\mu g/m^3$ increase in average $PM_{2.5}$ across the metropolitan statistical areas was associated with the following increases in mortality: all-cause 5% (95% CI: 2–11%); cardiopulmonary 9% (3–16%), and lung cancer 14% (4–23%) (see Table 24.2 of Reference [192]). The effects were approximately of the same magnitude when based on levels over the years 1979–1983 and 1999–2000. A similar finding with respect to period of exposure was reported for the Harvard Six Cities data [193].

Another cohort study that has contributed important data on the effects of long-term exposure to ambient PM aerosol is the Adventist Health Study of Smog (AHSMOG), which recruited subjects from the Adventist community in California [194,195]. This study had the advantage that cigarette smoking was virtually nonexistent as a result of religious prohibition and the fact that investigators had developed algorithms for the estimation of personal exposure of subjects to ambient PM [196]. Unlike the Harvard and ACS studies, AHSMOG reported exposure as the number of days on which PM_{10} was above 100 $\mu g/m^3$ as well as in terms of mean PM_{10}; results were comparable for both metrics (see Table 3 of Reference [194]). The initial publication from AHSMOG reported increased risk of death (1977–1992) from "any mention of non-malignant respiratory disease" for both sexes (1.28, 1.10, males and females, respectively, per interquartile range of PM_{10} of days [42.6 IQR] above 100 and 24.1 $\mu g/m^3$ for mean PM_{10}). Lung cancer risks were associated with PM only for males. Similar effects were observed for nonmalignant respiratory deaths in both sexes and lung cancer in males. Mean nitrogen dioxide and sulfur dioxide were associated with lung cancer in females. Thus, this study gives a less clear picture than the Harvard and ACS studies with regard to PM effects [189,190,192]. A later publication tried to partition the PM effects between $PM_{2.5}$ (partly estimated from airport visibility data) and PM_{10-25} [195]. Effects were largely confined to the fine fraction, with risk ratios similar to those for PM_{10}.

Subjects from the prospective study of diet in the Netherlands have been evaluated to determine the effect of living near roads with heavy traffic and mortality [197]. The general results from this study were summarized in Table 24.8. Mortality risk increased for those living near major roads (100 m from a freeway or 50 m from a major nonfreeway road) compared to those who did not (e.g., RR for cardiovascular death per 10 $\mu g/m^3$ increase in black smoke: near, 1.95 (1.09–3.51); not near, 1.34 (0.68–2.64) [197].

Thus, there is agreement among the prospective cohort studies with regard to associations between PM and mortality. The differences in the results in relation to gaseous pollutants remain to be explained. Given the difference in the summertime pollutant mixtures between California and more eastern portions of the United States [19], it is not likely that ozone is acting as a surrogate for PM-related pollutants as has been suggested for the eastern United States [198]. In addition, although both the Harvard study and the ACS studies pointed to sulfate as the potentially relevant component of the PM aerosol that is associated with mortality, neither of these studies, nor the AHSMOG study, conducted a comprehensive evaluation of this issue. Thus, the relevant physical and chemical components of the PM aerosol that are related to these effects remain virtually unexplained.

24.5.1.3 *Atopic allergy and asthma*

Over the past several decades, there has been a steady increase in the prevalence across all age groups, especially in the very young [109]. While there is a large body of data to support the fact that asthma symptoms are worsening on days with increased PM aerosol (see the section on Acute effects), there are few prospective studies that have provided data on the extent to which PM aerosol contributes to the onset of new asthma or the onset of related atopic allergy as might be expected from data on PM mechanisms discussed previously.

In 1992, the Medical Research Council (MRC) of Great Britain reported a 36 yr follow-up of a 1946 national birth cohort [199] (Table 24.12). Exposure was based on annual coal consumption during the first 11 yr of life. Although not a direct measure of PM aerosol, domestic coal consumption in the United Kingdom was a major contributor to PM through the time of the great London fog of 1954 [200]. "Air pollution attributable risk" for asthma/wheeze at age 36 yr was 6.7 and 8.8% for exposure between ages birth and 2 to 11 yr, respectively. This study is unique in that it tries to partition risk between air pollution,

Table 24.12 Association Between Ambient PM Aerosol and Onset of Asthma

Cohort	Follow-up Time	Exposure Metric	Findings	Reference
Medical Research Council, National Survey of Health and Development 1946 birth cohort — national sample	36 years multiple contacts	Annual coal consumption between birth and age 11 yr	Self-reported wheeze and asthma at age 36 yr: Attributable risk[a] for "air pollution" exposure between ages 0 and 2 yr = 6.7% Attributable risk[b] for "air pollution" exposure between ages 2 and 11 yr = 8.8%	(199)
Adventist Health Study of Smog Seventh Day Adventists who from various locations in California (90% southern California) enrolled in 1977	10 yr — 2 contacts (20 yr residential histories)	Estimated personal exposure to TSP, ozone, SO_4	Self-report of new physician diagnosis of asthma RR 2.9 (95% CI: 1.0–7.6) per 7 $\mu g/m^3$ — unit increase in mean annual SO_4 exposure[c] Effects not seen for new nonasthmatic onset obstructive airways or chronic bronchitis Unable to separate SO_4 association from that with O_3	(203)
Adventist Health Study of Smog	15 yr — 3 contacts	Estimated personal exposure to SO_4, PM_{10}	Self-report of new physician diagnosis of asthma Only association observed for 8-h average ozone for females	(204)
Children's Health Study Children up to 16 from 12 southern California communities	Up to 5 yr-annual contact	PM_{10}, $PM_{2.5}$, NO_2, ozone; Classified as high/low O_3 and PM based on 4-yr means	New onset self-report of doctor diagnosed asthma Associations limited to children who reported participation in three or more outdoor sports High PM communities:[d] RR=2.0 (95% CI: 1.1–3.6); low PM communities: 1.7 (0.9–3.2) High O_3 communities: RR=3.3 (1.9–5.8); low O_3 communities: RR=0.8 (0.4–1.6)	(206)

[a] Estimate adjusted for history of childhood lower respiratory illness, parental history of bronchitis, low socioeconomic status. No data given related to maternal smoking during pregnancy or during postnatal and childhood years.

[b] Adjusted for same risk factors as for birth to 2-yr exposure as well as cigarette smoking at age 36-yr. Attributable risk for smoking = 39.7% for comparison.

[c] Adjusted for years lived with a smoker, history of past lower respiratory illness before age 16, sex, age and education.

[d] The same communities were ranked as high for PM_{10}, $PM_{2.5}$, and NO_2. 4-year medians: LOW — PM_{10} ($\mu g/m^3$)=25.1, $PM_{2.5}$=7.6; High — PM_{10}=39.7, $PM_{2.5}$=21.4. All models adjusted for SED, family history of allergy and/or asthma, maternal smoking, body mass index.

smoking, and history of childhood respiratory illness in a single population in which all of these exposures were measured repeatedly over a long period of time. The major limitation of this study was the lack of data on maternal smoking during pregnancy and childhood, both of which have been shown to be important risk factors for asthma [201,202]. An early report from the Adventist Health Study reported an association between estimated personal exposure to sulfate over a 20 yr period and the new report of a doctor diagnosis of asthma that was specific for asthma [203] (Table 24.12). However, a subsequent report, based on an additional 5 yr of observation failed to confirm these findings [204]. In this latter study, new onset of asthma was associated only with mean ozone concentrations in females. Earlier reports from this cohort noted that correlations between ozone and PM aerosol (measured as TSP early on the cohort) made it difficult to distinguish PM and ozone-related health effects [205]. A recent report from the Children's Heath Study based on up to 5 yr of follow-up of children 9–16 yr from 12 communities in southern California failed to find a clear association between PM aerosol and the onset of new asthma [206] (Table 24.12). An association with ozone was observed for children who participated in three or more outdoor sports (Table 24.12). Other reports based on a follow-up of this cohort identified important indoor and familial factors as additional determinants of new onset asthma [207,208], but no attribution of risk was attempted between these latter exposures and ambient PM or other ambient pollutants. Although the mechanisms by which PM aerosol could contribute to the onset of asthma in susceptible individuals is compelling, the data in Table 24.12 are not so compelling. As the MRC study indicates, if PM aerosol and/or other ambient or indoor pollutants do contribute to the risk of the asthma etiology, the effect is likely small and will be difficult to sort out from other important exposures such as allergens and fungi. A further complication relates to the observation that the "epidemic" of asthma may have peaked, and asthma prevalence is declining (Figure 24.21). If this is the case, and the mechanisms related to PM aerosol discussed above are operative, then the focus of epidemiological studies will have to shift more intensely to identification of the susceptible profiles with large populations.

A series of studies from Germany indicates that the connection between exposure to PM aerosol and the occurrence of asthma and atopic allergy may involve different mechanisms or a different time course in populations. With the fall of the East German Republic (1989) and the unification of Germany (1990), there was a sharp decline in levels of SO_2 (precursor of secondary sulfate aerosol) [209]. Early studies indicated that the prevalence of atopic diseases and asthma were *less* frequent in the old East Germany Republic and that symptoms of bronchitis were more frequent [210]. Surveys carried out several years after reunification observed that prevalences of atopic allergy and positive prick skin tests were now similar in former East (Dresden) and West German (Munich) cities, but that the prevalence of asthma and reactive airways (an asthma-associated phenotype) were still more common in the western city [209]. However, at least one study has raised the question that diagnostic bias, in part, may be playing a role in the increased reported prevalence of atopic disease in the areas of former East Germany [211]. Cross-sectional surveys of children were conducted in two formerly highly polluted areas (combustion of brown coal with high sulfur content and heavy metal containing dust from smelters) and one area of minimal pollution in areas of former East Germany in 1992–1993 and 1995–1996. No changes were noted in the self-reported prevalence of asthma or atopic disease. However, physician-diagnosed allergy increased without a concomitant change in the prevalence of specific IgE to a variety of common aeroallergens. A decrease in physician-diagnosed bronchitis was also observed. These results conflict with a study of similar design carried out in Leipzig in 1991–1992 and 1995–1996 in which the prevalence of sensitization followed that of reported symptoms [212]. Whether explanation of the differences is due to technical problems with the skin testing device in Leipzig, as claimed by one set of

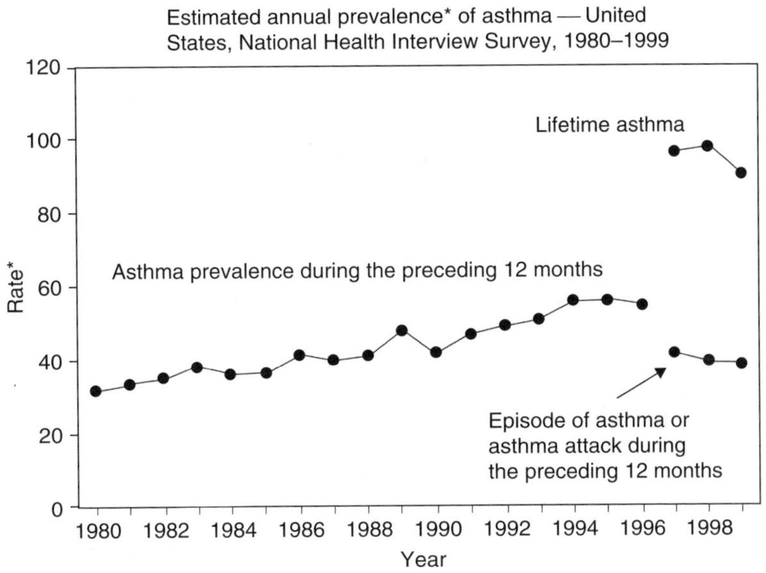

Figure 24.21 Annual self-reported prevalence of current asthma based on the National Health Interview Survey [109]. *Note*: Format of questions changed after 1996. Data demonstrate plateau in reported prevalence of asthma in the preceding 12 months (current asthma) and report of asthma at any time during life.

investigators [211], or is due to differences in air pollutant composition, population differences or chance cannot be determined from the data. In aggregate, these studies point to the difficulty in linking human health outcomes with the PM aerosol as would be expected from the data on mechanisms.

24.5.1.4 Cancer

As noted in Figure 24.20, PM aerosol has been associated with an increased risk of cancer. By and large, this excess risk has been noted for cancer of the lung [189,192]. The specific components of PM that are responsible for this excess risk have not been elaborated fully. However, several exhaustive reviews have been conducted related to the role of PM aerosol from diesel engines in the induction of cancer [213,214]. EPA provided a model for the hypothetical pathways that may be involved in carcinogenesis (Figure 24.22) and is based on concepts presented in Section 24.3. Most of the evidence to a carcinogenetic potential in humans is derived from occupational studies. EPA reviewed 22 such studies (Section 7.2 of Reference [214]) and estimated the pool relative risk at between 1.33 and 1.47. Given the much higher levels of exposure that occur in occupational environments relative to ambient environments, it is difficult to assess the risk beyond that which has been presented for lung cancer in the epidemiological studies cited previously. EPA has attempted to assess general population risk based on the conversion of exposures observed in occupational settings, for which risk estimates are available, to those experienced by the general population (Figure 24.23). Based on a variety of assumptions related to the data in Figure 24.23 (see pages 8–13 to 816 of Reference [214]), EPA estimated that the lifetime risk of lung cancer from exposure to ambient diesel PM could vary between 10^{-6} and 10^{-4}, although a zero risk could occur due to an unmeasured threshold.

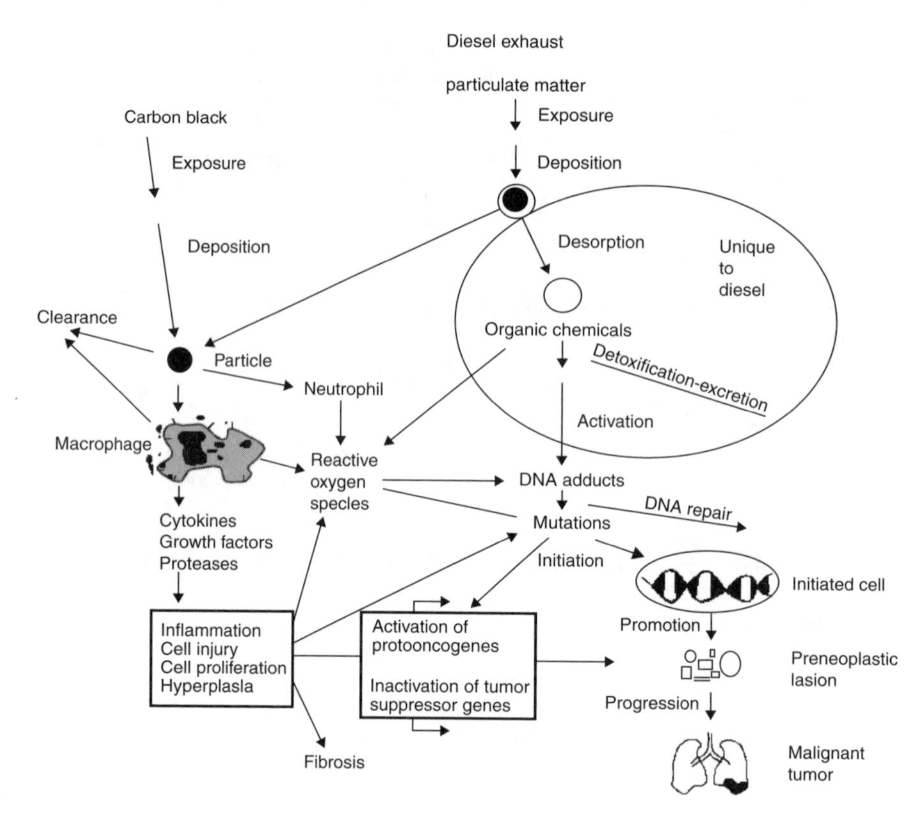

Figure 24.22 Postulated mechanism for carcinogenesis related to exposure to high-level diesel exhaust concentrations in rats (Figure 7.3 from U.S. Environment Protection Agency [214].

24.5.2 Health effects associated with acute and subacute exposures to PM

There is an enormous body of research that documents the association between short-term changes, usually measured over the week before a defined health outcome, and health effects. Effects on mortality, worsening of asthma and respiratory symptoms, and effects on lung function have been studied most widely. Perforce, this review is limited to a representative sample of studies that have demonstrated health effects and some of the critiques that have been levied against the validity of the associations. In addition, this section covers only the data related to the associations between daily changes in air pollution and daily mortality for several reasons: (1) mortality and years of life lost is an endpoint of major public health significance; and (2) the most extensive analyses on the effects of study design and statistical analysis techniques on estimates of effect have been carried out in relation to the mortality data. Excellent summaries of the large database related to worsening of asthma and other acute respiratory illnesses can be found in a number of the U.S. EPA references

24.5.2.1 Mortality

For obvious reasons, the effects of short-term increases in ambient PM have received more attention than any other PM aerosol-related health effect. The most famous example of the mortality associated with acute increases in PM aerosol is the London fog episode in December, 1952, which was responsible for both acute and subacute morality that accounted for approximately 13,000 excess deaths from the start of the fog through March 1953 [215]. Bell and Davis [215] estimated that a 0.10 ppm increase in weekly SO_2 (surrogate for the PM aerosol) was associated with a relative risk (RR) of mortality of 1.31 (95%

Table 8–1. DPM exposure margins (ratio of occupational ÷ environmental exposures)

Occpational group	Estimated occupational exposure/concentration - - - - - - - - - - Environmental equivalent[a]	Exposure margin ratio – average environmental Exposure for 0.8 μg/m^3 of environmental exposure[b]	Exposure margin ratio – average environmental Exposure for 4.0 μg/m^3 of environmental exposure[b]
public transit workers	15–98 μg/m^3 - - - - - - - - - - 3–21 μg/m^3	4–26	0.8–5
U.S. railroad workers	39–191 μg/m^3 - - - - - - - - - - 8–40 μg/m^3	10–50	2–10
Fork lift operators	7–403 μg/m^3 - - - - - - - - - - 1–85 μg/m^3	2–106	0.37–21
High-end boundary estimate	1200 μg/m^3 - - - - - - - - - - 252 μg/m^3	315	63

[a] Equivalent environmental exposure = occupational exposure × 0.21 (See Chapter 2, Section 2.4.3.1); some values are rounded.
[b] 0.8 μg/m^3 = average 1990 nationwide on-road exposure estimate from HAPEM model; the companion rural estimate is 0.5 μg/m^3, and 0.4 μg/m^3 is a high-end estimate. The 1996 nation wide average is 0.7 μg/m^3, the companion rural estimate is 0.2 μg/m^3, however, a high-end estimate is not available for 1996. See chapter 2, Sections 2.4.3.2.1 and 2.4.3.2.2.

Figure 24.23 Table 8 from U.S. Environment Protection Agency [214] that presents exposures to diesel PM from occupational studies and conversion to approximate levels in the ambient environment. The average and high-end estimates of observed ambient diesel PM are 0.8 and 4μ/m^3, respectively (see footnote b in figure).

CI 1.11–1.36) (Figure 24.24). Subsequent studies of mortality in London found associations between British smoke levels [216] and acid aerosol levels [217] and mortality. In 1994, Schwartz [218] summarized mortality TSP/British smoke/coefficient of haze associations from 13 studies, all of which used the same statistical methodology and reported a relative risk of death per 100 μg/m^3 increase in daily TSP that ranged from 1.06 to 1.08. A detailed analysis of the causes of death in Philadelphia, PA, reveals increased RRs for deaths from chronic obstructive lung disease (1.25 for high TSP vs. low TSP days) and pneumonia (1.13 for high vs. low days). Risk of death from cardiovascular disease was increased, but these deaths were often accompanied by underlying respiratory conditions [219]. Other studies have confirmed the increased susceptibility of patients with chronic obstructive lung disease (COPD) [220] and other underlying respiratory diseases to nonrespiratory mortality from daily changes in air pollution [221]. A recent study that evaluated deaths in New York City from 1985 to 1994, found that risk of death per interquartile range of PM_{10} was greater for persons whose death certificates indicated an underlying respiratory disease [221]. This was particularly true for those aged 75 yr and older, in whom a similar increased risk was also observed for cancer deaths. Numerous other studies have supported these findings. Table 24.13 summarizes representative examples of studies based

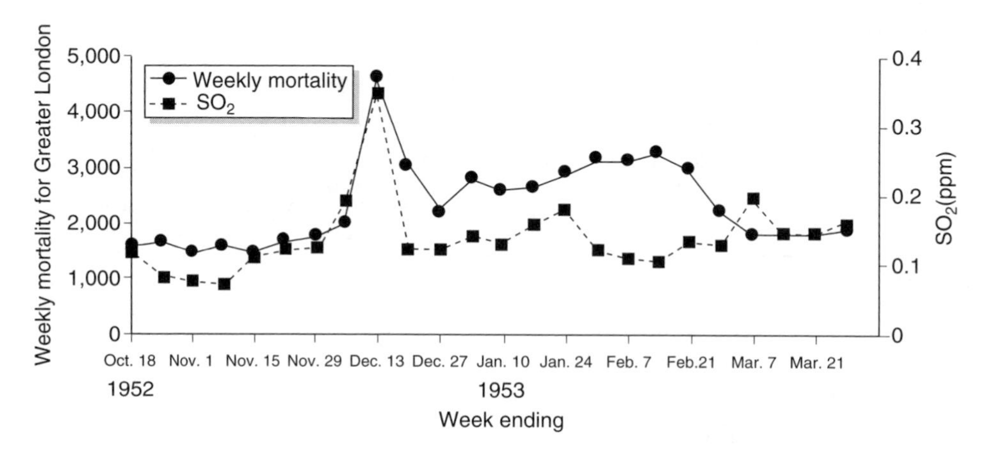

Figure 24.24 Relationship between London fog pollution measured as SO_2 between December 5 and 9, 1952 and mortality. Note that mortality remained elevated through February 1953. Average SO_2 during the time of the fog was 0.57 ppm (24-h U.S. NAAQS for SO_2=0.03 ppm) Bell and Davis [215].

on investigations of single cities or areas (a more detailed discussion of the multicity studies in U.S. and Europe follow). Studies from around the world (based on time series and case crossover) have generally reported associations between increased daily PM aerosol (or surrogates such as SO_2 and NO_2) and daily nonaccidental mortality, both from all causes and particularly from cardiovascular and respiratory causes. Although Schwartz and colleagues, in time series studies, reported that out-of-hospital deaths had a stronger association than in-hospital deaths, a case-crossover study from the Seattle area did not find any association between out-of-hospital deaths due to cardiac arrest and daily changes in any measure of PM aerosol [222] (Table 24.13). Two of the studies in Table 24.13 are of particular note. Clancy et al. [223] reported 10 and 15% decreases in daily cardiovascular and respiratory deaths in relation to daily changes in air pollution in the years that followed the 1990 ban on the sale of soft coal in Dublin, Ireland. These decreases in daily mortality were associated with decreases in average black smoke levels in 1984–1990 of 50 to 15 $\mu g/m^3$ in the years 1990–1996 (36% reduction) [223]. Vidal and colleagues studied daily changes in mortality in relation to ambient pollutants in an environment with low average PM_{10} levels (see footnote in Table 24.13). These investigators found an association between respiratory mortality and NO_2 and SO_2; PM_{10} was only associated with all-cause mortality at a 2-day lag. The explanation for why pollutants strongly associated with the PM aerosol seemed to have more specific effects on mortality was not explained by the author. A study from Seoul, Korea, points to the difficulties of ascribing effects to PM aerosol [224] (Table 24.13). In this study, the PM_{10} association was a 1.5% increase/interquartile interval; however, when O_3 was included with PM in the regression model, the PM effect was observed (2.7% increase) when ambient O_3 concentrations were above 13 ppb (Table 3 of Reference [224] (Table 24.13). This suggests that the overall chemical milieu is relevant to the interpretation of the magnitude of PM–mortality associations. Finally, a study in Mexico City found that over all lags and averages, the RR of death per 10 $\mu g/m^3$ of PM_{10} was greater for persons outside of hospitals at the time of death compared to deaths for persons in hospitals at the time of death [225]. Since hospitals usually have controlled environments, this supports a role for the ambient environment.

Subsequent to the publication of the time series studies noted in Table 24.15 and many other time series not cited in the table or this chapter, a problem was noted with the software that had been used by many of the studies to control the confounding effects of long-term trends in ambient PM aerosol, meteorological, and other time-dependent confounding

Table 24.13 Representative Sample of Time Series Studies that Relate Daily Changes in Measures of PM Aerosol to Daily Changes in Non-Accidental Mortality

General Population-Based Time Series Studies					
Study Population	Years	Pollutants[a]	Methods	Results	Reference
Santa Clara County, CA, U.S.	1980–1986	Winter COH	Multiple polynomial models; PR[2]	Results: % increase per unit change in COH (2 d lag) Respiratory 11.1 Cancer 1.0 Circulatory 9.4	
Philadelphia, PA, U.S.	1972–1982	TSP, SO_2	PR, GEE[b]	RR[2] for 100 μg/m^3 (mean of days 0 and 1) <65 yr = 1.03 65+ years = 1.1 associations for COPD[1], CVD,CVD, less clear for pneumonia	[299]
Buffalo/Rochester, NY, U.S.	1988–1990	H^+, SO_4^{2-}, SO_2, CO, NO_2, O_3, PM_{10} (predicted from COH and SO_4 for 5/6 days)	PR	RR/IQR[2] for respiratory/CVD mortality PM_{10} (unfilled) 1.048 1.034 COH 1.016 1.014 SO_4^2 1.024 1.011 O_3 1.037 1.009	[300]
Coachella Valley, CA, U.S.[c]	1989–1998	PM_{10}, CO, NO_2, O_3, $PM_{2.5}$ (measured directly last 2.5 yr), $PM_{10-2.5}$(predicted)	PR	RR/IQR for respiratory/CVD mortality PM_{10} (lag 0) 1.03 (1.01–1.05[d]) $PM_{10-2.5}$ (lag 0) 1.02 (1.01–1.04) $PM_{2.5}$ (lag 4) 1.03 (0.98–1.09) All $PM_{2.5}$ RR include 1 in sensitivity analyses; other PM exclude 1	[133]
Santiago, Chile	1989–1991	PM_{10}, NO_2, SO_2, O_3	OLS[2] PR	RR/115 μg/m^3 change in PM_{10} [133] Respiratory 1.15 (1.08–1.23) CVD 1.09 (1.04–1.14) Age 65+ years 1.11 (1.07–1.14) Male 1.11 (1.07–1.14) Female 1.06 (1.02–1.10) PM association robust to inclusion of other pollutants	

(Continued)

Table 24.13 *(Continued)*

Study Population	Years	Pollutants[a]	Methods	Results	Reference
Mexico City	1990–1992	TSP, PM_{10} SO_2, CO, O_3	PR	RR for 100 $\mu g/m^3$ change in TSP in regressions with SO_2, O_3 Age >65 yr 1.06 (1.03–1.09) Respiratory 1.10 (1.01–1.18) Cardiovascular 1.05 (1.01–1.10) 95% CI for estimates for SO_2 and O_3 include 1 in all models	[119]
Sydney, Australia	1989–1993	$PM_{0.01-2}$, NO_2, O_3	PR, GEE	% increase for 10th– 90th percentile change in PM_{10} in models with O_3 and NO_2 Cardiovascular 2.1% (0.3–5.0) Respiratory 0.7% (−5.5, 7.5) No associations with O_3 and NO_2	[302]
Dublin, Ireland	1984–1997		PR with interrupted time series	% decrease in deaths/10^3 person years before and after the 1990 ban on soft coal Cardiovascular 10.3 (12.6–8.0) Respiratory 15.5 (19.1, 11.6) Age <60 yr 7.9 (12.0, 3.6) Age 75+ years 4.5 (6.7–2.3)	[223]
Vancouver, Canada[e]	1994–1996	PM_{10}, CO, NO_2, SO_2, O_3	PR, GAM	Largest increases were for O_3 for respiratory mortality in summer, SO_2 for respiratory mortality in winter, and NO_2 for CVD deaths in winter. PM_{10} at lag 2 d showed association with total mortality	[303]
Seoul, Korea	1995–1998	PM_{10}, NO_2, CO, O_3	PR, GAM	Percentage change in mortality from stroke/interquartile increase in PM_{10} 1.5% (1.3–1.8) in single pollutant model −1.2%: PM_{10} effect for O_3 concentrations < median for O_3 (13 ppb) and 2.7% for O_3 above median O_3 concentration (correlation between O_3 and PM_{10}= −0.3)	[224]

Case-Crossover Studies[f]					
Philadelphia, PA, U.S.	1973–1980	TSP, no other pollutants evaluated	CLR[2]	Adjusted odds ratio (OR)/100 $\mu g/m^3$ increment in 48 h TSP = 1.06 >65 yr 1.07 (1.04–1.11) CVD 1.06 (1.02–1.11) Pneumonia 1.08 (0.92–1.26)	[305]
Seoul, Korea	1991–1995	TSP, SO_2, O_3	CLR	RR/100 $\mu g/m^3$ increase in 3-d moving average TSP = 1.01 (0.99–1.03) SO_2 (50 ppb) 1.05 (1.02–1.08) O_3 1.02 (0.99–1.05) RR for SO_2 from PR = 1.08 (1.06–1.10)	[306]
Seattle and King County, WA, U.S.	1988–1994	PM_{10}, $PM_{0.02-1.4}$, CO, SO_2, O_3 (summer)	CDL	Death associated with out-of-hospital cardiac arrest—no association with any pollutant	[222]
Time Series Studies of Association Between Exposure to Mobile Sources and Daily Mortality					
Amsterdam, Netherlands	1987–1998	Traffic counts, Black smoke (BS), PM_{10}, SO_2, CO, NO_2, NO, O_3	PR, GAM	Higher pollutant levels at traffic-influenced monitoring sites. Effects greatest when traffic-influence site data are applied to persons who lived on roads with >10^4 vehicles/day; most consistent for BS RR/100 $\mu g/m^3$ BS (lag 1), 1.89 (1.20, 2.95)	[287]

[a] All studies include some variables to control for meteorological effects, season and, for some, day of the week effects. SO_2, CO, and NO_2 are noted since they often are surrogates for fixed (SO_2) or mobile (CO, NO_2) source PM aerosol. O_3 is included, since in some areas (e.g., eastern U.S.) photochemical smog is highly correlated with ambient PM aerosol levels.

[b] TS=time series; PR=Poisson regression; GEE = generalized estimating equations; GAM=generalized additive models; OLS=ordinary least-squares regression; CLR=conditional logistic regression; COH = coefficient of haze; RR=relative risk; IQR=interquartile (25th – 75th) range; COPD=chronic obstructive lung disease; CVD=cardiovascular diseases.

[c] This is a follow-up study to that in Reference [301], and uses a longer time series and adds data on $PM_{2.5}$ to those of PM_{10}.

[d] 95% confidence intervals.

[e] This study is included since pollutant levels are quite low. Maximum PM_{10}=33.9 $\mu g/m^3$, 90th percentile=22.8 $\mu g/m^3$.

[f] Case-crossover designs are matched designs in which each subject serves as his/her own control. Unlike the population-based time series studies, case-crossover designs provide control for individual covariates and, with proper sampling, control of temporal confounding [222, 304].

Table 24.14 Multicity Time Series Studies of the Effect of Daily Changes in PM Aerosol on Daily Mortality

Study and References	Locations and Pollutants	Methods
The National Morbidity, Morbidity and Air Pollution Study (NMMAPS) *Initial analyses*: [307–309] *Reanalyses after discovery of the GAM problem*:[a] [238,310]	90 largest cities in the U.S. for 1987–1994. Subanalysis of 20 of the largest in this group [309] PM_{10}, O_3, CO, SO_2, NO_2 Meteorological data	Three-stage regional model 1. Within-city variability estimated with log-linear semi-parametric model 2. Within-region variability of the true regression coefficient with weighted second-stage regression 3. Between-region variability in the true regional regression coefficients Heterogeneity across cities and regions explored
Air Pollution and Health: a European Approach (APHEA) [228, pp. 311–315]	European Cities APHEA1 — 15 cities, Black smoke, PM_{10}, SO_2, NO_2, O_3 APHEA2 — 29 cities Black smoke, PM_{10}, SO_2, NO_2, O_3	Individual city analysis 1. Standard procedures for Poisson regression Between-city analyses 2. Assume city-specific means normally distributed around overall mean 3. Evaluate effect modifiers

[a] The initial analyses used the default convergence criteria in the GAM algorithm in S-PLUS. These default criteria did not lead to convergence and resulted in standard errors of parameter estimates that were too small. Subsequent reanalyses were carried out with more strict convergence criteria in S-PLUS and with a generalized linear model approach with natural and penalized splines. These issues are explored in detail in the references in the table and are summarized in Reference number [226].

factors [226] (see footnote in Table 24.14 for details). A number of these studies were reanalyzed, and, although, some of the effect estimates were reduced, the overall conclusion of an association between daily changes in PM aerosol and daily mortality was not altered (the same was observed for associations with hospital admission, which are not covered in this chapter) [227].

Given the heterogeneity of risk estimates that have been derived across studies carried out in individual cities, two large studies were undertaken in the United States and in Europe (Table 24.14), which attempted to study multiple cities with a broad range of PM aerosol environments using uniform statistical methodological approaches and to explore sources of heterogeneity between the estimates derived from individual cities. Since both studies were reanalyzed to account for the software problem noted above, only the reanalyzed data are summarized. The basic structure of each study is summarized in Table 24.15.

NMMAPS estimated that, for each 10 $\mu g/m^3$ increase in daily PM_{10}, daily mortality increased 0.7% (posterior SE 0.06), 0.21% (0.06), and 0.10% (0.06) for lags of 0, 1, and 2 d, respectively (Figure 24.25, top panel). At lag 1, the effects were largest for cardiorespiratory diseases (Figure 24.25, middle panel), and the lag 1 PM_{10} effect was found to be robust to the inclusion of other pollutants in the statistical models (Figure 24.25, bottom panel). There was evidence for geographic heterogeneity, with effect estimates being greatest in the northeast and least in the northwest (Figure 24.26a). City-specific estimates for all-cause and cardiorespiratory deaths showed heterogeneity; however, within regions, there seems to be relative homogeneity (Figure 24.26B — note circled areas for southern California and the northeast). This heterogeneity likely represents a combination of differences in the

Table 24.15 Association Between a 10 $\mu g/m^3$ Increment in PM_{10} (Average of Lag 0.1)[a] and Percentage Increase in Total Daily Mortality — APHEA Study [228]

GAM[b] with default convergence criterion	GAM with strict convergence criterion	GLM with natural spline	GLM with penalized spline
0.7% (0.6–0.8%)[c]	0.7% (0.5–0.7%)	0.4% (0.3–0.6%)	0.6% A(0.4–0.7%)

[a] Not directly stated but surmised from Table 2 in Reference [314].

[b] GAM=generalized additive model; GLM=generalized linear model. All results based on fixed-effects models. Results similar with random-effects models.

[c] Authors calculations from coefficients in Table 1 of Reference [228].

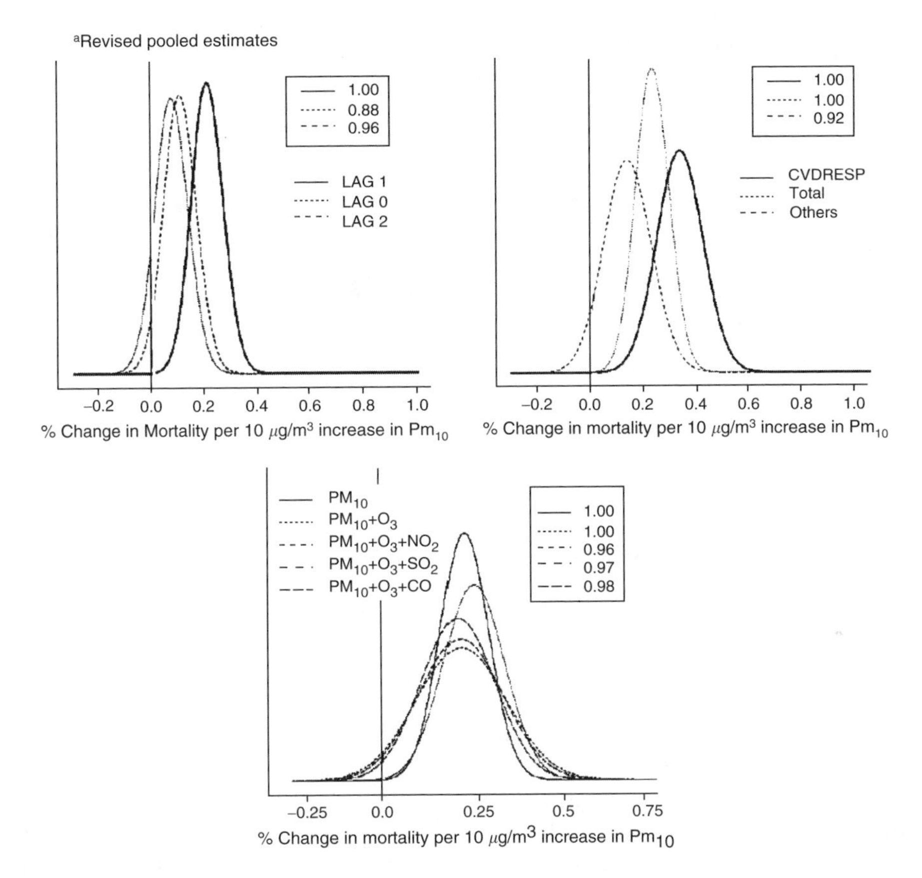

Figure 24.25 Results from the revised NMMAPS analyses [310]. Top: Marginal posterior distribution of the national average effect of 10 $\mu g/m^3$ increase in PM_{10} on total, nonaccidental mortality based on different lags. Middle: Comparative marginal posterior distribution of the national average effect of 10 $\mu g/m^3$ increase in PM_{10} at lag 1 on total, cardiorespiratory (CVDRESP) and "other" mortality. Bottom: Marginal posterior distribution of effect of 10 $\mu g/m^3$ increase in PM_{10} at lag 1in models with other pollutants. Box at upper right includes posterior probabilities that PM_{10} effects are greater than 0. Based on pooled city-specific estimates in a two-stage hierarchical model.

ambient pollutant mixtures between regions as well as demographic differences. In the original analyses, the percentage of adults without a high school diploma in each city had an independent effect on daily mortality. This was not included in the reanalysis. The APHEA project reported percentage increases in total daily mortality of approximately 0.7% per 10 $\mu g/m^3$ increase in PM_{10} (average of lag 0,1 — see footnote a, Table 24.15) for 21

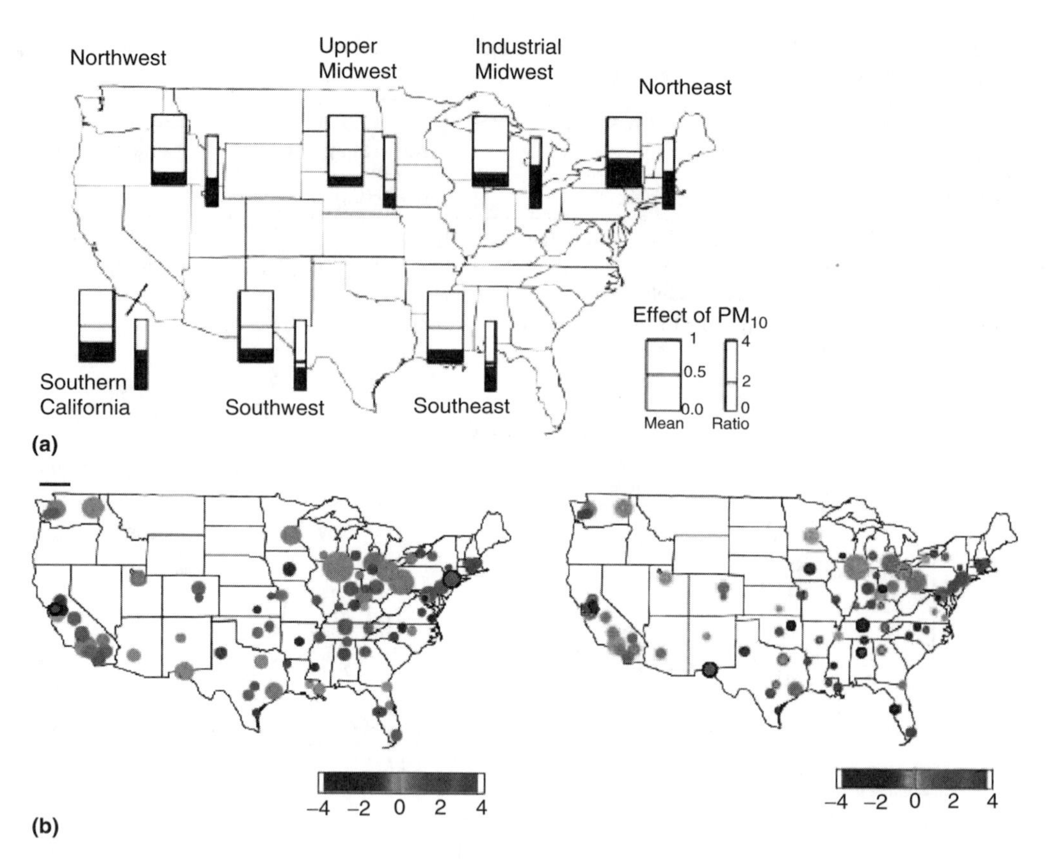

Figure 24.26 (**A**) Regional differences in estimated % change in daily total mortality per 10 $\mu g/m^3$ increase in PM_{10}. The height of the shaded area in the box on the left is the region-specific estimate. Shade area in the box on the right is the *t*-ratio (posterior mean divided by posterior standard deviation). *Source*: Dominici et al. [310]. (**B**) Within city total (left panel) and cardiovascular (right panel) mortality from NMMAPS. Color bars define ranges of % changes, and the sizes of the circles are proportional to the precision of the effect estimates. *Source*: Dominici et al. [325].

cities where such data were available (Table 24.15) [228]. The results were relatively robust to the types of modeling strategy chosen. These results are somewhat lower than that expected with a similar average of lags for NMMAPS (Figure 24.25, top panel).

A number of studies have attempted to identify specific components of mass or particle number as the source of increased risk for daily mortality (Table 24.16). Two separate reports from the Harvard Six Cities Study found that only fine PM was associated with increases in daily mortality; no association was observed for coarse PM or a "crustal" component based on source attribution [121,128]. A similar conclusion was reported in a Spokane, WA, study in which the contribution of coarse PM was evaluated with an indicator for days with dust storms [126]. A meta-analysis based on 19 U.S.-based studies came to the same conclusions [229]. In contrast, studies from California, Mexico City, and Santiago, Chile, all reported that daily changes in coarse PM aerosol had as large or larger effects on daily mortality than did fine PM aerosol (Table 24.16). The extent of differences in the composition of coarse PM (concentrations of iron, bioaerosol components, and a "tail" of combustion product PM that extends into the coarse range) were not evaluated by any of these studies. Based on the available data, it does not appear appropriate to extrapolate the relative contributions of fine and coarse PM to the associations with mortality from one location to another.

Table 24.16 Selected Time Series Studies that Evaluate the Relationship Between Changes in Specific Mass Components of PM Aerosol and Daily Mortality

Study Population	Years	Pollutants and Components	Methods	Results	Reference
Studies that Report Predominant Effects for Fine Component of Mass Fraction					
Spokane, WA, U.S.	1989–1996	Days with dust storms as surrogate for increased coarse PM, PM_{10}	PR[a] with indicator for days with dust storms	RR for days after dust storms 1.01 (0.87–1.17)	[126]
Wasatch Front, UT	1985–1995	PM_{10} ($PM_{2.5}$ ~ 70–90% of PM_{10}, $PM_{1.0}$), windblown dust episodes surrogate for $PM_{10-2.5}$	PR, GAM	Stagnant air episodes characterized by high concentrations of primary and secondary combustion-source PM more associated with mortality than wind-blown dust episodes characterized by PM with larger contribution from crustal elements	[316]
Harvard Six Cities, U.S.	1976–1987	PM_{10},[b] coarse PM, $PM_{2.5}$, H^+, SO_4^{2-}	PR, GAM	Combined % increase mortality over all cities/10 $\mu g/m^3$ increase on same day PM metric PM2.5 1.3% (0.9–1.7) Coarse PM 0.4% (−0.2, 0.9) (Results for 2 cities for course PM show comparable effects with $PM_{2.5}$	[128]
Harvard Six Cities, U.S.	1979–1988	Sources based on 15 elements, converted to five "factors"	PR, GAM	% increase in mortality/10 $\mu g/m^3$ increase in mass concentration from specific source Mobile sources (fine PM) 3.4 % (1.7–5.2) Crustal sources −2.3% (−5.8, 1.2)	[121]
Meta-analysis based on 19 studies from U.S.	Studies span 1973-90	PM_{10}, TSP, SO_2, CO, NO_2, O_3, $PM_{2.5}/PM_{10}$ ratio for 14 U.S. cities	Random effects and empirical Bayes summaries	% change in mortality/10 $\mu g/m^3$ increase in PM_{10} (ratio used directly) PM_{10} 0.67 (0.46–0.88) $PM_{2.5}/PM_{10}$ ratio as grouped variable: 0.31% (ratio <0.57), 0.68 (ratio 0.57–0.64), 0.81 (ratio >0.65) (95% CI excludes 0 for the highest two groups)	[229]

(Continued)

Table 24.16 (*Continued*)

Study Population	Years	Pollutants and Components	Methods	Results	Reference
Studies that Report Predominant Effects for Coarse Component of Mass Fraction					
Mexico City, Mexico	1993–1995	PM_{10}, SO_2, CO, NO_2, O_3,	PR	% increase in mortality/10 $\mu g/m^3$ in 5-d mean with and without adjustment for O_3 and NO_2 $PM_{2.5}$ % changes range from 1.25%–1.49% — all 95% CI include 0 change $PM_{10-2.5}$% changes range from 4.07%–4.28% — lower bounds of 95% CI >2%	[317]
Coachella Valley, CA, U.S.[c]	1989–1998	PM_{10}, CO, NO_2, O_3, $PM_{2.5}$ (measured directly last 2.5 yr) $PM_{10-2.5}$ (predicted)	PR	RR/IQR for respiratory/CVD mortality PM_{10} (lag 0) 1.03 (1.01–1.05[d]) $PM_{10-2.5}$ (lag 0) 1.02 (1.01–1.04) $PM_{2.5}$ (lag 4) 1.03 (0.98–1.09) All $PM_{2.5}$ RR include 1 in sensitivity analyses; other PM exclude 1	[133]
Santiago, Chile	1988–1996	$PM_{10-2.5}$, $PM_{2.5}$, SO_2, NO_2,CO, O_3	PR, GLM, and GAM	RR of death for increase in 2 d mean $PM_{2.5}$ 1.06 winter, 1.06 winter $PM_{10-2.5}$ 0.99 winter, 1.07 summer 95% CI for all RR >1 exclude one in single and all 2-pollutant models	[318]
Studies that Report Predominant Effects for Ultrafine Fraction of PM Aerosol					
Erfurt, Germany	1995–1998	Particle number, PM mass fractions	PR, GLM, and GAM	See Figure 25. Authors interpret results as showing particle number effects. Commentary by funder raises questions about whether the number and mass effects can be separated	[138]

[a] See footnotes in Table 24.13.

[b] PM_{15} was measured until 1984. Coarse PM until 1985 was $PM_{15-2.5}$.

[c] This is a follow-up study to that in Reference [301] and uses a longer time series and adds data on $PM_{2.5}$ to those of PM_{10}.

[d] 95% confidence intervals.

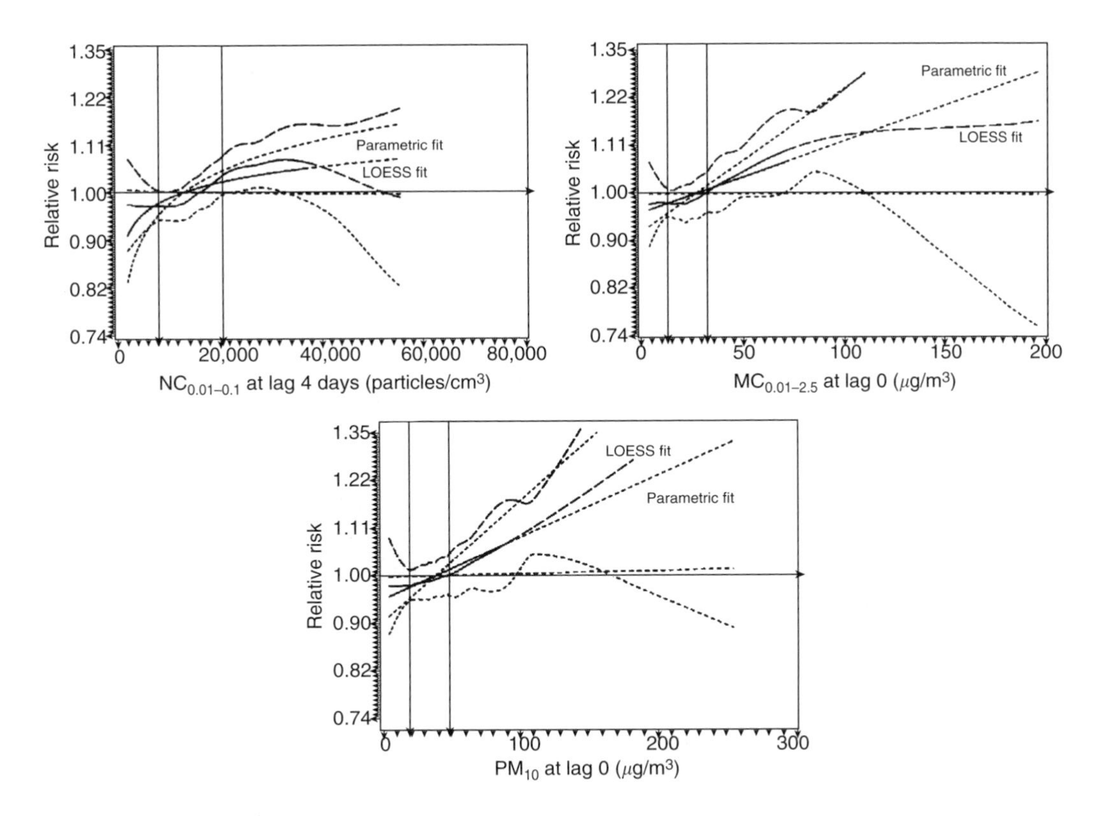

Figure 24.27 Exposure response curves for particle number count for particles between 0.1 and 1 μm (top curve), mass for particles between 0.1 and 2.5 μm (middle curve), and PM_{10} (lower curve). *Source*: Wichmann et al. [138].

A large study was undertaken in Germany to evaluate the relative strength of associations between changes in daily mortality and changes in particle number in the ultrafine range and mass fractions [138]. The authors concluded that the associations were driven by particle number effects. However, inspection of the data (Figure 24.27) does not provide a difference in the exposure–response relationships between particle number, fine mass, and total PM_{10} mass. A commentary by the reviewers of the report concluded that the evidence did not favor one component over another (begins on p. 93 of Reference [138]).

Among the studies that have reported associations between daily changes in PM and daily mortality, there has been a general consistency that the effects are greater in the elderly (usually defined as age 65 and older) and among individuals with selected underlying disease (e.g., diabetes [study of hospital admissions; [230]] congestive heart failure [231]). As noted previously, persons with underlying respiratory diseases seem to have an increased risk of cardiovascular deaths.

The form of the exposure-response relationship between daily changes in PM aerosol and daily mortality has been investigated extensively (e.g., see References [232–237]). In general, the data have been more consistent with a no-threshold model than models with thresholds, although alternative models with thresholds still remain a consideration (Figure 13-6 from Reference [28]). The NMMAPS study evaluated data from the 20 largest U.S. cities to assess the most likely exposure–response relationships between daily changes in PM aerosol and daily mortality [238]. A summary analysis for total mortality suggests that a no-threshold model is most consistent with the data for PM_{10} (Figure 24.28). A more detailed analysis, based on the distribution of posterior probability of a

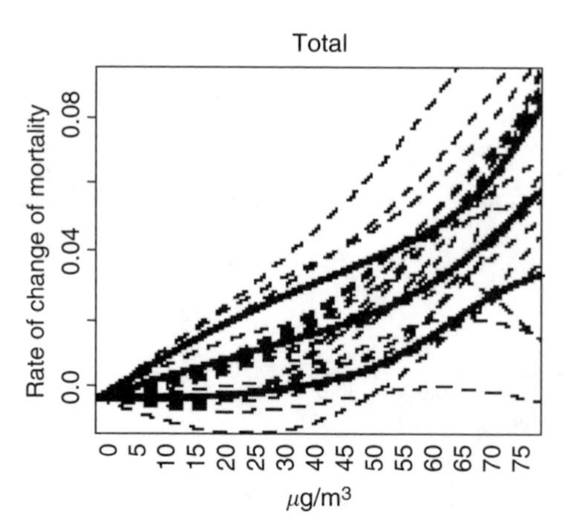

Figure 24.28 PM_{10} total mortality exposure–response curve for the mean of the current and previous day's PM_{10} concentrations for the 20 largest U.S. cities. Thick lines are the mean curve and 95% credible interval. Dashed lines are Bayesian estimates of the city-specific exposure–response curves. *Source* Dominici et al. [238].

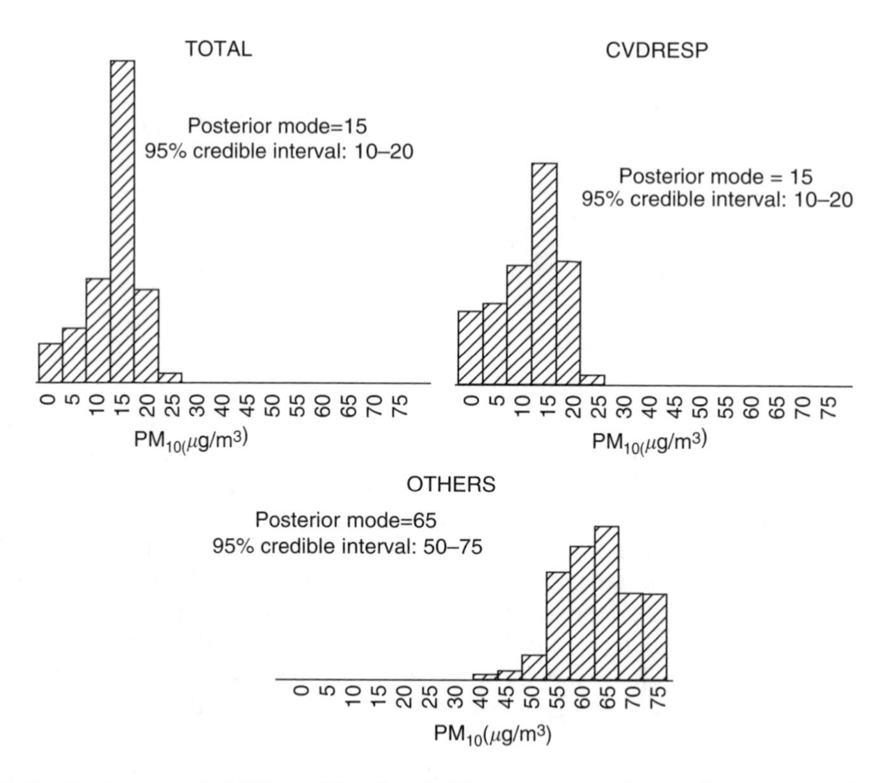

Figure 24.29 Posterior probabilities of the thresholds for groups of mortality based on the mean of the current and previous day's PM_{10} concentration for the 20 largest U.S. cities. Adapted from Figure 4 and Table 3 of Dominici et al. [238].

threshold, suggested that a threshold was not likely for cardiovascular mortality, but was consistent for the data for "other" causes of death (Figure 24.29). In the case of all-cause and cardiovascular mortality, any likely threshold was below the Federal annual PM_{10} annual 24 h standard of 50 $\mu g/m^3$ in force at the time over which the PM data were evaluated (Figure 24.29). In a commentary that accompanied the publication of the NMMAPS analysis, Pope [237] reviewed the various factors that could influence the detection of threshold (statistical methods, publication bias, measurement error) and presented a graphical summary of some of the more important time series studies that have addressed exposure–response relationships (Figure 24.30). He concluded that the weight

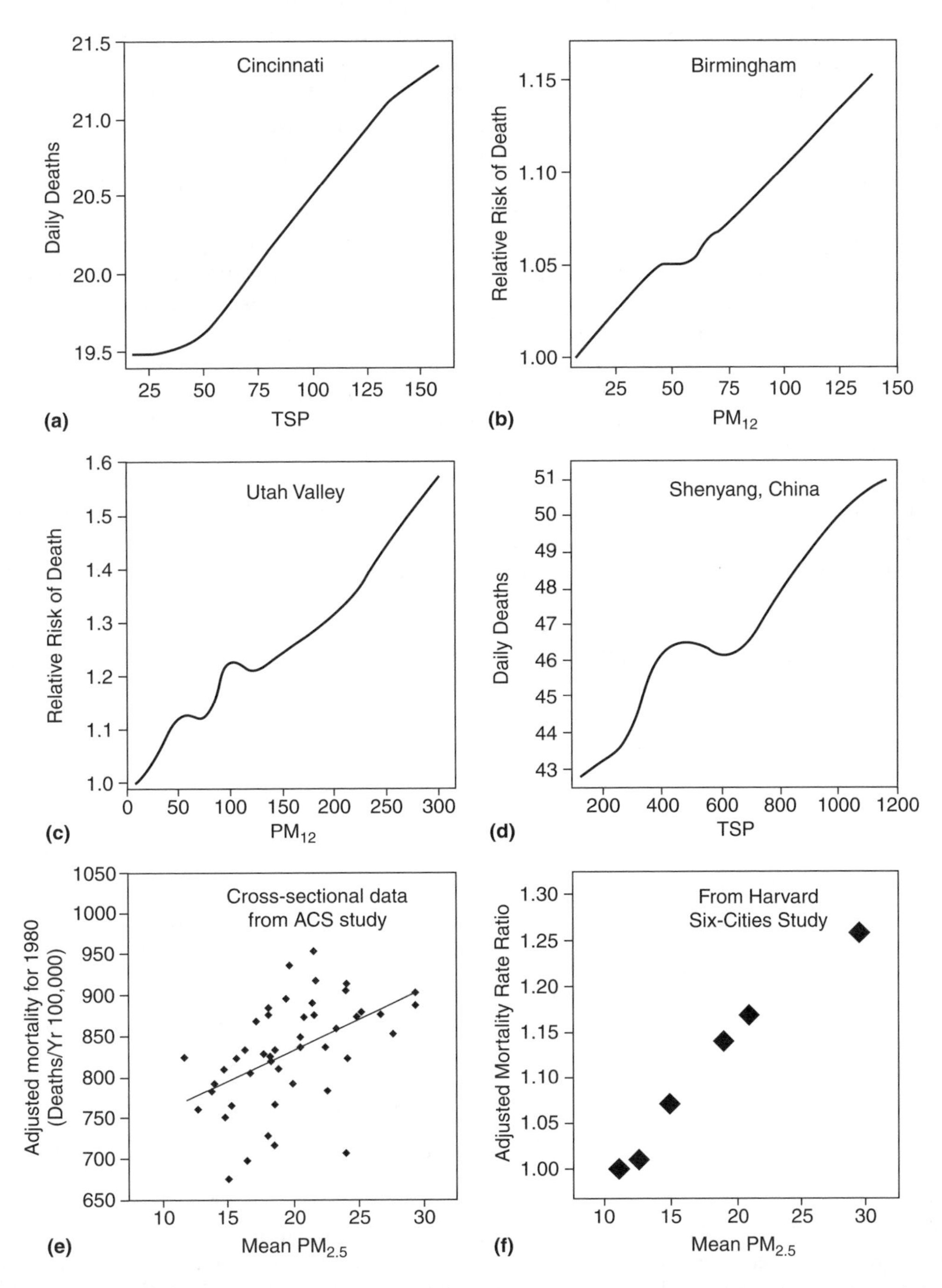

Figure 24.30 (From Figure 2 of Pope [237]. Plots of exposure–response relations from selected studies. (**A–D**) nonparametric smooth curves of adjusted daily deaths or adjusted relative risk of death for selected daily time series mortality studies.

of evidence "further indicates that assumptions or scientific priors of no-effects threshold levels for PM are not well supported by the empiric evidence" [237]. Other investigators, although in the minority, raise the argument that measurement errors preclude accurate specification of the exposure–response relationship [239].

In the end, the overall health impact of the associations between daily changes in PM aerosol and daily mortality depends upon the extent to which life expectancy is shortened by exposure [240]. If the people who are dying were those whose deaths were advanced by only a few days (a phenomenon that has been termed "harvesting" or mortality displacement [241]), there is general agreement (see Reference [240] for an alternative view) that the daily increases in mortality with increases in daily PM are not due simply to harvesting [241–245]. Schwartz carried out analyses based on smoothing windows of 15, 30, 45, and 60 d and observed that the percentage increases in mortality (1979–1986) per 10 $\mu g/m^3$ increase in $PM_{2.5}$ in Boston peaked at 15, 60, 60 d for COPD, pneumonia, and ischemic heart disease (IHD) deaths, respectively (Figure 24.31) [242]. COPD showed evidence of harvesting at the 60-d averaging window, and pneumonia showed evidence of short-term harvesting, followed by increased percentage increases (Figure 24.32). These data imply that COPD deaths are being brought forward by about 2 months, while deaths from pneumonia and IHD are not due to harvesting and may reflect "enrichment" of the at-risk pool as a consequence of persistent exposure to increased average levels of PM [242]. An analysis of the APHEA project, based on distributed lag models applied to ten cities, failed to find strong evidence for mortality displacement for total daily mortality (Figure 24.33) [244]. Moreover, the effect estimate for mortality for exposures 11–60 d before death was more than two thirds as large as that for 10 d just prior to death (0.688 ± 0.261 vs. 0.922 ± 0.184, respectively — both estimates are $\times 10^3$). Similar results were observed by Dominici et al. [245] in a study of daily mortality in four U.S. cities between 1987 and 1994. (Figure 24.34). Results were similar for frequency domain and time-scale estimates. In response to a critique of their approach [246], these authors evaluated 12 mortality rates (four cities and three categories of mortality) at time scales less

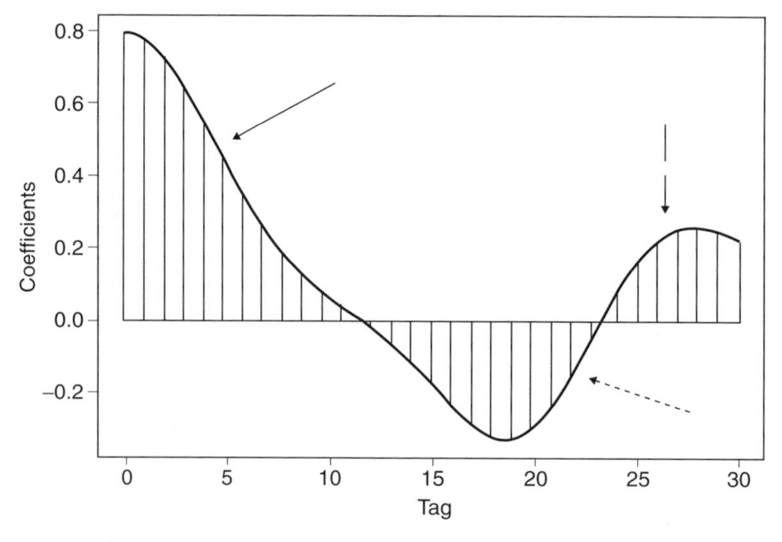

Figure 24.31 Schematic of mortality displacement. If frail individuals have the time of their deaths advanced by a few days due to increases in air pollution (solid arrow), the reduction of the at-risk pool will lead to a decrease in the number of deaths over ensuing days (dotted arrow). In the absence of any further increases, the hazard of death will return to its usual level of daily variability (dashed arrow). Adapted from Zanobetti et al. [244].

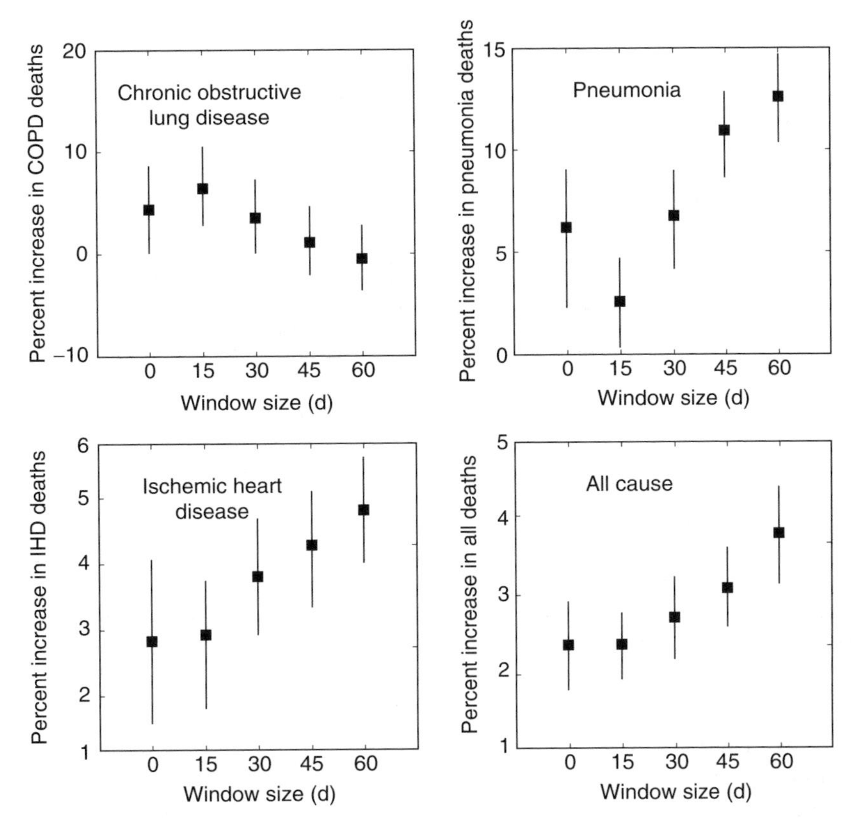

Figure 24.32 Percentage increase in class-specific mortality per 10 $\mu g/m^3$ increase in $PM_{2.5}$ in Boston, MA, for different smoothing windows of air $PM_{2.5}$. The figure shows increased percentage changes with longer $PM_{2.5}$ averaging times. Chronic obstructive lung disease shows some evidence of mortality displacement, with the percent change in death falling below 0 at a 60-d cycle length (see Figure 24.31). Pneumonia shows a trend to short-term harvesting, but larger effects with longer averaging times. Ischemic heart disease shows no change with the 15-day window and increasing percentage changes with progressively longer averaging times. Adapted from Schwartz (242).

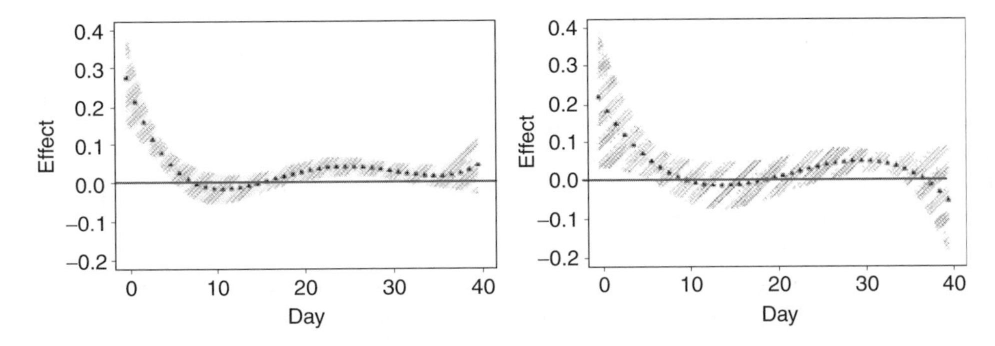

Figure 24.33 Shape of the association between PM_{10} and daily deaths 4th-degree distributed lag model (left panel) and a cubic-degree distributed lag model (right panel) in ten cities of the APHEA-2 project, 1990–1997. Both fit with a random effect for city. In neither case is there strong evidence for mortality displacement (see Figure 24.31), since deaths fall to near 0 but then rise to a second peak of longer duration than the first peak. Adapted from Zanobetti et al. [244].

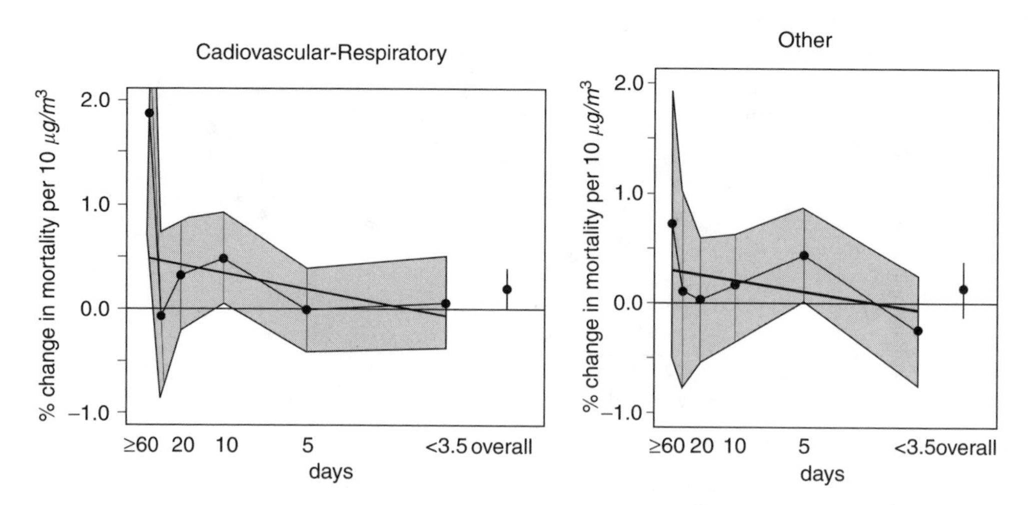

Figure 24.34 Pooled estimates of log relative rates of mortality at different time scales for cardiovascular-respiratory (left panel) and other causes of mortality (right panel) for four U.S. cities (Pittsburgh, PA; Minneapolis, MN; Seattle, WA; Chicago, IL), 1987–1994. Shaded regions are ±2 standard errors of the estimates. Percentage increases in mortality were greater at longer than shorter exposure intervals before death. *Source*: Dominici et al. (245).

than and greater than 5 d (Figure 24.35) [247]. In most cases, the estimates at greater than 5 d are greater than those at less than 5 d (Figure 24.35). While most of the data cited indicate that effects of ambient PM on daily mortality are largest for longer time scales, they do not negate the fact that shorter time scales are also associated with increased mortality. Nonetheless, the short-term associations do not appear to be the result of mortality displacement.

Although this chapter does not cover the extensive body of data related to the associations between PM and hospital admissions for respiratory and cardiovascular disease, these data have been used to support the causal connection between changes in ambient PM and daily mortality. Bates [143] argued that if the association was likely to be causal, then there should be increased risks for hospitalization for cardiorespiratory diseases. Such data have been compiled and were used to bolster the argument for the setting of the PM standard in 1996 (Figure 24.36).

24.5.2.2 *Physiological mechanisms related to the association of daily changes in PM and daily mortality*

Given that mortality from cardiovascular and respiratory diseases seem most closely associated with daily changes in ambient PM, number of studies have been undertaken to define the physiological pathways through which the proinflammatory, pro-oxidant effects of PM aerosol could result in the observed excess mortality.

Cardiovascular mortality: Inflammation is thought to be an important component of the pathophysiology of atherosclerotic cardiovascular diseases [248]. The proinflammatory properties of the PM aerosol have been discussed in previous sections. In addition to any role played by inflammation in the sudden worsening of cardiovascular disease that leads to death, several other general mechanisms have emerged as being possibly relevant to the excess daily deaths from cardiovascular disease:[4] alterations in heart rate variability and

[4]Mechanisms that could underlie cancer or disease related to inflammation have been discussed in earlier sections.

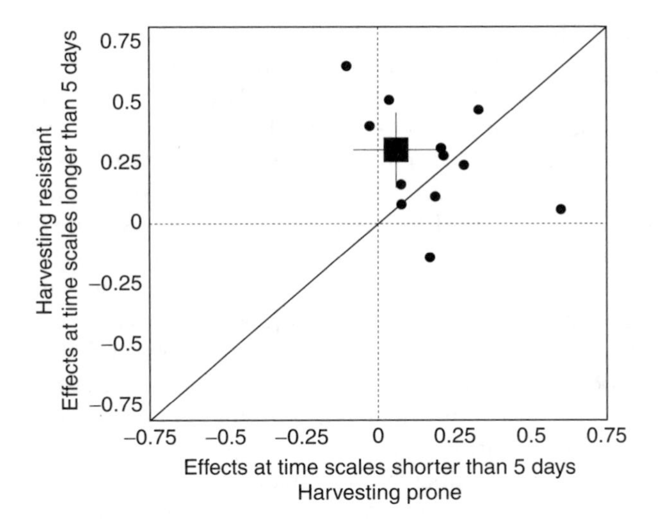

Figure 24.35 Estimated relative rates of mortality due to PM at time scales shorter than 5 d (harvesting-prone) vs. relative rate estimates obtained at time scales longer than 5 d (harvesting-resistant) from four U.S. cities (Pittsburgh, PA; Minneapolis, MN; Seattle, WA; Chicago, Il) and three mortality outcomes (total, cardiovascular-respiratory and other causes), 1987–1994. A large square is placed at the averages of the 12 estimates. *Source*: Dominici et al. [245].

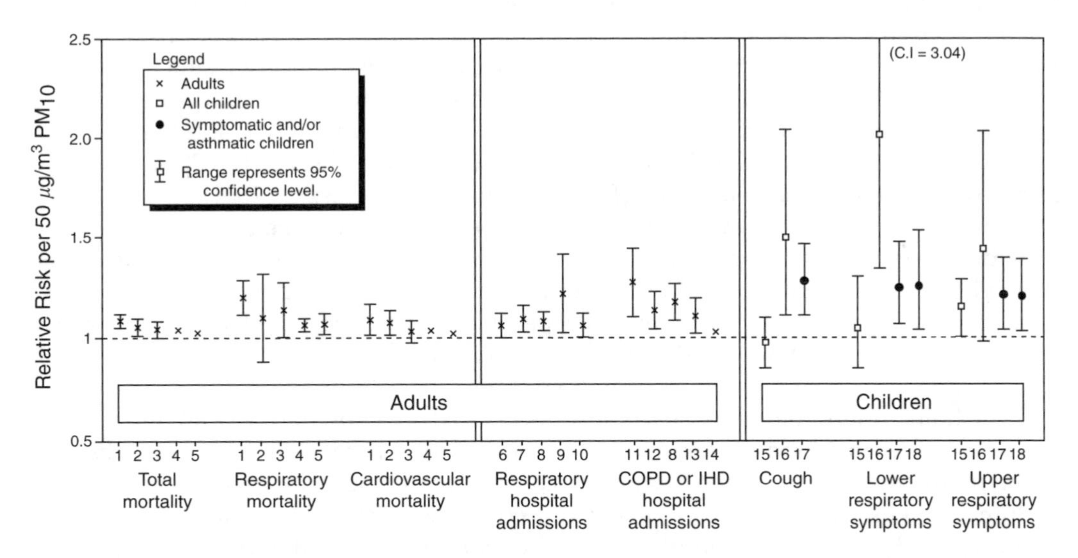

Figure 24.36 RRs of various health outcomes per 50 $\mu g/m^3$ increase in PM_{10} for adults and children. Each data point and error bar represents a separate study. *Source*: Federal Register [326].

increased cardiac arrhythmias, induction of a procoagulant state, and alteration of endothelial cell function.

The normal heart rate is quite variable as a result of control by the sympathetic and parasympathetic nervous systems [249]. Numerous studies have demonstrated that reduced heart rate variability is a risk factor for mortality after myocardial infarction (heart attack) and a general risk factor for mortality in the elderly [250,251]. A number of studies have been published that have found associations between increased levels of daily PM and reduced high-frequency heart rate variability (due to vagus nerve control) [252–257]. Gold et al. [254] evaluated changes in heart rate variability with 5 min of controlled outdoor exercise in subjects aged 53–87 yr. These investigators observed an association between increased 4-h mean pre-exercise $PM_{2.5}$ concentration and decreased heart rate variability. A 221-d study in an elderly population in a retirement center observed increased odds (3.1, 95% CI: 1.4–6.6) of decreased heart rate variability on days during indoor $PM_{2.5} > 15\ \mu g/m^3$ compared to days with PM levels $\leq 15\ \mu g/m^3$ [253]. A study in 20 healthy workers (nine smokers) from an occupational cohort (ages 21–58 yr) who wore personal PM monitors for 24 hs showed a 1.4% in heart rate variability per 100 $\mu g/m^3$ increase in the 3-h moving average $PM_{2.5}$ mean [256]. A study by the same investigators in 39 boilermakers who wore personal PM monitors observed that decreases in heart variability were associated with the lead and vanadium content of the $PM_{2.5}$ aerosol (vanadium, nickel, copper, chromium, lead, and manganese were tested) [257]. $PM_{2.5}$ mass did not show any association at any of the lags that were evaluated (see Table 4 of Reference [257]). Either through decreases in heart rate variability or direct toxic effects on the conduction system of the heart, PM aerosol could also trigger cardiac arrhythmias that could increase the risk of death [258]. Support for this possibility in humans comes from a study of patients with implanted defibrillators [258]. The odds of increases in defibrillator discharges were associated with increases in $PM_{2.5}$, black carbon, NO_2, and O_3. The largest and most precisely estimated effects were seen for NO_2 (see Table 4 in Reference [258]).

Acute heart attacks are thought to be initiated by the formation of a clot that obstructs the flow of blood in a blood vessel damaged by the atherosclerotic process, and anticoagulants are used routinely to reduce the risk of heart attacks [259]. Alterations in the coagulable state of blood have been suggested as one mechanism through which PM aerosol could increase the risk of death [137]. At least two studies have provided support for this hypothesis. Following a 1985 4-d episode of increased TSP and SO_2 in southern Germany, an increased risk of extreme values of plasma viscosity was observed in 3256 participants in a prospective randomized study of therapy for cardiac disease [260]. The results from this study were supported by those from a study of 112 subjects aged 60 yr and older who provided repeated blood specimens over a period of 18 months (maximum of 12 specimens) [261]. Estimated personal exposure to PM_{10} over the 3 d prior to blood sampling produced changes in a set of hematological parameters that suggested sequestration of red blood cells in the circulation [261]. Such a phenomenon could increase the risk of clot formation in the presence of a hypercoagulable state.

The lining cells of blood vessels (endothelial cells) are known to play an important role in the regulation of blood flow and vascular tone. Studies in rats have observed alterations in the production of endothelin 1, a potent constrictor of blood vessels that is produced by endothelial cells after exposure to concentrated ambient PM (CAPS) and O_3 in combination [117] and vasoconstriction of small pulmonary arteries with CAPS alone [262]. The relevance of these observations to humans was demonstrated with controlled 2-h exposure of 25 healthy adults to a mixture of CAPS and O_3 compared to exposure to filtered air [263]. Brachial artery vasoconstriction was observed after the CAPS+O_3 exposure but not with filtered air.

Thus, while the definitive physiological mechanisms that might underlie the association between increases in daily cardiovascular mortality with daily increase in PM aerosol are not known definitively, what is known is supportive of the fact that the association is likely to be causal and not due to some unmeasured confounding factors or some subtle problem with the statistical analysis approaches that have been employed by epidemiological studies. A similar case can be made for deaths from chronic lung disease, based on the expected results of the pro-inflammatory properties of the ambient PM aerosol.

24.6 Conclusions

This chapter has focused largely on the effects of the nonradioactive, nonoccupational ambient PM aerosol and its likely effect on human health. The chapter has been representative, rather than exhaustive, in its synthesis of current data. Only selected areas related to human health have been discussed. The most exhaustive review of the material presented herein can be found in the descriptive sections of the most recent public release of the draft document for PM standard being circulated for review by the U.S. EPA [264].

While there is little controversy about the effects of the bioaerosol component of the ambient PM aerosol on human health, there has been more skepticism about the effects of the anthropogenic component of the PM aerosol on human health, as clearly indicated by epidemiological studies. However, there now appears a sufficient body of basic biological and physiological mechanistic research data that makes it far more likely that the observed associations between the anthropogenic component of the PM aerosol and human health are causal than are the methodological issues related to epidemiological study designs, statistical analysis, and issues related to exposure measurement error.

References

1. Donaldson, K., Beswick, P.H., and Gilmour, P.S., Free radical activity associated with the surface of particles: a unifying factor in determining biological activity?, *Toxicol. Lett.*, 88, 293–298, 1996.
2. U.S. Environmental Protection Agency, Air Quality Criteria for Particulate Matter — Third External Review Draft, EPA 600/P-99/002aC, Report No.: EPA 600/P-99/002aC, Research Triangle Park, 2001.
3. Neas, L.M., Dockery, D.W., Burge, H., Koutrakis, P., and Speizer, F.E., Fungus spores, air pollutants and other determinants of peak expiratory flow, *Am. J. Epidemiol.*, 143, 797–807, 1996.
4. Delfino, R.J., Zeiger, R.S., Seltzer, J.M., Street, D.H., Matteucci, R.M., Anderson, P.R. et al., The effect of outdoor fungal spore concentrations on daily asthma severity, *Environ. Health Perspect.*, 105, 622–635, 1997.
5. Downs, S.H., Mitakakis, T.Z., Marks, G.B., Car, N.G., Belousova, E.G., Leuppi, J.D. et al., Clinical importance of Alternaria exposure in children, *Am. J. Respir. Crit. Care Med.*, 164, 455–459, 2001.
6. Soukup, J.M. and Becker, S., Human alveolar macrophage responses to air pollution particulates are associated with insoluble components of coarse material, including particulate endotoxin, *Toxicol. Appl. Pharmacol.*, 171, 20–26, 2001.
7. Schwartz, D.A., Does inhalation of endotoxin cause asthma? *Am. J. Respir. Crit. Care Med.*, 163, 305–306, 2001.
8. Monn, C. and Becker S., Cytotoxicity and induction of proinflammatory cytokines from human monocytes exposed to fine ($PM_{2.5}$) and course particles ($PM_{10-2.5}$) in outdoor and indoor air, *Toxicol. Appl. Pharmacol.*, 155, 245–252, 1999.
9. Monn, C., Naef, R., and Koller, T., Reactions of macrophages exposed to particles, <10 microm., *Environ. Res.*, 91, 35–44, 2003.
10. Wallace, L., Indoor particles: a review, *J. Air Waste Manag. Assoc.*, 46, 98–126, 1999.
11. Abt, E., Suh, H.H., Catalano, P., and Koutrakis, P., Relative contribution of outdoor and indoor particles sources to indoor concentrations, *Environ. Sci. Technol.*, 34, 3579–3587, 2002.
12. Leaderer, B.P., Naeher, L., Jankun, T., Balenger, K., Holford, T.R., Toth, C. et al., Indoor, outdoor, and regional summer and winter concentrations of PM_{10}, $PM_{2.5}$, SO_4^{-2}, H^+, NH_4^+, NO_3^-, NH_3^-, and nitrous acid in homes with and without kerosene space heaters, *Environ. Health Perspect.*, 107, 223–231, 1999.
13. Patterson, E. and Eatough, D.J., Indoor/outdoor relationships for ambient $PM_{2.5}$ and associated pollutants: epidemiological implications in Lindon, Utah, *J. Air Waste Manage. Assoc.*, 50, 103–110, 2000.

14. Wainman, T., Zhang, J., Weschler C.J., and Lioy, P.J., Ozone and limonene in indoor air: a source of submicron particle exposure, *Environ. Health Perspect.*, 108, 1139–1145, 2000.
15. Spengler, J.D., Long-term measurements of respirable sulfates and particles inside and outside homes, *Atmos. Environ.*, 15, 23–30, 1981.
16. National Research Council, *Environmental Tobacco Smoke: Measuring Exposures and Assessing Health Effects*, National Academy Press, Washington, DC, 1986.
17. National Cancer Institute, Monograph 10: Health Effects of Exposure to Environmental Tobacco Smoke, National Institutes of Health-National Cancer Institute, 2003, http://cancercontrol.cancer.gov/tcrb/monographs/10/m10_complete.pdf.
18. Abt, E., Suh, H.H., Allen, G., and Koutrakis, P., Characterization of indoor particle sources: a study conducted in the metropolitan Boston area, *Environ. Health Perspect.*, 108, 35–44, 2002.
19. U.S. Environmental Protection Agency, Air Quality Criteria for Particulate Matter, Report No: EPA/600/P-95-001bF, Washington, DC, April 1996.
20. Miller, F.J., Dosimetry of particles in laboratory animals and humans in relationship to issues surrounding lung overload and human health risk assessment, *Inhal. Toxicol.*, 12, 19–57, 2000.
21. Brain, J.D. and Valberg, P.A., Deposition of aerosol in the respiratory tract, *Am. Rev. Respir. Dis.*, 120, 1325–1373, 1979.
22. Yeates, D.B. and Mortensen, J., Deposition and clearance, in Murray, J.F., Nadel, J.A., Mason, R.J., Boushey, H.A., Eds., *Textbook of Respiratory Medicine*, W.B. Saunders Co., New York, 2000, pp. 349–384.
23. Churg, A. and Vidal, S., Carinal and tubular airway particle concentrations in the large airways of non-smokers in the general population: evidence of high particle concentration at airway carinas, *Occup. Environ. Med.*, 53, 553–558, 1996.
24. Brauer, M., Avila-Casado, C., Fortoul, T.I., Vedal, S., Stevens, B., and Churg, A., Air pollution and retained particles in the lung, *Environ. Health Perspect.*, 109, 1039–1043, 2001.
25. Schlesinger, R.B., Deposition and clearance of inhaled particles, in McClellan, R.O. and Henderson, R.F., Eds., *Concepts in Inhalation Toxicology*, Taylor & Francis, Washington, DC, 1995, pp. 191–224.
26. Pinkerton, K.E., Green, F.H., Saiki, C., Vallyathan, V., Plopper, C.G., Gopal, V. et al., Distribution of particulate matter and tissue remodeling in the human lung, *Environ. Health Perspect.*, 108, 1063–1069, 2000.
27. Kim, C.S., Lewars, G.A., and Sackner, M.A., Measurement of total lung aerosol deposition as an index of lung abnormality, *J. Appl. Physiol.*, 64, 1527–1536, 1988.
28. U.S. Environmental Protection Agency, Air Quality Criteria for Particulate Matter Volume II, in Air Quality Criteria for Particulate Matter Volumes I–III, EPA/600/P-95-001bF, National Center for Environmental Assessment Office of Research and Development, U.S. EPA, Washington, DC, 1996.
29. Wanner, A., Salathe, M., and O'Riordan, T.G., Mucociliary clearance in the airways, *Am. J. Respir. Crit. Care Med.*, 154(6 Part 1), 1868–1902, 1996.
30. Nemmar, A., Hoet, P.H., Vanquickenborne, B., Dinsdale, D., Thomeer, M., Hoylaerts, M.F. et al., Passage of inhaled particles into the blood circulation in humans, *Circulation* 105, 411–414, 2002.
31. Ilowite, J.S., Bennett, W.D., Sheetz, M.S., Groth, M.L., and Nierman, D.M., Permeability of the bronchial mucosa to 99mTc-DTPA in asthma, *Am. Rev. Respir. Dis.*, 139, 1139–1143, 1989.
32. Koenig, J.Q., Larson, T.V., Hanley, Q.S., Robelledo, V., Dumler, K., Checkoway, H. et al., Pulmonary function changes in children associated with fine particulate matter, *Environ Res.*, 63, 26–38, 1993.
33. Daigle, C.C., Chalupa, D.C., Gibb, F.R., Morrow, P.E., Oberdorster, G., Utell, M.J. et al., Ultrafine particle deposition in humans during rest and exercise, *Inhal. Toxicol.*, 15, 539–552, 2003.
34. Ward, P.A., Warren, J.S., and Johnson, K.J., Oxygen radicals, inflammation, and tissue injury, *Free Rad. Biol. Med.*, 5, 403–408, 1988.
35. Poli, G., Introduction — serial review: reactive oxygen and nitrogen in inflammation(1,2), *Free Rad. Biol. Med.*, 33, 301–302, 2002.

36. Nauseef, W.M. and Clark, R.A., Granulocyte phagocytosis, in Mandell, G.L., Bennett, J.E., and Dolin, R., Eds., *Principles and Practices of Infectious Diseases*, 5th ed., Churchill Livingston, New York, 2000, pp. 89–112.
37. Goldsmith, C.A., Frevert, C., Imrich, A., Sioutas, C., and Kobzik, L., Alveolar macrophage interaction with air pollution particulates, *Environ. Health Perspect.*, 105 (Suppl. 5), 1191–1195, 1997.
38. Li, N., Sioutas, C., Cho, A., Schmitz, D., Misra, C., Sempf, J. et al., Ultrafine particulate pollutants induce oxidative stress and mitochondrial damage, *Environ. Health Perspect.*, 111, 455–460, 2003.
39. Gilmour, P.S., Brown, D.M., Lindsay, T.G., Beswick, P.H., MacNee, W., and Donaldson, K., Adverse health-effects of PM(10) particles — involvement of iron in the generation of hydroxyl radicals, *Occup. Environ. Med.*, 53, 817–822, 1996.
40. Chang, L-Y. and Crapo, J.D., Inhibition of airway inflammation and hyperreactivity by an antioxidant mimetic, *Free Rad. Biol. Med.*, 33, 379–386, 2002.
41. McCord, J.M., Human disease, free radicals, and the oxidant/antioxidant balance, *Clin. Biochem.*, 26, 351–357, 1993.
42. Nauss, K.M., and the HEI Diesel Working Group, Critical issues in assessing the carcinogenicity of diesel exhaust: a synthesis of current knowledge, in Institute H.E., Ed., *Diesel Exhaust: A Critical Analysis of Emissions, Exposure and Health Effects*, Health Effects Institute, Charlestown, 1995, pp. 13–61.
43. Schroeder, W.H., Dobson, M., Kane, D.M., and Johnson, N.D., Toxic trace elements associated with airborne particulate matter: a review, *J. Air Pollut. Control Assoc.*, 37, 1267–1285, 1987.
44. Ghio, A.J. and Samet, J.M., Metals and air pollution particles, in Holgate, S.T., Samet, J.M., Koren, H.S., and Maynard, R.L., Eds., *Air Pollution and Health*, Academic Press, New York, 1999, pp. 635–651.
45. Ghio, A.J., Meng, H.H., Hatch, G.E., and Costa, D.L., Luminol-enhanced chemiluminescence after *in vitro* exposures of rat alveolar macrophages to oil fly ash is metal dependent, *Inhal. Toxicol.*, 9, 255–271, 1997.
46. Ghio, A.J., Stonehuerner, J., Pritchard, R.J., Piantodosi, C.A., Dreher, K.L., and Costa, D.L., Humic-like substances in air pollution particulates correlate with concentrations of transition metals and oxidant generation, *Inhal. Toxicol.*, 8, 479–494, 1996.
47. Pritchard, R.J., Ghio, A.J., Lehmann, J.R., Winsett, D.W., Tepper, J.S., Park, P. et al., Oxidant generation and lung injury after particulate air pollutant exposure increase with concentration of associated metals, *Inhal. Toxicol.*, 8, 457–477, 1996.
48. Dreher, K.L., Jaskot, R.H., Lehmann, J.R., Richards, J.H., McGee, J.K., Ghio, A.J. et al., Soluble transition metals mediate residual oil fly ash induced acute lung injury, *J. Toxicol. Environ. Health.*, 50, 285–305, 1997.
49. Dye, J.A., Lehmann, J.R., McGee, J.K., Winsett, D.W., Ledbetter, A.D., Everitt, J.I. et al., Acute pulmonary toxicity of particulate matter filter extracts in rats: coherence with epidemiologic studies in Utah Valley residents. *Environ. Health Perspect.*, 109 (Suppl. 3), 395–403, 2001.
50. Frampton, M.W., Ghio, A.J., Samet, J.M., Carson, J.L., Carter, J.D., and Devlin, R.B., Effects of aqueous extracts of PM(10) filters from the Utah valley on human airway epithelial cells. *Am. J. Physiol.*, 277 (5 Part 1), L960–L967, 1999.
51. Aust, A.E., Ball, J.C., Hu, A.A., Lighty, J.S., Smith, K.R., Straccia, A.M. et al., Particle characteristics responsible for effects on human lung epithelial cells, *Res. Rep. Health Effect Inst.*, 110, 1–65, 2002, discussion 67–76.
52. Li, N., Wang, M., Oberley, T.D., Sempf, J.M., and Nel, A.E., Comparison of the pro-oxidative and proinflammatory effects of organic diesel particle chemical in bronchial epithelial cells and macrophages, *J. Immunol.*, 169, 4531–4541, 2002.
53. Shi, T., Knaapen, A.M., Begerow, J., Birmili, W., Borm, P.J., and Schins, R.P., Temporal variation of hydroxyl radical generation and 8-hydroxy-2'-deoxyguanosine formation by coarse and fine particulate matter, *Occup. Environ. Med.*, 60, 315–321, 2003.
54. Nel, A.E., Diaz-Sanchez, D. and Li, N., The role of particulate pollutants in pulmonary inflammation and asthma: evidence for the involvement of organic chemicals and oxidative stress, *Curr. Opin. Pulm. Med.*, 7, 20–26, 2001.

55. Li, N., Venkatesan, M.I., Miguel, A., Kaplan, R., Gujuluva, C, Alam, J. et al., Induction of heme oxygenase-1 expression in macrophages by diesel exhaust particle chemicals and quinones via the antioxidant-responsive element, *J. Immunol.*, 165, 3393–3401, 2000.
56. Li, N., Kim, S., Wang, M., Froines, J., Sioutas, C., and Nel, A., Use of a stratified oxidative stress model to study the biological effects of ambient concentrated and diesel exhaust particulate matter, *Inhal. Toxicol.*, 14, 459–486, 2002.
57. Nel, A.E., Diaz-Sanchez, D., Ng, D., Hiura, T., and Saxon, A., Enhancement of allergic inflammation by the interaction between diesel exhaust particles and the immune system, *J. Allergy Clin. Immunol.*, 102, 539–554, 1998.
58. Hiura, T.S., Kaszubowski, M.P., Li, N., and Nel, A.E., Chemicals in diesel exhaust particles generate reactive oxygen radicals and induce apoptosis in macrophages, *J. Immunol.*, 163, 5582–5591, 1999.
59. Hiura, T.S., Li, N., Kaplan, R., Horwitz, M., Seagrave J.C., and Nel, A.E., The role of a mitochondrial pathway in the induction of apoptosis by chemicals extracted from diesel exhaust particles, *J. Immunol.*, 165, 2703–2711, 2000.
60. Hashimoto, S., Gon, Y., Takeshita, I., Matsumoto, K., Jibiki, I., Takizawa, H. et al., Diesel exhaust particles activate p38 MAP kiinase to produce interleukin 8 and RANTES by human bronchial epithelial cells and *N*-acetylcysteine attenuates p38 kinase activation, *Am. J. Respir. Crit. Care. Med.*, 161, 280–285, 2000.
61. Baulig, A., Garlatti, M., Bonvallot, V., Marchand, A., Barouki, R., Marano, F. et al., Involvement of reactive oxygen species in the metabolic pathways triggered by diesel exhaust particles in human airway epithelial cells, *Am. J. Physiol. Lung Cell Mol. Physiol.*, 285, L671–9, 2003.
62. Gilmour, P.S., Donaldson, K., and MacNee, W., Overview of antioxidant pathways in relation to the effects of air pollution, *Eur. Respir. Mon*, 21, 241–261, 2002.
63. van der Vliet, A. and Cross, C.E., Oxidants, nitrosants, and the lung, *Am. J. Med.*, 109, 398–421, 2000.
64. Hoidal, J.R., Reactive oxygen species and cell signaling, *Am. J. Respir. Cell Mol. Biol.*, 25, 661–663, 2001.
65. Vargas, L., Patino, P.J., Montoya, F., Vanegas, A.C., Echavarria, A., and Garcia, de Olarte, D., A study of granulocyte respiratory burst in patients with allergic bronchial asthma, *Inflammation*, 22, 45–54, 1998.
66. Barnes, P.J., Chung, K.F., and Page, C.P., Inflammatory mediators of asthma: update, *Pharmacol. Rev.*, 50, 515–596, 1998.
67. Ma, J.Y. and Ma, J.K., The dual effect of the particulate and organic components of diesel exhaust particles on the alteration of pulmonary immune/inflammatory responses and metabolic enzymes, *J. Environ. Sci. Health Part C Environ. Carcinog. Ecotoxicol. Rev.*, 20, 117–147, 2002.
68. Shukla, A., Timblin, C., BeruBe, K., Gordon, T., McKinney, W., Driscoll, K. et al., Inhaled particulate matter causes expression of nuclear factor (NF)-kappaB-related genes and oxidant-dependent NF-kappaB activation *in vitro*, *Am. J. Respir. Cell Mol. Biol.*, 23, 182–187, 2000.
69. Brown, D.M., Stone, V., Findlay, P., MacNee, W., and Donaldson, K., Increased inflammation and intracellular calcium caused by ultrafine carbon black is independent of transition metals or other soluble components, *Occup. Environ. Med.*, 57, 685–691, 2000.
70. Donaldson, K., Brown, D., Clouter, A., Duffin, R., MacNee, W., Renwick, L. et al., The pulmonary toxicology of ultrafine particles, *J. Aerosol. Med.*, 15, 213–520, 2002.
71. O'Grady, N.P., Preas, H.L., Pugin, J., Fiuza, C., Tropea, M., Reda, D. et al., Local inflammatory responses following bronchial endotoxin instillation in humans, *Am. J. Respir. Crit. Care Med.*, 163, 1591–1598, 2001.
72. Becker, S., Soukup, J.M., Gilmour, M.I., and Devlin, R.B., Stimulation of human and rat alveolar macrophages by urban air particulates: effects of oxidant radical generation and cytokine production, *Toxicol. Appl. Pharmacol.*, 141, 637–648, 1996.
73. Monick, M.M. and Hunninghake, G.W., Activation of second messenger pathways in alveolar macrophages by endotoxin, *Eur. Respir. J.*, 20, 210–222, 2002.
74. Whitekus, M.J., Li, N., Zhang, M., Wang, M., Horwitz, M.A., Nelson, S.K. et al., Thiol antioxidants inhibit the adjuvant effects of aerosolized diesel exhaust particles in a murine model for ovalbumin sensitization, *J. Immunol.*, 168, 2560–2567, 2002.

75. Diaz-Sanchez, D., Garcia, M.P., Wang, M., Jyrala, M., Saxon, A., Nasal challenge with diesel exhaust particles can induce sensitization to a neoallergen in the human mucosa, *J. Allergy Clin. Immunol.*, 104, 1183–1188, 1999.
76. Diaz-Sanchez, D., Dotson, A.R., Takenaka, H., and Saxon, A., Diesel exhaust particles induce local IgE production in vivo and alter the pattern of IgE messenger RNA isoforms, *J. Clin. Invest.*, 94, 1417–1425, 1994.
77. Diaz-Sanchez, D., Tsien, A., Fleming J., and Saxon, A., Combined diesel exhaust particulate and ragweed allergen challenge markedly enhances human *in vivo* nasal ragweed-specific IgE and skews cytokine production to a T helper cell 2-type pattern, *J. Immunol.*, 158, 2406–2413, 1997.
78. Diaz-Sanchez, D., The role of diesel exhaust particles and their associated polyaromatic hydrocarbons in the induction of allergic airway disease, *Allergy*, 52:52, 1997.
79. Pandya, R.J., Solomon, G., Kinner, A., and Balmes, J.R., Diesel exhaust and asthma: hypotheses and molecular mechanisms of action, *Environ. Health Perspect.*, 110 (Suppl. 1), 103–112, 2002.
80. Lambert, A.L., Dong, W., Winsett, D.W., Selgrade, M.K., and Gilmour, M.I., Residual oil fly ash exposure enhances allergic sensitization to house dust mite, *Toxicol. Appl. Pharmacol.*, 158, 269–277, 1999.
81. Lambert, A.L., Dong, W., Selgrade, M.K., and Gilmour, M.I., Enhanced allergic sensitization by residual oil fly ash particles is mediated by soluble metal constituents, *Toxicol. Appl. Pharmacol.*, 165, 84–93, 2000.
82. Kumagai, Y., Arimoto, T., Shinyashiki, M., Shimojo, N., Nakai Y., Yoshikawa, T. et al., Generation of reactive oxygen species during interaction of diesel exhaust particle components with NADPH-cytochrome P450 reductase and involvement of the bioactivation in the DNA damage, *Free Rad. Biol. Med.*, 22, 479–487, 1997.
83. Seagrave, J., McDonald, J.D., Gigliotti, A.P., Nikula, K.J., Seilkop, S.K., Gurevich, M. et al., Mutagenicity and *in vivo* toxicity of combined particulate and semivolatile organic fractions of gasoline and diesel engine emissions, *Toxicol. Sci.*, 70, 212–226, 2002.
84. Moller, P., Daneshvar, B., Loft, S., Wallin, H., Poulsen, H.E., Autrup, H. et al., Oxidative DNA damage in vitamin C-supplemented guinea pigs after intratracheal instillation of diesel exhaust particles, *Toxicol. Appl. Pharmacol.*, 189, 39–44, 2003.
85. van Eeden, S.F., Tan, W.C., Suwa, T., Mukae, H., Terashima, T., Fujii, T. et al., Cytokines involved in the systemic inflammatory response induced by exposure to particulate matter air pollutants (PM10), *Am. J. Respir. Crit. Care Med.*, 164, 826–830, 2001.
86. Jimenez, L.A., Drost, E.M., Gilmour, P.S., Rahman, I., Antonicelli, F., Ritchie, H. et al., PM(10)-exposed macrophages stimulate a proinflammatory response in lung epithelial cells via TNF-alpha, *Am. J. Physiol. Lung Cell. Mol. Physiol.*, 282, L237–L248, 2002.
87. Kennedy, T., Ghio, A.J., Reed, W., Samet, J., and Zagorski, J., Quay, J. et al., Copper-dependent inflammation and nuclear factor-kappaB activation by particulate air pollution, *Am. J. Respir. Cell Mol. Biol.*, 19, 366–378, 1998.
88. Quay, J.L., Reed, W., Samet, J., and Devlin, R.B., Air pollution particles induce IL-6 gene expression in human airway epithelial cells via NF-kappaB activation, *Am. J. Respir. Cell Mol. Biol.*, 19, 98–106, 1998.
89. Fujii, T., Hayashi, S., Hogg, J.C., Vincent, R., and van Eeden, S.F., Particulate matter induces cytokine expression in human bronchila epithelial cells, *Am. J. Respir. Cell Mol. Biol.*, 25, 265–271, 2001.
90. Salvi, S.S., Nordenhall, C., Blomberg, A., Rudell, B., Pourazar, J., Kelly, F.J. et al., Acute exposure to diesel exhaust increases IL-8 and GRO-alpha production in healthy human airways, *Am. J. Respir. Crit. Care Med.*, 161(2 Part 1), 550–557, 2000.
91. Nightingale, J.A., Maggs, R., Cullinan, P., Donnelly, L.E., Rogers, D.F., Kinnersley, R. et al., Airway inflammation after controlled exposure to diesel exhaust particulates, *Am. J. Respir. Crit. Care Med.*, 162, 161–166, 2000.
92. Mukae, H., Hogg, J.C., English, D., Vincent, R., and van Eeden, S.F., Phagocytosis of particulate air pollutants by human alveolar macrophages stimulates the bone marrow, *Am. J. Physiol. Lung Cell Mol. Physiol.*, 279, L924–L931, 2000.

93. Mukae, H., Vincent, R., Quinlan, K., English, D., Hards, J., Hogg, J.C. et al., The effect of repeated exposure to particulate air pollution (PM10) on the bone marrow. *Am. J. Respir. Crit. Care Med.*, 163, 201–209, 2001.
94. Gardner, S.Y., Lehmann, J.R., Costa, D.L., Oil fly ash-induced elevation of plasma fibrinogen levels in rats, *Toxicol Sci.*, 56, 175–180, 2000.
95. Tan, W.C., Qiu, D., Liam, B.L., Ng, T.P., Lee, S.H., van Eeden, S.F. et al., The human bone marrow response to acute air pollution caused by forest fires, *Am. J. Respir. Crit. Care Med.*, 161(4 Part 1), 1213–1217, 2000.
96. Gong, H., Jr., Linn, W.S., Sioutas, C., Terrell, S.L., Clark, K.W., Anderson, K.R. et al., Controlled exposures of healthy and asthmatic volunteers to concentrated ambient fine particles in Los Angeles. *Inhal. Toxicol.*, 15, 305–325, 2003.
97. Veronesi, B. and Oortgiesen, M., Neurogenic inflammation and particulate matter (PM) airway pollutants, *NeuroToxicology*, 22, 795–810, 2001.
98. Tsien, A., Diaz-Sanchez, D., Ma, J., and Saxon, A., The organic component of diesel exhaust particles and phenanthrene, a major polyaromatic hydrocarbon constituent, enhances IgE production by IgE-secreting EBV-transformed human B cells *in vitro*, *Toxicol. Appl. Pharmacol.*, 142, 256–263, 1997.
99. Gavett, S.H., Madison, S.L., Stevens, M.A., and Costa, D.L., Residual oil fly ash amplifies allergic cytokines, airway responsiveness, and inflammation in mice, *Am. J. Respir. Crit. Care Med.*, 160, 1897–1904, 1999.
100. Yang, H.M., Antonini, J.M., Barger, M.W., Butterworth, L., Roberts, B.R., Ma, J.K. et al., Diesel exhaust particles suppress macrophage function and slow the pulmonary clearance of *Listeria monocytogenes* in rats, *Environ. Health Perspect.*, 109, 515–521, 2001.
101. Yin, X.J., Schafer, R., Ma, J.Y., Antonini, J.M., Weissman, D.D., Siegel, P.D. et al., Alteration of pulmonary immunity to *Listeria monocytogenes* by diesel exhaust particles (DEPs). I. Effects of DEPs on early pulmonary responses, *Environ. Health Perspect.*, 110, 1105–1111, 2002.
102. Yin, X.J., Schafer, R., Ma, J.Y., Antonini, J.M., Roberts, J.R., Weissman, D.N. et al., Alteration of pulmonary immunity to *Listeria monocytogenes* by diesel exhaust particles (DEPs). II. Effects of DEPs on T-cell-mediated immune responses in rats, *Environ. Health Perspect.*, 111, 524–530, 2003.
103. Yang, H.M., Barger, M.W., Castranova, V., Ma, J.K., Yang, J.J., and Ma, J.Y., Effects of diesel exhaust particles (DEP), carbon black, and silica on macrophage responses to lipopolysaccharide: evidence of DEP suppression of macrophage activity, *J. Toxicol. Environ. Health. A.*, 58, 261–278, 1999.
104. Kay, A.B., T cells as orchestrators of the asthmatic response, *Ciba Found. Symp.*, 206, 56–67, 1997, discussion 67–70, 106–10.
105. Becker, S. and Soukup, J.M., Exposure to urban air particulates alters the macrophage-mediated inflammatory response to respiratory viral infection, *J. Toxicol. Environ. Health A*, 57, 445–457, 1999.
106. Myatt, T.A. and Milton, D.K., Endotoxins, in Spencer, J.P., Samet, J.M., and McCarthy, J.F., Eds., *Indoor Air Quality Handbook*. McGraw-Hill, New York, 2000, pp. 42.1–42.14.
107. Koppelman, G.H., Reijmerink, N.E., Colin Stine, O., Howard, T.D., Whittaker, P.A., Meyers, DA. et al., Association of a promoter polymorphism of the CD14 gene and atopy, *Am. J. Respir. Crit. Care Med.*, 163, 965–969, 2001.
108. Barton, G.M. and Medzhitov, R., Toll-like receptor signaling pathways, *Science*, 300, 1524–1525, 2003.
109. Centers for Disease Control and Prevention, Surveillance for asthma – United States, 1980–1999, *MMWR*, 51, 1–14, 2002.
110. Salvi, S.S., Babu, K.S., and Holgate, S.T., Is asthma really due to a polarized T cell response toward a helper T cell type 2 phenotype? *Am. J. Respir. Crit. Care Med.*, 164(8 Part 1), 1343–1346, 2001.
111. Magnan, A.O., Mely, L.G., Camilla, C.A., Badier, M.M., Montero-Julian, F.A., Guillot, C.M. et al., Assessment of the Th1/Th2 paradigm in whole blood in atopy and asthma. Increased IFN-gamma-producing CD8(+) T cells in asthma, *Am. J. Respir. Crit. Care Med.*, 161, 1790–1796, 2000.

112. Holt, P.G., OK, P., Holt, B.J., Upham, J.W., Baron-Hay, M.J., Suphioglu, C. et al., T-cell "priming" against environmental allergens in human neonates: sequential deletion of food antigen reactivity during infancy with concomitant expansion of responses to ubiquitous inhalant allergens, *Pediatr. Allergy Immunol.*, 6, 85–90, 1995.
113. Holt, P.G., Yabuhara, A., Prescott, S., Venaille, T., Macaubas, C., Holt, B.J. et al., Allergen recognition in the origin of asthma, in Chadwick, D.J. and Cardew, G., Eds., *The Rising Trends in Asthma: Ciba Foundation Symposium 206*, John Wiley & Sons, New York, 1997, pp. 35–55.
114. Nel, A.E., T-cell activation through the antigen receptor. Part 1: signaling components, signaling pathways, and signal integration at the T-cell antigen receptor synapse, *J. Allergy. Clin. Immunol.*, 109, 758–770, 2002.
115. Nel, A.E. and Slaughter, N., T-cell activation through the antigen receptor. Part 2: role of signaling cascades in T-cell differentiation, anergy, immune senescence, and development of immunotherapy, *J. Allergy Clin. Immunol.*, 109, 901–915, 2002.
116. Madden, M.C., Richards, J.H., Dailey, L.A., Hatch, G.E., and Ghio, A.J., Effect of ozone on diesel exhaust particle toxicity in rat lung, *Toxicol. Appl. Pharmacol.*, 168, 140–148, 2000.
117. Bouthillier, L., Vincent, R., Goegan, P., Adamson, I.Y., Bjarnason, S., Stewart, M. et al., Acute effects of inhaled urban particles and ozone: lung morphology, macrophage activity, and plasma endothelin-1, *Am. J. Pathol.*, 153, 1873–1884, 1998.
118. Alfaro-Moreno, E., Martinez, L., Garcia-Cuellar, C., Bonner, J.C., Murray, J.C., Rosas, I. et al., Biologic effects induced *in vitro* by PM_{10} from three different zones of Mexico City, *Environ. Health Perspect.*, 110, 715–720, 2002.
119. Borja-Aburto, V.H., Loomis, D.P., Bangdiwala, S.I., Shy, C.M., Rascon-Pacheco, RA., Ozone, suspended particulates, and daily mortality in Mexico City, *Am. J. Epidemiol.*, 145, 258–268, 1997.
120. Burnett, R.T., Cakmak, S., Brook, J.R., and Krewski, D., The role of particulate size and chemistry in the association between summertime ambient air pollution and hospitalization for cardiorespiratory diseases, *Environ. Health Perspect.*, 105, 614–620, 1997.
121. Laden, F., Neas, L.M., Dockery, D.W., and Schwartz, J., Association of fine particulate matter from different sources with daily mortality in six U.S. cities, *Environ. Health Perspect.*, 108, 941–947, 2000.
122. Brunekreef, B., Janssen, A.H., de Hartog, J., Harssema, H., Knape, M., and van Vliet, P., Air pollution from truck traffic and lung function in children living near motorways, *Epidemiology*, 8, 298–303, 1997.
123. van Vliet, P., Knape, M., de Hartog, J., Janssen, N., Harssema, H., and Brunekreef, B., Motor vehicle exhaust and chronic respiratory symptoms in children living near freeways, *Environ. Res.*, 74, 122–132, 1997.
124. Studnicka, M., Hackl, E., Pischinger, J., Fangmeyer, C., Haschke, N., Kuhr, J. et al., Traffic-related NO_2 and the prevalence of asthma and respiratory symptoms in seven year olds, *Eur. Respir. J.*, 10, 2275–2278, 1997.
125. Schwartz, J., Dockery, D.W., and Neas, L.M., Is daily mortality associated specifically with fine particles, *J. Air Waste Manage. Assoc.*, 46, 927–939, 1996.
126. Schwartz, J., Norris, G., Larson, T., Sheppard, L., Claiborne, C., and Koenig, J., Episodes of high coarse particle concentrations are not associated with increased mortality, *Environ. Health Perspect.*, 107, 339–342, 1999.
127. Schwartz, J. and Neas, L.M., Fine particles are more strongly associated than coarse particles with acute respiratory health effects in schoolchildren, *Epidemiology*, 11, 6–10, 2000.
128. Klemm, R.J., Mason, R.M., Jr., Heilig, C.M., Neas, L.M., and Dockery, D.W., Is daily mortality associated specifically with fine particles? Data reconstruction and replication of analyses, *J. Air Waste Manage. Assoc.*, 50, 1215–1222, 2000.
129. Ostro, B.D., Lipsett, M.J., Wiener, M.B., and Selner, J.C., Asthmatic responses to airborne acid aerosols, *Am. J. Publ. Health.*, 81, 694–702, 1991.
130. Thurston, G.D., Lippmann, M., Scott, M.B., and Fine, J.M., Summertime haze air pollution and children with asthma, *Am. J. Respir. Crit. Care Med.*, 155, 654–660, 1997.
131. Neas, L.M., Dockery, D.W., Koutrakis, P., and Speizer, F.E., Fine particles and peak flow in children: acidity versus mass, *Epidemiology*, 10, 550–553, 1999.

132. Spengler, J.D., Wilson, R., Emissions, dispersion, and concentrations of particles, in Wilson, R. and Spengler, J.D., Eds, *Particles in Our Air*, Harvard University Press, Cambridge, 1996, pp. 41–62.
133. Ostro, B.D., Broadwin, R., and Lipsett, M.J., Coarse and fine particles and daily mortality in the Coachella Valley, California: a follow-up study, *J. Expos. Anal. Environ. Epidemiol.*, 10, 412–419, 2000.
134. Mar, T.F., Norris, G.A., Koenig, J.Q., and Larson, T.V., Associations between air pollution and mortality in Phoenix, 1995–1997, *Environ. Health Perspect.*, 108, 347–353, 2000.
135. Loomis, D., Sizing up air pollution research, *Epidemiology*, 11, 2–4, 2000.
136. Lin, M., Chen, Y., Burnett, R.T., Villeneuve, P.J., and Krewski, D., The influence of ambient coarse particulate matter on asthma hospitalization in children: case-crossover and time-series analyses, *Environ. Health Perspect.*, 110, 575–581, 2002.
137. Seaton S., MacNee, W., Donaldson, K., and Godden, D., Particulate air pollution and acute health effects, *Lancet*, 345, 176–178, 1995.
138. Wichmann, H.E., Spix, C., Tuch, T., Wolke, G., Peters, A., Heinrich, J. et al., Daily mortality and fine and ultrafine particles in Erfurt, Germany Part I: role of particle number and particle mass, *Health Effects Inst. Res. Rep.*, 98, 5–86, 2000; discussion 87–94.
139. Peters, A., Wichmann, H.E., Tuch, T., Heinrich, J., and Heyder, J., Respiratory health effects are associated with the number of ultrafine particles, *Am. J. Respir. Crit. Care Med.*, 155, 1376–1383, 1997.
140. Pekkanen, J., Timonen, K.L., Ruuskanen, J., Reponen, A., and Mirme, A., Effects of ultrafine and fine particles in urban air on peak expiratory flow among children with asthmatic symptoms, *Environ. Res.*, 74, 24–33, 1997.
141. de Hartog, J.J., Hoek, G., Peters, A., Timonen, K.L., Ibald-Mulli, A., Brunekreef, B. et al., Effects of fine and ultrafine particles on cardiorespiratory symptoms in elderly subjects with coronary heart disease, *Am. J. Epidemiol.*, 157, 613–623, 2003.
142. Kunzli, N., Kaiser, R., Medina, S., Studnicka, M., Chanel, O., Filliger, P. et al., Public-health impact of outdoor and traffic-related air pollution: a European assessment, *Lancet*, 356, 795–801, 2000.
143. Bates, D., Health indices of the adverse effects of air pollution: the question of coherence, *Environ. Res.*, 59, 336–349, 1992.
144. Zhu, Y., Hinds, W.C., Kim, S., Shen, S., and Sioutas, C., Study of ultrafine particles near a major highway with heavy-duty diesel traffic, *Aitmos. Environ.*, 36, 4375–4383, 2002.
145. Levy, J.I., Houseman, E.A., Ryan, L., Richardson, D., and Spengler, J.D., Particle concentrations in urban microenvironments, *Environ. Health Perspect.*, 108, 1051–1057, 2000.
146. Ezzati, M. and Kammen, D.M., The health impacts of exposure to indoor air pollution from solid fuels in developing countries: knowledge, gaps, and data needs, *Environ. Health Perspect.*, 110, 1057–1068, 2002.
147. Wallace, L., Correlations of personal exposure to particles with outdoor measurements: a review of recent studies, *Aerosol Sci. Technol.*, 32, 15–25, 2000.
148. National Cancer Institute, Health Effects of Environmental Tobacco Smoke: The Report of the California Environmental Protection Agency, NIH 99-4645, Report No.: NIH 99–4645, National Cancer Institute, Bethesda, 1999.
149. Park, J.H., Spiegelman, D.L., Burge, H.A., Gold, D.R., Chew, G.L., and Milton, D.K., Longitudinal study of dust and airborne endotoxin in the home, *Environ. Health Perspect.*, 108, 1023–1028, 2000.
150. Park, J.H., Spiegelman, D.L., Gold, D.R., Burge, H.A., and Milton, D.K., Predictors of airborne endotoxin in the home, *Environ. Health Perspect.*, 109, 859–864, 2001.
151. Gereda, J.E., Leung, D.Y., Thatayatikom, A., Streib, J.E., Price, M.R., Klinnert, M.D. et al., Relation between house-dust endotoxin exposure, type 1 T-cell development, and allergen sensitisation in infants at high risk of asthma, *Lancet*, 355, 1680–1683, 2000.
152. Castellan, R.M., Cotton dust, in Harber, P., Schenker, M.B., and Balmes, J.R., Eds., *Occupational and Environmental Respiratory Disease*, Mosby, New York, 1995, pp. 401–419.
153. Vogelzang, P.F., van der Gulden, J.W., Folgering, H., Kolk, J.J., Heederik, D., Preller, L. et al., Endotoxin exposure as a major determinant of lung function decline in pig farmers, *Am. J. Respir. Crit. Care Med.*, 157, 15–18, 1998.

154. Park, J.H., Gold, D.R., Spiegelman, D.L., Burge, H.A., and Milton, D.K., House dust endotoxin and wheeze in the first year of life, *Am. J. Respir. Crit. Care Med.*, 163, 322–328, 2001.
155. Michel, O., Ginanni, R., Duchateau, J., Vertongen, F., Le Bon, B., and Sergysels, R., Domestic endotoxin exposure and clinical severity of asthma, *Clin. Exp. Allergy*, 21, 441–448, 1991.
156. Michel, O., Kips, J., Duchateau, J., Vertongen, F., Robert, L., Collet, H. et al., Severity of asthma is related to endotoxin in house dust, *Am. J. Respir. Crit. Care Med.*, 154, 1641–1646, 1996.
157. Braun-Fahrlander, C., Riedler, J., Herz, U., Eder, W., Waser, M., and Grize, L. et al., Environmental exposure to endotoxin and its relation to asthma in school-age children, *N. Engl. J. Med.*, 347, 869–877, 2002.
158. Baldini, M., Lohman, I.C., Halonen, M., Erickson, R.P., Holt, P.G., and Martinez, F.D., A polymorphism* in the 5′ flanking region of the CD14 gene is associated with circulating soluble CD14 levels and with total serum immunoglobulin E, *Am. J. Respir. Cell Mol. Biol.*, 20, 976–983, 1999.
159. Haider, S.C., Sommerfeld, A., Baldini, M., Martinez, F., Wahn, U., and Nickel, R., Evaluation of the CD14 C-159T polymorphism in the German Multicenter Allergy Study cohort, *Clin. Exp. Allergy*, 33, 166–169, 2003.
160. Jones, C.A., Holloway, J.A., Popplewell, E.J., Diaper, N.D., Holloway, J.W., Vance, G.H. et al., Reduced soluble CD14 levels in amniotic fluid and breast milk are associated with the subsequent development of atopy, eczema, or both, *J. Allergy Clin. Immunol.*, 109, 858–866, 2002.
161. Holmlund, U., Hoglind, A., Larsson, A.K., and Nilsson, C., Sverremark Ekstrom, E., CD14 and development of atopic disease at 2 years of age in children with atopic or non-atopic mothers, *Clin. Exp. Allergy*, 33, 455–463, 2003.
162. Downs, J., Zuidhof, A., Doekes, G., van den Zee, A., Wouters, I., Boezen, H.M. et al. (1→3)-b-D-glucan and endotoxiin in house dust and peak flow variability in children, *Am. J. Respir. Crit. Care Med.*, 162, 1348–1354, 2000.
163. Behrendt, H., Becker, W.M., Fritzsch, C., Silwa-Tomczok, W., Friedrichs, K.H., and Ring, J., Air pollution and allergy: experimental studies on modulation of allergen release from pollen by air pollutants, *Int. Arch. Allergy Immunol.*, 113, 69–74, 1997.
164. Knox, R.B., Suphioglu, C., Taylor, P., Desai, R., Watson, H.C., Peng, J.L. et al., Major grass pollen allergen Lol p 1 binds to diesel exhaust particle: implications for asthma and air pollution, *Clin. Exp. Allergy*, 27, 246–251, 1997.
165. Behrendt, H., Becker, W.M., Friedrichs, K.H., Darsow, U., and Tomingas, R., Interaction between aeroallergens and airborne particulate matter, *Int. Arch. Allergy Immunol.*, 99, 425–428, 1992.
166. D'Amato, G., Liccardi, G., D'Amato, M., and Cazzola, M., Outdoor air pollution, climatic changes and allergic bronchial asthma, *Eur. Respir. J.*, 20, 763–776, 2002.
167. Emberlin, J., Interaction between air pollution and aeroallergens, *Clin. Exp. Allergy*, 25, 33–39, 1995.
168. Emberlin, J., The effects of air pollution on allergic plants, *Eur. Resp. Rev.*, 53, 164–167, 1995.
169. D'Amato, G., Liccardi, G., D'Amato, M., and Cazzola, M., The role of outdoor air pollution and climatic changes on the rising trends in respiratory allergy, *Respir. Med.*, 95, 606–611, 2001.
170. Rossi, O.V.J., Kinula, V.L., Tienari, J., Huhti, E., Association of severe asthma attacks with weather, pollen and air pollutants, *Thorax*, 48, 244–248, 1993.
171. Garty, B.Z., Kosman, E., Ganori, E., Berger, V., Garty, L., Wietzen, T. et al., Emergency room visits of asthmatic children, relation to air pollution, weather and airborne allergens, *Ann. Allergy Asthma Immunol.*, 81, 563–570, 1998.
172. Anderson, H.R., Ponce de Leon, A., Bland, J.M., Bower, J.S., Emberlin, J., and Strachan, D.P., Air pollution, pollens, and daily admissions for asthma in London 1987– 92, *Thorax*, 53, 842–848, 1998.
173. Lewis, S.A., Corden, J.M., Forster, G.E., and Newland, M., Combined effects of aerobiological pollutants, chemical pollutants and meteorological conditions on asthma admission and A & E attendances in Derbyshire, UK, 1993–96, *Clin. Exp. Allergy*, 30, 1724–1732, 2000.
174. Miyamoto, T., Epidemiology of pollutant-induced airway disease in Japan, *Allergy*, 52 (Suppl. 38), 30–34, 1997.

175. Brunekreef, B., Hoek, G., Fischer, P., and Spieksma, F.T.M., Relation between airborne pollen concentrations and daily cardiovascular and respiratory disease mortality, *Lancet*, 355, 1517–1518, 2000.
176. Wyler, C., Braun-Fahrlander, C., Kunzli, N., Schindler, C., Ackermann-Liebrich, U., Perruchoud, A.P. et al., Exposure to motor vehicle traffic and allergic sensitization, The Swiss Study on Air Pollution and Lung Diseases in Adults (SAPALDIA) Team, *Epidemiology*, 11, 450–456, 2000.
177. Bobak, M. and Leon, D.A., Air pollution and infant mortality in the Czech Republic, 1986–88, *Lancet*, 340, 1010–1014, 1992.
178. Woodruff, T., Grillo, J., and Schoendorf, K., The relationship between selected causes of postneonatal infant mortality and particulate air pollution in the United States, *Environ. Health Perspect.*, 105, 608–612, 1997.
179. Bobak, M. and Leon, D.A., The effect of air pollution on infant mortality appears specific for respiratory causes. *Epidemiology*, 10, 666–670, 1999.
180. Pereira, L.A.A., Loomis, D., Conceicoa, G.M.S., Braga, A.L.F., Arcas, R.M., and Kishi, J.S. et al., Association between air pollution and intrauterine mortality in Sao Paulo, Brazil, *Environ. Health Perspect.*, 106, 325–329, 1998.
181. Loomis, D., Castillejos, M., Gold, D.R., McDonnell, W., and Borja-Aburto, V.H., Air pollution and infant mortality in Mexico City, *Epidemiology*, 10, 118–123, 1999.
182. Dejmek, J., Solansky, I., Benes, I., Lenicek, J., and Sram, R.J., The impact of polycyclic aromatic hydrocarbons and fine particles on pregnancy outcome, *Environ. Health Perspect.*, 108, 1159–1164, 2000.
183. Ritz, B., Yu, F., Chapa, G., and Fruin, S., Effect of air pollution on preterm birth among children born in Southern California between 1989 and 1993, *Epidemiology*, 11, 502–511, 2000.
184. Wang, X., Ding, J., Ryan, L., and Xu, X., Association between air pollution and low birth weight: a community-based study, *Environ. Health Perspect.*, 105, 514–520, 1997.
185. Dejmek, J., Jelinek, R., Solansky, I., Benes, I., and Sram, R.J., Fecundability and parental exposure to ambient sulfur dioxide, *Environ. Health Perspect.*, 108, 647–654, 2000.
186. Perera, F.P., Rauh, V., Tsai, W.Y., Kinney, P., Camann, D., Barr, D. et al., Effects of transplacental exposure to environmental pollutants on birth outcomes in a multiethnic population, *Environ. Health Perspect.*, 111, 201–205, 2003.
187. Chay, K.Y. and Greenstone, M., The impact of air pollution on infant mortality: evidence from geographic variation in pollution shocks induced by a recession, *Q. J. Econ.*, 118, 1121–1167, 2003.
188. Mortimer, K.M., Tager, I.B., Dockery, D.W., Neas, L.M., Redline, S., The effect of ozone on inner-city children with asthma: identification of susceptible subgroups, *Am. J. Respir. Crit. Care Med.*, 162, 1838–1845, 2000.
189. Dockery, D.W., Pope, Ad, Xu, X., Spengler, J.D., Ware, J.H., Fay, M.E. et al., An association between air pollution and mortality in six U.S. cities [see comments], *N. Engl. J. Med.*, 329, 1753–1759, 1993.
190. Pope, C.A., Thun, M.J., Namboodiri, M.M., Dockery, D.W., Evans, J.S., Speizer, F.E., et al. Particle air pollution as a predictor of mortality in a prospective study of U.S. adults, *Am. J. Respir. Crit. Care Med.*, 151, 669–674, 1995.
191. Health Effects Institute, *Special Report: Reanalysis of the Harvard Six Cities Study and the American Cancer Society of Particulate Air Pollution and Mortality*, Health Effects Institute, Charlestown, July 2000.
192. Pope, C.A., 3rd, Burnett, R.T., Thun, M.J., Calle, E.E., Krewski, D., Ito, K. et al., Lung cancer, cardiopulmonary mortality, and long-term exposure to fine particulate air pollution, *JAMA*, 287, 1132–1141, 2002.
193. Villeneuve, P.J., Goldberg, M.S., Krewski, D., Burnett, R.T., and Chen, Y., Fine particulate air pollution and all-cause mortality within the Harvard Six-Cities Study: variations in risk by period of exposure, *Ann. Epidemiol.*, 12, 568–576, 2002.
194. Abbey, D.E., Nishino, N., McDonnell, W.F., Burchette, R.J., Knutsen, S.F., Lawrence Beeson, W. et al., Long-term inhalable particles and other air pollutants related to mortality in nonsmokers, *Am. J. Respir. Crit. Care Med.*, 159, 373–382, 1999.

195. McDonnell, W.F., Nishino-Ishikawa, N., Petersen, F.F., Chen, L.H., Abbey, D.E., Relationships of mortality with the fine and coarse fractions of long-term ambient PM_{10} concentrations in nonsmokers, *J. Expos. Anal. Environ. Epidemiol.*, 10, 427–436, 2000.
196. Abbey, D.E., Moore, J., Petersen, F., and Beeson, L., Estimating cumulative ambient concentrations of air pollutants: description and precision of methods used for an epidemiological study, *Arch. Environ. Health*, 46, 281–287, 1991.
197. Hoek, G., Brunekreef, B., Goldbohm, S., Fischer, P., and van den Brandt, P.A., Association between mortality and indicators of traffic-related air pollution in the Netherlands: a cohort study, *Lancet*, 360, 1203–1209, 2002.
198. Sarnat, J.A., Schwartz, J., Catalano, P.J., and Suh, H.H., Gaseous pollutants in particulate matter epidemiology: confounders or surrogates? *Environ. Health Perspect.*, 109, 1053–1061, 2001.
199. Mann, S.L., Wadsworth, M.E.J., and Colley, J.R.T., Accumulation of factors influencing respiratory illness in members of a national birth cohort and their offspring, *J. Epidemiol. Commun. Health*, 46, 286–292, 1992.
200. Bates, D.V., A half century later: recollections of the London fog, *Environ. Health Perspect.*, 110, A735, 2002.
201. Gilliland, F.D., Li, Y.F., and Peters, J.M., Effects of maternal smoking during pregnancy and environmental tobacco smoke on asthma and wheezing in children, *Am. J. Respir. Crit. Care Med.*, 163, 429–436, 2001.
202. Kunzli, N., Schwartz, J., Stutz, E.Z., Ackermann-Liebrich, U., and Leuenberger, P., Association of environmental tobacco smoke at work and forced expiratory lung function among never smoking asthmatics and non-asthmatics, The SAPALDIA-Team. Swiss Study on Air Pollution and Lung Disease in Adults, *Soz Praventivmed.*, 45, 208–217, 2000.
203. Abbey, D.E., Petersen, F.F., Mills, P.K., and Kittle, L., Chronic respiratory disease associated with long term ambient concentrations of sulfates and other air pollutants, *J. Expos. Anal. Environ. Epidemiol.*, 3, 99–115, 1993.
204. McDonnell, W.F., Abbey, D.E., Nishino, N., and Lebowitz, M.D., Long-term ambient ozone concentration and the incidence of asthma in non-smoking adults: The Ahsmog Study, *Environ. Res.*, 80, 110–121, 1999.
205. Abbey, D., Petersen, F., Mills, P., and Beeson, W., Long-term ambient concentrations of total suspended particulates, ozone, and sulfur dioxide and respiratory symptoms in a nonsmoking population, *Arch. Environ. Health*, 48, 33–46, 1993.
206. McConnell, R., Berhane, K., Gilliland, F., London, S.J., Islam, T., Gauderman, W.J. et al., Asthma in exercising children exposed to ozone: a cohort study, *Lancet*, 359, 386–391, 2002.
207. London, S.J., James Gauderman, W., Avol, E., Rappaport, E.B., and Peters, J.M., Family history and the risk of early-onset persistent, early-onset transient, and late-onset asthma, *Epidemiology*, 12, 577–283, 2001.
208. McConnell, R., Berhane, K., Gilliland, F., Islam, T., Gauderman, W.J., London, S.J. et al., Indoor risk factors for asthma in a prospective study of adolescents, *Epidemiology*, 13, 288–295, 2002.
209. Weiland, S.K., von Mutius, E., Hirsch, T., Duhme, H., Fritzsch, C., Werner, B. et al., Prevalence of respiratory and atopic disorders among children in the East and West of Germany five years after unification, *Eur. Respir. J.*, 14, 862–870, 1999.
210. Von Mutius, E., Martinez, F.D., Fritzsch, C., Nicolai, T., Roell, G., and Thiemann, H-H., Prevalence of asthma and atopy in two areas of West and East Germany, *Am. J. Respir. Crit. Care Med.*, 149, 358–364, 1994.
211. Heinrich, J., Hoelscher, B., Jacob, B., Wjst, M., and Wichmann, H.E., Trends in allergies among children in a region of former East Germany between 1992–1993 and 1995–1996. *Eur. J. Med. Res.*, 4, 107–113, 1999.
212. von Mutius, E., Weiland, S.K., Fritzsch, C., Duhme, H., and Keil, U., Increasing prevalence of hay fever and atopy among children in Leipzig, East Germany, *Lancet*, 351, 862–826, 1998.
213. Health Effects Institute, *Diesel Emissions and Lung Cancer: Epidemiology and Quantitative Risk Assessment — A Special Report of the Institute's Diesel Epidemiology Expert Panel*, Health Effects Institute, Charlestown, 1999.

214. U.S. Environmental Protection Agency, Health Assessment Document for Diesel Engine Exhaust, Report No.: EPA/600/8-90/057F, National Center for Environmental Assessment, Washington, DC, 2002.
215. Bell, M.L. and Davis, D.L., Reassessment of the lethal London fog of 1952: novel indicators of acute and chronic consequences of acute exposure to air pollution, *Environ. Health Perspect.*, 109 (Suppl. 3), 389–394, 2001.
216. Schwartz, J. and Marcus, A., Mortality and air pollution in London: a time series analysis, *Am. J. Epidemiol.*, 131, 185–194, 1990.
217. Thurston, G.D., Ito, K., Lippmann, M., and Hayes, C., Reexamination of London, England, mortality in relation to exposure to acidic aerosols during the 1963–1972 winters, *Environ. Health Perspect.*, 79, 73–82, 1989.
218. Schwartz, J., Air pollution and daily mortality: a review and meta analysis, *Environ. Res.*, 64, 36–52, 1994.
219. Schwartz, J., What are people dying of on high air pollution days? *Environ. Res.*, 64, 26–35, 1994.
220. Sunyer, J., Schwartz, J., Tobias, A., Macfarlane, D., Garcia, J., and Anto, J.M., Patients with chronic obstructive pulmonary disease are at increased risk of death associated with urban particle air pollution: a case-crossover analysis, *Am. J. Epidemiol.*, 151, 50–56, 2000.
221. De Leon, S.F., Thurston, G.D., and Ito, K., Contribution of respiratory disease to nonrespiratory mortality associations with air pollution, *Am. J. Respir. Crit. Care Med.*, 167, 1117–1123, 2003.
222. Levy, D., Sheppard, L., Checkoway, H., Kaufman, J., Lumley, T., Koenig, J. et al., A case-crossover analysis of particulate matter air pollution and out-of-hospital primary cardiac arrest, *Epidemiology*, 12, 193–199, 2001.
223. Clancy, L., Goodman, P., Sinclair, H., and Dockery, D.W., Effect of air pollution control on death rates in Dublin, Ireland: an intervention study, *Lancet*, 360, 1210–1214, 2002.
224. Hong, Y.C., Lee, J.T., Kim, H., Ha, E.-H., Schwartz, J., and Christiani, D.C., Effects of air pollutants on acute stroke mortaltiy, *Environ. Health Perspect.*, 110, 187–191, 2002.
225. Tellez-Rojo, M.M., Romieu, I., Velasco, S.R., Lezana, M.-A., and Hernandez, M.M., Daily respiratory mortality and PM_{10} pollution in Mexico City: importance of considering place of death, *Eur. Respir. J.*, 16, 391–396, 2000.
226. Colburn, K.A. and Johnson, P.R.S., Air pollution concerns not changed by S-PLUS, *Science*, 299, 665–666, 2003.
227. A Special Panel of the Health Review Committee of the Health Effects Institute, Commentary on revised analyses of selected studies, in *Special Report: Revised Analyses of Time-Series Studies of Air Pollution and Health*, Health Effects Institute, Ed, Health Effects Institute, Charlestown, MA, 2003, pp. 255–291.
228. Katsouyanni, K., Touloumi, G., Samoli, E., Petasakis, Y., Analitis, A., Le Tertre, A. et al., Sensitivity analysis of various models of short-term effects of ambient particles on total mortality in 29 cities in APHEA2, in *Special Report: Revised Analyses of Time-Series Studies of Air Pollution and Health*, Health Effects Institute, Ed, Health Effects Institute, Charlestown, MA, 2003, pp. 157–164.
229. Levy, J.I., Hammitt, J.K., and Spengler, J.D., Estimating the mortality impacts from particulate matter: what can be learned from between study variability, *Environ. Health Perspect.*, 108, 109–117, 2000.
230. Zanobetti, A. and Schwartz, J., Are diabetics more susceptible to the health effects of airborne particles? *Am. J. Respir. Crit. Care Med.*, 164, 831–833, 2001.
231. Goldberg, M.S., Burnett, R.T., and Bailar, J.C., 3rd, Tamblyn, R., Ernst, P., Flegel, K. et al., Identification of persons with cardiorespiratory conditions who are at risk of dying from the acute effects of ambient air particles, *Environ. Health Perspect.*, 109 (Suppl 4), 487–494, 2001.
232. Schwartz, J. and Zanobetti, A., Using meta-smoothing to estimate dose–response relationship trends across multiple studies, with application to air pollution and daily death, *Epidemiology*, 11, 666–672, 2000.
233. Schwartz, J., Assessing confounding, effect modification, and threshold in the association between ambient particles and daily deaths, *Environ. Health Perspect.*, 108, 563–568, 2000.

234. Schwartz, J., Ballester, F., Saez, M., Perez-Hoyos, S., Bellido, J., Cambra, K. et al., The concentration–response relation between air pollution and daily deaths, *Environ. Health Perspect.*, 109, 1001–1006, 2001.
235. Schwartz, J., Laden, F., and Zanobetti, A., The concentration–response relation between $PM_{2.5}$ and daily deaths, *Environ. Health Perspect.*, 110, 1025–1029, 2002.
236. Daniels, M., Dominici, F., Samet, J.M., and Zeger, S.L., Estimating particulate matter–mortality dose–response curves and threshold levels: an analysis of daily time-series for the 20 largest US cities, *Am. J. Epidemiol.*, 152, 397–406, 2000.
237. Pope, C.A., Invited commentary: Particulate matter–mortality exposure–response relations and threshold, *Am. J. Epidemiol.*, 152, 407–412, 2000.
238. Dominici, F., Daniels, M., McDermott, A., Zeger, S.L., and Samet, J.L., Shape of the exposure–response relation and mortality displacement in the NMMAPS study, in *Special Report: Revised Analyses of Time-Series Studies of Air Pollution and Health*, Health Effects Institute, Ed, Health Effects Institute, Charlestown, MA, 2003, pp. 91–96.
239. Lipfert, F.W. and Wyzga, R.E., Air pollution and mortality: the implications of uncertainties in regression modeling and exposure measurement, *J. Air Waste Manage. Assoc.*, 47, 517–523, 1997.
240. Murray, C.J. and Nelson, R.N., State-space modeling of the relationship between air quality and mortality, *J. Air Waste Manage. Assoc.*, 50, 1075–1080, 2000.
241. Zeger, S.L., Dominici, F., and Samet, J.M., Harvesting-resistant estimates of air pollution effects on mortality, *Ann. Epidemiol.*, 10, 171–175, 1999.
242. Schwartz, J., Harvesting and long term exposure effects in the relation between air pollution and mortality, *Am. J. Epidemiol.*, 151, 440–448, 2000.
243. Schwartz, J., Is there harvesting in the association of airborne particles with daily deaths and hospital admissions, *Epidemiology*, 12, 55–61, 2001.
244. Zanobetti, A., Schwartz, J., Samoli, E., Gryparis, A., Touloumi, G., Atkinson, R. et al., The temporal pattern of mortality responses to air pollution: a multicity assessment of mortality displacement, *Epidemiology*, 13, 87–93, 2002.
245. Dominici, F., McDermott, A., Zeger, S.L., and Samet, J.M., Airborne particulate matter and mortality: timescale effects in four US cities, *Am. J. Epidemiol.*, 157, 1055–1065, 2003.
246. Smith, R.L., Invited commentary: timescale-dependent mortality effects of air pollution, *Am. J. Epidemiol.*, 157, 1066–1070, 2003.
247. Dominici, F., McDermott, A., Zeger, S.L., and Samet, J.M., Response to Dr. Smith: timescale-dependent mortality effects of air pollution, *Am. J. Epidemiol.*, 157, 1071–1073, 2003.
248. Fan, J. and Watanabe, T., Inflammatory reactions in the pathogenesis of atherosclerosis, *J. Atheroscler. Thromb.*, 10, 63–71, 2003.
249. Task Force of the European Society of Cardiology, Heart rate variability: standards of measurement, physiological interpretation and clinical use, *Circulation*, 93, 1043–1065, 1996.
250. Tsuji, H., Venditti, F.J., Jr., Manders, E.S., Evans, J.C., Larson, M.G., Feldman, C.L. et al., Reduced heart rate variability and mortality risk in an elderly cohort: the Framingham Heart Study, *Circulation*, 90, 878–883, 1994.
251. Tsuji, H., Larson, M.G., Venditti, F.J., Jr., Manders, E.S., Evans, J.C., Feldman, C.L. et al., Impact of reduced heart rate variability on risk for cardiac events: the Framingham Heart Study, *Circulation*, 94, 2850–2855, 1996.
252. Pope, C.A., 3rd, Verrier, R.L., Lovett, E.G., Larson, A.C., Raizenne, M.E., Kanner, R.E. et al., Heart rate variability associated with particulate air pollution, *Am. Heart J.*, 138, 890–899, 1999.
253. Liao, D., Creason, J., Shy, C.M., Williams, R., Watts, R., and Zweidiner, R., Daily variation of particulate air pollution and poor cardiac autonomic function in the elderly, *Environ. Health Perspect.*, 107, 521–525, 1999.
254. Gold, D.R., Litonjua, A., Schwartz, J., Lovett, E., Larson, A., Nearing, B. et al., Ambient pollution and heart rate variability, *Circulation*, 101, 1267–1273, 2000.
255. Magari, S.R., Hauser, R., Schwartz, J., Williams, P.L., Smith, T.J., and Christiani, D.C., Association of heart rate variability with occupational and environmental exposure to particulate air pollution, *Circulation*, 104, 986–991, 2001.

256. Magari, S.R., Schwartz, J., Williams, P.L., Hauser, R., Smith, T.J., Christiani, D.C., The association between personal measurements of environmental exposure to particulates and heart rate variability, *Epidemiology*, 13, 305–310, 2002.
257. Magari, S.R., Schwartz, J., Williams, P.L., Hauser, R., Smith, T.J., Christiani, D.C., The associatioin of particulate air metal concentrations with heart rate variability, *Environ. Health Perspect.*, 110, 875–880, 2002.
258. Peters, A., Liu, E., Verrier, R.L., Schwartz, J., Gold, D.R., Milttleman, M. et al., Air pollution and incidence of cardiac arrhythmias, *Epidemiology*, 11, aa–17, 2000.
259. Gaspoz, J.M., Coxson, P.G., Goldman, P.A., Williams, L.W., Kuntz, K.M., Hunink, M.G. et al., Cost effectiveness of aspirin, clopidogrel, or both for secondary prevention of coronary heart disease, *N. Engl. J. Med.*, 346, 1800–1806, 2002.
260. Peters, A., Doring, A., Wichmann, H.-E., and Koenig, W., Increased plasma viscosity during an air pollution episode: a link to mortality, *Lancet*, 349, 1582–1587, 1997.
261. Seaton, A., Soutar, A., Crawford, V., Elton, R., McNerlan, S., Cherrie, J. et al., Particulate air pollution and the blood, *Thorax*, 54, 1027–1032, 1999.
262. Batalha, J.R., Saldiva, P.H., Clarke, R.W., Coull, B.A., Stearns, R.C., Lawrence, J. et al. Concentrated ambient air particles induce vasoconstriction of small pulmonary arteries in rats, *Environ. Health Perspect.*, 110, 1191–1197, 2002.
263. Brook, R.D., Brook, J.R., Urch, B., Vincent, R., Rajagopalan, S., and Silverman, F., Inhalation of fine particulate air pollution and ozone causes acute arterial vasoconstriction in healthy adults, *Circulation*, 105, 1534–1536, 2002.
264. U.S. Environmental Protection Agency USEPA, Fourth External Review Draft of Air Quality Criteria for Particulate Matter (June, 2003) — Volume II, US EPA, Report No.: EPA/600/P-99/002aD, Research Triangle Park, NC, 2003.
265. Wilson, W.E. and Suh, H.H., Fine particles and coarse particles: concentration relationships relevant to epidemiologic studies, *J. Air Waste Manage. Assoc.*, 47, 1238–1249, 1997.
266. Health Effects Institute, *Understanding the Health Effects of the Particulate Matter Mix: Progress and Next Steps*, Perspectives, Health Effects Institute, Charlestown, April, 2002.
267. McDonald, B. and Ouyang, M., Air cleaning — particles, in Spengler, J.D., Samet, J.M., McCarthy, J.F., Eds, *Indoor Air Quality Handbook*, McGraw-Hill, New York, 2000, pp. 9.1–9.28.
268. Institute of Medicine Committee on the Assessment of Asthma and Indoor Air, *Clearing the Air: Asthma and Indoor Air Exposures*, National Academy Press, Washington, DC, 2000.
269. Huttunen, K., Hyvarinen, A., Nevalainen, A., Komulainen, H., and Hirvonen, M.R., Production of proinflammatory mediators by indoor air bacteria and fungal spores in mouse and human cell lines, *Environ. Health Perspect.*, 111, 85–92, 2003.
270. Wilson, S.R. and Spengler, J.D., Emissions, Dispersion, and concentration of particles, in: Wilson, S.R., Spengler, J.D., Eds, *Particles in Our Air: Concentrations and Health Effects*, Harvard University Press, Cambridge, MA, 1996, pp. 41–62.
271. Murray, J.F., *The Normal Lung*, W.B. Saunders Co, Philadelphia, 1986.
272. Jaques, P.A. and Kim, C.S., Measurement of total lung deposition of inhaled ultrafine particles in healthy men and women, *Inhal. Toxicol.*, 12, 715–731, 2000.
273. Kim, C.S. and Kang, T.C., Comparative measurement of lung deposition of inhaled fine particles in normal subjects and patients with obstructive airway disease, *Am. J. Respir. Crit. Care Med.*, 155, 899–905, 1997.
274. Mathew, O.P. and Sant'Ambrogio, G., Development of upper airway reflexes, in Cherniack, V., Mellins, R.B., Eds, *Basic Mechanisms of Pediatric Respiratory Disease: Cellular and Integrative*, Philadelphia, BC Decker, 1991, pp. 55–71.
275. Mandell, G.L., Bennett, J.E., and Dolin, R., Eds, *Principles and Practice of Infectious Diseases*, Churchill Livingstone, New York, 2000.
276. Diaz-Sanchez, D., Zhang, K., Nutman, T.B., and Saxon, A., Differential regulation of alternative 3′ splicing of epsilon messenger RNA variants, *J. Immunol.*, 155, 1930–1941, 1995.
277. Diaz-Sanchez, D., Tsien, A., Casillas, A., Dotson, A.R., and Saxton, A., Enhanced nasal cytokine production in human beings after *in vivo* challenge with diesel exhaust particles, *J. Allergy. Clin. Immunol.*, 98, 114–123, 1996.

278. Diaz-Sanchez, D., Jyrala, M., Ng, D., Nel, A., and Saxon, A., *In vivo* nasal challenge with diesel exhaust particles enhances expression of the CC chemokines rantes, MIP-1alpha, and MCP-3 in humans, *Clin Immunol.*, 97, 140–145, 2000.
279. Wjst, M., Reitmeir, P., Dold, S., Wulff, A., Nicolai, T., and von Loeffelholz-Colberg, E.F. et al., Road traffic and adverse effects on respiratory health in children, *B. Med. J.*, 307, 596–600, 1993.
280. Kunzli, N. and Tager, I.B., The semi-individual study in air pollution epidemiology: a valid design as compared to ecologic studies, *Environ. Health Perspect.*, 105, 1078–1083, 1997.
281. Edwards, J., Walters, S., and Griffiths, R.K., Hospital admissions for asthma in preschool children: relationship to major roads in Birmingham, United Kingdom, *Arch. Environ. Health*, 49, 223–227, 1994.
282. Oosterlee, A., Drijver, M., Lebret, E., and Brunekreef, B., Chronic respiratory symptoms in children and adults living along streets with high traffic density, *Occup. Environ. Med.*, 53, 241–247, 1996.
283. English, P., Neutra, R., Scalf, R., Sullivan, M., Waller, L., and Zhu, L., Examining associations between childhood asthma and traffic flow using a geographic information system, *Environ. Health Perspect.*, 107, 761–767, 1999.
284. Wilkinson, P., Elliott, P., Grundy, C., Shaddick, G., Thakrar, B., Walls, P. et al., Case–control study of hospital admission with asthma in children aged 5– 14 years: relation with road traffic in north west London, *Thorax*, 54, 1070–1074, 1999.
285. Venn, A., Lewis, S., Cooper, M., Hubbard, R., Hill, I., Boddy, R. et al., Local road traffic activity and the prevalence, severity, and persistence of wheeze in school children: combined cross sectional and longitudinal study, *Occup. Environ. Med.*, 57, 152–158, 2000.
286. Venn, A.J., Lewis, S.A., Cooper, M., Hubbard, R., and Britton, J., Living near a main road and the risk of wheezing illness in children, *Am. J. Respir. Crit. Care Med.*, 164, 2177–2180, 2001.
287. Roemer, W.H. and van Wijnen, J.H., Daily mortality and air pollution along busy streets in Amsterdam, 1987–1998, *Epidemiology*, 12, 649–653, 2001.
288. Lin, S., Munsie, J.P., Hwang, S.A., Fitzgerald, E., and Cayo, M.R., Childhood asthma hospitalization and residential exposure to state route traffic, *Environ. Res.*, 88, 73–81, 2002.
289. Brauer, M., Hoek, G., Van Vliet, P., Meliefste, K., Fischer, P.H., Wijga, A. et al., Air pollution from traffic and the development of respiratory infections and asthmatic and allergic symptoms in children, *Am. J. Respir. Crit. Care Med.*, 166, 1092–1098, 2002.
290. Wilhelm, M. and Ritz, B., Residential proximity to traffic and adverse birth outcomes in Los Angeles county, California, 1994–1996, *Environ. Health Perspect.*, 111, 207–216, 2003.
291. Dolk, H., Pattenden, S., Vrijheid, M., Thakrar, B., and Armstrong, B., Perinatal and infant mortality and low birth weight among residents near cokeworks in Great Britain, *Arch. Environ. Health*, 55, 26–30, 2000.
292. Bobak, M. and Leon, D.A., Pregnancy outcomes and outdoor air pollution: an ecological study in districts of the Czech Republic 1986–8, *Occup. Environ. Med.*, 56, 539–543, 1999.
293. Dejmek, J., Selevan, S.G., Benes, I., Solansky, I., and Sram, R.J., Fetal growth and maternal exposure to particulate matter during pregnancy, *Environ. Health Perspect.*, 107, 475–480, 1999.
294. Rogers, J.F., Thompson, S.J., Addy, C.L., McKeown, R.E., Cowen, D.J., and Decoufle, P., Association of very low birth weight with exposures to environmental sulfur dioxide and total suspended particulates, *Am. J. Epidemiol.*, 151, 602–613, 2000.
295. Bobak, M., Outdoor air pollution, low birth weight, and prematurity, *Environ. Health Perspect.*, 108, 173–176, 2000.
296. Ha, E.H., Hong, Y.C., Lee, B.E., Woo, B.H., Schwartz, J., and Christiani, D.C., Is air pollution a risk factor for low birth weight in Seoul? *Epidemiology*, 12, 643–648, 2001.
297. Ritz, B., Yu, F., Fruin, S., Chapa, G., Shaw, G.M., and Harris, J.A., Ambient air pollution and risk of birth defects in Southern California, *Am. J. Epidemiol.*, 155, 17–25, 2002.
298. Xu, X., Ding, H., and Wang, X., Acute effects of total suspended particles and sulfur dioxides on preterm delivery: a community-based cohort study, *Arch. Environ. Health*, 50, 407–415, 1995.
299. Schwartz, J. and Dockery, D.W., Increased mortality in Philadelphia associated with daily air pollution concentrations, *Am. Rev. Respir. Dis.*, 145, 600–604, 1992.

300. Gwynn, R.C., Burnett, R.T., and Thurston, G.D., A time-series analysis of acidic particulate matter and daily mortality and morbidity in the Buffalo, New York, region, *Environ. Health Perspect.*, 108, 125–133, 2000.
301. Ostro, B.D., Hurley, S., and Lipsett, M.J., Air pollution and daily mortality in the Coachella Valley, California: a study of PM_{10} dominated by coarse particles, *Environ. Res.*, 81, 231–238, 1999.
302. Morgan, G., Corbett, S., Wlodarczyk, J., and Lewis, P., Air pollution and daily mortality in Sydney, Australia, 1989 through 1993, *Am. J. Publ. Health*, 88, 759–764, 1988.
303. Vedal, S., Brauer, M., White, R., and Petkau, J., Air pollution and daily mortality in a city with low levels of pollution, *Environ. Health Perspect.*, 111, 45–51, 2003.
304. Bateson, T.F. and Schwartz, J., Control of seasonal variation and time trends in case-crossover studies of the acute effects of environmental exposures, *Epidemiology*, 10, 539–544, 1999.
305. Neas, L.M., Schwartz, J., and Dockery, D., A case-crossover analysis of air pollution and mortality in Philadelphia, *Environ. Health Perspect.*, 107, 629–631, 1999.
306. Lee, J.T., Shin, D., and Chung, Y., Air pollution and daily mortality in Seoul and Ulsan, Korea, *Environ. Health Perspect.*, 107, 149–154, 1999.
307. Samet, J.M., Dominici, F., Zeger, S.L., Schwartz, J., and Dockery, D.W., The National Morbidity, Mortality, and Air Pollution Study. Part I: Methods and methodologic issues, *Res. Rep. Health. Effect Inst.*, 94 (Part 1), 2000.
308. Samet, J.M., Zeger, S.L., Dominici, F., Curriero, F., Coursac, I., Dockery, D.W. et al., The National Morbidity, Mortality, and Air Pollution Study. Part II: Morbidity and mortality from air pollution in the United States, *Res. Repair. Health Effect Inst.*, 94(Part 2) 2000.
309. Samet, J.M., Dominici, F., Curriero, F.C., Coursac, I., and Zeger, S.L., Fine particulate air pollution and mortality in 20 U.S. cities, 1987–1994, *N. Engl. J. Med.*, 343, 1742–1749, 2000.
310. Dominici, F., McDermott, A., Daniels, D., Zeger, S.L., Samet, J.L., Mortality among residents of 90 cities, in *Special Report: Revised Analyses of Time-Series Studies of Air Pollution and Health*, Health Effects Institute, Ed, Health Effects Institute, Charlestown, MA, 2003, pp. 9–24.
311. Touloumi, G., Katsouyanni, K., Zmirou, D., Schwartz, J., Spix, C., Ponce de Leon, A. et al., Short-term effects of ambient oxidant exposure on mortality: a combined analysis of the APHEA project, *Am. J. Epidemiol.*, 146, 177–185, 1997.
312. Katsouyanni, K., Touloumi, G., Spix, C., Schwartz, J., Baldacci, F., Medina, S. et al., Short term effects of ambient sulphur dioxide and particulate matter on mortality in 12 European cities: results from time series data from the APHEA project, *Br. Med. J.*, 314, 1658–1663, 1997.
313. Zmirou, D., Schwartz, J., Saez, M., Zanobetti, A., Wojtyniak, B., Touloumi, G. et al., Time-series analysis of air pollution and cause-specific mortality, *Epidemiology*, 9, 495–503, 1998.
314. Katsouyanni, K., Touloumi, G., Samoli, E., Gryparis, A., Le Tertre, A., Monopolis Y. et al., Confounding and effect modification in the short-term-effects of ambient particles on total mortality: results from 29 European cities within the APHEA2 project, *Epidemiology*, 12, 521–531, 2000.
315. Samoli, E., Schwartz, J., Wojtyniak, B., Touloumi, G., Spix, C., Baldacci, F. et al., Investigation regional differences in short-term effects of air pollution on daily mortality in the APHEA project: a sensitivity analysis for controlling long-term trends and seasonality, *Environ. Health Perspect.*, 109, 349–353, 2001.
316. Pope, C.A., 3rd, Hill, R.W., and Villegas, G.M., Particulate air pollution and daily mortality on Utah's Wasatch Front, *Environ. Health Perspect.*, 107, 567–573, 1999.
317. Loomis, D., Castillejos, M., Borja-Aburto, V.H., Dockery, D.W., Stronger effects of coarse particles in Mexico City, in Phalen R., and Bell, Y., Eds, Proceedings of the Third Colloquium Particulate Air Pollution and Human Health, Durham, NC, 1999.
318. Cifuentes, L.A., Vega, J., Kopfer, K., and Lave, L.B., Effect of the fine fraction of particulate matter versus the coarse mass and other pollutants on daily mortality in Santiago, Chile, *J. Air Waste Manage. Assoc.*, 50, 1287–1298, 2000.
319. Casarett, L.J., Toxicololgy of the respiratory system, in Casarett, L.J. and Doull, J., Eds, *Toxicology: The Basic Science of Poisons.*, MacMillan Publishing Co., New York, 1975, pp. 201–224.
320. Lippmann, M. and Schlesinger, R.B., Interspecies comparisons of particle deposition and mucociliary clearance in tracheobronchial airways, *J. Toxicol. Environ. Health*, 13, 441–469, 1984.

321. Morrow, P.E., Task Group on Lung Dynamics: deposition and retention models for internal dosimetry of the human respiratory tract, *Health Phys.*, 12, 173–207, 1966.
322. International Commission on Radiological Protection, Human respiratory tract model for radiological protection, A report of a Task Group of the International Commission on Radiological Protection, *Ann. ICRP*, 24, 1–482, 1994.
323. Kay, A.B., Allergy and allergic diseases, *N. Engl. J. Med.*, 344, 30–37, 2001.
324. Busse, W.W., and Lemanske, R.F., Jr., *Asthma, N. Engl. J. Med.*, 344, 350–362, 2001.
325. Dominici, F., McDermott, A., Zeger, S.L., and Samet, J.M., National maps of the effects of particulate matter on mortality: exploring geographic variation, *Environ. Health Perspect.*, 111, 39–43, 2003.
326. Federal Register, National Ambient Air Quality Standard for Particulate Matter, Report No.: 40 CRF Part 50, National Archives and Records Administration, Washington, DC, December 1996.

Index

A

AAAR, *see* American Association for Aerosol Research
Accumulation mode, 32, 48, 54, 55, 57–58, 206
ACGIH, *see* American Conference of Governmental Industrial Hygienists
Actinon, 361, 364, 365
  basic equations, 390–395
Actinon decay products
  characteristics, 390
  generation, 558–559
  radioactive markers, 414
Administrative controls, 598
Adventist Health Study of Smog (AHSMOG), 656, 658
Aeroallergens, 639; *see also* Allergens
Aerodynamic diameter, 25, 77, 80
Aerodynamic particle sizer, 257
Aeroionizer, 439
Aeroions, 439
Aerosol Laboratory of the All–Union Institute of Physico–Technical and Radiotechnical Institute (VNIIFTRI), 346
Aerosol standard, 350, 557
Aerosol time–of–flight mass spectrometry (ATOFMS), 611
Aerosols; *see also* Bioaerosols; Indoor aerosols; Inhaled aerosols
  artificial radioactive, 563–565
  composition, 6–9
  concentration measurement, 1–3
  definition, 19
  indoor, 624–625
  outdoor, 622–623
  particle size, 20, 417–420
  stability, 418
Aethalometer, 41
AHSMOG, *see* Adventist Health Study of Smog
Air conditioners, 386
Air fresheners, 197–198, 202
Air sampling system (ASS), 444, 451, 452
Airflow, 82–86, 115–117
  bifurcations and branching networks, 85
  boundary conditions, 116
  computational fluid dynamics (CFD), 117
  curved tubes, 85
  equations, 115–116
  idealized tubes, 83–85
  indoor, 215–216
  laminar vs. turbulent 82–83, 84, 88–91, 117
  computational fluid–particle dynamics (CFPD), 119–120
  steady vs. unsteady, 82
  velocity profiles, 116–117
Airway epithelium, 315
Aitkin modes, 610
Allen, G., 43
Allergens, 300–302, 316, 319, 639
  dose, 317–318
  indoor, 322
  size, 307
Allergies, 294–295, 633, 635, 656–659
  diesel exhaust 603–604
Alpha particles, 348, 364, 379, 445
  counting, 374
American Association for Aerosol Research (AAAR), 11, 12
American Conference of Governmental Industrial Hygienists (ACGIH), 229, 320
Analytical filters, 552, 553
Andersen impactors, 244, 308
Angina, 277
Annual limit of intake (ALI), 460
Anthrax, 305
Anticholinergics, 268
Antigens, 329
Antiinflammatory agents, 268
Antimicrobials, 268
Artificial radioactive aerosols, 563–565
  concentration measurement, 406–417
Asbestos, 227, 229, 239, 247, 248
Asbestosis, 227
Asgharian, B., 91, 179–180
Aspergillosis, 268
Aspiration efficiency, 256
Aspirator condenser method, 421
Aspirators, 536, 549
ASS, *see* Air sampling system
Asthma, 132, 318, 319, 651, 656–659
  allergies, 293
  diesel exhaust, 601–602, 603–604
  endotoxins, 318, 648
  immunobiology, 639
  nebulizer, 274
  treatment, 160, 267
  triggers, 308
Atmospheric monitoring, 553–554
ATOFMS, *see* Aerosol time–of–flight mass spectrometry

Atomic emission spectroscopy, 246
Awards, 12–13

B

Bacteria, 297–299, 313
  dose, 320
  gram–negative, 295, 320, 328
  gram–positive, 305, 328
  negative, 639
  size, 307–308
Bacterial depsipeptide, 329
Balashazy, I., 107
Bateman equation, 366
Beeckmans, J.M., 86
Beeswax, *see* candles
BET method, 254
Beta gauge, 39
Beta–particles, 377, 378
Beta–spectrometry, 381–382, 387, 389, 390
Bimodal distribution, 207
Bioaerosols, 32
  concentration, 323
  defined, 292
  health effects, 293–301, 622, 647
  measurement and analysis, 321–330
  respiratory dosimetry, 314–321
  size distributions, 306–314
  sources and transmission, 304–305
Biological agent, 292
Biological assays, 328–329
Biotechnological agents, 268–269
Bluff body bias, 67–69
BMRC, *see* British Medical Research Council
Boehringer Ingelheim, 268
Boltzmann equilibrium charge distribution, 27
Bolus technique, 5–6
Breath–hold time, 178
Breath–holding method, 5
Breathing patterns, 125–126
Breathing rates, 177, 178, 182, 191, 231
  bronchial dose, 576
  miners, 351, 441, 444, 453–459, 473
Breathing zone, 63–69, 233
  bluff body bias, 67–69
  concentration measurements, 105, 450–453
  definition, 64–67
  size distributions, 244, 417
Breathing zone exposures (BZEs); *see also* Exposure assessment sampling, 62, 69–71
British Medical Research Council (BMRC), 229
Bronchial dosimeter, 386, 447
Bronchiectasis, 294
Bronchitis, 132, 658
Brownian motion, 26–27, 29, 81, 89, 107
Building envelopes, 194–195, 202, 215
Building materials, 195, 211
Buoyancy effects, 247
Butane, 270
Button sampler, 236
Byssinosis, 228
BZEs, *see* Breathing zone exposures

C

CAMMs, *see* Continuous aerosol mass monitors
CAMs, *see* Continuous air monitors
Cancer, 269, 659; *see also* Lung cancer
Candles, 197–198, 201
Carbon monoxide poisoning, 202
Carcinogenesis, 634, 659, 660; *see also* Cancer; Lung cancer
Cardiac arrhythmias, 678
Cardiorespiratory diseases, 108
Cardiovascular diseases, 676–678
Career Achievement Award (ISAM), 12
Cascade impactors, 2, 244, 245, 277–279, 324
Case–control studies, 160, 161, 162–165, 166, 168
Cass, G., 196
Cassette tape sampler, 311
Cast studies
  particle deposition measurements, 134, 137
CATHIA sampler, 238
CFC, *see* Chlorofluorocarbon
CFD, *see* Computational fluid dynamics
CFPD, *see* Computational fluid-particle dynamics
Charcoal canisters, 362
Charcoal technique, 385
Chemical assays, 329–330, 329
Chemical properties, 30–32
Chernobyl, 446, 462, 483, 517–538, 552
  aerosol characteristics, 520–526, 529–531
  aerosols of the "Shelter," 531–537
  ejection of radionuclides, 518–519
  forest fires, 531, 517
  gaseous components, 526–529
  global transfer, 519–520
  sampling devices, 520
Chlorofluorocarbon (CFC) propellants, 270; *see also* Inhalers
Chronic obstructive pulmonary disease (COPD), 131–132, 169, 170, 177–178, 268, 294, 319, 661, 674
CIP–10 sampler, 236, 237, 242
Classical risk paradigm, 61
Cleaning, 197–198, 203–204,
Clearance, 128–131, 135, 158, 182, 281, 315, 462, 629–632
  bronchial dose, 576
  deposition patterning, 135
  diesel exhaust, 638
  macrophages, 130–131, 130
  modeling, 178–180, 178
  mucociliary 128–130, 631
  radon, 359
Cloud, 32
Cloud motion, 92–93
CNC, *see* Condensation nucleus counter
Coagulation, 29
Coarse aerosols (CA), 417
Coarse mode, 32, 48, 206

Coarse particles, 37–38, 40
Coaxial electrostatic precipitator, 235
Cohen, D., 3, 8, 165–166, 167
Cohort studies, 161, 162
Combustion, 229
Combustion sources, 196–205, 610, 611; *see also* Indoor aerosols
 air fresheners, 202
 candles, 201
 cleaning, 203–204
 cooking, 196, 624
 fireplace, 199–200
 health effects, 647
 incense, 201–202
 pesticides, 202
 renovations, 204
 tobacco smoke, 200–201, 624
 unvented kerosene heaters (UKHs), 202
 walking, 202–203
Computational fluid dynamics (CFD), 85–86, 117
 modeling, 124
 simulation, 173
 techniques, 177
Computational fluid–particle dynamics (CFPD), 119–120
 modeling, 118, 121, 122, 140
Concentrations, 61, 62–63, 66; *see also* Dose; Epidemiological studies
 bioaerosols, 323
 expression, 395
 health effects, 160
 indoor, 194, 195
 long–term temporal variability, 164
 mass, 21, 229
 optical particle counters, 250
 photometers, 251
 urban, 192
Condensation, 29–30, 209, 536
 reactions, 609
Condensation aerosol generator (La Mer–generator), 7
Condensation nucleus counter (CNC), 609
Condensation particle counter, 253
Confocal microscopy, 5
Confounding variables, 164
Conical inhalable sampler (CIS), 237
Conifuge, 9
Continuous aerosol mass monitor (CAMM), 39, 45
Continuous air monitors (CAMs), 564–565
Continuous sulfate analyzers, 43
Continuum regime, 81
Cooking, 196, 197–198, 624
COPD, *see* Chronic obstructive pulmonary disease
Corticosteroids, 267–268
Coughing, 178, 304; *see also* Clearance
Count median diameter (CMD), 23
Count–nephelometric method, 438
Count–weight method, 438
Cowled sampler, 234
CR–39 plastic, 445
Cross–validation, 173
Crystals, 410–411, 412, 480
Crystal scintillation, 414; *see also* Spectrometry
Cultivation–based methods, 327–328
Cunningham slip correction factor, 24, 78, 79, 82
Cyclone samplers, 311, 312
Cyclones, 323, 325
Cystic fibrosis (CF), 132, 177, 268, 274, 276
Cystic fibrosis transport receptor (CFTR), 268
Cytotoxic agents, 316

D

David Sinclair Award, 13
Defibrillators, 678
Department of Environmental Medicine at New York University, 10
Deposition, 86–93, 117
 cloud motion, 92–93
 coefficient, 105
 diffusion, 431–434
 diffusion, 89–90
 efficiency, 5, 86–87, 88, 125
 electrostatic charge, 91–92
 indoor, 208, 213, 215, 216–217
 inertial impaction, 86–87
 interception, 90–91
 lung, 612
 measurement, 280–281
 modeling, 172, 175, 177, 182
 radioactive markers, 109
 regional, 4–6
 respiratory tract, 626–629
 respiratory tract, 630
 sedimentation, 88–89
 site specific, 177
 ultrafine particles, 106–107
Deposition modeling
 advantages, 122
 age, 133
 breathing patterns, 125–126
 clearance, 131
 compared with actual data, 139–140, 143–144
 computational fluid dynamics (CFD), 124
 computational fluid-particle dynamics (CFPD), 119–120, 121, 122, 140, 144
 deterministic models, 117, 118–119, 121, 122, 140–144
 disease, 132–133
 empirical models, 117, 118–119, 121–122
 extrathoracic, 140
 measurements, 134–139
 mode of respiration, 125
 predictions, 133–134
 pulmonary, 142
 respiratory system environment, 126–128
 respiratory system morphology, 122–124
 selecting a model, 120–122
 stochastic models, 118–119, 121, 122, 123, 140
 tracheobronchial, 140–142
 uncertainties, 140
Deterministic models, 117, 118–119, 121, 122, 140–144

Diabetes, 269
Diesel engines, 56–57, 602
Diesel exhaust, 47–49, 172, 321, 601–604, 642
  asthma and allergies, 603–604
  characteristics of emissions, 602
  inflammation, 636, 637–638
  lung cancer, 603
  oxidative damage, 634, 635
Diesel fuels, 602
Diesel particles
  formation, 49
  measurements, 49–56
  sampling issues, 49–50
  size distributions, 52, 53, 54, 57
  structure and composition, 52–56
Differential mobility analyzer (DMA), 2, 463, 608–609
Diffusion, 89–90, 626
  barrier, 578
  deposition, 431–434
  filtration, 544–545
  indoor, 208
  properties, 26–27
  rate, 207
  ultrafine particles, 608, 612
Diffusion batteries, 396, 397, 398, 406, 420, 421, 422, 426, 428, 430, 431, 561
Diffusion method, 421, 425, 438, 561
  errors, 434–436
Dimethylether, 270
Direct measurement, 464–483, 490–491; *see also* Direct reading instruments; Dosimetry; Measurement
  assessment of uncertainities, 467–468
  correction for shift of equilibrium, 468–472
  errors, 476–480
  model measurement, 474–475
  parametric variations, 472–474
  phantom measurements, 475–476
  portable instruments, 480
  radioactive markers, 481–483
  theory, 465–467
Direct method, 357, 358
Direct reading instruments
  applications, 255–257
  condensation particle counter, 253
  light scattering, 248–252
  pressure drop sensor, 253–254
  surface area, 254–255
  tapered element oscillating microbalance (TEOM), 252–253
Discretization, 117, 121
Diseases, 131–133, 304
  cardiovascular, 676
  hypersensitivity, 294–295
  infectious, 293–294
  ischemic heart disease (IHD), 674
  respiratory system, 131–133
  skin, 276
Disk pulverizer, 560
Dispersion
  dry powder inhalers (DPIs), 272
Distribution
  count, 22–23
  lognormal, 21–22, 23
  moment, 22
  number, 31
  volume or mass, 22–23, 31
Diurnal patterns
  bioaerosols, 313
DMA, *see* Differential mobility analyzer
DNA aerosols, 268–269
Dorr–Oliver cyclone, 241–242
Dose, 101–110, 158, 226; *see also* Dosimetry; Epidemiological studies
  allergens, 317–318
  bacteria, 320
  concentration safety standards, 106
  definition, 105
  endotoxin, 319
  environmental dosimetry, 102–104
  fungi, 320
  glucans, 318–319
  health effects, 158–160
  infectious, 316–317
  nanometer particles, 106–109
  nonuniformity of deposition, 106–107
  particle size distribution, 417
  radioactive markers, 109
  radon decay products, 354–358
  radon, 363–364
  uncertainties, 105–106
  vs. exposure, 104–105, 180–182
Dosimetry; *see also* Direct measurement; Dose
  accuracy, 492
  breathing rate, 453–459
  breathing zone, 450–453
  dose assessment, 459–464
  environmental, 102–104
  medical aerosols, 281–283
  microdosimetry, 136–137
  radon decay products, 447–448
  radon, 347, 445–447, 574–577
  unattached fraction measurements, 450
  uncertainties, 346–348, 441–445
DPIs, *see* Dry powder inhalers
Droplet method, 414
Drugs
  macrophage clearance 130–131
  mucociliary clearance, 128
Dry powder inhalers (DPIs), 266, 272–273, 282, 283
Dry powder aerosols, 268
Dust, 32
Dyspnea, 228

E

EAA, *see* European aerosol assembly
Ecological fallacy, 161, 166, 171
Ecological studies, 160–161, 162, 165–166, 167, 168, 171
Electret–passive environment radon monitors (E–PERMs), 362
Electrical properties, 27, 439–441

Electroaerosols, 439
Electron microscopy, 421
Electrostatic charge, 91–92, 247, 306, 419
  contamination in radiosynthesis laboratories, 598
  filtration, 545–546
Electrostatic collecting system, 483
Electrostatic effects, 315
Electrostatic precipitators (ESPs), 39, 439, 440
Emphysema, 6, 131–132
Empirical models, 117, 118–119, 121–122
Endocytosis, 316
Endotoxins, 295, 304, 305, 316, 322, 328, 612, 635, 638
  chemical analysis, 330
  dose, 319
  health effects, 623, 648
  inflammation, 639
Engineering controls, 598
Entrance effect, 431, 434
Environmental dosimetry, 102–104
Environmental Protection Agency (EPA), US, 36, 38, 40, 44, 45, 106, 191, 133, 247, 574, 587
Environmental Radiation Ambient Monitoring System (ERAMS), 587
Enzyme–linked immunosorbent assay (ELISA), 329
EPA, *see* Environmental Protection Agency
Epidemiological studies, 158, 160–171, 182, 490
  aerosol exposure, 168–169
  fine particle inhalation, 169–171
  miners, 441, 442, 509–513
  radon case–control studies, 162–165, 166
  radon ecological studies, 165–166, 167
  radon risk estimates, 166–168
Epiphanometer, 254, 400, 401
Equilibrium factor, 161
Equivalent equilibrium volume activity (EEVA), 504
Ergosterol, 330
Ergotamine tartrate, 269
Errors
  model misspecification, 172, 173
  parameter, 173, 174
ESPs, *see* Electrostatic precipitators
European Aerosol Assembly (EAA), 11, 12
European Committee for Standardization (CEN), 230
Evaporation, 30
Evaporators, 596
Exhaust emissions, see Diesel exhaust; Traffic
Exhaust filters, 55–57
Exhaust–boxes, 598
Exposure, 102, 104–105, 158; *see also* Epidemiological studies; Exposure assessment
  aerosols, 168–169
  ambient vs. personal, 169–170
  error, 165
  health effects, 159–160, 165
  indoor, 191
  infection, 321
  personal, 248
  radon, 167, 169
  regulations, 232
  vs. dose, 180–182
Exposure assessment; *see also* Exposure
  bioaerosols, 316–321
  bluff body bias, 67–69
  breathing zone, 63–69, 70
  concentrations vs. personal exposures, 62–63
  concentrations, 61, 66
  health, 642
  personal exposures, 66–67
  sampling by contaminant type, 70–71
Exposure metrics, 228–229
Express method, 372–373, 447
Extrathoracic simulations, 140

F

FDMS, *see* Filter dynamics measurement systems
Federal Reference Method (FRM), 36, 37, 38, 40
Fibers, 128, 130, 543; *see also* Filtration
  asbestos, 8, 227
  deposition, 90–91
  hollow, 91
  sampling, 239
Fibrinogen, 637
Fick's first law of diffusion, 27
Filters; *see also* Filtration
  analytical filters, 549, 550, 552, 553
  Chernobyl Shelter, 535
  fiber, 548–549
  holders, 234
  radon decay products, 325, 367, 506
  sampling, 312, 506
  technology, 256
Filter dynamics measurement systems (FDMS), 39–40
Filtration, 541–554; *see also* Filters
  analytical filters AFA, 549, 550
  atmospheric monitoring, 553–554
  definitions, 542–544
  desorption of volatile substances, 553
  fiber filters FP, 548–549
  gaseous compounds, 552–553
  mechanisms, 544–546
  most penetrating size, 546–547
  nonstationary, 547–548
  particle size measurement, 549–552
  pressure drop, 547
Filtration ability of lungs (FAL), 103, 456
Findeisen, W., 86
Fine aerosols (FA), 38–40, 417
  composition, 438–439
  generation, 436–438
  inhalation, 169–171
Fine particle fraction (FPF), 272
Fire
  Chernobyl, 531, 534, 536, 538
Fireplaces 197–198, 199–200
Fluid dynamics, *see* Airflow
Fluid flow modeling, 82; *see also* Airflow
Fluorophores, 326
Forward–marching approach, 215
Fraser, D.A., 90–91

Fraunhofer Institute of Toxicology and Aerosol Research (Fh–ITA), 11
Free–molecule regime, 81
FRM, *see* Federal Reference Method
Fuchs' coagulation theory, 401
Fuel–containing materials (FCMs), 532
Fuel sulfur, 52, 54–55
Fumes, 32, 253
Fume hoods, 597
Fungi, 228, 299, 300–301, 306, 313, 318, 322, 327, 622
  dose, 320
  size 307–308, 307

G

Gamma ray detector system, 4
Gamma scintigraphy 135–136, 281, 282
Gas chromatography (GC), 329
Gases
  Chernobyl, 526–529
  detection with filters, 552–553
  pollutants, 169
  radon, 577–578
  thoron, 577–578
Geiger counter, 597
Generators
  radiation, 558
Genes, 268–269, 268
Geometric diameters, 89, 90
Glove–box, 598
Glucans, 295, 316
  dose, 318–319
Gram–negative bacteria, 295, 320, 328
Gram–positive bacteria, 305, 328
Gravimetric analysis, 246–247; *see also* Sample analysis
Gravitational effect, 546
Gravitational settling, 324
Growth processes, 29–30
GSP inhalable sampler, 237

H

Haber's law, 104
Hamilton, A., 226
Harris, R.L., 90–91
Harvard School of Public Health (HSPH), 43
Harvard Six Cities Study, 668
Harvesting, 674
Hatch–Choate equations, 23
Haze, 32
Health effects, 47, 619–622; *see also* Epidemiological studies
  acute exposure and mortality, 660–676
  aerosol characteristics, 622–645
  allergy and asthma, 656–659
  ambient particulate matter, 640–649
  cancer, 659
  cardiovascular mortality, 676–678
  clearance, 629–632
  concentrations, 160
  deposition, 626–629
  dose, 158–160
  exposure and birth outcomes, 649–651, 652
  exposure, 159–160
  induction of inflammation, 635–639
  long–term exposure and mortality, 651–656
  oxidative damage, 634–635
  toxicity, 633–634
HFC, *see* Hydrofluorocarbon
High–purity germanium detector, 482
Hoffman, W., 3, 107, 175
HSPH, *see* Harvard School of Public Health
Human Body Spectrometer, 474
Hungary
  radon exposure, 446
Hydrofluorocarbon (HFC) popellants, 270; *see also* Inhalers
Hygroscopic aerosols, 9
Hygroscopic growth, 126–128, 129
Hygroscopicity, 79–81, 306, 462
  measurements, 54–55
Hypersensitivity diseases, 294–295; *see also* Allergies

I

IAEA, *see* International Atomic Energy Agency
ICRP, *see* International Commission of Radiologiocal Protection
IDS, *see* Individual dosimetric system
IEC, *see* International Electrotechnical Commission
Image charge, 91
Immunoassays, 329
Impaction, 208, 626
Impactors, 309, 310, 311, 312, 536
  Andersen, 244, 308
  cascade, 2, 244, 245, 277–279, 324
  comparison with filters, 551–552
  inertial, 9
  whirling–arm, 325
Impingers, 233, 246, 311, 323
  liquid, 325
  twin, 279–280
IMPROVE sites, 41
Incense, 197–198, 201–202
Individual dosimetric system (IDS), 444, 451, 452, 453
Indoor aerosols, 205–212; *see also* combustion sources
  allergens, 322
  model of behavior, 213–215
  particle formation, 209–211
  particle motion, 208
  size distributions, 205–208
  surface area, 211–212
  transformation processes, 213, 214
  transport, 215–218
Industrial revolution, 226, 227
Inertia
  filtration, 544, 545, 551
Inertial sampling, 324–325; *see also* Measurement
Inertial impaction, 86–87, 323
Inertial impactor, 9
Inertial precipitation, 438

Inertial properties, 26
Infection
  agents, 296, 321
  diseases, 293–294
  dose, 316–317
  transmission, 304–305
Inflammation, 316, 635–639
  agents, 302–303
  diseases, 295, 304, 676
  endotoxins, 648
Influenza, 293, 321
Inhalable samplers, 236–237
Inhalable fraction, 21
Inhalation fever, 228
Inhalation modeling, 172–182, 460
  breathing rate, 178, 157
  dose vs. exposure, 180–182
  interpersonal variability, 174–175
  intersubject morphometric variability, 175–177
  lung clearance, 178–180
  motivation, 172
  respiratory tract morphology, 175
  systematic morphometric variability, 177–178
Inhalation rates, 102
Inhaled aerosols, 76–81; *see also* Aerosols; Bioaerosols; Indoor aerosols
  aerodynamic diameter, 77, 80
  hygroscopicity, 79–81
  nonspherical particles, 78
  radioactive and toxic, 6
  relaxation time, 77, 88
  slip correction, 78
  Stokes's law 76–77, 77–79
  terminal settling velocity, 78
Inhalers; *see also* nebulizers
  dry powder (DPIs), 266, 272–273, 282, 283
  pressurized metered dose (pMDIs), 266, 269–272, 282, 283
Inlet systems, 63, 64, 65, 66, 68
Institute de Protection et de Surete Nucleaire (IPSN), 563
Institute of Biophysical Radiation Research, Frankfurt, 10
Institute of the Research Center for Environment and Health, Neuherberg, 11
Insulin, 269
Intake, 102, 103, 105
Interception, 90–91
  filtration, 544, 545
International Aerosol Fellow Award, 13
International Atomic Energy Agency (IAEA), 562
International Commission of Radiologiocal Protection (ICRP), 406, 460
International Commission of Radiologiocal Protection (ICRP) dosimetry model, 4
International Electrotechnical Commission (IEC), 563
International Radon Metrology Program (IRMP), 562
International Society for Aerosols in Medicine (ISAM), 11
International Standards Organization (ISO), 230
Interstitial transfer, 108
IOM, *see* United Kingdom Institute of Occupational Medicine
Ion chromatography (IC), 41, 43
Ionization chamber, 410
IRMP, *see* International Radon Metrology Program
Iron, 610
ISAM, *see* International Society for Aerosols in Medicine
Ischemic heart disease (IHD), 674

J

Journals, 11–12
Juraj Ferrin Award, 13

K

Kelvin effect, 30
Kerosene heaters 197–198, 202
Kinetic properties, 24–30
Knocking off effect, 402–406

L

Laboratories, 9–11
  radiosynthesis, 595–599
LAL assay, 328
Laminar flow, 208
  vs. turbulent flow 82–83, 84, 88–91, 117
Landahl, H.D., 86
Langmuir adsorption theory, 210
Laser diffraction, 280
Latin Hypercube sampling, 174
Lead, 201, 228
Leakage testing, 255, 256
Leuprolide acetate, 269
Levoglucosan, 199
Light scattering, 135, 395, 396
  instruments, 248–252
Limonene, 195, 202, 203
Liposome, 275
Liquid impingers, 325
Liquid chromatographic mass spectrometers (LCMS), 596
Liquid scintillation counting (LSC), 597
London fog, 660–661, 662
Long–term temporal variability, 164
Lovelace organization, 10
Lubin, J.H., 166, 167
  asbestos, 227
  diesel exhaust, 603
  irradiation, 354–358, 444
Lung cancer, 160, 161–162, 163, 345, 358, 656; *see also* Radon; Radon decay products; Thoron; Thoron decay products
  miners, 442, 453, 483–487, 509–513
  radon, 165–166, 167, 169, 363–364, 446, 574
  risk projections, 579–581
  thoron, 574
Lung counter, 5–6
Lungs, 85, 91, 92, 101–102, 119, 122–124; *see also* Weibel model
  deposition of ultrafine particles, 612
  functions, 315

Lungs, (continued)
  injury, 615, 634
  modeling, 4, 6, 177, 178, 179
  particle retention, 627
Lyophilization, 596, 597

M

Macrophages, 315, 319
  ingestion of particles, 631
Magnetic particles, 8
Magnetopneumography (MPG), 8
Man–made vitreous fibers (MMVFs), 228, 229
Marple personal cascade impactor, 244
Martonen (name), 80, 87, 89, 93, 117, 119, 124, 140
Mass median diameter (MMD), 23
Mass size distribution
  indoor, 206–207
Maximum exposure limits (MELs), 232
McBride, S.J., 63
MDIs, *see* Metered dose inhalers
Measles, 317
Measurement; *see also* Direct Measurement; Sampling
  bioaerosols, 323–324
  biological assays, 328–329
  chemical analyses, 329–330
  condensation particle counter, 253
  continuous mass, 38–40
  deposition, 280–281
  direct reading instruments, 248–258
  dosimetry, 281–283
  filter–based, 36–40
  general aerosol samplers, 233–235
  genetic assays, 330
  immunoassays, 329
  inhalable samplers, 236–237
  light scattering instruments, 248–252
  mass, 36–38
  multifraction sampling, 242–246
  particle size distribution, 417–420
  particle size using fiber filters, 549–552
  particle size, 270–280
  particulate constituent (fixed site), 40–44
  personal exposure, 248
  pressure drop sensor, 253–254
  respirable samplers, 241–242
  samplers, 232–233
  surface area, 254–255
  surrogates and indicators, 321–323
  tapered element oscillating microbalance (TEOM), 252–253
  thoracic samplers, 237–240
Mechanical properties, 24–27
Medical aerosols, 8; *see also* inhalers; nebulizers
  future prospects, 266
  history, 266
  research, 9
  size and chemical properties, 6–7
  therapeutic agents, 266–269
Medical Research Council (MRC) of Great Britain, 656
Melandri, C., 92
MEM, *see* Microenvironmental exposure measurement
Mercer Award, 12
Mesh, 117, 121
Mesothelioma, 227
Metalworking fluids (MWFs), 305
Metered dose inhalers (MDIs), 266, 269–272, 282, 283
Method of multilayer filters (MMF), 550, 551
Methoxyphenols, 199
Mexico city, 641
Microbes, 197–198
Microbial volatile organic compounds (MVOCs), 292
Microdosimetry, 136–137
Microenvironment, 62
Microenvironmental exposure measurement (MEM) 62–63, 68; *see also* Exposure assessment
Micrometer–sized aerosols, 32
Microorganisms, 293
Microscopy, 280, 326–327, 549
  confocal, 5
  electron, 247, 421
Microtubules, *see* Fibers, hollow
Mie scattering regime, 249
Mie theory, 28, 420, 437
Migraine headaches, 269
Miners; *see also* Tadjikistan; Russia
  data comparisons, 490
  direct measurement, 464–465
  distribution of radon decay, 448–450
  dosimetry and risk assessment, 490–493
  geography, 350–351
  lung cancer mortality, 483–487
  lung sickness data, 487–490
  measuring ultrafine aerosols (UFAs), 422
  radon concentration, 570
  radon decay measurements, 447–448
  sampling, 451–452
  uncertainities in dose assessment, 441–445
  use of respirators, 513–514
  working conditions, 351–352
Mites, 197–198, 317
MMFs, *see* Method of multilayer filters
MMVFs, *see* Man–made vitreous fibers
Mode of respiration, 125
Modeling, 356–357; *see also* Inhalation modeling
Models
  deterministic, 117, 118–119, 121, 122, 140–144
  empirical, 117, 118–119, 121–122
  ICRP dosimetry, 4
  stochastic, 118–119, 121, 122, 123, 140, 175, 176, 179, 180
Molecular genetic assays, 330
Monday Morning Asthma, 228
Monitors
  continuous, 41–44
  filter–based integrated, 40–41
Monodisperse aerosols, 7, 10, 32
Monodispersity factor, 418
Monte Carlo simulation, 174, 180

Morphology
intersubject variability, 175–177
lung, 179
respiratory tract, 175
systematic variability, 177
Most penetrating size, 546–547
Mouth breathing, 182, 315, 627
Mucociliary clearance, 4–6, 178, 631; *see also* Clearance
Mucosal immune factors, 317
Multifraction samplers, 242–246
Multiorifice uniform deposit impactor (MOUDI), 244
MVOCs, *see* Microbial volatile organic compounds
Mycobacterial infections, 227
Mycotoxins, 330

N

NAAQS, *see* United States National Ambient Air Quality Standard
Nanometer particles, 32, 49, 106–109, 607; *see also* Ultrafine aerosols (UFAs)
National Institute of Environmental Medicine in Stockholm, 4
Natural gas, 446
Navier–Stokes equations, 81, 115
Nazaroff, W.W., 191
Nebulizers, 274–276, 282–283; *see also* Inhalers
Negative bacteria, 639
Nephelometers, 424, 430; *see also* Photometers
Nicotine, 201; *see also* Tobacco smoke
Niosomes, 275
Nitric oxide (NO), 602
Nitroglycerin, 277
Nitrous acid (HONO), 202
Nonemanating reference standard, 381
Nonlinear least–squares approach, 246
Nonuniform rational B–splines (NURBS), 123
Nose–breathing, 182, 315, 627
No–slip condition, 116
Nuclear accident, 416; *see also* Chernobyl
Nuclear fallout, *see* Radionuclides
Nuclear power plants, 386
Nuclear track film, 578
Nucleation, 206, 209, 609–610
Nuclei mode, 32, 48, 50, 51, 52, 54, 55, 57–58, 609
Null batteries, 428
Nylasorb filter, 41

O

Occupational exposure standards (OES), 232
Occupational exposures, 226, 229; *see also* Miners
ODTS, *see* Organic dust toxic syndrome
Optical particle counters (OPCs), 2, 249–251; *see also* Light scattering
Optical particle size analysis, 280
Optical properties, 27–28
Orbicules, 327
Organic dust toxic syndrome (ODTS), 295, 304
Organizations, 11–12
Overloading, 180
Oxidative damage, 634–635, 636
Ozone, 624
asthma, 658
depletion, 270
health effects, 641
Ozyeyen equation, 432

P

PAEC, *see* Potential alpha energy concentration
PAHs, *see* Polycyclic aromatic hydrocarbons
Pankow's nomenclature, 209
Paraffin, *see* Candles
Particles; *see also* Aerosols; Deposition
density, 23–24
dynamics, 81–86, 115–117
indoor formation, 209–211
indoor motion, 208
overload, 603
properties, 21–24
shape, 24, 419
size, 21–23, 31, 32
PAS–6 inhalable sampler, 237
Passive badge samplers, 64, 70
PCM, *see* Phase contrast microscopy
PCR, *see* Polymerase chain reaction
PDA, *see* Phase doppler anemometry
Peclet number, 544
PEMs, *see* Personal exposure measures
Penetration 230–231, 542
Perfluoroalkyl sulfonamides, 195
Permissible exposure limits (PELs), 232
Personal aerosol sampler (PAS), 443
Personal cascade impactor sampler (PCIS), 244
Personal cloud, 191, 204
Personal dosimeters, 492
Personal dust monitor (PDM), 252
Personal exposure measures (PEMs) 62–63, 66, 68, 69, 70; *see also* Exposure assessment
Personal inhalable dust spectrometer (PIDS), 244, 245
PERSPEC, 242
Pesticides, 197–198, 202
Petryanov's filters, *see* Filtration, fiber filters FP
PET, *see* Positron emission tomography
Pets, 197–198
Phagocytosis, 8, 178, 639; *see also* Clearance, macrophages
Phantom measurements, 475–476
Pharmaceutical aerosols, *see* Medical aerosols
Pharmacokinetic methods, 282
Phase contrast microscopy (PCM), 247, 248
Phase doppler anemometry (PDA), 280
Phase partitioning, 209–211
Photometers, 251–252; *see also* Light scattering
PIDS, *see* Personal inhalable dust spectrometer
Pinene, 202, 203
Planar gamma cameras, 135–136
PLM, *see* Polarized light microscopy
Plug flow, 83
Plutonium, 586
inhalation estimates, 592–593
measurements, 588–590

Plutonium, (continued)
  respirable, 591
  resuspension, 591
PM standards, 36–38, 45
Pneumoconiosis, 227; *see also* Silicosis
Pneumonia, 268, 321, 674
Poiseuille flow, 431, 434
Polarized light microscopy (PLM), 247
Pollen, 313, 318, 321, 322, 327, 622
  aerodynamic diameter, 306
  geography, 314
  health effects, 648–649
  size, 308
Polycyclic aromatic hydrocarbons (PAHs), 42, 602, 634, 651
Polydisperse aerosols, 32, 257
Polymerase chain reaction (PCR) assays, 330
Polytetrafluoroethylene (PTFE), 108
Polyurethane foams (PUF), 239
Population mobility, 164
Porton Down, UK, 9
Positive matrix factorization (PMF), 204
Positron emission tomography (PET), 136, 281
Potential alpha energy concentration (PAEC), 370, 373, 401
Pressure drop, 543
  filtration, 547, 548
  sensor, 253–254
Primary particles, 33
Progressive massive fibrosis, 227
Propane, 270
Propellant systems 270–271, 272; *see also* Inhalers
Proximity effect, 63
Pulmonary research, 1–4, 10

Q

Quantal infection, 317

R

Radiation safety techniques, 598
Radioactive aerosol standards, 413–414
  currently applicable standards, 562–566
  U.S.S.R. standard, 557–561
Radioactive aerosols, *see* Chernobyl; Russia; Miners; Radionuclides; Radiosynthesis laboratories; Radon; Radon decay products
Radioactive markers, 109
Radioassay program, 482
Radiometers, 504
  testing, 565–566
Radionuclides, 585–593
  concentrations at Chernobyl, 524
  contribution to gamma radiation, 526
  ejection from Chernobyl reactor, 518–519, 521–522
  global transfer from Chernobyl, 519–520
  imaging, 135–136
  inhalation estimates, 592–593
  long–term monitoring, 592
  plutonium in TSP, 588–590
  respirable concentrations, 591
  resuspension, 590–591
  sampling, 586–587
  similarities and differences, 586
  thorium in TSP, 587–580
  uranium in TSP, 587–588
Radiosynthesis laboratories, 595–599
  air effluents, 597
  control of contamination, 598–599
  radiation safety guidelines, 596
  sources of airborne radioactivity, 596–597
  surface contamination, 597
Radon, 161–162, 536; *see also* Radon decay products
  air radon concentration monitoring, 361–362
  bronchial lung dose, 574–577
  case–control studies, 162–165, 166
  concentration distribution measurement, 362–363
  drinking water, 571, 573–574, 577
  ecological studies, 165–166, 167
  environmental concentrations, 570
  health effects, 169
  indoor concentrations, 570, 579
  lung cancer, 363–364, 579–581
  measurements, 350, 445–447
  mines, 345–346
  outdoor concentrations, 570–571, 578
  overview, 569
  radon and thoron gas, 577–578
  risk estimates, 166–168
  standards, 562–563
  stratospheric concentrations, 571–573
  studies on animals, 358–361
Radon decay products, 407–408, 409, 413, 416; *see also* Miners; Radioactive markers; Radon
  absorption of alpha–radiation, 377–381
  basic equations, 366–370
  breathing rate and deposition, 453–459
  characteristics, 364–366
  continuous air monitors (CAMs), 564–565
  direct measurement, 466, 468
  distribution of concentration, 448–450
  dose distribution of mines, 513–514
  dosimetry, 354–358, 442–447, 459–464
  equilibrium factor, 386–387
  errors in measurement, 387–390, 476–480
  generation, 558–559
  indoor concentration distributions, 503–504
  indoor exposure, 512
  lung cancer, 363, 486, 509–513, 574
  lung irradiation, 354–358
  lung sickness, 487–490
  measurement, 346–348, 353, 370–373, 440, 447–448
  portable measurement instrument, 480
  progeny concentration measurement, 385–386
  radioactive markers, 481–483
  sampling breathing zone, 504–508
  scintillation chamber count rate measurement, 383–385
  spectrometry, 373–374, 381–382, 559–560
  standards, 562–563
  studies on animals, 358–361
  Tadjikistan case study, 352–354
  unattached fraction 386–387, 395–406

Radon progeny, *see* Radon decay products
Rayleigh scattering, 28, 249
Real–Time Ambient Monitoring System (RAMS), 39
Recoil effect, 402–406
Recommended exposure limits (RELs), 232
Regularization method, 246
Relative limit values (RLVs)
  endotoxin, 320
Relaxation time, 77, 88
Renovations, 197–198, 204
Research areas, 158
Residual oil fly ash (ROFA), 634, 637
RESPICON sampler, 243
Respirable samplers, 241–242
Respirable dust, 591
Respirable dust dosimeter (RDD), 253–254
Respirable fraction (RF), 21, 282, 591
Respirators, 232, 599
  miners, 513–514
  testing, 255–256
Respiratory illness, 131–133, 305, 615
Respiratory dosimetry, 314–315
Respiratory infections, *see* Infections
Respiratory syncytial virus, 639
Respiratory system
  clearance, 128–131
  disease, 131–133
  environment, 126–128
  morphology, 122–124
Respiratory tract, 175, 626–629; *see also* lungs
Resuspension, 529, 531, 586
  indoor, 213, 214, 217–218
  radionuclides, 590–591
Resuspension factor (RF), 590
Retention, 102
  medical aerosols, 281
ROFA, *see* Residual oil fly ash
Rupprecht and Patashnick Inc., 38
Russia, 503–514
  Krasnokamensk mines, 513–514
  Lermontov mines, 509–513
  radon decay products, 503–504
  sampling breathing zone, 504–508

S

Safety standards, 106
Saltation, 590
Sample analysis, 246–248
Samplers, 232–233
  cowled sampler, 234
  general, 233–235
  inhalable, 236–237, 239, 242, 243, 256–257
  multifraction, 242–246
  passive badge, 64, 70
  respirable, 241–242
  testing, 256–257
  thoracic, 237–240, 256–257
Sampling; *see also* Measurement
  breathing zone, 69–70
  Chernobyl, 520
  contaminant type, 70–71
  cultivation–based methods, 327–328
  electrostatic precipitation, 325–326
  filtration, 312, 325, 551
  impactor vs. filter, 552
  inertial sampling, 324–325
  inhalable conventions, 231
  light microscopy, 326–327
  personal vs. fixed, 450–451
  radionuclides, 508, 586–587
  radon decay products, 367, 504–508
  radon, 361–363
  respirable convention, 231
  size–selective sampling, 21, 22, 229–232, 233, 240
  thoracic convention, 230, 231, 232
  thoron decay products, 504–508
Saxony
  lung cancer risk, 579
Scaling–law–based techniques, 177
Scanning electron microscopy (SEM), 247, 248
Scanning mobility particle size analyzer (SMPS), 608
Scintillation detectors, 374
Seasonal patterns
  bioaerosols, 313
Secondary aerosol formation, 202
Secondary organic aerosols (SOAs), 193
Secondary particles, 33
Sedimentation, 3, 88–89, 438, 561, 626
SEM, *see* Scanning electron microscopy
Semicontinuous elements in aerosol system (SEAS), 43
SFM, *see* Sorption filter materials
Shelter, Chernobyl, 518, 519, 520, 525, 530, 531–532, 538
  aerosol concentration, 532–534
  aerosol dispersity inside, 535–536
  aerosol transport into the atmosphere, 534–535
  nearby radioactive aerosols, 536–537
  types of aerosols, 532
Shields, H.C., 190, 195
Short–term exposure limits (STELs), 232
Silicosis, 227, 229, 485
Sinclair–LaMer generator, 2; *see also* Spinning disc generator
Single breath recovery method, 5
Single fiber capture coefficient, 543
Single particle counters, 2–3
Single–photon emission computed tomography (SPECT), 136, 138, 281
Size distributions, 55–56, 205–208
  bioaerosols, 306–307
  breathing zone, 244
  fine aerosols, 438
  functions, 206
Size–selective sampling, 21, 22, 229–232, 233, 240
Skin diseases, 276
Sip correction, 24, 25
Slip–flow regime, 81–82
Smog, 33
Smoluchowski Award, 13
Sneezing, 304

Sorption filter materials FP (SFM), 520, 552, 553
Sorption/desorption, 209–211, 211–212, 377–381, 632
Space charge, 91
Spark generator, 8, 11
Spark–ignition (SI) size distributions, 48
Spatial variability, 164
Special aerosol sources (SAS), 565–566
SPECT, *see* Single–photon emission computed tomography
Spectrometry, 395, 480, 559–560, 611
 alpha–, 381–382, 386, 387, 389, 390, 393, 396, 408, 410, 422, 462
 beta–, 381–382, 387, 389, 390
 mass, 246
Spectroscopical methods, 365
Spinning disc generator, 2, 10
Spiral centrifuge, 11
Spray, 33
Standard aerosol sampler (SAS), 443
Standard density spheres, 23
Standards, 36–38, 45; *see also* Radioactive aerosol standards
Steroids, 275
Stochastic models, 118–119, 121, 122, 123, 140, 175, 176, 179, 180
Stokes's law, 24, 76–77, 77–79, 81, 82
Stokes's regime, *see* Continuum regime
Subepithelial fibrosis, 132
Submicrometer size, 33
Sunset Laboratory, 42
Supersaturation, 209
Supersites, 44
Surface area measurement, 254–255
Surface area size distribution, 206
Surface creep, 590
Surfactants, 270
Suspension, 590
Sweden
 radon exposure, 446
Swift, D., 2, 7

T

Tadjikistan mines, 347, 350, 351, 442, 448, 455, 575
 breathing rates, 458
 case study, 352–354
 comparisons, 490
 dosimetry and risk assessment, 490–493
 lung sickness, 487
 thoron decay, 480
Tandem differential mobility analyzers (TDMAs), 54
Tapered element oscillating microbalance (TEOM), 39, 40 252–253
Teflon filter, 41
TEM, *see* Transmission electron microscopy
TEOM, *see* Tapered element oscillating microbalance
Terminal settling velocity, 77, 78, 92
Therapeutic aerosols, see Medical aerosols
Thermal desorption particle beam mass spectrometer (TDPBMS), 54
Thermistor detectors, 428
Thermodynamic diameter, 607–608
Thoracic samplers, 237–240, 256–257
Thoracic fraction, 21
Thorium, 462, 483, 586; *see also* Radionuclides
 inhalation estimates, 592–593
 measurements, 587–588
 respirable, 591
 resuspension, 591
Thoron, 361, 362, 365, 386, 536; *see also* Radon; Radon decay products; Thoron decay products
 basic equations, 390–395
 bronchial lung dose, 574–577
 environmental concentrations, 560
 overview, 569
 radon and thoron gas, 577–578
 stratospheric concentrations, 571–573
Thoron decay products, 393–394, 408, 413, 416, 480, 484
 characteristics, 390
 generation, 558–559
 radioactive markers, 414
 spectrometry, 560
Three tiered–monitoring program, 44
Three–count method, 373
Threshold limit values (TLVs), 232
Throughput, 282
Time series studies, 161, 169
 acute exposure and mortality, 662–666, 669–670
Time–of–flight (TOF) particle size analyzers, 280
Tobacco smoke, 200–201, 321, 624, 632
 birth outcomes, 651
 chemical composition, 197–198
 cloud motion, 92
 lung cancer, 166, 485
 lung deposition, 463, 574
Tobramycin, 268
Topical sprays, 276–277
Total deposition, 1–4, 7
Toxicity, 226, 612–615, 633–634
 oxidative damage, 634–635
Tracheobronchial simulations, 140–142
Traffic, 192, 193, 321; *see also* Diesel exhaust
 health effects, 642–647, 656
Transformation processes, 213, 214
Transition metals, 634, 635
Transition regime, 81
Transmission electron microscopy (TEM), 247, 248
Trimodal size distribution, 31, 32, 48
Tritiation process, 596
Tuberculosis, 294, 317
Twin impingers, 279–280
Tyndall spectrometry, 2

U

Ultrafine aerosols (UFAs), 32, 417; *see also* Nanometer particles
 ambient, 607–609
 composition, 610–612
 electrical parameters, 440–441
 formation, 609–610

generating, 422–427, 428–431
health effects, 642
indoors, 194
lung deposition, 612
measurement errors, 430
measurements, 420–422
method for investigating, 427–428
oxidative damage, 635
toxicology, 612–615
Ultrafine condensation particle counter (UCPC), 50
Ultraviolet aerodynamic particle size spectrometer (UV–APS), 326
Unattached activity, 406
Unattached fraction, 395–406, 455, 482
concentration measurements, 401–402
correlation with aerosol concentration, 395–401
measurements, 450
recoil nuclei, 402–406
United Kingdom
mining, 442
radon concentrations, 362
United Kingdom Mines Research Establishment (MRE), 241
United Kingdom Institute of Occupational Medicine (IOM) inhalable sampler, 236, 239, 242, 243
United States
radon concentrations, 362
United States National Ambient Air Quality Standard (NAAQS), 36, 37, 38, 40
University of Minnesota mobile emissions laboratory (MEL), 50
Uranium, 364, 586; *see also* Miners; Radionuclides; Russia
inhalation estimates, 592–593
measurements, 587–588
respirable, 591
resuspension, 591

V

Variability
interpersonal, 174
intersubject morphometric, 175–177
systematic morphometric, 177
Vector model, 3
Vehicle exhaust, *see* Diesel exhaust; Traffic
Velocity, 24–26
profiles, 83, 85, 116–117
Ventilation, 125–126
mines, 351, 355, 448, 465, 473
radon, 364, 387
Vermiculite, 227
Vertical elutriator, 237
Video monitoring, 257
VNIIFTRI, *see* Aerosol Laboratory of the All–Union Institute of Physico–Technical and Radiotechnical Institute
Volatile organic compounds (VOCs), 193
Volatility experiments, 54

W

Waiting method, 365
Walking, 197–198, 202–203
Weapons testing, 589
Weibel, E.R., 122, 123
Weibel model, 3. *see also* Lungs, modeling
Weschler, C.J., 190, 195, 211, 212
Whirling–arm impactors, 325
Whitby Award, 13
White light optical counter, 2
Whole–body counters, 482
Wipe–testing, 597
Wolkoff, B., 190
Wood, *see* Fireplaces
World Health Organization (WHO), 294

X

XRF analysis, 41

Y

Young Investigator Award, 13
Yu, C.P., 91

Z

Zeolite, 227